Water Management and Climate Smart Agriculture

Proceedings of National Seminar on Water Management and Climate Smart Agriculture

(13th – 14th February, 2015)

Vol – I

Vulnerability of Agriculture to Climatic Aberrations

R. Subbaiah

G. V. Prajapati

Centre of Excellence on Soil & Water Management
Research, Testing & Training Centre
Junagadh Agricultural University, Junagadh, Gujarat, India

GYAN

ISBN: 978-81-212-1315-8

Price: ₹ 4500 (set of 3 Vols)
₹ 1300 (Vol–I)

First Published, 2015

Published by

Gyan Publishing House
5, Ansari Road, Daryaganj,
New Delhi-110002
Phones: 9811692060
E-mail: books@gyanbooks.com

Printed at: G. Print Process, Delhi

Cataloging in Publication Data—DK
Courtesy: D.K. Agencies (P) Ltd. <docinfo@dkagencies.com>

National Seminar on Water Management and Climate Smart Agriculture (2015 : Junagadh Agricultural University)
Proceedings of National Seminar on Water Management and Climate Smart Agriculture (13th-14th February, 2015) / R. Subbaiah, G.V. Prajapati.
3 volumes cm
National seminar organized by Junagadh Agricultural University.
Includes bibliographical references.
Contents: vol. I. Vulnerability of agriculture to climatic aberrations — vol. II. Climate change impact on water cycle — vol. III. Adaption of climatic resilient water management and agriculture.
ISBN 9788121213158

1. Water-supply, Agricultural—India—Management—Congresses. 2. Climatic changes—India—Congresses. 3. Crops and climate—India—Congresses. 4. Agriculture—Environmental aspects—India—Congresses. 5. Agriculture—Economic aspects—India—Congresses. I. Subbaiah, R., editor. II. Prajapati, G. V., editor. III. Junagadh Agricultural University.organizer. IV. Title.

S494.5.W3N38 2015 DDC 333.9130954 23

Vice Chancellor
Junagadh Agricultural University
Junagadh-362 001
Ph. :(o) 0285-2671784
Fax :(o) 0285-2672004
E-mail : vc@jau.in
website : www.jau.in

Message

Agriculture faces the challenge of fulfilling ever-expanding global food demand, in the context of changing climatic conditions that negatively affect food production. Higher temperatures, shifting seasons and rainfall patterns, more frequent and extreme floods and droughts, pests and diseases, greenhouse gas emissions, depleted soil fertility and stress on water resources wreak havoc on food production and farmers' livelihoods.

Stable and Climatic Smart Agricultural production and ecologically sound resource conservation technologies that improve input use efficiency, conserve and protect our natural resources need to be aggressively promoted to increase production and better manage and sustain the planet's biodiversity. Climate Smart Agriculture creates an opportunity to assess and design agricultural production models that help smallholder producers adapt to and mitigate the effects of changing climates, while improving their own household food security. Our scientists must therefore work intensively to accurately assess the felt needs of our country and develop new methods, new technologies and new knowledge to build resilience to mitigate the negative impacts of climate change, as well as contribute to reducing greenhouse gas emissions through practicing climate-smart agriculture, both in policies and practices and through judicious utilization of natural reserves

I sincerely hope that the concerted efforts of this amalgamation of country's leading agricultural scientists and water management specialists will take stock of the alarming effect of climatic drift on agriculture and quantity and quality of water resources and come up with tangible and feasible eco-technological interventions and solutions to free our agriculture of the veritable maladies and unleash the potential of the country in a truly bio secure India.

I congratulate the concentrated efforts of Centre of Excellence on Soil and Water Management of Research, Testing and Training Centre of this University to organize this National Seminar on Water Management and Climate Smart Agriculture in a most proficient manner.

I wish the National Seminar a grand success.

Place : Junagadh **(A. R. Pathak)**

Preface

Agriculture represents a core part of the Indian economy and provides food and livelihood activities to much of the Indian population. Climate is a direct input into the agricultural production process, the agricultural sector has been a natural focus for research. Agriculture, particularly in India with nearly 60% rainfed area, has been a highly risky venture with vagaries of monsoon besides the interplay of other abiotic and biotic factors. Climate change is set to compound the daunting complex challenges already being faced by agriculture through affecting crop yields (both positively and negatively), as well as the types of crops that can be grown in certain areas, by impacting agricultural inputs such as water for irrigation, amounts of solar radiation that affect plant growth, as well as the prevalence of pests. The predicted changes to agriculture vary greatly by region and crop. Climate change and climate variability also affecting irrigated agriculture, installed power capacity, environmental flows in the dry season and higher flows during the wet season, thereby causing severe droughts and floods in urban and rural areas.

Due to the complex interaction of climate impacts, combined with varying irrigation techniques, regional factors, and differences in crops, the detailed impacts of these factors need to be investigated and concerted efforts are required for mitigation and adaptation strategies for gearing up our efforts to evolve climate-resilient crop varieties, cropping patterns and water management practices for ensuring food security, enhancing livelihood opportunities and contributing to economic stability at the national level. Keeping this in view the Junagadh agricultural University organized a National seminar on Water management and climate Smart Agriculture during Feb 13-14, 2015 to address the above challenges. The proceedings is a compendium of scientific papers covering different dimensions of issues associated with climate change and sustainable food security and ways to mitigate the problem. I am sure, is the need of the hour, as it offers lot of pragmatic guidelines to the policy makers to improve Indian agriculture, besides serving as a very useful reference work for academicians and researchers engaged in studies on agricultural development.

I am extremely indebted to Dr. A. R. Pathak Hon Vice Chancellor, JAU, Junagadh, who had to bear a heavy responsibility and concern in directing this event to a successful end. I am also thankful to Dr. A. Y. Desai, Director of Research, JAU and Dr. A. M. Parakhia, DEE, JAU, Junagadh and Dr. N. K. Gontia, Principal C.A.E.T, JAU, Junagadh for their continuous support.

We will be grateful to Rajiv Kumar Gupta IAS, Chairman, Ground Water Supply Sewerage Board and Climate Change Department, Government of Gujarat, ICAR, New Delhi, Gujarat Council on Science & Technology, Department of Science and Technology, Gandhinagar, Monsanto Holdings Pvt. Ltd., Mumbai for sparing generous amount towards organizing this National gathering.

Junagadh
Date: 05/02/2015

(R Subbaiah)
Research Scientist (Agril. Engg)
&Organizing Secretary

Contents

1

The Challenges of Climatic Drift on Productivity of Wheat in India

Indu Sharma, Ravish Chatrath, Satish Kumar and R Sendhil

Abstract

Agriculture is the mainstay and a source of livelihood for millions of farmers, and a significant fraction of the gross domestic product of Indian economy. It has been one of the main subjects of analysis to understand the impacts of both climatic variability and climatic change. It is projected that by the end of 21st century, the mean annual temperature will increase by 3-5°C and the maximum increase is expected to occur in northern India. Wheat is likely to be negatively impacted due to terminal heat stress. Further, uncertain movement in the weather variables shall force the use of scarce resources like water beyond its optimal level resulting in groundwater depletion. India is the second largest producer of wheat and during 2013-14 harvested about 96 million tons of wheat, beating its previous year record. Since we have highly variable wheat growing environments, there is a possibility of wheat being affected for all the predicted climatic concerns in one or the other zone. In the case of temperature, there is a 0.6°C rise in the last 100 years and it is projected to rise by 3.5-5°C by 2100. The implications on wheat of present climate change can be: (1) Erratic behaviour of the ambient temperature – variable spells of cold or warm temperatures during crop cycle, (2) Simultaneous drought and flooding conditions, (3) Poor seed and grain quality, (4) Altered pest and disease dynamics and (5) Fluctuating yield. However, in spite of all these factors and impact of climate change across the globe, Indian wheat production has been increasing for last 4-5 years continuously. This continuous increase in yields can be attributed to several management practices such as timely control of diseases and pests and slight changes in climatic factors. With global wheat yields showing a decreasing trend development of climate smart wheat is need of the hour.

Introduction

Agriculture is the mainstay and a source of livelihood for millions of farmers, and a significant fraction of the gross domestic product of Indian economy. It has been one of the main subjects of analysis to understand the impacts of both climatic variability and climatic change. Volatility in the weather variables, under present and future climate scenarios is a topic of concern and interest for a variety of reasons among researchers and policy-makers. The impact of fluctuations in climatic variables and its immediate consequences in agriculture sector are equally important as that of variations in the overall mean climatic variables. Individual crop growth processes at different stages have a different effect owing to the climate change. For instance, a seasonal increase in the temperature is expected to raise the crop developmental rate resulting in an early harvest. Heat stress poses several negative effects on crop production. Alternatively, the quantum of increased rainfall in the dry regions is expected to increase the crop duration resulting in higher yields. Nevertheless, the importance of weather variables and climatic factors in getting higher yield cannot be undermined as agriculture practised in different forms at diverse locations remains highly vulnerable to the change in climate.

Inter alia, climate change is one of the biggest environmental threats affecting crop production and food security of global population. The Fourth Assessment Report of Intergovernmental Panel on Climate Change (IPCC) states that global warming is unequivocal and is expected to increase the evapo-transpiration of crops, inconsistency in the quantum of rainfall and the frequency of drought. Further, the future tropical cyclones are likely to be more intense with larger peak wind speeds and heavy precipitation. The emerging weather conditions are expected to reduce the crop yields directly and indirectly encourage the weeds and pests incidence. Further, changes in the rainfall pattern increase the likelihood of yield loss in the short-run leading to overall production decline in the long-run. Studies on inter-annual and long-term variability of rainfall indicated a significant variation in rainfall with a high degree of correlation between the rainfall and agricultural production. India is highly prone to the impact of climate change due to its typical geographical location as well its high dependency on agriculture for national food security. IPCC projections for India have been 0.5 to 1.2°C rise in temperature by 2020, 0.88 to 3.16°C by 2050 and 1.56 to 5.44°C by 2080 depending on the future development scenario. Overall, the temperature rise is likely to be much higher during the *Rabi* (November to April) season in comparison to *Kharif* (June to October). It is projected that by the end of 21st century, the mean annual temperature will increase by 3-5°C and the maximum increase is expected to occur in northern India. Extreme climatic events cause enormous losses to the standing crops affecting the food security of the country. For instance, Rice, a *Kharif* season crop is expected to be impacted more by the variability in rainfall while wheat, a cold loving plant which is grown in the *Rabi* season will be affected by the volatility in the minimum temperature. Wheat is likely to be negatively impacted due to terminal heat stress while rice to be impacted both by temperature and water availability. Further, uncertain movement in the weather variables shall force the use of scarce resources like water beyond its optimal level resulting in groundwater depletion. The concern is, India, a self-sufficient nation may sooner turn to a food insecure country unless adaptation measures are implemented.

Effects of weather variables on crop production can be either beneficial or detrimental. Farmers always expect the mercy against detrimental effect of climate change or seldom derive the benefit. Hence, farmers have to adapt their practices according to the changes in solar radiation, temperature, relative humidity and precipitation produce significant changes in crop yield, crop mix, cropping system, scheduling of field operations and grain moisture content so as to improve their livelihood. Inherently, crop production practices are manageable and farmers always have a set of possible adaptations strategies which either reduce the likely future yield losses on account of climate change or shall improve the regional yield levels where it benefits. It can be concluded that crop production responds to physiological changes as well as management practices. The present paper enlighten and provide new insights about the relationship between climate change and wheat yield giving more attention on the adaptation strategies to reduce the negative impact of climate change.

Indian Wheat Scenario and Climate

India is the second largest producer of wheat and during 2013-14 harvested about 96 million tons of wheat, beating its previous year's record. The attainable yield levels indicate that India is sitting over a potentially large harvestable yield. Wheat growing India is divided in to six main wheat growing zones. The cooler North Western Plains Zone (NWPZ) has about 10.5 million hectares under wheat. The Northern Hills Zone (NHZ) in Himalaya with varying valleys, slopes and elevation has more than a million hectare under long duration wheat. The cool and humid North Eastern Plains Zone (NEPZ) has about 10.5 million hectares. Central Zone (CZ) and Peninsular Zone (PZ) that have short winter and are relatively warm have about 5.5 and 1.2 million hectares respectively under wheat. Relatively the Southern Hills Zone (SHZ) accounts for less than one per cent area. India uses nearly three million tons of wheat to seed the nearly 30.0 million hectares. Wheat is sown in November and is harvested by March/mid- April and almost 90 per cent of the wheat sown gets some irrigation and fertiliser. Since we have highly variable wheat growing environments, there is a possibility of wheat being affected for all the proposed climatic concerns in one or the other zone.

Climate change impacts on crop yields especially wheat are variable in different areas. It may lead to increase in some areas where as it will decrease in other. The crop yield can be increased with irrigation application and precipitation increase during the crop growth; meanwhile, crop yield is more sensitive to the precipitation than temperature. If water availability is reduced in the future, soil of high water holding capacity will be better to reduce the frequency of drought and improve the crop yield. With climate change, the growing period will reduce, and the planting date also needs to change for higher crop production. Climate change can decrease the crop rotation period, so farmers need to consider crop varieties, sowing dates, crop densities and fertilisation levels when planting crops. The positive effects of climate change on agriculture are concerned with the atmospheric CO_2 concentration, crop growth period increases in higher latitudes; the negative effects include the increasing incidence of pests and diseases, and soil degradation owing to climate change. This has urged scientists to develop more crop varieties suiting the changing climate and degrading soil in order to obtain sufficient yield for the increasing population. Quantification of climate uncertainty is an important indicator for crop yield variation in future climate scenarios.

Factors Responsible for Climatic Change/drift

Climate change is manifesting itself in many ways across the country. Among the indicators, while long-term rainfall data analysis shows no clear trend of change, regional variations as well as increased rainfall during summer and reduced number of rainy days can be noticed. In the case of temperature, there is a 0.6°C rise in the last 100 years and it is projected to rise by 3.5-5°C by 2100. The carbon dioxide concentration is increasing by 1.9 ppm each year and is expected to reach 550 ppm by 2050 and 700 ppm by 2100 (IPCC, 2007). Extreme events like frequency of heat and cold waves, droughts and floods have been observed in the last decade. The sea level has risen by 2.5 mm every year since 1950 while the Himalayan glaciers are retreating. These are all symptomatic of climate change.

Atmospheric GHG and the mean sea level: The changes in climate parameters are being felt globally in the form of changes in temperature and rainfall pattern. One of the factors responsible for global warming is the global atmospheric concentration of carbon dioxide. A greenhouse gas (GHG), the value of atmospheric carbon dioxide has increased from a 280 ppm to 387 ppm in 2010 due to factors that are mainly industrial. Other GHG such as methane and nitrous oxides have also increased considerably. This has led to the warming of the climate system by 0.74°C between 1906 and 2005 (IPCC, 2007b). This has further led to an increase in global average sea water level by 1.8 mm per year from 1961 to 2003. In recent years this rate was faster making the affects of climate change more visible.

Rise in temperature: The frequent changes in temperature and rainfall patterns have become more prominent over the last decade. The rise in global temperature and inconsistent rainfall has led to an increased frequency of droughts as well as heavy precipitation events over many regions. Cold days, cold nights and frost events have become less frequent,while hot days, hot nights and heat waves have become more frequent. The IPCC (2007) projected that temperature increase by the end of this century is expected to be in the range 1.8 to 4.0°C. For the Indian region (South Asia), the IPCC projected 0.5 to 1.2°C rise in temperature by 2020, 0.88 to 3.16°C by 2050 and 1.56 to 5.44°C by 2080, depending on the future development scenario (IPCC, 2007b).

In context of Indian wheat the rise in average temperatures across the growing season is threatening to reduce the duration of the crop. High initial temperatures are not beneficial for seed germination and vigour whereas the terminal heat stress causes a loss in the yield by reducing the grain filling duration (Aggarwal, 2008).

Variable rainfall: Untimely and variable rainfall is also a concern as drought like conditions as well as flooding will lead to loss in production. Overall, the temperature rise is likely to be much higher during the winter (*Rabi*) rather than in the rainy season (*Kharif*). It is projected that by the end of the 21st century, rainfall over India will increase by 10-12 per cent and the mean annual temperature by 3-5°C. The warming is more pronounced over land areas with a maximum increase over northern India.These environmental changes are likely to increase the pressure on Indian agriculture, in addition to the on-going stresses of yield stagnation, land-use, competition for land, water and other resources and globalisation. It is estimated that by 2020, food grain requirement would be almost 30-50 per cent more than the current demand (Paroda and Kumar, 2000). This will have to be produced from the same or even the shrinking land resource due to increasing competition for land and other resources by the non-agricultural sector.

Visualising the Climate Change Impact on Wheat Crop

The studies about wheat production affected by climate change are mainly concerned with future CO_2 concentrations. Ortiz *et al.*(2008) discussed how wheat can adapt to climate change in Indo-Gangetic Plains for 2050s and suggested that global warming is beneficial for wheat crop production in some regions, but may reduce productivity in critical temperature areas, so it is urgent to develop some heat-tolerant wheat germplasm to mitigate climate change. Similarly the rainfall patterns may change leading to drought in some areas whereas flooding in others. Recent rains in Utrakhand and Kashmir are an example of uneven distribution of rainfall. Thus, emphasis needs to be paid for developing wheat cultivars tolerant to drought as well as water logging conditions. Similarly changes in both temperature and rainfall/humidity may lead to change in the insect/pest and disease dynamics for wheat crop. All these changes will lead to poor seed yield and also deteriorate quality of the grain. The implications of present climate change can be: (1) Erratic behaviour of the ambient temperature – variable spells of cold or warm temperatures during crop cycle, (2) Simultaneous drought and flooding conditions, (3) Poor seed and grain quality (4) Altered pest and disease dynamics and (5) Fluctuating yield.

Erratic behaviour of the ambient temperature and rainfall: Throughout the 21st century, India is projected to experience warming above the global mean. A warming trend has been observed along the west coast, in central India, the interior peninsula and Northeast India. A single factor of climate change like temperature can have a catastrophic effect on crop yield. Temperature increases of 1°C, 2°C and 3°C in Punjab, would reduce the grain yield of rice by 5.4 per cent, 7.4 per cent and 25.1 per cent respectively (Aggarwal *et al.*, 2006). However, cooling trends have been observed in northwest India and parts of South India (Gbetibouo *et al.*, 2005). Rainfall data for the last 100 years at the all-India level do not show any significant trend. However, at the regional level, increasing monsoon seasonal rainfall has been found along the west coast, northern Andhra Pradesh and north-western India, while a trend of decreasing monsoon seasonal rainfall has been observed over eastern Madhya Pradesh, NE India and some parts of Gujarat and Kerala. The possible changes in temperature, precipitation, concentration of CO_2, CH_4, nitrous oxide (N_2O) and O_3 are expected to have significant impact on crop growth. However two to three for last years the central India has been receiving good rainfall during wheat season thus accounting for higher wheat yields. The unpredictable nature of climate change thus makes it difficult to assess the change and its management.

Poor seed and grain quality: Weather wheat varieties of hot climate lack in grain or product quality remains an obvious anxiety in the global warming era. Escalated CO_2 levels do affect wheat quality (Conroy, 1994; Wu *et al.*, 2004 and Erda *et al.*, 2005) whereas some reports have also shown uncertainty in this regard (Wrigley, 2006 and Hogy *et al.*, 2008). Studies on global warming have envisaged effect on grain quality in India as well (Houghton 1991 and Ortiz *et al.*, 2008). However, wheat grown under hot climate of central India has always been rated quality superior (Mohan *et al.*, 2011). Grain quality data of ten year period (2005-14) in All India Coordinated Wheat and Barley Improvement Project (AICW and BIP) however indicate that grain and product quality in genotypes of the north-eastern plains, central and peninsular zones was no way inferior to NWPZ, zone of most congenial wheat growth environment. In fact, wheat from hot climates registered superiority in several quality components like test weight, kernel weight, flour recovery and wet gluten content. End-product quality, especially of bread and *Chapati,* also holds better in hot regions, may it be north-eastern or central-peninsular parts of the country.

Reduction in crop duration has been envisaged as major concern of global warming in the eastern Indo-Gangetic Plains (Ortiz *et al.*, 2008). Effect of reduced crop duration on wheat quality was mitigated by comparing grain quality of wheats raised under timely and late sown production conditions in AICW and BIP (Period: 2005-14). Comparison within two types of wheat was made in two regions, (i) Indo-Gangetic Plains covering NWPZ and NEPZ and (ii) central and peninsular India comprising CZ and PZ. Investigations reflected that reduction in grain weight is most prominent change in short duration wheats (Mohan *et al.,* 2011). Consequently, decline in kernel and test weights and elevation in grain protein content and gluten harvest occur in each region (Table 2). Short duration wheats have exhibited improvement in bread loaf volume in AICW and BIP. Magnitude of such change varies in different zone as difference was small but significant in Indo-Gangetic and significantly large in central-peninsular India. Gluten strength and gluten quality also appeared better in late sown wheats of central-peninsular India. Mohan and Gupta (2013) highlighted differences in frequency of high yielding bread wheat genotypes with 5+10 and 2+12 High Molecular Weight Glutenin Subunits (HMW-GS) banding patterns at *Glu D1* under Indian conditions. They observed that 2+12 type wheats are more frequent in the stressed environments suggesting preponderance of this category of wheat under global warming. It could lower strength and quality of gluten as dough quality is adversely in 2+12 wheats. But, Mohan and Gupta (2013a) observed that 2+12 wheats have more gluten content in comparison to 5+10 as gluten per unit protein is higher in this category of wheat. Since gluten content is the main quality component in bread making (Mohan *et al.,* 2013b), they opined that higher gluten content in 2+12 wheats can amend the decline caused by lowering of gluten quality. They have envisaged shift in the route to product quality under elevated temperature conditions which might need special selection criteria for identifying product superior varieties of 2+12 category wheats.

Altered pest and disease dynamics: Plant diseases are one of the important factors which have a direct impact on global agricultural productivity and climate change will further aggravate the situation (IPCC, 2007). Combined infestation of pests and diseases in plants could result up to 82 per cent losses in attainable yield in case of cotton and over 50 per cent losses for other major crops and if we combine these losses with post-harvest spoilage and deterioration in quality; these losses become critical particularly for resource poor regions of the world (Oerke, 2006). Further, plant diseases are estimated to cause yield reduction of almost 20 per cent in the principal food and cash crops worldwide (Thind 2012). Local weather conditions such as rain, temperature, sunshine and wind in combination with locally adapted plant varieties, cropping systems and soil conditions can maximise food production as long as plant diseases can be controlled. Changes in temperature and precipitation regimes due to climate change may alter the growth stage, development rate and pathogenicity of infectious agents, and the physiology and resistance of the host plant (Chakraborty *et al.,* 1998 and Chakraborty and Datta, 2003). A change in temperature could directly affect the spread of infectious diseases and their survival between seasons. There are indications of increased aggressiveness at higher temperatures of stripe rust isolates (*Pucciniastriiformis*), suggesting that rust fungi can adapt to and benefit from higher temperatures (Mboup 2012). Climate change is also reported to cause a shift in the geographical distribution of host pathogens (Mina and Sinha 2008). A change in temperature may favour the development of different dormant pathogens, which could induce an epidemic. Increase in temperature with sufficient soil moisture may increase evapo-transpiration resulting in humid microclimate in crops and may lead to incidence of diseases favoured under these conditions (Mcelrone *et al.,* 2005). Diseases such as common bunt (*Tilletia caries*) and Karnal bunt (*Tilletia indica*) in wheat can be of importance under changing climatic conditions in regions with low productivity if proper seed treatment is not followed in this crop (Oerke, 2006). Temperature can also affect disease resistance in plants, thus affecting the incidence and severity of the diseases. Temperature sensitivity to resistance has been reported for leaf rust (*Puccinia recondita*) in wheat (Gregory *et al.,* 2009). In India, wheat, rice and potato are important crops. Analysis of the last 30 years historical weather data from different locations of Punjab has indicated that significant changes have occurred in the weather causing early warming in February. Such climatic changes will influence wheat crop, which is of importance in the region. Temperature affects the growth of the crop, host–pathogen interactions and will alter the susceptibility window. Temperature rise above normal and increased humidity will predispose the crop to severe brown rust (*P. recondita*) infection and abet facultative pathogens. In wheat belt of India in Punjab, while change in temperature and humidity will reduce the importance of yellow rust (*P. striiformis*) and Karnal bunt (*T. indica*); the importance of leaf rust, foliar blights, *Fusarium* head blight and stem rust may increase in the future, particularly in the absence of resistance in wheat cultivars (Kaur *et al.,* 2008) . The incidence if pre-harvest sprouting has also increased in the eastern parts of the country. On the other hand, the importance of leaf rust, foliar blights *Fusarium* head blight and stem rust may increase in Punjab in the future. In plant diseases, there are some pathogens of major crops which have a huge potential to cause losses in those crops. In wheat, *Sr*31 stem rust resistance has been effective in cultivars for over 30 years, which can be overcome by new races of *Pucciniagraminis f. sp. tritici* like *Ug*99. According to estimates, *Ug*99 race of the stem rust can result in up to 10 per cent yield loss in Asia alone, amounting to US$ 1–2 billion per year (Duveiller *et al.,* 2007).

Climate change and wheat production: Understanding how different climate factors interact and impact food production (Savary *et al.,* 2012) is essential when reaching decisions on how to adapt to the effects of climate change. To implement such strategies the contribution of various climate variables on crop yields need to be separated and quantified. Wheat yields are projected to be affected by climate change in India. The variable wheat growing zones represent different factors of changing climate which finally may lead to reduced wheat production. The North Western Plains Zone (NWPZ), which is the main wheat growing and producing area, is prone to changes in both temperature and rainfall. Wheat crop in this region is affected by both initial and terminal heat stresses. Heat stress at the grain filling stage can lead to severe loss in wheat yields in this zone. Recent changes in the rainfall pattern are also going to impact wheat production. There is also threat of stripe rust which in epidemic form can cause 100 per cent loss in yield. The North Eastern Plains Zone (NEPZ), is the second largest zone and mostly rainfed. This zone is mainly affected by leaf rust, leaf blight diseases and is also prone to heat stress throughout growing season. Similarly the Central Zone (CZ) and Peninsular Zone (PZ) are mostly rainfed areas and the wheat crop is under heat stress throughout the wheat season. Stem rust is a major threat in these areas. Also the crop duration is a major factor here. Any reduction in growth period will reduce the wheat yields in CZ and PZ.

However in spite of all these factors and impact of climate change across the globe, Indian wheat production has been increasing for last three to four years continuously. This continuous increase in yields can be attributed to several management practices such as timely control of diseases and pests and slight changes in climatic factors. Temperature and rainfall occurrences at a particular stage of wheat crop can lead to increase or decrease in yield. For example, warmer temperatures at tillering will advance anthesis dates, but cooler temperatures during grain filling will delay maturity dates, resulting in higher yields. These warmer-to-cooler and cooler to warmer thermal variations create disparity even though the overall mean temperature remains normal.

This increase in wheat production of India is, however, contrary to the predicted yields for cereal crops like rice and wheat. Sinha and Swaminathan (1991) showed that an increase of 2°C in temperature could decrease the rice yield by about 0.75 ton/ha in the high yield areas; and a 0.5°C increase in winter temperature would reduce wheat yield by 0.45 ton/ha. Rao and Sinha (1994) showed that wheat yields could decrease between 28 to 68 per cent without considering the CO_2 fertilisation effects; and would range between +4 to -34 per cent after considering CO_2 fertilisation effects. Aggarwal and Sinha (1993) using WTGROWS model showed that a 2°C temperature rise would decrease wheat yields in most places. Saseendran *et al.*, (2000) showed that for every one degree rise in temperature the decline in rice yield would be about 6 per cent. All these studies focused only on agronomic impacts of climate change. Negative impacts of climate change are projected to be less severe in low-emission scenarios than in high-emission scenarios. The magnitude of uncertainty varies spatially and increases with time. Differences in sowing time are one of the major reasons for variable impacts on yield. Late-sown areas are projected to suffer more than the timely-sown ones. Considerable spatial variation in impacts is projected. Warmer central and south-central regions of

India may be more affected. Despite CO_2 fertilization benefits in future climate, wheat yield is projected to be reduced in areas with mean seasonal maximum and minimum temperatures in excess of 27°C and 13°C, respectively.

Strategies for Adapting Wheat to Climate Change

Climate change is a reality now and its affects have been showing on Indian agriculture. With global predictions of reduction in wheat yields due to temperature rise getting realised in some European countries, there is need for preparedness to this challenge. The diverse agro-climatic zones of wheat in India will make the combating strategies a herculean task. Taking into stride the increasing wheat production in India we should also embark on developing climate smart strategies to deal to the unpredictable nature of changing climate. Simple adaptation options, such as change in sowing times, and increased and efficient use of inputs, could not only offset yield reduction, but could also improve yields. Converting late-sown areas into timely-sown regions could further significantly improve yield even with the existing varieties in the near future. However, some regions may still remain vulnerable despite the adaptation interventions considered. Therefore, emphasis should be for intensive, innovative and location-specific adaptations to improve wheat productivity in the future climate.

Conservation Agriculture (CA): Conservation agriculture practices such as zero-till in particular have been promoted for their potential to mitigate climate change. The potential for zero-till alone to mitigate climate change by sequestering carbon may be less than previously thought, but other aspects of CA have mitigation potential. The water and soil conserving effects of CA help to stabilise yields against weather extremes. Often, CA increases average yields in the long term. CA increases soil water content by increasing infiltration and reducing runoff and evaporation. Increased infiltration improves water use efficiency and buffers crops against drought. Mulch cover also buffers the soil against temperature extremes. Residue retention increases the potential for carbon sequestration by increasing biomass inputs to the soil. Permanent soil cover with crop residues or mulch provides a constant source of fresh organic material. Rotating with a leguminous crop helps balance the carbon-to-nitrogen ratio of crop residues, which allows nitrogen from decaying surface residues to be released slowly and serve as a source for the following crop. Increasing soil carbon only mitigates climate change if it represents an additional net transfer of carbon from atmosphere to land. The most common cropping system in the Indo-Gangetic Plain of India is a rice-wheat rotation, which generally requires intensive tillage before rice is planted. However, farmers in the Northwest Indo-Gangetic Plain have begun adopting a zero-tillage system for the wheat crop. Wheat planting is done with a tractor-drawn zero-till seed drill that plants seeds directly into unploughed fields, sometimes applying fertiliser at the same time, which eliminates the need for multiple tractor passes. Zero-till thus reduces the turnaround time between rice and wheat crops and allows timelier planting of wheat. On-farm trials have shown yield gains between 1 per cent and 15 per cent (0.05 and 0.63 tons per ha), due to timelier planting and decreased emergence of an herbicide-resistant weed that is a particular problem in the area, even though other weeds were more abundant.

Climate smart agriculture: Development of weather smart agriculture by devising techniques for growing wheat in niche areas can help mitigate the climate change impact. Similarly input smart wheat production methods can also help in enhancing wheat production. Techniques for saving inputs like water and methods for applying fertilisers at a particular stage of the crop can reduce the stress on climate change.

Physiological interventions: Physiological interventions such as hormonal sprays and application of Potash and Silica based fertilisers has been recommended for mitigating the effect of temperature on wheat crop.

Stress Tolerant/resistant varieties: Development of heat and drought tolerant wheat cultivars is the most economical and eco-friendly method. At the same time breeding wheat lines for resistance to the predicted pest/insect and diseases can be a way forward. However, the unpredictability of the climate change across time and space makes it a difficult option.

Shuttle breeding: Shuttle breeding is a method of wheat improvement under which the breeding material is screened/evaluated at different locations with variable climatic conditions. This will help in developing stable genotypes which will be less affected by the climatic adversaries.

Conclusion

Climate change is projected to affect wheat yields across the globe. The affect can be negative as well as positive. However, most of the predictions have shown a reduction in wheat yields across the globe. In Indian scenario till now we have been benefitting from this change. There is an increasing trend in wheat production of the country for last four to five years. This increase can however be attributed equally to the management strategies and favourable climatic conditions. The global realisation of climate change and its visualisation in the form of decreasing global yields presses for development of climate smart wheats. Conservation agricultural practices at present have been shown to be effective in mitigating the impact of climate changes. However, development of stress tolerant/resistant wheat cultivars is the best way forward.

References

Aggarwal, P.K. and S.K. Sinha (1993). Effect of probable increase in carbon dioxide and temperature on productivity of wheat in India. Journal of Agricultural Meteorology, 48: 811-814.

Aggarwal, P.K., N. Kalra, and S. Chander, *et al.*, (2006). InfoCrop: A dynamic simulation model for the assessment of crop yields, losses due to pests, and environmental impact of agro-ecosystems in tropical environments. I. Model description. AgricSyst, 89:1–25.

Aggarwal, P.K. (2008). Global Climate Change and Indian Agriculture: Impacts, adaptation and mitigation. Indian Journal of Agricultural Sciences, 78(10): 911 - 919.

Chakraborty, S. and S. Datta (2003). How will plant pathogens adapt to host plant resistance at elevated CO_2 under a changing climate? New Phytologist, 159: 733–742.

Chakraborty, S. *et al.*, (1998). Potential impact of climate change on plant diseases of economic significance to Australia. Australas Plant Pathology, 27: 15–35.

Conroy, J.P., S. Seneweera, A.S. Basra, G. Rogers and B. Nissen-Wooller (1994). Influence of rising atmospheric CO_2 concentrations and temperature on growth, yield and grain quality of cereal crops. Australian Journal of Plant Physiology 21(6): 741-758.

Duveiller, E., R.P. Singh and J.M. Nicol (2007). The challenges of maintaining wheat productivity: pests, diseases, and potential epidemics.Euphytica 157, 417–430.

Erda, L., X. Wei, J. Hui, X. Yinlong, L. Yue, B. Liping and X. Liyong (2005). Climate change impact on crop yield and quality with CO_2 fertilisation in China. Philosophical Transitions, Royal Society London, Biological Science 360 (1463): 2149-2154.

Gbetibouo, G. and R. Hassan (2005). Economic impact of climate change on major South African field crops: A Ricardian approach. Glob Planet Change 47:143–152.

Gregory, P.J., S.N. Johnson, A.C. Newton and J.S.I. Ingram (2009). Integrating pests and pathogens into the climate change/food security debate. Journal of Experimental Botany 60: 2827–2838.

Hogy, P. and A. Fangmeier (2008). Effect of elevated atmosphere CO_2 on grain quality of wheat. Journal of Cereal Science 48: 580-591.

Houghton, J.T. (1991). Scientific assessment of climate change: summary of the IIPC working group I report. In: J Jager and HL Ferguson (eds.) Climate change: science, impact and policy. Cambridge University: 23-40.

Intergovernmental Panel on Climate Change, Climate Change (IPCC), The Fourth IPCC Assessment Report, Cambridge University Press, Cambridge, UK, 2007.

IPCC (2007). IPCC Fourth Assessment Report: Climate Change 2007 (AR4). Geneva: IPCC.

IPCC (2007b). Summary for Policymakers. In: Climate Change 2007: Impacts, Adaptation and Vulnerability. Contribution of Working Group II to the Fourth Assessment Report of the Intergovernmental Panel on Climate Change, M.L. Parry, O.F. Canziani, J.P. Palutikof, P.J. van der Linden and C.E. Hanson, Eds., Cambridge University Press, Cambridge, UK, 7-22.

Kaur, S., L. Dhaliwal and P. Kaur (2008). Impact of climate change on wheat disease scenario in Punjab. Journal of Research 45(3): 161–170.

Mboup, M., B. Bahri, M. Leconte, C. De Vallavieille-Pope,O. Kaltz and J. Enjalbert (2012). Genetic structure and local adaptation of European wheat yellow rust populations: the role of temperature-specific adaptation. Evolution Applications 5: 341–352.

McElrone, A.J., C.D. Reid, K.A. Hoye, E. Hart and R.B. Jackson (2005). Elevated CO_2 reduces disease incidence and severity of ared maple fungal pathogen via changes in host physiology and leaf chemistry. Global Change Biology 11: 1828–1836.

Mina, U. and P. Sinha (2008). Effects of climate change on plant pathogens. Environ. News 14(4): 6–10.

Mohan, D. and R.K. Gupta (2013a). An analysis to visualise influence of global warming on wheat grain quality with high molecular weight glutenin subunits at *Glu D1* Locus. Molecular Plant Breeding 5(7): 36-42.

Mohan, D. and R.K. Gupta (2011). Harness value-addition in bread wheat (*Triticum aestivum* L.) through genotype and location specificity in highly productive north western Indo-Gangetic plains. Indian Journal of Agricultural Sciences 81(5): 433-437.

Mohan, D. and R.K. Gupta (2013b). Analysing grain properties of Indian bread-wheat cultivars for defining route to end-product quality and key attributes for selection. Indian Journal of Genetics and Plant Breeding 73(4): 1-8.

Oerke, E.C. (2006). Crop losses to pests. Journal of Agricultural Science 144: 31–43.

Ortiz, R., K.D. Sayre, B. Govaerts, R. Gupta, G.V. Subbarao, T. Ban, D. Hodson, J.M. Dixon, J. Ivan Ortiz-Monasterio and M. Reynolds (2008). Climate change: Can wheat beat the heat? Agriculture, Ecosystems and Environment 126(1-2): 46-58.

Paroda, R.S. and P. Kumar (2000). Food production and demand in South Asia. Agr Econ Res:1-25.

Rao, A. and A.K. Sinha. (1994). Climate changes and agriculture. Nature 437: 102- 109.

Saseendran, R.M., I.M. Smith and P.A. Matson (2000). Ecological and evolutionary responses to climate change. Science 284: 1943-1947.

Savary, S., A. Ficke, J.N. Aubertot and C. Hollier (2012). Crop losses due to diseases and their implications for global food production losses and food security. Food Security 4: 519–537.

Sinha, A.K. and M.S. Swaminathan (1991). Long-term climate variability and changes. Journal of Indian Geographical Union 7(3): 125-134.

Thind, T.S. (2012). Fungicides in crop health security. Indian Phytopathology 65(2): 109–115.

Wrigley, C. (2006). Global warming and wheat quality. Cereal Foods World 5(1): 34-36.

Wu, D., G. Wang, Y. Bai and J. Liao (2004). Effects of elevated CO_2 concentration on growth, water use, yield and grain quality of wheat under two soil water levels. Agriculture, Ecosystems and Environment 104(3):493-507.

Table 1: Grain quality in wheats of different zones.

Characteristics	NWPZ (124)	NEPZ (129)	CZ (88)	PZ (81)	NHZ (81)
Bread loaf volume (cc)	550	556	557	580	539
Chapati score	7.52	7.61	7.82	7.65	7.35
Flour recovery (%)	68.8	69.8	69.9	70.3	64.3
Grain appearance score	5.8	5.8	6.5	6.1	6.0
Grain protein (%)	12.1	11.5	11.7	12.3	10.8
Test weight (kg/hl)	77.9	77.8	81.5	80.6	79.2
Kernel weight (mg)	37.5	38.6	42.1	40.4	40.6
Sedimentation value (ml)	43	44	41	44	42
Wet gluten content (%)	30.8	29.4	31.6	32.7	26.9
Gluten index	61	62	53	60	60
Grain hardness index	77	75	75	76	68
Yellow pigments (ppm)	3.37	3.32	2.63	2.76	3.60

Figure in parenthesis denotes total genotypes evaluated for the zone under irrigated conditions

Table2: Wheat quality in timely and late sown wheats of different region

Parameters	Indo-Gangetic Plains		Central-Peninsular India	
	Timely sown (133)	Late sown (120)	Timely sown (71)	Late sown (98)
Bread loaf volume (cc)	561	565*	561	572**
Chapati score	7.57	7.56	7.73	7.75
Flour recovery (%)	69.3	69.1	70.0	70.1
Grain appearance score	5.9	5.8	6.3	6.3
Grain protein (%)	11.6	12.0**	11.5	12.4**
Test weight (kg/hl)	78.5**	77.3	81.5**	80.8
Kernel weight (mg)	39.5**	36.9	43.2	39.9
Sedimentation value (ml)	44	44	40	44**
Wet gluten content (%)	29.4	30.9**	30.8	33.1**
Gluten index	62	61	53	58**
Grain hardness index	76	75	76	75
Yellow pigments (ppm)	3.39	3.29	2.73	2.67

2

Challenges of Climatic Change on Productivity of Groundnut

K. Chakraborty, D. Bhaduriand Radhakrishnan T.

Abstract

Climate change is one of the most important factors limiting agriculture in modern era. The change in climate over the years has usually been thought of as a natural phenomenon. However, its rapid and 'unnatural' changes, and the spatial and temporal variabilities have assumed greater importance in recent years over the world. As India is being considered among the affected countries of the world, so it is very much likely to face climatic challenges especially in agricultural sector. Climate change has both the climatic and biological effects of elevated CO_2 and temperature on plant and soil, representing the most important constituents of the agricultural ecosystem. Rising CO_2 promotes photosynthetic carbon assimilation and adversely affects the photo-respiratory activity with alteration in stomatal behaviour. Higher level of atmospheric CO_2 is likely to affect C_3 and C_4 plants differentially, through enhancing the growth rates of all plant species, but much higher enhancement in C_3 plants. On the contrary, rise in temperature may have negative effect on plants as a whole and its increase beyond critical limit would results in retardation of growth, physiological development and suppression metabolic activities. The C_4 plants, being adapted to warmer environment may show less harmful effects than C_3 plants. Thus, a mutual benefit to both types of plants may nullify the effect of climate change as far as the plant processes are concerned. Being a C_3 crop groundnut is also very much vulnerable to both the elevation of atmospheric CO_2 level as well as increase in air temperature. As both plant and soil processes are interrelated having a cumulative impact on agriculture scenario; simultaneous changes in soil processes may have direct relation under changing *climate scenario with respect to soil carbon fractions, nutrient cycling and microbial activities target towards alteration in soil quality, soil water balance, salinity etc.*

Introduction

Climate change is now the single gravest threat in the current century. Its impact is being intensely felt by the countries which largely depend on agriculture, and India being recognised among the worst-hit countries of the world. In this review, both the climatic and biological effects of elevated CO_2 and temperature on plant and soil, representing the most important constituents of the agricultural ecosystem have been discussed. Rising CO_2 promotes photosynthetic carbon assimilation and adversely affects the photo-respiratory activity with alteration in stomatal behaviour. Higher level of atmospheric CO_2 is likely to affect C_3 and C_4 plants differentially. Atmospheric CO_2 enrichment enhances the growth rates of all plant species but enhancement is much significant in C_3 plants. The C_4 plant system represents an evolutionary adaptation to low CO_2 environment with lower CO_2 compensation point. On the contrary, rise in temperature may have negative effect plants as whole and its increase beyond critical limit would results in retardation of growth, physiological development and suppression metabolic activities. The C_4 plants, being adapted to warmer environment may show less harmful effects than C_3 plants. Thus, a mutual benefit to both types of plants may nullify the effect of climate change as far as the plant processes are concerned.

The change in climate over the years has usually been thought of as a natural phenomenon. These rapid and unnatural changes off late have been attributed mostly to anthropogenic factors, in sharp contrast to the natural course of climate change. People have released chlorofluorocarbons (CFCs) into the atmosphere, thereby degrading stratospheric ozone and increasing biologically harmful UV radiation that reaches the earth's surface. Through mining and combustion of fossil fuels, deforestation, maintenance of livestock herds, and even through crop cultivation, people have released enormous quantities of CO_2, CH_4, and other "greenhouse" gases into the atmosphere. There is growing evidence of increase in atmospheric CO_2 concentration from a base value of 280 to a level of 379 μ mol mol^{-1} (IPCC, 2007). During the last 12 years, its increase has been 1.9 μ mol mol^{-1} year^{-1} and is predicted to be as high as 543 μ mol mol^{-1} by the middle of this century. Accordingly, as a consequence of this rise in CO_2, a warmer earth (rise in earth's average temperature from 1.8 to 4.0°C; IPCC, 2011) is predicted. This might possibly exerts a significant impact on agricultural productivity. Being sub-tropical climatic conditions Uttar Pradesh and other north and central Indian states face extreme weather throughout the year, which is characterised by extremely hot and dry summer, humid monsoon and chilling winter period. Hence, these parts of our country is vulnerable to changing climatic conditions viz. increase in mean day/night temperature, erratic rainfall and of course increasing levels of atmospheric CO_2 levels.

The international effort to address climate change began since two decade with the implementation of the new programme called United Nation's Framework Convention on Climate Change (UNFCC) in 1992. The objective of the convention was stabilisation of greenhouse gas concentration in the atmosphere. Anthropogenic activities are creating continuous changes on global climate and ecosystem. Hence, emissions of greenhouse gases including CO_2 and others are increasing day by day. The rate of global climate change and warming expected over the next century is more than that occurred in the past 10,000 years (Reddy and Hodges, 2000). Global climate change will have immense effects and consequences on natural and agricultural ecosystems. Agricultural productivity is particularly vulnerable to the disruption of global climate changes. Increasing concentration of the gases such as atmospheric carbon dioxide [CO_2], methane [CH_4], nitrous oxide [N_2O], and sulphur dioxide [SO_2] have direct impact on crops, and on the other hand these are also critical in increasing air temperature (Table 1) (Sharma and Bazaz, 2012). The increase in the CO_2 concentration of the atmosphere is responsible for global warming as it absorbs long wave heat radiations from earth's surface and may also affect cloudiness and precipitation. This is referred to as climatic effect. There is considerable uncertainty associated with the changes in the temperature of earth's atmosphere with the elevation of these atmospheric gases.

Table 1: Rise in greenhouse gas concentration in atmosphere as influenced by anthropogenic activities (Source: IPCC Reports)

Major GHGs	*CO_2*	*CH_4*	*N_2O*	*CFC-12*
Pre-industrial atmospheric concentration	280 ppmv	0.70 ppmv	280 ppbv	0
Current concentration	379 ppmv	1.72 ppmv	320 ppbv	500 pptv
Current Annual increase (per cent)	0.5% (1.5–1.8 ppmv)	0.8% (0.013 ppmv)	0.25% (0.75 ppbv)	4% (18-20 pptv)
Atmospheric life time (years)	50-200	12-17	150	102
Global warming potential relative to CO_2	1	24.5	320	10000

Plants have been directly affected by climate change, especially by rising atmospheric CO_2 and temperature. Higher level of atmospheric CO_2 is likely to affect C_3, and C_4 plants differentially. Atmospheric CO_2 enrichment typically enhances the growth rates of all plant species but enhancement is very significant in case of C_3 plants. C_4 system represents an evolutionary adaptation to low CO_2 environment and also has lower CO_2 compensation point. But, till now our knowledge of the climate effects of the rising CO_2 and other greenhouse gases in the atmosphere is inadequate for initiating any global attempt to formulate effective mitigation strategies. If the climate does change, some warming could be tolerated, and may even be beneficial with no reductions in food production. A warming trend may encourage farmer adaptations and favour the introduction of new technologies and cultural practices. So, it is very much important to gather as much knowledge as possible in this regard to combat against the future climate change and develop effective mitigation and adaptation strategies.

Most of the research on plant responses to high CO_2 and temperature have been conducted under laboratory in green houses or controlled field conditions (Bazzaz and Garbutt, 1988) and the technologies used based on the single component approach, have been successfully applied to study the responses of plants to increased CO_2 and temperature separately. Results obtained with isolated plants growing under controlled growth conditions, life phase synchrony, absence of organismic interactions and provision of luxurious amounts of mineral nutrients are the major limitation for the application of results of a large number of climate change studies. Thus, there is need to address the recent developments of impact of climate change on various plant and soil processes which could determine the fate of agricultural scenario in near future.

Groundnut (*Arachishypogaea* L.) is one of the major oilseed and food crops grown in subtropical and tropical regions of the world. It is grown in different rainfall and temperature regimes on a variety of soils. Being a C_3 crop, higher temperatures and other climatic factors may affect its productivity and to some extent its distribution (Kumar *et al.*, 2012). The possible impact of climate change on physiology and growth of groundnut would be as follows:

Germination and Seedling Establishment

As the optimum temperature for germination the optimum temperature range is from 25-30°C and higher temperature (upto 35°C) during this stage may lead to faster germination and seedling establishment (Awal and Ikeda, 2002; Prasad *et al.*, 2006).

Vegetative Development

Differential response among Spanish and Virginia genotypes to heat stress may be observed, as both habit groups differ with respect to their base temperature values. The plants grown in comparatively warmer soil produced more leaves on their branches than on the main axis. This would lead to better initial plant vigor for establishment by capturing more light and CO_2. Under higher CO_2 condition both leaf area and leaf weight is expected to increase which would directly affect the plant architecture and light use efficiency (Awal and Ikeda, 2002).

Canopy Expansion and Growth

In higher temperature regime leaf growth is much faster which increase SLA of the plant. Both high CO_2 and temperature is expected to increase the leaf area in groundnut. Under drought conditions, elevated CO_2 increases canopy development in terms of higher LAI. Elevated CO_2 benefits the crop growth under both water limiting and non-limiting conditions; however, the relative benefits are more under water limiting conditions (Clifford *et al.,*1993). Studies showed elevated CO_2 accelerates leaf senescence in groundnut (Hardy and Havelka,1977).

Stomatal Conductance and Transpiration

Stomatal conductance and transpiration rates increase with the increase in temperature and decreases with the increase in CO_2 concentration. Elevated CO_2 decreases stomatal frequency on both the surfaces of leaves up under irrigated as well as drought conditions (Clifford *et al.,*1995).

Photosynthesis

With increase in mean daily temperature upto 30-35°C, the net photosynthesis rate increases in groundnut, beyond that rate of photosynthesis is likely to decrease. Increase in atmospheric levels of CO_2 is likely to increase net photosynthesis rate in groundnut as current levels of atmospheric CO_2 do not saturate Rubisco. Interactive effect of both elevated CO_2 and temperature in future climatic condition may nullify each other's effect. Both elevated CO_2 and temperature may increase the growth rate (both RGR and NAR) in groundnut (Rao, 1999).

Reproductive Development

Temperature plays most crucial role in reproductive development in any crop if it is otherwise not limited by any other external factor. Increase in temperature may lead to early flower initiation and onset of reproductive phase by shortening the vegetative phase. Number of flower and also per week flower production is likely to be increased with elevated temperature. As number of flower and number peg production in groundnut is highly correlated, hence increase in temperature below the detrimental level would increase number of pods per plant. Both hot days and warm nights reduce the pollen production and its viability in groundnut (Prasad *et al.,*1999). Elevated CO_2 is likely to have positive effect on reproductive development as under elevated CO_2 condition both availability and supply of net assimilate for reproductive development will be more.

Total Dry Matter, Pod and Seed Yield

Studies showed both pod and seed development is delayed beyond the optimal growing temperature in groundnut. Elevated CO_2 have little effect on crop duration but the size and dry matter accumulation in pods was found more under elevated CO_2 condition (Bagnall *et al.*, 1991).

Harvest Index and Shelling Percentage

Studies showed that in groundnut pod and seed harvest indices at harvest maturity were significantly affected by temperature, but not by CO_2. Severe reduction in harvest index was also recorded when the crop faces high temperature stress throughout its growth stage. Partitioning of biomass towards reproductive parts tends to reduce with increase in daily air temperature beyond certain limit. High temperature decreases the partitioning of dry matter to seeds which results in low shelling percentage (Prasad *et al.*, 2003).

Yield Loss due to Temperature

About 90 per cent of the world's peanut production occurs in the tropical and semi-arid tropical regions, which are characterised by high temperature and low or erratic rainfall. In the tropics, most of the crops are near their maximum temperature tolerance; therefore, crop yields may decrease even with minimal increases in temperature. Both short- and long-term exposure to air and soil temperatures above optimum can cause up to 50 per cent yield loss in peanut. Day temperature more than 34°C decreased fruit-set and resulted in fewer numbers of pods. Decreased fruit-set at high temperatures was mainly due to poor pollen viability, reduced pollen production and poor pollen tube growth, all of which lead to poor fertilisation of owers. Increasing daytime temperature from 26–30°C to 34–36°C signicantly reduced the number of subterranean pegs and pods, seed size and seed yield by 30–50 per cent (Cox, 1979; Ong, 1984).

Overall temperature plays major role in determining crop duration in groundnut. The length of the crop growth period is determined by growing degree days. So, rise in daily mean temperature will lead to reduction in crop duration which in turn will shorten the crop development stage (variable vegetative stage) and thus will result in decrease in crop yield. As per IPCC projection the mean global temperature could rise between 1.8–4.5°C by 2050 which can reduce groundnut yield in the range of 30-35 per cent if existing cultivars are grown.

Recommendation for Cultivation of Crop in view of Climate Change

As temperature rise is obvious in future climatic conditions, development and cultivation of thermo-tolerant varieties needed. Selection of responsive genotypes to higher levels of atmospheric CO_2 is required to increase the crop productivity. Cultivation of short duration crop with high per day productivity would be more remunerative for farmers with erratic and uncertain pattern of rainfall with change in climate. Development of crop with high initial growth rate and better water use efficiency. Efficient crop and nutrient management practices need to developed in view of the differential response under high temperature and elevated CO_2 conditions to meet up the crop demand in future climate.

Effects of Climate Change on Soil Processes

Soil provides mechanical support and providing essential nutrient ions to them to plant roots, and also it is an abode of several biological creatures. Various physical, chemical, and biological processes are continuously happening within the soil in favour of plant growth. However, in a changing climatic situation some of the soil-forming factors and processes are assumed to be altered. Climate change, be it elevation of atmospheric CO_2 concentration or fluctuation of temperature or integration of both can alter the indigenous either or all soil processes. In some areas, warmer conditions are expected, in others, lower temperatures or no change at all; precipitation and wind velocity can also behave differently. Soil, an indispensible unit of this ecosystem, should react under this drastic behaviour of climate. Long-lasting shifts in soil moisture and temperature regimes in view of increasing atmospheric CO_2 concentration and temperature indifferent parts of the globe are matter of concern. Chemical and biological processes are more prone to these climatic changes with an obvious impact on soil fertility status. Rising temperature can hasten the rate of organic matter decomposition, release of organic carbon from the soil into the atmosphere and thereby accelerate the status of organic matter by carbon sequestration.

Major factors of climate change that will surely have an impact in soil quality are: increasing temperatures, elevated CO_2 concentration, variability in the amount, intensity and distribution of rainfall, and increasing atmospheric N deposition and land use change (Allen *et al.,* 2011). Though several indicators of different agro-ecosystem were considered, but key indicators affected by climate change include aggregate stability (Salvador *et al.,* 2008), soil organic matter (Richter *et al.,* 2007; Kuzyakovand Gavrichkova, 2010), carbon and nitrogen cycling (Weil and Magdoff, 2004), microbial biomass and activity (Rinnan *et al.,* 2007; Ritz *et al.,* 2009), and microbial fauna and flora diversity (Briones *et al.,* 2009) using controlled environment and long-term research experiments.

Thus, the overall the major impacts on soil health resulting due to climate change would be in the organic matter supply, declining SOC level, temperature regimes, hydrology and salinity. Increase in soil temperature will decline soil available-N through processes such as mineralisation, volatilisation and denitrification.

Soil Carbon Stock

The much-talked effect of climate change is concerned with the soil organic C stock. The effect of rising global temperatures on soils has already begun to thaw permafrost, the stored carbon and organic nutrients become available for microbial decomposition, by releasing CO_2 into the atmosphere. Other climatic-driven factors such as flooding, drought, freezing etc. and better soil management (for example, conservation tillage) could also modify the soil carbon stocks. As the N concentration in plant tissues is widely reported to reduce under elevated CO_2, the C:N ratio of the crop residues is increased, which may further reduce their rate of decomposition, hence nutrient supply to plants in the long-term and possibility to cause more C sequestration in soil (Viswanath *et al.*, 2010). Whereas high temperature also may play role here by changing soil microbial activity, by accelerated rates of microbial decomposition, thus increasing CO_2 emitted by soil respiration leads to large soil carbon losses, thus prevails the uncertainty in climate change predictions for soil C-stock (Briones *et al.*, 2004).

So, it is essential to the development of best management practices to enhance carbon sequestration in soils by managing optimum levels of soil organic matter through crop rotation, fertility maintenance, use of organic manures and inorganic fertilisers, tillage methods, along with other cropping system components.

Nutrient Cycling

Though increasing atmospheric CO_2 concentrations are expected to intensify plant growth and yield, the positive effect of CO_2 fertilisation on crop yield is achieved only when the soil is supplemented by adequate soil available nutrients. This issue could be of great concern to Indian soils where most of the states are low in N and P availability. Even an altered synergism or antagonism also controls the macro- and micronutrients (N, P, Zn, Fe etc.) availability in soil under changing climate scenario. Increased plant's P requirement under the rising atmospheric CO_2 and temperature caused by higher uptake of P despite the reduced yield (Manoj-Kumar *et al.*, 2012). On the other hand, soil N-dynamics was also affected under elevated CO_2 and temperature as NO_3^--N, potential nitrification activity, microbial biomass-N and population of denitrifiers were reduced (Rakshit *et al.*, 2012).

Microbial Processes

Microbial processes, often dependent on environmental factors (temperature, moisture, and nutrient availability) are likely to be affected by climate change. However, the effects are multiple and complex and often difficult to measure. Enhanced microbial activity observed under elevated CO_2 due mainly to increased carbon flux in the rhizosphere (IPCC, 2007). A greater root mass and root exudation, caused by more litter or crop residues, increases mycorrhizal colonisation and activity of other rhizosphere or soil-inhabited microorganisms (including symbiotic and root-zone N_2 fixers). On the other hand, higher soil microbial-cum-root activity would imparts higher CO_2 partial pressure in soil air and CO_2 activity in soil water, hence increased rates of plant nutrient release (e.g., K, Mg, micronutrients) from weathering of soil minerals (Brinkman and Sombroek, 1996). Enhancement of soil respiration rates and plant responses attributed to an interactive effect between warming, decline in soil moisture, limitations in soil nutrient supply for plants, decline in the availability of labile C sources for soil microbes (Arft, 1999). Soil biodiversity and its functions (enzyme activity) are expected to change under conditions of elevated CO_2 and changed moisture and temperature regimes.

References

Allen, D.E., B.P. Singh and Dalal, R.C. 2011. "Soil Health Indicators under Climate Change: A Review of Current Knowledge", In: Singh BP *et al.*, (eds) Soil Health and Climate Change, Soil Biology 29, Springer-Verlag, Berlin Heidelberg, pp. 25-45.

Arft, A.M. 1999. "Response patterns of tundra plant species to experimental warming: a meta-analysis of the International Tundra Experiment", Ecological Monograph, Vol. 69, pp. 491–511.

Awal, M.A. and T. Ikeda. 2002. "Effects of changes in soil temperature on seedling emergence and phenological development in field-grown stands of peanut (*Arachishypogaea*)", Environmental and Experimental Botany, Vol. 47, pp. 101-113.

Bagnall, D.J. and R.W. King. 1991. "Response of peanut (*Arachishypogaea*) to temperature, photoperiod, and irradiance. 2. Effect on peg and pod development", Field Crops Research, Vol. 26, pp. 279-293.

Bazzaz, F.A. and K. Garbutt. 1988. "The response in annuals in competitive neighbourhoods: effects of elevated CO_2", Ecology, Vol. 69, pp. 937-946.

Brinkman, R. and W.G. Sombroek. 1996. "The effects of global change on soil conditions in relation to plant growth and food production", In: Bazzaz F, Sombroek W. (eds) Global Change and Agricultural Production. FAO of the United Nations. John Wiley and Sons, New York. Available at: www.fao.org/docrep/W5183E/w5183e05.htm

Briones, M.J.I., N.J. Ostle, N.P. McNamara and J. Poskitt. 2009. "Functional shifts of grassland soil communities in response to soil warming", Soil Biology and Biochemistry, Vol. 41, pp. 315–322.

Briones, M.J.I., N.J. Ostle and J. Poskitt. 2004. "Influence of warming and enchytraeid activities on soil CO_2 and CH_4 fluxes", Soil Biology and Biochemistry, Vol. 36, pp. 1851-1859.

Clifford, S.C. *et al.,* 1995. "The effect of elevated atmospheric CO_2 and drought on stomatal frequency in groundnut (*Arachishypogaea* L.)", Journal of Experimental Botany, Vol. 46, pp. 847-852.

Clifford, S.C. *et al.,* 1993. "The effects of elevated atmospheric carbon dioxide and water stress on light interception, dry matter production and yield in stands of groundnut (*Arachishypogaea* L.)", Journal of Experimental Botany, Vol. 44, pp. 1763-1770.

Cox, F.R. 1979. "Effect of temperature treatment on peanut vegetative and fruit growth", Peanut Science, Vol. 6, pp.114-117.

Hardy, R.W.F. and U.D. Havelka 1977. "Possible routes to increase the conversion of solar energy to food and feed by grain legumes and cereal grains (crop production): CO_2 and N fixation, foliar fertilization, and assimilate partitioning", "In Biological solar energy conversion" A. Mitsui *et.al.,* (ed.), Academic Press, New York, pp. 299-322.

IPCC, 2007. "Climate Change 2007: Synthesis Report", Available at:http://www.ipcc.ch/publications_and_data/publications_ipcc_fourth_ assessment_report _synthesis_report.htm

IPCC, 2011. "Renewable Energy Sources and Climate Change Mitigation—IPCC Special report", 2011.

Kuzyakov, Y. and O. Gavrichkova. 2010. "Time lag between photosynthesis and carbon dioxide efflux from soil: a review of mechanisms and controls", Global Change Biology, Vol.16, pp. 3386–3406.

Manoj-Kumar *et al.,* 2012. "Effects of elevated atmospheric CO_2 and temperature on phosphorus efficiency of wheat (*Triticumaestivum* L.) grown in an Inceptisol of subtropical India", Plant, Soil and Environment, Vol. 58(5), pp. 230–235.

Ong, C.K. 1984. "The influence of temperature and water deficits on the partitioning of dry matter in groundnut (*Arachishypogaea* L.)", Journal of Experimental Botany, Vol. 35, pp. 746-755.

Prasad, P.V.V., P.Q. Craufurd and R.J. Summerfield. 1999. "Fruit number in relation to pollen production and viability in groundnut exposed to short episodes of heat stress", Annals of Botany, Vol. 84, pp. 381-386.

Prasad, P.V.V., K.J. Boote, L.H. Allen Jr. and J.M.G. Thomas. 2003. "Super-optimal temperatures are detrimental to peanut (*Arachishypogaea* L.) reproductive processes and yield at both ambient and elevated carbon dioxide", Global Change Biology, Vol. 9, pp. 1775-1787.

Prasad, P.V.V. *et al.,* 2006. "Influence of soil temperature on seedling emergence and early growth of peanut cultivars in field conditions", Journal of Agronomy and Crop Science, Vol. 192, pp.168-177.

Rakshit, R. *et al.,* 2012. "Effect of elevated CO_2 and temperature on nitrogen dynamics and microbial activity in soil during wheat growth" Journal of Agronomy and Crop Science, Vol. 198, pp. 452-465.

Rao, K.V. 1999. "The combined effect of elevated CO_2 levels and temperature on growth characteristics of groundnut (*Arachishypogaea* L.)", Indian Journal of Plant Physiology,Vol. 4, pp. 297-301.

Reddy, K.R. and H.F. Hodges. 2000. "Climate Change and Global Crop Productivity", CABI Publishing, New York.

Richter, D.D. *et al.,* 2007. "Long-term soil experiments: keys to managing Earth's rapidly changing ecosystems", Soil Science Society of America Journal, Vol. 71, pp. 266–279.

Rinnan, R., A. Michelsen, E. Baath and S. Jonasson 2007. "Fifteen years of climate change manipulations alter soil microbial communities in a subarctic heath ecosystem", Global Change Biology, Vol. 13, pp. 28–39.

Ritz, K., H.I.J. Black, C.D. Campbell and J.A. Harris. 2009. "Wood C Selecting biological indicators for monitoring soils: a framework for balancing scientific and technical opinion to assist policy development", Ecological Indicator, Vol. 9, pp.1212–1221.

Salvador Sanchis, M.P. *et al.,* 2008. "Climate effects on soil erodibility", Earth Surface Processes and Landforms, Vol. 33, pp.1082–1097.

Sharma, S.K. and A.B. Bazaz. 2012. "Sustainable Management of Biodiversity in the Context of Climate Change-Issues, Challenges and Response", Proceedings of National Academy of Sciences, India, Section B Biological Sciences, Vol. 82, pp. 251–260.

Kumar, U., P. Singh and K.J. Boote. 2012. "Chapter two – Effect of Climate Change Factors on Processes of Crop Growth and Development and Yield of Groundnut (*Arachishypogaea* L.)", Advances in Agronomy. Vol. 116, pp. 41–69.

Viswanath, T., D. Pal, and T.J. Purakayastha. 2010. "Elevated CO_2 reduces rate of decomposition of rice and wheat residues in soil", Agriculture, Ecosystem and Environment, Vol. 139, pp. 557–564.

Weil, R.R. and F. Magdoff 2004. "Significance of soil organic matter to soil quality and health" In: Soil organic matter in sustainable agriculture, Weil RR, Magdoff F (eds). CRC press, Florida, pp. 1–43.

3

Identification of Early Maturing High Yielding Mutants in Black Gram [*Vigna Mungo* (L.) Hepper]

Mohan Dadarwal

Abstract

*"Identification of Early Maturing High Yielding Mutants in Black Gram [*Vigna Mungo *(L.) Hepper]" study was carried out with the effect of EMS, DMS and their post treatment with respective modifiers IAA and GA on different traits in black gram variety Barkha. The experiment was carried out in compact family block design during* Kharif, *2005 at Experimental field of Department of Plant Breeding and Genetics, Rajasthan College of Agriculture, Udaipur.EMS 0.4 + 3 x 10^{-3} M IAA induced early maturity with higher yield and yield contributing traits. High coefficient of variation and range was observed for all the treatments. On the basis of yield parameters there were eight progenies identified which were high yielder with most of the yield contributing traits and eighteen progenies were earlier in maturity than control.Variability parameters viz., GCV, PCV, Heritability and Genetic advance were calculated for all the families. Family EMS 0.4 per cent and DMS 0.04 per cent with modifiers exhibited high values for variability parameters for yield and yield contributing traits.*

Key Words: Ethylmethane sulphonate (EMS), Dimethyl sulphonate (DMS), Indol Acetic Acid (IAA) and Gibberellic Acid (GA), GCV, PCV, Heritability and Genetic advance.

Introduction

Black gram is the fourth important pulse crop in India which occupies about 12 per cent of the total pulse area and contributing about eight per cent to the total pulse production . Inspite of its high nutritional value its productivity is very low.

Barkha variety of black gram is well adapted to the climatic conditions of southern Rajasthan. It is liked by the farmers due to its bold, black shining grains but it is late in maturity and faces terminal drought. This strongly thrust the importance to induce earliness is that it can be well adopted. Genetic improvement of a crop primarily depends upon extent of genetic variability present in the population. Experimentally, induced mutations provide an important source of variability. The practical utility of induced mutations for the improvement of quantitatively inherited characters in urd bean is well recognised.

Therefore, present investigation is to induce variability in black gram through mutation breeding by using chemicals mutagens like Ethylmethane sulphonate (EMS) and Dimethyl sulphonate (DMS) and their modifiers viz., Indol Acetic Acid (IAA) and Gibberellic Acid (GA) to isolate early maturing and high yielding genotype with shining bold grains from M_2 generation of Barkha, so that, it can very well fit in the cropping pattern under rainfed conditions.

Material and Methods

Seed of Barkha variety was treated with chemical mutagens like Etylmethane sulphonate (EMS), Dimethyle sulphonate (DMS) and their combinations with different plant growth regulators like Indol Acetic Acid (IAA) and Gibberellic acid (GA). Normal appearing competitive ten M_1 plants (progenies) from each selected treatment (families) as below were advanced to M_2 generation.

(1) Control
(2) EMS 0.4 per cent
(3) EMS 0.6 per cent
(4) DMS 0.04 per cent
(5) DMS 0.06 per cent
(6) EMS 0.4 + 3×10^{-3} M IAA
(7) EMS 0.4 + 3×10^{-4} M IAA
(8) EMS 0.6 + 3×10^{-3} M IAA
(9) EMS 0.6 + 3×10^{-4} M IAA
(10) DMS 0.04 + 3×10^{-3} M GA
(11) DMS 0.04 + 3×10^{-4} M GA

Experimental Methodology

Normal appearing plants from eleven treatments (M_1 generation) which exhibited relatively high fertility were selected to advance in M_2 generation. The M_2 generation was raised in compact family block design with three replications. Each treatment was taken as family and within treatment the individual plant progeny was taken as progenies. All the progenies within family were randomised, likewise the family blocks were also randomised. Progenies were sown in single row plot of three meter length and distance between plant to plant and row to row (30 × 10 cm) was maintained.

Procedure of Observations

Observations were recorded on ten randomly selected plants in each progeny of all the families except days to 50 per cent flowering and days 75 per cent maturity, which were observed on whole plot basis.

Statistical Methods

The analysis of variance for each of the family was done separately according to Panse and Sukhatme (1978). The components of variance and their expectations are given as following.

Analysis of variance for differences between families and within families (between progenies)

Source	*d.f.*	*S.S.*	*M.S.*	*EMS*
Replication	r – 1	SSr	MSr	
Families	n –1	SSn	MSn	$\sigma^2_e + rp\sigma^2_F$
Progenies	p - 1	SSp	MSp	$\sigma^2_e + r\sigma^2_p$
Error		SSe	MSe	σ^2_e
Total (T)		SS(T)		

where, r, n, p = replications, families, progenies in a family and error, respectively.

(i) **Critical difference for difference between families :**

$$CD = \sqrt{\frac{2MSe}{Pr}} \times \text{t value at 5\% level of significance of error d.f.}$$

(ii) **Critical difference for differences between progenies within family :**

$$CD = \sqrt{\frac{2MSe}{r}} \times \text{t value at 5\% level of significance of error d.f.}$$

(iii) **Coefficient of variation:**

$$CV = \frac{\sqrt{Mse}}{\bar{X}} \times 100$$

Variability parameters

Following genetic variability parameters calculated in all those families which revealed significant progeny differences.

(i) Genotypic Coefficient of Variation (GCV) : The magnitude of genetic variation existing in a character was worked out by the formula given by Burton (1952).

$$GCV = \frac{\sqrt{Vg}}{\bar{X}} \times 100$$

where,

$$V_g = \frac{MSp - MSe}{r}$$

where,

V_g = Genotypic variance
$\bar{X}$ = Mean of the character under study
MSp = Mean square due to progenies
MSe = Error variance
r = Number of replications.

(ii) Phenotypic Coefficient of Variation (PCV) : The magnitude of phenotypic variation existing in a character was estimated by using the formula suggested by Burton (1952).

$$PCV = \frac{\sqrt{V_{ph}}}{\bar{X}} \times 100$$

$$V_{ph} = V_g + MSe$$

where,

V_{ph} = Phenotypic variance
$\bar{X}$ = Mean of the character under study
V_g = Genotypic variance
MSe = Error variance

(iii) Heritability : Heritability in broad sense was estimated by using following formula proposed by Burton and Davane (1953), Jhonson *et al.* (1955) and Hanson *et al.* (1956).

$$H = \frac{V_g}{V_{ph}} \times 100$$

where,

V_g = Genotypic variance
V_{ph} = Phenotypic variance

(iv) Genetic gain : It is the genetic advance expressed as per cent of mean:

$$\text{Genetic gain} = \frac{GA \times 100}{\bar{X}}$$

Where genetic advance (GA) was estimated using the formula suggested by Robinson *et al.* (1949), Johnson *et al* (1955).

$$GA = k\,[V_g/V_{ph}] \times \sqrt{V_{ph}}$$

where,

V_g = Genotypic variance

V_{ph} = Phenotypic variance

$\overline{X}$ = Mean of the character under study

K = Selection differential (constant) at 5% selection intensity (Allard, 1960) i.e. 2.06

Result and Discussion

Analysis of variance revealed that highly significant differences existed between the families for all the traits.

(i) Days to 50 per cent flowering: Progenies differences were significant in all the families except control and EMS 0.6 per cent. On the basis of mean family EMS 0.4 +3 × 10^{-4} M IAA (37.77) was significantly earlier in flowering than control (38.53). Maximum range for days to 50 per cent flowering was recorded in family EMS 0.4 + 3 × 10^{-3} M IAA (35.67-41.00) against the control (38.33–39.00). Maximum coefficient of variation was observed in family EMS 0.4 + 3 × 10^{-3} M IAA (7.70) than control (1.26).

(ii) Days to 75 per cent maturity: Progeny differences were significant in all the families except control. Family EMS 0.4 + 3 × 10^{-3} M IAA (78.83) and Family EMS 0.4 + 3 × 10^{-4} M IAA (78.90) were significantly earlier in maturity than control (79.77). Highest range for days to 75 per cent maturity was observed in family EMS 0.6 + 3 × 10^{-3} M IAA (76.00 – 82.33) against control (79.00–80.33). Coefficient of variation in control was 0.97 per cent which increased to 4.95 in family EMS 0.4 + 3 × 10^{-3} M IAA. Eighteen progenies exhibited early maturity than control. The variability parameters recorded for all the families were low in magnitude except EMS 0.4 + 3 × 10^{-3} M IAA (GCV = 2.69, PCV = 2.74, Heritability = 96.59, Genetic gain = 5.45) (Table 1).

(iii) Plant height (cm): Progeny differences were significant in all the families except control. Maximum plant height was observed in family EMS 0.4 +3 × 10^{-3} (28.60) than control (27.40). Wide range was observed in family DMS 0.06 per cent (20.14–32.12) than control (26.93–27.86). Coefficient of variation in control was 2.29 per cent which increased in all the families. Highest coefficient of variation was observed in family EMS 0.4 + 3 × 10^{-4} M IAA (18.91). Thirty four progenies exhibited significant increase in plant height than control in all the mutagenic families.

(iv) Number of primary branches per plant: Progeny differences were significant in all the families except control. Maximum number of primary branches per plant was observed in family EMS 0.4 + 3 × 10^{-3} M IAA (2.80) than control (2.56). Highest range for this trait was observed in family DMS 0.04 + 3 × 10^{-3} MGA (1.87–3.14) as compared to (2.36 – 2.69) control. Coefficient of variation was maximum in family DMS 0.04 +3 × 10^{-4} M GA (37.08). Higher number of primary branches per plant was observed in twenty seven progenies over all the families.

Table 1: Estimates of variability parameters in Barkha variety of black gram in M_2 generation

Characters/ families	*GV per cent*	*PV per cent*	*GCV per cent*	*PCV per cent*	*H (in broad sense) per cent*	*GG per cent*
Days to 75 per cent maturity						
Control	0.11	0.32	0.42	0.71	34.60	0.50
EMS 0.4 per cent	1.52	3.12	1.55	2.22	48.66	2.22
EMS 0.6 per cent	0.55	1.17	0.92	1.34	47.04	1.30
DMS 0.04 per cent	1.77	2.19	1.66	1.84	80.90	3.07
DMS 0.06 per cent	2.57	2.82	1.99	2.08	91.20	3.91
EMS 0.4 + 3 × 10^{-3} M IAA	**4.51**	**4.67**	**2.69**	**2.74**	**96.59**	**5.45**
EMS 0.4 + 3 × 10^{-4} M IAA	3.45	3.83	2.36	2.48	90.23	4.61
EMS 0.6 + 3 × 10^{-3} M IAA	2.76	3.17	2.09	2.23	87.15	4.01
EMS 0.6 + 3 × 10^{-4} M IAA	2.84	3.48	2.12	2.35	81.57	3.94
DMS 0.04 + 3 × 10^{-3} M GA	2.29	2.74	1.91	2.09	83.40	3.59
Seed yield per plant						
Control	0.00	0.01	0.00	0.00	0.00	0.00
EMS 0.4 per cent	0.06	0.08	6.47	7.55	73.47	11.43
EMS 0.6 per cent	0.00	0.01	1.63	3.57	20.98	1.54
DMS 0.04 per cent	0.02	0.03	4.28	4.77	80.66	7.92
DMS 0.06 per cent	0.01	0.03	0.00	0.00	0.00	0.00
EMS 0.4 + 3 × 10^{-3} M IAA	0.03	0.04	4.41	5.11	74.48	7.84
EMS 0.4 + 3 × 10^{-4} M IAA	**0.06**	**0.06**	**6.86**	**7.06**	**94.46**	**13.73**
EMS 0.6 + 3 × 10^{-3} M IAA	0.03	0.04	4.87	5.76	71.45	8.47
EMS 0.6 + 3 × 10^{-4} M IAA	0.00	0.03	0.00	0.00	0.00	0.00
DMS 0.04 + 3 × 10^{-3} M GA	**0.03**	**0.04**	**5.44**	**5.63**	**93.39**	**10.83**
DMS 0.04 + 3 × 10^{-4} M GA	0.03	0.04	5.46	5.90	85.78	10.42

(v) Number of pods per plant: Statistically significant differences between progeny were recorded in all the families except control, EMS 0.6 per cent, DMS 0.04 per cent and DMS 0.06 per cent. None of the families exhibited positive shift in number of pods per plant as compared to control (15.86). Wide range for number of pods per plant was observed in family DMS 0.04 + 3 × 10^{-3} MGA (13.60 – 18.23) against control (15.67 – 16.03). Coefficient of variation was 1.20 per cent in control. It was increased at all the mutagenic treatments and highest in family EMS 0.4 per cent (19.40). Among all the families six progenies exhibited higher number of pods per plant than control.

(vi) 100-seed weight (g): No significant differences between progenies were observed for 100-seed yield than control (4.37). Coefficient of variation was 1.82 per cent in control and it was increased in all the families. Range for seed yield per plant was observed narrow in all the families than control. None of the progeny was found superior than control for 100-seed weight.

(vii) Seed yield per plant (g): Progeny differences were found significant in all the families except control, EMS 0.6 per cent and DMS 0.06 per cent. Grain yield was higher in two families viz., EMS 0.4 per cent (3.78) and family EMS 0.4 + 3 × 10^{-3} M IAA (3.70) than control (3.60). Coefficient of variation was 1.96 per cent in control which increased in all the mutagenic treatment and it ranged from 1.96 per cent (control) to 12.64 per cent (EMS 0.4 + 3 × 10^{-4} M IAA). Wide range for seed yield per plant was observed in all the families than control (3.53 – 3.65). In all the families only eight progenies were statistically superior for grains yield per plant than control. Maximum value of variability parameters was observed in family EMS 0.4 + 3 × 10^{-4} M IAA (GCV = 6.86, PCV = 7.06, Heritability = 94.46 and Genetic gain = 13.73) followed by family DMS 0.04 + 3 × 10^{-3} M GA (GCV=5.44, PCV = 5.63, Heritability = 93.39 and Genetic gain = 10.83) (Table 1).

(viii) Biological yield per plant (g): For biological yield per plant progenies differences were significant in all the families except control. Biological yield was significantly higher in all the families than control (16.39). Highest biological yield was observed in family DMS 0.06 per cent (21.53). Increase in range was observed in all the families than control (16.08–17.17). Higher coefficient of variation was observed in all the treatments than control (4.53). Maximum coefficient of variation (22.69) was observed in DMS 0.04 per cent. Among all the families thirty five progenies were superior for biological yield per plant over control.

(ix) Harvest index (per cent): For harvest index progeny differences were significant in all the families except control. None of the family exhibited significant increase in harvest index than control (18.03), though increased in range was exhibited by all the families. Highest coefficient of variation was observed in DMS 0.04 + 3 × 10^{-3} M GA (22.26) than control (4.35). One progeny among all the families was found superior than control.

(x) Seed Protein content (per cent): Significant progeny differences were observed in all the families except control and DMS 0.04 + 3 × 10^{-3} M GA. Seed Protein content was significantly higher in family EMS 0.4 per cent (24.67), EMS 0.6 + 3 × 10^{-3} M IAA (24.94) and EMS 0.6 + 3 × 10^{-4} M IAA (24.78) than control (24.45). Narrow range for protein content was observed in all the families. Coefficient of variation was highest in EMS 0.4 + 3 × 10^{-3} M IAA (7.10) than control (2.72). Ten progenies among all the families were significantly superior to control for protein content.

Analysis of variance of M_2 generation revealed significant inter-family differences for all the traits. This included that different loci/genes have been affected in the progenies in respect of all traits (Mahala *et al.*, 2010). Progeny differences were significant in all most all the families for all the traits except 100-seed weight.

The mean values of different characters shifted either to positive or negative direction from the control due to mutagenic treatments. EMS 0.4 + 3 × 10^{-3} M IAA treatment exhibited early maturity with higher number of primary branches per plant, plant height and seed yield. EMS 0.4 per cent was also efficient treatment for seed yield and seed protein content. High estimate of coefficient of variation with increased range for most of the traits indicate induction of polygenic mutants. Similar observations were recorded by Prema Manapure *et al.* (1998) and Solanki and Joshi (2000).

The coefficient of variation and range were higher in all the families as compared to their respective control for all the traits. These results are in broad conformity with those of earlier researchers Solanki and Sharma (2002).

The genotypic and phenotypic coefficient of variation gave an idea about the relative magnitude of variability. Gap between genotypic coefficient of variation and phenotypic coefficient of variation was narrow which reflected lesser degree of environmental influence on the genotypic variability. Similar trend were also observed by Hipziba and Subramanium (1994) in black gram.

EMS 0.4 per cent with IAA induced high genotypic and phenotypic coefficient of variations for most of the yield traits.The efficiency with which genotypic variability can be exploited for improvement through selection depends on its heritability (Burton 1952). A character possessing higher GCV with high heritability can be easily improved by selection. In present investigation high heritability was recorded for seed protein content, biological yield and harvest index in almost all the families. These results were in accordance with the findings of Vandana and Dubey (1992) in black gram.

In present study high value of genetic gain was observed in almost all the families for biological yield per plant and harvest index. Variability coupled with high heritability and greater expected genetic gain is an induction of greater role of additive gene effects in genetic control of the given character. Character with greater contribution and additive gene effect can be easily improved through selection (Panse, 1978). Higher genotypic variability along with high heritability and greater genetic gain was recorded in family EMS 0.4 + 3 × 10^{-3} M IAA, EMS 0.4 +3 × 10^{-4} M IAA and DMS 0.04 + 3 × 10^{-3} M GA for grains yield and most of the traits. These results are in agreement with those reported by Hipziba and Subramaniam (1994).

Conclusion

Based on the results summarised above it is concluded that Family EMS 0.4 + 3 × 10^{-3} M IAA, EMS 0.4 +3 × 10^{-4} M IAA and DMS 0.04 + 3 × 10^{-3} MGA exhibited higher values for variability parameters for most of the traits studied. EMS 0.4 treatment along with modifier was found to be the most efficient treatment, it induced early maturity along with high mean for yield and yield contributing traits. Range and variance was also increased for most of traits. It also exhibited high values of variability parameters indicating polygenic mutants. In all there were eight progenies identified which was higher yielder with most of the yield contributing traits and eighteen progenies were early in maturity than control. Therefore, these treatments and progenies should be carried further in M_3 generation to isolate early maturing, high yielding lines.

References

Arulabalachandran, D. and L. Mullainathan (2009).Changes in quantitative traits of black gram (*Vigna mungo* L. Hepper) induced by EMS in M_2 generations. J. Phytol., 1:230-235.

Burton, G.W. (1952). Quantitative inheritance in grasses. *Proceedings of 6th International Grassland Congress*. 1 : 227-283.

Burton, G.W. and Devane, E.M. (1953) Estimation of heritability in tall Fescus. *Agronomy Journal*. 45 : 478-481.

C.Thilagavathi, L. Mullainathan, (2011) Influence of physical and chemical mutagens on quantitative characters of (*Vigna mungo* (L.) Hepper), International multidisciplinary research journal. 1/1:06-08.

D. Arulbalachandran, L. Mullainathan, S. Velu and C. Thilagavathi, (2010) Genetic variability, heritability and genetic advance of quantitative traits in black gram by effects of mutation in field trial. African journal of Biotechnology. 9 (19):2731-2735.

Gautam, A.S. and Mittal, R.K. (1998)Induced mutations in black gram [*Vigna mungo* (L.) Hepper.]. *Crop Research.* 16 : 344-348.

Goyal, S. and Khan S. (2010) Induced mutagenesis in urdbean [*Vigna mungo* (L.) Hepper] : A Review. Int. J. Bot., 6:196-206.

Hipziba, S.J. and Subramaniam, M. (1994) Gamma rays induced variability in black gram (*Vigna mungo* L. Hepper). *Annals of Agricultural Research.* 15: 512-515.

Jain, S.K. and Khandelwal Vikas (2009) Induced polygenic variability in black gram [*Vigna mungo* (L.) Hepper]. *Indian Journal of Genetics and Plant Breeding.* 69(1): 72-75.

Johnson, H.W., Robinson, H.F. and Comstock, R.E. (1955) Genotypic and phenotypic correlations in soybean and their implication in selection. *Agronomy Journal* 47 : 477-482.

Kharkwal, M.C. and Shu Q.Y. (2009) Role of induced mutations in world food security. In: Q.Y. Shu(ed.)induced mutations in the Genomics Era. Food and Agriculture Organisation of the United Nations, Rome. 33-38.

Kumar, Y. and Mishra, V.K. (1999) Effect of gamma rays and diethylsulphate on germination, growth, fertility and yield in *Vigna radiata* L. (Wilczek). *Annals of Agricultural Research.* 20 : 144-147.

Kundu, S.K. and Singh, D.P. (1981) EMS induced variability in black gram. *Crop Improvement* 8 : 71-72.

Mahla, H.R., Shekhawat A. and Kumar D. (2010) A study on EMS and Gama rays mutagenesis of cluster bean [*Cymopsis tetragonoloba* (L.)Taub]. Plant mutations Reports, 2:28-32.

Panse, V.G. and Sukhatme, P.V. (1978) Statistical method for agriculture workers. ICAR, New Delhi.

Prema, Manapure, Shanti Patil and Bhoyar, H.S. (1998). Induced mutagenic changes in quantitative characters of black gram (*V. mungo* L. Hepper). *Journal of Soils and Crops.* 8 : 204-206.

Solanki, I.S. and Joshi, K.R. (2000) Induced genetic variability in blackgram. *Madras Agricultural Journal.* 12 : 412-415.

Solanki, I.S. and Sharma, B. (2002) Induced polygenic variability in different groups of mutagenic damage in lentil (*Lens culinaris* Medik). *Indian Journal of Genetics and Plant Breeding.* 62 : 135-139.

Vandana and Dubey, D.K. (1992) Heritability and genetic advance in induced mutants of faba bean (*Vicia faba* L.). *Indian Journal of Genetics and Plant Breeding.* 46 : 143-145.

Vanniarajan, C., Vivekanandan, P. and Ramalingam, J. (1996) Induced variability for quantitative characters in black gram. *Crop Research.* 11 : 341-346.

4

Salt Stress and Its Management in Fruit Crops

V. B. Shinde, S. S. Wandre, M.G. Bhoyar and M.V. Ramdevputra

Abstract

In the present review paper, an attempt has been made to study the effects of salt stress and different techniques used for its management in fruit crops. Approximately, 10 per cent of world's 7000 million hectares arable land surface consists of saline or sodic soils. About 230 million hectares of irrigated land, 45 million hectares were salt-affected soils; and of the almost 1500 million hectares of dry land agriculture, 32 million were salt-affected soils. These soils contain high pH, poor infiltration rate and permeability, poor physical condition which offer high mechanical impedance to the growth of roots. Salt stress have a pronounced effect on leaf, stem, shoot, root, relative water content, photosynthetic rate and chlorophyll pigment contents of fruit plants (Citrus, date palm, etc.). Maximum reduction occurred in germination per cent, seedling growth and survival of seedlings of fruit crops under salt stress condition (Mango, aonla, etc.). Use of rootstock ameliorates the adverse effect of salt stress. It enhances the vegetative growth along with number of leaves (Olour-mango, etc.). In sodic soil, mulching of orchard gives positive effect on plant growth and minimises weed population in orchard (Aonla-black polythene mulch, etc.). Few varieties of different fruit crops show more vigorous growth with early flowering in saline conditions (Banarasi-ber, etc.). Application of magnetised water under saline soil condition improves nutrient availability at root zone of fruit plants in comparison to non-magnetised water (Apple, citrus, olive, etc.).

Introduction

Salt Stress is osmotic forces exerted on plants when they are growing in a salt marsh or under other excessively saline conditions. It is nothing but the salt concentration is high enough to lower the water potential. The process of increasing the salt content is known as 'Salination'. Salinity remains one of men's oldest environment and horticultural problems and challenging to scientist for production and productivity in our country. Generally the salt concentration is natural source to soil and water. If water with high concentration of 100 mg salts per liter, add to the soil, metric ton of salt is added to the soil per hectare of applied water. Normally the salinity problems are acute in the arid-semiarid regions of the country. There is a lot amount of marginal lands with a high degree of natural salinity. Salt-affected soils may inhibit seed germination, retard plant growth and cause irrigation difficulties. Approximately, 10 per cent of world's 7000 million hectares arable land surface consists of saline or sodic soils. About 230 million hectares of irrigated land, 45 million hectares (19.5 per cent) were salt-affected soils; and of the almost 1500 million hectares of dry land agriculture, 32 million (2.1 per cent) were salt-affected soils (F.A.O., 2014). Saline soils normally occur in Gujarat, West Bengal, Rajasthan, Maharashtra, Orissa and Andhra Pradesh. These soils contain high pH, poor infiltration rate and poor permeability, poor physical condition which offer high mechanical impedance to the growth of roots. Highly accumulation of sodium salt causes the sodic soil. Sodic soil found extensively in the states of Uttar Pradesh, Gujarat and Maharashtra (C.S.S.R.I., 2012).

Impact of Salt Stress on Plants

When concentration of sodium salt is increased:

- They affect the plant growth both physically and nutritionally.
- Leaf burn and necrotic symptoms were observed in mango and grapes.
- Yield losses in fruits are generally more because of more effect of specific toxicity than osmotic effect in fruit crops.

When concentration of the salt exceeds threshold value:

- Growth rate and size of the plant progressively decreased.
- Plant becomes stunted with restricted lateral shoot level.
- Reduction in size of the leaves and fruits.
- Decrease in fresh and dry weight of different plant parts
- Reduction on number of leaves resulted to low yield.
- Generally leaves become thicker than normal.
- The top growth suppressed more than the root growth.

Physiological Effects of Salt Stress

1. Osmotic effects: The reduction of the sum of metric and osmotic potential reduced the water uptake and so the presence of excess salts causes the water deficiency and which leads to physiological drought.

2. Specific ion effects or toxicity: Injurious effects on the plants may be due to high total soluble salts or toxic concentration of ions. Ions like Na^+ and Cl^- causes specific toxicity to fruit crops. The yield losses of fruits and nuts are generally greater due to specific ion toxicity rather than due to osmotic effect.

3. Nutritional imbalance: The uptake of ions affected due to differences in membrane permeability, transport kinetics and selectivity. The K^+ concentration in plant tissue is reduced as the Na^+ concentration is increased. The uptake of Ca^{2+} may decreases and reduces the activity of Ca^{2+}. There are antagonistic effects like high Na^+ and Cl^-; causes deficiency of Ca^{2+}, Mg^{2+} and K^+.

4. Effect on germination: Generally plants are more sensitive to salinity during germination and early seedling growth due to increase in the osmotic pressure of the soil solution or toxicity to the embryo or the growing seedling.

Prasad *et al.* (1991) observed that maximum seed germination, seedling growth and survival of aonla seedlings were recorded in non-sodic soil.

Fruit crops grades for their germination, Date palm: tolerant (5-10 dS/m); Fig, grapes, guava and ber: medium tolerant (3-5 dS/m); Citrus, apple and peach: sensitive (1.5-3 dS/m).

5. Growth and development: Salinity in soil is harmful at all stages of growth and development of the plant. The growth stunting and retardant is the most common effect of salt stress to salt affected plant. All growth and yield attributing parameters are reduced.

Ramoliya and Pandey (2010) revealed that salt stress had a pronounced effect on leaf, stem, shoot and root growth of date palm plant. Varu and Barad (2010) reported that germination, plant height, root length and survival of mango seedlings were significantly decreased with increase in salinity stress. Murkute *et al.* (2006) found that minimum reduction in plant height; stem diameter and number of leaves per branch of 'Karna Khatta' citrus plant were observed in non-saline condition. Rao and Singh (2006) revealed that as salt stress increased there is increase in reduction in plant height, plant spread and stem diameter of Aonla cv. NA-7 plants were observed.

6. Effect on metabolism: The salinity tends to alter the nutritional imbalance, which resulted in changes in metabolic activities leading to increases in free amino acid, proline, total organic acids and hydrolytic enzymes. Sever salinity resulted in death due to loss of ionic control in root and chlorosis, necrosis and wilting. Srivastav *et al.* (2009) reported that minimum mortality in 'Kurukkan' mango seedling at nursery stage was found in non-saline soil condition.

7. Effect on photosynthesis

- Reduction in photosynthetic areas and increase in resistant to CO_2 diffusion
- Structural damage to chloroplasts

- Inhibition of light and dark reaction
- Reduction in transport of photosynthetic assimilates
- Chlorophyll content of the leaves are decreases
- Adverse effects to activity of photosynthetic enzymes

Singh *et al.* (2014) observed that minimum reduction in relative water and chlorophyll pigment content in citrus were observed in non-saline condition, but maximum proline content was observed in 50 mM salinity. Sharma *et al.* (2011) found that minimum reduction occurred in relative water content, photosynthetic rate and chlorophyll pigments contents of non-salinised 'Karna Khatta' citrus seedlings compared to salinise.

8. Effect on plant hormones: The plant hormones also altered that the level of endogenous cytokinin is decreased whereas the level of ABA increased to some plants. So the salinity induced symptoms also due to hormonal imbalance.

Management of Salt Stress

1. Natural salinity resistance

Tolerance mechanism: The salts are accepted first by plants but their adverse effects are nullified by some natural mechanism and protect itself against the salt. There is two possible ways which are:

- Excreting or secreting the absorbed salts into the vacuole.
- Tolerate the ion balance strain by requiring protoplasmic substances.
- The halophytes, which markedly increase their cell sap concentration by storing salts.

2. Crop Improvements

(A) Breeding aspects against salt stress: Selection of tolerant varieties is the most important tools. However, it reported that the important tolerance were not phenomenal perhaps because most of fruit crop varieties from narrow genetic base and were propagated by vegetative. Fruit crops can be evaluated by the capacity of the seed to germinate or cuttings to root in the salt affected soil or the capacity of particular fruit crops to survive in the saline media or the relative performance in saline and non-saline soils with respect to reduction in growth. Eg. citrus, grape, etc.

(B) Genetic Engineering: In genetic engineering, genetic manipulation to by-pass the sexual cycle is done to establish an individual during a new combination of inherited properties. Eg. Kiwifruit, citrus, etc. It presently follows two major approaches.

- The cellular approached involving in vitro culturing of cells and the hybridization of somatic cells.

- The molecular approach involves the direct manipulation of nucleic acid.

Balamohan, *et al.* (2001) concluded *Banarasi* variety of ber grows well in sodic soil with maximum plant height, number of branches per tree, crop canopy and minimum days to first flowering as compared to other varieties. Balamohan *et al.* (2002) revealed that *Mridula* variety of pomegranate grows well in sodic soil with maximum plant height, number of branches per tree, crop canopy and minimum days to first flowering as compared to other varieties.

3. Soil and Water Management

(A) Leaching of salts through drainage system: Leaching is the process in which extra water is added to a field and allowed to soak through the soil and drain away underground. A common method of leaching is to pond the water in basins over the entire field. The amount of leaching water that enters the soil by surface flooding determines how much salt is removed from the soil.

(B) Chemical: Most saline soils need chemical amendments to restore their productivity of the many suitable available amendments; gypsum, sulphur and sulphuric acid are the most common. Application of acid is effective in reclaiming saline soils by lowering ESP (exchangeable sodium percentage), SAR (sodium adsorption ratio) and pH. Dayal, *et al.* (2014) observed that minimum reduction in mango plant height and leaf numbers under salinized condition was found in graft with local seedlings and Olour rootstocks, respectively. Maximum dry weight of shoot and root was recorded in Olour rootstocks and local seedlings, respectively.

(C) Scrapping: Scrapping of salt layer, land leveling, deep ploughing, sub-soiling, sanding (i.e. adding sand), flushing with good quality and enough quantity of water and drainage also help in improving soil condition.

(D) Green manuring and FYM: Green manuring through leguminous crops and application of farm yard manure not only provide organic matter and other nutrient, but also make the soil porous for aeration and moisture absorption and enhance soil microorganisms, thus improving the overall condition of the soil.

(E) Halophytes: Research has identified more than 1,500 plant species that have high levels of tolerance to saline soils; these are called halophytes. Some of these are able to withstand salt concentrations in excess of those found in sea-water. Eg. Date palm (Kutch-Gujarat), etc.

(F) Mulching: During periods of high evapotranspiration between the two irrigations and during periods of fallow there is a tendency for the leached salts to return to the soil surface. Mulching reduce evaporation from the soil surface will help to control root zone salinity. Mulching material: Black polythene, straw, grasses, rice husk, sugarcane trash, etc. Rao and Pathak (1996) indicated that orchard mulched by black polythene leads to maximum increase in height and girth of aonla trees in sodic soil. It also reduced the weed population.

(4) Use of Rootstock

Rootstocks differ widely in absorption and transport of toxic ions and develop the salt resistivity in fruit crops either by exclusion or accumulation mechanism. Rootstock is the most important and easier tools against salinity problems in fruit crops.

- Citrus: Rough lemon, Rangpur lime, Sour orange, Sweet orange, Ridge apple
- Mango: *Kurukkan*
- Apple: Merton-793
- Pear: Oregon-249
- Walnut: Paradox
- Grape: Salt creek, Dogridge, 1613
- Cherry: Mazzard (Prunus avium)

Modern Methods

1. Skimming Well Technology

Skimming well is any technique employed with an intention to extract relatively freshwater from the upper zone of the fresh-saline aquifer. The skimming wells are low discharge (less than 28 lps.) cluster of wells drawing groundwater from relatively shallow depth. Skimming wells are generally designed for irrigation (Saeed *et al.*, 2002) or drinking water supply (Rao *et al.*, 2006; 2007) purposes.

The types of skimming wells:

- Conventional single strainer wells
- Multi-strainers wells
- Scavenger wells
- Radial collector wells
- Dug wells

Benefits

- Improved quality of pumping groundwater
- Availability of locally manufactured material

- Availability of local expertise for drilling, installation and maintenance
- Shallow water table which helps in use of centrifugal pump units
- Technically simple systems
- Economically viable
- Improves soil health and crop yields

2. Doruvu Technology

A dug out conical pits locally called Doruvu in Andhra Pradesh and Kottai in Tamil Nadu, is used to skim fresh water floating on the saline water. Each Doruvu occupy an area of about 200 m^2. The water collected from each Doruvu is just sufficient to irrigate 800 m^2. As such 10-12 Doruvu are needed for one hectare area. An alternative to traditional Doruvu, AICRP center at Bapatla developed a popularly known Improved Doruvu Technology. In this set up, flow of water 1.8-2.0 m below the ground surface in collectors embedded is collected in a sump. This water is pumped and used to irrigate crops using sprinklers/drips.

3. Magnetized Water

Water never travels one molecule at a time. It travels in clusters. Many clusters are oversized, carrying toxins and pollutants inside. So they can't penetrate the cellular membrane of the cell. Oversized clusters make water and nutrients bypass the plant. When magnetic resonance is added to water, the oversized cluster breaks up and is reconstructed into smaller, bioavailable units with six molecules per structure. These clusters easily enter the cell membrane. Poor hydration allows salts and pollutants to affect plant health. Full hydration means healthier, more productive plants. Selim (2008) indicated that maximum nutrient availability in saline soil at root zone of apple, citrus and olive plants recorded at plants irrigated by magnetized water as compared to non-magnetized water.

Expenses: Approximately, Rs 50,000 per hectare expenses occurs during the installment of magnetic systems. Magnetic devices do not need technical maintenance and do not require electrical energy. Life of devices is minimum 25 years.

Benefits

- Magnetic field further breaks down minerals into smaller particles making them more bio-available to the plant cells.
- Maximum hydration of healthy water with greater uptake of minerals results in greater yields, larger and better end product.
- It allows a reduction of amount of water needed, fertilizer and pesticides.

- Since the magnetic structuring breaks all minerals into smaller particles, salt in the soil is broken down by the structured water, causing it to sink deep into the soil, away from plant roots and wash away.
- The desalinization happens quickly over a season, creating much healthier plants, greater yields and better final product.

Effect on Plants

- High quality, bigger fruit; fewer inferior grade fruit.
- Quantity of fruits.
- Time for collecting the harvest.
- Difference in quantity and quality of harvest.
- All plants developed improved stress resistance.
- Percentage of crops standing.
- Development of root system.

Effects on Soil Desalination

- Speed of filtration.
- Magnetic water dissolves salts effectively.
- Magnetic water removes nitrate and sulphates from the soil.

4. Bio-drainage

Tree species such as *Eucalyptus Spp.*, *Acacia nilotica*, *Dalbergia sissoo* and *Casurina equisetifolia* are quite efficient in intercepting the canal seepage. Planting of 2-4 rows of these trees, 5 m away from the canal could intercept more than 80 per cent of the seepage and relieve water logging problem along the canals.

Conclusion

From the forgoing discussion it can be concluded that:

1. Salt stress had a pronounced effect on leaf, stem, shoot, root, relative water content, photosynthetic rate and chlorophyll pigment contents of fruit plants (Citrus, date palm, etc.).
2. Maximum reduction occurred in germination, seedling growth and survival of seedlings of fruits crops under salt stress (Mango, aonla, etc.).
3. Use of rootstock ameliorates the adverse effect of salt stress. It enhances the vegetative growth along with number of leaves (Olour-mango, etc.).

4. In sodic soil, mulching of orchard gives positive effect on plant growth and minimizes weed population in orchard (Aonla- black polythene mulch, etc.).
5. Few varieties of different fruit crops show more vigorous growth with early flowering in saline conditions (*Banarasi*- ber, etc.).
6. Application of magnetized water under saline soil condition improves nutrient availability at root zone of fruit plants in comparison to non-magnetized water (Apple, citrus, olive, etc.).

References

Anonymous (2014). Food and Agriculture Organization, Rome, Italy.

Anonymous (2012). Central Soil Salinity Research Institute, Karnal, Haryana.

Balamohan, T.N.; Arulmozhiyan, R.; Sivakumar, K.C. and Velmurugan, S. (2002). "Study on the performance of amla cultivars under sodic soils". South Indian Hort., 50: 509-512.

Balamohan, T.N., Nayaki, D.A.; Ajagopalan, R.R. and Sivanantham, M. (2001). "Effect of sodic soil on popular cultivars of ber (*Zizyphus mauritiana* Lam.)". South Indian Hort., 49 (special): 382-383.

Dayal, V.; Dubey, A.K.; Awasthi, O.P.; Pandey, R. and Dahuja, A. (2014). "Growth, lipid peroxidation, antioxidant enzymes and nutrient accumulation in Amrapali mango (*Mangifera indica L*) grafted on different rootstocks under NaCl stress". Pl. Knowledge J., 3(1): 15-22.

Murkute, A.A.; Sharma, S. and Singh, S.K. (2006). "Studies on salt stress tolerance of citrus rootstock genotypes with arbuscular mycorrhizal fungi". Hort. Sci. (Prague), 33(2): 70-76.

Prasad, J.; Singh and Kumar, R. (1991). M.Sc. (Agri.) thesis submitted to Narendra Dev University of Agriculture & Technology, Faizabad, Uttar Pradesh.

Ramoliya, P.J. and Pandey, A.N. (2010). "Soil salinity and water status affect growth of Phoenix dactylifera seedlings". New Zealand J. Crop Hortil. Sci., 31(4): 345-353.

Rao, V.K. and Pathak R.K. (1996). "Response of sodicity and salinity levels on vegetative growth and nutrient uptake of aonla genotypes". Indian J. Hort., 53(4): 251-254.

Selim, M.M. (2008). The 3rd international conference on water resources and arid environments, Cairo, Egypt.

Sharma, D.; Dubey, A.K.; Srivastav, M.; Singh, A.K.; Sairam, R.K.; Pandey, R.N.; Dahuja, A. and Kaur, C. (2011). "Effect of Putrescine and Paclobutrazol on Growth, Physiochemical Parameters, and Nutrient Acquisition of Salt-sensitive Citrus Rootstock Karna khatta (Citrus karna Raf.) under NaCl Stress". J. Pl. Growth Regul., 30: 301-311.

Singh, A.; Prakash, J.; Srivastav, M.; Singh, S.K.; Awasthi, O.P.; Singh, A.K.; Chaudhari, S.K. and Sharma, D.K. (2014). "Physiological and biochemical responses of citrus rootstocks under salinity stress". Indian J. Hort., 71(2): 162-167.

Srivastav, M.; Dubey, A.K.; Singh, A.K.; Singh, R.; Pandey, R.N. and Deshmukh, P.S. (2009). "Effect of salt stress on mortality, reduction in root growth and distribution of mineral nutrients in *Kurukkan* mango at nursery stage". Indian J. Hort., 66(1): 26-34.

Varu, D.K. and Barad, A.V. (2010). "Standardization of mango rootstock for mitigating salt stress". Indian J. Hort., 67 (Special): 79-83.

5

Genetic Divergence Analyses in Rice (*Oryza Sativa* L.) Germplasm Lines

Harsh Vardhan Singh Shekhawat

Abstract

This study was carried out with 96 genotypes of rice representing diversity for yield and yield attributing traits. The main objective of the study was the estimation of nature and magnitude of variability; to estimate inter se genetic distance among the genotypes and grouping them into different clusters. In order to accomplish it divergence analysis (using Tocher's method) was carried out in order to assess the genetic divergence amongst the 96 germplasm lines. In this analysis germplasm lines arranged into nine clusters of which cluster (1) was found to be the largest containing 35 genotypes followed by clusters (3), (2), (4), (9) and (6) which comprises of 19, 17, 10, 7 and 5 genotypes respectively. Clusters (5), (7) and (8) comprises of one genotype each. Maximum intra cluster distance [10.92] was observed in cluster (6), containing five genotypes. Likewise, maximum inter cluster distance (75.73) was observed between clusters (6) and (9) indicating suitability of germplasm lines of these clusters for hybridization. In the study character grain yield per plant showed highest per cent contribution towards genetic divergence followed by tillers per plant, effective tillers per plant and grains per panicle.
Keywords: *Divergence, Tocher's method, Cluster, Cluster distance, Hybridization.*

Introduction

Rice belongs to genus ORYZA of Gramineae family. The genus oryza includes 24 species, of which 22 are wild and 2 namely, *Oryza sativa* and *Oryza glaberrima* are cultivated. *O.sativa* is diploid species having 24 chromosomes. Rice being the staple food for more than 70 per cent of Indian population occupies a pivotal place in Indian agriculture. Rice is grown, particularly in

India in a wide range of agro climatic situations, ranging from the high altitude Himalayan valleys to the tropical coastal areas of Kerala. The principal systems of rice cultivation followed in India are 'dry', 'semidry' and 'wet'. However, 'wet' system is the one which is widely followed and most productive. This is practiced in areas of adequate supply of water either by rainfall or irrigation. The forthcoming trend in rice is that, much more rice will be needed to be produced on less land with less labour and less water along with sustainable achievement and protection to environment. At the current rate of population growth of 1.8 per cent, the rice requirement of the country is estimated to be around 140 million tonnes by 2020. For achieving this production target in the next few decades, it has been advocated that hybrid rice may be of greater hope as it has hypothetical to produce about 20-30 per cent higher yield than the pure line varieties. The important issue is how to balance the need for ever greater food production against very real concerns about protecting natural resources and the environment for the generations to come. Biological diversity and genetic resources have great role to play in achieving sustainable agricultural development and the ecological health of the earth. Plant breeding programme with diverse genetic base could sustain a high level of crop yield. The narrow genetic base of semi-dwarf varieties is likely to make them vulnerable to different biotic and abiotic stresses. Therefore, to meet the ever-increasing demand of food grains, for higher production emphasis should be given to the genetic improvement of the existing varieties of rice. The major thrust area for such genetic improvement has been known to lay importance on selecting efficient breeding system and identifying potential genotypes in hybridization programme. This would depend to large extent on the knowledge of the genetic system controlling the various characters. With this backdrop the present investigation was undertaken to estimate the genetic distance amongst the genotypes and to identify the trait contributing towards maximum diversity.

Materials and Methods

The present investigation was carried out at agricultural research farm, Institute of Agricultural Sciences, Banaras Hindu University, Varanasi. The genotypes under study were a set of 96 diverse rice germplasm lines from IRRI. All the 96 genotypes were grown in *Kharif* season of 2010. The crop was directly sown under the irrigated conditions on 7th June 2010. Ninety six genotypes were grown in a completely randomized Block Design with two replications. Each plot consisted of two rows of 1.5 m length. Spacing was maintained at 15×10 cm excluding the induced irrigation; the normal recommended agronomic practices were followed. Fertilizers were applied at the rate of 120 kg N, 80 kg P_2O_5 and 80 kg K_2O per hectare. Five plants were randomly selected from each entry in each replication and their means were used for the statistical analysis.

Result and Discussion

In this investigation in order to study genetic divergence 96 germplasm lines were subjected to divergence analysis using Tocher's method. In the study nine clusters were obtained. Table (1). As per the findings a highest of 35

Table 1

Cluster	*Genotypes as Per Tocher's Method*	*Number*
1.	7, 13, 14, 16, 17, 20, 21, 22, 23, 29, 32, 36, 38, 39, 41, 44, 45, 46, 47, 49, 50, 54, 55, 57, 58, 61, 64, 71, 72, 77, 81, 82, 83, 84, 95	35
2.	5, 12, 19, 25, 31, 34, 42, 48, 56, 59, 60, 62, 67, 70, 89, 90, 94	17
3.	4, 18, 24, 27, 28, 33, 40, 52, 68, 69, 73, 76, 79, 80, 85, 86, 87, 88, 96	19
4.	3, 6, 8, 9, 15, 43, 51, 53, 74, 98	10
5.	93	1
6.	30, 37, 75, 91, 97	5
7.	92	1
8.	26	1
9.	1, 2, 10, 63, 65, 66, 99	7

genotypes clustered in cluster (1) followed by clusters (3), (2), (4), (9) and (6) having 19, 17, 10, 7 and 5 genotypes respectively.Through the cluster diagram D^2 values and D values are extracted as shown in Table (2). Among the clusters highest intra cluster distance of 10.92 (D values) has been found for Cluster 6 which comprises of LINES (30, 37, 75, 91 and 97). Cluster 6 is followed by Cluster 9 (10.13) which comprises of LINES (1, 2, 10, 63, 65, 66 and 69).

Similarly, lowest intra cluster distances [0] have been found for Clusters (5), (7) and (8) respectively which itself reveals that each of these clusters are formed of single genotype, namely LINE 93 (Cluster 5), LINE 92 (Cluster 7) and LINE 26 (Cluster 8). The highest inter cluster distance (75.53) through Tocher method has been found between Cluster 6 (LINES 30, 37, 75, 91 and 97) and Cluster 9 (LINES 1, 2, 10, 63, 65, 66 and 69) followed by inter cluster distances (61.00), (58.36) and (55.06) between clusters 4 and 6, 7 and 9 and 8 and 9 respectively. High inter cluster distance would mean high phenotypic diversity for genotypes contained in clusters which would be mainly because of genetic diversity when environment influence is small. Therefore, the genotypes selected from distant clusters would be expected to yield wide spectrum of variability on hybridization in segregating generations. Mean values of different traits for different clusters have been presented in Table (10). The highest mean values for different traits are observed to fall in different clusters. Cluster (5), (6) and (7) exhibited highest mean values for most of the traits. Cluster (5) exhibited highest values for total tillers per plant, test weight and grain L/B ratio. Cluster (6) recorded highest values for effective tillers per plant

Table 2: Average Intra and Inter cluster D^2 values (D values) among 9 clusters of 96 rice genotype

	1 Cluster	*2 Cluster*	*3 Cluster*	*4 Cluster*	*5 Cluster*	*6 Cluster*	*7 Cluster*	*8 Cluster*	*9 Cluster*
1 Cluster	30.28 (5.50)	209.40 (14.47)	197.50 (14.05)	753.86 (27.46)	69.22 (8.32)	1196.01 (34.58)	375.32 (19.37)	265.03 (16.28)	1786.30 (42.26)
2 Cluster		33.21 (5.76)	715.47 (26.75)	217.20 (14.74)	268.64 (16.39)	2272.39 (47.67)	952.22 (30.86)	796.83 (28.23)	870.24 (29.49)
3 Cluster			42.83 (6.54)	1604.06 (40.05)	207.14 (14.39)	512.03 (22.63)	159.63 (12.63)	96.15 (9.81)	3000.12 (54.77)
4 Cluster				35.79 (5.98)	856.34 (29.26)	3721.42 (61.00)	1911.03 (43.72)	1676.24 (40.94)	285.26 (16.89)
5 Cluster					0.00 (0.00)	1192.67 (34.54)	251.78 (15.87)	377.60 (19.43)	1950.28 (44.16)
6 Cluster						119.16 (10.92)	584.33 (24.17)	528.29 (22.98)	5704.46 (75.53)
7 Cluster							0.00 (0.00)	329.10 (18.14)	3405.83 (58.36)
8 Cluster								0.00 (0.00)	3032.11 (55.06)
9 Cluster									102.61 (10.13)

and grian yield and cluster (7) expressed highest mean values for chlorophyll content, panicle length and grains per panicle. Only maximum cluster mean value for plant height was exhibited by cluster (3).

Genotypes contained in clusters exhibiting high cluster mean for specific traits may be selected from the above mentioned clusters for breeding programme. Lowest mean values for most of the traits except chlorophyll content are exhibited by Cluster 1. Therefore, genotypes in this cluster are expected to be contrasting for most of the traits against the genotypes exhibiting higher cluster mean. Based on the information from cluster means contrasting genotypes for different traits may be utilised for making divergent crosses. Lowest mean value for chlorophyll content is recorded from Cluster 6.

A perusal of cluster means indicated that their existed considerable differences in mean value of the different traits. Distribution of maximum and minimum mean values for different traits in distinct cluster indicated the traits contributing to the total divergence. In this regard, grain yield per plant contributed maximum to the total divergence followed by tillers per plant, effective tillers per plant and grains per panicle. More than 80 percent contribution towards total divergence was mainly because of these four characters. In their findings Ahmed and Borah (1999) found out that tiller number, panicles per plant, grains per panicle, grain fertility and grain yield accounted for the major portion of genetic divergence; while Banumathy *et al.* (2010) reported maximum contribution was made by grain yield followed by days to 50 per cent flowering, total grains per panicle and plant height; and Vennila *et al.* (2011) in their studies accounted maximum contribution to number of grains per panicle, plant height, grain length and grain breadth.

The major contributing character towards genetic divergence was found to be grain yield per plant (28.52 per cent) followed by effective tillers per plant (19.59 per cent), tillers per plant (18.24 per cent) and grains per panicle (17.20 per cent). In Tocher's method of clustering maximum intra cluster distance was observed for cluster 6 containing five genotypes followed by cluster 9 and cluster 3. Cluster 5, 7 and 8 each contained single genotype namely, LINE 26, LINE 92 and LINE 93. All the single genotype clusters showed large inter cluster distances with cluster 9. Cluster 1 was the largest with respect to number and genotypes it accommodated 35 genotypes and exhibited low intra cluster distance indicating that a large number of genotypes were phenotypically similar with respect to number of traits in the present study. The maximum inter cluster distance was observed between cluster 6 and cluster 9. This indicates that the hybridization between the genotypes of cluster 6 and cluster 9 would yield desirable segregates with accumulation of favourable genes in the segregating generations. Similar was suggested by Sarma, Richharia and Agarwal (1996). Following cluster 6 and 9 were the clusters 4 and 6, cluster 7 and 9 and cluster 8 and 9 respectively. The greater the distance between two clusters, wider the expected genetic distance between their genotypes. The genetic distance between the parents largely governs the variability spectrum generated in the segregating generation and also the F1 heterosis. Similar suggestions were given by Babu *et al.*(2003), Chauhan and Singh (2003).

References

Ahmed, T. and Borah, P. (1999). Genetic diversity in glutinous rice germplasm of Assam. *Oryza* 36(1): 74-75.

Arun Sharma, Gupta, K.R., and Rakesh Kumar. (2008). Genetic divergence in Basmati rice (*Oryza sativa*) under irrigated ecosystem. *Crop improvement*. 35(1): 8-10.

Awan, M.A. 1999: Mutation breeding for crop improvement: A review. *Proc. Pak. Acad. Sci.*, 36: 65-73.

Babu, S., Anbumalaramati, J., Yogameenakshi, P., Sheebha, A. and Rangaswami (2003). Genetic diversity studies in rice (*Oryza sativa* L.). *Crop Res*. 25(2): 280-286.

Banumathy, S. Manimaran, R. Sheeba, A. Manivannan, N. Ramya, B. Kumar, D. Ramasubramanian, G. V. (2010). Genetic diversity analysis of rice germplasm lines for yield attributing traits. *Electronic Journal of Plant Breeding*. **1**(4): 500-504.

Baruah, R.K.S.M., Das, G.R. and Hazarika, M.H. (1994). Diversity of *Ahu* rice of Assam and its characterization. *Oryza* 31:163-168.

Chauhan, J.S. and Singh, K.H. (2003). Genetic divergence analysis using quality traits in upland rice (*Oryza sativa* L.). *Annals of Agric. Res.* 24(3): 673-675.

Padmaja, D. Radhika, K.Rao, L.V.S. Padma, V. (2010). Studies on genetic divergence in rice (*Oryza sativa L.)* germplasm. *Crop Research (*Hisar). 40(1/3): 117-121.

Parimalan, R., Joel, A. J. and Vanniarajan, C. (2008). Genotypes clustering in rice (*Oryza sativa* L.). *Crop Improvement*. 35(2):115-118.

Sarma, M.K., Richharia, A.K. and Agrawal, R.K. (1996). Variability, heritability, genetic advance and genetic divergence in upland rice. *IRRN* 21(1):25-26.

Vennila, S. Anbuselvam, Y. Palaniraja, K. (2011). D^2 analysis of rice germplasm for some quantitative and quality traits. *Electronic journal of Plant Breeding*.2*:3, 392-396.*

Eswaran, R. (2012). Genetic divergence analysis for certain yield traits in rice (*Oryza sativa* L.). *International journal of recent scientific research*. 3 (6):486-488.

Ahmed, A., Singh,V.K., Sharma, V. and Paswan, S.K. (2014). Genetic divergence analysis in rice (*Oryza sativa* L.) germplasm in sodic soil condition. *The Ecoscan* 6:51-54.

6

Reliability of Single Vs Ensemble Model Simulations to Study the Impacts of Climate Change on Agriculture: CLIVAGRI Project Experience

Raveendra H. Patil (Ravi Patil)

Abstract

With the growing threats of climate change research efforts are being done to identify or develop adaptation strategies for the projected future climates using both the climate manipulative field experiments (e.g., FACE studies) and dynamic multi-process crop simulation models. Of these two methods, in the last two decades our dependence on process-based models has heavily increased to assess the impacts of CC on global food production. While these models were initially developed for a given locality and to simulate crop growth at plot or field level and in decision support system, later the use of same models were extended across regions and at multiple scales to assess the impacts of CC and identify adaptation strategies. This is done because crop models are being considered as efficient and effective tools, and modelers / modelling teams achieved reasonable success as they re-calibrated the models using newer datasets and applied the models across environments and production systems. This could be seen in the sudden increase in publications during the last 10-15 years.

However, with so much at stake under changing climate, the question now being asked is: are our current pool of models and modelling approaches take into account all the complex interactions between crop, soil, environment and management as well as non-linearity in crop response to climatic factors? The answer certainly seems NO based on the recent multi-model inter-comparison studies. One of the reasons is most of the models currently being used were developed using control environment experiments and old crop cultivars prevalent during 1970s and 1980s, thus models in all probability, may not have incorporated the latest know-how on crop physiology and response to changing environment, thus may not represent modern crop cultivars and management practices properly. Furthermore, most of the modelling studies carried out until recently have failed to quantify the uncertainty of individual models, which is questioning the robustness of models as well as the reliability of model outputs in decision-making process. There is a growing voice amongst the modelling community that the models need to be updated to reflect new research in crop physiology, agronomy and soil science. This presentation will go over number of case studies taken up in recent past under COST-734 modelling project entitled "Climatic Variability and its Impacts on Agriculture" (CLIVAGRI) funded by European Union between 2009 and 2012 to prove a hypothesis that multi-model outputs are more reliable than single model outputs in assessing the impacts of climate change on agriculture.

Keywords: *climate change, impacts, adaptation, crop models, model inter-comparison.*

Introduction and Rationale

In recent past process-based crop simulation models (CSM) as decision support tools are increasingly being used in decision-making processes for different purposes (prediction, projections, policy, preparedness etc.) and planning in the field of agriculture at different levels (field, farm, landscape, region etc.). Particularly in the context of changing climate and global warming dependence on models is increasing to understand/predict the impacts on crops under future projected climates and accordingly prepare or develop adaptation strategies at various scales for regional and global food security. Most CSMs currently used are mechanistic as they explain not only the relationship between parameters (climate, soil, input and management) and simulated variables (e.g., yield), but also the mechanism of the various underlying processes (Porter and Semenov, 2005). Therefore, process-based mechanistic CSMs are being considered as an efficient and effective tools in climate change impact assessment studies, and in the decade or so, modellers/ modelling teams across the world have achieved reasonable success as they keep on re-calibrated and validating CSMs using newer datasets for improvement and applied the models across environments and production systems. This could be seen in the sudden increase in publications during the last 10-15 years (White *et al.*, 2011).

Climate change is scientifically proven fact and a reality. Its threats and impacts are global in nature, more so on agriculture and its allied activities.

Therefore, adaptation to climate change is must for the sustainable development of agriculture and global food security, but is a costly affair. United Nations Framework Convention on Climate Change estimates that in the field of agriculture, forestry and fisheries sectors some US$14 billion per annum will be required by 2030 to cope with the adverse impacts of climate change. Although experts think this figure could actually be twice or thrice greater (Fankhauser S, 2009). Information about the potential impacts of climate change on food production and supply is essentially required by the agricultural scientists and policymakers to plan for the coming decades. The United Nations Food and Agriculture Organisation estimates that current agricultural productivity needs to be increased by 70 per cent to meet the growing food demand expected by 2050.

However, with so much at stake in agriculture under changing climate the models we use should quantify complex interactions between crop-soil-environment accurately and realistically to support agricultural production, management and policy decisions. In this context, the question now being asked is: are our current pool of models and modelling approaches take into account all the complex interactions between crop, soil, environment and management, each varying so much at every scale (e.g., field, farm, region etc.) as well as non-linear response of crops to climatic variables (e.g., temperature)? The answer certainly seems to be no based on the recent multi-model inter-comparison studies undertaken within CLIVAGRI project in Europe. Single model outputs tend to simulate with large uncertainties (Rotter *et al.,* 2011). This paper intends to highlight this issue and caution overenthusiastic model users in jumping on to unrealistic conclusions based on single model outputs. One of the reasons for the unacceptably high uncertainty with model outputs is most of the models currently being used were developed using control environment experiments and old crop cultivars that existed during 1970s and 1980s, thus models in all probability, may not have incorporated the latest know-how on crop physiology and the response patterns and rates to changing environment (e.g., rising temperature, heat and moisture stress, solar radiation etc.), thus may not represent modern crop cultivars and management practices properly (Rotter *et al.,* 2011).

Another potential reason for high uncertainty could be that the CSMs when first developed they were meant to be evaluated and used at the field scale, hence they were not meant for scaling-up to be used at larger area *viz.*, landscape, regional, national or global scale. Despite this, in the last decade or so with the growing concern of global warming, hundreds' of studies have been carried out across the world to assess the impacts of changes in climatic parameters (temperature, rainfall) and climate variability (extreme events) on crops and agro-ecosystems, and tried to develop adaptation strategies both at the field scale to regional and national scale using the same CSMs (Parry *et al.,* 2005; White *et al.,* 2011). While these efforts have been useful to improve our understanding on the impacts of climate change, but as these CSMs' parameters and processes were basically structured and developed to suit at field level, scaling up or applying at larger scale to assess climate impacts and to develop adaptation options may have introduced greater uncertainties due to the issues related CSMs parameters and because of each CSM's basic structure (Palosuo *et al.,* 2011). Larger the uncertainty in CSM output meant the larger error in the output leading to lower reliability while using at decision-making process.

For example, field experiments in the last decade or so have shown that in crops like wheat, rice and maize when temperatures go above thresholds, say for instance, >34–36°C especially during pre- and post-anthesis period cause sharp reduction in grain setting and final grain yield. Most process-based models do not account for this as most models either have set Tmax of a given crop below this range or do not include the process occurring in the plant at or beyond this range of temperatures. So current models tend to overestimate projected future yields in regions semi-arid and arid regions) already experiencing more frequent hot days, beyond optimum and/or maximum range, during the growing season. Similarly, it is also known that increased atmospheric CO_2 concentrations help reduce crop water loss through the stomata, but are not accurately represented in many models. Further, many models lack explicit details about photosynthesis, hence cannot account for the interaction between atmospheric CO_2 concentrations, higher temperature and water use pattern, thus may again overestimate the effects of droughts period under elevated atmospheric CO_2 concentrations (Rotter *et al.,* 2011). Therefore, these issues need to be either incorporated into the models or taken into account while interpreting the model outputs in decision-making process.

Why go for Multi-model Simulation?

One of the basic steps in model development and improvement is to calibrate the chosen CSM thoroughly for the given locality/condition and optimise model parameters (soil, crop, management modules) before used as decision support system (DSS) tool. This step is often not been done thoroughly and are not been evaluated rigorously to check if the chosen CSM captures the effect of climatic variability on crop performance vis-à-vis physiological processes, phenology and yield both under the conditions for which the CSM was developed or for the new conditions where CSM was applied. Further, it is also the known fact that most model users (not model developers) are not fully familiar with detail of each process and its mechanism, and potential limitations and specificities for their proper application. Furthermore, most of the modelling studies carried out until recently have failed to identify and quantify the uncertainty of individual models, which is questioning the robustness of models as well as the reliability of model outputs in decision-making process when used at larger scale. There is a growing voice amongst the modelling community that the models need to be updated to reflect new research in crop physiology, crop improvement, agronomy and soil science.

Owing to the above mentioned complexity many models do not agree among themselves on how crops respond to projected changes in climate. For instance, Jamieson *et al.* (1998) in New Zealand compared five different models for winter wheat crop. They reported deviation in simulated wheat yields by up to 20 per cent, and the tested models differed more in estimating other yield attributing variables (e.g., above-ground biomass). Similar observations on wheat and potato have also been reported in different models. Therefore, back in 2002, many a modellers suggested to take up further work to identify and quantify the reasons for these large inter-model variations in their simulations.

Rosenzweig and Wilbanks, (2010) in their recent work underlined the urgency to take up inter-comparison modelling studies. In this project hypothesis was multi-model ensemble simulations provide a range of possible outcomes, valuable insight into each CSM's simulation in comparison to others, and help identify importance of different physical and biological processes included into the models. These days with fast increasing computing and storage capacity crop simulations modelling has become less challenging computationally than climate modelling, hence the project partners believed multi-model ensemble approach would be more useful and reliable in impact assessment than single model simulation. As such there is no such thing as 'one best model' applicable to all environments, conditions and purposes, deriving and using multi-model simulation output averages seems to be a promising practice (Rotter *et al.*, 2011; Asseng *et al.*, 2013).

Comparison of different modelling approaches and CSMs can help us identify the uncertainties / errors coming from each CSM with regard to crop growth and yield predictions, and uncertainty related to CSM structure / mechanism itself, although quantification of the latter is very difficult to achieve (Chatfield, 1995). Despite that multi- model comparison help identify process(s) of CSM that produce systematic errors and obviously require further changes for improvement (Palosuo *et al.*, 2011). Realising the importance of this process quite a few studies have been attempted in the past, and have compared different mechanistic CSMs with regard to their performance in predicting crop yield and yield variability in response to climate change and other environmental variables (Porter *et al.*, 1993; Goudriaan *et al.*, 1994; Wolf *et al.*, 1996; Meinke *et al.*, 1998; Jamieson *et al.*, 1998; Eitzinger *et al.*, 2004; Kersebaum *et al.*, 2007). But majority of these studies at the most included two or couple of CSMs for the comparison, and for more than a decade no attempt had been made at the European or global level to compare considering most of the major easily accessible and widely used CSMs, particularly on cereal crops of global importance.

CLIVAGRI Modelling Project

In the context of above discussion and rationale CLIVAGRI Project: Impacts of Climate Change and Variability on European Agriculture (COST 734) funded by the European Union under European Cooperation in the field Science and Technical Research (www.cost.esf.org) between 2009 and 2012 undertook a large scale inter-comparison modelling study with an aim to examine and compare how different process-based CSMs perform at the field scale with limited calibration when applied to larger areas representing multiple sites across Europe and for the Europe's dominant cereal crops (winter wheat, spring barley and maize).

This presentation will go over the studies taken up under COST-734 CLIVAGRI modelling project to indicate if single- or multi-model outputs are more reliable in assessing the impacts of climate change on agriculture. In this project some eight to twelve crop models were run for different cereal crops: winter wheat (Palosuo *et al.*, 2011, Eitzinger *et al.*, 2012), spring barley (Rotter *et al.*, 2012; Salo *et al.*, 2015 in publication process) and maize (Eitzinger *et al.*, 2012) and for large number of growing seasons (max. 49 seasons) at eight different study sites across Europe: Czech Republic, Denmark, Germany, Slovakia and Turkey.

Three scientific papers have already been published in high impact international journals (Palosuo *et al.*, 2011; Rotter *et al.*, 2012; Eitzinger *et al.*, 2012) and fourth one is in review process with the Journal of Agriculture Sciences, Cambridge (Salo *et al.*, 2015). For instance, a study on winter wheat for across European climates published by Palosuo *et al.* (2011) used eight models: APES, CROPSYST, DAISY, DSSAT, FASSET, HERMES, STICS and WOFOST. Second study on spring barley again across European sites published by Rotter *et al.* (2012) used nine models: APES, CROPSYST, DAISY, DSSAT-CERES, FASSET, HERMES, STICS, WOFOST and MONICA. Third study on winter wheat and maize by Eitzinger *et al.* (2012) for two Austrian sites used only seven models: CROPSYST, DSSAT, FASSET, HERMES, EPIC, WOFOST and AQUACROP. Whereas the paper submitted by Salo *et al.* (2015) studies response of different models under N limited environments for Finnish conditions on spring barley used eleven models: APSIM, DSSAT-CERES, CROPSYST, COUP, DAISY, EPIC, FASSET, HERMES, MONICA, STICS and WOFOST.

Material and Methods

The total CSMs included in the inter-comparison were twelve: APES, APSIM, AQUACROP, COUP, CROPSYST, DAISY, DSSAT-CERES, EPIC, FASSET, HERMES, MONICA, STICS, WOFOST and the details of these models can be obtained from the references listed in Table-1. For the inter-comparison of models previously unpublished experimental data from different chosen locations/site across Europe on winter wheat, spring barley and/or maize were used. The production systems at different sites represented diverse agro-ecological conditions of Europe. All the model users within the partner modelling groups performed "blind model runs" based on minimum data set.

Table 1: Models used in CLIVAGRI project and the reference(s) respective model's descriptions

Model	*Applications for cereals*
APSIM	Keating *et al.*, 2003 and Anwar *et al.*, 2009
CERES	Hlavinka *et al.*, 2010, Eitzinger *et al.*, 2004
CROPSYST	Donatelli *et al.*, 1997, Stokle *et al*, 2003
DAISY	Hansen *et al.*, 1990, Abrahamsen and Hansen, 2000
EPIC	Strauss *et al.*, 2012

Therefore, data required by the models actually excluded many a sites from being included in this project and finally only limited the number of sites were selected. Further, from the finally chosen sites some more data was excluded for the seasons during which yields were reported to be severely affected by pests, diseases or lodging, in spite of plant protection measures.

Main Findings

This project work clearly suggested that application of CSMs simulation models with very limited calibration leads to a high degree of uncertainty especially is used to assess climate impacts on yield and yield variability. The mean model predictions were found to be much more closer to the observed yields than individual model's outputs (Figure 1) and further it is also proved in another recent study as part of AgMIP modelling that to reduce variability in CSM outputs multi-model mean output is better (Figure 2). This proves the hypothesis that multi-model estimates are more reliable than single model simulations. Individual models seem to perform well for a specific regions and agro-ecological conditions where they have been developed and used again and again after repeated calibrations. In this project all the models used were calibrated only for phenology *i.e.,* limited calibration and found that crop models need thorough calibration for the all the important parameters before they can be applied with confidence. As expected, couple of models performed better than others in estimating grain yield and other crop variables or at certain locations, but none performed well for all the variables and across locations, hence no model could be termed robust and accurate across different environments and for different crop cultivars. Therefore, this project's outcome strongly support the argument for ensemble crop modelling.

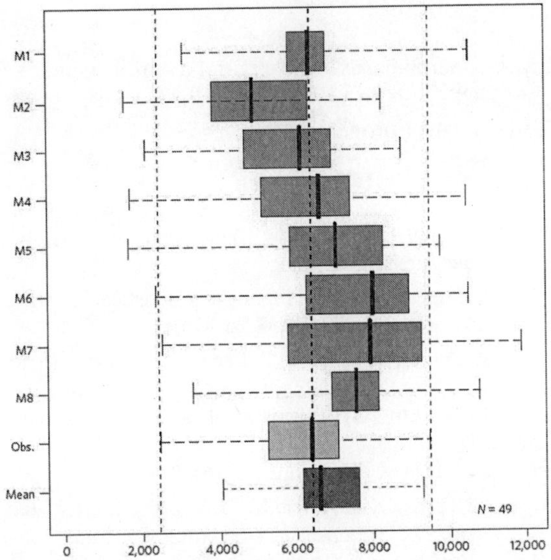

***Fig. 1:* Simulated grain yield of winter wheat by eight different models (M1 to M8) in comparison with the observed grain yield from the experimental field (Obs) and mean of all the eight CSMs grain yield outputs (Mean) (Source: Rotter et al., 2011).**

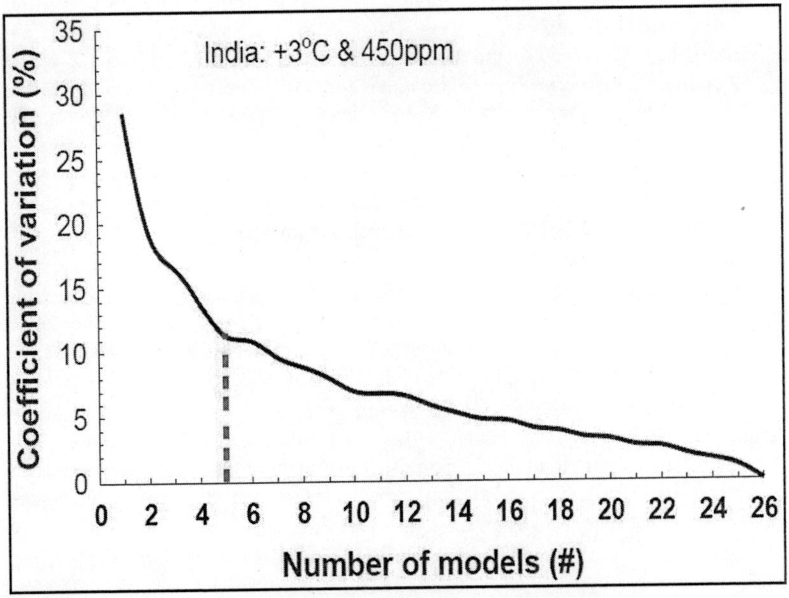

***Fig. 2:* Coefficient of variation (CV in per cent) of simulated yield responses to combined changes in temperature and atmopheric CO_2 concentrations as a function of number of average model responses for India. Vertical dotted line shows number of models used (five) in this particular study for India (Source: Asseng et al., 2013).**

Acknowledgement

This study was undertaken as a co-operative project under European COST734 "Impacts of Climate Change and Variability on European Agriculture – CLIVAGRI" funded by the European Union.

References

Asseng *et al.*, 2013. "Uncertainty in simulating wheat yields under climate change". *Nature Climate Change*, 3 (9): 827-832.

Chatfield, C., 1995. "Model uncertainty, data mining and statistical inference". *Journal of the Royal Statistical Society: Series A*, 158: 419–466.

Eitzinger, J., S. Thaler, E. Schmid, F. Strauss, R. Ferrise, M. Moriondo, M. Bindi, T. Palosuo, R. Rotter, K.C. Kersebaum, J.E. Olesen, R.H. Patil, L. Saylan, B. Caldag, and O. Caylak, 2013. "Sensitivities of crop models to extreme weather conditions during flowering period demonstrated for maize and winter wheat in Austria". *The Journal of Agricultural Science,* Cambridge, 151(6): 813-835.

Eitzinger, J., M. Trnka, J. Hosch, Z. Zalud, and M. Dubrovsky, 2004. "Comparison of CERES, WOFOST and SWAP models in simulating soil water content during growing season under different soil conditions". *Ecological Modelling*, 171: 223–246.

Fankhauser, S, 2009. "The costs of adaptation". *Working paper-8 submitted to the ESRC Center for Climate Change Economics and Policy*, UK (http://www.cccep.ac.uk/), pp. 12.

Goudriaan, J., S.C. van de Geijn, and J.S.I. Ingram, 1994. "GCTE Focus 3 Wheat modelling and experimental data comparison workshop report", *Lunteren, The Netherlands, November 1993. GCTE Focus 3 Office, University of Oxford, Oxford, UK.*

Jamieson, P.D., J.R. Porter, J. Goudriaan, J.T. Ritchie, H. van Keulen, and W. Stol, 1998. "A comparison of the models AFRCWHEAT2, CERES-Wheat, Sirius, SUCROS2 and SWHEAT with measurements from wheat grown under drought". *Field Crops Research*, 55: 23–44.

Kersebaum, K.C., J.M. Hecker, W. Mirschel, and M. Wegehenkel, 2007. "Modelling water and nutrient dynamics in soil–crop systems: a comparison of simulation models applied on common data sets". In: Kersebaum, K.C., Hecker, J.M., Mirschel, W., Wegehenkel, M. (Eds.), *Modelling Water and Nutrient Dynamics in Soil Crop Systems. Springer, Dordrecht*, pp. 1–17.

Meinke, H., R. Rabbinge, G. Hammer, H. van Keulen, and P.D. Jamieson, 1998. "Improving wheat simulation capabilities in Australia from a cropping systems perspective II. Testing simulation capabilities of wheat growth". *European Journal of Agronomy*, 8: 83–99.

Palosuo, T., K.C. Kersebeaum, C. Angulo, P. Hlavinka, M. Moriondo, J.E. Olesen, R.H. Patil, F. Ruget, C. Rumbaur, J. Takac, M. Trnka, M. Bindi, B. Caldag, F. Ewert, R. Ferrise, W. Mirschel, L. Saylan, B. Siska, and R. Rotter, 2011. "Simulation of winter wheat yield and its variability in different climates of Europe. A comparison of eight crop growth models". *European Journal of Agronomy*, 35: 103-114.

Parry, M., C. Rosenzweig, and M. Livermore, 2005. "Climate change, global food supply and risk of hunger". *Philosophical. Transactions of the Royal Society of London/B 360*, 2125–2138.

Porter, J.R., P.D. Jamieson, and D.R. Wilson, 1993. "Comparison of the wheat simulation models AFRWHEAT2, CERES-Wheat and SWHEAT for non-limiting conditions of crop growth". *Field Crops Research*, 33: 131–157.

Porter, J.R., and M.A. Semenov, 2005. "Crop responses to climatic variation". *Philosophical Transactions of the Royal Society of London/B 360*, 2021–2035.

Rosenzweig, C., and T.J. Wilbanks, 2010. "The state of climate change vulnerability, impacts, and adaptation research: strengthening knowledge base and community". *Climate Change*, 100, 103–106.

Rotter, R.P., T.R. Carter, J.E. Olesen and J.R. Porter, 2011. "Crop-climate models need an overhaul". *Nature Climate Change*, 1: 175-178.

Rotter, R., T. Palosuo, K.C. Kersebaum, C. Angulo, M. Bindi, F. Ewert, R. Ferrise, P. Hlavinka, M. Moriondo, C. Nendel, R.H. Patil, F. Ruget, J. Takac, M. Trnka, and J.E. Olesen, 2012. "Simulation of spring barley yield in different climatic zones of Northern and Central Europe. A comparison of nine crop models". *Field Crops Research*, 133: 23-36.

Salo, T., T. Palosuo, R.P. Rotter, K.C. Kersebaum, C. Nendel, C. Angulo, F. Ewert, M. Bindi, P. Calanca, T. Klein, M. Moriondo, R. Ferrise, J.E. Olesen, R.H. Patil, F. Ruget, J. Takac, P. Hlavinka, and M. Trnka, (in press). "Comparing the performance of eleven agro-ecosystems models in simulating yield response to nitrogen fertilization" (*Journal of Agriculture Sciences,* Cambridge, UK).

White, J.W, G. Hoogenboom, B.A. Kimball, and G.W. Wall, 2011. "Methodologies for simulating impacts of climate change on crop production". *Field Crops Research*, 124: 357–68.

Wolf, J., L.G. Evans, M.A. Semenov, H. Eckersten, and A. Iglesias, 1996. "Comparison of wheat simulation models under climate change. 1. Model calibration and sensitivity analyses". *Climate Research*, 7: 253–270.

7

Nano Particle Production – For Climate Smart Agriculture

P. S. Pandit

Abstract

The nano-particle may be the atom, molecules, structure or compound, which is having at least one dimension bellow 100 nm size. This nano-particle behaves separately, accurately, smartly, aggressively, more specifically compared to normal particle of same atom or compound. There are two type of nano particles; naturally occurred and engineered. The material specific property like, high surface to volume ratio has generated space at the bottom size of scale to utilise this technology. In field of agriculture and allied-science, nano-technology could be utilise to produce metabolically more absorbing fertiliser compounds like; nano-rock-phosphate, nano-potassium; efficient micronutrient supplement materials like; nano-zinc powder, nano-magnesium dust, nano-copper powder. As far as instrumentation side is concern, carbon or metal based nano-particle could combine with sensors and utilised in the precision agriculture. Speed, accuracy and range of sensor could be improved if it utilise with nano-particle. Nanotechnology could also utilise to develop biosensor-based tools for detecting spoilage of food and drug, chemical trace detection in water and blood as well as environmental remediation. There are three methods for production of nano-particle viz; physical, chemical and biological. Each method has its own advantages and limitations. Some of the successful nano-particle technology developed were; silicate nano-particle, Clay-based nano-composite, Nano-composite KU2-2601 plastic film, Kodak nano-biological-antibody composite film etc. There are very tremendous scopes for nano-particle for climate smart agriculture but, due to new field, scientists still not confidant about nano-particle behaviour in the ecosystem. It may diffuse, absorb, leached or react with existing matter or media is not defined. Reaction of unknown or newly engineered nano-particles with human metabolic physiology or present production system may generate great disaster for livesphere. With the precise precaution, the uses of nano-particles have wide scope in smart agriculture and climate control so it became the upcoming area of researcher and educationalist for precision and smart agriculture.

Nano-Particle

The word 'nano' has originated from a Greek language. It means 'draft' and so technically, the word 'nano' means, 10^{-9} or one billionth of part. The symbol of 'nano' is 'n'. It is now used as prefix of scale e.g., nano-meter (nm=10^{-9} m) (Anonymous, 2002). The nano-scale is not just another step toward miniaturization but a qualitative new scale of the material. The nano-particles (NPs) may be defined as the materials included; atom, molecules, structure or compound, that are bellow 100 nm size at least at one dimension. This means that nano materials can be three dimensional particles like; sphere, cuboids, pyramid etc., or two dimensional particles like; ultra thin film, cylinder, etc. or one dimensional like; thin rode, thread, etc. (Lijima, 2008). The nano-particles (NPs) might be defined based on size at which fundamental properties differ from those of the corresponding bulk materials. (Banfield and Zhang, 2001). This nano-particle behaves separately, aggressively, more specifically compared to bulk materials (Kundu, 2014).

The manipulation or self-assembly of individual atoms, molecules or molecular clusters into structures for creation of materials and devices with new or vastly different properties. In history also, the trend toward the small began with the miniaturisation of macro techniques, which led to the now well-established field of micro-technology, electronic, optical and mechanical micro-technologies have all profited from the smaller, smarter and less costly sensor-based technology that resulted from work to ICs, fiber optics, micro lenses and micro-electro-mechanical system (Biswas, 2014). The most of present traditional technology works on the high volume to surface ratio, where reaction efficiency, mass, duration and activation energy are high. If we transform entire reaction on high surface to volume ratio, the reaction efficiency, mass, duration and activation energy are low. Thus, the properties of NPs depend on high surface to volume ratio. The nano-material specific property like, high surface to volume ratio has generated the space at the bottom size of scale to utilise this technology. This ratio has provided the use of NPs in the form of technology (Adhikari, 2014).

There are two type of nano particles, (i) naturally occurred and (ii) engineered. In naturally occurred nano-particle, one has to bring down the particle size bellow 100 nm using external source of energy through magnetron, burr mill, electromagnetic chemical distillation, magneto-optical oscillation etc.

In this the structure of the particle cannot be designed-constructed pure material like; Nano-phosphate, nano-zink, nano-gold, nano-silver etc. (Kundu, 2014). A number of NPs are commercially available in the form of dry powder or liquid dispersions e.g., SiO_2, TiO_2, Al_2O_3, Fe_3O_4, ZnO, Cu_2O_3 and ZrO_2, In_2O_3 SnO_2, Sb_2O_3-SnO_2, $BaTiO_3$, metal oxide, ceramics and silicate, magnetic and materials, lyposomes, dendrimers, emulsions, nano-carbon tube CNT), nano-carbon balls (CNB), nano-carbon plates (CNP) (Singh, 2014, Hochella, *et. al.* 2008 and Consales, *et. al.* 2007).

Table 1: Size Dimensions of Different Natural Materials

Name	*Size*
Molecules	< 1 nm
Colloids	1-1000 nm
Nanoparticles	1-100 nm
Virus	10-100 nm
Red blood cell	2000-5000 nm
Bacteria	250-1000 nm
Tissue cell	1000-10000 nm
Width of human hair	80000 nm
Intracellular spaces in seed coat parenchyma	<10000 nm

Nano-Sensor and Technology

Nanotechnology is emerging out as the sixth revolutionary technological field in the current era after the Industrial Revolution of mid-1700s, Nuclear Revolution of the 1940s, The Green Revolution of 1960s, Information Technology Revolutions of 1980s and Biotechnology Revolution of 1990s. As per the SCI for the period 1999-2008, India occupied the 10th position among the most active ten countries involved in the field of nanotechnology (Kundu, 2014). Nanotechnology is defined as the understanding and control of matter at dimensions of roughly 1-100 nm, where unique physical properties make novel applications possible (Singh, 2014).

The British Standard Institution (BSI, 2005) and American Society for Testing and Materials (ASTM, 2006) defined nanotechnology as 'Design, characterisation, production and application of structure, devices and systems controlling shape, size and composition at the nanoscale. (http:/ nanotechweb.org) The instrumentation and arrangement of nano-particle to use as specific toll of science and engineering is known as nano-technology. It is technology, which deals with nano size particle and sensors. This technology could be work from the top to bottom, which means reduce the size of the smallest structure to the nanoscale e.g., photonic application in nano-engineering or the bottom up, which involves manipulating individual atom and molecules into nano-structures and more closely resembles chemistry and biology. The nanomaterial specific property like, high surface to volume ratio has generated the space at the bottom size of scale to utilise this technology. This particular property will allow developing nano-sensors (Vo-Dinh, *et. al.*, 2001).

The sensors are devices that respond to chemical, physical and biological stimuli, resulting in the production of a signal that provides a quantitative measurement of the stimuli. These are the fundamental elements of all the machines that gather data, require feedback for their operation of are required to provide a Human Machine Interface. Purpose specific sensors that are observable by instruments have been developed to enhance the scope and range of instruments (Padey, 2014). It can be engineered or produce using various technique of unit operations of engineering and science. Consider nano-wire sensors that direct chemicals and biologics, nano-sensors placed in blood cells to detect early radiation damage in astronauts and nano-shells that detect and destroy tumors. Similarly, nanotechnology has the potential to revolutionise the agricultural and food industry with new tools for the molecular treatment of diseases, rapid diseases detection, enhancing the abilities of patents of absorb nutrients etc. Smart sensors and delivery systems will help the agricultural industry combat viruses and other crop pathogens. Nanostructured catalysis will be available which will increase the efficiency of pesticide and herbicide. It is also allowing lower dose to use. Nanotechnology will also protect the environment indirectly through the use of alternative energy supplies, and filters or reduce pollution and clean-up existing pollution (Singh, 2014, Scott, 2007, Kizma, 2007, www.nature.com).

In the field of agriculture and allied science, the nano-technology could be utilise to produce metabolically more absorbing fertilizer compounds like; nano-rock-phosphate, nano-potassium etc., efficient micronutrient supplement materials like; nano- zinc powder, nano- magnesium dust, nano-copper powder etc. As far as instrumentation side is concern, nano-particle could combine with sensors and utilised in the precision agriculture. The speed, accuracy and range of the sensor could be improved if it utilise with nano-particle. Nanotechnology could also utilise to develop biosensor based sensor for detecting spoilage of food and drug, chemical trace detection in water and blood as well as environmental remediation (Kundu, 2014, Singh, 2014 and http://link.springer.com).

Nanomaterials can also be engineered to actively interact with a pollutant and decompose it into less toxic species. This technology will be useful to trap and measure the individual protein or molecules with smallest quantity of reagents. It can generate the signals in presence of bacterial or ethylene gas. Development of electronic tongue and electronic nose has involved the nano science in the identification of complex and minute difference in taste and fragrance. Use of silicate nano-particle in packaging has enhanced the shelf life of produce by maintaining selective gases. Clay-based nano-composite plastic beer container is lighter and stronger than glass as well as less likely to shatter with shelf life up to 26 weeks. Nano-composite KU2-2601 plastic film is lighter, stronger and more heat resistive than normal packing file available for retorted pouches. Kodak nano-biological antibody composite film has specific properties to absorb the Oxygen from packaging materials and thus, impending food deterioration (http://iopscience.iop.org and Handy, *et. al.* 2008).

Production of Nano Particle

Several techniques are available for the synthesis of nano-particles using three types of precursors, namely; solid, liquid and vapour. These techniques of synthesis of nano particles for use in various fields could be classified into following groups are the ways (Kundu, 2014).

Physical Method: Generally solid precursors are used which are directly converted to nano-size by two ways.

- ***Mechanical Method:*** This is generally done for about several hours of size reduction in the high energy ball mill. But the distribution of size and shape are uncontrolled. The mechanical grinding one can attain particle in the nano-scale, but it has some limitations such as (i) the ground mixture of nano-particles are in different shape and size and (ii) the distribution of size and shape are uncontrolled. However, this technique can suitably be used in agriculture for making non-soluble mineral sources more useful as sources of various plant nutrient and amendments in soil.
- ***Vapourisation Method:*** This involves four techniques:

  Inert Gas Condensation: In this solid precursor material is heated to evaporate it into a background gas and then it is mixed with a cold gas to reduce the temperature. This method is well suited for production of metal nano-particles, since many metals evaporate at reasonable rate at attainable temperature. Due to temperature limitation, synthesis of metal-based nanoparticles are suitable in this mothods.

  - ***Pulse Laser Ablation:*** In this method pulsed laser is used to vaporise a plume of material that is tightly confined, both spatially and temporally. Laser ablation can vaporise materials that cannot readily be evaporated. Quantity is limiting factor in this method.
  - ***Spark Discharge Generation:*** Another means of vapourising metal is to change electrodes made of the metals to be vapourised in presence of an inert gas background until the breakdown voltage is reached. The arc produced across the electrodes then vaporises a small amount of metal, which is to be cooled down further. Here also, quantity is limiting factor in this method.
  - ***Ion Sputtering:*** One more means of vapourising a solid is via sputtering with a beam of inert gas ions.

Chemical Method: The liquid and vapour precursor are allowed to undergo chemical reaction for achieving the super saturation that is required to induce homogeneous nucleation of particles. The different chemical methods are used for preparation of nano-particles depending upon the nature of the base material and size and shape of the desired nano-particles. In this class, the methods are as follows:

- ***Simple Chemcial Reactions:*** Number of nano-particles can be prepared in the laboratory by employing simple chemical reaction followed by autoclaving at high temperature at specific time. e.g. Cd, Se nano-belts are prepared by mixing aqueous solution of Cadmium nitrate (0.01 mol) and Na_2SeO_3 in presence of NH_4OH and NaOH followed by incubation at of about 180°C for 10 hour.
- ***Microemulsion Method:*** Micro-emulsion is a combination of surfactant (e.g. dedecyl dimethylamino oxide and acetyl alcohol), organic acid (e.g. hydrocarbon or olefins) and a water soluble salt (e.g. titanium chloride, titanium tetraisopropoxide, titanium tetra butoxide). Dissolution of surfactant in organic acid solvent leads to the formation of spherical particle called micro-emulsion and it is used to produce TiO_2 nanoparticle.
- ***Sol-Gel Method:*** Particle size smaller than 50 nm that are suspended in solution are called sol. Aggregation of these particles together leads to formation of long chain of particles. With this mechanism, the solvent is imprisoned in structure of chain and a complete homogenious mixture is obtained which is known as gel. Reactants like metal alkaloid of metal chloride are commonly used in sol-gel method, which eventually leads to production of metal oxide. Sol-gel occurs in three stages; (i) Hydrolysis, (ii) Condensation and Polymerisation and (iii) Particle growth.
- ***Precipitation Methods:*** In synthesis of TiO_2 nano-particles method reactants such as titanium tetra chloride and titanium tetra butoxide are usually used. Titanium tetra chloride and titanium tetra butoxide are dissolved in the solvent which is usually water and by adding a basic solution the desired product precipitates.
- ***Chemical Vapour Synthesis:*** In this method, vapour phase precursors are brought to hot-wall reactor under conditions that favour nucleation of particle in vapour phase.
- ***Spray Pyrolysis:*** Rather than delivering the nano-particle precursors in to hot reactor as vapour, one can use a nebulizer to directly injected very small droplets of precursors solution. This has been called spray pyrolysis, aerosol decomposition synthesis, droplet-to-particle conversion, etc.

- ***Laser Pyrolysis/Photothermal Synthesis:*** An alternate means of heating the precursors to induce reaction and homogeneous nucleation is absorption of laser energy. Compared to heating the gases in the furnace, this allows highly localised heating and rapid cooling. The silicone particles are prepared in laboratory by laser pyrolysis of silane.
- ***Thermal Plasma Synthesis:*** Another means of providing the energy to induce the reactions that lead to super-saturation and particle nucleation is to inject the precursors into the thermal plasma. This generally decomposes them fully into atom, which can be then react or condense to form particles when cooled by mixing with cool gas or expansion through nozzle.
- ***Flame Synthesis:*** Rather than supplying energy externally to induce reaction and particle nucleation, one can carry out the particle synthesis within a flame, so that the heat needed is produced in situ by the combustion reactions.
- ***Flame Spray Pyrolysis:*** Rather than injecting vapour precursors into the flame, one can directly spray liquid precursors into it. This process is generally called flame spray pyrolysis. This method allows use of precursors that do not have sufficiently high vapour pressure to be delivered as a vapour.

Biological Method: Higher plants and number of microbes are employed for the synthesis of nano-particles. The development of reliable and eco-friendly biological processes of production of nano-particle is still a challenging task. The use of biological system for synthesis of nano-particles is new area of interest as this method could be more eco-friendly compared to others like physical and chemical. Among the various biological systems, plant provides an easy and safe green route for the synthesis of various metal nanoparticles. When aqueous solution of gold chloride is mixed with root extract of 'Ashwagandha', ruby red colour appears, indication the formation of nano-gold particle. It is believed that, the reduction of metal ion occurs due to the presence of alkaloids and terpenoids present in the root, e.g. Plant extract of dry leaves of *Mellia azedarch* and fruit of *Ficus racemosa*, rich in tannic acid and terpenoids, were found effective in reducing silver nitrate solution to give silver nano-particles. Copper biomineralisation with some wetland plants that transformed copper into metallic nanoparticles at soil-root interface with the help of endo-mycorrhizal fungi.

Application of Nanotechnology for Climate Smart Agriculture

Sensor/Biosensor, incorporated with new technologies in molecular biology, microfluidics and nanomaterials, have diverse application in agricultural production, food processing and environmental monitoring for in-field, on-line and real-time detection of pesticide, pathogens, toxic materials, protein, antibiotics, odours creating bacteria, microbes etc. in soil, air, water, food, plants and animals (Buffle, 2006).

Sensor based upon the Surface Plasmon Resonance (SPR) exhibited by metal nanoparticles. Metal nanoparticles with different composition and size can be tailored to scatter light of different wavelengths according to their characteristics SPR. This sensor is capable of forecasting the possible occurrence of soil diseases on the basis of differential oxygen consumption in the respiration of favourable and unfavourable microbes (Singh, 2014).

An inexpensive sensor-based wireless nanotechnology, which composed of Micromachined Micro Electron Mechanical System (MEMS) cantilever beams coated with a water sensitive nanopolymer for moisture detection and reaction at root zone in real time (Consales, *et. al.*, 2007).

A simple cheap optical sensor for detecting and quantifying the presence of pesticides in water and humic acid was reported by Anderson, *et. al.*, 2005. The use of porous silicon (PSi) and allowed capillary permeation of liquid into its pores. PSi nanostructure is highly sensitive to reflective index of a liquid within its pores which affects its optical properties.

A disposable electrochemical immunosensor for detection of Vibrio Parahaemolyticus (VP) based on the screen-printed electrode (SPE) coated with agarose/nano-Au membrane and horseradish peroxidase (HRP) labeled VP antibody (HRP-anti-VP) has been developed in the form of tiny devices, which they leveraged to detect a range of contaminations, from molecules to whole bacteria, in food and water (Zhao *et. al.* 2007).

Nanocantilevers piezoelectric could be crucially designed a small ultra-small sensors for the quick detection of viruses, bacteria and other targeted pathogens. The electronic pathogen detector or electronic nose, detects the specific pathogen or gas by changing the resistant of nanowires or structure when molecular pressure of the specific matter exists (Singh, 2014).

Mechanically developed rock phosphate have shown that the absorbance of low grade P is high through root in solution culture compared to normal P fertilizer without interaction of soil Ca, Fe and Al. (Adhikari, T., 2014)

The nano-size level gluconite/waste mica, dolomite, magnetite and pyrite could be utilise as good a source of K, Ca, Mg and Fe as well as S respectively for agriculture with reduced cost and without imparing damage to environment (Kundu, 2014).

TiO_2 and ZnO can successfully be utilise in various plant protection and food processing technologies. SnO_2 nano-particle have many good humidity sensing property which can be adopted for making nano-sensors for humidity, evaporation rate and soil moisture measurement.

The life of storage batteries can be enhanced considerably by making the separator plates made of nano-powder of metal. The separator plates made of nano-powder are homogeneous porous structure and thereby store more energy (Consales, *et. al.*, 2007).

Swarms of specially designed nano-machines could be released to upper atmosphere to destroy the CFCs and built new molecular of ozone. Similarly, specialised nano-machines could be used to destroy the excess CO_2 in the lower atmosphere and slow down the rate of climate change (Singh, 2014).

The use of zero-valent (FeO) iron nanoparticles for the remediation of contaminated groundwater and soil compared to iron powder. When exposed to air, iron oxidizes easily to rust which are non-toxic. When oxidize around contaminants such as trichloroethylene (TCE), Carbon tetrachloride, dioxind or PCBs, these organic molecules are broken down into simple, far less toxic carbon compound. The iron-nanoparticles are 10-1000 times more reactive compared to iron powder with less cost. Ahead of that, bimetallic iron/ palladium nanoparticle slurry can be injected to the contaminated plume where treatment is needed. The nanoparticles are not changed by soil acidity, temperature, or nutrient level, so they can remains in suspension maintaining their oxidative properties for extended period of time to establish an in-situ treatment zone (Handy, *et. al.*, 2008)

The edges of the blades of the agricultural tools or implements can be treated with TiO_2 to increase the life by reducing the rusting. The nanoparticles of tin or iron are coated simply by high pressure spraying with base solution to fix up the fishers and cracks of edges generated during forging and heating operations. This will not only reduces the production cost but also provide high quality material (Kuzma, 2007). Use of nano-particle in paint industry is not hidden to the world, the life of paint increase and also reduce the volume of epoxy. The nanoparticle emulsion based paint can remains on agricultural equipments for long period of time compared to normal one (Singh, 2014).

In spite of tremendous scope in the field of nanotechnology, scientists are still not confident about nano-particle behaviour in the ecosystem i.e. atmosphere, aquatic environment, terrestrial environment and into plant. It may diffuse, absorb, leached out or react with existing matter or media is not defined. Reaction of unknown nano-particle with human metabolic physiology may generate great disaster in humanity. Hence, it is necessary to be precautious while dealing with new technology (Biswas, 2014; Banfield and Zhang, 2001).

References

Adhikari, T., 2014. Mechanical Synthesis of Nano Rock Phosphates and its Application to Crops. Technical Bulletin No. CIAE/IDED/2014/447. Pub. CIAE, pp. 81-84.

Anderson, D.G., D. Putnam, E.B. Lavik, T. A. Mahmood and Langer. G, 2005. Biomaterial microarrays Ripid, microscale screening of polymer-cell interaction. Biomaterials. Vol. 26, pp. 4892-4897.

Anonymous, 2002. Small Wonders, Endless Frontiers: A Review of the National Nanotechnology Institute. National Academy Press, 2002.

Banfield, J.F. and Zhang, H., 2001. Nanoparticle in the environment. Pub.-Mineralogical Society of America, WDC. Ch-1, pp. 1-58.

Biswas, A.K., 2014. Application of Nanotechnology for environmental Remediation. Technical Bulletin No. CIAE/IDED/2014/447. Pub.CIAE: pp. 167-170.

Buffle, J., 2006. The key role of environmental colloids/nanoparticles for the sustainability of life. Environ. Chem. 3, pp. 155-158.

Consales, M., Crescitelli, A., Campopiano, S., Cutolo, A., Penza, M., Aversa, P., Giodano, M. and Cusano, A., 2007. Chemical detection in water by single-walled carbon nanotubes-based optical fiber sensors. IEEE sensors J, Vol.7 (7), pp. 1004-1007.

Handy, R.D., Owen, R., Valsami-Jones, E., 2008. The ecotoxicology of nano-particle and nano-materials: Current status, knowledge gaps, challenges and future needs. Ecotoxicology. 17, pp. 315-325.

Hochella, M.F., Jr. Lower, S.K., Maurica, P.A., Penn, R.L., Sahai, N., Sparks, D.L. and Twining, B.S., 2008. Nanominerals, minerals nanoparticles and Earth chemistry. Science. 21, pp. 1631-1635.

Kundu, S., 2014. Synthesis of Nano Particles and Their Application in Natural Resource Management. Technical Bulletin No. CIAE/IDED/2014/447. Pub.CIAE, pp.140-142.

Kundu, S., 2014. Conceptual Understanding of Nano-Science and Nano-Technology. Technical Bulletin No. CIAE/IDED/2014/447. Pub.CIAE: pp. 143-145.

Kuzma, J., 2007. Moving forward responsibly: Oversight of the nanotechnology-biology interface. Journal of Nanoparticle Research. Vol. 9, pp.165-182.

Lijima, S., 2008. Helical Microtubes of graphitic carbon, Nature. Vol. 354(6348), pp. 56-58.

Pandey, S.K., 2014. Physics of Nano-Particle Based Sensors. Technical Bulletin No. CIAE/IDED/2014/447. Pub.CIAE, pp. 258-261.

Scott, N.R., 2007. Nanoscience in Veterinary Medicine. Veterinary Research Communications. Vol. 31(suppl.), pp. 139-144.

Singh, C.D., 2014. Use of Nano-sensors and Instrumentation in Precision Agriculture. Technical Bulletin No. CIAE/IDED/2014/447. Pub.CIAE, pp. 75-80.

Ullah Ruh and Dutta, Joydeep, 2008. Photocatalytic degradation of organic dyes with manganese-dopped ZnO nanoparticles. Journal of Hazardous Materials, Vol. 156, pp. 194-200.

Vo-Dinh, T. Cullum, B.M. and Stokes, D.L., 2001. Nanosensors and Biochips: Frontiers in Biomolecular Diagnostics, Sensor and Actuators. B-74(2).

http://iopscience.iop.org/0957-4484/23/45/455106

http://link.springer.com/article/10.1186/1556-276x-6-225/fulltext.html

http://nanotchweb.org/cws/article/tech/48783

http://www.nature.com/lsa/journal/v1/n10/abs/lsa201234a.html

Zhao, Y., Wang, B., Feng, W. and Bai, C. 2005. Nanoscience and Nanotechnologies - Nanotoxicology: Toxicological and Biological Activities of Nanomaterials. Encyclopedia of Life Support Systems (EOLSS) web portal.

8

Impact of Different Salinity Levels of Irrigation Water on Castor Genotypes and Soil Properties

Shalini Kumari and M.S. Solanki

Abstract

A pot experiment was conducted in a coastal saline region of Saurashtra of Gujarat state at Junagadh district during kharif-2008. This experiment was conducted in mini plot of (1.20 × 0.90 m) containing medium black calcareous clay soil to evaluate the salt tolerance of three genotypes of castor (C_1: GCH-4, C_2: GCH-6 and C_3: JI-273) using four levels of saline irrigation water (S_0: Control, S_1: 4, S_2: 8 and S_3:12 EC ds/m). The results showed that irrespective of varieties, as the levels of salinity increases the castor seed and stalk yield decreases significantly. Similarly, irrespective of varieties as the levels of salinity increases soil properties i.e. EC, pH and ESP increased significantly in both the soil depth i.e. 0-15cm and 15-30 cm. Irrespective of the salinity levels, variety JI-273 was found significantly superior in term of seed (0.467 g/plot) and stalk (0.739 g/plot) yield of castor as compared to rest of two varieties i.e. GCH-4 and GCH-6. Different varieties did not show any significant effect on soil properties i.e., EC, pH and ESP.

***Key Words**: Castor genotypes, salinity, salt tolerance.*

Introduction

Castor is one of the important non-edible oilseed cash crops in Saurashtra region of Gujarat. Salinity is one of the important constrain which limits crop production by increasing osmotic pressure of soil solution. However, farmers grew castor crop in the region having saline irrigation water for benefit. Hence, it is worthwhile to have information related to salt tolerance of different varieties of castor crop. There is no any new release varieties identified for salt tolerance so far in Saurashtra region. Keeping this in view a micro plot trial was undertaken to study the effect of different levels of saline irrigation water on castor genotypes and on soil properties.

Materials and Methods

An experiment was conducted in mini plot of 1.2 x 0.9 m, containing medium black calcareous clay soil (Typic Ustochrepts) during kharif-2013 to evaluate the salt tolerance of three genotypes of castor (C_1: GCH-4, C_2: GCH-6 and C_3: JI-273) using four levels of saline irrigation water (S_0: Control, S_1: 4, S_2: 8 and S_3:12 EC ds/m) in a completely randomised block design with four replications. The saline water of 4.0, 8.0 and 12.0 EC ds/m were prepared by dissolving sulphate and chloride salt of Na, Ca and Mg so as Na : Ca : Mg proportion was maintained at 5 : 1 : 2 and of Cl : SO_4 at 4 : 1. Three healthy seed were sown in each pot (90 cm spacing) and they were covered with the soil. Finally two seedlings/plot were allowed to grow up to maturity. Forty litres of saline irrigation water as per the treatments were applied uniformly in each plot in order to attain slightly higher moisture than that of field capacity. Twelve irrigations of above solution were applied as and when required. The experimental soil was clayey in texture, having pH-7.9, $EC_{2.5}$-0.51 ds/m, free lime-165 g/kg, O.C- 6.6 g/kg, Olsen's available P-14.4 kg/ha. The half dose N and full dose of P_2O_5 (75.0: 50.0 kg N and P_2O_5/ha) were applied as basal in the form of urea and DAP and remaining half dose of N was applied as split at 45 days after sowing of castor seeds in the form of urea. The castor seed and stalk yield was recorded at harvest. The soil samples were collected after harvest of the crop to a depth of 0-15 and 15-30 cm and were analysed using standard methods as described by Richards (1954).

Results and Discussion

The data presented in (Table 1) clearly indicated that soil properties and yield of different castor genotypes significantly influenced by different salinity levels. Irrespective of varieties, as the levels of salinity increases, the castor seed and stalk yield decreased significantly. The highest castor seed and stalk yield were recorded significantly in the lowest salinity level S_0 (control) with values of 0.460 and 0.651 g/plot, respectively, but it was at par with salinity level S_1 (EC 4 ds/m). Similarly observations were also made by Patel *et al.* (1992) for different genotype of groundnut.

Similarly, irrespective of varieties, as the levels of salinity increases, soil properties i.e EC, pH and ESP increased significantly in both the soil depth *i.e.* 0-15 and 15-30 cm. The significantly highest values of EC 1.368 and 1.118 ds/m, pH of 8.513 and 8.429 and ESP of 20.49 and 19.31 were noted in the depth of 0-15cm and 15-30 cm, respectively under S3 salinity level. Irrespective of the salinity levels, variety C_3 (JI-273) was found significantly superior in term of seed (0.739 g/plot) yield of castor as compared to rest of two other varieties

Table 1: Effect of salinity levels and varieties on yield of castor and on soil properties

Treatments	Yield kg/plot		Soil Properties					
	Seed	Stalk	$EC_{2.5}$ (ds/m)		pH		ESP	
			0-15	15-30	0-15	15-30	0-15	15-30
Salinity level (ds/m)								
S_0:Control	0.460	0.651	0.523	0.496	8.164	8.164	9.838	7.063
S_1:4 EC	0.410	0.608	0.746	0.673	8.266	8.235	12.347	10.319
S_2:8 EC	0.376	0.552	1.053	0.854	8.425	8.323	15.991	16.194
S_3:12 EC	0.307	0.509	1.368	1.118	8.513	8.429	20.490	19.31
S.Em.±	0.027	0.032	0.044	0.037	0.030	0.021	0.456	0.535
C.D.5%	0.078	0.090	0.127	0.107	0.085	0.061	1.307	1.536
Variety								
C_1:GCH-4	0.349	0.503	0.954	0.784	8.358	8.279	14.088	13.409
C_2:GCH-6	0.349	0.508	0.881	0.771	8.338	8.284	14.901	12.798
C_3:JI-273	0.467	0.730	0.932	0.801	8.352	8.301	15.011	13.457
S.Em±	0.024	0.027	0.038	0.032	0.026	0.018	0.395	0.463
C.D.5%	0.067	0.078	NS	NS	NS	NS	NS	NS

(GCH-4 and GCH-6). Different varieties did not show any significantly effect on soil properties i.e. EC, pH and ESP.

Interaction Effects

Interaction effect of different salinity levels and varieties (Table 2) was found significant only in stalk yield of castor. The combination of S_1C_3 produced the highest stalk yield of 0.865 g/plot and it was at par with the combination of S_0C_3.

Table 2: Interaction effect of salinity levels and varieties on stalk yield of castor

Salinity	Variety		
	C_1:GCH-4	C_2:GCH-6	C_3:JI-273
S_0:Control	0.570	0.558	0.825
S_1:4 EC	0.405	0.555	0.865
S_2:8 EC	0.488	0.485	0.683
S_3;12 EC	0.548	0.433	0.548
S.Em±	0.055	-	-
C.D.5%	0.157	-	-

Conclusion

The foregoing discussion revealed that irrespective of varieties, as the levels of salinity increases, the castor seed and stalk yield decreases significantly but reverse was true in case of soil properties i.e. EC, pH and ESP which increased significantly in both the soil depth i.e. 0-15 cm and 15-30 cm. The variety JI-273 was found salt tolerance at all salinity levels. Different varieties did not show any significant effect on soil properties i.e. EC, pH and ESP.

References

Anonynious, 1995. Annual progress report, Castor 1994-95. Directorate of Oilsecds Research, Hyderabad.

Arangarsaran, V., Palanippan, S.P. and Chelliah, S. (1999). Response of castor to azospirillum and phosphobacteria inoculation. J. Oilseeds Res., 16(1): 121-122.

Joshi, R S., Bhuvn N.P. and Patel S.A. 1980. Response GAUCH-I to irrigation and fertilizer on loamy sand soil in arid zone areas of Gujurat. Gujarat Agricultural Untversity, Navsari S(20): 73-76.

Mathukia, R.K. and Modhwadia, M.M. (1993). Response of castor to nitrogen and phosphorus. Indian J. Agron., 38(1): 152-153.

Narkheda, B.N., Patil A.B. and Deokar A.B. 1994. Varietal response to plant densities and nitrogen application in castor. Journal of Oilseeds Research l(2): 109-1 14.

Patel, M.S., Gundalia, J.D. and Polara, K.B. (1992). Evaluation of salt tolerance of different groundnut (*Arachis hypogaea* L.) genotypes. Gujarat Agricultural University Research Journal.18: (1), 17-23.

Patel, I.S., Jadhav, N.J., Patel, J.C. and Patel, J.J. (1999). Effect of different irrigation treatments on seed yield of castor, WUE and economics. Soil and Water Management, Gujarat Agricultural University, Navsari Research Bulletin, 10: 147-148.

Patel, M.K. (1997). Effect of nitrogen and phosphorus on yield and chemical composition of castor variety GAUCH 1 under rainfed and irrigated cond (Agri.) thesis, Gujarat Agricultural University, Sarda Krushinagar.

Qadir, G. (2006). Morpho Genetic expression of sunflower under varied temperature and moisture regimes. A Ph.D thesis University of Arid Agriculture, Rawalpindi, Pakistan.

Richards, L.A. (1954). "Diagnosis and improvement of saline and alkali soils", Hand book No, 60. Oxford and IBH Pub. Co Pvt. Ltd., New Delhi.

Rojas-Baros, P., A, de Haro and j. M. Fernandez martinez. (2005), Inhabitance of high oleic/low ricin oleic acid content in the seed oil of castor mutant OLE-1.Crop Science 45:157-162.

Sreedhar, C. (2001). Influence seedling time on castor genotypes under irrigated conditions. M.Sc. (Ag) thesis submitted to Acharya N.G. Ranga Agricultural University, Hyderabad.

Sudhakar, C. and Rao, V.P. (1998). Water use and yield response of pigeonpea, castor, wheat and soybean under different irrigation schedules. J. Res. ANGRAU, 26(1): 1-8.

Subba Reddy, G., Gangadhar Rao D., Venkateswaralu S. and Maruthi V. 1996. Drought management options for rainfed castor in Alfisols. Journal of oilseeds research 13(2): 200-207.

Thadoda, N.K., Sukhadia N.M., Malavia D. and Moradia A.M. 1996. Response of castor (*Ricinus cormminus* L.) GCH-4 to planting geometry and nitrogen fertilization under rainfed condition. Gujarat Agricultural University Research Journal 21(2): 85-87.

Turkhed, B.B., Gangasaran and Rajat De 1982. Response of castor varieties to dates of seeding under dryland conditions of Northwest India. Indian Journal of Agricultural Sciences 16(4): 209-214.

Umadevi, M., Santhaiah V., Singa Rao M. Prasad Rao A. and Rama Rao S. 1991. Effect of conservation tillage practices and nitrogen levels on nutrient uptake by rainfed castor (*Ricinus comminus* L) in alfisols. Journal of Research Andhra Pradesh Agricultural University 19(3): 123-131.

Vijay, K.B. (1992). Effect of irrigation, nitrogen and plant density on yield attributes and yield of castor varieties. Indian Journal of Agronomy 37(l): 203-205.

Vijay Kumar, B. and Shiva Shankar M. (1992). Response of castor varieties to irrigation, nitrogen and plant density. Journal of Research APAU 20(1): 25-26.

Wali, R.M., Palled B., Kalaghatagi S.B., Babalud H.B. and Megeri S.N. Response of castor genotypes to irrigation and nitrogen, Journal of Maharashtra Agricultural University 16 (2) : 262-263.

William, G., M.A.H. Gregory and S. Sugalski, (2006). Toxicity, Plants-Castor Bean and Jequirity Bean, J.emed. 42. No (280), pp.1627-1630.

Williams, J.H. and D.L. Kittock.(1969). Management factors influencing viability of castor bean (*Ricinus communis* L.) Seed *Agron J.*, 61: 954-958.

Zimmerman, L.H. (1958). Castor bean: A new oil crop for mechanized production. Adv. Agron, (10): 258-289.

9

Effect of Irrigation and Foliar Fertilisation on Yield of Chickpea (*Cicer Arietinum* L.)

J.K. Nandaniya, D.S. Hirpara, G. S. Sutaria and V.D. Vora

Abstract

*The experiment entitled effect of irrigation and foliar fertilisation on yield of chickpea (*Cicer arietinum *L.) was conducted during* rabi *season of the year 2012-13 at the Instructional Farm, Department of Agronomy, College of Agriculture, Junagadh Agricultural University, Junagadh. Total ten treatment combinations consisting of two irrigation schedules based on day interval (I_1=0, 20, 50 and 80 DAS) (I_2=0, 20, 35, 50, 65 and 80 DAS) and five foliar spray of inorganic chemicals (F_1 = 2 % KNO_3, F_2 = 2 % urea, F_3 = 2 % thiourea, F_4 = 1 % $ZnSO_4$, F_5 = water spray) were tested in a split plot design with four replications. The results indicated that scheduling irrigation at I_2 level recorded significantly higher values of almost all the growth characters, yield attributes, quality parameters as well as seed and stover yields of chickpea over irrigation at I_1 level. Thus, from the result of present study, it is indicated that, chickpea (cv. GG-1)) irrigated with six irrigations (two common, remaining four at 35, 50, 65, 80 DAS) along with the foliar application of two per cent urea recorded higher yield and maximum net realisation on clayey soils of South Saurashtra Agro climatic conditions.*

Key words: *Chickpea, (Cicer arietinum* L*), Irrigation, Foliar fertilisation, Yield, Economic.*

Introduction

Chickpea is one of the important pulse. India is the largest producer of chickpea in the world having an area of 8.95 million hectares and 7.77 million tonnes of production. (Nadarajan, 2011). The major inputs that determine productivity are varieties, irrigation, nutrients etc. Among them irrigation water is one of the costlier and scare input therefore, it's economic and efficient utilisation becomes quite imperative. Thus, precise information on water requirement by chickpea is essential and inevitable. The concept of foliar feeding implies that nutrient applied to leaves and other above ground plant parts are absorbed and taken up by plants. Accordingly, nutrients needed by plants can theoretically be fed through plant leaves. Moreover, several research workers revealed that foliar feeding is more efficient than soil fertilisation. Precise information regarding appropriate schedule of irrigation for chickpea crop is very limited. Further, to improve yield and yield attribute of chickpea, without deteriorating the soil fertility and productivity it is imperative to study the effect of foliar fertilisation. Keeping in view the above considerations, comprehensive research programme planned to study the judicious use of irrigation, which may provide information not only on water use by crop but also when to irrigate and to study the effect of irrigation and foliar fertilisation on yield of chickpea (*Cicer arietinum* L.) at the Junagadh Agricultural University, Junagadh during *rabi* 2012-13.

Materials and Methods

The experiment was conducted in D-7 plot of Instructional Farm, Department of Agronomy, Junagadh Agricultural University, Junagadh during *rabi* 2012-13. The soil of the experimental plot was clayey in texture and slightly alkaline in reaction with pH 8 and EC 0.56 dS m^{-1}. The soil was low in available nitrogen, medium in available phosphorus and potash. An improved variety GG-1 (Gujarat Gram-1) was selected which yields about 29.43 and 17.32 per cent higher than check *viz*., Dahod Yellow and ICCC-4, respectively, under irrigated condition. The treatments comprised main factor two level of irrigation I_1 = 4 irrigation (at 0, 20, 50, 80 DAS) and I_2 = 6 irrigation (at 0, 20, 35, 50, 65, 80 DAS) while in case of sub factor foliar fertilisation five treatments F_1 = KNO_3 2 %, F_2 = Urea 2 % , F_3 = Thiourea 2 %, F_4 = Zinc sulphate 1 % and F_5 = Water spray.

Note: (1) Common basal application of recommended dose of fertiliser i.e. 20-40-0 (NPK kg ha^{-1}) at sowing. (2) Irrigation given as per treatments schedule. (3) Spray should be done at flowering and pods formation stage. (4) Irrigation was applied to the depth of 50 mm at each irrigation. Chickpea variety Gujarat gram-1 was used for sowing manually with seed rate of 60 kg ha^{-1} on 11th November during 2012. The seeds were placed at 3-5 cm depth, keeping spacing of 45×10 cm and covered with the soil.

Results and Discussion

Effect of irrigation growth attributes of chickpea

Water is the key input in chickpea crop production. However, both water deficit and excess water condition adversely affect the plant growth, development and productivity. Although the scheduling of irrigation in chickpea crop is affected by number of factor like climate and weather condition, ability of root system to withstand moisture stress without any adverse effect on yield, method of crop establishment and method of irrigation etc. but the basic principle is maintaining adequate moisture stress could result in proportionately high losses in yield. The growth parameters *viz.*, plant height and number of branches per plant (Table 1) recorded significantly

Table 1: Growth Attributes as Influenced by Irrigation Levels and Foliar Fertilisation

Treatments	*Plant height (cm)*	*No. of branches per plant*
Irrigation: (I)		
I_1: (4 irrigation)	33.49	11.26
I_2: (6 irrigation)	37.11	12.33
S.Em. ±	0.65	0.20
C.D. at 5%	2.93	0.91
C.V. %	8.25	7.64
Foliar fertilization (F)		
F_1: 2% KNO_3	35.30	11.88
F_2: 2% urea	37.11	12.29
F_3: 2% thiourea	35.72	12.20
F_4: 1% $ZnSO_4$	35.25	11.36
F_5: Water spray	33.13	11.24
S.Em. ±	0.61	0.28
C.D. at 5%	1.77	0.83
C.V. %	4.87	6.78

higher values when irrigated with I_2, which received six irrigations (two common, remaining four at 35, 50, 65, 80 DAS). This might be due to the availability of adequate moisture at all critical stages of growth and development and obviously contributed to luxurious growth. Availability of optimum soil moisture regime along with higher growth rate might boost the uptake of nutrients at various growth and developmental stages resulting in increased plant height and number of branches per plant. The results obtained by Brij Bihari *et al.* (1992), Sharma (1994), Arya *et al.* (2005), Reddy and Ahlawat (1998) and Rajput *et al.* (1999) are corroborative with the above results.

Effect of Foliar Fertilisation

Effect of Foliar Fertilisation on Growth Attributes

Foliar fertilisation significantly increased the growth parameters *viz.*, plant height at harvest and number of branches per plant (Table 1). Significantly higher plant height and number of branches per plant was recorded by foliar application of 2 per cent urea (F_2). The increase in plant height and number of branches per plant at harvest might be attributed to the beneficial and favourable effect of foliar application of urea, thiourea, zinc and potassium and their role in the synthesis of IAA, metabolism of auxins, biological activity, stimulating effect on photosynthetic pigments and enzyme activity which in turn encourage vegetative growth of plants (Michail *et al.*, 2004). The results of this investigation are in consonance with the findings of Virendra and Sharma (2005), Solanki and Sahu (2007), Sarkar and Pal (2006), Thalooth *et al.* (2006) and Yadav *et al.* (2010).

Effect of Irrigation on Yield Attributes and Yield

Increasing number of irrigation significantly increased the yield attributing characters *viz.*, number of pods per plant, seed and stover yield per hectare and test weight (Table 2). These might be due to the increased dry matter

Table 2: Yield Attributes and Yield as Influenced by Irrigation Levels and Foliar Fertilisation

Treatments	*Number of pods plant^{-1}*	*Seed yield plant^{-1} (g)*	*Seed yield (kg ha^{-1})*	*Stover yield plant^{-1} (g)*	*Stover yield (kg ha^{-1})*	*Test weight (g)*
Irrigation: (I)						
I_1: (4 irrigation)	63.88	11.06	1843	14.73	2410	14.49
I_2: (6 irrigation)	70.93	12.40	2032	16.31	2687	16.35
S.Em. ±	1.25	0.22	36	0.33	61	0.25
C.D. at 5 %	5.64	0.98	160	1.47	274	1.12
C.V. %	8.31	8.27	8.23	9.39	10.69	7.21
Foliar fertilisation (F)						
F_1: 2 % KNO_3	67.63	11.50	1918	14.91	2443	15.70
F_2: 2 % urea	70.44	12.38	2057	16.85	2776	16.18
F_3: 2 % thiourea	70.13	12.33	2027	16.63	2737	16.11
F_4: 1 % $ZnSO_4$	65.69	11.38	1867	15.00	2457	14.98
F_5: Water spray	63.13	11.06	1819	14.22	2329	14.15
S.Em. ±	1.79	0.32	50	0.49	85	0.35
C.D. at 5 %	5.23	0.94	146	1.43	248	1.02
C.V. %	7.51	7.79	7.32	8.92	9.45	6.42

production because of higher of photosynthetic rate which contributing to increase in pods per plant (Reddy and Ahlawat, 1998). Further, availability of adequate moisture might be resulted in better translocation and partitioning of these photosynthate from source to sink and increased the number of pods per plant and thus test weight. The data presented in Table 2 revealed that seed and stover yield per hectare were increased with increasing number of irrigation. The higher values in seed and stover yields in treatment I_2. The probable reason for the potential increase in seed and stover yields with increasing level of irrigation was because of contribution of growth and yield attributes to the yields through efficient synthesis and conversion of photosynthesis into dry matter and ultimately in yields.

Effect of foliar fertilisation on yield attributes and yield. Significant increase in yield attributes *viz*., number of pods per plant, seed yield per plant, stover yield per plant and test weight (Table 2) were recorded with the foliar fertilisation over control. It might be due to its involvement in various physiological and biochemical processes along with N-fixation process, which may enhance the dry matter accumulation playing key role in higher photosynthetic rate resulting in increased yield attributes and finally yield in pulse crops (Palsaniya and Ahlawat, 2009). Its effect is further more pronounced in test weight as it plays vital role in improving the vegetative structure for nutrient absorption and provides strong sink strength through the development of reproductive structures as well as production of assimilates to fill economically important sink. A perusal of data on yield presented in Table 2 revealed that foliar fertilisation significantly increased the seed yield in 2 % urea (F_2) being at par with 2% thiourea (F_3). Significantly higher stover yield was recorded with the application of 2% urea (F_2) being at par with 2% thiourea (F_3). This potential increase of seed and stover yields with foliar fertilisation might be due to the contribution of growth and yield attributes. All those factors discussed above collectively boosting the yields under the influence of various foliar fertilisation in chickpea. These findings are corroborating with the reports of Hafiz *et al.* (2004), Siavashi *et al.* (2004), Yadav *et al.* (2010), Pathak *et al.* (2012), Solanki and Sahu (2007), Burman *et al.* (2007), Mathur *et al.* (2006) and Poonia *et al.* (2005).

Economic

The economic evaluation of different treatments revealed that the maximum gross realisation Rs 58,240 ha^{-1}, net realisation Rs 30,871 ha^{-1} and B : C ratio 2.13 were obtained in treatment I_2. Data further showed that treatment 2% urea (F_2) recorded maximum gross realisation Rs 58,984 ha^{-1}, net realisation Rs 32,404 ha^{-1} and B : C ratio 2.22. (Table 3) These results are in total agreement with those obtained by Palsaniya and Ahlawat (2009).

Table 3: Economic along with benefit cost ratio of chickpea as influenced by irrigation levels and foliar fertilisation

Treatments	*Gross realisation (₹ ha^{-1})*	*Total expenditure (₹ ha^{-1})*	*Net realisation (₹ ha^{-1})*	*Benefit: cost ratio*
Irrigation (I)				
I_1	52809	26804	26005	1.97
I_2	58240	27369	30871	2.13
Foliar fertilisation (F)				
F_1	54926	27527	27399	2.00
F_2	58984	26580	32404	2.22
F_3	58125	28364	29761	2.05
F_4	53505	26861	26644	1.99
F_5	52097	26104	25993	2.00

References

Arya, K.L., L. Kumar, K.K. Singh and B.L. Kushwaha. 2005. "Effect of fertilizers and tillage management in rice — chickpea cropping system under varying irrigation schedules", Indian Journal of Agronomy, Vol. 50, pp. 256-259.

Bihari, B., H.S. Kushwaha and M.S. Vaidya. 1992. "Response of chickpea to irrigation and fertilisation", Indian Journal of Agronomy, Vol. 37, pp. 110-111.

Burman, U.,G. Kumar, Balvinder and S. Kathju 2007. "Interactive effects of phosphorus, nitrogen and thiourea on clusterbean under rainfed conditions of the Indian arid zone", Journal of Plant Nutrition and Soil Science, Vol. 170, pp. 803-810.

Hafiz, S.I. 2004. Response of chickpea crop to biofertilization and foliar spraying with zinc under different levels of N and P fertilization in newly reclaimed sandy soils. Annals of Agricultural Science, Moshtohor. Vol. 42, pp. 933-948.

Mathur, N., J. Singh, S. Bohra, A. Bohra and A. Vyas. 2006. "Improved productivity of mung bean by application of thiourea under arid condition", World Journal of Agricultural Sciences, Vol. 2, pp. 185-187.

Michail, T.T., W. Walter, G. Astrid, G. Walter, S.J. Dieter, Maria and M. Domingo (2004). "A survey of foliar mineral nutrient concentrations of *Pinus canariensis* at field plots in Tenerife" Forest Ecology and Management, Vol. 189, pp. 49-55.

Nadarajan, N. 2011. "In Vision 2030", Indian Institute of Pulses Research", Kanpur, pp. vi-vii.

Palsaniya, D.R. and I.P.S. Ahlawat. 2009. "Sulphur management in pigeonpea (*Cajanus cujan*) wheat (*Triticum aestivum*) cropping system", Indian Journal of Agronomy, Vol. 54, pp. 272-277.

Pathak, G.C., B. Gupta and N. Pandey. 2012. "Improving reproductive efficiency of chickpea by foliar application of zinc", *Brazilian Journal of Plant Physiology*, Vol. 24, pp. 173-180.

Poonia, K.L. 2005. "Response of cluster bean to thiourea foliar application under rainfed conditions", Arid Legumes for Sustainable Agriculture and Trade, Vol. 1, pp. 185-186.

Rajput, A.S., N.K. Choubey, G.K. Shrivastava, S.S. Khola and R.S. Tripathi. 1999. "Effect of irrigation schedules and weed management practices on performance of late sown chickpea in vertisols" Indian Journal of Weed Science, Vol. 31, pp. 255-257.

Reddy, N.R.N. and I.P.S. Ahlawat. 1998. "Response of chickpea genotypes to irrigation and fertilizers under late sown conditions", Indian Journal of Agronomy, Vol. 43, pp. 95-101.

Sarkar, R.K. and P.K. Pal. 2006. "Effect of pre-sowing seed treatment and foliar spray of nitrate salts on growth and yield of greengram" Indian Journal of Agricultural Sciences, Vol.76, pp. 62-65.

Sharma, R.S. 1994. "Response of chickpea to irrigation and fertility levels under double cropping systems" Indian Journal of Agronomy, Vol. 39, pp. 310-311.

Siavashi, K., R. Soleymani and M.J. Malakouti. 2004. "Effect of zinc sulfate application times and methods on seed yield and protein content of chickpea in rainfed conditions" Iranian Journal of Soil and Waters Sciences, Vol. 18, pp. 5-7.

Solanki, N.S. and M.P. Sahu. 2007. "Productivity and P-use efficiency of cluster bean as influenced by bio-regulators and phosphorus", Indian Journal of Agronomy, Vol. 52, pp. 143-147.

Thalooth, A.T., M.M. Tawfik and H.M. Mohamed. 2006. "A comparative study on the effect of foliar application of zinc, potassium and magnesium on growth, yield and some chemical constituents of mung bean plants grown under water stress condition", World Journal of Agricultural Sciences, Vol. 2, pp. 37-46.

Virendra, S., S.K. Sharma. 2005. "Response of mothbean to zinc and thiourea under dryland condition", Annals of Agricultural Research, 26(2): pp. 313-314.

Yadav, M.K., B. Singh, A.K. Singh, G. Mahajan, R. Kumar, M.K. Singh and S.R. Balai. 2010. "Response of chickpea to sowing methods and zinc sulphate levels under rainfed condition of Eastern Uttar Pradesh", Environment and Ecology, Vol. 228, pp. 1652-1654.

10

Climate Change Compliant Crop: Pearl Millet

C.J. Dangaria, K.D. Mungra and P.R. Padhar

Abstract

Pearl millet [Pennisetumglaucum *(L.)*] *is primarily grown for food and dry fodder purposes in India and Africa. It has certain adaptive traits/mechanisms which are helpful to plant to withstand against changed climatic and edaphic parameters. An important aspect of the ability of pearl millet to survive under high stress is its root system, maturity duration, leaf structure and certain other traits. High temperatures have relevance at both seedling and reproductive stages of pearl millet. The pearl millet is one of the crop from which you can get economic yield under > 42°C temperature situation. On contrarily in Gujarat and Maharashtra it is also cultivated as a cool season crop during post-rainy season. In future, pearl millet is likely to play a greater role not only providing food and nutritional security but will serve as a climatic variability resilient crop. Selection criterion for the improvement of pearl millet against major climatic variability has been suggested. Pearl millet would also be an excellent genomic resource for isolation of candidate genes responsible for tolerance to climatic and edaphic stresses for accelerating further genetic improvement of this crop.*

Introduction

Agriculture is not only sensitive to climate change but is also one of the major drivers for climate change. Scientific evidence about the seriousness of the climate threat to agriculture is now unambiguous, but the exact magnitude is uncertain because of the complex interaction and feedback processes in the ecosystem and the economy.

Pearl millet [*Pennisetumglaucum* (L.) R. Br.] is a C4 plant having a very high photosynthetic efficiency and dry matter production capacity. It is usually grown under the most adverse agro-climatic conditions where other crops like sorghum and maize fail to produce economic yields. Primarily a tropical plant, pearl millet is often referred to as the "Camel", because of its exceptional ability to tolerate drought. Even with minimal rainfall millet will typically still produce reasonable yields. Besides, pearl millet has a remarkable ability to respond to favourable environments because of its short developmental stages and capacity for high growth rate, thus making it an excellent crop for short growing seasons under improved crop management.

Pearl millet is cultivated on about 30 million ha in more than 30 countries of five continents viz., Asia, Africa, North America, South America and Australia. Though the majority of crop area is in Asia (10 million ha) and Africa (about 18 million ha), pearl millet cultivation is being expanded in some of the non-traditional areas, with Brazil having the largest area (about two million ha). It is also being experimented as a grain and forage crop in the USA, Canada, Mexico, the West Asia and North Africa (WANA), and Central Asia.

In India, pearl millet is the third most widely cultivated food crop after rice and wheat. It is grown on nine million ha with an average productivity of 1,000 kg ha^{-1}. The major pearl millet growing states are Rajasthan, Maharashtra, Gujarat, Uttar Pradesh and Haryana (Fig. 1). Most of pearl millet in India is grown in rainy (*kharif*) season (June–September). It is also being increasingly cultivated during the summer season (February–May) in parts of Gujarat, Rajasthan and Uttar Pradesh; and during the post-rainy (*rabi*) season (November–February) at a small scale in Maharashtra and Gujarat.

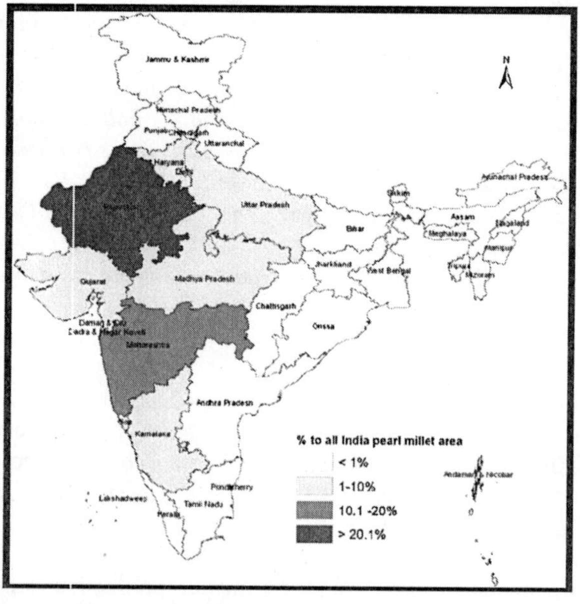

Fig. 1. Distribution of pearl millet in India (www.agricoop.nic.in).

Pearl millet is primarily grown for food and dry fodder. Its grains are highly nutritious with high levels of metabolisable energy and protein, have high densities of iron and zinc, and more balanced amino acid profile than maize or sorghum (Rai *et al.*, 2008). Grains are mainly used for human consumption in the form of diverse food, mostly as leavened and unleavened flat breads and porridges. Dry stover of pearl millet is a major component of livestock ration during the dry period of year (Jordan *et al.*, 2012).

Defence System in Pearl Millet against Climatic Vulnerability and its Research Prioritisation

Drought

Deep Root Penetration: An important aspect of the ability of pearl millet to survive under high stress is its root system. Pearl millet roots can penetrate up to 180 cm deep (Mangat *et al.*, 1999). This deep root penetration may help millet species to exploit soil water more effectively and therefore overcome drought stress. Pearl millet root systems also have the ability to penetrate through hard clay pans in the lower soils.

The adventitious roots start appearing six to seven days after seedling emergence at the basal nodes of the stem. These grow rapidly and form a root system of secondary and tertiary roots which are the principle route of absorption of water and nutrients during the major part of the life cycle of the plant. Crown roots develop from the lower nodes near the soil surface approximately 30-40 days after seedling emergence. The crown roots above the soil surface thicken and support the plant, preventing it from lodging.

Leaf structure: Stomata are present on both leaf surfaces. The colour of the leaves vary from light green to yellow to deep purple. The maximum leaf area (LAI) occurs at the time of 50 per cent flowering, when the majority of the tillers have produced leaves. After flowering there is a decline in leaf area during which time the leaves begin senescing. At physiological maturity only the upper 3-4 leaves may be green on the main stem (Mangat*et al.*, 1999).

Stomatal sensitivity to evaporative demand is dependent upon leaf age and leaf area of the crop. This suggests that the degree to which water use is controlled by stomata and leaf area is influenced ontogeny so as to optimise crop water use for growth (Winkel *et al.*, 2001). It appears that stomata tend to remain open even under high levels of moisture stress. The implication is that millet does not tend to conserve moisture but rather transpires freely as long as the root system can supply the water it needs (Wallace *et el.*, 1993). However, leaves will begin to senesce, thus reducing the LAI of the plant canopy. Stomatal regulation and leaf senescence are not mutually exclusive; stomatal conductance decreases at leaf area increases. Conversely, a reduction in transpiring area increases stomatal conductance in the remaining leaves.

A great deal of work has been done on understanding the response of pearl millet to moisture stress at various growth stages that is related to its adaptation to drought stress conditions. It has been conclusively established that effect of water stress depends on the developmental stage during which the crop is subjected to stress. Water stress during vegetative growth may have little adverse effect on grain yield of pearl millet but it has been shown to increase the number of panicles per plant (Bidinger *et al.,* 1987a and Mahalakshmi and Bidinger, 1985). The apparent excess production of tillers provides potential compensation for damaged main shoot or primary tillers (Van Oosterom *et al.,* 2003 and Van Oosterom *et al.,* 2006). High tillering and asynchrony of tillering contribute to adaptation to drought stress during the vegetative growth phase (Mahalakshmi *et al.*, 1987 and Siband, 1983). Seed setting that determines the number of grains per panicle is usually less affected if terminal stress occurs after flowering (Bidinger *et al.,* 1987a and Mahalakshmi *et al.*, 1987). These understandings have helped to prioritise the research in dealing with management of drought stress.

Several efforts have been made to identify traits that can be used as selection criteria in breeding drought-tolerant genotypes. Most research has concentrated on the identification of physiological parameters like dehydration tolerance (Henson *et al.,* 1982), dehydration avoidance (Bidinger and Witcombe, 1989), growth maintenance through stability of cellular membrane (Blum and Ebercon, 1981), osmotic adjustment (Turner and Jones, 1980), desiccation and heat tolerance, leaf gas exchange rate (Kholova *et al.*, 2010 and Vadez and Sinclair, 2001) and radiation reflectance (Bidinger and Witcombe, 1989).

On the other hand, morphological characters that can be measured easily appeal most to plant breeders for use as selection criteria. A rapid development of the crop in the initial stages i.e. early vigour has been correlated with drought tolerance as measured by time taken for wilting initiation and permanent wilting in pearl millet (Manga and Yadav, 1995).

Early flowering, the most important factor determining yield under terminal water stress (Bidinger *et al.*, 1987b and Fussel *et al.*, 1991), is recognised as another drought mitigation mechanism, although its advantage is due to drought escape rather than due to drought tolerance. Large genetic variability for earliness is widely available in pearl millet (Rai *et al.*, 1996 and Rattunde *et al.*, 1989b) and simple selection has been successful under most circumstances (Rattunde *et al.*, 1989a).

Panicle threshing percentage or panicle harvest index (PNHI) is another mechanism proposed for improving tolerance to terminal drought that indicates the plants' ability to set and fill grain under water-limiting conditions and it integrates the effects of assimilation and translocation under water stress. A wide range of variability in panicle is observed in pearl millet (Fig. 2). This variability is a result of plasticity mechanism of pearl millet to adapt the changed climatic and edaphic parameters. Research has indicated that it usually explains a large proportion of the variation among genotypes for grain yield under terminal drought stress (Bidinger and Mahalakshmi, 1993; Bidinger and Witcombe, 1989 and Fussel *et al.*, 1991).

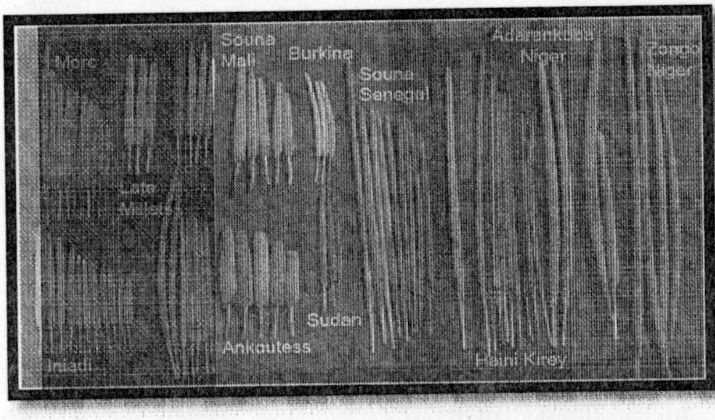

Fig. 2: Diversity of pearl millets showing different ranges of adaptation (Cultivating an optimal range of different pearl millet cultivars in a region enhances resilience to climate variability).

Low tillering and large panicles are commonly being used as selection criteria in pearl millet breeding. Selecting for these traits results in higher grain yield per panicle (Bidinger *et al.*, 1987b and Yadav, 1994) which is important yield components in pearl millet (Bidinger *et al.*, 1993; Solanki *et al.*, 2001; Yadav *et al.*, 1994 and Yadav *et al.*, 2001). These traits are frequently assessed visually, under both drought and non-drought conditions. However, their specific contribution to improved grain yield and stability under terminal drought condition has not been quantified.

Stay-green trait has been widely considered as potentially useful selection criterion for drought tolerance in sorghum (Jordan *et al.*, 2012). However, its effect, both in the magnitude and the direction, has been shown to be significantly influenced by the genetic background and the environment (Jordan *et al.*, 2012). Breeding lines with stay-green trait have also been identified in pearl millet, and their multilocational evaluation has been initiated. It would be critical to evaluate the level of expression and stability of this trait in diverse genetic backgrounds before starting its utilisation in stay green breeding programme.

Because of intrinsic difficulties in breeding for drought adaptation (Barker *et al.,* 2005 and Blum, 1988), this field has become a prime focus for molecular marker-assisted breeding. Genetic mapping for drought tolerance has targeted terminal drought. Several major QTLs have been identified that have significant effects on pearl millet grain and stover yields in terminal drought stress environments (Bidinger *et al.,* 2005; Bidinger *et al.,* 2007; Yadav *et al.,* 2002 and Yadav *et al.,* 2004).

Temperature

Heat Tolerance/High Temperature

High temperatures have relevance at both seedling and reproductive stages of pearl millet. The maximum air temperatures around 43°C are observed in some years in parts of India, especially in the north-western arid zone during the rainy season. The soil surface temperatures during germination may reach 60–62°C in this zone. Pearl millet seedlings are most vulnerable to high temperatures during the first 10 days of sowing (Peacock *et al.*, 1993 and Stomph, 1990). Rapid screening procedures for seedling survival under these high temperatures have been shown and genotypes with tolerance to high temperature levels have been identified (Peacock *et al.*, 1993 and Soman and Peacock, 1985).

There is a growing interest in pearl millet cultivation as irrigated summer season (February–June) crop in parts of Gujarat, Rajasthan and Uttar Pradesh where high temperatures (4°C) are of common occurrence during flowering. Such high air temperatures coinciding with flowering in this region can cause spikelet sterility, leading to drastic reductions in grain yield. Several heat tolerance genotypes beyond 42°C temperature resistance have been developed by ICRISAT and has made significant contribution to the development of reproductive stage heat- tolerance in pearl millet adapted to hot climate (Gupta *et. al.*, 2015). However, a few hybrids which specifically target this environment for hybrid development, have been found having good seed set and high grain yield (4–5 t ha^{-1}). The pearl millet area under summer season crop has been on rise due to having irrigation facility and little option with farmers of this region due to high temperature during (>43°C frequently) summer season. Hence expansion of breeding activities related to flowering-period heat tolerance at ICRISAT and in some of the public and private sectors field facilities in the target region. As a result of this, large genetic variation for heat tolerance at reproductive stage have been identified.

Low Temperature

Study of the winter season adaptation of pearl millet in Gujarat and Maharashtra (Mula *et al.*, 2009), showed increasing interest of farmers in this cropping system because of the importance of the crop in sustaining farmers' household requirements of nutritious food and fodder. A major constraint faced by farmers in these areas is the changing growing conditions such as low temperature of 10–15°C, which reduces plant growth, delays maturity, and leads to a reduced biomass yield. The existence of genetic variation for this trait can effectively address this issue. The farmers' practice of adjusting sowing time largely overcomes the problem of poor seed setting at low temperature. However, genetic improvement of this trait is needed to allow flexible sowing time. Employing other cultural management practices, such as crop rotation juxtaposed with the adherence to a cropping calendar of pearl millet, yields economic benefits to farmers. These attributes of pearl millet combined with high water-use efficiency under water limiting environments make it a potential replacement crop in the global scenario of climate change with bright prospect for developing hybrids suited to winter season environments.

Day Length/Photoperiod

Day length, or photoperiod, is a critical control in the initiation of the reproductive phase of the millet in many pearl millet cultivars. Photosensitive cultivars are grown as long season crops while non-photoperiodic cultivars are grown as short season crops (The Syngenta Foundation for Sustainable Agriculture, 2006). Because of photoperiod sensitivity, the growth cycle of local millet cultivars changes greatly with sowing date.

Solar Radiation

Pearl millet is a C_4 type plant, i.e., it has a high photosynthetic efficiency, particularly under high temperatures conditions because of reduced photorespiration (OMM, 1993).

Low Soil Fertility

Soils in the regions where pearl millet is cultivated are often poor in fertility as they contain low amount of organic matter (0.05–0.40 %) because of low vegetation cover, coarse texture of soils and prevailing high temperatures (Kumar *et al.* 2009). Soils also contain low-to-medium levels of available phosphorous (10–25 kg ha^{-1}). This issue has mainly been addressed through nutrient management. Possibilities of genetic enhancement for nutrient-use efficiency have been increasingly being explored in a few crops (Hash *et al.*, 2002).

Future Prospects

The increasing urbanisation and industrialisation will put heavy pressure on water resources for agricultural use. Further, the climate change will also lead to variable water scarcity, increase in air and edaphic temperatures and soil salinity. Pearl millet being the most water use efficient major cereal the most heat-tolerant cereal. It is also the most salinity-tolerant cereal next only to barley. These attributes of pearl millet put it in a unique advantageous position to address the multiple environmental challenges, which are likely to be of increasingly serious proportion in the future. Pearl millet, being a C4 species with high photosynthetic potential and biomass production ability, has a greater role to play in arid and semi-arid regions to bridge the gap between demand and supply of dry fodder. Pearl millet is also an excellent genomic resource for isolation of candidate genes responsible for tolerance to climatic and edaphic stresses and their possible deployment in other crops.

References

Barker, T., H. Campos, M. Cooper, D. Dolan, G. Edmeades, J. Habben, J. Schussler, D. Wright and C. Zinselmeir (2005). Improving drought tolerance in maize. Plant Breed Rev, 25:173–226.

Bidinger, F.R. and V. Mahalakshmi (1993). Selection for drought tolerance. In: Cereals programme, ICRISAT 1993. Annual report. ICRISAT Patancheru, India 1993:57–59.

Bidinger, F.R. and J.R. Witcombe (1989). Evaluation of specific drought avoidance traits as selection criteria for improvement of drought resistance. In: Baker FGW (ed) Drought resistance in cereals. CAB International, Wallingford, pp. 151–164.

Bidinger, F.R., V. Mahalakshmi and G.D.P. Rao (1987a). Assessment of drought resistance in pearl millet [*Pennisetumamericanum* (L.) Leeke]. I. Factors affecting yield under stress. Aust J Agric Res, 38:37–48.

Bidinger, F.R., V. Mahalakshmi and G.D.P. Rao (1987b). Assessment of drought resistance in pearl millet [*Pennisetumamericanum* (L.) Leeke]. II. Estimation of genotype response to stress. Aust J Agric Res, 38:49–59.

Bidinger, F.R., G. Alagarswamy and K.N. Rai (1993). Use of grain number components as selection criteria in pearl millet. Crop Improv, 20:21–26.

Bidinger, F.R., R. Serraj, S.M.H. Rizvi, C. Howarth, R.S. Yadav and C.T. Hash (2005). Field evaluation of drought tolerance QTL effects on phenotype and adaptation in pearl millet [*Pennisetumglaucum* (L.) R. Br.] top cross hybrids. Field Crops Res, 94:14–32.

Bidinger, F.R., T. Nepolean, C.T. Hash, R.S. Yadav and C.J. Howarth (2007). Quantitative trait loci for grain yield in pearl millet under variable post flowering moisture conditions. Crop Sci, 47:969–980.

Blum, A. (1988). Plant breeding for stress environments. CRC Press, Boca Raton.

Blum, A. and A. Ebercon (1981). Cell membrane stability as a measure of drought and heat tolerance in wheat. Crop Sci, 21:43–47.

Fussel, L.K., F.R. Bidinger and P. Bieler (1991). Crop physiology and breeding for drought tolerance: research and development. Field Crops Res, 27:183–199.

Gupta, S.K., K.N. Rai, Piara Singh, V.L. Ameta, Suresk K. Gupta, A.K. Jaylekha, R.S. Mahala, S. Pareek, M.L. Swami and Y.S. Verma (2015). Seed set variability under high temperatures during flowering period in pearl millet (*Pennisetumglaucum* L. (R.) Br.). Field Crop Res., 171:41-53.

Hash, C.T., R.E. Schaffert and J.M. Peacock (2002). Prospects for using conventional techniques and molecular biological tools to enhance performance of 'orphan' crop plants on soils low in available phosphorus. Plant Soil, 245:135–146.

Henson, I.E., V. Mahalakshmi, F.R. Bidinger and G. Alagarswamy (1982). Osmotic adjustment to water stress in pearl millet [*Pennisetumamericanum* (L.) Leeke] under field conditions. Plant Cell Env, 5:147–154.

Jordan, D.R., C.H. Hunt, A.W. Cruickshank, A.K. Borrell and R.G. Hanzell (2012). The relationship between the stay-green trait and grain yield in elite sorghum hybrids grown in a range of environments. Crop Sci, 52:1153–1161.

Kholova, J., C.T. Hash, P. Lava Kumar, R.S. Yadav, M. Kholova and V. Vadez (2010). Terminal drought-tolerant pearl millet [*Pennisetumglaucum* (L.) R. Br.] have high leaf ABA and limit transpiration at high vapour pressure deficit. J Exp Bot, 61:1431–1440.

Kumar, P., J.C. Tarafder, D.K. Painuli, P. Raina, M.P. Singh, R.K. Beniwal, M.L. Soni, M. Kumar, P. Santra and M. Shamsudin (2009). Variability in arid soils characteristics. In: Kar A, B.K. Garg, M.P. Singh and S. Kathju (eds). Trends in arid zone research in India. Central Arid Zone Research Institute, Jodhpur, pp. 78–112.

Mahalakshmi, V. and F.R. Bidinger (1985). Flowering response of pearl millet to water stress during panicle development. Ann Appl Biol, 106:571–578.

Mahalakshmi, V., F.R. Bidinger and D.S.Raju (1987). Effect of timing of water deficit on pearl millet (*Pennisetumamericanum*). Field Crop Res, 15:327–339.

Manga, V.K. and O.P. Yadav (1995). Effect of seed size on development traits and ability to tolerate drought in pearl millet. J Arid Env, 29:169–172.

Mula, R.P., K.N. Rai, C.J. Dangaria and M.P. Kulkarni (2009). Pearl millet as a postrainy cool season crop: case studies from Gujarat and Maharashtra, India. J SAT Agric Res, 7:1–7. (www.ejournal.icrisat.org).

Mangat, B.K., R.K. Maiti and I.S. Kharval (1999). Pearl Millet Biology. In Khairwal, I.S., K.N. Rai, D.J. Andrews and G. Harinarayana. (1999). Pearl Millet Breeding. Science Publishers, Inc. USA.

OMM, (1993). L agrometeorologie du mil [*Pennisetumglaucoma* (L) Leeke], Rapport CMAg No. 56.

Peacock, J.M., P. Soman, R. Jayachandran, A.U. Rani, C.J. Howarth and A. Thomas (1993). Effect of high soil temperature on survival in pearl millet. ExpAgric, 29:215–225.

Rai, K.N., K. Anand Kumar, D.J. Andrews, S.C. Gupta and B. Ouendeba (1996). Breeding pearl millet for grain yield and stability. In: Proceedings of international conference on genetic improvement of sorghum pearl millet, Lubbock, Texas, 23–27 Sept. INTSORMIL, Lincoln, pp. 71–83.

Rai, K.N., C.L.L. Gowda, B.V.S. Reddy and S. Sehgal (2008). The potential of sorghum and pearl millet in alternative and health food uses. Compr Rev Food Sci Food Saf, 7:340–352.

Rattunde, H.F., P. Singh and J.R. Witcombe (1989a). Feasibility of mass selection in pearl millet. Crop Sci, 29:1423–1427.

Rattunde, H.F., J.R. Witcombe and K.J. Frey (1989b). Structure of variation among morphological and physiological traits in three pearl millet composites. Euphytica, 43:233–244.

Siband, P. (1983). Essai d'analyse du fonctionnement du mil (*Pennisetumtyphoides*) en zone Sahelienne. L'Agron Trop, 38:27–36.

Solanki, Y.P.S., I.S. Khairwal and F.R. Bidinger (2001). Panicle surface area as a selection criterion for grain yield in pearl millet (*Pennisetumglaucum*). Indian J Agric Sci, 71:450–454.

Soman, P. and J.M. Peacock (1985). A laboratory technique to screen seedling emergence of sorghum and pearl millet at high soil temperature. ExpAgric, 21:335–341.

Stomph, T.J. (1990). Seedling establishment in pearl millet [*Pennisetumglaucum* (L) R. Br.]: the influence of genotype, physiological seed quality soil temperature and soil water. Ph.D. Thesis, University of Reading, Reading.

The Syngenta Foundation for Sustainable Agriculture (2006). Harnessing Modern Science in Africa to Sustain Sorghum and Pearl Millet Production for Resource poor Farmers. www.syngentafoundation.com/millet.html.

Turner, N.C. and M.M. Jones (1980). Turgor maintenance by osmotic adjustment: a review and evaluation. In: Turner NC, Kramer PJ(eds) Adaptation of plant to water and high temperature stress. Wiley Interscience, New York, pp. 87–103.

Vadez, V. and T.R. Sinclair (2001). Leaf ureide degradation and N_2 fixation tolerance to water deficit in soybean. J Exp Bot, 52:153–159.

Van Oosterom, E.J., F.R. Bidinger and R.E. Weltzien (2003). A yield architecture framework to explain adaptation of pearl millet to environmental stress. Field Crops Res, 80:33–56.

Van Oosterom, E.J., E. Weltzien, O.P. Yadav and F.R. Bidinger (2006). Grain yield components of pearl millet under optimum conditions can be used to identify germplasm with adaptation to arid zones. Field Crops Res, 96:407–421.

Wallace, J.L., C.R. Lloyd and M.V.K. Sivakumar. (1993). Measurement of soil, plant and total evaporation from millet in Niger. Agril and Forest Meteo, 63:149-169.

Winkel, T., W. Payne and J. F. Renno (2001). Ontogeny modifies the effects of water stress on stomatal control, leaf area duration and biomass partitioning of *Pennisetumglaucum*. New Phytologist, 149(1): 71-82.

Yadav, O.P. (1994). Genetic divergence in pearl millet accessions of Indian and exotic origin. Indian J Genet Plant Breed, 54:89–93.

Yadav, O.P., Manga VK, Saxena MBL (1994). Ontogenetic approach to grain production in pearl millet (*Pennisetumglaucum*) based on path-coefficient analysis. Indian J Agric Sci, 64:233–236.

Yadav, O.P., R.E. Weltzien, and D.C., Bhandari DC (2001). Genetic variation and trait relationship in pearl millet landraces from Rajasthan. Indian J Genet Plant Breed. 61:322–326.

Yadav, R.S., C.T. Hash, F.R. Bidinger, G.P. Cavan and C.J. Howarth (2002). Quantitative trait loci associated with traits determining grain and stover yield in pearl millet under terminal drought stress conditions. TheorAppl Genet, 104:67–83.

Yadav, R.S., C.T. Hash, F.R. Bidinger, K.M. Devos and C.J. Howarth (2004). Genomic regions associated with grain yield and aspects of post-flowering drought tolerance in pearl millet across stress environments and testers background. Euphytica, 136:265–277.

11

Enhancement of Water Productivity under Rice Ecosystem: A Review

Fakeerappa Arabhanvi, Amit M. Pujar and Shruti M. Yogikoppa

Abstract

Rice is one of the main source of food for more than half of the world's population and its cultivation secures a livelihood for more than two billion people. In Asia, more than two billion people are getting 60-70 per cent of their energy requirement from rice and its derived products. In India, rice occupies an area of about 42.7 m.ha with an average production of 105 m.t and average productivity of 2.45 t/ha (2012-13). To sustain the present food self-sufficiency and to meet future food requirements, achieving the higher productivity per unit area is most essential. In India, only 55 per cent of the rice area is irrigated that accounts for 75 per cent of the rice production in the world (Bouman, 2001). Rice is a very water-intensive crop. It is estimated that to produce one kilogram of rice requires 3,000-5,000 litres of water. About two or three times more water is needed for rice cultivation than other irrigated crops. Major constraint in rice production in India is the lack of suitable crop management practices, availability of irrigation facilities and improper water management practices leads lower water use efficiency in rice ecosystem. Hence, there is a need to develop modern technologies and production systems that can produce more rice with less use of water depending upon different rice cultivation methods viz., transplanted rice, direct sown rice, alternate wetting and drying method, aerobic rice and system of rice intensification. However, increasing water scarcity, a shifting trend towards less water demanding crops against rice is noticed in most part of the India and this warrants alternate methods of rice cultivation that aims at higher water and crop productivity. For instance, cultivation of rice through system of rice intensification increase the grain yield by five to seven per cent besides saving of water by 12 to 15 per cent over conventional method of rice cultivation under wetland ecosystem.

Introduction

Rice is a staple food in most of the Indian states and plays a major role in Indian economy. In Asia, more than two billion people are getting 60-70 per cent of their energy requirement from rice and its derived products. In India, rice occupies an area of about 42.7 m.ha with an average production of 105 m.t and average productivity of 2.45 t/ha (2012-13). Water is the most crucial input resource for agricultural production. Globally, agriculture alone accounts for more than 80 per cent of all freshwater used by humans, most of that is for crop production (Morison *et al.*, 2008). Water productivity is defined as crop production per unit amount of water used or in agricultural production systems is focused on producing more food with the same water resources or producing the same amount of food with less water resources. Rice cultivation is a very water-intensive crop. It is estimated that to produce one kilogram of rice requires 3,000-5,000 litres of water.

About two or three times more water is needed for rice cultivation than other irrigated crops. Major constraint in rice production in India is the lack of suitable crop management practices, availability of irrigation facilities and improper water management practices leads lower water use efficiency in rice ecosystem. Hence, there is a need to develop modern technologies and production systems that can produce more rice with less use of water depending upon different rice cultivation methods *viz.*, aerobic rice, system of rice intensification, direct sown rice and alternate wetting and drying method. However, increasing water scarcity, a shifting trend towards less water demanding crops against rice is noticed in most part of the India and this warrants alternate methods of rice cultivation that aims at higher water and crop productivity.

For instance, Aerobic rice have shown that water requirement in aerobic rice were more than 50 per cent lower (only 470-650 mm) and water productivities were 64-88 per cent higher than the lowland rice (Amudha *et al.*, 2009). Aerobic way of growing rice saves water by eliminating continuous seepage and percolation, reducing evaporation and eliminating wet land preparation (Castaneda *et al.*, 2002). With respect to water productivity and yield, system of rice intensification method of rice cultivation recorded the higher water productivity (0.43 and 0.47 kg/m^3) and higher grain yield (6014 and 6682 kg/ha), followed by alternate wetting and drying and aerobic rice cultivation during both the seasons (*summer and kharif*) observed by Geethalakshmi *et al.* (2008). The work of various scientists discussed above indicated that the enhancement of water productivity play significant role to enhance crop productivity in rice ecosystem.

Aerobic Rice

Aerobic rice cultivation is an approach in rice cultivation is to grow the crop like an irrigated upland crop such as wheat or maize. Unlike lowland rice, it is grown in unpuddled, non-saturated (*i.e.,* aerobic system) soil without flooded water. It is a new concept of reducing water requirement for rice in which rice is grown like an upland crop with high inputs and supplementary irrigations when rainfall is insufficient reported by Bouman, (2001). A new development in water saving technology is the concept of "aerobic rice". Aerobic rice is a new way of cultivating rice that requires less water than lowland rice. It entails the growing of rice in aerobic soil, with the use of external inputs such as supplementary irrigation and fertilisers and aiming at higher yields observed by Wang *et al.* (2002).

Aerobic way of growing rice saves water by eliminating continuous seepage and percolation, reducing evaporation and eliminating wet land preparation reported by Castaneda *et al.* (2002). Aerobic rice saves 73 per cent of irrigation water for land preparation and 56 per cent during the crop growth period reported by Castaneda *et al.* (2002). Gupta *et al.* (2004) found that Aerobic rice production system eliminates continuous seepage and percolation losses, greatly reduces evaporation as no standing water is present at any time during the cropping season, and effectively uses the rainfall and thus helps in enhancing water productivity and comparison of water requirement of lowland flooded rice and aerobic rice system clearly shows that aerobic rice system can save about 45 per cent of water. Rainfall is insufficient to sustain lowland rice production (estimated to require about 1200-1500 mm) but sufficient for aerobic rice (about 800 mm).

Water is scarce during the first part of the growing season (requiring irrigation) but floods occur in the second part and favourable uplands with access to supplementary irrigation reported by Bouman, (2001). In aerobic rice systems, fields remain unsaturated throughout the season studied by Bouman *et al.* (2004). Bouman *et al.* (2002) estimated that water requirement for aerobic condition by growing two elite aerobic rice genotypes and one popular lowland variety both under flooded and aerobic conditions. The results of their study has shown that under aerobic condition, the combined amount of rainfall and irrigation water from sowing to harvest varied from 470 to 650 mm, compared with 1350 to 1400 mm in lowland rice and water productivities. Wang and Tang (2002) observed that the water use efficiencies of aerobic varieties under aerobic conditions were 164-188 per cent higher than that of the lowland variety under lowland conditions. The returns on water use were 50-100 per cent higher than those in lowland rice.

Moreover, the use of labour was more than 50 per cent lower than in lowland rice because of the absence of puddling and transplanting as well as less frequent irrigation application and possibilities of mechanised sowing and harvesting in aerobic rice observed by Bouman *et al.* (2002). Ravendra Singh *et al.* (2010) revealed that aerobic rice is a new concept to decrease water requirements in rice production and is highly suitable for irrigated lowland rice with insufficient rainfall and favourable uplands with access to supplementary irrigation. As reduction in water use was more pronounced than that in grain yield, water productivity under aerobic cultivation increased by 22.4 – 45.0 per cent compared to flooded condition. Therefore, researchers are looking for ways and means to decrease water use in rice production and increase the water use efficiency.

SRI Method of Rice

System of rice intensification (SRI) which is relatively a new methodology involves a set of practices for plant, soil, water and nutrient management. It is a revolutionary technology in the sense that it tries to change traditional practices especially with respect to water management that existed for thousands of years. The future of country's rice production will depend heavily on developing and adopting strategies and practices that use irrigation water efficiently at the farm level. SRI is one such method which has a potential to produce more rice with less water. Until 1990, the impression was that rice yields better only under flooded conditions. Recent reports from International Water Management Institute (IWMI), Colombo, however suggest that continuous submergence is not essential for obtaining higher rice yields.

Further, experiences from studies on SRI in China and Sri Lanka during the last decade have conclusively demonstrated that unflooded soil is ideal for rice plant to grow well and yield better than under conventional method of continuous shallow submergence. Hence, Krishna *et al.* (2008) and Udyakumar (2005) reported an enhanced tillering, early flowering, better yield and grain quality in SRI plots as compared to traditional methods due to increased conversion of tillers into productive tillers/flowering panicles with the adoption of SRI management reported by Gani *et al.* (2002). This favourable influence might be due to efficient utilisation of resources and less inter-and-intraspace competition under SRI management, which may be assigned as the reason for superiority in such yield attributes of rice and consequently increased yield reported by Dass and Chandra (2013).

Geethalakshmi *et al.* (2008) found that with respect to water productivity, SRI method of rice cultivation registered the higher water productivity (0.43 and 0.47 kg/m^3), followed by AWD and aerobic rice cultivation during both the seasons. Similarly, SRI methods saves of 25 per cent irrigation water without any effect on yield components compared to conventional transplanting reported by Chowdhary *et al.* (2005). Thiyagarajan *et al.* (2002) reported that water saving of 50 per cent in SRI over the traditional flooding without any adverse effect on grain yield. Total water productivity of the SRI was 29 per cent higher as compared to conventional method reported by Thiyagarajan *et al.* (2002). Similarly, Manajunath *et al.* (2010) reported that modified SRI method and recommended SRI method where lower water requirement and higher water use efficiencies when compared to normal method of cultivation. These results were authenticated by Bali *et al.* (1995) revealed that grain yield was higher with five week/old seedlings and water use efficiency. Thiagarajan *et al.* (2002) reported that limited irrigation of two cm depth after surface crack development recorded higher water productivity (0.732 kg/m^3) with 56 per cent saving in irrigation water (IW) compared to conventional practice of five cm continuous flooding (0.398 kg/m^3). Thiruneelakandan Subbulakshmi (2013) reported that higher water productivity in SRI methods (0.514 kg/m^3), followed by conventional practices (0.259 kg/m^3).

Direct Seeded Rice

Dry-seeded rice technology offers a significant opportunity for conserving irrigation water by using rainfall more effectively. In transplanted and wet-seeded rice systems, farmers normally wait for delivery of canal water before they start land soaking. In dry-seeded rice, land preparation is done with dry or moist soil conditions and is started using early monsoon rainfall. Crop emergence and early growth also occur in the early part of the monsoon, and only later, when canal water is available, is the crop irrigated as needed. Tabbal *et al.* (2001) demonstrated the feasibility of dry-seeded rice in wet season irrigated areas in the Philippines. Cabangon *et al.* (2001) reported that dry-seeded rice significantly increased water productivity with respect to irrigation water over wet-seeded and transplanted rice in the Muda Irrigation Scheme, Malaysia.

However, it was also observed that all three crop establishment practices had similar total water input and water productivity with respect to total water input. An additional advantage of dry seeding is the early establishment of the crop which may allow farmers to grow an extra crop after harvest on residual soil moisture (My *et al.* 1995; Saleh *et al.* 1995) or using saved irrigation water. In purely rainfed systems, early establishment and harvest of dry-seeded rice allows the rice plants to escape any late season drought and hence improve the yield and its reliability. Water-saving technologies that improve productivity and income will be easily accepted by farmers. Direct seeded rice (DSR) gave higher water productivity than transplanted rice observed by Gill *et al*, (2013).

Alternate Wetting and Drying Irrigation (AWD) Method in Rice

Alternate wetting and drying (AWD) is commonly used as a water-saving practice in many parts of the world for more than a decade. It is the most efficient method of utilising the water through irrigation. Cabangon *et al.* (1999) observed that in this system, the soil is allowed to dry for a few days within irrigation events depending on plant developmental stages. Some success has been reported as far as yield and water demand is concerned Lu *et al.*, (2000).

However, unproductive water losses could not be totally prevented by AWD. Since, the water consumption is still high in AWD. Therefore, soils need to be submerged at least during the irrigation period. Savings irrigation water in the AWD treatments were 53–87 mm (13–16 per cent) compared with the continuously submerged regime. Rice grain yields ranged from 7.2 to 8.7 t ha^{-1} and were not markedly affected by the water regimes. Water productivity was significantly higher in the AWD regime than in the continuously submerged regime (Belder *et al.* 2004). Yield penalty was commonly observed under AWD compared with continuously flood-irrigated (CF) rice (Bouman and Tuong, 2001). In general, AWD method of irrigation increased water productivity with respect to total water input because the yield reduction was smaller than the amount of water saved. Variety has a large influence on the grain yield of AWD (Bhushana *et al.,* 2007). The development of water-saving and drought-resistance rice (WDR) is another strategy to produce more rice with less water (Serraj *et al.*, 2011). WDR aims to produce the same yield as paddy rice with much less water consumption (50 per cent water saving compared with the normal paddy rice) under irrigated conditions.

At the same time, WDR should have the ability of drought tolerance to minimise yield loss under water-limited conditions. Several field studies were conducted to determine grain yield, water saving, water productivity and drought tolerance of HY3. HY3 demonstrated an ability of drought tolerance. Water saving and relatively high yield was also achieved in HY3. However, these studies did not compare HY3 with high-yielding varieties for the continuously flood-irrigated rice system such as—super hybrid rice varieties. China's super hybrid varieties have increased rice yield potential by 8–15 per cent compared with ordinary hybrid and inbred varieties. In recent years, super hybrid rice varieties have occupied about 20 per cent rice planting areas in China. However, the high yielding super hybrid rice varieties was often achieved when water was amply supplied (Bueno *et al.* 2010). Ultimately, it was found that practicing the AWD method of irrigation in rice results in higher yield and water use efficiency by consuming lesser amount of water.

Raised Beds System in Rice

Raised beds system has greater advantage as far as the efficient utilisation of water is concerned. Puddling induces high bulk density, high soil strength and low permeability in subsurface layers (Kukal and Aggarwal, 2003). These factors restrict root development, water and nutrient use from the soil profile by wheat sown after rice. The development of hardpan also leads to aeration stress in wheat crop at the time of the first irrigation and this problem is predominant in the region where rice–wheat system is being practiced. Various technologies for water saving in rice like direct seeding, ground cover system, alternate wetting and drying, direct seeding and transplanting on beds (soil saturation culture), etc. are being tested. The latter one, i.e. transplanting of rice on beds omits puddling and hence avoids the detrimental effects of puddling. In this case rice is grown on raised beds and irrigation is applied in furrows between the beds. Several studies found that water saving associated with plant installation in beds, water management (continuously flooded condition or intermittent irrigation) is often poorly reported. This is an important consideration in assessing whether the raised beds saved irrigation water because of their particular geometry or whether the water saving was the result of applied intermittent irrigations which can also be applied to flat land (Kukal and Aggarwal, 2003).

Sandhu *et al.* (2012) observed that transplanting of rice seedlings on slopes of freshly constructed beds resulted in 15 per cent saving of irrigation water as compared to puddled plots (conventional method used by farmers) without any significant reduction in grain yield of rice. Irrigation water can also be saved in puddled transplanted rice by applying irrigation three days after disappearance of ponded water as compared to recommended practice of applying irrigation two days after disappearance of ponded water and this practice does not leads to any significant reduction in grain yield. However, beds are to be irrigated two days after disappearance of ponded water (Sandhu *et al.,* 2012). Maynard, 1991, observed that the practice of raised beds for the production of irrigated non-rice crops was pioneered in the heavy clay soils of the rice-growing region in Australia in the late 1970s and for irrigated wheat in the rice–wheat system of the Indo-Gangetic plains during the 1990s, inspired by the success of beds for wheat–maize systems in Mexico. Potential agronomic advantages of beds include improved soil structure due to reduced compaction through controlled trafficking, and reduced water logging and timely machinery operations due to better surface drainage. Beds also provide the opportunity for mechanical weed control and improved fertiliser placement. On the other way, the potential benefits of beds for wheat production in the Indo-Gangetic plains have been known for some time, evaluation of beds for rice and permanent beds in rice–wheat system systems commenced more recently (Sayre and Hobbs).

Conclusions

Cultivation of rice through system of rice intensification increase the grain yield by five to seven per cent besides saving of water by 12 to 15 per cent over conventional method of rice cultivation under wetland ecosystem. Practicing of aerobic rice cultivation has increased water productivity on an account of 22.4-45.0 per cent as compared to flooded condition. Ultimately, it was found that practicing the Alternate wetting and drying (AWD) method of irrigation in rice not only increases the yield but also resulted in higher water use efficiency by consuming lesser amount of water.

References

Abdullha, 2003. "The potential of aerobic rice to reduce water use in water-scarce irrigated lowlands in the tropics". In Water-Wise Rice Production.

Amudha, K., K. Thiyagarajan and N. Sakthivel. 2009. "Aerobic rice: A review", Agricultural Review., Vol. 30, pp. 145 – 149.

Anchal Dass and Shiva Dhar. 2014. "Irrigation management for improving productivity nutrient uptake and water-use efficiency in system of rice intensification": A review Annual Agriculture Research. New Series Vol. 35, pp. 107-122.

Bali, A. S., K.N. Singh., A.S Balli., B. A. Ganai and B. Hasan. 1995. "Effect of transplanting date, number and age of seedlings on rice" Indian Journal of Agronomy. Vol. 90, pp. 504-507.

Belder, P., B.A.M., Bouman, R. Canbagon, L.Guaon, T.P. Tuong. 2004. "Effect of water-saving irrigation on rice yield and water use in typical lowland conditions in Asia". Agricultural Water Management, Vol. 65, pp. 193–210.

Bhushan, L., J.K. Ladha, R.K. Gupta, S. Singh, M. Gathala, H. Pathaka, 2006. "Saving of water and labor in a rice–wheat system with no-tillage and direct seeding technologies". Agronomy Journal, Vol.99, pp. 1288–1296.

Bouman, B.A.M., R.M. Lampayan and T.P. Tuong. 2002 and 2004. "Water management in irrigated rice: Book".

Bouman, B.A.M., R.M. Lampayan and T.P. Tuong. 2002. "Proc. 12th International Soil Conservation Organization Conference". Tsinghua University, Beijing, China.

Bouman, B.A.M. and T.P. Tuong, 2001. "Field water management to save water and increase it productivity in irrigated rice". Agricultural Water Management, Vol. 49, pp. 11–30.

Bouman, B.A.M., H. Hengsdijk, B. Hardy, P. S. Bindraban, T. P. Tuong, and J. K. Ladha. 2003. "Proceedings of a thematic workshop on water-wise rice production" at IRRI Headquarters in Los Banos, Philippines.

Bueno, C.S., M. Buccort, N. Kobayashi, K. Inubhushi, T. Lafarge, 2010. Water productivity of Contrasting rice genotypes grown under water-saving conditions in the tropics and investigation of morphological traits for adaptation. Agricultural Water Management, Vol. 98, pp. 241–250.

Cabangon, R., T.P. Tuong, E.B. Tiak, N.B. Abdullah. 2001. "Increasing water productivity in rice cultivation: impact of large-scale adoption of direct seeding in the muda irrigation system. In Direct Seeding in Asian Rice Systems: Strategic Research Issues and Opportunities. Proceedings of an International workshop, on Direct Seeding in Asia, Bangkok, Thailand, 25-28 January 2000. International Rice Research Institute, Makati City, Philippines.

Cabangon, R.J., T.P. Tuong, N.B. Abdullha. 2002. Comparing water input and water productivity of transplanted and direct seeded rice production systems. Agricultural Water Management, Vol.57, pp. 11–13.

Castaneda, R.J., T.P. Tuong, N.B. Abdullha (2002). "Proc. of the International Workshop on Water Wise Rice Production". IRRI, Los Banos, Philippines.

Choudhury, B.U., A.K. Singh and B.A.M. Bouman. 2005. "Effect of establishment techniques on yield, crop water relationship in rice and wheat. Transitions in Agriculture for Enhancing Water Productivity", Proceedings of an International Symposium held in Killikulam, Tamil Nadu, India.

Dass, A. and Chandra, S. 2013. "Irrigation, spacing and cultivar effects on net photosynthetic rate, dry matter partitioning and productivity of rice under system of rice intensification in mollisols of northern India". Experimental. Agriculture. Vol. 49, pp. 504-523.

Gani, A. 2001. "Improving water-use efficiency for sustainable rice production systems. Water saving rice production systems". In: Proc. International workshop on Water Saving Rice Production systems. Nanjing University, China.

Geethalakshmi, T. Ramesh, A. Palamuthirsolai and A. Lakshmanan. 2008. "Agronomic evaluation of rice cultivation systems for water and grain Productivity" Agro Climate Research Centre, Tamil Nadu Agricultural University, Coimbatore 641 003.

Gill, J.S, S.W. Sohan and R.S. Gill. 2014. "Direct seeded rice: An alternative rice establishment technique in north-west India – A review" International Journal of Advanced Research (2014), Vol. 2 (3). pp. 375-386

Gupta, A.K., S. Kumar and J. Sai Prasad. 2014. "Efficient use of water for maximum production of aerobic rice (*Oryza sativa)"*, Journal of Progressive Agriculture, Vol. 5, pp. 52.

Krishna, A, N.K. Biradar Patil and B.B. Channappagoudar. 2008. "Influence of system of rice intensification (SRI) cultivation on seed yield and quality". Karnataka Journal Agricultural Sciences, Vol. 2, pp. 369-372.

Kukal, S.S. and G.C. Aggarwal, 2003. Puddling depth and intensity effects in rice–wheat system on a sandy loam soil. Development of subsurface compaction. Soil Tillage Research, Vol. 72, pp. 1–8.

Lu, K., T. Kasharaj and M. Sudhee. 2000. "The effects of irrigation regimes on the water use, dry matter production and physiological responses of paddy rice", Plant Soil, Vol. 223, pp. 207–216.

Manjunatha, B.N., R. Basavarajappa and B.T. Pujari. 2010. "Effect of age of seedlings on growth, yield and water requirement by different system of rice intensification" Karnataka Journal of Agriculture Sciences. Vol. 23, pp. 231-234.

Maynard, M., 1991. Permanent beds—Their potential role in soil management for the future. Farm Newsletter Large Area, Vol.137, pp. 14–18.

Morison, J.I.L., N.R. Baker Mullineaux and W.J. Davies. 2008. "Improving water use in crop production". Philosophical Transactions of the Royal Society of London B. Biological Sciences 363(1491), 639-658.

Ravender, S., D.K. Kundu and K.K. Bandyopadhyay. 2010. "Enhancing Agricultural Productivity through enhanced Water", Journal of Agricultural Physics Use Efficiency, Vol. 10, pp. 1-15.

Sandhu, S.S, S.S. Mahal, K.K. Vashist, G.S. Buttar, A.S. Brar, M. Singh. 2012. Crop and water productivity of bed transplanted rice as influenced by various levels of nitrogen and irrigation in northwest India. Agricultural Water Management, Vol. 104, pp. 32–39.

Sayre, K.D. and P.R. Hobbs. 2004. The raised-bed system of cultivation for irrigated production conditions. In - Sustainable Agriculture and the International Rice–Wheat System" (R. Lal, P. R. Hobbs, N. Uphoff, and D. O. Hansen, Eds.). pp. 337–355.

Serraj, R., L. Kenneth, Mc. Nally, I. Loedin, A. Kohli., S.M. Haefele, G. Atlin and A. Kumar. 2011. Drought resistance improvement in rice: an integrated genetic and resource management strategy. Plant Production Science, Vol.14, pp. 1–14.

Shi, Y. C., 1999. Way of exploitation: Biological water saving. Sci. Technol. Rev. 10, 3–5 Lu, J., T. Ookawa, T. Hirasawa, 2000, The effects of irrigation regimes on the water use, dry matter production and physiological responses of paddy rice. Plant Soil, Vol. 223, pp. 207–216.

Tabbal, D.F., R.M. Lampayan, and S.I. Bhuiyan. 1992. "Water-efficient irrigation technique for rice. Soil and water engineering for paddy field management". Proceedings of the International Workshop on Soil and Water Engineering for Paddy Field Management, 28-30 January 1992, Asian Institute of Technology, Bangkok, Thailand. pp. 146-159.

Thiruneelakandan, R. and Subbulakshmi, G. 2014. "Potential of the System of Rice Intensification (*Sri*) for Cuddalore District" Chemistry Science Review Letter, Vol. 3, pp. 10-17.

Thiyagarajan, T.M., Velu, V., Ramasamy, S., Durgadevi, D., Govidarajan, K., Priyadarshini, R., Sudhalakshmi, C., Kuamr, K.S., Nish, P.T., Gayathry, G., Hengsdijk, H. and Bindraban, P.S. 2002. "Effects of SRI practices on hybrid rice performance in Tamil Nadu, India". In: Water Wise Rice Production. IRRI, pp. 119-127.

Udyakumar, 2005. "Studies on system of rice intensification (SRI) for seed yield and seed quality". M.Sc. (Agri.) Thesis, Acharya N.G. Ranga Agricultural University, Hyderabad (India), pp. 146.

Wang, H. and S.Tang, 2002. In: Proc. "Upland Rice Research Consortium Meeting", IRRI, Los Banos, Philippines.

Wang, H., S.Tang and Tang, M. 2002. "Proc. of the International Workshop on Water Wise Rice Production". IRRI, Los Banos, Philippines.

Wang, H.Q. and S.X. Tang. 2000. "Upland rice production in China, Its past, today and future". Paper presented at the Aerobic Rice Workshop, pp. 7-8 September, IRRI, Los Baños, Philippines.

12

Effect of Fertiliser and Crop Geometry on Productivity and Economics of Cluster Bean Varieties under Climate Change Condition

Satyakumari Sharma, Rahul Sharma and H.M. Bhuva

Abstract

A field experiment was conducted on sandy loam soil at the Rajmata Vijyaraje Sindhiya Krishi Vishwa Vidhyalaya, Gwalior during kharif seasons of 2009-10. To study the assess the effect of fertiliser and crop geometry on productivity and economics of cluster bean. Results revealed that significantly higher plant height branches/ plant and length of pod, biomass/plant number of pod/plant, pod length and protein yield and maximum net accounting to Rs 20,999 along with highest benefit: cost ratio of 3.29 with variety 'HG-563 (c)' with 45 spacing and at least 50 per cent of recommended doses of fertilisers (10 nitrogen + 20 phosphorus kg ha^{-1}). Plant population at 15 DAS and maturity, plant height and branches per plant at 30 DAS and protein content in seed remained unaffected due to futility levels, row spacing and some promising genotypes of cluster bean. But, branches per plant at 60, 90 DAS and at harvest remained unaffected due to varietal levels and plant height at 60, 90 DAS and at harvest remained unaffected due to row spacing.

Key words:*Cluster bean, Cyamopsistetragonoloba, Fertiliser, Yield and Variety.*

Introduction

Pulse are important food grain crop because of higher nutritive value and occupy a unique position in ever known system of farming as a main, catch, green manure and intercrop. Among the different pulses, cluster bean locally known as guar considered as an important pulse crop and important crop of northern M.P., because of its gum content (28-32 per cent) which is being utilised in various industries and potential for exports, has added to the demand of cluster bean especially to the USA bowing demand, production technology of Cluster bean needs to be standardised. Cluster bean can be grown successfully on poor fertile soil and areas of erratic rainfall. Use of improved varieties in addition to adequate nutrient management has been reported effective in improving the productivity of cluster bean in arid zone (Singh *et al.*, 2004). Therefore, there is a need to supply required nutrients for obtaining higher yield of cluster bean, which may be cheaper and renewable. Though cluster bean is an important crop of semi-arid zone of M.P., but its productivity is low mainly due to imbalanced fertilisation. Application of recommended dose of fertiliser (20 kg N + 40 kg P_2O_5/ha) to guar gave significantly higher yield than 50 per cent recommended fertiliser dose (10 kg N + 20 kg P_2O_5/ha) (Rajput, 2002). Proper plant population is the main key for higher yield of cluster bean. Higher yield and quality cannot be achieved without proper row spacing and plant stand. Row spacing of 45 cm recorded significantly higher seed yield, more branches, pods/plant, seeds/pod and test weight as compare to 30 cm row spacing (Anonymous, 2005-06).

Material and Methods

An experiment was conducted at the Research Farm, Rajmata Vijyaraje Sindhiya Krishi Vishwa Vidyalaya, Gwalior (Madhya Pradesh) during *kharif* 2009-10. The experiment site is situated at 26.1°N latitude and 74.4°E longitude with an altitude of 211.5 m above the mean sea level. The area has typically sub-tropical climate with extreme hot about 46°C in summer and minimum temperature below 1°C in the winter season. The total rainfall received during the crop growth period was 533.7 mm during 2009-10. The maximum and minimum temperature during the crop growth period ranged between 21.6°C to 43.9°C and 6.3°C to 29.5°C during 2009-10. The soil of the experimental plot was sandy loam and slightly alkaline in reaction (pH 7.9) as well as low in available nitrogen (178 kg/ha.), medium in available phosphorus (12.6 kg/ha.) and medium in available potash (281 kg/ha.). The experiment comprising of sixteen treatment combinations consisting two levels of fertility (10-20 and 20-40 kg N-P_2O_5/ha), two levels of row spacing (30 cm and 45 cm) and four promising genotypes (HG-04- 875, RGR-6, RGR-7 and HG-563), were laid out in randomized complete block design with three replications. The cluster bean was sown at 30 cm or 45 cm spacing using seed rate of 20 kg ha^{-1}and harvested in October, 2009. The entire dose of fertiliser in form of DAP and SSP was applied as basal at sowing. Following formula were used to calculate harvest index (HI).

$$\text{HI}(\%) = \frac{\text{Economic yield (kg ha}^{-1})}{\text{Biological (kg ha}^{-1})}$$

Results and Discussion

Effect of Fertiliser

The growth and yield parameters, *viz.,* number of branches/plant and length of pods was significantly influenced by fertility levels. Significantly superior plant height at 60 and 90 DAS, number of branches per plant at 60, 90 DAS and maturity characters were increased dose of fertiliser from observed with application of 20-40 kg N-P_2O_5/ha over application of 10-20 kg N P_2O_5/ha. This might have resulted due to better and timely availability of N and P_2O_5 for their utilisation by plant. However, both the fertility levels were statistically comparable in respect of plant height, biomass/plant, number of pods/plant, number of seeds/pod, 100-seed weight, seed weight/plant, stalk weight/plant, seed yield, stalk yield and harvest index. This might be ascribed to the fact that clusterbean, being a leguminous crop, fixes atmospheric N and thus requires only a starter dose of N. Moreover, medium status of P in experimental soil might be responsible for non-response of the crop to applied P. Similar results were also reported by Rajput (2002), Gupta *et al.* (2006) and Rathore *et al.* (2007).

Effect of Row Spacing

Crop sown at 45 cm row spacing produced significantly higher number of branches/plant than 30 cm row spacing at all crop growth stages except 30 DAS. Row spacing at 45 cm resulted in significantly more biomass/plant, number of pods/plant, number of seeds/pod, seed weight/plant and stalk yield over 30 cm row spacing. While harvest index significantly decreased at 45 cm row spacing over 30 cm row spacing. Crop sown at 45 cm row to row distance increased the stalk yield per hectare significantly by 13.03 per cent over 30 cm row spacing. Harvest index significantly decreased from 28.26 per cent at 30 cm row spacing to 26.01 per cent at 45 cm row spacing. However, plant height, pod length, 100-seed weight, stalk weight/plant and seed yield remained unchanged due to row spacings. This indicates that the genotypes tested in the experiment have been specially bred for gum purpose having determinate growth habit, which can be grown at closer spacing of 30 cm rather than wider spacing of 45 cm suitable for vegetable cluster bean. Meena *et al.* (2004), Patel *et al.* (2005) and Burman *et al.* (2009) also reported analogous results.

Effect of Genotypes

Cluster bean genotypes differed significantly in plant height at all the stages of observation except at 30 DAS, while in case of number of branches/ plant, it did not differ significantly at any stage of the crop growth. HG-563 (c) followed by RGr-6 had significantly taller plant than other varieties. HG-563 (c) also produced significantly higher biomass/plant than other varieties.

Differences among varieties with respect to yield attributing characters viz., number of pods/plant, 100-seed weight, seed yield, stalk yield were significant. HG-563 (c) followed by RGr-6 had highest number of pods per plant, seed and stalks yields/plant, although RGr-6 followed by RGr-7 had maximum seed index. However, number of branches/plant, pod length and number of seeds per pod, protein content in seed remained almost equal in all the genotypes. Parallel results were also obtained by Singh *et al.* (2004) and Choudhary *et al.* (2005).

Effect on Economics

Although maximum net profit amounting to Rs 20999 along with highest B;C ratio of 3.29 was computed with $F_2S_2V_4$ followed by $F_2S_1V_4$ (Rs 20605/ha and 3.25) and $F_1S_1V_4$ (Rs 19697/ha and 3.03) and minimum net profit of Rs 11471 was worked out when the variety "RGr-7" was sown at 30 cm row spacing with 50 per cent NP (10 N + 20 P_2O_5 kg/ha).

Conclusion

The results indicate that application of variety 'HG-563' at 45 cm row spacing and fertilising with at least 10-20 kg N-P_2O_5/ha gave higher production and economics from cluster bean can be secured under the agro-climatic conditions of Gwalior region.

Table 1: Effect of different treatments on growth parameters of cluster bean varieties

Treatments	*Plant stand/m row length*		*Plant height (cm)*				*Biomass/ plant (g)*
	15 DAS	*Harvest*	*30 DAS*	*60 DAS*	*90 DAS*	*Harvest*	
Fertiliser levels							
F_1	12.39	11.24	20.56	74.71	78.05	81.06	7.57
F_2	12.62	11.50	0.13	0.29	3.77	77.12	7.38
S.Em ±	0.25	0.28	0.28	0.98	1.00	1.70	0.13
C.D. at 5%	NS	NS	NS	2.83	2.90	NS	NS
Row spacing							
S_1	12.38	11.21	20.47	71.93	75.34	76.88	7.23
S_2	12.63	11.53	20.21	73.06	76.47	81.30	7.73
S.Em ±	0.25	0.28	0.28	0.98	1.00	1.70	0.13
C.D. at 5%	NS	NS	NS	NS	NS	NS	0.39
Varieties							
V_1	12.37	11.26	20.53	70.99	74.40	77.72	7.20
V_2	12.47	11.21	20.10	73.28	76.72	77.52	7.57
V_3	12.79	11.51	20.37	69.64	72.92	72.20	6.64
V_4	12.39	11.50	20.38	76.09	79.60	88.92	8.49
S.Em ±	0.36	0.39	0.40	1.39	1.42	2.41	0.19
C.D. at 5%	NS	NS	NS	4.00	4.10	6.95	0.55

Table 2: Effect of different treatments on growth and quality parameters of cluster bean varieties

Treatments	*Branches/plant*				*Protein content in seeds (per cent)*	*Protein yield (kg/ha)*
	30 DAS	*90 DAS*	*60 DAS*	*Harvest*		
Fertiliser levels						
F_1	1.46	4.92	6.58	6.65	33.11	3.50
F_2	1.17	4.00	5.95	6.16	32.84	3.44
S.Em ±	0.13	0.19	0.21	0.14	0.40	0.045
C.D. at 5%	NS	0.54	0.62	0.40	NS	NS
Row spacing						
S_1	1.21	4.13	5.90	6.19	32.88	3.43
S_2	1.42	4.79	6.63	6.62	33.07	3.51
S.Em ±	0.13	0.19	0.21	0.14	0.40	0.045
C.D. at 5%	NS	0.54	0.62	0.40	NS	NS
Varieties						
V_1	1.25	4.25	5.85	6.08	33.15	3.25
V_2	1.42	4.50	6.34	6.25	32.86	3.34
V_3	1.08	4.33	6.24	6.38	33.48	3.16
V_4	1.50	4.75	6.63	6.83	33.41	4.13
S.Em ±	0.19	0.26	0.30	0.20	0.57	0.063
C.D. at 5%	NS	NS	NS	NS	NS	0.180

Table 3: Effect of different treatments on yield parameters of clusterbean varieties

Treatments	*Pods/ plant*	*Pod length (cm)*	*Seeds/ pod*	*100 seed weight (g)*	*Yield (g/plant)*		*Yield (g/plant)*		*HI (per cent)*
					Seed	*Stalk*	*Seed*	*Stalk*	
Fertiliser levels									
F_1	43.62	5.59	9.15	3.569	4.78	2.80	1068.63	2878.38	27.22
F_2	42.13	5.31	9.08	3.608	4.75	2.63	1033.92	2817.88	27.04
S.Em ±	0.85	0.04	0.10	0.021	0.12	0.06	15.81	35.95	0.33
C.D. at 5%	NS	0.10	NS	NS	NS	NS	NS	NS	NS
Row spacing									
S_1	40.33	5.40	8.91	3.591	4.58	2.65	1042.25	2673.88	28.26
S_2	45.41	5.49	9.33	3.586	4.94	2.79	1060.29	3022.38	26.01
S.Em ±	0.85	0.04	0.10	0.021	0.12	0.06	15.81	35.95	0.33
C.D. at 5%	2.45	NS	0.28	NS	0.33	NS	NS	103.82	0.96
Varieties									
V_1	39.50	5.49	9.22	3.476	4.67	2.53	989.50	2760.67	26.43
V_2	46.32	5.42	9.05	3.699	4.79	2.79	1003.67	2769.67	27.19
V_3	39.02	5.39	8.97	3.638	4.09	2.56	942.08	2682.25	26.04
V_4	46.65	5.49	9.23	3.542	5.50	2.99	1269.83	3179.92	28.88
S.Em ±	1.20	0.05	0.14	0.030	0.16	0.08	22.36	50.84	0.47
C.D. at 5%	3.46	NS	NS	0.090	0.47	0.24	64.58	146.83	1.36

Table 4: Effect of different treatments combination on economics of clusterbean

Treatments	*Total income (Rs/ha)*	*Total cost (Rs./ha))*	*Net income (Rs/ha)*	*B:C ratio*
$F_1S_1V_1$	21790	9708	12082	2.24
$F_1S_1V_2$	22277	9708	12569	2.29
$F_1S_1V_3$	23900	9708	14192	2.46
$\mathbf{F_1S_1V_4}$	29405	9708	19697	3.03
$F_1S_2V_1$	26083	9708	16375	2.69
$F_1S_2V_2$	25865	9708	16157	2.66
$F_1S_2V_3$	21489	9708	11781	2.21
$\mathbf{F_1S_2V_4}$	28784	9708	19076	2.96
$F_2S_1V_1$	23095	9156	13939	2.52
$F_2S_1V_2$	23277	9156	14121	2.54
$F_2S_1V_3$	20627	9156	11471	2.25
$\mathbf{F_2S_1V_4}$	29761	9156	20605	3.25
$F_2S_2V_1$	21630	9156	12473	2.36
$F_2S_2V_2$	22444	9156	13287	2.45
$F_2S_2V_3$	22252	9156	13096	2.43
$\mathbf{F_2S_2V_4}$	30155	9156	20999	3:29

References

Burman, Uday, Garg, B.K. and Kathju, S. (2009). Effect of phosphorus application on clusterbean under different intensities of water stress. *Journal of Plant Nutrition*, 32(4): 668-680.

Burman, Uday, Garg, Balvinder Kumar and Kathju, Shyam (2009). Interactive effects of phosphorus, nitrogen, and thiourea on cluster bean (*Cyamopsistetragonoloba* L.) under rainfed conditions of the Indian arid zone. *Journal of Agronomy and Crop Science*, 181(4): 209-214.

Choudhary, S.P.S., Singh, R.V., Singh, N.P., Saini, D.D. and Khedar, O.P. (2005). Genetic improvement of guar (*Cyamopsistetragonoloba*) at a glance. *Proc.* National Symposium on "Advances in Forage Research and Sustainable Animal Production" held at CCSHAU, Hisar, pp. 19-20.

Gupta, A., Sharma, V.K., Sharma, G.D. and Chopra, P. (2006). Effect of bio-fertiliser and phosphorus level on yield attributes yield and quality of urdbean (*Vignamungo*). *Indian Journal of Agronomy*, 9(12): 142-144.

Meena, N. L., Naagar, K. C. (2004). Effect of phosphorus, sulphur and phosphate solubilising bacteria on yield components, yield and quality of clusterbean [*Cyamopsistetragonoloba* (L.) Taub.], *Legume Research*, 27(1): 27-31.

Patel, I.C., Patel, M.M., Patel A.G. and Tikka, S.B.S. (2005). Response of *kharif* guar to fertilizers and row spacing under rainfed conditions. *Indian Journal of Pulses Research*, 18(2): 246-247.

Rajput, R.S. (2002). Response of clusterbean (*Cyamopsistetragonoloba* L. Taub) varieties to varying fertility levels and row spacing in northern Madhya Pradesh. M.Sc. (Agri.) thesis submitted to JNKVV, Jabalpur.

Rathore, V.S., Singh, J.P., Soni, M.L. and Beniwal, R.K. (2007). Effect of nutrient management on growth, productivity and nutrient uptake of rainfed clusterbean (*Cyamopsistetragololoba*) in arid region. *Indian J. Agric. Sci.*, 77(6): 349-53.

Singh, Raj, Singh, Bhagvan and Saxena, Anurag. (2004). Response of clusterbean genotypes to nutrient management under arid condition of Rajasthan. *J. Arid Legumes,* 1(1): 32-34.

13

Effect of Rainfall Variability and Supplemental Irrigation in Rainfed Groundnut under Scarce Rainfall Zone of Andhra Pradesh

K. Madhusudhan Reddy, Y. Pavan Kumar Reddy, B. Sahadeva Reddy and Ch. Appa Rao

Abstract

The main objective of this study is to assess the mean annual rainfall behaviour and also to basify the influence on the groundnut production and productivity in the Anantapuram district of Andhra Pradesh. The annual rainfall had significantly positive influence on the production and productivity of Groundnut in the Anantapuram district of Andhra Pradesh A trend analysis of groundnut production and productivity showed that the key to realising stable yields of groundnut in Anantapuram is the distribution of rainfall rather than total. In areas like Anantapuram Farm Pond Technology should be encouraged to provide the supplemental irrigation at critical stages are highly sensitive to dry spells realise the better yields in the situations like Anantapuram district where erratic and uneven distribution of rainfall to realise better yields.

Introduction

Rain fed agriculture is predominant in Anantapuram district, one of the most drought prone district in the country and it is the Single largest groundnut growing belt in the state growing annually an area of 5-8.5 lakh ha of groundnut. During *Kharif* season where the yield is severely limited by water deficits during crop growth due to unpredictable rainfall, high evaporative demands and low water holding capacity of soils. Pod yields fluctuate widely from 550-1100 Kg/ha. The crop is generally sown between 15th June to

15th July, depending on the normal onset of monsoon. However, the erratic onset of monsoon and commencement of sowing rains, sometimes force the farmers to sow the crop late in the season. Yield variation in rain fed groundnut can be attributed to rainfall variability *i.e.*, amount and distribution of rainfall. The effectiveness of rainfall in crop production depends mainly on commencement of sow ingrains and amount and distribution of rainfall during the season, as water deficit is a major constraint in peanut production, especially during the critical period of pod set which results in reduced pegging (Chandrika *et al.*, 2008 and Sahu *et al.*, 2004).

Stresses can occur at any stage of plant growth and development, thus illustrating the dynamic nature of crop plants and their productivity. Among heabiotic stresses, deficiency of soil water constitutes primary limitation to crop productivity in many regions of the world. The yield is sensitive to weather, soil as well as management practices. Uncertainties in rainfall and other environmental hazards in dry land farming cause large year to year fluctuation in groundnut production and productivity. Yield arrivals showed wide variability because its response to seasonal weather variables like rainfall, temperature, sunshine, relative humidity and evaporation etc. The unreliability in the rainfall is posing serious problem in groundnut production. The study on influence of weather variables on production and productivity of groundnut is greatly useful to know the impact weather and climate change on groundnut production.

The objective of this study is to ascertain the influence of rainfall distribution and its amount on the performance of groundnut production, productivity and seasonal rainfall behaviour on groundnut production in the Anantapuram district of Andhra Pradesh.

Methodology

Anantapuram district lies in the Rayalaseema region of Andhra Pradesh between 13°40' and 15°15' northern Latitude and 76°50' and 78°30' eastern Longitude with a total geographic area of the district is 19130 Km^2 with an annual average rainfall of the district is 572 mm. Red soils are dominant in the district and occupy 66 per cent of total geographical area. The depth of majority soils ranges from extremely shallow (10-25 cm) to shallow (25-50 cm). The soils have low inorganic carbon and available nitrogen, medium to high in available phosphorous and available potassium content. About 40 per cent soils are below critical level of zinc and 33 per cent soils are below critical level for iron. The district has 90 per cent area under rainfed conditions. Groundnut is the principal *Kharif* crop grown over an area of about eight lakh ha. The monthly rainfall data of 15 years (1999-2014) was taken from Chief Planning Officer, Anantapuram and Groundnut productivity, production was taken from Joint Director of Agriculture, Anantapuram to assess the annual rainfall behaviour and its influence on productivity of groundnut.

Results and Discussion

A trend analysis of groundnut production and productivity showed that a substantial increase in area over years was with slight deviations in some years. On the other hand, productivity did not show any consistent trend (Fig.1). During 2007-08, productivity of about 1207 kg/ha which highlights the fact that there is a possibility of realising more than one ton per ha productivity of groundnut in Anantapuram. On the other hand, extremely low yields were also recorded during 2006-07 and 2008-09. When compared, during 2006-07, annual rainfall were 408 mm, whereas in 2008-09, they were 714 mm which again emphasises that the key to realising stable yields of groundnut in Anantapuram is the distribution of rainfall rather than total. The peg penetration and pod development stages are highly sensitive to dry spells and hence distribution is more important (Table: 1).

Table 1. Rainfall and productivity of groundnut in Anantapuram district

S.No	*Year*	*Actual Rainfall (mm)*	*Productivity (Kg/ha)*
1.	1999-00	520.6	344.5
2.	2000-01	612.2	1010.8
3.	2001-02	702.1	465.4
4.	2002-03	290.3	381.4
5.	2003-04	523.2	269.8
6.	2004-05	433.5	714.8
7.	2005-06	791.2	355.8
8.	2006-07	408.0	079.4
9.	2007-08	816.0	1206.7
10.	2008-09	714.0	74.9
11.	2009-10	615.6	211.3
12.	2010-11	639.3	531.4
13.	2011-12	410.0	114.0
14.	2012-13	445.0	400.0
15.	2013-14	273.0	236.0

The district level yield data were plotted against seasonal rainfall to find out optimum levels of seasonal rainfall required for obtaining higher yields. A rainfall of 40mm appears to be critical for obtaining minimal yield of around 400 kg/ha. However some years, despite high seasonal rainfall, lower yields were realised which signifies improper distribution of rainfall.

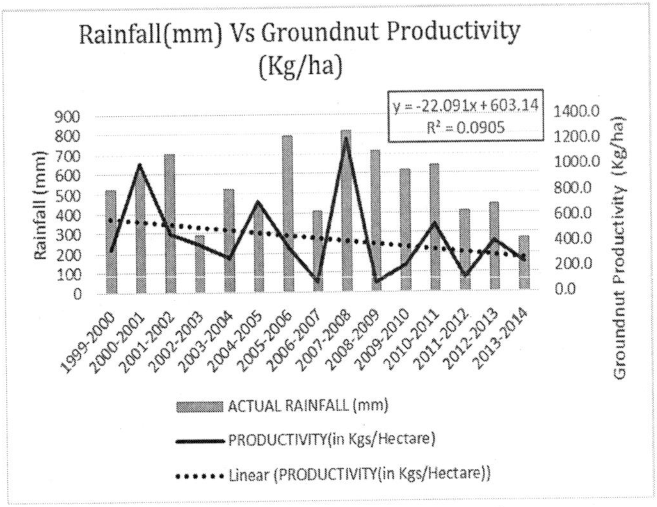

Fig. 1. Trend Analysis of rainfall and productivity of Anantapuram district, A.P.

The annual rainfall had significantly positive influence on the production and productivity of Groundnut in the Anantapuram district of Andhra Pradesh (r = 0.80 and 0.09).

Table 2. Correlation between rainfall and production, productivity in groundnut of Anantapuram district

	Actual Rainfall (mm)
Production (In '000' Tonnes)	0.1347
Productivity (In Kgs/ha)	0.1132

From the above it is clearly depicting that the yield of rainfed groundnut and the extent of yield reduction depends upon the amount and distribution of the rainfall.

Hence, there is a need to adopter in water harvesting techniques not only to store water for supplemental irrigation at the peg penetration and pod development stages but also to improve the yield levels.

Farm pond of 250 m³ capacity with side slopes of 1.5:1 is sufficient for catchment area of 5-10 ha depending upon slope of the field. Studies carried out at Agricultural Research Station, Anantapuram from 2005-14 revealed that supplemental irrigation of 10 mm with water stored in farm pond through sprinkler irrigation at times of moisture stress increased pod, haulm yield and shelling percentage of groundnut by 25-30 per cent, 13.1 per cent and 8.3 per cent respectively. Farm pond technology should be encouraged to provide supplemental irrigation at critical stages which are highly sensitive to dry spells to realise the better yields in the situations like Anantapuram district.

Policy Intervention

In such situations farm pond technology should be encouraged to store water for supplemental irrigation at the peg penetration and pod development stages which are highly sensitive to dry spells. Farmers should be provided with low cost diesel motors or solar power motors on subsidy to lift the water for supplemental irrigation from farm ponds. Farmers who adopted farm pond technology should be given incentives.

Acknowledgements

The authors are highly grateful to Chief Planning Officer, Anantapuram and Joint Director of Agriculture, Anantapuram for providing related data.

References

Chandrika, V., Parameswari, P. and Sreenivas, G. 2008. "Effect of sowing time and rainfall distribution on yield of rain fed groundnut (*Arachishypogaea* L.) in southern agro-climatic zone of Andhra Pradesh", Legume Research, Vol. 31(1), pp. 54-56.

CRIDA. 2012. A Report of the ICAR Expert Team on Agricultural Situation in Anantapuram district, Andhra Pradesh, CRIDA, 2012.

Sahu, D.D., Golakiya, B.A. and Patoliya, B.M. 2004. "Impact of rainfall on the yield of rain fed groundnut" *J. Agrometeorology,* Vol .6(2):249-253.

14

Development of Weather Based Forewarning Modules for the Management of Fruit Flies in Mango Orchard

D. M. Jethva, V. R. Virani and J. B. Bhut

Abstract

Mango, an important fruit crop of South Saurashtra region is attacked by many insect pests during the year. Among the pests, fruit fly, Bactroceradorsalis (Hendel) is a major pest of mango and cause economic losses. The importance of fruit flies has gone very high due to the major constraint in the exportation of fresh mango fruits. The study required orchard and laboratory studies, which were conducted on the commercial variety Kesar, at Department of Entomology, Junagadh Agricultural University, Junagadh, India, during the year 2005 to 2011. Pooled results showed that the populations of the trapped fruit flies in methyl eugenol trap showed that the insect was active throughout the year, except severe winter months. The maximum catches of fruit flies per trap were ranged from 200 to 690 during 14th to 36th standard week. Pooled correlation of fruit fly catches in methyl eugenol traps was found to be positive and significant with minimum temperature, morning and afternoon relative humidity, wind speed and rains. Bright sunshine hours showed significant negative correlation with fruit flies trapped. Rest of weather parameters did not showed significant positive or negative correlation of co-efficient with adult trapped. The multiple linear regression analysis indicated that the influence of all the weather parameters was highly significant and it was 84 per cent (R^2=0.84461). Thus, it can be recommended that the activity of fruit flies in mango orchards seemed to be related positively with high humidity (80-90 per cent RH) and 24 to 26°C minimum temperature during the month of April to September (14th to 36th standard week). This information could help to take the management strategies *in an opportune and accurate way, when the fruit flies populations surpassing its innocuous level numbers.*

Introduction

Mango (*Mangiferaindica* Linnaeus) is one of the important commercial fruits ranked first in India. It is called "King of fruits" due to its wide adaptability, attractive colour, excellent taste, exotic flavour, and exemplary nutritive value, richness in variety, attractive appearance and popularity among the masses. It is a rich source of vitamin A and C. Fruit flies are the most important pest causing enormous damage to mango. Above 128 species of fruit flies has been recorded, out of these, eight species are found infesting mango fruits in India. Management of fruit flies remains a challenging task as they are polyphagous with overlapping generations and there is limited scope for the use of chemical insecticides because of the residual problems. Hence, development of an integrated management package with emphasis on non-insecticidal approach becomes imperative. As a pre-requisite to this, studies on monitoring of fruit fly populations and their relationship with weather parameters were made in mango orchard with development of forewarning systems against this pest are presented in this paper.

Materials and Methods

Field cum laboratory experiments were made in the mango orchard of the Junagadh Agricultural University, Junagadh from the year 2005 to 2011. A total of 10 methyl eugenol traps were set up 50 m apart from each other to avoid the interference of the traps. Each trap had four holes of 20 mm on four sides and a cotton were charged with 0.4 ml methyl eugenol and one ml dichlorvos was placed in a loop of iron wire. All the traps were serviced with these chemicals at monthly and weekly intervals, respectively. Fruit flies collected from each trap were pooled for each standard week throughout the study period, identified and recorded. Correlations were worked out to find the relationship with the weather parameters *viz.,* minimum and maximum temperature, morning and afternoon relative humidity, evaporation, wind speed, rainfall and rainy days.

Results and Discussion

The populations of the trapped fruit flies (mean of 2005-06 to 2011-12), were recorded on weekly basis (Table 1).Three species of fruit flies *viz., Bactroceradosalis* Hendel, *B. correcta* and *B. zonata* were recorded in methyl traps. The catches of fruit flies in methyl eugenol traps showed that the insect was active throughout the year, except severe winter months. The maximum catches of fruit flies per trap were ranged from 200 to 690 during 14th to 36th standard weeks (April to September) and minimum catches were observed during 1st to 9th and 41st to 52nd standard weeks. Present results are in close agreement with the reports of Suresh Babu and Viraktamath (2003), who reported that the peak catches of *B. dorsalis* during 21st, 23rd and 46th standard weeks and a low population from 1st to 20th standard weeks. The pooled result indicated that the level of population of adult flies went on

Table 1: Number of moth catches and correlation matrix of mango fruit fly with weekly mean abiotic factors in mango (Pooled of 2005-06 to 2011-12)

Std. Week No.	*Means of 2005-06 to 2011-12*									
	No. of Fruit flies trapped/trap	*Temp. C0*		*R.H. per cent*		*W.S.*	*B.S.H*	*EVAPO.*	*Rain*	*Rainy*
		Max	*Min*	*I*	*II*	*(kmph)*	*(hrs)*	*(mm)*	*(mm)*	*Days*
1	12.71	28.14	11.61	65.86	32.29	5.29	7.76	4.53	0.00	0.00
2	20.00	29.14	11.90	70.29	31.00	5.70	8.14	4.64	0.00	0.00
3	23.57	29.50	12.69	68.57	27.14	4.29	8.36	4.54	0.00	0.00
4	33.57	29.31	12.64	62.57	23.57	4.63	8.31	4.97	0.00	0.00
5	48.29	30.51	13.04	67.57	25.71	4.57	9.06	5.14	0.37	0.14
6	41.00	30.69	14.49	71.00	27.14	4.63	8.71	4.93	0.00	0.00
7	48.29	31.29	15.14	63.86	26.43	5.41	9.11	5.93	0.86	0.14
8	51.29	33.10	15.56	57.57	19.29	5.26	9.04	6.54	0.00	0.00
9	85.00	35.23	17.46	62.14	19.43	5.36	9.59	7.31	0.00	0.00
10	121.00	34.31	18.61	65.43	19.57	5.74	9.31	7.26	0.80	0.14
11	147.29	37.14	19.14	57.43	16.86	5.43	9.90	8.13	0.00	0.00
12	160.29	37.53	20.80	68.57	18.86	6.09	10.00	8.64	0.00	0.00
13	188.57	38.06	21.51	60.43	17.57	6.11	9.99	9.11	0.00	0.00
14	200.43	38.94	22.10	58.29	16.43	6.74	9.43	9.73	3.57	0.14
15	221.29	39.27	22.26	64.57	16.29	6.23	9.39	9.34	0.00	0.00
16	222.71	40.10	23.71	66.71	18.29	6.84	9.83	9.79	0.00	0.00
17	191.71	40.34	23.96	74.57	20.00	7.03	10.24	10.03	0.00	0.00
18	165.57	39.51	24.87	79.71	29.29	7.86	9.73	9.43	0.14	0.00
19	182.00	38.77	25.34	78.57	33.14	8.71	9.87	9.49	0.00	0.00
20	178.71	38.01	26.11	79.57	39.00	9.83	10.16	9.44	0.00	0.00
21	172.71	38.57	26.99	79.00	40.43	10.79	8.81	9.46	0.00	0.00
22	216.43	38.06	26.64	76.00	42.29	9.96	7.94	9.09	11.93	0.43

Cont...

Cont...

23	200.43	37.84	26.77	77.00	42.00	10.30	6.96	8.16	3.33	0.29
24	266.86	36.34	26.49	80.29	51.57	9.93	5.23	7.44	41.19	1.29
25	549.14	36.17	26.76	81.57	54.86	9.91	3.09	6.14	36.44	1.29
26	690.00	32.83	26.40	86.86	69.57	9.57	1.29	4.09	70.74	3.14
27	633.43	31.50	25.76	88.57	75.00	8.64	0.57	3.27	119.61	3.57
28	657.71	31.53	25.37	89.71	75.14	9.37	0.46	3.63	43.10	2.86
29	557.43	32.36	24.89	89.00	70.43	7.47	0.99	3.90	81.89	2.86
30	507.71	30.97	25.06	92.14	78.71	7.60	1.54	3.21	123.71	4.29
31	425.71	29.70	24.61	93.43	82.57	7.13	0.77	2.36	160.76	4.86
32	404.29	29.50	24.80	93.14	82.00	9.04	0.34	2.69	82.36	3.57
33	412.00	29.87	24.64	91.71	77.71	7.26	0.93	2.60	44.19	2.86
34	422.00	31.16	24.50	90.29	70.57	5.67	2.37	3.40	22.57	2.00
35	374.71	31.17	24.34	91.00	74.57	5.30	2.31	3.04	67.51	3.29
36	266.71	31.81	24.73	89.57	70.14	5.61	3.23	3.71	64.41	2.43
37	187.43	31.46	24.30	88.29	63.71	5.39	4.21	3.60	131.06	2.71
38	146.43	30.97	23.99	88.14	61.43	5.06	5.44	3.54	57.76	2.71
39	128.57	32.46	23.63	86.00	59.71	4.29	7.60	4.27	14.34	0.71
40	125.71	34.34	23.53	82.29	47.00	4.23	8.69	4.76	1.06	0.14
41	94.43	34.90	23.04	80.29	48.14	4.47	9.10	5.06	7.89	0.29
42	72.57	36.46	21.46	77.00	40.29	3.46	8.79	5.26	0.00	0.00
43	61.86	35.73	19.19	67.86	26.29	3.07	9.01	5.04	0.00	0.00
44	68.00	35.89	19.43	65.14	26.86	3.61	8.56	4.93	0.00	0.00
45	78.00	35.51	19.60	68.43	31.14	3.00	8.26	4.63	1.46	0.14
46	71.00	33.61	17.99	70.29	32.00	3.50	8.09	4.33	1.36	0.14
47	53.00	32.93	16.13	69.00	31.43	3.07	8.11	4.09	8.11	0.43

Cont...

Cont...

48	49.29	31.86	16.61	63.86	31.29	4.30	7.74	4.53	0.26	0.00
49	36.43	31.83	15.81	71.00	30.86	4.20	7.97	4.54	0.00	0.00
50	19.86	29.97	13.39	63.14	26.57	4.83	8.53	4.73	0.00	0.00
51	20.86	30.41	13.64	68.71	30.57	4.34	7.60	4.16	0.00	0.00
52	18.14	29.10	12.00	71.14	32.29	4.23	8.41	4.17	0.00	0.00
Mean Correlation co-efficient		**-0.0209**	**0.7107**	**0.7071**	**0.7569**	**0.6481**	**-0.8324**	**-0.2192**	**0.7001**	**0.8013**

*, ** Significant at P=0.05 (0.412) and at P=0.01 (0.479), respectively

N = 52, Max.: Maximum temperature; Min. : Minimum temperature; RH: Relative humidity; WS: Wind Speed; BSH: Bright sun shine hour: Evap : Evaporation.

increasing and reached to its peak of 690 fruit flies per trap at a minimum temperature of 26.40°C, morning relative humidity 86.86 per cent, afternoon relative humidity 69.57 per cent, wind speed 9.57 km/hr, bright sunshine hour 1.29 hrs, 70.74 mm rains in 26th standard week. Then, adult population showed a declined trend and thereafter continued in the field up to harvesting of mango fruit. Present results also corroborate the findings of Kumar *et al.* (1997) from Gujarat.

Pooled correlation of fruit fly catches in methyl eugenol traps was found to be positive and significant with minimum temperature, morning and afternoon relative humidity, wind speed and rain and bright sunshine hours showed significant negative correlation with fruit flies trapped. Rest of weather parameters did not showed significant positive or negative correlation of co-efficient with adult trapped.These results with respect to *B. dorsalis* corroborate the findings of Gupta and Bhatia (2000), who observed a significant positive correlation between trap catches of *B. dorsalis* and maximum and minimum temperature. More or less similar findings was also observed by Jalaluddin *et al.* (2001), who observed the population of *B. correcta* having a significant positive correlation with mean maximum temperature and minimum temperature. Results with respect to *B. zonata* are in agreement with the reports of Suresh Babu (2002) who observed that *B. zonata* population had a high significant positive correlation with maximum temperature.Verghese and Sudha Devi (1998) observed a positive correlation of *B. dorsalis* population with maximum temperature and minimum temperature. Similarly, Agarwal and Kumar (1999) reported positive relationship between trap catches of *B. zonata* and maximum temperature. This slight variation was attributed to change in the fruiting period of host plants around the experimental site.

The multiple linear regression analysis indicated that the influence of all the weather parameters was highly significant and it was 71% (R^2=0.71463) in 2005-06, 43% (R^2=0.43114) in 2006-07, 65% (R^2=0. 0.65458) in 2007-08, 65% (R^2=0.65932) in 2008-09, 67% (R^2=0.67342) in 2009-10, 55% (R^2=0.55589) in 2010-11, 69% (R^2=0.69593) in 2011-12 and 84% (R^2=0.84461) in mean result of years on number of fruit fly catches in methyl eugenol trap.

Conclusion

The populations of the trapped fruit flies in methyl eugenol showed that the insect was active throughout the year, except severe winter months. The highest catches of fruit flies per trap was observed during April to September (14th to 36th standard weeks) and it was further informed that the activity of fruit flies in mango orchards seemed to be related positively with high humidity (80-90 % RH) and 24 to 26°C minimum temperature during the month of April to September (14th to 36th standard weeks). The minimum temperature, morning and evening relative humidity, wind speed, rainfall showed positive effect and bright sun shine hours adversely affected the fruit flies catches in traps. The multiple linear regression analysis indicated that the influence of all the weather parameters was high (84 per cent).

Table 2: Correlations coefficient between number of fruit fly catches in different year (2005-06 to 2011-12) and weather parameters

Year	*Temp. °C*		*R.H. per cent*		*W.S.*	*B.S.H*	*EVAPO.*	*Rain*	*Rainy*
	Max	*Min*	*I*	*II*	*(kmph)*	*(hrs)*	*(mm)*	*(mm)*	*Days*
2005-06	0.0944	0.5574	0.3672	0.4526	0.4451	-0.5779	-0.0108	0.0452	0.1362
2006-07	-0.0386	0.3887	0.4012	0.3905	0.1380	-0.4239	-0.1118	0.1116	0.2064
2007-08	-0.0876	0.4405	0.4274	0.4478	-0.0013	-0.3511	-0.1096	0.1580	0.5365
2008-09	-0.0431	0.4904	0.4155	0.5236	0.4735	-0.6525	-0.2867	0.1215	0.1491
2009-10	0.1182	0.6453	0.3963	0.4574	0.6692	-0.5281	0.0726	0.3398	0.3485
2010-11	0.1554	0.4450	0.2561	0.2299	0.2075	-0.3283	0.0149	0.2455	0.1908
2011-12	-0.0199	0.5131	0.4744	0.5574	0.5910	-0.7607	-0.1871	0.3837	0.4858
Mean correlation co-efficient of 2005-06 to 2011-12	-0.0209	0.7107	0.7071	0.7569	0.6481	-0.8324	-0.2192	0.7001	0.8013

*, ** Significant at P=0.05 (0.412) and at P=0.01 (0.479), respectively

N= 52, Max.: Maximum temperature; Min.: Minimum temperature; RH: Relative humidity; WS: Wind Speed; BSH: Bright sun shine hour: Evap: Evaporation.

Table 3: Multiple linear regression analysis between weather parameters and number of fruit flies catches in different year

Year	*Regression equation*	*R^2 value*
2005-06	Y=781.35+5.3302(X1)-4.0135(X2)+0.7937(X3)+1.7993(X4)-64.7674(X5)-140.9169(X6)+115.2733(X7)-1.2343(X8)-34.0856(X9)	**0.71463**
2006-07	Y=717.32-18.5372(X1)+9.6692(X2)+1.4108(X3)+2.0747(X4)-44.0354(X5)-51.5163(X6)+64.8811(X7)-0.5507(X8)-31.2089(X9)	**0.43114**
2007-08	Y=182.25-6.9102(X1)+10.6341(X2)+0.2982(X3)+3.7582(X4)-22.9998(X5)-30.6051(X6)+30.8460(X7)+0.0774(X8)-42.4823(X9)	**0.65458**
2008-09	Y=-179.80+28.1132(X1)-1.5340(X2)-1.3850(X3)-0.5585(X4)+ 45.9773 (X5)-56.3774(X6)-49.7916(X7)+0.1170(X8)-65.6688(X9)	**0.65932**
2009-10	Y=-563.48+19.8943(X1)-3.3816(X2)5.6784(X3)-3.7813(X4)+31.0737(X5)-59.5858(X6)+18.4665(X7)+0.5342(X8)-13.8796(X9)	**0.67342**
2010-11	Y=-1306.60-48.12(X1)+56.55(X2)+9.06(X3)-14.76(X4)-2.39(X5)-109.48(X6)+22.44(X7)+4.63(X8)-209.39(X9)	**0.55589**
2011-12	Y=-357.78+34.1682(X1)-11.6808(X2)+4.2970(X3)-6.1353(X4)+52.3023(X5)-69.0817(X6)-36.4748(X7)-1.0534(X8)+63.5246(X9)	**0.69593**
Mean	Y=-358.37+25.5361(X1)-2.9875(X2)-1.9832(X3)+2.5884(X4)+21.5407(X5)-36.5909(X6)-13.0385(X7)-0.4188(X8)+8.9871(X9)	**0.84461**

X1: Maximum temperature, X2: Minimum temperature, X3: Relative humidity morning, X4: Relative humidity evening, X5: Mean sunshine Hrs, X6: Mean wind speed Km/hrs X7: Mean Eva. Op. mm/day.

References

Agarwal, M. L. and Kumar, P., 1999. "Effect of weather parameters on population dynamics of peach fruit fly, *Bactrocerazonata* (Saunders)". Entomonon, 24 : 81-84.

Gupta, D. and Bhatia, R., 2000. Population fluctuations of fruit flies, Bactrocera spp. in submountainous mango and guava orchards. Journal of Applied Horticulture, Lucknow, 2:47-49.

Jalaluddin, S. M., Natarajan, K. and Sadakathulla, S., 2001. Population fluctuation of the guava fruit fly, *Bactroceracorrecta* (Bezzi) in relation to hosts and abiotic factors. Journal of Experimental Zoology, India, 4 : 323 - 327.

Kumar, S., Patel, C.B. and Bhatt, R.I., 1997. Studies on seasonal cyclicity of *Bactroceracorrectus* Bezzi in mango and sapota orchards using methyl eugenol trap. Gujarat Agricultural University Research Journal, 22: 68-74.

Suresh Babu, K., 2002. Species complex and population dynamics of fruit flies (Diptera: Tephritidae) on some fruits and vegetable crops. M.Sc. (Agri.) thesis, University of Agricultural Sciences, Dharwad, p. 82.

Suresh Babu, K. and Viraktamath, S., 2003. Species diversity and population dynamics of fruit flies (Diptera: Tephritidae) on mango in Northern Karnataka. Pest Management and Economic Zoology, 11: 103-110.

Verghese, A. and Sudhadevi, K., 1998, Relation between trap catch of *Bactroceradorsalis* Hendel and abiotic factors. In: Proceedings of First National Symposium on Pest Management in Horticultural Crops, Indian Institute of Horticulture Research, Bangalore, pp.15-18.

15

Salt Tolerance in Tropical Fruit Crops – A Review

Akula Venu, A.V. Barad, D.V. Delvadia, M.V. Ramdevputra and P. Parameshwar

Abstract

Soil salinity is one of the major environmental problems affecting agricultural production in arid and semi-arid regions of the world, both in irrigated and dry land agriculture. It is necessary to determine the environmental factors under which fruit crops give higher yields and better quality. The salt tolerance of fruit species is important because the cash value of fruits is usually high compared to field crops. The problem of salinity is characterised by an excess of inorganic salts and is common in the arid and semi-arid lands, where it has been naturally formed under the prevailing climatic conditions and due to higher rates of evapotranspiration and lack of leaching water. Although more frequent in arid lands, salt-affected soils are also present in areas where salinity is caused by poor quality of irrigation water. Saline soil induces physiological and metabolic disturbances in plants, affecting development, growth, yield, and quality of plants. Plants affects adversely as a result of salinity, growth inhibition, toxicity and injury, metabolic changes, osmotic adjustment and nutritional imbalance. In general, salt stress decreases the photosynthesis and respiration rate of plants. They affect the plant growth both physically and nutritionally also such fruit crops are mango and grapes show leaf burn and necrotic symptoms under sodic conditions. Growing salt tolerant crops and adopting appropriate soil and water management practices to minimise salinity damage are the main components of successful management in salt affected soils.

Introduction

A wide range of environmental stresses such as drought, salinity, alkalinity and pathogen infection are harmful to the plants. Salt stress in soil is one of the major stresses especially in arid and semi-arid regions and can severely limit plant growth and productivity (Parvaiz and Satyawati, 2008). About 23 per cent of the world's cultivated lands is saline and 37 per cent is sodic (Khan and Duke, 2001).In India, soil salinity is causing concern in over seven million hectares and the potential soil salinity hazard is threatening another 27 per cent of land area.

Under the situations of both salinity and sodicity, plant growth is suppressed. Fruit crops are adversely affected by soil salinity and sodicity because of lowered osmotic potentials and specific ion toxicities. Citrus decline in Punjab, Murkute, *et al.* (2005). Poor mango and guava production in Uttar Pradesh, failure in productivity in arid tracts of Rajasthan and Gujarat, and the problems in grape production in Maharashtra and Andhra Pradesh are related to salinity hazards.

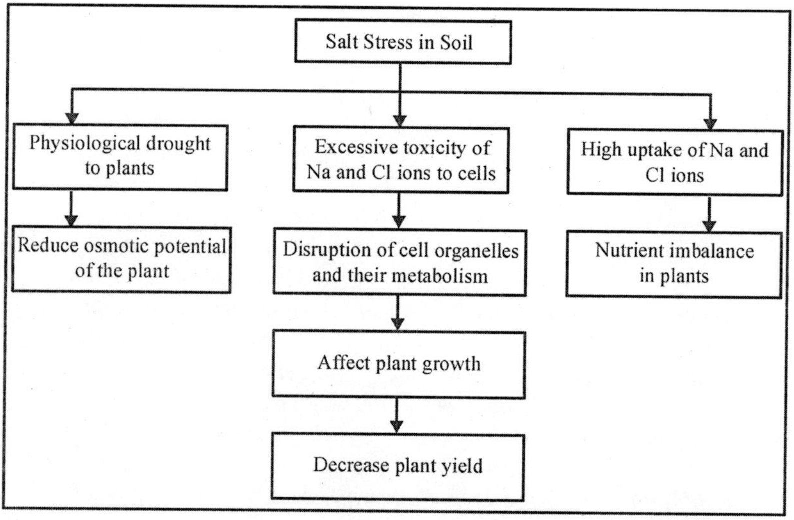

Fig. 1. Effects of salt stress on plants.

Source: Sabbagh *et al.* (2014).

Salt Tolerance

Tolerance to salinity level by specific salts like sodium and chloride at which initial yield declines and yield decreases per unit increase in salinity beyond threshold salinity values are more useful in fruit crops than general ratings and groupings of the plants. The ratings will be valuable guides in predicting how one crop or variety may fare relative to another on saline and sodic soils.

Table 1: General salt tolerance of fruit crops

Sensitive (0-4 dSm^{-1})	*Moderately tolerant (4-6 dSm^{-1})*	*Tolerant (6-8 dSm^{-1})*	*Highly tolerant (8-16 dSm^{-1})*
Grapefruit	Peach	Sapota	*Ber*
Orange	Papaya	Pomegranate	*Jamun*
Lemon	Banana	Fig	Date palm
Lime	Grape	Olive	
Avocado	Mango	Guava	
Pear	Custard apple	*Aonla*	
Apple	*Phalsa*	*Bael*	
Cherry			
Plum			

Sourse: Datta (2013).

Effects of Salt Stress

When the salt concentration exceeds threshold value, growth rate and size of the plant progressively decrease with increase in salinity and the plants become stunted showing restricted lateral shoot development, reduction in size of leaves and fruits, decrease in fresh and dry weight of different plant parts, reduced leaf number resulting in diminished yield. Many a time, leaves may be thicker than normal. The intensity varies with the concentration and type of salinity and plants. The top growth may be suppressed more than root growth and vice versa in different species. For instance, sodicity affects the soil physical properties which adversely restricts root growth.

Woody tree species like mango and grapes show leaf burn and necrotic symptoms under sodic conditions. Chlorosis is not always a characteristic symptom of salt injury. But in calcareous soils associated with salt accumulation, lime-induced chlorosis may be frequently observed in plants. Yield losses in fruits are generally greater because of specific ion toxicity than due to osmotic effects.

Table 2: Effects of various degrees of salinity on plant growth

Salinity class	*Conductivity (dSm^{-1})*	*Effect on plant growth*
Non-saline	0-2	Negligible effect
Slightly saline	2-4	Very sensitive *crops like* orange, grapefruit and lemon are affected
Moderately saline	4-8	Only a few fruit crops can thrive
Strongly saline	8-16	Only tolerant fruits like date palm and *ber* can yield satis factorily
Very strongly *saline*	>16	Only a few tolerant plants and grasses can grow

Source: Amacher *et al.* (2000).

Growth Inhibition

In mango rootstocks, plant height, leaf number and area were drastically reduced due to saline water (5 dSm^{-1}) application. The fruit yield and mean leaf area of mango were reduced due to chlorine pollution in the industrial belt around Vadodara in Gujarat (Vijayan and Bedi, 1989).

In Anab-e-Shahi, Selection-7 and Bangalore Purple grape, reduced dry weight of roots and shoots due to chloride, sulphate and carbonate salinity. Bud sprouting was delayed by one to three weeks, one week delay in bud burst observed in Anab-e-Shahi and Perlette and that further sprouting was reduced by 27 per cent due to sodicity in the soil.

In pomegranate also reduction in stem diameter, plant height and number of leaves even at low ESP levels showing growth depression of 20 per cent at 25 ESP. In Panipat Selection and Banipur Selection cullivars, reduction in growth was to the extent of 48 and 60 per cent respectively. Increasing salinity was also found to reduce shoot dry weight.

Some fruit crops can be grown successfully in saline and sodic situations. In ber was observed decrease in plant height, stem diameter, leaf area, shoot dry matter and yield at a soil salinity level of 14.2 dSm^{-1}. However, water salinity (9 dSm^{-1}) caused 10 to 20 per cent yield reduction in Gola ber (Jain *et al.*, 1988) and sodic conditions (40 ESP) adversely affected root length and stem and root weight due to poor physical conditions of the soil. Similarly, little reduction in growth has been noticed due to sodic conditions in aonla and bael and due to salinity in jamun.

Metabolic Changes

Salinity has been observed to alter the hormonal balance of plants. The resultant nutritional imbalance also adversely affects plant metabolism. Thus, disturbances in growth and development altered normal metabolic process leading to increased free amino acids, proline, total organic acids and hydrolytic enzymes in guava.The salt stress either accelerated the degradation of protein or impaired the incorporation of amino acid into protein or both. In guava, salt stress had pronounced effect specifically on RNA and generally on nucleic acid contents. The increased respiration was probably at the cost of restricted growth when the carbohydrates accumulated more rapidly without being utilised for formation of new cells and tissues. Salinity resulted in the destruction of cell organelles capable of large scale protein synthesis leading to amino acid accumulation. Less chlorophyll resulted in lower photosynthesis.

Adaptation to salinity thus resulted in depletion of energy pools needed for growth in these fruit crops. At 0.15 per cent salt concentration in soil, proline accumulated fast in grape leaves. The level of endogenous cytokinins was considerably low in root exudates of salt-stressed plants. Rate of photosynthesis decreased while that of respiration got stimulated with increase in salt concentration. The changes in hormone levels following salination may well be an immediate after effect of such treatments but a rather ephemeral one that may disappear in most cases as soon as osmotic adjustment is affected and water stress is released. These hormonal responses may mediate the adjustment of plants to salinity by reducing stomatal aperture and water loss. In case the stress continues, it may end up in death of the plant. Pandey (1982) came to the conclusion that growth was retarded in grape due to diversion of nutrients towards respiring tissues damaged by the salts. He recorded damage to the root tips and leaf cells within thirty days of salination.

Adverse effects of high salinity on growth of jamun were related to decrease in chlorophyll content and increased respiration and reduced photosynthesis. In bael, observed interveinal chlorosis with increase in sodicity as a result of reduced chlorophyll. The metabolic changes due to salinity mostly take place even before the injury symptoms are clearly seen on the plant.

Nutritional Imbalance

Nutritional imbalance may be caused by an excess of a particular ion, i.e., cations like sodium, magnesium and calcium, and anions like chlorides, sulphates and nitrates.

Calcium and magnesium nutrition in mango plants is equally affected at higher concentration of salts. Deficiency of phosphorus showing a purplish pigment along the margins of leaves and that of magnesium as a green wedge down the central part and bronze colour between the veins have been described as a result of high levels of sodium sulphate.

In sweet orange budded on rough lemon, that the leaf potassium contents were markedly depressed by high salinity levels (4 to 6 dSm^{-1}). The effect was reverse on calcium content. Among micronutrients, copper and boron decreased as zinc and iron increased with increased salinity. On different rootstocks, macronutrient contents also varied in sweet orange scion. Reduction in N, P, K, Ca, Mg and Zn contents were observed in citrus under higher salinity levels irrespective of the rootstock.

Under sodic situations, increasing exchangeable sodium per cent (ESP) from 24 to 85 decreased calcium, manganese and potassium contents in guava leaf (Singh *et al.* 1983). Similarly, Mehta (1982) noticed adverse mineral nutrition in young ber plants at ESP 40 and above and reduction in potassium content in ber leaves with increasing levels of salinity. Jamun also had low nitrogen content in the root at high salinity level.

Thus, nutritional disorders occur in fruit trees when the soluble salt concentration or exchangeable sodium in soil exceeds certain limits. Potassium and calcium nutrition are seriously impaired, e.g., potassium nutrition in Thompson Seedless grape by saline irrigation waters (EC 2.5 and 5 dSm^{-1}) and by excess of sodium in banana grown in Krishna district and Nuzwid region of

Andhra Pradesh. K/Na ratios changed from high to low with increasing salinity. At low sodium value, sodium was beneficial as a nutrient, but at high levels of sodium, it adversely affected calcium absorption.

Management

Tolerant Fruits and Cultivars

Ber: Dhankar *et al.* (2006) noticed satisfactory growth in ber up to 21 dS m^{-1}. Once established, growing ber in a saline soil was not difficult. In fact, Dahiya and Dhankar (2004) reported appreciable reduction in growth beyond 14 dSm^{-1}. Even in nursery stage, ber seedlings withstood high salinity and saline waters with 6.5 dSm^{-1} electrical conductivity. Cultivar Sev survived in salinity up to 8-10 months without adverse effect on growth. Performance of Gola ber with use of saline water (EC 9.0 dSm^{-1}) was good with 10-20 per cent marginal decrease in yield (Jain *et al.*, 1988).

Citrus: Citrus fruits are the most sensitive to salinity. Even 500 ppm of sodium chloride was found to be harmful. Tolerance of root stocks was mainly influenced by differential uptake and accumulation of salts. Cleoptera mandarin and Rangapur lime could tolerate 6.8 dSm^{-1} salinity in the soil. Death of root stocks at 9.9 dSm^{-1} and Jatti Khatti and Bhadri lemon were in the most sensitive (2.3 dSm^{-1}). Thus, tolerant root stocks have the capacity to exclude chloride uptake from the system. It has, therefore, been suggested that sweet oranges should not be irrigated with waters having an electrical conductivity more than 2500 μ Siemens m^{-1}.

Grapes: Grape is rather sensitive to chloride and tolerant to other salts. All vines died at 3500 ppm total salt concentration and identified Bhokri and Gold as salt tolerant and Anab-e- Shahi and Karachi Gulabi as sensitive. The upper salt tolerance limit for grape was found to be 0.3 per cent. Soils have containing more than 0.3 per cent salts should not be selected for grape cultivation. Kandari and Perlette as sensitive, Himrod as medium tolerant and Alamwick, Thompson Seedless and Beauty Seedless as resistant. Observations suggest that Pusa seedless is more tolerant than other varieties. Grape as moderately is more tolerant both to salinity and sodicity. Use of root stocks like Dogridge and 1613 can exclude salts and can increase tolerance to even higher levels of salinity.

Mango: Jindal *et al.* (1990) found Langra as the most sensitive variety. Many mango seedlings withstood 4 dSm^{-1} salinity. Dashehari and Chusa varieties accumulated less chlorides and sulphates and hence withstood higher levels of salinity. However, mango cannot be considered as highly tolerant. Among mango root stocks, Bappakai and Olour tolerated 4.5 dSm^{-1} salinity in irrigation water and were rated moderately tolerant.

Other Fruits: Other fruits likes jamun can be grown successfully up to 16 dSm^{-1} soil salinity. Pomegranate and Sapota have been identified for moderate salt conditions. Date palm has been recommended for growing in Jaisalmeer, Bikaner, Puggal, Anupgarh and Ganganagar regions of Rajasthan. Aonla and Bael are considered suitable up to salinity levels of 8 dSm^{-1} and sodicity levels of 25 ESP.

Varietal Selections

Varietal difference also occur in salt tolerance. Significant varietal differences have been observed in mango, citrus, grapes and ber. However, the reported tolerances were not phenomenal perhaps because most of these varieties developed from a narrow genetic base and were propagated by vegetative means.

Use of Rootstocks

Rootstocks differ widely in absorption and transport of toxic ions. Rangapur lime and Cleopatra mandarin in citrus, Bappakai and Olour in mango and Dogridge and 1613 in grape were found to tolerant higher levels of salt either by exclusion or accumulation mechanisms.

Soil Management

Soil and land management should be done carefully to level the land such that one can achieve more uniform water application. During irrigation, sodic soils are prone to surface crusting. Practices that maintain high organic matter levels will help in maintenance of good tilth. More frequent and light irrigations may be applied to soften crusts.

Application of Amendments

Excess salinity in the root zone can be effectively controlled by leaching and draining of soluble salts either by surface or sub-surface drainage. The best means of controlling soil and water salinity is by adequate leaching. Reclamation steps for sodic soil include bunding and land leveling, provision of irrigation water, drainage and application of amendment. Gypsum is the standard amendment for reclamation of alkali soils. Spreading and mixing adequate quantity of gypsum based on gypsum requirement in the field and then irrigating with good quality water washes away the by-products through the drainage system.

Other Methods

Increasing the number of plants per unit area to account for the mortality rate, mulching of soils and growing the tree in the least saline furrow are some of the techniques to be followed. Frequent irrigations, adequate leaching, drainage, maintain water table depth, improving irrigation efficiency by drip, pitcher irrigation and uniform water distribution and green manuring also help to reclaim such soils.

Conclusion

The approximation of global salt-affected area is one billion hectares, which correspond to about seven per cent of the earth's continental extent or about 20 times the size of a country such as France. Salt stress causes huge losses of fruit productivity worldwide due to lack of relevant information about on salt stress on fruit crops. However, effect of salt stress on fruit crops drastically reduce fruit yield, quality, growth inhibition and nutritional imbalance observe in fruit plants. Therefore its need to be overcoming salinity stress quickly and fully implemented with conventional and modern methods likes that flood irrigations, tillage, collecting wide diversity of germplasm and application of amendments. Although salt resistance in plants is multigenic and thus complex, breeding and transgenic approaches to improve salinity resistance could contribute to enhancing crop production over millions of ha of salt-affected areas worldwide.

References

Amacher, J.K., Koenig, R., Kitchen, B., (2000). Salinity and Plant Tolerance. Electronic Publishing. AG-SO-03.

Dahiya, S.S. and Dhankar, O.P. (2004). Studies on salt tolerance of ber. Haryana J. Hort. Sci., 10: 53-58.

Datta, S. (2013). Impact of climate change in Indian horticulture - A review. Intl J. Sci.Envi. and Tech. 2(4), 661–671.

Jdate, G.M., Kokate, A.S. and Chaudhari, K.G. (1989). Effect of different chloride containing potasic fertilizers on physical quality parameters of grapevine. *Maharashtra J. Hort.*, 5: 81-85.

Jain, B.L. and Dass, H.C. (1988). Effect of saline water on performance of saplings of jujube (*Zizyphus mauritiana*), Indian cherry (Cordia dichotoma var. Wallichu) and Pomegranate (Punica granatum) at nursery stage. *Indian J. Agric. Sci.*, 58: 203-7.

Jindal, P.C., Singh, J.P. and Gupta, O.P. (1990). Salt tolerance of Mango seedling for salt tolerance. *Haryana J. Hort. Sci.* 4: 112-15.

Joolka, N.K. (2003). Studies on the effect of chloride and sulphite salts on growth and leaf composition in grape (V. vinefera). M.Sc. thesis, HAU, Hissar.

Khan, M.A., Duke N.C. 2001. Halophytes- A resource for the future. Wetland Ecology and Management, 6:455-456.

Makhija, M., Dhankar, O.P. and Singhrot, R.S. (1980). Effect of soil salinity levels on growth and leaf mineral composition of guava. *Haryana J. Hort. Sci.* 9:21-25.

Mahta, P.K. (1982). Studies on the effect of soil sodicity and boron on seed germination, growth and mineral composition of young ber (Zizyphus species) plants. *Thesis abstrt.*, 8:47-48.

Murkute, A., Sharma, S., and Singh, S.K. (2005). Citrus in terms of Soil and Water Salinity: A review. J. Sci. and Indu. Res. 64: 393-402.

Pandey, R.M. (1982). Physiological studies on salt tolerance in graps. Var. perlitte. Research Report, ICAR Fruit Research workshop (Nagpur), 414-16.

Parvaiz, A., Satyawati S. (2008). Salt stress and phyto biochemical responses of plants. Plant Soil. J. 54(3): 89-99.

Sabbagh, E., Lakzayi, M., Keshtehgar, A. and Rigi, K. (2014). The effect of salt stress on respiration, PSII function, chlorophyll, carbohydrate and nitrogen content in crop plants. Intl J Farm & Alli Sci., 3 (9): 988-993.

Singh, S.B., Chhabra, R. and Abrol, I.P. (1983). Effect of exchangable sodium on the growth and mineral composition of jujube and guava. *Indian J. Agri. Sci.* 53: 446-50.

Vijayan, R. and Bedi, S.J. (1989). Effect of chlorine pollution on three fruit tree species at ranoli near Baroda, India. *Environ. Pollution.* 57: 97-102.

16

Effect of Cyclic Water Stress and Potassium on Dry Matter Yield and Nutrients Uptake by Summer Groundnut

H.L. Sakarvadia, K.B. Parmar and K.B. Polara

Abstract

A field experiment was conducted in medium black calcareous soil (Typic Ustochrepts*) with groundnut var. GG-2 in summer to study the effect of cyclic water stress at individual four stages i.e. flowering (SF), pegging (Spg), pod formation (Spf) and pod development (Spd) as well as at any two and three stages in all possible combinations along with one control of normal irrigation and three levels of potassium (0, 2.0 % K_2O as foliar spray and 60 kg K_2O ha^{1} as soil application). The normal irrigation treatment gave the highest pod and total biomass yield and dry matter and nutrients (N, P, K) accumulation at all the stages of growth. Water stress at any one or more phenophases reduced pod and total biomass yield, dry matter and nutrients accumulation. Water stress at pegging or its association with any one or two phenophase(s) reduced the accumulation of dry matter and nutrients, while pod yield was most adversely affected due to water stress during pod development stage or its association with any other phenophase. The accumulation of all the nutrients increased with advancement of crop growth except K in which the maximum accumulation was found at 78 DAS. Foliar application of potassium only increased its uptake, while soil application increased N and K uptake by the crop at all the growth stages.*

***Key words**: Growth stages, water stress, potassium, dry matter yield and nutrients uptake.*

Introduction

Groundnut (*Arachishypogaea* L.) is an important crop accounting for over 50 per cent of the production of oil seeds in the country. It is the principal crop of Gujarat state grown under erratic and scanty rainfall condition. Many factors influenced the crop yield, among which water supply is considered as one of the basic factor. Water affects plant growth by affecting internal physiological process and conditions. However, water stress at critical stages in plant growth caused more injury than at other stage (Kramer, 1969). The magnitude of reduction in growth due to water stress varied with the phenophase at which the stress was imposed (Patel and Golakiya, 1988; Golakiya and Patel, 1988). Water deficit not only reduced the total amount of growth, but it also changes the pattern of growth by controlling the availability of nutrients to the plants. Potassium has been known to help the plant achieving water economy under situation of moisture deficit and therefore it has been specially identified as a nutrient with drought buffering capacity.

Materials and Methods

A field trial was conducted at Instructional Farm, Junagadh Agricultural University, Junagadh with Groundnut Var. GG-2 as the test crop during summer season. The soil of experimental site was clayey (*Typic Ustochrepts*) with a pH 8.2 and EC 0.36 dSm^{-1}. The available nutrients status was medium for $KMnO_4$-N (247 kg ha^{-1}) and Olsen – P_2O_5 (31 kg ha^{-1}) and high NH_4OAc – K_2O (340 kg ha^{-1}). The experiment was laid out in a split plot design with four replications. The water stress treatments at different growth stages were taken in main plots. These 15 water stress treatments, consisting of water stress at no, one and/or two, and/or three growth stages, were applied to the crop. The details of these treatments are given in Table-1. Potassium was applied in sub plot. The three levels of potassium K_0 = Control, K_f= Foliar spray of K @ 2% K_2O as KCl at 32 and 56 DAS and Ks = soil application of K @ 60 kg K_2O ha^{-1} as KCl at sowing. Plant samples were collected at flowering (30 DAS), pegging (46 DAS), pod formation (62 DAS) and pod development (78 DAS) from one square meter area and at harvest from net plot to know the pattern of accumulation of dry matter and nutrients as well as for yield. Plant sample were analysed for N content by kjeldahl's method, whereas, P and K were from the di-acid mixture standard methods (Jackson, 1973).

Results and Discussion

The dry matter yield was increased with advancement of crop growth.The maximum dry matter yield at all the stages of growth and pod yield was obtained in normal irrigation treatment. The dry matter yield was significantly reduced at 46, 62 and 78 DAS and at harvest by water stress treatments over that obtained in the respective unstressed control treatment (Table-2).

Table 1: Details of main plot treatments

Sl. No.	*Symbol*	*Period* (DAS)*	*Description*
1	S_0	—	Normal irrigation (no water stress)
Single cycle water stress			
2	S_F	23-47	Water stress at flowering
3	Spg	33-55	Water stress at pegging
4	Spf	57-79	Water stress at pod formation
5	Spd	73-95	Water stress at pod development
			Double cycle water stress
6	$S_F$pg	23-55	Water stress at flowering and pegging
7	$S_F$pf	23-47 and 57-79	Water stress at flowering and pod formation
8	$S_F$pd	23-47 and 73-95	Water stress at flowering and pod development
9	Spgpf	33-79	Water stress at pegging and pod formation
10	Fpgpd	33-55 and 73-95	Water stress at pegging and pod development
11	Spfpd	57-95	Water stress at pod formation and pod development
Triple cycle water stress			
13	$S_F$pgpf	23-79	Water stress at flowering, pegging and pod formation
14	$S_F$pgpd	23-55 and 73-95	Water stress at flowering, pegging and pod development
15	$S_F$pfpd	23-47 and 57-95	Water stress at flowering, pod formation and pod development
16	Spgpfpd	33-95	Water stress at pegging, pod formation and pod development

* Period indicates the duration in which irrigation was not applied

The minimum dry matter yield (672 kg ha^{-1}) was obtained due to water stress during flowering at 46 DAS, due to double cycle water stress treatment during flowering to pegging (1096 kg ha^{-1}) at 62 DAS, due to triple cycle water stress treatment ($S_F$pgpf) during flowering to pod formation (1750 kg ha^{-1}) at 78 DAS. At harvest, the total dry matter yield of the crop was most adversely affected due to water stress imposed during pegging (3749 kg ha^{-1}), pegging to pod formation (3442 kg ha^{-1}) and flowering to pegging and pod development stages (2719 kg ha^{-1}) among single, double and triple cycle water stress treatments, respectively (Table-2). The continuous imposition of water stress during pegging to pod development stage resulted in minimum (559 kg ha^{-1}) pod yield. The maximum reduction in pod yield was 55.7 per cent under treatment Spd, 60.6 per cent under Spfpd and 75.2 per cent under Spgpfpd treatment among single, double and triple cycles of water stress treatments, respectively over that of control, i.e., no water stress. As water content in soil decreased, water deficit in plant become more severe and breakdown of RNA,

DNA and protein increased and finally photosynthesis and growth ceases (Kramer, 1969). The increase in magnitude of adverse effect of water stress on dry matter accumulation with increase in its duration is in agreement with the work reported by Pallas *et al.* (1979). The application of potassium either in soil or on foliage and interaction effect of water stress and potassium were found non-significant on dry matter accumulation by crop at all the growth stages and on pod yield of groundnut.

Table 2: Effect of potassium on total biomass and pod yield of groundnut under different cycles of water stress

Treatment	*Total biomass yield (kg ha^{-1})*					
	30 DAS	*46 DAS*	*62 DAS*	*78 DAS*	*At harvest*	*Pod yield*
Water stress						
S_0	459	1193	2570	3903	7153	2251
S_F	453	672	1540	2420	4438	1427
Spg	460	887	1245	1990	3749	1179
Spf	150	1201	2501	2540	4282	1068
Spd	462	1191	2575	3893	4433	996
$S_F$pg	459	681	1096	1876	3524	1004
$S_F$pf	457	678	1614	2243	3989	1020
$S_F$pd	452	675	1549	2412	3974	926
Spgpf	460	894	1242	1954	3442	903
Fpgpd	449	898	1237	1991	3462	913
Spfpd	461	1198	2520	2553	4020	886
$S_F$pgpf	453	668	1096	1750	2802	575
$S_F$pgpd	456	673	1105	1870	2719	785
$S_F$pfpd	458	678	1530	2243	3302	808
Spgpfpd	468	895	1234	1946	2790	559
S.Em.±	22.6	44.3	78.5	100.5	197.6	52.4
C.D. at 5%	NS	126.6	224.4	287.2	564.0	149.4
Potassium						
K_0	455	871	1649	2369	3928	1017
K_f	457	872	1650	2371	3936	1019
Ks	459	873	1652	2374	3950	1023
S.Em.±	4.0	8.3	13.5	15.4	13.2	2.4
C.D. at 5%	NS	NS	NS	NS	NS	NS
Interaction	NS	NS	NS	NS	NS	NS

The uptake of all the nutrients (N, P, K) by crop was significantly reduced due to water stress at 46, 62 and 78 DAS and at harvest over that obtained in respective unstressed control treatment. The minimum uptake of nitrogen (24.78, 37.15, 44.78 and 81.31 kg ha^{-1}), phosphorus (1.900, 3.318, 4.675 and 7.368 kg ha^{-1}) and potassium (25.15, 40.71, 46.91 and 47.72 kg ha^{-1}) were recorded under treatments of S_F at 46 DAS, $S_F$pg at 62 DAS, $S_F$pgpf at 78 DAS and at harvest, respectively. On an average single, double and triple cycle of water stress resulted in respective uptake of nitrogen 114.7, 101.7 and 89.79 kg ha^{-1}, phosphorus 10.130, 9.457 and 8.089 kg ha^{-1} and potassium 60.69, 56.37 and

Table 3: Uptake of nutrients (N, P, K) by groundnut crop at different growth stages as influenced by the water stress and potassium

Treatments	*Nitrogen (kg ha⁻¹)*					*Phosphorus (kg ha⁻¹)*					*Potassium (kg ha⁻¹)*				
	Days after sowing				*At harvest*	*Days after sowing*				*At harvest*	*Days after sowing*				*At harvest*
	30	46	62	78		30	46	62	78		30	46	62	78	
Water stress															
S_0	14.6	36.3	66.7	85.2	168	1.14	2.78	5.93	8.65	15.9	15.5	39.1	72.7	87.6	82.2
S_F	14.5	24.5	49.0	63.9	127	1.17	1.87	4.28	6.29	10.73	15.7	24.9	53.6	67.4	60.2
Spg	15.3	31.1	39.7	53.4	105	1.18	2.28	3.54	5.70	9.14	15.7	30.9	45.4	56.0	54.2
Spf	14.3	37.3	66.0	66.6	114	1.11	2.80	5.75	6.81	10.15	15.1	40.2	70.8	72.8	63.3
Spd	14.8	37.0	67.4	88.4	114	1.16	2.86	6.01	8.63	10.49	15.6	39.9	70.1	88.2	64.4
$S_F$pg	15.6	24.5	36.9	50.9	101	1.22	1.93	3.30	5.32	8.72	16.4	25.4	40.5	53.5	52.8
$S_F$pf	15.5	25.0	51.4	59.3	109	1.22	1.96	4.58	6.21	9.63	16.0	25.0	56.8	64.1	59.2
$S_F$pd	15.3	24.8	49.3	63.7	105	1.18	1.86	4.40	6.48	9.47	16.0	25.2	54.3	67.3	59.9
Spgpf	14.8	31.3	39.9	51.0	94	1.16	2.24	3.56	5.50	8.71	15.6	31.0	43.9	54.2	52.1
Fpgpd	14.8	31.6	39.3	52.8	94	1.16	2.40	3.51	5.55	8.55	15.6	31.3	43.1	55.7	52.3
Spfpd	15.0	36.7	68.2	68.5	107	1.18	2.87	6.11	7.00	9.87	15.8	40.2	70.3	69.8	62.0
$S_F$pgpf	15.6	24.7	37.2	44.8	81	1.21	1.87	3.34	4.68	7.59	16.4	24.9	40.8	46.9	47.7
$S_F$pgpd	15.1	24.8	37.4	50.4	93	1.18	1.92	3.32	5.12	8.36	15.9	25.2	40.8	53.2	51.6
$S_F$pfpd	15.4	25.2	49.9	60.9	103	1.20	1.90	4.40	6.25	9.04	16.2	25.5	54.3	64.2	57.4
Spgpfpd	15.3	31.9	40.4	52.4	82	1.19	2.33	3.58	5.42	7.37	16.1	31.4	43.7	54.3	48.3
S.Em.±	0.9	1.4	2.7	3.5	5	0.07	0.12	0.28	0.34	0.52	0.9	1.5	2.9	3.0	2.6
C.D. at 5%	NS	4.1	7.7	9.9	15	NS	0.33	0.81	0.98	1.49	NS	4.2	8.4	8.5	7.3
Potassium															
K_0	14.7	28.6	48.3	56.9	101	1.19	2.38	4.39	6.24	9.55	15.3	29.3	52.2	60.1	55.3
K_f	14.8	30.0	49.4	61.4	106	1.20	2.30	4.34	6.18	9.58	15.4	31.0	54.4	63.7	58.0
Ks	15.8	30.8	50.2	63.7	112	1.14	2.09	4.43	6.28	9.62	16.8	31.9	56.0	66.2	60.7
S.Em.±	0.2	0.3	0.5	0.7	0.8	0.01	0.02	0.05	0.08	0.08	0.2	0.3	0.6	0.6	0.4
C.D. at 5%	0.5	0.8	1.4	1.9	2.4	0.04	0.06	NS	NS	NS	0.5	0.8	1.8	1.6	1.0
Interaction	NS	NS	NS	NS	NS	NS	NS	NS	NS	NS	NS	NS	NS	NS	NS

51.25 kg ha^{-1}. The favourable effect of moisture content increases solubility of nutrients (Russell and Barbar, 1960), conductivity of the soil (Tisdall and Nelson, 1975), availability of the nutrients to the roots and ultimately uptake of nutrients by the plant.

Soil application of potassium significantly increased N uptake by the crop over that of control at all the growth stages (Table-3). However its foliar application increased N uptake by the crop only at 46 and 78 DAS and at harvest which indicates synergistic effect of K on N utilisation. Lakshaminarasimhan and Surendran, (1978) also reported higher uptake of N by groundnut plant due to K application. While the soil application of potassium significantly decreased P uptake over that of respective control and foliar spray at 30 and 46 DAS (Table-3). However, its foliar application significantly decreased P uptake over that of control only at 46 DAS. Specific inhibitory effect of K^+ on phosphate transport have been demonstrated by several workers (Emmert, 1971, Jones *et al.,* 1977). Application of potassium either in soil or on foliage significantly increased its uptake by the crop over that of control at all the growth stages (Lakshaminarasimhan and Surendran, 1978). The interaction effect of water stress and potassium on N, P and K uptake was found non-significant at all the growth stages.

References

Emmert, F.H., 1971. Centripetal ion penetration as a causal agent of ion interactions in foliar tissue. In Recent Advances in Plant Nutrition (Ed. Samish, R.M.) Vol. II, pp. 477-489.

Golakiya, B.A. and M.S. Patel 1988. Role of potassium in counteracting the effect of cyclic drought on groundnut. J. Potassium Res., Vol. 4(4), pp.163-167.

Jackson, M.L.,1973. Soil Chemical Analysis. Prentice Hall of India Pvt. Ltd., New Delhi.

Jones, G.D., J.A. Lutz, and T.J. Smith, 1977. Effect of phosphorus and potassium on soybean nodules and seed yield. Agron. J., Vol. 69, pp.1003-1006.

Kramer, P.J.,1969. Plant and Soil-Water Relationship. A modern synthesis. McCraw Hill, Inc., New York.

Lakshaminarasimhan, C.R. and R. Surendran, 1978. Influence of split application of potash on a groundnut variety for table purpose. *Indian Potash J.* **3(1):** 15-18.

Pallas, J.E., J.R. Stansell and T.J. Koske, 1979. Effect of drought on florunner peanuts. Agron. J.Vol. 71(5), pp. 853-857.

Patel, M.S. and B.A. Golakiya, 1988. Effects of water stress on yield attributes and yield of groundnut. Indian J. Agric. Sci.Vol. 58(10), pp. 704-705.

Russell, R.S. and D.A. Barbar, 1960. The relationship between salt uptake and absorption of water by intact plants. Ann. Rev. Plant Physiology, Vol. 11, pp. 127-140.

Tisdall, S.L. and W.L. Nelson, 1975. Soil Fertility and Fertilizer (3rd Ed.). The Macmillan Co., New York, pp. 189-242.

17

Response of Drilled Rabi Fennel (*Foeniculumvulgare* Mill.) to Irrigation based on IW: CPE Ratios and Varying Level of Nitrogen

Mukesh Kumar Joshi and B.S. Patel

Abstract

A field experiment was conducted on loamy sand soil of Agronomy Instructional farm,Chimanbhai Patel College of Agriculture, Gujarat Agricultural University, Sardar Krushinagar, During rabi season of 2000-2001.This experiment was laid out in split plot design with four levels of irrigation based on IW: CPE ratios (0.60, 0.75, 0.90 and 1.05) as main plot treatment and four levels of nitrogen (60, 80,100 and 120 kg N/ha) as sub-plot treatment with four replications. The highest seed yield of 1576 kg/ha was recorded under irrigation scheduled at an IW:CPE ratio of 1.05 which was closely followed by the IW:CPE ratio of 0.90 (1539 kg/ha), but both these treatments gave significantly higher seed yield than obtained under IW:CPE ratios of 0.75 and 0.60. Likewise stover and total biomass production followed the same pattern of seed yield. Seed, stover as well as total biomass production increased with increase in nitrogen levels from 60 to 120 kg/ha, but increase between 80,100 and 120 kg N/ha was not significant.

Introduction

Fennel (*Foeniculum vulgare* Mill.) commonly known as Saunf is a native of Southern Europe and Mediterranean region, belonging to Umbelliferae. From ancient time, India is known as a land of spices and world's largest producer, consumer and exporter of the spices. India produced almost 75 per cent of the total world spices of production. The area production and productivity of funnel in India during the year 1999-2000 were 17,561 hectares, 18,937 tonnes and 1,078 kg/ha respectively. In India, it is cultivated in Gujarat, Rajasthan, Punjab and Haryana. Among these states Gujarat ranks first in area (12,400 ha), production (16,300 tonnes) and productivity (1348 kg/ha) contributing about 86 per cent of the country's production. (DOA, 2001) Fennel is a cold weather crop and grows well under dry and mild cold climate. Mainly it is cultivated as transplanted crop in Gujarat and Rajasthan. However, during recent past, experimental evidences have proved good scope of *rabi* drilled fennel in Gujarat (Mehta *et.al.*, 1990).

In India, the productivity of fennel crop is low as compare to other fennel growing countries. This is particularly because of lack of high yielding varieties and proper agro-techniques including irrigations water as well as fertility management for *rabi* fennel crop.Yield potential for *rabi* fennel crop is about 28 q/ha,while the average yield of the crop is about 13 q/ha.Thus, there is a wide gap between potential and average yield.

Irrigation water is scare and costly input in North-Gujarat.Therefore, its economic and scientific utilisation become imperative.Thus, precise information on water requirement of funnel is essential and inevitable. Recently irrigation is being scheduled on the basis of climatological approach (IW: CPE ratios), which is now considered as most scientific. Since it integrates all the weather parameters giving them natural weightage in a given climate water plant continuum (Parihar *et.el.*,1974). Because of this reason scheduling of irrigation based on IW:CPE ratios is likely to result not only in better fennel production, but also to increase water use efficiency.

It is an established fact that there is a positive correlation between fertiliser use and agricultural production. Among the primary nutrients nitrogen is very important as it directly influences plant growth and dry matter production of the crop.

Materials and Methods

A field experiment was conducted at Agronomy Instructional Farm, Chimanbhai Patel College of Agriculture, Gujarat Agricultural University, Sardar Krushinagar, during *rabi* season of 2000-2001. This experiment was laid out in split plot design with four levels of irrigation based on IW: CPE ratios (0.60, 0.75, 0.90 and 1.05) as main plot treatment and four levels of nitrogen (60, 80,100 and 120 kgN/ha) as sub-plot treatment with four replications. The experimental field was loamy sand in texture with pH 7.85. The soil was low in organic carbon (0.17%), and available nitrogen (160 kg/ha) and medium in available phosphorus (37.2 kg/ha) and high in available potash (273 kg/ha). Electrical conductivity was very low showing that soil was free from any kind of salinity hazards.

Results and Discussion

Effect of Irrigation

The results presented in Table: 1 revealed that the effect of irrigation treatments on seed yield, stover yield and total biomass production was significant. Seed yield obtained with treatment application of irrigation at an IW: CPE ratios of 1.05 (1576 kg/ha) and 0.90 (1539 kg/ha) were at par and significantly higher over treatment of irrigation at 0.75 (1382 kg/ha) and 0.60 (1086 kg/ha) IW: CPE ratios. Seed yield obtained under IW: CPE ratio of 0.75 was significantly higher than that obtain under IW: CPE ratio of 0.60.

IW: CPE ratio of 0.60. The similar trend was obtained for stover yield. The stover yield increased with increase in IW: CPE ratios from 0.60 to 1.05. Significantly the highest stover yield was obtained under IW: CPE ratio of 1.05 (2533 kg/ha) but was at par with that recorded under 0.90 IW: CPE ratio (2488 kg/ha). Total biomass production recorded under IW: CPE ratios of 1.05 and 0.90 were at par and significantly higher over rest of the IW: CPE ratios.

Effect of Nitrogen

The effect of nitrogen levels on seed yield was significant. The seed yield increased with increase in nitrogen levels from 60 to 120 kg/ha. The seed yield obtained with the application of 120,100 and 80 kg N/ha (1541, 1503 and 1455 kg/ha) respectively, were at par, but significantly higher over 60 kgN/ha (1084 kg/ha). Similarly stover yield and total biomass yield was also increased with increase in nitrogen levels from 60 to 120 kg/ha. Data presented in Table 1 indicates that the significant increase in stover yield and total biomass production was only up to 80 kgN/ha.

Table 1: Seed yield, stover yield and total biomass production kg/ha as influenced by different IW: CPE ratios and nitrogen levels

Treatments	*Seed yield*	*Stover yield*	*Total Biomass production*
IW:CPE Ratios			
I1:0.60	1086	1775	2861
I2:0.75	1382	2244	3626
I3:0.90	1539	2488	4027
I4:1.05	1576	2533	4109
S.Em±	46	74	120
C.D. at 5%	148	237	385
C.V.%	13	13	13
Nitrogen levels (Kg/ha)			
N1:60	1084	1775	2859
N2:80	1455	2355	3810

(contd...)

N3:100	1503	2427	3930
N4:120	1541	2483	4024
S.Em±	35	56	91
C.D. at 5%	100	160	261
C.V.%	10	10	10
Interaction			
S.Em±	70	112	181
C.D. at 5%	NS	NS	NS

Conclusion

Based on findings and economics, it is concluded that the economically potential production from fennel crop can be secured by scheduling 11 irrigations based on IW: CPE ratios of 0.90 and fertilising the crop with 80 kgN/ha on loamy sand soils of North Gujarat.

References

Directorate of Agriculture, 2001. District-wise area, production and yield per hectare and important food and non-food crops in Gujarat state. Directorate of Agriculture, Gujarat State, Ahmedabad.

Mehta, K.G.; Patel, P.G.; and Patel, I.D.1990. A new avenue of fennel cultivation in Gujarat. *Indian Coca Areca nut and Spice Journal* 13(4): 139-141.

Parihar, S.S.; Gajari, P.R. and Narang, R.S.1974. Scheduling irrigation to wheat using pan evaporation. *Indian Journal of Agricultural Sciences* 44 (9):567-571.

18

Effect of Cutting Management and Nitrogen Levels on the Growth and Yield of Forage Chicory (*ChicoriumIntybus* L.) Under North Gujarat Conditions

Mukesh Kumar Joshi and B.S. Patel

Abstract

The studies entitled "Effect of cutting management and nitrogen levels on the growth and yield of forage chicory (Chicorium intybus L.) under North Gujarat Conditions" were carried out on loamy sand soil at Chimanbhai Patel College of Agriculture, Sardarkrushinagar Dantiwada Agricultural University, Sardarkrushinagar during rabi *2004-05 and 2005-06. The experiment was laid out in randomised block design (Factorial) with twenty seven treatment combinations {Comprising of three levels of each first cut days after sowing (50, 60 and 70 days after sowing), and cutting interval days after first cut (20, 30 and 40 days after first cut) and three levels of Nitrogen (15, 30 and 45 kg N/ha) applied as top dressed after harvest of each cut} replicated thrice. During both the years first cut taken at 50 days after sowing recorded the maximum green forage yields i.e.,1138.8 q. The results on green forage yield for cutting interval days after first cut indicated that cutting interval of 20 days (1131 q/ha) and 30 days (1143 q/ha) being at par with each other but, superior over cutting interval of 40 days after first cut. Application of Nitrogen as top dress indicated that fertilising the chicory crop with 45 kgN/ha after harvest of each cut produced highest green forage (1170 q/ha) yield.*

Introduction

Indian economy is primarily agricultural-based where animal health is very important. To establish and improve the animal production, a critical factor in sustainable agriculture the availability of quality forage crops and grasses and their production needs urgent attention.

Gujarat state has a total animal population of 18.44 million heads and the total forage production is 20.0 million tonnes against the requirement of 49.2 million tonnes. Thus a gap of 29.2 million tonnes exists between the demand and supply of fodder, which is ought to further widen due to further rise in the livestock population and diversion of more area to grain and cash crop. The total area under forage crops in the state is 0.8 million hectares (i.e. 6.4 per cent of the total cultivated area), thus the state is not only short of quantity good quality of fodder too (Patel, 2005). The area under forage crops shall have to be raised from existing 6.4 per cent to 10 per cent to meet fodder requirement of total livestock population. But, it seems difficult because of limited irrigation facilities and receipt of higher profit from contemporary food and cash crops. However, there is a scope for introduction of multicut high yielding forage crops/species like chicory and its input management.

Forage chicory, an herbaceous perennial of the family Asteraceae, is suitable for well drained or moderately drained soils having medium to high fertility and a pH of greater 5.5 or greater. Unlike most of the forage crops, it is herb rather than either a grass or a legume. The vigorous growth of chicory in winter and its apparent tolerance of soil water deficits due to relatively taproot system and responsiveness to management inputs and comparable or exceeding nutritive value than the commonly used species offer an ideal combination of traits suited to the needs of growing livestock (Jung *et el.*,1996). A weed locally known as Kasani (Wild chicory) comes up in Berseem in North India was studied for its palatability, nutritive value and found comparable with berseem (Shukla,1970)

Forage chicory a non-leguminous crop, propogated through seeds has different growth habit and high leaf-stem ratio than Lucerne lacks information on its production technology.

Cutting management is the crucial aspect for getting higher yield in forage chicory. The time of first cut after sowing is important to obtain maximum numbers of cuts as well as green forage at each cut. Cutting frequency is an other important aspect for multicut forage crops, which not only provide information about the regeneration potential of the crop but also growth peak and yield too.

Materials and Methods

A field experiment was conducted during *rabi* 2004-05 and 2005-06, respectively at the Agronomy Instructional Farm, Chimanbhai Patel College of

Agriculture, Sardarkrushinagar Dantiwada Agricultural University, Sardarkrushinagar. The experiment was laid out in randomised block design (Factorial) with twenty seven treatment combinations (Comprising of three levels of each first cut days after sowing (50,60 and 70 days after sowing), and cutting interval days after first cut (20, 30 and 40 days after first cut) and three levels of Nitrogen (15,30 and 45 kg N/ha) applied as top dressed after harvest of each cut} replicated thrice. The experimental field was loamy sand in texture with pH 7.6. The soil was low in organic carbon (0.18 per cent), and available nitrogen (182 kg/ha) and medium in available phosphorus (43 kg/ha) and high in available potash (210 kg/ha). Electrical conductivity was very low showing that soil was free from any kind of salinity hazards. The sowing during first season was done on 6th November-2004 while on second season sowing was done on 16th November-2005.

Results and Discussion

1.Yields

1.1 Effect of First Cut days after Sowing

Pooled data presented in Table 1 revealed that significant effect of first cut days after sowing on green forage yield of chicory was observed. At first cut, data showed that first cut taken at 70 days after sowing produced significantly higher green forage yield than that recorded under the treatments first cut at 50 or 60 days after sowing.

Table 1: Cut-wise green forage yield of chicory as influenced by first cut days after sowing, cutting interval days after first cut and levels of Nitrogen applied as top dress after harvest of each cut in pooled data

Treatments	*Cut wise green forage yield (q/ha)*						
	1st	*2nd*	*3rd*	*4th*	*5th*	*6th*	*Total*
First cut: Days after sowing (F)							
F1:50 DAS	109.4	237.7	318.4	278.9	225.9	120.6	1138.8
F2:60 DAS	124.4	235.7	306.4	252.6	189.4	103.3	1085.2
F3:70 DAS	143.5	227.0	296.8	240.2	149.8	62.1	1033.2
S.Em±	1.4	5.0	3.2	3.4	—	—	10.7
C.D. at 5%	3.8	NS	8.9	9.5	—	—	29.5
Cutting Interval: Days after first cut ©							
C1:20 Days	124.8	180.9	246.9	270.4	199.2	95.4	1130.6
C2:30 Days	126.4	225.1	324.1	288.1	177.4	—	1142.9
C3:40 days	126.2	294.3	350.6	213.2	—	—	983.7
S.Em±	1.4	6.9	8.6	3.4	—	—	10.7
C.D. at 5%	NS	41.8	52.0	9.5	—	—	29.5

(Contd...)

Levels of Nitrogen applied as top dress after harvest of each cut (N/ha)							
N1:15 kg	125.3	201.4	276.9	245.8	65.2	79.5	992.4
N1:30 kg	126.0	236.1	310.0	258.1	192.8	97.6	1094.5
N1:45 kg	126.0	262.8	334.7	267.8	207.0	109.0	1170.3
S.Em±	1.4	2.5	3.2	3.4	—	—	10.7
C.D. at 5%	NS	6.8	8.9	9.4	—	—	29.5
Interaction	NS	NS	NS	NS	—	—	NS
C.V.%	8.0	7.7	7.7	9.8	—	—	7.2

At third and fourth cut yield taken under first cut at 50 days after sowing was significantly higher than that obtained under first cut at 60 or 70 days after sowing.

The total green forage yield of all cuts revealed that first cut taken at 50 days after sowing produced significantly the highest green forage yield than that obtained from first cut taken at 60 and 70 days after sowing.

1.2 Effect of Cutting Interval days after First Cut

The data on green forage yield of chicory revealed that cutting interval days after first cut had significant effect at second, third and fourth cut and total of all the cuts.

At second cut, treatment cutting interval of 40 days after first cut produced significantly higher green forage yield than that recorded with the cutting intervals of 20 or 30 days after first cut. At third cut cutting interval of 30 and 40 days after first cut yielded at par green forage but significantly higher than that obtained under treatment cutting interval of 20 days after first cut. At fourth cut, treatment cutting interval of 30 days after first cut yielded significantly the highest green forage yield over cutting intervals of 20 or 40 days after first cut. At fifth cut, treatment cutting interval of 20 days after first cut produced maximum green forage than that obtained under cutting interval of 30 days after first cut. The data on total green forage yield revealed that treatment cutting intervals of 20 and 30 days after first cut yielded at par total yield but, significantly higher than that obtained under treatment cutting interval of 40 days after first cut.

1.3 Effect of Levels of Nitrogen applied as top dress after harvest of each cut

The green forage yield of chicory significantly influenced by different levels of nitrogen applied as top dress after harvest of each cut. The green forage yield obtained at all six individual cuts and total green forage yield obtained under treatment application of 45 kgN/ha was significantly higher than that obtained under treatment application of 15 or 30 kg N/ha.

References

Jung, G.A.; Shaffer, J.A; Varga, G.A and Everhart, J.R (1996). Performance of "Grassland funa" chicory at different management levels. *Agronomy Journal* 88, pp.131-136.

Shukla, S.K. (1970). Nutritional studies on *kasani* (*Chicorium intybus* L.) chemical composition and nutritive values of *kasani* for sheep. *Indian Journal of Dairy Science*, 22, pp.123-170.

Patel, D.M. (2005). Yield and quality of forage chicory (*Chicorium intybus L.*) as influenced by nitrogen. Ph.D Thesis (unpublished), Submitted to Gujarat Agricultural University Sadarkrushinagar.

19

Study on Summer Sesame Response to Different Sowing Time and Irrigation Regimes

K.N. Sondarva, H.D. Rank and P.S. Jayswal

Abstract

A field experiment was conducted to study the summer sesame response to different sowing time and irrigation regimes. It was observed that the seasonal depth of irrigation decreased with delay in sowing from 1st February and the growing days requirement decreased with delaying sowing after 1st February. The highest and lowest sesame grain yield of 1131.59 kg/ha and 555.20 kg/ha was observed for the thermal windows of 16th February and 1st February respectively. The grain yield increased rapidly by delaying the sowing from 1st Feb to 21st Feb., then after it decreased slowly and continously. The highest grain yield of 991.27 kg/ha was found under drip irrigation at 3-days interval which was higher by the tune of 10.33 per cent, 17.32 per cent and 20.86 per cent as compared to that of under four, Five days under drip and seven-days under surface irrigation respectively. The vegetative stage was found the most sensitive stage to thermal regimes followed by establishment stage, flowering stage, ripening stage and reproductive stage. Result showed that the sesame yield response to seasonal as well as stage-wise thermal heat units could be described well by the quadratic model and irrigation interval by the linear model, which shows that the applied water was less than the crop evapotranspiration or the optimal water requirements and yet there is a scope for increasing the yield by increased water application.

Introduction

Water resources of a country constitute one of its vital assets. India occupies only 329 M-ha geographical area, which forms 2.4 per cent of the world's land area; it supports over 15 per cent of the world's population. The population of India as on 31 March 2011 stood at 1,210,193,422 persons. Thus, India supports more than one-sixth of world population, 1/50th of world's land and 1/25th of world's water resources. Drip irrigation is one of the best and latest methods for efficient utilisation of irrigation water. It is an efficient method of application of water in which, the water is applied at low rate over long period of time at frequent intervals with low-pressure delivery system, in order to avoid water stress to the plant.

Drip irrigation provides high water use efficiency, higher crop yield, less labour requirement and relatively low operating cost, less weed growth, less insect and pest attacks, shorter growing season and earlier harvest of the crop. The soil moisture in the upper root zone is evacuated mainly due to soil evaporation and the water stored in the lower portion can be utilised efficiently by plant. To conserve water for a longer period and to reduce evaporation mulching is used. Mulching is the application of any plant residues or other materials to cover the top soil surface.

Temperature is an important weather parameter that affects plant growth, development and yield. Photosynthesis produces the sources of assimilates which plants use for growth and development. Temperature and radiation influence the rate of photosynthesis. However, plants also have an obligatory development in time, which must be met if the photosynthetic assimilates are to be converted into economically useful yields of satisfactory quantity and quality.

Temperature (and day-length in case of photosensitive crops) influences the developmental sequence of crop growth in relation to crop phonology. Evolutionary changes that have occurred in the biochemical and physical characteristics of photosynthesis have resulted in a large variation between crops in both their optimum temperature requirements and the responses of photosynthesis to changes in temperature, radiation, and composition of the atmosphere. Objective : (a) To assess the effect of thermal regimes during different growth stages of sesame crop on yield and development of mathematical model. (b) To assess the effect of irrigation interval on crop yield. and (c) To study the techno-economic of sesame crop grown under drip irrigation on yield.

Materials and Methods

A field experiment was conducted at instructional farm of soil and water engineering, CAET, JAU, Junagadh during summer season (Feb-May)-2012 with three factorial strip plot design. The experiment comprising of 24 treatment combinations shown in Table 1, were laid out in strip plot design with four replications.

Table 1: Treatments Details

Treatment	Treatment details		
	Date of sowing	*Irrigation Interval (days)*	*Mulching*
W_1I_1 M	01.02.2012	3 by drip irrigation	Wheat straw @ 5 tonne/ha
W_1I_1 NM			No mulch (NM)
W_1 I_2 M		4 by drip irrigation	Wheat straw @ 5 tonne/ha
W_1 I_1 NM			No mulch
W_1 I_3 M		5 by drip irrigation	Wheat straw @ 5 tonne/ha
W_1 I_3 NM			No mulch
W_1 I_4 NM		7 by surface irrigation	No mulch
W_2 I_1 M	16.02.2012	3 by drip irrigation	Wheat straw @ 5 tonne/ha
W_2 I_1 NM			No mulch
W_2 I_2 M		4 by drip irrigation	Wheat straw @ 5 tonne/ha
W_2 I_2 NM			No mulch
W_2 I_3 M		5 by drip irrigation	Wheat straw @ 5 tonne/ha
W_2 I_3 NM			No mulch
W_2 I_4 NM		7 by surface irrigation	No mulch
W_3 I_1 M	01.03.2012	3 by drip irrigation	Wheat straw @ 5 tonne/ha
W_3 I_1 NM			No mulch
W_3 I_2 M		4 by drip irrigation	Wheat straw @ 5 tonne/ha
W_3 I_2 NM			No mulch
W_3 I_3 M		5 by drip irrigation	Wheat straw @ 5 tonne/ha
W_3 I_3 NM			No mulch
W_3 I_4 NM		7 by surface irrigation	No mulch
W_4 I_1 M	16.03.2012	3 by drip irrigation	Wheat straw @ 5 tonne/ha
W_4 I_1 NM			No mulch
W_4 I_2M		4 by drip irrigation	Wheat straw @ 5 tonne/ha
W_4 I_2 NM			No mulch
W_4 I_3 M		5 by drip irrigation	Wheat straw @ 5 tonne/ha
W_4 I_3 NM			No mulch
W_4 I_4 NM		7 by surface irrigation	No mulch

The treatment combination of four levels of thermal regime (four dates of sowing 1st February, 16th February, 1st March and 16th March, 2012) and three levels of irrigation interval viz, three days, four days, five days with drip irrigation and seven days with surface irrigation without mulch as common to all treatment.

Sesame yield response to different sowing date

Various observations of the water consumption, durations of flowering, pod development and maturity, and sesame yield were observed and recorded. Daily maximum and minimum temperature were utilised for calculating the available heat units. The base temperature (T_b) was taken as 8.0°C (Ramankutty N., 2002). The following expression was used to calculate the heat units. (Matt Johnston *et al.*, 2011)

$$HU = [T_{max} + T_{min}]/2 - T_b \quad (1) \qquad ...(1)$$

where, HU is the heat units (degree-days/day); T_{max} is the maximum temperature of the day (°C); and T_{min} is the minimum temperature of the day (°C), T_b is the base temperature for the sesame crop (°C).

The water use efficiency (kg/ha-mm) and heat use efficiency (kg/ha/degree-days) were computed by dividing the seed sesame yield (kg/ha) by the seasonal water consumption and accumulated seasonal heat units respectively.

The mathematical models for the sesame yield response to seasonal heat unit availability (Eq. 2) and sesame yield response heat unit availability during growth periods (Eq. 3) was developed through regression analysis. The mathematical models for crop yield response to heat units were fitted. Various forms of the mathematical model were tried and the model having the highest goodness of fit (R^2) was proposed.

$$Y = f(THU) \qquad ...(2)$$

$$Y = f(thu_i) \qquad ...(3)$$

where, Y is sesame yield (kg/ha); THU is total seasonal thermal heat unit availability (degree-days); thu_i, is heat unit availability (degree-days) during the ith growth stage; and 'i' is an index for the growth period/stage (i = 0, 1, 2, 3 and 4 for establishment, vegetative development, flowering, pod maturity and pod ripening stage respectively.)

Sesame yield response to irrigation regimes

The seasonal yield production functions for border and drip irrigation have been empirically determined for field conditions by below equation.

$$Y = f(W) \qquad ...(4)$$

where, Y is the sesame yield and W is the seasonal water consumption of the sesame crop. The regression analysis was carried out to test the various mathematical frames and the best model describing the experimental data for drip irrigation was proposed.

Economics

The economics of sesame cultivation includes the total cost and total return per unit area. The total cost of cultivation was computed using the following expression:

$$TCC = CCC + CI \quad ...(5)$$

where, TCC is the total cost of sesame cultivation (Rs/ha/season); CCC is the common cost of sesame cultivation (Rs/ha/season); CI is cost of irrigation (Rs./ha/season)

$$CI = FCI + VCI \quad ...(6)$$

where FCI is the fix cost of irrigation (Rs/ha/season); VCI is the variable cost of irrigation (Rs/ha/season).

Result and Discussions

Effects of Date of Sowing on Growing Days and Seasonal Heat Units Requirements

The highest growing days requirements was observed as 109 days for thermal window-1st February with highest inputs of seasonal thermal heat unit 2152 degree-days as shown in Fig. 1. The lowest growing days requirement were observed as 74 days for the sesame sown at 16th March with lowest thermal heat units of 1722 degree-days. The estimated crop water requirements increased with delayed sowing from 1st February to 16th February and then after decreased. The reason behind the increase in crop water requirement up to 16th February was due to increase in mean daily temperature.

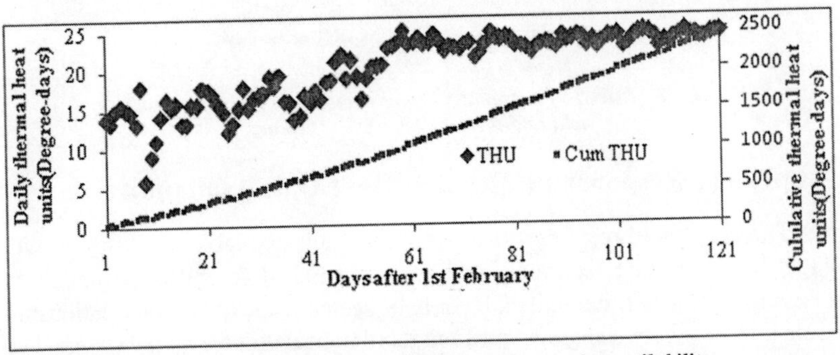

Fig. 1: Daily and cumulative Thermal heat units availability from 1st-February.

Effects of Date of Sowing on Crop Yield

The effects of seasonal thermal heat units' availabilities on the sesame grain yield as shown in Table 2 and thermal heat use efficiency were observed as presented in Fig. 2 below. The crop yield decreased for the thermal windows later than 16th Feb. The pod yield of 828.10 kg/ha was found under thermal windows of 16th March as the crop matured rapidly due to higher rate of thermal heat unit availability per day. The highest grain yield under thermal windows of 16th Feb. shows that the crop production can be optimum if the daily thermal heat units availability are around 15, 18, 23 and 24 degree-days/day during the establishment, vegetative, flowering-reproductive and ripening stages respectively.

Table 2: Effect of Thermal Window on Grain Yield

Treatments	*Grain Yield (kg/ha)*
Thermal window	
W_1=1st February	555.20
W_2=16th February	1131.59
W_3= 1st March	1084.72
W_4=16th March	828.09
S.Em. ±	33.36
C.D. (0.05)	106.75
C.V. (%)	18.16

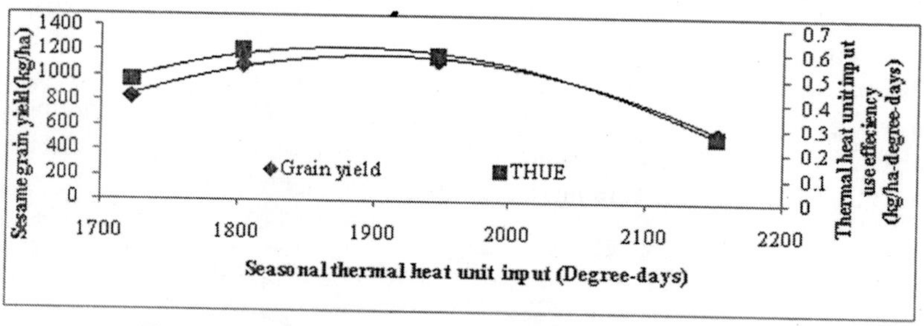

Fig. 2: Effects of seasonal thermal heat units on sesame grain yield and thermal heat use efficiency.

Crop Yield Response to Thermal Heat Unit Availability

The crop yield response to thermal heat unit availability during growth stages are presented separately in Fig. 3 and Fig. 4 for different irrigation interval with and without mulch. It could be seen that the quadratic relationship could be found between grain yield and thermal heat unit for all the stages. The yield decreased with increase in thermal heat unit availability up to certain level for establishment and vegetative stages. This indicated that these two stages require the lower thermal heat units. The reason behind the yield increase after reaching the minimum yield level could not be identified. It was found that the increase in thermal heat unit was due to increase in growing days requirement because of lower rate of thermal heat unit availability per day, which had slow down the physiological growth. The vegetative stage was found most sensitive followed by establishment stage, flowering stage, ripening stage and reproductive stage.

Crop Response to Seasonal Depth of Irrigation

Crop yield response to seasonal depth of irrigation under different thermal windows with mulch and no mulch are given in Table 3. It could be seen that the linear relationship could found between grain yield and seasonal depth of irrigation for all thermal window. The yield decreased with decrease in seasonal depth of irrigation and water applied was lower than the optimal water requirements and yet there is a scope for increasing the yield by increased water application.

Table 3: Crop Yield Response to Irrigation Interval

Irrigation interval	*Grain Yield (kg/ha)*
I_1=3 days	991.27
I_2=4 days	888.86
I_3=5 days	819.56
Control = surface at 7 days	784.46
S.Em. ±	31.92
C.D. (0.05)	110.47
C.V. (per cent)	20.06

The similar results obtained by Ucan Kenan (2007) reported that the amount of irrigation water applied significantly affected the seed yield of sesame. Significant higher grain yield was recorded with higher water quantities treatments. This supports the results of Foroud *et al.* (1993), Balasubramanigan and Dharmalingam (1996), they found that increase in yield was directly related to increased number of irrigations.

Effects of Thermal Regime on Crop Yield

It shows that the highest and lowest sesame grain yield of 1131.59 kg/ha and 555.20 kg/ha was observed for the thermal windows of 16th February and 1st February respectively. The crop yield decreased for the thermal windows later than 16th Feb. The pod yield of 828.10 kg/ha was found under thermal windows of 16th March as the crop matured rapidly due to higher rate of thermal heat unit availability per day. The highest grain yield under thermal windows of 16th February shows that the crop production can be optimum if the daily thermal heat unit availability is around 15, 18, 23 and 24 degree-days/

day during the establishment, vegetative, flowering-reproductive and ripening stages respectively.

Crop Yield Response to thermal Heat Unit

Among the various mathematical frames tried, the second order polynomial (quadratic) relationships were found best fitted as presented in Table 4. The values of the goodness of fit shows that the sesame yield response to seasonal thermal heat units can be described well by the quadratic model. The

Table 4: Optimized Yield for Different Irrigation Interval

Irrigation interval (Days)	*Mulch*	*Model*	R^2	*Optimized thermal heat units, (degree-days)*	*Optimized Yield (kg/ha)*
3	M	$y = -0.0094\ (THU_s)^2 + 35.58\ (THU_s) - 32507$	0.999	1892.55	1161.52
	NM	$y = -0.0077\ (THU_s)^2 + 29.20\ (THU_s) - 26433$	0.977	1896.10	1250.12
4	M	$y = -0.0097\ (THU_s)^2 + 37.28\ (THU_s) - 34547$	0.982	1921.65	1272.55
	NM	$y = -0.0093\ (THU_s)^2 + 35.44\ (THU_s) - 32796$	0.991	1905.38	967.27
5	M	$y = -0.0113\ (THU_s)^2 + 43.09\ (THUs) - 39856$	0.996	1906.64	1222.50
	NM	$y = -0.0117\ (THU_s)^2 + 44.46\ (THU_s) - 41076$	0.998	1900.00	1161.00
Where, THU_s = Seasonal thermal heat unit (degree-days), Y = yield (kg/ha)					

developed model showed that for the three days and four days irrigation interval, the sesame grain yield increased more rapidly under mulch as compared to no mulch for the lower values of thermal heat units but for the higher values of thermal heat units, the yield decreased more rapidly under mulch as compared to no mulch. However, the opposite was found for the models of five days irrigation interval. Therefore, for the lower values of the thermal regimes, the mulch helps to maintain the required warmth micro climate for the soil micro organisms.

Crop Yield Response to Irrigation Regimes

The observed yield and seasonal irrigation water application were used to develop the model. The regression analysis was carried out using the yield data as output and seasonal irrigation water application as input. The developed model for various thermal windows and different irrigation interval under drip with and without mulch were as presented in Table 5 and 6. The developed models were linear in nature, which were in contradictory with the results reported by Grimes and EI-Zik (1990), Wanjura *et al.* (2002), and Rank (2006) for the cotton crop.

A substantial body of previous work had been reviewed by Vaux and Pruit (1983) and ET production function was derived for a variety of agricultural crops. They concluded that the yield of different crops gave as linear function of seasonal consumptive use. The benefit cost ratio (BCR) of summer sesame

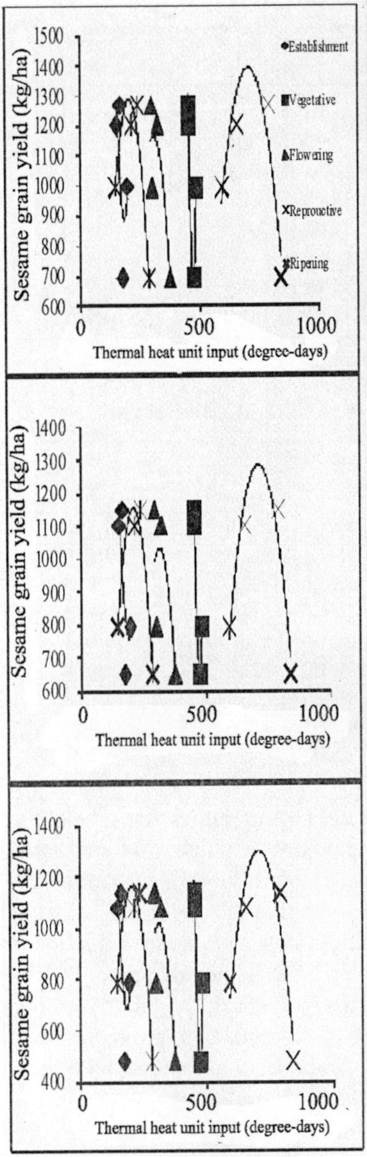

Fig. 3: Sesame Grain Yield Response to Thermal Heat Units Inputs During Different Growth Stages for 3 Day, 4 Day and 5 Day Irrigation Interval with Mulch.

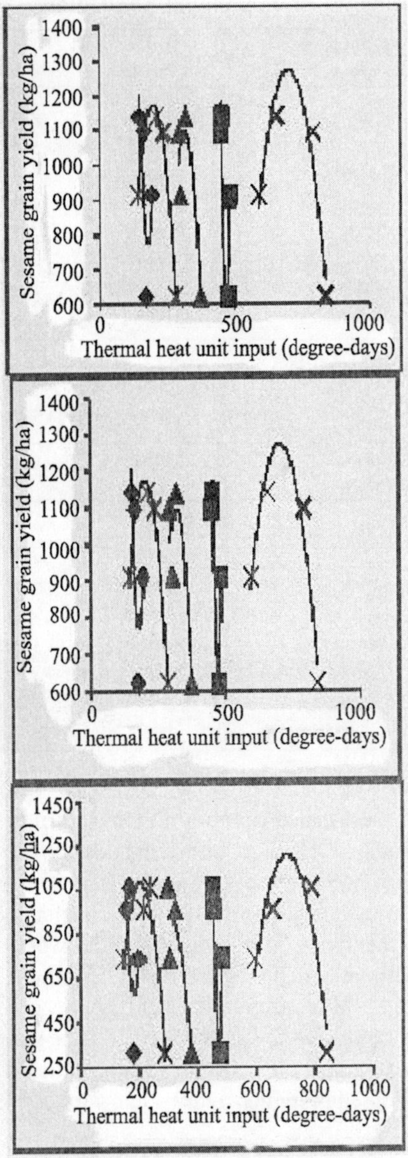

Fig. 4: Sesame Grain Yield Response to Thermal Heat Units Inputs During Different Growth Stages for 3 Day, 4 Day and 5 Day Irrigation Interval Under No Mulch.

Table 5: Sesame Yield Response to Seasonal Water Application

Thermal window	*Mulch level*	*Model*	*R^2*
1-Feb	M	y = 3.484W – 724.9	0.997
	NM	y = 5.209W – 1489	0.999
16-Feb	M	y = 1.062W + 807.8	0.985
	NM	y = 0.248W + 989.7	0.993
1-Mar	M	y = 1.317W + 704.2	0.980
	NM	y = 1.868W + 439.2	0.999
16-Mar	M	y = 2.939W + 111.5	0.985
	NM	y = 2.433W + 181.2	0.991

where, W = Optimised seasonal depth of irrigation (mm), y = yield (kg/ha)

Table 6: Sesame yield response to irrigation interval

Thermal window	*Mulch*	*Model*	*R^2 value*
1-Feb	M	y = –105.7I + 818.9	0.890
	NM	y = –154.9I + 812.8	0.857
16-Feb	M	y = –65.49I + 1316	0.813
	NM	y = –16.42I + 1110	0.941
1-Mar	M	y = –61.19I + 1250	0.871
	NM	y = –90.17I + 1221	0.958
16-Mar	M	y = –104.9I + 1067	0.780
	NM	y = –87.91I + 974.6	0.804

where, I = Irrigation interval (days), y = yield (kg/ha)

was estimated as ratio of gross return to total cost of cultivation. The highest and lowest benefit cost ratio under drip irrigation with mulch was found as 4.07 and 1.59 respectively under irrigation treatment 14th Feb. with three days irrigation with drip and mulch and 1st Feb. with five days with drip and mulch respectively. While highest and lowest benefit cost ratio under irrigation but without mulch was found as 4.63 and 1.13 under irrigation treatment W_3I_4 and W_1I_3NM respectively. The higher benefit cost ratio of drip irrigation was found as compared to surface method. The mulch also helped to improve the benefit cost ratio. All drip treatments either alone or in combination with mulch resulted in higher returns than surface method of irrigation.

Conclusions

Based on the data observations, analysis and its interpretations, the following conclusions could be drawn from the present investigation.

- It was revealed that if the daily thermal heat unit availability is lower, the more number of growing days are required to mature the crop with less total thermal heat unit requirements.

- The highest grain yield of 991.27 kg/ha was found under irrigation by drip at three-days interval which was higher by the tune of 10.33 per cent, 17.32 per cent and 20.86 per cent as compared to that of under four, five days under drip and 7-days under surface irrigation respectively. The grain yield decreased with increase in irrigation interval.
- The sesame yield response to seasonal thermal heat units as well as to stage wise thermal heat unit's availabilities could be described well by the quadratic model.
- The linear form of the model for the yield response to irrigation interval shows that the yield decreases with increase in irrigation interval.
- The highest and lowest benefit cost ratio under drip irrigation with mulch was found as 4.07 and 1.59 respectively under irrigation treatment 14th Feb. with three days irrigation with drip and mulch and 1st Feb. with five days with drip and mulch respectively.

References

Balasubramaniyan, P. and V. Dharmalingam, 1996. Influence of irrigation and N levels on summer sesame. Sesame and Safflower Newsletter, Vol.11, pp. 45- 49.

Foroud, N., H.H. Mundel, G. Saindon and T. Entz, 1993. Effect of level and timing of moisture stress on soya beans yield, protein, and oil response, Field Crop Res., Vol. 33 (3), pp. 195- 205.

Grimes, D.W. and El-Zik,1990. Cotton. In: B. A. Stewart and D.R. Lielson (Eds). Irrigation of Agriculture Crops, Agronomy Monograph No. 30, ASA, CSSA and SSSA, Ladison, Wiscosin.

Matt Johnston, R. Licker, J. Fole, Hollowa, N.D. Mueller, C. Barford and C. Kucharik, 2011. Closing the gap: global potential for increasing biofuel production through agricultural intensification, Environ. Res. Lett. 6 (July-September -2011) 034028.

Ramankutty, N., 2002. The global distribution of cultivable lands: current patterns and sensitivity to possible climate change, Glob. Ecol. Biogeography, Vol. 11, pp. 377-392.

Rank, H.D., 2006. Effect of irrigation system and moisture regimes on physiological water response function of cotton. Ph.D Thesis, Junagadh Agricultural University, Junagadh.

Ucan, K., F. Killi, C. Gencoglan and H. Merdun, 2007. Effect of irrigation frequency and amount on water use efficiency and yield of sesame (*Sesamum indicum*, L.) under field conditions, Field Crops Res., Vol.101, pp. 249-258.

Vaux, H.J. and W.O. Pruitt, 1983. Crop water production functions. In: Hillel, D. (Ed). Advances in irrigation,Vol. 2, pp. 61-97. Academic Press, New York.

Wanjura, D.F., D.R. Upchurch, J.R. Mahan, and J.J. Burke, 2002. Cotton yield and applied water relationships under drip irrigation, Agric. Water Manage, Vol. 55, pp. 217-237.

20

Response of Garlic to Sulphur Sources under Sodic Soil Conditions

Pamu Swetha and M.S. Solanki

Abstract

A field experiment was conducted on sodic soil during Rabi-2008-09 at Dry Farming Research Station, Junagadh Agril. University, Jamkhabhalia, Dist: Jamnagar grown cv. G.G.-3 of garlic crop. The results revealed that the T_3: (RD + S_{20}AS) gave significantly higher bulb yield (9173.5 kg ha^{-1}), bulb girth (2.77cm) and uptake of K (16.1 kg ha^{-1}) over their full recommended dose. While, T_5: (RD +S_{20} Gypsum) gave significantly higher no. of cloves per bulb (16.2), sulphur in bulb (0.78 per cent), uptake of sulphur (15.97 kg ha^{-1}), available sulphur (19.4 kg ha^{-1}) and also gave significantly lower soil ESP (9.6). Treatment T_2: (RD) + S_{10} AS) shown significantly lower EC (1.02 dSm^{-1}) over rest of the treatments.

Key words: *Sulphur sources, Garlic yield, AS, Gypsum, Sodic soil*

Introduction

Garlic is an important vegetable crop commercially grown over a large area in Gujarat state. Garlic has higher nutritive value than other bulb crops and hence also finds medicinal usages. Sulphur (S) is now recognised as the fourth major plant nutrient, along with nitrogen (N), phosphorus (P) and potassium (K). Sulphur is an important nutrient for garlic, which is essential for the growth and development of garlic plant and bulb/bulbils production.The availability of phosphorus and sulphur in soil and their uptake by plants are fairly proportionate and application of these nutrients together affects the availability of each other and thereby growth and yield of crops (shinde *et al.*, 1979). Bulb crops respond well to P and Sapplication (Rohal and Kalra, 1986; Singh *et al.*, 1995). Garlic crop requires sulphur equivalent of phosphorus. Moreover, sulphur deficiencies have been reported in Saurashtra region (Jetpara, 2006). Generally farmers do not know the economical source and importance of sulphur for garlic production. Sulphur is also beneficial for reclamation of sodic soil. Keeping this in view, a field experiment will be conducted study the effect of various sources of sulphur on bulb yield of garlic in sodic soil of Saurashtra. This experiment will be conducted at Dry Farming Research Station, Junagadh Agril. University, Jamkhabhalia, Dist: Jamnagar. The soils of the state are highly acidic and low in available P and 5 contents (Prasad, 1981).

Materials and Methods

A field experiment was conducted on sodic soil during *Rabi*-2008-09 in RBD with four replications at Dry Farming Research Station, J.A.U., Khambhalia (Jamnagar) North Saurashtra Agroclimatic zone, grown cv.G.G. -3 of garlic crop. The treatment comprise T_1: Recommended Dose (RD) (50:50:50 kg N: P_2O_5: K_2O ha^{-1}) P and K through DAP and MOP, T_2: (RD) + (S_{10} AS) remaining N through urea, T_3: RD + (S_{20} AS) remaining N through urea, T_4: RD + (S_{10} Gyp), T_5: RD + (S_{20} Gyp). Garlic crop (var. G.G. -3) was sown at spacing of 15 x 10 cm having seed rate of 600 kg bulbils ha^{-1}. Crop was fertilised as per the treatments.

Results and Discussions

Effect on Yield and Yield Attributes

The experimental results (Table 1) revealed the bulb yield, bulb girth, no. of cloves per bulb were significantly influenced by levels and sources of

Table -1: Effect of levels and sources of sulphur on yield and yield attributes of garlic

Treatments	Yield (kg ha^{-1})		Yield attributes			
	Bulb	Fodder	Plant height (cm)	Bulb girth (cm)	Bulb height (cm)	No. of cloves per bulb
T_1:(RD) (50:50:50 kg N: P_2O_5: K_2O ha^{-1})	8102.0	1004.5	34.4	2.35	2.46	12.8
T_2: (RD) + (S_{10} AS)	9301.0	1150.3	34.9	2.59	2.65	14.6
T_3: RD + (S_{20} AS)	9673.5	1199.3	35.3	2.77	2.68	15.9
T_4: RD + (S_{10} Gyp)	9187.5	1182.8	33.6	2.69	2.74	15.3
T_5: RD + (S_{20} Gyp)	9333.3	1231.5	32.6	2.68	2.63	16.2
S. Em±	2972	79.10	1.39	0.084	0.08	0.58
C. D. @ 5%	**915.9**	**NS**	**NS**	**0.26**	**NS**	**1.80**
C.V. %	6.52	13.71	8.15	6.46	6.07	7.81

sulphur, while fodder yield, plant height and bulb height were remain unaffected. The bulb yield (9173.5 kg ha^{-1}) and bulb girth (2.77 cm) were significantly higher with treatment RD + S20 AS(T_3) which was remain at par with T_2, T_4 and

T_5, while no. of cloves per bulb (16.2) was significantly higher with treatment T_5 i.e. RD + (S_{20} Gyp) and it was remain at par with T_2, T_3 and T_4. Application of P and S significantly increased the P uptake by garlic bulb (Table 2) which might be due to increase in soil available P andits higher absorption by bulbs. Similar increase in P uptake by applied P was also reported byVerma *et al.* (1996). The interaction between P and S was positive for P uptake. P and S are absorbed as anions by plants, so if anyone of them is available in higher amount, the available pool for other will also increase (Joshi and Seth, 1975).

Table 2 : Effect of levels and sources of sulphur on N, P, K and S content (per cent) in garlic.

Treatments	*Bulb*				*Straw*			
	N	*P*	*K*	*S*	*N*	*P*	*K*	*S*
T_1 (RD) (50:50:50 kg N: P_2O_5: K_2O ha^{-1})	2.26	0.442	0.75	0.56	0.66	0.182	2.72	0.24
T_2: (RD) + (S_{10} AS)	2.32	0.433	0.76	0.60	0.62	0.180	2.85	0.25
T_3: RD + (S_{20} AS)	2.29	0.437	0.76	0.61	0.65	0.180	2.78	0.26
T_4: RD + (S_{10} Gyp)	2.20	0.436	0.75	0.72	0.67	1.183	2.81	0.29
T_5: RD + (S_{20} Gyp)	2.18	0.454	0.76	0.78	0.67	0.181	2.80	0.30
S. E.m±	0.13	0.02	0.02	0.04	0.04	0.011	0.167	0.015
C. D. @5%	**NS**	**NS**	**NS**	**0.13**	**NS**	**NS**	**NS**	**NS**
C.V. %	11.58	7.98	5.36	12.52	12.48	12.14	11.93	10.91

Effect on Content of Nutrients

The concentration of N, P, K and S in straw and bulb of garlic did not influenced significantly by various levels and sources of sulphur, except S content in bulb (Table 3. 3. 2). The application of RD + S_{20} Gyp (T_5) gave significantly higher values of sulphur (0.78 per cent) in bulb, which was found at par with T_4 treatment. Sulphur uptake increased significantly at all the levels of S application, which might be the result of rapid absorption and translocation of it by the plant with adequate supply of S to the soil in combination with increase in bulb yield. Similar result has also been reported by Singh *et al.* (1995). The bulb and straw samples collected at harvest were analysed for total N, P, K and S by micro Kjeldahl, vanadomolybdate yellow colour, flame photometric and turbidimetric methods, respectively (Tandon, 1993).

Effect on Uptake of Nutrients

The data (Table 3) indicated that the uptake of potassium and sulphur by garlic bulb were significantly influenced under various levels and sources of sulphur, while uptake of nitrogen, phosphorus, potassium and sulphur by straw as well as nitrogen and phosphorus by garlic bulb were found non-significant. The application of RD + S20 AS(T_3) recorded significantly higher uptake of K (16.1 kg ha^{-1}) which was found at par with T_5, T_4 and T_2, while significantly higher uptake of S (15.97 kg ha^{-1}) by garlic bulb was found with the treatment T_5 (RD + S_{20} Gyp) which was found at par with T_4. Application of P and S significantly increased the P uptake by garlic bulb (Table 2) which might be due to increase in soil available P and its higher absorption by bulbs. Similar increase in P uptake by applied P was also reported by Verma *et al.* (1996).

Table 3: Effect of levels and sources of sulphur on N, P, K and S uptake (kg ha^{-1}) by garlic

Treatments	*Bulb*				*Straw*			
	N	*P*	*K*	*S*	*N*	*P*	*K*	*S*
T_1:(RD) (50:50:50 kg N: P_2O_5: K_2O ha^{-1})	40.0	7.9	13.4	10.02	6.7	1.8	27.31	2.46
T_2: (RD) + (S_{10} AS)	47.5	8.9	15.5	12.26	7.2	2.1	33.09	2.86
T_3: RD + (S_{20} AS)	48.8	9.3	16.1	13.02	7.8	2.2	33.31	3.12
T_4: RD + (S_{10} Gyp)	44.6	8.8	15.3	14.41	8.0	2.2	33.43	3.44
T_5: RD + (S_{20} Gyp)	44.4	9.3	15.6	15.97	8.2	2.2	34.41	3.66
S. Em±	2.392	0.515	0.567	0.622	0.715	0.219	3.25	0.28
C. D. @ 5%	**NS**	**NS**	**1.746**	**1.916**	**NS**	**NS**	**NS**	**NS**
C.V. %	10.61	11.64	7.490	9.47	18.9	20.84	20.14	17.75

Effect on Soil Properties and Nutrient Availability

Soil analysis data (Table 4) indicated that the soil EC and ESP values were significantly influenced by various levels and sources of sulphur treatments. The lowest value of EC (1.02 dS m^{-1}) was observed with treatment T_2 (RD + S_{10} AS) which was remain at par with T_1 and T_3, while soil ESP (9.6) was recorded significantly lower with treatment T_5 (RD + S_{20} Gyp).

The treatment T_5 (RD + S_{20} Gyp) showed significantly higher values of available S (19.4 kg ha^{-1}) over rest of the treatments. The available nitrogen,

Table 4: Effect of levels and sources of sulphur on soil chemical properties and available nutrients after harvest of garlic

Treatments	*Soil chemical properties*			*Available nutrients (kg/ ha)*			
	EC (dSm^{-1})	pH	ESP	N	P_2O_5	K_2O	S
T_1:(RD) (50:50:50 kg N: P_2O_5: K_2O ha^{-1})	1.04	7.57	11.9	156	50.3	613	11.1
T_2: (RD) + (S_{10} AS)	1.02	7.62	11.8	162	51.2	578	12.6
T_3: RD + (S_{20} AS)	1.04	7.59	11.9	164	56.8	617	14.2
T_4: RD + (S_{10} Gyp)	1.28	7.55	11.3	164	58.1	612	17.8
T_5: RD + (S_{20} Gyp)	1.46	7.54	9.6	161	55.5	604	19.4
S. E.m±	0.081	0.054	0.57	7.53	3.04	33.54	2.52
C. D. @ 5%	**0.249**	**NS**	**1.78**	**NS**	**NS**	**NS**	**0.818**
C.V. %	13.86	1.44	10.32	9.33	11.20	11.09	10.89

phosphorus and potassium were remaining unaffected by the treatments after harvest of the crop. The available S content increased with P application and viceversa suggesting synergism between S and P. Similar results in acid soil has been reported byMajumdar andVenkatesh (2001).

Conclusion

From the above results, it can be concluded that under sodic soil condition recommended dose along with S_{20} AS gave significantly higher bulb yield (9173.5 kg ha^{-1}), bulb girth (2.77 cm) and uptake of K (16.1 kg ha^{-1}) while, RD along with S_{20} Gypsum gave significantly higher no. of cloves per bulb (16.2), sulphur in bulb (0.78 per cent), uptake of sulphur (15.97 kg ha^{-1}), available sulphur (19.4 kg ha^{-1}) and also gave significantly lower soil ESP (9.6) and RD along with S_{10} AS gave significantly lower EC (1.02 dSm^{-1}) over rest of the treatments. Increasing levels of phosphorus and sulphur significantly increased the bulb and straw yield of garlic (Table 1). This result confirms the findings of Singh *et al.* (1995) and Rohal and Kalra (1986). The significant increase in S uptake by P application could be due to higher S absorption as a result of greater root proliferation and increase in the activity of S oxidizing bacteria (Singh *et al.,* 1986).

References

Dubey, S.K., 2000. "Effectiveness of rock phosphate and superphosphate amended with phosphate solubilising microorganisms in soybean grown on vertisols". J. Indian Soc. Soil Sci., Vol. 48, pp. 71-75.

Jetpara, P. I. Sakarvadia, H. L. Parmar, K.B., 2009. Status and depletion of sulphur in the soils of Saurashtra region of Gujarat. Asian J. of Soil Sci., Vol. 4(1), pp. 156-157.

Joshi, D. C. and Seth, S.P., 1975. "Effect of sulphur and phosphorus application on soil characteristics, nutrient uptake and yield of wheat crop". J. Indian Soc. Soil Sci., Vol. 23, pp. 217-221.

Francisco, D. and Rodriguez, O., 1998. "Effects of sulphur, phosphogypsum and polymaleicacid on the physical properties of a saline, sodic soil and the bulb growth in onions (*Allium cepa* L.)". Suelos Ecuatoriales, Vol. 28, pp. 71-179.

Majumdar, B. and Venkatesh, M.S., 2001. "Influence of phosphorus and sulphur on residual nutrient availability and dry matter yield of soybean in acid Alfisol of Meghalaya". Indian J. Hill Farming, Vol. 14, pp. 33-38.

Majumdar, B.,Venkatesh, M.S., Lal, B., Singh, C.C., (1998). "Response of garlic (*Alluvium sativum* L.) to lime application in acidic soil". Ind. J. of Hill Farming.Vol. 11(1/2), pp.111-112.

Meghalaya, Rohal, S.V. and Kalra, G.S., 1986. "Effect of different levels of nitrogen, phosphorus and sulphur on dry weight of leaves, bulb diameter and pungency (volatile sulphur) of onion bulbs". Indian J. Agric. Res., Vol. 20, pp. 47-50.

Page, A.L., Miller, R.H. and Keeney, D.R., 1982. "Methods of Soil Analysis Part II". American Society of Agronomy Inc. and Soil Sci. Society of America Inc. Madison, Wisconsin, USA.

Prasad, R.N., 1981. Soil science research in NEH region. In: Singh, G.B. (Ed.) Proceedings of Workshop on Agricultural Research in NEH Region, (pp. 89-101), ICAR Research Complex for NEH Region, Shillong.

Shinde, D.A., Soni, B.K., Jha, M.N. and Sinha, S.B., 1979. "Response of soybean to phosphorus and sulphur in black soil". Bull. Indian Soc. Soil Sci., Vol. 12, pp. 498-504.

Shainberg, I.M.E., Sumner, W.P., Miller, M.P.W., Farina, M.A., Pavan, and M.V. Fey, 1989. "Use of gypsum on soils". Advances in Soil Science, Vol. 9, pp. 1-11.

Singh, V., Mehta, M.S. and Singh B., 1986. "Individual and interaction effect of sulphur, phosphorus and molybdenum in mustard". J. Indian Soc. Soil Sci., Vol. 39, pp. 535-538.

Singh, V., Singh, A. and Mehta, V.S., 1995. "Effect of S sources and levels on yield and uptake of nutrients by garlic". Fert. News., Vol. 40, pp. 47-49.

Tandon, H.L.S., 1993. "Methods of Analysis of Soil, Plant, Water and 14. Fertilizers". Fertilizer Development and Consultation Organization, New Delhi.

Verma, D.P., Sharma, B.R., Chadha, A.P.S., Bajpai, H.K. and Bhaduria, U.P.S., 1996. "Response of garlic to nitrogen, phosphorus and potassium levels". Advances in Plant Sciences, Vol. 9, pp. 37-41.

21

Yield and Yield Attributes as Influenced by Terminal Water Stress and Benzyl Adenine in Pearl Millet [*Pennisetum glaucum* (L.) R.Br.] Genotypes

G.V. Marviya and D.N. Vakharia

Abstract

Pearl millet [Pennisetum glaucum (L.) R.Br.] is a monocot species belonging to the family Poaceae. *As being arid and semi-arid crop, pearl millet is the component of dry land cropping system usually grown extensively on the soil with depleted fertility. The cultivation of pearl millet in summer season is popular in Gujarat state. It demonstrates highest level of tolerance to abiotic stresses such as drought and heat but when it is exposed to water stress specifically at flowering to maturity causes drastic reduction in yield and yield attributes. A field experiment was conducted to study the effect of benzyl adenine (BA) soaked seed treatment (50 mg BA/L) and terminal water stress on grain yield and yield attributing parameters of four pearl millet genotypes (drought tolerant i.e. J-2340 and 95444B and drought sensitive i.e. J-2290 and JMS-101B) during summer-2008 season with split-split plot design at Millet Research Station, Junagadh Agricultural University, Jamnagar. Among the parameters studied, terminal water stress condition significantly declined grain yield, dry matter yield, tillers/plant, threshing per cent and harvest index as compared to irrigated condition. Genotypes also varied significantly with respect to all these parameters. Presoaking seed treatment did not show any significant difference except tillers/plant which was maximum with BA soaked seed treatment. Genotype J-2340 consisted higher grain yield and dry matter yield with maximum number of tillers/plant, while threshing per cent and harvest index observed maximum with genotype*

95444B. On exposure of terminal stress condition, the maximum reduction in grain yield was by 3.5 fold with the drought sensitive genotype J-2290 while minimum reduction was by 2.1 fold in the drought tolerant genotype 95444B. The terminal water stress condition caused decreased the number of tillers per plant, threshing per cent and harvest index in all the genotypes however, the greater reduction was observed in the drought sensitive genotypes and less was with drought tolerant genotypes.

Introduction

Pearl millet [*Pennisetum glaucum* (L.) R. Br.] is locally known as bajra, and also known as bulrush millet, cat tail or spiked millet which is a monocot species belonging to the family Poaceae and sub family Penicedae. It demonstrates highest level of tolerance to abiotic stresses such as drought and heat but terminal water stress at flowering to maturity causes drastic reduction in yield and yield attributes.The importance of escape and tolerance, however, depends on the timing and intensity of stress occurrence (Oosterom *et al.*, 1995). The cultivation of pearl millet in summer season is popular in Gujarat state which occupied 3.30 lakh hectares area with production of 8.11 lakh tonnes and very high yield potentiality of 2458 kg ha^{-1}(Anonymous., 2013).

Apart from Gujarat, it is also grown in Maharashtra, Rajasthan, Uttar Pradesh and peninsular India under summer condition for grain purpose.The optimum and maximum temperatures for pearl millet growth are 17°C and 35°C. High soil temperatures during germination and emergence affect seedling stand establishment, while high air temperatures during vegetative growth and grain formation increase evapo-transpiration. (Konate, 1984).

Normal crop development and plant health depend on environmental conditions adequate to meet the crop species requirements. If these conditions deviate considerably from the optimum range of adaptation of the plant species, the crop experiences abiotic stresses. The improvement of yield and yield stability under abiotic stress conditions is the main target. Drought, heat, and poor soil fertility and structure are the most important abiotic stress factor for pearl millet. They usually occur in combination, leading to the low average grain yield of this crop (Andrews and Bramel-Cox, 1993).

Benzyl adenine is a synthetic cytokinin that elicits plant growth and development responses, setting blossoms and stimulating fruit richness by stimulating cell division. In wheat, it is observed by Sunita *et al.*, (2012) that drought stress severely reduced the photosynthetic attributes, water status and chlorophyll content which were significantly improved by foliar application of putrescine/benzyladenine as well as yield and yield attributes were also exhibited significantly higher values in putrescine/benzyladenine treated plants over control. The information regarding the effect of moisture deficit stress at terminal stage and benzyl adenine (BA) soaked seed treatment on yield and yield attributes in summer pearl millet is lacking therefore, the present experiment was conducted.

Materials and Methods

Seeds of four pearl millet parental lines with different sensitivity to water deficit stress viz., J-2340 (Tolerant inbred line), J-2290 (Sensitive inbred line), 95444 B (Tolerant male sterile line) and JMS-101 B (Sensitive male sterile line) were received from Millet Research Station, Junagadh Agricultural University, Jamnagar and the experiment was conducted under field conditions during summer 2008 at same research station. A split-split plot design consisted two main plots; I_1- Irrigated (Irrigation was given at regular 7-8 days interval) and I_2- Terminal stress (irrigation stopped from flowering to maturity) and each main plot was sub-divided into four sub-plots; with Genotypes (V_1- J-2340, V_2-J-2290, V_3- 95444 B and V_4-JMS-101 B) and further in three sub-plots with pre-sowing seed soaking treatments (C_1- Control (Without soaking), C_2- Distilled Water soaked for 6 hours and C_3- Benzyl adenine soaked (50 mg/L) for 6 hours. A split-split plot design with twenty four treatment combinations (12 for Irrigated and 12 for Terminal water stress) and three replications was carried out containing total seventy-two plots. Recommended Agronomical package of practices for summer pearl millet was followed.

Each replication was divided in main plots and each main plot was subdivided into nine sub-plots and randomized. Cultural operations were carried out as per the package of practices recommended for pearl millet cultivation. From each net plot, plants were harvested and various phenological and post harvest parameters including yield and yield attributing parameters were recorded. Data were analysed statistically using standard statistical procedures as described by Snedecor and Cochran (1967). The threshing per cent and harvest index were calculated by using following formulas.

$$\text{Threshing (\%)} = \frac{\text{Grain Yield}}{\text{Earhead Yield}} \times 100$$

$$\text{Harvest Index (HI \%)} = \frac{\text{Grain Yield}}{\text{Biological Yield}} \times 100$$

Results and Discussion

1. Number of tillers/plant

Table 1 represents the data on the influence of terminal water stress and BA on tillers/plant. On imposing the terminal stress condition, tillers per plant was decreased from 4.09 to 2.78 and the reduction was by 32.03 per cent. Genotypes varied significantly for the mean tillers/plant and the genotype

Table 1: Influence of terminal water stress and benzyl adenine on number of tillers/plant of pearl millet

Treatments/ Genotypes	Irrigated (I_1)			Mean V	Terminal stress (I_2)			Mean V
	Control	Water soaked	BA soaked		Control	Water soaked	BA soaked	
	(C_1)	(C_2)	(C_3)		(C_1)	(C_2)	(C_3)	
J-2340 (V_1)	4.53	4.47	4.53	4.51	3.47	3.20	3.47	3.38
J-2290 (V_2)	3.47	3.73	3.60	3.60	2.07	2.20	2.27	2.18
95444 B (V_3)	4.27	4.07	4.33	4.22	3.07	3.13	3.27	3.16
JMS-101B (V_4)	3.93	4.13	4.07	4.04	2.13	2.40	2.67	2.40
Mean C	4.05	4.10	4.13		2.68	2.73	2.92	
Mean I	4.09				2.78			
	I	V	IXV	C	IXC	VXC	IXVXC	
S.Em. ±	0.014	0.058	0.081	0.028	0.039	0.055	0.078	
C.D. @ 5%	0.083	0.177	0.251	0.080	NS	0.159	NS	
C.V. (%)	Main Plot	2.38	Sub Plot	7.11		Sub-Sub Plot	3.94	

J-2340 possessed the maximum (3.94) tillers while, it was minimum with the genotype J-2290 (2.89). Overall effect of presoaking treatments were also found significant and the BA soaked seed treatment possessed the highest number of tillers/plant which was the lowest with the control. Interaction effects of IXVwere significant and the highest number was recorded by the genotype J-2340 which was the lowest in the genotype J-2290 under both the conditions. In the terminal water stress condition, the maximum reduction in tillers was observed in the drought sensitive genotypes J-2290 (39.44 per cent) and JMS-101 B (40.59 per cent) which was less in the drought tolerant genotypes J-2340 and 95444B by 25.06 per cent and 25.12 per cent, respectively. Combined effects of the IXC found non-significant however, numerically the BA soaked treatment contained the maximum tillers per plant under both the conditions. Interaction effects of VXC were found significant. In general, the drought tolerant genotypes J-2340 and 95444B contained more number of tillers/plant, while it was less in the drought susceptible genotype J-2290 and JMS-101 B. The interaction effects of IXVXC were found non-significant. Maracchi *et al.* (1993) suggested that yield reductions in the field resulting from water stress during the reproductive stage are linked to the late development of unproductive tillers. Joshi *et al.* (2006) reported that tiller number is considered one of the important yield contributing attributes in pearl millet. Patil and Jadeja (2006) also reported that under stress conditions, grain yield per plant was positively and significantly associated with total number of tillers per plant and number of productive tillers per plant.

2. Grain Yield

The data on grain yield are presented in Table 2. The significant and drastic reduction in grain yield was observed from 2017 kg/ha in irrigated to 791 kg/ha by 2.6 fold in the terminal stress condition. Genotypes also varied significantly, the highest mean grain yield (1598 kg/ha) was with the genotype J-2340 and JMS-101 B was the minimum (1240 kg/ha) yielder. On exposure of terminal stress the maximum reduction was recorded by 3.5 fold in drought sensitive genotype J-2290 and the minimum (2.1 fold) reduction in the drought tolerant genotype 95444 B. The presoaking treatments were found to be non-significant, however, the BA soaked treatment visualized the maximum grain yield. Combined effects of irrigation and the genotype were significant. Under the irrigated condition, the genotype V_1 gave the highest yield (2232 kg/ha) and the genotype $V_2$was at par (2221 kg/ha) with the V_1. The non-significant IXC interaction effects were recorded for the grain yield, however, the BA soaked treatment gave the highest grain yield 2035 kg/ha and 843 kg/ha for both the I_1 and I_2 condition, respectively. Combined effects of VXC were also found to be non-significant. The overall IXVXC interaction effects were non-significant. Under I_2 condition, the reduction in grain yield was noticed in all the treatments. The genotype J-2290 resulted minimum (625 kg/ha) grain yield

Table 2: Influence of terminal water stress and benzyl adenine on grain yield (kg/ha) of pearl millet

Treatments/ Genotypes	Irrigated (I_1)			Mean V	Terminal stress (I_2)			Mean V
	Control	Water soaked	BA soaked		Control	Water soaked	BA soaked	
	(C_1)	(C_2)	(C_3)		(C_1)	(C_2)	(C_3)	
J-2340 (V_1)	2215.7	2259.3	2222.2	2232.4	888.9	981.5	1018.5	963.0
J-2290 (V_2)	2129.6	2231.5	2300.9	2220.7	625.0	638.9	662.0	642.0
95444 B (V_3)	1838.0	1726.9	1925.9	1830.2	745.4	893.5	949.1	862.7
JMS-101B V_4)	1893.5	1768.5	1689.8	1784.0	689.8	657.4	740.7	696.0
Mean C	2019.2	1996.5	2034.7		737.3	792.8	842.6	
Mean I	2016.8				790.9			
	I	V	IXV	C	IXC	VXC	IXVXC	
S.Em. ±	40.04	36.27	51.30	24.72	34.96	49.44	69.92	
C.D. @ 5%	243.67	111.78	158.08	NS	NS	NS	NS	
C.V. (%)	Main Plot	17.11	Sub Plot	10.96		Sub-Sub Plot	8.63	

under control treatment and the BA soaked treatment of the genotype J-2340 recorded the highest (1019 kg/ha). Grain yield of pearl millet and its components were severely reduced by the terminal stress (Bidinger *et al.*, 1987; Winkel *et al.*, 1997; Joshi *et al.*, 2005) and it reduces up to 20 per cent (Mahalakshmi *et al.*,1993).

3. Dry Matter Yield

The data on dry matter yield are sited in Table 3. In the irrigated condition, the mean value for dry matter yield was significantly reduced by 42.83 per cent from 8395 kg/ha to 4799 kg/ha in the terminal stress condition. For the genotypes, mean dry matter yield varied significantly and was the highest with the genotype J-2340 (7670 kg/ha) and it was at par with the genotype JMS-101B (7153 kg/ha), while the genotype 95444 B (4799 kg/ha) was the lowest yielder. The mean dry matter yield for the presoaking treatments did not differ significantly. Interaction effects of IXV varied significantly. Under the irrigated condition, the genotype J-2340 possessed the highest dry matter yield while, JMS-101 B was the highest yielder in the terminal stress condition. The IXC integrated effects were significant. In the terminal stress condition, the highest dry matter was with the BA soaked treatment and control had the lowest. The VXC interaction effects were found significant. The BA soaked treatment of the genotype J-2340 resulted the highest (7940 kg/ha) dry matter yield. The significant IXVXC interaction effects were recorded. In the BA soaked treatment of the genotype J-2340 and JMS-101B gave the highest (10324 and 5880 kg/ha) dry matter yield under the irrigated and terminal stress condition, respectively. Water use efficiency based on dry matter and grain yields decreased with water stress in pearl millet (Ibrahim *et al.*, 1995). Under stress conditions, grain yield per plant was positively and significantly associated with total biomass accumulation per plant, Patil and Jadeja, (2006).

4. Threshing per cent

The threshing per cent data are presented in Table 4. The significant reduction of 37.47 per cent from 50.6 per cent in the irrigated condition to 31.6 per cent was recorded under the terminal stress condition.The mean threshing per cent of genotypes varied significantly and it ranged from 33.6 per cent (JMS-101 B) to 47.7 per cent (95444 B). The presoaking treatments found nonsignificant. Combined IXV effects were significant. The genotype 95444 B consisted highest (54.9 per cent and 40.6 per cent) and JMS-101 B had the lowest (44.6 per cent and 22.5 per cent) threshing per cent in both the irrigated and terminal stress conditions, respectively. The reduction in threshing per cent due to terminal water stress was the greater in the drought sensitive genotypes JMS-101 B (49.5 per cent) and J-2290 (48.4 per cent) while less in the drought tolerant genotypes J-2340 (27.6 per cent) and 95444B (26.1 per cent). The IXC and VXC interaction effects were non-significant.The overall IXVXC

Table 3: Influence of terminal water stress and benzyl adenine on total dry matter yield (kg/ha) of pearl millet

Treatments/ Genotypes	*Irrigated (I_1)*			*Mean V*	*Terminal stress (I_2)*			*Mean V*
	Control	*Water soaked*	*BA soaked*		*Control*	*Water soaked*	*BA soaked*	
	(C_1)	*(C_2)*	*(C_3)*		*(C_1)*	*(C_2)*	*(C_3)*	
J-2340 (V_1)	10046.3	9444.4	10324.1	9938.3	5277.8	5370.4	5555.6	5401.2
J-2290 (V_2)	9814.8	8287.0	8888.9	8996.9	4398.1	4722.2	4490.7	4537.0
95444 B (V_3)	6388.9	5972.2	5648.1	6003.1	3287.0	3750.0	3750.0	3595.7
JMS-101B (V_4)	8148.1	8842.6	8935.2	8642.0	5463.0	5648.1	5879.6	5663.6
Mean C	8599.5	8136.6	8449.1		4606.5	4872.7	4919.0	
Mean I		8395.1				4799.4		
	I	V	IXV	C	IXC	VXC	IXVXC	
S.Em. ±	118.82	177.08	250.43	65.05	91.99	130.09	183.98	
C.D. @ 5%	723.05	545.68	771.70	NS	265.12	374.94	530.25	
C.V. (%)	Main Plot	10.81	Sub Plot	11.39		Sub-Sub Plot	4.83	

Table 4: Influence of terminal water stress and benzyl adenine on threshing per cent of pearl millet

Treatments/ Genotypes	*Irrigated (I_1)*			*Mean V*	*Terminal stress (I_2)*			*Mean V*
	Control	*Water soaked*	*BA soaked*		*Control*	*Water soaked*	*BA soaked*	
	(C_1)	*(C_2)*	*(C_3)*		*(C_1)*	*(C_2)*	*(C_3)*	
J-2340 (V_1)	50.38	50.83	48.00	49.74	33.68	36.55	37.29	36.01
J-2290 (V_2)	46.46	54.77	57.13	52.97	27.55	27.06	27.50	27.36
95444 B (V_3)	50.90	52.54	60.29	54.91	41.28	39.39	39.42	40.56
JMS-101B (V_4)	49.88	42.92	41.01	44.59	24.03	20.88	22.54	22.53
Mean C	49.53	50.34	51.79		31.88	31.20	31.76	
Mean I	50.55				31.61			
	I	**V**	**IXV**	**C**	**IXC**	**VXC**	**IXVXC**	
S.Em. ±	1.000	1.171	1.657	0.895	1.266	1.791	2.533	
C.D. @ 5%	6.086	3.610	5.105	NS	NS	NS	7.300	
C.V. (%)	Main Plot	14.61	Sub Plot	12.10		Sub-Sub Plot	10.68	

interaction effects were found significant. In the irrigated condition, the genotype 95444 showed maximum (60.3 per cent) value for threshing per cent with the BA soaked treatment and in the terminal stress it was with the control of the genotype 95444 B (41.3 per cent). Joshi *et al.*, (2005) evaluated pearl millet lines under drought stress and observed that the male sterile line 95444B and inbred line J-2340 recorded the highest threshing percentage.

5. Harvest Index (HI)

The data on harvest index are sited in Table 5. The reduction in harvest index was significant and it was by 30.77 per cent from 24.7 per cent in irrigated to 17.1 per cent in the terminal stress condition. The mean value for harvest index of genotypes varied significantly which was ranged from 16.5 per cent (JMS-101 B) to 27.4 per cent (95444 B). The presoaking treatments were found non significant. The IXV interaction effects were significant. In both the irrigated and terminal stress condition, the highest harvest index was observed with the genotype 95444 B (30.8 per cent and 24.1 per cent per cent) and the lowest harvest index was recorded with the genotype JMS-101 B (20.7 per cent and 12.3 per cent), respectively. In the terminal water stress condition the drought tolerant genotype J-2340 (17.9 per cent) and 95444 B (24.1 per cent) had higher harvest index as compared to drought sensitive genotype JMS-101 B (14.2 per cent) and J-2290 (12.3 per cent). Interaction effect of I and C were non-significant, however, the BA soaked treatment showed the highest value in both the conditions. Joint effects V and C were found significant and the BA soaked treatment had both the minimum HI with the genotype JMS-101 B (15.7 per cent) and the maximum HI with the genotype 95444 B (29.8 per cent). The interaction effects of I, V and C were also found significant. The BA soaked treatment was resulted the highest HI 34.1 per cent and 25.3 per cent for the genotype 95444 B under the irrigated and terminal stress condition, respectively while, the lowest under irrigated condition was with the BA soaked treatment of the genotype JMS-101B (18.9 per cent) and the same genotype showed the lowest HI (11.6 per cent) with the water soaked treatment under the terminal stress condition. Yadav *et al.* (2003) reported that the harvest index was associated with drought tolerance in the pearl millet. Joshi *et al.* (2005) evaluated pearl millet lines under drought stress and found that the male sterile line 95444B and the inbred line J-2340 recorded the highest harvest index.

Conclusion

Pearl millet is being comparatively abiotic stress tolerant, even though in the present study the terminal water stress created at flowering caused great impact on yield and yield attributes. The presowing seed soaking treatment of benzyl adenine was given to aware about its role in compensating water stress effect. The terminal stress condition caused drastic and significant reduction

Table 5: Influence of terminal water stress and benzyl adenine on Harvest Index (HI %)of pearl millet

Treatments/ Genotypes	*Irrigated (I_1)*			*Mean V*	*Terminal stress (I_2)*			*Mean V*
	Control	*Water soaked*	*BA soaked*		*Control*	*Water soaked*	*BA soaked*	
	(C_1)	*(C_2)*	*(C_3)*		*(C_1)*	*(C_2)*	*(C_3)*	
J-2340 (V_1)	22.06	23.92	21.52	22.49	16.84	18.28	18.33	17.85
J-2290 (V_2)	21.70	26.93	25.89	24.83	14.21	13.53	14.74	14.16
95444 B (V_3)	28.77	28.91	34.10	30.76	22.68	23.83	25.31	24.12
JMS-101B (V_4)	23.24	20.00	18.91	20.72	12.63	11.64	12.60	12.28
Mean C	23.98	25.00	25.12		16.61	16.93	17.76	
Mean I	24.70				17.10			
	I	**V**	**IXV**	**C**	**IXC**	**VXC**	**IXVXC**	
S.Em. ±	0.238	0.650	0.919	0.737	0.528	0.747	1.056	
C.D. @ 5%	1.447	2.002	2.831	NS	NS	2.152	3.043	
C.V. (%)	Main Plot	6.83	Sub Plot	13.19		Sub-Sub Plot	8.75	

in all the attribute like the number of tiller plant by 32.03 per cent, grain yield by 2.6 fold, dry matter yield by 42.83%, threshing per cent by 37.47 per cent and harvest index by 30.77 per cent as compared to irrigated condition.Considering genotypes, all these parameters varied significantly and the best performer were the genotype J-2340 for tillers/plant, grain yield and dry matter yield, while the genotype 95444B for the threshing per cent and harvest index which proves the tolerance of drought tolerant genotypes. The presowing seed soaking treatments did not show any significant influence except number of tillers/plant which was highest with the BA soaked seed treatment and this treatment also visualized the maximum grain yield, higher dry matter production and threshing per cent. The IXV interaction was found significant for all these parameters. On exposure of terminal stress condition, the grain yield reduced by 3.5 fold with the drought sensitive genotype J-2290 and the minimum reduction was by 2.1 fold in the drought tolerant genotype 95444 B. The mean threshing per cent and harvest index of genotype JMS-101 B was the lowest and the highest in 95444 B. In the terminal stress condition, the maximum dry matter yield was recorded with the genotype JMS-101 B. Overall interaction effect IXVXC varied significantly for dry matter yield, threshing per cent and harvest index while nonsignificant for grain yield and tillers per plant. The performance of drought tolerant genotypes was better under terminal water stress condition compared to the drought sensitive genotypes.The positive effect of benzyl adenine also proved up to some extent in compensating terminal water stress.

References

Andrews, D.J. and P.J. Bramel-Cox. 1993. Breeding cultivars for sustainable crop production in low-input dry land agriculture in the tropics. Proceedings International Crop Science, pp. 211-223.

Anonymous, 2013. Crop-wise area, production and yield of food and non-food crops in Gujarat State for the year 2009-10 to 2011-12. Directorate of Agriculture, Gujarat State, Gandhinagar.

Bidinger, F.R., V. Mahalakshmi and G.D.P. Rao. 1987. Assessment of drought resistance in pearl millet (*Pennisetum americanum* (L) Leeke). 1. Factors affecting yields under stress. Australian Journal of Agricultural Research, 38: 37-48.

Ibrahim, Y.M., V. Marcarian and A.K. Dobrenz. 1995. Pearl millet response to different irrigation water levels: II. Porometer parameters, photosynthesis, and water use efficiency. Emirates Journal of Agricultural Sciences, 7: 20-38.

Joshi, A.K., G.V. Marviya and C.J. Dangaria.2005. Identification of drought tolerant inbred lines of pearl millet. International Sorghum and Millets Newsletter, 46: 100-102.

Joshi, A.K., G.V. Marviya, J.N. Pandya, K.V. Pethani, K.V. Buhecha and C.J. Dangaria. 2006. The share of main and side tillers to grain yield in pearl millet [*Pennisetum glaucum* (L.) R.Br.] Third National Seminar on Millets: Research and Development Future Policy Options in India, 2: 131-140.

Konate, M. 1984. Climate of the sorghum and millet cultivation zones of the semi-arid tropical regions of West Africa. Proceedings International Symposium. Agrometeorology of Sorghum and Millet in the Semi-arid Tropics, pp. 101-113.

Mahalakshmi, V., S. Sivaramakrishnan and F.R. Bidinger 1993. Contribution of pre and post anthesis photosynthates to grains in pearl millet under water deficit. Journal of Agronomy and Crop Science, 170: 91-96.

Maracchi, G., L.Bacci, C. Cantini and M.T. Haimanot 1993. Effects of water deficit in vegetative and post-flowering phases of pearl millet grown in controlled environment. Agricoltura Mediterranea, 123: 65-71.

Oosterom,E.J. van,V. Mahalakshmi, G.K. Arya, H.R. Dave, B.D. Gothwal, A.K. Joshi, P. Joshi, R.L. Kapoor, P. Sagar, M.B.L. Saxena, D.L. Singhania and K.L. Vyas. 1995. Effect of yield potential, drought escape and drought tolerance on yield of pearl millet (*Pennisetum glaucum*) in different stress environments. Indian Journal of Agricultural Sciences, 65(9): 629-635.

Patil, H.E. and G.C. Jadeja. 2006. Phenotypic and genotypic correlations of various characters of pearl millet [*Pennisetum glaucum* (L.) R. Br.] under terminal water stress environment. Advances in Plant Sciences, 19: 621-625.

Snedecor, G.W. and W.G. Cochran. 1967. Statistical methods. 6th Ed. Oxford and IBH Publishing Co. Calcatta.

Sunita Gupta, Vishnu Prakash Agarwal, and N. K. Gupta (2012). Efficacy of putrescine and benzyladenine on photosynthesis and productivity in relation to drought tolerance in wheat (*Triticum aestivum* L.). Physiology and Molecular Biology of Plants, 18(4): 331–336.

Winkel, T., J.F. Renno and W.A. Payne. 1997. Effect of the timing of water deficit on growth, phenology and yield of pearl millet (*Pennisetum glaucum* (L.) R. Br.) growth in Sahelian conditions. Journal of experiment botany, 48: 1001-1009.

Yadav, R.S., F.R. Bidinger, C.T. Hash, Y.P. Yadav, O.P. Yadav, S.K. Bhatnagar and C.J. Howarth. 2003. Mapping and characterisation of QTL × E interactions for traits determining grain and stover yield in pearl millet. Theor Appl Genet., 106: 512–520.

22

Effect of Weather Parameters on Abundance of Shoot and Fruit Borer, *Leucinodesorbonalis Guenee* Infesting Brinjal

A.V. Khanpara, P.J. Kalsariya and L.K. Chejara

Abstract

Investigation were carried out on population dynamics of shoot and fruit borer, Leucinodesorbonalis Guenee on brinjal at Instructional farm, College of Agriculture, Junagadh Agricultural University, Junagadh during Rabi season 2012-2013. Studies on population dynamics of shoot and fruit borer revealed that the male moth catches (12/trap) in the pheromone traps commenced during 4th week after transplanting (WAT) of brinjal plant and reached at different peaks in 8th, 10th and 13th WAT with population density of 55, 61 and 82 moth/trap, respectively. Thereafter, moth population declined fastly. Highly significant positive correlation showed between trap catches of male moths of L. orbonalis in brinjal and maximum temperature, mean bright sunshine hours and rate of evaporation, while significantly negative correlation with relative humidity i.e., morning, evening and mean relative humidity during rabi *season of 2012-13. The weather parameters* viz., *temperature maximum 27.50 to 41.70°C and minimum 25.50 to 6.50°C along with relative humidity of 86 to 32 per cent, bright sunshine hours 6.10 to 11.10 per day and wind speed of 9.60 to 2.50 km/hr found most congenial for fast multiplication of L. orbonalis on brinjal. Shoot damage strated 4th WAT with 2.38 per cent and increased gradually reaching to a peak level of 34.86 per cent during 1st week of March. Thereafter the percentage of damaged shoots was decreased when the crop reached to fruiting stage. Significant positive correlation was observed between per cent damaged* *shoots and mean bright sunshine hours, while significant negative correlation was found with relative humidity i.e., morning, evening and mean relative humidity. Fruit damage appeared in 7th WAT with 15.63 per cent. Thereafter, it increased continuously and reached its maximum 35.25 per cent damaged fruit in 13th WAT. Thereafter the percentage of damaged fruits was decreased. Highly significantly negative correlation exhibited between percentage of damaged fruits and mean relative humidity, while significantly negative correlation was found with relative humidity i.e., morning, evening and significant` positive correlation with mean bright sunshine hours and rate of evaporation.*

Introduction

Brinjal, *solanummelongena* Linnaeus, is an important vegetable crop, also known as a "King of vegetables" cultivated since ages. The name brinjal is popular in Indian subcontinents and is derived from Arabic and Sanskrit whereas the name eggplant has been derived from the shape of the fruit of some varieties, which are white and resemble in shape to chicken eggs. It is also called aubergine (French word) in Europe. Brinjal is used in Ayurvedic medicine for curing the diabetes, aphrodisiac, cardio tonic, laxative and reliever of inflammation. Shoot and fruit borer, *Leucinodesorbanalis* Guenee is the key pest throughout Asia (Allam *et al.*, 1982). In India, this pest has a countrywide distribution and has been categorised as the most destructive and the most serious pest causing huge amount of losses of brinjal (Patil, 1990). In early stage of the crop growth, larva bores into the shoots resulting in drooping, withering and drying of the affected shoots. During the reproductive stage, tiny larva bores into the flower buds and fruits, the bored hole are invariably plugged with excreta. It is also reported that there will be reduction in vitamin C content to an extent of 68 per cent in the infested fruit (Hemi, 1955).

Materials and Methods

The population dynamics of shoot and fruit borer, *Leucinodesorbonalis* Guenee on brinjal was monitored and the impact of abiotic factors on the population fluctuation of shoot and fruit borer on brinjal was studied. The crop variety GJB-3 was grown in plot size of 20 m × 20 m keeping 90 cm × 60 cm spacing between row to row and plant to plant. All the recommended agronomical practices were followed to raise the good crop. Seed treatment was given with imidacloprid @ 5 g/kg seed to prevent the incidence of sucking pests of brinjal. The observations on shoot and fruit damage were recorded from five randomly selected plants each from nine quadrate (1.8 m × 2.4 m). One pheromone trap was installed in the field and observations on the moth catches per week were recorded.

Correlation study: In order to study the effect of weather parameters on the population of shoot and fruit borer, correlation between the various weather parameters and pest incidence was worked out with the help of statistically analysis software (SAS 9.3).

Observation: Observation on number of damaged shoots and fruit per plant were recorded at weekly interval after appearance of the pest till harvest of the crop. Five randomly selected plants from nine quadrate (1.8 m × 2.4 m). The data, thus obtained were converted into percentage fruits infestation and were also correlated with weather parameters.

Results and Discussion

Population Build-up

The data presented in Table 1 revealed that the male moth catches (12/trap) in the pheromone traps commenced from 4th week after transplanting (WAT) of brinjal plants. The moth catches increased during four weeks after commencement and 1st peak of population density level of 20 moths/trap was recorded during 9th WAT. The moth catches increased in each successive week and reached its highest peak level with 82 moths in 13th WAT. Thereafter the number of male moths/trap catches in pheromone traps was decreased.

The moth catches of 15 to 82/trap recorded from 2nd week of January (3rd standard week) to 1st week of March (10th standard week) being maximum in 1st week of March. The data given in table 1 indicated that the male moth commenced in pheromone trap from 4th week after transplanting of the crop and in the same week the damage to shoot also recorded. Further it was seen that as the number of moth in the traps increased the damaged shoots also increased. Similarly as the moth catches in pheromone traps increased during fruiting stage of brinjal crop, percentage of damaged fruit also increased.

Per cent Shoot Damage

The data on percentage of damaged shoots in different weeks in brinjal crop during *rabi* season of 2012-13 are presented in Table 1. The observations recorded during*rabi* season indicated that shoots damage stared 4th week after transplanting (WAT) (1st standard week) with a level of 2.38 per cent and increased gradually reaching to a peak level of 34.86 per cent during 1st week of March (10th standard week). Theafter the percentage of damaged shoots was decreased and onward when the crop reached to fruiting stage. Patnaik (2000) observed the peak infestation on shoots occurred during March (68.0 per cent) and August (29.2 per cent).

Per cent Fruit Damage

The data on percentage of damaged fruits recorded at each picking are presented in Table 1. The data revealed that damaged fruits ranged between

Table 1: Weekly catches of male moths of *L. orbonalis* in pheromone traps and its damage to shoots and fruits during *rabi* 2012

Standard week	*Week after transplanting*	*Date of observation*	*No. of moths catches in pheromone trap*	*Damage (per cent)*	
				Shoot	*Fruit*
50	1	10/12/12	0	0.00	0.00
51	2	17/12/12	0	0.00	0.00
52	3	24/12/12	0	0.00	0.00
1	4	31/12/12	12	2.38	0.00
2	5	07/01/13	6	5.66	0.00
3	6	14/01/13	15	10.77	0.00
4	7	21/01/13	27	10.53	15.63
5	8	28/01/13	55	23.46	16.00
6	9	04/02/13	20	15.12	13.24
7	10	11/02/13	61	27.17	21.37
8	11	18/02/13	29	16.49	17.36
9	12	25/02/13	70	30.10	23.73
10	13	04/03/13	82	34.86	35.25
11	14	11/03/13	78	25.44	25.47
12	15	18/03/13	71	26.67	34.00
13	16	25/03/13	50	16.41	12.00
14	17	01/04/13	53	18.18	18.84
15	18	08/04/13	40	9.42	11.39
16	19	15/04/13	38	12.68	8.76
17	20	22/04/13	25	7.48	8.33
18	21	29/04/13	34	5.92	8.16

8.16 to 35.25 per cent in brinjal crop. Fruit damage was started during 7th week after transplanting (WAT) (*i.e.,* from 1st picking). The fruits damage increased in text three weeks after commencement and reached at level of 17.36 per cent during 11th WAT. Again the damaged fruits increased in each successive week and reached at highest damage fruits 35.25 in 13th WAT. However, fruit damaged decreased in the following week and again it was increased and reached at 34.00 per cent in 15th WAT. Thereafter the percentage of damaged fruit was decreased. Thus, the result indicated that percentage of damaged fruits was maximum 35.25 per cent during 13th WAT (10th standard week) in *rabi* season.

Correlation Studies

Studies on effects of various weather parameters on population of pheromone trap catches of *L. orbonalis,* per cent shoot damage and per cent fruit damage on brinjalwere carried out and the correlation co-efficient values presented in Table-2 indicated that weekly pheromone trap catches of male moths of *L. orbonalis* exhibited a highly significant positive correlation with maximum temperature (r = 0.552), mean bright sunshine hours (r = 0.565) and rate of evaporation (r = 0.574). While significantly negative correlation of pheromone trap catches exhibited with relative humidity *i.e.,* morning (r = - 0.469), evening (r = - 0.537) and mean (r = - 0.535, while minimum temperature (r = 0.165), mean temperature (r = 0.369) and wind speed (r = 0.111) showed positive non-significant correlation, during *rabi* season of 2012-13. Shukla and Khatri (2010) indicated that the maximum and minimum temperature and abundance of moth showed a positive correlation.

The percentage of damaged shoots exhibited significant positive correlation with mean bright sunshine hours (r = 0.466), while significant negative correlation exhibited with relative humidity *i.e.*, morning (r = - 0.446), evening (r = - 0.440) and mean relative humidity (r = - 0.481). While maximum temperature (r = 0.342), mean temperature (r = 0.130), rate of evaporation (r = 0.310), minimum temperature (r = - 0.071) and wind speed (r = - 0.126) showed positive and negative non-significant correlation, respectively. Matur *et al.* (2012) observed that the infestation had positive correlation with maximum temperature and significant negative correlation with mean relative humidity.

The percentage of damaged fruits exhibited highly significant negative correlation with mean relative humidity (r = - 0.560), while significant negative correlation exhibited with relative humidity *i.e.*, morning (r = - 0.523), evening (r = - 0.506) and significant positive correlation with mean bright sunshine hours (r = 0.446) and rate of evaporation (r = 0.468). While maximum temperature (r = 0.421), minimum temperature (r = 0.070), mean temperature (r = 0.250) and wind speed (r = - 0.036) showed non-significant correlation, respectively. Sobharani *et al.* (2012) observed that the minimum temperature during the *kharif* was positively correlated with shoot and fruit incidence, while the maximum temperature during *rabi* had positive correlation with fruit incidence.

Summary and Conclusion

Through present findings it is concluded that the male moth catches (12/trap) in the pheromone traps commenced during 4th week after transplanting (WAT) of brinjal plant and reached at different peaks in 8th, 10th and 13th

Table 2: Correlation matrix of the relationship between weather parameters and population of pheromone trap catches of *L. orbonalis*, per cent shoot damage and per cent fruit damage on brinjal during *rabi* 2012

	Temperature (°C)			*Relative humidity (per cent)*			*Wind speed (km/hr)*	*Mean bright sun shine (hours)*	*Evapora-tion (mm)*
	Max.	*Min.*	*Mean*	*Morning*	*Evening*	*mean*			
Population of pheromone trap catches of *L. orbonalis*	0.552**	0.165	0.369	-0.469*	-0.537*	-0.535	0.111	0.565**	0.574**
Per cent shoot damage	0.342	-0.071	0.130	-0.446*	-0.440*	-0.481*	-0.126	0.466*	0.310
Per cent fruit damage	0.421	0.070	0.250	-0.523*	-0.506*	-0.560**	-0.036	0.481*	0.468*

WAT with population density of 55, 61 and 82 moth/trap, respectively. Thereafter, moth population declined fastly. Highly significant positive correlation showed between trap catches of male moths of *L. orbonalis* in brinjal and maximum temperature, mean bright sunshine hours and rate of evaporation, while significantly negative correlation with relative humidity *i.e.*, morning, evening and mean relative humidity during *rabi* season of 2012-13. The weather parameters *viz.*, temperature maximum 27.50 to 41.70°C and minimum 25.50 to 6.50Table 2: Correlation matrix of the relationship between weather parameters and population of pheromone trap catches of *L. orbonalis*, per cent shoot damage and per cent fruit damage on brinjal during *rabi* 2012C along with relative humidity of 86 to 32 per cent, bright sunshine hours 6.10 to 11.10 per day and wind speed of 9.60 to 2.50 km/hr found most congenial for fast multiplication of *L. orbonalis* on brinjal. Shoot damage strated 4th WAT with 2.38 per cent and increased gradually reaching to a peak level of 34.86 per cent during 1st week of March. Thereafter the percentage of damaged shoots was decreased when the crop reached to fruiting stage. Significant positive correlation was observed between per cent damaged shoots and mean bright sunshine hours, while significant negative correlation was found with relative humidity *i.e.,* morning, evening and mean relative humidity. Fruit damage appeared in 7th WAT with 15.63 per cent. Thereafter, it increased continuously and reached its maximum 35.25 per cent damaged fruit in 13th WAT. Thereafter the percentage of damaged fruits was decreased. Highly significantly negative correlation exhibited between percentage of damaged fruits and mean relative humidity, while significantly negative correlation was found with relative humidity *i.e.*, morning, evening and significant positive correlation with mean bright sunshine hours and rate of evaporation.

References

Allam, M. A.; Rao, P.K. and Rao, B. H.K. (1982). Chemical control of brinjal shoot and fruit borer, *Leucinodes orbonalis* Guenee with newer insecticides. *Entomology*, 7: 133-135.

Hemi, M.A. (1995). Effect of borer attack on the vitamin 'C' content of brinjal. *Pakistan Journal of Health*, 4: 223-224.

Mathur, A.; Singh, N.P.; Meena, M. and Singh S. (2012). Seasonal incidence and effect of abiotic factors on population dynamics of major insect pests on brinjal crop. *Journal of Environmental Research and Development*, 7(1A): 431-435.

Patil, P. D. (1990). Technique for mass rearing of the brinjal shoot and fruit borer, *Leucinodesorbonalis* Guenee. *Journal of Entomological Research*, 14: 164-172.

Shukla Anju and Khatri, S.N. (2010). Incidence and abundance of brinjal shoot and fruit borer *Leucinodes orbonalis* Guenee, *The bioscan an inter-national quarterly journal of life science*, 5(2): 305-308.

23

Forecasting Adult Populations of Gram Pod Borer *Helicoverpa Armigera* (Hübner) on Chickpea using Pheromone Traps

Divya, C., D.M. Jethva and M.V. Ramdevputra

Abstract

Chickpea is valuable pulse crop and cultivated sporadically in North Saurashtra region where irrigation facility is limited. This crop is attacked by the number of insect pests. Among them, the gram pod borer, Helicoverpa armigera (Hübner) is the most important and economic pest of gram. The experiment was conducted on farmers field near Grassland Research Station, JAU, Dhari during the year 2006 to 2010. Pooled results showed that the populations of the trapped moths showed that the insect was active throughout the year, except severe winter months. The maximum catches of moths per trap per week were ranged from 63 to 73 moths during 2nd to 4th standard week. The highest catches of adult moths per trap at a maximum temperature of 31.50°C, minimum temperature of 13.46°C and relative humidity of 61.00 per cent and mean evaporation of 5.50 mm per day in 3rd standard week. Pooled correlation of moth catches in pheromone traps was found to be negative and significant with maximum and minimum temperature and mean evaporation. The multiple linear regression analysis indicated that the influence of all the weather parameters was high (80 per cent) and it was found that maximum temperature and mean evaporation had significant negative impact on the number of moth catches (one unit increase in maximum temperature and mean evaporation, there will be 1.74 and 4.97 unit decrease in population, respectively). Thus, it can be recommended that the activity of gram pod borer seemed to be related negative with maximum and minimum temperature and mean evaporation. This information could help the farmers in development of management strategies like installation of sex pheromone trap as tool of IPM in chickpea reduces the hazardous insecticides load, providing low cost bio-intensive management strategy for gram pod borer.

Key words: *Chick pea, Helicoverpa armigera, Forecasting modules, sex pheromone trap.*

Introduction

Chickpea is valuable pulse crop and cultivated sporadically in North Saurashtra region where irrigation facility is limited. Now, the area under this crop is increasing day by day with increasing irrigation facility. This crop is attacked by the number of pests. Among them, the gram pod borer (*Helicoverpa armigera* H.) is the important and economic pest of gram though, it is phytophagous. Indiscrimate use of chemical insecticides in chickpea ecosystem creates the adverse effects resulting in to residue in grains, bioagent destruction and environmental pollution. The use of sex pheromones for monitoring insect pests is of recent origin. It has been reported to be very useful for determining seasonal activity of pests by several workers (Tamhankar *et al.*, 1989; Singh and Sachan, 1991; Patil *et al.*, 1992). Integration of sex pheromone trap as tool of IPM for chickpea reduces the hazardous insecticides load, providing low cost bio-intensive management strategy for *H. armigera* in chickpea. As a pre-requisite to this, studies on monitoring of gram pod borer populations and their relationship with weather parameters were made in chickpea with development of forecasting systems against this pest are presented in this paper.

Materials and Methods

A sex pheromone trap was installed in the chickpea field at Grassland Research Station, Dhari during the year 2006 to 2011. The pheromone trap comprising a single plastic funnel was used and five pheromone traps were installed at one meter above the ground level in one hectare of land and each pheromone lure was replaced with a new one after an exposure of 20 days. Number of male catches in traps was recorded at weekly (std. week) intervals throughout the crop season. Correlation between male catches and various weather parameters was worked out. Meteorological data was obtained from local weather station.

Results and Discussions

The population of the trapped moths (pooled of 2006-07 to 2010-11), recorded on weekly basis (Table 1). The catches of gram pod borer moths in pheromone traps showed that the insect was active throughout the crop period, except severe winter months. Pooled results showed that the adult moth of *H. armigera* was recorded for the first time in 43rd standard week under observation. The level of population of adult male moth went on increasing and reached to its peak of 73.08 adult moths per trap at a maximum temperature of 31.50°C, minimum temperature of 13.46°C, RH 61.00 per cent and mean evaporation of 5.50 mm/day in 3rd standard week. Then, male adult population was showed declined trends and thereafter continued in the field up to harvest of experimental crop. Pooled correlation of moth catches in pheromone traps was found to be negative and significant with maximum (r = - 0.46599) and minimum temperature (r = - 0.36559) and mean evaporation (r = - 0.50807). Rest of weather parameters did not showed significant positive or negative correlation of co-efficient with male adult trapped moth.

Table 1: Number of moth catches and correlation matrix of *Helicoverpa armigera* with weekly mean abiotic factors in chickpea (Pooled of 2006-07 to 2010-11)

Std week No.	*Pooled of 2006-07 to 2010-11*							
	No. of pod borer/trap (5 trap average)	*Temperature*		*Humidity (per cent)*		*Mean sun-shine Hrs*	*Mean wind speed Km/hrs*	*Mean Eva. Op. mm/day*
		Max.	Mini.	I	II			
43	34.12	35.50	18.82	66.60	23.84	8.72	3.76	6.36
44	30.84	35.16	18.44	62.20	23.72	8.94	3.68	6.52
45	34.84	34.58	19.40	69.20	32.46	8.00	3.68	5.84
46	34.24	32.64	17.42	71.40	32.64	7.34	4.02	4.66
47	45.24	31.64	15.62	66.60	30.40	8.10	4.00	4.68
48	47.80	31.28	15.32	64.20	24.50	8.16	4.72	4.64
49	45.68	31.80	15.52	64.00	27.62	8.32	4.46	4.58
50	53.36	32.16	12.34	64.80	25.22	8.60	4.22	4.54
51	53.28	30.34	14.28	65.40	26.32	7.64	4.30	4.34
52	54.84	30.04	12.90	64.60	24.20	8.60	3.90	4.14
1	57.56	28.52	11.76	58.40	32.40	8.80	5.14	3.40
2	63.64	29.30	12.04	51.20	29.80	8.94	5.02	3.74
3	70.48	30.44	12.80	59.80	27.60	8.60	5.04	4.98
4	73.08	31.50	13.46	61.00	28.40	8.76	4.66	5.50
5	38.48	32.16	14.40	65.80	25.60	9.14	5.12	5.78
6	25.68	31.52	14.94	72.60	28.00	8.56	5.44	5.66
7	0.00	31.90	14.96	62.00	24.40	9.12	6.04	6.20
8	0.00	34.04	16.26	58.00	21.80	7.92	6.04	7.14
9	0.00	35.80	17.48	57.40	18.00	9.40	6.94	8.18

Table 2: Correlations coefficient between number of moths catches of *Helicoverpa armigera* in different year (2006-07 to 2010-11)

Year	Temperature		Humidity (per cent)		Mean sunshine Hrs	Mean wind speed Km/hrs	Mean Eva. Op. mm/day
	Max.	Mini.	I	II			
2006-07	-0.59795	-0.6209	-0.01191	0.477403	-0.17702	-0.41402	-0.58903
2007-08	-0.35516	-0.60372	-0.42299	0.239534	0.063542	-0.31295	-0.53678
2008-09	**-0.6419**	-0.33448	0.577849	0.171893	-0.58771	-0.27659	**-0.80465**
2009-10	**-0.67387**	-0.60346	0.266067	0.306047	0.184767	-0.01303	-0.51355
2010-11	-0.58853	-0.43961	-0.41950	0.112608	0.076027	-0.43107	-0.63056
Pooled	**-0.46599**	**-0.36559**	0.11106	0.13191	-0.04223	-0.15927	**-0.50807**

Individual year: Significant at P=0.05 (0.641) and P=0.01 (0.724) (N=19)
Pooled: Significant at P=0.05 (0.315) and at P=0.01 (0.368) (N=95)
Max.: Maximum temperature; **Min.:** Minimum temperature; **RH:** Relative humidity.

Table 3: Multiple linear regression analysis between weather parameters and number of moths catches in different year.

Year	Regression equation	R^2 value
2006-07	Y=336.2515 **-3.35037** (X1)-3.62416(X2)-0.56744(X3)+ 0.920562 (X4)-9.17064(X5)-6.12076(X6)**-0.239349**(X7)	0.573129
2007-08	Y=170.7485**-0.35541**(X1)-4.68309(X2)-0.93364(X3)+ 1.309306 (X4)-3.96663(X5)-9.30753(X6)**-1.03676**(X7)	0.699214
2008-09	Y=-102.866-**3.940165**(X1)-1.697684(X2)-1.224924(X3)+ 0.618627(X4) 1.04391(X5)-4.368059(X6)-**20.9352** (X7)	0.787756
2009-10	Y=46.39876-**1.59481**(X1)-1.80985(X2)-0.147911(X3)+ 0.309382(X4) 0.740603(X5)-0.28406(X6)-**3.311008**(X7)	0.698584
2010-11	Y=-166.726-**10.06322**(X1)-5.10508(X2)-0.88251(X3)+ 2.619212 (X4)-6.095611(X5)-1.126605(X6)-**20.4266**(X7)	0.659587
Pooled	Y=154.1169-2.5322(X1)-1.2502(X2)+0.3919(X3)-0.1534 (X4)+ 0.1661(X5)-1.6296(X6)-5.2922(X7)	0.313993

X1: Maximum temperature, **X2:** Minimum temperature, **X3:** Relative humidity morning **X4:** Relative humidity evening, **X5:** Mean sunshine Hrs, **X6:** Mean wind speed Km/hrs **X7:** Mean Eva. Op. mm/day.

The multiple linear regression analysis indicated that the influence of all the weather parameters was highly significant and it was 57% (R^2=0.573129) in

2006-07, 69% (R^2=0.699214) in 2007-08, 78% (R^2=0.787756) in 2008-09, 69% (R^2= 0.698584) in 2009-10, 65% (R^2= 0.659587) in 2010-11 and 80% (R^2= 0.806091) in pooled result of years on number of moth catches in pheromone traps. The pooled regression equation further indicated that the maximum temperature had significant negative impact on the number of moth catches i.e., for one unit increase in maximum temperature, there will be 3.35, 0.35, 3.94, 1.59, 10.06 and 1.74 unit decreased moth population in 2006-07, 2007-08, 2008-09, 2009-10, 2010-11 and pooled result of all these years, respectively. The results obtained by Chatar, V. P. *et al* (2010) and Kanhere, R.D. *et al* (2013) support the findings.

Conclusion

The populations of the trapped moths of *Helicoverpa armigera* (H.) in pheromone traps showed that the insect was active throughout the crop period, except severe winter months. The highest catches of adult moths per trap at a maximum temperature of 31.50°C, minimum temperature of 13.46°C and RH 61.00 per cent and mean evaporation 5.50 mm per day in 3rd standard week. The maximum temperature and mean evaporation adversely affected the moth catches in pheromone traps. The multiple linear regression analysis indicated that the influence of all the weather parameters was high (80 per cent) and it was found that maximum temperature and mean evaporation had significant negative impact on the number of moth catches (one unit increase in maximum temperature and mean evaporation, there will be 1.74 and 4.97 unit decrease in population, respectively).

References

Chatar, V.P., Raghvani, K.L., Joshi, M.D., Ghadge, S.G., Deshmukh, S.G. and Dalave, S.K. (2010). Population dynamics of pod borer, *Helicovrpa armigera* (Hubner) infesting chickpea. *International Journal of Plant Protection*, Vol. 3, No. 1: 65-67.

Kanhere, R.D., V.N. Patel, P.S. Umbarkar and A.M. Kakde (2013). Impact of weather parameters on population of pod borer, Helicoverpa armigera (Hubner) infesting cowpea. *Insect Environment,* Vol. 19(2): 96-97.

Patil, B.V., Nandihalli, B.S., Hugar, P. and Somashekar, 1992. Influence of weather parameters on pheromone trap catches of cotton bollworms. *Karnataka Journal of Agricultural Sciences,* 5: 346-350.

Singh, K.N and Sachan, G.C. 1991. Assessment of the use of sex pheromone traps in the management of *Spodoptera litura. Indian J. Proc.* 21(1):7-13.

Tamhankar, A.J., Guthi, K.K. and Rahalkar, G.W. 1989. Responsiveness of *Earaias vitella* and *Earias insulana* males to their female sex pheromone. *Insect Scie. Appli.*,10(5):625-630.

24

Impact of Climate Change on Population Dynamics of Pest Complex of Soybean

L.K. Chejara and A.V. Khanpara

Abstract

Investigations were carried out on impact of climate change on population dynamics pest complex of soybean [Glycine max (L.) Merrill.] at Instructional Farm, College of Agriculture, Junagadh Agricultural University, Junagadh (Gujarat) during the kharif-2013. Studies on population dynamics of pest complex of soybean revealed that the whitefly (Bemisiatabaci Gennadius) commenced from third week of September (1.9 whiteflies per three leaves per plant) and peak level of 5.6 whiteflies per three leaves per plant was observed during second week of October. None of the weather parameter was found significantly correlated (r = 0.5324 at 5 per cent) with population of whitefly but the significant negative correlation between whitefly population and evaporation rate (r = 0.6614 at 1 per cent) on soybean. The infestation of jassid (Empoascakerri pruthi) initiated from the second week of September (0.16 jassid per three leaves per plant) and peak level (3.78 jassid per three leaves per plant) in second week of October was observed. None of the weather parameters was positive significantly correlated with the jassid population. Incidence of aphid, (Aphis glycine Koch) commenced from third week of September (0.01 aphid index) and peak level (3.57 aphid index) was observed during second week of October. None of the weather parameter was found significantly correlated (r = 0.5324 at 5 per cent) with population of aphid but significant negative correlation between aphid population and evaporation rate (r = 0.5324 at 5 per cent level). Incidence of tobacco leaf eating caterpillar (Spodopteralitura Fabricius) commenced from third week of September. The first peak level of 5.7 larvae per plant during second week of October and second peak level of 3.1 larvae per plant was observed during second week of November. All the weather parameters showed non-significant correlation with the population of S. *lituraduring kharif-2013.*

Introduction

Soybean [*Glycine max* (L.) Merrill.] is an important leguminous crop. Native of soybean is in Asia and the first known records, however, indicate that soybean emerged as a domesticated crop around the eleventh century B.C. in China (Hymowitz, 1970), where it was cultivated for its oil since centuries and they named it as a "yellow jewel" which feeds China's entire population. From nutritional point of view, soybean contains 43.20 per cent protein and 20.00 per cent of edible oil. Soybean is a good source of phosphorus and lecithin. It also contains good amount of potassium, sulphur and Vit. E. Soybean protein is mainly rich in amino acids like leucine, methionine and threonine that the human body requires. For vegetarians, it is known as "poor man's meat". Owing to the absence of sugar content, it is considered to be very suitable diet for diabetic patients. These characteristics have made soybean to fit well in sustainable agriculture. During the late sixties and early seventies, the soybean crop was considered to be comparatively safe crop as regards to insect pest attack. According to Bhattacharya and Rathore (1979), 60 to 100 species of insects are found to attack the soybean at various stages in Uttar Pradesh. Mundhe (1980) reported about 16 insect species on soybean crop in Maharashtra. Sucking pests *viz.,*white fly, jassid, aphid and foliage feeders leaf eating caterpillar are the major insect pests of soybean in Gujarat state.

Materials and Methods

The seasonal incidence of pest complex of soybean was monitored and the impact of abiotic factors like maximum temperature, minimum temperature, relative humidity morning, relative humidity evening, sunshine hours, evaporation rate, rainfall and rainy days on the population fluctuation of pest complex of soybean was studied. The crop was grown in plot size of 20 m × 20 m keeping 45 cm × 15 cm spacing between row to row and plant to plant. The incidence of pest complex was recorded on five randomly selected plants from each quadrate of 1 m × 1 m at weekly interval on standard week basis. The crop under the experiment was kept free from pesticides throughout the season. The observations were recorded on fifty randomly selected plants at weekly interval from entire area, starting from one week after germination till harvest of the crop.

Observations Recorded

The observations were recorded by examining the plant thoroughly at weekly interval from the first week after sowing till harvest of the crop.

Observation of sucking pests *viz*., whitefly and jassid were recorded at weekly interval by observing three leaves (upper, middle and lower) of each five tagged plant per plot. Observations on aphid were recorded by counting aphid index per plant. To record the observations of foliage on five plants per plot were selected randomly and tagged. Numbers of larvae of foliage feeders were recorded on the five tagged plant per plot at weekly interval and continued upto harvest. Mean population of larvae per plant *S. litura* were worked out separately.

Results and Discussion

The data presented in Table-1 indicated that the population of whitefly (1.9 whiteflies per three leaves per plant) appeared from third week of September and remained throughout the crop period. The pest population increased gradually and reached to the peak at of 5.6 whiteflies per three leaves per plant during second week of October. Thereafter the population was declined gradually during the successive weeks and reached to zero level at the time of harvest.The frequent ups and downs of the pest population observed during crop growth period might be due to sudden change in rainfall pattern. Similar observations were also made by Bhattacharjee (1976). The population of jassid was initiated from the second week of September with 0.16 jassid per three leaves per plant. Initially the population of this pest was increased slowly and reached the peak level (3.78 jassid per three leaves per plant) in second week of October. However, the population of jassid observed throughout the crop period. The population was declined gradually during the successive weeks and reached to zero level at the time of harvest. Maximum population of jassid on soybean was during first and second week of October during *kharif* season of 2013.

Incidence of aphid was started from third week of September (0.01 aphid index) which increased during each successive week and reached at peak level (3.57 aphid index) during second week of October. Thereafter, the pest population suddenly declined at minimum level of 0.10 aphid index during first week of December and disappeared during second week of December. According to Donaldson *et al.* (2007), the pest population of aphid started from the first week i.e. vegetative stage and remained upto maturity stage of crop thereafter, aphid population decreased.

The population of *S. litura* (1.1 larvae per plant) commenced from third week of September. The pest population increased constantly during successive weeks and reached at first peak level of 5.7 larvae per plant during second week of October. Later on, second peak was observed with a peak level of 3.1 larvae per plant during second week of November. After reaching a peak the pest population declined gradually and disappeared during second week of December at the time of maturity of the crop. Sojitra (1990) reported two peaks of *S. litura* during *kharif* season on soybean.

Table 1: Seasonal incidence of whitefly, *B. tabaci,* jassid, *E. kerri*, aphid, *A. glycines* and tobacco leaf eating caterpillar, *S. litura* on soybean during *kharif*- 2013

Std. week	*Month and Week*	*Mean no. of whitefly / three leaves / plant*	*Mean no. of jassid / three leaves / plant*	*Aphid Index*	*Mean no. of tobacco leaf eating caterpillar larvae/plant*	*Temperature (°C)*		*Relative humidity (per cent)*		*Sunshine (hr.)*	*Evaporation rate (mm)*	*Rainfall (mm)*	*Rainy days*
						Max.	*Min.*	*Morning*	*Evening*				
36	**Sep-I**	0.0	0.00	0.00	0.0	32.3	24.2	89	56	6.4	5.0	0	0
37	**Sep-II**	0.0	0.16	0.00	0.0	34.0	24.9	83	63	7.0	4.8	67.4	2
38	**Sep-III**	1.9	1.70	0.01	1.1	34.1	25.4	86	68	6.3	4.4	19.4	1
39	**Sep-IV**	3.4	2.70	1.52	2.9	28.7	25.3	97	86	2.1	2.0	248.2	4
40	**Oct-I**	4.2	3.00	2.86	3.8	32.4	25.1	76	70	6.6	2.7	89.8	3
41	**Oct-II**	5.6	3.78	3.57	5.7	32.6	25.6	86	63	5.7	3.1	0	0
42	**Oct-III**	4.7	2.70	1.78	3.7	35.0	24.2	87	41	8.5	4.7	0	0
43	**Oct-IV**	3.5	2.10	1.18	2.6	35.2	20.6	79	32	9.6	5.1	0	0
44	**Nov-I**	2.3	1.60	0.90	2.2	34.1	19.0	78	34	9.5	4.5	0	0
45	**Nov-II**	1.2	1.40	0.70	3.1	33.4	19.6	68	34	9.0	5.2	0	0
46	**Nov-III**	1.9	1.20	0.60	2.9	31.5	16.0	67	36	8.8	4.8	0	0
47	**Nov-IV**	1.1	1.10	0.20	2.8	32.9	15.1	77	31	9.3	4.6	0	0
48	**Dec-I**	0.0	0.88	0.10	1.1	32.8	18.8	58	34	8.7	5.6	0	0
49	**Dec-II**	0.0	0.00	0.00	0.0	31.5	14.0	77	33	9.0	4.3	0	0

Correlation Studies

Studies on effects of various weather parameters on the fluctuation of pest complex of soybean were carried out and the data is presented in Table-2. The correlation matrix indicated that non-significant positive correlation exhibited between population of whitefly and minimum temperature, morning relative humidity, evening relative humidity, rainfall and rainy days. The significant negative correlation between whitefly population and evaporation rate (r = 0.6614 at 1 per cent) and maximum temperature, sunshine hours were non-significant negatively correlated. Thus none of the weather parameter was found significantly correlated (r = 0.5324 at 5 per cent) with population of whitefly on soybean during *kharif* season of 2013. Joshi (2007) reported that whiteflies population exhibited non-significant positive correlation with morning relative humidity, evening relative humidity. Correlation between whitefly population and maximum temperature, sunshine hours were negatively non-significant.

The non-significant positive correlation exhibited between population of jassid and minimum temperature, morning relative humidity, evening relative humidity, rainfall and rainy days. The significant negative correlation between jassid population and evaporation rate (r = 0.5324 at 5 per cent level) and maximum temperature, sunshine hours were negatively non-significant on soybean. Joshi (2007) reported that jassid population exhibited non-significant positive correlation with minimum temperature, morning relative humidity and evening relative humidity. Correlation between jassid population and maximum temperature was negatively non-significant.

The non-significant positive correlation exhibited between population of aphid and minimum temperature, morning relative humidity, evening relative humidity, rainfall and rainy days. The significant negative correlation between aphid population and evaporation rate (r = 0.5324 at 5 per cent level). The correlation between pest population and maximum temperature, sunshine hours was negatively non-significant. According to Vinay (2010), correlation between aphid population and morning relative humidity, evening relative humidity were positive non-significant and the correlation between aphid population and maximum temperature, evaporation rate and sunshine hours were negative non-significant on soybean during *kharif* season of 2009.

The non-significant positive correlation exhibited between population of *S. litura* and minimum temperature, morning relative humidity, evening relative humidity, rainfall and rainy days. The correlation between pest population and maximum temperature, sunshine hours and evaporation rate was negatively non-significant. Vinay (2010) reported that among the different weather parameters morning relative humidity and evening relative humidity had positive non-significant correlation with population of *S. litura* and the correlation between *S. litura* population and maximum temperature and evaporation rate was negative non-significant.

Table 2: Correlation matrix of the relationship between weather parameters population of whitefly, *B. tabaci*, jassid, *E. kerri*, aphid, *A. glycines* and tobacco leaf eating caterpillar, *S. litura* on soybean during *kharif*-2013

Insect pest	*Temperature (°C)*		*Relative humidity (per cent)*		*Sunshine (hr.)*	*Evaporation rate (mm)*	*Rainfall (mm)*	*Rainy days*
	Max.	*Min.*	*Morning*	*Evening*				
Whitefly	–0.0349	0.4776	0.3366	0.4049	–0.3592	–0.6852**	0.2623	0.2903
Jassid	–0.0097	0.4963	0.2945	0.3871	–0.3688	–0.6345*	0.2940	0.2934
Aphid	–0.0709	0.4705	0.2549	0.3703	–0.3363	–0.6607*	0.2241	0.2606
Tobacco leaf eating caterpillar	–0.0130	0.2088	0.0622	0.1035	–0.1271	–0.4641	0.0765	0.0456

*Significant (r = 0.5324) at 5 per cent level

**Significant (r = 0.6614) at 1 per cent level n = 14

Summary and Conclusion

Through present findings it is concluded that the whitefly (*Bemisiatabaci* Gennadius) commenced from third week of September (1.9 whiteflies per three leaves per plant) and peak level of 5.6 whiteflies per three leaves per plant was observed during second week of October. None of the weather parameter was found significantly correlated (r = 0.5324 at 5 per cent) with population of whitefly but the significant negative correlation between whitefly population and evaporation rate (r = 0.6614 at 1 per cent) on soybean. The infestation of jassid (*Empoascakerri* pruthi) initiated from the second week of September (0.16 jassid per three leaves per plant) and peak level (3.78 jassid per three leaves per plant) in second week of October was observed. None of the weather parameters was positive significantly correlated with the jassid population. Incidence of aphid, (*Aphis glycine* Koch) commenced from third week of September (0.01 aphid index) and peak level (3.57 aphid index) was observed during second week of October. None of the weather parameter was found significantly correlated (r = 0.5324 at 5 per cent) with population of aphid but significant negative correlation between aphid population and evaporation rate (r = 0.5324 at 5 per cent level). Incidence of tobacco leaf eating caterpillar (*Spodopteralitura* Fabricius) commenced from third week of September. The first peak level of 5.7 larvae per plant during second week of October and second peak level of 3.1 larvae per plant was observed during second week of November. All the weather parameters showed non-significant correlation with the population of *S. litura* during *kharif*-2013.

References

Bhattacharjee, N.S. 1976. Control of the yellow mosaic virus of "moong" in soybean. *Entomologists News Letter*, 6 (11/12): 64-65.

Bhattacharya, A.K. and Rathore, Y.S. 1979. Soybean insect problem in India. pp. 291-303. In Proceeding of World Soybean Research Conference-II (Ed Cobin, C. T.), March 26-29, 1979, North Carolina State University, Raleigh, U. S. A., [Fide: R. A.E.(A), 69(9): 665].

Donaldson, J.R.; Myers, S.W. and Gratton, C. 2007. Density dependent responses of soybean aphid (*Aphis glycines* Matsumura) population to generalist predators in mid to late season soybean fields. *Biological-Control*, 43 (1): 111-118.

Hymowitz, T. 1970. On the domestication of soybean. *Economic Botany*, 24: 408-421.

Joshi, M.D. 2007. Seasonal incidence, varietal screening, yield losses and efficacy of ecofriendly pesticides against major insect pest of soybean [*Glycine max* (L.) Merrill]. M. Sc. (Agri.). Thesis submitted to Junagadh Agricultural University, Junagadh. 112 pp.

Mundhe, D.R. 1980. Insect pest complex on soybean, *Glycine max* (L.) Merrill in Marathawada region. *Journal of Marathawada Agricultural Unversity,* 5(3): 259-260.

Sojitra, B.P. 1990. Population dynamics of major soybean pest's along with bionomics of grey semilooper, *Amynaocta* (Guenee) on different hosts and its chemical control. M.Sc. (Agri.) Thesis. GAU, Sardar Krishinagar.

Vinay Kumar, M.M. 2010. Population dynamic and chemical control of pest complex of soybean with newer insecticides. M.Sc. (Agri.) Thesis JAU, Junagadh.

25

Interrelationships and Path Coefficients for Twelve Components of Wheat (*Triticumaestivum* L.) in Late Shown Condition under Normal and Limited Irrigated Condition

Raiyani G.D., Bhatiya V.J., Patel K.D., Hadiya A.M. and Lalwani H.H.

Abstract

A set of 40 genotype of wheat were used to estimate genetic variability, interrelationships and path coefficients for 12 components. The estimates of genotypic correlation were higher than the corresponding phenotypic correlation. The grain yield per plant had significant and positive association, at genotypic and phenotypic levels, with biological yield per plant, grain weight per main spike, harvest index, number of grain per main spike, ear length, number of productive tillers per plant and 100 grain weight under normal irrigation while biological yield per plant, number of productive tillers per plant, grain weight per main spike, harvest index, days to 50 per cent flowering, number of grain per main spike and days to maturity under limited irrigated condition. The path coefficient analysis revealed high positive direct effects of biological yield per plant, harvest index, grain filling period, number of grain per main spike and days to 50 per cent flowering towards the grain yield per plant under normal irrigated condition. In case of limited irrigated condition, high positive direct effects of days to maturity, biological yield per plant, harvest index and plant height towards the grain yield per plant. The most of the characters contributed indirectly to grain yield through biological yield per plant under both condition. Based on correlation and path analysis, biological yield per plant, number of grain per main spike and harvest index

were identified as the most important components of grain yield per plant under normal as well as limited irrigated condition.
Key words: *Wheat, correlation, path analysis, yields components.*

Introduction

Wheat (*Triticumaestivum* L.) is the staple food for a large part of the world population including India. India accounts an area, production and a productivity of 29.9 million h, 93.9 million metric tonnes and 3140 kg/h, respectively (Anonymous, 2012). Grain yield is a complex trait and highly influenced by many genetic factors and environmental fluctuations. In plant breeding programme, direct selection for yield as such could be misleading. A successful selection depends upon the information on the genetic variability and association of morpho-agronomic traits with grain yield. Correlation studies along with path analysis provide a better understanding of the association of different characters with grain yield. Path coefficient analysis separates the direct effects from the indirect effects through other related characters by partitioning the correlation coefficient (Dixet and Dubey, 1984). The study of associations among various traits is useful to breeders in selecting genotypes possessing groups of desired traits. Hence, in this investigation, exotic as well as local genotypes were used and an attempt was made to generate information on inheritance, relationships of yield and its components and their implication in selection of better genotypes of wheat for the development or improvement of cultivars and germplasm as well.

Materials and Methods

The experimental material consisted of 40 diverse genotypes of wheat (*Triticumaestivum* L.) representing different geographic origins. The pure seeds of these genotypes were obtained from the Research Scientist (Wheat), Wheat Research Station, Junagadh Agricultural University, Junagadh. Forty genotypes of wheat were sown at late under normal irrigation and limited irrigation condition in a Randomised Block Design with three replications during *Rabi* 2013-2014. Under limited irrigated condition, irrigation was skipped after anthesis growth stage. Each entry was accommodated in a single row of 2.0 m length with a spacing of 22.5 cm. The recommended agronomical practices and plant protection measures were followed for the successful raising of the crop. Five competitive plants per genotype in each replication were selected randomly and observations were recorded on different characters and their averages were used for statistical analysis except days to 50 per cent flowering and days to maturity while taken as plot basis. This data were used to determine the genotypic and phenotypic correlation coefficients between all possible pairs of the characters from the genotype means through the method given by Al-Jibouri *et al.* (1958) and the path coefficient analysis as described by Dewey and Lu (1959).

Results and Discussion

The genotypes differed significantly for all the 12 characters indicating the presence of wind variation in the material evaluated. The values of phenotypic coefficient of variation were higher than genotypic coefficient of variation indicating the influence of environmental factors under both the conditions. The highest genotypic coefficient of variation was observed under both normal as well as limited irrigated condition for grain weight per main spike followed by number of grain per main spike, grain yield per plant, number of productive tillers per plant, biological yield per plant. High magnitude of GCV indicated the presence of wide variation for the characters under study to allow further improvement by selection of the individual traits. High estimates of GCV in wheat have been reported for grain weight per main spike, grain yield per plant and number of productive tillers per plant by Singh *et al.* (2013) while high estimates of GCV was reported for biological yield per plant and number of grains per spike by Sharma *et al.* (2006). Johnson *et al.* (1995) suggested that the heritability and genetic advance when calculated together would be more useful in predicting the resultant effects of selection. In the present study, the estimates of high heritability coupled with high genetic advance were observed for grain yield per plant, grain weight per main spike, biological yield per plant, number productive tillers per plant and number of grains per main spike under both normal as well as limited irrigated condition which may be attributed to the preponderance of additive gene action and possess high selective value and thus, selection pressure could profitably be applied on these characters for their rationale improvement (Panse, 1957).

The association analysis revealed that, in general, the values of genotypic correlation were higher than their phenotypic correlation under both the irrigation condition (Table 1 and Table 2) indicating the inherent association among the traits. Similar were the finding of Iftikhar *et al.* (2012) and Singh *et al.* (2012). In the present investigation, grain yield per plant had found significant and positively correlated with biological yield per plant, grain weight per main spike, harvest index, number of grain per main spike, number of productive tillers per plant and 100 grain weight at both the genotypic and phenotypic levels under both normal as well as limited irrigated condition. Such positive interrelationship between grain yield and these attributes have also been reported in wheat by several researchers (Singh *et al.*, 2012 andIftikhar *et al.*, 2012). Thus, these attributes can serve as marker characters for yield improvement in wheat. Grain yield per plant was found to be significant and positively correlated with ear length under normal irrigated condition and significant and positively correlated with days to 50 per cent flowering and days to maturity under limited irrigated condition. It was positive and significantly correlated with plant height at genotypic level under normal condition.Similar results were found for ear length by Singh *et al.* (2012) and for days to 50 per cent flowering and days to maturity by Monpara *et al.* (2009).

Table 1: Genotypic (r_g) and phenotypic (r_p) correlation coefficients among 12 characters in normal irrigated wheat under late shown condition

Characters		Days to 50 per cent flowering	Days to maturity	Grain Filling Period (Days)	Plant height (cm)	Ear length (cm)	No. of productive tillers per plant	Number of grains per main spike	Grain weight per main spike (g)	Biological yield per plant (g)	Harvest index (per cent)	100 grain weight (g)
Grain yield per plant (g)	r_g	0.1438	0.0498	-0.1441	0.2153*	0.5114**	0.4484**	0.5836**	0.7220**	0.9571**	0.5936**	0.2564**
	r_p	0.1119	0.0402	-0.0995	0.1532	0.3526**	0.4052**	0.4860**	0.5764**	0.8635**	0.4387**	0.2197*
Days to 50% flowering (days)	r_g		0.5037**	-0.7674**	0.0321	0.1400	0.2961**	0.1255	0.0467	0.2506**	-0.1915*	-0.0930
	r_p		0.4833**	-0.7530**	0.0325	0.1173	0.2780**	0.1168	0.0470	0.2015*	-0.1118	-0.0910
Days to maturity (days)	r_g			0.1677	-0.1718	0.1075	0.3390**	-0.0463	-0.1373	0.0987	-0.1458	-0.3814**
	r_p			0.2011*	-0.1503	0.1088	0.3081**	-0.0383	-0.1324	0.0943	-0.0944	-0.3585**
Grain Filling Period (Days)	r_g				-0.1793*	-0.0849	-0.0959	-0.1756	-0.1528	-0.2255*	0.0919	-0.1732
	r_p				-0.1645	-0.0658	-0.0925	-0.1609	-0.1486	-0.1636	0.0609	-0.1639
Plant height (cm)	r_g					0.4751**	-0.2068*	0.2967**	0.2857**	0.2118*	0.1371	0.0650
	r_p					0.4366**	-0.1922*	0.2964**	0.2831**	0.1358	0.0771	0.0583
Ear length (cm)	r_g						0.0192	0.6456**	0.6227**	0.5197**	0.2462**	0.0071
	r_p						0.0185	0.5458**	0.5543**	0.3487**	0.0858	0.0291

Cont...

Number of productive tillers per plant	r_g							-0.0102	-0.1013	0.5328**	-0.0241	- 0.2118**
	r_p							-0.0005	-0.0995	0.4391**	0.0226	-0.1842*
Number of grains per main spike	r_g								0.8825**	0.5460**	0.3904**	-0.0762
	r_p								0.8376**	0.4138**	0.2383**	-0.0683
Grain weight per main spike (g)	r_g									0.6557**	0.5454**	0.3206**
	r_p									0.5042**	0.2507**	0.3170**
Biological yield per plant (g)	r_g										0.3366**	0.1658
	r_p										-0.0651	0.1473
Harvest index (%)	r_g											0.4069**
	r_p											0.1866*

*, ** Significant at five per cent and one per cent levels, respectively.

Table 2: Genotypic (r_g) and phenotypic (r_p) correlation coefficients among 12 characters in limited irrigated wheat under late shown condition

Characters		*Days to 50% flowering*	*Days to maturity*	*Grain Filling Period (Days)*	*Plant height (cm)*	*Ear length (cm)*	*No. of productive tillers per plant*	*Number of grains per main spike*	*Grain weight per main spike (g)*	*Biological yield per plant (g)*	*Harvest index (%)*	*100 grain weight (g)*
Grain yield per plant (g)	r_g	0.4732**	0.3133**	-0.1879*	-0.0061	0.1590	0.7596**	0.4186**	0.6533**	0.9272**	0.6390**	0.2086*
	r_p	0.4109**	0.2745**	-0.1526	0.0120	0.1441	0.6719**	0.3742**	0.5715**	0.8766**	0.5674**	0.1980*
Days to 50% flowering (days)	r_g		0.5114**	-0.5443**	0.1196	0.2879**	0.3411**	0.0507	0.2150*	0.4509**	0.2996**	0.0640
	r_p		0.4884**	-0.5439**	0.1007	0.2653**	0.3337**	0.0550	0.2057*	0.3843**	0.2071*	0.0683
Days to maturity (days)	r_g			0.4425**	0.2979**	0.3539**	0.2393**	0.2723**	0.3460**	0.3282**	0.1047	-0.1774
	r_p			0.4666**	0.2790**	0.3299**	0.2109*	0.2566**	0.3128**	0.2817**	0.0739	-0.1585
Grain Filling Period (Days)	r_g				0.1660	0.0451	-0.1224	0.2129**	0.1134	-0.1502	-0.2104*	-0.2400**
	r_p				0.1662	0.0483	-0.1355	0.1910*	0.0923	-0.1186	-0.1388	-0.2217*
Plant height (cm)	r_g					0.4826**	-0.1915*	0.1941*	0.2213*	0.1228	-0.3468**	0.0381
	r_p					0.4311**	-0.1876*	0.2010*	0.2105*	0.1147	-0.2156*	0.0502
Ear length (cm)	r_g						-0.0321	0.5610**	0.3488**	0.3379**	-0.3337**	-0.3499**
	r_p						-0.0260	0.5069**	0.3255**	0.2869**	-0.2066*	-0.2746**
No of productive tillers per plant	r_g							0.0107	0.0830	0.7547**	0.4206**	0.1464
	r_p							-0.0143	0.0731	0.6818**	0.2463**	0.1495

Cont...

No. of grains per main spike	r_g								0.7462**	0.5238**	-0.0536	-0.3811**
	r_p								0.7269**	0.4392**	0.0028	-0.3585**
Grain wt. per main spike (g)	r_g									0.5481**	0.4976**	0.1550
	r_p									0.4793**	0.3259**	0.1573
Biological yield per plant (g)	r_g										0.3138**	-0.0216
	r_p										0.1141	0.0058
Harvest index (%)	r_g											0.6054**
	r_p											0.4032**

*, ** Significant at five per cent and one per cent levels, respectively.

The days to 50 per cent flowering, which had significant and positive association with days to maturity at genotypic and phenotypic levels, is an important component in identifying and deciding the duration of the wheat crop.The number of productive tillers per plant exhibited positive and significant association with biological yield per plant under normal irrigated condition but it exhibited positive and significant association with biological yield per plant and harvest index under limited irrigated condition at both genotypic and phenotypic levels. Ear length showed highly significant and positive correlation with plant height at both the condition. The number of grains per spike also had positive and significant association under both the condition with plant height and ear length. Similar relationship has been reported earlier by Sidharthan and Malik (2006). Significant and positive association was also observed between number of grains per spike and grain weight per spike under both normal as well as limited irrigated condition, which is akin to the results reported by Singh *et al.* (1985). Grain weight per spike exhibited positive and significant association with 100 grain weight under normal irrigated condition, which strengthen the earlier report made by Sharma *et al.* (2004).

The results obtained for direct and indirect effects of different characters on grain yield are presented in Table 3 for normal irrigated condition and table 4 for limited irrigated condition. The path coefficient analysis revealed that biological yield per plant and harvest index exhibited high and positive direct effects on grain yield per plant under both irrigation condition. Thus, these characters turned out to be the major components of grain yield. Similar results were obtained by Singh *et al.* (2012) and Fellahi *et al.* (2013). The character 100 grain weight exerted positive direct effect on grain yield under normal irrigated condition. Positive direct effect of 100 grain weight was also reported by Iftikhar *et al.* (2012). The character ear length and number of productive tillers per plant had positive direct effect on grain yield under limited irrigated condition. Similar results were obtained by Singh *et al.* (2012) for ear length and Rangare *et al.* (2010) for number of productive tillers per plant.The characters days to maturity, plant height, ear length and number of productive tillers per plant showed moderate to low and negative direct effect but, their positive correlation with grain yield was primarily formed through its indirect contribution from biological yield per plant under normal irrigated condition but, the characters days to 50 per cent flowering, 100 grain weight and grain weight per main spike showed high to moderate and negative direct effect but, their positive correlation with grain yield was primarily formed through its indirect contribution from number of productive tillers per plant under limited irrigated condition. The residual effect was of very low magnitude under both conditions suggesting that the majority of the yield attributes have been included in the path analysis.

Considering the genetic variability parameters, interrelationships and path analysis for grain yield and its components, an ideal plant type in wheat

Table 3: Genotypic path coefficient analysis showing direct (diagonal) and indirect effects of different characters on total grain yield per plant in normal irrigated wheat under late shown condition

Characters	*Days to 50 per cent flowering*	*Days to maturity*	*Grain filling period (days)*	*Plant height (cm)*	*Ear length (cm)*	*No. of productive tiller per plant*	*Number of grain per main spike*	*Grain Weight per main spike (g)*	*Biological yield per plant (g)*	*Harvest index (per cent)*	*100 grain weight (g)*	*Genotypic correlation with grain yield per plant*
Days to 50% flowering (days)	**0.0606**	0.0305	-0.0465	0.0019	0.0085	0.0179	0.0076	0.0028	0.0152	-0.0116	-0.0056	0.1438
Days to Maturity (days)	-0.0146	**-0.0291**	-0.0049	0.0050	-0.0031	-0.0099	0.0013	0.0040	-0.0029	0.0042	0.0111	0.0498
Grain filling period (days)	-0.0599	0.0131	**0.0781**	-0.0140	-0.0066	-0.0075	-0.0137	-0.0119	-0.0176	0.0072	-0.0135	-0.1441
Plant height (cm)	-0.0002	0.0010	0.0011	**-0.0059**	-0.0028	0.0012	-0.0017	-0.0017	-0.0012	-0.0008	-0.0004	0.2153**
Ear length (cm)	-0.0026	-0.0020	0.0016	-0.0088	**-0.0185**	-0.0004	-0.0120	-0.0115	-0.0096	-0.0046	-0.0001	0.5114**
No. of productive tiller per plant	-0.0107	-0.0122	0.0035	0.0075	-0.0007	**-0.0361**	0.0004	0.0037	-0.0192	0.0009	0.0077	0.4484**
Number of grain per main spike	0.0085	-0.0031	-0.0119	0.0201	0.0438	-0.0007	**0.0679**	0.0599	0.0371	0.0265	-0.0052	0.5836**
Grain Weight per main spike (g)	-0.0044	0.0130	0.0144	-0.0270	-0.0588	0.0096	-0.0833	**-0.0944**	-0.0619	-0.0515	-0.0303	0.7220**
Biological yield per plant (g)	0.2280	0.0898	-0.2052	0.1928	0.4731	0.4850	0.4970	0.5968	**0.9102**	0.3063	0.1509	0.9571**
Harvest index (per cent)	-0.0595	-0.0453	0.0285	0.0426	0.0765	-0.0075	0.1212	0.1694	0.1045	**0.3106**	0.1264	0.5936**
100 grain weight (g)	-0.0014	-0.0059	-0.0027	0.0010	0.0001	-0.0033	-0.0012	0.0050	0.0026	0.0063	**0.0155**	0.2564**

$R^2 = 0.9921$, Residual effect = 0.0891

*, ** Significant at five per cent and one per cent levels, respectively.

will be the one with biological yield per plant, grain weight per spike, number of grains per spike, number of productive tillers per plant, 100 grain weight and harvest index accompanied with early flowering and maturity under both normal as well as limited irrigated condition. Therefore, more emphasis should be given to those components while making selection for higher yield in wheat.

References

Al-Jibouri, H.A.; Miller, P.A. and Robinson, H.F. 1958. Genotypic and environmental variances in upland cotton cross of interspecific origin. *Agron. J.*, 50: 633-635.

Anonymous, 2012. Progress report of All India Co-ordinated Wheat and Barly Improvement Project. Directorate of Wheat Research, Karnal, India.

Dewey, D.R. and Lu, K.H. 1959. A correlation and path coefficient analysis of component of crested wheat grass seed production. *Agron. J.*, 51: 515-518.

Dixet, P. and Dubey, D. K. 1984. Path analysis in lentil (*Lens culinaris* Med.). *Lens Newsletter,* 11: 15-17.

Fellahi, Z.; Hannachi, A.; Bouzerzour, H. and Boutekrabt, A. 2013. Correlation between traits and path analysis coefficient for grain yield and other quantitative traits in bread wheat under semi-arid conditions. *J. of Agric. and Sustainability,* 3: 16-26.

Iftikhar, R.; Khaliq, I.; Ijaz, M. and Rashid, M.A.R. 2012. Association analysis of grain yield and its components in spring wheat (*Triticumaestivum* L.). *American-Eurasian J. Agric. and Environ. Sci.,* 12: 389-392.

Johnson, H.W.; Robinson, H.F. and Comstock, R.E. 1955. Genotypic correlations in soybean and their implications in selection. *Agron. J.*, 47: 477-483.

Monpara, B.A. 2009. Relationship of durum wheat yield to agronomical and physiological growth parameters. *Int. J. Agric. Sci.*, 5: 399-402.

Panse, V.G. 1957. Genetics of quantitative traits in relation to plant breeding. *Indian J. Genet.*, 17: 318-328.

Rangare, N.R.; Krupakar, A.; Kumar, A. and Singh, S. 2010. Character association and component analysis in wheat (*Triticumaestivum* L.). *Electronic J. Pl. Breed.*, 1: 231-238.

Sharma, P.K.; Kumar, P. and Kumar, N. 2004. Character association analysis in different height groups of wheat (*Triticumspp*). *Prog. Agric.*, 4: 167-171.

Sharma, V.; Pawar, I.S. and Munjal, R. 2006. Variability parameters, correlation and path coefficient for yield and its component and quality traits in bread wheat. *Natl J. Pl. Improv.*, 8: 153-155.

Sidharthan, B. and Malik, S.K. 2006. Association analysis in wheat. *Int. J. Agric. Sci.*, 2: 427-428.

Singh, A.K.; Singh, S.B.; Singh, A.P. and Sharma, A.K. 2012. Genetic variability, character association and path analysis for seed yield and its component characters in wheat (*Triticumaestivum* L.) under rainfed environment. *Indian J. Agric. Res.*, 46: 48-53.

Singh, B.J.; Verma, A.; Prakash, S.; Patidar, I.; Prakash, T.L.; Saiprasad, S.V.; Singh, A. and Mishra, A.N. 2013. Variability and interrelationship analysis in bread wheat under moisture stress conditions. *J. Wheat Res.,* 5: 27-34.

Singh, H.B.; Sharma, J.K. and Sethi, G.S. 1985. Correlation among different traits and their relative contribution to grain yield in wheat. *Himachal J. Agric. Res.*, 11: 86-91.

26

Selection Indices for Yield Improvement in Wheat (*Triticumaestivum* L.) in Normal and Limited Irrigation Condition under Late Sown Condition

Komal Patel, Raiyani G.D., Bhatiya V.J. Lalwani H.H., and Hadiya A.M.

Abstract

Sixty-three selection indices, involving grain yield per plant and five yield components, were constructed using the discriminant function technique. The efficiency of selection increased with the inclusion of more number of characters in the index. The selection index based on six characters viz., grain yield per plant, grain weight per main spike, number of grain per main spike, ear length, biological yield per plant and harvest index under normal condition and grain yield per plant, days to maturity, number of productive tillers per plant, number of grain per main spike, biological yield per plant and harvest index under limited irrigated condition exhibited maximum gain and relative efficiency. It is expected that grain yield could be improved if due consideration is given to these traits in future breeding programme of wheat.

Key words: *Wheat, discriminant function, relative efficiency, selection indices.*

Introduction

Wheat (*Triticumaestivum* L.) is the staple food for a large part of the world population including India. India accounts an area, production and a productivity of 29.9 million h, 93.9 million metric tonnes and 3140 kg/h, respectively (Anonymous, 2012). Yield is governed by a polygenic system and is highly influenced by the fluctuations in the environment. Hence, selection of plant based directly on yield would not be very reliable in many cases. The effectiveness of component approach to selection breeding is well appreciated.

An application of discriminant function developed by Smith (1936) helps to identify important combination of yield components useful for selection by formulating suitable selection indices. Therefore, the objective of the present study was to construct and assess the efficiency of selection indices in wheat.

Materials and Methods

The experimental material consisted of 40 diverse genotypes of wheat (*Triticumaestivum* L.) were sown at late under normal irrigation and limited irrigation condition in a Randomised Block Design with three replications during *Rabi* 2013-2014 at Wheat Research Station, Junagadh Agricultural University, Junagadh. Under limited irrigated condition, irrigation was skipped after anthesis growth stage. Each entry was accommodated in a single row of 2.0 m length with a spacing of22.5 cm. Five competitive plants per genotype in each replication were selected randomly and observations were recorded on different characters and their averages were used for statistical analysis except days to 50 per cent flowering and days to maturity while taken as plot basis. For constructing the selection indices, the characters which had highly significant with grain yield per plant were considered. In this context, the grain yield per plant (X_1) along with five components *viz.* grain weight per main spike (X_2), number of grain per main spike (X_3), ear length (X_4), biological yield per plant (X_5) and harvest index (X_6) under normal irrigated condition while the grain yield per plant (X_1) along with five components *viz.* days to maturity (X_2), number of productive tillers per plant (X_3), number of grain per main spike (X_4), biological yield per plant (X_5) and harvest index (X_6) under limited irrigated condition were identified and considered. Sixty-three selection indices were constructed in all possible combinations of the five yield contributing characters and grain yield per plant in both the environment. Their respective genetic advance was calculated and relative efficiency of different discriminant functions in relation to straight selection for grain yield was compared, assuming the efficiency of selection for seed yield as 100 per cent.

Results and Discussion

Selection indices for grain yield and other characters were constructed and examined to identify their relative efficiency in the selection of superior genotypes. The results on selection indices, discriminant function, expected genetic gain and relative efficiency are presented in Table 1 for normal and Table 2 for limited irrigated condition. Hazel and Lush (1943) showed that the selection based on such an index is more efficient than selecting individually for the various characters. The basis for the development of the selection indices has been provided by Smith (1936), Hazel (1943) and Robinson *et al.* (1951). Hazel and Lush (1943) stated that the superiority of selection based on index increases with an increase in the number of characters under selection and Mcvetty and Evans (1980) and Esheghi *et al.* (2011) also suggested that

Table 1. Selection index, discriminant function, expected genetic advance in grain yield and relative efficiency from the use of different selection indices in normal irrigated wheat under late shown condition

Sl. No.	*Selection Index advance*	*Discriminant function (per cent)*	*Expected genetic*	*Relative efficiency*
1	**2**	**3**	**4**	**5**
1.	X1 Grain yield per plant	0.742X1	2.21	100.00
2.	X2 Grain weight per main spike	0.945X2	0.83	37.55
3.	X3 Number of grain per main spike	0.910X3	16.90	764.70
4.	X4 Ear length	0.788X4	2.25	101.81
5.	X5 Biological yield per plant	0.685X5	2.75	124.43
6.	X6 Harvest index	0.249X6	3.53	159.72
7.	X1.X2	0.616X1 + 1.793X2	3.02	136.79
8.	X1.X3	0.973X1 + 0.911X3	18.50	837.23
9.	X1.X4	0.815X1 + 0.896X4	4.10	185.56
10.	X1.X5	1.084X1 + 0.498X5	5.05	228.39
11.	X1.X6	1.644X1 + 0.152X6	6.14	277.84
12.	X2.X3	4.878X2 + 0.756X3	17.78	804.29
13.	X2.X4	1.410X2 + 0.713X4	2.91	131.80
14.	X2.X5	2.030X2 +0.577X5	3.56	161.09
15.	X2.X6	4.432X2 + 0.197X6	5.32	240.76
16.	X3.X4	0.897X3 +1.160X4	18.68	845.05

Cont...

1	2	3	4	5
17.	X3.X5	0.921X3 + 0.918X5	18.93	856.64
18.	X3.X6	1.020X3 + 0.174X6	19.69	890.80
19.	X4.X5	0.974X4 + 0.716X5	4.65	210.16
20.	X4.X6	1.223X4 + 0.250X6	5.19	234.80
21.	X5.X6	1.249X5 + 0.311X6	6.46	292.29
22.	X1.X2.X3	0.667X1 + 5.905X2 + 0.740X3	19.42	878.86
23.	X1.X2.X4	0.697X1 + 2.011X2 + 0.762X4	4.88	220.79
24.	X1.X2.X5	0.818X1 + 2.692X2 + 0.487X5	5.89	266.68
25.	X1.X2.X6	1.031X1 + 4.651X2 + 0.152X6	7.43	336.34
26.	X1.X3.X4	0.997X1 + 0.891X3 + 1.223X4	20.32	919.39
27.	X1.X3.X5	0.745X1 + 0.941X3 + 0.910X5	20.68	935.92
28.	X1.X3.X6	1.858X1 + 0.971X3 + 0.072X6	21.62	978.36
29.	X1.X4.X5	1.115X1 + 1.036X4 + 0.493X5	6.81	308.12
30.	X1.X4.X6	1.654X1 + 1.064X4 + 0.154X6	7.70	348.53
31.	X1.X5.X6	4.372X1 + -1.133X5 + -0.08X6	8.67	392.10
32.	X2.X3.X4	4.686X2 + 0.764X3 + 1.001X4	19.52	883.37
33.	X2.X3.X5	5.665X2 + 0.753X3 + 0.732X5	19.84	897.62
34.	X2.X3.X6	10.394X2 + 0.654X3 + 0.144X6	21.00	950.15
35.	X2.X4.X5	2.205X2 + 0.806X4 + 0.630X5	5.42	245.14
36.	X2.X4.X6	5.247X2 + 0.519X4 + 0.197X6	6.77	306.52
37.	X2.X5.X6	4.735X2 + 0.828X5 + 0.247X6	7.76	350.90
38.	X3.X4.X5	0.902X3 + 1.227X4 + 0.922X5	20.76	939.47
39.	X3.X4.X6	0.996X3 + 1.271X4 + 0.182X6	21.45	970.57

Cont...

1	2	3	4	5
40.	X3.X5.X6	0.978X3 + 1.359X5 + 0.252X6	21.97	994.08
41.	X4.X5.X6	1.123X4 + 1.243X5 + 0.314X6	8.11	366.82
42.	X1.X2.X3.X4	0.734X1 + 5.521X2 +0.745X3 + 1.063X4	21.20	959.32
43.	X1.X2.X3.X5	0.445X1 + 6.826X2 +0.738X3 + 0.853X5	21.65	979.39
44.	X1.X2.X3.X6	1.222X1 + 10.513X2 + 0.641X3 + 0.087X6	22.86	1034.17
45.	X1.X2.X4.X5	0.875X1 + 2.734X2 + 0.864X4+0.504X5	7.64	345.73
46.	X1.X2.X4.X6	1.137X1 + 5.272X2 + 0.527X4 + 0.146X6	8.95	404.79
47.	X1.X2.X5.X6	4.409X1 + 5.631X2 + -1.681X5 + -0.165X6	10.03	453.86
48.	X1.X3.X4.X5	0.808X1 + 0.912X3 + 1.328X4 + 0.866X5	22.54	1019.80
49.	X1.X3.X4.X6	1.875X1 + 0.953X3 + 1.188X4 + 0.078X6	23.40	1058.75
50.	X1.X3.X5.X6	7.420X1 + 1.007X3 + -3.106X5 + -0.468X6	23.90	1081.35
51.	X1.X4.X5.X6	4.132X1 + 1.280X4 + -1.020X5 + -0.057X6	10.32	466.97
52.	X2.X3.X4.X5	5.305X2 + 0.757X3 + 1.080X4 + 0.767X5	21.63	978.90
53.	X2.X3.X4.X6	10.339X2 + 0.667X3 + 0.862X4 + 0.148X6	22.69	1026.74
54.	X2.X3.X5.X6	10.534X2 + 0.648X3 + 0.935X5 + 0.200X6	23.20	1049.72
55.	X2.X4.X5.X6	5.227X2 + 0.595X4 + 0.900X5 + 0.252X6	9.32	421.62
56.	X3.X4.X5.X6	0.955X3 + 1.253X4 + 1.366X5 + 0.259X6	23.77	1075.61
57.	X1.X2.X3.X4.X5	0.525X1 + 6.368X2 + 0.740X3 + 1.141X4 + 0.838X6	23.45	1061.18
58.	X1.X2.X3.X4.X6	1.324X1 + 10.308X2 + 0.655X3 + 0.861X4 + 0.075X6	24.57	1111.60
59.	X1.X2.X3.X5.X6	6.649X1 + 11.516X2 + 0.643X3 + -3.062X5 + -0.445X6	25.18	1139.16
60.	X1.X2.X4.X5.X6	2.159X1 + 5.986X2 + 0.648X4 + -0.112X5 + 0.082X6	11.56	523.17
61.	X1.X3.X4.X5.X6	7.131X1 + 0.976X3 + 1.359X4 + -2.917X5 + -0.430X6	25.70	1163.03
62.	X2.X3.X4.X5.X6	10.265X2 + 0.658X3 + 0.935X4 + 0.987X5 + 0.208X6	24.93	1128.19
63.	X1.X2.X3.X4.X5.X6	6.347X1 + 11.187X2 + 0.649X3 + 1.008X4 + -2.814X5 + -0.405X6	26.91	1217.45

Table 2. Selection index, discriminant function, expected genetic advance in grain yield and relative efficiency from the use of different selection indices in limited irrigated wheat under late shown condition

Sl. No.	*Selection Index*	*Discriminant function*	*Expected genetic advance*	*Relative efficiency (per cent)*
1	**2**	**3**	**4**	**5**
1.	X1 Grain yield per plant	0.784X1	1.56	100.00
2.	X2 days to maturity	0.924X2	5.94	380.77
3.	X3 number of productive tillers per plant	0.911X3	1.07	68.59
4.	X4 Number of grain per main spike	0.912X4	15.37	985.26
5.	X5 Biological yield per plant	0.764X5	2.98	191.03
6.	X6 Harvest index	0.469X6	5.52	353.85
7.	X1.X2	0.814X1 + 0.938X2	6.65	426.50
8.	X1.X3	0.674X1 + 1.232X3	2.55	163.40
9.	X1.X4	0.881X1 + 0.915X4	16.16	1035.99
10.	X1.X5	1.071X1 + 0.651X5	4.53	290.07
11.	X1.X6	1.887X1 + 0.354X6	7.02	449.74
12.	X2.X3	0.924X2 + 1.047X3	6.31	404.41
13.	X2.X4	0.966X2 + 0.913X4	18.01	1154.30

Cont...

1	2	3	4	5
14.	X2.X5	0.959X2 + 0.773X5	7.58	486.14
15.	X2.X6	0.987X2 + 0.467X6	8.69	557.33
16.	X3.X4	0.864X3 + 0.912X4	15.42	988.38
17.	X3.X5	1.575X3 + 0.631X5	3.95	253.42
18.	X3.X6	2.602X3 + 0.433X6	6.55	419.89
19.	X4.X5	0.921X4 + 0.906X5	17.37	1113.44
20.	X4.X6	0.886X4 + 0.416X6	15.74	1009.00
21.	X5.X6	1.163X5 + 0.488X6	7.71	494.12
22.	X1.X2.X3	0.642X1 + 0.939X2 + 1.387X3	7.20	461.45
23.	X1.X2.X4	0.881X1 + 0.972X2 + 0.915X4	18.86	1209.01
24.	X1.X2.X5	1.046X1 + 0.972X2 + 0.666X5	8.70	557.88
25.	X1.X2.X6	1.952X1 + 0.920X2 + 0.348X6	10.05	644.49
26.	X1.X3.X4	0.891X1 + 0.938X3 + 0.914X4	16.29	1044.35
27.	X1.X3.X5	0.953X1 + 1.836X3 + 0.517X5	5.51	353.38
28.	X1.X3.X6	1.172X1 + 2.462X3 + 0.390X6	7.90	506.72
29.	X1.X4.X5	0.360X1 + 0.928X4 + 1.098X5	18.33	1175.18
30.	X1.X4.X6	2.811X1 + 0.805X4 + 0.212X6	17.16	1099.86

Cont...

1	2	3	4	5
31.	X1.X5.X6	7.146X1 + -1.493X5 +-0.033X6	9.26	593.55
32.	X2.X3.X4	0.970X2 + 0.947X3 + 0.912X4	18.14	1162.79
33.	X2.X3.X5	0.962X2 + 1.713X3 + 0.611X5	8.24	527.96
34.	X2.X3.X6	0.927X2 + 2.742X3 + 0.430X6	9.58	614.42
35.	X2.X4.X5	0.973X2 + 0.921X4 + 0.910X5	20.08	1287.02
36.	X2.X4.X6	1.060X2 + 0.878X4 + 0.411X6	18.60	1192.19
37.	X2.X5.X6	0.951X2 + 1.177X5 + 0.487X6	10.86	696.44
38.	X3.X4.X5	0.872X3 + 0.920X4 + 0.920X5	17.56	1125.70
39.	X3.X4.X6	2.700X3 + 0.885X4 + 0.377X6	16.15	1035.13
40.	X3.X5.X6	2.881X3 + 0.772X5 + 0.459X6	8.73	559.50
41.	X4.X5.X6	0.837X4 + 1.489X5 + 0.422X6	18.33	1175.03
42.	X1.X2.X3.X4	0.813X1 + 0.974X2 + 1.105X3 + 0.918X4	19.06	1221.97
43.	X1.X2.X3.X5	2.212X1 + 1.036X2 + 1.597X3 + -0.106X5	9.53	610.60
44.	X1.X2.X3.X6	1.165X1 + 0.922X2 + 2.616X3 + 0.387X6	10.87	696.61
45.	X1.X2.X4.X5	0.327X1 + 0.989X2 + 0.926X4 + 1.113X5	21.07	1350.62
46.	X1.X2.X4.X6	2.812X1 + 0.981X2 + 0.804X4 + 0.210X6	19.97	1280.35
47.	X1.X2.X5.X6	7.090X1 + 0.950X2 + -1.456X5 + -0.030X6	12.29	787.55

Cont...

1	2	3	4	5
48.	X1.X3.X4.X5	0.322X1 + 1.434X3 + 0.940X4 + 0.989X5	18.59	1191.52
49.	X1.X3.X4.X6	2.704X1 + 1.100X3 + 0.809X4 + 0.222X6	17.50	1121.57
50.	X1.X3.X5.X6	7.831X1 + 3.362X3 + -2.269X5 + -0.124X6	10.33	662.26
51.	X1.X4.X5.X6	8.072X1 + 0.831X4 + -1.546X5 + -0.173X6	19.63	1258.08
52.	X2.X3.X4.X5	0.976X2 + 1.034X3 + 0.923X4 + 0.881X5	20.33	1303.05
53.	X2.X3.X4.X6	0.996X2 + 2.763X3 + 0.883X4 + 0.373X6	19.04	1220.40
54.	X2.X3.X5.X6	0.951X2 + 3.033X3 + 0.756X5 + 0.456X6	11.77	754.76
55.	X2.X4.X5.X6	1.002X2 + 0.835X4 + 1.483X5 + 0.419X6	21.15	1355.94
56.	X3.X4.X5.X6	1.862X3 + 0.859X4 + 1.257X5 + 0.412X6	18.73	1200.55
57.	X1.X2.X3.X4.X5	0.260X1 + 0.986X2 + 1.606X3 + 0.943X4 + 0.973X6	21.38	1370.51
58.	X1.X2.X3.X4.X6	2.577X1 + 0.982X2 + 1.293X3 + 0.814X4 + 0.230X6	20.35	1304.66
59.	X1.X2.X3.X5.X6	7.832X1 + 0.947X2 + 3.514X3 + -2.284X5 + -0.128X6	13.26	850.11
60.	X1.X2.X4.X5.X6	5.266X1 + 1.006X2 + 0.834X4 + -0.386X5 + 0.059X6	22.43	1437.63
61.	X1.X3.X4.X5.X6	8.114X1 + 2.402X3 + 0.866X4 + -1.928X5 + -0.195X6	20.09	1287.58
62.	X2.X3.X4.X5.X6	0.994X2 + 2.027X3 + 0.861X4 + 1.214X5 + 0.408X6	21.58	1383.48
63.	X1.X2.X3.X4.X5.X6	7.993X1 + 0.986X2 + 2.572X3 + 0.869X4 + -1.919X5 + -0.190X6	22.92	1469.01

the selection index to be superior to direct selection in wheat. In the present study also the expected genetic advance and relative efficiency assessed for different indices increased considerably when selection was based on two or more characters. The maximum genetic advance (GA) and relative efficiency (RI) in single character discriminant function was 16.90 g and 764.70 per cent, in normal and 15.37 g and 985.26 per cent under limited irrigation, respectively for number of grain per main spike which however, increased upto 19.69 g and 890.80 per cent respectively in two character combinations under normal irrigated condition while increased upto 18.01 g and 1154.30 per cent respectively in two character combinations under limited irrigated condition and 21.97 g and 994.08 per cent respectively in three characters combination under normal condition but 20.08 g and 1287.02 per cent respectively in three character combination under limited irrigated condition. Thus, there was an increase in the genetic gain as well as relative efficiency with an increase in the character combinations.

In four character combinations, the highest genetic advance and relative efficiency were 23.90 g and 1081.35 per cent, respectively under normal irrigated condition while four character combinations, the highest genetic advance and relative efficiency were 21.15 g and 1355.94 per cent, respectively under limited irrigated condition. Whereas, the maximum genetic advance and relative efficiency in five character combination was 25.70 g and 1163.03 per cent respectively under normal irrigated condition while five character combinations were 22.43 g and 1437.63 per cent respectively under limited irrigated condition. Ferdous *et al*. (2010) and Kemelew Muhe (2011) were also with the same opinion that an increase in characters result in an increase in genetic gain and that the selection indices improve the efficiency than the straight selection for grain yield alone. Further, it was observed that the straight selection for grain yield was not that much rewarding (GA=2.21 g, RI=100 per cent) as it was through its component like grain weight per main spike (GA=0.83 g, RI=37.55 per cent), number of grain per main spike (GA=16.90 g, RI=764.70 per cent), ear length (GA=2.25 g, RI=101.81 per cent), biological yield per plant (GA=2.75 g, RI=124.43 per cent), harvest index (GA=3.53 g, RI=159.72 per cent) and or in their combinations under normal irrigation while, under limited irrigation, it was observed that the straight selection for yield was not that much rewarding (GA=1.56 g, RI=100 per cent) as it was through its component like days to maturity (GA=5.94 g, RI=380.77 per cent), number of productive tiller per plant (GA=1.07 g, RI=68.59 per cent), number of grain per main spike (GA=15.37 g, RI=985.26 per cent), biological yield per plant (GA=2.98 g, RI=191.03 per cent), harvest index (GA=5.52 g, RI=353.85 per cent) and or in their combinations.

The maximum efficiency in selection for grain yield was exhibited by a discriminant function involving grain yield per plant, grain weight per main spike, number of grain per main spike, ear length, biological yield per plant and harvest index which had a genetic advance and relative efficiency of 26.91 g and 1217.45 per cent, respectively followed by an index of five characters (grain yield per plant, number of grain per main spike, ear length, biological yield per plant and harvest index) with the 25.70 g genetic advance and 1163.03 per cent relative efficiency under normal irrigated condition. While under limited irrigated condition, the maximum efficiency in selection for grain yield was exhibited by a discriminant function involving grain yield per plant, days to maturity, number of productive tillers per plant, number of grain per main spike, biological yield per plant and harvest index which had a genetic advance and relative efficiency of 22.92 g and 1469.01 per cent, respectively followed by an index of five characters (grain yield per plant, days to maturity, number of grain per main spike, biological yield per plant and harvest index) with the 22.43 g genetics advance and 1437.63 per cent relative efficiency. High efficiency in selection based on grain yield per plant, ear length, number of grain per main spike, biological yield per plant and harvest index under normal irrigation condition while under limited irrigation condition, the high efficiency in selection based on grain yield per plant, days to maturity, number of grain per main spike, biological yield per spike and harvest index or in combination of all these five characters. Singh and Diwivedi (1999) suggested that number of effective tillers per plant, number of grains per spike, grain weight per spike, biological yield per plant and harvest index to be included in selection criteria for improvement of grain yield in wheat.

The present study showed consistent increase in the relative efficiency of the succeeding index with simultaneous inclusion of each character. Therefore, improvements of grain yield through these selection indices are suggested. However, in practice, the plant breeder might be interested in maximum gain with minimum number of characters. In such a case, selection index consisting of four traits *viz.*, grain yield per plant, number of grain per spike, biological yield per plant and harvest index at both the condition could be advantageously exploited in the wheat breeding programmes. The present study also revealed that the discriminant function method of making selection in plant appears to be the most useful than the straight selection for grain yield alone and hence, due weightage should be given to the important selection indices while making selection for grain yield advancement in wheat breeding programme.

References

Anonymous, 2012. Progress report of All India Co-ordinated Wheat and Barly Improvement Project. Directorate of Wheat Research, Karnal, India.

Esheghi, R.; Javid, O. and Samira, S. 2011. Genetic gain through selection indices in hulless barley. *Int. J. Agric. and Biol.*, 13: 191-197.

Ferdous, M.; Nath, U.K. and Islam, A. 2011. Genetic divergence and genetic gain in bread wheat through selection practices. *J. Bangladesh Agril. Univ.*, 9: 1-4.

Hazel, L.N. 1943. The genetic basis for constructing selection indices. *Genetics*, 28: 476-490.

Hazel, L.N. and Lush, J.L. 1943. The efficiency of three method of selection. *J. Hered.*, 33: 393-399.

Kemelew, M. 2011. Selection index in durum wheat *(Triticumturgidum var. durum)* variety development. *Academic J. of Pl. Sci.*, 4: 77-83.

McVetty, P.B. and Evans, L.E. 1980. Breeding Methodology in Wheat. II. Productivity, harvest index and height measured on F_2 spaced plants for yield selection in spring wheat. *Crop Sci.*, 20: 587-589.

Robinson, H.F.; Comstock, R.E. and Harvey, P.H. 1951. Genotypie and phenotypic correlation in corn and their implication in selection. *Agron. J.*, 43: 282-287.

Singh, S.P. and Diwivedi, V.K. 1999. Character association and path analysis in wheat (*Triticumaestivum* L.). *New Botanist*. 26: 135-140.

Smith, H. F. 1936. A discriminant function for plant selection. *Ann. Eugen.*, 7: 240-250.

27

Genetic Variability, Correlation and Path Analysis in Bread Wheat (*Triticumaestivum* L.) Genotypes under Limited Water for Timely Sown Condition

Hadiya A.M., Parnaliya, J.B., Raiyani G.D., Komal Patel and K.H. Dabhi

Abstract

A set of 40 genotype of wheat were used to estimate genetic variability, correlation coefficients and path coefficients for 12 components. A wide range of variation was observed for important yield components. The estimates of heritability were observed high for grain weight per main spike, ear length, plant height, number of grains per main spike, days to maturity, days to 50 per cent flowering, 100 grain weight, grain filling period, number of productive tillers per plant, grain yield per plant and biological yield per plant. Values of genetic advance expressed as percentage of mean were found high for number of productive tillers per plant, grain yield per plant and biological yield per plant. The estimates of genotypic correlation were higher than the corresponding phenotypic correlation. The grain yield per plant showed significant and positive correlation with biological yield per plant, number of productive tillers per plant, harvest index, number of grains per main spike, grain filling period, grain weight per main spike and days to maturity at genotypic and phenotypic levels. The path coefficient analysis revealed high positive direct effects of biological yield per plant, harvest index and grain weight per main spike towards the grain yield per plant. The most of the characters contributed indirectly to grain yield through biological yield per plant. Based on correlation and path analysis, biological yield per plant, number of productive tillers per plant and harvest index was identified as the most important components of grain yield per plant.

Key words: *Wheat, variability, correlation, path analysis.*

Introduction

Wheat is the second most important staple food next to rice, consumed by nearly 35 per cent of the world population and providing 20 per cent of the total food calories. It is the most widely cultivated food crop of the world. India accounts an area, production and a productivity of 29.9 million ha, 93.9 million metric tonnes and 3140 kg/ha, respectively (Anonymous, 2012). Grain yield is a complex trait and highly influenced by many genetic factors and environmental fluctuations. In plant breeding programme, direct selection for yield as such could be misleading. A successful selection depends upon the information on the genetic variability and association of morpho-agronomic traits with grain yield. Correlation studies along with path analysis provide a better understanding of the association of different characters with grain yield. Path coefficient analysis separates the direct effects from the indirect effects through other related characters by partitioning the correlation coefficient (Dixet and Dubey, 1984). The study of associations among various traits is useful to breeders in selecting genotypes possessing groups of desired traits. Hence, in this investigation, exotic as well as local genotypes were used and an attempt was made to generate information on inheritance, relationships of yield and its components and their implication in selection of better genotypes of wheat for the development or improvement of cultivars and germplasm as well.

Materials and Methods

The experimental material consisted of 40 diverse genotypes of wheat (*Triticumaestivum* L.) representing different geographic origins. The pure seeds of these genotypes were obtained from the Research Scientist (Wheat), Wheat Research Station, Junagadh Agricultural University, Junagadh. Forty genotypes of wheat were sown at normal time under limited irrigation (stopping irrigation 65 days after sowing) condition in a Randomised Block Design with three replications during *Rabi* 2013-2014. Each entry was accommodated in a single row of 2.0 m length with a spacing of 22.5 cm. The recommended agronomical practices and plant protection measures were followed for the successful raising of the crop. Five competitive plants per genotype in each replication were selected randomly and observations were recorded on different characters and their averages were used for statistical analysis except days to 50 per cent flowering and days to maturity while taken as plot basis. The phenotypic coefficient of variation (PCV) and genotypic coefficient of variation (GCV) was estimated as per the formula suggested by Burton (1952). Heritability in broad sense and genetic advance at five per cent selection intensity was calculated by using the formula suggested by Allard (1960). This data were also used to determine the genotypic and phenotypic correlation coefficients between all possible pairs of the characters from the genotype means through the method given by Al-Jibouri*et al.* (1958) and the path coefficient analysis as described by Dewey and Lu (1959).

Results and Discussion

The genotypes differed significantly for all the 12 characters included in the study indicated the presence of wide variation in the material evaluated. The results on genetic variability parameters are presented in Table 1. The values of phenotypic coefficient of variation were higher than genotypic coefficient of variation indicating the influence of environmental factors. The high value of GCV was observed for grain yield per plant followed by number of tillers per plant and biological yield per plant which was earlier reported by Singh *et al.* (2012) and Talebi and Fayyaz (2012). High magnitude of GCV indicated the presence of wide variation for the characters under study which allow further improvement by selection of the individual traits. High heritability (broad sense) values observed for days to 50 per cent flowering, grain weight per main spike, number of grains per main spike, days to maturity, ear length, plant height, grain filling period, 100 grain weight, number of productive tillers per plant, biological yield per plant and grain yield per plant. High magnitude of heritability has been reported for days to 50 per cent flowering (Singh *et al.*, 2012), grain weight per spike (Jagshorani, 1995), number of grains per spike (Jagshorani, 1995), days to maturity (Singh *et al.*, 2012), 100 grain weight (Jagshorani, 1995), number of productive tillers per plant, biological yield per plant and grain yield per plant (Jagshorani, 1995 and Singh *et al.*, 2012). The genetic advance expressed as percentage of mean was the high for grain yield per plant followed by number of productive tillers per plant and biological yield per plant which was earlier reported by Jagshorani (1995). In the present study, high estimates of heritability coupled with high genetic advance expressed as percentage of mean were observed for number of productive tillers per plant, grain yield per plant and biological yield per plant which may be attributed to the preponderance of additive gene action and possess high selective value and thus, selection pressure could profitably be applied on these characters for their rationale improvement (Panse, 1957).

The association analysis revealed that, in general, the values of genotypic correlation were higher than their phenotypic correlation (Table 2) indicating the inherent association among the traits. Similar were the finding of Iftikhar *et al.* (2012) and Singh *et al.* (2012). In the present investigation, grain yield per plant was found to be significantly and positively correlated with biological yield per plant, number of productive tillers per plant, harvest index, number of grains per spike, grain filling period, grain weight per main spike and days to

Table 1. Phenotypic and genotypic coefficients of variation, heritability and genetic advance for various characters in wheat under limited water condition

Characters	*Range of variation*	*Coefficient of range (per cent)*	*Mean*	*Phenotypic coefficient of variation (PCV) (per cent)*	*Genotypic coefficient of variation (GCV) (per cent)*	*Heritability in broad sense (per cent)*	*Genetic advance*	*Genetic advance expressed as percentage of mean*
Days to 50% flowering	46.00-69.00	20.00	56.07	12.44	12.29	97.63	14.03	25.02
Days to maturity	89.33-106.67	8.85	99.45	5.23	5.13	96.36	10.32	10.37
Grain Filling Period (Days)	31.33-51.67	24.51	43.38	10.64	10.25	92.82	8.83	20.34
Plant height (cm)	50.37-77.37	21.14	67.71	9.44	9.21	95.24	12.54	18.52
Number of productive tillers per plant	1.80-5.20	48.57	3.28	25.41	23.56	85.96	1.47	45.00
Ear length (cm)	5.99-10.70	28.22	8.46	15.02	14.72	95.98	2.51	29.71
Number of grains per main spike	23.33-65.53	47.49	46.65	18.57	18.27	96.73	17.26	37.01
Grain weight per main spike (g)	1.15-2.62	38.99	1.64	18.96	18.66	96.84	0.62	37.83
Grain yield per plant (g)	2.40-9.83	60.75	5.45	30.93	27.03	76.41	2.65	48.68
Biological yield per plant (g)	6.53-18.73	48.33	12.10	24.86	22.23	80.00	4.96	40.96
Harvest index (%)	31.28-52.59	25.41	44.72	13.99	11.49	67.45	8.69	19.44
100-grain weight (g)	2.46-5.07	34.66	3.59	15.85	15.21	92.04	1.08	30.06

Table 2. Genotypic (r_g) and phenotypic (r_p) correlation coefficients among 12 characters in wheat

Characters		Days to 50% flowering	Days to maturity	Grain Filling Period (Days)	Plant height (cm)	Number of productive tillers per plant	Ear length (cm)	Number of grains per main spike	Grain weight per main spike (g)	Biological yield per plant (g)	Harvest Index (per cent)	100-grain weight (g)
Grain yield per plant (g)	r_g	-0.1242	0.2295*	0.4558**	0.0130	0.8151**	-0.0443	0.4782**	0.3893**	0.8926**	0.6169**	0.0573
	r_p	-0.0906	0.2029*	0.3654**	0.0062	0.7446**	-0.0374	0.4318**	0.3250**	0.8771**	0.6325**	0.0650
Days to 50 per cent flowering	r_g		0.7645**	-0.6731**	0.4821**	0.0338	-0.0645	0.1531	-0.1423	0.0591	-0.4149**	-0.4834**
	r_p		0.7498**	-0.6671**	0.4675**	0.0355	-0.0582	0.1549	-0.1419	0.0617	-0.3192**	-0.4616**
Days to maturity	r_g			-0.0378	0.5404**	0.3260**	0.0410	0.3519**	-0.1422	0.4106**	-0.2633**	-0.6526**
	r_p			-0.0074	0.5181**	0.3047**	0.0428	0.3409**	-0.1461	0.3680**	-0.2143*	-0.6134**
Grain Filling Period (Days)	r_g				-0.1274	0.3216**	0.1470	0.1664	0.0574	0.3794**	0.3410**	0.0007
	r_p				-0.1232	0.2895**	0.1362	0.1498	0.0500	0.3210**	0.2411**	0.0071
Plant height (cm)	r_g					-0.0369	0.3436**	0.2384**	0.0497	0.1956*	-0.3431**	-0.3835**
	r_p					-0.0486	0.3326**	0.2336*	0.0506	0.1669	-0.2811**	-0.3589**
Number of productive tillers per plant	r_g						-0.2027*	0.2473**	0.0354	0.9381**	0.1709	-0.1469
	r_p						-0.1744	0.228*	0.0212	0.8601**	0.1717	-0.1253
Ear length (cm)	r_g							0.1643	0.4175**	-0.0123	-0.1265	-0.016
	r_p							0.1661	0.4015**	-0.0094	-0.0985	-0.0097
Number of grains per main spike	r_g								0.4444**	0.4232**	0.3220**	-0.5424**
	r_p								0.4284**	0.3731**	0.2985**	-0.5039**
Grain weight per main spike (g)	r_g									0.2073*	0.5309**	0.3623**
	r_p									0.1755	0.4227**	0.3433**
Biological yield per plant (g)	r_g										0.2153*	-0.1841*
	r_p										0.2045*	-0.1408
Harvest index (%)	r_g											0.4885**
	r_p											0.3943**

*, ** Significant at five per cent and one per cent levels, respectively

Table 3. Genotypic path coefficient analysis showing direct (diagonal) and indirect effects of different characters on grain yield per plant in wheat.

Characters	*Days to maturity*	*Grain Filling period (Days)*	*Number of productive tillers per plant*	*Number of grains per main spike*	*Grain weight per main spike (g)*	*Biological yield per plant (g)*	*Harvest index (%)*	*100 grain weight*	*Genotypic correlation with grain yield per plant*
Days to maturity	**-0.1712**	0.0065	-0.0558	-0.0602	0.0243	-0.0703	0.0451	0.1117	0.2295*
Grain Filling period (Days)	0.0043	**-0.1128**	-0.0363	-0.0188	-0.0065	-0.0428	-0.0385	-0.0001	0.4558**
Number of productive tillers per plant	-0.1595	-0.1573	**-0.4892**	-0.121	-0.0173	-0.4589	-0.0836	0.0719	0.8151**
Number of grains per main spike	0.2456	-0.1161	-0.1726	**-0.698**	-0.3101	-0.2954	-0.2247	0.3786	0.4782**
Grain weight per main spike (g)	-0.0317	0.0128	0.0079	0.099	**0.2227**	0.0462	0.1182	0.0807	0.3892**
Biological yield per plant (g)	0.5695	0.5263	1.3013	0.5871	0.2876	**1.3872**	0.2986	-0.2554	0.8926**
Harvest index (per cent)	-0.2293	0.2969	0.1488	0.2803	0.4623	0.1874	**0.8707**	0.4253	0.6169**
100 grain weight (g)	0.4929	-0.0005	0.111	0.4097	-0.2737	0.1391	-0.3689	**-0.7553**	0.0573

*, ** significant at five per cent and one per cent levels, respectively

Residual effect, R = 0.0665

maturity at both the genotypic and phenotypic levels. The positive genotypic association has been reported between grain yield per plant and biological yield per plant, number of productive tillers per plant, harvest index, number of grains per main spike by Singh *et al.* (2012) and Gelalcha and Hanchinal (2013). The grain yield per plant has positive correlation with plant height has been reported by Iftikhar *et al.* (2012). The 100 grain weight was positively correlated with grain yield per plant which confirms the earlier findings of Monpara (2009). The days to 50 per cent flowering showed positive and significant correlation with days to maturity and plant height and positive and significant genotypic and phenotypic association was found for days to maturity with biological yield per plant and number of grains per main spike which was conformed with earlier finding of Sharma *et al.* (2006). Number of productive tillers per plant was found to be positively correlated with biological yield per plant (Sharma *et al.*, 2006 and Munir *et al.*, 2007) and number of grains per spike (Munir *et al.*, 2007) at both genotypic and phenotypic levels.Number of grains per main spike exhibited significant and positive correlation with grain weight per main spike (Munir*et al.*, 2007 and Talebi and Fayyaz, 2012), biological yield per plant (Sharma *et al.*, 2006) and harvest index (Talebi and Fayyaz, 2012) at both genotypic and phenotypic levels.The present results on correlation coefficients thus, revealed that the biological yield per plant, grain weight per spike, number of grains per spike, number of productive tillers per plant, 100 grain weight and harvest index were the most important traits and may contribute considerably towards higher grain yield. The interrelationship among yield components would help in increasing the yield levels and therefore, more emphasis should be given to these components while selecting better types in wheat.

The results obtained for direct and indirect effects of different characters on grain yield are presented in Table 3. In the present study, path coefficient analysis revealed that the grain weight per main spike, biological yield per plant and harvest index exhibited high and positive direct effects on grain yield per plant. Thus, these characters turned-out to be the major components of grain yield. The characters like days to maturity, grain filling period, number of grains per spike exerted low and negative direct effects on grain yield per plant. Singh *et al.* (2012) and Gelalcha and Hanchinal (2013) reported high and positive direct effect of biological yield per plant and harvest index on grain yield per plant in wheat. The negative direct effect of days to maturity on grain yield has been reported earlier by Singh *et al.* (2012). The number of tillers per plant and 100 grain weight showed negative direct effect on grain yield as reported earlier by Singh *et al.* (2012). The number of grains per main spike also exhibited negative direct effect on grain yield per plant has as reported by Talebi and Fayyaz (2012). The direct effect of biological yield per plant on grain yield was high and positive. This character also contributed indirectly by exhibiting positive effect *via* days to maturity, grain filling period, number of productive tillers per plant, number of grains per spike, grain weight per main spike and harvest index. The character harvest index showed positive direct effect and also contributed indirectly by grain filling period, number of productive tillers per plant, number of grains per main spike, grain weight per main spike, biological yield per plant and 100 grain weight. The character days to maturity exhibited high negative direct effect but contributed indirectly by exerting positive indirect effect via 100 seed weight.

Considering the genetic variability parameters, interrelationships and path analysis for grain yield and its components, an ideal plant type in wheat will be the one with biological yield per plant, grain weight per spike, number of grains per spike, number of productive tillers per plant and harvest index accompanied with early flowering and maturity under limited irrigated condition. Therefore, more emphasis should be given to those components while making selection for higher yield in wheat.

References

Al-Jibouri, H.A.; Miller, P.A. and Robinson, H.F. 1958. Genotypic and environmental variances in upland cotton cross of interspecific origin. *Agron. J.*, 50: 633-635.

Allard, R.W. 1960. Principles of Plant Breeding. John Willey and Sons, New York.

Anonymous, 2012. Progress report of All India Co-ordinated Wheat and Barly Improvement Project. Directorate of Wheat Research, Karnal, India.

Burton, G.W.; and Dewane, E. H. 1952. Estimating heritability in tall Fescues (*Festucaallamidiaceae*) from replicated clonal material. *Agri. J.*, 45: 1476-1481.

Dewey, D.R. and Lu, K.H. 1959. A correlation and path coefficient analysis of component of crested wheat grass seed production. *Agron. J.*, 51: 515-518.

Dixet, P. and Dubey, D.K. 1984. Path analysis in lentil (*Lens culinaris* Med.). *Lens Newsletter,* 11: 15-17.

Gelalcha, S. and Hanchinal, R.R. 2013. Correlation and path analysis in yield and yield components in spring bread wheat *(Triticumaestivum* L.) genotypes under irrigated condition in Southern India. *African J. of Agril. Res.*, 8(24): 3186-3192.

Iftikhar, R.; Khaliq, I.; Ijaz, M. and Rashid, M.A.R. 2012. Association analysis of grain yield and its components in spring wheat (*Triticumaestivum* L.). *American-Eurasian J. Agric. and Environ. Sci.*, 12: 389-392.

Jagshorani. 1995. Estimation of variability parameters and path coefficients for some quantitative characters in hill wheat. *Madras Agric. J.*, 82: 441-444.

Monpara, B.A. 2009. Relationship of durum wheat yield to agronomical and physiological growth parameters. *Int. J. Agric. Sci.*, 5: 399-402.

Munir, D.; Chowdhry, M.A. and Malik, T.A. 2007. Correlation studies among yield and its components in bread wheat under drought conditions. *Intern. J. Agril. and Biol.*, 9(2): 287–290.

Panse, V.G. 1957. Genetics of quantitative traits in relation to plant breeding. *Indian J. Genet.*, 17: 318-328.

Sharma, V.; Pawar, I.S. and Munjal, R. 2006. Variability parameters, correlation and path coefficient for yield and its component and quality traits in bread wheat. *Natl J. Pl. Improv.*, 8: 153-155.

Singh, A.K.; Singh, S.B.; Singh, A.P. and Sharma, A.K. 2012. Genetic variability, character association and path analysis for seed yield and its component characters in wheat (*Triticumaestivum* L.) under rainfed environment. *Indian J. Agric. Res.*, 46: 48-53.

Talebi, R. and Fayyaz, F. 2012. Estimation of heritability and genetic parameters associated with agronomic traits of bread wheat (*Triticumaestivum* L.) under two constructing water regimes. *J. Applied Biol. Sci.*, 6 (3): 35-39.

28

Influence of Weather Parameters on Major Insect Pests of Mustard

H.N. Patel, M.R. Patel and A.V. Khanpara

Abstract

A study was carried out on the population dynamics of pest complex and predator-parasite of aphid on mustard in relation to weather parameters at College Farm, N.M. College of Agriculture, Navsari under south Gujarat, India during the year 2005-06. Incidence of aphid (Lipaphis erysimi Kalt.) started from November with a peak during January whereas mustard sawfly (Athalia lugens proxima Klug.) started form November and reached to a peak during December. The population of natural enemy, syrphid fly started from December and reached to a peak during January whereas Diaeretiella rapae M. Intosh started from December and reached to a peak at the time of harvest of crop.

Introduction

Among the different oilseed crops in country, mustard and rapeseed have assumed significance in the national economy by occupying second position in area and production next to groundnut. In India, oilseeds are grown in an area of 23 to 32 million hectares with total production of 21.16 million tons having productivity of 907 kg/ha (Sarkar *et al.*, 2003). Among the oilseeds, mustard, *Brassica juncea* (L.) is one of the important oilseed crops of India which occupies an area of 71.2 lac hectares with total production of 7.64 million tons and productivity of 1073 kg/ha during 2004-2005 (Anon., 2005). In India, major mustard and rapeseed producing states are Uttar Pradesh, Rajasthan, Madhya Pradesh, Harayana, Gujarat, Assam, West Bangal and Punjab. In Gujarat, it is cultivated in about 2.69 lac hectares, total production of 3.95 lac tons with productivity of 1469 kg/ha during 2003-2004 (Anon., 2003).

Mustard is a native of China. Mustard is also known as *Sarson, Toria* and *Taramira.* Mustard seed contains about 40 per cent oil. The seeds are also used in preparing pickles, flavouring vegetable and curries. Mustard oil is mainly used for cooking, frying and in pickles. Oil is used in preparing vegetable ghee, hair oil, medicines, soaps, lubricating oil and in tanning industries. The major constraints affecting yield in mustard are insect pests causing injury to different parts right from seedling to maturity stage of this crop. In India, about 24 insects have been reported on this crop up to 1970 at different stages of growth (Rai, 1976). However, with increase in cropping intensity and the changing cropping patterns under different agroclimatic condition, the insect pests and diseases complex of this crop has also changed a great deal. According to Bakhetia and Sekhon (1989), 38 insect pests are now known to be associated with different stages of mustard and rapeseed crops in India. In Gujarat, aphid, sawfly, leaf webber and painted bug are important (Kapadia *et al.*, 1980).

Materials and Methods

A variety of mustard (Varuna) was grown in 400 sq. m. The experimental area was kept free from insecticidal spray throughout the crop season in order to record the incidence of insect pests and predator-parasite. Population of major insect pests and predator-parasite was recorded at weekly interval from 30 randomly selected and tagged plants starting from one week after sowing till the harvest of the crop.

Data on maximum, minimum and average temperature and morning, evening and average relative humidity recorded at the meteorological observatory, College Farm, N.A.U., Navsari were used to study the effect of weather parameters on population of various insect pests *viz.*, aphid, mustard sawfly and painted bug. The simple correlation coefficient was worked out.

Results and Discussion

The results (Table-1) revealed that aphid, *Lipaphis erysimi* Kalt. population commenced from 4th week after sowing (WAS) *i.e.* the third week of November with 0.13 aphid index/plant. Further, the mean aphid index indicated that the activity of this pest increased steadily and reach to the peak level of 3.70 aphid index/plant at 12th week after sowing coinciding with the 2nd week of January, then aphid index gradually decreased at the time of harvesting of the crop.

The population and per cent damaged leaves of mustard sawfly, *Athalia lugens proxima* Klug. started from 2nd week after sowing *i.e.* the 1st week of November (0.50 larvae/plant and 3.69 per cent damaged leaves, respectively) and reached to a pick level (3.73 larvae/ plant and 30.67 per cent damaged leaves, respectively) during the 8th week after sowing coincided with 3rd week of December. Thereafter, the mustard sawfly larvae and per cent damaged leaves was gradually declined and reached to low level (0.26 larvae/plant and 2.89 per cent damaged leaves, respectively) at the time of harvesting.

Table 1. Population Dynamics of Insect Pests and Predator-Parasite of Aphid on Mustard During Winter, 2005-06.

Met. week	*Date*	*WAS*	*Insect-pests*			*Predator-parasite*	
			Aphid index /plant	*Mustard sawfly larval population/plant*	*Per cent damaged leaves by mustard sawfly*	*Syrphid fly larvae/plant*	*Per cent parasitisation of D. rapae*
43	28/10/2005	1	0.00	0.00	0.00	0.00	0.00
44	04/11/2005	2	0.00	0.50	3.69	0.00	0.00
45	11/11/2005	3	0.00	0.70	5.83	0.00	0.00
46	18/11/2005	4	0.13	1.26	10.59	0.00	0.00
47	25/11/2005	5	0.43	1.40	15.36	0.00	0.00
48	02/12/2005	6	1.35	2.33	24.33	0.00	0.00
49	09/12/2005	7	2.06	3.60	29.45	0.00	0.00
50	16/12/2005	8	2.66	3.73	30.67	0.00	0.00
51	23/12/2005	9	3.26	2.67	25.83	0.00	0.97
52	30/12/2005	10	3.36	2.53	20.03	0.47	2.16
1	06/01/2006	11	3.56	2.17	15.83	0.83	6.33
2	13/01/2006	12	3.70	1.76	9.35	1.71	19.43
3	20/01/2006	13	2.90	0.73	5.06	1.81	35.58
4	27/01/2006	14	1.70	0.26	2.89	2.17	49.56
5	03/01/2006	15	0.90	0.00	0.00	1.90	79.24

Met. Week : meteorological standard week, **WAS :** week after sowing.

Table 2. Correlation Matrix of The Relationship Between Weather Parameters and Population of Major Insect Pests of Mustard.

Sr. No.	*Weather parameter*	*Standard week*	*Aphid*	*Mustard sawfly*	*Syrphid fly*	*Diaeretiella rapae MC, Intosh.*
1.	Minimum Temperature°C	Preceding	-0.706*	-0.184	-0.566*	-0.442
		Same	-0.751*	-0.278	-0.454	-0.306
2.	Maximum Temperature°C	Preceding	-0.634*	-0.134	-0.657*	-0.581*
		Same	-0.730*	-0.375	-0.390	-0.162
3.	Average Temperature°C	Preceding	-0.736*	-0.179	-0.651*	-0.537*
		Same	-0.811*	-0.349	-0.466	-0.627
4.	Morning Relative Humidity (per cent)	Preceding	-0.022	-0.354	0.057	-0.137
		Same	0.253	-0.026	-0.064	-0.174
5.	Evening Relative Humidity (per cent)	Preceding	-0.327	-0.361	-0.186	-0.213
		Same	-0.084	-0.117	-0.143	-0.140
6.	Average Relative Humidity (per cent)	Preceding	-0.193	-0.392	-0.072	-0.192
		Same	0.107	-0.078	-0.114	-0.178
7.	Wind Velocity (km/hr)	Preceding	0.394	-0.050	0.475	0.420
		Same	0.708*	0.384	0.334	0.033
8.	Sunshine Hours	Preceding	-0.314	-0.472	0.164	0.289
		Same	-0.322	-0.557*	0.332	0.365

Significant at five per cent level ($r = \pm 0.512$).

The population of syrphid fly (0.47 larvae/plant) started from ten week after sowing (WAS) *i.e.* 4th week of December and reach to a peak level (2.17 larvae/plant) during 14th week after sowing coincided with 4th week of January. Thereafter, the syrphid fly population declined and reached to 1.90 larvae per plant at the time of harvest.

The population of *Diaeretiella rapae* M. Intosh was started from the 9th week after sowing *i.e.,* the 4th week of December (0.97 per cent/plant). The population increased slowly and reached the peak level at the time of harvest (79.24 per cent/plant).

The Correlation matrix of the relationship between weather parameters and population of major insect pests of must (Table-2) revealed that the correlation coefficient for relationship between aphid index and weather parameter. The results revealed that the aphid index had significant negative correlation with maximum temperature (r = – 0.634), minimum temperature (r = –0.706) and average temperature (r = – 0.736). The different weather parameters had no marked effect on population fluctuation of mustard sawfly. The syrphid fly population exhibited significant negative correlation with minimum temperature (r = – 0.566), maximum temperature (r = – 0.657) and average temperature (r = – 0.651). Significant negative correlation between maximum temperature (r = – 0.581) as well as average temperature (r = – 0.537) and population of *D. rapae*.

References

Anonymous (2003). "District wise area, production and yield per hectare of important food and non-food crops in Gujarat state". Krishi Bhavan, Paldi, Ahmedabad, pp. 10-15.

Anonymous (2005). "District wise area, production and yield per hectare of important food and non-food crops in Gujarat state". Krishi Bhavan, Paldi, Ahmedabad. pp. 12-15.

Bakhetia, D.R.C. and B.S. Sekhon (1989). Insect pests and their management in rapeseed mustard. *Journal of Oilseeds Research* 6 : 269-299.

Kapadia, M.N., R.K. Bharodia and V. J. Vora (1980). Biology and larval and post-larval development of *Athalia lugens proxima* Klug. on different host plants. *Gujarat Agricultural University Research Journal* 6 (1): 13-16.

Rai, B.K. (1976). "Pest of oilseed crops in India and their control". ICAR, New Delhi, pp. 121.

Sarkar, A., A.R. Singh, B.J. Gangwar, S.K. Sharma and R.L. Yadav (2003). "Technology transfer of oilseed based cropping system through frontline demonstration in India". PDCSR Technological Bull. No.: 7.

29

Simulating the Crop Yield Response to Climate

H.M. Bhatu, H.D. Rank, R.J. Patel, D.M. Paradava, J.H. Kelaiya, and Sarika S. Wandre

Abstract

The intensive and extensive research efforts are required to increase or even to maintain the agricultural production in the changing climate. The global food crisis and food security for the ever-increasing population pressure can be another concern for the agricultural development. In fact, the huge budgetary allocation towards field experimentation for agricultural research is very difficult for the developing countries like India. However, many research results are possible by model simulation without field experiment. In the present effort, two major crops of Saurashtra region of Gujarat, like wheat and summer groundnut crop yield could be simulated for various climatic inputs using the "Wageningen model" and "Agro ecological zone model". Also, the optimal period of sowing and optimal evapotranspiration were determined for both crops simulating the ET-yield behaviours. The optimal period of sowing for wheat crop was found in first week of November. The potential yield of summer groundnut is increasing with delay in sowing from first week of January to onwards. However, having the risk of early monsoon in first week of June, it was proposed to sow the groundnut during the third week of February. The yield simulated by "Agro ecological zone model" was found comparable to actual yield data observed in field. The data of wheat yield by Agroecological zone model and ET by Blaney-Criddle method could not be assembled in proper mathematical frame. However, the wheat yield by Wageningen model and ET could be found assembled in polynomial form. The optimal ET for maximum wheat yield was found as 37 cm. The mathematical relationships between summer groundnut yield and ET was found polynomial in nature. The optimal ET for maximum yield of summer groundnut crop was found as 44 cm.

Introduction

The climate change is witnessed by us since last many years through natural events occurring on the earth. The agriculture is one of the most important sectors which can be notably affected by the climate change. The climatic parameters have the essential and major role in crop yield production. The green house gases have caused global warming. The increase in atmospheric temperature can increase or decrease the yield depending on the genotype of the crops. The agriculture can help to consume the carbon dioxide of the atmosphere which is one of the pollution gasses. The evapotranspiration from the agriculture can decrease the atmospheric temperature. Therefore, the agriculture can play a great role in mitigating the global warming.

Therefore, the research is required to assess the effect of the climatic parameters on the crop yields and water requirements. However, due to limited budgetary allocation, India's investment on agricultural research is very less as compared to developed countries, which is 2.5 per cent of their agricultural income. According to one estimate, in India, every rupee spend on agricultural research, returns 25 rupees. Due to changing climate, there is a lot of scope in the agricultural research. Their results can be useful in enhancement of the agricultural productions. However, India is not spending the huge amount of the expenditure towards the research due to negligence of agricultural sector. However, by model simulation, many research results are possible without any field experiment. The yield of any crop depends largely on climatic conditions prevailing in the area. The effect of climatic parameters can be assembled in the mathematical frame and the simulation models can determine the optimal input parameters without any investment and other resources required for the field experiment. Also, in the field experiment, there are huge possibilities of error in the outputs due to difficulties in the control. So, in the present study, an attempt was made to assess the effects of the climatic parameters on the yield and ET (water requirements) of wheat and groundnut crop and the relationship between the ET and the yield of crops. The study was made for simulating the wheat and groundnut yield response to climatic inputs using "Wageningen" and "Agro ecological zone" models and determining the optimal date of sowing and optimal evapotranspiration for yield maximisation.

Materials and Methods

Study area Description

Location: The study and prediction of crop yield was done at college of agricultural engineering and technology, Gujarat Agricultural University,

Junagadh. It is located at 21.5°N latitude and 70.58° E longitude with an altitude of 60 m above mean sea level. Maximum and minimum temperature, relative humidity, sunshine hours and Pan evaporation records of the past 15 years (1985-2000) were obtained from the meteorological observatory, Sagdividi farm, Gujarat Agricultural University, Junagadh. Meteorological data was procured from meteorological department of Agricultural College, Junagadh.

Climate: The climate area is subtropical and semi-arid type with an average annual rainfall of 900 mm and average pan evaporation of 6.41 mm/day. May is the hottest month with mean weekly pan evaporation 10.95 cm and mean monthly temperature varying between 35°C to 45°C and January is the coldest month with mean monthly minimum temperature varying between 7°C and 10°C as observed from the 15 yrs data. Maximum yield (Y_m) can be calculated for different climatic conditions. The methods enable quantification of production potential of different areas and thereby identify the most suitable areas for production for a given crop. The complexity of the relationships between many parameters makes the derivation of the methods complicated. However, there use is not, provided the essential data are' available. Computation techniques are given for two selected methods.

Wageningen method: The so-called linear model is to determine the dry matter production adequately. To calculate 'experiment' yield (Y_{me}) the steps needed are: (a) Calculate gross dry matter production of a standard crop (Yo), (b) Apply correction for climate (E_{tm} / ($e_a - e_d$), (c) Apply correction for crop species (K). (d) Apply correction for temperature (c_t). (e) Apply correction for harvested part (c_h).

Calculation of gross dry matter production of standard crop (Y_0): To calculate Y_o in kg/ha/day for a given location, the method of De Wit (1965) was used. This method is based on the level of incoming active short wave radiation for standard conditions

$$Y_0 = F\,y_0 + (1-F)\,y_c \qquad ...(1)$$

where, Y_o = Gross dry matter production of a standard crop, kg/ha /day; F = Fraction of the day time the sky is clouded, or F = (R_{se} – 0.5 R_s) / 0.8 R_{se}. where R_{se} is the maximum active incoming short-wave radiation on clear days in cal/cm²/day (De Wit, 1965) and Rs is the actual measured incoming short wave radiation in cal/cm²/day. Y_o = Gross dry matter production rate of a standard crop for a given location on a completely overcast day, kg! ha/day (De Wit, 1965). yc = Gross dry matter production rate of a standard crop for a given location on a clear (cloudless) day, kg/ha/day (De Wit, 1965)

$$R_s = (0.25 + 0.50\,n/N)\,R_a \qquad ...(2)$$

When, R_a is the extra terrestrial radiation in mm/day (De Wit, 1965), *N* is the maximum possible sunshine duration in hours/day. n is the actual measured sunshine duration in hours/day.

Correction for effect of climate = ($Et_m/(e_a - e_d)$): In addition to radiation, the rates of crop growth for a given climate is closely related to the ratio mean maximum evapotranspiration (E_{tm}) in mm/day and the vapour pressure deficit ($e_a - e_d$) in mean daily mbar over, the total growing period.

Correction for crop species (K): To relate gross dry matter production of a standard crop (Y_o) to gross dry matter production of wheat (Y_o. $Et_m/(e_a - e_d)$, empirically derived crop constant (K) are used with a value of 0.65 for winter wheat.

Correction for temperature (c_t): The correction were made according to empirically derived values using mean temperature over the total growing period.

Correction for harvested part (c_h): In general, only a part of the total dry matter is harvested. The ratio between net total dry matter and harvested yield is given by the harvest index (c_h) for high-producing varieties under irrigation. For wheat it is taken between 0.3 to 0.4.

In summary, the production (Y_{me}, kg/ha/period) under experimental conditions of a high-producing, climatically adapted variety, grown under optimum climatic condition is:

$$Y_{me} = 0.65\, C_h C_t G Y_0 E_{tm}/(e_a - e_d) \qquad ...(3)$$

The Agro-ecological Zone Method

The methodology to calculate crop production was developed to suit the assessment on a continental basis. However, the method can also be applied to a degree of detail required to suit specific locations. For a given climate, the possible potential yield is calculated for a standard crop using the concept of De Wit (1965) using radiation data; for agricultural crops, corrections are made for the genetically controlled growth processes of the crop under the given climate. It is assumed that the climatic requirements of the crop are met and that water, nutrients, salinity, pests and diseases do not affect crop growth and potential yield (Y_{mp}).

Under actual farming conditions, yield losses will occur due to adverse climatic conditions over short periods, limited water and nutrient supply, and problematic farm operations including land preparation, weeding and harvesting. These constraints are complex and it is difficult to quantify their effect on yield. However, when compared to actual farmers' yields; the calculated potential yield (Y_{mp}) will give an indication of the efficiency in agricultural production.

Calculation of gross dry Matter Production of a standard crop (yo): From Wageningen method, Y_o in kg/ha/day is calculated for a given location applying the concept of De Wit (1965):

$$Y_0 = F y_0 + (1 - F) Y_c \qquad ...(4)$$

where: Y_0 = Gross dry matter production of a standard crop, kg/ha/day. F = fraction of the daytime the sky is clouded, fraction; or F =R_{se} – 0.5 R_s) / 0.8 R_{se} where R_{se} is the maximum active incoming short wave radiation on clear days in cal/cm²/day (Doorenbos and Kassam, 1979). and R_s is the actual measured incoming short wave radiation in cal/cm 2/day. Rs ~an also be calculated from measured sunshine duration data (n) in hours/day. y_0 = gross dry matter production rate of a standard crop for a given location on a completely overcast day, kg/ha/day (De Wit, 1965). y_c = gross dry matter production rate of a standard crop for a given location on a clear (cloudless) day, kg/ha/day (De Wit, 1965).

Correction for crop species and temperature: The gross dry matter production is crop species and temperature dependent. The production rate (y_m) can be larger or smaller than 20 kg/ha/hour as assumed for the standard crop. Production rates (y_m) in kg/ha/hour for group of crops are given in Doorenbosand Kassam (1979). Applying De Wit's concept, the value of y_0 and y_c in kg/ha/ day can be adjusted for different crop groups:

(a) When y_m > 20 kg/ha/hour :

$$Y_0 = F(0.8 + 0.01y_m)Y_0 + (1 - F)(0.5 + 0.025y_m)y_c \qquad \text{...(5a)}$$

(b) When y_m < 20 kg/ha/hour

$$Y_0 = F(0.25y_m)Y_0 + (1 - F)(0.5y_m)y_c \qquad \text{...(5b)}$$

Correction for Crop Development over time and leaf area (cL): In relation to the maximum growth rate during the middle of the total growing period, crop growth will be small at the start and the end of the growing period or the average rate over the growing period is about 50 per cent of the rate during the maximum growth. Also for the standard crop an active leaf area of five times the ground surface is assumed (LAI = 5). When leaf area is smaller a correction must be applied and when greater than five the effect is small as per Doorenbos and Kassam (1979).

Correction for net dry matter production (cN): To maintain dry matter production, energy is required by the plant for the within plant growth processes, (also called respiration). Only the remaining energy can be used to produce new growth, which is about 0.6 for cool (mean temp.< 20°C) and 0.5 for warm (mean temp.> 20°C) conditions, or c_N = 0.5-0.6.

Correction for harvested Part (c_H): In general, only a part of the total dry matter such as grain, sugar or oil is harvested. The ratio between net total dry matter and harvested yield is given by the harvest index (c_H) for high-producing varieties under irrigation (Doorenbosand Kassam, 1979). In summary, potential yield (Y_{mp}, in kg/ha/period) of a high-producing, climatically adapted variety grown under constraint-free conditions over a growing period of G days is,

(a) When $y_m > 20$ kg/ha/hour

$$Y_{mp} = C_L C_N C_H G[F(0.8+0.01y_m)y_0+(1-F)(0.05+0.025\,y_m)y_c] \quad ..(6a)$$

(b) When $y_m < 20$ kg/ha/hour

$$Y_{mp} = C_L C_N C_H G[F(0.8+0.01y_m)y_0+(1-F)(0.5\,y_m)y_c] \quad ...(6b)$$

where, c_L = correction for crop development and leaf area index (Doorenbosand Kassam, 1979). c_N = correction for dry matter production, 0.6 for cool and 0.5 for warm conditions. c_H = correction for harvest index (Doorenbos and Kassam, 1979). G = total growing period (days). F = fraction of the daytime the sky is clouded. y_m= maximum leaf gross dry matter production rate of a crop for a given climate, kg/ha//day. y_o= gross dry matter production of a standard crop for a given location on a completely overcast (clouded) day, kg/ha/day (Doorenbos and Kassam,1979). y_c= gross dry matter production rate of a standard crop for a given location on a clear (cloudless) day, kg/ha/day (Doorenbos and Kassam, 1979).

Maximum Evapotranspiration (ET_m)

Climate is one of the most important factors determining the crop water requirements needed for unrestricted optimum growth and yield. Crop water requirements are normally expressed by the rate of evapotranspiration (ET) in mm/day or mm/period. The level of ET has been shown to be related to the evaporative demand of the air. The evaporative demand can be expressed as the reference evapotraspiration (Et_o) which when calculated, predicts the effect of climate on the level of crop evapotranspiration Et_o represents the rate of evapotranspiration of an extended surface of an 8 to 15 cm tall green grass. cover, actively growing completely shading the ground and not short of water.

Empirically-determined crop coefficients (k_c) can be used to relate Et_o to maximum crop evapotranspiration (Et_m) when water supply fully meets the water requirements of the crop. The value of k_c varies with crop, development stage of the crop, and to some extent with wind speed and humidity. For most crops, the k_c value increases from a low value at time of crop emergence to a maximum value during the period when the crop reaches full development, and declines as the crop matures.

The various methods available to calculate ET are the Penman, Blaney-Criddle, radioactive, Menman-Monteninth and Pan evaporation methods. However, due to limitations of climatic data availability for Junagadh station, the most commonly used Blaney-Criddle method was adopted for predicting the ET for wheat and groundnut crop. Blaney and Criddle (1950) observed that, the amount of water consumptively used by crops during their growing season was closely co-relates with mean monthly temperature and daylight hours. The relation crop developed by Blaney and Criddle as below:

$$ET_m = \sum_{i=1}^{n} K_i\{[(4.6t + 81.3)P]/100\} \quad ...(7)$$

where, ET_m = evapotranspiration in cm during season; K_i = empirical consumptive use factor for the ith month of year period. t = mean monthly temperature, in °C, of ith month. P = i^{th} months daylight hours expresses as per cent of daylight hours of the year as per given in Doorenbosand Kassam (1979) for given location. i = An index for the month. The Blaney-Criddle formula has generally given sufficiently accurate estimates of seasonal consumptive use owing to the inclusion of locally developed crop coefficient factor (k). Basically, however, it assumes that the consumptive use of water is dependent only on temperature and day length, which is not fully true. Crop water requirements have been found to vary widely between climates having similar air temperature but different humidity and wind conditions. The consumptive use crop coefficient K will, therefore, vary not only with crop but also with climatic conditions.

Results and Discussion

Crop requires cardinal levels of environmental factors like air, soil, temperature, day length, humidity, quality and intensity of light etc. Cool weather during vegetative development and warm weather for maturity deemed ideal for wheat. However, the climatological parameters vary spatially and temporally. Too early or too delay in crop sowing is not congenial for wheat cultivars development. For attaining optimum yield, the date of sowing should be selected in such a way to match the climatic condition with crop needs.

The groundnut crop during *kharif* season is very common in this area. However, due to insect and pest problems with uncertainty in rainfall, the yield is very low. The groundnut crop yield is very high if sown in summer season. However, the water resources are very limited during the summer season. The date of sowing plays a very important role in water management planning. If it is sown in January, the length of growing period is high due to low temperature and less sunshine hours. If it is sown in later periods, the length of growing period is less but the water requirement is increased. Also there is a risk of early starting monsoon before harvesting can be completed. So there is a need of determining the optimal and suitable date of sowing for the summer groundnut crop.

Impact of Date of Sowing

Wageningen Method

Wageningen method was proposed for predicting the maximum yield of alfalfa, maize, sorghum and wheat. The mathematical relationship was given to convert dry matter production to yield of marketable product depending on water shortages during different crop growth periods. The maximum yield, for a given area, the reference yield level obtainable under a high standard of crop and water management where water and nutrient are not limited and pests and diseases are minimal.

The gross dry matter production of standard crop in kg/ha/day for Junagadh location was calculated (De Wit, 1965). It was based on the level of incoming active short wave radiation for the standard conditions. The corrections for effect of climate, crop species, temperature, and harvested part were made. The impact date of sowing on the maximum yield was assessed by this method. The date of sowing was considered from 1 Nov. to onwards at regular interval of one week. The considered dates of sowing were 1 Nov, 8 Nov, 15, Nov, 22 Nov and 29 Nov. Due to unavailability of climatic data in proper forms, the maximum yield for all the date of sowing was calculated over for the year 1995-96, 1996-97,1997-98, 1998-1999 and 1999-2000. The predicted data showed that the yield was lowest during 1995-96 and highest during 1997-98 for all the date of sowing. It indicates that the climatic conditions favourable for optimum crop growth like humidity, temperature and sunshine hours mere optimum during 1997-98. The optimal date of sowing during 95-96, 96-97, 97-98, 98-99 and 99-2000 was found as 8 Nov, 15 Nov, 29 Nov, 8 Nov and 8 Nov respectively. The averaged data showed that the optimal date of sowing for the wheat crop for this region is 8 Nov. The average of maximum yield for the five years were-2181.199, 2226.696, 2168.563, 2055.824, 2054.761 kg/ha respectively for the 1 Nov, 8 Nov, 15 Nov, 22 Nov and 29 Nov date of sowing. The effect of date of sowing on predicted yield by Wageningen method was depicted in Fig. 1.

Agro-Ecological Zone Method

The methodology to calculate crop production was developed to suit the assessment on a continental basis. However, the method can also be applied to a degree of detail required to suit specific locations. For a given climate, the possible potential yield is calculated for a standard crop using the concept of De Wit (1965) using radiation data; for agricultural crops, corrections are made for the genetically controlled growth process of the crop under the given climate. It was assumed that the climatic requirements of the crop are met and that water, nutrients, salinity, pests, and diseases do not affect crop growth and potential yield.

Under actual farming conditions, yield losses will occur due to adverse climatic conditions over short periods, limited water and nutrient supply, and problematic farm operations, including land preparation, weeding and harvesting. There constituents are complex and it is difficult to quantify their effect on yield. However, when compared to actual farmer's yields, the calculated potential yield gives the efficiency in agricultural productions.

This method was used to calculate the potential yield of wheat and summer groundnut crop for the different date of sowing the gross dry matter production of a standard crop was calculated and after that the corrections for the crop species, and temperature, crop development over time and leaf area, net dry matter production and nutrients part were applied.

(A) Wheat crop : The highest predicted potential yield was found as 4428.24 kg/ha, 4718.73 kg/ha, 4467.19 kg/ha, 4432.85 kg/ha and 4517.13 kg/ha respectively during the year 1995-96, 1996-97, 1997-98, 1998-99, and 1999-2000. During all the years, the highest potential yield was found for the date of sowing as 1 Nov. the optimal date of sowing was found as 8 Nov under the Wageningen method. The reason behind this result is difference in the considerations of different climatic parameters. In Agro-climatic zone method, the relative humidity and vapour pressure deficits are not considered while those are incorporated in the Wageningen method. The impact of date of sowing on the potential yield is depicted in Fig 2. It was found that the potential yield was decreased with delaying in the sowing after 1 Nov. so it can be judged that the cooler and humid weather may be favourable for the initial

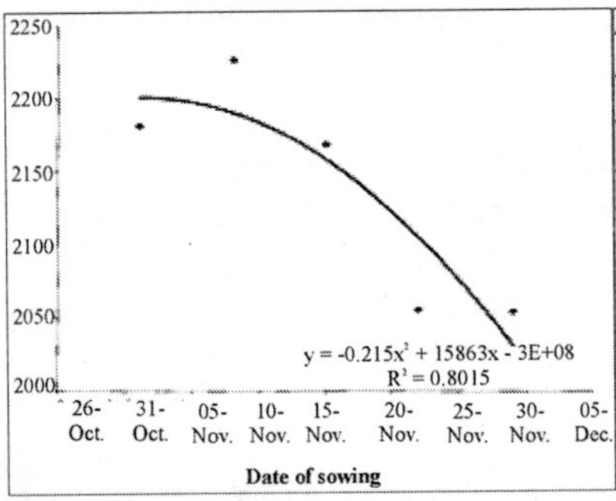

Fig1. Effects of date of sowing on wheat yield predicted by Wageningen method.

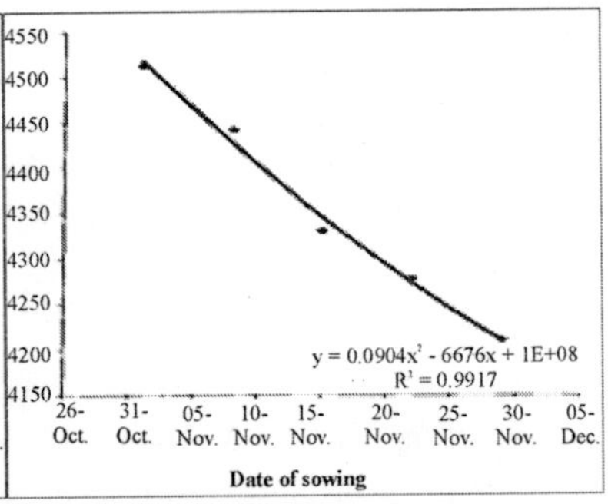

Fig 2. Effect of date of sowing on the wheat yield by AEZ model.

crop growth store of crop and warmer and dry weather during latter stages. Also, it can be said that if the wheat crop sown latter than 1 Nov, the temperature will be higher during initial stages and excessive during latter stages than optimally required. The average potential yield of wheat for the sowing on 1 Nov, 8 Nov, 15 Nov, 22 Nov and 29 Nov were found as 4512.827, 4443.182, 4331.991, 4278.719, 4211.137 kg/ha respectively.

(B) Summer Groundnut Crop: The potential yield for summer groundnut crop was predicted considerably the different date of sowing as 18 Jan, 25 Jan, 1 Feb, 8 Feb and 15 Feb. during all the years, the highest potential yield was observed for the date of sowing as 15 Feb. The potential yield was found to increase with delaying the date of sowing from 18 Jan. the potential yield might further increase if sown latter than 15 Feb. However, in case of sowing latter than 15 Feb, the harvesting of crop will be in the first week of June. However, in this region, there is a probability of 0.1 of starting of monsoon in the first half of the June. The highest potential yield predicted for the summer 1996, 1997, 1998, 1999 and 2000 seasons were found as 4805.53, 4591.73, 4781.91, 4921.32 and 4888.99 kg/ha respectively, on the early date of sowing as 15 Feb. The average of potential yield predicted for the summer 1996 to summer 2000 were found as 4465.61, 4532.143, 4637.928, 4722.066, 4797.895 kg/ha respectively for the date of sowing as 18 Jan, 25 Jan, 1 Feb, 8 Feb and 15 Feb.

Potential Yield and ET

Potential yield of a crop is defined as the harvested yield of a high producing variety, will adapted to the given growing environment, including the time a available to a reach maturity, under conditions where water, and nutrients and pests and diseases do not limit yield. Climatic factors, which determine potential yield, are temperature, radiation and length of total growing season in addition to any specific temperature, and day length requirement for crop development. Crop growth and yield are affected by the total radiation received during the growing period. At a given radiation and temperature, crop differ in the response to how much of the total radiation received can be converted into growth and yield. In addition to climatic requirements, consideration must be given to crop water requirements needed for unrestricted growth and yield. Crop water requirements are normally expressed by the rate of evapotransipiration (Et) in mm/day or cm/per day or cm per period.

(A) Wheat Crop: The highest potential yield predicted was 4918.66 kg/ha during the year 1986-87 with water requirements of 38.91 cm/growing season during the year 1986-87 yield was maximum it indicates that the climatic conditions were favourable for optimal growth. Consequently during 1987-88 second highest yield was obtained as 4785.22 kg/1m and water requirement was 40.67 cm/growing period. The lowest yield was obtained during the years 1999-2000. It was due to unfavourable climatic conditions. The potential yield obtained was 4243.316 kg/ha with water requirement of 37.34 cm/growing period. Generally as the yield increases water requirement also increases. The lowest water requirement was observed during the year 1989-90 with the yield 4523.53 kg/ha as the mean monthly temperature and sunshine were very favourable. Fig. 4 depicts the relationship between potential yield and water requirement for wheat crop.

(B) Groundnut Crop: As in wheat crop the highest yield was observed during the year 1986-87. Similarly same trend was noticed and the potential yield predicted in the year 1986 was 4694.95 kg/ha and the water requirement for this year was 39.7 cm/growing period. Second highest yield was obtained in the year 1999 as 4693.95 with water requirement of 42.52 cm/growing period. The maximum water consumption was observed to be 46.73 cm/growing period with yield of 4678.72 kg/ha (third highest). The water consumption was increased due to increase in mean monthly temperature. The lowest yield was

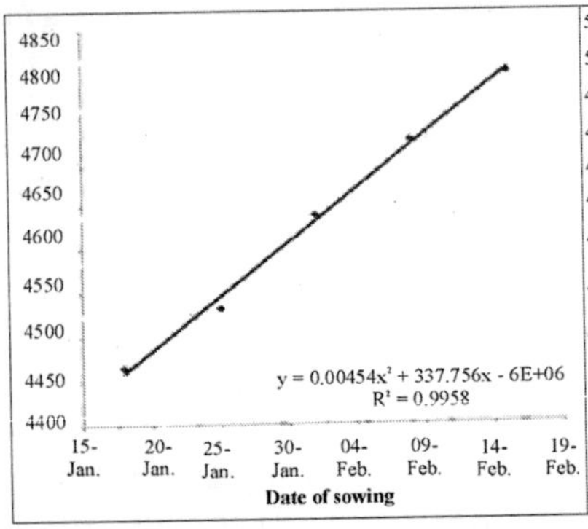

Fig 3. Effect of date of sowing on the groundnut yield by AEZ model.

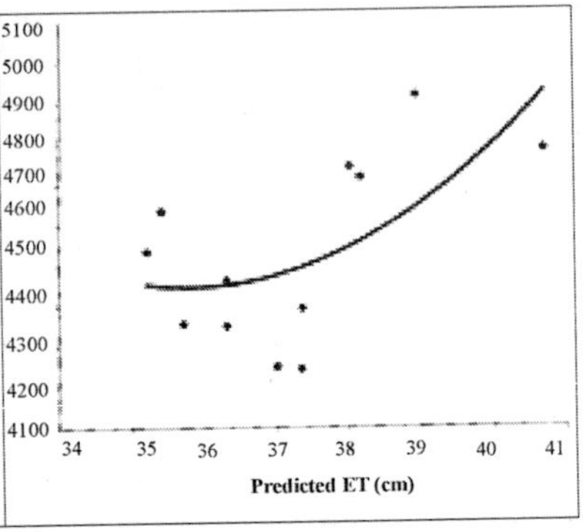

Fig 4. Relationship between predicted ET and yield of wheat predicted by AEZ model.

obtained during the year 1993 and a sharp decrease in the yield was noticed. The yield obtained during the above mentioned year was 4146.58 kg/ha with water requirement 38.1 cm/growing period. The relationship between predicted ET and groundnut yield by Agro Ecological Zone method was shown in Fig. 5.

(C) Wheat (Wageningen Method): The same trend was also observed under this method i.e. highest yield in the year 1986-87. The predicted yield was 2692.174 kg/1m. That clearly depicts that during 1986-87 the climate was most favourable. The water requirement for this duration was 38.91 cm/ growing period. The second highest yield was noticed in the year 1998-99 as 2408.035 kg/ha with water requirement of 36.3 cm/growing period. The lowest yield was observed during 1987-88 with highest ET. This is due to unfavourable climatic parameters. This method covers a widerange of climatic conditions. The possible production potential for given climate is calculated for a standard crop by concept of De Wit (1965), using radiation and evapotranspiration data for application to Agricultural crops. The corrections are required using crop dependent constants and expressions of the effect of temperature, growth

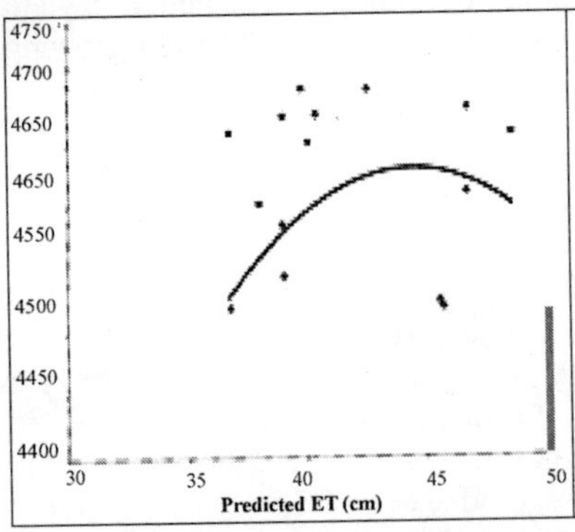

Fig 5. Relationships between predicted ET and groundnut yield AEZ model.

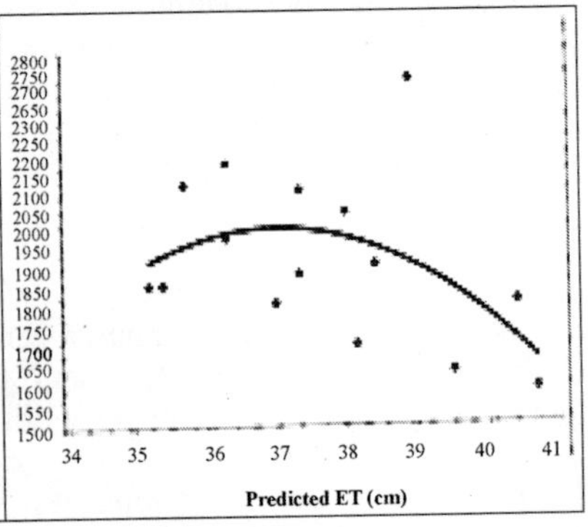

Fig 6. Relationships between predicted ET and yield of wheat by Wageningen.

efficiency and for the harvested part on final yield. So this method gives the actual yield which is obtainable in experimental plot provided the necessary climatic data are available. The relationship between predicted yield by Wageningen method and ET was shown in Fig. 6.

Conclusion

The analysis of the results so obtained leads to conclude the following conclusions.

(i) The optimal period of sowing for the wheat crop is the first week of November. (ii) The potential yield of summer groundnut crop is increasing with delay in sowing from first January. However, having the risk of early monsoon in first week of June, it is recommend the sowing during the third week of February. (iii) The average potential yield of the wheat crop predicted by the Wageningen and Agro ecological zone method is 2074.767 and 4401.994 kg/ha respectively.

References

De wit, C.T. 1965. Photosynthesis of leaf canopies. Agricultural Research Report. Wageningen, pp. 663- 1-57.

Doorenbos, J. and A.H. Kassam, 1979. Yield response to water. Irrigation and Drainage paper No. 33, FAO, Rome.

Doorenbos, J. and W.O. Pruitt 1977. Crop Water Requirements. Irrigation and Drainage Paper No. 24, FAO, Rome, Italy.

30

Effect of Weather Parameters on Occurrence of Jassid Infesting Blackgram

A.R. Rupavatiya, A.V. Khanpara and H.N. Patel

Abstract

The present study embodies population dynamics of jassid, Empoasca kerri Pruthi infesting blackgram. The investigations were carried out on population dynamics of sucking pest jassid of blackgram during summer 2014 at Junagadh Agricultural University, Junagadh. The population of jassid was started from 3rd WAS i.e., the second week of February (0.8 jassid/three leaves). The infestation of this pest reached the peak level (15.2 jassid/three leaves) at 10th WAS coinciding with first week of April. The sunshine hour (r=0.877), maximum temperature (r=0.919) and minimum temperature (r=0.833) were positively significantly correlated.

Introduction

Pulses play an important role in human nutrition, pre-dominantly in a vegetarian country like India, where pulses are major sources of protein. Among legumes, blackgram and green gram are two major pulses next to red gram in India. Blackgram (*Vigna mungo* L.) is one of the important pulses of India. In India, blackgram is grown mainly for grain purpose. In addition, it is widely used as nutritive fodder especially for milch animals. Blackgram is commonly grown in both the *kharif* and summer seasons, in India. India and Pakistan are the two prominent blackgram growing countries in Asian continent (Thakur, 1975). In Gujarat, it covers an area of about 1032 hectares with production of 720 MT and productivity of 698 kg/ha (Anon., 2012).

Among various constraints responsible for miserable low yield of such an important pulse crop, the loss due to insect pests is considered to be important. It is unfortunate fact that every year we lost 15-20 per cent pulses due to ravages of pest complex infesting pulses (Lal and Sachan, 1987). In blackgram, the avoidable loss in yield due to insect pests was recorded to be 34.7 per cent (Saxena, 1983). According to Pandey *et al.* (1991), yield of blackgram was reduced to about 2.75 q/ha i.e. 33.51 per cent due to insect pests.

In India, about 18 species of insect pests damage the blackgram (Singh and Singh, 1977). While, according to Nayar *et al.* (1976), the crop is attacked by more than 20 insect pest species. As many as 30 insect pests of different groups appear in succession at different stages of crop growth of blackgram (Dhuri and Singh, 1983). Among these, only a few of them are considered to be major pests of blackgram. The important insect pests attacking this crop are as jassid, thrips, whitefly, aphid etc. This insect pests cause considerable damage to various parts viz. buds, flowers, pods and seeds of blackgram resulting in poor quality of crop and loss in yield.

Materials and Methods

The population study of jassid was carried out on blackgram variety GU-1 during Summer 2014. The crop was grown in a plot size of 10.0 m × 10.0 m at the spacing of 45 cm × 10 cm. All other agronomical practices were followed as per the scientific recommendations. The crop under the experiment was kept free from the insecticides throughout the season. Ten quadrates, 1.0 m × 1.0 m, were made in crop area from which five plants were randomly selected and tagged.

The population of jassid nymphs were recorded early in the morning at weekly interval from the tagged five plants from appearance of the pests till harvest. Three trifoliate leaf (upper, middle and lower) of each plant were observed for recording the pest population. Mean pest population was worked out and the data were correlated with weather parameters.

Results and Discussion

The result presented in Table-1 showed that the population of jassid was started from 3rd WAS i.e., the second week of February (0.8 jassid/three leaves). The infestation of the pest reached the peak level (15.2 jassid/three leaves) at 10th WAS coinciding with first week of April. Thereafter, the infestation declined to (3.4 jassid/three leaves) at 15th WAS.

According to Borah (1995), jassid appeared in the first week after germination in summer greengram. Population continued to build up throughout the crop growth period and reached at peak (11.4 nymph/5 plants) during fourth week of April. Thus, the present findings are more or less in agreement with the result reported by earlier workers.

The correlation coefficient for incidence of jassid with weather parameters revealed that (Table-2) sunshine hour (r = 0.877), maximum temperature (r = 0.919) and minimum temperature (r = 0.833), positive significant correlation, while morning relative humidity (r = 0.278) had exhibited positive

Table 1: Population of Jassid on Blackgram During Summer 2014

Week after sowing (WAS)	*Month week*	*Meteorological*	*Jassid population/three leaves/plant*
3	February	7	0.8
4		8	1.6
5		9	3.6
6	March	10	5
7		11	6.6
8		12	9.8
9		13	13.2
10	April	14	15.2
11		15	11.6
12		16	12.2
13		17	7.2
14		18	5.4
15	May	19	3.4

Table 2: Correlation between weather parameters and population of jassid on blackgram during summer 2014

Sucking pests	*Temperature (°C)*		*Relative humidity (per cent)*		*Sunshine hours*
	Max.	*Min.*	*Morning*	*Evening*	
Jassid	0.919**	0.833**	0.278	–0.539	0.877**

non-significant correlation. Whereas, evening relative humidity (r=-0.539) had negative non-significant correlation with jassid population. In past, Poonia (2005) studied the impact of various abiotic factors on population buildup of *E. kerri* on mungbean crop. It was found that the pest showed positive correlation with maximum temperature, minimum temperature and morning relative humidity.

References

Anonymous (2012). Directorate of Agriculture, Gujarat State, Gandhinagar.

Thakur, C. (1975). Scientific crop production, Vol I. Metropolitan Book Co. Pvt. Ltd., New Delhi p. 637.

Saxena, H.P. (1983). Losses in blackgram due to insect pests (special issue on crop losses due to insect pests). *Indian Journal of Entomology.* 2: 294-297.

Pandey, S.N., R. Singh, V.K. Sharma and P.W. Kanwat (1991). Losses due to insect pests in some *Kharif* pulses. *Indian Journal of Entomology.* 53(4): 629-631.

Poonia, F.S. (2005). Population dynamics of *E. kerri*, a jassid pest on mungbean crop in arid regions of Rajasthan. *Arid legumes for sustainable agriculture and trade.* 1:179-180.

Singh, K.M. and R.N. Singh (1977). Succession of insect pests in greengram and blackgram under dry land conditions at Delhi, *Indian Journal of Entomology.* 39(4):365-370.

Nayar, K.K., T.N. Anonthakrishnan and B.V. David (1976). General and Applied Entomology, Tata McGraw Hill Pub. Ltd., New Delhi. p. 589.

Dhuri, A.V. and K.M. Singh (1983). Pest complex and succession of insect pests in blackgram *Vigna mungo* (L) Hepper. *Indian Journal of Entomology.* 45(4): 396-401.

Borah, R.K. (1995). Insect pest complex in summer greengram (*Vigna radiata* L.). *Annals of Agricultural Research.* 16(1), 91-92.

31

Effect of Weather Parameters on Incidence of Leaf Miner Infesting Soybean

B.A. Ghadage, A.V. Khanpara and B.V. Sureja

Abstract

The present study were carried out on population dynamics leaf miner, Aproaerema modicella (Deventer) on soybean [Glycine max (L.) Merrill] at Instructional Farm, College of Agriculture, Junagadh Agricultural University, Junagadh during kharif season of 2010. A study on population dynamics of leaf miner, A. modicella on soybean variety Gujarat Soybean-1 indicated that the incidence of leaf miner commenced from third week after sowing i.e., fourth week of July with 1.30 larvae/plant and remained active throughout the crop season and peaked with 8.56 larvae/plant during eighth week after sowing i.e., first week of September then after it was declined and observed up to 12th week after sowing i.e. first week of October. Weather parameters correlated with the pest population, the minimum temperature (r = 0.2893), rainfall (r = 0.3573), rainy days (r = 0.3583), morning relative humidity (r = 0.3953) and evening relative humidity (r = 0.4527) which exhibited non-significant positive correlation during kharif-2010. The weather parameters i.e., maximum temperature (r = - 0.4572) and bright sunshine hours (r = - 0.4246) were non-significant negative correlation with the population of leaf miner, A. modicella.

Introduction

Soybean [*Glycine max* (L.) Merrill.] is an important leguminous crop. It was cultivated for its oil since centuries and they named it as a "yellow jewel" which feeds China's entire population. Soybean was introduced in India in

1870-80 (Andole, 1984). It was introduced in Gujarat by ex-Baroda State as early as 1934. In India, area under soybean cultivation was 9.52 million hectares and productivity was 1040 kg per ha with total production of 9.90 million tonnes (Anonymous, 2008). In Gujarat, the area under soybean was 68.0 hectares and the productivity was 382 kg per ha with total production of 26.0 tonnes (Anonymous, 2007). The national production of soybean has not been able to meet its domestic requirement due to the poor average yield. Among the various factors responsible for the poor yield, non-replacement of old cultivars with new improved ones and insect pest damage are important (Anonymous, 1988). Such an important pulse crop suffers a lot due to the attack of number of insect-pests (Lal *et al.*, 1981). Out of the various factors attributed to the low yield, one of the main factors is the economic losses caused by insect-pests varied from 10 to 15 per cent (Bindra, 1968).

According to Bhattacharya and Rathore (1979), 60 to 100 species of insects are found to attack the soybean at various stages in Uttar Pradesh. Among the various insect pests infesting soybean, leaf miner, *Aproaerema modicella* (Deventer) is considered as major pest affecting the yield and quality of soybean in Gujarat State (Lathiya, 1985; Anonymous, 1989; Sojitra, 1990; Vyas, 1996 and Joshi, 2007). Lathiya (1985) noted 22.74 per cent yield loss in soybean due to leaf miner, *Aproaerema modicella* Deventer. Leaf miner causes 30-50 per cent losses in grain yield and 75 per cent loss in the leaflets (Shetgar and Thombre, 1984).

To avoid the yield losses caused by this destructive pest and to encourage the cultivation of soybean on large area, to increase the production and productivity of Gujarat as well as India, all our efforts are needed to tackle this major insect pest by knowing the peak period of infestation and relation with weather parameters under field conditions. In view of the importance of the soybean crop and seriousness of this pest, it becomes necessary to have comprehensive information on population dynamics of leaf miner, *A. modicella* infesting soybean under Junagadh conditions are needed.

Materials and Methods

In order to study the population dynamics of leaf miner (*A. modicella*) on soybean crop, variety G. Soybean-1 was sown at Instructional Farm, College of Agriculture, Junagadh Agricultural University, Junagadh during *kharif*-2010. The crop was grown in plot size of 27 m × 15 m keeping 45 cm × 15 cm spacing between row to row and plant to plant. All the agronomical practices were adopted as per the scientific recommendations. The crop under the experiment kept without spraying of insecticides throughout the season. The area divided into ten quadrates. Five plants selected randomly from each quadrate and tagged them for recording the observations. Method for recording observations are as under:

1. The observations were taken from entire plant which was randomly selected and tagged for the observation.
2. Observation of number of larvae of leaf miner started from appearance of the pest till maturity of the crop.
3. Larval counts recorded from five plants randomly selected in each quadrate at an interval of seven days.
4. Mean population of larvae of leaf miner *A. modicella* larvae per plant was worked out. The data thus obtained, were statistically analysed.

The effect of weather parameters viz., maximum temperature, minimum temperature, morning relative humidity, evening relative humidity, sunshine hours, rainfall and rainy days on population of leaf miner pest, the simple correlation coefficients were worked out. The weekly meteorological data were obtained from Meteorological Observatory, Instructional Farm, Junagadh Agricultural University, Junagadh.

Results and Discussion

Weekly observations were recorded on the larval population of *A. modicella* on soybean during *kharif*-2010 are presented in Table 1 and Fig. 1.

The larval population commenced from 31st standard week (last week of July) with 1.30 larvae per plant. Then the larval population increased from 2.54 to 7.34 larvae per plant during 32nd to 35th standard week (first week of August to last week of August). Then after, the pest population attains a peak level of 8.56 larvae per plant during 36th standard week (first week of September). Then, the larval population of *A. modicella* declined slowly up to 38th standard week (last week of September) and then drastically reduced from 39th standard week (last week of September) and disappeared during 41st standard week (second week of October) at the time of maturity of the crop.

According to Lathiya (1985), incidence of leaf miner population on soybean continued throughout the crop period. Behera *et al.* (1989) noticed that larval activity commenced on soybean in the third week of July. Chattopadhyay *et al.* (2002) stated that the peak activity of soybean leaf miner, *A. modicella* was found during July to September. Rana (2005) noted that incidence of leaf miner, *A. modicella* started from July and two peaks observed during August and September. Joshi and Patel (2010) stated that *A. modicella* incidence initiated in soybean from third week after sowing i.e. from first week of August and reached to peak level 6.9 larvae per plant during 10th week after sowing i.e. third week of September.

The correlation study between the incidence of soybean leaf miner, *A. modicella* and weather parameters was carried out during *kharif*-2010 and the results are presented in Table 2. None of the weather parameters was found significantly correlated with the pest population. The correlation coefficient values between population of leaf miner and minimum temperature ($r = 0.2893$),

Table 1: Population of leaf miner, *A. modicella* on soybean during *kharif*-2010

Standard Week	*Week after sowing*	*Mean no. of leaf miner larvae/plant*	*Temperature (°C)*		*Relative humidity (per cent)*		*Bright Sunshine (hr.)*	*Wind velocity (km/hr)*	*Rainfall (mm)*	*Rainy days*
			Maxi	*Mini.*	*Morning*	*Evening*				
29	1	0.00	31.3	25.3	94.4	84.7	0.8	4.0	53.3	6
30	2	0.00	29.2	24.7	96.1	82.3	1.3	7.2	113.1	6
31	3	1.30	29.6	24.5	96.4	88.9	1.2	5.4	198.4	6
32	4	2.54	29.6	24.8	94.9	85.7	1.1	5.7	80.6	5
33	5	5.28	30.8	25.3	91.0	77.9	1.5	4.6	61.8	4
34	6	4.58	31.4	25.2	91.6	76.6	3.0	5.3	15.1	3
35	7	7.34	30.1	24.2	94.6	92.4	0.8	4.3	244.6	7
36	8	8.56	29.4	24.7	95.9	88.7	0.4	5.7	117.2	5
37	9	7.64	29.5	24.4	95.0	80.6	0.1	5.6	72.2	5
38	10	5.98	33.4	23.0	87.7	58.1	8.1	3.2	4.8	1
39	11	2.44	34.1	22.7	84.9	63.3	7.9	3.3	46.2	1
40	12	1.58	36.5	22.3	80.1	36.9	9.1	3.2	0	0
41	13	0.00	34.3	22.3	87.6	53.7	7.8	4.2	0	0
42	14	0.00	35.4	23.8	85.4	49.9	7.6	4.2	0	0
43	15	0.00	34.6	22.7	72.3	40.1	8.2	3.5	0	0

Table 2. Correlation between weather parameters and larval population of leaf miner, *A. modicella* on soybean during *kharif*–2010

Pest	*Temperature °C*		*Relative humidity (per cent)*		*Bright Sunshine (hr.)*	*Wind velocity (km/hr)*	*Rainfall (mm)*	*Rainy days*
	Max.	Mini.	Morning	Evening				
Leaf miner	-0.4572	0.2893	0.3953	0.4527	-0.4246	0.1348	0.3573	0.3583

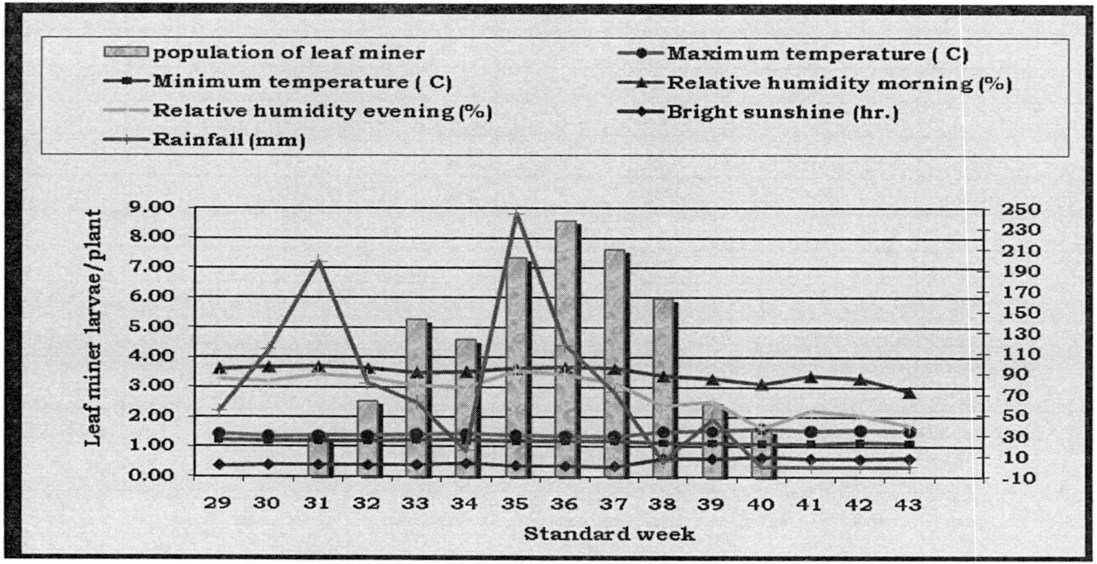

Fig. 1: Population of leaf miner, A. modicella *infesting soybean during kharif-2010.*

rainfall (r = 0.3573), rainy days (r = 0.3583), morning relative humidity (r = 0.3953) and evening relative humidity (r = 0.4527) exhibited non-significant positive correlation during *kharif*-2010. The maximum temperature and bright sunshine hour had non-significant negative relationship with A. modicella which recorded the correlation coefficients to the tune of – 0.4572 and – 0.4246, respectively.

Earlier, Vyas (1996) reported that the leaf miner population had highly significant negative correlation with maximum temperature and sunshine hours, where as it had highly significant positive correlation with morning and afternoon relative humidity. Rana (2005) observed that the negative non-significant correlation existed between leaf miner population and maximum temperature and sunshine hours, while positive non-significant correlation between pest population and minimum temperature, morning and evening relative humidity. Joshi and Patel (2010) stated that none of the weather parameters was found significantly correlated with the leaf miner population during *kharif*-2006.

References

Andole, V.C. (1984). Annual report, All India Coordinated Research Project on Soybean, MAU, Parbhani, pp. 58-61.

Anonymous (1988). Chairman's address, Project Co-ordinator's Report of AICRP on soybean (Indore, India). pp. 1-2.

Anonymous (1989). Annual Research Report (Oilseeds Entomology). Main Oilseeds Research Station, G.A.U., Junagadh, pp. 44.

Anonymous (2007). Agriculture, Centre for Monitoring Indian Economy 2010, pp. 199.

Anonymous (2008). Agriculture Statistics at a Glance 2009, pp. 121.

Behera, P.K., H.P. Patnaik and B. Senapati (1989). Leaf miner, *Bilobata* (*Stomopteryx*) *subsecivella* Zeller (Gelechiidae: Lepidoptera) incidence on soybean cultivars in Northern Orissa. *Orissa Journal of Agricultural Research.* 2 (3-4): 227-229.

Bhattacharya, A.K. and Y.S. Rathore (1979). Soybean insect problem in India. pp. 291-303. In Proceedings of World Soybean Research Conference -II (Ed Corbin, C.T.), March 26-29, 1979, North Carolina State University, Raleigh, U. S. A., [Fide: R. A. E. (A), 69 (9): 665].

Bindra, O.S. (1968). Insect pests of pulse crops. *Indian farming*, 17 (11): 12-15.

Chattopadhyay, N., R.P. Samui, S.N. Wadekar, A.R. Singh and N.G. Kumar (2002). Prediction of the incidence of soybean leaf miner based on weather parameters. *Indian Journal of Entomology.* 64 (3): 358-367.

Joshi, M.D. (2007). Seasonal incidence, varietal screening, yield losses and efficacy of eco-friendly pesticides against major pests of soybean. *M. Sc. (Agri.) Thesis,* J.A.U., Junagadh, pp. 95.

Joshi, M.D. and V. N. Patel (2010). Seasonal incidence of leaf miner *Aproaerema modicella* Deventer on soybean. *Agriculture Science Digest.* 30 (3): 212 – 214.

Lal, S.S., C.P. Yadav and C.A.R. Dias (1981). Insect pests of pulse crops and their management. Pesticide Annual, pp. 66-67.

Lathiya, B.T. (1985). Comparative bionomics and insecticidal control of leaf miner, *Aproaerema modicella* Deventer (Gelechiidae: Lepidoptera) infesting groundnut. *M. Sc. (Agri.). Thesis.* G.A.U., Sardar Krishinagar, pp. 70.

Rana, K.B. (2005). Population dynamics, varietal screening and bio-efficacy of newer insecticides against pest complex of soybean [Glycin max (L.) Merrill]. *M. Sc. (Agri.) Thesis,* N.A.U., Navsari, pp. 89.

Shetgar, S.S. and U.T. Thombre (1984). Occurrence of natural enemies on soybean leaf miner and relative susceptibility of some soybean varieties to its attack. *Journal of Maharashtra Agricultural University*. 9 (2): 218-219.

Sojitra, B.P. (1990). Population dynamics of major soybean pest's along with bionomics of grey semilooper, *Amyna octo* (Guenee) on different hosts and its chemical control. *M. Sc. (Agri.) Thesis,* G.A.U., Sardar Krishinagar, pp. 117.

Vyas, J.A. (1996). Population dynamics and chemical control of pest complex of soybean along with crop loss estimation. *M. Sc. (Agri.) Thesis,* G.A.U., Junagadh, pp. 104.

32

Effect of Weather Parameters on Incidence of Jassid Infesting Bt Cotton

D.K. Makwana, B.B. Kabaria and A.V. Khanpara

Abstract

The present investigation was carried out on seasonal incidence of jassid infesting Bt cotton during Kharif *2011-2012. The experiment on seasonal incidence of jassid infesting Bt cotton crop was conducted at Instructional Farm, JAU, Junagadh. The incidence of jassid on Bt cotton commenced during the second week of August and reached at a peak in the fourth week of September. The pest was found to be active from second week of August to fourth week of December. Jassid population had highly significant positive correlation with maximum temperature (r = 0.8447), morning relative humidity (r = 0.7123) and afternoon relative humidity (r = 0.3710), while non-significant negative correlation occurred with minimum temperature (r = -0.8802), rainy days (r = -0.5827).*

Introduction

Cotton crop is known as "White gold" and is a premier commercial crop in India. India is a unique among the cotton growing countries of the world in which four species of cotton viz, *Gossypium hirsutum, G. barbadense, G. arboreum* and *G. herbaceum* are grown commercially under diversified ecosystem. In India, it is cultivated in 110.00 million hectares with 325.00 lakh bale production and an average productivity was 503 kg/ha during 2010-11. India accounts for approximately 23 per cent of the world cotton area but the average productivity of cotton is markedly low at about 521 kg/ha as compared to world average of 765 kg/ha. Gujarat stands the second in respect of area 26.20 million ha and the first in respect of production 106.82 lakh bale with the productivity of 693 kg/ha during 2010-11 (Anonymous, 2011). The average productivity of cotton in India is lower than Brazil, China, USA and Pakistan in the world. The major factor for low productivity of cotton in India is the fluctuation in monsoon. In addition to cultivation on marginal lands and the complexities of pest management are also important. Among the different constraints that limit the production and productivity of cotton crop in India, the insect pests are considered to be the most serious.

In India, 160 species of insect pests have been reported to attack the cotton crop right from time of germination till the final harvesting of cotton (Agarwal, 1978). The total loss due to cotton pests (jassid and bollworm complex) in north, central and south zone was estimated as 52.80, 46.53 and 47.56 per cent, respectively (Basu *et. al.*, 1990). Dhawan *et al.* (1988) reported that jassid, *Amrasca biguttula biguttula* is one of the most serious sucking pests of cotton in India causing reduction in yield to an extent of 20 per cent.

Materials and Methods

The Bt hybrid cotton cultivar KDCHH-441 was grown with the plot size of 18 m × 18 m at spacing of 120 cm × 45 cm during *Kharif* season of the year 2011-12. All the recommended agronomical practices were followed to raise the crop. The plot was kept unsprayed throughout the crop seasons. Plot was divided in to ten quadrates. Five plants were selected randomly from each quadrate and observed at weekly intervals to record incidence of jassid from three leaves, one each from top, middle and bottom portions of cotton plant. Recording of the observations was started with the initiation of infestation of the jassid upto the completion of the crop season. Mean pest population was worked out and the data were analysed statistically for correlation study.

The effect of weather parameters viz., maximum temperature, minimum temperature, morning relative humidity, evening relative humidity, sunshine hours, rainfall and rainy days on population of jassid pest, the simple correlation coefficients were worked out. The weekly meteorological data were obtained from Meteorological Observatory, Instructional Farm, Junagadh Agricultural University, Junagadh.

Results and Discussion

The incidence of the pest started appearing during five week after sowing i.e. 2nd week of August with a low level of 0.8 jassid per three leaves per plant (Table 1). The pest population increased gradually and reached to a peak level of 16.4 jassid per three leaves per plant during 16th week after sowing i.e., fourth week of October. Later on, the population of the pest gradually declined. From the results on population of jassid on cotton, it can be seen that the pest was active on cotton from August and gradually increased after fifth week of sowing.

Patel and Rote (1995) stated that the jassid population on cotton ranged between 0.59 in August and 2.78 in first fortnight of October, then after it declined. The incidence of jassid started from 30th std. week and continued up to end of season. Severe infestation of jassid was observed from 38th to

Table 1. Seasonal Incidence of Jassid, *A. biguttula biguttula* on Bt cotton KDCHH-441 during *Kharif* - 2011-12

Weeks after sowing	*Standard Week*	*Date of Observation*	*Population of jassid/3 leaves/plant*
5	32	12/08/11	0.8
6	33	19/08/11	1.4
7	34	26/08/11	2.8
8	35	02/09/11	1.6
9	36	09/09/11	2.6
10	37	16/09/11	3.2
11	38	23/09/11	4.6
12	39	30/09/11	4.8
13	40	07/10/11	7.2
14	41	14/10/11	9.6
15	42	21/10/11	10.8
16	43	28/10/11	16.4
17	44	04/11/11	11.2
18	45	11/11/11	9.4
19	46	18/11/11	7.2
20	47	25/11/11	4.8
21	48	02/12/11	3.4
22	49	09/12/11	2.4
23	50	16/12/11	2.2
24	51	23/12/11	1.4
25	52	30/12/11	1.2

44th std. week with maximum population (10.8 jassid/3 leaves) during 43rd std. week on cotton (Anonymous, 2007). The peak population of jassid on cotton was reported in 43th std week (first fortnight of October) by Patel and Rote (1995), in the last week of July to mid September by Gupta *et. al*. (1998) and in 49th std week by Anonymous (2007). While in present study it was in 43th std week. The variation in time of peak population of jassid might be due to variation in location and weather condition.

The jassid population had exhibited highly significantly positive correlation with maximum temperature (r = 0.8447), morning relative humidity (r = 0.7123), evening relative humidity (r = 0.3710) (Table 2). Whereas, it was non-significantly negative between pest population and minimum temperature (r = -0.8802), rainfall (r = -0.5339), rainy days (r = -0.5827) and bright sunshine hours (r = -0.7953).

Table 2. Correlation co-efficient between population of jassid, *A. biguttulla biguttulla* of cotton and weather parameters during *Kharif* 2011-12

Sl. No.	*Name of sucking pests*	*Weather parameters*									
		Temperature (°C)			*Relative humidity (per cent)*			*Wind speed (Km/Hr)*	*Mean bright sunshine hours*	*Rainfall (mm)*	*Rainy days*
		Max.	Min.	Ave.	Morn.	Even.	Ave.				
1	Jassid	0.8447**	-0.8802	0.5342	0.7123	0.3710	0.9768	0.6245	-0.7953	-0.5339	-0.5827

*Significant at 0.05 level Jassid (n = 21) (r = 0.433)

**Significant at 0.01 level Jassid (n = 21) (r = 0.549)

Patel *et al.* (1997) reported significant relationship between jassid population level and maximum temperature (r = 0.76) as well as hours of bright sunshine (r = 0.82) on okra. Singh *et al.* (2002) reported that population of cowpea jassid positively correlated with maximum temperature and negatively correlated with relative humidity and minimum temperature.

References

Agarwal, R.A. (1978). Cotton insect pests and their control, *Richer harvest.* 1: 22-29.

Anonymous (2007). Population dynamics of key insect pests of cotton. Annual Report, AICCIP Group Meeting, during April 9-11, Ludhiana, pp.7-9.

Anonymous (2011). Crop wise area, production and productivity of Gujarat state. Directorate of Economics and Statistics, Dept. of Agriculture and Cooperation, pp. 28-29.

Basu, A.K., S. Jayaraj, S. Uthemasamy, M. Gopalan and R.J. Rabindra (1990). Action plan for containing *Heliothis* in cotton in Andhra Pradesh. Eds: Proc. National workshop on *Heliothis* management Feb.18-19, Coimbatore, pp. 297-300.

Dhawan, A.K., A.S. Sidhu and G.S. Simwat (1988). Assessment of avoidable loss in cotton (*Gossypium hirsutum* and *G. arboreum*) due to sucking pests and boll worms. *Indian Journal of Agricultural Science.* 58(4): 290-292.

Gupta, G.P., G.K. Mahapatro, S.K. Kundu and Roshan Lal (1998). Impact of abiotic factors on population dynamics of whitefly in cotton. *Indian Journal of Entomology.* 60(3): 293-296.

Patel, I.S. and N.B. Rote (1995). Seasonal incidence of sucking pest complex of cotton under rainfed condition of South Gujarat. *GAU Research Journal.* 21(1): 127-129.

Patel, K.I., J.R. Patel, D.B. Jayani, A.M. Shekh and N.C. Patel (1997). Effect of seasonal weather on incidence and development of major pests of okra (Abelmoschus esculentus). *Indian Journal of Agricultural Science.* 67(5): 181-183.

Singh, A.K., Santosh Kumar and V. Pandey (2002). Effect of meteorological parameters on the population build-up of sap feeders on cowpea. *Shashpa.* 9(2): 149-152.

Effect of Water Stress on Flowering, Fruit Set and Yield in *mrig bahar* Guava (*Psidium guajava* L.) under Saurashtra Climatic Condition cv. Bhavnagar Red

Tejendra Singh, D.K. Varu, Akula Venu and Ajay Karna

Abstract

Present investigation carried out in department of Horticulture, Junagadh Agricultural University, Junagadh. Effect of water stress on plant growth, flowering behaviour and yield was studied in 'Bhavnagar Red' guava. Stress was exerted on tree were withholding irrigation on three date an interval of 30 days commencing from 1st March and ending on 1st May and non-stress control tree were irrigated at an interval of 10-15 Days. The result found that withholding irrigation at 1st March (W_1) was observed for earliest flowering (131.83), Maximum number of flower (6.08) and fruit set per shoot (3.74) during kharif season. The reverse trend was found in summer season. Days tofruit set was non-significantly noted with time withholding irrigation. Highest fruit retention (60.08 per cent) and (72.08 per cent) were observed with regular irrigation (W_0) during summer and kharif season, respectively. Maximum fruit weight (121.75 g), fruit length (7.28 cm) and fruit width (7.18 cm) were observed with withholding irrigation at 1st March (W_1). Withholding irrigation at 1st March gave significantly highest number of fruit (623.33), yield kg per plant (87.83 kg) as well as ton per hectare (24.40 ton) in winter and total fruit yield While kharif *season, maximum number of fruit/plant (342.17) and fruit yield (36 kg/pl) were recorded in regular irrigation (W_0).*

Introduction

Guava (*Psidium guajava L.*) is one of most important subtropical and tropical fruit crops of the world and is a potential horticultural crop of India. It claims to be the fifth most important fruit crop in area and production after banana, mango, citrus and papaya in India. (Anon, 2013). Its fruit is called poor man's apple, though the fruit is neither poor its nutritive value nor in commercial value. Guava (*Psidium guajava* L.) belongs to the family of myrtaceae. Guava fruit is berry, edible portion of fruit is thalamus and pericarp.The guava is the native of tropical America from where it was introduced during 17th century in India by the Portuguese people. Guava was considered to be one of the most exquisite and nutritionally valuable remunerative crops. Guava fruits are used for both eating afresh and also for processing. As soon as, one got accustomed to its penetrating aroma, it became the most delicious and the most fascinating fruit for consumers and it also excelled most of the other fruit trees in respects of its productivity, hardiness, adaptability and its vitamin C content. Besides, its high nutritive value, it yields a heavy crop every year and it also gave handsome economic returns involving very little inputs. This had prompted several Indian farmers to take up guava cultivation on a commercial scale.

Winter season excellent quality in both the size and quality of guava fruit are found better than the other season. This necessitates for developing effective crops regulating technique in guava for manipulating winter season crop a major one. Several methods have been tried to induce new vegetative growth during rainy season so that bumper crop is obtained in subsequent winter season coordination of the fruiting cycle can help in maintaining fruit supplies during winter month.

Withholding irrigation is economic tools of crop regulation because not required skill work than plant growth regulator and chemical spray that's laborious and uneconomical. The chemical method for regulating flowering some time causes harmful effect on production, if the chemical not sprayed appropriate stage and recommended concentration (Dubey *et al.,* 2002). Thereby, growers failed to identify full bloom stage for foliar spray. Withholding irrigation easily effort by farmer and easily applicable. And reduce extra irrigation cost in summer season.

Guava can withstand prolonged period of drought (Shigeura and Bullock 1983). The metabolic activity of the plant was better maintained, even at low relative water content in guava. Withholding irrigation condition significantly decreases yield during rainy season (Singh *et al.,* 1997) best growth and earliest flowering, fruit maturity, largest fruit and highest yield in winter season. For obtaining winter crop (Cheema *et al.,* 1954) recommended withholding irrigation during summer month. The cropping pattern in guava (to reduce rainy season yield) can be regulated to some extent by means of withholding irrigation. Withholding irrigation is favourable for *bahar* regulation of guava. It produces a significant reduction in fruit set and yield during the rainy season with subsequent increase during the winter season.

Material and Method

The experiment was carried out on 18th year old guava tree of Bhavnagar red during 2013-14 at the Fruit Research Station, Madhadi Baugh, Department of Horticulture, College of Agriculture, Junagadh Agricultural University, Junagadh during 2013-14. The soil was slightly calcarious having medium fertility status. The experiment was a randomised block design with four levels of time of withholding irrigation i.e. regular irrigation (W_0), withholding irrigation at 1st March (W_1), withholding irrigation at 1st April (W_2) and withholding irrigation at 1st May (W_3). Each treatment was replicated three times in a Randomised Block Design. In each treatment, there were single tree. The irrigation was stopped as per treatment. Drip irrigation system was used for irrigation and quality of water was normal, having 6-7 pH. Similarly, trees were subjected to water stress by withholding irrigation during March to mid-June. At the mid-June the stressed trees were watered to field capacity with *mansoon* rainfall. The observation on days to flowering, number of flower per shoot, days to fruit set, number of fruit set, fruit retention and fruit yield. Days to flowering counted from 1st March to 50 per cent blooming in both season.

Result and Discussion

Flowering Fruiting Attribute

Withholding irrigation at 1st March observed earliest days to flowering (131.83). In summer season earliest flowering was observed with regular irrigation treatment (W_1) compare to withholding irrigation treatment. While, earliest flowering in *kharif* season (131.83) was recorded in withholding irrigation at 1st March (W_1) compare to regular irrigation (W_0). This fact was in accordance with the findings of Singh *et al.* (1993) in guava. In summer season no flowering was recorded in withholding irrigation at 1st March (W_1). Whereas, maximum flower per shoot (2.97) was observed in regular irrigation (W_0). Subsequently in *kharif* season significantly increases number of flower per shoot with withholding irrigation at 1st March (W_1) than regular irrigation treatment. These results clearly show that the cropping pattern of guava can easily be manipulated to obtain a desired flowering as well as a good flowering during *kharif.* Withholding irrigation at 1st March create water stress conditions that inhibit and distinct vegetative growth result in a shift in carbohydrate partitioning and a diversion of soluble assimilates to stem apices. The elevated carbohydrate status in bud, together with a floral stimulus, results in floral induction of crop phase. This fact was in accordance with the findings of Li-WenQing (2011) and Kumar *et. al* (2013).

Minimum number of fruit set per shoot *i.e.* not fruit set (0.00) were recorded in withholding irrigation at 1st March (W_1) and maximum in regular irrigation/control (W_0) during summer season, whereas, the reverse trend was

Table-1: Effect of time of withholding irrigation on flowering and fruit set during *kharif* and winter season

Treatments	*Plant height (m)*	*Plant Spread*		*Days to flowering*		*Number of flower/shoot*		*Number of fruit set/shoot*		*Fruit retention (per cent)*	
		EW	*NS*	*Summer*	*Kharif*	*Summer*	*Kharif*	*Summer*	*Kharif*	*Summer*	*Kharif*
W_0	5.07	5.87	5.68	60.67	151.67	2.97	2.37	1.93	1.75	60.08	72.08
W_1	5.28	6.48	5.98	0.00	131.83	0.00	6.08	0.00	3.74	0.00	55.00
W_2	5.39	6.18	6.20	0.00	136.33	0.73	5.74	0.45	3.33	9.42	59.67
W_3	5.56	6.54	6.16	63.67	141.83	2.89	2.94	1.93	2.09	50.28	69.68
S.Em.±	0.21	0.21	0.18	1.24	1.94	0.09	0.14	0.06	0.11	1.49	1.74
C.D. at 5 %	NS	NS	NS	3.58	5.59	0.26	0.40	0.16	0.32	4.30	5.03

Note: Days to flowering counted from 1st March.

Table 2: Effect of time of withholding irrigation and pruning on number of fruit and yield during *kharif* and winter season

Treatment	*Number of fruit/pl*			*Fruit yield (kg/pl)*			*Fruit yield (t/ha)*		
	Kharif	*Winter*	*Total*	*Kharif*	*Winter*	*Total*	*Kharif*	*Winter*	*Total*
W_0	342.17	209.58	551.75	32.11	21.76	53.87	8.92	6.04	14.96
W_1	0.00	623.33	623.33	0.00	87.83	87.83	0.00	24.40	24.40
W_2	39.83	579.42	619.25	4.38	79.35	83.73	1.22	22.04	23.26
W_3	327.67	216.50	544.17	36.08	26.06	62.14	10.02	7.24	17.26
S.Em.±	7.67	14.13	15.72	0.74	2.24	2.42	0.21	0.62	0.67
C.D. at 5 %	22.16	40.80	45.39	2.15	6.47	6.98	0.60	1.80	1.94

found in *kharif* season. Maximum fruit set per shoot (3.74) were noted in withholding irrigation at 1st March (W_1) and minimum fruit set per shoot (1.75) in regular irrigation (W_0). This might to be due to the fact that reduce fruit set in summer season due to withholding irrigation, facilitates a better balance of carbohydrates in the water stress tree enabling them to bear more number of fruits in the following *kharif*. These observations agreed with the finding of Singh *et al.* (1993).

The significantly maximum fruit retention during sumer (60.08 per cent) and *kharif* season (72.08 per cent) were observed in the regular irrigation/ control (W_0) treatment. The regular irrigation helped to stimulate fruit development in summer season than other treatment. But in case of *kharif* season, the maximum fruit retention was also observed under regular irrigation (W_0) because of less number of fruit set per shoot that decreases the competition for food source. Whereas in other treatment the number of fruit set per shoot was very high (W_1, W_2 and W_3) which enhance the competition among them for food source.

Yield Attribute

Significant maximum number of fruits per plant (623.33) was recorded in withholding irrigation at 1st March (W_1) during winter season. Subsequently, decreases number of fruit per plant (0.0) during *kharif* season i.e. no fruit per plant was recorded in withholding irrigation in 1st March (W_1). The similar trend was also observed in total number of fruit per plant. The improvement in number of fruit during winter due to stress may be attributed to less competition between vegetative organs during summer, leading to build up of carbohydrate reserve prior to flowering for winter crop. Hence more number of fruit retention and fruit set leads to more number of fruit. The result of present study are in conformity with those of Singh *et al.* (1997) in guava.

The yield per plant as well as ton per hectare was recorded fail to produce *kharif* season crop. Subsequently, increase higher yield per plant (87.83 kg) and per hectare (24.40 t/ha) in winter season. The similar trend of winter was observed for total fruit yield (87.83 kg) and per hectare (24.40 t/ha) was obtained with withholding irrigation at 1st March as compare to other treatment. When the water-deficient plant were receive the *Mansoon* rainfall (mid-June), there was induction of vegetative flush that enabled flower bud formation, resulting maximum yield during winter season with withholding irrigation treatment. The improvement in winter yield efficiency due to stress may be attributed to less competition between vegetative organs during summer, leading to build up of carbohydrate reserve prior to flowering for winter crop. The findings are found in consonance those of Singh *et al.* (1997) in guava.

Conclusions

From the foregoing discussion it can be concluded the withholding irrigation at 1st March were found to be effective for reduction the days to flowering, more number of flower, fruit set per shoot and highest fruit retention percentage. It was also found effective for yield and yield attribute number of fruit/tree as well as fruit yield. The treatment W_1 became effective to shifted the crop load from *kharif* to winter season in *mrig bahar*.

References

Anonymous, 2013. Indian horticulture database. National Horticulture Board, Department of Agriculture, India.

Dina, O.M. Ali, Abdelhalim R.A. and Elrakha B.B. 2014. Physicochemical and nutritional value of red and white guava cultivars grown in Sudan. *Journal of Agri-Food and Applied Sciences*, 2(2): 27-30.

Lal, S. 1992. Response of guava (Psidium guajava L.) cv. Sardar to spacing and pruning intensities. Unpublished Thesis, Ph.D. (Horticulture). G.B. Pant University of Agriculture and Technology, Pantnagar, 126.

Dubey, A.K.; Singh, D.B.; Barche, S.; Singh, A. and Dalal, M. 2002. Deblossming in summer season flowering in guava. *Indian Hort.*, 4: 35-36.

Shigeura, G.T. and Bullock, R.M. (1983). In: Hawaii – history and production. Uni. Hawwaii, College of Trop. Agric. and Human Resources Extension Series, 035.

Singh, Gorakh; Rajan, S.; Pandey, D. and Singh, A.K. 1997. Effect of soil moisture stress on water relation by plant and cropping behavior in guava (Psidium guajava L.). *Indian Journal of Agricultural Science*, 67 (7): 303-306.

Cheema, G.S.; Bhat, S.S. and Naik, K.C. 1954. Commercial fruits of India with special reference to Western India. *Macmillan & Co.* 253-254.

Singh, U.P.; Sharma, R.K. and Hoda, M.N. 1993. Regulation of cropping in guava cv. Allahabad Safeda. *Horticultural Journal*, 6(1): 1-5.

Kumar, M.; Ponnuswami, V.; Jeyakumar, P.; Richard K.R. and Saraswathy, S. 2013. Studies on influence of season for biochemical parameters in mango cultivars. *African Journal of Agricultural Research*, 8 (49) : 6394- 6400.

Li-WenQing Kang ShaoJie Zhang MingYue. 2011. Effects of drought on Citrus reticulate 'Shatangju' nutrition status and its relations with flowering. *Acta Horticulturae Sinica*, 38(11): 2085-2091.

34

Influence of Weather Parameters on Progress of Alternaria Leaf Spot of Cotton (*Gossypiumhirsutum* L.)

V.V. Rajani, M.G. Valu and M.D. Khanpara

Abstract

Alternaria leaf spot of cotton caused by Alternariamacrospora Zimm. is one of the major foliar disease occurring regularly. The pathogen infects the cotton leaves and spreads very fast under the most favourable climatic conditions. The meteorological data (average of five years) viz., maximum and minimum temperature, relative humidity (morning and evening), rainfall (mm) and rainy days were used for correlation studies. The correlation analysis indicated that intensity of Alternaria leaf spot was negatively correlated with all the weather parameters. The disease initiated in general 67-77 days after sowing (36 SW) and peak development stage of the disease (PDI 4.42) took place 145-155 days after sowing (47 SW) in the temperature range 16.3°C (minimum) to 32.5°C (maximum), relative humidity of 34.4 per cent (evening) to 70.2 per cent (morning) and rainfall of 1.6 mm within seven days prior to the day of peak development of disease. These parameters can be considered as critical parameters for maximum rate of increase of Alternaria leaf spot of cotton. Therefore, it is advisable to undertake plant protection measures against Alternaria leaf spot of cotton when weather factors prevail in aforesaid range in later growth stage (140-150 DAS or 47 SW) of the crop.

Introduction

Diseases under favourable weather conditions cause serious loss to cotton production. Among them, one of the most important foliar disease like Alternaria leaf spot caused by *Alternariamacrospora* Zimm. occurring regularly in Maharashtra (Rane and Patel, 1956), Gujarat, Karnataka and Andhra Pradesh. The pathogen infects the cotton leaves and spreads very fast under the most favourable climatic conditions. The climate is decided by abiotic factors like temperature, humidity, rainfall and rainy days, etc. Rao and Murti (1984) and Padaganur *et al.* (1989) reported 3.06 to 12.56 per cent loss in seed cotton yield of cv. Jaydhar due to Alternaria leaf spot. Whereas, the losses estimated by Chattanwar *et al.* (2002) were 16 to 40 per cent with 42 to 45 per cent disease intensity realising the role played by the meteorological factors in the disease development. The research work carried out on existing weather parameters in relation to disease development was directed only towards the relation of weather factors with the occurrence of the disease. If the intervention between disease and weather were properly understood, the outbreak of the disease can be forecasted well in advance and also fungicidal spray programme can be formulated. The present investigation was undertaken at Cotton Research Station, Junagadh Agricultural University, Junagadh to study the relation of different weather parameters on development of Alternaria leaf spot of cotton and to forecast the occurrence of disease and to suggest the control measures.

The experiment was conducted during 2009 to 2013 as non-replicated trial. The susceptible cotton variety G. Cot-10 was sown on an area of 40 x 30 m^2 at a distance of 120 x 45 cm^2 in irrigated medium black soil. The crop was sown in the month of June. All the recommended cultural practices were implemented. The crop was covered regularly for natural incidence and development of Alternaria leaf spot disease in the field. After appearance of disease, observations were recorded at weekly interval for progress of the disease on 25 randomly selected plants in the plot by adopting 0 – 4 grade scale. The average maximum disease intensity was 19.34 per cent. The meteorological data such as maximum and minimum temperature, relative humidity (morning and evening), rainfall (mm) and rainy days recorded at Instructional Farm, College of Agriculture, Junagadh Agricultural University, Junagadh was used for correlation studies. The relationship between meteorological and biological data was worked out by Karl Pearson's correlation coefficient and the incremental values of per cent disease intensity (average of five years) were used for correlation studies.

The analysed data (Table-1) regarding correlation analysis studies indicated that intensity of Alternaria leaf spot was negatively correlated with all the weather parameters. The significant negative correlation (-0.870) was observed between increase in alternaria leaf spot intensity and minimum temperature indicating its true relationship with disease development. The significant negative correlation also observed between increase of Alternaria leaf spot severity with morning humidity, evening humidity and rainfall. However, maximum temperature and rainy days had negative, but non-significant correlation. These results are in accordance with the findings of Perane *et al.* (2007).

Table 1: Correlation value for Alternaria leaf spot intensity in cotton with weather parameters under Junagadh condition (2009-2013)

Sr. No.	*Meteorological parameters*	*Disease correlation matrix ALS*
1.	Temperature (Maximum)	–0.366
2.	Temperature (Minimum)	–0.870
3.	Relative Humidity (Morning)	–0.719
4.	Relative Humidity (Evening)	–0.783
5.	Rainfall (mm)	–0.628
6.	Rainy days	–0.017

Significance level, if correlation r = > at 5 % = 0.433.

Table 2: Influence of weather parameters on occurrence and development of Alternaria leaf spot of cotton (average of years 2009, 2010, 2011, 2012 and 2013)

Std. week	*Period*	*Temp. (°C)*		*RH (%)*		*Rainfall (mm)*	*Rainy Days*	*Av. PDI*	*PDI Incremental value*
		Max.	*Min.*	*Morn.*	*Even.*				
36	3.9-9.9	31.2	24.9	92.1	73.9	59.4	3	0.30	0.22
37	10.9-16.9	31.5	24.6	89.9	71.0	35.9	2.8	0.74	0.44
38	17.9-23.9	33.2	24.1	85.3	58.3	10.1	1.4	1.16	0.42
39	24.9-30.9	32.8	23.9	85.9	60.5	52.1	1	1.64	0.48
40	1.10-7.10	35.1	24.3	78.6	49.9	18.0	0.8	1.98	0.34
41	8.10-14.10	34.8	23.7	78.2	45.2	0.0	0	2.40	0.42
42	15.10-21.10	36.2	22.8	76.7	35.0	0.0	0	3.42	1.02
43	22.10-28.10	35.9	19.8	68.2	28.6	0.0	0	4.34	0.92
44	29.10-4.11	35.8	19.4	62.6	27.4	0.0	0	4.84	0.50
45	5.11-11.11	35.0	19.7	65.9	33.1	1.3	0.4	6.76	1.92
46	12.11-18.11	33.0	18.6	69.9	36.5	0.3	0.2	9.58	2.82
47	19.11-25.11	32.5	16.3	70.2	34.4	1.6	0.4	14.00	4.42
48	26.11-2.12	32.1	17.1	66.2	31.2	0.0	0	17.34	3.34
49	3.12-9.12	32.2	15.7	71.5	30.4	0.0	0	17.38	0.04
50	10.12-16.12	30.4	13.5	70.4	27.3	0.0	0	18.54	1.16
51	17.12-23.12	30.7	13.4	70.8	28.3	0.0	0	19.34	0.80

Alternaria leaf spot disease occurred at in general 67-77 days after sowing (36 SW) when maximum and minimum temperature, morning and evening relative humidity, rainfall and rainy days were 31.2°C, 24.9°C, 92.1 per cent, 73.9 per cent, 59.4 mm and 3.0 days, respectively. This exhibited that the disease appeared in the temperature range 24.9°C to 31.2°C, relative humidity range 73.9 per cent to 92.1 per cent with moderate rainfall 59.4 mm (Table-2). The results are in agreement with the report of Chattanwar *et al.* (2002) and

Perane *et al.* (2007). The peak development of the Alternaria leaf spot (PDI 4.42) took place 145-155 days after sowing (47 SW) in the temperature range16.3°C (minimum) to 32.5°C (maximum), relative humidity of 34.4 per cent (evening) to 70.2 per cent (morning) and rainfall of 1.6 mm within seven days prior to the day of peak development of disease. Hence, these can be considered as critical parameters for maximum rate of increase of Alternaria leaf spot of cotton. The foregoing results are in accordance with the findings of Bhaskaran and Shanmuganan (1973). Analysed results indicated that none of the meteorological parameter can independently influence the Alternaria leaf spot intensity. Hence, the results of multiple regression analysis could throw more light in establishing the influence of various weather factors on Alternaria leaf spot severity in cotton.

Conclusion

Foregoing results clearly indicated that low rains (1.6 mm) coupled with moderate temperature (16 to 32°C) and moderate to high humidity (34 to 70 per cent) were highly favourable for maximum Alternaria leaf spot development. Therefore, it is advisable to undertake plant protection measures against Alternaria leaf spot of cotton when weather factors prevail in aforesaid range in later growth stage (140-150 DAS or 47 SW) of the crop.

References

Bhaskaran, R. and Shanmugan, P. (1973). *Madras Agric. J.* 60: 646-647.

Chattanwar, S.N., Hegde, Prakash, Hiremath, S.V., Khadi, B.M. and Narula, A. M. (2002). *J. Cotton Res. and Dev.* 16: 233-234.

Govind Rao and Dakshina Murti V. (1984). *Andhra. Agric. J.* 11: 272-275.

Padaganur, G.M., Hiremath, R.V. and Basavaraj, M.K. (1989). *ISCI Journal*, September 1989, pp. 144-145.

Perane, R.R., Pawar, A.P., Gaikwad, A.P. and Navale, P.A. (2007). Pl. Dis. Res. 22(1): 84-86.

Rane, M.S. and Patel, M.K. (1956). Indian Phytopath. 9:106-113.

35

Impact of Weather Factor on Incidence of Inflorescence Pests of Mango

J.B. Bhut and D.M. Jethva

Abstract

*The experiment on impact of weather factor on incidence of inflorescence pests of mango was carried out at Department of Entomology, Junagadh Agricultural University, Junagadh during 2013. The incidence of all the pests was found higher during initiation of new flowering January to March. The hopper population found more or less during the year except second fortnight of April to second fortnight of July. The maximum activities of flower bug was found at first fortnight of February, while thrips per leaf and thrips per inflorescence were found maximum at second fortnight of February. The correlation analysis showed that morning relative humidity had negatively correlated with hopper population and it showed highly negative correlation with flower bug, thrips per leaf and thrips per inflorescence. The minimum temperature and evening relative humidity had significant negative correlated with flower bug and thrips per inflorescence population, while it showed highly negative correlation with thrips per leaf populations (P= -0.7788** and -0.7579**). The bright sunshine hours and rainy days showed significantly positive (P= 0.6137*) and negative (P= -0.5455*) relationship with thrips per leaf population.*

Introduction

Mango (*Mangifera indica* Linnaeus) is national fruit of India and known as "King of fruits" due to its wide adaptability, attractive colour, excellent taste, exotic flavour, and exemplary nutritive value, richness in variety, attractive appearance and popularity among the masses. It is considered as a major fruit in India. This fruit tree suffers regularly a colossal loss due to ravages of pests, a serious threat to mango industry. The crop is attacked by about 492 species of insects, 17 species of mites and 26 species of nematodes in the world level. Of these, 188 species of insects have been reported from India (Tandon and Verghese, 1985). Among all pests, mango hopper *Amritodus atkinsoni* (Lethiery), thrips *Frankliniella schultzei* Bangnall and flower bug *Campylomma sp.* are abundantly observed damaging the mango inflorescences. Now-a-days the pest scenario has changed with changing in climatic condition hence study on impact of inflorescence pests of mango is very important.

Materials and Methods

Field experiments were conducted during 2013 Department of entomology, Junagadh agricultural University, Junagadh on impact of weather factor on incidence of inflorescence pests of mango. Ten mango trees of kesar variety of same age and growth were randomly selected to record inflorescence pest population. The activity of inflorescence pests recorded at fortnightly interval throughout the year starting from January to December 2013. The adult hopper population count was made by observing ten inflorescences randomly selected on each trees. Similarly the hopper population count was also made on ten sprouted twigs when available on each tree. For recording thrips population randomly selected ten twigs or inflorescence on each trees and counting number of thrips per leaves and thrips per inflorescence. Number of flower bugs also counted per inflorescences on each trees. The pests data was correlated with different weather parameters *i.e.* evaporation (EP), bright sunshine hours (BSS), maximum temperature (MaxT), minimum temperature (MinT), mean temperature (MT), Morning RH_1, Evening, RH_2 mean relative humidity (MRH), Evaporation (Eo), wind speed (WS) and rainfall (RF) and Rainy days through correlation analysis. The data on various weather parameters were collected from the Meteorological Department, Junagadh agricultural University, Junagadh.

Results and Discussion

Mango Hopper

The data (Table 1) indicated that the population of mango hoppers was recorded more or less throughout the year except second fortnight of April to July. It increased by first fortnight of January (1.75 hoppers/inflorescence) with initiation of flowering and reached at peak in second fortnight of February (28.56 hoppers/inflorescence) thereafter, it decreased gradually from the April to July. After these month, population of this pest raised but at low level with starting of new vegetative growth during August to October with maximum activities at first fortnight of September (5.40 hoppers/twig). The correlation study (Table 2) shows that morning relative humidity negatively (p = - 5998*) correlated with hopper population. Anithakumari *et al.,* 2009 reported that hoppers correlated negatively and significantly with morning relative humidity (r = -0.549, -0.581) and evening relative humidity (r = -0.658, -0.688) in two cultivars *viz.,* Baneshan and Dashehari. Kudagamage *et al.* (2001) reported that the population of hopper begins to increase in February with a peak in March–April. Talpur *et al.* (2002) from Pakistan noticed the peak populations of mango hopper on Almas (24.23/shoot) and Sindhri (25.66/shoot) varieties on 23 and 30 March, respectively. Jhala *et al.* (1989) reported that peak incidence of hoppers are seen during full bloom stage of the crop. Varshneya and Rana (2007) noted that maximum and minimum temperature positively affected the hopper population, whereas relative humidity had negative effect, but rainfall showed no significant effect, as it was fluctuating. This difference might be due to different location and variation in climatic condition.

Flower Bug

The population of flower bug (Table 1) was observed ranging from 1.85 to 6.00 bugs/inflorescence during first fortnight of January to second fortnight of March. The maximum population of bugs found during first fortnight of February (6.00 bugs/inflorescence). The minimum temperature and evening relative humidity was negatively correlated, while morning relative humidity was highly negatively correlated (p = -0.7717**) (Table 2) with flower bug population. Anon. (2012) reported that the higher population of flower bug was found during early February-12 (8.00 bug/inflo.).

Thrips/leaf

The thrips population was found between 0.23 to 5.40 per leaf during study period. The maximum population of thrips on leaf was observed at second fortnight of February (5.40 thrips/leaf). The correlation study shows that Minimum Temperature Morning relative humidity, Evening relative humidity was highly negatively correlated with pest population, while Bright Sunshine Hours was positively correlated (p = 0.6137*) with the thrips population.

Thrips/inflorescence

The thrips population found during December to March with maximum population in second fortnight of February (20.40 thrips/ inflorescence). The correlation study show that the minimum temperature and Evening relative humidity was negatively while, morning relative humidity was highly negatively (p = -0.7628**) correlated with thrips population. Aliakbarpour and Salmah (2010) reported the peak activities of thrips found two times first at December to February and second at August to September in new flower stage. Anon. (2012) reported that the activities of thrips was higher during it was late February (14.60 thrips/inflo.).

Table: 1 Population dynamics of inflorescence pests of mango

Month	*Fort night*	*No. of hopper per Inflo.*	*No. of Flower bug/inflo.*	*No. of Thrips/leaf*	*No, of Thrips/inflo.*
JAN	I	1.75	1.85	2.50	4.05
	II	5.20	3.65	1.60	8.35
FEB	I	25.95	6.00	2.60	19.50
	II	28.56	4.30	5.40	20.40
MARCH	I	10.25	3.50	3.38	13.75
	II	5.50	2.60	2.60	7.05
APRIL	I	4.80	0.00	1.50	0.00
	II	0.00	0.00	0.81	0.00
MAY	I	0.00	0.00	0.32	0.00
	II	0.00	0.00	0.23	0.00
JUNE	I	0.00	0.00	0.43	0.00
	II	0.00	0.00	0.02	0.00
JULY	I	0.00	0.00	0.00	0.00
	II	0.00	0.00	0.00	0.00
AUG	I	2.15	0.00	0.24	0.00
	II	3.80	0.00	0.00	0.00
SEP	I	5.40	0.00	0.00	0.00
	II	3.48	0.00	0.00	0.00
OCT	I	1.80	0.00	0.00	0.00
	II	0.50	0.00	0.00	0.00
NOV	I	0.90	0.00	2.65	0.00
	II	0.00	0.00	2.80	0.00
DEC	I	0.00	0.00	1.50	1.20
	II	1.20	0.00	2.90	1.45

Table: 2 Correlation co-efficient between weather parameters and inflorescence pests of mango

Weather parameters	No. of hopper per inflo. or twig	No. of Flower bug/inflo.	No. of Thrips/leaf	No, of Thrips/inflo.
Maximum Temp., (MaxT)°C	0.0064	0.0180	0.1184	0.0576
Minimum Temperature, (MinT) °C	-0.4030	-0.5959*	-0.7788**	-0.5642*
Mean Temperature, (MT) °C	-0.2912	-0.4280	-0.5096	-0.3824
Morning RH_1 (%)	-0.5998*	-0.7717**	-0.8141**	-0.7628**
Evening, RH_2 (%)	-0.4270	-0.5875*	-0.7579**	-0.5756*
Wind Speed, (WS) $kmhr^{-1}$	-0.0973	-0.0986	-0.2793	-0.0999
Bright Sunshine Hours, (BSS) $hrday^{-1}$	0.2674	0.3730	0.6137*	0.3794
Evaporation (Eo) mm	0.1824	0.2453	0.3663	0.2664
Rainfall, (RF) mm	-0.2350	-0.2899	-0.4378	-0.2825
Rainy days	-0.2596	-0.3491	-0.5455*	-0.3401

*significant at 5%, **significant at 1%

Conclusion

The population of inflorescence pests of mango found more at new flowering stage during December to February. The maximum activities of hopper, flower bug, thrips per leaf and inflorescence found at first and second fortnight of February. The correlation analysis showed that morning relative humidity had negatively correlated with hopper population and it showed highly negative correlation with flower bug, thrips per leaf and thrips per inflorescence. The minimum temperature and evening relative humidity had significant negative correlated with flower bug and thrips per inflorescence population, while it showed highly negative correlation with thrips per leaf populations. The bright sunshine hours showed significantly positive relationship with thrips per leaf population.

References

Aliakbarpour and Salmah (2010). Species composition and population dynamics of thrips (Thysanoptera) in mango orchards of northern peninsular Malaysia. *Environ Entomol.*, 39 (5):1409-19.

Anithakumari, D., Lakshmi, B.K.M., Reddy, S.G. and Reddy, L.M. (2009). Influence of abiotic factors on the incidence of hopper and chemical control strategies in mango. *Karnataka J. Agric. Sci.*, 22 (3): 601-602.

Anonymous (2012). Annual Report, Department of Entomology, JAU. Junagadh. p. 14.

Jhala, R.C., Shah, A.H., Patel, Z.P. and Patel, R.L. (1989). Studies on population dynamics of mango hopper and scope of off-seasonal spraying in integrated pest management programme. *Acta Hort.*, 231 (3): 597-601.

Kudagamage, C.Z., Rajapakse, H.L., Rajapakse, R.H., Ratnasekara, D.A., (2001). The population dynamics and insecticidal control of three leaf hoppers, *Amritodus brevistylus* Leth., *Idioscopus niveosparus* Leth. and *Idioscopus clypealis* Leth. (Homoptera: Cicadellidae) in mango. *J. Ento. Res.*, 25 (2): 121-125.

Talpur, M.A., Khuhro, R.D. and Nizamani, I.A. (2002). Phenological relationship between mango hoppers, *Idioscopus* spp. and mango inflorescence/fruit. *Pakistan J. Appl. Sci.*, 2 (5): 533-535.

Tandon, P.L. and Verghese, A. (1985). World list of insect, mite and other pests of mango. Technical Document, No. 5, IIHR, Bangalore.

Varshneya, A. and Rana, K.S. (2007). Effect of some abiotic factors on population buildup of *Idioscopus clypealis* (Lethierry) in western Uttar Pradesh. *J. Environ. Bio.*, 29 (5): 811-812.

36

Relationship of Meteorological Parameters on Progress of Bacterial Leaf Blight of Cotton (*Gossypium hirsutum* L.)

M.G. Valu, V.V. Rajani and M.D. Khanpara

Abstract

Bacterial leaf blight of cotton caused by Xanthomonasaxonopodis pv malvacearum is major foliar disease occurring regularly. The pathogen infects the cotton leaves and spreads very fast under the most favourable climatic conditions. Meteorological data (average of five years) viz., maximum and minimum temperature, relative humidity (morning and evening), rainfall (mm) and rainy days were used for correlation studies. The correlation analysis indicated that intensity of Bacterial leaf blight was negatively correlated with all the weather parameters except maximum temperature. The disease initiated in general 74-84 days after sowing (37 SW) and the peak development of Bacterial leaf blight (PDI 1.56) took place 130-140 days after sowing (45 SW) in the temperature range 19.7°C (minimum) to 35.0°C (maximum), relative humidity of 33.1 per cent (evening) to 65.9 per cent (morning) and rainfall of 1.3 mm within seven days prior to the day of peak development of disease. These parameters can be considered as critical parameters for maximum rate of increase of Bacterial leaf blight of cotton. Therefore, it is advisable to undertake plant protection measures against Bacterial leaf blight of cotton when weather factors prevail in aforesaid range in later growth stage (125-135 DAS or 45 SW) of the crop.

Introduction

Bacterial leaf blight of cotton caused by *Xanthomonasaxonopodis* pv *malvacearum*. The disease is under favourable condition cause serious loss to cotton production. The seed cotton yield losses vary with host cultivar, age, host nutrition and environmental factors. Weather factors are considered important in development and spread of this disease (Diekmann and Bogyo, 1992). Appearance of this disease is quite erratic and no relevant information on disease-weather relationship is available. The research work carried out on existing weather parameters in relation to disease development was directed only towards the relation of weather factors with the occurrence of the disease. If the intervention between disease and weather were properly understood, the outbreak of the disease can be forecasted well in advance and also fungicidal spray programme can be formulated. The present investigation was undertaken at Cotton Research Station, Junagadh Agricultural University, Junagadh to study the relation of different weather parameters on development of Bacterial leaf blight of cotton and to forecast the occurrence of disease and to suggest the control measures.

The experiment was conducted during 2009-2013 as non-replicated trial. The susceptible cotton variety G.Cot-10 was sown on an area of 45×35 m^2 at a distance of 120×45 cm^2 in irrigated medium black soil. The crop was sown in the month of June. All the recommended cultural practices were implemented. The crop was covered regularly for natural incidence and development of Bacterial leaf blight disease in the field. After appearance of disease, observations were recorded at weekly interval for progress of the disease on 25 randomly selected plants in the plot by adopting 0-4 grade scale. The average maximum disease intensity was 7.18 per cent. The meteorological data such as maximum and minimum temperature, relative humidity (morning and evening), rainfall (mm) and rainy days recorded at Department of Meteorology, College of Agriculture, Junagadh Agricultural University, Junagadh was used for correlation studies. The relationship between meteorological and biological data was worked out by Karl Pearson's correlation coefficient and the incremental values of percent disease intensity (average of five years) were used for correlation studies.

The analysed data (Table-1) regarding correlation analysis studies indicated that intensity of Bacterial leaf blight was negatively correlated with all the weather parameters except maximum temperature. The significant negative correlation (-0.672) was observed between increase in bacterial leaf blight intensity and relative humidity (evening) indicating its true relationship with disease development. The significant negative correlation also observed between increase of Bacterial leaf blight severity with morning humidity, minimum temperature and rainfall. However, maximum temperature had positive and rainy days had negative, but non-significant correlation. These results are in accordance with the findings of Venkateswarlu *et al.* (2006).

Bacterial leaf blight disease occurred at in general 74-84 days after sowing (37 SW) when maximum and minimum temperature, morning and evening relative humidity, rainfall and rainy days were 31.5°C, 24.6°C, 89.9 per cent, 71.0 per cent, 35.9 mm and 2.8 days, respectively. This exhibited that the disease appeared in the temperature range 24.6°C to 31.5°C, relative humidity range 71.0 per cent to 89.9 per cent with moderate rainfall 35.9 mm (Table-2). The results are in agreement with the report Verma and Singh (1971), Sheoraj *et al.* (1991) and Dang *et al.* (1994). The peak development of Bacterial leaf blight (PDI 1.56) took place 130-140 days after sowing (45 SW) in the temperature range 19.7°C (minimum) to 35.0 0C (maximum), relative humidity of 33.1% (evening) to 65.9% (morning) and rainfall of 1.3 mm within seven days prior to the day of peak development of disease. Hence, these can be

Table 1: Correlation value of Bacterial leaf blight intensity in cotton with weather parameters under Junagadh condition (2009-2013)

Sl. No.	*Meteorological parameters*	*Disease correlation matrix BLB*
1.	Temperature (Maximum)	0.074
2.	Temperature (Minimum)	-0.568
3.	Relative Humidity (Morning)	-0.545
4.	Relative Humidity (Evening)	-0.673
5.	Rainfall (mm)	-0.602
6.	Rainy days	-0.017

Significant level, if correlation r = > at 5 % = 0.043.

Table 2: Influence of weather parameters on occurrence and development of Bacterial leaf blight of cotton (average of years 2009, 2010 2011, 2012 & 2013).

Std. week	*Period*	*Temp. (°C)*		*RH (per cent)*		*Rainfall (mm)*	*Rainy Days*	*Av. PDI*	*PDI Incremental value*
		Max.	*Min.*	*Morn.*	*Even.*				
37	10.9-16.9	31.5	24.6	89.9	71.0	35.9	2.8	0.14	0.06
38	17.9-23.9	33.2	24.1	85.3	58.3	10.1	1.4	0.24	0.10
39	24.9-30.9	32.8	23.9	85.9	60.5	52.1	1.0	0.42	0.18
40	1.10-7.10	35.1	24.3	78.6	49.9	18.0	0.8	0.59	0.17
41	8.10-14.10	34.8	23.7	78.2	45.2	0.0	0.0	0.80	0.21
42	15.10-21.10	36.2	22.8	76.7	35.0	0.0	0.0	1.22	0.42
43	22.10-28.10	35.9	19.8	68.2	28.6	0.0	0.0	2.06	0.84
44	29.10-4.11	35.8	19.4	62.6	27.4	0.0	0.0	3.26	1.20
45	5.11-11.11	35.0	19.7	65.9	33.1	1.3	0.4	4.82	1.56
46	12.11-18.11	33.0	18.6	69.9	36.5	0.3	0.2	5.32	0.50
47	19.11-25.11	32.5	16.3	70.2	34.4	1.6	0.4	6.06	0.74
48	26.11-2.12	32.1	17.1	66.2	31.2	0.0	0.0	6.30	0.24
49	3.12-9.12	32.2	15.7	71.5	30.4	0.0	0.0	7.18	0.88

Conclusion

Foregoing results clearly indicated that low rains (1.3 mm) coupled w ith moderate temperature (19 to 35 °C) and moderate to high humidity (33 to 66 per cent) were highly favourable for maximum Bacterial leaf blight development. Therefore, it is advisable to undertake plant protection measures against Bacterial leaf blight of cotton when weather factors prevail in aforesaid range in later growth stage (125-135 DAS or 45 SW) of the crop.

References

Dang, J.K.; Sangwan, M.S.; Chauhan, M.S. and Pal Vijal (1994). Epidemiological studies on angular leaf spot. *J. Cotton Res. Dev.* 8: 175-78.

Diekmann, M. and Bogyo, T.P. (1992). Distribution of bacterial leaf blight depending on climatological factors. *Z. Pfl. Pfl.*99: 127-136.

Sheoraj, M.K.; Ajmera, B.D.; Harula, A.M. and Chauhan, M.S. (1991). Outbreak of bacterial blight of cotton -cause and future strategies for its management. *J. Cotton Res. Dev.* 5: 35-42.

Venkateswarlu, D.; Srinivas, T. and Rao, H.N. (2006). Epidemiological studies on bacterial blight of cotton caused by *Xanthomonasaxonopodis J. Cotton Res. Dev.* 20(2): 270-272.

Verma, J.P. and Singh, R.P. (1971). Epidemiology and control of diseases of cotton. *Pro. of National Sci.Aca. B*37: 317-26.

37

Climate Change Impact on Pulse Beetle *Callosobruchus Maculatus* (Fabricius) Comparative Biology on Different Stored Pulses

Chudasama J.A., Sharma, Satyakumari, Acharya M.F. and Sagarka N.B.

Abstract

An experiment was conducted at Junagadh Agricultural University, Junagadh to study climate change impact on pulse beetle comparative biology on different stored pulses. Biology of C. maculatus was made on different three hosts viz., *cowpea (local available variety of cowpea), chickpea variety GG-2 and soybean Gujarat soybean 2 at an average temperature of 30 ± 1°C under laboratory conditions. The mean incubation period in different hosts ranged from 2.52 to 8.16 days being minimum in cowpea 2.52 ± 0.81 with a range of 1 to 4 days, while maximum in soybean with an average of 8.16 ± 0.71 and range of 7-9 days. The mean duration larval + pupal periods ranged from 19.66 to 38.29 days in different hosts. Developmental period varied from 22.14 to 46.53 days in different hosts. The total developmental period varied from 20 to 24 days with an average of 22.14 ± 1.02 days in cowpea, an average or 34.67 ± 1.09 with a range of 31-37 days in chickpea and in soybean, it was 46.53 ± 1.64 with a range of 39 to 50 days. Significant differences were observed in the longevity of male and female beetles on various hosts. The longevity of male beetles varied from 6.60 to 9.72 days being minimum in chickpea and maximum on cowpea, but in case of female it was varied from 7.20 to 9.28 days being minimum in soybean and maximum on cowpea. The data on fecundity revealed that maximum number of eggs laid on cowpea (80.08) followed by chickpea (70.56) and soybean (58.80).*

Key words: biology, Callosobruchus maculatus, cowpea, chickpea, soybean

Pulses form an important part of Indian cuisine. The pulses are a large family and various species are capable or surviving in very different climates and soils. Pulses are not only important source of protein but also essential adjuncts to predominantly cereal-based diet and enhance the biological value of protein consumed. Thus, pulses play an important role in human nutrition in a predominantly vegetarian country like India.

During 2011-12, the total pulse production in India was 17088.9000 tonnes from 24462.2000 ha area and in Gujarat, the total pulse production was 780000 tonnes from 957000 ha area (Anon., 2013). In India, 12.5 million tonnes of edible legumes are produced every year and nearly 18.6 per cent of cowpea alone is damaged by bruchids during storage (Agarwal *et al.*, 1988). Caswell (1981) reported a loss of approximately 50 per cent of cowpeas in storage for three or four months due to infestation by *Callosobruchus maculatus* (Fab.). Raina (1970) reported that among five known species of *Callosobruchus* from India, three species *viz.*, *Callosobruchus chinesis* (L.), *C. maculatus* and *Callosobruchus analis* (Fab.) are important pests of stored pulses.

Females lay single eggs on the surface of grains sheath in field or on dried seeds in stores of grain legumes. The hatched larvae bored the seed below the lower surface of eggs and the egg shell remains glued to the bean. The larval developmental period is different by physical conditions. Larvae remained inside the seed and the appearance of a capped exit hole on the seed showed the pupal stage (Singh and Pandey, 2001). The adults do not need food or water, and can reproduce immediately after emergence.

Material and Methods

Studies on biology of *C. maculatus* on different hosts were conducted at the Storage Entomology and Packaging Research Laboratory, Department of Entomology, College of Agriculture, Junagadh Agricultural University, Junagadh during the year of 2013-14.

A stock culture of the bruchid beetle *C. maculatus* was maintained on used hosts seed *viz.*, chickpea, cowpea and soybean at normal room temperature in the laboratory before start of experiment. One to two days old 50 pairs of the beetle obtained from the initial culture were released in each of the wide mouth cylindrical glass jar measuring 20 × 15 cm, containing 250 g disinfested seeds of chickpea, cowpea and soybean. The jar was covered with muslin cloth and fastened with rubber band. After two days, all the adult insects were removed. Subsequently, the adults emerged from this culture were used for further mass multiplication.

25 pairs of one to two days old adults of *C. maculatus* were released for egg laying in glass jars (21 cm × 15 cm) containing 100 g chickpea, cowpea and soybean grains. The grains containing the eggs were collected on next day morning. In order to facilitate the observations, one egg was kept on each grain, while others were removed with the help of a needle. Such 100 grains

were kept individually in plastic vials (6.5 cm × 2.5 cm) under laboratory conditions.

Observations on hatching of eggs, larval + pupal period and total development period were recorded. Observations were taken daily in the morning. The adults emerged on the day were paired and released in individual plastic vials (6.5 cm × 2.5 cm) containing fifty grains of chickpea, cowpea and soybean grains. The grains with eggs were replaced daily by healthy grains and the number of eggs laid by an individual female was recorded till all the females died. Thus, hatching period, larval + pupal period, total development period, longevity of males and females, sex-ratio, pre-oviposition period, oviposition period, post-oviposition period and fecundity of females were worked out.

Results and Discussion

The results of the present study revealed the obvious effects of pulses on fecundity and development of *C. maculatus*. The freshly egg laid by female of *C. maculatus* was small, oval and whitish and shiny. Generally, singly eggs were laid on the surface of the seeds. The data on fecundity revealed that the mean number of eggs laid on different hosts ranged from 58.80 to 80.08 eggs (Table 1) being maximum on cowpea (80.08) followed by chickpea (70.56) and soybean (58.80). The minimum number of eggs (58.80) was laid on soybean. Raina (1970) found that females of *C. maculatus* laid an average of 128 eggs ranging from 109 to 157 eggs over a period of nine days. Singh and Rina Kumari (2000) reported an average of 70 eggs laid by female of *C. chinensis*. Sharma *et al.* (2007) observed that the single female of *C. chinensis* laid 65-72 eggs with an average 59.5 on soybean. The present findings were in agreement with previous studies by Bhargava *et al.* (2008) who recorded the maximum fecundity in cowpea (75.60 ± 6.87) and minimum on soybean (47.40 ± 6.26) eggs.

The mean incubation period in different hosts ranged from 2.52 to 8.16 day being minimum in cowpea (2.52 ± 0.81 days) while maximum in soybean (8.16 ± 0.71days). Raina (1970) recorded four days incubation period of *C. maculatus*. These results were similar to those reported by Zalavadia (1971). In present results, the incubation period on chickpea was observed an average of 5.91 ± 0.62 days with a range of 5 -7 days which was almost similar to be reported by Vyas (2004) for *C. chinensis*. Patel *et al.* (2005) observed that the incubation period of *C. chinensis* was 4.10 days on lentil. Patil (2007) reported that the average incubation period was 5.03 ± 0.11 days on chickpea. The incubation period of *C. chinensis* on cowpea (4.40 ± 0.54 days) and soybean (7.20 ± 0.44 days) were reported by Bhargava *et al.* (2008) and Borude *et al.* (2012).

The first instar larva bears a large spine on either side of the first abdominal segment and groups of smaller spines dorsally on the tergal plate of the pronatum. These spines appeared to help the larva in getting a hold on the

Table 1: Biological Parameters of *C. Maculatus* on Different Varieties of Pulses

Biological parameters	*Cowpea (Local variety)*	*Chickpea (GG 2)*	*Soybean (Gujarat soybean 2)*
Incubation period (days)	1 – 4 (2.52 ± 0.81)	5 – 7 (5.91 ± 0.62)	7 – 9 (8.16 ± 0.71)
Larval + pupal period (days)	18 – 23 (19.66 ± 1.31)	26 – 29 (27.73 ± 0.91)	31 – 42 (38.29 ± 1.47)
Total developmental period (days)	20 – 24 (22.14 ± 1.02)	31 – 37 (34.67 ± 1.09)	39 – 50 (46.53 ± 1.64)
Adult sex ratio: (Male: Female)	1:1.10	1:0.86	1:0.79
Longevity: Male (days)	8 – 12 (9.72 ± 1.54)	5 – 9 (6.60 ± 1.22)	5 – 9 (6.64 ± 1.19)
Female (days)	7 – 11 (9.28 ± 1.31)	6 – 11 (7.68 ± 1.25)	5 – 8 (5.88 ± 0.97)
Pre-oviposition period (hours)	3 – 7 (4.76 ± 1.27)	4 – 8 (5.32 ± 1.31)	4 – 8 (5.04 ± 1.21)
Oviposition period (days)	5 – 10 (7.36 ± 1.70)	5 – 10 (7.08 ± 1.29)	5 – 9 (6.76 ± 1.13)
Post-oviposition period (days)	1 – 3 (1.80 ± 0.65)	1 – 3 (2.20 ± 0.58)	1 – 2 (1.28 ± 0.46)
Fecundity	65 – 95 (80.08 ± 8.50)	64 – 76 (70.56 ± 3.69)	50 – 66 (58.80 ± 4.80)

egg shell while it used its mandibles to cut through the seedcoat. Frass produced by larva filled the egg shell and gave it a white appearance. After boring vertically into seed, the larva turns at right angles and pushes forward horizontally feeding continuously on the cotyledons. The larva undergoes four moults before pupation. Before pupation it chews a circular hole near the seed coat till only a thin layer of testa is left intact. This gives appearance of a dark spot. So, it indicates the pupation (Raina, 1970). In the present findings, the mean larval + pupal periods ranged from 19.66 to 38.29 days in different hosts. The combined larval + pupal period of *C. maculatus* was 20 days reported by Raina (1970). The mean incubation period of *C. chinensis* was 18 – 20 days noted by Singh and Rina Kumari (2000). These findings are also corroborated with those of Bhargava *et al.* (2008), Shivanna *et al.* (2011) and Borude *et al.* (2012) who reported significant variation in the larval + pupal periods of the *Callosobruchus spp.* in different pulses.

Developmental period varied from 22.14 to 46.53 days in different hosts. Complete developmental period was 24 days ranging from 23 to 27 days observed by Raina (1970). These results were in agreement with the previous studies by Zalavadia (1971) and Srivastav and Pant (1989). The complete developmental period of *C. maculates* was comparatively the shortest on the cowpea (average 22.14 ± 1.02 days) and longest on soybean (average 46.53 ± 1.64 days). This was in accordance of the findings of Wijenayake and Karunaratne (1999). Vyas (2004) reported that the mean duration of larval + pupal stage was 16.11 ± 0.88 days and the total developmental period from egg to adult was 20.09 ± 0.91 days. Similar results were also reported by Patel *et al.* (2005) and Sharma *et al.* (2007). This supports the present findings in this study. In this study, legume seeds which had the highest as well as lowest mean egg count correspondingly had the shortest and longest developmental period were reported by Bhargava *et al.* (2008), Kazemi *et al.* (2009), Swella and Mushobozy (2009), Shivanna *et al.* (2011), Malaikozhundun and Raj (2012) and Borude *et al.* (2012).

The emerging adults were sexed and the ratio determined. The ratio of male and female was quite different on these hosts. In cowpea, the ratio was 1:1.10. It was 1:0.86 and 1:0.79 in chickpea and soybean, respectively. Raina (1970) noted that male and female occurred in almost equal numbers with male having a slight majority. The ratio was 7:6. Similar observation regarding *C. chinensis* on mung bean and cowpea were reported by Singh and Rina Kumari (2000). Mandal and Konar (2006) stated that sex ratio was more in favour of male than female.

Significant differences were observed in the longevity of male and female beetles on various hosts. The longevity of male beetles varied from 6.60 to 9.72 days being minimum in chickpea and maximum on cowpea, but in case of female it was varied from 7.20 to 9.28 days being minimum in soybean and maximum on cowpea. Raina (1970) observed that males of *C.maculatus* lived for 7-10 days and females 6-10 days with an average of 8.2 and 7.6 days, respectively. The longevity of male and female of *C. maculatus* was reported similar by Zalavadia (1971). Almost similar observations in respect to male and female of *Callosobruchus spp.* longevity were also reported by Vyas (2004), Patil (2007), Sharma *et al.* (2007), Bhargava *et al.* (2008) and Borude *et al.* (2012).

In present finding, the highest pre-oviposition period in chickpea (5.32 ± 1.31 hours with a range of 4 - 8 hours) and the lowest pre-oviposition period in cowpea (4.76 ± 1.27 hours with a range of 3 -7 hours). Similar observations were obtained in oviposition and post-oviposition period. Almost similar findings were reported for *Callosobruchus spp.* on different pulses by Zalavadia (1971), Vyas (2004) and Patil (2007).

During the development, the highest mortality from egg to adult stage was observed in soybean 23 per cent and the lowest mortality was noticed in cowpea (14 per cent). It was observed that in the seed from which adults did not emerge, that most of the mortality occurred in the egg and early larval stages. Development responses of the cowpea beetle varied among different pulses. Adult emergence was highest in cowpea, followed by chickpea. Significantly, high developmental mortality of cowpea beetle was seen in soybean. There were less number of adults emerged from soybean because of the presence of the soybean cysteine protease inhibitor, soyacystatin N (scN) in soybean, which affected mainly during and second instar larval development (Koiwa *et al.*, 1998).

A number of physical and chemical factors of seeds appear to be responsible for the reduction of the development of the cowpea beetle in some pulses. The ability of the larva to penetrate the seed coat appears to be influenced by the physical properties of the seed coat such as thickness, hardness and roughness (Wijenayake and Karunaratne, 1999). In addition, factors such as poor nutrition, presence of growth deterrents or toxic substances may have been responsible for the death of all or most of the larvae after penetration into the cotyledons of soybean. Also compactness of the seed cotyledon may cause difficulties for the larvae to feed resulting in their death due to starvation.

Conclusion

The results indicate that pre-emergence application of pendimethalin 30% EC @ 0.900 kg ha-followed by post-emergence application of imazethapyr @ 75 g ha^{-1} or pre-emergence application of pendimethalin 30% EC @ 0.900 kg ha^{-1} followed by one hand weeding in row zone and interculturing in inter row zone at 40 DAS, or one hand weeding at 20 DAS followed by one interculturing at 40 DAS can be used to get higher productivity in *kharif* groundnut in Saurashtra region of Gujarat.

References

Agarwal, A.; Lal, S. and Gupta, K.C. 1988. Natural products as protectants of pulse against pulse beetles. *Bulletin of Grain Technology*. 26: 154-164.

Anonymous. 2013. Directorate of Economics and Statistics, Department of Agriculture and Cooperation (Fourth advance estimates of production of foodgrains).

Bhargava, M.C.; Choudhary, R.K. and Yadav, S.R. 2008. Biology and host preference of pulse Beetle (*Callosobruchus chinesis* L.) on different pulses. *Journal of Maharashtra Agriculture University*. 33(1): 44-46.

Borude, D.M.; Sonkamble, M.M. and Mutkule, D.S. 2012. Screening of different pulses for biology of *Callosobruchus chinesis* Linn. *Journal of Eco-friendly Agriculture*. 7(1): 98-99.

Caswell, G.H. 1981. Damage to stored cowpeas in Northern Nigeria. *Samaru Journal of Agricultural Research*. 1: 11-19.

Kazemi, F.; Talebi, A.; Fathipour, Y. and Farahani, S. 2009. A comparative study on the effect of four leguminous species on biological and population growth parameters of *Callosobruchus maculatus* (Fab.) (Col.: Bruchidae). *Advances in Environmental Biology.* 3(3): 226-232.

Koiwa, H.; Shade, R.E.; Zhu-Salzman, K.; Subramanian, L; Murdock, L.; Neilsen, S.S.; Bressan, R.A. and Hasegawa, P. M. 1998. Phage displayed selection can differentiate insecticidal activity of soybean cystatins. *The Plant Journal*, 14(3): 371-379.

Malaikozhundun, B. and Raj, S.T. 2012. A study on the developmental biology of *Callosobruchus maculatus* (Fab.) in different pulses. *Legume Research.* 35(2): 159-163.

Mandal, S. and Konar, A. 2006. A study on the biology of pulse beetle, (*C. chinesis*) infesting green gram, *Vigna radiata* L. *Legume Research.* 29(2): 134-136.

Patel, V.K.; Chaudhuri, N. and Senapati, S.K. 2005. Biology of pulse beetle (*C. chinesis*) as influenced by feeding of different grain pulses. *Agriculture Science Digest.* 25(4): 254-256.

Patil, D.R. 2007. Population build-up, varietal screening and management of pulse beetle (*callosobruchus chinensis* L.) through botanical materials in stored chickpea. M. Sc. thesis (Unpublished). Junagadh Agricultural University, Junagadh.

Raina, A.K. 1970. *Callosobruchus* spp. infesting stored pulses (grain legumes) in India and comparative study of their biology. *Indian Journal of Entomology.* 32(4): 303-310.

Sharma, M.; Pareek, B.L. and Sharma, K.P. 2007. Studies on the biology of pulse beetle, *Callosobruchus maculatus* (F.) (Coleoptera: bruchidae). *Soybean Research.* 5: 38-42.

Shivanna, B.K.; Ramamurthy, B.N.; Gangadhara, N.B.; Gayathri, S. and Krishna, N. B. 2011. Varietal screening of cowpea against pulse beetles, *Callosobruchus maculatus* (Fab.) and *C. analis. International Journal of Science and Nature.* 2(2): 245-247.

Singh, S.C. and Rina Kumari, 2000. A study of the biology of *Callosobruchus chinensis* (L.) infesting stored pulses (grain legumes) in India. *Indian Journal of Entomology.* 62(4): 319-322.

Singh, V.N. and Pandey, N.D. 2001. Growth and development of *Callosobruchus chinensis* Linn. on different varieties. *Indian Journal of Entomology.* 63(2): 182-185.

Srivastava, K.M. and Pant, J.C. 1989. Growth and development response of *C. maculatus* (Fab.) to different pulses. *Indian Journal of Entomology.* 51(3): 269-272.

Swella, G.B. and Mushobozy, M.K.D. 2009. Comparative susceptibility of different legume seeds to infestation by cowpea bruchid *Callosobruchus maculatus* (F.). *Plant Protection Science.* 45(1):19-24.

Vyas, H. H. 2004. Biology, varietal screening and management of *C. chinensisi* (Linneus) in stored cowpea. M. Sc. Thesis (Unpublished). Junagadh Agricultural University, Junagadh.

Wijenayake, D.U.S. and Karunaratne, M.M.S.C. 1999. Ovipositional preference and development of the cowpea beetle *Callosobruchus chinensis* on different stored pulses. *Vidyodaya Journal of Science*. 8: 135-147.

Zalavadia, R.V. 1971. Some studies on bionomics and control of pulse beetle (*C. maculatus* (Fabricius) under storage condition of Saurashtra area. M.Sc. thesis (Unpublished). Gujarat Agricultural University, Junagadh Campus, Junagadh.

38

Population Dynamics of *Aphis gossypii* Glover on Coriander in Relation to Biotic and Abiotic Factors

M.K. Kanani, Aheibam Ranila, J.B. Bhut and P.K. Borad

Abstract

Field experiment was conducted on population dynamics of Aphis gossypii Glover on coriander (Gujarat coriander 2) in relation to biotic and abiotic factors during winter season of 2011-12 and 2012-13. Incidence of aphid on coriander started from third week of December (51st SMW) and continued till first week of March (10th SMW) with a peak level during fifth week of January (5th SMW). Minimum and mean temperature had highly significant negative association with aphid incidence during first year, whereas non-significant negative correlation with the pest activity during second year. The morning vapour pressure had significant negative relationship with A. gossypii during first year and non-significant negative association during second year. Remaining all the tested meteorological parameters showed non-significant relationship with aphid population on coriander crop during both the years. The biotic factors, viz., coccinellids and syrphids as well as their cumulative numbers showed highly significant positive correlation with aphid population during both the years.

Key words: A. gossypii, population dynamics, biotic – abiotic factors, coriander.

Introduction

Coriander (*Coriandrum sativum* Linnaeus) commonly known as 'Dhania' is a very important seed spice which is used both as green vegetables as well as spice and medicine. The consumption of this spice crop is almost round the year in almost all parts of the country. In view of its increasing importance as medicine, its demand is increasing both in domestic and exports market. The sucking insect pests are the limiting factors for higher production and good quality seeds. *Aphis gossypii* Glover was recorded as a pest of coriander for the first time under Gujarat conditions (Ghetiya and Butani, 1995) and plays major role in making huge quantitative and qualitative losses. It has been found to attack about 220 host plants belonging to 46 families throughout the world (Roy and Behura, 1983). Both the nymphs and adults feeds by sucking the cell sap from inflorescence/ umbel. Due to their fast multiplication, aphid covers the entire surface of apical shoots. It exudates honeydew, which favour the growth of sooty mould and inhibit photosynthetic activity of the plants. Owing to the high rate of reproduction of this pest and continuous desapping of the flowers, grain formation is very much reduced or shriveled seeds are formed. While in less infested one the seeds are deformed resulting in poor yield. In addition to this they are known to be the vectors of various diseases. (Kenndy *et al.*,1962). Keeping these points in view, the present investigation was carried out to study the relationship between aphid population and biotic as well as abiotic factors in coriander crop.

Materials and Methods

The field experiment on activity of *A. gossypii* on coriander was carried out at Anand Agricultural University, Anand during *rabi*, 2011-12 and 2012-13. The coriander (Gujarat Coriander 2) crop was grown during first week of November in an area of 15 × 10 m at 30 × 15 cm spacing in both the years by following recommended agronomical practices. The coriander crop under experiment was kept free from application of any insecticides. For recording observations, coriander plots were divided into six equal sectors. From each sector, five plants were selected randomly and tagged with label for recording the observations. On each selected plant, three branches were randomly selected and aphid was counted from five cm shoot of each branch at weekly interval starting from one week after germination till harvest of the crop. Observations on coccinellids (adult + grub) and syrphid larvae were also recorded from the whole tagged plants. Data, thus, obtained were correlated with biotic and abiotic factors to determine its effect on fluctuation of *A. gossypii.* In order to study the instantaneous effect of weather parameters on population fluctuation of *A. gossypii* the data of physical factors of environment were analysed for correlation coefficient values.

Result and Discussion

The activity of *A. gossypii* was started from third week of December (51th SMW) and continued till first week of March (10th SMW) during both the year of experimentation (Table 1). After appearance it was increased approximately at a rate of two times in subsequent four weeks *i.e.* fourth week of December to third week of January during 2011-12 and 2012-13. The increasing rate was

Table 1: Population fluctuation of *A. gossypii* on coriander

Standard Meteorological Week	*Month and Week*		*No. of aphids/3 branches*	
			2011-12	*2012-13*
50	December	2	0	0
51		3	4	8
52		4	13	15
1	January	1	28	30
2		2	59	60
3		3	120	130
4		4	180	190
5		5	220	240
6	February	1	160	200
7		2	120	150
8		3	70	100
9		4	30	65
10	March	1	12	20
11		2	0	0

approximately 1.5 times during fourth and fifth week of January (fourth and fifth SMW). The activity of this pest was at maximum level on fifth week of January in both the years. During February, pest activity was started to decline at a rate of 0.25 to 0.50 time in both the years. The pest was disappears on coriander crop from second week of March during both the years. The higher activity of the pest was noticed during third week of January to 3rd week of February in 2011-12 and 2012-13.

The correlation coefficient results indicated that both the predators (Coccinellids and Syrphid fly) were significantly positive associated with aphid incidence on coriander during first and second year of experimentation. Similarly, cumulative number of predators had also significantly positive relationship with aphid population on coriander (Table 2).

The abiotic factors like maximum temperature exhibited significantly negative correlation with aphid incidence on coriander during first and second year of study (Table 2). The negative relationship between maximum temperature and *A. gossypii* in coriander was noted by Ghetiya (1992). The minimum and mean temperature had highly significant negative association with aphid incidence during first year, during second year. Slosser *et al.* (1998) observed negative association between temperature (maximum, minimum and mean) and *A. gossypii* population on cotton.

The morning vapour pressure had significant negative relationship with *A. gossypii* during first year and non-significant negative association during second year. The negative correlation of vapour pressure with *A. gossypii* on okra was observed by Bhoi (2008). Wind speed had positive association with

Table 2: Correlation coefficient (r) between *A. gossypii* incidence and biotic as well as abiotic factors on coriander

Factors		*Correlation coefficient (r)*	
		2011-12	2012-13
Biotic factors			
	Coccinellids	0.970**	0.966**
	Syrphids	0.930**	0.938**
	Total	0.964**	0.969**
Abiotic factors			
	Bright Sun Shine (hrs)	0.247	-0.102
	Wind Speed (kmph)	0.218	0.440
Temperature (°C)	Maximum	-0.656*	-0.554*
	Minimum	-0.738**	-0.257
	Mean	-0.702**	-0.450
Relative humidity (per cent)	Morning	-0.139	0.077
	Evening	-0.131	0.412
	Mean	-0.011	0.247
Vapour pressure (mm)	Morning	-0.643*	-0.308
	Evening	-0.314	0.041
	Mean	-0.511	-0.136
Vapour pressure deficit (mm)	Morning	-0.183	-0.170
	Evening	-0.391	-0.517
	Mean	-0.340	-0.377

*Significant at five per cent level (r = 0.553), **Significant at one per cent level (r = 0.681)

A. gossypii activity on coriander crop during both the years but non-significant. The mean vapour pressure, morning, evening and mean vapour pressure deficit were negatively correlated with *A. gossypii* population but non-significant in the first and second year. The bright sunshine hours had positive and negative relationship with aphid incidence on coriander during first and second year of study, respectively but non-significant. Ghetiya (1992), Devi *et al.* (2002) and Dabhi (2008) reported that bright sunshine hours had positive relationship with *A. gossypii* on coriander, brinjal and okra, respectively. Morning, evening and mean relative humidity and evening vapour pressure were negatively and positively associated with aphid population during first and second year, respectively. Ghetiya (1992) reported that morning and evening relative humidity had negative and positive association with *A. gossypii* on coriander, respectively.

Conclusion

Population dynamics of *A. gossypii* studied during the year 2011-12 and 2012-13 revealed that the pest incidence was commenced from 51th standard week (third week of December) and continued till 10th standard week (first week of March) on coriander crop during both the years. The higher incidence of *A. gossypii* was observed during 3rd standard week (3rd week of January) to 7th standard week (2nd week of February) in first (120-220 aphids/3 branches) and second (130-240 aphids/3 branches) year of study with a peak level in 5th standard week (5th week of January). The pest was disappeared from 11th standard week (2nd week of March) in coriander crops during both the years. The biotic factors, viz., coccinellids and syrphids as well as their cumulative numbers showed highly significant positive correlation with aphid population during both the years. Among the abiotic factors minimum and mean temperature had highly significant negative association with aphid incidence during first year, whereas non-significant negative correlation with the pest activity during second year. The morning vapour pressure had significant negative relationship with *A. gossypii* during first year and non-significant negative association during second year.

References

Bhoi, S.R. (2008). Population dynamics, varietal susceptibility and management of sucking pests of okra. M.Sc. (Agri.) thesis submitted to AAU, Anand.

Dabhi, M.V. (2008). Succession, population dynamics and management of major insect pests of okra (*Abelmoschus esculentus* (L) Moench). M. Sc. (Agri.) thesis submitted to AAU, Anand.

Devi, M.N., Singh, T.K. and Devi, C. (2002). Field density of *Aphis gossypii* in relation to predatory and biotic factors. *Uttar Pradesh J. Zool.*, 22 (1): 67-71.

Ghetiya, L.V. (1992). Bionomics, population dynamics and chemical control of aphid (*Aphis gossypii* Glover) on coriander. M. Sc. (Agri.) thesis submitted to GAU, Junagadh.

Ghetiya, L.V. and Butani, P.G. (1995). New record of *Aphis gossypii* Glover as a pest of coriander (*Coriandrum sativum* Linnaeus) in Gujarat. *GAU Res. J.*, 20 (2): 179-185.

Kennedy, J.S., Day, M.F. and Eastop, V.F. (1962). A Conspectus of Aphids as Vectors of Plant Viruses. Commonwealth Institute of Entomology, London, UK. 1-114.

Roy, D.K. and Behura, B.K. (1983). Notes on host plants, feeding behaviour, infestation and ant attendance of cotton aphids, *Aphis gossypii* Glover. *J. Bombay Nat. Hist. Soc.*, 80 (3): 654-656.

Singh, N.N. and Rai, V. N. (1994). Effect of abiotic factors on the development of aphid (*Lipaphis erysimi* Kalt Bach.) population. *Indian J. Ent.*, 56 (1): 99-103.

Slosser, J.E., Pinchak, W.E. and Rummel, D.R. (1998). Biotic and abiotic regulation of *Aphis gossypii* Glover in West Texas dryland cotton. South-west Entomologist, 23 (1): 31-65.

39

Population Dynamics of Sucking Pests and Their Correlation with Weather Parameters in Cluster Bean, *Cyamopsis Tetragonaloba* L.

R.L. Kalasariya, H.G. Kanara, M.F. Acharya and D.M. Jethva

Abstract

*Results revealed that the incidence of aphid (*Aphis craccivora *Koch), jassid (Emposcakerri Pruthi) and thrips (Megaleurothripsdistalis Karny) were appeared on the Cluster bean crop from the first week of March with 1.2 aphid nymph / leaf, 0.9 jassid/leaf and 0.8 thrips/flower, respectively, while mite observed first time during third week of March (20.25 mite/leaf). The highest population of aphid (2.8 nymphs/leaf) was attained during second week of March, while peak of jassid (5.9 jassid/leaf), thrips (9.8 thrips/flower) and mite (56.5 nymphs/leaf) was observed during fourth week of April. The correlation studies indicated that population of aphid exhibited significant negative correlation with wind speed (r = -0.660*), minimum (r = -0.680*) and mean temperature (r = -0.660*). Population of jassid favours the evaporation (r = 0.775**) and maximum temperature (r = 0.718), while it was found negative correlation with evening (r = -0.582*) and means relative humidity (r = -0.590*). Correlation studies indicated that thrips exhibited a significant positive correlation with maximum temperature (r = 0.597*), whereas found negative in case of morning (r = -0.630*) and mean relative humidity (r = -0.616*). In case of mite, there was significant positive correlation with evaporation rate (r = 0.725**), maximum (r = 0.733**) and mean temperature (r = 0.603*).*

Introduction

Cluster bean (*Cyamopsistetragonaloba* Linn.) is one of the important leguminous vegetable crop due to its nutritional and industrial values. About 70 per cent of India's production comes from Rajasthan. India produces 9.0 lakh tohnes of cluster bean annually. India contributes 80 per cent share in the world's total production (Anonymous, 2012a). In Gujarat State, it is mainly cultivated in Ahmedabad, Vadodara, Gandhinagar, Kheda, Surat, Navsari, Valsad, Kutch, Banaskantha, Mehsana and Sabarkantha. The estimated area under Cluster bean crop for vegetable purpose in Gujarat is 1359 hectares with a production of 568 MT. (Anonymous, 2012b). Cluster bean being an important vegetable crop is found to be attacked by aphid, *Aphis craccivora* Koch, jassid, *Empoascakerri* Pruthi, thrips, *Megaleurothripsdistalis* (Karny) and Mite, *Polyphagotarsonemuslatus* (Banks) (Butani and Jotwani, 1984; Whistler and Hymolvitz, 1979).

They have been observed as key pests of Cluster bean in this area. Excessive use of chemical insecticides not only causes the economical restrain on farmers but also produces the harmful side effects on the environment as well as human being. The best way to overcome this situation is to destroy the pest at its initial stage of the life cycle. This is possible if timely prediction of the occurrence of the pest can be made. Available scientific literature shows that not much information is available especially on seasonal incidence and influence of various environmental factors on the fluctuation on major insect pests on Cluster bean crop in semi-arid region conditions of Gujarat. Hence, a region oriented study on sucking pest's population dynamics would give an idea about peak period of their activity and may be helpful in developing pest management strategy.

Material and Methods

In order to study the population dynamics of aphid (*Aphis craccivora* Koch), jassid (*Emposcakerri* Pruthi), thrips (*Megaleurothripsdistalis* Karny) and mites (*Polyphagotarsonemuslatus* Banks) in field, the Cluster bean variety "Pusa Navbahar" which recommended for this region was used for the experiment. The experiment was conducted at Gujarat, College of Agriculture Farm, Junagadh Agricultural University, Junagadh during 2009-10. To study the population dynamics of major insect pests of Cluster bean were done by counting the population of insect pests on ten randomly selected plants in plots 20 m x 20 m divided into ten quadrates of 2.0 m x 2.0 m, having 60 cm x 20 cm spacing between row to row and plant to plant. All the other agronomical practices were adopted as per the scientific recommendations. The crop under the experiment was maintained without employing any plant protection measures. The observation of major insect pests was recorded at weekly intervals during morning hours between 6.30 am to 8.30 pm.

The population of aphid, jassid and thrips were recorded from three leaves one each from the upper, middle and lower position of ten randomly selected plants. The numerical count method described by Heathcote (1972) was used to record the population of aphid. The population was counted only on three leaves as per the method suggested by Satpathy (1973). The aphid population was expressed on per plant basis. The density of *P. latus* (eggs and mobile stages) was recorded under stereo binocular microscope on 2 cm × 2 cm leaf bit area.

The weekly meteorological data on temperature, relative humidity, wind speed, bright sunshine hours and evaporation rate were recorded during the experimental period. Correlation was worked out between insect pest's population and abiotic factors.

Results and Discussion

Aphid

The results presented in Table 1 showed that the aphid population initiated from the first week after sowing (first week of March) with 1.2 nymph/ leaf and remained active throughout the crop period. The pest population increased gradually and reached to the peak level of 2.8 aphid/leaf during the second-week after sowing coinciding with the second week of March. Later on, it started to decline slightly. The second peak of population (1.50 nymph / leaf) was observed in the sixth-week after sowing coinciding with the second week of April.

Table 1: Incidence of Major Insect and Mite Pests of Cluster Bean Crop

Sl. No.	*Date of observations*	*Week after sowing*	*Standard week*	*Mean No. of aphids nymphs/leaf*	*Mean No. of jassids/leaf*	*Mean No. of thrips /leaf or flower*	*Mean No. of mites/leaf*
1.	09/03/09	1	10	1.20	0.90	0.80	0.00
2.	16/03/09	2	11	2.80	1.50	0.90	0.00
3.	23/03/09	3	12	2.50	1.40	0.50	20.25
4.	30/03/09	4	13	1.80	3.99	4.60	30.25
5.	06/04/09	5	14	0.90	4.35	5.60	36.20
6.	13/04/09	6	15	1.50	3.68	6.02	45.50
7.	20/04/09	7	16	0.90	5.26	9.32	46.80
8.	27/04/09	8	17	0.60	5.90	9.80	56.50
9.	04/05/09	9	18	1.60	3.10	0.80	45.25
10.	11/05/09	10	19	0.50	2.50	0.95	40.25
11.	18/05/09	11	20	0.90	0.03	0.50	25.25
12.	25/05/09	12	21	0.00	0.00	0.00	0.0

However, the aphid population observed throughout the crop period. The population declined gradually during the successive weeks and reached to zero level at the time of harvest. The correlation matrix (Table 2) indicated that population of aphid exhibited significant negative correlation with minimum temperature (r = -0.680*) and wind speed (r = -0.660*). However, the pest showed negatively non-significant correlation with maximum temperature, evaporation rate, morning and evening relative humidity, whereas non-significant positive correlation was observed with the pest population and mean bright sunshine hours. The finding confirmed the results obtained by Defrawi *et al.* (2000) who reported that the population of aphid activity with highest counts during the third week of March. The present findings also supported by Helal *et al.* (1997) who reported the weather parameters (temperature and relative humidity) did not significantly affect the aphid population. Thus, present finding are in conformity with the above research workers.

Jassid

The results (Table 1) revealed that population of jassid (0.9 nymph / leaf) was appeared from the first-week after sowing (first week of March). Then after jassid population increased gradually and reached at first peak level of 4.35 nymph / leaf during the fifth week after sowing (first week of April). Later on it decline slightly. The second peak of population (5.90 nymph/leaf) was observed in the eighth week after sowing (fourth week of April). Then after population declined gradually during the successive weeks and reached to zero level at the time of harvest.

The correlation study (Table 2) clearly indicated that population of jassid exhibited significant positive correlation with maximum temperature (r = 0.718) and evaporation rate (r = 0.775**), while negative significant correlation exhibited between pest population and evening (r = -0.582*) and mean relative humidity (r = -0.590*), whereas incidence of jassid should positive non-significant correlation with minimum and mean temperature and mean bright sunshine hours, while negative non-significant correlation was observed in morning relative humidity and wind speed, while Gupta and Singh (2006) reported maximum population of jassid in the fourth week of April. This might be due to variable climatic conditions of that particular region and time of cultivation that particular crop.

Thrips

The results (Table 1) indicated that the population of thrips (0.8 nymph / flower) appeared from the first-week after sowing (second week of March) and remained active throughout the crop period. The pest population increased gradually and reached to the peak level of 9.80 nymph / flower during the eighth week after sowing (fourth week of April). The population declined gradually during the successive weeks and reached to zero level at the time of harvest, The results on effects of the various weather parameters on the fluctuation of thrips population (Table 2) indicated that population of thrips exhibited significant positive correlation with maximum temperature (r = 0.597*) and evaporation rate (r = 0.702**) and significant negative correlation with morning (r = -0.630) and mean relative humidity (r = -0.616*), while positively non-significant correlations exhibited between pest population and minimum and mean temperature and mean bright sunshine hours, whereas correlation between thrips population and evening relative humidity and wind speed was negatively non-significant. Pithava (1996) revealed that thrips appeared after fifth week of sowing and remained active throughout the crop season. The pest showed two peaks of its population, 3.80 and 8.12 thrips/flower during the seventh and tenth week of sowing, respectively. Slight variation in this regard might be due to some biotic and abiotic factors. Yadav and Singh (2006) observed the incidence of thrips during summer season.

Mite

The result (Table 1) indicated that the population of mite (20.5 mite/leaf) appeared from the third-week after sowing (fourth week of March). Then after population of mite increased slowly from the third-week after sowing and reached to a peak level (56.5 mite/leaf) in the eighth-week after sowing (fourth week of April). Later on, it started to decline slightly. However, the mite population remained throughout the crop period. The population declined gradually during the successive weeks and reached to zero level at the time of harvest. The correlation co-efficient values presented in Table 2 indicated the significant positive correlation with maximum (r = 0.733**) and mean temperature (r = 0.603*) and evaporation rate (r = 0.725**) and non-significant negative correlation with morning, evening and mean relative humidity and wind speed while positively non-significant correlation existed between pest population and minimum temperature and mean bright sunshine hours. The present finding concord with the finding of Ahuja (2000) who reported that the incidence of *P. latus* started appearing after one month of sowing in August and continued until harvest and peak population of this pest was recorded around (28.7 mite/leaf). Jovicich *et al.* (2004) reported that mite infestations occurred at seedling developmental stages and increased and developed more rapidly as seedling older that was infested. Seedling damage was greater in those infested at younger developmental stage.

Table 2: Correlation Between Key Abiotic Factors and Insect and Mite Pests Population on Cluster Bean During Summer 2009-10

Name of insect /pests	*Year*	*Weather parameters*								
		Temperature (°C)			*Relative humidity (%)*			*Wind speed (km/hr)*	*Mean bright sun shine hours*	*Evapor ation rate (mm)*
		Maximum	**Minimum**	**Mean**	**Morning**	**Evening**	**Mean**			
Aphid	2009-10	-0.308	-0.680*	-0.660*	-0.215	-0.438	-0.344	-0.660*	0.105	-0.424
Jassid	2009-10	0.718**	0.010	0.360	-0.545	-0.582*	-0.590*	-0.454	0.166	0.775**
Thrips	2009-10	0.597*	0.049	0.328	-0.630*	-0.543	-0.616*	-0.295	0.110	0.702*
Mite	2009-10	0.733**	0.341	0.603*	-0.087	-0.140	-0.117	-0.087	0.254	0.725**

*Significant at five per cent level (r = -0.576) n = 12

**Significant at one per cent level (r = 0.708)

References

Ahuja, D.B. (2000). Seasonal occurrence of insect pests in *Sesamum indicum* Linn. *J. Ento. Res.* 13(1-2): 116-120.

Anonymous, (2012a). Directorate of Economics and Statistics, Department of Agriculture and Cooperation. Area, production and yield of principal crops. 99.

Anonymous, (2012b). Directorate of Agriculture. Gujarat state, Gandhinagar. District wise Area, production and yield of 2009 to 2012. 23-24.

Butani, D.K. and Jotwani, M.G. (1984). Insect in vegetables, Periodical Expert Book Agency, Delhi, 356.

Defrawi, G.M.; Emam, A.K.; Marzouk, I. A. and Rizkalla, L. (2000). Population dynamics and seasonal distribution of *A. craccivora* (Koch) and associated natural enemies in relation to virus disease incidence in fababean fields. *Egyptian J. Agril. Res.* 78(2): 627-641.

Gupta, P.K. and Singh, J. (2006). Population studies on insect pest of greengram. *Ind. J. Ento.* 55(1):45-51.

Heathcoate, G.C. (1972). Evaluating aphid population on plants. *In: Aphid Technology (Ed. H.V. Van Emden.) Academic Press,* New York,105-145.

Helal, H.A.; EL-Khouly, A.S.; Khalafalla, E.M.E.; Metwally, M. M. and EL-Mozaien, A. B. (1997). Preliminary Studies on the population of legume aphid, *Aphis craccivora* (Koch.) and jassid, *Empoasca* spp. On different fababean varieties and breeding lines under the field conditions of Kafr El-Sheikh goverenorate, Egypt. Egyptian *J. Agril. Res.* 75(4): 951-959.

Jovicich, E.; Cantliffe, D.J.; Osborne, L.S. and Stoffella, P.J. (2004). Mite population and damage caused by broad mite, *P. latus* (Banks) infesting bell pepper (*Capsicum annuum* L.) at different seedling developmental stages. *Acta Hort.* 659(1): 339-344.

Pithava, B.B. (1996). Population dynamics, varietal screening and chemical control of pest complex of green gram. *Vigna radiate* (L.) Wilczek. M.Sc. (Agri.) Thesis. Gujarat Agricultural University, Sardar Krushinagar.

Satpathy, J.M. (1973). Field tested with granulated insecticides for the control of *L. orbonalis*. *Indian J. Agric. Sci.* 43(12): 1081-1086.

Whistler, R.L. and Hymolvitz, T. (1979). Guar agronomy, production, industrial use and nutrition. *Prude Univ. Press*. India USA, 89.

Yadav, D.K. and Singh, S.K. (2006). Forecast model of major insect pests of mung bean. *Annals Pl. Prot. Sci.* 14(2): 323-328.

40

Effect of Abiotic and Biotic Factors on Population Build-Up of Pest of Cluster Bean Through Path Co-efficient Analysis (*Aphis Craccivora* Koch)

T.K. Chandravadiya, D.J. Vadjaand and R.L. Kalasariya

Abstract

Cluster bean is one of the most important leguminous crop. The efforts were made to quantify the effect of weather parameters and bio-agents on population fluctuation of aphid. The experiment was conducted at Junagadh Agricultural University, Junagadh during kharif 2013-14. Results revealed that the direct effect on aphid population was noticed negatively very high (-1.6441) and positively high (1.1880 and 1.0259) through mean temperature, evening relative humidity and mean bright sunshine hours, respectively. Residual value indicated that all these biotic and abiotic factors had a total combined effect of 95.95 per cent on population build-up of cluster bean aphid. It indicated that only 4.05 per cent effect was through other unknown factors existing in the ecosystem.*

Introduction

Cluster bean (*Cyamopsistetragonaloba* Linn.) is one of the important leguminous vegetable cropdue to its nutritional and industrial values. About 70 per cent of India's production comes from Rajasthan. India produces 9.0 lakh tonnes of cluster bean annually. India contributes 80 per cent share in the world's total production (Anonymous, 2012a). In Gujarat State, it is mainly cultivated in Ahmedabad, Vadodara, Gandhinagar, Kheda, Surat, Navsari, Valsad, Kutch, Banaskantha, Mehsana and Sabarkantha. The estimated area under Cluster bean crop for vegetable purpose in Gujarat is 1359 hectares with a production of 568 MT. (Anonymous, 2012b). Cluster bean being an important vegetable crop is found to be attacked by aphid, *Aphis craccivora* Koch (Butani and Jotwani, 1984; Whistler and Hymolvitz, 1979). They have been observed as key pest of Cluster bean in this area. Excessive use of chemical insecticides not only causes the economical restrain on farmers but also produces the harmful side effects on the environment as well as human being. The best way to overcome this situation is to destroy the pest at its initial stage of the life cycle. This is possible if timely prediction of the occurrence of the pest can be made. Available scientific literature shows that not much information is available especially on seasonal incidence and influence of various environmental factors on the fluctuation on major insect pests on Cluster bean crop in semi-arid region conditions of Gujarat. Hence, a region oriented study on sucking pest's population dynamics would give an idea about peak period of their activity and may be helpful in developing pest management strategy.

Material and Methods

In order to study the population dynamics of aphid (*Aphis craccivora* Koch) in field, the Cluster bean variety "Pusa Navbahar" which recommended for this region was used for the experiment. The experiment was conducted at Gujarat, College of Agriculture Farm, Junagadh Agricultural University, Junagadh during *kharif* 2013-14. To study the population dynamics of aphid on Cluster bean were done by counting the population of *A. craccivora* on ten randomly selected plants in plots 20 m × 20 m divided into ten quadrates of 2.0 m × 2.0 m, having 60 cm × 20 cm spacing between row to row and plant to plant. All the other agronomical practices were adopted as per the scientific recommendations. The crop under the experiment was maintained without employing any plant protection measures. The observation of major insect pests was recorded at weekly intervals during morning hours between 6.30 am to 8.30 am. The population of aphid was recorded from three leaves one each from the upper, middle and lower position of ten randomly selected plants. The numerical count method described by Heathcote (1972) was used to record the population of aphid. The population was counted only on three leaves as per the method suggested by Satpathy (1973). The aphid population was expressed on per plant basis.

Correlation Study

In order to study the instantaneous effect of weather parameters on population fluctuation of various insect pests, the data of physical factors of environment *viz.*, maximum (MaxT) and minimum (MinT) temperature, morning (RH_1) and evening (RH_2) relative humidity, bright sunshine (BSS) and wind speed (WS) were correlated. Week-wise data on various parameters recorded by Department of Meteorology, College of Agriculture, Junagadh Agricultural University, Junagadh during *kharif* season of the year 2013-2014 and same were used to work out the correlationship with the incidence of the aphid.

With a view to study the impact of different weather parameters on pest incidence, a simple correlation between pest population and weather parameters was worked out by the method suggested by Panse and Sukhatme (1985).

Path Co-efficient Analysis

The association of insect pests population and different biotic and abiotic factors was ascertained by computing correlation co-efficient which gave strength and direction of association and provided the mutual association between the two characters (weekly average insect pest population and any one of the biotic or abiotic factor). The correlation, however, did not reflect actual contribution of individual biotic or abiotic factor on insect pest population and, therefore, it becomes difficult to determine the influence of a particular biotic and abiotic factor on the expressivity of dependent variable (weekly average insect pest population per plant), where confounding effects was encountered.

The path co-efficient analysis on the other hand, describes direct and indirect effects of each of the variable and gives the path in which an independent variable is affecting the dependent variable in a given set of independent variables (Nageswar Rao, 1983).

During the present investigation, the effect (direct as well as indirect) of ten different biotic and abiotic parameters *i.e.* maximum temperature (X_1), minimum temperature (X_2), mean temperature (X_3), relative humidity at morning (X_4), at evening (X_5), average relative humidity (X_6), sunshine hours (X_7), wind velocity (X_8), evaporation (X_9) and natural enemy (X_{10}) on the population build-up of mustard insect pests was studied with the help of path co-efficient analysis.

A path co-efficient was obtained by solving simultaneous equation which represents the basic relationship between correlation and path co-efficient of the formula given below.

$$R_{iy} = P_{1y} . r_{i1} + P_{2y} . r_{i2} + \ldots\ldots + P_i - 1_y . r_{i,i-1} + \ldots\ldots + P_{iy} + P_i + 1_y . r_{i,i+1} + \ldots\ldots + P_{ky} . r_{ik} .$$
for $i = 1, 2, \ldots\ldots k$.

where,

$P_{1y,}\ P_{2y} \ldots\ldots P_{ky}$ are the co-efficients in the linear relation and are known as path co-efficients: $r_{i1}, r_{i2}, \ldots\ldots r_{ik}$ are simple correlation co-efficient among independent variables (biotic or abiotic factors) and r_{iy} is the simple correlation co-efficient between i[th] independent variable X_1 and dependent variable Y (average insect pests population per plant). P_{iy} is the direct effect of X_i on Y and $P_{1y} . r_{i1} \ldots\ldots P_{i-1,y} .r_{i,}$ i-2' Pi + 1, $y_{ri,}$ i + 1.......... Pkyrik are indirect on X_i on Y through $X_{1,}\ X_{2,}\ X_3 \ldots\ X_{i-1'}\ X_{i+1'} \ldots\ldots X_k$' . Therefore, the simple correlation co-efficient (Total effect) between X_i and Y is the sum of direct and indirect effects of X_i on Y. The residual variation which is the variation due to uncontrolled causes in dependent characters (insect pests population) was estimated by the following relationship.

$$\text{Residual Effect } (P_{RY}) = \sqrt{1 - (P_{1y}.r_{1y} + P_{2y}.r_{2y} + \ldots\ldots\ldots + P_{ky}.r_{ky})}$$

The weekly weather parameters were obtained from meteorological observatory of Central Experimental Research Station, Junagadh Agricultural University, Junagadh, in order to study the effect of weather parameters on the population of mustard insect-pests, correlation between various weather parameters and pest incidence was worked out with the help of statistically analysis software (SAS 9.3 and SPSS 18).

Results and Discussion

Path co-efficient analysis of various abiotic factors express the direct and indirect influence on population build-up to insect pests. Direct and indirect effects of various key abiotic factors on the aphid population were worked out by applying path co-efficient analysis and the results thus obtained are presented in Table 1.

The maximum temperature was found to be positive and low direct effects (0.2007) on aphid population. The minimum temperature having significant negative association (r = -0.6221*) with aphid population but had a positively low direct effect (0.2975) on aphid population. Indirect effects of both these parameters were also found to have positively moderate to low on aphid population through almost all other biotic and abiotic factors.

Direct effect of morning and evening relative humidity were observed positively low (0.0162) and high (1.1880) respectively, indicating a favourable individual influence on the population build-up of aphid. The evening relative humidity having a high positive direct influence (1.1880) was however found to have more favourable in population build-up of the aphid as compared to morning relative humidity. This parameter exhibited positive moderate to low indirect effects on aphid population through almost all the other biotic and abiotic factors except mean temperature, mean relative humidity, mean bright sunshine hours and bio-agents.

The mean bright sunshine hours was found to be positively high direct effect (1.0259), indicating a favourable individual influence on the population build-up of aphid. This parameter exhibited negative high to low indirect effects on aphid population through almost all the other biotic and abiotic factors except maximum temperature, mean temperature, mean relative humidity and bio-agents.

The direct effect of wind speed was found to be moderate negative (-0.8141), indicating an unfavourable individual influence on the population build-up of aphid. This parameter exhibited negative moderate to low indirect effects on aphid population through almost all the other biotic and abiotic factors except mean temperature, mean relative humidity and mean bright

Table 1: Direct and Indirect Effect of Key Biotic and Abiotic Factors on Population Build-up of Aphid, *A. craccivora*

Biotic and abiotic factors	*Temperature (°C)*			*Relative humidity (per cent)*			*Mean bright sunshine hours*	*Wind speed (km/hr)*	*Evaporation (mm)*	*Bio-agents*	*Correlation with aphid population*
	Max.	*Mini.*	*Mean*	*Morning*	*Evening*	*Mean*					
Maximum temperature	**0.2007**	0.2178	-1.4524	0.0081	0.3732	-0.2486	0.1532	0.5315	0.0924	0.0131	-0.1109
Minimum temperature	0.1469	**0.2975**	-1.5881	0.0058	0.9076	-0.3217	-0.5186	0.3724	0.0929	-0.0169	**-0.6221***
Mean temperature	0.1773	0.2873	**-1.6441**	0.0071	0.7661	-0.3157	-0.2986	0.4580	0.0990	-0.0066	-0.4701
Morning relative humidity	0.1004	0.1069	-0.7191	**0.0162**	0.4638	-0.4336	-0.0325	0.7292	0.0331	-0.0130	0.2513
Evening relative humidity	0.0631	0.2273	-1.0602	0.0063	**1.1880**	-0.3946	-0.8800	0.3966	0.0774	-0.0236	-0.3997
Mean relative humidity	0.1002	0.1922	-1.0424	0.0141	0.9414	**-0.4979**	-0.4914	0.6948	0.0633	-0.0212	-0.0468
Mean bright sunshine hours	0.0300	-0.1504	0.4785	-0.0005	-1.019	0.2385	**1.0259**	-0.0413	-0.0404	0.0309	0.5521
Wind speed (km/hr)	-0.1310	-0.1361	0.9251	-0.0145	-0.5788	0.425	0.052	**-0.8141**	-0.0598	-0.0043	-0.3365
Evaporation (mm)	-0.1435	-0.2138	1.2595	-0.0041	-0.7122	0.244	0.3205	-0.3771	**-0.1292**	-0.0118	0.2322
Bio-agents	0.0448	-0.0854	0.1854	-0.0036	-0.4779	0.1799	0.5402	0.0603	0.0260	**0.0587**	0.5284

* Significant at 5 per cent level $(r = -0.576)$ $n = 12$

** Significant at 1 per cent level $(r = 0.708)$

Residual = 0.0405

sunshine hours. The evaporation had a negative low direct effect (-0.1292) on aphid population. This parameter exhibited negative moderate to low indirect effects on aphid population through almost all the other biotic and abiotic factors except mean temperature, mean relative humidity and mean bright sunshine hours. The bio-agents exhibited positive and low direct effect (0.0587) on population of aphid. This parameter exhibited positive moderate to low indirect effects on aphid population through almost all the other biotic and abiotic factors except maximum temperature, minimum temperature, morning relative humidity and evening relative humidity.

Combined Effect of Biotic and Abiotic Parameters on Population Build-up of Aphid

Residual value presented in Table 1 indicated that all these biotic and abiotic factors had a total combined effect of 95.95 per cent on population build-up of cluster bean aphid. It indicated that only 4.05 per cent effect was through other unknown factors existing in the ecosystem.

Conclusion

Among all the weather factors, the maximum and minimum temperature, morning relative humidity and bio-agents found to be positive and very low direct effect on aphid population, indicating that increase in maximum and minimum temperature and morning relative humidity would favourable for fast multiplication and dispersal of aphid population but negative direct effect of mean temperature, mean relative humidity, wind speed and evaporation on aphid population was found decrease the aphid in cluster bean. The aphid population in cluster bean is affected greatly by different factors that fluctuate at various crop growth stages.

References

Anonymous, (2012a). Directorate of Economics and statistics, Department of Agriculture and cooperation.Area, production and yield of principal crops, 99.

Anonymous, (2012b). Directorate of Agriculture. Gujarat state, Gandhinagar. District wise Area, production and yield of 2009 to 2012, 23-24.

Butani, D.K. and Jotwani, M.G. (1984). Insect in vegetables, Periodical Expert Book Agency, Delhi, 356.

Heathcoate, G.C. (1972). Evaluating aphid population on plants. *In: Aphid Technology (Ed. H.V. Van Emden.) Academic Press,* New York, 105-145.

Nagesar Rao, G. (1983). Statistics for agriculture sciences. Oxford and IBH publication. Co. New Delhi, 376.

Panse, V.G. and Sukhatme, P.V. (1985). Statistical methods for agricultural workers. ICAR., New Delhi, 381.

Satpathy, J.M. (1973). Field tested with granulated insecticides for the control of *L. orbonalis*. *Indian J. Agric. Sci.* 43(12): 1081-1086.

Whistler, R.L. and Hymolvitz, T. (1979). Guar agronomy, production, industrial use and nutrition. *Prude Univ. Press*. India USA, 89.

41

Effect of Farm Yard Manure and Potassium on Yield of Summer Groundnut under Different Cycles of Water Stress

K.B. Parmar, H.L. Sakarvadia, J.M. Modhvadia and Vinotha B.

Abstract

A field experiment was conducted in medium black calcareous soil (Typic Ustochrepts) with groundnut var. GG-2 in summer to study the effect of FYM and potassium under cyclic water stress at individual four stages i.e., flowering (SF), pegging (Spg), pod formation (Spf) and pod development (Spd) as well as at any two and three stages in all possible combinations along with one control of normal irrigation. Water stress at pegging drastically reduced haulm yield, while pod yield at pod development stage. As water stress at these phenophases is critical to haulm and pod yield, inclusion of any one of these with any other phenophase caused great reduction in the yield of respective parameter. In stress cycles, the consecutive stress was more damaging than the intermittent cycles. Under management treatments, the application of FYM @25 tha^{-1} gave significantly higher pod and haulm yields of groundnut than remaining treatments.

Introduction

Groundnut (*Arachis hypogaea* L.) is an important crop accounting for over 50 per cent of the production of oil seeds in the country. It is the principle crop of Gujarat state grown under erratic and scanty rainfall condition. Many factors influenced the crop yield, among which water supply is considered as one of the basic factor. Water affects plant growth by affecting internal physiological process and conditions. However, water stress at critical stages in plant growth caused more injury than at other stage (Kramer, 1969). The magnitude of reduction in growth due to water stress varied with the phenophase at which the stress was imposed (Golakiya and Patel, 1988). Water deficit not only reduced the total amount of growth, but it also changes the pattern of growth by controlling the availability of nutrients to the plants. Farm yard manure makes more available water to plant through increasing available water holding capacity of the soil by improving its physical properties. Potassium has been known to help the plant achieving water economy under situation of moisture deficit and therefore it has been specially identified as a nutrient with drought buffering capacity.

Materials and Methods

A field trial was conducted at Instructional Farm, Junagadh Agricultural University, Junagadh with Groundnut Var. GG-2 as the test crop during summer season. The soil of experimental site was clayey (*Typic Ustochrepts*) with a pH 8.2 and EC 0.36 dSm^{-1}. The available nutrients status was medium for $KMnO_4$-N (247 kg ha^{-1}) and Olsen – P_2O_5 (31 kg ha^{-1}) and high NH_4OAc – K_2O (340 kg ha^{-1}). The experiment was laid out in a split plot design with four replications. The water stress treatments at different growth stages were taken in main plots. These 15 water stress treatments, consisting of water stress at no, one and/or two, and/or three growth stages, were applied to the crop (Table-1), while sub plot treatments consist of F- FYM @ 25 tha^{-1}, Kf-Foliar spray of K @ 2% K_2O as KCl at 32 and 56 DAS, Ks- soil application of K @ 60 kg K_2O ha^{-1} as KCl at sowing and C-Control. Recommended dose of N and P_2O_5 (25.00 and 50.00 kg ha^{-1}) was applied to all plots of experiment. Pod and haulm yield were recorded from individual plots at harvest of crop.

Table 1: Details of main plot treatments

Sl.No.	*Symbol*	*Period* (DAS)*	*Description*
1.	S_0	—	Normal irrigation (no water stress)
Single cycle water stress			
2.	S_F	23-47	Water stress at flowering
3.	Spg	33-55	Water stress at pegging
4.	Spf	57-79	Water stress at pod formation
5.	Spd	73-95	Water stress at pod development
Double cycle water stress			
6.	$S_F$pg	23-55	Water stress at flowering and pegging
7.	$S_F$pf	23-47 and 57-79	Water stress at flowering and pod formation
8.	$S_F$pd	23-47 and 73-95	Water stress at flowering and pod development
9.	Spgpf	33-79	Water stress at pegging and pod formation

10.	Fpgpd	33-55 and 73-95	Water stress at pegging and pod development
11.	Spfpd	57-95	Water stress at pod formation and pod development
Triple cycle water stress			
12.	$S_F$pgpf	23-79	Water stress at flowering, pegging and pod formation
13.	$S_F$pgpd	23-55 and 73-95	Water stress at flowering, pegging and pod development
14.	$S_F$pfpd	23-47 and 57-95	Water stress at flowering, pod formation and pod development
15.	Spgpfpd	33-95	Water stress at pegging, pod formation and pod development

*Period indicates the duration in which irrigation was not applied.

Results and Discussion

Effect of Water Stress

A perusal of the date (Table 2) revealed that different water stress treatments significantly influenced the pod yield. In general, pod yield tended to decrease with increase in number of cycles of water stress. The maximum pod yield of 2302 kg ha^{-1} was recorded under S_1 and it was significantly superior to that in rest of the treatment. Significantly lowest pod yield was recorded under S_{15} (592 kg ha^{-1}).

In case of single water stress cycle imposed at different growth phases the adverse effect of water stress was significantly, least when imposed at flowering stage but it was maximum when at pod development phase (Sg). The effect of water stress at this stage was statistically at par with that produced due to imposition of water stress at pod formation stage (S_4), but significantly lower than S_3 (Table 2). The magnitude of decrease in pod under S_2 was 35.7 per cent and that under S_5 was 53.9 per cent over that of control.

Among the treatments S_6 to S_{11} covering double cycle water stress, the imposition of water stress at any two of the four growth phases almost produced equal effect as no significant differences in the pod yield were recorded among these treatments (Table 2). However, the maximum reduction of 58.6 per cent in the yield was found under treatment S_{11} over that of S_1 and the minimum of 53.1 per cent under S_7.

When the crop was subjected to triple cycle water stress, the imposition of water stress in three continuous phases (S_{12} and S_{15}) significantly produced lower pod yield (608 and 592 kg ha^{-1}) when compared with the treatments with intermittent water stress (S_{13} & S_{14}) (851 and 875 kg ha^{-1}). The per cent decreased in the yield due to the continuous water stress treatments ranged from 73.5 to 74.5 over that stress treatments.

Maximum pod yield reduction was registered under the stress at pod development stage as also reported by Patel and Golakiya, 1988.

Effect of FYM-potassium Treatments

The data (Table 2) revealed that among the FYM-potassium treatments, FYM gave significantly highest pod yield of 1247 kg ha^{-1}, while significantly lowest yield of (1016 kg ha^{-1}) was produced under control. The application of FYM can partly attributed to increased water status and biophysical activity of

Table 2: Pod Yield of Groundnut as Influenced by Water Stress and FYM-Potassium

Water Stress	*Pod yield (kg ha^{-1})*				*Mean*
	C	*FYM*	*Ks*	*Kf*	
S_0-S_1	2243	2459 (9.11)	2257	2251	2302(-)
S_F- S_2	1422	1642(15.4)	1432	1427	1480(35.7)
Spg- S_3	1177	1406(19.4)	1181	1179	1236(46.3)
Spf- S_4	1066	1288(20.8)	1068	1068	1123(51.2)
Spd- S_5	994	1255(26.2)	996	988	1061(53.9)
$S_F$pg- S_6	999	1238(23.9)	1006	1006	1063(53.8)
$S_F$pf- S_7	1016	1251(23.1)	1024	1019	1078(53.1)
$S_F$pd- S_8	922	1176(27.5)	929	927	989(57.0)
Spgpf- S_9	899	1155(28.4)	906	903	966(58.0)
Fpgpd- S_{10}	911	1061(16.4)	914	913	975(57.6)
Spfpd- S_{11}	883	1149(30.1)	888	887	952(58.6)
$S_F$pgpf- S_{12}	572	708(23.7)	576	575	608(73.5)
$S_F$pgpd- S_{13}	783	1049(34.0)	816	756	851(63.0)
$S_F$pfpd - S_{14}	805	1068(32.6)	810	807	875(61.9)
Spgpfpd- S_{15}	555	693(24.8)	562	557	592(74.5)
Mean	**1016**	**1247**	**1024**	**1018**	
	Stress(S)	**Management (M)**			**S × M**
C.D. at 5%	160	10			40.00
CV %	20.8	4.7			4.7

Figures in parenthesis show per cent increase or decrease in pod and haulm yield over that of respective control.

plant (Bennet *et.al.*,1981). The treatment Ks and Kf were at par with control, suggesting that K application either in soil or on foliage did not produce any beneficial effect.

Interaction Effect

The interaction effect among water stress, FYM-potassium treatments was significant on pod yield (Table 2). For a given water stress treatment, application of FYM significantly increased pod yield over control but the magnitude of increase due to FYM for various water stress treatments varied from 9.11 (S_1F) to 34.0 per cent (S_{13}F). Significantly maximum pod yield of 2459 kg ha^{-1} was recorded in FYM treated normally irrigated plots (S_1F), early growth phase/phases produced pronounced adverse effect on vegetative growth than that involving later growth phase/phases. This effect was reflected on haulm yield. In case of single cycle water stress, the higher pod yield of 1642 kg ha^{-1} was recorded when the crop was imposed stress at pegging stage (S_2F). When the crop was imposed to double water stress cycle, the higher pod yield was recorded under treatment S_7F (1251 kg ha^{-1}), it was at par with (S_6F). In case of triple cycle water stress, the higher pod yield was recorded under treatment S_{14}F (1068 kg ha^{-1}) and it was statistically at par with that obtained in S_{13}F (1049 kg ha^{-1}). The trend of the effect of FYM for different cycles of water stress was not consistent. It more depended on duration of water stress and phase of growth. The application of K in any form did not produce any significant effect on pod yield at any level of water stress treatment.

Haulm Yield

Effect of Water Stress

The data (Table 3) revealed that haulm yield was significantly decreased due to different water stress treatments. Significantly highest haulm yield of 4932 kg ha^{-1} was recorded in S_1, whereas lowest yield of 2311 kg ha^{-1} in S_{12}. The extent of decrease in the haulm yield under treatment S_{12} over that of S_1 was 53.1 per cent. Water stress which may reduce the cell size and leaf expansion and transport of photosynthates. (Ghosh *et al.*,1997) The maximum adverse effect of stress on haulm yield at pegging could be due to the vigorous vegetative growth and to transition of the crop from vegetative into reproductive phase.

Effect of FYM-potassium Treatments

The data (Table 3) revealed the FYM application produced favourable effect on haulm yield, but potassium treatments failed in doing so. The application of FYM established its superiority by recording significantly the highest haulm yield of 3338 kg ha^{-1}.

Table 3: Haulm yield of groundnut as influenced by water stress and FYM-Potassium

Water Stress	*Pod yield (kg ha^{-1})*				*Mean*
	C	*FYM*	*Ks*	*Kf*	
S_0S_1	4888	5021(2.7)	4914	4903	4932(-)
S_F- S_2	3000	3246(8.2)	3020	3011	3069(37.8)
Spg- S_3	2566	2857(11.3)	2574	2569	2642(46.4)
Spf- S_4	3211	3514(9.4)	3218	3213	3289(33.3)
Spd- S_5	3433	4024(17.2)	3436	3441	3583(27.4)
$S_F$pg- S_6	2511	2857(15.2)	2522	2524	2612(47.0)
$S_F$pf- S_7	2955	3385(14.5)	2966	2961	3067(37.8)
$S_F$pd- S_8	2972	3517(18.3)	3050	3045	3146(36.2)
Spgpf- S_9	2533	3015(19.0)	2545	2539	2658(46.1)
Fpgpd- S_{10}	2544	3005(18.2)	2552	2550	2662(46.0)
Spfpd- S_{11}	3127	3777(20.7)	3138	3135	3294(33.2)
$S_F$pgpf- S_{12}	2222	2562(15.3)	2230	2278	2311(53.1)
$S_F$pgpd- S_{13}	2911	3612(24.1)	2976	2938	3109(36.0)
$S_F$pfpd - S_{14}	2488	3057(22.8)	2624	2493	2666(45.9)
Spgpfpd- S_{15}	2223	2584(16.2)	2237	2232	2319(52.9)
Mean	**2906**	**3338**	**2933**	**2919**	
	Stress(S)	**Management(M)**			**S × M**
C.D. at 5%	415	33			130
CV %	19.2	5.1			5.1

Interacting Effect

Interaction effect of water stress, with FYM and potassium was significant on haulm yield. The interaction was significant due to FYM and not to potassium. For a given water stress treatment, application of FYM significantly increased haulm yield but the magnitude of increase varied widely ranging from 2.7 in S_1F to 24.0 per cent S_{13}F. The trend of the effect of FYM on various water stress treatments on haulm yield was almost same as was found in case of pod yield. Significantly highest haulm yield of 5021 kg ha^{-1} was obtained with S_1F and the lowest with $S_{12}C$. Application of K either in soil or in foliage did not increase haulm yield under any water stress treatment.

References

Bennett, J.M., Boote, K.J. and Hammond, L.C., 1981. "Alteration in the components of peanut leaf water potential during desiccation", *J. of Exp. Bot.,* Vol. 32, pp. 1035-1043.

Golakiya, B.A. and Patel, M.S., 1988. "Role of potassium in countracting the effect of cyclic drought on groundnut", *J. Potassium Res.,* Vol. 4(4), pp. 163-167.

Ghosh, P.K., Mathur, R.K., Ravindra, V. and Gor, H.K.., 1997. "Dry matter accumulation, nitrogen uptake and their partitioning pattern in Virginia groundnut (*Arachis hypogaea* L.)", *Indian J. of Pl. Physiol.,* Vol. 2(3), pp. 234-236.

Jackson, M.L., 1973. "Soil Chemical Analysis", Prentice Hall of India Pvt. Ltd., New Delhi.

Kramer, P.J., 1969. "Plant and Soil-Water Relationship", A modern synthesis, McCraw Hill, Inc., New York.

Patel, M.S. and Golakiya, B.A., 1988. "Effects of water stress on yield attributes and yield of groundnut". *Indian J. of Agric. Sci.,* Vol. 58(10), pp. 704-705.

42

Climate Change Impact on Crop Productivity and Economics under Irrigation, Nitrogen and Phosphorus Levels of Pearl Millet (*Pennisetumglaucum* L.)

H.M. Bhuva, B.C. Bochalya, Satyakumari Sharma and R.M. Javia

Abstract

A field experiment was conducted during rabiseasons of 1998 on Research farm at Junagadh Agricultural University, Junagadh to study effect of various level of irrigation, nitrogen and phosphorus on productivity and economics of pearl millet. The result revealed that application of IW/CPE of 0.7 recorded significantly higher growth, yield attributes and yields viz., *plant height, LAI, number of internodes, total and effective tiller, ear head thickness, earhead length, grain yield per plant, grain and straw yields and quality parameter viz., protein content and soil moisture parameters viz., consumption use of water, water use efficiency and economic returns was at par with application of IW/CPE ratios of 0.9. Although highest net monetary return and B:C was recorded in with application of IW/CPE of 0.7 closely followed by IW/CPE ratios of 0.9.*

Key words: Pearlmillet, Irrigation, Fertilizer, water use efficiency and Yield.

Introduction

Pearl millet [*Pennisetumglaucum* (L.) R. Br. emend. Stuntz] is an important drought hardy millet crop basically cultivated in the arid and semi-arid areas on light textured sandy soils. It is a dual purpose crop, grains are used for human consumption and its fodder as cattle feed. Pearl millet grains are eaten cooked like rice or 'chapattis' are prepared out of flour like wheat or maize flour and also used as feed for poultry and green fodder or dry *karbi* for cattle. India is the largest producer of pearl millet with an annual production of 10.05 million tonnes from an area of 8.69 million ha and productivity of 1156 kg ha^{-1}(AICMIP, 2014). Rajasthan, Maharashtra, Gujarat, Uttar Pradesh and Haryana are the major pearl millet growing states. Scientific scheduling of irrigation is a technique for determining the quantity of irrigation water, which also aim at optimising crop yield with the maximum water use efficiency and ensures minimum deterioration of the soil and crop. The present investigation was undertaken with the objective of finding suitability of different irrigation, nitrogen and phosphorus levels and their doses for pearl millet so that practically convenient and economically feasible strategy for scheduling irrigation for *rabi* pearl millet can be developed.

Material and Methods

An experimental was conducted at the Instructional Farm, Junagadh Agricultural University, Junagadh (Gujarat) for *rabi* season of the years 1998-1999. The experiment site is situated at 21.5°N latitude and 70.5°E longitude with an altitude of 60 m above the mean sea level on the western side at the foothill of famous mount 'Girnar'. The area has typically sub-tropical climate with fairly cool and dry winter, hot and dry summer, and warm and moderately humid rainy season. The maximum and minimum temperature during the crop growth period ranged between 27.3°C to 34.6°C and 7.8°C to 21.9°C during 1998-1999. The soil was medium clayey in texture and slightly alkaline in reaction with pH (8.10), OC (0.59 per cent) and EC (0.32 dS m^{-1}), low in available N (261.2 kg ha^{-1}), and medium in available P (32.12 kg ha^{-1}), available K (295.1 kg ha^{-1}) and moisture content at field capacity (28.02 per cent) and moisture content at permanent wilting point (13.81 per cent). Total 24 treatments *viz*., main plot treatments were 0.3 IW/CPE ratio, 0.5 IW/CPE ratio, 0.7 IW/CPE ratio 0.9 IW/CPE ratio and sub plot treatments nitrogen 100 kg ha^{-1}, 120 kg ha^{-1} and 140 kg ha^{-1} and phosphorus 40 kg ha^{-1} and 60 kg ha^{-1} tested in split plot design keeping four levels of irrigation in main plot and six treatment combinations comprising of three levels of nitrogen and two levels of phosphorus, in sub plot with four replications. 'Bajra-316' was sown at 45 × 15 cm spacing with seed rate of 3.75 kg ha^{-1}.

Results and Discussion

Effect on Growth attributes

The results revealed that the growth parameters, except plant population, were significantly influenced due to different irrigation and nitrogen levels (Table 1). Among the different treatment IW/CPE ratio of 0.9 over 0.3 and 0.5 IW/CPE ratio were most effective to significantly increase plant height (at 30 and 60 DAS and harvest) and number of total tiller/plant. The higher value in

Table 1. Effect of different levels of irrigation, nitrogen and phosphorus on growth parameters

Treatment	Plant population/net plot		Plant height (cm)			LAI at 70 DAS	No. of internode/ main shoot
	Initial	Final	30 DAS	60 DAS	Harvest		
Irrigation (IW/CPE ratio)							
I_1(0.3)	110.79	107.42	32.67	88.71	109.75	2.35	6.10
I_2(0.5)	111.75	108.83	35.04	95.58	115.42	3.52	6.30
I_3(0.7)	113.75	111.21	38.92	102.25	122.17	4.25	6.39
I_4(0.9)	114.17	111.33	40.13	103.42	123.67	4.31	6.43
SEm±	1.22	1.60	0.81	1.79	1.69	0.09	0.08
C.D. at 5%	NS	NS	2.60	5.73	5.39	0.28	0.25
C.V.%	5.29	7.16	10.84	8.99	7.01	11.91	6.32
N kg ha^{-1}							
N_1(100)	111.31	108.09	34.50	94.75	115.25	3.24	6.15
N_2(120)	112.38	109.25	37.14	98.38	118.59	3.75	6.39
N_3(140)	114.16	111.75	38.38	99.34	119.41	3.83	6.40
SEm±	0.91	1.05	0.43	0.73	1.00	0.07	0.06
C.D. at 5%	NS	NS	1.23	2.05	2.83	0.19	0.17
C.V.%	4.55	5.44	6.69	4.21	4.80	10.42	5.30
P kg ha^{-1}							
P_1(40)	112.08	109.15	36.44	97.15	116.96	3.57	6.29
P_2(60)	113.15	110.25	36.94	97.83	118.54	3.65	6.33
SEm±	0.74	0.86	0.35	0.59	0.82	0.05	0.05
C.D. at 5%	NS	NS	NS	NS	NS	NS	NS
C.V.%	4.55	5.44	6.69	4.21	4.80	10.42	5.30
Interaction	-	-	-	-	-	-	-

this respect might be due to better cell division and greater cell elongation owing to adequate moisture availability in the soil under higher irrigation levels. LAI increased with each successive increase in level of irrigation. Higher levels of irrigation application had led to excessive vegetative growth in term of higher LAI by maintaining optimum turgid conditions which would be beneficial for easy running of physiological process, growth and development of plant. Water is involved as a transporting agent in many of these processes.

Thus, higher dose of nitrogen improved cell division which resulted in greater growth parameters. Significantly higher LAI, number of internodes/ main shoot, number of tillers/plant were recorded under application of 140 N kg

Table 2. Effect of different levels of irrigation, nitrogen and phosphorus on yield attributing and yields

Treatment	No. of tiller/plant		Earhead length	Earhead thickness	Grain yield/ plant	Test weight	Yield kg ha^{-1}		Grain stover ratio
	Total	Effective					Grain	Straw	
Irrigation (IW/CPE ratio)									
I_1(0.3)	3.20	2.13	14.85	1.92	18.09	7.25	1680	4941	0.339
I_2(0.5)	3.71	2.50	15.90	2.09	20.94	7.65	2242	5970	0.377
I_3(0.7)	4.56	3.18	17.69	2.27	22.71	8.04	2607	6449	0.405
I_4(0.9)	4.81	3.21	18.23	2.32	23.00	8.07	2633	6561	0.403
SEm±	0.11	0.07	0.34	0.04	0.41	0.12	51.62	148.16	0.008
C.D. at 5%	0.35	0.24	1.09	0.14	1.31	0.38	165.12	473.94	0.024
C.V.%	13.17	13.21	10.01	9.85	9.43	7.53	11.04	12.14	9.67
N kg ha^{-1}									
N_1 (100)	3.72	2.51	15.61	2.04	19.88	7.48	1992	5554	0.358
N_2 (120)	4.17	2.83	16.94	2.18	21.59	7.81	2429	6178	0.393
N_3 (140)	4.31	2.92	17.45	2.24	22.08	7.95	2450	6208	0.394
SEm±	0.08	0.04	0.20	0.02	0.25	0.06	32.88	120.74	0.007
C.D. at 5%	0.24	0.12	0.56	0.07	0.69	0.17	93.01	341.51	0.018
C.V.%	11.62	8.64	6.74	3.75	6.55	4.50	8.12	11.42	9.55
P kg ha^{-1}									
P_1(40)	3.99	2.74	16.60	2.14	21.15	7.72	2288	6001	0.381
P_2(60)	4.14	2.76	16.73	2.16	21.22	7.78	2293	5959	0.384
SEm±	0.07	0.03	0.16	0.01	0.20	0.05	26.85	98.59	0.005
C.D. at 5%	NS	NS	NS	NS	NS	NS	NS	NS	NS
C.V.%	11.62	8.64	6.74	3.75	6.55	4.50	8.12	11.42	9.55
Interaction	-	I*N	-	I*N	-	-	-	-	-

ha^{-1} over 100 N kg ha^{-1}, but statistically at par with 120 N kg ha^{-1}, however growth parameters were not significantly influenced due to different phosphorus levels. The non-response to higher phosphorus levels 60 P kg ha^{-1} could be due to medium available phosphorus status of the soil.

Effect on Yield Attributes and Yield

The results revealed that the yield attributes were significantly influenced due to different irrigation, nitrogen and phosphorus levels (Table 2). Among the different treatment IW/CPE ratio of 0.9 and 0.7 over lower IW/CPE ratio of 0.5 and 0.3 were most effective to significantly increase effective tiller, ear head thickness, ear head length, grain yield per plant, grain and straw yields that might be due adequate supply of water under which plant become physiologically more active and also more nutrient availability might have been increased and ultimately resulted in improved the growth and development of sink. Adequate supply of water has stimulating effect on the progressive development in ear head thickness. The magnitude of increase in grain yield with 0.9, 0.7, 0.5 over 0.3 IW/CPE was the tune of 56.75, 55.17, 33.44 per cent, respectively. The reduction of grain yield under lower level of irrigation i.e. IW/CPE ratio of 0.3 could be explained on the basis of internal water status in relation to different physiological processes taking place in plant body based on functional behaviour of water in relation to the processes of photosynthesis taking place in plant. The increase in stover yield with 0.9, 0.7, 0.5 over 0.3 IW/CPE was the tune of 32.79, 30.51 and 20.82 per cent, respectively.

Significantly higher number of effective tillers/plant, ear head length and thickness, grain yield/plant and test weight were recorded under application of 120 N kg ha^{-1} over 100 N kg ha^{-1}, but statistically at par with 140 N kg ha^{-1}, however yield attributes were not significantly influenced due to different phosphorus levels. Relative constancy in grain and stover yields with an increase in phosphorus levels 40 P kg ha^{-1} to 60 P kg ha^{-1} indicate that application of phosphorus more than 40 P kg ha^{-1} is not helpful for further increase in grain and stover yields of pearl millet.

Effect on Quality Parameters

The results revealed that the protein per cent, consumption use of water and water use efficiency, except phosphorus levels significantly influenced due to different irrigation and nitrogen levels (Table 3). Among the different treatment IW/CPE ratio of 0.9 and 0.7 IW/CPE ratio were most effective to significantly increase protein per cent over IW/CPE ratio of 0.3. However, the per cent increase in protein content due to 140 N kg ha^{-1} over 120 N kg ha^{-1} and 100 N kg ha^{-1}, respectively. Application of phosphorus more than 40 P kg ha^{-1} did not improve pearl millet grain protein content as phosphorus fertilisation failed due to medium available phosphorus status of the soil.

Effect on Soil Moisture Parameters

Water used by pearl millet increased significantly with each increment in irrigation levels (Table 3). Scheduling irrigation at IW/CPE ratio of 0.9 recorded significantly the highest consumptive use of water over rest of lower ratios. The per cent increase in water use due to IW/CPE ratio of 0.9 was 9.95, 16.92 and 63.02 over IW/CPE ratio of 0.7, 0.5 and 0.3, respectively. The reason for higher yield water use under higher IW/CPE ratio of 0.9 is attributed to the fact that this ratio received the maximum irrigations each of 50 mm depth with a total of 300 mm water. However, the per cent increase in consumption use of water due to 140 N kg ha^{-1} and 120 N kg ha^{-1} as compare to 100 N kg ha^{-1} due to significant improvement in growth parameters, which demand more water in transpiration process. Application of phosphorus more than 40 P kg ha^{-1} did not improve pearl millet grain protein content as phosphorus fertilisation failed due to medium available phosphorus status of the soil.

The result pertaining to water use efficiency showed that different

Table 3. Effect of different levels of irrigation, nitrogen and phosphorus on quality and soil moisture parameters

Treatment	Protein (per cent)	Consumption use of water	Water use efficiency	Gross return	Cost of cultivation	Net return	B:C ratio
Irrigation (IW/CPE ratio)							
I_1 (0.3)	10.85	148.50	11.30	12621	6780	5841	0.86
I_2 (0.5)	11.27	207.04	10.78	16437	7373	9064	1.23
I_3 (0.7)	11.53	220.17	11.75	18793	7669	11124	1.45
I_4 (0.9)	11.63	242.08	10.88	19014	7965	11049	1.39
SEm±	0.16	1.29	0.28				
C.D. at 5%	0.50	4.12	0.89				
C.V.%	6.81	3.09	12.13				
N kg ha^{-1}							
N_1 (100)	11.06	200.63	9.91	14771	7255	7516	1.04
N_2 (120)	11.37	205.81	11.74	17619	7447	10172	1.36
N_3 (140)	11.53	206.91	11.89	17760	7639	10121	1.32
SEm±	0.08	0.57	0.17				
C.D. at 5%	0.23	1.62	0.49				
C.V.%	4.15	1.59	8.69				
P kg ha^{-1}							
P_1 (40)	11.31	203.96	11.22	16719	7227	9492	1.31
P_2 (60)	11.33	204.94	11.13	16714	7666	9048	1.18
SEm±	0.07	0.47	0.14				
C.D. at 5%	NS	NS	NS				
C.V.%	4.15	1.59	8.69				
Interaction	-	-	-				

treatment appreciably significantly affected water use efficiency. Significantly higher water use efficiency was obtained under IW/CPE ratio of 0.7 over 0.5, but it was statistically at par with lower ratios of IW/CPE ratio of 0.3. under the highest levels of irrigation lower water use efficiency was attributed to increase in consumption use of water. However, the per cent increase in water use efficiency due to 140 N kg ha^{-1} and 120 N kg ha^{-1} over 100 N kg ha^{-1}.

Effect on Economics

The result revealed that treatment combination of irrigation scheduling at 0.9 IW/CPE ratio along with 140 N kg ha^{-1} and 60 P kg ha^{-1} gave the highest grain yield and gross return (Table 3). However, the maximum net return was accrued under treatment combination of irrigation scheduling at 0.9 IW/CPE ratio along with 120 N kg ha^{-1} and 40 P kg ha^{-1} could be attributed to saving of extra cost of irrigation and fertiliser inputs.

Conclusion

Based on the results of one year experimentation, it can be concluded that potential production and profit from *rabi* pearl millet can be secured by scheduling irrigation at IW/CPE of 0.7 with 50 mm depth which required five irrigation ad fertilising the crop with 120 N kg ha^{-1} and 40 P kg ha^{-1} on clayed soil in South Saurashtra region of Gujarat.

References

AICPMIP, 2014. Annual report of all India Coordinated Pearl millet improvement Project (ICAR) University of Mysore, pp. 15.

Balyan, J.S. (1996). Response of pearl millet (*Pennisetumglaucum*) to applied nitrogen and stover management on succeeding wheat (*Triticumaestivum*). *Indian Journal of Agronomy*, 41(3):364-367.

Karnavat, B.B., Patel, B.A. and Kalaria, K.K. (1991). Response of pearl millet to depth and schedules of irrigations. *Indian Journal of Agronomy*, 36 :272.

Khan, M.H. and Reddy, S.N. (1997). Response of legumes and bajra to phosphorus application. *Journal of Research Andhra Pradesh Agricultural University*, 25(1): 1-8.

Khistaria, M.K., Vadaria, K.N., Vadodaria, J.R. and Akbari, K.N. (1998). Fertilizer management in groundnut-pearl millet cropping system under rainfed conditions. *Indian Journal of Agronomy*, 43(4): 591-595.

Kumar, S. and Chauhan, S.P.S. (1999). Response of intercropping (pearl millet + moong) to tillage practices and nitrogen. *Crop Research (Hisar)*, 17(3):307-308.

Yadav, N.D. (1998). Note on performance of pearl millet entries with nitrogen in rainfed conditions of Western Rajasthan. *Current Agriculture*, 22(1-2): 123-124.

43

Hydroponic Techniques in Floriculture

A.L. Regar, B.V. Thumar, S. Bishnoi, N.D. Polara and T.C. Saini

Abstract

Hydroponics as an agricultural production system includes those activities, which have the potential to displace conventional soil based horticultural production. Commercial hydroponics covers nearly 50000 ha globally. Main countries are Netherland, UK, Spain, Australia, USA, NZ and Italy. In order to expand the scope for development in floriculture in India there is a penury for hi-tech production systems. It is not only a solution for problematic soils but it also helps to improve the quality and quantity of floriculture produce. To overcome constraints like water availability, small land holdings, disease and pest incidence, problematic soils etc. It is more efficient use of water and nutrients, controls the environment and abridges use of pesticides and herbicides. It is part of soilless culture growing plants in nutrient solution with or without the use of artificial medium. It uses less than 1/10th-1/15th of the water used in soil cultivation. It can reduce irrigation water usage by 70-90 per cent by recycling the run-off water. In arid countries water can be shipped in and stored and when this is recycled the growing of crops is made possible. It can be argued that hydroponics is "The next logical step" after traditional agriculture. We can confidently say that application of hydroponics has a great scope in Indian floriculture industry. The hydroponics low cost device and procedure are recommended for the rose gardening in soil stress area for both urban and city growers. The vegetative and reproductive characteristics in gladiolus was found highest when used in nutrafin fertiliser.

Introduction

Hydroponic is a new production system. The word "Hydroponic" is derived from Greek words 'Hydros' (water) and 'Ponos' (working). In order to expand the scope for development in floriculture in India there is a need for Hi-tech production systems. To overcome constraints like water availability, small land holdings, disease and pest incidence, problematic soils etc. hydroponic is the best suited high-tech technique.

Hydroponics is a part of soilless culture growing plants in nutrient solution with or without the use of artificial medium. Simply it means working with water. Hydroponics in its true sense strictly means use of only water or solution culture. The interest of growers in soilless culture is rapidly increasing in the recent years. Soilless culture is commonly referred to Hydroponics or cultivation of flower crops without soil or mixed with some substrates.

Hydroponics was practiced many centuries ago in Amazon, Babylon, Egypt, China and India. The "Hanging Garden of Babylon" and the Aztec's floating farms were actually prototypes of hydroponic systems. Later, when plant physiologists started to grow plants with specific nutrients for experimental purposes, they gave the name "nutriculture." In 1929, Dr. William F. Gericke of the University of California succeeded in growing tomato vines of 7.5 m height in nutrient solutions. During 1960s and 70s, commercial hydroponics farms were developed in Abu Dhabi, California, Denmark, Germany, Holland, Iran, Italy, Japan, Russian Federation and other places. During 1980s, many automated and computerised hydroponics farms were established around the world. During1990s home hydroponics kits became popular.

Classification of Hydroponic/Soilless Culture

A. Solution Culture or Liquid Hydroponics

1. Circulating methods
   - (a) Nutrient film technique
   - (b) Deep flow technique

(2) Non-circulating method
   - (a) Root dipping technique
   - (b) Floating technique
   - (c) Capillary action technique

B. Solid media culture (Aggregate systems)

- (a) Hanging bag technique
- (b) Grow bag technique
- (c) Trench or trough technique
- (d) Pot technique

C. Aeroponics

- (a) Root mist technique
- (b) Fog feed technique

(a) Nutrient Film Technique (NFT)

NFT is a true hydroponics system where the plant roots are directly exposed to nutrient solution. A thin film (0.5 mm) of nutrient solution flows through channels. The channel is made of flexible sheet.

The seedlings with little growing medium are placed at the centre of the sheet and both edges are drawn to the base of the seedlings and clipped together to prevent evaporation and to exclude light. The growing medium absorbs nutrient solution for young plants and when the plants grow the roots form a mat in the channels.

The nutrient solution is pumped to the higher end of each channel and flows by gravity to the lower end wetting the root mat. At lower end of the channels nutrient solution gets collected and flows to the nutrient solution tank.

The solution is monitored for salt concentration before recycling. Some growers replace the nutrient solution every week with fresh solution.

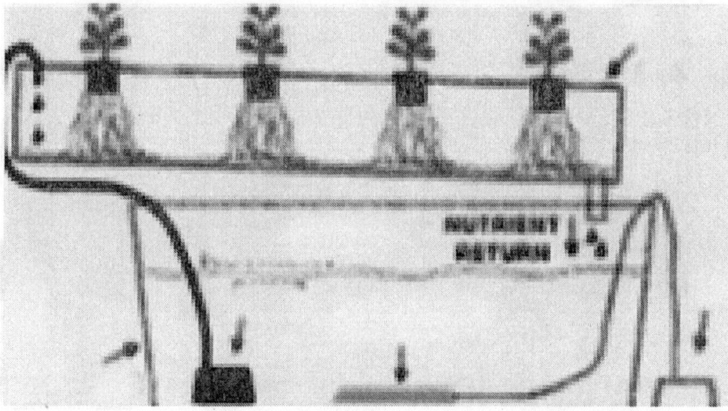

Fig. 1: Nutrient Film Systems.

(b) Deep Flow Technique (DFT)

In this method two types pipes are used:

1. **Single plane pipe**
    - As the name implies, 2-3 cm deep nutrient solution flows through 10 cm diameter PVC pipes to which plastic net pots with plants are fitted.
    - The plastic pots contain planting materials and their bottoms touch the nutrient solution that flows in the pipes.
    - The PVC pipes may be arranged in one plane or in *zig-zag* shape depending on the types of crops grown.

Fig. 2: A Single Plane Pipe System DFT.

2. Zig-zag Shape Pipe

The *zig-zag* system utilises the space efficiently but suitable for low growing crops.

Fig. 3: A Zig-zag Pipe System DFT.

Non-circulating Method

The nutrient solution is not circulated but used only once. When its nutrient concentration decreases or pH or EC changes, it is replaced.

In this technique, plants are grown in small pots filled with little growing medium. The pots are placed in such a way that lower 2–3 cm of the pots is submerged in the nutrient solution. Some roots are dipped in the solution while others hang in the air above the solution for nutrient and air absorption, respectively. This technique is easy and can be developed using easily available materials. This 'low tech' growing method is inexpensive to construct and needs little maintenance. Importantly this technique does not require expensive items such as electricity, water pump, channels, etc.

Fig. 4: Root Dipping Technique.

Floating Technique

Fig. 5: Floating Technique.

This is similar to box method but shallow containers can be used. Plants established in small pots are fixed to a Styrofoam sheet or any other light plate an allowed to float on the nutrient solution filled in the container and solution is artificially aerated.

Capillary Action Technique

Planting pots of different sizes and shapes with holes at the bottom are used. Fill these pots with an inert medium and plant seedlings in that inert medium. These pots are placed in shallow containers filled with the nutrient solution. Nutrient solution reaches inert medium by capillary action.

Materials and Methods

Three commercial rose cultivars namely, Tajmahal, Paradise and Calcutta were collected from the local nursery. The saplings of these cultivars were of 45 days old, developed through budding on *Rosa indica* stock plant and grown in a rooting media of rice husk and FYM (1:1). For each cultivar five saplings on an average of 15 cm. length were taken. They were transferred in equal sized Single Plant Culture System devices of hydroponics.

Fig. 6: Capillary Action Technique.

Aeration is very important in this technique. Therefore, old coir dust mixed with sand or gravel can be used. This technique is suitable for ornamental, flower and indoor plants.

This work was conducted during the 2006 and 2007 seasons, under controlled glass house conditions. The nutrient film technique (NFT) system was designed by using plastic tubing of 4.0 inches diameter and 4.0 meterslength with upper holesevery 20 cm with 8.0 cm diameters. They were fixed with a gentle slope of 1.0%. Accordingly, the nutrient solution flowed under the influence of gravity. The nutrient solution tank was a white plastictank of 50 L capacity. The nutrient solution was supplied from PVC pipesof 0.5 inch diameters starting at the highest end to allow the flow of thenutrient solution. The used nutrient solutions were re-collected within thenutrient tank and recycled again by the emerged pump to the upper side.The immersed plastic pump was used to avoid salt corrosion, and a pumpedsolution was run continuously with a given flow rate of 120 L $hour^{-1}$. Gladiolus corms of about 3 cm diameter wide and 20 gm weight were bought from Taylor Corms Co (Aberdeen, Scotland) and placed in the NFT pipes. The corms were placed in holes of the plastic pipes, with each hole containing two bulbs. As such, the four plastic pipes contained 160 corms. Three commercial fertilizer formulas (Signral, 20-20-20, Nasr Co, Egypt; Nutrafin 23-33-24, Marinedepot, CA, USA; and Heavy harvest Bloom, Hydroempire, Cambridge, England) were utilised as hydroponic growing solutions and were studied and compared to both Hoagland A solutions (Lorenz and Maynard, 1980), which were considered the control. The NFT unit was represented by four pipes, each containing 40 corms, consequently all pipes contained 160 bulbs and occupied 1.6 m^2 (100 corms m–2). All used nutrient solutions were changed weekly to avoid changes in pH and electrical conductivity (EC). Every solution was adjusted to a pH of 6.0–6.5 and the EC to 2–3 dS m–1. The analysis of the used tap water was as follows: nitrogen (N) 0.0125 ppm, phosphorous (P) 8.09 ppm, potassium (K) 5.08 ppm, and sodium (Na) 8.09 ppm. The EC was 1.24 dSm"1, total soluble salts 793.6 ppm, and the pH 7.66. At the end of the vegetative stage the following data were recorded: number of leaves plant–1, leaf length (cm), leaf width (cm), leaves area plant–1 (cm^2), leaves fresh weight plant–1 (gm), Chlorophyll A and B, and carotenoids according to Wettestein (1957). At the flowering stage the following data were recorded: spike length (cm), number of florets spike–1, and spike fresh and dry weight (gm). Samples of vegetative leaves at the beginning of flowering were dried at 70% C and used for chemical analysis.

After the leaves were dried on the plant, the bulbs were collected, dried, and the following data were recorded: fresh and dry weights of the corm (gm) and the number of cormlet corm–1. Samples of the bulbs were dried at 70% C and the leaves were subjected to chemical analysis as follows: total nitrogen percentages in the leaves and bulbs were determined according to Naguib (1969). Total phosphorus percentages in the leaves and corms were determined according to Troug and Mayer (1939). Total potassium percentages in the leaves and bulbs were determined according to Jackson (1970). Total carbohydrate percentages in leaves and bulbs were determined according to the method described by Dubois *et al.* (1956).

Results

Three cultivars of rose could be grown well in the Single Plant Culture System of hydroponics as constructed above. From the estimation of different growth parameters of the plants at every 15 days intervals it could also be revealed that all the cultivars showed a normal and steady growth in this hydroponics system. The total amount of different component of nutrients applied to the cultivars during 165 days growth period following the standard of Hoagland and Arnon [10] is given in Table 1. Neither any deficiency symptom due to unavailability of nutrition nor any adverse effect due to over-nutrition was noticed among the plants due to the application of the given amount of nutrition. There was also variation in plant growth parameters and water requirement among the three cultivars.

Data in Table 1 indicate that there was no significant difference in leaf number plants–1, leaf length, or leaf width due to different nutrient solution types. However, a slight increase was observed when Hoagland's solution was used. On the other hand, the other characters, i.e., leaf fresh and dry weights per plant, root length, root fresh and dry weights per plant, corm fresh and dry weights, and number of cormlets corm–1 showed significant differences. The comparison of Hoagland's solution and commercial fertilizers indicated that the Nutrafin and Singeral resulted in the highest values for leaf number plant–1 compared to Hoagland's solution. Leaf width had the highest values with Nutrafin. Nevertheless, leaf fresh weight plant–1 wassignificantly different.

Discussion

Calcutta than the other two cultivars. Number of shoot branches was almost similar for the cvs. Tajmahal and Paradise but the lengths of the shoot branches were more in Tajmahal than Paradise Number of axillary bud initiations was also the highest in the cv. Calcutta (6±0.70) followed by cv Paradise (5±0.73) and the cv. Tajmahal (4±0.24) during 165 days in hydroponics culture. In the correlation analysis it was also observed that the two growth parameters e.g. shoot branch length per plant and the number of shoot branches had significant correlation with the initiation of axillary buds of the plants throughout the period of observation.

Hydroponic nutrient solution costs and preparation are one of the main reasons for obstruction practicing the NFT by public. Referring to the effect of Hoagland solutions and commercial fertilizers on the vegetative growth, it was clear that the effect of Nutrafin was more pronounced by significant increase than Hoagland solution for leaf length plant–1, leaf dry weight, root length, root fresh and dry weight, corm dry weight and number of cormlets corm–1. On the other side, the increase in flowering characters was insignificant. These results can be explained on the base that Nutrafin solution contains two forms of nitrogen (nitrate and ammonia) whereas Hoagland A solution contains nitrate form only. In this regard, Jones (1997) mentioned that some ammonium may by desirable in the nutrient solution as it stimulates the uptake of nitrate and has been shown that as little as five per cent of total N in solution as ammonium in a flowing nutrient solution system is sufficient. Moreover, selecting the proper ratio of ammonium to nitrate in the nutrient solution should be considered in plant species. Also in this regard, Mohammed (2001) showed the enhancing effect of ammonia and nitrate in the solution than nitrate alone on vegetative growth of Ficushawaii and Ocimumcanum. Herein, it could be concluded that the effect of Nutrafin was similar to Heavy harvest bloom solution for obtaining the best results of vegetative growth and flowering. From economical point of view Nutrafin and Singral solutions are preferred since it prepared from commercial fertilisers that have low costs compared to Hoagland solutions which should be prepared from pure salts. On the other side, Singeral compound commercial fertiliser came in the second order after Nutrafin. This research considers steps in knowing the efficiency of using commercial fertilisers instead of using the high-cost hydroponic ready prepared solutions. However, this question needs to be answered with different plants under different situations.

References

Cooper, A. 1979. Nutrient Film *Technique for Growing Crops.* London: Grower Books.

Cooper, A. 1996. The A B C of NFT, Nutrient Film Technique. Narrabeen, Australia: Casper Publications.

Dubois, M., K.A. Gilles, J. Hamillon, P.A. Reber, and F. Smith. 1956. Colorimetric method for determination of sugars and related substances. Analytical chemistry, 28: 350–355.

FAO. 1990. Soilless Culture for Horticulture Crop Production. Rome: Food and Agriculture Organisation of the United Nations.

Jackson, M.L. 1970. Soil Chemical Analysis. Englewood Cliffs, NJ: Prentice Hall.

Jones, J.B. 1997. Hydroponics, A Practical Guide Soilless Grower. Boca Raton, FL: St. Luci Press.

Lorenz, O.A., and D.N. Maynard. 1980. Knott's Hand Book for Vegetable Growers. New York: John Wiley and Sons.

Mohammed, H.A.H. 2001. Study on the use of nutrient film technique for growing some ornamental, medicinal and aromatic plants. *Ph.D. Thesis,* Al-Azhar University, Cairo, Egypt.

Naguib, M.I. 1969. Colorimetric determination of nitrogen components of plant tissues. Journal of the Faculty of Science, Cairo University 34: 16–24.

Snedecor, G.W. and W.G. Cochran. 1980. Statistical Methods, 7th ed. Ames, IA: Iowa State University.

Troug, E., and A.H. Mayer, 1939. Improvement in the denies colorimetric method for phosphoric and arsenic. Industrial and Engineering Chemistry, Analytical Edition 1: 136–139.

Wettestein, D. 1957. Chlorophyll-letale under submikroskopis-che form Wechsel der plastiden Exptl [Chlorophyl analysis of plant plastides]. Cell Research 12: 427–433.

Bhattachrya, S.K. and L.C. Dey, 2003. Rose In: *Advanced commercial floriculture,* Aaviskar Pub. Rajasthan, 1: 207-228.

Raviv, M. and T.J. Blom, 2001. The effects of water availability on photo-synthesis and productivity of soilless-grown cut roses. *Scientia Horticulturae*, 88: 257-276.

Samartzidis, C., T. Awada, E. Maloupa, K. Radoglou and A.H.I. Constantinidou, 2005. Rose productivity and physiological responses to different substrates for soil-less culture. *Scientia Horticulturae*, 106: 203-212.

Schwarz, D., R. Grosch and W. Gross, 2004. Water quality for hydroponics: nutrients, bacteria and algae in rainwater ponds. Acta Horticulturae, 644: 533-540.

Gericke, W.F., 1937. Hydroponics crop production in liquid culture. Sci., 85: 177-78.

Jensen, H.M., 1997. Hydroponics. Hort Sci., 32: 1018-1021.

Das, A. and D.S. Majhi, 2009. Low cost Hydroponics devices use of harvested water for vegetables and flower cultivation. J. Appl. Sci, 15:1015-1055.

List of Contributors

Authors	Page No.
H.D. Rank, Research Engineer, AICRP on Irrigation Water Management, College of Agricultural Engineering and Technology, Junagadh,	165, 252
H.G. Kanara, Department of Entomology, College of Agriculture, Junagadh Agricultural University, Junagadh - 362 001.	314
H.L. Sakarvadia, Associate Research Scientist, Directorate of Research, Junagadh Agricultural University, Junagadh-362001.	150, 327
H.M. Bhatu, M.Tech student, College of Agricultural Engineering and Technology, Junagadh.	252
H.M. Bhuva, Department of Agronomy, College of Agriculture, J.A.U., Junagadh (Gujarat).	119
H.M. Bhuva, Vigyan Kendra, J.A.U, Junagadh, Nana Kandhansar (Chotila).	334
H.N. Patel, Department of Agricultural Entomology, N.M. College of Agriculture, Navsari Agricultural University, Navsari-396 450, Gujarat, India.	247, 264
Hadiya A.M., Department of Genetics and Plant Breeding, College of Agriculture, Junagadh Agricultural University, Junagadh, 362 001.	215, 226, 238
Harsh Vardhan Singh Shekhawat, Department of Plant breeding and genetics, SKRAU, Bikaner-334006.	60
Indu Sharma, Indian Institute of Wheat and Barley Research, Karnal – 132 001.	15
J.B. Bhut, Biocontrol Research Laboratory, Department of Entomology, Junagadh Agriculture University, Junagadh 362001.	132, 291, 309
J.H. Kelaiya, Senior Research Fellow (NICRA), College of Agricultural Engineering and Technology, Junagadh.	252
J.K. Nandaniya, Farming Research Station, AICRP for Dryland Agriculture, Junagadh Agricultural University, Targhadia (Dist: Rajkot) 360 003-India.	92
J.M. Modhvadia Associate Research Scientist, Directorate of Research, Junagadh Agricultural University, Junagadh-362001.	327
K. Chakraborty, ICAR-Directorate of Groundnut Research, Ivnagar Road, Junagadh – 362 001.	29
K. Madhusudhan Reddy, Scientist, ANGRAU, Anantapuramuamu, A.P.	127
K.B. Parmar, Associate Research Scientist, Directorate of Research, Junagadh Agricultural University, Junagadh-362001.	150, 327
K.B. Polara, Department of Agricultural Chemistry and Soil science, Junagadh Agricultural University, Junagadh-362001.	150
K.D. Mungra, Associate Research Scientist (Plant Breeding). 12 *Proceedings of National Seminar on Water Management*	99
K.H. Dabhi, Department of Genetics and Plant Breeding, College of Agriculture, Junagadh Agricultural University, Junagadh.	238

Authors	*Page No.*
P.S. Jayswal, Assistant Professor, College of Agriculture, Junagadh Agricultural University, Amreli.	165
P.S. Pandit, Assistant Professor, Centre of Excellence on Post Harvest Technology, Navsari Agricultural University, Navsari, Gujarat.	76
Pamu Swetha, Department of Agricultural Chemistry and Soil Science, Junagadh Agricultural University, Junagadh-362001(Gujarat).	176
Parnaliya, J.B., Department of Genetics and Plant Breeding, College of Agriculture, Junagadh Agricultural University, Junagadh.	238
Patel K.D., Department of Genetics and Plant Breeding, College of Agriculture, Junagadh Agricultural University, Junagadh.	215
R Sendhil, Indian Institute of Wheat and Barley Research, Karnal – 132 001.	15
R.J. Patel, Assistant Professor, College of Agricultural Engineering and Technology, Junagadh.	252
R.L. Kalasariya, Department of Entomology, College of Agriculture, Junagadh Agricultural University, Junagadh - 362 001.	314, 321
R.M. Javia, Krishi, Vigyan Kendra, J.A.U, Junagadh, Nana Kandhansar (Chotila).	334
Rahul Sharma, Department of Agronomy, College of Agriculture, J.A.U., Junagadh (Gujarat).	119
Raiyani G. D., Department of Genetics and Plant Breeding, College of Agriculture, Junagadh Agricultural University, Junagadh 362 001.	215, 226, 238
Raveendra H. Patil (Ravi Patil), Agronomist, Department of Agronomy, College of Agriculture, Dharwad, University of Agricultural Sciences, Dharwad, Karnataka, PIN: 580 005, India.	66
Ravish Chatrath, Indian Institute of Wheat and Barley Research, Karnal – 132 001.	15
S. Bishnoi, M.Sc. Horti. (Floriculture), Department of Horticulture, College of Agriculture, JAU, Junagadh-362001 (GJ)	341
S.S. Wandre, Ph.D. Scholar, Department of Soil and Water Engineering, CAET, JAU, Junagadh-362001 (GJ).	50
Sagarka N.B., Department of Entomology, College of Agriculture, J.A.U., Junagadh (Gujarat).	301
Sarika S Wandre, Ph.D Student, College of Agricultural Engineering and Technology, Junagadh.	252
Satish Kumar, Indian Institute of Wheat and Barley Research, Karnal – 132 001.	15
Satyakumari Sharma, Department of Agronomy, College of Agriculture, J.A.U., Junagadh (Gujarat).	119, 301, 334
Shalini kumara, Department of Agriculture Chemistry and Soil Science, Junagadh Agricultural University, Junagadh, (Gujarat)-362001.	87

Water Management and Climate Smart Agriculture

Proceedings of National Seminar on Water Management and Climate Smart Agriculture

(13th – 14th February, 2015)

Vol – II

Climate Change Impact on Water Cycle

R. Subbaiah

G. V. Prajapati

Centre of Excellence on Soil & Water Management
Research, Testing & Training Centre
Junagadh Agricultural University, Junagadh, Gujarat, India

GYAN

ISBN: 978-81-212-1316-5

Price: ₹ 4500 (set of 3 Vols)
₹ 1500 (Vol–II)

First Published, 2015

Published by

Gyan Publishing House
5, Ansari Road, Daryaganj,
New Delhi-110002
Phones: 9811692060
E-mail: books@gyanbooks.com

Printed at: G. Print Process, Delhi

Cataloging in Publication Data—DK
Courtesy: D.K. Agencies (P) Ltd. <docinfo@dkagencies.com>

National Seminar on Water Management and Climate Smart Agriculture (2015 : Junagadh Agricultural University)
Proceedings of National Seminar on Water Management and Climate Smart Agriculture (13th-14th February, 2015) / R. Subbaiah, G.V. Prajapati.
3 volumes cm
National seminar organized by Junagadh Agricultural University.
Includes bibliographical references.
Contents: vol. I. Vulnerability of agriculture to climatic aberrations — vol. II. Climate change impact on water cycle — vol. III. Adaption of climatic resilient water management and agriculture.
ISBN 9788121213165

1. Water-supply, Agricultural—India—Management—Congresses. 2. Climatic changes—India—Congresses. 3. Crops and climate—India—Congresses. 4. Agriculture—Environmental aspects—India—Congresses. 5. Agriculture—Economic aspects—India—Congresses. I. Subbaiah, R., editor. II. Prajapati, G. V., editor. III. Junagadh Agricultural University.organizer. IV. Title.

S494.5.W3N38 2015 DDC 333.9130954 23

Contents

1

21st Century Climate Change Scenario of Haridwar District in Uttarakhand

S.K. Dubey, S.K. Tripathi, G. Pranuthi

Abstract

The daily observed weather data (rainfall, maximum temperature and minimum temperature) was obtained from the Department of Water Resources Development and Management, Indian Institute of Technology Roorkee (29°52'N and 77°53'E) for the period 1979-2013 for validation and calibration of PRECIS RCM data for the same period. The PRECIS RCM data for 2014-2090 was also down loaded and bias corrected using linear correlation factor developed for 1979-2009. The bias corrected data was also statistically analysed. Mann-Kendall test and Theil Sen's Slope test was also performed. Analysis of the bias corrected PRECIS RCM data for the period 2014-2090 revealed that the Roorkee station in general record increases rainfall (@ 5.83 mm/year), maximum temperature (@ 0.05°C/year) and minimum temperature (@ 0.04°C/year). The normal rainfall during the forecasted period was recorded as 1174 mm/year. It is also observed that out of the 77 years forecast, 50 years will receive normal rainfall, 13 years will be excessive rainfall (>25 per cent of Normal) whereas 14 years will receive deficit rainfall years (<25 per cent of Normal). However, temporal variation in rainfall was quite remarkable. In the last quarter of the 20th century, rainfall increased @ 15.3 mm/year. The change in maximum and minimum temperatures was not so remarkable.
Keywords: *PRECIS, Climate change, Bias correction, Mann Kendall Trend test, Sen's Slope.*

Introduction

Increased concentration of greenhouse gases (GHG) due to anthropogenic reasons has led to climate change and has emerged in the world as a key environmental and economic concern. The increasing GHG concentrations are resulting in the increasing temperature and thereby influencing the weather pattern (Yu *et al.*, 2010; Ahsan *et al.*, 2011). The Intergovernmental Panel on Climate Change (IPCC) in its fourth assessment report (IPCC-2007) states that the world over industrialisation has caused the planet to warm by about 3°C in the 21st century. This will effect directly and indirectly the agricultural productivity. (FAO, 2007; Roudier *et al.*, 2011). Indian agriculture contributing about 20 per cent to the national economy will not remain untouched in the changed weather scenario (Anonymous, 2011).

Uttarakhand with eleven mountainous and two plane, (Haridwar and Udham Singh Nagar) districts was formed in 2000. In order to increase the per capita income of the state, industrial development was undertaken even on the agricultural lands of Haridwar and Udham Singh Nagar due to geographic limitations of mountainous region.

The PRECIS RCM developed by the UK's Hadley Centre for IPCC for the climate prediction and research programme of IPCC, is widely used worldwide to forecast future weather scenario (Fowler *et al.*, 2007; Grotch and Mac Cracken, 1991; IPCC, 2007; Maraun *et. al.*, 2010). Undoubtedly the actual observations are different from the forecasted values. Therefore, it is minimised using bias correction by linear correction approach (Lenderink *et al.*, 2007) and non-linear correction approach (Leander and Buishand, 2007) so that both (actual and forecasted) values get statistically acceptable. Rice, wheat and sugarcane are the main crops in the plain districts of Uttarakhand. Tripathi *et al* (2007, 2008a and 2008b)) analysing the weather data of 1979-2008 reported that the average temperature is rising @ 0.025°C/yr in Roorkee (Haridwar District). Some evil effects of industrial development on agriculture and animal husbandry is being realised by the farmers. However, the a biotic cause needs to be ascertained.

Therefore, in this background the study entitled "21st Century Climate Change Scenario of Haridwar district in Uttarakhand" was undertaken.

Methodology

PRECIS Regional Climate Model data of Rainfall, maximum temperature (Tmax) and minimum temperature (Tmin) was downloaded in binary format at the spatial resolution of 0.44° × 0.44° (50 Km × 50 Km). This data is converted into *netcdf* format using FORTRAN program and thereafter the daily data of rainfall, Tmax and Tmin was extracted for Roorkee location (29° 52′ N and 77° 54'E) for the period 01/01/1979-31/12/2090 using Arc GIS 9.3.

In the PRECIS model daily weather data is available for 360 days/year only considering 30 days per month. Therefore the downloaded data were adjusted to match the Gregorian calendar days/month taking average of the last five days in the respective months and accounting the same for 31st day in January,

March, May, July, August, October and December. No adjustment was made for those months having only 30 days/month. February is either 28 or 29 days/month which is less than 30 days therefore one or two days were deleted from the data of 30 days.

Daily weather data (Rainfall, Tmax and Tmin) recorded at the Agromet observatory of the Department of Water Resources Development and Management, Indian Institute of Technology Roorkee was obtained for the period 01/01/1979 – 31/12/2013. PRECIS RCM weather data (Rainfall, Tmax and Tmin) for Uttarakhand was obtained from IITM Chennai for the period 01/01/1979-31/12/2013. The daily PRECIS RCM data for the period 1979-2009 was calibrated doing bias correction. PRECIS RCM data were validated for the period 2010-2013. Thereafter data was forecasted for the period 2014-2090. Bias correction procedure is described in the forthcoming paragraphs.

Bias correction of PRECIS RCM data

Following steps were adopted in the bias correction:

First Step: Correction for Standard Deviation (Leander and Buishand, 2007)

Observed and PRECIS RCM down loaded data (daily/monthly) for the period 1979-2009 was taken. Correction for Standard Deviation of PRECIS RCM data was done (Leander and Buishand, 2007) using equation 1 as given below:

$$P_{SDcorr} = \overline{P}_{obs} + \left(\frac{\sigma(P_{obs})}{\sigma(P_{rcm})}\right) \times \left(P_i - \overline{P_{obs}}\right) + \left(\overline{P_{obs}} - \overline{P_{rcm}}\right) \quad ...(1)$$

where,

P_{SDcorr} = Bias Corrected PRECIS RCM precipitation or temperature (daily/monthly/yearly) data.

$\overline{P}_{obs}$ = Observed precipitation or temperature (daily/monthly/yearly) data.

$\overline{P}_{rcm}$ = PRECIS RCM precipitation or temperature (daily/monthly/yearly) data.

$\sigma(P_{obs})$ = SD of the observed precipitation or temperature (daily/monthly/yearly) data.

$\sigma(P_{rcm})$ = SD of the PRECIS RCM precipitation or temperature (daily/monthly/yearly) data.

P_i = PRECIS RCM precipitation or temperature (*i*th daily/monthly/yearly) data.

Correlation coefficient (r) between observed and SD corrected PRECIS RCM data was calculated. If the r value calculated was > 0.5 then, correction for mean was calculated (Hay *et al*, 2002) using equation 2 as given below:

Second Step: Correction for Mean (Hay et al., 2002)

$$P_{mean\ corr} = P_{SDcorr} \times \left(\frac{\overline{P_{obs}}}{P_{rcm}} \right) \qquad ...(2)$$

where,

P_{SDcorr} = Bias Corrected PRECIS RCM precipitation or temperature (daily/monthly/yearly) data

$\overline{P}_{obs}$ = Oserved precipitation or temperature (daily/monthly/yearly) data

P_{rcm} = PRECIS RCM precipitation or temperature (daily/monthly/yearly) data

In the step first and second there is chance of generating negative values which is not desirable especially in case of precipitation and even in temperatures.

Third Step: Correction for Biases

In order to check above mentioned situation, following Bias correction conditions were applied on mean corrected PRECIS RCM data:

1. The precipitation data of a given day was programmed in such a way that the negative values were replaced with zero.
2. If the Tmax (°C) of bias corrected value exceeds the observed value of a given day then the bias corrected value was replaced by the observed value.
3. If the Tmin (°C) bias corrected value was lower than the observed value for a given day then the bias corrected value was replaced by the observed value.

Forth Step: Forecasting for 2014-2090

$$P_{SDcorr\ F} = \overline{P}_{obs\ N} + \left(\frac{\sigma\left(P_{obs\ N}\right)}{\sigma\left(P_{rcm\ F}\right)} \right) \times \left(P_{iF} - \overline{P_{obs\ N}}\right) + \left(\overline{P_{obs\ N}} - \overline{P_{rcm\ F}}\right) \qquad ...(3)$$

where,

$P_{SDcorrF}$ = Bias Corrected PRECIS RCM precipitation or temperature (daily/monthly/yearly) data

$\overline{P}_{obs\ N}$ = Observed precipitation or temperature (daily/monthly/yearly) data

$\overline{P}_{rcm\ F}$ = PRECIS RCM precipitation or temperature (daily/monthly/yearly) data

$\sigma\left(P_{obs\ N}\right)$ = SD of the observed precipitation or temperature (daily/monthly/yearly) data

$\sigma\left(P_{rcm\ F}\right)$ = SD of the PRECIS RCM precipitation or temperature (daily/monthly/yearly) data.

P_{iF} = PRECIS RCM precipitation or temperature (*i*th daily/monthly/yearly) data.

Fifth Step: Correction for Mean of Forecasted data during 2014-2090

$$P_{meancorr} = P_{SDcorr\ F} \times \left(\frac{\overline{P_{obs\ N}}}{\overline{P}_{rmm\ F}} \right) \qquad ...(4)$$

where,

$P_{SDcorr\ F}$ = Bias Corrected PRECIS RCM precipitation or temperature (daily/monthly/yearly) data

$\overline{P}_{obs\ N}$ = Observed precipitation or temperature (daily/monthly/yearly) data

$\overline{P}_{rcm\ F}$ = PRECIS RCM precipitation or temperature (daily/monthly/yearly) data

Calibration was done by using observed and PRECIS RCM data (daily/monthly) for the period 1979-2009. This data was standard deviation corrected using equation 1. Correlation coefficient (r) between observed and SD corrected data (equation 1) was calculated. If the r value calculated was > 0.5 then a correction for mean was done by developing a multiplying factor $\left(\overline{P}_{obs} / \overline{P}_{rcm}\right)$ from equation 2. The outcome of the factor multiplied by SD corrected PRECIS RCM data was mean corrected PRECIS RCM data. The mean PRECIS RCM data was bias corrected by adopting the biasness conditions given in step. 3 of correction for biases. Observed and bias corrected data were statistically tested.

Climate Change Pattern Analysis

In order to assess the trend and its magnitude of slope in the trend line Mann–Kendall Trend test [Mann, (1945), Kendall, (1975)] and Theil Sen's Slope [Theil, (1950) and Sen, (1968)] was analysed form the bias corrected PRECIS RCM weather data of2014-2090.

Results and Discussion

Results obtained are discussed in the forthcoming paragraphs:

(1) Rainfall

Data given in Table 1 recorded the average annual rainfall during 2014-2090 as 1174 mm/year which is about 125 mm/year above the current normal annual rainfall. In general rainfall increased @ 1.3 mm/year. Fig. 1 shows that the rate of rise in rainfall will be slow during 2014-2070 but increase abruptly @15.3 mm/year during 2070-2090. This is peculiar trend arising and needs to be further investigated using other available weather models.

Table1: Drought and Excessive rain years forecasted by bias corrected PRECIS RCM data during 2014-2090 at Roorkee

Deficit/Drought (<25% N (1174 mm))		*Excessive rain (>25% N (1174 mm))*	
Years	*Rainfall (mm)*	*Years*	*Rainfall (mm)*
2015	774.1	2039	1540.0
2022	836.8	2040	1501.3
2026	833.4	2049	1514.1
2027	862.3	2063	1470.8
2030	793.6	2067	1576.5
2031	763.5	2071	1671.7
2033	863.2	2079	1672.5
2038	816.6	2080	1625.0
2043	757.8	2081	1473.8
2047	769.2	2082	1695.5
2051	841.0	2085	1483.4
2060	809.1	2086	1923.7
2062	805.8	2087	1476.1
2083	822.2.		

In the nature, occurrence of normal, deficit and excess rainfall is a part of the weather game. Therefore, deficit (drought) years recording <25 per cent less rain whereas excess years recording >25 per cent higher rain were identified and respective years along with their corresponding rainfall are presented in Table-1. It is interesting to note that, out of 77 years (2014-2090) of forecast, 50 years will receive normal rain, 14 years will receive deficit rain and 13 years will receive excess rain. Highest rainfall of 1923.7 mm will be received during 2086 whereas the lowest rainfall of 774.1 mm will be received during 2015.

PRECIS RCM model run weather data results although has their own limitations but IPCC (2007) has accepted it as one of the best out of all the available sources of weather data. Therefore, this information is very helpful in advance planning to save our crops against the vagaries of nature. This could also be very good information for agricultural planners and managers of the state for better prepareness against any unwonted incident. Further studies to forecast rainfall in other stations of Uttarakhand state for the same period could also be tried to ascertain the facts. Pranuthi *et al* (2014) has reported only 60 per cent parity in the actual and model predicted rainfall results for Haridwar.

Rainfall data presented in Fig.1, 2 and 3 showed that rainfall trend up to 2070 recorded a rise @1.3 mm/year but afterward this made a remarkable change and recorded rise @ 15.3 mm/year. This indicate that the last quarter of the century will encounter heavy rainfall events. Those events could be distributed even in the NE (winter) monsoon. This could be a location specific rainfall character and reasons for such a changed behaviour is a matter of further scrutiny.

Monthly rainfall trend analysis employing Mann Kendall's and Theil Sen's slope method as given in Table 2 showed that rainfall increased significantly during July and August both between 2014-2090 @ about 3.2 mm/year. In general the overall increase in rainfall will be @5.8 mm/year. This

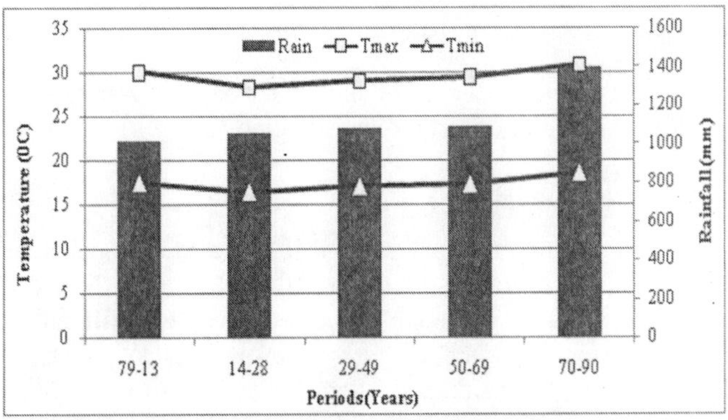

Fig. 1: Rainfall and Temperature trend during different periods (1979-2090) at Roorkee.

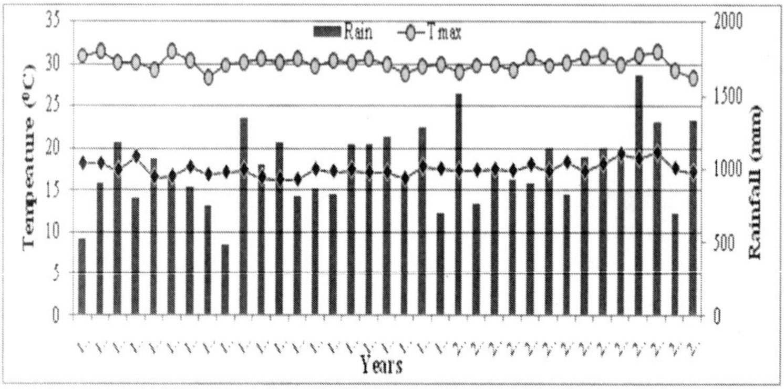

Fig. 2: Observed Annual Rainfall (mm), Max and Min. Temperature (°C) at Roorkee during 1979-2013.

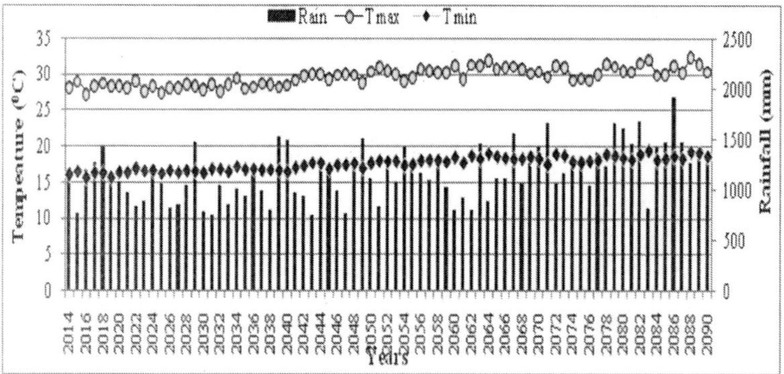

Fig. 3: Forecasted Annual Rainfall (mm), Max & Min. Temperature (°C) at Roorkee during 2014-2090.

response could; however, be pronounced during the last quarter of the century which is a matter of further investigation.

(2) Maximum temperature

Data presented in Fig. 1 showed that the average annual maximum temperature will record a little dip during 2014-2028 and will rise gradually afterward to become slightly above the current level. It is difficult to ascribe any definite reason for such a pattern however is a matter of further investigation. Mann Kendall's and Theil Sen's slope of monthly maximum temperature during 1979 and 2013 (Table 2) recorded significant increase during February and March and decrease during June and July. Forecast for 2014-2090 (Table 3) presented a different scenario. All the months, January to

December, recorded significantly increased maximum temperature. This confirms a significant change in maximum temperature regime to occur especially during the last quarter of the century.

Table 2: Mann Kendall statistics and Thiel Sen's slope estimated for rainfall, maximum and minimum temperature for the projected period (2014 - 2090)

Month	*Rainfall (2014 - 2090)*			*MaxT (2014 - 2090)*			*MinT (2014 - 2090)*		
	Z	*Sig*	*Sen's Slope*	*Z*	*Sig*	*Sen's Slope*	*Z*	*Sig*	*Sen's Slope*
Jan	-1.05		-0.041	3.58	***	0.051	4.74	***	0.032
Feb	0.79		0.012	3.03	**	0.041	5.57	***	0.043
Mar	0.69		0.019	2.23	*	0.025	4.41	***	0.030
Apr	-1.99	*	-0.065	5.01	***	0.050	4.25	***	0.030
May	0.05		0.000	5.31	***	0.045	4.74	***	0.027
Jun	0.78		0.370	2.97	**	0.039	4.26	***	0.028
Jul	4.40	***	3.310	5.98	***	0.041	9.00	***	0.042
Aug	4.68	***	3.161	7.20	***	0.045	9.43	***	0.038
Sep	-1.23		-0.632	7.02	***	0.048	8.40	***	0.042
Oct	-1.81	+	-0.001	6.60	***	0.058	6.88	***	0.061
Nov	-0.68		0.000	4.63	***	0.039	5.39	***	0.044
Dec	0.62		0.004	4.15	***	0.050	5.99	***	0.040
Annual	3.97	***	5.828	7.47	***	0.046	9.50	***	0.039

Note: Bold values indicate significance of trend at any level of significance.
*** Significant trend at 0.001 level of significance, **Significant trend at 0.01 level of significance,*Significant trend at 0.05 level of significance, +Significant trend at 0.1 level of significance.

(3) Minimum Temperature

Data presented in Fig. 1 showed that the average annual minimum temperature will record a little dip during 2014-2028 and will rise gradually afterward to become slightly above the current level. This trend is at par with that of the maximum temperature. The exact reason is a matter of further investigation. Mann Kendall's and Theil Sen's slope of monthly minimum temperature during 1979 and 2013 (Table 2) recorded significant increase during February and March, May and June as well as the September and October. However, the forecast for 2014-2090 (Table 3) presented a different scenario. All the months, January to December, recorded significantly increased

Table 3: Total/ Average annual Forecast of Rainfall (mm), Maximum Temperature (°C) and Minimum Temperature (°C) for the period 2014-2090

S.No.	Years	Total Rain (mm)	Average Tmax (°C)	Average Tmin (°C)	S.No.	Years	Total Rain (mm)	Average Tmax (°C)	Average Tmin (°C)	S.No.	Years	Total Rain (mm)	Average Tmax (°C)	Average Tmin (°C)
1	2014	1201.2	28.2	16.1	27	2040	1501.3	28.6	16.6	53	2066	1113.2	31.1	18.4
2	2015	774.1	29.1	16.6	28	2041	982.4	29.3	17.1	54	2067	1576.5	31.3	18.3
3	2016	1201.0	27.2	15.7	29	2042	949.0	30.0	17.4	55	2068	1070.1	30.9	18.3
4	2017	1278.5	28.5	16.4	30	2043	757.8	30.1	17.8	56	2069	1289.5	30.3	18.6
5	2018	1427.0	28.8	16.3	31	2044	1261.1	30.2	17.7	57	2070	1433.7	30.5	18.4
6	2019	1087.7	28.5	15.7	32	2045	1171.4	29.4	16.9	58	2071	1671.7	29.8	17.6
7	2020	1079.4	28.4	16.4	33	2046	996.1	30.0	17.5	59	2072	1070.4	31.4	19.0
8	2021	983.2	28.3	16.4	34	2047	769.2	30.1	17.5	60	2073	1170.3	31.1	18.8
9	2022	836.8	29.1	17.0	35	2048	1245.2	30.0	17.7	61	2074	1337.8	29.3	17.9
10	2023	904.3	27.7	16.6	36	2049	1514.1	28.9	17.0	62	2075	1352.7	29.5	17.8
11	2024	1157.8	28.4	16.7	37	2050	1121.1	30.5	17.8	63	2076	1044.9	29.3	18.0
12	2025	1061.5	27.5	16.3	38	2051	841.0	31.1	18.0	64	2077	1380.9	30.2	18.2
13	2026	833.4	28.3	16.7	39	2052	1317.9	30.7	18.0	65	2078	1240.8	31.7	18.9
14	2027	862.3	28.3	16.4	40	2053	1092.9	30.1	18.0	66	2079	1672.5	31.2	18.8
15	2028	1050.6	28.8	16.7	41	2054	1428.0	29.2	17.4	67	2080	1625.0	30.7	18.4
16	2029	1488.7	28.5	16.6	42	2055	1205.8	29.7	17.5	68	2081	1473.8	30.5	18.2
17	2030	793.6	28.0	16.4	43	2056	1170.9	31.0	18.1	69	2082	1695.5	31.8	19.0
18	2031	763.5	28.7	16.9	44	2057	1101.8	30.6	18.1	70	2083	822.2	32.2	19.5
19	2032	1058.1	27.8	16.9	45	2058	1314.3	30.4	18.1	71	2084	1431.6	30.0	18.2
20	2033	863.2	28.7	16.6	46	2059	1033.3	30.4	18.0	72	2085	1483.4	30.2	18.3
21	2034	1021.3	29.5	17.2	47	2060	809.1	31.3	18.5	73	2086	1923.7	31.3	18.7
22	2035	951.2	28.1	16.8	48	2061	927.7	29.4	17.8	74	2087	1476.1	30.4	18.4
23	2036	1186.0	28.3	16.9	49	2062	805.8	31.5	18.8	75	2088	1282.7	32.6	19.3
24	2037	995.6	28.8	16.8	50	2063	1470.8	31.3	18.3	76	2089	1400.6	31.5	19.2
25	2038	816.6	28.8	16.8	51	2064	891.7	32.1	19.1	77	2090	1356.0	30.5	18.7
26	2039	1540.0	28.3	16.7	52	2065	1112.9	30.9	18.7	**Total/Avg**		**1174.1**	**29.8**	**17.6**

minimum temperature. This confirms a significant change in minimum temperature regime to occur especially during the last quarter of the century.

Summary and Conclusion

Weather (Rainfall, Tmax, Tmin) condition is undoubtedly going to be appreciably different in 21st century from that of the 20th century but will be clearly visualised in the last quarter of the 21st century. Actual observations of 1979-2013 showed a significant increase in Rainfall during August; significant increase in Tmax during February and March whereas significant increase in Tmin during February to June and September, October. Forecasted weather data (bias corrected PRECIS RCM) during 2014-2090 showed a significant increase in Rainfall during July and August; significant increase in Tmax and Tmin both during January to December. Highest rainfall (1923.7 mm) in 2086 may present an alarming situation for most of the crops grown and water conducting systems in the district. Precautionary measure for such event is, therefore suggested.

References

Ahsan, M.N, M.F. Ahmed, M.H. Bappy, M.N. Hasan, and N. Nahar, 2011. "Climate change induced vulnerability on living standard a study on south western coastal region of Bangladesh". J. Innov. Dev. Strategy, Vol. 5(3) pp. 24-28.

Anonymous. 2011. "Agricultural Statistics of Uttarakhand". Published by Directorate of Agriculture Uttarakhand, Dehradun, India. 2.

FAO. 2007. "Livelihood adaptation to climate variability and changes in drought-prone areas of Bangladesh", Rome, Italy.

Fowler, H. J., S. Blenkinsopp and C. Tebaldi 2007. "Linking climate change modelling to impacts studies: Recent advances in downscaling techniques for hydrological modeling". Int. J. Climatol, Vol. 27, pp. 1547–1578.

Grotch, S.L., and M.C. MacCracken, 1991. "The use of general circulation models to predict regional climatic change". Journal of Climate, Vol. 4(3) pp. 286-303.

IPCC. 2007. "Climate Change impacts, adaptation and vulnerability": contribution of Working Group II to the fourth assessment report of the Inter-governmental Panel on Climate Change, Cambridge University Press, Cambridge, UK.

Kendall, M.G. 1975. "Rank Correlation Methods", 4th edition, Charles Griffin, London.

Leander, R., and T. A. Buishand. 2007. "Resampling of regional climate model output for the simulation of Extreme River flows". J Hydrol., Vol. 332(3) pp. 487-496.

Lenderink, G., A., Buishand, and van W. Deursen, 2007. "Estimates of future discharges of the river Rhine using two scenario methodologies: direct versus delta approach". Hydrol. Earth Syst. Sci., Vol. 11 pp.1145–1159.

Mann, H.B. 1945. "Non-parametric tests against trend" Econometrica, Vol. 13 pp. 163-171.

Maraun, D., F., A.M. Wetterhall, R.E., Chandler, E.J., Kendon, M., Widmann, and I. ThieleEich. 2010. "Precipitation downscaling under climate change": Recent developments to bridge the gap between dynamical models and the end user. Rev Geophys, Vol. 48(3).

Pranuthi, G. S.K. Tripathi and S.K. Dubey. 2014 "Climate change assessment using PRECIS-RCM data for Hardwar district of Uttarakhand". Paper presented in International Annual Meeting Grand Challenges and Great Solutions organised by ASA, CSSA and SSSA at Convention Centre, Long Beach, CA from Nov. 2-5, 2014.

Roudier, P., B. Sultan, P. Quirion, and A. Berg. 2011. "The impact of future climate change on West African crop yields": What does the recent literature say? Global Environ. Change, Vol. 21(3) pp. 1073-1083.

Sen, P.K. 1968. "Estimates of the regression coefficient based on Kendall's tau". J. Am. Stat. Assoc. Vol. 63, pp.1379–1389.

Theil, H. 1950. "A rank invariant method of linear and polynomial regression analysis Part 3". Nederl.Akad.Wetensch. Proc. Vol. 53, pp. 1397–1412.

Tripathi, S.K., Manvedra Singh and Ashish Pandey 2007. "Agro climatic Variability Analysis of Roorkee, Uttarakhand". J. Indian Water. Resour. Soc. Vol. 3 and 4 pp. 24-30.

Tripathi, S.K., Manvendra Singh and Ashish Pandey 2008a. "Climate change at Roorkee (Uttarakhand): A case study". J. Indian Water. Resour. Soc. Vol. 28 (1) pp. 19-28.

Tripathi, S.K., Manvendra Singh and Ashish Pandey 2008b. "Analysis of Agroclimatic Extremes for Crop Cultivation and Diversification". J. Indian Water. Resour. Soc. Vol. 28 (4) pp. 34-42.

Yu, J.Y., and S. Tae Kim. 2010. "Three evolution patterns of Central-Pacific El Niño". Geophys. Res. Lett., Vol. 37, pp. 3-6.

2

Long Term Temperature Trends around Six Cities in North East India Using Mann-Kendall and Sen's Slope Technique

S.R. Kumar

Abstract

In the present study, the non-parametric Mann-Kendall test is used to test for significant trend in atmospheric temperature in North East India through randomly selected six locations. The selected locations are; (1) Agartala, (2) Cooch Behar, (3) Guwahati, (4) Dibrugarh, (5) Imphal and (6) Cherrapunji, with an elevation range from 12.8 to 1313 meters above mean sea level. The Mann-Kendall method does not give the rate of change, so to estimate the slope of an existing trend that is the change in temperature (°C) per year, the nonparametric Sen's slope method is used. The trend analysis has been carried out using twelve monthly series of mean monthly maximum and minimum temperature series for the selected cities of North East (NE) India.

Introduction

In recent years, investigations on climate change due to the buildup of greenhouse gases have mushroomed, and climate-related issues have started dominating and influencing policy decisions at various levels of governance in both developed and developing nations. Therefore, global warming due to the anthropogenic-driven emissions of greenhouse gases, land-use and land-cover changes has emerged as one of the important environmental issues of the 21st century (Jhajharia *et al.*, 2011). The global mean surface temperature has increased by 0.89 [0.69 to 1.08] °C over the period 1901–2012 which is mainly attributed to anthropogenic activities (IPCC, 2013). This unprecedented increase is expected to have severe impact on global hydrological systems, ecosystems, sea level, crop production and related processes. The impact would be particularly severe in the tropical areas, which mainly consist of developing countries, including India.

Trend detection is an active area of interest for both hydrology and climatology in order to investigate climate changes scenarios and enhance climate impact research. Statistical tools commonly used to detect significant trends in climatic and hydrological time series is either or both of the non-parametric test such as Mann-Kendall or Spearman's rank correlation and the parametric test such as student's t-test. The non-parametric test is considered better because it is a function of the ranks of observation and it displays much insensitivity to outliers unlike the parametric counterpart.

The Mann-Kendall method does not give the rate of change, so to estimate the slope of an existing trend that is the change in temperature (°C) per year, the nonparametric Sen's method is used. This test is applied in cases the trend is assumed to be linear, depicting the quantification of change per unit time.

These methods offer many advantages that have made them useful in analysing atmospheric data. Missing values are allowed and the data need not conform to any particular distribution. Besides, the Sen's method is not greatly affected by single data errors or outliers.

Mann-Kendall Analysis

The Mann-Kendall test, which was originally devised by Mann (1945) and further derived by Kendall (1975) as a non-parametric approach, has been widely used in different fields of research for the detection of trend in time series. These tests are widely used in environmental science because they are simple, robust and can cope with missing values and values below a detection limit. Therefore, it was found to be an effective tool for identifying trends in hydrology and water resources (e.g. Burn, 1994; Xu *et al.*, 2003; Ludwig *et al.*, 2004). One benefit of this test is that the data need not conform to any particular distribution. Mann-Kendall test has the advantage of not assuming any distribution form for the data and has similar power as its parametric competitors. Therefore, it is highly recommended for general use by the World Meteorological Organisation.

Mann-Kendall Statistic (S)

The non-parametric Mann-Kendall test, which is commonly used for hydrologic data analysis, can be used to detect trends that are monotonic but not necessarily linear. The null hypothesis in the Mann-Kendall test is that the data are independent and randomly ordered. The Mann-Kendall test does not require the assumption of normality, and only indicates the direction but not the magnitude of significant trends (McBean and Motiee, 2008).

The Mann-Kendall test is a non-parametric test for identifying trends in time series data. The test compares the relative magnitudes of sample data rather than the data values themselves (Gilbert, 1987). One benefit of this test is that the data need not conform to any particular distribution. Moreover, data reported as non-detects can be included by assigning them a common value that is smaller than the smallest measured value in the data set. The data values are evaluated as an ordered time series. Each data value is compared to all subsequent data values. The initial value of the Mann-Kendall statistic, S, is assumed to be 0 (*e.g.*, no trend). If a data value from a later time period is higher than a data value from an earlier time period, S is incremented by 1. On the other hand, if the data value from a later time period is lower than a data value sampled earlier, S is decremented by 1. The net result of all such increments and decrements yields the final value of S.

Given a data set of *n* size, whereby n is not less than 10, for a mean of zero and a standard deviation of 1, and assuming that the time series is independent, then the Mann-Kendall statistic *S* is described as:

Let $x_1, x_2, \ldots x_n$ represent n data points where x_j represents the data point at time j. Then the Mann-Kendall statistic (S) is given by,

$$S = \sum_{k=1}^{n-1} \sum_{j=k+1}^{n} sign\left(x_j - x_k\right)$$

where,

$$\begin{aligned} sign\ (x_j - x_k) &= \text{if } x_j - x_k > 0 \\ &= 0 \text{ if } x_j - x_k = 0 \\ &= -1 \text{ if } x_j - x_k < 0 \end{aligned}$$

sign() is the signum function. A very high positive value of S is an indicator of an increasing trend, and a very low negative value indicates a decreasing trend. However, it is necessary to compute the probability associated with S and the sample size, *n*, to statistically quantify the significance of the trend (Khambhammettu, 2005).

Probability Associated with M-K Statistic

Kendall (1975, p. 55) describes a normal-approximation test that could be used for datasets with more than 10 values, provided there are not many tied values within the data set. The test procedure is as follows:

- Calculated S as described above.

- Calculated the variance of S, VAR(S), by the following equation:

$$VAR\ (S) = \frac{1}{18}[n(n-1)(2n+5)]$$

For situation where ties occur, then $VAR(S)$ is extended to the form:

$$VAR\ (S) = \frac{1}{18}\left[n(n-1)(2n+5)-\sum_{p-1}^{g}t_p(t_p-1)(2t_p+5)\right]$$

where n is the number of data points (larger than10), g is the number of tied groups (a tied group is a set of sample data having the same value), and t_p is the number of data points in the pth group.

- Computed a normalised test statistic Z_c as follows (Partal and Kahya, 2006):

$$Z_c = \begin{cases} \frac{(S-1)}{\sqrt{var(S)}} & \ldots \text{ if } S>0 \\ 0 & \ldots \text{ if } S=0 \\ \frac{(S+1)}{\sqrt{var(S)}} & \ldots \text{ if } S<0 \end{cases}$$

- Computed the probability associated with this normalised test statistic. The probability density function *(pdf)* for a normal distribution with a mean of 0 and a standard deviation of 1 is given by the following equation:

$$f(z_c) = \frac{1}{\sqrt{2\pi}}e^{\frac{z^2}{2}}$$

The value of *pdf* should be equal to 1.000 or relatively very close to 1.000 to have a high chance of a significant change occurring.

- Decided on a probability level of significance (*95% typically*). The hypothesis of no trend is rejected if $|Z_c| > Z_{1-\frac{\alpha}{2}}$, where Z is taken from the standard normal distribution table and α is level of significance. The values of Mann-Kendall test statistic are computed and it may be seen that, if the value lies within the limits (–)1.96 and (+)1.96, i.e.,

$-Z_{1-\frac{\alpha}{2}} \leq Z_c \leq +Z_{1-\frac{\alpha}{2}}$, then the null hypothesis of no trend can be accepted at the 5% (α) level of significance using a two-tailed test. Otherwise, the null hypothesis can be rejected and the alternative hypothesis can be accepted at the significant level of a, i.e., 5%. If $Z_c >$ +1.96, there is 'increasing' trend; and if $Z_c <$ -1.96, there is 'decreasing' trend. If the computed probability is less than the level of significance, there is 'no trend'.

Because it is based on ranks, this test evaluates the strength of a trend and not a rate of temporal change.

Sen's slope estimator

One of the most useful parametric models to detect trend is the "Simple Linear Regression" model. However, the method of linear regression requires the assumptions, normality of residuals (McBean and Motiee, 2008). Viessman *et al.*, (1989) has noted that many hydrologic variables exhibit a marked right skewness partly due to the influence of natural phenomena, and not follow a normal distribution. This methods offer many advantages that have made them useful in analysing atmospheric data. Missing values are allowed and the data need not conform to any particular distribution.

The Mann-Kendall method does not give the rate of change, so to estimate the slope of an existing trend that is the change in temperature (°C) per year, the nonparametric Sen's method (Salmi *et al.*, 2002) given below is used. This test is applied in cases the trend is assumed to be linear, depicting the quantification of change per unit time (Sen. 1968; Gilbert. 1987). De Lima *et al.* (2005) point out the utility of this estimator in cases there are missing values or other gaps in the data, as it remains unaffected from outliers or gross errors (Salmi *et al.*, 2002). Sen's slope is the median of all possible pair-wise slopes (Burn *et al.*, 2002), and assumes the trend is linear and is calculated by:

$$f(t) = Qt + \beta$$

where Q is slope and β is a constant indicating the vertical axe crossing of the slope line and Q is obtained by calculating the slopes of all data pairs:

$$Q_i = (x_j - x_k)/(j-k)$$

where $j > k$ (Salmi *et al.*, 2002).

If there are n values x_j in the time series there will be as many as $N = n(n-1)/2$ slope estimates Q_i. When there is only one datum in each time period, then $N = n(n-1)/2$, where n corresponds to the number of time periods. The N values of slopes are ranked from the smallest to largest and if N is odd, Sen's estimator of slope is calculated as $Q_{median} = Q_{(N+1)/2}$. On the other hand, in case that N is even, the estimator arises from $Q_{median} = [Q_{N/2} + Q_{(N+2)/2}]/2$.

The Sen's estimator of slope is the median of these N values of Q_i. To obtain an estimate of β in above equation, the n values of differences $x_i - Qt_i$ are calculated. The median of these values gives an estimate of β (Salmi *et al.*, 2002).

Serial Correlation Correction

Autocorrelation refers to the correlation of a time series with its own past and future values. It has been found that positive serial correlation inflates the variance of the MK statistic S and hence increases the possibility of rejecting the null hypothesis of no trend (Von Storch, 1995). In order to reduce the impact of serial correlation, it is common to pre-whiten the time series by removing serial correlation from the series. For this purpose the new time series as proposed by Kumar *et al.* (2009) and Jhajharia *et al.* (2011) can be obtained as:

$$x_i' = x_i - (Q * i)$$

where Q is Sen's estimator (Sen, 1968). The value of r_l (significance of lag-1 serial correlation) of the new time series is first computed and later used to determine the residual series as:

$$y' = x' - r_{1*}x_{i-1}'$$

The value of Q_*i was added again to the residual data set as:

$$y_i = y_i' + (Q * i)$$

The y_i series was subjected to trend analysis. If the absolute value of r_l was less than the significance level value, then rainfall time series were subjected to the original MK test. Otherwise, the effect of serial correlation was removed from the time series by pre-whitening prior to applying the MK test in the series as mentioned above.

Data Collection

Monthly mean data of maximum and minimum temperature record for the selected cities (Fig.1 and Table 1) were obtained from the Indian Meteorological Department (IMD) website. IMD is the agency responsible for the measurement, control, and storage of the meteorological data in India.

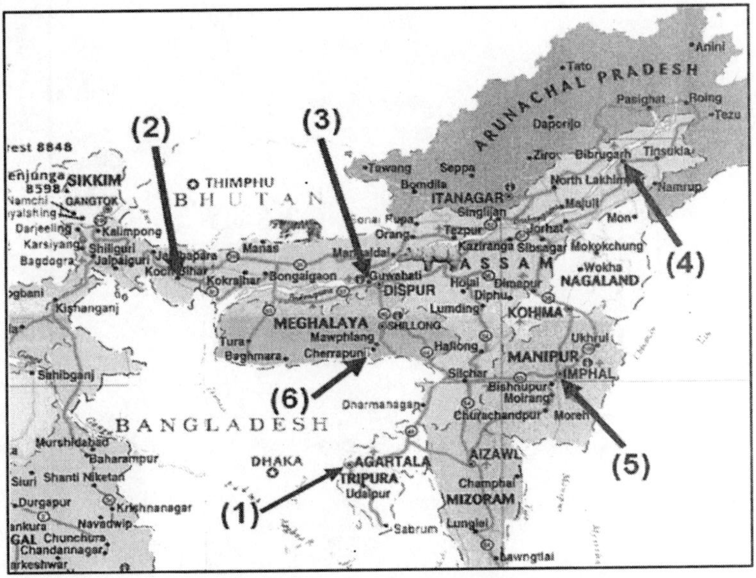

Fig. 1: Study Area with Selected Locations.

Table 1: Details of Temperature Measurement Stations in NE Region of India

Sl. No.	*Location*	*Latitude*	*Longitude (N)*	*Elevation (E)*	*State (MSL-m)*	*Data Availability*
1.	Agartala	23°	91°	12.8	Tripura	48 year
2.	Cooch Behar	26°13'	89°17'	42	West Bengal	100 year
3.	Guwahati	26°11'	91°54'	54	Assam	50 year
4.	Dibrugarh	27°29'	94°54'	111	Assam	49 year
5.	Imphal	24°44'	93°58'	774	Manipur	47 year
6.	Cherrapunji	25°15'	91°44'	1313	Meghalaya	98 year

Trend Analysis

The trend analysis was done in two phases for each month of selected locations. First the presence of a monotonic increasing or decreasing trend was tested with the nonparametric Mann-Kendall test and secondly the slope of a linear trend is estimated with the non parametric Sen's slope estimator. The result of the Mann-Kendal and Sen's slope analysis are presented in Table 2. For the five tested significance levels the following symbols are used in the analysis:

Table 2: Mann-Kendall Statistic and Sen's Trend Slope of the Studied Temperature Stations

Time series	*Mann-Kendall Trend*				*Sen's slope and constant*		*Mann-Kendall Trend*			*Sen's slope and constant*		
	Test Z value		*pdf*	*Significance*	*Sen's Slope Q (°C/year)*	*Constant β*	*Test Z value*		*pdf*	*Significance*	*Sen's Slope Q (°C/year)*	*Constant β*
Agartala-Jan	-0.87	↓	0.19	$	-0.008	25.82	1.16	↑	0.88	$	0.018	9.56
Feb	-1.93	↓	0.03	+	-0.030	29.00	0.41	↑	0.66	$	0.007	13.34
Mar	-2.74	↓	0.00	**	-0.041	33.61	0.37	↑	0.64	$	0.007	18.51
Apr	-2.95	↓	0.00	**	-0.056	34.77	-0.37	↓	0.36	$	-0.003	22.29
May	-1.40	↓	0.08	$	-0.023	33.12	-1.93	↓	0.03	+	-0.025	24.00
Jun	2.96	↑	1.00	**	0.032	31.11	2.32	↑	0.99	*	0.019	24.30
Jul	1.29	↑	0.90	$	0.009	31.10	1.35	↑	0.91	$	0.008	24.68
Aug	3.17	↑	1.00	**	0.024	31.22	2.18	↑	0.99	*	0.011	24.70
Sept	-0.04	↓	0.48	$	0.000	31.70	0.82	↑	0.79	$	0.005	24.37
Oct	1.92	↑	0.97	+	0.015	30.79	0.14	↑	0.56	$	0.000	21.90
Nov	3.14	↑	1.00	**	0.027	28.76	2.10	↑	0.98	*	0.038	15.78
Dec	0.38	↑	0.65	$	0.000	26.50	0.71	↑	0.76	$	0.009	11.16
Cooch Behar-Jan	0.26	↑	0.6033	$	0.00000	23.55	-3.54	↓	0.0002	***	-0.01846	10.65
Feb	0.37	↑	0.6444	$	0.00211	25.58	-2.23	↓	0.0130	*	-0.01485	12.38
Mar	0.96	↑	0.8324	$	0.00619	29.65	-1.02	↓	0.1549	$	-0.00430	15.88
Apr	0.46	↑	0.6766	$	0.00408	31.49	-1.19	↓	0.1172	$	-0.00526	20.58
May	0.31	↑	0.6209	$	0.00156	30.90	-2.54	↓	0.0056	*	-0.01045	22.73
Jun	2.33	↑	0.9900	*	0.00899	30.55	-2.32	↓	0.0101	*	-0.00758	24.52

Cont...

Cont...

Jul	-0.37	↓	0.3569	$	0.00000	31.30	-3.74	↓	0.0001	***	-0.01250	25.76
Aug	2.69	↑	0.9964	**	0.01154	31.20	-3.41	↓	0.0003	***	-0.01176	25.72
Sept	-1.11	↓	0.1333	$	-0.00340	31.40	-4.58	↓	0.0000	***	-0.01826	25.39
Oct	1.14	↑	0.8733	$	0.00400	30.35	-3.28	↓	0.0005	**	-0.02000	22.08
Nov	3.70	↑	0.9999	***	0.01449	27.28	-2.15	↓	0.0158	*	-0.01368	16.12
Dec	2.42	↑	0.9923	*	0.01099	24.51	-1.90	↓	0.0288	+	-0.01282	11.86
Guwahati-Jan	-0.04	↓	0.4833	$	0.00000	23.80	3.08	↑	0.9990	**	0.02857	9.55
Feb	-1.76	↓	0.0393	+	-0.02727	26.88	2.75	↑	0.9971	**	0.03226	11.06
Mar	-1.16	↓	0.1238	$	-0.02000	30.52	1.12	↑	0.8692	$	0.00805	15.38
Apr	-0.76	↓	0.2228	$	-0.01176	31.10	0.00	↑	0.5000	$	0.00000	20.10
May	0.81	↑	0.7917	$	0.00976	30.96	0.67	↑	0.7489	$	0.00385	22.39
Jun	2.27	↑	0.9882	*	0.01538	31.22	1.65	↑	0.9509	+	0.00857	24.63
Jul	0.82	↑	0.7942	$	0.00526	31.67	0.36	↑	0.6388	$	0.00000	25.50
Aug	1.03	↑	0.8482	$	0.00646	32.00	-0.27	↓	0.3940	$	0.00000	25.50
Sept	-1.18	↓	0.1181	$	-0.00755	31.73	-0.24	↓	0.4042	$	0.00000	24.60
Oct	1.21	↑	0.8870	$	0.00909	30.06	1.00	↑	0.8417	$	0.00876	21.82
Nov	1.99	↑	0.9765	*	0.01625	27.19	3.46	↑	0.9997	***	0.03463	15.69
Dec	0.49	↑	0.6889	$	0.00352	24.55	1.75	↑	0.9603	+	0.01835	11.41
Dibrugarh-Jan	1.25	↑	0.8948	$	0.01319	22.16	0.87	↑	0.8083	$	0.01429	8.54
Feb	-0.11	↓	0.4553	$	0.00000	23.90	1.41	↑	0.9205	$	0.01379	11.86
Mar	-2.18	↓	0.0145	*	-0.03162	27.06	2.75	↑	0.9970	**	0.02466	15.11

Cont...

Cont...

Apr	-1.74	↓	0.0407	+	-0.02500	28.53	1.23	↑	0.8900	$	0.00833	18.61
May	2.00	↑	0.9774	*	0.02692	28.89	2.60	↑	0.9954	**	0.01719	21.27
Jun	0.89	↑	0.8130	$	0.01000	30.79	2.82	↑	0.9976	**	0.01429	23.69
Jul	-0.67	↓	0.2529	$	-0.00563	31.21	2.78	↑	0.9973	**	0.01111	24.28
Aug	1.18	↑	0.8815	$	0.01342	31.45	3.44	↑	0.9997	***	0.01379	24.47
Sept	-1.50	↓	0.0664	$	-0.01231	30.89	2.23	↑	0.9873	*	0.00960	23.55
Oct	3.30	↑	0.9995	***	0.03158	28.73	1.77	↑	0.9618	+	0.01818	20.16
Nov	2.80	↑	0.9975	**	0.02952	26.50	3.15	↑	0.9992	**	0.04558	13.37
Dec	2.47	↑	0.9932	*	0.02957	23.27	1.61	↑	0.9458	$	0.01968	9.33
Imphal-Jan	1.72	↑	0.9572	+	0.02414	21.02	1.86	↑	0.9688	+	0.03280	3.17
Feb	0.46	↑	0.6777	$	0.00690	23.34	1.85	↑	0.9680	+	0.03846	6.14
Mar	-0.26	↓	0.3967	$	-0.00526	27.01	3.43	↑	0.9997	***	0.05714	9.78
Apr	-0.95	↓	0.1707	$	-0.01868	28.92	1.65	↑	0.9510	+	0.01968	15.09
May	0.65	↑	0.7416	$	0.00785	28.71	0.74	↑	0.7718	$	0.01026	18.31
Jun	3.28	↑	0.9995	**	0.03101	28.40	1.59	↑	0.9438	$	0.01376	20.95
Jul	1.99	↑	0.9768	*	0.01111	28.73	1.00	↑	0.8417	$	0.00678	21.64
Aug	2.91	↑	0.9982	**	0.02000	28.66	0.48	↑	0.6846	$	0.00333	21.54
Sept	0.74	↑	0.7707	$	0.00506	28.70	0.27	↑	0.6045	$	0.00000	20.60
Oct	3.00	↑	0.9986	**	0.03333	27.37	0.14	↑	0.5565	$	0.00000	17.20
Nov	2.46	↑	0.9931	*	0.02500	24.64	1.99	↑	0.9769	*	0.04514	9.89
Dec	1.83	↑	0.9667	+	0.02376	21.68	0.70	↑	0.7596	$	0.01127	4.83

Cont...

Cont...

Cherrapunji-Jan	0.53	↑	0.7016	$	0.00185	15.66	-3.23	↓	0.0006	**	-0.01282	7.99
Feb	2.74	↑	0.9969	**	0.01304	16.56	-1.66	↓	0.0480	+	-0.00635	9.08
Mar	1.27	↑	0.8988	$	0.00571	20.17	-1.16	↓	0.1236	$	-0.00455	12.80
Apr	3.21	↑	0.9993	**	0.01347	21.21	-0.48	↓	0.3161	$	-0.00164	15.04
May	1.50	↑	0.9335	$	0.00435	22.15	0.09	↑	0.5344	$	0.00000	16.30
Jun	2.69	↑	0.9964	**	0.00725	22.05	-0.50	↓	0.3087	$	0.00000	17.90
Jul	-0.41	↓	0.3425	$	0.00000	22.20	-1.99	↓	0.0234	*	-0.00417	18.53
Aug	3.80	↑	0.9999	***	0.01111	22.22	-1.23	↓	0.1086	$	-0.00169	18.39
Sept	0.54	↑	0.7050	$	0.00000	22.90	-2.92	↓	0.0018	**	-0.00588	18.22
Oct	4.72	↑	1.0000	***	0.01304	21.93	-0.94	↓	0.1727	$	-0.00306	15.86
Nov	3.84	↑	0.9999	***	0.01304	19.43	-0.26	↓	0.3993	$	0.00000	12.00
Dec	3.52	↑	0.9998	***	0.01286	16.56	-0.63	↓	0.2639	$	-0.00227	8.74

Results for mean monthly maximum temperature Results for mean monthly minimum temperature.

↓ Downward Trend ↑ Upward Trend

*** if trend at [illegible] = 0.001 level of significance

** if trend at [illegible] = 0.01 level of significance

* if trend at [illegible] = 0.05 level of significance

+ if trend at [illegible] = 0.1 level of significance

$ the significance level is greater than 0.1

Developed Sen's model equations for mean monthly maximum and minimum temperature (°C) is presented through general equation format as:

$$f(\text{year}_{\text{month}}) = Q^*(\text{year}_{\text{month}} - \text{first data year}_{\text{month}}) + \alpha$$

Conclusion

Temperature is one of the climatic factors that can indicate climate change. Trend detection in temperature time series is crucial for planning and designing regional water resources management. The non-parametric test was used to detect monotonic trends, and the Sen's slope estimator was used to develop model for the variable. The outcome could help to inject inputs for proper planning in the region.

Acknowledgement

Author is thankful to Director, NIH for kind permission to publish the paper.

References

Burn, D.H. (1994). Hydrologic effects of climate change in west-central Canada. *Journal of Hydrology*, 160(1), pp. 53–70.

Burn, D.H., Elnur, M.A.H. (2002). Detection of hydrologic trends and variability. *Journal of Hydrology*, 255(1), pp. 107–122.

De Lima, M.I.O., Marques, A.C.P., De Lima, J.L.M.P., Coelho M.F.E.S. (2005). Precipitation trends in Mainland Portugal in the period 1941-2000. The Fourth Inter-Celtic Colloquium on Hydrology and Management of Water Resources, Guinmaraes, Portugal.

Gilbert, R.O. (1987). Statistical methods for environmental pollution monitoring. Van Nostrand Reinhold, New York.

IPCC (2013). Climate Change 2013—The Physical Science Basis. Cambridge University Press, UK.

Jhajharia, D., Dinpashoh, Y., Kahya, E., Singh, V.P., Fakheri-Fard, A. (2011). Trends in reference evapotranspiration in the humid region of northeast India. *Jr. Hydrol. Process.* (2011). Online publication DOI: 10.1002/hy, p. 8140.

Kendall, M.G. (1975). Rank correlation methods, 4th ed. Charles Griffin, London.

Khambhammettu, P. (2005). Mann-Kendall Analysis, Annual Groundwater Monitoring Report of Hydro Geologic Inc, Fort Ord, California Olofintoye, O.O. USEP : *Journal of Research Information in Civil Engineering*,Vol.7, No. 2, 2010, Sule, B.F. 48.

Kumar, R., Bhakar, S.R., Jhajharia, D. and Durbude, D.G. (2009). Performance evaluation of synthetic envelopes in Indira Gandhi Command Area, Rajasthan, *ISH Journal of Hydraulic Engineering*, 15(2), 1-15.

Ludwig, W., Serrat, P., Cesmat, L., Garcia-Esteves, J. (2004). Evaluating the impact of the recent temperature increase on the hydrology of the Têt River (southern France). *Journal of Hydrology*, 289(1): pp. 204–221.

Mann, H.B. (1945). Non parametric test against trend. *Econometrica*, 13, pp. 245-259.

McBean, E., Motiee, H. (2008). Assessment of Impact of Climate Change on Water Resources: A long term analysis of the Great Lakes of North America, *Hydrology and Earth System Sciences*, Vol. 12, pp. 239–255.

Partal, T., Kahya, E. (2006). Trend analysis in Turkish precipitation data. *Hydrological Processes*; 20: 2011-2026.

Salmi, T., Määttä, A., Anttila, P., Ruoho-Airola T. and Amnell T. (2002). Detecting Trends of Annual Values of Atmospheric Pollutants by the Mann–Kendall Test and Sen's Slope Estimates -The Excel Template Application MAKESENS. Finnish Meteorological Institute, Pub. on Air Quality No. 31. Helsinki, Finland.

Sen, P.K. (1968). Estimates of the regression coefficient based on Kendall's tau. *J. American Statistical Association*; 63: 1379-1389.

Viessman, W., Krapp, J.W. and Harbough, T.E. (1989). Introduction to Hydrology, Third ed., Harper and Row Publishers Inc., New York.

Von Storch, H. (1995). Misuses of statistical analysis in climate research. In: *Analysis of Climate Variability: Applications of Statistical Techniques* (ed. by H. von Storch and A. Navara), 11–26. Springer-Verlag, Berlin, Germany.

Xu, Z.X., Takeuchi, K., Ishidaira, H. (2003). Monotonic trend and step changes in Japanese precipitation. *Journal of Hydrology*, 279(3), pp. 144–150.

3

Joint Event and Continuous Hydrological Modeling Using HEC-HMS- A Suitable Strategy for Improving Modeling Efficiency

G.L. Chunale, S.D. Gorantiwar, G.U. Satpute

Abstract

The present paper deals with joint event and continuous hydrological modelling using HEC-HMS model and its application to small agricultural watershed under Sub-montane zone of Maharashtra. Event model reveals how a basin responds to an individual rainfall event, whereas the continuous model synthesises hydrologic processes and responses of watershed to a number of rainfall events and their cumulative effects over a longer period. The SCS-CN and Soil Moisture Accounting (SMA) process in event and continuous models use dissimilar procedure for simulating surface runoff, infiltration, and other related hydrologic processes, but their basic derivations provide some way to estimate the involved parameters in continuous model. The model parameters viz. initial abstraction (Ia = 22.70 mm) and recession constants (k = 0.766) so optimised for event model at finer time step of 5 min for 18 rainfall-runoff events (2004-2008) and composite curve number (CNi = 63) so computed from known hydrological soil-cover complex were used in continuous modelling operated at coarser scale (1-hr). Six parameters in continuous model optimised using hourly gauged data (2004-2008) was used for verification studies. The quantitative evaluation shows better agreements between the simulated and observed runoff in both the modelling. Small computational time steps of 5-min and 1-hr, respectively used for both event and continuous models. are suitable for detail understanding of hydrologic processes and responses of small watershed. Finer time step (5-min) in event model enables parameter optimisation process in continuous model; which sequentially improves the modelling efficiency. The flood hydrographs simulated by both models are realistic and have a closer match with observed ones in terms of magnitude, shape and symmetry. The study suggests that the joint event and continuous models at finer and coarser time step supported with intensive data can be the effective strategy for improving the overall modelling capabilities.

Introduction

The event based hydrologic modelling reveals how a basin responds to an individual rainfall event, whereas the continuous hydrologic modelling synthesises hydrologic processes and responses of the watershed to a number of rainfall events and their cumulative effects over a longer time period that includes both wet and dry conditions (Chu and Steinman, 2009). The hydrologic modelling and its associated calibration and verification processes often require a large set of spatial and temporal data on topography, land use/ covers, soils, rainfall, flow monitoring data etc., which is in practice, not available or the quality of such data is often an issue one needs to cope with. Most of the watersheds in India are ungauged, where such long-term monitoring data may not be available, or may not be "continuous," or may not have sufficient resolution (small time-interval data). Many times one has to compromise the overall modeling quality because of insufficient high-resolution data for developing, calibrating and validating the model. Under such circumstances, it is critical to develop an effective modelling strategy that will not only take full advantage of the available data but will also improve modeling efficiency.

Present study therefore deals with developing suitable strategy of combining event based modeling at fine scale with the coarse-scale continuous hydrologic modelling using HEC-HMS (USACE-HEC 2006) and its application to the small agricultural watershed, which comes under Sub-montane zone (M.S.). Event based hydrologic modeling when operated at finer time step is useful for better understanding of underlying hydrologic processes and identifying the relevant parameters. This kind of strategy has been already used by Chu and Steinmen (2009) successfully in Mona Lake watershed in west Michigan. The parameters that are well calibrated and validated in fine scale event model will help to improve the quality continuous modelling.

The latest version of HEC-HMS (v 3.4.0) has been used for both kinds of modelling in the present study. The HEC-HMS is a physically based semi-distributed watershed model that incorporates both event based and continuous rainfall-runoff simulations. It is public domain software developed by U.S. Army Corps of Engineers (USACE-HEC, 2006). All the methods included in the model use deterministic techniques, where the parameter values are fixed and same for every simulation. It is designed to simulate the precipitation-runoff processes of watershed systems in a wide range of geographic areas.

The strategy of combining both event and continuous modelling was applied to small agricultural watershed (12.1 ha), which is gauged with autographic instruments such as automatic water stage level recorder (AWSLR) and self-recording rain gauge (SRRG).The watershed represents the typical situation of Sub-montane Zone of Maharashtra for its undulating topography with small ridges and valleys having land slopes ranging from 5.5 to 9.0 per cents (Fig. 1). It is located at Zonal Agricultural Research Station, Shenda park farm, Kolhapur (M.S.). It receives average annual rainfall of 1015.3 mm in 65 rainy days of which monsoon season contributes to more than 80 per cent. The soils are mostly reddish brown to greyish black in colour tending towards lateritic in nature with varying depth and texture due to geologic and climatic changes (Anon, 1996). The soils are well drained with moderately low runoff potential (Hydrologic Soil Group B).

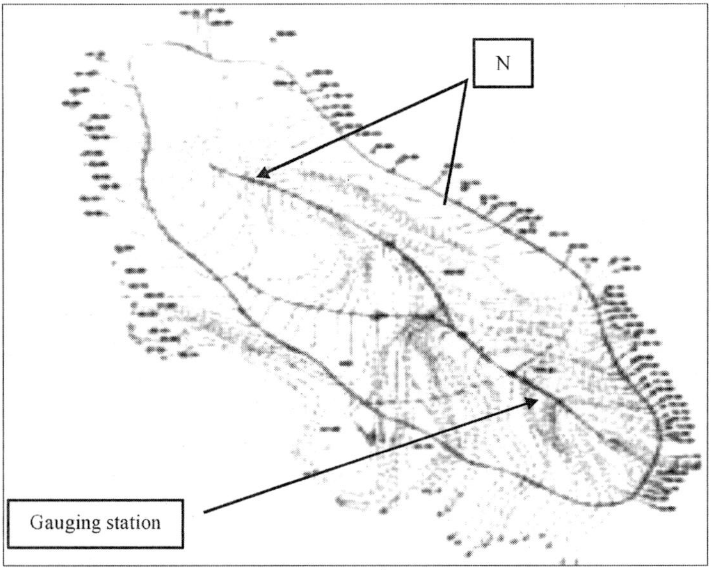

Fig. 1: Shenda park watershed at ZARS, Kolhapur (M.S.).

Methods and Model Development

Event based HEC-HMS model

The event model was formulated using SCS-CN (USDA, 1986), Clark transform (Clark, 1945) and recession base flow (Chow *et al.,* 1988) methods and operated at fine time scale of five-min for simulating different runoff components from the single sub-basin (Fig. 2). The SCS-CN is an empirical, event based rainfall-runoff model. The dimensionless curve number (CN) takes into account, in a lumped way, the effects of land use/cover, soil types, and hydrologic conditions on surface runoff, and relates direct surface runoff with rainfall.

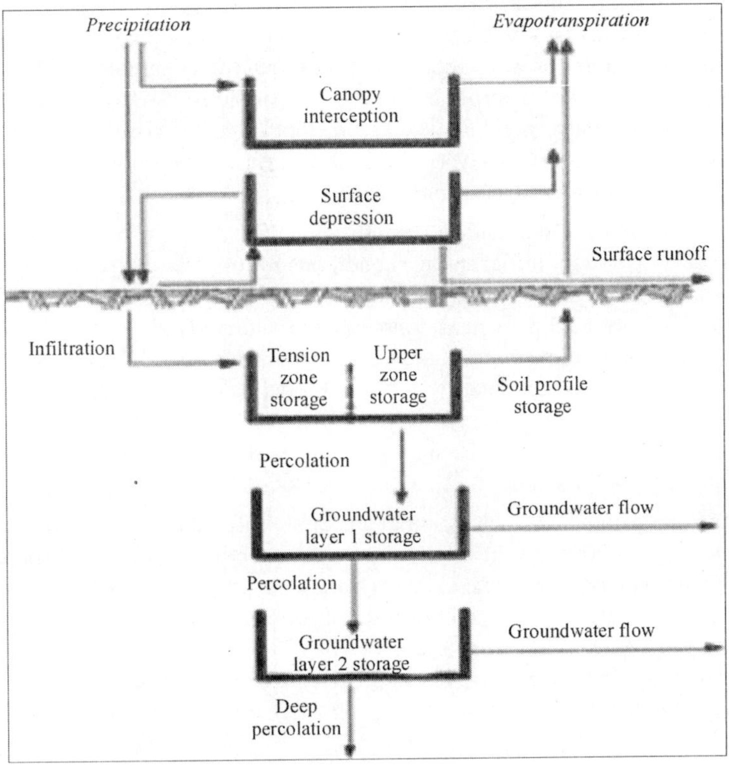

Fig. 2: Conceptual schematic of SMA algorithm (Bennett, 1998).

The Clark transforms method transforms rainfall excess to direct runoff at the outlet of sub-basin, whereas the recession method utilises an exponential recession model to represent base-flow from a sub-basin. Two model parameters *viz.* initial abstraction (Ia) and recession constant (k) were optimised using 18 observed rainfall-runoff events (2004-2008) and the optimal values were decided based on Implicit Stochastic Optimisation (ISO) strategy. Other parameters such as composite curve number (CNi), Clark time of concentration (t_c) and storage coefficient (R) were computed from known hydrological soil-cover complex and physiographic characteristics (Kirpich, 1940; Hoggan, 1997; Bedient and Huber, 1992 and Chunale *et al.*, 2001).

Continuous modeling using SMA process of HEC-HMS

Continuous model was formulated using soil moisture accounting (SMA) process (Bennett, 1998), Clark transform (Clark, 1945) and exponential recession base flow (Chow *et al.,* 1988) methods of HEC-HMS and operated at coarser time step of one hour for *kharif* season of five years from 20-25 May to 31 October (2004-2008). The SMA method is basically a lumped bucket-type model that represents a sub-basin with well-linked storage layers/buckets accounting for canopy interception, surface depression storage, infiltration, evapo-transpiration, as well as soil-water and groundwater percolation (Fig. 2). This method has been incorporated in HEC-HMS (USACE-HEC, 2006). The SMA process is also precipitation loss method used for simulating rainfall excesses and losses in five layers *viz.* canopy, surface depressions, soil profile and two groundwater layers continuously.

There are in all 14 parameters and five initial conditions to be specified for the model operation. Initial storage conditions for these layers were assumed to be constant during simulation. Only upper GW layer was considered for shallow subsurface flow such as drainage from saturated hill slopes and lower GW layer, representing more extensive aquifer and hydraulically connected to stream, was omitted for small watershed. Model parameters such as storage capacities of upper and lower tension zone in the soil and maximum infiltration rate were derived based on the similarities between event based SCS-CN and continuous SMA method.

Model parameters such as initial abstraction (Ia) and recession constant (k) optimised for the event model were used in continuous model, whereas remaining six model parameters such as storage capacities of canopy vegetation and surface depression, percolation rates of soil and upper groundwater layer, its storage capacity and storage coefficient, which cannot be determined or estimated by direct methods, were optimised using observed rainfall-runoff data (2004-2008) and based on ISO strategy, the optimal model parameters were decided.

SCS-CN and SMA methods in HEC-HMS

Although SCS-CN and SMA methods utilise dissimilar methods for simulating surface runoff, infiltration, and other related hydrologic processes, the below given derivations provide a way to estimate the involved parameters in continuous model. The SCS-CN model can be expressed as per USDA (1986).

$$R = P_e^2/(P_e + S) \quad \ldots(1)$$

$$P_e = P - I_a \quad \ldots(2)$$

$$I_a = \square * S \qquad \ldots(3)$$

$$S = (25400/CN)-254 \qquad \ldots(4)$$

where, R = cumulative runoff; P = cumulative rainfall; P_e = cumulative effective rainfall.

(P_e>0; otherwise, R = 0), S = maximum potential retention; Ia = initial abstraction (all losses-interception, surface depression and infiltration prior to runoff); Ia= initial abstraction coefficient. Using model parameters (CN = 63, Ia = 22.70 mm and Ia = 0.152) optimised and computed for event modelling, the Eq. 1 becomes,

$$R = (P-22.70)^2/(P+126.47) \qquad \ldots(5)$$

According to the SMA method in HEC-HMS, rainfall contributes first to the canopy interception (C_s), and then rainwater is available for infiltration, which is determined by infiltration capacity and soil storage (S_s). Any excess rainwater sequentially fills the surface depression storage (S_d) and eventually becomes surface runoff. The potential infiltration rate is given by USACE-HEC (2000).

$$i(t) = i_m * (S_{sd}(t)/S_s) \qquad \ldots(6)$$

$$S_s(t) = S_{s\,max}(t) \qquad \ldots(7)$$

Where, i(t) = potential infiltration rate at time t (the actual infiltration rate also depends on the water available for infiltration at time t); i_m= maximum infiltration rate; S_{smax}= maximum soil water storage; S_s = soil water storage at time t; and $S_{sd}(t)$ = soil water storage deficit at time t. The infiltration rate equals zero when $S_{sd}(t) = 0$ [i.e., $S_s(t) = S_{smax}$] and reaches maximum rate i_m when $S_{sd}(t) = S_{smax}$ [i.e., $S_s(t) = 0$].

$$I_a = C_s + S_d + F_0 \qquad \ldots(8)$$

In which, $F_0 = i_0 t_0$, where F_0= cumulative infiltration before surface runoff starts; i_0= average infiltration rate before surface runoff starts; and t_0 = initial time. Substituting the expression of I_a in Eq. 3 and 4 into Eq. 8, we have,

$$F_0 = Ia\,(25400/CN-254)-C_s-S_d \qquad \ldots(9)$$

Substituting Ia = 0.152 and CN = 63;

$$F_0 = 22.70-C_s-S_d \qquad \ldots(10)$$

For entire event, the total runoff is expressed as,

$$R = P - C_s - S_d - F_0 - F_r \qquad \text{...(11)}$$

i.e.,

$$R = P_e - F_r \qquad \text{...(12)}$$

$$F_r = (P - 22.70) - R \qquad \text{...(13)}$$

The cumulative infiltration (F) over entire rainfall event time (T) is given as,

$$F = F_0 + F_r \qquad \text{...(14)}$$

The average infiltration rate (i) is given as :

$$i = F/T \qquad \text{...(15)}$$

Hydrologic data monitoring and field data collection

The required hydro-meteorological data (2004-2008), physiographic and soil-water characteristics of watershed, land use/cover, soil types and hydrologic conditions were obtained from Zonal Agricultural Research Station, Kolhapur (M.S.). The daily rainfall-runoff charts were analysed for five minutes and one hour time steps for event based and continuous hydrological modeling, respectively.

Error functions for goodness-of-fit

Various error functions *viz.* root mean square error (RMSE), Nash-Sutcliffe efficiency (NSE) as suggested by ASCE (1993), Normalised Objective Function (NOF) by Ibbitt and O'Donnell (1971) and per cent errors were used to quantify the goodness-of-fit between simulated and observed runoff components during verification run.

Analysis of results

The optimal model parameters (Ia = 22.70 and k = 0.766)were decided on the basis of ISO strategy. These parameters were further verified for runoff simulation process. The results presented in Table 1, the event model showed better predictive performance for simulation of runoff volume and runoff peak based on minimum overall per cent errors (PEV = –8.97% and PEP = –7.66%). The findings are further confirmed by overall lower values of RMSE (0.016) and NOF (0.426) with higher values of NSE (0.82). Thus, the HEC-HMS model has achieved good agreement between simulated and observed runoff for event based simulation. Event model at finer time step of five-min simulates realistic flood hydrographs having closer agreement with observed ones in terms of shape, magnitude and symmetry and given the least RMSE = 0.002 and NOF = 0.147 and highest NSE = 0.96 for event E-6 (Fig. 3).

Table 1. Predictive performance of event based HEC-HMS model (2004-2008)

Period	*Year 2004-2008*
Number of Events	E-1 to E-18
Simulation time step	05 min
Rainfall, mm	41.2 to 205.9 mm (Av. 68.1 mm)
Percent Error for runoff volume (PEV)	-1.31 to-28.60 % (Av. -8.97 %)
Percent Error for runoff peak (PEP)	1.33 to-26.06 % (Av. -7.66 %)
RMSE	0.002 to 0.039 (Av. 0.015)
NSE	0.66 to 0.94 (Av. 0.82)
NOF	0.147 to 0.701 (Av. 0.426)

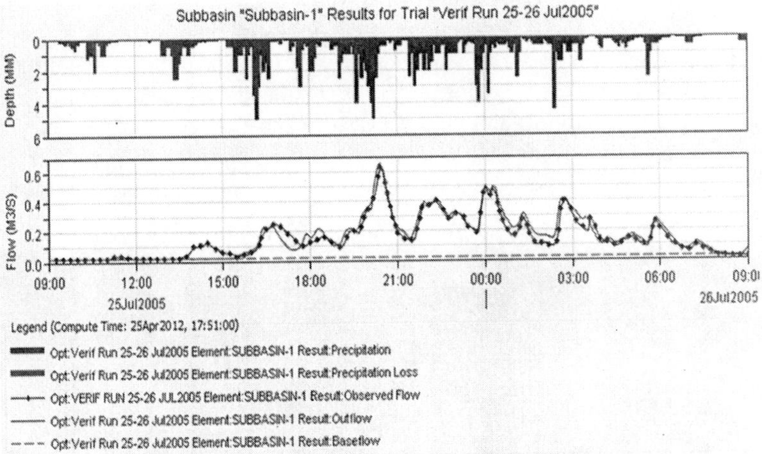

Fig. 3: Simulated and observed flood hydrographs for event.

The optimised parameters (Ia = 22.70 mm and k = 0.766) were further used in continuous model. The verification study indicates that various runoff components (e.g. surface, sub-surface and total runoff, flood peaks and their timings) simulated by the model have closer agreement with the observed ones. The over prediction caused in runoff volume by 22.30 per cent for the year 2005 may be due to excessive rains, however minimum overall percentage of errors over the years (PEV = 4.45 % and PEPF= +3.90%) indicate better predictive performance of continuous model for runoff volume and peaks (Table 2).

Table 2. Predictive performance of continuous HEC-HMS model (2004-2008)

Period	*Year 2004-2008*
Simulation period	10 May-31 October
Simulation time step	01 Hr
Seasonal rainfall, mm	959.3 to 1770.6 mm (Av. 1235.2 mm)
Percent Error for runoff vol. (PEV)	-0.22 to +22.30% (Av. +4.45 %)
Percent Error for runoff peak (PEP)	+0.29 to +6.08% (Av. +3.90 %)
RMSE	0.004 to 0.013 (Av. 0.006)
NSE	0.70 to 0.89 (Av. 0.84)
NOF	1.67 to 2.64 (Av. 2.12)

Similarly the findings are further confirmed by qualitative evaluation based on lower RMSE (0.006 cumec) and NOF (2.12) with higher NSE (0.84). The model even at coarser time step of one-hr simulates realistic flood hydrographs with better agreement with observed ones in terms of magnitude, shape and symmetry (Fig. 4).The model has simulated 19.71 per cent as excess rainfall i.e. surface runoff and 9.53 per cent as sub-surface runoff in delayed interflow from the given watershed and remaining portion 70.76 per cent is subjected to various losses. This kind of water budgeting information is useful for watershed based water resource planning and management.

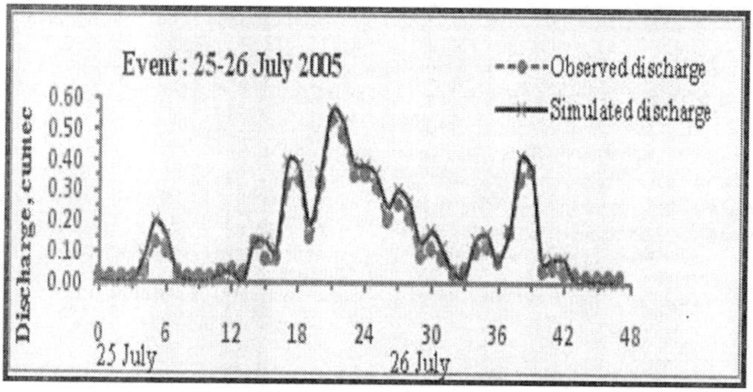

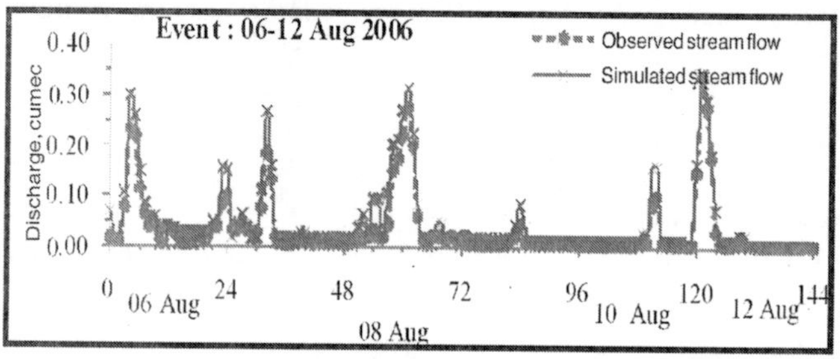

(Contd...)

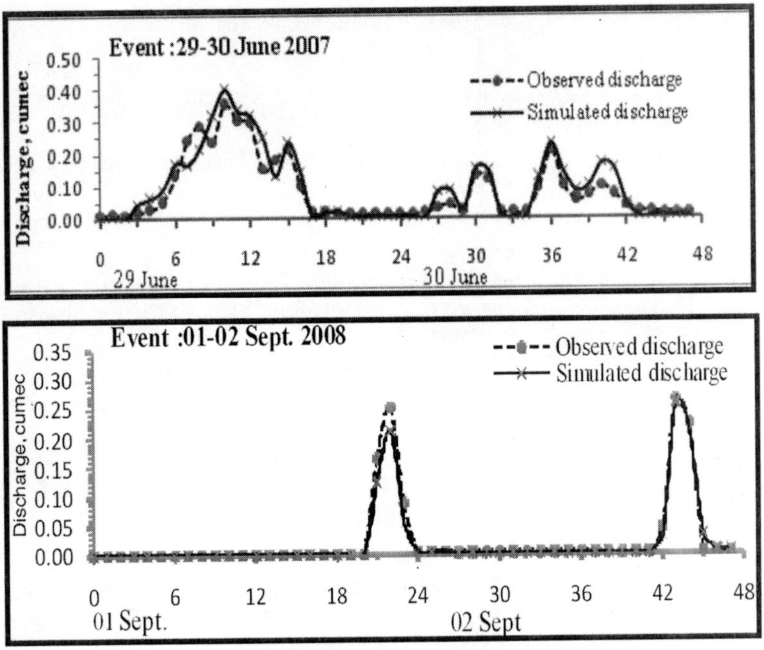

Fig. 4: Simulated and observed flood hydrographs in continuous modeling for some major events (Year 2005-2008).

Summary and Conclusions

Both event and continuous models using HEC-HMS, respectively at fine and coarser time steps have better applicability and better predictive performance for small agricultural watersheds falling under Sub-montane Zone of Maharashtra. The hydrologic processes and hydrologic responses simulated by these models provide a way for watershed based water resource planning and management. The flood hydrographs simulated by both models are realistic and have closer match with observed ones in terms of magnitude, shape and symmetry. Similarly the model parameters so optimised for event model can improve the quality of coarser scale continuous modeling. Thus, the combination of event and continuous models can be the suitable strategy for improving the overall modeling efficiency with available data.

Acknowledgments

The authors are very thankful to the Zonal Agricultural Research Station, Sub-montane zone, Kolhapur (M.S.) for providing necessary data and contributions to various aspects of this research.

References

Anonymous. 1996. A Report presented in RRC Meeting in Agricultural Chemistry and Soil Science, Department of Agril. Chemistry and Soil Science, Mahatma Phule Krishi Vidyapeeth, Rahuri (M.S.).

ASCE Task Committee. 1993. Criteria for evaluation of watershed models. *J. Irrig. Drain. Eng.* (ASCE) 119(3): pp. 429–442.

Bedient, P.B., and W.C. Huber. 1992. Hydrology and Floodplain Analysis. Reading, Massachusetts, Addison-Wesley.

Bennett, T.H. 1998. Development and application of a continuous soil moisture accounting algorithm for the HEC-HMS. MS thesis, Dept. of Civil and Environ. Engr., Uni. of California, Davis.

Burger, C.M., Kolditz, O., H.J. Fowler and S. Blenkinsop. 2007. Future climate scenarios and rainfall-runoff modeling in the Upper Gallego Catchment (Spain). *Environ. Pollution* (148): pp. 842–854.

Chow, V.T., D.R. Maidment, and L.W. Mays. 1988. *Applied Hydrology*. New York: McGraw-Hill.

Chu, X. and A. Steinman. 2009.. Event and Continuous Hydrologic Modeling with HEC-HMS. *J. Irrig. and Drain. Engg.* Vol. 135 (1): pp. 119-124.

Chunale, G. L., A.A. Atre, G.B. Bangal and S.M. Kareppa. 2001. On the initial abstraction of SCS Curve Number model for various storm sizes for Sub-humid region in Maharashtra-I, *Indian J. Soil Cons.* Vol. 29(3): pp. 248-255.

Chunale, G.L. 2013. Optimisation of rainwater harvesting storage structure based on hydrologic modeling. A Ph.D. thesis submitted to the Dept. of Soil and Water Conservation Engineering, Dr. PDKV, Akola.

Clark, C.O. 1945. Storage and the unit hydrograph. *Trans. Am. Soc. Civ. Eng.*, Vol. 110, (2261): pp. 1419-1446.

Hogan, D.H. 1997. Computer-Assisted Floodplain Hydrology and Hydraulics. New York: McGraw-Hill.

Ibbitt, R.P., and O'Donnell, T. 1971. Fitting methods for conceptual catchment models. *J. Hydr. Div.*, Vol. 97(9): pp. 1331-1342.

Kirpich, Z.P. 1940. Time of concentration of small agricultural watersheds. *Civil Engineering*. Vol.10 (6): p. 362.

USACE-HEC. 2006. Hydrologic Modeling System, HEC-HMS (3.0.1.) User's Manual, US Army Corps of Engrs. Hydrol. Engg. Center (HEC), Davis, USA.

USDA. 1986. "Urban Hydrology for Small Watersheds." *Technical Release 55*, Natural Resources Conservation Services (NRCS), Washington, D.C.

4

Condensed Disaggregation Procedure for Simulating Monthly Rainfall

R. Subbaiah, R.N. Das and R.N. Yasaswi

Abstract

Disaggregation models downscale the annual or seasonal rainfall among sub-periods and divide the aggregate rainfall among sub-basins. The two basic forms of disaggregation modelling are temporal and spatial. The focus of this study was application of disaggregation model in the temporal domain within a basin. The lane's condensed disaggregation model was used to disaggregate the annual rainfall series of Junagadh to monthly values to avoid the estimation of excessive number of parameters. Out of many combinations of ARMA models selected, ARMA (1,0) generated the annual rainfall series to reasonable accuracy. The lane's condensed disaggregation model succeeded in preserving additivity, statistical moments and correlation among annual series with monthly values, but failed to maintain the cross correlations among monthly series at different lags. Some negative monthly values were also generated without affecting the statistical moments.

Key words: *Condensed disaggregation, Lane Model, moment preservation, simulation.*

Introduction

Very often a hydrological stochastic process must be studied in different time scales. Relative paucity of rainfall data at lower time-scales coupled with insufficient rain gauge network demands the generation of the hydrologic sequences for the design of water resource system. Two basic approaches have been taken in the generation of monthly or seasonal flow series. Periodic autoregressive moving average extensions generate monthly or seasonal flows directly, which can be summed to obtain annual flows. Such models may not capture the distribution and persistence of annual totals. An alternative is to generate annual flows is disaggregation approaches to obtain finer time scale rainfall series. The goal of the downscaling procedure is to generate time series of downscaled values at any specific site such that they exhibit patterns, statistical characteristics, temporal dependences, and distributional properties that are statistically indistinguishable from those of the observations. Two approaches exist in disaggregation scheme include parametric and non-parametric procedures. Both parametric and non-parametric methods have been widely used for stochastic simulation and forecasting of hydro-climatic processes, the former over the past 40 years, the latter more recently during the last decade or so. Although both types of approaches have relative advantages and disadvantages, one approach or the other may be more appropriate for a specific situation.

Parametric stochastic models have long been used for disaggregation, from annual to monthly timescales. These methods are iterative in nature and thus, computationally intensive besides requiring assumptions of linearity. Unlike the linear approach where a single linear model is fit to the entire data, the non-parametric methods involve "local" functional fitting. The function is fit to a small number of neighbours at each point. This approach has the ability to capture any arbitrary features (non-linearities, non-normal, etc.) exhibited by the data. Some non-parametric approaches still have some shortcomings common to parametric schemes, such as generation of negative flows in some cases, lack of preservation of cross-boundary correlations and generation of flow sequences that can be very similar to the observations (Rajagopalan *et al.*, 2010; Lee *et al.*, 2010). A general review of models under each category with their advantages and disadvantages is presented in Rajagopalan *et al.*, (2010). Hoshi and Burges (1980) highlighted that Lane's condensed disaggregation model (Parametric type) was expressed as the best overall model due to its minimum number of parameters and recommended it for general use. Therefore, in the present study the lane's condensed disaggregation model is used to disaggregate seasonal rainfall series to monthly series. The disaggregated values are very useful for planning ungauged watersheds.

Materials and Methods

The Lane (1982) condensed disaggregation procedure is expressed as,

$$Y_\tau = A_\tau X + B_\tau \varepsilon + C_\tau Y_{\tau-1} \qquad ...(1)$$

where X is column matrix of annual values, Y_τ is the column vector of monthly series which sum to X, ε is the noise vector, and A, B, and C are parameter matrixes C. The matrix C preserves the correlations between current year values with past year values and also preserves variances and lag-one covariance among the seasonal values. The *A, B* and *C* are estimated as: (Salas *et al,* 1988)

$$\hat{A}_\tau = \left[\frac{S_{YX}(\tau,\tau) - S_{YY}(\tau,\tau-1)S_{YY}^{-1}(\tau-1,\tau-1)S_{YX}(\tau-1,\tau)}{S_{XX}(\tau,\tau) - S_{XY}(\tau,\tau-1)S_{YY}^{-1}(\tau-1,\tau-1)S_{YX}(\tau-1,\tau)}\right] \quad ...(2)$$

$$\hat{C}_\tau = \left[\frac{S_{YY}(\tau,\tau-1) - \hat{A}_\tau S_{XY}(\tau,\tau-1)}{S_{YY}(\tau-1,\tau-1)}\right] \quad ...(3)$$

$$\hat{B}_\tau \hat{B}_\tau^T = S_{YY}(\tau,\tau) - \hat{A}_\tau S_{XY}(\tau,\tau) - \hat{C}_\tau S_{YY}(\tau-1,\tau) \quad ...(4)$$

$$S_{YY}(\tau,\tau) = \frac{1}{N-1}\sum_{v=1}^{N}\left[y_{v,\tau}x_{V,\tau}^T\right] \quad ...(5)$$

$$S_{YX}(\tau,\tau) = \frac{1}{N-1}\sum_{v=1}^{N}\left[y_{v,\tau}x_{V,\tau}^T\right] \quad ...(6)$$

$$S_{YY}(\tau,\tau-1) = \frac{1}{N-1}\sum_{v=1}^{N}\left[y_{v,\tau}x_{V,\tau}^T\right] \quad ...(7)$$

$$S_{YY}(\tau,\tau-1) = \frac{1}{N-1}\sum_{V=1}^{N}\left[y_{V,\tau}y_{V,\tau-1}^T\right] \quad ...(8)$$

$$S_{YX}(\tau-1,\tau) = \frac{1}{N-1}\sum_{v=1}^{N}\left[y_{v,\tau}x_{V,\tau}^T\right] \quad ...(9)$$

S_{uv} = Covariance of the general vectors *u* and the transpose of *v*. All the vectors contain elements with zero mean and unit variance. To achieve this, the data are typically transformed to a normal distribution by appropriate transforms before the model is fit. The model simulation proceeds as follows: (i) the seasonal rainfall is generated from an appropriate model or equivalent dataset; (ii) the simulated seasonal value is then disaggregated to monthly series using the condensed Lane's model. The simulated monthly series are back transformed to the original space.

Seasonal series generation model

For a staged disaggregation scheme, the input (annual or seasonal value for temporal disaggregation) to the disaggregation model has to be generated by another stochastic model. The ARMA (p, q) model can be expressed as,

$$X_t = \sum_{i=1}^{p} \phi i X_{t-i} + \varepsilon_t - \sum_{j=1}^{q} \theta i \varepsilon_{t-j} \qquad ...(10)$$

where X_t represents the rainfall for year t, which is normally distributed with mean zero and variance $\sigma^2(X)$, ε_t is the uncorrelated normally distributed noise term with mean zero and variance $\sigma^2(\varepsilon)$, $\{\Phi_1,\ldots,\Phi_p\}$ are the autoregressive parameters and $\{\theta_1,\ldots,\theta_q\}$ are the moving average parameters. Multivariate ARMA models of order less than or equal to (2, 2) are commonly sufficient to model multisite flow series (Bras and Rodriguez-Iturbe, 1985). The parameters of the each of the model structure were estimated using the approximate least squares estimation procedure (Box and Jenkins, 1976).

Storage related statistics

The storage, drought, and surplus related statistics are particularly important in modelling time series for simulation studies of reservoir systems. Such characteristics are generally functions of the variance and auto-covariance structure of a time series. Consider the time series y_i , i = 1, ..., N and a subsample y_1, ..., y_n with $n < N$. Form the sequence of partial sums S_i as,

$$S_i = S_{i-1} + (y_i - \bar{y}_n) \quad i=1,2,3...n \qquad ...(11)$$

where $S_0 = 0$ and $\bar{y}_n$ is the sample mean of y_1 , ..., y_n. Then, the adjusted range R_n^* and the rescaled adjusted range R_n^{**} can be calculated by,

$$R_n^* = \max(S_0, S_1,...S_n) - \min(S_0, S_1,...S_n) \qquad ...(12)$$

$$R_n^{**} = \frac{R_n^*}{\sigma_n} \qquad ...(13)$$

where S_n = standard deviation of y_1 , ..., y_n. The Hurst coefficient for a series is estimated by,

$$K = \frac{\ln(R_n^{**})}{\ln(n/2)} \text{ for } n > 1 \qquad ...(14)$$

The storage capacity is calculated using Ripple mass curve method (Loucks *et al.*, 1981).

Drought and surplus related statistics

For any series y_i $i = 1, ..., N$, the demand level d may be defined as $\alpha * \overline{y}$, $0 < \alpha < 1$ (for $\alpha = 1, d = \overline{y}$). A deficit occurs when $y_i < d$ consecutively during one or more years until $y_i > d$ again. Such a deficit can be defined by its duration L, by its magnitude M, and by its intensity $I = M/L$. Assume that m deficits occur in a given hydrologic sample, then the maximum deficit duration (longest drought or maximum run-length) is given by,

$$L_n^* = \max(L_1, L_2 \ldots L_m) \quad ...(15)$$

and the maximum deficit magnitude (maximum run-sum) is defined by,

$$M_n^* = \max(M_1, M_2 \ldots M_m) \quad ...(16)$$

The longest drought duration and the maximum deficit magnitude are estimated for both annual and seasonal series. Surplus related statistics are simply the opposite of drought related statistics.

Statistical criteria

The fitted annual model must be tested to determine whether the model complies with the model assumptions and whether the model is capable of reproducing the historical statistical properties of the data at hand. The statistical criteria adopted are the corrected Akaike information criterion, AIC, (Hurvich and Tsai, 1989, Brockwell and Davis, 1996) RMSE, Goodness of fit and the Schwarz information criterion, SIC (Hurvich and Tsai, 1993, Shumway and Stoffer, 2000) also often referred to as the Bayesian information criterion. The selection rule for the above criteria is to select the model with minimum AIC, SIC and RMSE and maximum goodness of fit value. The statistical properties like mean, standard deviation, coefficient of variation, lag zero and lag one correlation coefficient and the correlation structure between the annual and monthly series for generated monthly series from the disaggregated models were compared with the historical series.

Results and Discussion

Rainfall variability

The daily historical rainfall data for the past 54 (1957-2010) years were collected from Agro-meteorological observatory, Junagadh Agricultural

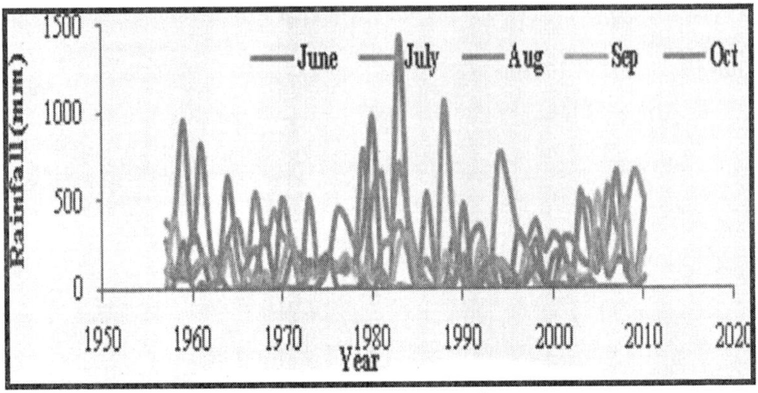

Fig.1: Monthly rainfall during various years.

University, Junagadh to determine the monthly and yearly variation of rainfall data. The monthly rainfall received during different years is plotted in Fig. 1. Large variation in among the monthly rainfall series was recorded. January to May and November to December received scanty amount of rainfall (one per cent of total rainfall). The statistical properties of monthly and annual observed rainfall are presented in Table 1. The total rainfall is confined between June to October. The average rainfall for June, July, August, September October, seasonal and annual series was respectively 183.79 mm, 357.23 mm, 206.3 mm, 127.68 mm and 33.18 mm, 871.23 mm, 928.30 mm. Nearly 93.85 per cent of yearly rainfall is confined to monsoonal months. Large variation of rainfall from year to year was observed.

The standard deviation for June, July, Aug, Sep, Oct, and Seasonal series was 231.3 mm, 230.5 mm, 172.5 mm, 119.7 mm, 55.10 mm and 427.9 mm respectively (Table 1). The skewness of monthly series varied from 0.8948 to 3.613. The seasonal series recorded a skewness of 1.561.The lag one and lag two auto correlation function for historical series were estimated to be -0.023 and -0.0247. The cross correlations for monthly and annual rainfall series were also estimated and are presented in Fig. 2. In June the third and 18th lag period showed correlation. In July 10th lag showed negative correlation whereas 13th and 16th lag showed same value of positive correlation. Correlation structure was poor for August and October series. The September month showed good correlation at 14th lag. Annual series showed good correlation at 4 and 24 lag periods.

Seasonal generation model

ARMA model of various orders were fitted to the seasonal rainfall series. The parameters for ARMA for various orders are presented in Table 2. The model with lowest values of AIC, SIC and minimum noise variance was

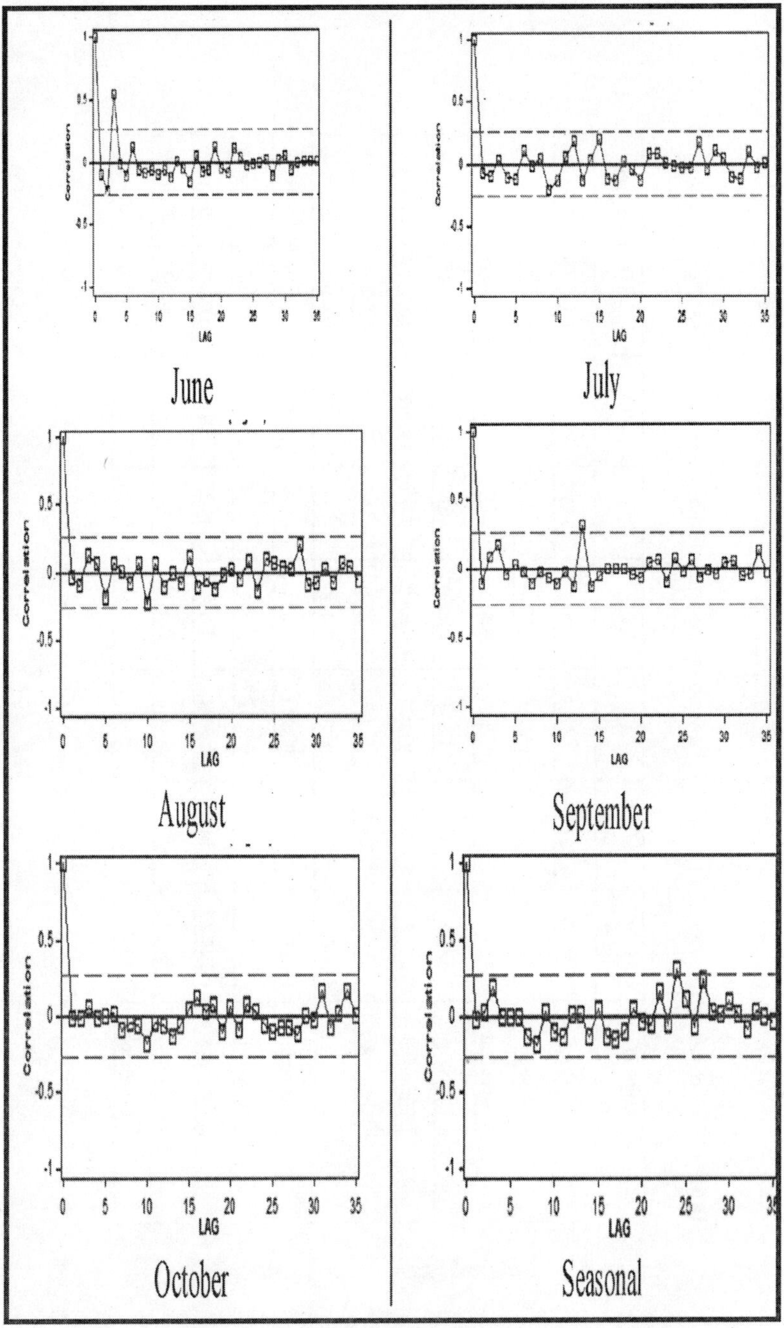

***Fig. 2:** Cross Correlation coefficient for monthly and seasonal rainfall series.*

Table 1: Statistical properties of monthly and annual generated and historical rainfall series of Junagadh

Time scale	*Type of series*	*Mean (mm)*	*St Dev (mm)*	*CV*	*Skew*	*Min (mm)*	*Max (mm)*	*acf(1)*	*acf(2)*
Annual	Historical	908.2	427.9	0.4711	1.561	138.2	2790	-0.02343	0.02479
	Generated	884.8	425.2	0.4787	1.1046	-57.6	1808	-0.05496	-0.04003
June	Historical	183.8	231.3	1.259	3.613	2.3	1432	-0.09480	-0.02388
	Generated	175.6	229.6	1.364	1.8534	-67.1	1081	-0.03857	0.007298
July	Historical	357.2	230.5	0.6452	0.8948	30.60	1056	-0.06397	-0.09721
	Generated	352	223.3	0.6366	0.3976	-16.8	898.7	-0.02217	-0.03465
Aug	Historical	206.3	172.5	0.8361	1.478	2	787.3	-0.03121	-0.09956
	Generated	194.6	161.3	0.8294	1.023	-65.94	587.5	-0.0154	-0.07536
Sep	Historical	127.7	119.7	0.9372	1.550	0	532.9	-0.11	0.08685
	Generated	123.2	114.6	0.9331	1.023	-19.87	395.6	0.02143	-0.03489
Oct	Historical	33.19	55.10	1.661	2.408	0	265.3	-0.02027	-0.01994
	Generated	30.73	57.3	1.94	1.5634	-81.23	180.9	0.06181	-0.6623

Table 2: ARMA model estimates for annual transformed rainfall series of Junagadh

ARMA (p q)		*AR parameters*		*MA parameters*		*AIC*	*SIC*	*White Noise Variance*
p	*q*	ϕ_1	ϕ_2	θ_1	θ_2			
1	0	-0.024515				58.204	57.958	0.998225
2	0	-0.024024	0.025496			60.415	61.913	0.998805
0	1			0.023784		58.206	57.960	0.999458
0	2			0.029480	-0.031991	60.382	61.880	0.998179
1	1	-0.448015		-0.423991		60.514	62.012	1.00064
1	2	-0.040451		0	0	62.798	65.949	0.999667
2	1	-0.026956	0.023479	0		62.752	65.903	0.998817
2	2	-0.031962	0.015502	0	0	65.194	69.900	0.998956

Table 3. Lane's condensed disaggregation model estimates for various months

Matrix	*Seasonal Months*				
	June	*July*	*August*	*Sep*	*Oct*
A	0.350375	0.498631	0.310751	0.145338	0.018161
B	175.274	162.278	134.454	104.27	54.6307
C	0.330768	-0.514512	-0.397553	-0.0504546	-0.0519517
G	30720.9	26334.2	18078	10872.2	2984.51

Table 4: Drought characteristics of monthly and annual generated and rainfall series of Junagadh

Time scale	*Type of series*	*Longest Drought*	*Max Deficit*	*Longest Surplus*	*Max Surplus*	*Storage Capacity*	*Rescaled Range*	*Hurst Coefficient*
Annual	Historical	5	1500	6	2005	3296	7.704	0.6195
	Generated	5.6	2314	6	1755	4289	7.973	0.6255
June	Historical	6	709.5	2	1248	1049	8.580	0.6522
	Generated	4.7	1120	4.7	923.7	1939	7.418	0.6061
July	Historical	5	841.9	3	698.8	1753	7.607	0.6157
	Generated	5.7	1185	4.7	1034	2066	8.828	0.6558
Aug	Historical	7	746.9	3	581	1244	7.213	0.5995
	Generated	5.5	858.7	4.3	602.8	1454	7.502	0.6048
Sep	Historical	7	476.5	4	898.3	1160	9.694	0.6892
	Generated	6	554.7	5.3	724.7	948.3	8.897	0.6579
Oct	Historical	11	277.5	3	236.7	406.7	7.380	0.6065
	Generated	8.1	277,8	4.2	268.5	487.6	8.219	0.6443

selected in the present study. Lowest AIC, SIC and minimum variance was observed in ARMA (10). Hence, the lag-one autoregressive model, $AR(1)$ was selected for generation of the annual rainfall series for utilisation in the disaggregation model. The comparison between the historical and generated inflow series are presented in Fig. 3. In some places either higher or lower values were generated. The maximum value generated by the model was found to be lesser than historical series. These values however could not affect the statistical moments of the annual series. The model under predicted the properties like mean, standard deviation and skewness of the historical seasonal series (Table 1). The mean, standard deviation and skewness of the generated the series were less than the historical series respectively by 2.57 per cent, 0.63 per cent, and 29.23 per cent. A good correlation (r = 0.9835) was observed between the historical and synthetic seasonal rainfall series. The ACF of annual historical and generated series was also plotted in Fig. 4. The correlation of ACF at different lags was found to be 0.7763. The model slightly preserved the serial correlation among the seasonal series at different lags. The model failed to maintain cross correlations at different lags.

Disaggregated monthly series

The transformed and normalised seasonal generated series were taken as input to the disaggregation models to overcome the difficulty of non-normality in the historical series. The disaggregated series had zero mean and unit variance and were in normalised form. The Lane's condensed disaggregation model was adopted to down scale the seasonal series to monthly ones. The A, B, C and G matrix values for different months are presented in Table 3. The value A, B, C and G vary from 0.018 to 0.498, 54.63 to 175.27, -0.5195 to 0.33 and 10872.2 to 30720.9 respectively.

The sums of the generated monthly rainfall totals were compared with the observed seasonal totals for checking the preservation of additivity. The monthly synthetic series generated using Lane's condensed disaggregation model is presented in Fig. 3. A good correlation between historical and synthetic series was observed for June (r = 0.9346), July (r = 0.9268), August (r = 0.9528), September (r = 0.9639), and October (r = 0.9192). The RMSE, AIC, SIC and goodness of fit for the generated series were calculated after comparison with historical series (Table 4). The RMSE was varying from 23.24 to 48.79. Low RMSE was observed for July and highest for October. This may be due more number of series appeared in the October. Low AIC, SIC and goodness of fit were observed for July and August months and high values for June, September and October months. The correlation between annual series with historical and generated monthly series was calculated (Table 5). Good correlation structure was observed between seasonal and monthly historical series. The correlation for historical series was varying from 0.086 to 0.647. Low

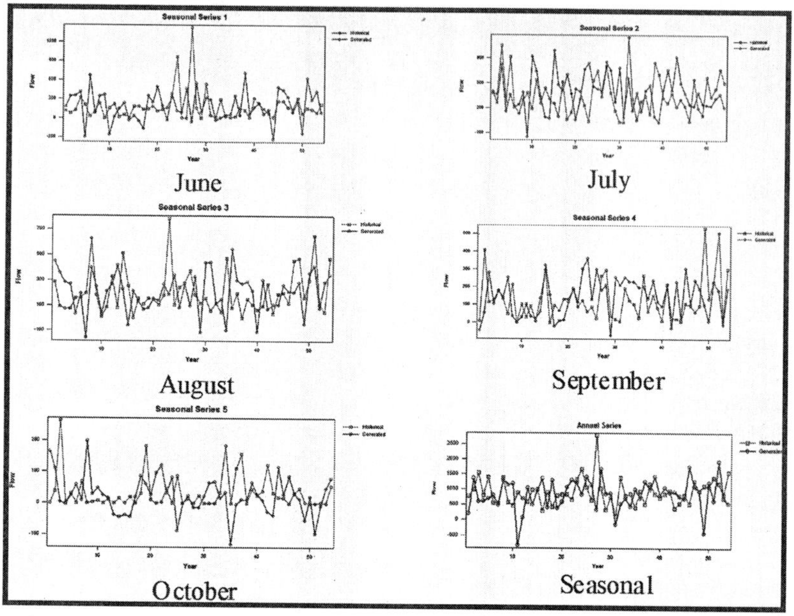

Fig. 3. Predicted vs observed monthly and seasonal rainfall.

Table 5: Correlation between historical and synthetic monthly rainfall with annual series of Junagadh

Season	*Historical*	*Generated*	*Std Deviation*
June	0.647881	0.636596	0.087209
July	0.591129	0.537167	0.102817
Aug	0.456873	0.437185	0.104117
Sep	0.486428	0.508849	0.087836
Oct	0.086138	0.055077	0.106069

and high correlation was observed for October and June for historical series. Same trend was prevailed for the generated series. The correlation values were lower for most of the months.

The ACF of the historical and synthetic series generated is presented in Fig. 4. The correlation structure of simulated and observed monthly series were well preserved for most of the months at different lags. However, the correlation between the first month of a year and the last month of the preceding year is not preserved. The models capability was also analysed in preserving the statistical moments of the historical series. The statistical properties of synthetic and historical monthly series are presented in Table 1. The statistical moments for the simulated and observed monthly series are well preserved for most of the months. The maximum and minimum flow statistics are also reasonably well simulated for most of the months. However, one can observe some discrepancies in the simulated maxima and minima values, indicating that the extreme values need some consideration within the proposed methodology. Extrapolation beyond the maximum historic flow occurs more

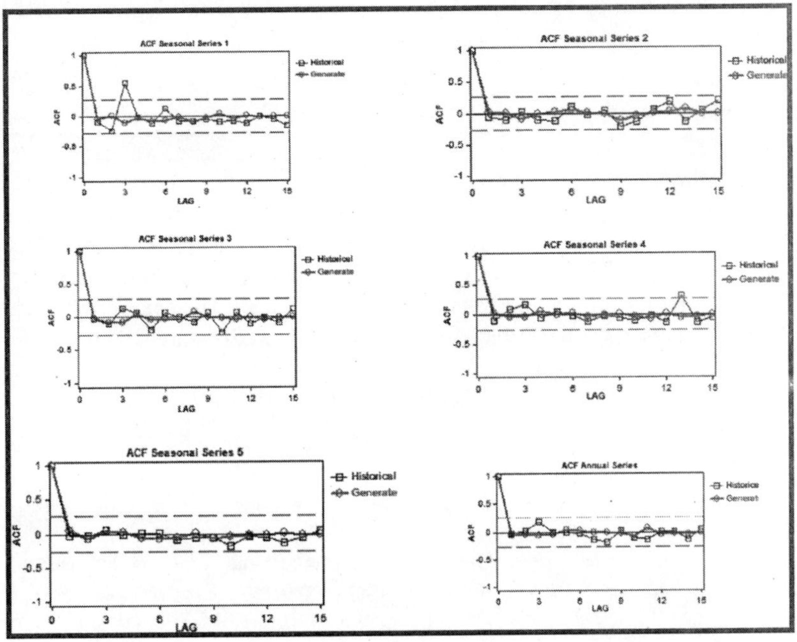

Fig. 4: ***Auto correlation of historical and generated monthly series.***

extensively for some months, while other months (Jul and Oct) display limited to no extrapolation. Negative rainfall series were generated, mostly in dry months, but this had no impact on statistical moments.

Storage, drought and surplus related statistics

The storage, drought and surplus related statistics were calculated using the above described procedure for historical monthly and annual series as well as for generated series (Table 6). In annual historical series five and six years longest drought and longest surplus was observed. The maximum deficit and surplus was observed to be 1500 and 2005 respectively. The storage capacity was 3296 and the Hurst coefficient was 0.6195. The ARMA (1,0) model preserved the Hurst coefficient and longest surplus characteristics. Model over predicted longest drought, max deficit and storage capacity and under predicted surplus. This depicts the conservative estimates of the model. The Hurst coefficient and rescaled range values for all months were preserved.

Table 6: Statistical criteria of the forecasted series for the monsoonal months

Month	*RMSE*	*AIC*	*SIC*	R^2
June	37.8264	4.8264	3.8583	0.67943
July	23.2436	2.2807	2.9366	0.71235
August	29.3060	3.1983	3.1472	0.81264
September	33.9674	4.5993	3.8125	0.65437
October	48.796	5.1367	4.2284	0.68357

Longest drought was under predicted by the disaggregation model except for July. Max deficit was over predicted except for October. Longest surplus was over predicted for all the months. Maximum surplus was under predicted for June and September and over predicted for remaining months. Storage capacity was over predicted for all the months except September. The conservativeness of the model is not reflected in the Lane's condensed disaggregation model.

References

Box, G.E.E and Jenkins, G.M. (1976): Time Series Analysis: Forecasting and Control, Holden-Day, San Franscisco, Cal.

Bras, R.L. and Rodriguez-Iturbe, I. (1985). Random functions and hydrology, Addison-Wesley, Reading, Massachusetts.

Brockwell, P.J. and Davis, R.A. (1996). Introduction to Time Series and Forecasting. Springer Texts in Statistics. Springer-Verlag, New York.

Hoshi, K. and Burges, S.J. (1980). Disaggregation of stream-flow volumes. *Journal of Hydraulics Division*, ASCE 106 (6):1127-38.

Hurvich, C.M. and Tsai, C.L. (1993). A corrected Akaike information criterion for vector autoregressive model selection. *Journal of Time Series Anals* 14: 271–279.

Lane, W.L. (1982). Corrected parameter estimates for disaggregation scheme. In: Statistical Analysis of Rainfall and Runoff (Eds. V.P. Singh) Water Resources Publication, Littleton, Colorado.

Loucks, D.P., Stedinger, J.R. and Haith, D.A. (1981). Water Resources Systems Planning and Analysis, Prentice-Hall, Englewood Cliffs, New Jersy.

Lee, T., Salas, J.D. and Prairie J. (2010). An enhanced nonparametric streamflow disaggregation model with genetic algorithm, *Water Resources Research*, 46 (8):210-216.

Rajagopalan, B., Salas, J.D. and Lall, U. (2010). Stochastic methods for modelling precipitation and streamflow. In: Advances in Data-based Approaches for Hydrologic Modeling and Forecasting (Eds. B. Sivakumar and R. Berndtsson) World Scientific, Singapore, 17-52.

Salas, J.D. (1993). Analysis and Modelling of Hydrologic Time Series, In: Handbook of Hydrology (Eds D.R. Maidment) Chapter: 19, McGraw-Hill, Inc., New York 1-72.

Shumway, R.H. and Stoffer, D.S. (2000). Time Series Analysis and Its Applications. Springer Texts in Statistics. Springer-Verlag edition, New York.

5

Artificial Neural Networks for Rainfall Forecasting

P.M. Kyada, Kumar Pravendra and R.A. Kachhadiya

Abstract

Accurate rainfall forecasting is very necessary for water resource management. Recently, several modelling approaches have been investigated to perform such forecasting task. In the present study, possibility of forecasting rainfall in Junagadh has been analysed through feed forward artificial neural network models. The 30 years data has been used for training and testing the ANN networks. In formulating the ANN based Predictive model, single and double hidden layers network have been constructed. The performance of models have been evaluated using Correlation coefficient, Mean Square Error, Normalised Mean Square Error, Akaike's information criterion, Coefficient of Efficiency and volumetric error, and two best suitable models (7-12-13-1 and 4-6-4-1) have been selected from case I and II for rainfall forecasting in Junagadh. Based on the performance evaluation of the models, two models found suitable for prediction of daily rainfall for the study area.

Keywords: *Artificial Neural Network; Feed forward algorithm; Rainfall Prediction; Back propagation algorithm.*

Introduction

The climate change is expected to affect precipitation and water availability. Gujarat mainland region receives an average annual rainfall of 800 to 2000 mm, while Saurashtra has an average annual rainfall of 400 to 800 mm. The incidence and distribution of rainfall, particularly in Saurashtra and Kutch regions and in the northern part of Water scarcity and droughts are not new phenomena for the people of Saurashtra. They have always lived with them and thus have treated water as a highly scarce and precious resource. The above facts lead to conclude that the artificial neural network is the most efficient rainfall forecasting methods for bringing the more agricultural productions from the limited land and water resources. Artificial Neural Networks (ANNs) are non-linear mapping structures based on the functions of human brain. They are powerful tools for modelling, especially when the underlying data relationship is unknown. The artificial neural network is the most efficient rainfall forecasting methods for bringing the more agricultural productions from the limited land and water resources.

Different scientist over the globe have developed stochastic weather models which are basically statistical models that can be used as random number generators whose output resembles the weather data to which they have been fit (Wilks, 1998). Guhathakurta (2006) implemented ANN technique to predict rainfall over a state (Kerala) of India. Hu (1964) initiated the implementation of ANN, an important soft computing methodology in weather prediction. Michaelides *et al.* (1995) compared the performance of ANN with multiple linear regressions in estimating missing rainfall data over Cyprus. Singh *et al.* (2012) has developed radial basis neural network (RBNN) for Nagwa watershed for simulating monthly surface runoff and sediment yield. Abbot and Marohasy (2012) used artificial intelligence to monthly and seasonal rainfall forecasting in Queensland, Australia. It was assessed by inputting recognised climate indices, monthly historical rainfall data, and atmospheric temperatures into a prototype stand-alone, dynamic, recurrent, time-delay, artificial neural network. Chauhan and Shrivastava (2012) developed ANN models for estimation of reference crop evapotranspiration with climate data required for Penman-Monteith (P-M) method, to test artificial neural networks (ANNs) for estimating reference evapotranspiration (ET_0) with limited climate data (ET0) and compares the performance of ANNs with P-M method. Nastos *et al.* (2013) developed predictive models in order to forecast rain intensity (mm/day) in Athens, Greece, using Artificial Neural Networks (ANN) models. The objectives of the present study were to formulate Artificial Neural Network model with reliable accuracy and validate formulated models for the study area.

Materials and Methods

The main objective of this study is to develop Artificial Neural Networks models for forecasting of monsoon rainfall. This chapter deals with the location and climate of study area, collection of meteorological data, methodology adopted for rainfall modelling using artificial neural networks models. Procedure used for calibration and validation of the model and various criteria for evaluating performance of the models discussed here.

Descriptions of Area

The study area falls in the District of Junagadh (Gujarat), India. The latitude and longitude of the study area is 21.5°N and 70.1°E, respectively. The elevation of gauging station is 60 m above the mean see level. The climate of area is subtropical and semi-arid type with an average annual rainfall of 900 mm and average annual pan evaporation of 6.41 mm/day. May is the hottest month with mean monthly temperature varying between 35°C to 45°C and mean monthly minimum temperature varying between 7°C and 10°C as observed from the 10 years data collected by meteorological observatory, Krushigadh, JAU, Junagadh.

Data Acquisition

The daily meteorological data of 30 years (1979 to 1981, 1984 to 1989 and 1991 to 2011) were collected from meteorological observatory of Krushigadh, Junagadh Agricultural University. The Observatory is situated at Junagadh in the "Saurastra" region of Gujarat state of India.

Development of Models for study area

The forecasting of rainfall is very complex and non-linear process, which is affected by many factors which inter-related, some of the parameters includes as vapour pressure, relative humidity, wind velocity, dry and wet bulb temperature and mean temperature etc. Models can identify and learn correlated patterns between input data sets and target values. After training, Models can be used to forecast the outcome of new independent data. In this present study Artificial Neural Networks models have been developed for forecasting rainfall on daily basis.

Artificial Neural Networks (ANNs)

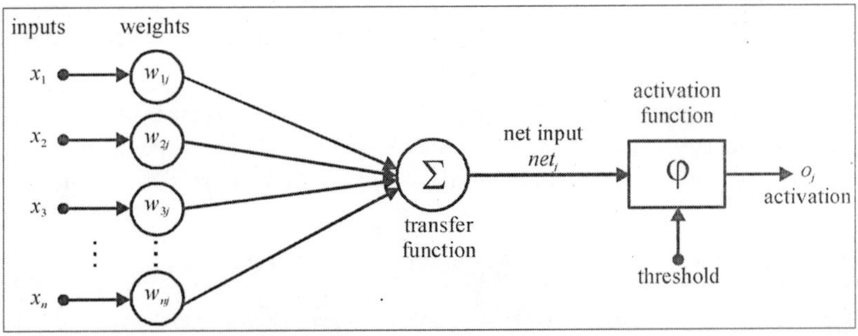

Fig. 1: Mathematical representation of neural network.

Artificial neural networks (ANNs) are inspired by the structure of human brain that is well suited for complicated task such as rainfall prediction, rainfall-runoff modeling, river flow modeling etc., in hydrologic systems. This approach is based on the human brain and it is faster compared with its conventional compatriots, flexible in the range of problems it can solve, and highly adaptive to the newer environments. The process of training is an important aspect, and the performance of an ANN is dependent on successful training. The training procedure involves the adjustment of connection between weights and threshold values for each of the nodes. The Fig. 1 shows mathematical representation of ANN model.

Back-Propagation Training Algorithm

Back-propagation training algorithm is the most commonly used supervised algorithm for training the multihidden layer ANN. In back-propagation ANN, information is processed in the forward direction from the input layer to the hidden layer(s) and then to output layer. The objective of a back-propagation network is to find the weights that approximate target values of output with a selected accuracy. It requires a continuous, differentiable and non-linear function on the ANN to compute output from each neuron.

The input data are multiplied by the initial weights, then the weights inputs are added by simple summation to yield the net input to each neuron.

$$\text{Net} = \sum_{i=1}^{N} V_{ji} X_i \qquad ...(1)$$

Where, X_i = input to any neuron, V_{ji} = weighted matrix from j^{th} layer to i^{th} layer, N = number of inputs and Net = net for j^{th} neuron.

Development of ANN model

In this method, there is a combinations prepared as input of the ANN model. The following two cases of the data sets were taken in account for the modelling of ANN for rainfall prediction.

Case I: In this case N = 149 days in year and M=30 years (1979-1981, 1984-1989 and 1991-2011). Let the observed values of vapour pressure, relative humidity, wind velocity, temperature and rainfall represented as $VP_{i,j}$, $RH_{i,j}$, $V_{i,j}$, $T_{i,j}$ and $P_{i,j}$ respectively for j^{th} day of the ith year (i = 1,2,....,M and j = 1,2,...,N).

The functional form of the models can be represented as:

$$P_{i,j} = f(P_{i,j-3}, P_{i,j-2}, P_{i,j-1}, VP_{i,j-1}, RH_{i,j-1}, V_{i,j-1}, T_{i,j-1}) \qquad ...(2)$$

It can be seen from eq. 2, there are seven numbers of input and one number of output in this case. Denoting input variables by X_1, X_2,...., X_7 and output by Y, the year-wise arrangement of input and output variables is shown in Table 3.3. Thus each row of Table 3.2 will represent an input-output pair for the development of the artificial neural network model.

Case II: In this case the observed time series of vapour pressure, relative humidity, wind velocity, temperature and rainfall of previous days are taken as the input variables and current day rainfall as the output variable, considering rainfall occurrence period starting from 1st June to 30th October only. In this case $N = 152$ days in year and M=30 years (1979-1981, 1984-1989 and 1991-2011). Let the observed values of vapour pressure, relative humidity, wind velocity, temperature and rainfall represented as $VP_{i,j}$, $RH_{i,j}$, $V_{i,j}$, $T_{i,j}$ and $P_{i,j}$ respectively for j^{th} day of the ith year ($i = 1,2,\ldots, M$ and $j = 1,2,\ldots,N$).

The functional form of the models can be represented as:

$$P_{i,j} = f(VP_{i,j-1}, RH_{i,j-1}, V_{i,j-1}, T_{i,j-1}) \tag{3}$$

It can be seen from eq. 3.15, there are four numbers of input and one number of output in this case. Denoting input variables by $X_1, X_2, \ldots, X_4$ and output by Y, the year wise arrangement of input and output variables is shown in Table 3.3. Thus each row of Table 3.2 will represent an input-output pair for the development of the artificial neural network model.

Results and Discussion

Performance Evaluation of Model

Qualitative and quantitative evaluation of model is an important task to assess their capability or potential of developed model in simulation of actual circumstances. In the present study the following qualitative and quantitative performance indices were applied to verify the applicability of developed ANN model.

The performance of the models was evaluated qualitatively and quantitatively by visual observation and employing various statistical indices viz. correlation coefficient, mean square error, normalised mean square error, Akaike's information criterion, co-efficient of efficiency and volumetric error. The qualitative performance of the single and double hidden layer models (7-30-1), (4-12-1), (7-12-13-1) and (4-6-4-1) are shown in the Fig. 2, 3, 4 and 5 respectively for the testing periods. In this study, the acceptable limits for the correlation coefficient, mean square error have been considered to be above 75 per cent, and less than 0.01 respectively for quantitative performance of the models. For the comparison of different models, the Mean square error, Correlation coefficient and Akaike's information criterion of different model are given in Table 2.

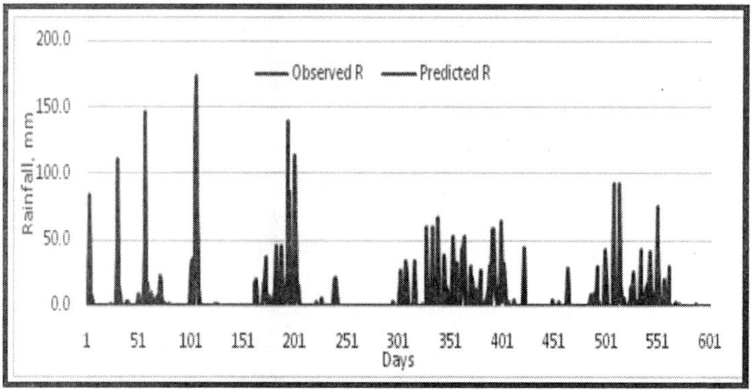

Fig. 2: Observed and Predicted daily rainfall using ANN model (7-30-1) during testing period.

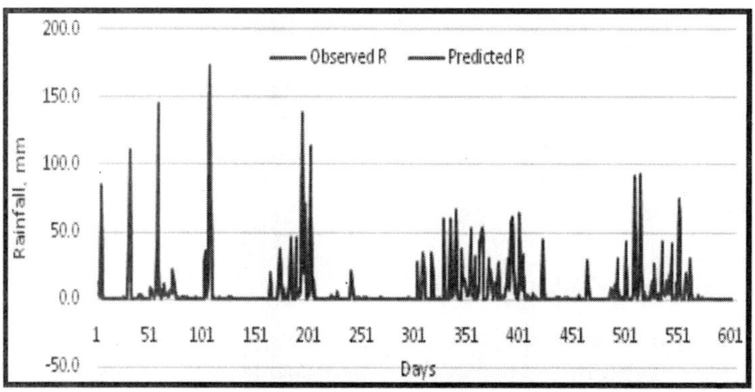

Fig. 3: Observed and predicted daily rainfall using ANN model (7-12-13-1) during testing period.

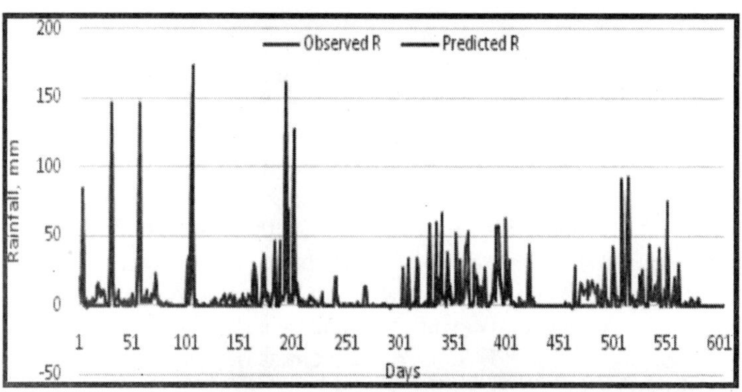

Fig. 4: Observed and predicted daily rainfall using ANN model (4-12-1) during testing period.

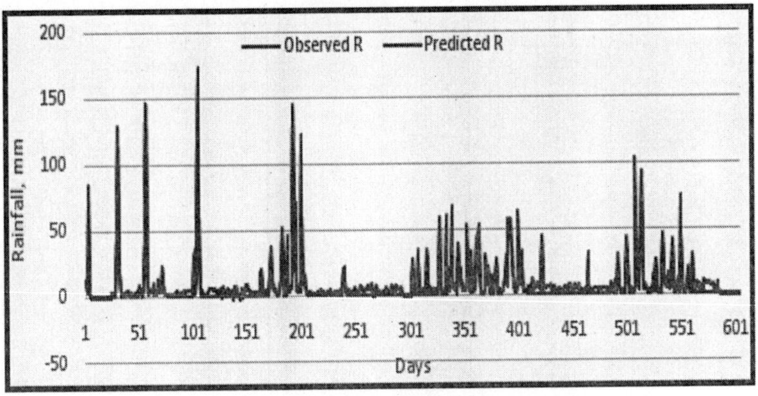

Fig. 5: Observed and predicted daily rainfall using ANN model (4-6-4-1) during testing period.

Quantitative evaluation

For better appreciation of the model, the predictive effectiveness of ANN model is judged on the basis of performance indicators. To judge the predictive capability of the developed model, Correlation Coefficient, Mean Square Error (MSE), Normalised Mean Square Error (NMSE), Akaike's Information Criterion (AIC), Coefficient of Efficiency (CE) and Volumetric Error (EV) were employed as shown in Table 1. The Y_j is observed values, Y_{ej} is the predicted values, $\overline{Y}$ is the mean of observed values, n is the number of observations, *P* is the number of output processing elements, *N* is the number of observation in the data sets, MSE is mean squared error, d_{ij} is the desired output for observation *i* at processing element *j, E* is the sum-square-error, *k* is the number of parameters.

Correlation coefficient (CC)

The values of correlation coefficient for artificial neural networks models were computed. Based on ANN model, the values of correlation coefficient for Junagadh testing of single and double (7-30-1, 4-12-1, 7-12-13-1 and 4-6-4-1) hidden layers networks (2008-2011) of periods are 80.00, 89.00, 89.00 and 95.00 per cent respectively. The higher values of correlation coefficient for testing periods show good agreement between observed and predicted values of rainfall. According to the correlation coefficient, the model 4-6-4-1 has better accuracy than the other selected models.

Mean square error (MSE)

The mean square error (MSE) values between observed and predicted values of rainfall based on developed models of rainfall for single and double hidden layers (7-30-1, 4-12-1, 7-12-13-1 and 4-6-4-1) for testing periods are

Table 1: Performance Indicators

Indicator	*Equation*
Mean Square Error (MSE)	$MSE = \frac{\sum_{j=1}^{n}\left(Y_j - Y_{ej}\right)^2}{n}$
Normalised Mean Square Error (NMSE)	$MSE = \frac{PNMSE}{\sum_{j=0}^{P}\frac{N\sum_{i=1}^{N}d_{ij}^{2}-\left(\sum_{i=1}^{N}d_{ij}\right)^{2}}{N}}$
Correlation Coefficient (CC)	$CC = \frac{\sum_{j=1}^{n}\left\{\left(Y_j - \overline{Y}\right)\left(Y_{ej} - \overline{Y_{ej}}\right)\right\}}{\sum_{j=1}^{n}\left(Y_j - \overline{Y}\right)^2\sum_{j=1}^{n}\left(Y_{ej} - \overline{Y_{ej}}\right)^2}\times 100$
Akaike's Information Criterion (AIC)	$CC = 2k + n\ln\left(\frac{2E}{n}\right)$
Coefficient of Efficiency (CE)	$CE = \left(1 - \frac{\sum_{i=1}^{n}\left(Y_j - Y_{ej}\right)^2}{\sum_{i=1}^{n}\left(Y_j - \overline{Y}\right)^2}\right)\times 100$
Volumetric Error (EV)	$EV = \left[\frac{\sum_{j=1}^{n}\left(Y_{ej} - Y_j\right)}{\sum_{j=1}^{n}\left(Y_j\right)}\right]\times 100$

Table 2: Quantitative performance evaluation of developed models during testing for the best chosen network for ANN

Performance indies	*Single Hidden Layer ANN Network*		*Double Hidden Layer ANN Network*	
	7-30-1	*4-12-1*	*7-12-13-1*	*4-6-4-1*
MSE	0.0015	0.0006	0.0008	0.0003
NMSE	0.43	0.37	0.40	0.29
CC	0.80	0.89	0.89	0.95
AIC	-3275.82	-4250.50	-3472.79	-4691.61
CE	69.7	83.48	82.5	87.1
EV	42.14	17.34	24.7	13.54

0.0015, 0.0006, 0.0008 and 0.0003 respectively.The lowest MSE shows the higher performance of the model and the model 4-6-4-1 has lowest MSE.

Normalised mean square error (NMSE)

The normalised mean square error (NMSE), between observed and predicted values for developed single and double hidden layers models were determined and values so obtained are given in Table 2. Based on ANNs model, the values of Normalised mean square error of developed models (7-30-1, 4-12-1, 7-12-13-1 and 4-6-4-1) for testing period (2008 to 2011) are 0.43, 0.37, 0.40 and 0.29 respectively. Among all selected models, the model 4-6-4-1 has the lowest NMSE.

Akaike's information criterion

The Akaike's information criterion (AIC) is used to measure the trade off between training performance and network size. The goal is to minimise this term to produce a network with the best generalisation. The values of akaike's information criterion of developed models (7-30-1, 4-12-1, 7-12-13-1 and 4-6-4-1) for testing periods are -3275.82, -4250.50, -3472.79 and -4691.61 respectively. It can be seen from the performance of the models, the model 4-6-4-1 has greater accuracy as compared to the other models.

Coefficient of efficiency

The coefficient of efficiency of developed single and double hidden layer models for rainfall between the observed and predicted values of rainfall were computed and are shown in Table 2. Applying the ANN model, the values of coefficient of efficiency of models (7-30-1, 4-12-1, 7-12-13-1 and 4-6-4-1) for testing periods are 69.7, 83.48, 82.5 and 87.1 % respectively. The model 4-6-4-1 has higher value of CE. The higher value of CE show the good association between observed and predicted rainfall.

Volumetric error

The volumetric error of the model was also assessed by another measure i.e. volumetric error and is given in Table 2. Using the ANN models, the values of volumetric error for single and double hidden layers models (7-30-1, 4-12-1, 7-12-13-1 and 4-6-4-1) for testing period are 32.14, 14.34, 14.70 and 13.54 % respectively. It is observed that the model 4-6-4-1 has lower EV and it shows the higher performance of the model.

Conclusions

In this work, an attempt has been made to train and validate the artificial neural networks models for monsoon season of Junagadh, Gujarat, India. In order to test these models, the actual rainfall data was collected from Junagadh Agricultural University, Junagadh, Gujarat, India for the period of 2008 to 2011. Two different combinations of the input parameters have been used for predict the rainfall. Four different single and double hidden layers models of artificial neural networks having a less mean square error and high correlation coefficient. The model 4-6-4-1 has lowest MSE and higher CC. The model 4-6-4-1 has been selected from the four models. It has been proved that ANNs provide better accurate forecasting of rainfall. These models based on the individuals parameters like vapour pressure, relative humidity, wind velocity and mean temperature.

References

Abbot, J., and Marohasy, J. 2012. Application of artificial neural networks to rainfall forecasting in Queensland, Australia. Advance in Atmospheric sciences. 29:717–730.

Banik, S., Anwer, M., Khan, A. F. M. K., Rouf, R. A. and Chanchary, F.H. 2009. Forecasting Bangladeshi monsoon rainfall using neural network and genetic algorithm approaches. International Technology Management Review, 2:1-18.

Chauhan, S. and Shrivastava, R.K. 2012. Estimating Reference Evapo-transpiration Using neural computing Technique, Journal of Indian Water Resources Society, 32:22-30.

De, S.S. 2009. Artificial neural network based prediction of maximum and minimum temperature in the summer monsoon months over India. Applied Physics Res.,1(2):37-44.

El-Shafie, A.H., El-Shafie, A., El Mazoghi, H.G. and Taha, M.R., 2011. Artificial Neural Network technique for rainfall forecasting applied to Alexandria, Egypt. International Journal of the Physical Sciences, 6(6):1306-1316.

Guhathakurta, P. 2006. Long-range monsoon rainfall prediction of 2005 for the districts and sub-division Kerala with artificial neural network, Current Science, 90:773-779.

Hu, M.J.C. 1964. Application of ADALINE system to weather forecasting, Technical Report, Stanford Election, Stanford, CA.

Jeong, C., Shin, J., Kim, T. and Heo, J.H. 2012. Monthly Precipitation Forecasting with a Neuro-Fuzzy Model. Water Resources Management, Doi:10.1007/s11269-012-0157-3.

Kumar, A., Singh, M.P., Ghosh, S. and Anand, A. 2012. Weather forecasting model using Artificial Neural Network, Procedia Technology, 4:311-318.

Michaelides, S.C., C.C. Neocleous and C.N. Schizas, 1995. Artificial neural networks and multiple linear regressions in estimating missing rainfall data, Proc. DSP95 International Conference, "Digital Signal Processing", Limassol, Cyprus, 668-673.

Nastos, P.T., Moustris, K.P., Larissi, I.K. and Paliatsos, A. G. 2013. Rain intensity forecast using Artificial Neural Networks in Athens, Greece. Atmospheric Research. 199:153-160.

Singh, A., Mohd., I., Isaac, R.K. and Denis, D.M., 2012. Hydrological Process Modeling using RBNN – A Neural Network Computing Technique, Journal of Agricultural Engineering, 49(2):27-33.

Toth,E.,Brath,A. and Montanari, A. 2000. Comparison of short-term rainfall prediction models for real-time flood forecasting. J. Hydrology, 239:132–147.

Zurada, M. Z. 1992. Introduction to artificial neural systems; St. Paul New York, West Pub. Co.

6

Modelling and Simulating the Climate Change Impact on Water Resources System in Ozat Basin using SWAT

H.D. Rank, J.H. Kelaiya, D.M. Pardava, R.J. Patel, H.V. Parmar, H.H. Mashru and P.B. Vekariya

Abstract

The study was undertaken for Ozat river basin (3176.24 km² area) having four watersheds.The satellite digital images of SRTM DEM having 90 m resolution, land use/land cover and soil map was used in this study. The downscaled global climate model simulations (CMIP3) data (1979-2013) of total 10 weather grid points' were used for the analysis. The four components of the water resources system namely rainfall, runoff, evapotranspiration and groundwater recharge were simulated. The climate change impacts were assessed by the trend analysis of the time series of rainfall, runoff, reference evapotranspiraion and groundwater recharge using the standard methods as described by Gilbert (1987), Kendall and Gibbons (1990) and Brauner (1997) along with best fit trend analysis.The SWAT simulated area weighted spatial and temporal average rainfall, runoff, evapotranspiration, reference evapotranspiration and groundwater recharge in the Ozat basin are 968 mm, 170 mm, 304 mm, 663 mm and 163 mm respectively which may change at 145.2, 43.1, 14.1, -14.6 and 44.6 mm/decade due to climate change impacts. The monsoon rainfall in the basin may increase at 1.75 per cent/year. The surface and ground water resources of the basin is 540 mCM and 517MCM which may increase at 2.5 per cent/ year (80 MCM/year) and 2.7 per cent/year (87 MCM/year) due to the climate change impacts.The all of four water resources system components i.e. rainfall, runoff, evepotranspiration and groundwater recharge are found increasing in the Ozat basin may be because of climate change.

Key words: *climate change, impact assessment, water resources.*

Introduction

Climate projections developed for India for the 2050s, indicate an increase in the average temperature by 2-4°C during that period, an overall decrease in the number of rainy days by more than 15 days in western and central India and an increase by 5-10 days near the foothills of Himalaya and in North-East India. The projections also indicate an overall increase in the rainy day intensity by 1-4 mm/day except for small areas in northwest India where the rainfall intensities may decrease by 1 mm/day. As many as 99 districts, spread over 14 states, were identified by the Central Water Commission as drought prone in the country. Most of the drought prone areas so identified are concentrated in the states of Rajasthan, Karnataka, Andhra Pradesh, Maharashtra and Gujarat. It is projected that by 2050s the quantity of surface run off due to climate change would decrease (Gosain *et al.,* 2006). This may be due to increase in evapotranspiration on account of increased temperatures or variation in the distribution of the rainfall. Climate change has had an effect on the monsoons too. India is heavily dependent on the monsoon to meet its agricultural and water needs, and also for protecting and propagating its rich biodiversity. Subtle changes have already been noted in the monsoon rain patterns by scientists at IIT, Delhi. They also warn that by the 2050s, India will experience a decline in its summer rainfall, which accounts for almost 70 per cent of the total annual rainfall and is crucial to agriculture. Many parts of peninsular India, especially the Western Ghats, are likely to experience a 5–10 per cent increase in total precipitation (IPCC 2007); however, this increase is likely to be accompanied by greater temporal variability. The trend may reflect a continuation of some past trends. Increased frequency of extremely wet rainy seasons (Gosain and Rao, 2007) is also likely to mean increased runoff. According to Milly *et al.* (2005), compared to 1900–1970, most of India is likely to experience 5–20 per cent increase in annual runoff during 2041–60.

Prediction of water availability in space and time has been a challenging task for fraternity working in the area of water resources. The problem has become more complicated due to changing climatic conditions across the globe. The recent past observations revealed the fact that extremes (floods and droughts) have become more frequent, and threat for human civilisation. The basic and applied research in the area of climate change and its impact on water resources are being taken as the top priority by government and nongovernment organisations in the country.

Since the TAR there have been many studies related to trends in runoff flows during the 20th century at scales ranging from catchment to global. Some of these studies have detected significant trends in some indicators of river flow, and some have demonstrated statistically significant links with trends in temperature or precipitation; but no globally homogeneous trend has been reported. Many studies, however, have found no trends, or have been unable to separate the effects of variations in temperature and precipitation from the effects of human interventions in the catchment, such as land-use change and reservoir construction. At the global scale, there is evidence of a broadly coherent pattern of change in annual runoff, with some regions experiencing an increase for China, Finland, coterminous USA, particularly at higher latitudes, and others a decrease, for example in parts of West Africa, southern Europe and southern Latin America (Bates *et al.*, 2008). Milly *et al.* (2005) claimed a four per cent increase in global total runoff per 1°C rise in temperature during the 20th century, with regional variation around this trend. Higher temperatures and increased variability of precipitation would, in general, lead to increased irrigation water demand, even if the total precipitation during the growing season remains the same. The impact of climate change on optimal growing periods, and on yield-maximising irrigation water use, has been modelled assuming no change in either irrigated area and/or climate variability (Döll and Siebert, 2002). Applying the IPCC SRES A2 and B2 scenarios as interpreted by two climate models, it was projected that the net irrigation requirements of China and India, the countries with the largest irrigated areas worldwide, would change by +2 per cent to +15 per cent in the case of China, and by –6 per cent to +5 per cent in the case of India, by 2020, depending on emissions scenarios and climate model (Döll and Siebert, 2002). Different climate models project different worldwide changes in net irrigation requirements, with estimated increases ranging from 1–3 per cent by the 2020s and 2–7 per cent by the 2070s. The largest global-scale increases in net irrigation requirements result from a climate scenario based on the B2 emissions scenario.

Most national and international studies focus on climate change impacts to surface water resources. Climate change effects on groundwater resources have not been systematically assessed, despite many regions being greatly dependent on groundwater for irrigation, industrial use (including mining) and urban water supply. To support a sustainable groundwater management, it is necessary to assess the renewable groundwater resources, i.e. the long-term average annual groundwater recharge. Assessing the impacts of climate change on the future renewable groundwater resources under various scenarios can help to identify regions with significantly changing groundwater resources and thus inform on the planning of possible measures for climate change adaptation. The groundwater recharge depends on many factors such as land use pattern, slopes, soil types, hydro-geological and hydro-meteorological parameters. Also the effect of the climate change on rainfall i.e. rainfall amount and its temporal and spatial distribution can ultimately alter the groundwater recharge. The most recent Intergovernmental Panel on Climate Change (IPCC) report noted a comparative lack of studies addressing the effects of climate change on groundwater (IPCC, 2007). This challenge has been taken up by the ground water research community and has resulted in an increasing focus of study on this topic in recent years (Green *et al.*, 2011;Taylor, 2013), although there still remains many aspects of the climate change effects on groundwater that have not been well studied. The most direct method in which climate change will impact groundwater resources is by modifying the renewable portion of groundwater storage through changes in recharge. Knowledge of future recharge rates is desirable in order to promote proactive management of groundwater, as historical observations may not be an appropriate basis for management under a future climate. The High Plains Aquifer is a very important source of water for the United States. The High Plains Aquifer is ranked first in terms of volume of groundwater extracted in the USA; for the year 2000, 16 km^3 was extracted, of which 97 per cent was for irrigation (Maupin and Barber, 2005). This irrigation water underpins a market value of agricultural products of $35 billion or more than 10 per cent of the national total (National Agricultural Statistics Services, 2011).

The climate change impact assessment on the water resources system components at the basin scales is the utmost need in the era of changing climate. Keeping in view the said challenges, the present study was planned. It can be helpful to adapt the appropriate strategies against the adverse impact of climate change.

Methodology

Study area

The study was undertaken for Ozat river basin (3176.24 km^2) having four watersheds as shown in Fig. 1. Ozat river originates from near Visavadar and meets in a Arebian sea having total length of 169 km. It is situated between latitude of 21°N to 22°N and longitude of 70°E to 71°E. The climate of the project area can be classified as tropical and sub-tropical. January is the coldest month with mean monthly temperature varying from 4°C and 15°C and maximum monthly temperature varies between 40°C and 46°C in the month of May.

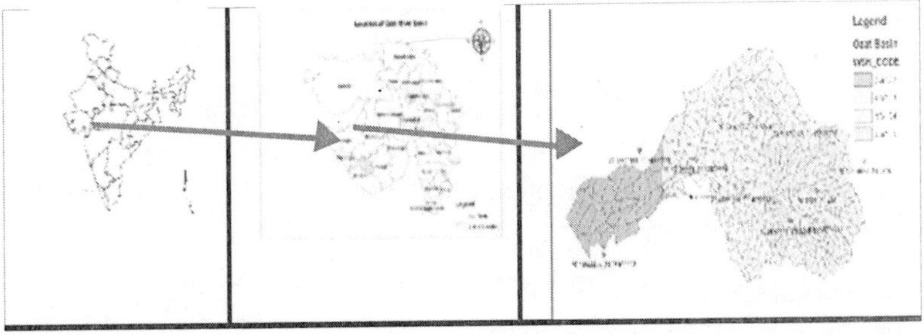

Fig. 1: Location map of Ozat river basin.

Simulation of Climate change impact on water resources components

The software namely, remote sensing and GIS software-Arc GIS V10.1, PCI GEOMETICA V10.1, Arc SWAT 2012 and WGEN maker 4.1 were used for the present SWAT simulation studies. The Soil and Water Assessment Tool (SWAT) model is a medium to large-scale river basin model that was developed to predict the impact of climate change on water resources system components of the basin over long periods of time (Neitsch *et al.,* 2005). SWAT is a physically based, spatially semi-distributed and computationally efficient model that can be used to simulate a single basin or a system of multiple basins that are hydrologically connected (Luzio *et al.,* 2002). It is a continuous time series model with a GIS interface and that uses readily available input data. SWAT has proven to be an effective model for river basin studies under different environmental and climatic conditions (Arnold and Allen, 1996). Input climate variables required by SWAT are precipitation, maximum and minimum air temperature, solar radiation, wind speed and relative humidity. The satellite images of SRTM DEM having 90 m resolution, land use/land cover and soil map was used in this study. The downscaled global climate model simulations (CMIP3) data (1979-2013) of total 10 weather grid points' were used for the analysis.

The SWAT simulated runoff, evapotranspiration and groundwater recharge were also estimated using other methods for the comparison. The runoff was estimated by the SCS curve number method. The groundwater recharge was estimated by Krishna Rao (1970) empirical model and water table fluctuation methods. The evapotranspiration was estimated on daily time scale using FAO-56 Penmen-Monteinth method (Allen *et al*., 1998).

The time series of seasonal and annual rainfall, runoff, evapotranspiration and groundwater recharge by different methods (1979-2013) for each of four watersheds and basin were analysed separately. The climate change impacts were assessed by the trend analysis of the time series of rainfall, runoff, reference evapotranspiration and groundwater recharge using the standard methods as described by Gilbert (1987), Kendall and Gibbons (1990) and Brauner (1997) along with best fit trend analysis.

Results and Discussions

The trend statistics of time series of rainfall along with SWAT simulate drunoff, reference evapotranspiration and groundwater recharge and estimated by other standard methods are given in Table-1 and 2.

Rainfall and runoff

The area weighted average monsoon rainfall from Ozat basin was found as 968 mm. The Mann-Kendall and Sen's statistics showed that monsoon rainfall

Table 1: Trend analysis of SWAT simulated water resources system components for Ozat river

Water resources System components during monsoon season	*Man Kendall(Z)*	*Sen's Slope*	*Lower and Upper limit of Sen's slope at 1 per cent*		*Lower and Upper limit of Sen's slope at 5 per cent*		*Conf. Level*	*Best fit trend slope*	*R^2*	*Mean*	*CV*
		mm/yr	mm/yr		mm/yr		(per cent)	mm/yr		mm	Per cent
Runoff (mm)	2.27**	5.15	-0.80	11.417	0.52	9.67	98.8	4.31	0.07	170	98.57
Groundwater recharge (mm)	2.61***	4.63	0.08	9.729	0.88	8.50	99.6	4.46	0.19	163	64.87
ET_c (mm)	1.76**	1.52	-0.59	3.696	-0.22	3.02	96.1	1.41	0.11	304	14.44
ET_o (mm)	-1.08NS	-1.21	-5.20	1.910	-4.35	0.98	86.0	-1.46	0.04	663	11.35

Table 2 : Trend analysis of Seasonal/annual water resources system components for Ozat river basin

Watershed/ Basin	Season/ period	Man Kendall (Z)	Sen's Slope	Lower and Upper limit of Sen's slope at 1 per cent		Lower and Upper limit of Sen's slope at 5 per cent		Conf. Level	Best fit trend slope	R^2	Mean	CV
			mm/yr	mm/yr		mm/yr		(per cent)	mm/yr		mm	Per cent
Rainfall(mm)												
5G1C2	Monsoon	2.24**	20.78	-3.46	43.46	1.51	35.99	98.76	18.31	0.15	1153	41.65
5G1C3		2.22**	17.51	-3.37	34.39	1.61	28.43	98.66	12.30	0.12	779	47.17
5G1C4		1.90**	17.53	-4.12	32.98	-0.34	30.40	97.15	12.61	0.10	980	41.97
5G1C5		1.99**	14.86	-4.90	36.38	0.20	30.62	97.66	14.86	0.14	957	42.67
Ozat basin		2.33***	15.96	-2.30	36.89	3.26	30.43	99.01	14.52	0.13	968	42.10
Runoff(mm) estimated by SCS-CN method												
5G1C2	Monsoon	1.73**	8.88	-2.84	20.7	-0.85	17.45	95.8	8.04	0.12	346	69.02
5G1C3		2.07**	4.05	-1.15	10.4	0.13	8.29	98.1	3.28	0.04	148	110.3
5G1C4		2.16**	5.14	-1.21	10.7	0.30	9.36	98.6	3.60	0.03	216	95.21
5G1C5		1.99**	5.44	-1.81	13.2	0.16	11.4	97.6	3.70	0.04	189	104.8
Ozat basin		2.02**	6.63	-1.91	14.7	0.18	11.9	97.8	4.66	0.06	225	85.80
Reference evapotranspiration (mm)												
5G1C2	Winter	1.76**	0.926	-0.440	2.402	-0.12	1.978	96.09	0.77	0.08	722	3.96
	Summer	1.33*	0.506	-0.427	1.477	-0.22	1.257	90.91	0.58	0.09	806	2.40
	Monsoon	0.00 NS	-0.009	-2.235	1.758	-1.52	1.261	50.00	-0.08	0.00	586	7.12
	Annual	1.62*	1.645	-1.209	4.457	-0.34	3.549	94.73	1.27	0.05	2115	2.91
5G1C3	Winter	0.31NS	0.204	-1.735	2.262	-1.26	1.620	62.26	0.27	0.00	714	5.50
	Summer	0.28NS	0.198	-0.908	2.095	-0.70	1.447	61.18	0.56	0.03	986	3.54
	Monsoon	- 0.80NS	-0.777	-4.375	2.138	-3.49	1.499	78.68	-0.91	0.02	533	12.38
	Annual	0.14NS	0.235	-5.097	4.492	-3.67	3.576	55.65	-0.08	0.00	2234	4.61

Cont...

5G1C4	Winter	0.76NS	0.818	-1.590	3.496	-0.99	2.704	77.70	0.41	0.01	738	5.31
	Summer	1.74**	0.962	-0.410	2.901	-0.12	2.480	95.86	0.77	0.06	964	3.39
	Monsoon	0.41NS	0.457	-3.209	2.992	-2.48	2.336	65.74	-0.73	0.02	541	11.04
	Annual	1.22NS	2.387	-3.077	7.312	-2.49	5.254	88.81	0.46	0.00	2244	4.15
5G1C5	Winter	0.73NS	0.586	-1.583	3.110	-0.99	2.420	76.72	-0.11	0.00	711	5.65
	Summer	1.12NS	0.824	-0.836	3.229	-0.51	2.640	86.84	0.15	0.00	998	4.06
	Monsoon	0.28NS	0.434	-3.210	2.929	-2.80	2.516	60.86	-1.04	0.03	498	12.62
	Annual	1.22NS	2.272	-3.659	6.964	-2.15	5.079	88.81	-1.00	0.01	2209	4.83
Ozat basin	Winter	0.76NS	0.743	-1.269	3.235	-0.77	2.508	77.70	0.34	0.01	721	4.92
	Summer	1.31*	0.854	-0.383	2.721	-0.24	2.321	90.55	0.52	0.03	939	3.31
	Monsoon	0.31NS	0.459	-3.066	3.102	-2.20	2.290	62.10	-0.69	0.02	539	10.55
	Annual	1.28*	2.031	-2.836	6.811	-1.90	4.842	89.99	0.16	0.00	2200	4.03
Groundwater recharge by Krishna-Rao empirical model (mm)												
5G1C2	Monsoon	2.27**	5.95	-0.718	11.63	0.604	9.87	98.84	5.31	0.17	194.7	68.25
5G1C3		2.22**	3.43	-0.304	7.66	0.098	6.51	98.67	2.93	0.12	97.8	89.35
5G1C4		1.99**	3.83	-1.216	8.94	0.049	7.89	97.66	3.14	0.10	144.4	71.75
5G1C5		1.90**	4.03	-1.156	8.53	-0.078	7.75	97.14	3.66	0.13	138.6	73.97
basin		2.22**	3.98	-0.647	9.414	0.702	7.51	98.66	3.76	0.14	143.9	72.25
Groundwater recharge by water table fluctuation (mm)												
5G1C2	Monsoon	1.52*	0.082	-0.014	0.260	0.000	0.222	93.63	0.10	0.05	48.64	104.1
5G1C3		0.27NS	0.000	-0.147	0.174	-0.091	0.127	60.72	0.00	0.00	64.54	73.4
5G1C4		0.08NS	0.000	-0.123	0.168	-0.089	0.122	53.03	0.01	0.00	65.19	71.1
5G1C5		0.16NS	0.000	-0.202	0.300	-0.125	0.214	56.18	-0.01	0.00	65.02	78.3
basin		0.51NS	0.009	-0.102	0.186	-0.055	0.145	69.52	0.03	0.00	61.26	75.8
****,***,**,* = indicate significant at 0.1 per cent, 1 per cent, 5 per cent, 10 per cent level respectively, NS= Non Significant												

was found significantly increasing at 145.2 mm/decade. The SWAT simulated area weighted runoff and estimated by SCS-CN method was observed as 170 and 225 mm respectively. It can be seen in Fig. 1 that the runoff estimated by SCS-CN method is higher than the estimated by SWAT as the former estimated the runoff considering five per cent slope of the basin while majority basin area is having slope within one per cent only. The monsoon runoff is increasing at 46.6 and 43.1 mm/decade as per SCS-CN and SWAT model respectively may be due to the climate change impacts. The result is also supported by the slope of best fitted line (Fig. 2) and Sen's slope method except watershed 5G1C2. The runoff Sen's slope was observed as 66.3 mm per decade for the entire river basin which was significant. The rainfall and runoff was found increasing significantly in the all watersheds 5G1C2, C3, C4 and C5. The slope of the best fit trend lineofrainfall was observed as 18.31, 12.30, 12.61 and 14.86 mm per year (Fig. 2) while that of estimated by Sen's method was found as 20.78, 17.51, 17.53

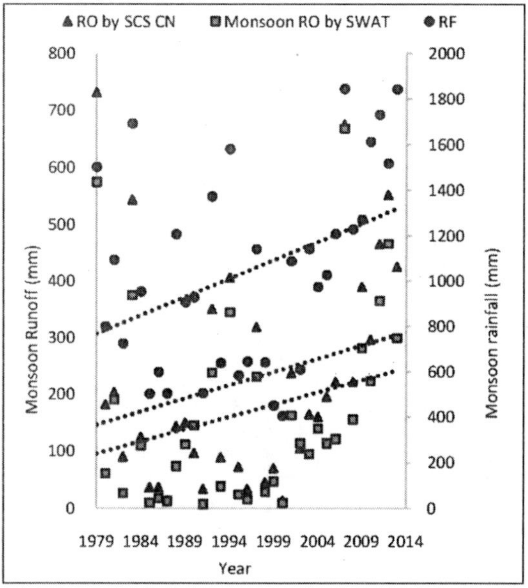

Fig. 2 : Comparison of SWAT simulated and SCS-CN runoff.

and 14.86 per year respectively (Table 2). The runoff was also found significantly increasing all four watersheds and basin. The close agreement was found between the slopes of the best fit trend line and estimated by Sen's method. The Sen's slope was found significant for the watershed 5G1C2, C3, C5 and basin. The result showed that there will be significant change in rainfall and runoff for the entire basin due to climate change impact. The rainfall and runoff data showed that the highest rainfall and runoff are not in the same year may be due to change in the temporal distributions during monsoon. This result obtained for the Ozat basin was supported by the findings reported by Singh (1999) stating that an increasing linear trend was observed in the rainfall for the Luni river basin of north-west arid India. Rupakumar and Ashrit (2001) have projected 13 per cent increase in monsoon rainfall in India using ECHAM4 model.

Evapotranspiration

The SWAT simulated monsoon evapotranspiration and reference transpiration was found as 304 mm and 663 mm which was found changing respectively at 14.1 (significant) and -14.6 mm/decade (Fig. 3). The reference evapotranspiration during monsoon season estimated by FAO-56 PM method was found as 540 mm. The crop evapotranspiration is in increasing trend even though the insignificant decreasing trend of reference evapotranspiration. The reason may be the reduction in stress due to the increasing trend of the rainfall in the basin. During the winter season (1-8 and 43-52 standard week), no significant change trend in reference evapotranspiration could be found during all standard weeks except standard week of 4, 7 and 47. During summer season, the significant decreasing trend was found in weekly reference evapotranspiration (ET_o) during the standard week 20, 23 and 24 while the significant increasing trend during the standard week 9, 12, 13 and 17. During monsoon season (27-42), it was found significantly decreasing during the standard week 28 and 39-41.

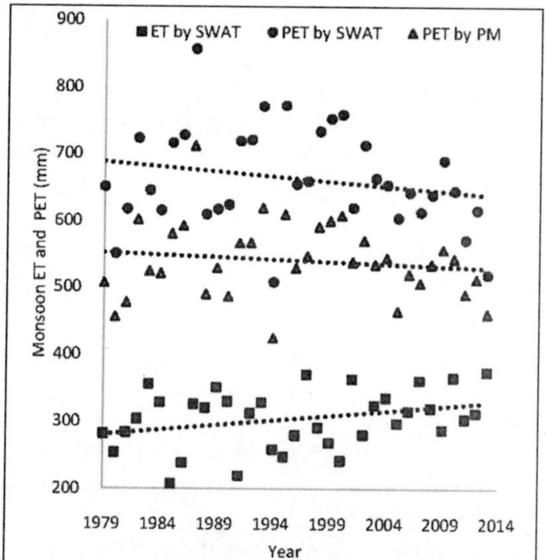

Fig. 3 : Comparison of SWAT simulated ET and PET with FAO-56 PM method.

The monthly reference evapotranspiration (ET_0) were found increasing significantly during March and April only.

The seasonal reference evapotranspiration (ET_0) was found increasing insignificantly in Winter (16th October to 15th February) and monsoon (16th June to 15th October) seasons while significantly in Summer (16th February to 15th June) season of the year in Ozat basin (Table 2 and Fig. 3).

The annual reference evapotranspiration (ET_0) was found significantly increasing for the entire river basin. However, the Sen's slope was not found significant for any of four watersheds and basin. The annual reference evapo-transpiration in the Ozat basin was found increasing significantly at 0.16 mm/year due to climate change. The increase in the annual reference evapotranspiration is due to increase in temperature and solar radiation.

Groundwater resources

The area weighted mean of the SWAT simulated groundwater recharge was 163 mm which was comparable to 143 mm found by Krishna Rao. However, the groundwater recharge estimated by groundwater table fluctuations method was found as 61.6 mm which is too low. The groundwater recharge by water table fluctuation may be underestimated. The reason was lack of data of spatial aquifer properties for the entire basin. Methods for estimating groundwater recharge based on groundwater level fluctuations are prone to large uncertainties due to the uncertainty in values for specific yield (S_y) (Healy *et al.*, 2002).

The groundwater recharge estimated by the water table fluctuation (WTF) method can be lower than the estimated by SWAT model using water balance approach as time is required for the pressure front from increased deep drainage to move downward through the unsaturated zone. In the present study, water table level data of pre-monsoon and post-monsoon was used. So while observing the water table level of post-monsoon, the part of recharged water might not had been reached to the groundwater. So, the groundwater recharge by WTF method could be lower than that of by SWAT model using water balance approach.

Analysis of water-level fluctuations can, however, be useful for determining the magnitude of long-term changes in recharge. Changes in climate or land use are not reflected immediately at the water table. This time delay is related to the recharge rate, the soil-water content, and the depth to the water table (Jolly *et al.* 1989). Allison *et al.* (1990) observed that 70 years after the change in land use at their site, groundwater levels were raising only where the depth to the water table was less than 40 m. presumably the pressure front was still being propagated through the saturated zone in areas with a deeper water table.

The groundwater recharge estimated by SWAT and Krishna Rao model Groundwater level fluctuation were found increasing at 44.6 and 37.6 mm/ decade (Table1-2 and Fig. 4). It was found that the groundwater recharge by Krishna-Rao model was significantly increasing for all of four watersheds and

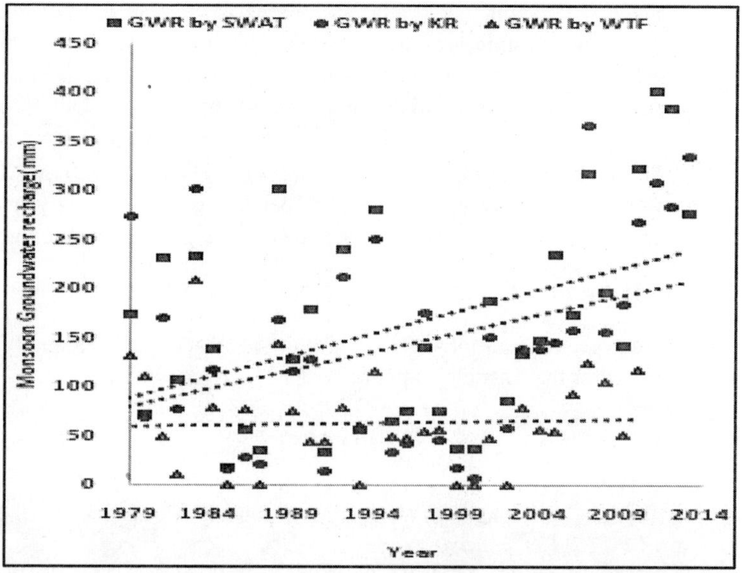

Fig. 4: Comparison of SWAT simulated groundwater recharge with other methods.

entire basin. The Sen's slope statistics also supported the Mann-Kendall statistics. The Sen's slope was found as 39.8 mm/decade for the entire river basin and which was significant. The groundwater recharge in watersheds 5G1C2, C3, C4, C5 and basin may be increased at 5.31, 2.93, 3.14, 3.66 and 3.76 mm/year respectively.

There was a close agreement between the Sen's slope and the slope of the best fitted trend line. In fact it can be also be reflected in Fig. 4. The result showed that there will be significant change in groundwater resources due to increasing trend of groundwater recharge in river basin. However, a chalk aquifer in Belgium showed for some of the future climate scenarios drops in groundwater level of up to seven metres (Brouye'ŕe *et al.*, 2004; Woldeamlak *et al.*, 2007). For the UK, particularly in the southern part a reduction in groundwater recharge is expected in spite of the increased rainfall (Arnell, 1998). The groundwater recharge by WTF method was found increasing in four watersheds and basin but significantly only in watershed- 5G1C2. The Sen's slope statistics also support the Mann-Kendall statistics.

Conclusions

- The all of four water resources system components i.e. rainfall, runoff, evapotranspiration and groundwater recharge are found increasing in the Ozat basin may be because of climate change.
- The SWAT simulated spatial and temporal average rainfall, runoff, evapotranspiration, reference evapotranspiration and groundwater recharge in the Ozat basin are 968 mm, 170 mm, 304 mm, 663 mm and 163 mm respectively which may change at 169.1, 43.1, 14.1, -14.6 and 44.6 mm/decade.
- The surface and ground water resources of the basin is 540 mCM and 517 MCM which may increase at 2.5 per cent/year (80 MCM/year) and 2.7 per cent/year (87 MCM/year) due to the climate change impacts.

Acknowledgements

The authors are thankful to the Directorate of Water Management, ICAR, Bhubaneswar for the financial supports under NICRA project.

References

Allen, R.G., L.S. Pereira., D. Raes and M. Smith. 1998. "Crop Evapotranspiration, Guideline for Computing Crop Water Requirements", FAO Irrigation and Drainage Paper 56. FAO Rome, Italy.

Allison, G.B., P.G. Cook, S.R. Barnett, G.R. Walker, I.D. Jolly and M.W. Hughes.1990. "Land clearance and river salinization in the western Murray Basin, Australia", J. Hydrol. 9:1–20.

Arnell, N.W., 1998. "Climate change and water resources in Britain", Climatic Change, Vol. 39, No. 1, pp. 83-110.

Arnold, J.G. and P.M. Allen.1996. "Estimating hydrological budgets for three Illinois watersheds", J. Hydrol.176: 57-77.

Bates, B.C., Z.W. Kundzewicz, S. Wu and J. P. Palutikofeds. 2008. "Climate change and water", Technical Paper of the Intergovernmental Panel on Climate Change, IPCC Secretariat, Geneva, Switzerland.

Brauner, J.S., 1997. "Nonparametric estimation of slope—Sen's method". In Environmental sampling and monitoring primer. (Eds. Gallagher and Daniel). (online at URL http://www.cee.vt.edu).

Brouye're, S., G. Carabin and A. Dassargues. 2004. "Climate change impacts on groundwater resources: modeled deficits in a chalky aquifer, Geer Basin, Belgium", Hydrogeology Journal, Vol. 12, No. 2, pp. 123-34.

Döll, P. and S. Siebert. 2002. "Global modeling of irrigation water requirements", Water Resour. Res., 38(4).

Gilbert, R.O., 1987. "Statistical methods for environmental pollution monitoring", Van Nostrand Reinhold, NY.

Gosain, A.K., S. Rao and D. Basuray. 2006. "Climate change impact assessment on hydrology of Indian River basins", Current Science, 90(3).

Gosain, A.K. and S. Rao. 2007. "Impact assessment of climate change on water resources of two river systems in India", Jalvigyan Sameeksha. 22: 21.

Green, G., M. Gibbs and C. Wood. 2011. "Impacts of Climate Change on Water Resources", Phase 3 Vol. 1: Northern and Yorke Natural Resources Management Region, DFW Technical Report 2011/03, Gov. South Aus., Dept. Water, Adelaide.

Jolly, I.D., P. G. Cook, G. B. Allison and M. W. Hughes. 1989. "Simultaneous water and solute movement through an unsaturated soil following an increase in recharge", J. Hydrol.111:391–396.

Healy, R.W., P.G. Cook and B.R. Scanlon. 2002. "Choosing appropriate techniques for quantifying groundwater recharge", Hydrogeology Journal, 10:18–39.

IPCC, 2007. "Summary of Policymakers of Synthesis Report of the IPCC", Fourth Assessment Report [M]. Cambridge, UK.

Krishna Rao, 1970. In: Kumar C.P., 1996. Assessment of Ground Water Potential. All India Seminar on small watershed development, organized by IAH, W.B. Center, 15 February, 1996, Calcutta.

Kendall, M. and J.D. Gibbons. 1990. "Rank Correlation Methods", 5th Eds. Oxford University Press, NY. p. 272.

Luzio, D.M., R. Srinivasan and G.A. Jeffrey. 2002. "Integration of watershed tools and swat model into basins", J. Ame. Water Reso. Asso. (JAWRA). 38(4):1127-1141.

Maupin, M. A., and N. L. Barber. 2005. "Estimated withdrawals from principal aquifers in the US", Circular-1279, 52pp., U.S. Geol. Surv., Reston, Va.

Milly, P.C.D., K.A. Dunne and A.V. Vecchia. 2005. "Global pattern of trends in stream flow and water availability in a changing climate nature", Vol. 438, pp. 347–350.

Natural Resources Conservation Service. 2006. USDA, US Gen. Soil Map. Available at http://soildatamart.nrcs.usda.gov.

Neitsch, S.L., J.G. Arnold, J.R. Kiniry, R. Srinivasan and J. R.Williams. 2005. "Soil and Water Assessment Tool Input/Output File Documentation", Ver. 2005. Available at:www.brc.tamus.edu/swat/doc.html.

Rupakumar, K. and R.G. Ashrit. 2001. "Regional Aspects of global climate change simulations-Validation and assessment of climate response over Indian monsoon region to transient increase of greenhouse gases and sulphate aerosols", Mausam, Vol. 52, pp. 229–244.

Singh, R.S., K.D. Sharma and A. S. Faroda. 1999. "Climate change and its impact on drought and floods in Luni river basin of north-west arid India", Journal of Agrometeorology, 1(2): 99-111.

Taylor, R.G., 2013. "Ground water and climate change", Nature Clim. Change, 3(4), 322–329, doi:10.1038/nclimate1744.

Woldeamlak, S.T., O. Batelaan and F. de Smedt. 2007. "Effects of climate change on the groundwater system in Grote-Nete catchment, Belgium", Hydrogeology Journal, Vol. 15. No. 5, pp. 891-901.

7

Development of Short Period Rainfall Forecasting Models Using Artificial Neural Network

M.A. Sojitra, R.C. Purohit, Mahesh Kothari and K.K. Yadav

Abstract

Present research paper described the application of Artificial Neural Networks (ANNs) in the rainfall forecasting. Research shows ability of ANNs for daily rainfall forecasting of Udaipur, India. Two different combinations of data sets of weather parameters, one day lag and previous moving average week were taken as Case I and Case II respectively to generate nonlinear relationship. Different single and multi-hidden layer ANNs generated by increasing and decreasing of hidden layer(s) and Processing Element (PE) by trial and error method. Developed models are selected mainly two basics criteria least Mean Square Error and high Correlation Coefficient. Two single hidden layer(s) and two multi hidden layer(s) were selected for two input dataset. Models were trained and tested using last 30 (1979-2008) years and five (2009-13) years of weather parameter respectively. Result shows multi-hidden layer model ANN Model (7-4-1-1) of Case II has good Correlation Coefficient (0.93) and least Mean Square Error (0.001) was selected best among four models. It clearly revealed that monsoon depends on long-term of weather parameters. It also unveils that there donot necessary to higher number of Processing Element (PE) and higher number of hidden layer(s) gave good result. Sensitivity analysis revealed wet bulb temperature is most sensible parameter followed by mean temperature, dry bulb temperature, relative humidity, evaporation, rainfall, and wind velocity.

Introduction

Water is essential for all life. Rainfall is vital resources of water. It is also one of the prime requirements for agriculture, industrial production, domestic and recreation. Components of precipitation, resolved into soil moisture and groundwater, are the prerequisites for biomass production and social development in dry areas. The world's total water resources are estimated at 1.36×10^8 M ha-m. About 97.2 per cent of these world water resources salt water mainly in oceans, and only 2.8 per cent is available as freshwater at any time on the planet earth. Out of this 2.8 per cent of fresh water, about 2.2 per cent is available as surface water and 0.6 per cent as ground water. Even out of this 2.2 per cent of surface water, 2.15 per cent is fresh water in glaciers and icecaps and only of the order of 0.01 per cent is available in lakes and streams, the remaining 0.04 per cent being in other forms. Out of 0.6 per cent of stored ground water, only about 0.25 per cent can be economically extracted with the present drilling technology (Raghunath, 2006). So, rainfall is cheap and prime source of fresh water. As per UNICEF (2013) estimates 3.4 bn population will be faced water scarcity in 2025; nearly 40 per cent of the world population will be faced water scarce in 2050. About 20 per cent of the world's aquifer will be depleted in 2050. (UNICEF, 2013). It shows serious issue of water.

India occupies only 3.28 million sq. km geographical area, which is 2.4 per cent of the world's land area supports over 17 per cent of the world's population with 4 per cent of world's water resources. India also has a livestock population of 500 million, which is about 20 per cent of the world's total livestock population. More than half of these are cattle, forming the backbone of Indian agriculture. Indian agriculture shows 14.2 per cent share of the total GDP. It shows Water is essential for growth of agricultural and allied sectors. India's present average annual rainfall is 119 cm. Owing to physiographic factors; rainfall in India is highly variable. For example, in 2008, rainfall measured from about 500 mm in east and west Rajasthan to 3,798 mm in coastal Karnataka (UNICEF, 2013). Guhathakurta and Rajeevan (2006) reported contribution of June, July and September rainfall to annual rainfall is decreasing for few sub-divisions while contribution of August rainfall is increasing in few other subdivisions. It would be gear job if practices of rainfall forecasting carried before it occurrence.

Artificial Neural Networks (ANNs) are biologically inspired computer programmes designed to simulate the way in which the human brain processes information. ANNs gather their knowledge by detecting the patterns and relationships in data and learn (or are trained) through experience, not from programming. An ANN is formed from hundreds of single units, artificial neurons or processing elements (PE), connected with coefficients (weights), which constitute the neural structure and are organised in layers (Agatonovic-Kustrin and Beresford, 2000). Rainfall is forecasting is tough and non-linear process. Perceived strengths of ANNs are the capability for representing complex, non-linear relationships as well as being able to model interaction effects (Cameron *et al.*, 1999). ANNs were applied for flood forecasting, evaporation study, rainfall runoff relationship, tide forecasting, ground water level forecasting, River stage forecasting, stream flow forecasting, drought forecasting and spring discharge. Adamowski and Chan (2011) used to wavelet neural network conjunction model for ground water level forecasting. ANNS have been widely used rainfall or precipitation forecasting in monsoon season (Changsamjeong *et al.*, 2012; Ranjan Nayak *et al.*, 2013). Luk *et al.*(2000) have developed rainfall forecasting model using ANNs. Mekanik *et al.* (2013) have used multi-regression and ANNs for long-term rainfall forecasting using large scale climate modes. Moustris *et al.* (2011) has developed precipitation forecasting for Greece region. Toth *et al.* (2000) studied comparison of short-term rainfall prediction models for real time flood forecasting. BPNN of ANNs was used to rainfall forecasting (Vamsidhar *et al.*, 2010). Back propagation algorithm gave better result for river stage forecasting (Thirumalaiash and Deo, 1998).

Rainfall is main source of water for Indian agriculture where 68 per cent (86 Mha) of a total land are under rainfed area (Amarsinghe and Sharma, 2009). On an average, the rainfed regions of the country such as Western Rajasthan, eastern Rajasthan, Gujarat, western Uttar Pradesh, Tamil Nadu, Kashmir and Andhra Pradesh are most vulnerable to droughts, suffering once in every three years (Singh *et al.*, 2009). It would be great job if rainfall forecasting carried out, it would be help to farmer community, local bodies and nation for making future policy. It also helps disaster management for flood prone area. It help to farmer community to pre-sowing of seeds and reduced to economics loss. Here one approached carried out of rainfall forecasting model of Udaipur might be helpful to farming community and local municipal bodies to aware about rainfall occurrence for the same year.

Material and Methodology

This chapter contains study area, data acquisition and material and methodology were adopted to achieve main objective of develop artificial neural network. Comparative study of developed ANNs models and sensitivity analysis.

Description of the study area

Location

The study area is the Udaipur city of Rajasthan, India. The latitude and longitude of the study area is 24°58' N and 73°68' E, respectively. The elevation of gauging station is at 598 m above the mean sea level.

Climate

Udaipur city has particularly a tropical climate. The three main seasons, summer, monsoon and winter respectively. Udaipur is the only place in Rajasthan that has quite moderate climate throughout the year. An average annual rainfall of Udaipur is 617 (1979-2013) mm. Average annual pan evaporation is 5.1 mm/day (2002-12). May is the hottest month with mean monthly maximum temperature 39.8°C (2002-12) and January is the coolest month with mean monthly minimum temperature 4.9 °C (2002-2012). Highest mean monthly pan evaporation shows in the month of July 10.6 mm/day (2002-2012).

Data acquisition and pre-analysis

The depth of rainfall and its distribution in the temporal and spatial dimensions depends on many variables, such pressure, temperature, and wind speed and direction (Luk *et al.,* 2001). The collected daily meteorological data (i.e. Wet bulb Temperature, Dry bulb Temperature, Relative Humidity, Wind Velocity, Mean Temperature, Evaporation and Rainfall) of the period from the month May 25th to October 30th for the year 1979 to 2013 were subjected to Pre-analysis and formulation of the data base. Thus, the total number of data for each year's period comes out to be 153. Data from 1979-2008 were used formulate and validate and data from 2009-13 were used forecasting and check the error.

ANN architecture

A neural network is characterised by its architecture that present the pattern of connection between nodes. The architecture of an ANN is designed by weights between neurons, a transfer function the controls the generation of output in a neuron, and learning laws that define the relative importance of weights for input to a neuron. The architecture of ANN is classified into two types: single hidden layer and multi hidden layer. Fig. 1 and Fig. 2 are Single Hidden layer and Multi Hidden Layer respectively.

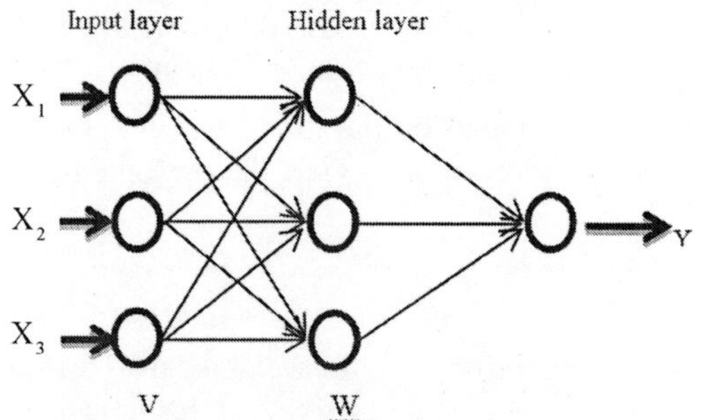

Fig. 1. Single Hidden layer Neural Network.

As shown in Fig. 1, X_1 and X_2 represent the input of the network, and the connection between the neurons, represented by lines, is quantified by their weights, which are shown in the V_{ji} and W_{kj}, Y is the output from single hidden layer ANN.

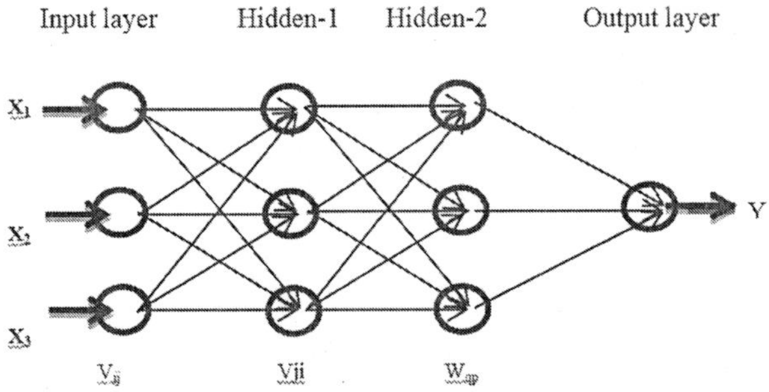

Fig. 2. Multi Hidden layer(s) Neural Network.

In the Fig. 2, Inputs are shown by X_1, X_2 and X_3 and V_{ji} represents the connection weight from the jth node in the preceding layer to i^{th} node. 'Y' is the observed output of the network.

Back-propagation training algorithm

Back-propagation training algorithm is the most commonly used supervised algorithm for training the multi-hidden layer ANN. An ANN which used back-propagation algorithm for its training is also called back-propagation ANN.

In back-propagation ANN, information is processed in the forward direction from the input layer to the hidden layer(s) and then to output layer. The objective of a back-propagation network is to find the weights that approximate target values of output with a selected accuracy. The least mean square error method, along with the generalised delta rule, is used to optimize the networkweights in back-propagation networks. The gradient descent method along with the chain rule of derivatives is employed to modify the network weights. It requires a continuous, differentiable and non-linear function on the ANN to compute output from each neuron.

The input data are multiplied by the initial weights, then the weights inputs are added by simple summation to yield the net input to each neuron.

$$Net = \sum_{i=1}^{N} V_{ji} X_i \quad ...(1)$$

where, X_i is the input to any neuron, V_{ji} is the weighted matrix from j^{th} layer to i^{th} layer, N is the number of inputs and Net is the net for j^{th} neuron.

The net of neuron is passed through an activation or transfer function to produce output form a neuron:

$$O = \frac{1}{1+\exp(-Net)} \qquad ...(2)$$

where, O is the output signal form i^{th} neuron.

After error between the output of the network and the target output are computed at the end of each forward pass, and is summed over as follows:

$$E = \sum_{i=1}^{N} \frac{1}{2}(O_i - D_i)^{2.} \qquad ...(3)$$

where, E is the Total error, O_i is the observed output and D_i is the target output.

The weight values are originally initialised randomly for all the connection weights in the network. During the back-propagation of error signal at output neuron, the weights are modified according to the following equations:

$$V_{ji}(n+1) = V_{ji}(n) + \Delta V_{ji}(n) \qquad ...(4)$$

$$\Delta V_{ji}(n) = \eta(\delta_i)(O_j) + \alpha \Delta V_{ji}(n-1) \qquad (5)$$

where, $V_{ji}(n)$ is the change in weight V_{ji} at n^{th} iteration, $\Delta V_{ji}(n-1)$ is the change in weight V_{ji} at n-1th iteration, $V_{ji}(n)$ is the value of weight V_{ji} at n^{th} iteration, $V_{ji}(n+1)$ is the updated value of weight V_{ji} at *(n^{th})* iteration, O_j is the output from j^{th} neuron in the output layer, α is the momentum constant and η is the learning constant. The value of δ_i for output neuron is given by,

$$\delta_i = O_i(1-O_i)(D_i - O_i) \qquad ...(6)$$

where, O_i is the output from the network, D_i is the target value of the output, is the error signal term of the output layer.

In the output layer, the target outputs are known, in the hidden layers, target values are not known. Therefore, the back-propagation algorithm uses the sum of the error signals of all the neurons of the succeeding layers to calculate error signals of any neuron '*j*' in the hidden layer.

$$\delta_i = O_i(1-O_i)\sum_p \delta_p W_{qp} \qquad ...(7)$$

where, p runs over all the neurons in the subsequent layers and in the error signal term corresponding to subsequent layer of p. The value of is then substituted in the equation 5. This procedure is repeated until the selected accuracy is achieved.

Activation function

The output from a neuron is calculated through the use of an activation function. The basic characteristics of the sigmoid function are that it is continuous, differentiable and is monotonically increasing. The sigmoid function is shown in Fig. 3.

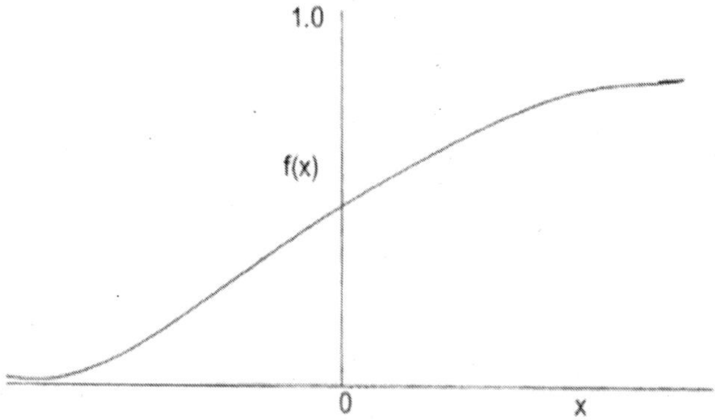

Fig. 3. Sigmoid function.

The sigmoid function can be represented by the following equation.

$$f(x) = \frac{1}{1+\exp(-ax)} \quad \ldots(8)$$

Where, a is the slope parameter.

The output from sigmoid function is always bounded between 0 to 1 and input to the function can vary between $-\infty$ to $+\infty$.

Selection of architecture

The most popular and successful technique for selecting the appropriate number and size(s) of the hidden layer(s) is trial and error method. A number of networks with one or two hidden layers are trained with different combinations of hidden neurons and a network is selected the yields the minimum Mean Square Error (MSE) and maximum Correlation Coefficient (CC). It is important that the size of the network should be as small as possible.

Development of ANNs Model

Single and multi-hidden layer(s) Neural Network Case I and Case II

In this case the observed time series of Wet bulb Temperature, Dry bulb Temperature, Relative Humidity, Mean Temperature, Evaporation, Wind Velocity and Rainfall of previous days and previous moving average week were taken as the input variables Case I and Case II respectively and current day rainfall as the output variable for both Case I and Case II for single and multi-hidden layer(s). Levenberg–Marquardt learning algorithm and sigmoid logistic activation function have been used for prediction of annual precipitation (Azadi and Sepaskhah, 2011).

Let the observed values of wet bulb temperature, dry bulb temperature, mean temperature, relative humidity, wind velocity, evaporation and rainfall represented as $Tw_{i,j,}$ $Td_{i}jT_{i,j,}$ $RH_{i,j,}$ $V_{i,j,}$ $E_{i,j,}$ and $P_{i,j}$ respectively for j^{th} day of the ith year ($i = 1, 2, \ldots, M$ and $j = 1, 2, \ldots, N$).

The functional form of the models can be represented as:

$$P_{i,j} = f(Tw_{i,j-1}, Td_{i,j-1}, T_{i,j-1}, RH_{i,j-1}, V_{i,j-1}, E_{i,j-1}, P_{i,j-1}) \quad \ldots(9)$$

Sensitivity Analysis

Sensitivity analysis were carried by adding and removing weather parameter one by one then model were formulated and validate for the best identified model ANN (7-4-1-1).

Performance and Evaluation of Models

Performance evaluations of the models were carried through visual observation based on the graphical comparison between observed and predicted values. There were also statistical and hydrological indices used for testing the goodness of fit for comparison between observed and predicted values of rainfall.

Results and Discussion

Fig. 4 to Fig. 7 shows Observed and Predicted Rainfall of ANNs network of 7-12-1, 7-6-6-1, 7-11-1 and 7-4-1-1 Case I and Case II of single hidden layer(s) and multi-hidden layer(s) respectively during the testing period (2009-2013). It is observed from the Fig. 4 to Fig. 7 that there is a close agreement between the predicted and observed rainfall. Developed models give satisfactory results the basis of qualitative evaluation during testing period. Therefore, qualitative performance during training has been found satisfactory.

There were so many artificial neural network generated for Case I and Case II out of them only four model selected of each case having highest CC and lower MSE value given in Table 1.Value of different statistics indices Mean Square Error (MSE), Normalise Mean Square Error (NMSE), Correlation Coefficient (CC), Akaike's Information Criterion (AIC) and Minimum Description Length (MDL), Per cent error (% error) and value of Hydrologic indices Volumetric error and Coefficient of Efficiency (CE) of different ANNs were given in the Table 1.

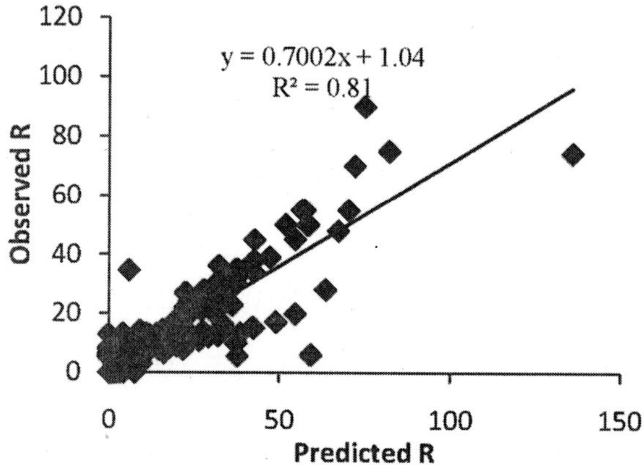

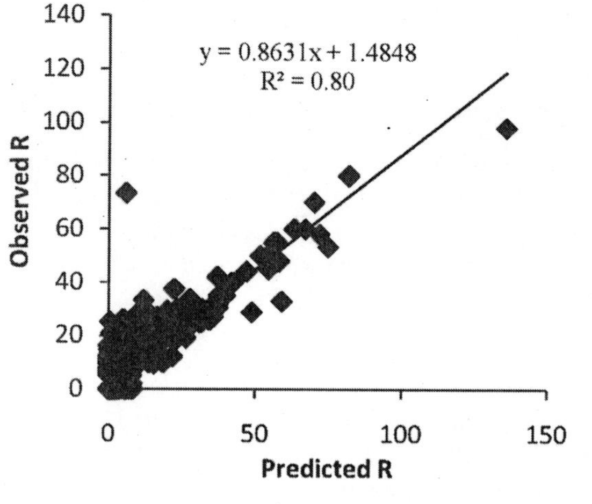

Fig. 4 and 5: Observed R and Predicted R of Developed Model 7-12-1 and 7-6-6-1 for Testing Period.

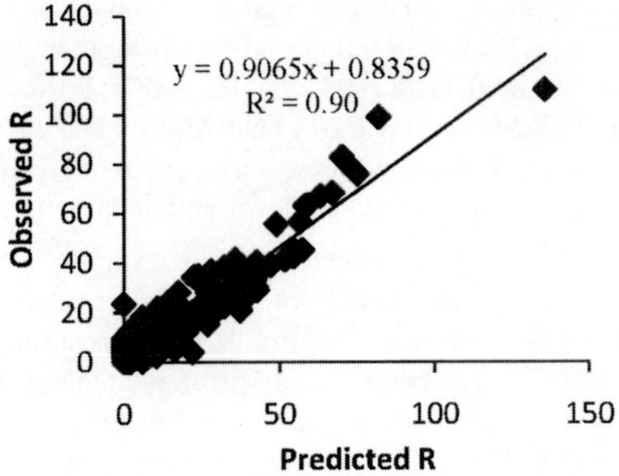

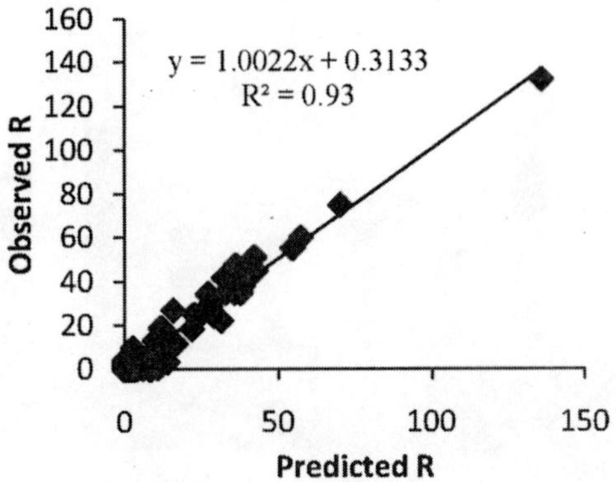

Fig. 6 and 7: Observed R and Predicted R of Developed Model 7-11-1 and 7-4-4-1 for Testing Period.

Table 1: Performance evaluations of developed ANNs Models during training and testing period

Performance indies	*Case I (One day ahead)*				*Case II (Moving average week)*			
	Training		*Testing*		*Training*		*Testing*	
	7-12-1	7-6-6-1	7-12-1	7-6-6-1	7-11-1	7-4-1-1	7-11-1	7-4-1-1
MSE	0.009	0.004	0.003	0.001	0.007	0.005	0.002	0.001
NMSE	0.62	2.69	0.54	0.68	0.15	0.35	0.39	0.29
CC	0.84	0.82	0.81	0.80	0.92	0.94	0.90	0.93
AIC	-31715	-24845	-4819	-4965	-30625	-27114	-4914	-5096
MDL	-31545	-24480	-4754	-4847	-30444	-27003	-4762	-4923
per cent ERROR	52.15	24.14	36.18	22.12	66.87	51.40	39.69	36.09
CE	83.55	85.60	80.65	79.75	92.59	96.74	86.62	88.45
EV	26.15	28.94	17.14	18.05	15.40	13.06	8.42	5.34

Qualitative performance was carried through Statistic Indices and Hydrologic Indices. Statistics indices and hydrologic indices shows network 7-4-1-1 shows best model (CC 0.93 and MSE 0.001) followed by 7-11-1 (CC= 0.90 and MSE=0.002), 7-12-1 (CC=0.81 and MSE=0.003) and 7-6-6-1 (CC=0.80 and MSE= 0.001). In 7-4-1-1 shows seven inputs, four Neurons for first hidden layer, one neuron for second hidden layer and one output respectively.

ANN model 7-4-1-1 was selected for the sensitivity analysis with highest CC value. Sensitivity was carried out on the basis performance indices for ANN model 7-4-1-1 were given in Fig. 8. On the basis of statistics indices correlation coefficient and Mean Square Error and other parameter shows Wet bulb Temperature is most sensible parameter followed by Mean Temperature, Relative Humidity, Dry bulb Temperature, Evaporation, Rainfall and Wind Velocity.

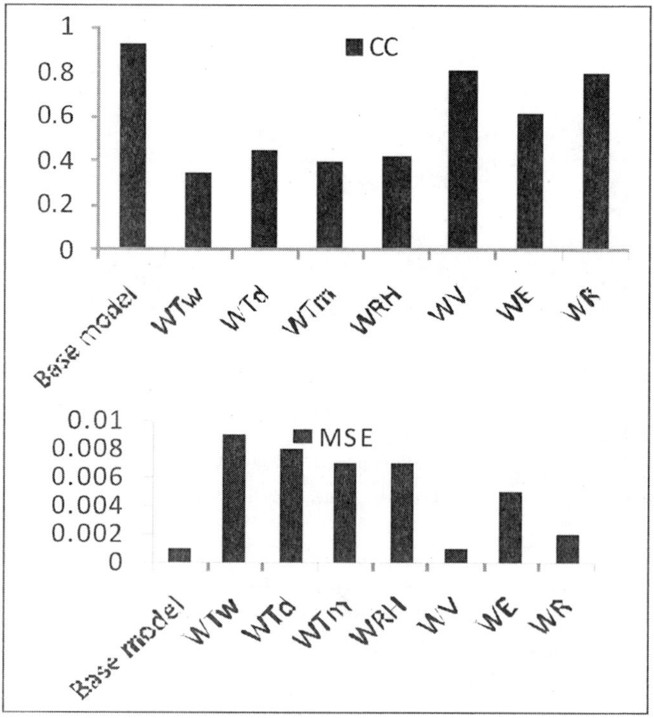

Fig. 8. Performance of sensitivity analysis of base ANNs (7-4-1-1) model.

Conclusions

Based on the performance indices of Artificial Neural Networks (ANNs) 7-4-1-1, 7-11-1 show better than networks, 7-6-6-1, 7-12-1 of single and multi-hidden layer of Case II and Case I respectively. It clearly indicates rainfall forecasting depends on long-term weather parameter. It also noted that higher number of Processing Element (PE) and higher number of hidden layer do not give always best result. Selection of network carried out through trial and error process with lower value of MSE and higher value Correlation Coefficient. Sensitivity analysis of Artificial Neural Network 7-4-1-1 revealed that Wet bulb Temperature is most sensible parameter followed by Mean Temperature, Relative Humidity, Dry bulb Temperature, Evaporation, Rainfall and Wind velocity. Visual observation, statistical and hydrological indices were showed that ANN (7-4-1-1) can be useful for rainfall forecasting for Udaipur.

References

Adamowski, J. and Chan, H.F. (2011). A wavelet neural network conjunction model for ground water level forecasting. *J. Hydrology,* 407:28–40.

Agatonovic-Kustrin, S. and Beresford, R. (2000). Basic concepts of artificial neural network (ANN) modeling and its application in pharmaceutical research. Journal of Pharmaceutical and Biomedical Analysis. 22(5):717-727.

Amarsinghe, U. and Sharma, B. (2009). Water PS Productivity of Food Grains in India Exploring Potential Improvements. Strategic Analysis of National River Linking Project (NRLP) of India: Series 4, pp. 13-54. New Delhi: IWM.

Azadi, S. and Sepaskhah, A. R. (2011). Annual Precipitation forecast for west, southwest, and south provinces of Iran using artificial neural networks. *Theor. Appl. Climatol.,* 109:175–189.

Cameron, M., Zealand Donald, H. Burn, Slobodan P. and Simonovic (1999). Short term stream flow forecasting using artificial neural networks, *Journal of Hydrology,* 214:32–48.

Changsam Jeong, Ju-Young Shin, Taesoon Kim and Jun-Haneg Heo (2012). Monthly Precipitation Forecasting with a Neuro-Fuzzy Model. *Water Resour. Manage.,* DOI 10.1007/s11269-012-0157-3.

Deepak Ranjan Nayak, Amitav Mahapatra and Pranati Mish (2013). A survey on Rainfall Prediction using Artificial Neural Network.*International Journal of Computer Applications,* 72,163, pp. 2-41.

H M Raghunath (2006). Hydrology: Principles, Analys and Design, New Age International Publication, New Delhi.

Luk, K. C., Ball, J. E. and Sharma, A.(2000). A study of optimal model lag and spatial inputs to artificial neural network for rainfall forecasting. *J. Hydrology,* 227: 56–65.

Mekanik, F., Imteaz, M.A., Gato-Trinidad, S. and Elmahdi, A.(2013). Multiple regression and Artificial Neural Network for long-term rainfall forecasting using large scale climate modes, *Journal of Hydrology,* 503:11–21.

Moustris, K. P., Larissi, I. K. Nastos, P. T. and Paliatsos, A. G. (2011). Precipitation Forecast Using Artificial Neural Networks in Specific Regions of Greece. *Water Resource Management,* 25:1979-1993.

Singh, Sahadeva, Singh, V.P. and Raghuvanshi, M.S. (2009). Weeds threat to rainfed crops National Symposium on Weed Threat to Environment, Biodiversity and

Agriculture Productivity, August 2-3, 2009, Tamil Nadu Agricultural University, Coimbatore (Tamil Nadu) pp. 19-20.

Toth, E., Brath, A. and Montanari, A. (2000). Comparison of short-term rainfall prediction models for real-time flood forecasting. *J. Hydrology,* 239:132–147.

Thirumalaiash, K. and Deo, M. C. (1998). Riverstage forecasting using neural networks. *J. Hydrologic Engineering,* 3(1):26-32.

Vamsidhar, E., Varma, K. V. S. R. P., Sankara Rao, P. and Satapati, R., (2010). Prediction of Rainfall Using Back propagation Neural Network Model. *International Journal on Computer Science and Engineering,* 2(4):1119-1121.

UNICEF (2013) *Water in India: Situation and Prospects.* Reports available at. http://www.unicef.org/india/Final_Report.pdf.

8

Rainfall Forecast Model for Agriculture Water Allocation Planning in Madhya Pradesh

Arun Jhajharia, Shrikant and Sumeet Meshram

Abstract

This paper deals an empirical study of time series modelling for crop planning in M.P. The forecast molds has been develop for forecasting the rains so that farmers may plan their cultivation of crops accordingly. In this paper, Box–Jenkins ARIMA Time Series was considered for modelling using thirty years rainfall data of Jabalpur. For stationarity check in the data set augmented Dicky Fuller (ADF) test was applied. Root Mean Square Error (RMSE), Akaike Information Criteria (AIC) and Bayesian Information Criteria (BIC) were applied to identify the best model. The performance of fitted model was checked using plotting actual to fitted value. Adjusted R^2, RSE and MAPE. The ARIMA (1, 0, 1) model performed best among other ARIMA family.

Introduction

Rainwater is an important water resource, particularly for paddy cultivation, which requires high water consumption compared to other crop water requirements. The rainfall is concern with a time series which is a sequence of observation ordered in time. Therefore time series modelling is essential for the forecast of rainfall so that the farmer can allocate the resources accordingly.

The study area, Jabalpur is centrally located district in the State and comes under Kymore Plateau and Satpura Hills agroclimatic zone (Zone-VII). The district has 1393 villages and spread in 5,19,757 ha geographical area. The climate of the district is congenial for horticultural crops, with the annual rainfall of 1358 mm. The minimum temperature of 4°C is recorded in January and maximum 47°C in May. The district comes under the paddy-wheat zone, but arhar, soybean, maize and mustard are also grown predominantly. Rainfed cultivation is the major practice among the farmers. Therefore, an accurate rainfall forecasting is helpful to planning of crop pattern.

In time series modelling, the prediction of values for the future periods is based on the pattern of past values of the variable under study. Hence by forecasting we can predict the upcoming drought and flood areas. Various statistical approaches viz. regression, time series and stochastic approaches are in vogue for arriving at crop forecast as well as rainfall forecast.

This study is to develop a Statistical model which may be able to forecast annual rainfall in Jabalpur (M.P.). It is expected that the proposed model could provide forecast of annual rainfall with acceptable accuracy which may be able to fulfill the requirement of agricultural water allocation.

In early 1970's BOX and JENKINS pioneered in evolving methodologies for time series modelling in univariate case often referred to as Univariate Box–Jenkins ARIMA modelling. In course of time various organisations/worker in India and abroad have done modelling/forecasting based on time series data using different methodologies viz. time series decomposition model, exponential smoothing models, ARIMA models etc. Recently many researchers like: Yurekli and Ozturk (2003), Gilbert (2005), Yurekli and Kurunc (2006) worked in this field. The Autoregressive Integrated Moving Average (ARIMA) models, or Box-Jenkins methodology, are a class of linear models that is capable of representing stationary as well as non-stationary time series.

Therefore, a study has been taken to develop suitable time series model for forecasting with the following objectives:

1. To identify the model.
2. To estimate the parameters involved in model.
3. To diagnose the model

Material and Methods

Data collection: In order to study the appropriate model for rainfall forecasting, rainfall data of Jabalpur be used for time series analysis or time series modelling. Thirty years (1981-2014) annual rainfall data of Jabalpur were obtained from Department of Physics and Agrometeorology, College of Agricultural Engineering, JNKVV, Jabalpur. Rainfall data considered as time series. Mathematically, a time series is defined by the functional relationship,

$$Y_t = f(t)$$

where Y_t is the value of phenomenon (or variable) under consideration at time t. In this study annual rainfall value at different time t, considered as Y_t where value of t, to be given in yearly, 1981 to 2014.

Methods

Stationarity: To fit a appropriate model for any time series its essential to check about stationarity of series.

A Model-Building Strategy: Finding appropriate models for time series is a non-trivial task. We will develop a multistep model-building strategy espoused so well by Box and Jenkins is as follows.

Linear Stationary Models

General linear processes: A general linear process, $\{Y_t\}$, is one that can be represented as a weighted linear combination of present and past white noise term as

$$Y_t = e_t + \psi_1 e_{t-1} + \psi_2 e_{t-2} - \psi_q e_{t-q} \ldots\ldots \qquad \ldots(1)$$

Autoregressive processes: Autoregressive processes are as their name suggest- regressions on themselves. Specifically, a pth order autoregressive process viz. *AR (p)*, $\{Y_t\}$ satisfies the equation,

$$Y_t = \phi_1 Y_{t-1} + \phi_2 Y_{t-2} + \ldots + \phi_p Y_{t-p} + e_t \qquad \ldots(2)$$

where we now use the symbols ϕ_1, ϕ_2, ...,ϕ_p for the finite set of weight parameters.

Now we can write (2) in the equivalent form,

$$(1 - \phi_1 B - \phi_2 B^2 - \ldots - \phi_p B_p) Y_t = e_t$$

or
$$\Phi(B) Y_t = e_t \qquad \ldots(3)$$

where *B,* is a backshift operator.

Moving Average Processes: Where only finite numbers of the ψ -weights are non-zero, we have what is called a moving average process. In this case, we change notation somewhat and write,

$$Y_t = e_t - \theta e_{t-1} - \theta_2 e_{t-2} - \ldots - \theta_q e_{t-q} \qquad \ldots(4)$$

We call such a series a moving average of order q and abbreviate the name to MA (q).

Where we now use the symbols $-\theta_1$, $-\theta_q$ for the finite set of weight parameters.

We can also write (4) in the equivalent form,

$$Y_t = (1 - \theta_1 B - \theta_2 B^2 - \ldots - \theta_q B^q)_{et}$$

or $$Y_t = \theta(B)\, e_t \qquad \ldots(5)$$

Mixed autoregressive-moving average processes (ARMA Model): If we assume that the series is partly autoregressive and partly moving average, we obtain a quite general time series model. In general, if,

$$Y_t = \phi_1 Y_{t-1} + \phi_2 Y_{t-2} + \ldots + \phi_p Y_{t-p} + e_t - \theta_1 e_{t-1} - \theta_2 e_{t-2} \ldots - \theta_q e_{t-q}$$

or $$\Phi(B) Y_t = \theta(B_t) e_t \qquad \ldots(6)$$

is called the mixed autoregressive-moving average process of order (p, q),which we sometimes abbreviated as ARMA(p,q), where $\Phi(B)$ and $\theta(B)$ are the polynomials of degree p and q in B. An ARMA(p,q) process $\Phi(B)Y_t = \theta(B)e\text{-}_t$ will define a stationary process provided that the characteristic equation $\Phi(B) = 0$ has all its roots lying outside the unit circle.

Model for non-stationary Time series: Any time series without a constant mean and variance over time is non-stationary. When a time series is non-stationary then stationary should be achieved by differencing the series.

ARIMA Model: A time series $\{Y_t\}$ is said to follow an Integrated autoregressive moving average model if the dth difference $W_t = \nabla^d Y_t$ is a stationary ARMA process. If $\{W_t\}$ follows an ARMA (p, q) model, we say that $\{Y_t\}$ is an ARIMA (p, d, q) process.

Let us consider the model,

$$\varphi(B) Y_t = \theta(\mathrm{B})\, e_t \qquad \ldots(7)$$

where (B) is a non-stationary autoregressive operator.

Then we can express the model eq.(7) in the form,

$$\varphi(B) Y_t = \phi(B)(1-B)^d Y_t = \theta(B) e_t \qquad \ldots(8)$$

where $\phi(B)$ is a stationary autoregressive operator.

$$\phi(B)\nabla^d = \theta(\mathrm{B}) e_t \qquad \ldots(9)$$

Equivalently, the process is defined by the two equations,

$$\phi(B) W_t = \theta(B) e_t \qquad \ldots(10)$$

and $$W_t = \nabla^d Y_t \qquad \ldots(11)$$

Consider then an ARIMA (p, 1, q) process. With $W_t = Y_t - Y_{t-1}$, we have,

$$W_t = \phi_1 W_{t-1} + \phi_2 W_{t-2} + ... + \phi_p W_{t-p} + e_t - \theta_1 e_{t-1} - \theta_2 e_{t-2} - ... - \theta_q e_{t-q} \qquad ...(12)$$

Or, in terms of the observed series,

$$Y_t - Y_{t-1} = \phi_1 (Y_{t-1} - Y_{t-2}) + \phi_2 (Y_{t-2} - Y_{t-3}) + ... + \phi_p (Y_{t-p} - Y_{t-p-1}) + e_t - \theta_1 e_{t-1} - \theta_2 e_{t-2} ... - \phi_q e_{t-q}$$

Which we may rewrite as,

$$Y_t = (1+\phi_1) Y_{t-1} + (\phi_2 - \phi_1) Y_{t-2} + (\phi_3 - \phi_2) Y_{t-3} + ... (\phi_p - \phi_{p-1}) Y_{t-p} - \phi_p Y_{t-p-1} + e_t - \theta_1 e_{t-1} - \theta_2 e_{t-2} - ... - \theta_q e_{t-q} \qquad ...(13)$$

We call this the difference equation form of the model.

General form of the Autoregressive Integrated Moving Average process: A rather general form of the model (3.2.9) by adding a constant term θ_0. Thus, a rather general form of the model that we shall use to describe time series is the autoregressive integrated moving average process;

$$\varphi(B) Y_t = \phi(B) \nabla^d Y_t = \theta_0 + \varphi(B) e_t \qquad ...(14)$$

where, $\phi(B) = 1 - \phi_1 B - \phi_2 B^2 - ... - \phi_p B^p$

$\theta(B) = 1 - \theta_1 B - \theta_2 B^2 - ... - \theta_q B^q$

we rewrite eq.(13) as follows,

$$\phi(B) \nabla^d Y_t = \theta_0 + \phi(B) e_t \qquad ...(15)$$

Model Identification: To identify an appropriate subclass of models from the general ARIMA family,

$$\phi(B) \nabla^d Y_t = \theta_0 + \theta(B) e_t \qquad ...(16)$$

which may be used to represent a given time series. The approach will be as follows:

1. To difference Y_t as many times as is needed to produce stationary, hopefully reducing the process under study to the mixed autoregressive moving average process,

$$\phi(B) W_t = \theta_0 + \phi(B) e_t$$

where, $W_t = (1-B)^d Y_t = \nabla^d Y_t$

2. To identify the resulting ARMA process.

Model Estimation

Least squares estimation

For autoregressive models, we introduce a possibly non-zero mean ì, into our stationary models and treat it as another parameter to be estimated by least squares.

Consider the first-order case where,

$$Y_t - \mu = \phi(Y_{t-1} - \mu) + e_t \qquad \ldots(17)$$

Since only $Y_1, Y_2, \ldots, Y_n$ are observed, we can only sum from t = 2 to t = n.

Let $$S_c(\phi, \mu)^2 = \sum_{t=1}^{n} [(Y_t - \mu) - \phi(Y_{t-1}\mu)]^2 \qquad \ldots(18)$$

Consider the equation $\partial S_c / \partial\mu = 0$. We have,

$$\partial S_c / \partial\mu = \sum_{t=1}^{n} (Y_t - \mu) - \phi(Y_t - 1 - \mu) = 0$$

or, simplifying and solving for μ.

$$\mu = \frac{1}{(n-1)(1-\phi)}\left[\sum_{t=2}^{n} Y_t - \phi \sum_{t=2}^{n} Y_{t-1}\right]$$

Now, for large *n*,

$$\frac{1}{(n-1)}\sum_{t=2}^{n} Y_t \approx \frac{1}{n-1}\sum_{t=2}^{n} Y_{t-1} \approx \bar{Y}$$

Thus, regardless of the value of ô, Equation (17) reduce t,

$$\hat{\mu} \approx \frac{1}{1-\phi}\left(\bar{Y} - \phi\bar{Y}\right) = \bar{Y} \qquad \ldots(19)$$

Consider now the minimisation of $S_c(\phi, \bar{Y})$ with respect to ϕ we have,

$$\frac{\partial S_c(\phi, \bar{Y})}{\partial\phi} = \sum_{t=2}^{n} 2\left[(Y_t - \bar{Y}) - \phi(Y_{t-1} - \bar{Y})\right](Y_{t-1} - \bar{Y})$$

Setting this equal to zero and solving for ϕ yields,

$$\hat{\phi} = \frac{\sum_{t=2}^{n}(Y_t - \bar{Y})(Y_{t-1} - \bar{Y})}{\sum_{t=2}^{n}(Y_{t-1} - \bar{Y})^2}$$

For moving average models,

Consider now the least-squares estimation of *è* in the MA(1) model:

$$Y_t = e_t - \theta e_{t-1} \qquad \text{...(20)}$$

The invertible MA (1) model can be expressed as,

$$Y_t = -\theta Y_{t-1} - \theta^2 Y_{t-2} - \theta^3 Y_{t-3} - \text{....} + e_t$$

Consider evaluating $S_c(\theta)$ for a single given value of θ. The only Y's we have available are our observed series $Y_1, Y_2, Y_{3\ldots}\ Y_n$. Rewrite Equation (19) as,

$$e_t = Y_t + \theta e_{t-1} \qquad \text{...(21)}$$

Model Diagnostics: In the model-building process, if an ARIMA (p, d, q) model is chosen (based on the ACFs and PACFs), some checks on the model adequacy are required. A residual analysis is usually based on the fact that the residuals of an adequate model should be approximately white noise. Therefore, checking the significance of the residual autocorrelations and comparing with approximate two standard error bounds, i.e., $\pm 2\sqrt{n}$ are needed.

One important test to diagnose ARIMA model fit is the Ljung–Box test test can be defined as follows.

H_0: The data are independently distributed (i.e. the correlations in the population from which the sample is taken are 0, so that any observed correlations in the data result from randomness of the sampling process).

H_a: The data are not independently distributed.

The test statistic is:

$$Q = n(n+2)\sum_{k=1}^{h}\frac{\hat{\rho}_k^2}{n-k}$$

where n is the sample size, $\hat{\rho}_k$ is the sample autocorrelation at lag k, and h is the number of lags being tested.

The Ljung–Box test is commonly used in autoregressive integrated moving average (ARIMA) modeling.

Results

To develop a appropriate model we select here time series analysis or modelling. Based on Box-Jenkins procedure, the beginning step is identification by using autocorrelation function and partial autocorrelation function plots, for this first, to check stationarity of observed rainfall time series. For stationary process mean and variance of observed series should be constant through over the time shown in (Figs. 1 and 2) and also by plotting collerogram for sample ACF at different time lag which shown in (Fig. 3.), we can check stationarity of time series, mean and variance constant through over the time then the series can be modeled as a stationary process, as well as autocorrelation function of observed series (Fig. 3) quickly decay to zero there for series is a stationary process. Thus, there is not necessary to make series stationary, and the order of differencing considered d = 0 for ARIMA model, we can say it ARMA (*p, q*) model.

Model which has maximum value of R^2 and minimum value BIC (Bayesian Information Criterion). The models and corresponding statistics are given in Table 1. Here ARIMA (1,0,1) having maximum value of R-square and minimum value of BIC. So the most suitable model is ARIMA (1,0,1).

Model estimation and verification: Model parameters of ARIMA (1, 0, 1) were estimated using SPSS package. Results of estimation. are reported in Table 2.

The model verification is concerned with checking the residuals of the model to see if they contain any systematic pattern which still can be removed to improve on the chosen ARIMA. This is done through examining the autocorrelations and partial-autocorrelations of the residuals of various orders. For this purpose, the various correlations up to 16 lags were computed and the same along with their significance which is tested by Box-Ljung test are provided in Table 4.

As the results indicate, none of these correlations is significantly different from zero at a reasonable level. This proves that the selected ARIMA model is an appropriate model. The ACF and PACF of the residuals (Figs. 5 and 6) also indicate 'good fit' of the model.

So the fitted ARIMA model.

$$Y_t = 0.460Y_{t-1} + e_t - 0.998e_{t-1} \qquad \text{...(4.1.1)}$$

Forecasting with ARIMA model: ARIMA models are developed basically to forecast the corresponding variable. There are two kinds of forecasts: sample period forecasts and post-sample period forecasts. The former are used to develop confidence in the model and the latter to generate genuine forecasts for use in planning and other purposes. The ARIMA model can be used to yield both these kinds of forecasts.

Sample period forecasts: The sample period forecasts are obtained simply by plugging the actual values of the explanatory variables in the estimated eq. (1).The explanatory variables here are the lagged values of Y_t and the estimated lagged errors. The so obtained values for Y_t together with the actual values of Y_t are shown in Table 4.

To judge the forecasting ability of the fitted ARIMA model, important measures of the sample period forecasts' accuracy were computed. The Mean Absolute Percentage Error (MAPE) for annual rainfall turns out to be 20.303. This measure indicates that the forecasting inaccuracy is low.

Post-sample forecasts

The principal objective of developing an ARIMA model for a variable is to generate post-sample period forecasts for that variable. This is done through using equation (1). The forecasts for annual rainfall during 2012 to 2015 are given in lower part of table.

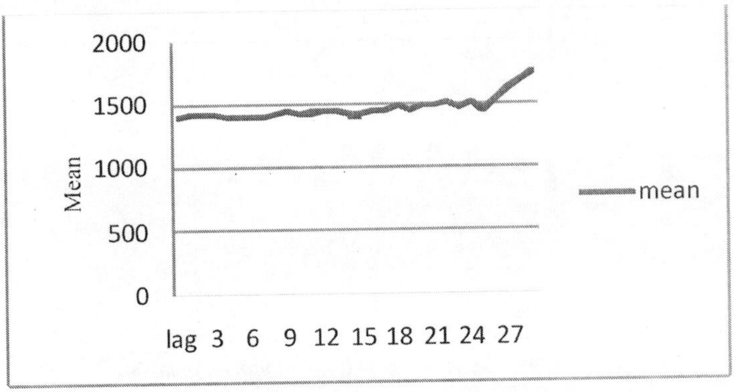

Fig.1: Mean at different time lag of observed rainfall series

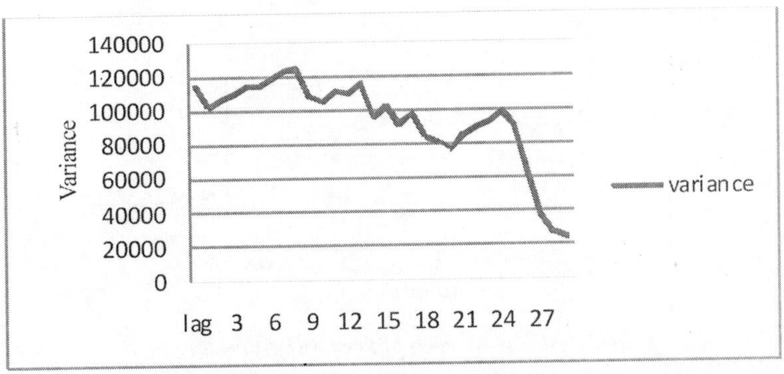

Fig.2: Variance at different time lag of observed rainfall series.

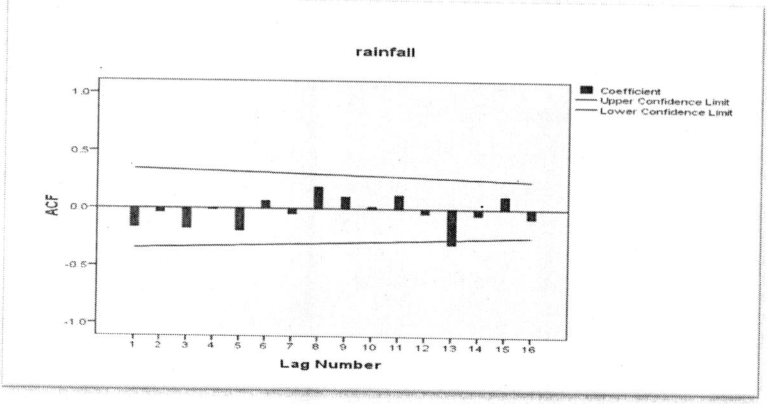

Fig. 3: ACF of observed time series at different time lag.

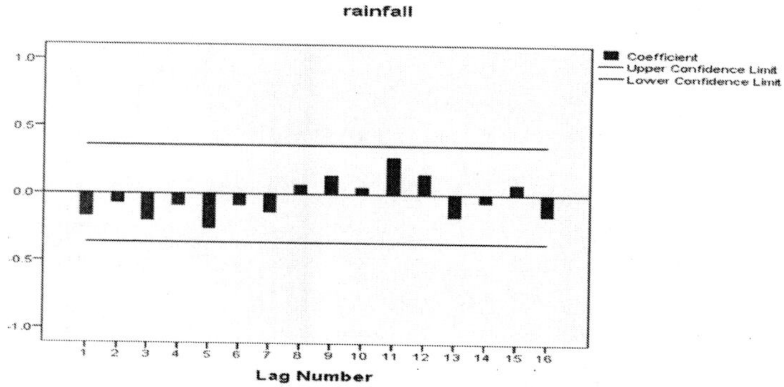

Fig. 4: PACF of observed time series at different time lag.

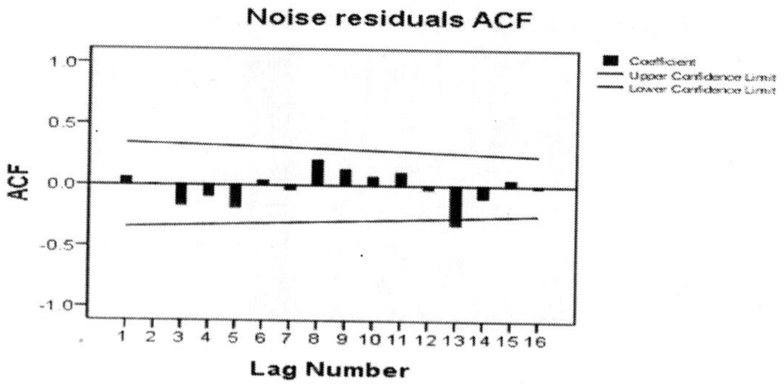

Fig. 5: ACF of residuals of fitted ARIMA model.

Table 1: ARIMA models and corresponding statistics

Model	*R square*	*BIC*
ARIMA (1,0,1)	0.152	11.83
ARIMA(1,0,2)	0.131	12.055

Table 2: Estimates of ARIMA Model

		Estimate	*SE*	*T*	*Sig.*
Constant	1.386E3	23.343	59.361	0.000	
AR	Lag 1 (φ)	0.460	0.276	1.667	0.107
MA	Lag 1 (θ)	0.998	10.267	0.097	0.923

Table 3: ARIMA Model Fit Statistics

Model Fit statistics							*Ljung-Box Q(18)*		
R-squared	*RMSE*	*MAPE*	*MAE*	*MaxAPE*	*MaxAE*	*Normalised BIC*	*Statistics*	*df*	*Sig.*
0.152	322.334	20.303	239.952	95.591	711.673	11.883	14.838	6	0.537

Table 4: Actual and Estimated annual rainfall and 95 per cent confidence limit

Year Rainfall (mm/year)	*Observed Rainfall (mm/year)*	*Forecasted*	*Noise residuals*	*Lower CL*	*Upper CL*
1981	731.2	1385.63	-654.43	649.05	2122.22
1982	1456.5	1562.29	-105.79	853.04	2271.53
1983	1473.5	1501.54	-28.04	808.5	2194.58
1984	1532.2	1449.19	83	766.89	2131.5
1985	1732	1382.45	349.54	707.77	2057.13
1986	1314.1	1240.84	73.25	571.87	1909.81
1987	1358	1287.87	70.12	623.33	1952.42
1988	1074.1	1310.02	-235.92	649.02	1971.03
1989	744.5	1456.17	-711.67	798.05	2114.28
1990	1824	1741.48	82.51	1085.77	2397.19
1991	1370	1511.32	-141.32	857.65	2165
1992	1065.8	1509.45	-443.65	857.52	2161.39
1993	1486.2	1651.94	-165.74	1001.52	2302.37
1994	2083.2	1587.1	496.09	937.99	2236.2
1995	1417.2	1240.22	176.97	592.28	1888.16
1996	918.8	1233.18	-314.38	586.27	1880.09
1997	1483.5	1468.39	15.1	822.41	2114.38
1998	943.3	1416.33	-473.03	771.18	2061.48
1999	1806.8	1632.27	174.52	987.87	2276.67
2000	1109	1412.95	-303.95	769.24	2056.66
2001	1509.3	1548.9	-39.6	905.81	2191.99
2002	1274.8	1480.47	-205.67	837.95	2122.99

(Contd...)

(Contd...)

2003	1747.4	1531.98	215.41	889.98	2173.97
2004	1242.3	1345.06	-102.76	703.56	1986.57
2005	1853.4	1418.6	434.79	777.54	2059.65
2006	1038	1181.78	-143.78	541.14	1822.42
2007	1188.5	1364.44	-175.94	724.19	2004.69
2008	1398.7	1464.94	-66.24	825.05	2104.83
2009	1540	1455.73	84.26	816.18	2095.28
2010	1635.9	1375.06	260.83	735.83	2014.29
2011	1858	1247.94	610.05	609	1886.87
2012	1235.4	1011.05	540.6	372.4	1649.71
2013	1665.9	1213.28	203.97	496.31	1930.24
2014	1776.2	1306.33	187.35	573.85	2038.8
2015		1349.14		613.42	2084.86
2016		1408.22		289.43	2076.84
2017		1397.86		456.75	2022.3
2018		1376.43		554.6	2111.29

Conclusions

1. Rainfall data of Jabalpur was found as stationary time series which needs not any differencing or transformation to make stationary.
2. ARIMA (1, 0, 1) and ARIMA (1, 0, 2) or we can say stationary model ARMA (1,1) and ARMA (1, 2) found to be a appropriate model for forecasting rainfall in Jabalpur.
3. After comparing both models, ARMA (1,1) model found to be a good model for fitting on rainfall data of Jabalpur.
4. Graph of ACF of residuals and PACF of residuals indicates that choosing model is an appropriate model for forecasting rainfall of Jabalpur.
5. Value of MAPE is found to be in range of good accuracy of forecasting by model.

References

Box, G.E.P., and G. M. Jenkins. 1970. Time series analysis: forecasting and control. Holden Day, San Francisco, CA.

Gilbert, Kenneth (2005). An ARIMA Supply Chain Model. Management Science, 51 (2): 305-310.

Yurekli, Kadri and Ozturk, Fazil (2003) Stochastic Modeling of Annual Maximum and Minimum Stream flow of Kelkit Stream. Water International, 28(4): 433 – 441.

Yurekli K., Kurunc A. (2006). Simulating Agricultural Drought Periods Based on Daily Rainfall and Crop Water Consumption. Journal of Arid Environments, 67(4): 629-64.

9

Trend Analysis of Climatological Parameters using Mann-Kendal Test: A Case Study of Vadodara District, Gujarat, India

I.S. Risaldar, T.M.V. Suryanarayana and Falguni. P. Parekh

Abstract

Climate change may refer to a change in average weather conditions, or in the time variation of weather around longer-term average conditions. The impact of climate change on the various meteorological parameters has received a great deal of attention by scholars worldwide. The means for detecting climate trends and variability are time series analysis based on data sets e.g. Temperature, precipitation, wind speed and relative humidity. In this study the focus is on detecting trends in mean monthly maximum temperature, mean monthly minimum temperature, mean monthly precipitation, mean monthly wind speed and mean monthly relative humidity for the Vadodara district in the state of Gujarat. For this study, Mann-Kendall test was run at five per cent significance level on time series data for Vadodara district for the period of thirty seven years starting from 1969 to 2005. Mann-Kendall test is carried out using software XLSTAT to analyse trend in given data. The resultant Mann-Kendall test statistic (Z_s) indicates how strong the trend is and whether it is increasing or decreasing. A high positive value (Z_s) statistic indicates an increasing trend while a low negative value indicates a decreasing trend in the time series of random variables. In this study maximum temperature shows highest increase in December while highest decrease in July. The minimum temperature has shown an increasing trend in all the months and a significant increase is observed in the month of March. The relative humidity is showing an increasing trend in all the months except the *months of June, July and August. The wind speed is decreasing for all the months and a significant decrease is observed in June. Also, Sen's slope estimator has been used along with Mann-Kendall test statistic (Z_s). A positive value of (Q) indicates an upward trend and negative value indicates (Q) downward trend. The results are same with both the tests.*

Key words: *Climate change, Mann-Kendall Test, Sen's Slope Estimator.*

Introduction

The physical conditions in the environment and atmosphere such as precipitation, temperature, wind pressure and humidity, which have direct or indirect consequences upon the biosphere, are termed as weather while the pattern of weather in a region over a period of time is called climate (Cunningham *et al.*, 2005). Climate is the long-term regime of the atmospheric variables or the composite of the day to day to values of the weather elements for a given place or area for a given period. The period of averaging may be several months, seasons, years or even centuries. Drastic changes have taken place in the climate of the earth over the ages. Taken over a long period the trends of climate change are more pronounced and it shows variations according to the choice of period these changes are referred to as climate change. The Intergovernmental Panel on Climate Change (IPCC) estimates that the global mean surface temperature has increased 0.6 ± 0.2 °C since 1861, and predicts an increase of 2 to 4 °C over the next 100 years (IPCC, 2007).

Climate change and global warming are commonly detected throughout studies of variability of climatic parameters such as rainfall, temperature, runoff and groundwater. The changes in regional temperature and precipitation have important implications for all aspects of the hydrologic cycle. Variations in these parameters determine the amount of water that reaches the surface, evaporates or transpires back to the atmosphere, becomes stored as snow or ice, infiltrates into the groundwater system, runs off the land, and ultimately becomes base flow to streams and river.

One of the best ways of understanding how climate may change in future is to examine how it has changed in the past based upon long-term observational records. Changes in precipitation patterns affect water availability and runoff directly, while changes in temperature, radiation and humidity have an effect on evapotranspiration. Hence in this study an attempt has been made to detect the trends in various climatological factors like monthly maximum temperature, monthly minimum temperature, monthly relative humidity and monthly wind speed for Vadodara district from the year 1969 to 2005 using Mann-Kendall test and Sen's slope estimator.

Yadav *et al.* (2014) have indicated an increasing trend of annual rainfall in Almora, Bageshwar, Nainital, Pauri Garhwal, Pithoragarh, Udhamsingh Nagar and Uttarkashi; and a dereasing trend of annual rainfall for Chamoli,Champawat, Rudraprayag and Tehri Garhwal district of Uttarakhand state, India in a study conducted from year 1971 to 2011. Shah *et al.* (2012) have observed that the rainfall is decreasing, minimum temperature, mean temperature and relative humidity are increasing and maximum temperature has no change for the Mann-Kendall test applied to Meteorological data of Peshawar (1981-2010) district, Pakistan. Babar *et al* have studied the rainfall data of 40 years from 1971 to 2001 in Nethravathi river basin and have indicated an increasing trend for month of September while decreasing trend for June, July and August months. Safari *et al* have conducted a study for fifty two years starting from 1958 to 2010 for the state of Rwanda, Africa on mean annual temperature and have indicated a significant change in temperature from year 1977.

Study Area and Data Collection

The area selected for this study is Vadodara district in state of Gujarat, India. The district covers an area of 7794 km^2 and lies between 22.18° N Latitude and 73.12 °E Longitude. The district has a sub-tropical climate with moderate to high summer, cold winter and medium rainfall and with moderate humidity.

The climatological data required for the study was collected from Indian Meteorological Department (IMD), Pune and State Water Data Centre, Gandhinagar for the years starting from 1969 to 2005.

Methodology

There are several approaches for detecting the trend in the time series. These approaches can be either parametric or non-parametric. Parametric methods assumed the data should normally distributed and free from outliers. On the other hand, non-parametric methods are free from such assumptions. The most popularly used non-parametric tests for detecting trend in the time series is the Mann-Kendall (MK) test. It is widely used for different climatic variables. For this study, the modified Mann-Kendall test was run at five per cent significance level on time series data from the year 1969 to 2005 for various parameters like maximum temperature, minimum temperature, wind speed, relative humidity and sunshine hours. Also, Sen's slope estimator (Q) has been used along with Mann-Kendall test statistic (Z_s).

Mann–Kendall test

Mann-Kendall test is a statistical test widely used for the analysis of trend in climatologic and in hydrologic time series. There are two advantages of using this test. First, it is a non-parametric test and does not require the data to be normally distributed. Second, the test has low sensitivity to abrupt breaks due to inhomogeneous time series. According to this test, the null hypothesis H_0 assumes that there is no trend (the data is independent and randomly ordered) and this is tested against the alternative hypothesis H_1, which assumes that there is a trend. The computational procedure for the Mann-Kendall test considers the time series of n data points and T_i and T_j as two subsets of data where $i = 1, 2, 3,\ldots, n$-1 and $j = i$+1, i+2, i+3, …, n. The data values are evaluated as an ordered time series. Each data value is compared with all subsequent data values. If a data value from a later time period is higher than a data value from an earlier time period, the statistic S is incremented by 1. On the other hand, if the data value from a later time period is lower than a data value sampled earlier, S is decremented by 1. The net result of all such increments and decrements yields the final value of S. The Mann-Kendall S Statistic is computed as follows:

$$S = \sum_{i=1}^{n-1}\sum_{j=i+1}^{n} \operatorname{sign}\left(T_j - T_i\right)$$

$$\operatorname{sign}\left(T_j - T_i\right) = \begin{cases} 1 \text{ if } T_j - T_i > 0 \\ 0 \text{ if } T_j - T_j = 0 \\ -1 \text{ if } T_j - T_i < 0 \end{cases}$$

where T_j and T_i are the annual values in years j and i, $j > i$, respectively.

If n < 10, the value of |S| is compared directly to the theoretical distribution of S derived by Mann and Kendall. The two tailed test is used. At certain probability level H_0 is rejected in favor of H_1 if the absolute value of S equals or exceeds a specified value $S_{\alpha/2}$, where $S_{\alpha/2}$ is the smallest S which has the probability less than $\alpha/2$ to appear in case of no trend. A positive value of S indicates an upward (downward) trend and negative value of S indicates a downward trend.

For $n \geq 10$, the statistic S is approximately normally distributed with the mean and variance. The variance (σ^2) for the S-statistic is defined by:

$$\sigma^2 = \frac{n(n-1)(2n+5) - \Sigma t_i (i)(i-1)(2i+5)}{18}$$

in which t_i denotes the number of ties to extent i. The summation term in the numerator is used only if the data series contains tied values. The standard test statistic Z_s is calculated as follows:

$$Z_s = \begin{cases} \dfrac{s-1}{\sigma} \text{ for } S > 0 \\ 0 \text{ for } S = 0 \\ \dfrac{s+1}{\varsigma} \text{ for } S < 0 \end{cases}$$

The test statistic Z_s is used a measure of significance of trend.

Sen's slope estimator

The method of calculating the Sen's slope estimator requires a time series of equally spaced data. The magnitude of trend is predicted by the Sen's estimator. Here, the slope (T_i) of all data pairs is computed as,

$$T_i = \frac{X_j - X_k}{j-k}$$

Where:

X_j = data measurement at time j

X_k = data measurement at time k

j = time after time k,

X_j and X_k constitute the pairs of observations identified by place in the series. The median of these N values of T_i is represented as Sen's estimator of slope which is given as:

$$Q_i = \begin{cases} T_{\frac{N+1}{2}} N \text{ is odd} \\ \dfrac{1}{2}\left(T_{\frac{N}{2}} + T_{\frac{N+2}{2}} N \text{ is even}\right) \end{cases}$$

Results and Discussions

The trends over the study period of 1969-2005 of all the meteorological parameters are observed in all months of the study period with the help of Mann-Kendall non-parametric test and Sen's Slope estimator.

Mean monthly maximum temperature

The Mann-Kendall test statistic (Z_s) and Sen's Slope estimator (Q) of monthly maximum temperature of all months are shown in Table-1 and Table-2

Table 1: Mann-Kendal Trend Statistic (Zs) for mean monthly maximum and minimum temperature, mean monthly relative humidity and mean monthly wind speed from years 1969-2005

	Maximum Temperature		*Minimum Temperature*		*Relative humidity*		*Wind Speed*	
Month	Z_s	*Remark*	Z_s	*Remark*	Z_s	*Remark*	Z_s	*Remark*
January	114.0	↑	220.0	↑	264.0	↑	-246.0	↓
February	123.0	↑	146.0	↑	265.0	↑	-329.0	↓
March	39.0	↑	230.0	↑	266.0	↑	-254.0	↓
April	-23.0	↓	107.0	↑	219.0	↑	-315.0	↓
May	6.0	↑	111.0	↑	22.0	↑	-386.0	↓
June	36.000	↑	142.0	↑	-31.0	↓	-441.0	↓
July	-42.0	↓	149.0	↑	-50.0	↓	-426.0	↓
August	116.0	↑	165.0	↑	-116.0	↓	-418.0	↓
September	33.0	↑	224.0	↑	26.0	↑	-339.0	↓
October	39.0	↑	156.0	↑	121.0	↑	-232.0	↓
November	189.0	↑	7.0	↑	119.0	↑	-295.0	↓
December	235.0	↑	31.0	↑	193.0	↑	-228.0	↓

Where Z_s is the Mann-Kendall Test Statistic. ↑ indicates an increasing trend. ↓ indicates a decreasing trend.

Table 2: Sen's Slope Estimator (Q) for mean monthly maximum and minimum temperature, mean monthly relative humidity and mean monthly wind speed from years 1969-2005

	Maximum Temperature		*Minimum Temperature*		*Relative humidity*		*Wind Speed*	
Month	*Q*	*Remark*	*Q*	*Remark*	*Q*	*Remark*	*Q*	*Remark*
January	0.025	↑	0.051	↑	0.264	↑	-0.08	↓
February	0.029	↑	0.042	↑	0.275	↑	-0.106	↓
March	0.01	↑	0.045	↑	0.295	↑	-0.057	↓
April	-0.004	↓	0.022	↑	0.203	↑	-0.083	↓
May	0.002	↑	0.014	↑	0.03	↑	-0.194	↓
June	0.011	↑	0.019	↑	-0.049	↓	-0.305	↓
July	-0.005	↓	0.014	↑	-0.097	↓	-0.221	↓
August	0.025	↑	0.021	↑	-0.155	↓	-0.167	↓
September	0.009	↑	0.037	↑	0.043	↑	-0.1	↓
October	0.012	↑	0.031	↑	0.254	↑	-0.044	↓
November	0.05	↑	0.001	↑	0.189	↑	-0.072	↓
December	0.042	↑	0.011	↑	0.185	↑	-0.078	↓

Where Q_s is the Mann-Kendall Test Statistic. ↑ indicates an increasing trend. ↓ indicates a decreasing trend.

below. The increasing rates of maximum temperature are shown for the months of January, February, March, May, June, August, September, October, November and December. The decreasing trend is observed in April and July. However, December month shows higher increasing trend (Figure-1) amongst all the months and July month shows highest decreasing trend (Figure-2).

Mean monthly minimum temperature

The Mann-Kendall test statistic (Z_s) and Sen's Slope estimator (Q) of monthly minimum temperature of all months are shown in Table-1 and Table-2. The increasing rates of minimum temperature are shown for the months of January, February, March, April, May, June, July, August, September, October, November and December. The decreasing trend is observed in none of the months. However March month shows highest increasing trend (Figure-3) amongst all the months.

Mean monthly relative humidity

The Mann-Kendall test statistic (Z_s) and Sen's Slope estimator (Q) of monthly relative humidity of all months are shown in Table-1 and Table-2. The increasing rates of relative humidity are shown for the months of January,

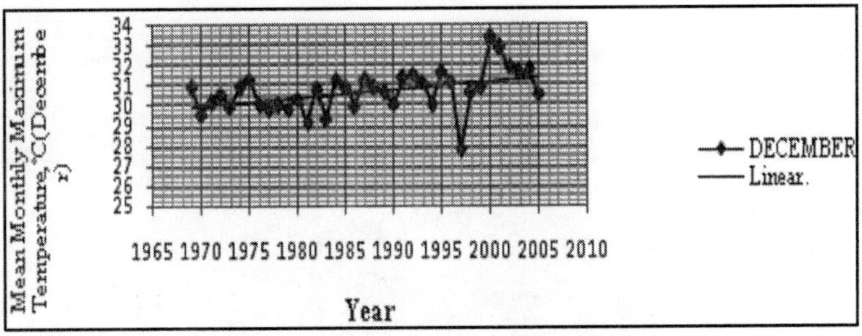

Fig. 1: Trends for Mean Monthly Maximum Temperature, °C (December 1969-2005)

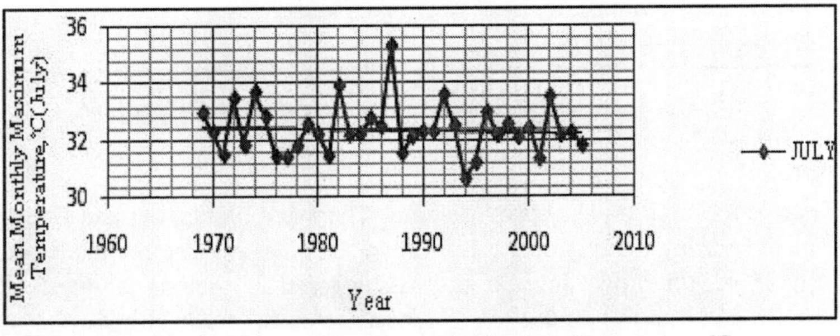

Fig. 2: Trends for Mean Monthly Maximum Temperature, °C (July 1969-2005)

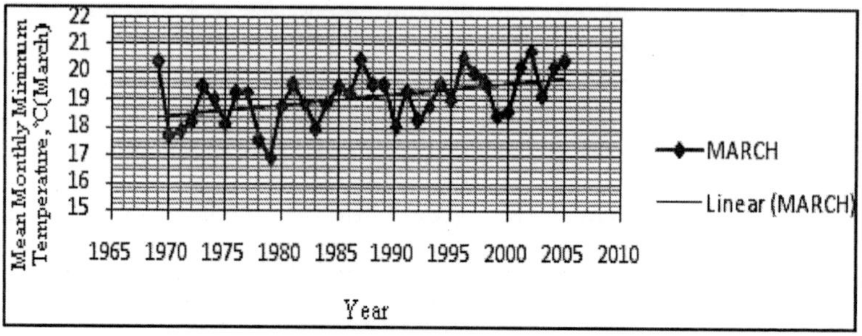

Fig. 3: Trends for Mean Monthly Minimum Temperature, °C (March 1969-2005).

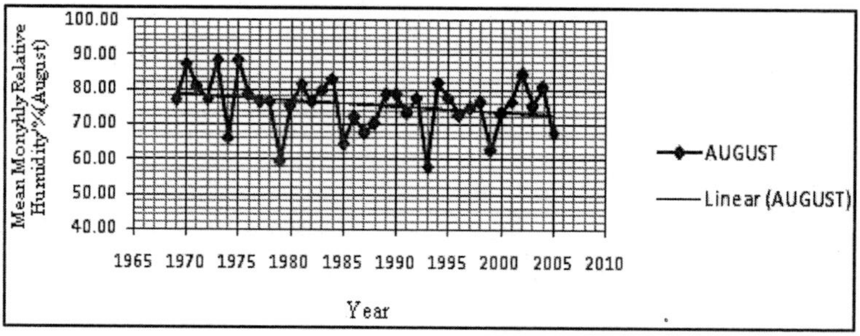

Fig. 4: Trends for Mean Monthly Relative Humidity, per cent (August, 1969-2005).

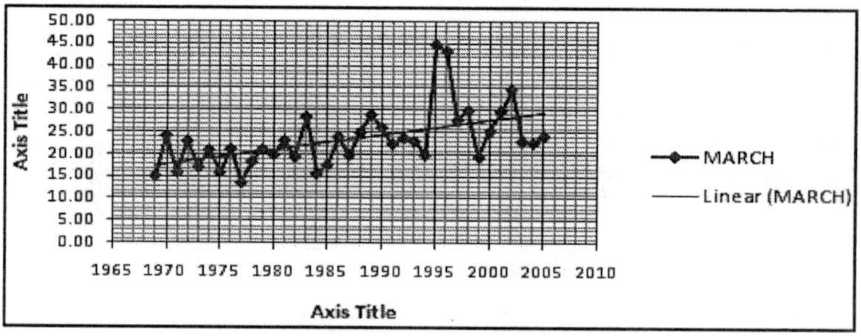

Fig. 5: Trends for Mean Monthly Relative Humidity, per cent (March, 1969-2005)

February, March, April, May, September, October, November and December. The decreasing trend is observed in June, July and August months. However, August month shows highest decreasing trend (Figure-4) amongst all the months and March month shows highest increasing trend (Figure-5).

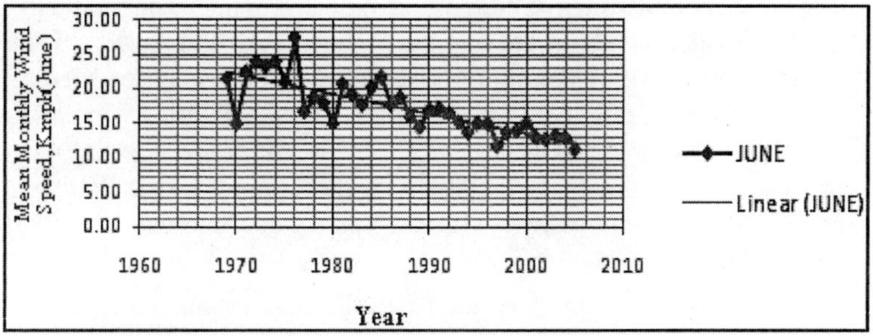

Fig. 6: Trends for Mean Monthly Wind Speed, kmph (June, 1969-2005).

Mean monthly Wind Speed

The Mann-Kendall test statistic (Z_s) and Sen's Slope estimator (Q) of monthly wind speed of all months are shown in Tables 1 and 2. The decreasing rates of wind speed are shown for the months of January, February, March, April, May, June, July, August, September, October, November and December. However, June month shows highest decreasing trend (Figure-6) amongst all the months.

Conclusions

The analysis of various climatological factors like monthly maximum temperature, monthly minimum temperature, monthly relative humidity and monthly wind speed for Vadodara district from the year 1969 to 2005 using Mann-Kendall test and Sen's slope estimator have been carried out. Pattern of trends as discussed in the results leads to a conclusion that climate in showing some changes in Vadodara district. The changes in climate are observed as follows:

- The mean monthly maximum temperature has also shown an increasing trend in all the months except April and July. The highest increase in temperature is observed in the month of December with the Mann-Kendall Statistic having value as high as Z_s= 235 and Sen's slope Estimator having value Q= 0.042. A significant decrease is observed in the month of July with the Mann-Kendall Statistic having value as low as Z_s= -42 and Sen's slope Estimator having value Q= -0.005.
- There is increasing trend for mean monthly minimum temperature in all the months. However, a significant increase in temperature is observed in the month of June with the Mann-Kendall Statistic having value as high as Z_s= 230 and Sen's slope Estimator having value Q=0.045 .

- The mean monthly relative humidity is showing an increasing trend in all the months except in the months of June, July and August. The highest increase in temperature is observed in the month of March with the Mann-Kendall Statistic having value as high as Z_s= 266 and Sen's slope Estimator having value Q= 0.295, the highest decrease is observed in the month of August with the Mann-Kendall Statistic having value as low as Z_s= -116 and Sen's slope Estimator having value Q= -0.155
- The mean monthly wind speed is decreasing for all the months. However, a significant decrease in wind speed is observed in the month of June with the Mann-Kendall Statistic having value as low as Z_s= -441 and Sen's slope Estimator having value Q= -0.305

It indicates that it will have significant effect on yield of *kharif* crops as rainfall depends on relative humidity. Hence farmers need to adopt new techniques for irrigation and farming practices to get better yield. Also, attempts should be made to reduce the emissions of green house gases as are the main reason for changing climate.

References

Rai, R.K., Upadhyay, A. and Ojha, C.S. P. (2010), "Temporal Variability of Climatic Parameters of Yamuna River Basin: Spatial Analysis of Persistence, Trend and Periodicity", The Open Hydrology Journal, 4, 184-210.

Babar, S, Ramesh H. 2013. "Analysis of South West Monsoon Rainfall Trend using Statistical Techniques Over Nethravathi Basin" International Journal of Advanced Technology in Civil Engineering, ISSN: 2231 –5721, Volume-2, Issue-1.

Shah, S, Sobia, N, Khan, A., Zia-Ur-Rahman1 and Shah, M. 2012 "Trends and Variability in Climate Parameters Of Peshawar District" sci., Tech. And Dev., 31 (4) , pp. 341-347.

Climate Change 2007 – "The Physical Science Basis", Contribution of Working Group I to the Fourth Assessment Report of the IPCC (ISBN 978 0521 88009-1 Hardback; 978 0521 70596-7 Paperback).

Cunningham, W.P., Cunningham, M.A. and B. Saigo. 2005. "Environmental Sciences: A Global Concern." 8th ed. New York: Published by McGraw-Hill: 16-316.

Yadav, R, Tripathi, S.K., G. Pranuthi and Dubey, S.K. 2014. "Trend Analysis by Mann-Kendall Test for precipitation and temperature for thirteen districts of Uttarakhand", Journal of Agrometeorology 16(2):164-171 (December).

Safari, B. (2012), "Trend Analysis of the Mean Annual Temperature in Rwanda during the Last Fifty Two Years", Journal of Environmental Protection, 3, 538-551 (June).

10

Application of Artificial Neural Networks to Rainfall Disaggregation

R.N. Yasaswi, R.N. Das and R. Subbaiah

Abstract

Disaggregation models provide an algorithm to divide annual or seasonal flows among sub-periods and dividing aggregate flows among sub-basins. Feed forward neural network with back propagation algorithm was applied for disaggregating the seasonal rainfall data into shorter time intervals (months) for Junagadh rainfall. A hyperbolic tangent function is used as transfer functions in hidden layers. The identity function was used as transfer function in the output neuron. Learning rate and momentum was taken as 0.1 and 0.9. All ANNs were trained to minimise the Mean Squared Error (MSE). The gradient descent technique was used. ANN architecture of 1-5-5 was recommended for disaggregation. The average relative for training and testing set was observed to be 0.688 and 0.932 respectively. Low RMSE, AIC, SIC and R^2 *was obtained for all the months for the selected architecture. The correlation coefficient between the observed and computed series for June, July, August September and October was estimated to be 0.642, 0.687, 0.713, 0.722, and 0.654 respectively.*

Key Words: *Covariance, Disaggregation, model, multilayer, neural networks, temporal.*

Introduction

Disaggregation is a mathematical technique for downscaling coarser temporal or spatial levels into finer levels. Disaggregation models were introduced in hydrology by Valencia and Schaake (1973). This model fails to preserve the seasonal covariance between the first season of current year and the last season of the past year. Hoshi and Burges (1979) attempted to overcome the shortcoming of the VS model by simultaneously disaggregating two successive annual events. However, their disaggregation scheme introduces some distortion between the summed inverse transformed individual seasonal flows and the annual amount from which they are disaggregated. Mejia and Rousselle (1976) overcome this constraint by adding a term to the Valencia and Schaake model. However, this model has an inconsistent structure and, thus, fails to preserve all the correlations as anticipated if using the conventional parameter estimation method (Lane, 1982; Stedinger and Vogel, 1984; Lin, 1990a, 1990b). Both the above models have very large number of parameters. Lane (1980; 1982) developed a condensed model by simplifying the original disaggregation model, to decrease the number of parameters, by explicitly modelling fewer of the correlations among the seasonal flows. This model fails to preserve the additivity property. Alternate condensed models which preserve additivity property (Stedinger and Peri, 1982; Grygier and Stedinger, 1988) generate monthly flows that would add up to the previously generated annual flows.

To overcome the trauma in the above models, one of the most successful and brilliant classes of method is an artificial neural network that recently have come into the limelight [Jang *et al,* 1997]. This paradigm mimics human intelligence. Artificial Neural Networks (ANN) is mathematical paradigm, which tries to represent low-level intelligence in natural organisms. ANN shows greater performances in comparison to conventional approaches based on statistical analysis. A detailed review is presented by Kumar (2010). In the present study the application of multi-layered feed forward network which is soft computing technologies is applied to find appropriate network architecture and a set of parameters which can best model the disaggregation of rainfall series from yearly to monthly series.

Materials and Methods

The ANN type adopted in this study is the static multilayer feed forward network. It consists of one input layer, one hidden layer and one output layer. Each layer composed of a number of processing nodes called neurons with connections linking the nodes in successive layers. The input layer converts input values into an output from each node without doing any processing on them.

$$O_i^I = X_i \quad \forall i = 1, 2, \ldots\ldots n \qquad \ldots(1)$$

Here, O_i^I = output from node i in input layer and X_i= input value from node i. The processing nodes in hidden layer will receive the output from node i in input layer from each node and perform a weighted sum of these values via:

$$net_j^h = \sum_{i=1}^{n} O_i^h W_{ij}^h + b_j^h \quad \forall j = 1, 2, \ldots\ldots m \qquad \ldots(2)$$

This weighted sum of values for each node in hidden layer will be filtered out by a particular activation function f to find the output from node j in hidden layer via $O_j^h = f(net_j^h)$. An additional value, so called bias (b), is commonly added to this value. The processing nodes in output layer will receive O_j^h from each node in hidden layer and perform a weighted sum of these values via:

$$net_k^o = \sum_{j=1}^{m} O_j^h \, W_{jk}^o + b_k^o \quad \forall k = 1,2,......p \qquad ...(3)$$

This weighted sum of values for each node in output layer will be filtered out by a particular activation function g to find the output from node k in output layer via $O_k^o = g\,(net_k^o)$. The output from each node in output layer will be compared with the corresponding observed value (t_k) and weights between input layer and hidden layer (w_{ij}) as well as weights between hidden layer and output layer (w_{jk}) will be adjusted to minimise sum of square of residuals. In the present study the selected activation functions for hidden and output layers were the hyperbolic tangent function ($f(x) = \tanh(x)$) and identity functions ($g(x) = X$) respectively. All ANNs were trained to minimise the Mean Squared Error (MSE). The merits of using a good algorithm are threefold: (1) better accuracy leads to better ANN performance, (2) faster convergence leads to smaller calculation times, and (3) lower spread in the performance makes it easier and more honest to evaluate and compare ANNs. Unfortunately, few algorithms are able to combine these three merits. MLP is trained by Back-Propagation (BP) algorithm to adjust the weights in an iterative manner i.e., the weight adjustments are summed over all the training patterns in an epoch and then the weights are adjusted. New weights are a function of the derivatives and previous weights, for example

$$\Delta W_{ij}^t = \eta \, W_{ij}^{t-1} + \alpha \, \Delta W_{ij}^{t-1} \qquad ...(4)$$

$$W_{ij}^t = W_{ij}^{t-1} + \Delta W_{ij}^{t-1} \qquad ...(5)$$

where, η is the learning parameter and α is the momentum. In this study the initial learning rate and momentum was taken as 0.1 and 0.9. Use of BP in network training is very slow, tedious and the convergence to the optimum parameter set could be oscillatory.

Increasing the number of parameters of an ANN by adding hidden neurons or layers, complicates network training. A few hidden nodes result in a high training and generalisation errors due to under-fitting, whereas too many hidden nodes result in low training error but give higher estimation error due to over-fitting. The procedure started with two nodes in the hidden layer initially and number of nodes in the hidden layer increased to 50 with step size of one at each trial. In this way, the optimal number of hidden nodes is estimated by estimating the error training for several networks. Based on extensive testing, De Voseand Rientjes (2005) found that the optimal number of hidden neurons was usually approximately around the square root of the number of input neurons. However, Swingler (1996) and Berry and Linoff (1997) pointed out that the networks rarely require hidden nodes more than twice the number of input nodes for the MLP. ANNs with one hidden layer are considered to offer enough complexity to accurately simulate the dynamic and non-linear properties system. Preliminary test results showed that ANNs indeed outperform the networks with two hidden layers. The optimal size of the hidden layer was found by systematically increasing the number of hidden neurons until the network performance on the test set no longer improved significantly.

The available data are commonly split in two separate data sets: (1) the training set, and (2) the validation set. The first provides the data on which an ANN is trained. The latter data sets are used to validate the performance of a trained ANN. In this study 60 per cent data was taken towards training set and the remaining 40 per cent towards the validation set. All ANNs were trained to minimise the Mean Squared Error (MSE). The merits of using a good algorithm are threefold: (1) better accuracy leads to better ANN performance, (2) faster convergence leads to smaller calculation times, and (3) lower spread in the performance makes it easier and more honest to evaluate and compare ANNs. Unfortunately, few algorithms are able to combine these three merits. The most popular algorithms are gradient descent techniques (e.g., back propagation algorithm) and Newtonian optimisation techniques (Levenberg–Marquardt algorithm). The gradient descent technique was used. Batch training approach was used for training the ANNs. Lower learning rate results in slow learning of the networks, whereas higher training rate makes the weights and objective (error) function diverge, so there is no training at all. Use of momentum term reduces the training time by enhancing the stability of the training process and keeps the training process going in the same general direction. In this study the initial learning rate and momentum was taken as 0.1 and 0.9.

Statistical Criteria

The fitted ANN architecture must be tested to determine whether the model capable of reproducing the historical statistical properties of the data at hand. The measures for evaluating model performance that are used in this paper are the Root Mean Square Error (RMSE), the Nash–Sutcliffe coefficient of efficiency, R^2(Nash and Sutcliffe, 1970), and the Persistence Index, *PI* (Kitanidis and Bras, 1980). The R^2 is formulated as:

$$R^2 = 1 - \frac{F}{F_o} \qquad \ldots(6)$$

$$F = \sum_{i=1}^{N} (Q_o - Q_c)^2 \qquad \ldots(7)$$

$$F_o = \sum_{i=1}^{N} (Q_o - \bar{Q})^2 \qquad \ldots(8)$$

where N is the total number of data elements, Q_o and Q_c are observed and computed monthly rainfall at i^{th} interval respectively is the mean value of runoff over time. The other statistical criteria adopted are the corrected Akaike information criterion (AIC) and the Schwarz information criterion (SIC) also often referred to as the Bayesian information criterion. The AIC is given by (Hurvich and Tsai, 1989; Brockwell and Davis, 1996):

$$AIC = n \ln \sigma^2(\varepsilon) + n + \frac{2(k+1)}{n-k-2} \qquad \ldots(9)$$

where n is the size of the sample used for fitting, k is the number of parameters excluding constant terms, and $\sigma^2(\varepsilon)$ is the maximum likelihood estimate of the residual variance (biased). The AIC statistic is efficient but not consistent and is good for small samples but tends to overfit for large samples and large k. The SIC is given by (Hurvich and Tsai, 1993; Shumway and Stoffer, 2000):

$$SIC = n \ln \sigma^2(\varepsilon) + n + k \ln n \qquad (10)$$

where n, k and are defined in the same way as for the AIC statistic. In general the SIC is good for large samples, but tends to underfit for small samples. The selection rule for the above criteria is to select the model with minimum AIC and minimum SIC and minimum white noise variance. The generated monthly series were compared with the historical ones using statistical properties like mean, standard deviation, coefficient of variation, lag zero and lag one correlation coefficient and the correlation structure between the annual and monthly series for disaggregation models.

Results and Discussion

The proposed procedure was applied to the monthly rainfall series of Junagadh. The daily historical rainfall data for the past 54 (1957-2010) years was collected from Agro-meteorological observatory, Junagadh Agricultural University, Junagadh to determine the monthly and yearly variation of rainfall data. The daily rainfall amounts received were converted into monthly, seasonally and yearly amounts. The monthly and yearly rainfall received during different years is plotted in Fig 1. Large variation in among the monthly rainfall series was recorded. The months of January to May and November to December received scanty amount of rainfall. The statistical properties of monthly rainfall observed are presented in Fig. 2. The total rainfall is confined

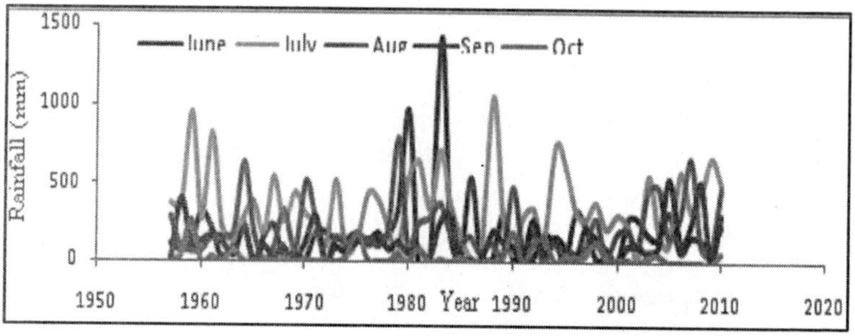

Fig. 1: Monthly rainfall during various years.

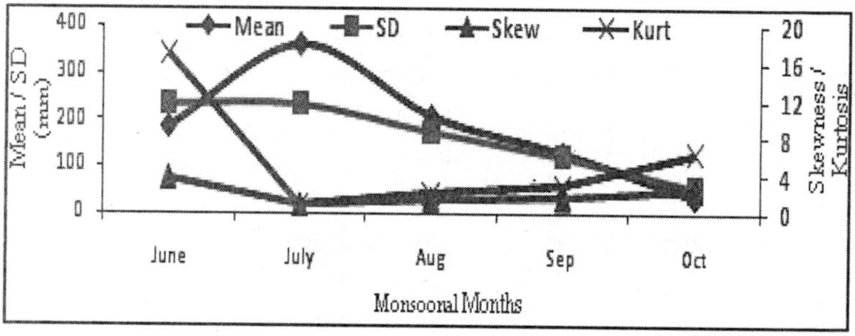

Fig. 2: Statistical parameters of monsoonal rainfall months.

between June to October. The average rainfall for June, July, August, September October, seasonal and annual series was respectively 183.79 mm, 357.23 mm, 206.3 mm, 127.68 mm and 33.18 mm, 871.23 mm, 928.30 mm. This showed that nearly 93.85 per cent of yearly rainfall is confined to these monsoon months. Large variation in monthly rainfall from year to year was observed (Fig. 1). The yearly rainfall was varying from 144 mm during 1987 to 2800 mm during 1983. After 1958 the highest rainfall exceeding 2000 mm occurred again in the year 1983. The average rainfall was found to 925.56 mm based on 54 years of record.

The standard deviation, CV, Skewness and Kurtosis for monthly series is depicted in Fig. 2. The standard deviation for June, July, Aug, Sep, Oct, seasonal and annual series was 233.52 mm, 232.65 mm, 174.09 mm, 127.68 mm, 55.62 mm, respectively. Lowest coefficient of variation was observed during July to September and highest during June and October and non-monsoon months. Lowest coefficient of variation during July indicates the wettest of month (Fig. 2). Lowest skewness and kurtosis were observed in August and highest during remaining monsoonal periods.

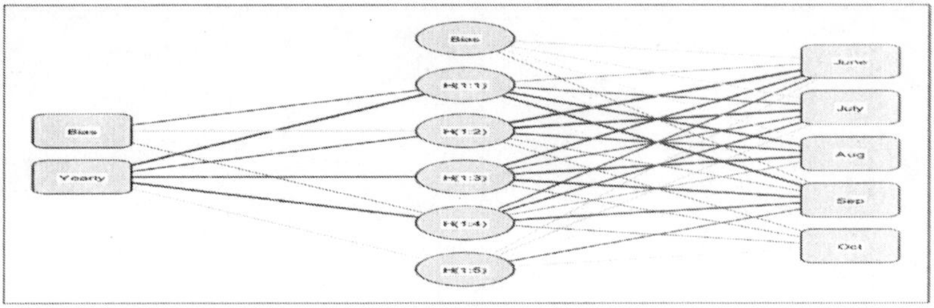

Fig. 3: ANN architecture.

The ANN architecture is shown in Fig. 3. Many ANN architectures were tried for rainfall disaggregation. The number of hidden neurons as well as the layers was also increased. The results show that there is indeed a point at which the performance no longer increases (five hidden neurons). Addition of second layer did not improve the results significantly. Of the several network configurations tried results from the 1-5-5 architecture are presented here due to limitation of space. Note that the 95 per cent confidence bounds widen as the number of hidden neurons increases. This implies that the training algorithm is less likely to find optima as the dimensionality of the parameter space increases. After the network is trained the resulting weights (Table 1) are frozen and are used for disaggregating seasonal rainfall data of the remaining 21 years.

Table 1. Parameter estimates for ANN (1-5-5) architecture

Predictor		*Predicted*									
		Hidden Layer 1					*Output Layer*				
		H(1:1)	*H(1:2)*	*H(1:3)*	*H(1:4)*	*H(1:5)*	*June*	*July*	*Aug*	*Sep*	*Oct*
Input Layer	(Bias)	-.249	.771	.139	-.046	.078					
	Yearly	-.638	-.321	-.344	-.560	.400					
Hidden Layer 1	(Bias)						.721	.435	.202	-.050	.023
	H(1:1)						-.178	-.317	-.517	-.504	.065
	H(1:2)						-1.059	-.736	-.356	-.033	-.051
	H(1:3)						-.479	-.288	-.702	-.564	-.198
	H(1:4)						-.281	-.342	-.073	-.471	-.077
	H(1:5)						.328	.504	.163	-.202	.262

The scatter plot of observed and computed values for June, July, August, September and October were shown in Fig. 4. The average relative error for the training set and testing set was 0.688 and 0.932 respectively. The relative error for June, July, August, September and October for the training set was found to be respectively 0.503, 0.602, 0.667, 0.695 and 0.973. Highest relative error was observed for October series as there are more than 70 per cent zero values were found. The relative error for the remaining months was at par and is within the

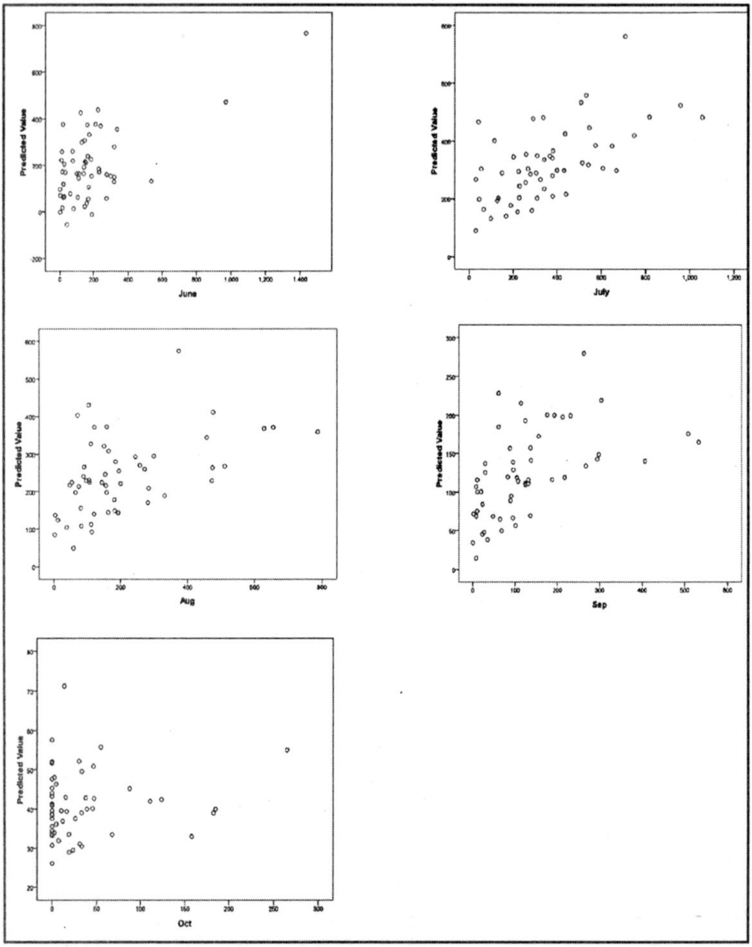

Fig. 4: Scatter plot of observed versus disaggregated rainfall series for monsoonal months Using ANN (1-5-5).

limits. The relative error for June, July, August, September and October for the testing set was found to be respectively 0.891, 0.774, 1.510, 0.760 and 1.258. Maximum relative error for August and October was observed. This may be due to high variability observed in the series. The error attained for the testing series was within the accepted limits.

The biggest drawback is that the model underestimates quite a number of moderate peak flows by upto 40 per cent. Low rainfall series are well simulated, even though these forecasts show more fluctuations than the observed flow pattern. This is most likely due to the model overestimating the effect of small rainfall events. The correlation coefficient between the observed and computed series for June, July, August and September and October was estimated to be 0.642, 0.687, 0.713, 0.722, and 0.654 respectively. The other criteria like RMSE, AIC, SIC, PI and R^2 was also estimated for June, July, August and September and are presented in Table 2. Higher RMSE was observed for the months of June, September and October compared to July and August. The synthetic series also preserved the statistical properties like mean and skewness to a great extent.

Table 2: Statistical criteria of the forecasted series for the monsoonal months

Month	RMSE	Skewness	AIC	SIC	R^2
June	37.8264	2.253	3.8264	2.8583	0.67943
July	23.2436	3.861	4.2807	3.9366	0.71235
August	29.3060	2.162	4.1983	3.8472	0.81264
September	33.9674	4.455	2.5993	2.1125	0.65437
October	48.796	4.765	5.1367	4.1284	0.68357

Plotting the residuals can tell a great deal about the model, including whether it was appropriate to use the model in practice, and whether the forecasting was done correctly. There are no odd fan or curved trends in the plot, the average of the residuals is zero, and the points are equally represented about the x-axis. This residual represents the difference between the observed response variable and the value predicted by the model.

Conclusion

Feed forward neural network with back propagation algorithm was applied for disaggregating the seasonal rainfall data into shorter time intervals (months). The algorithm was applied to Junagadh in Gujarat. Out of various architectures tried the 1-5-5 architecture was quite encouraging. The average relative for training and testing set was observed to be 0.688 and 0.932 respectively. Higher RMSE was observed for the months of June, September and October compared to July and August. The synthetic series also preserved the statistical properties like mean and skewness to a great extent.

References

Berry, M.J.A. and Linoff, G. 1997. Data mining techniques. Wiley, NY.

Brockwell, P. J. and Davis, R.A. 1996. Introduction to time series and forecasting. Springer Texts in Statistics. *Springer-Verlag*, first edition.

deVos, N. J. and Rientjes, T.H.M. 2005. Constraints of artificial neural networks for rainfall-runoff modelling: trade-offs in hydrological state representation and model evaluation, Hydrology and Earth System Sciences, 9: 111–126.

Grygier, J.C. and Stedinger, J.R. 1988. Condensed disaggregation procedures and conservation corrections for stochastic hydrology. *Water Resources Research*, Vol. 24, p. 1574.

Hoshi, K. and Burges, S. J. 1980. Disaggregation of streamflow volumes. *Journal of Hydraulics Division*, American Society of Civil Engineers, Vol. 106, No. HY6, p. 1127.

Hurvich, C.M. and Tsai, C.L. 1993. A corrected Akaike information criterion for vector autoregressive model selection. *Journal of Time Series Annuals*, Vol. 14, p . 271.

Kumar, M.,Raghuwanshi, N.S. and Singh, R. 2010. Artificial neural networks approach in evapotranspiration modeling: a review. Irrig Sci. DOI 10.1007/s00271-010-0230-8.

Lane, W.L. 1980. Applied stochastic techniques (LAST Computer Package), User Manual. *Division of Planning Technical Services, Bureau of Reclamation*, Denver, Colorado.

Lane, L.W. 1982. Corrected parameter estimates for disaggregation scheme, In: V.E. Singh (ed.), *Statistical analysis of rainfall and runoff,* Water Resources Publication, Littleton, Colorado.

Lane, W.L. and Fervert, D.K. 1988. Applied stochastic techniques, User's Manual. *Bureau of Reclamation, Engineering and Research Center*, Denver, Colorado.

Lin, G.F. 1990a. Corrected parameter estimates for staged disaggregation. *Journal of Chinese Institution of Engineers,* Vol. 13, p. 405.

Lin, G.F. 1990b. Parameter estimation for seasonal to sub-seasonal disaggregation. *Journal of Hydrology,* Vol.120, p. 65.

Loucks, D. E., Stedinger, J. R. and Haith, D. A. 1981. Water resources system planning and analysis. Prentice-Hall, Englewood Cliffs, NJ.

Mejia, J.M. and Rousselle, J. 1976. Disaggregation models in hydrology revisited. *Water Resources Research*,Vol. 12, p. 185.

Nash, J. E. and Sutcliffe, J. V. 1970. River flow forecasting through conceptual models; part- I – a discussion of principles, J. Hydrol., 10: 282–290.

Salas, J.D., Delleur, J. W.,Yevjevich, V. and Lane, W.L.1980. Applied modelling of hydrologic time series. *Water Resources Publications*, Littleton, Colorado.

Stedinger, J. R. and Vogel, R.M. 1984. Disaggregation procedures for generating serially correlated flow vector. *Water Resources Research*,Vol. 20, p. 47.

Stedinger, J.R. and Pei, D. 1982. An annual monthly stream flow model for incorporating parameter uncertainty into reservoir simulation. Time series methods in hydroscience, A.H. El-shaarawi and S.R. Esterby, eds., Elsevier, New York, N.K., 520-529.

Shumway, R.H. and Stoffer, D. S. 2000. Time series analysis and its applications. *Springer Texts in Statistics, Springer-Verlag*, first edition.

Swingler, K. 1996. Applying neural networks: a practical guide. Academic Press, London.

Valencia, R. D. and Schaake, J. C. 1973. Disaggregation processes in stochastic hydrology, *Water Resources Research*, Vol. 9, p. 580.

11

Drought Investigation using the Standardised Precipitation Index

R.N. Das, R.N. Yasaswi, R. Subbaiah, and S. Pandya

Abstract

Drought which is related to the time, the duration, and the place of its occurrence was assessed using Standard Precipitation Index. Extremely wet and extremely dry events were not noticed in the 54 years of record. The probability of nearly normal event is more than 53 per cent in monthly, seasonal and annual series. The probability of moderately wet event is ranging from 3.70 per cent to 18.51 per cent. Low probability was observed during October and high was noticed in seasonal series. No severe dry events are recorded in October. The probability of severe dry event was ranging from 5.55 per cent to 11.11 per cent. The probability of moderately dry event is 9.25 per cent and 11.11 per cent during July and August.
***Keywords**: Drought, extreme, gamma distribution, model, probability.*

Introduction

Climate change leads to increasing frequency of extreme weather events evident around the globe. Unusually high rainfall, which many scientists agree is due to climate change, is a significant cause of floods. On the other hand droughts are becoming longer, harder and more frequent. Assessing the frequency of drought in the changing climate is very much needed to reduce the crop failure. Standardised Precipitation Index (SPI) has found widespread application (McKee *et al*., 1993; Heim, 2000; Wilhite *et al*., 2000; Rossi and Cancelliere, 2002) for estimating droughts among different time periods and regions with different climatic conditions. The SPI performs better than other drought indices (Bonaccorso *et al*., 2003; Paulo and Pereira, 2006; Vicente-Serrano, 2006). The SPI does not require information about land surface conditions and is solely a function of the precipitation amount (Kim *et al.*, 2006). An evaluation of common indicators, according to six weighted evaluation criteria of performance (robustness, tractability, transparency, sophistication, extendibility, and dimensionality), indicates strengths of the SPI and Deciles over the PDSI (Keyantash and Dracup, 2002). Keeping this in view the drought assessment was done using SPI index.

Standardised Precipitation Index

Standard Precipitation Index was developed by McKee et al. (1993, 1995). The index is based on an equi-probability transformation of aggregated monthly precipitation into a standard normal variable. In practice, computation of the index requires fitting a probability distribution to precipitation series, computing the non-exceedence probability related to such aggregated values and defining the corresponding standard normal quantile as the SPI. McKee *et al.* (1993) assumed an aggregated precipitation gamma distributed and used a maximum likelihood method to estimate the parameters of the distribution. The gamma distribution is defined by its frequency or probability density function:

$$g(x) = \frac{1}{\beta^{\alpha}\Gamma(\alpha)} X^{\alpha-1} e^{-x/\beta} \qquad ...(1)$$

The alpha (α) and beta (β) parameter of the gamma probability density function are estimated for each station (Haan 1994).

$$\alpha = \frac{1}{4A}(1+\sqrt{1+\frac{4A}{3}}) \qquad ...(2)$$

$$\beta = \frac{\overline{X}}{\alpha} \qquad ...(3)$$

$$A = \ln(\bar{x}) - \sum \ln(x)/n \qquad ...[4]$$

where n is number of precipitation observations. The resulting parameters are then used to find the cumulative probability of an observed precipitation event for the given time scale for the station in question. The cumulative probability is given by:

$$G(x) = \int_0^x g(x)dx = \frac{1}{\beta^{\alpha}\Gamma(\alpha)}\int_0^x x^{\alpha-1}e^{-x/\beta}dx \qquad ...(5)$$

Let t = x/β the equation become the incomplete gamma function:

$$G(x) = \frac{1}{\Gamma(\alpha)}\int_0^t t^{\alpha-1}e^{-t}dt \qquad ...(6)$$

Since the gamma function is undefined for $x = 0$ and a precipitation distribution may contain zeros, the cumulative probability becomes:

$$H(x) = q + (1-q)\,G(x) \qquad \text{...(7)}$$

where, q is the probability of a zero. The cumulative probability $H(x)$ is then transformed to the standard normal random variable Z with mean zero and variance of one, which is the value of the SPI.

$$Z = \text{SPI} = -\left(t - \frac{c_0 + c_1 t + c_2 t^2}{1 + d_1 t + d_2 t^2 + d_3 t^3}\right) \qquad \text{...(8)}$$

for $0 < H(x) \leq 0.5$

$$Z = \text{SPI} = +\left(t - \frac{c_0 + c_1 t + c_2 t^2}{1 + d_1 t + d_2 t^2 + d_3 t^3}\right) \qquad \text{...(9)}$$

for $0.5 < H(x) \leq 1.0$

$$t = \sqrt{\ln\left[\frac{1}{(H(x))^2}\right]} \qquad \text{...(10)}$$

for $0 < H(x) \leq 0.5$

$$t = \sqrt{\ln\left[\frac{1}{(1-H(x))^2}\right]} \qquad \text{...(11)}$$

for $0.5 < H(x) \leq 1.0$

and $C_0 = 2.515517$; $C_1 = 0.802853$; $C_2 = 0.010328$; $d_1 = 1.432788$; $d_2 = 0.189269$; $d_3 = 0.001308$. Since the SPI is equal to the z-value of the normal distribution, McKee *et al.* (1993, 1995) proposed a seven-category classification for the SPI: extremely wet ($z > 2.0$), very wet (1.5 to 1.99), moderately wet (1.0 to 1.49), near normal (–0.99 to 0.99), moderately dry (–1.49 to 1.0), severely dry (–1.99 to 1.5), and extremely dry (<–2.0). Also, the probabilities "*P*", that the index lies within each class are also listed in Table 1. Since the present work focuses on forecasting drought conditions, the near normal and wet classes have been grouped into one class termed "Non-drought".

Testing the goodness of fit

McKee originally used an incomplete gamma distribution to calculate the SPI. The gamma probability distribution function is determined from a long-term data record by fitting the function to the data. To determine the suitability

Table 1: Wet and Drought Period Classification according to SPI Index

Type	*Index value*	*Class*	*Probability*	ΔP
Non Drought	SPI = 2.00	Extremely wet	0.977–1.00	0.023
	1.50 = SPI < 2.00	Very wet	0.933–0.977	0.044
	1.00 = SPI < 1.50	Moderately wet	0.841–0.933	0.092
	- 1.00 = SPI < 1.00	Near normal	0.159–0.841	0.682
Drought	- 1.50 = SPI < - 1.00	Moderate drought	0.067–0.159	0.092
	- 2.00 = SPI < - 1.50	Severe drought	0.023–0.067	0.044
	SPI < - 2.00	Extreme drought	0.000–0.023	0.023

of selected distribution, it is necessary to fit the observed distribution to a theoretical distribution. This is done by comparing the observed frequencies in the data to the expected frequencies of the theoretical distribution since certain types of variables follow specific distribution (Tilahun, 2006). The following goodness-of-fit tests viz. Kolmogorov-Smirnov test and Anderson-Darling test were used along with the chi-square test at α (0.01) level of significance for the determining the suitability of gamma probability distribution.

Results and Discussion

Rainfall Variability

The daily historical rainfall data for the past 54 (1957-2010) years was collected from Junagadh Agricultural University, Junagadh to determine the monthly and yearly variation of rainfall data. The monthly rainfall received during different years is plotted in Fig. 1. Large variation in among the monthly rainfall series was recorded. The yearly and seasonal variation of rainfall received from 1957 to 2010 is presented in Fig. 2. The months of January to May and November to December received scanty amount of rainfall. The statistical properties of monthly rainfall observed are presented in Fig. 3. The total rainfall is confined between June to October. The average rainfall for June, July, August, September, October, seasonal and annual series was respectively

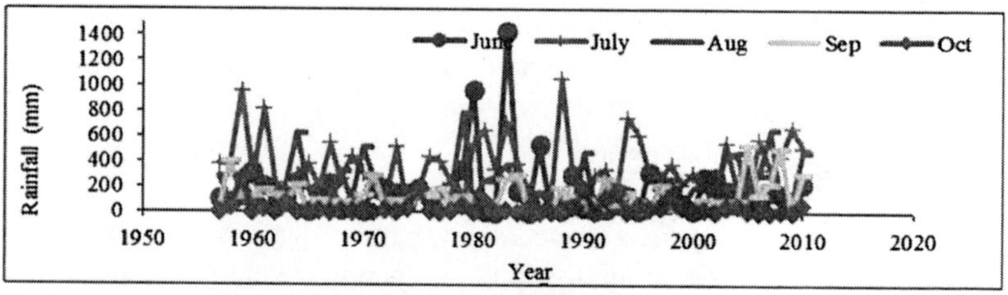

Fig. 1: Monthly rainfall received during various years in Junagadh.

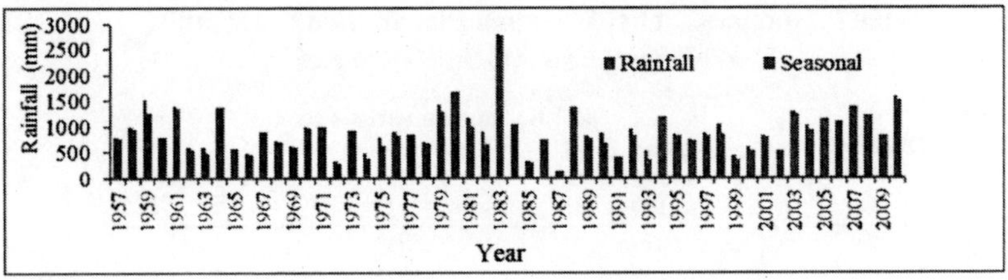

Fig. 2: Seasonal and annual rainfall variation

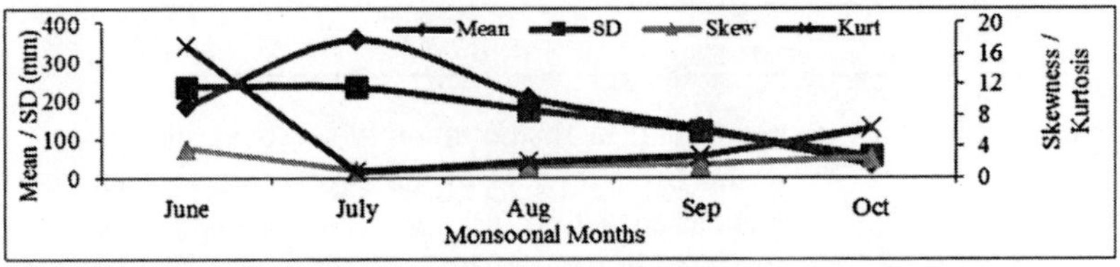

Fig. 3: Statistical parameters of monsoonal months rainfall.

183.79 mm, 357.23 mm, 206.3 mm, 127.68 mm and 33.18 mm, 871.23 mm, 928.30 mm. This showed that nearly 93.85 per cent of yearly rainfall is confined to these monsoon months. Large variation of rainfall from year to year was observed (Fig. 2). The yearly rainfall was varying from 144 mm during 1987 to 2800 mm during 1983. After 1958 the highest rainfall exceeding 2000 mm occurred again in the year 1983. The average rainfall was found to 925.56 mm based on 54 years of record (Fig. 3).

The standard deviation for monthly series is depicted in Fig. 3. The standard deviation for June, July, Aug, Sep, Oct, seasonal and annual series was 233.52 mm, 232.65 mm, 174.09 mm, 127.68 mm, 55.62 mm, respectively. Lowest coefficient of variation was observed during July to September and highest during June and October and non-monsoon months. Lowest coefficient of variation during July indicates the wettest of month (Fig. 3). Lowest skewness and kurtosis were observed in August and highest during remaining monsoonal periods. The independence was tested by the Wald-Wolfowitz test (Wald and Wolfowitz, 1943) and the Anderson test (Anderson, 1941). For most cases, they turned out to be non-significantly different from zero in the test suggested by Clarke (1973) (i.e. $\pm 2/(n)^{0.5}$) thereby suggesting that the weekly rainfall sequences are independently distributed.

Gamma Probability Distribution

Mathematically, the SPI is based on the cumulative probability of a given rainfall event occurring at a station. The historic rainfall data of the station is fitted to a gamma distribution. The suitability of gamma distribution is tested with KS test, Anderson Darling test and Chi Square test. The results of these tests are presented in Table 2. The statistics of the above tests indicate that the

Table 2: Goodness of fit of Gamma Distribution to Monthly, Yearly and Seasonal Rainfall Series

Duration	Statistic value in various tests		
	Kolmogorov Smirnov	Anderson-Darling	Chi-Square
June	0.1329	1.1713	3.776
July	0.0743	0.5243	2.227
August	0.1049	0.5815	3.017
September	0.0926	2.339	2.494
October	0.1889	2.1814	3.0098
Seasonal	0.0923	0.298	2.654
Yearly	0.0834	0.323	2.901

gamma distribution has been found to fit the monthly seasonal and annual precipitation series. The a and b parameters of the two parameter gamma distribution was estimated as per eq. 2 to 4 for monthly, yearly and seasonal rainfall amounts and are presented in Fig. 4. The value of alpha was varying

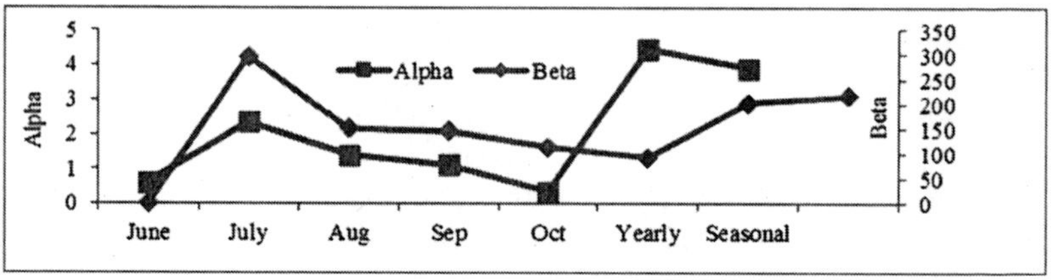

Fig. 4: Scale and Shape parameters of Gamma Distribution for rainfall of Junagadh.

from 0.356 to 3.895. The value of Beta was varying from 0 to 296.7. An increasing trend in the value of Alpha was observed during June to July and a decreasing trend was observed during July to October for the rainfall of Junagadh. A similar trend in the Beta value was seen for the monthly rainfall series of Junagadh. The alpha value for seasonal and yearly rainfall series was found to be respectively 3.895 and 4.458. The value of beta for seasonal and yearly rainfall series was found to be respectively 202.5 and 93.23.

Rounding off fractional values of return periods revealed that observed annual rainfall amounts ranging from 144 mm to 705 mm had one-year return period, and 720 mm to 920 mm had two-year return period. Thus at least 144 mm rainfall could be expected to occur every year and 720 mm in every alternate year. The corresponding values of rainfall of one-year and two-year return periods as estimated by gamma distribution were 204.5 mm and 820 mm respectively. The prediction analysis revealed 99 per cent chance of at least 204.5 mm rainfall every year. The observed annual amount exceeding 2000 mm had a return period of 22 years.

Z–score

The estimation of Z was carried and are presented in Fig. 5 to 12 respectively for June, July, August, September, October, Yearly and Seasonally. The z score indicate that there is no extremely wet or extremely dry event observed either in monthly series or seasonal or annual series. Positive SPI

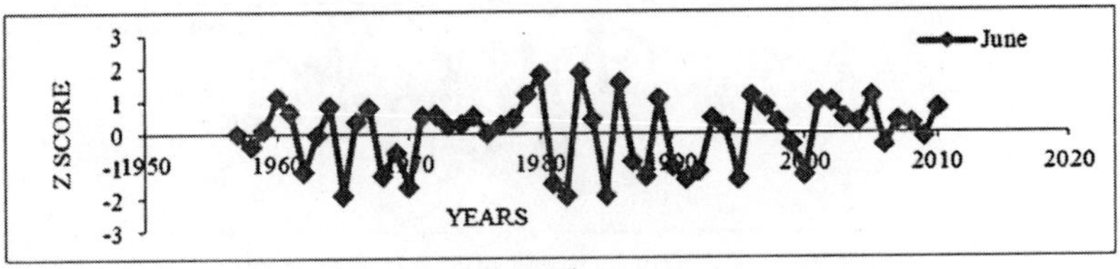

Fig. 5: Z Score for Different Years of Junagadh Rainfall.

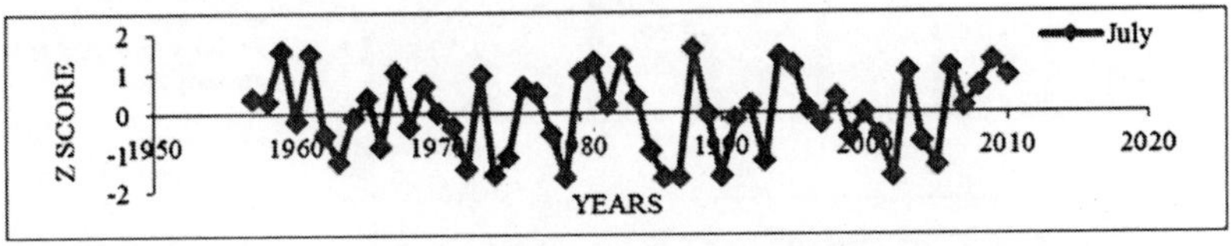

Fig. 6: Z Score for Different Years of Junagadh Rainfall.

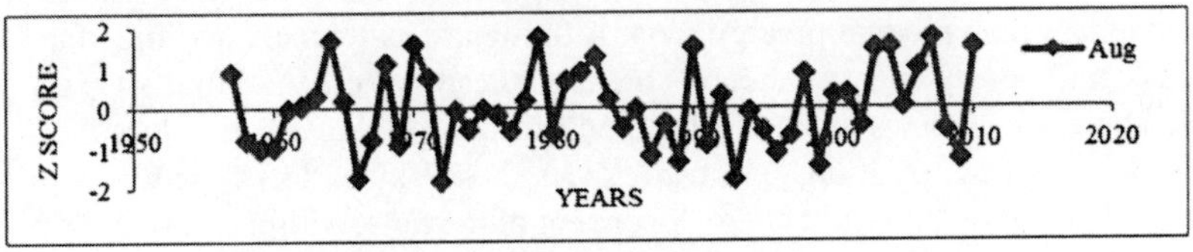

Fig. 7: Z Score for Different Years of Junagadh Rainfall.

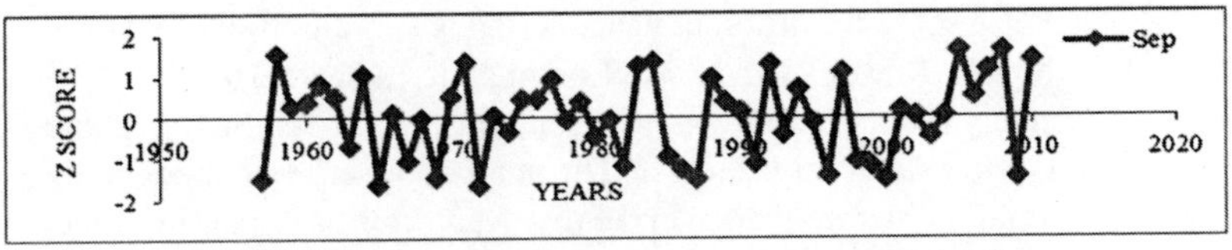

Fig. 8: Z Score for Different Years of Junagadh Rainfall.

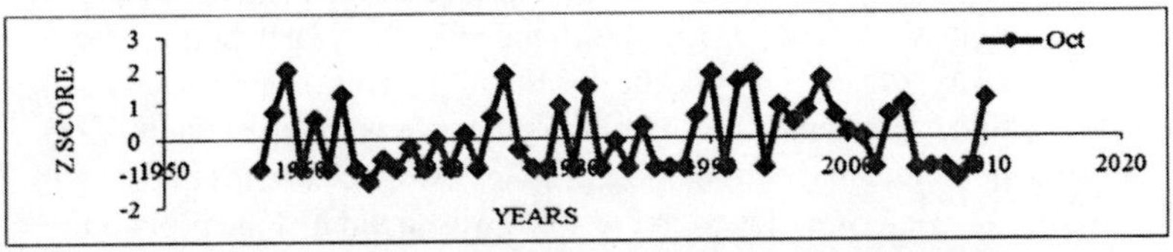

Fig. 9: Z Score for Different Years of Junagadh Rainfall.

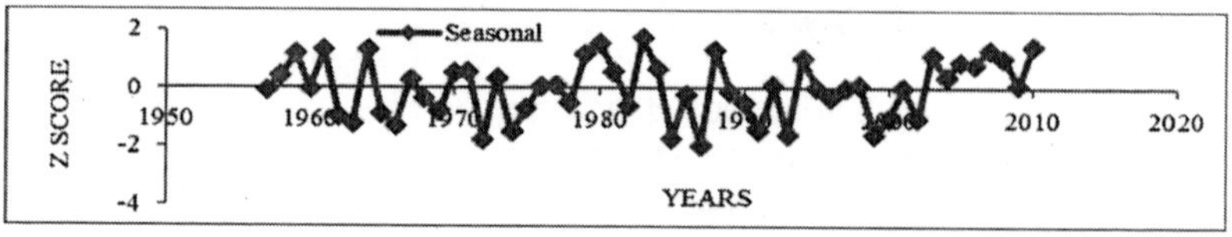

Fig. 10: Z Score for Different Years of Junagadh Rainfall.

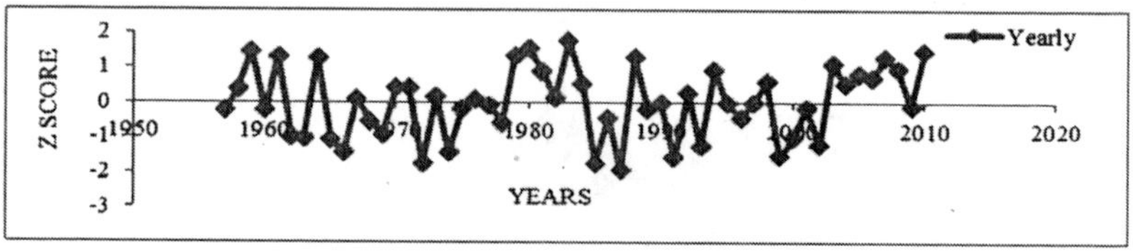

Fig. 11: Z Score for Different Years of Junagadh Rainfall.

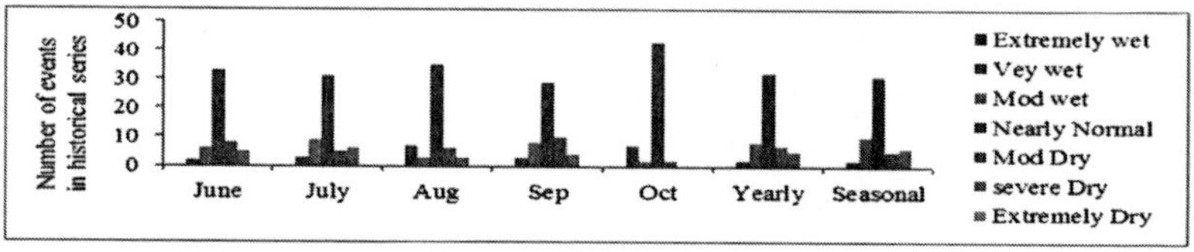

Fig. 12: Drought and non drought events during monthly seasonal and annual series.

values indicate greater than median precipitation, while negative values indicate less than median precipitation. A drought event occurs any time the SPI is continuously negative and reaches an intensity where the SPI is –1.0 or less. The event ends when the SPI becomes positive. Because SPI values fit a typical normal distribution, one can expect these values to be within one standard deviation approximately 68 per cent of the time, within two standard deviations 95 per cent of the time, and within three standard deviations 99 per cent of the time (Wilhite *et al.*, 2000).

The value of Z score for different years during is varying from -1.9414 to 1.8355, -1.66 to 1.61, -1.79 to 1.75, -1.65 to 1.62, -1.25 to 1.98, -1.93 to 1.74, -1.94 to 1.75 respectively for June, July, August, September, October, Seasonal and Yearly rainfall of Junagadh (Fig. 5 to 11). The number of very wet, moderately wet, nearly normal, moderately dry, severe dry events appearing during June, July, August, September, October, Seasonal and Yearly series are presented in Fig. 12. The years appearing under different types of drought and non-drought events are presented in Table 3. Light difference in the yearly and seasonal rainfall z score was noticed. This may be due to low difference in the yearly and seasonal rainfall and in some cases similar amount in both cases due to zero rainfall in the non-monsoon periods. The probability of occurrence of drought and non-drought events are plotted in Fig 13. It was found that the probability of nearly normal event is more than 53 per cent in monthly seasonal and annual series. The probability of moderately wet events is ranging from 3.70% to

Table 3: Recurrence of drought and non-drought events

Drought/ Non Drought events	*June*	*July*	*Aug*	*Sep*	*Oct*	*Seasonal*	*Yearly*
Extremely wet	0	0	0	0	0	0	0
Very Wet	84,86,	59,61,88,	64,70,79, 90,04,07, 10	58,05,08	59,75,81, 90,92,93, 98	80,83	80,83
Moderately wet	60,79,80, 89,96, 05	67,80,81, 83, 84, 95,03,06,09	68,83,03	64,71,83, 84,92,97, 07,10	63,10	59,61,64, 79,88,94, 03, 07,08, 10	59,61,64, 79,88,03, 07,10
Moderately Dry	62,68,88, 90,91, 92,95,00	63,72,75, 93,05	59,87,89, 96,99,09	67,69,82, 86,91,96, 98,99,00, 09	65,08	63,66,91, 00,02	63,65,66, 74,93,00, 02
Severely Dry	65,70,81, 82,85	74,79,86, 87,90, 02	66,72,93	57,65,72, 87		72,74,85, 87,93,99	72,85,87, 91,99
Extremely Dry	0	0	0	0	0	0	0

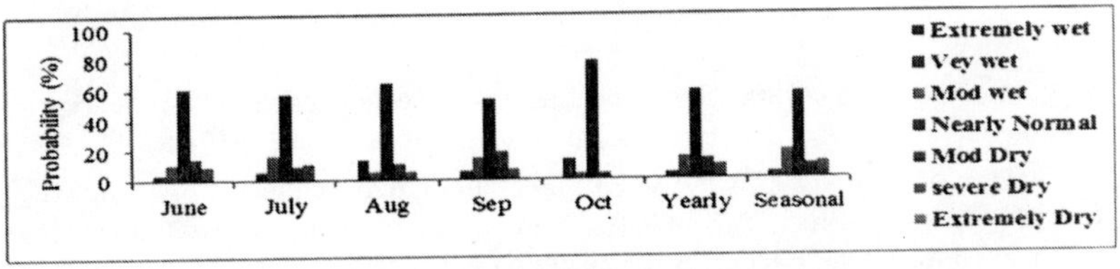

Fig. 13: Drought and non drought events probability during monthly seasonal and annual series.

18.51 per cent. Low probability was observed during October and high was obtained in seasonal series. No severe dry events are recorded in October. The probability of severe dry event was ranging from 5.55 per cent to 11.11 per cent. The probability of moderately dry event is 9.25 per cent and 11.11 per cent during July and August.

With minimum data simple indices such as the SPI can be used as the basis of a drought of an area. The procedure can be considered as the first step for assessing drought. The main advantage of the proposed procedure is its simplicity, its transparency and its universality. The results can be produced quickly, with low infrastructure and are understandable and compatible. Therefore, the proposed methodology can be the basis for drought estimation and management.

References

Anderson, R.L., (1941). Distribution of serial correlation coefficient. *Annals of Mathematical Statistics*, 8(2):13.

Bonaccorso, B., Bordi, I., Cancelliere, A., Rossi, G., Sutera, A., (2003). Spatial variability of drought: an analysis of the SPI in Sicily. *Water Resour Manag.*17(4): 273–296.

Clarke, R.T., (1973). Mathematical models in hydrology, *FAO Irrigation and Drainage Paper No.* 19, Rome, Italy.

Haan, C.T., (1994). Statistical methods in hydrology. Affiliated East-West Press Pvt. Ltd., *New Delhi*.

Heim, R.R., (2000). Drought indices: a review. In: Wilhite DA., (ed) *Drought: a global assessment.* Routledge, London and New York.

Keyantash, J., Dracup, J.A., (2002). The quantification of drought: an evaluation of drought indices. *Bull. Am. Meteor. Soc.* 83(8):1167–1180.

Kim, T.W., Valds, J.B., Nijssen, B., Roncayolo, D., (2006). Quantification of linkages between large-scale climatic patterns and precipitation in the Colorado River Basin. *J. Hydrol.* 321:173–186.

McKee, T,B., Doesken, N.J., Kleist, J. (1993). The relationship of drought frequency and duration to time scales. 8th Conference on *Applied Climatology. Am. Met. Soc., Boston*, MA, pp. 179–184.

McKee, T.B., Doesken, N.J., Kleist, J., (1995). Drought monitoring with multiple time scales. Proc 9th Conf. on *Applied Climatology, Am. Met. Soc., Boston*, MA, pp. 233–236.

Paulo, A.A., Pereira, L.S., (2006). Drought concepts and characterization. Comparing drought indices applied at local and regional scales. *Wate Int* 31(1):37-49.

Rossi, G., Cancelliere, A., (2002). Early warning of drought: development of a drought bulletin for Sicily. Proc. 2nd Int Conf. "New trends in water and environmental engineering for safety and life: eco-compatible solutions for aquatic environments". Capri, Italy, June 24–28, 2002, pp. 1–12.

Tilahun, K., (2006). The characterization of rainfall in the arid and semiarid regions of Ethiopia. *Water S.A.* 32(3): 429-436.

Vicente-Serrano, S.M., (2006). Differences in spatial patterns of drought on different time scales: an analysis of the Iberian Peninsula. *Water Resour. Manag.*, 20(1): 37–60.

Wald, A. and Wolfowitz, J., (1943). An exact test for randomness in the nonparametric case based on serial correlation, *Annals Math. Stat.* 14:378-388.

Wilhite, D. A., Hayes, M. J. and Svoboda, M. D.: 2000, Drought Monitoring and Assessment: Status and Trends in the United States. In Voght, JV and Somma, F. (eds). *Drought and Drought Mitigation in Europe*, Kluwer Academic Publishers, Dordrecht, pp. 149–160.

12

Forecasting Weekly Temperature with Seasonal ARIMA Model

P.A. Damor, R. Subbaiah, G.V. Prajapati, H.D. Rank and H.V. Parmar

Abstract

Climatic change projected an increase of 1.8 to 4 °C in temperature by the end of the century. The Autoregressive Integrated Moving Average (ARIMA) is used to analyse the fact above and also to forecast prevailing weekly maximum and minimum temperature of Junagadh during the coming years. Twenty seven years data (1984 to 2010) of T_{max} *and* T_{min} *were used. Seasonal ARIMA models of different orders were selected for data analysis. Based on ACF and PACF plot, AIC and SBC values ARIMA (4,0,0) (0,1,1)*$_{52}$ *and ARIMA (1,0,1) (2,1,1)*$_{52}$ *models suited best to forecast* T_{max} *and* T_{min}*. The results indicate that the seasonal ARIMA model provide reliable and satisfactory prediction for temperature on weekly basis.*

Introduction

Climatic change projected an increase of 1.8 °C to 4 °C in temperature by the end of the century. This will cause instability in biosphere including increase in crop water requirements, reduce crop growth and agricultural production. There is a dire need to forecast the temperature trend with climatic variability. Many methods and approaches for formulating forecasting models are available in the literature. This research exclusively deals with time series forecasting model, in particular, the Auto Regressive Integrated Moving Average (ARIMA). Chiew *et al* (1993) and Langu (1993) mentioned that ARIMA models can provide adequate estimates of monthly and annual yields in the water resources of the catchments.

Wide spectrum availability of computer software enable to work with large observations, simplify calculations on the part of time and allow simulations of various types of models with correspondent results. In comparison with decomposition analysis, Box-Jenkins method describes existent time series with one model that gets to the basis of historical data development, per quad decreases forecasting error possibility. Keeping this in view a study was undertaken to forecast weekly maximum and minimum temperature of Junagadh, Gujarat, using Statistical Analysis System (SAS) Software.

Materials and Methods

Junagadh geographically situated between latitude 21°31′23.29"N, longitude 70°27'17.90" E and at altitude of 86 meter above mean sea level in South Saurashtra region of Gujarat state. The climate of study area is subtropical and semi-arid type with an average rainfall of 929.81 mm. Rainfall is confined between mid-June and mid-October. Average annual pan evaporation is 5.6 mm/day. January is the coldest month with mean monthly temperature varying from 7°C to 15 °C and maximum monthly temperature is recorded in the month of May varying between 29.50 °C to 39.40 °C. Relative humidity vary between 45 per cent and 89 per cent. Wind speed vary between 2 to 9.70 km/h.

Seasonal ARIMA modelling

An autoregressive model of order *p* is conventionally classified as AR (*p*) and a moving average model with *q* terms is known as MA (*q*). A combined model that contains *p* AR-terms and *q* MA-terms is called an ARMA (*p, q*) model. To make a generally non-stationary time-series stationary time-shifted (by *d* lags, whereby in most cases *d* =1) differences are computed before further processing. Such a model is then classified as ARIMA (*p, d, q*), where the symbol "I" signifies "integrated". The general form of the above model describing the current value y_t of a time series by its own past is expressed as:

$$\phi_p(B)\phi_P(B_5)\nabla^d\nabla_s^D T_t = \Theta_Q(B^s)\phi_q(B)\varepsilon_t \qquad \ldots(1)$$

where $\phi_p(B) = 1-\phi_1 B - \ldots \phi_p B^p$ = Non-Seasonal autoregressive (AR) operator; $\theta_q(B) = 1-\theta_1 B - \ldots \theta_q B^q$ = Non-Seasonal moving average operator (MA) operator; $\phi_p(B^s) = 1-\phi_1 B^s - \ldots \phi_1 B^{sQ}$ = Seasonal auto regressive (SAR) operator; $\Theta_Q(B^s) = 1-\Theta_1 B^s - \ldots \Theta_1 B^{sQ}$ = Seasonal moving average operator (SMA). Here, B = backshift operator (i.e. $B^1 y_t = y_{t-1}$, $B^2 y_t = y_{t-2}$ and so on); s = the seasonal lag; ε_t = sequence of independent normal error variables with mean zero and variance σ^2; *p* and *q* are orders of non-seasonal auto-regression and moving average parameters respectively and *P* and *Q* are that of the seasonal auto regression and moving average parameter respectively; *d* and *D* denote the non-seasonal and seasonal differences respectively. The main stages in setting up a forecasting ARIMA model includes model identification, model parameters estimation and diagnostic checking for the identified model appropriateness for modelling and forecasting.

Classical Box-Jenkins model describe stationary time series. Thus, in order to tentatively identify Box-Jenkins model, verifying the time series for stationary is essential. Stationary models assume that the process remains in equilibrium about a constant mean level that is when the plotting shows that the data fluctuates around its constant mean. A cursory look at the graph of the data and structure of autocorrelation and partial correlation coefficients at various lags may provide clues for the presence of stationarity. If the model is found to be non-stationary, stationarity could be achieved mostly by differencing the series. The next step in the identification process is to find the initial values for the orders of seasonal and non-seasonal parameters, p, q, and P, Q. They could be obtained by looking for significant autocorrelation and partial autocorrelation coefficients.

After choosing the most appropriate model (step 1 above) the model parameters are estimated (step 2) by using the least square method. In this step, values of the parameters are chosen to make the Sum of the Squared Residuals (SSR) between the real data and the estimated values as small as possible. In general, non-linear estimation method is used to estimate the above identified parameters to maximise the likelihood (probability) of the observed series given the parameter values. The methodology uses the following criteria in parameter estimation:

(a) The estimation procedure stops when the change in all parameters estimate between iterations reaches a minimal change of 0.001.

(b) The parameters estimation procedure stops when the SSR between iterations reaches a minimal change of 0.0001.

In diagnose checking step (step three), the residuals from the fitted model shall be examined against adequacy. This is usually done by correlation analysis through the residual ACF plots and the goodness-of-fit test by means of Chi-square statistics χ^2. If most of the sample autocorrelation coefficient of the residuals are within the limits $\pm 1.96/\sqrt{N}$., where N is the number of observations upon which the model is based then the residuals are white noise indicating that the model is a good fit. Otherwise, the autocorrelations are white noise and the model is adequate to represent our time series. In the last phase, i.e forecasting we calculate point extrapolated prognosis of time series and eventually confidence lag of prognosis.

Evaluation Criteria

The other statistical criteria adopted in the study are:

1. ***Akaike Information Criterion*** **(AIC)**

The AIC is given by,

$$AIC = n \ln \sigma^2(\varepsilon) + n + \frac{2(k+1)}{n-k-2} \qquad ...(2)$$

where n is the size of the sample used for fitting, k is the number of parameters excluding constant terms, and $\sigma^2(\varepsilon)$ is the maximum likelihood estimate of the residual variance.

2. ***Schwarz information criterion*** (SIC)

The SIC is given by,

$$SIC = n \ln \sigma^2(\varepsilon) + n + k \ln n \qquad ...(3)$$

where n, k and are defined in the same way as for the AIC statistic.

3. *Mean absolute percentage error* (MAPE) :

$$\text{MAPE} = \frac{1}{N}\sum_{i=1}^{N}\left|\frac{X_c - O_c}{O_t}\right| \times 100 \qquad ...(4)$$

where X_t= forecast value at time t; O_t= actual value at time t; N= number of weeks considered for forecasting.

4. Mean absolute error (MAE)

$$\text{MAE} = \frac{1}{N}\sum_{i=1}^{N}|X_t - O_t| \qquad ...(5)$$

where X_t= forecast value at time t; O_t= actual value at time t; N= number of weeks considered for forecasting.

Results and Discussion

In this study the weekly data of maximum temperature and minimum temperature of 27 years (1984 to 2010) were collected from Agro meteorology Department of Junagadh Agricultural University, Junagadh and forecast was made for the three years (2011-2013) using seasonal ARIMA model.

ACF and PACF of T_{max} and T_{min}

Data of maximum temperature (T_{max}) and minimum temperature (T_{min}) was analysed using Statistical Analysis System (SAS) software. Auto correlation function (ACF) and Partial Auto correlation function (PACF) of the original time series of maximum (T_{max}) and minimum temperature (T_{min}) are shown in Fig. 1 and Fig. 2.

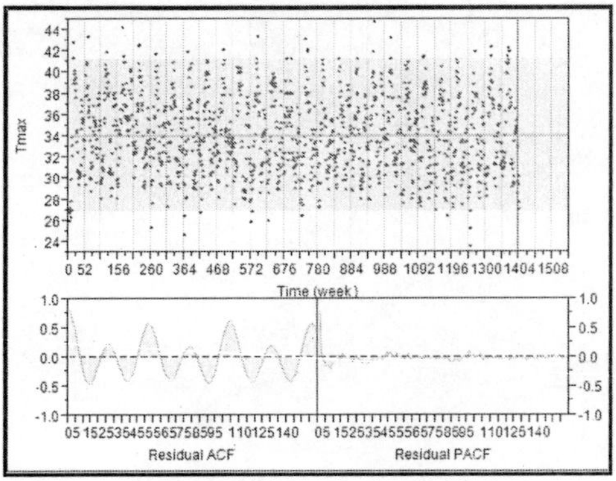

Fig. 1: ACF and PACF of Maximum temperature.

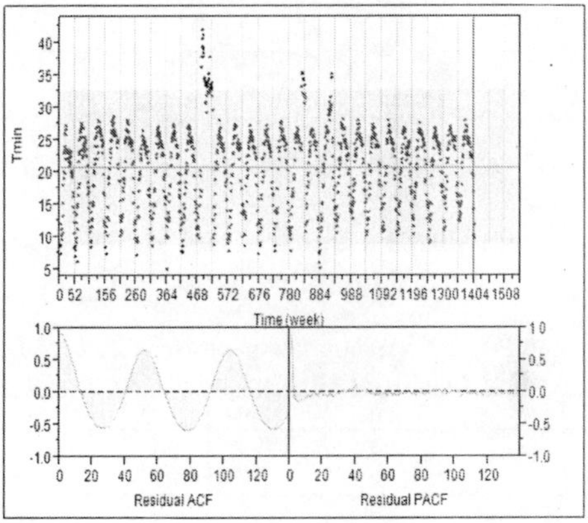

Fig. 2: ACF and PACF of Minimum temperature.

Model selection for T_{max} and T_{min}

The figures of ACF and PACF show that variables of T_{max} and T_{min} are periodic in nature. These functions behave similarly in their period cycles involving seasonal variations. Therefore given time series is periodic and involve seasonal variations. Based on these assumptions a period of 52 weeks i.e. one year is assumed for given time series of maximum temperature and minimum temperature. After making the stationeries, Seasonal ARIMA models of different orders were tried for the series of maximum temperature and minimum temperature. Based on the ACF and PACF chart, AIC value, MAPE and MAE value the model selection was made. Among those Seasonal ARIMA models (4, 0, 0) $(0, 1, 1)_{52}$ and (1, 0, 1) $(2, 1, 1)_{52}$ were found best for maximum temperature (T_{max}) and minimum temperature (T_{min}) respectively. The forecast was made for three years (2011-2013) data of maximum temperature (T_{max}) and minimum temperature (T_{min}) using the selected models for Junagadh region. The ACF and PACF chart of the selected models is shown in the Fig. 3 and Fig. 4. The parameters and statistics information of the selected models are shown in the Table 1 and Table 2.

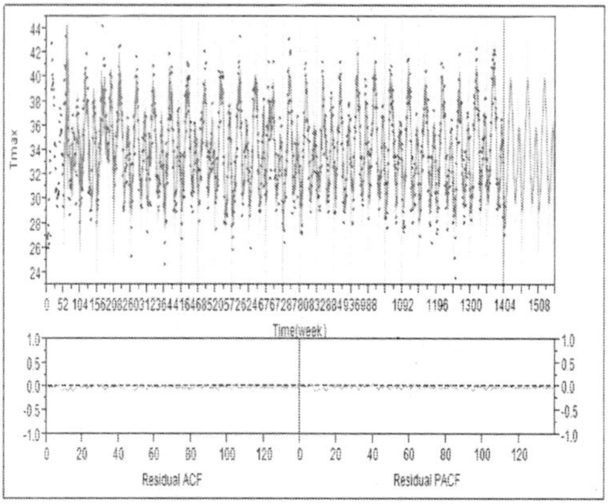

Fig. 3: ACF and PACF in ARIMA models (4, 0, 0) (0, 1, 1).

Table 1: Model Statistics, Parameters and Mathematical Expression for Maximum Temperature (T_{max})

Maximum Temperature (T_{max})						
ARIMA (4,0,0) (0,1,1)52 Parameter Estimates for T_{max}						
Term	F	Lag	Estimate	SE	t Ratio	Prob>|t|
AR 1,1	1	1	0.494	0.027	18.243	1.28E-66
AR 1,2	1	2	0.067	0.030	2.242	0.025
AR 1,3	1	3	0.022	0.029	0.755	0.449
AR 1,4	1	4	0.077	0.027	2.85	0.004
MA2, 52	2	52	0.900	0.021	41.62	4.6E-244
Intercept	1	0	-0.014	0.022	-0.65	0.510

Statistics of Model	
DF	1346
Sum of Squared Errors	4646.77
Variance Estimate	3.452
Standard Deviation	1.858
Akaike's 'A' Information Criterion (AIC)	5605.06
Schwarz's Bayesian Criterion (SBC)	5636.32
RSquare	0.70
Rsquare Adj	0.70
MAPE	4.41
MAE	1.48
-2LogLikelihood	5593.06

Mathematical Expression of the Model

$$(1-0.494B-0.067B^1-0.022B^1-0.077B^1)\ (1-B^1)Yt=-0.005+(1-0.9B^1)$$

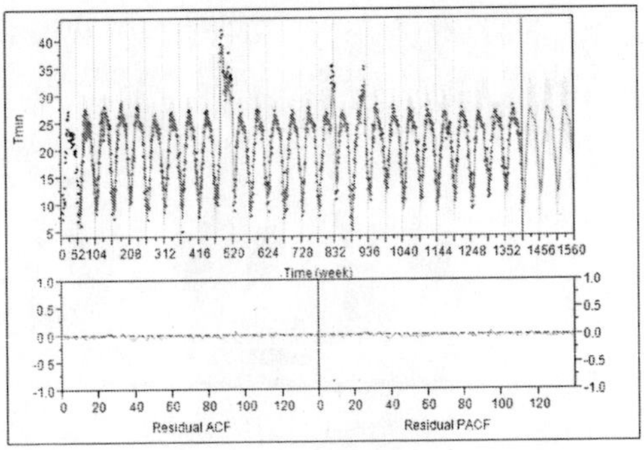

Fig. 4: ACF and PACF in ARIMA models (1, 0, 1) (2, 1, 1)

Table 2: Model Statistics, Parameters and Mathematical Expression for Minimum Temperature (T_{min})

Minimum Temperature (T_{min})						
ARIMA (1,0,1) (2,1,1)52 Parameter Estimates for T_{min}						
Term	F	Lag	Estimate	SE	t Ratio	Prob>|t|
AR1,1	1	1	0.913	0.013	69.66	0
AR2,52	2	52	-0.066	0.031	-2.097	0.036
AR2,104	2	104	-0.074	0.030	-2.39	0.016
MA1,1	1	1	0.339	0.029	11.40	8.19E-29
MA2,52	2	52	0.965	0.052	18.47	4.49E-68
Intercept	1	0	0.043	0.045	0.941	0.346
Statistics of Model						
DF						1346
Sum of Squared Errors						4829.28
Variance Estimate						3.58
Standard Deviation						1.89
Akaike's 'A' Information Criterion						5716.08
Schwarz's Bayesian Criterion						5747.34
R Square						0.89
R square Adj						0.89
MAPE						8.02
MAE						1.34
-2LogLikelihood						5704.08
Mathematical Expression of the Model						
$(1-0.913)(1+0.066B^{52}+0.074B^{104})(1-B^{52})Y_t=0.004+(1-0.339B)(1-0.965B^{52})$						

Note: F=Factor, SE = Standard error, Prob.= Probability.

Figure 5 and Figure 6 shows the comparison of actual and predicted values of maximum temperature (T_{max}) and minimum temperature (T_{min}). From the visual inspection of these graphs we note that predicted time series is very close to the actual data series which indicates that the ARIMA model is valuable tool for forecasting of the two meteorological variables maximum and minimum temperature.

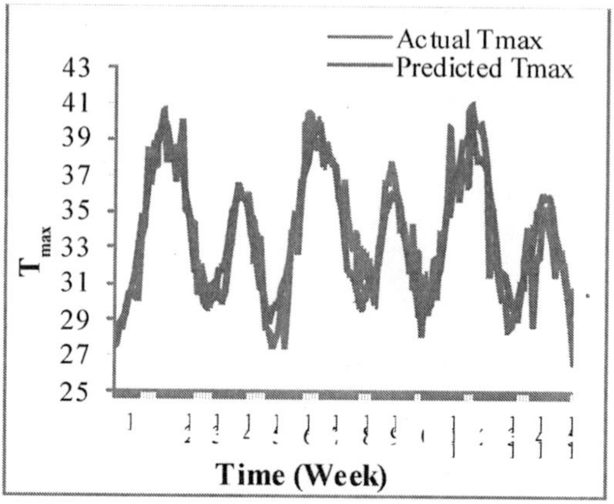

Fig. 5: Comparison of actual and forecasted maximum temperature (T_{max}).

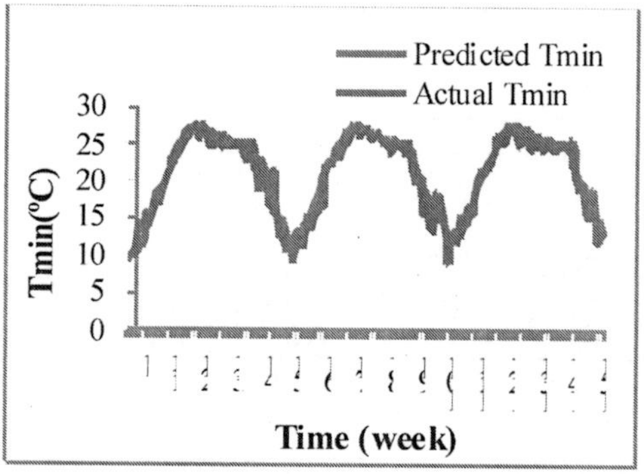

Fig. 6: Comparison of actual and forecasted maximum temperature (T_{min}).

Conclusion

Study indicates that the seasonal ARIMA model is available tool for forecasting the maximum temperature and minimum temperature of Junagadh region. The seasonal ARIMA model can predict minimum and maximum temperature with good accuracy as statistics of model indicate. Thus, the Box-Jenkins methodology can help decision makers to establish better strategies and to set up priorities for arming themselves against upcoming weather changes which may have effects last, but not to the least, on the water resources in the Junagadh region.

References

Bender, M. and Stohodan, S. 1994. Time-series modelling for long-range stream flow forecasting. Journal of Water Resources Planning and Management, ASCE, 120(6): 857-870.

Inderjeet, K., and Singh, S. M., (2008). Seasonal arima model for forecasting of monthly rainfall and temperature Journal of Environmental Research and Development, Vol. 3, No. 2.

Mohan, S. and Arumugam, N., (1995). Forecasting weekly reference evapotranspiration series. Hydrological Science. Vol. 40(6), 689-702.

Nury, A .H., Koch, M., and M.J.B., Alam (2013). Time Series Analysis and Forecasting of Temperatures in the Sylhet Division of Bangladesh.

Popale, P.G., and S.D. Gorantiwar (2014). Stochastic Generation and Forecasting of Weekly Rainfall for Rahuri Region International Journal of Innovative Research in Science, Engineering and Technology An ISO 3297: 2007 Certified Organization Volume 3, Special Issue 4, April 2014 Two days National Conference – VISHWATECH 2014.

Raghuwanshi, N.S. and W.W. Wallender, (2000). Forecasting daily evapotranspiration for a grass reference crop. Agricultural Engineering Journal, 9(1), 1-16.

Seyed, A. S., N. Shahidi, A. Liaghat, A. Sarraf, S.F. Vahdat, (2011). Modeling of Weather Parameters Using Stochastic Methods (ARIMA Model) (Case Study: Abadeh Region, Iran) International Conference on Environment and Industrial Innovation IPCBEE, Vol.12 (2011) IACSIT Press, Singapore.

Subbaiah, R. (2004). Stochastic Modeling of evapotranspiration. Journal of Agro-meteorology. Vol. 6(2), 153-161.

Trivedi, M. M. (2006). Application of ARIMA model for forecasting of Mahi River inflows. A Ph.D thesis submitted to Maharana Pratap University of Agriculture and Technology, Udaipur.

Wang, Y., W. Namaona, Traore, S. and Zhang, Z. (2009). Seasonal temperature based model for reference evapotranspiration estimation under semi-arid condition of Malawai. African Journal of Agricultural Research. Vol. 4(9), 878-886.

13

An Assessment of Potential Surface and Groundwater Resources of Bhadar River Basin of Gujarat State using SWAT Model

J.H. Kelaiya, D.M. Paradava, R.J. Patel, H.V. Parmar, H.H. Mashru, P.B. Vekariya and H.D. Rank

Abstract

The assessment of surface and ground water resources will help to prepare a future plan for ground water development and management planning for the basin. It will also be helpful for agricultural as well as other resources planning. The required shift in cropping pattern can also be judged. Therefore, keeping this in views, present study was undertaken for the Bhadar river basin having an area of 8593.46 km^2. The entire basin was found consisting of 26 sub-basins. The Soil and Water Assessment Tool (SWAT) having an interface with ArcView GIS software (AVSWAT2012) was used for the assessment of surface and groundwater resources of Bhadar river. Daily CFSR data (1979-2010) of the various stations falling in the basin with the respective longitude and latitude for the 32 years were fed into SWAT to build the model in GIS environment. The potential surface water resources in the Bhadar basin can be 560 MCM. These surface water potential can irrigate the 1117 sq.km. cropped area in a season which is 25 per cent of the cultivated area in the basin. The potential groundwater resources is 370 MCM which can command 740 sq.km cropped area for one season. The potential surface and ground water resources in the basin may be increased due to climate change impacts at 410 MCM (254 MCM surface and 156 MCM groundwater) per decade will can increase the irrigated agriculture at 82000 ha per decade.

Introduction

The precise assessment of surface and groundwater resources are required for the resources management planning on basin scale. A number of models have been developed to evaluate water resource system at both field and watershed scale. The widely used field scale models include CREAMS (Chemicals, Runoff, Erosion from Agricultural Management Systems) and watershed scale models include continuous daily time step model SWRRB (Simulator for Water Resources in Rural Basins). Expansion of SWRRB model's capacities to facilitate more subbasins and sophisticated routing structure resulted in a new watershed scale model SWAT (Soil and Water Assessment Tool). In recent years, SWAT model developed by Arnold *et al.,* (1998) has gained international acceptance as a robust interdisciplinary watershed modelling. SWAT is currently applied worldwide and considered as a versatile model that can be used to integrate multiple environmental processes, which support more effective watershed management and the development of better informed policy decision (Gassman *et al.,* 2005). But little have been published on the application of SWAT model for assessing surface and groundwater resources in the arid and semi-arid area catchments particularly in India. Similar studies in Tanzanian catchments using SWAT model (Birhanu, 2005; Ndomba *et al.,* 2005) which recommended further testing and customising SWAT model for different climatic conditions. Thus, this study examines the applications of SWAT model for estimating runoff and groundwater recharge.

Study area

Bhadar is one of the major rivers of Saurashtraand (Fig. 1) it drains about 1/7th of the area of Saurashtra. The Bhadar basin is the South Western basin and situated between 21°25' to 22°10′ North latitude and 69°45′ to 71°20′ East longitude. The river Bhadar originates at an elevation of 261 m above m.s.l. in Vaddi about 26 km North –West of Jasdan in Rajkot district and flows into Arabian sea at Navibandar (Porbandar).Thus from Jetpur to Porbandar, the River Bhadar fertile Rajkot, Jamnagar, Amreli and Junagarh districts of Saurashtra. The river Bhadar drains an area of 8593.46 km^2 out of which 706 km^2 in hilly and the rest in plain regions of Saurashtra.

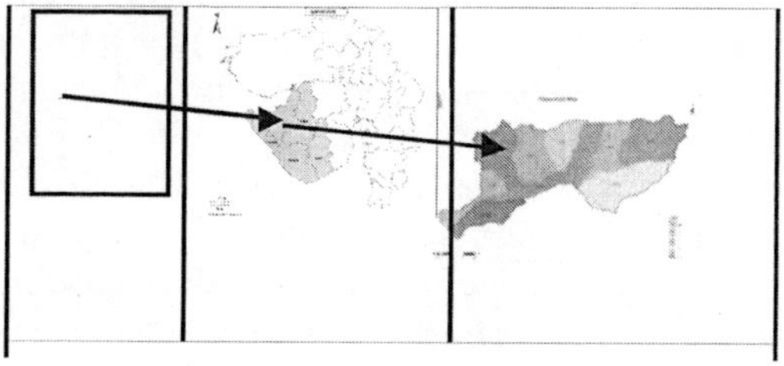

Fig. 1: Location map of Bhadar river Basin.

Data collection

The maps of India, Gujarat and watershed along with cadastral map of Bhadar basin and Satellite images of IRS P6 of sensor LISS IV and Awifs digital data, SRTM 90 m DEM along with land use and soil map were collected from BISAG, Gandhinagar. The daily weather data records (1979-2010) were collected from National Centers for Environmental Prediction (NCEP) Climate Forecast System Reanalysis (CFSR). The CFSR was designed and executed as a global, high resolution, coupled atmosphere-ocean-land surface-sea ice system to provide the best estimate of the state of these coupled domains over this period.

Methodology

SWAT is a basin-scale, continuous model that operates on a daily time step. It is designed to predict the runoff, groundwater recharge and flow in watersheds for various inputs of weather parameters and is capable for large, complex watersheds with variable land uses, elevations and soils. The model is physically based, computationally efficient and capable of continuous simulation over long periods. In SWAT, an entire basin/watershed is divided into sub-basins, which are then further subdivided into hydrologic response units (HRUs) on the basis of unique combinations of land use, soil and slope class. Climatic data can be input from measured records or generated using the weather generator (or any combination of the two). Hydrological computations are performed at the level of each HRU. They are summed to the subbasin level and routed through channels, ponds, wetlands or lakes to the watershed outlet. Hydrology in SWAT is based on the following water balance.

The water balance equation that represents the hydrologic cycle simulated in SWAT can be expressed mathematically as (Neitsch *et al.*, 2005):

$$SW_t = SW_0 + \sum_{i=1}^{t}\left(R_{day} - Q_{surf} - E_a - W_{seep} - Q_{gw}\right) \quad \ldots(1)$$

where, SW_t is the soil water content at time t (mm); SW_0 is the initial soil water content on the day i (mm); t is time (days); R_{days} is the amount of precipitation on the day i (mm); Q_{surf} is the amount of surface runoff on day i (mm); E_a is the amount of evapotranspiration on day i (mm); W_{seep} is the amount of water entering the vadoze zone from the soil profile on day i (mm); and Q_{gwis} the amount of return flow on day i (mm). Overland flow runoff volume is computed using the Natural Resources Conservation Service (NRCS) curve number method. Curve numbers are a function of hydrologic soil group, vegetation, land use, cultivation practice and antecedent moisture conditions. SWAT accounts for sediment contributions from overland runoff through the Modified Universal Soil Loss Equation.

The minimum required meteorological time series for SWAT simulations are daily precipitation, and daily maximum and minimum air temperature. Potential evapotranspiration (PET) is also required for simulation, but the model estimates it directly using one of several options. For the Champlain watershed model, SWAT estimates Penmann-Monteith PET using a statistical weather generator for inputs other than temperature and precipitation. The Penman-Monteith energy balance method of estimating PET was selected for this model. A total of 36 precipitation stations were identified for use in the basin model with a common period of record of January 1, 1979, to December 31, 2010. Temperature records are sparser; where these were missing, the temperature was taken from nearby stations with an elevation correction.

The spatial characteristics of the basin were furnished through digital images namely, SRTM 90 m DEM, land use and soil map. The temporal and spatial weather variability (1979-2010) were also fed. The 32-year times series of SWAT simulated runoff, groundwater recharge, evapotranspiration, reference evapotranspiration and stress coefficient were analysed.

Results and Discussion

Surface water

The daily weather data of the various stations falling in the basin with the respective longitude and latitude for the 32 years were fed into SWAT to build the model in GIS environment. The skewness and kurtosis coefficient of the spatial average runoff series was found as 0.89 and -0.09 respectively. Similarly for the rainfall, it was observed as 0.43 and -0.5 respectively. The values of the skewness and kurtosis coefficient indicated that both rainfall and runoff time series follows nearly normal distribution. Also the mean (751 mm) and median (768 mm) were nearly equal. The results on the rainfall and runoff relationships of simulated period are depicted in Fig. 2 and Table 1.

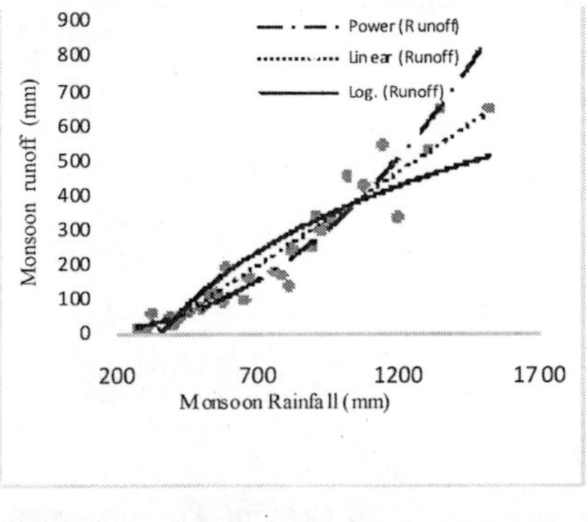

Fig. 2: Rainfall-runoff relationships for Bhadar basin using SWAT model.

The results showed that the power form of the model is the best model describing the rainfall runoff relationships followed by linear and logarithm from. The linear model showed that the harvesting the runoff for the surface water resources development are only possible if the seasonal rainfall is more than 176 mm. The average seasonal runoff for the harvesting through the surface water resources storage structures in the bazaar basin can be 225 mm for the rainfall of 751 mm. The Fig. 3 show the runoff amount for the various probability level. It showed that at 75 per cent dependability level, the runoff can be obtained by the tune of 65 mm.Therefore, the potential surface water resources in the Bhadar basin can be 560 MCM at 75 per cent dependability level. These surface water potential can irrigate the 1117 sq.km. cropped area in a season which is 25 per cent of the cultivated area in the basin. There lies a scope for the surface water resources development in the basin.

Table 1: Rainfall-Runoff models

SN	*Model*	R^2
A	Rainfall-Runoff relationships	
1	$R = 9E\text{-}05(P)^{2.1876}$	0.93
2	$R = 0.53(P) - 176$	0.91
3	$R = 348\ln(P) - 2043$	0.81

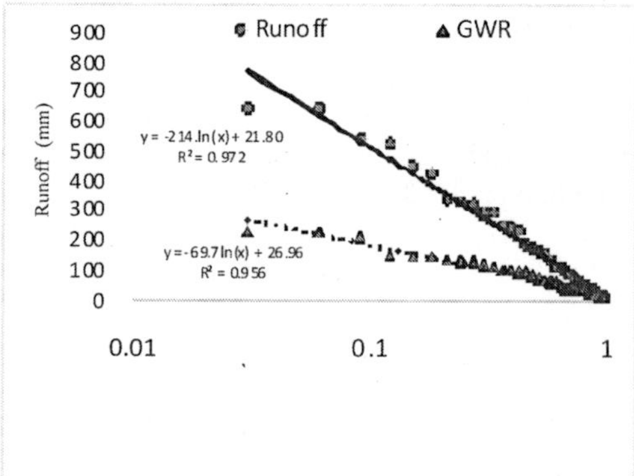

Fig. 3: Probability curve for the runoff from Bhadar basin.

Groundwater

The average groundwater recharge during the study period (1979-2010) was found as 93 mm from the rainfall of 751 mm. The median of the groundwater recharge (83 mm) was close to mean values. The skewness and kurtosis coefficient of the groundwater recharge was found as 0.8 and -0.01 indicating that times series of groundwater recharge in the basin is close to normal distribution. The relationship between rainfall and groundwater recharge is depicted in Fig. 4 and Table 2.

The results showed that the rainfall and SWAT simulated groundwater recharge relationships can be best described by power form of the model followed by linear and log form. The linear model indicates that the groundwater recharge starts only after rainfall exceeds 32.6 mm at 17 per cent of the rainfall excess. The groundwater recharge estimated by SWAT seems somewhat higher than that of by water table fluctuation method as the former does not exclude the moisture storage below field capacity in vadozzone (Kelaiya, 2014). The groundwater recharge at different probability level was found as depicted in Fig. 3. It shows that the groundwater recharge at 75 per cent confidence level was found as 43 mm. The potential groundwater resources in the Bhadar basin is 370 MCM which can command 740 sq.km. cropped area for one season at 75 per cent dependability level.

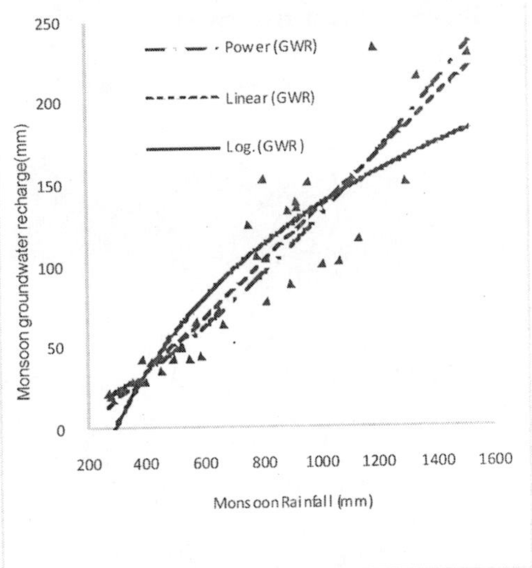

Fig. 4: Rainfall-groundwater recharge relationships for Bhadar basin using SWAT model.

Effect of rainfall on potential/crop evapotranspiration and moisture stress coefficient

The SWAT simulated reference evapotranspiration, crop evapotranspiration and stress coefficient for the Bhadar basin is depicted in Fig. 5. It can be seen that the reference evapotranspiration decrease with increase in rainfall while crop evapotranspiration increase with rainfall due to

Table 2: Rainfall-Groundwater recharge model

SN.	*Model*	R^2
B	Rainfall-Ground water recharge relationships	
1	$GWR = 0.006(P)^{1.4423}$	0.91
2	$GWR = 0.17(P) - 32.6$	0.83
3	$GWR = 112\ln(P) - 635$	0.77
F	Rainfall-adjusted crop coefficient	
1	$Kcadj = 0.0002(P) + 0.24$	0.59

reduction in moisture stress. Also, the adjusted crop coefficient (K_{cadj}) increased with reduction in moisture stress which is quite obvious. The mathematical relationships for the effect of rainfall on reference evapotranspiration (PET), crop evapotranspiration (ETc) and moisture stress coefficient (Ks) depicted in Table 3 also reflect the same. Similar results were found by Paghdal (2014) also for Ozat river basin.

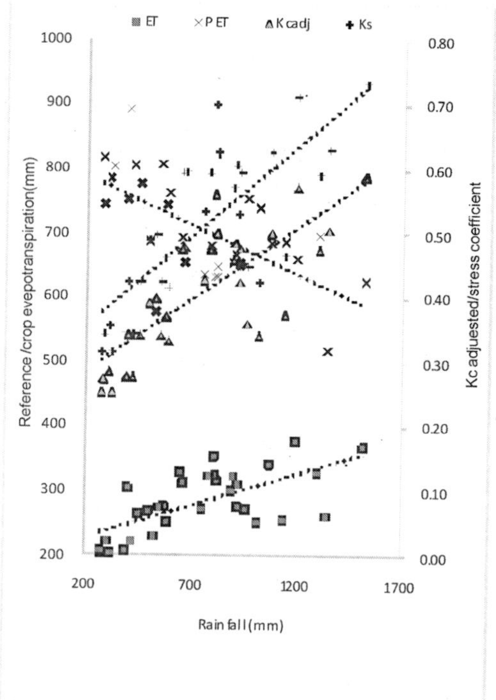

Fig. 5: Effect of rainfall in reference evapotranspiration, crop evapotranpiration adjusted crop coefficient and stress coefficient.

Climate change impacts on water resources

The climate change is now being a word wide fact, its impacts on the catchment hydrology is also detected in world including India (Bates *et al.*, 2008; Gosain and Rao, 2007). So, an effort was made to assess the climate

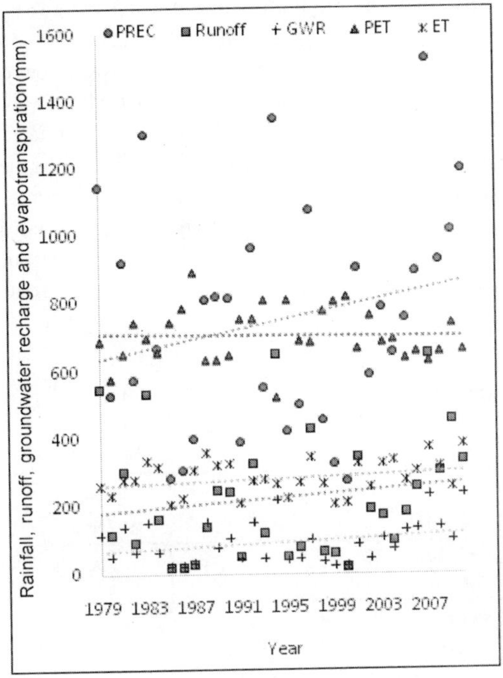

Fig 6. The trend of rainfall, runoff, and evapotranspiration.

Table 3

SN.	*Model*	R^2
C	Rainfall-Reference evapotranspiration	
1	$ET_o = -0.15(P) + 817$	0.41
D	Rainfall-Crop evapotranspiration	
1	$ET_c = 0.1(P) + 207$	0.44
E	Rainfall-Stress coefficient	
1	$Ks = 0.0003(P) + 0.3$	0.59
F	Rainfall-adjusted crop coefficient	
1	$Kcadj = 0.0002(P) + 0.24$	0.59

change impacts on catchment hydrology which directly affects the water resources components. The results of the trend analysis of rainfall, runoff, and potential/crop evapotranspiration is depicted in Table 4 and Fig. 6. Table 4: Trend statistics of rainfall, runoff, potential/crop evapotranspiration.Table 3: Rainfall- evapotranspiration and stress coefficient relationships.

The results shows that the potential surface water resources in the basin can be increased at 29.6 mm/decade (254 MCM per decade). The groundwater recharge is increasing at rate of 18.2 mm/decade (156 MCM per decade). The total water resources in the basin will increase 410 MCM/year which can increase the irrigated agriculture 82000 ha/decade. The increased in irrigated agriculture can enhance the carbon sequestration which ultimately help to reduce the warming. The potential evapotranspiration is decreasing may be because of increased relative humidity due to increasing rainfall in the basin. However, even though the decreasing potential evapotranspiration, the crop evapotranspiration is found increasing because of increasing trend of rainfall.

Table 4

Parameter	*best fit trend slope (mm/year)*	*Sense slope (mm/year)*
Rainfall(PREC)	7.4*	10.141
Run off(RO)	2.96	3.759
Groundwater recharge(GWR)	1.82*	1.783
Reference evapotranspiration (PET)	-0.23	-0.152
Crop evapotranspiration (ETc)	1.37	1.259
*significant		

Conclusions

The potential surface and ground water resources in Bhadar basin can be 560 MCM and 370 MCM which can irrigate 1860 sq.km. cropped area in a season. The climate change impacts is favourable for the surface and groundwater resources of the basin.

Acknowledgements

The authors are thankful to the Directorate of Water Management, ICAR, Bhubaneswar for the financial supports under NICRA project.

References

Arnold, J.G., R. Srinivasan, R.S. Muttiah, and J.R. Williams. 1998. "Large area hydrologic modelling and assessment part I:Model development", J. of Amer. Water Res. Assoc. 34(1): 73:89.

Bates, B.C., Z.W. Kundzewicz, S. Wu and J. P. Palutikofeds. 2008. "Climate change and water", Technical Paper of the Intergovernmental Panel on Climate Change, IPCC Secretariat, Geneva, Switzerland.

Birhanu, B.Z., 2005. "Application of SWAT model in simulating the available water resources in Pangani river basin upstream of NYM reservoir", A case study of 1DD1 catchment, M.Sc Thesis, University of Dar es Salaam, Tanzania, 2005.

Gassman, W.P., M. R. Reyes, C.H. Green and J.G. Arnold. 2005. "SWAT peer – reviewed literature", A review, Proceedings of the 3rd International SWAT Conference, Zurich, 2005.

Gosain, A.K. and S. Rao. 2007. "Impact assessment of climate change on water resources of two river systems in India", Jalvigyan Sameeksha. 22: 21.

Kelaiya, J H., 2014. "Impact of climate change on groundwater recharge in Bhadar basin", Unpub. M Tech Thesis, sub to JAU, Junagadh.

Ndomba, P.M., F.W. Mtalo and A. Killingtveit. 2005. "The Suitability of SWAT model in sediment yield modelling for ungauged catchments", A case study of Simiyu river basin catchments, Tanzania. Proceedings of the 3rd International SWAT conference, Zurich, 2005.

Neitsch, S.L., J. G. Arnold, J. R. Kiniry, R. Srinivasan and J. R.Williams. 2005. "Soil and Water Assessment Tool Input/Output File Documentation", Ver.2005. Available at: www.brc.tamus.edu/swat/doc.html.

Paghdal, A. M., 2014. "An assessment of surface and groundwater resources of Ozat river basin using remote sensing and GIS",Unpub. M Tech Thesis, sub to JAU, Junagadh.

14

A Systematic Month-Wise Trends of Temperature and Rainfall in Junagadh (Saurashtra Region) of Gujarat, India: A Study

Bhavin Ram, S.S. Chianchorkar, A.N. Kunapara, F.G. Sayyad and M.S. Khardiwar

Abstract

Climate change has brought in unexpected changes not only in India but all over the regions across the world. Emergence of global warming due to climate change is the new and most talked subject of today's world as it being the most threatening issue for very existence of life on the earth. One of the consequences of climate change is the alteration of rainfall patterns and increase in temperature. The temperature and rainfall trends are analysed for meteorological data of Junagadh (Saurashtra) in Gujarat, India over approximately last three decades stretching between years 1980 to 2011. This implies that in Junagadh (Saurashtra) the highest increase in MMAX temperature occurs in November (0.119°C) during last 35 years. The annual mean of monthly mean of maximum temperatures observed an increasing trend having an annual increase of 0.017°C per year. It is evident from above figures that monthly mean of TMRF have increased significantly for the months January, February, March, April, May, July, August, September, and decreasing trend in June, October, November and December for the Junagadh (Saurashtra). This implies that in Junagadh district the highest increase in TMRF occurs in September and has increased during the last 35 years. The highest decrease in TMRF occurs in November and decreased during 35 years. Annual mean of monthly mean of total mean rainfall observed an increasing trend having an increase of 0.35 mm per year. This implies that in

Junagadh (Saurashtra) annual TMRF has increased during the last 35 years. Annual TMRF shows increasing trend which is statistically insignificant at five per cent level of significance.

Introduction

Rainfall plays an important role in agricultural production. It has a profound influence on the growth, development and yields of a crop, incidence of pests and diseases, water needs and fertiliser requirements in terms of differences in nutrient mobilisation due to water stresses and timeliness and effectiveness of prophylactic and cultural operations on crops. However, it is possible to adapt to or mitigate the adverse effects if a forecast of the rainfall can be had in time. Nearly 70 per cent of total cultivated land in India depends on rainfall for assured crop yield. As far as agricultural planning is concerned, the importance of weather forecasting needs no over emphasis. Weather during crop seasons strongly influences the crop growth and development. The variation in crop productivity is mainly due to weather fluctuations. Weather and climate are the important factors determining the growth, development and yield of crops. Weather influences the agricultural operations and experimental treatments from sowing of a crop to the harvest and storage. Crop yields are the integrated result of external and internal processes that occur during the crop growing period. The external environment is the climate which regulates and the weather determines the growth and development and finally the yield of the crop. It is therefore, that the meteorological parameters are recorded in all agricultural experimental stations. The interpretation of experimental results in the light of weather conditions prevailing during the period of crop growth is essential. Weather observations are also necessary for accurate weather forecasting and comparison of forecasted weather.

Climate change has brought in unexpected changes not only in India but all over the regions across the world. Emergence of global warming due to climate change is the new and most talked subject of today's world as it being the most threatening issue for very existence of life on the earth. One of the consequences of climate change is the alteration of rainfall patterns and increase in temperature. According to Intergovernmental Panel on Climate Change (IPCC) reports, the surface temperature of the earth has risen by 0.6+ 0.2°C over the 20th century. Also in the last 50 years, the rise in temperature has been 0.13 + 0.07°C per decade. As the warming depends on emissions of GHGs in the atmosphere, the IPCC has projected a warming of about 0.2°C per decade. Further, surface air temperature could rise between 1.1°C to 6.4°C over 21st century. In case of India, the climate change expected to adversely affect its natural resources, forestry, agriculture, and change in precipitation, temperature, monsoon timing and extreme events (Fulekar, M.H., *et.al, 2010)*. Due to global warming, precipitation amount, type and timing are changing or are expected to change because of increased evaporation, especially in the tropics (Ritter). The pattern and amount of rainfall are among the most important factors that affects agricultural production. Agriculture is vital to India's economy and the livelihood of its people. Agriculture is contributing 21 per cent to the country's GDP, accounting for 115 of total export, employing 56.4 per cent of the total workforce, and supporting 600 million people directly and indirectly (Beena Shah, 2010).

Karl *et.al.* analysed monthly mean maximum and minimum temperatures from countries comprising 37 per cent of the global landmass and found that the minimum temperature increased over the period 1951-1990 by 0.84°C compared to only 0.28°C increase in maximum temperatures. Hasanean investigated the trends and periodicity of surface air temperatures series from eight meteorological stations in the east Mediterranean using different correlation tests. Shafiqur Rehman *et.al.* analysed extreme temperature trends for a meteorological data collection station in Jeddah, Saudi Arabia over four decades during 1970 and 2006. Del Rio *et.al.* presented the analysis of mean, minimum and maximum temperatures data from 171 stations in Spain on monthly, seasonal, and annual time scales and they observed that mean, minimum and maximum temperatures increases in all months of the year. Julius M. Huho *et.al.* examined the changing rainfall pattern during the main growing season (March, April and May) and associated effects on subsistence agriculture in Laikipia East District of Kenya and observed rainfall intensities declined in March but increased in April and May. Jestinos Mzezewa *et. al.* analysed 23 years (1983 to 2005) of rainfall data in order to study the basic statistical rainfall characteristics at the University of Venda ecotype and found that the distribution of daily rainfall was highly skewed with high frequency of occurrence of low rainfall events.

Data and Methodology

The data used in this paper are the monthly averages of total mean rainfall, minimum and maximum atmospheric temperatures during 1980-2011. The yearly averages were calculated from the monthly readings. The time series is made up of four components known as seasonal, trend, cyclical and irregular (Patterson). Trend is defined as the general movement of a series over an extended period of time or it is the long-term change in the dependent variable over a long period of time (Webber and Hawkins). Trend is determined by the relationship between the two variables as temperature and time, rainfall and time. The statistical methods such as regression analysis and coefficient of determination R^2 (Murray R. Spiegel, Larry J. Stephens,) are used. The magnitudes of the trends of increasing or decreasing maximum temperatures and total mean rainfall were derived and tested by the Mann-Kendall (M-K) trend test and slope of the regression line using the least squares method.

Study Area

The Junagadh city is located between latitudes 21° 31´N and 70° 49´ E. The city is a gateway to famous Gir Forest which is the natural habitat for the last existing population of Asiatic Lion in the wild. Apart from Gir, there is Girnar Ranges, Barda Hills and extensive grasslands known as Vidis, which also support a variety of wildlife especially avifauna. Junagadh has a tropical wet and dry climate, with three distinct seasons observed, a mild winter from November to February, a hot summer from March to June, and a monsoon from July to October. Junagadh faces adverse climatic conditions in the summer months with the temperature ranging from 28° Celsius to 38° Celsius. In the winter months, the temperature ranges from 10° Celsius to 25° Celsius. The city is drained by the south-west monsoon from June to September. It records a rainfall of 1000 to 1200 mm annually. Various factors such as its close proximity to the sea influence the weather of Junagadh. The latent winds from the sea affect the climatic conditions in the region (Beena Shah).

Results and Discussion

The coefficient of variation for MMAX temperature is highest in the month of January and it is observed as 6.839 per cent whereas it is lowest in the month of October and it is 3.587 per cent for the Junagadh (Saurashtra). This means that maximum temperature is most stable in the month of June (43° C) and least stable in the month of March (40.4°C) for the Juangadh (Saurashtra) show in Table. 1.

Table 1: Statistical Summary of Monthly Mean of MMAX Temperatures.

Month	*Mean`*	*Stnd. Devi. (per cent)*	*Mean Devi. (per cent)*	*Coeff.of Variance (per cent)*
Jan	33.5	2.291	1.446	6.839
Feb	36.6	1.905	1.269	5.201
Mar	40.4	1.822	1.369	4.508
Apr	42.3	1.528	1.014	3.592
May	42.9	2.007	1.360	4.681
Jun	43.0	1.974	1.323	4.596
Jul	35.4	1.674	1.145	4.731
Aug	33.0	1.662	1.049	5.033
Sept	36.1	1.531	1.079	4.243
Oct	38.3	1.375	0.813	3.587
Nov	36.4	1.590	1.102	4.368
Dec	34.4	1.907	1.385	5.546

Trend Analysis of Monthly Mean of Maximum Temperature (MMAX)

The trends of monthly mean of maximum temperatures over different years were obtained using linear regression best fit lines. The linear regression trends with their linear regression equations and coefficient of determinations for all the months from January to December are represented in (Fig.1) and summarised in Table. 2 below.

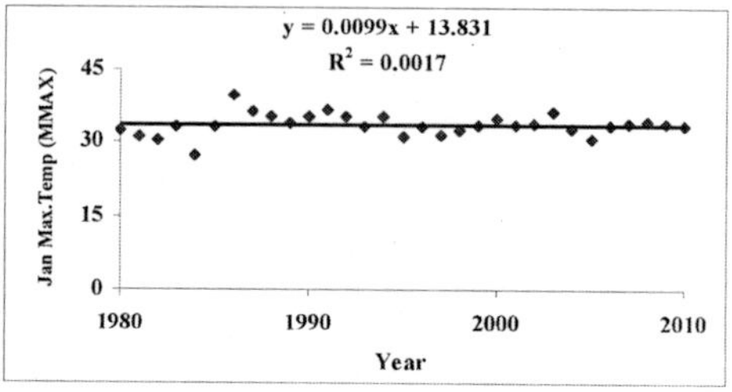

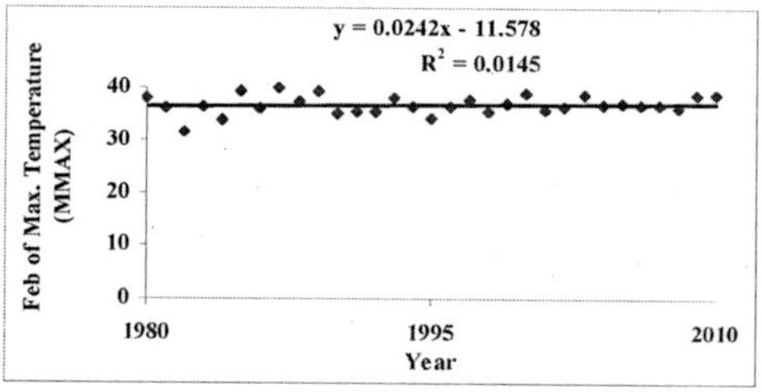

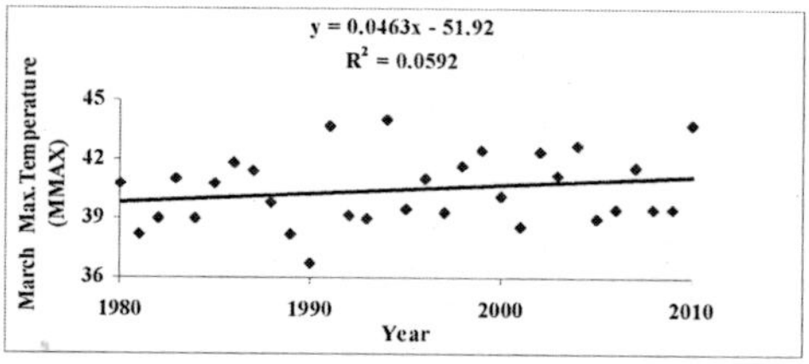

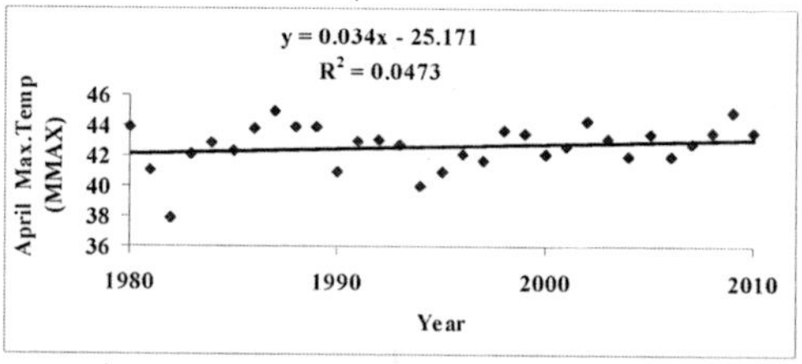

(*contd...*)

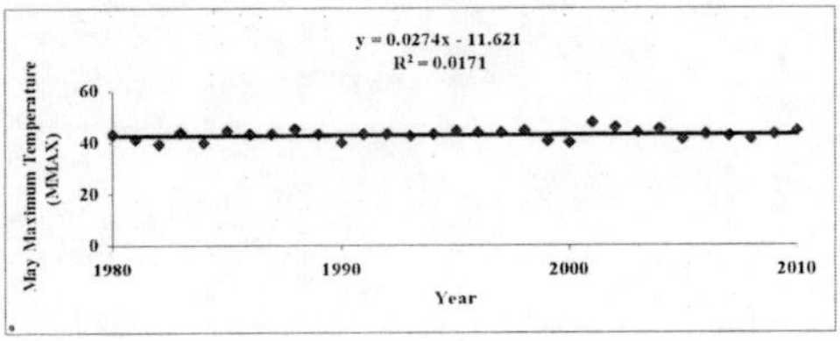
y = 0.0274x - 11.621
R² = 0.0171
May Maximum Temperature (MMAX)
60
40
20
0
1980
1990
2000
2010
Year

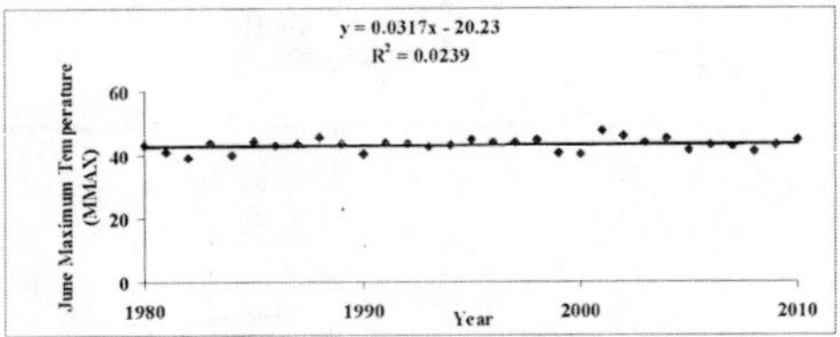
y = 0.0317x - 20.23
R² = 0.0239
June Maximum Temperature (MMAX)
60
40
20
0
1980
1990
2000
2010
Year

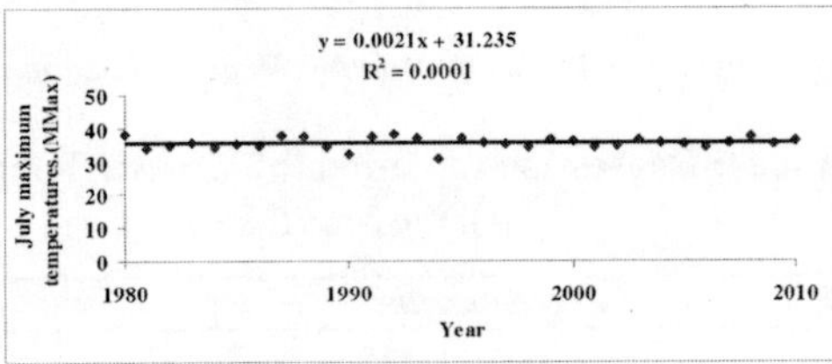
y = 0.0021x + 31.235
R² = 0.0001
July maximum temperatures.(MMax)
50
40
30
20
10
0
1980
1990
2000
2010
Year

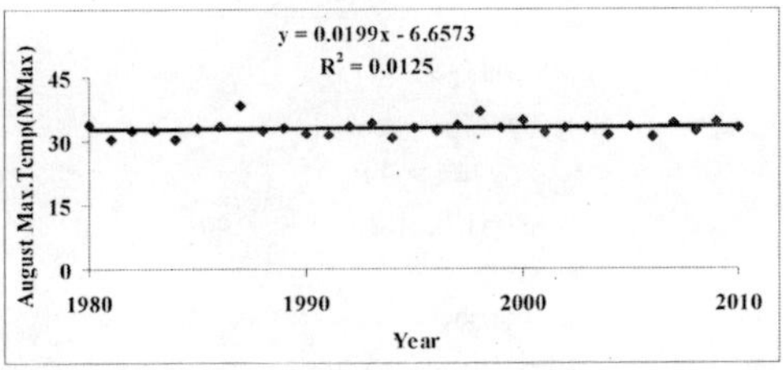
y = 0.0199x - 6.6573
R² = 0.0125
August Max.Temp(MMax)
45
30
15
0
1980
1990
2000
2010
Year

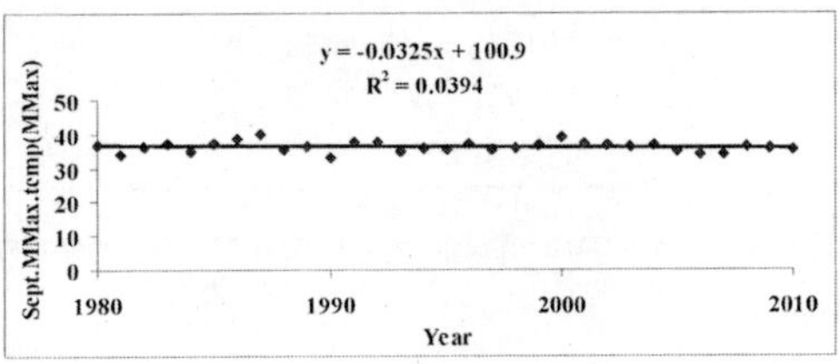
y = -0.0325x + 100.9
R² = 0.0394
Sept.MMax.temp(MMax)
50
40
30
20
10
0
1980
1990
2000
2010
Year

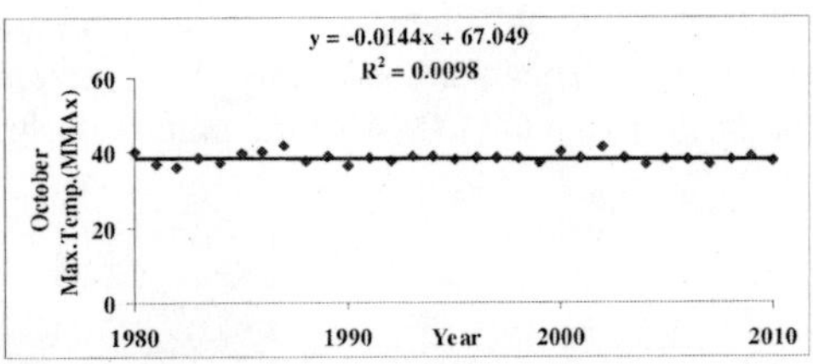
y = -0.0144x + 67.049
R² = 0.0098
October Max.Temp.(MMAx)
60
40
20
0
1980
1990
Year
2000
2010

(contd...)

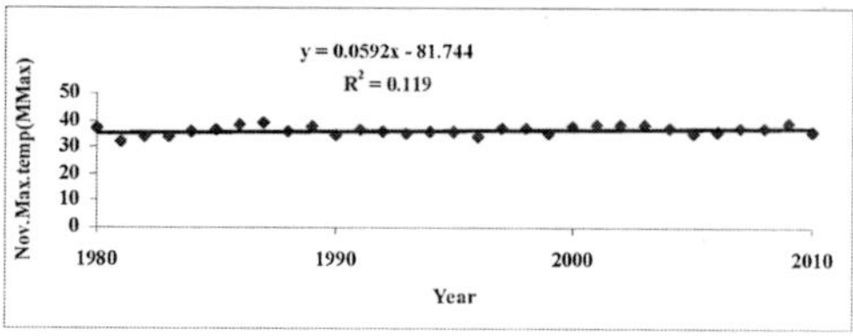

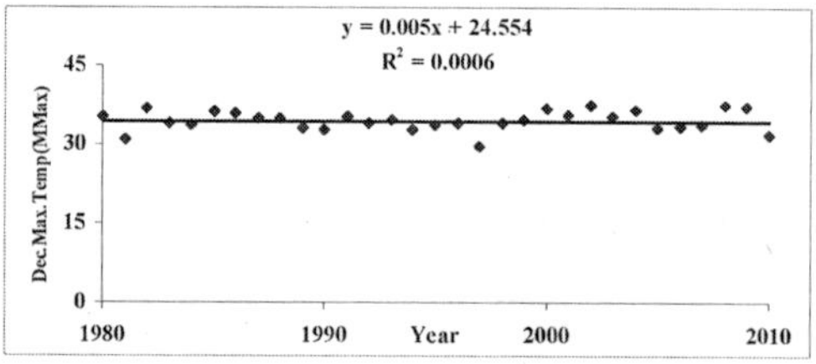

Fig.1: Linear Regression Trends of Monthly Mean of Maximum Temperatures.

Table 2. Linear Regression Equations of MMAX Temperatures for All the Months

Month	*Regression line*	*R^2*
Jan	y = 0.0099x + 13.831	0.0017
Feb	y = 0.0242x - 11.578	0.0145
Mar	y = 0.0463x - 51.92	0.0592
Apr	y = 0.034x - 25.171	0.0473
May	y = 0.0274x - 11.621	0.0171
Jun	y = 0.0317x - 20.23	0.0239
July	y = 0.0021x + 31.235	0.0001
Aug	y = 0.0199x - 6.6573	0.0125
Sep	y = -0.0325x + 100.9	0.0394
Oct	y = -0.0144x + 67.049	0.0098
Nov	y = 0.0592x - 81.744	0.119
Dec	y = 0.005x + 24.554	0.0006

Trend Analysis of Annual Mean of monthly maximum Temperature (MMAX)

The annual mean of monthly mean of maximum temperatures observed an increasing trend having an annual increase of 0.017°C per year, this implies that in Junagadh (Saurashtra) annual MMAX temperature during the last 35 years (Figure 2).

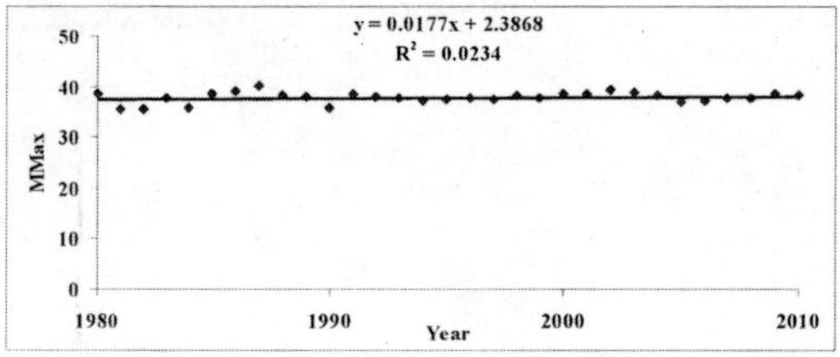

Fig. 2: Trend of Annual Mean of Monthly Maximum Temperature.

Table 3: Statistical Summary of Monthly Mean of Total Mean Rainfall (TMRF)

Month	*Mean*	*Stnd Devi (per cent)*	*Mean devi.*	*Coeff.of variation (per cent)*
Jan	1.74	8.865	1.619	508.774
Feb	0.05	0.219	0.0429	455.239
Mar	0.39	1.408	0.0334	360.841
Apr	2.24	8.125	1.92	362.870
May	5.75	13.209	8.229	229.716
Jun	213.1	268.23	153.016	125.848
July	361.3	236.17	137.912	65.369
Aug	217.4	163.65	128.351	75.252
Sep	361.10	241.22	136.853	66.803
Oct	38.78	63.755	41.6574	164.411
Nov	53.79	75.392	51.62394	140.148
Dec	0.11	0.410	0.14583	368.044

The coefficient of variation for TMRF observed highest in the month of January and it is 508.77 per cent whereas coefficient of variation is minimum for the month of July and it is 65.369 per cent for the Junagadh (Saurashtra) in Table 3. This shows that rainfall is more stable in the month of July and is more variable in the month of April for the Junagadh (Saurashtra).

Trend Analysis of Monthly Mean of Total Mean Rainfall (TMRF)

The trends of monthly mean of total mean rainfall over different years were obtained using linear regression best fit lines. The linear regression trends with their linear regression equations and coefficient of determinations for all the months from January to December are represented in (Figure 3) and summarised in Table 4. It is evident from above figures that monthly mean of TMRF have increased significantly for the months of January, February March, April, May, July, August, September, and decreasing trend in June, October, November and December for the Junagadh (Saurashtra). This implies that in Junagadh district the highest increase in TMRF occurs in September and has increased during the last 35 years. The highest decrease in TMRF occurs in November and decreased during 35years.

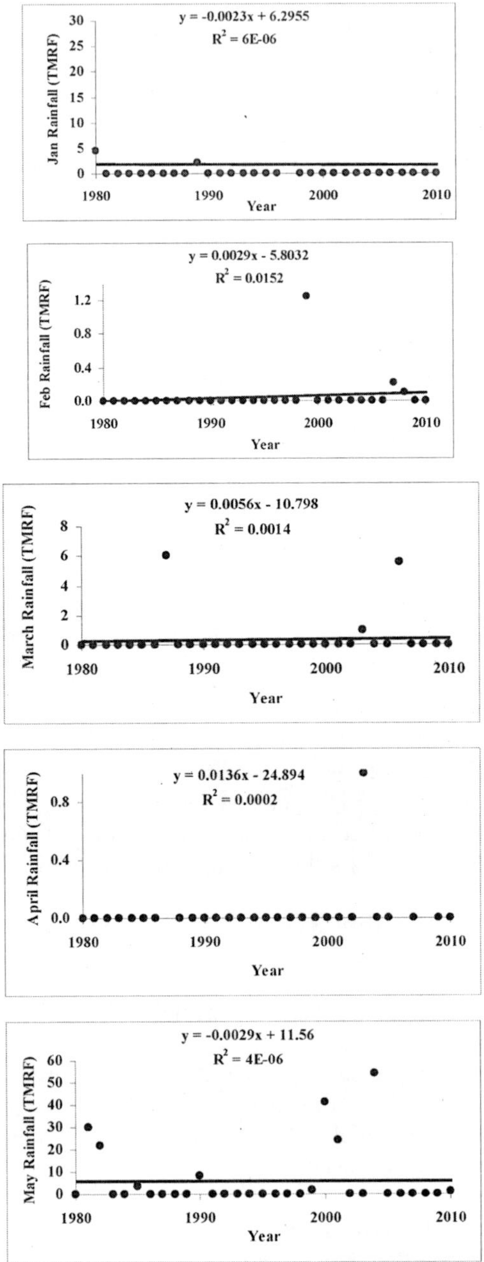

(contd...)

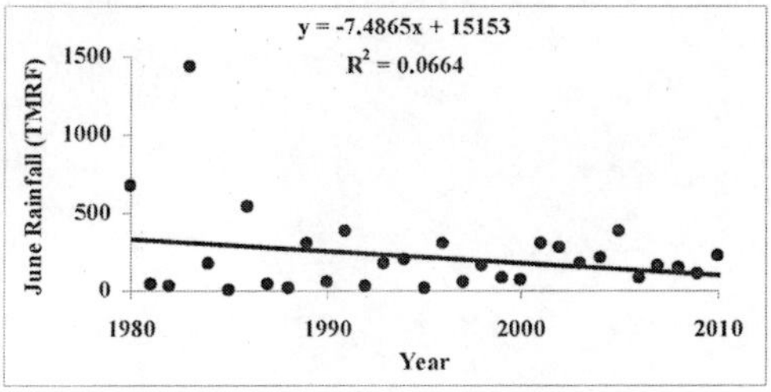
y = -7.4865x + 15153
R^2 = 0.0664
June Rainfall (TMRF)
Year

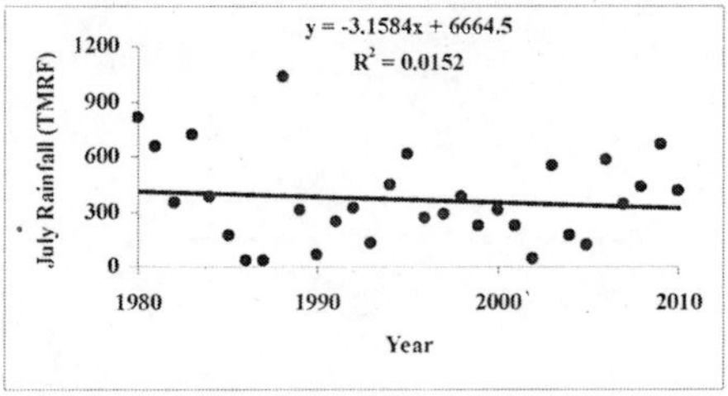
y = -3.1584x + 6664.5
R^2 = 0.0152
July Rainfall (TMRF)
Year

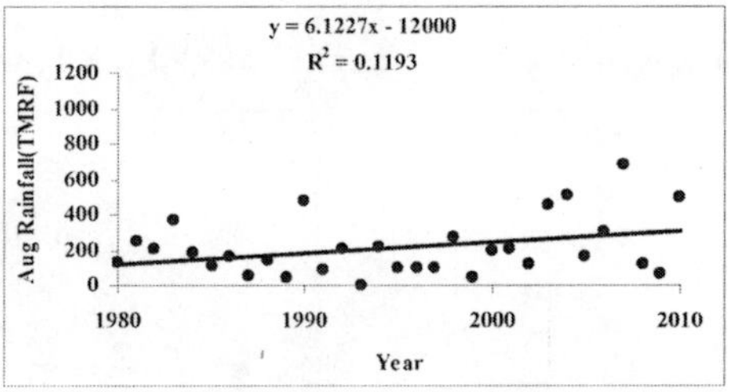
y = 6.1227x - 12000
R^2 = 0.1193
Aug Rainfall(TMRF)
Year

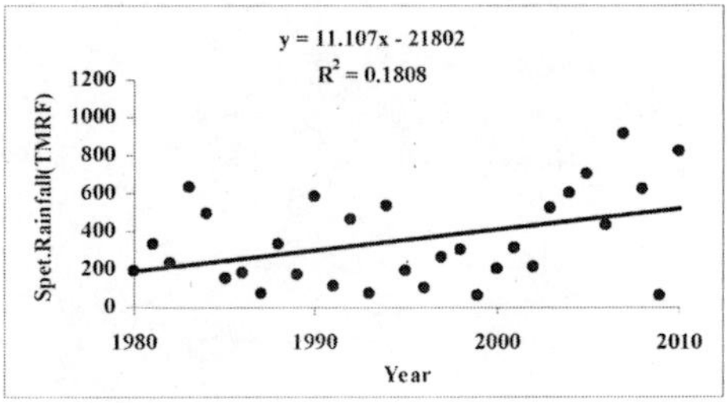
y = 11.107x - 21802
R^2 = 0.1808
Spet.Rainfall(TMRF)
Year

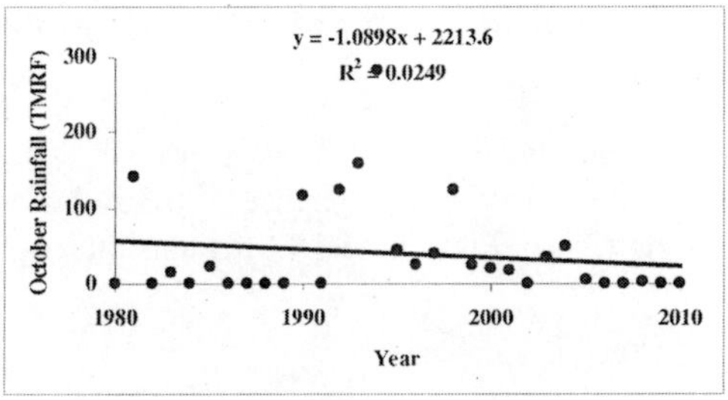
y = -1.0898x + 2213.6
October Rainfall (TMRF)
Year

(contd...)

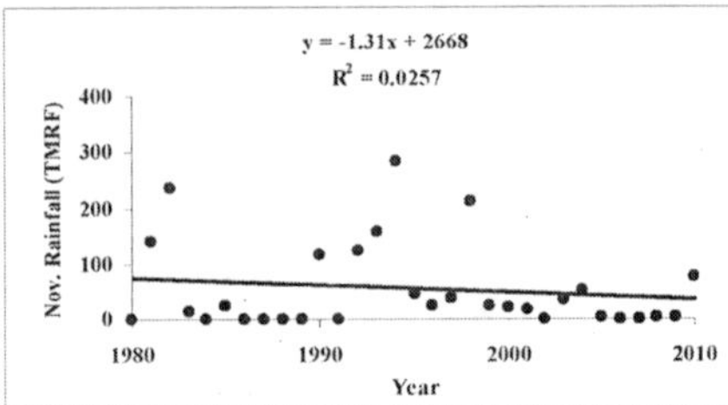

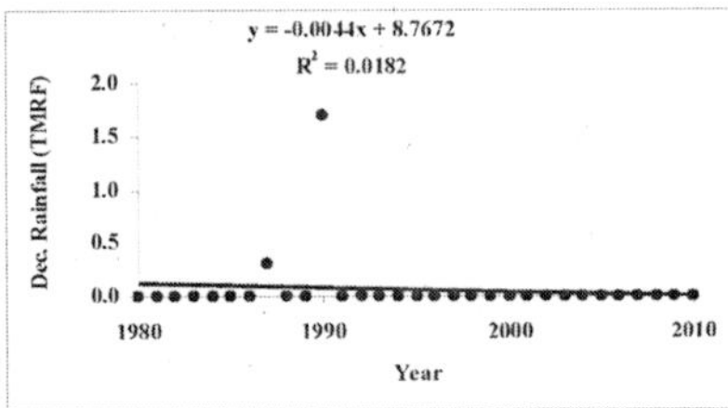

Fig. 3: Linear Regression Trends of Monthly Mean of Total Mean Rainfall.

Table 4: Linear Regression Equations of TMRF for All the Months

Month	*Regression line*	R^2
Jan	y = -0.0023x + 6.2955	6E-06
Feb	y = 0.0029x - 5.8032	0.0152
Mar	y = 0.0056x - 10.798	0.0014
Apr	y = 0.0136x - 24.894	0.0002
May	y = -0.0029x + 11.56	4E-06
Jun	y = -7.4865x + 15153	0.0664
July	y = -3.1584x + 6664.5	0.0152
Aug	y = 6.1227x - 12000	0.1193
Sep	y = 11.107x - 21802	0.1808
Oct	y = -1.0898x + 2213.6	0.0249
Nov	y = -1.31x + 2668	0.0257
Dec	y = -0.0044x + 8.7672	0.0182

Trend Analysis of Annual Mean of Total Mean Rainfall (TMRF)

From the (Figure 4), the annual mean of monthly mean of total mean rainfall observed an increasing trend having an increase of 0.35 mm per year. This implies that in Junagadh (Saurashtra) annual TMRF has increased during the last 35 years.

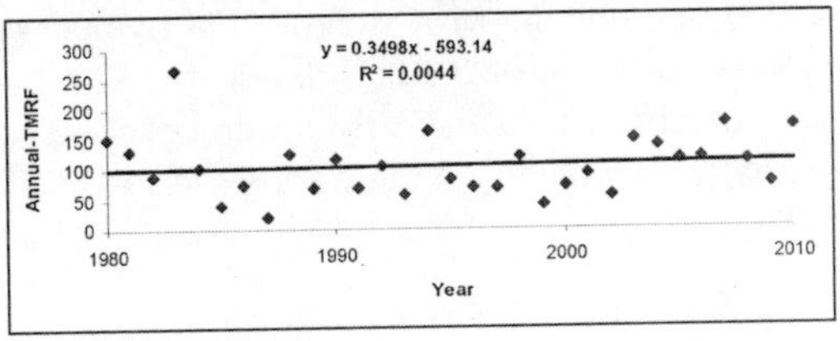

Fig. 4: Trend Analysis of Annual Mean of Total Mean Rainfall (TMRF).

The Mann-Kendall Test for Trend

The Mann-Kendall test is a non-parametric test for identifying trends in time series data. The test was suggested by Mann (1945) and has been extensively used with environmental time series (Hipel *et.al.*). The test compares the relative magnitudes of sample data rather than the data values them. One benefit of this test is that the data need not conform to any particular distribution. Let $X_1, X_2 .... X_n$ represents n data points where X_j represents the data point at time *j*. Then the Mann-Kendall statistic (*S*) is given by,

$S=\Sigma\Sigma$ sign $(X_j - X_k), j=2,3....n; k=1,2.....j-1$

Where: sign $(X_j-X_k) = 1$ if $X_j-X_k>$

$= 0$ if $X_j-X_k = 0$

$= -1$ if $X_j-X_k<0$

A very high positive value of S is an indicator of an increasing trend, and a very low negative value indicates a decreasing trend. However, it is necessary to compute the probability associated with *S* and the sample size, *n*, to statistically quantify the significance of the trend.

For a sample size >10, a normal approximations to the Mann-Kendall test may be used.

For this, variance of *S* is obtained as,

$V(S) = [n(n-1)(2n+5) - \Sigma tp(tp-1)(2tp+5)]/18, p=1,2.....q$

Where tp is the number of ties for the pth value and q is the number of tied values.

Then standardised statistical test is computed by:

$Z=S-1/\sqrt{V(S)}$ if $S>0$,

$=0$ if $S=0$,

$=S+1/\sqrt{V(S)}$ if $S<0$

For annual MMAX temperature, the value of S obtained as 208, a very high positive value indicating increasing trend and is statistically significant that there is enough evidence to determine an upward trend as shown in

(Fige. 2) which is confirmed by the M-K trend test at five per cent level of significance. For annual TMRF, the value of *S* obtained as 42, a positive value indicating increasing trend but is statistically insignificant that there is not enough evidence to determine there is an upward trend for TMRF shown in (Fig. 4) and is confirmed by the M-K trend test at five per cent level of significance.

Conclusion

It is observed that monthly mean of maximum (MMAX) temperatures have increased significantly for all the months except the month of July for which a very weak decrease in MMAX temperature is observed in Junagadh (Saurashtra). This implies that in Junagadh (Saurashtra) the highest increase in MMAX temperature occurs in November (0.119°C) during last 35 years. The annual mean of monthly mean of maximum temperatures observed an increasing trend having an annual increase of 0.017°C per year, as represented in (Fig. 2). It is evident from above figures that monthly mean of TMRF have increased significantly for the months January, February, March, April, May, July, August, September, and decreasing trend in June, October, November and December for the Junagadh (Saurashtra). This implies that in Junagadh district the highest increase in TMRF occurs in September and has increased during the last 35 years. The highest decrease in TMRF occurs in November and decreased during 35 years. Annual mean of monthly mean of total mean rainfall observed an increasing trend having an increase of 0.35 mm per year. This implies that in Junagadh (Saurashtra) annual TMRF has increased during the last 35 years annual TMRF.

Acknowledgement

We are thankful to Agricultural Meteorology Department, Junagadh Agricultural University, Janagadh for providing meteorological data of Junagadh for study of temperature, rainfall and trends.

References

Beena Shah, Global and National Concerns on Climate Change, University News, Vol. 48 No. 24, June 14-20, 2010, 15-23.

Deshmukh, D. T., Lunge, H. S., A Study of Temperature and Rainfall Trends in Buldana District of Vidarbha, India. International Journal of Scientific & Technology Research, volume 2, issue 2, February 2013 ISSN 2277-8616

Fulekar, M.H., Kale, R.K., "Impact of Climate Change: Indian Scenario," University News, Vol. 48 No. 24, June 14-20, 2010, 15-23.

H.M. Hasanean 2001. "Fluctuations of Surface Air Temperature in the East Mediterranean," Theoretical and Applied Climatology, Vol. 68, No.1-2, pp.75- 87.

Hipel, K.W. and McLeod, A.I., Time Series Modelling of Water Resources and Environmental Systems, Electronic reprint of our book orginally published in 1994.

IPCC Climate Change-A Synthesis Report of the IPCC, Technical Report, Inter-governmental Panel on Climate Change 2007.

Jestinos, Mzezewa, Titus Misi and Leon D van Rensburg (2009). Characterisation of rainfall at a semi-arid ecotope in the Limpopo Province (South Africa) and its implications for sustainable crop production, available on website www.wrc.org.za

Julious, M. Huho, Josephine, K.W. Ngaira, Harun, O, Ogindo and Nelly Masayi. The changing rainfall pattern and the associated impacts on subsistence agriculture in Laikipia East District, Kenya, Journal of Geography and Regional Planning, Vol. 5 (7), pp. 198-206, 4 April.

Murray, R. Spigel, Larry J. Stephens, SCHAUM Soutlines STATISTICS, Third Edition, Tata McGraw-Hill Edition, 2000.

Mann, H.B. 1945. Nonparametric tests against trend," Econometrica, 13, 245-259.

Patterson, P.E. Statistical Methods, Richard D. Irwin Inc, Homewood, IL,1987.

Ritter, ME (2006). "The physical environment: an introduction to physical Geography," available online at: http://www.uwsp.edu/geo/faculty/ritter/geog101/textbook/title_ page.html.

Shafiqur Rehman, Luai M. Al-Hadhrami, 2012. Extreme Temperature Trends on the West Coast of Saudi Arabia, Atmospheric and Climate Sciences, 2,351-361.doi:10.4236/acs.2012.23031 Published Online July 2012 (http://www.SciRP.org/journal/acs).

http://shodhganga.inflibnet.ac.in/bitstream/10603/9220/7/07_chapter %202.pdf

S.Del Rio, R.Fraile, L. Herrero and A.Penas, 2007. Analysis of Recent Trends in Mean Maximum and Minimum Temperatures in a Region of the NW of 73 IJSTR©2013 www.ijstr.org Spain (Castillay Leon), Theoretical and Applied Climatology, Vol.90, No.1-2, pp. 1-12. doi:10.1007/s00704-006-0278-9.

T.R. Karl, P.D. Janes, R.W. Knight, J. Kukla, N. Plummer, V. Razuvayev, K.P. Gallo, J. Lindesay, R.J. Charlson and T.C. Peterson, 1993. A Symmetric Trends of Daily Maximum and Minimum Temperatures: Empirical Evidence and Possible Causes, Bulletin of the American Mathematical Society, Vol.74, No. 6, pp.1007-1023.

Webber, J. and Hawkins, C. 1980. Statistical Analysis Applications to Business and Economics Harper and Row, New York.

www.jau.in/attachments%5CWeather%20Bulletins%5Creports% 5C2010.pdf

15

Development of Statistical Models for Forecasting of Gram Crop Yield for Amreli District of Gujarat State

N.J. Rankja and P.B. Maraviya

Abstract

Crop production and productivity depend on climate in general and weather in particular. The productivity of the crop depends on soil characteristics and weather conditions during season. The present study of forecasting of gram crop yield for Amreli district based on weekly average data of weather parameters like maximum and minimum temperature, morning and afternoon relative humidity, sunshine hours and past year total rainfall over a period of 30 years (1981-2010) for Amreli district. The approach used for forecasting yield was original weather variables and week-wise approach. The time trend was include as an explanatory variable. For this district, the model of 14 weeks period (week-wise approach) was selected. The model for Amreli district can be used for providing pre-harvest forecast two weeks before expected harvest. The result of the study shows that model selected for Amreli district for pre-harvest forecast explain more than 83 per cents variation.

Introduction

The overall growth of the Indian economy depends upon the performance of agriculture, which depends, quite a lot, upon the weather conditions every year and yield of most of the crops are fairly below the levels of these achieved in other countries. In Gujarat, gram occupied an area of 2.15 lack hectares with a production of 2.10 lack tonne with an average yield of 977 kg/ha which accounts 2.46 per cent of area and 2.80 per cent of production of the country, respectively (Singh, 2010). Crop production and productivity depend on climate in general and weather in particular. The total production of any crop depends on the actual acreage under cultivation and the weather conditions during the crop life period whereas the productivity of the crop depends on both the soil characteristics and weather conditions during the season. There are many weather variables which affect crop growth and development in different ways and at different times during the growth cycle of the crop. Weather has long been recognised as major control over growth and yield of crops.

The relationship between crop yield and weather parameters is generally carried out with the help of multiple regression models. These models are generally employed for making quantitative crop yield forecast on operational basis. The present study was undertaken to investigate the feasibility of estimating chickpea yields by regression analysis of Amreli district of Gujarat state majoring in area and production from the weather parameters and technological advances. The main objectives of the investigation are:

1. To study the effect of weather parameters on productivity of chickpea for Amreli district in Gujarat state.
2. To develop suitable statistical models for forecasting chickpea productivity of Amreli district in Gujarat state.

Materials and Methods

Considering the specific objectives under research study, Amreli district was selected based on last ten years average yield of chickpea. The area, production and productivity of chickpea crop for the period of past 30 years (1981-82 to 2010-11) were collected from Department of Economics, J.A.U., Junagadh. The weather data collected for the last 30 years Agromet Observatory of Junagadh and Amreli, J.A.U., Junagadh.

Weather Parameters

The corresponding data on weather parameters were collected from the weather stations or meteorological observatories situated in Amreli and Junagadh. For this study weekly average weather data were collected for the growing season of chickpea. The following variables were considered.

Y : Average chickpea yield of the district in kg/ha.

T : Time trend, year number included to correct for upward or downward trend in yield.

X_1 : Minimum temperature (°C)

X_2 : Maximum temperature (°C)

X_3 : Morning relative humidity (per cent)

X_4 : After noon relative humidity (per cent)

X_5 : Sunshine hours (hrs)

X_6 : Total rainfall of past year (mm)

The details on crop season of chickpea, meteorological standard weeks and physiological stages of chickpea crop are given in Table 1.

Table 1: Various physiological stages and meteorological standard week of chickpea crop

Sr. No.	*Stages*	*Crop duration (days)*	*Meteo. Week*	*Meteo. Period*
1.	Establishment	7 – 10	42 – 43	15th Oct. to 28th Oct.
2.	Vegetative	11 – 40	44 – 48	29th Nov. to 2nd Dec.
3.	Flowering	40 – 50	49 – 51	3rd Dec. to 23rd Dec.
4.	Maturity	95 – 120	52 – 05	24th Dec. to 28th Feb.

Source: As per personal inquiry in Pulse Research Station, J.A.U., Junagadh.

Analysis was carried out to formulate a relationship for predicting the chickpea yield (Y) of the district by investigating the influences of important weather factors (X_is) on the chickpea crop. Due to technological advancement, the time trend (T) assumed to be present, therefore time variable also considered as one of the independent variables in the study.

In week wise-approach, the weekly average data were used as per original scales. The data on weather variables were collected from 42nd meteorological standard week (MSW) of current year to seventh standard week of foreword year. With a view to assess the accuracy and capability of earlier forecasts at an interval of weeks, four models was fitted, considering up to 12, 13, 14 and 15 weeks after sowing during the crop period. The details of variables included in model up to 18 weeks crop period are given in Table 2. The time trend variable was included in this analysis as an explanatory variable.

The mathematical expression of this approach,

$$Y = A_0 + \sum_{i=1}^{p}\sum_{j=1}^{w} a_{ij} X_{ij} + bR + cT \quad \text{(Fisher, R.A. (1924).}$$

Where,

Y = Average chickpea yield of district in kg/ha

A_0 = Constant

X_{ij} = Observed value of ith weather variable in j^{th} week $i = 1, 2, \ldots p = 6$ and $j = 1, 2, \ldots w = 12, 13, 14, 15$

R = Total rainfall of past season of the district in mm.

T = Year number included to correct for the long-term upward or downward trend in yield ($T = 1, 2 \ldots t = 30$)

a_{ij}, b and c are partial regression coefficients associated with each X_{ij}, rainfall and time trend (T) respectively.

Table 2: Variables included in the model week-wise approach up to 18 weeks crop period

Meteo Std. week No.	*Crop week No.*	*Temperature*		*Relative humidity*		*Sunshine hour*	*Rain-fall*
		Max	*Min.*	*M*	*E*		
		X_{1j}	X_{2j}	X_{3j}	X_{4j}	X_{5j}	X_6
42	01	X_{101}	X_{201}	X_{301}	X_{401}	X_{501}	X_6
43	02	X_{102}	X_{202}	X_{302}	X_{402}	X_{502}	
44	03	X_{103}	X_{203}	X_{303}	X_{403}	X_{503}	
45	04	X_{104}	X_{204}	X_{304}	X_{404}	X_{504}	
46	05	X_{105}	X_{205}	X_{305}	X_{405}	X_{505}	
47	06	X_{106}	X_{206}	X_{306}	X_{406}	X_{506}	
48	07	X_{107}	X_{207}	X_{307}	X_{407}	X_{507}	
49	08	X_{108}	X_{208}	X_{308}	X_{408}	X_{508}	
50	09	X_{109}	X_{209}	X_{309}	X_{409}	X_{509}	
51	10	X_{110}	X_{210}	X_{310}	X_{410}	X_{510}	
52	11	X_{111}	X_{211}	X_{311}	X_{411}	X_{511}	
1	12	X_{112}	X_{212}	X_{312}	X_{412}	X_{512}	
2	13	X_{113}	X_{213}	X_{313}	X_{413}	X_{513}	
3	14	X_{114}	X_{214}	X_{314}	X_{414}	X_{514}	
4	15	X_{115}	X_{215}	X_{315}	X_{415}	X_{515}	
5	16	X_{116}	X_{216}	X_{316}	X_{416}	X_{516}	
6	17	X_{117}	X_{217}	X_{317}	X_{417}	X_{517}	
7	18	X_{118}	X_{218}	X_{318}	X_{418}	X_{518}	

T= Time trend Y= Yield

Selection of Variables

For selecting the best regression equation with significant independent variables among number of independent variables the stepwise regression procedure was adopted (Draper and Smith, 1966). The best regression equation for predictive purpose should be such as would include as many independent variables as possible, but at the same time, reject all those variables which do not contribute significantly to the prediction model. There is no unique statistical procedure to compromise between these two extremes, but stepwise regression analysis procedure which is generally applied under such circumstances and considered as the best selection procedure.

Simulated Forecast Model

Using these fitted prediction equations, simulated forecasts were obtained for subsequent years, which were not included in the model. The per cent deviations of these forecasts from the reported yields were worked out to assess the suitability of these equations as pre-harvest forecast models. Finally, the model which provided the early forecast and accounted more than 80 per cent of the total variation and which did not deviate much more (<20 per cent) from the observed yields of the district, were considered as pre-harvest forecasts models.

Results and Discussion

The result related to the 12 week crop period, represented in the Table 3 revealed that the coefficient of determination (R^2) varied from 54.09 to 62.01 per cent. In case of 13 and 15 week crop period, represented in the Table 4 and Table 5 revealed that the coefficient of determination (R^2) varied from 44.65 to 56.68 per cent and 64.66 to 68.83 per cent respectively.

The result related to the 14 week crop period, represented in the Table 6 revealed that the coefficient of determination (R^2) varied from 70.63 to 83.39 per cent and the deviation in predicted values varied from 2.51 to 34.68 per cent (Table 7). The fitted regression equations related to 14 week crop period, positive and significant effect of minimum temperature of 14th week (X_{114}) and sunshine hours of 12th week (X_{512}) were observed. The effect of maximum temperature of 7th and 12th week (X_{207}, X_{212}) and sunshine hours of 6th and 11th week (X_{506}, X_{511}) were negatively influenced on the yield of chickpea crop in Amreli district. The weeks correspond to vegetative and flowering stages of chickpea crop.

Table 3: Regression equations for 12 - week crop period of Amreli district

	Models for different years			
Variables in model	*Model-I (25 year)*	*Model-II (26 year)*	*Model-III (27 year)*	*Model-IV (28 year)*
Constant	2292.53	2796.52	3348.07	4278.04
T	8.30**	11.93**	12.90**	11.72**
X_{207}	-21.94**	-40.84**	-52.61**	-60.22**
X_{212}	-46.92**	-49.22**	-43.39**	-43.93**
X_{305}	2.47**	3.52**	4.38**	2.79**
X_{504}	11.18**	10.28**	2.92**	1.62**
X_{506}	49.60**	65.61**	22.03**	-49.72**
X_{511}	-58.95**	-67.90**	-7.28**	-51.52**
S.E.	177.70	180.20	196.52	227.09
R^2 (%)	**54.09**	**61.79**	**62.01**	**60.39**

*Significant at five per cent level. ** Significant at one per cent level.

Table 4: Regression equations for 13 - week crop period of Amreli district

	Models for different years			
Variables in model	*Model-I (25 year)*	*Model-II (26 year)*	*Model-III (27 year)*	*Model-IV (28 year)*
Constant	3076.33	3961.15	4419.58	4929.72
T	7.00**	11.37**	15.19**	19.58**
X207	-29.07**	-57.06**	-74.88**	-91.49**
X212	-57.84**	-61.23**	-60.91**	-59.71**
X305	5.73**	7.79**	9.79**	11.24**
X306	-4.05*	-4.99*	-5.98*	-8.24*
S.E.	179.04	186.78	198.08	220.71
R2 (%)	44.65	51.71	54.97	56.68

*Significant at five per cent level. ** Significant at one per cent level

Table 5: Regression equations for 15-week crop period of Amreli district

	Models for different years			
Variables in model	*Model-I (25 year)*	*Model-II (26 year)*	*Model-III (27 year)*	*Model-IV (28 year)*
Constant	3076.33	3961.15	4419.58	4929.72
T	7.00**	11.37**	15.19**	19.58**
X207	-29.07**	-57.06**	-74.88**	-91.49**
X212	-57.84**	-61.23**	-60.91**	-59.71**
X305	5.73**	7.79**	9.79**	11.24**
X306	-4.05*	-4.99*	-5.98*	-8.24*
S.E.	179.04	186.78	198.08	220.71
R2 (%)	44.65	51.71	54.97	56.68

*Significant at five per cent level. ** Significant at one per cent level

Table 6: Regression equations for 14 - week crop period of Amreli district

	Models for different years			
Variables in model	*Model-I (25 year)*	*Model-II (26 year)*	*Model-III (27 year)*	*Model-IV (28 year)*
Constant	2804.85	2740.46	2986.37	3464.14
T	7.21**	10.75**	11.27**	10.54**
X_{210}	-49.53**	-48.65**	-57.49**	-65.03**
X_{212}	-30.47**	-32.96**	-23.35**	-19.28**
X_{506}	23.01**	41.70**	-1.37**	-60.27**
X_{511}	-81.08**	-65.56**	-109.94**	-111.16**

(contd...)

X_{512}	62.19**	63.85**	88.39**	111.52**
S.E.	151.25	163.93	175.03	196.34
R^2 (%)	**64.66**	**66.52**	**68.19**	**68.83**

*Significant at five per cent level. ** Significant at one per cent level

Table 7: Simulated forecast values for 14 - week crop period of Amreli district (week-wise approach)

Variables in model	*Models for different years*			
	Model-I (25 year)	*Model-II (26 year)*	*Model-III (27 year)*	*Model-IV (28 year)*
Constant	2020.99	2370.86	2576.63	2821.08
T	4.98**	7.00**	6.84**	4.98**
X_{114}	41.46**	43.13**	48.02**	57.38**
X_{207}	-44.17**	-58.54**	-69.41**	-74.64**
X_{212}	-21.53**	-18.84**	-12.70**	-7.67**
X_{213}	-0.07	-1.59*	0.61*	0.85*
X_{302}	-1.17*	-2.27*	-2.59*	-3.63*
X_{305}	5.57**	6.41**	7.25**	7.09**
X_{506}	-2.53**	3.95**	-27.40**	-58.61**
X_{511}	-120.28**	-132.23**	-146.32**	-153.69**
X_{512}	81.36**	92.77**	113.81**	126.37**
S.E.	157.67	155.69	158.64	160.26
R^2 (%)	**70.63**	**76.51**	**79.37**	**83.39**

*Significant at five per cent level. ** Significant at one per cent level

The perusal of the results of Amreli district indicated that the approaches under study, the 14 weeks crop period model (using original weather variables, week-wise approach) could be suggested as a pre-harvest forecast model. The variation explained by this model was very high (> 83 per cent) and simulated forecast error was less than three per cent. This model could be utilised for pre-harvest forecast 2 weeks before expected harvesting period of chickpea crop.

The recommended model for Amreli district is,

$$Y = 2821.08 + 4.98^{**}\,T + 57.38^{**}\,X_{114} - 74.64^{**}\,X_{207} - 7.67^{**}\,X_{212} + 0.85^{*}\,X_{213} - 3.63^{*}\,X_{302} + 7.09\,X_{305} - 58.61^{**}\,X_{506} - 153.69^{**}\,X_{511} + 126.37^{**}\,X_{512}$$

(R^2 = 83.39 per cent)

A brief review of studies which have direct or indirect bearing in commensuration with the objectives of the present study.

Snedecor and Cocheren (1967) and Runge (1968) have used this model to study joint effect of temperature and rainfall on crop yield.

Shrivastava *et al.* (2002) studied on the model of CHICKPGRO for the evaluation of the importance of projected climate change on yield of popular chickpea cultivar on central and north-west India.

Berger *et al.* (2005) revealed that Low temperatures can cause flower and pod abortion. Under field conditions, chickpea delays pod set at mean temperatures below 16°C, and ceases podding below 12°C.

Dubey *et al.* (2011) studied Impact of climate change on pulse productivity and results indicated that there was greater impact of increase in average maximum temperature on yield reduction as compared to increase in minimum temperature and the temperature difference.

Shweta Singh and Suman Kumar (2012) concluded that Low temperature is favourable for vegetative phase and high temperature is favourable for reproductive phase of chickpea. The yield per hectare was higher during 2005-06 in comparison to 2006-07 due to favourable weather conditions in 2005-06.

References

Berger, J.D., Buck, R.P., Henzell, J.M. and Turner, N.C. (2005). Evolution in the genus Cicer – vernalization response and low temperature pod set in chickpea (*C. arietinum* L.) and its annual wild relatives. *Aust J. Agric. Res.* 56:1191–1200.

Dubey, Shantanu Kumar, Sah, Uma and Singh, S.K. (2011). Impact of climate change on pulse productivity and adaptation strategies as practiced by the pulse growers of Bundelkhand region of Uttar Pradesh. *Journal of Food Legumes*. 24 (3): 230-234.

Fisher, R. A. (1924). The influence of rainfall on the yield of wheat at Rothamsted Roy. Soc. (London) Phil. Transser B. 213:89-142.

Singh, N.P. (2010). Project Co- ordinators Report 2009-2010, Annual group meet Aug. 29-31, (2010). AICRP on chickpea, IIPR, Kanpur, pp. 30-32.

Singh, Shweta and Kumar, Suman. (2012). Crop Weather Relationship and its Impact on Growth and Yield of Chickpea (*Cicerarietinum* L.) of Uttarakhand, Department of Agrometeorology, G. B. Pant University of Agriculture and Technology, Pantnagar, Uttarakhand, India. U.S. Nagar – 263145.

Snedecore, G.W. and Cochran, W.G. (1967). Statistical Methods: Oxford and IBH Publishing Co. New Delhi. (6th Edi.).

Srivastava, A.K., Gupta, B.R.D. and Panda, B.C. (2002). Impact of climate change on chickpea yield in central and northwest India, Dept. of Geophysics, Bhu, Varanasi- 05.

16

Comparative Assessment of Meteorological Drought Indices for Amravati District

Chaitali V. Thakare, G.U. Satpute, and Suhas K. Upadhye

Abstract

Drought is a complex phenomenon that is difficult to accurately describe. A drought index can be used to quantify the moisture condition of a region and thereby to measure the severity of a drought and quantify the spatial extent of a drought event. In this study, monthly rainfall at selected taluka places in Amravati district of Maharashtra was analysed to estimate and compare Decile Drought Index (DDI), Effective Drought Index (EDI) and Standardised Precipitation Index (SPI) for identifying drought years in Amravati district using 22 years rainfall data (1991 - 2012). DDI identified maximum number of years in moderate drought category and widespread drought years during 1991, 1993, 1995, 1997, 2000-2004, 2008 and 2009. EDI identified maximum number of years in normal condition and only one year 1995 was found as widespread drought year. SPI identified maximum number of years in mild wet condition and the years 1991, 1995, 2000 and 2002 were observed to be widespread drought years in Amravati district. The three indices were then compared using three important assessment criteria of the seven criteria adopted by Ntale and Gan (2003) and Dabare (2007). The SPI was found to be superior over that of DDI and EDI because it describes all the major droughts occurred in Amravati district, higher positive values of Spearman rank correlation coefficient between year-wise crop yield data and climatic severity and more consistency with historical drought events. Since SPI satisfies all the three assessment criteria followed by DDI which fulfills only two criteria, SPI can be considered as the most suitable index for drought assessment in Amravati district.

Introduction

Precipitation deficiency, due to natural climatic variability in space and time, is the primary cause of drought. Drought is a complex phenomenon that is difficult to accurately describe. The most common tool for monitoring drought conditions is a drought index, which quantifies the moisture condition of a region and thereby to detect the onset and measure the severity of a drought event. A drought index can also be used to gain additional information on seasonal cycle and frequency of water deficiencies, frequency of various levels of irrigation requirements (Heim, 2002). The index of meteorological drought is used to measure how much precipitation for a given period of time has deviated from historically established norms. Although none of the major indices is inherently superior to the rest in all circumstances, some indices are better suited than others for certain uses (Ntale and Gan, 2003). The drought indices can be used to provide an early drought warning system (Lohani *et al.*, 1998), to calculate the probability of drought termination (Karl *et al.*, 1987), to determine drought assistance (Wilhite *et al.*, 1986), to predict crop yield (Kumar and Panu, 1997), to examine the spatial and temporal characteristics of drought, the severity of drought, and to make comparisons between different regions (Alley 1984, and Quiring *et al.*, 2007).

Standardised Precipitation Index (SPI), Decile Drought Index (DDI) and Effective Drought Index (EDI) are some of the widely used meteorological drought indices for drought quantification. These indices are solely based on precipitation data. Khetmalas (2006) estimated and compared four drought indices namely, aridity index, decile index, effective drought index and Palmer drought severity index for 19 years of climatological data at Akola. Gibbs and Maher (1967) developed decile drought index technique to avoid some of the weaknesses within percent of normal approach. George and Kalyansundaram (1969) used monthly rainfall deciles for assessing agricultural drought in Bihar state. Das *et al.* (2003) applied decile method to six stations viz., Delhi, Jaipur, Ahmedabad, Ahamadnagar, Jalgaon and Anantpur. Khetmalas (2006) analysed 19 years (1986-2004) of weekly rainfall data for Akola using decile index method using the procedure of Gibbs and Maher (1967) and drought years were identified according to the procedure adopted by Das *et al.* (2003). Hughes (2004) reviewed the Effective Drought Index (EDI) and suggested that EDI is applicable to drought monitoring over large regions. Kim *et al.* (2009) compared the performances of the Effective Drought Index (EDI) and 1, 3, 6, 9, 12, and 24 months Standardised Precipitation Indices (SPIs) for drought monitoring data accumulated over 200-year period from 1807 to 2006 for Seoul, Korea. McKee (1993) used 3, 6, 12, 24 and 48 months Standardised Precipitation Index (SPI) for 120 years of random monthly precipitation based on the annual cycle and expected variability of precipitation at Fort Collins, CO. Hayes *et al.* (1999) examined the 1996 drought in the Southern plains and southwestern

United States using Standardised Precipitation Index (SPI). A series of maps are used to illustrate how the SPI would have assisted in being able to detect the onset of the drought and monitor its progression. Bordi *et al.* (2001) demonstrated that SPI can be used as a tool in the historical reconstruction of drought events in locations in Italy and analysed the rainfall on Marche region.

Oza *et al.* (2002) used standardised precipitation index (SPI) for drought detection, quantification and agricultural impact assessment at a homogeneous rainfall region level (North-West India) and at meteorological subdivision level (West Rajasthan and Saurashtra-Kutch) using monthly rainfall data (1951-1999). Muna Elhag (2012) observed the impact of climate change on the sorghum production and sorghum productivity in Gedaref region in east Sudan using Standard Precipitation Index (SPI), rainy days and rainfall amount and distribution. The relation between the drought index and the crop parameters indicated that there was strong correlation with crop productivity.

Hall and Brown (2005) compared precipitation and drought indices related to fire activity and potential in the United States and used Spearman rank correlation between various indices and number of fire starts. Khetmalas (2006) compared the four drought indices viz. aridity index, decile index, effective drought index (EDI) and Palmer drought severity index (PDSI) by using seven assessment criteria suggested by Ntale and Gan (2003) for Akola station. Morid *et al.* (2006) compared the performance of seven indices for drought monitoring in the Tehran province of Iran. Dabare (2007) compared the decile drought (DDI) index, standardised precipitation index (SPI) and Bhalme-Mooley Drought Index (BMDI) by using seven assessment criteria suggested by Ntale and Gan (2003) for eastern Vidarbha. The SPI was found to be superior over that of decile index and BMDI because it describes all the major droughts occurred in eastern Vidarbha. The meteorological drought study has been carried out by different researchers by using three drought indices such as Decile drought index (DDI), Effective drought index (EDI) and Standardised precipitation index (SPI) at different locations. The present study was undertaken for comparative assessment of meteorological drought indices i.e., Decile drought index (DDI), Effective drought index (EDI) and Standardised precipitation index (SPI) for identifying drought years in Amravati district using 22 years rainfall data (1991 - 2012).

Materials and Methods

Location of Study Area

The study was conducted for Achalpur, Anjangaon, Amravati, Chandur Bazar, Chandur Railway, Chikhaldara, Dharani, Morshi and Warud tehsils of Amravati district.

Meteorological Data and Yield Data

The 22 years (1991-2012) rainfall data and the yield data of different dry land crops grown in the *kharif* and *rabi* season during 1991-2012 in selected taluka places in Amravati district was used.

Determination of Decile Drought Index

The Decile drought index for each monsoon month from June to September has been calculated and compared with actual rainfall of that month according to the criteria (Table 1) given by Gibbs and Maher (1967) to identify the severity of drought in that month.

Table 1: Classification of DDI (Gibbs and Maher, 1967)

Decile classes	*Drought severity*
Decile 1 (90 to 100 per cent)	Severe Drought
Decile 2 (80 to 90 per cent)	
Decile 3 (70 to 80 per cent)	Moderate Drought
Decile 4 (60 to 70 per cent)	
Decile 5 (50 to 60 per cent)	Mild Drought
Decile 6 (40 to 50 per cent)	
Decile 7 (30 to 40 per cent)	Normal Condition
Decile 8 (20 to 30 per cent)	
Decile 9 (10 to 20 per cent)	Above Normal Condition
Decile 10 (0 to 10 per cent)	

Determination of Effective Drought Index (EDI)

Effective drought index (EDI) was developed by Byun and Wilhite (1999). EDI values are standardised which allows drought severity at two or more locations to be compared with each other regardless of climatic differences between them. EDI roughly varies in the range from -2 to 2 (Table 2). It has thresholds indicating the range of wetness from extremely dry to extremely wet conditions.

Table 2: Classification of EDI (Byun and Wilhite, 1999)

Sr. No.	*Range of EDI value*	*Class*
1.	EDI > 2	Extremely wet
2.	$1.5 \leq$ EDI ≤ 1.99	Severely wet
3.	$1 \leq$ EDI ≤ 1.49	Moderately wet
4.	$0.7 \leq$ EDI ≤ 1.49	Mild wet
5.	$-0.69 \leq$ EDI $\leq .69$	Near normal

(contd...)

6.	-0.7 ≥ EDI ≥ - 0.99	Mild drought
7.	-1 ≥ EDI ≥ -1.49	Moderate drought
8.	-1.5 ≥ EDI ≥ -1.99	Severe drought
9.	EDI ≤ - 2	Extremely dry

Determination of Standardised Precipitation Index (SPI)

SPI developed by McKee *et al.* (1993) in Colorado State University is based on the probability distribution of precipitation and requires less input data and calculation. The SPI program (SPI_SL_6.exe) available in windows/ PC version was used to predict SPI values. In present study the SPI values were predicted at three months interval starting from July upto September. Drought classification by SPI value and corresponding event probabilities is as given in Table 3:

Table 3: Classification of SPI (McKee et al., 1993)

SPI value	*Category*	*Probability (per cent)*
2.0 or more	Extremely wet	2.3
1.50 to 1.99	Severely wet	4.4
1.0 to 1.49	Moderately wet	9.2
0 to 0.99	Mildly wet	34.1
0 to -0.99	Mild drought	34.1
-1.0 to -1.49	Moderately drought	9.2
-1.5 to -1.99	Severely drought	4.4
-2.0 and less	Extremely drought	2.3

Comparison of Drought Indices

Although there is no drought index that is inherently superior to others in all circumstances, some indices are better suited than others for certain regional application. Ideally, this study is planned to identify the most suitable index for characterising drought in Amravati district, but realistically no one index can perfectly track the dimensional variation of the climatic system. Ntale and Gan (2003) discussed and used the criteria that can be used to gauge the suitability of drought indices. In present study, Decile drought index, Effective drought index and Standardised precipitation drought index as different meteorological indices were compared according to the procedure followed by Dabare (2007) by identifying the following three assessment criteria to determine most appropriate drought index for monitoring meteorological drought in Amravati District.The drought years identified by different indices were analysed according to their severity class. The index, which is having more consistency with historical drought events, was characterised as good indicator of drought for this region. Due to the discrete variation of two

Table 4: Moderate (MO), Severe (S) and Extremely Drought (ED) Years Identified by Different Indices at Different Taluka Places in Amravati District (1991- 2012)

Taluka	*DDI*		*EDI*			*SPI*		
	MO	*SD*	*MO*	*SD*	*ED*	*MO*	*SD*	*ED*
Achalpur	91 93 96 97 00 01 02 03 04 05	95 08 09	95 96 01 02 03	04 08 09	2005	2000 01 09	1995 08	2002
Anjangaon	1992 93 95 00 01 03 09 11	1991	——	——	——	1991 2000	——	——
Amravati	1991 93 97 2000 02 03 09	1995 2004	2002	1995 97 2004	1996	——	2002	1995
Chandur Bazar	1991 93 96 99 01 02 03 04 08 09	1995 2000	1995 2001 02	1996	——	2001	1991	1995
Chandur Railway	1991 92 94 95 97 98 2000 01 02 04 08 09 12	1993	1995 96 2008	2009	——	1991 93 95 2002	2001 08	——
Chikhaldara	1991 92 99 2002 03 04 05 09	1995 00 01	2000	1995 2001	——	1991 00 02	1995	2001
Dharani	1997 2000 01 04 08 09 10	1991 95	2000	2009	——	1997 2000	——	——
Morshi	1991 93 96 97 98 2001 02 08 09	1995 2000	1997 2000 08	1996	——	1993 95	2000	2002
Warud	1996 97 98 2001 02 03 04 08 09	1991 2000	2002	——	——	2000 01 02	1991	——

variables (i.e. yield and drought year), Spearman rank correlation coefficient was chosen as a measure of how well years ranked by drought index value are compared to years ranked by yield of the area. For all drought indices, a positive index value indicates optimum or wetter than normal condition and negative index value imply dryer than normal conditions. Correlation between the drought years and yield of the year can range between -1 and 1 (Chandel, 1965). A positive correlation indicates a direct relationship between two variables (drought severity and crop yield). This means that a higher ranked value of one variable will be associated with a lower ranked value of the other variable (Hall and Brown, 2005). The three indices were checked with the well-known/major historic drought event occurred during 1991, 1995, 2000 and 2002 in Amravati district. The consistency of each index was studied and best index was chosen.

Results and Discussion

Drought Years Identified by Different Indices and Comparison of Drought Indices

The drought years identified by three drought indices in different taluka places in Amravati district during 1991–2012 are sorted out according to drought severity (Table 4). From Table 4, it is observed that, DDI identified maximum number of drought (18 years) in severe category followed by EDI (12 years) and SPI (9 years). Extreme category of drought years were identified by SPI and EDI methods and were found to be five years in different taluka places by SPI method and at two years in different taluka places by EDI method. The years 1991, 1995, 2000 and 2002 are identified as most widespread disastrous years since these years are shown as moderate, severe and extreme category drought by all the indices in most of the taluka places of Amravati district.

From Table 4 it is revealed that, SPI has identified all the drought years which were also traced by other two indices. EDI finds only two years as a extreme drought year. At Achalpur, Morshi and Warud, the years 1995, 2000 and 1991 respectively were identified as a severe drought years by both DDI and SPI methods, whereas EDI method, identified those years as moderate drought year which gives a false picture since these years were the historical drought years. Compared to EDI and SPI methods, DDI identified more number of drought years at different taluka places in Amravati district. At Anjangaon (2011), Chandur Railway (2012) and Dharani (2010) these years were identified as moderate drought by DDI, which is not true as it was a normal year according to other two indices. From above discussion for identification of drought years at different taluka places in Amravati district, it is observed that SPI and DDI methods had identified most of the drought years which were not observed by EDI method. However, DDI method has identified extra large

Table 5: Spearman Rank Correlation Between Various Crop Yield and Drought Severity by Different Drought Indices for Different Talukas in Amravati District

Taluka	*DDI*				*EDI*				*SPI*			
	Soybean	*Sorghum*	*Cotton*	*Pigeon pea*	*Soybean*	*Sorghum*	*Cotton*	*Pigeon pea*	*Soybean*	*Sorghum*	*Cotton*	*Pigeon pea*
Achalpur	0.70	0.16	-0.03	0.32	0.42	-0.09	-0.04	0.02	0.7	0.09	-0.17	-0.02
Anjangaon	0.30	0.24	-0.29	0.28	0.46	0.045	0.048	0.37	0.3	0.3	-0.02	0.23
Amravati	0.71	-0.15	-0.10	-0.07	0.49	-0.13	-0.13	-0.03	0.58	-0.30	0.1	-0.2
Chandur Bazar	0.47	-0.04	0.05	0.33	0.49	-0.29	0.08	0.15	0.47	0.01	0.12	0.20
Chandur Railway	0.46	-0.08	-0.08	-0.14	0.36	-0.18	-0.44	-0.36	0.49	-0.16	-0.14	-0.25
Chikhaldara	0.57	0.41	-0.01	0.28	0.59	0.41	-0.04	0.11	0.49	0.47	-0.12	0.18
Dharani	0.37	0.39	0.23	0.13	0.26	0.45	0.14	0.24	0.35	0.38	0.10	0.20
Morshi	0.58	-0.15	0.05	-0.03	0.37	-0.28	0.07	-0.09	0.40	-0.18	0.18	-0.13
Warud	0.47	-0.007	0.35	0.25	0.37	-0.01	0.34	0.30	0.47	0.17	0.28	0.32

number of year in moderate and severe category at different taluka places compare to SPI and EDI methods. From the above results, it can be concluded that SPI method identifies all the drought years consistently and distinctly over DDI and EDI method which proves the superiority of SPI method over the latter two drought indices.

Spearman Rank Correlation Coefficient

The drought severity during different years of study period identified by various drought indices viz., DDI, EDI, SPI were arranged according to their severity from most severe to most wet years and were given rank. Similarly, the yield data of different kharif crops like Soybean and Sorghum and *rabi* crops like Cotton and Pigeon pea were arranged from lowest yield to highest yield and were given rank. The ranks of these two variables were then compared for each year and Spearman rank correlation coefficient, for each index was estimated. The Spearman rank correlation coefficient values between various crops and different indices for different taluka places in Amravati district are presented in Table 5.

Result of Spearman rank correlation coefficient test (Table 5) indicates a positive relationship as well as negative relationship in case of all the three indices with different crop yield data. A positive correlation indicates a direct relationship between two variables (yield and drought severity) while a negative correlation indicates an inverse relationship (Chandel, 1965). Accordingly, even though, the positive correlation values in Table 5 are very low, they definitely indicate somewhat direct relationship between crop yield and the climatic conditions explained by corresponding indices. This shows that the selected drought indices in the present study are suitable for the purpose for which they were selected. However, the performance of different indices varies in identifying drought years and its severity.

From Table 5, it can be revealed that for Soybean crop, DDI gives highest positive correlation followed by SPI and EDI, for different taluka places in Amravati district.

For Sorghum, SPI gives highest positive correlation at six taluka places followed by DDI (four taluka places) and EDI (three taluka places) in Amravati district. At Amravati, Chandur Railway and Morshi all the three indices showed negative correlation for sorghum crop. At Warud, DDI and EDI gave negative correlation for sorghum crop. Also at Achalpur EDI showed negative correlation for sorghum crop.

For Cotton, the EDI and SPI gives same positive correlation followed by DDI. At Achalpur, Chandur Railway and Chikhaldara, all the three indices give negative correlation. At Anjangaon, EDI gives positive correlation whereas Decile and SPI gives negative correlation for cotton. However, SPI gives positive correlation for cotton at Amravati whereas DDI and EDI show negative correlation for the same crop.

For Pigeon pea, the DDI and EDI show positive correlation at six taluka places followed by SPI. At Amravati, Chandur Railway and Morshi all the three indices showed negative correlation while at Achalpur, DDI and EDI recorded positive correlation. From the above results of correlation between different rainfed crops yield data and drought years severity obtained by different indices, it can be concluded that the performance of DDI and SPI can be considered as better in identification of drought over EDI.

Analysis of Major Historical Droughts

The three indices used for drought characterisation in different taluka places of Amravati district have identified four major historical droughts in Amravati district viz., 1991, 1995, 2000 and 2002. The severity of these four drought years indicated by different indices is tabulated in Table 6. The Table 6 revealed that the severity of drought during 1991 identified by DDI and SPI was found to be matching with each other with slight variation in severity but EDI failed to identify the drought severity at four taluka places at all. During 1995, DDI and SPI identified droughts at eight and nine taluka places repectively while EDI identified the droughts at only five taluka places. Similarly, during 2000, DDI and SPI identified the drought at nine and eight talukas respectively while EDI identified the drought as low severity at six talukas only. During the historical drought year of 2002; both DDI and SPI had identified drought at all the talukas while EDI failed to identify drought at one taluka.

From above results, it can be concluded that DDI and SPI are showing more consistency with historical drought events indicating the superiority of these two indices over EDI in identifying proper severity of drought in the region.

Conclusion

The monthly rainfall at selected taluka places in Amravati district of Maharashtra was analysed to estimate and compare Decile drought index (DDI), Effective drought index (EDI) and Standardised precipitation index (SPI) for identifying drought years in Amravati district using 22 years rainfall data (1991 - 2012). The Decile drought index identified maximum number of years in moderate drought category and widespread drought years during 1991, 1993, 1995, 1997, 2000-2004, 2008 and 2009. Effective drought index identified maximum number of years in normal condition and only one year 1995 was found as widespread drought year. SPI identified maximum number of years in mild wet condition and the years 1991, 1995, 2000 and 2002 were observed to be

Table 6: Drought Severity Indicated by Various Indices in Historical Drought Years

Taluka	*1991*			*1995*			*2000*			*2002*		
	DDI	*EDI*	*SPI*	*DDI*	*EDI*	*SPI*	*DDI*	*EDI*	*SPI*	*DDI*	*EDI*	*SPI*
Achalpur	MoD	MIW	MIW	SD	MoD	SD	MoD	MID	MoD	MoD	MoD	ED
Anjangaon	SD	MID	MoD	MoD	Nor	MID	MoD	MID	MoD	MID	Nor	MID
Amravati	MoD	Nor	MID	SD	SD	ED	MoD	MoW	MIW	MoD	MoD	SD
Chandur	MoD	MID	SD	SD	MoD	ED	SD	MID	MID	MoD	MoD	MID
Chandur	MoD	MID	MoD	MoD	MoD	MoD	MoD	Nor	MID	MoD	MID	MoD
Chikhaldara	MoD	Nor	MoD	SD	SD	SD	SD	MoD	MoD	MoD	MID	MoD
Dharani	SD	Nor	MID	SD	Nor	MID	MoD	MoD	MoD	MID	MID	MID
Morshi	MoD	Nor	MID	SD	MID	MoD	SD	MoD	SD	MoD	MID	ED
Warud	SD	MID	SD	Nor	MoW	MID	SD	Nor	MoD	MoD	MoD	MoD

Mow – Moderate wet, MIW – Mild wet, Nor – Normal condition, MID – Mild drought, MoD- Moderate drought, SD – Severe drought, ED – Extremely dry.

widespread drought years in Amravati district. The three indices were then compared using the important three assessment criteria adopted by Ntale and Gan (2003) and Dabare (2007). The SPI was found to be superior over that of DDI and EDI because it describes all the major droughts occurred in Amravati district, higher positive values of Spearman rank correlation coefficient between year-wise crop yield data and climatic severity and more consistency with historical drought events. Since Standardised precipitation index (SPI) satisfies all the three assessment criteria followed by Decile drought index (DDI) which fulfills only two criteria, SPI can be considered as the most suitable index for drought assessment in Amravati district.

References

Bordi, I., S. Frigio, P. Parenti, A. Speranza and A. Sutera. 2001. "The analysis of the Standardized Precipitation Index in the Mediterranean area: large-scale patterns", Annali Di Geofisica, 44, pp. 965-978.

Byun, H.R. and D.A. Wilhite. 1999. "Objective quantification of drought severity and duration", J. Climate, 12, pp. 2747-2756.

Chandel, S.R.S. 1965. "A Handbook of Agricultural Statistics", Achal Prakashan Mandir, pp. 312-316.

Dabare, N.B. 2007. "Comparative study of meteorological drought indices in Eastern Vidarbha", M.Tech. Thesis (Unpub.) CAET, Dr Panjabrao Deshmukh Krishi Vidhyapeeth, Akola.

Das, H.P., P.A. Kore and V.N. Jadhav. 2003. "An effective method of identification of drought in Kharif season", Mausam, 54 (4), pp. 908-916.

George, C.J. and V. Kalyansundram. 1969. "A use of monthly rainfall deciles for assessing agricultural droughts in Bihar state", Scientific Report No. 96. IMD, Pune, pp. 1-20+9 annexures.

Gibbs, W.J. and J.V. Maher. (1967). "Rainfall deciles as drought indicators", Commonwealth Bureau of Meteorological, Bulletin No. 48, Melbourne, Australia.

Hall, L. H. and Timothy J. Brown. 2005. "A comparison of precipitation and drought indices related to fire activity and potential in U.S.", Desert Research Institute, Reno, Nevada. *http//ensoul/edu/ndmc/enigma indices/html*

Hayes, M.J., M.D. Svoboda, D.A. Wilhite and O.V. Vanyarkho. 1999. "Monitoring the 1996 drought using the Standardized Precipiation Index", Bulletin of American Meter. Soci. 80 (3), pp. 429-438.

Heim, R. R. 2002. "A review of twentieth-century drought indices used in the United States", Bulletin of American Meteorological Society, 31, pp. 1149-1165.

Hughes, D. A. 2004. "Review, automated estimation and analysis of drought indices in South Asia", *http://www.iwni.cgiar/org/drought/files/pdf/wp%2083.pdf.*

Karl, T., F. Quinlan and D.S. Ezell. 1987. "Drought termination and amelioration: Its climatological probability", J. of Climate and Applied Meteorology 26, pp. 1198-1209.

Khetmalas, S.E. 2006. "Comparative study of drought indices at Akola", M.Tech. Thesis (Unpub) CAET, Dr Panjabrao Deshmukh Krishi Vidhyapeeth, Akola.

Kim, D.W., H.R. Byun and K.S. Choi. 2009. "Evaluation, modification, and application of the Effective Drought Index to 200-Year drought climatology of Seoul, Korea", J. of Hydrology, DOI:10.1016/j.jhydrol.378:1-12.

Lohani, V.K., G.V. Loganathan, S. Mostaghimi. 1998. "Long-term analysis and short-term forecasting of dry spells by Palmer Drought Severity Index", Nordic Hydrology, 29, pp. 21-40.

McKee, T. B., N. J. Doesjen and J. Kelist. 1993. "The relationship of drought frequency and duration to time scales", 8th AMS Conference on Applied Climatol, Anaheim CA., American Meteorological Society, Boston, pp. 179-184, January 17-22,1993.

Morid, S., V. Smakhtin and M. Moghaddasi. 2006. "Comparison of seven meteorological indices for drought monitoring in Iran", Int. J. Climatol, 26, pp. 971–985.

Muna, Elhag, M. 2012. "Assessment of climate change impact on sorghum productivity in Sudan", Third RUFORUM Biennial Meeting, 24-28 September 2012, Entebbe, Uganda.

Ntale, H.K. and Thian Y. Gan. 2003. "Drought indices and their application to East Africa", International J. Climatology, 23, pp. 1335-1357.

Oza, S.R., J.S. Parihar and V.K. Dadhwal. 2002. "Evaluating use of Standardized precipitation index for drought assessment in North-West India", Proceeding of National Symposium, TROPMET-2002, New Delhi, 60-62 March, 2002, pp. 248-254.

Quiring, S. 2007. "Drought monitoring index for Texas", Final Technical Report submitted by the Texas A & M Research Foundation (RF - 468511) to the Texas Water Development Board.

Wilhite, D.A., N.J. Rosenberg and M.H. Glantz. 1986. "Improving federal reponse to drought", Journal of Climate and Applied Meteorology, 25, pp. 332-342.

17

Water Management and Climatic Smart Agriculture

Mausumi Raychaudhuri and S. K. Ambast

Abstract

Water resources are important to both society and ecosystems. We depend on a reliable, clean supply of drinking water to sustain our health. We also need water for agriculture, energy production, navigation, recreation, and manufacturing. Many of these uses put pressure on water resources, stresses that are likely to be exacerbated by climate change. In many areas, climate change is likely to increase water demand while shrinking water supplies. In some areas, water shortages will be less of a problem than increases in intensity of rainfall, runoff, flooding, enhanced soil erosion, or sea level rise. These effects can reduce the quality of water and can damage the infrastructure. Rising temperatures also likely to lower water quality in lakes through increased thermal stability and altered mixing patterns, resulting in reduced oxygen concentrations and an increased release of phosphorus from the sediments. This shifting balance in different parts of the country would challenge water managers to simultaneously meet the needs of growing communities, sensitive ecosystems, farmers, ranchers, energy producers, and manufacturers. Hence in the present effort some of water management interventions are advocated to combat the future crisis emanating from climatic change.

Introduction

Climate change has become one of the most severe developmental challenges for developing countries in the world. Its impact on natural resources affects the lives and livelihoods of people. Being dependent on natural resources Indian agriculture is more vulnerable to climate change and environmental shocks. According to India's Initial National Communication

(NATCOM) to United Nations Framework Convention on Climate Change (UNFCC), climate change is likely to affect the water balance in different parts of India adversely due to changes in precipitation and evapotranspiration and rising sea levels, leading to increased saline intrusion into coastal and island aquifers. The National Water Mission (NWM), a part of the National Action Plan on Climate Change (NAPCC), identifies the threat to water resources in India due to climate change in terms of the expected decline in the glaciers and snow-fields in the Himalayas; increased drought like situations due to the overall decrease in the number of rainy days over a major part of the country; increased flood events due to the overall increase in the rainy day intensity; effect on groundwater quality in alluvial aquifers due to increased flood and drought events; impact on groundwater recharge due to changes in precipitation and evapotranspiration; and increased saline intrusion of coastal and island aquifers due to rising sea levels (NWM, 2009). There is wide temporal and spatial variability of water resources over the country. Not only quantity but the quality of surface and ground water is also of concern due to discharge of large amount of untreated urban and industrial effluent into the water bodies. Thus, concerted effort is required to assess the effect of climate change on water resources, to develop techniques of reducing the impact of extreme hydrological events on crop growth and yield, and to find out suitable pathways to combat the adverse effect of climate change.

At present agriculture sector is the largest consumer of water (82.8 per cent) and continued to be the major consumer with an increased demand of 910 km^3 by 2025 against the total demand of water of 1093 km^3 (CWC, 2010). The stress on water resources in terms of quantity and quality is increasing and efficient water management is one of the important key to achieve climate smart agriculture in India. Developing strategies through efficient management of available on-farm water resources to enhance agricultural productivity on sustainable basis under climate change scenario is crucial in achieving climate resilient agriculture. Understanding the problem scientifically and critically and prioritising suitable available technologies in mitigating the causes and adapting adverse situations especially the climatic extremes is highly required.

Climatic extremes

India receives on an average 1182.2 mm rainfall which is the highest in the world for a country of comparable size. The rainfall over the country has large spatial as well as temporal variability from less than 13 cm over the western Rajasthan to 1141 cm at Mausiram in the Meghalaya and also bioclimate-wise from humid in the northeast (about 180 days rainfall in a year), to arid in Rajasthan (20 days rainfall in a year). About 877.2 mm of rainfall constituting 74.2 per cent of annual rainfall is received during south-west monsoon (June, July, August and September) and the coefficient of variation is higher during the months of November, December, January and February. The annual and monsoon rainfall of the country for the period 1901-2009 do not show any significant increasing or decreasing trend. Based on the index of percentage departure of rainfall from normal, West Rajasthan (34 cases) and Saurashtra and Kutch (31 cases), experienced highest occurrences of drought during 1875-2009. The adjoining Gujarat Region which mostly belongs to semi-arid climate also experienced high incidences of drought (28) during the period. One-sixth of the geographical area of the country (about 40 M ha) is drought-prone while 7.5 M ha of land is affected by floods around each year. Though flood and drought has become regular feature in the country still managing their impacts is difficult and claims lives and damages natural resources. Models that are used for flood forecasting and its influence zones are not rigorous enough due to lack of integration of hydrology and the weather forecasting systems. Modern site-specific forecasting systems with satellite based telemetry are required for overall basin-wise forecasting. This will help in managing the calamities by executing the contingency plans well in advance.

Water resources

India receives total precipitation, including snowfall, of around 4000 km^3 but the availability from surface water and replenishable groundwater is estimated as 1869 km^3. It has been estimated that only 1121.3 km^3 (690 km^3 from surface water and 431.3 km^3 from groundwater resources) can be put to beneficial use (Planning Commission, 2011) against the current use of 634 km^3 reflecting a surplus scenario (Table 1).

Table 1: Water resources of India

S. No.	Parameters	Quantity (BCM)	Percentage of precipitation
1.	Annual precipitation (including snowfall)	4000	100
2.	Evaporation + groundwater flow	2131	53.3
3.	Average surface runoff	1869	46.7
4.	Groundwater recharge	431.3	10.8
5.	Available water	2300.3	57.5
6.	Estimated utilisable water resources	1121.3	28.0
6. (i)	Estimated utilisable surface water	690	17.3
6. (ii)	Replenishable groundwater	431.3	10.8
7.	Current water use	634	15.9

Source: Narsimhan (2008), cited in Planning Commission (2010:427).

It is projected that with increase in population the per capita availability of water in the country as a whole will decrease from 1545 m^3/year in 2011 to 1,340 and 1,140 m^3/year in 2025 and 2050 respectively. At national level presently India is a water stress country approaching towards water scarcity by 2050. Scaling down to individual river basin levels per capita water availability varied from 260 m^3/year in the river basins such as Sabarmati, west flowing rivers in Kutch and Saurashtra including Luni, Cauvery, Pennar, etc. to 11,782 m^3/year in Brahmaputra and Barak basin. Among the twenty river basins/group of river basins the Ganga-Meghna-Brahmaputra basin covers a land area of 33 per cent and accounts for 60 per cent of India's water resources, while the catchment of rivers flowing west is three per cent and they account for 11 per cent of the country's water resources. The majority of the runoff is generated during the three-four months of the monsoon season causing recurring floods in Brahmaputra, Barak and Ganga. At the same time large areas in Rajasthan, Gujarat, Andhra Pradesh, Karnataka and Tamil Nadu get scanty rainfall and faces drought.

Water Management Issues

Indian Institute of Water Management (IIWM) formerly known as Directorate of Water Management is a premier institution in agricultural water management under Indian Council of Agricultural Research (ICAR), Department of Agriculture, Research and Education, Government of India. It caters the need of the farmers, researchers and policy-makers in managing water resources efficiently in agriculture. Issues are lack of water resource databases, existing wide gap between demand and supply under canal commands being short of optimal crop plan, too little conjunctive use plan of surface and groundwater, insufficient use of pressurised irrigation in high value crops for improved water-use efficiency and water productivity; over exploitation of groundwater, inadequate groundwater recharge structures, lack of recharge plans, inadequate water harvesting structures in high rainfall areas, decline in the quality of surface and groundwater, not enough treatment facilities to treat wastewater and recycle and so on. The problem of managing water resources efficiently is multifaceted and its magnitude is sky high. To deliver at the maximum level scientific knowledge is not enough and requires participatory approach and strong linkages between the stakeholders. Technologies are available but its implementation at the right time at right place is demanding.

Priorities

At National level the project of interlinking of rivers will divert water resources from surplus to deficit regions. This in turn will mitigate the hydrological extremes like flood and droughts in the basins to a certain extent. Infrastructure to be developed to store the excess runoff to adapt intermittent drought. Since independence a total storage capacity of 212.78 km^3 has been created in the country through major and medium projects and an additional 76.26 km^3 is under consideration (ADB, 2009). The irrigation efficiency in the canal command areas to be increased to adapt intermittent drought conditions in a larger way and the priorities are assessment of available water resources, secondary storage structure to store the excess runoff, optimal crop plan (cropping pattern and cropping sequence) based on crop water requirement for enhancing on-farm water use efficiency, conjunctive use of surface and groundwater, use of pressurised irrigation system, irrigation scheduling based on IW/ CPE and multiple use of water resources. Groundwater recharge may be enhanced in areas of over exploitation through delineation of potential zones and suitable recharge areas based on aquifer properties, construction of artificial recharge structures like check dams, percolation pond, recharge shaft, rainwater harvesting cum recharge structures, preparation of recharge plans for basin areas to maintain the stage of groundwater development within the safe category as per the guidelines of the CGWB, skimming technology to control sea water intrusion in the coastal aquifers and prevention of groundwater pollution from other sources. On the other hand in areas within the safe category of groundwater potential infrastructure may be developed and regularised electricity may be provided for sustainable development of groundwater leading towards climate resilient agriculture. Wastewater is growing, especially urban effluent and becoming an important resource to adapt climatic variability for peri-urban agriculture. Judicious use of wastewater in agriculture safeguarding the risk to ecology, environment and human health is quite challenging. Methodology, protocols and national policies need to be derived for its safe use in agriculture which will be a bold step towards climate smart agriculture.

Water Management at Regional Level

ICAR-IIWM has recommended specific water management and conservation practices for the entire country based on location specific research in collaboration with SAUs and other ICAR Institutes (Table 2).

Table 2. Region wise water management technologies

Region/States	Climatic variability and water management issues	Recommended water management and conservation practices
Northern region (Jammu and Kashmir, Punjab, Haryana, Himachal Pradesh, Uttarakhand)	Water scarcity, over-exploitation of groundwater, inadequate and erratic canal water supply, decline in groundwater quality,	Laser levelling, assessing the impact of climatic variability on groundwater, remediation of heavy metal contamination in groundwater, irrigation

	decreasing trend of stream discharges, production of high water demanding crops, drying of artesian wells, development of new water logged areas due to seepage along the canals, water quality deteriorated due to development of industries, high irrigation cost, and inadequate electricity supply (Uttarakhand).	scheduling based on IW/CPE, micro irrigation like drip and sprinkler methods, methods of irrigation like border strip method, ridges and furrow, furrow irrigated raised bed system, alternate furrow irrigation, SRI, bed planting, wet bed planting etc.
Western region (Gujarat, Rajasthan, Maharashtra)	Frequent drought, uncertainty in canal water supply (Rajasthan), overexploitation of groundwater (Rajasthan), groundwater affected by salinity and alkalinity.	Identification of suitable groundwater recharge sites, water harvesting cum groundwater recharge, aquifer properties, resisting sea water intrusion through skimming, establishment of rainfall runoff relationships, assessment of groundwater contamination, conjunctive use, irrigation scheduling based on IW/CPE, micro irrigation such as drip and sprinkler methods of irrigation, and alternate methods.
Central Region (Uttar Pradesh, Madhya Pradesh and Chhattisgarh)	Floods on river banks, water stagnation during *kharif*, poor and interrupted power supply, high irrigation cost (Uttar Pradesh), low water use efficiency (WUE), increased irrigation cost, deterioration of groundwater quality (Maharashtra).	Assessing potential of groundwater, groundwater recharge through rain water harvesting, performance evaluation of pumps, potential and classification of groundwater quality for irrigation, improved pumping systems, groundwater recharge through percolation tank, recharge shaft for recharge in Haveli fields, recharge filters, conjunctive use, efficient water use in canal command, participatory irrigation management, soil aquifer treatment, water quality assessment, ground water potential estimation, strategic plans for GW stress areas and GW safe areas.

Southern region (Andhra Pradesh, Karnataka, Kerala, Tamil Nadu)	Intermittent floods and drought condition during summer, salinisation of groundwater, inequality in water distribution, poorly maintained waterways and tanks, inadequate power supply, and decline in water quality due to pollution.	Irrigation scheduling based on IW/CPE, alternate furrow irrigation, border strip irrigation, drip and sprinkler irrigations, mulching, alternate wetting and drying, System of Rice Intensification (SRI), paired row system, sub-surface drip fertigation, recharge shaft, conjunctive use of surface and groundwater
Eastern region (Bihar, Odisha, West Bengal)	Frequent drought and flood, 20-30 per cent of potential is exploited in canal system (Bihar), Drainage problem, formation of hard pans, low WUE due to conveyance losses, lack of awareness among farmers about efficient management of water (Odisha) and poor groundwater quality in As affected areas and irrigation water scarcity in various crop stages	Groundwater development, conjunctive use, irrigation scheduling based on IW/CPE, SRI, drip and fertigation, drum seeded rice, paira cropping in rice (West Bengal).
North-eastern region (Assam, Arunachal Pradesh, Meghalaya, Manipur, Mizoram, Nagaland, Sikkim, Tripura)	Inadequate storage capacity, Water scarcity during summer, intermittent drought, flooded fertile lands, drainage problem in the valley, predominance of iron, arsenic and fluoride, poor adoption of irrigation principles due to increasing cost of pumping and lifting, and inadequate supply of electricity for the purpose of irrigation.	Rainwater harvesting, mulching, furrow liming, residue management and conservation tillage, intercropping in terrace, in-situ soil, moisture conservation, improved irrigation and water management through micro irrigation, pre-sowing irrigation, 2, 3 irrigations and seed hardening, drainage management.

for climate smart agriculture

Source: Compiled from AICRP on GWU, 2010, 2011, 2012, 2013 and AICRP on WM, 2010, 2011, 2012, 2013.

The water management technologies cited above were studied for most of the cereals, pulses, oilseeds, vegetables and fruits grown in different regions under different agroclimatic zones. The adoption of technologies varied widely over the region due to lack of awareness among the farmers and socio-economic conditions. It has been observed that the adoption is higher where water resources are limited as in the northern region followed by southern region; even the success rates are also higher. The water saving due to wider adoption of technologies ranged from 10 to 80 per cent and the yield increased to a maximum by 200 per cent.

Conclusion

Climate change is scientifically accepted and its impact on water resources in all its forms in terms of quantity and quality are adversely affected. Immense pressure on foodgrain production from irrigated lands exercises huge stress on available water resources. Infrastructure for conservation of water, more equitable distribution of water resources over the country, groundwater recharge and development, increasing water use efficiency and irrigation efficiency through integrated water resource management, recycling and safe use of wastewater in agriculture needs to be focussed to achieve climate smart agriculture. The Institute is working hand in hand with National Water Mission in achieving the goal of increasing water use efficiency by 20 per cent. A sincere effort to develop a comprehensive database on water resources over the region and location specific studies on impact of climate change on water resources is the need of the hour. The available technologies need to be prioritised and implemented for immediate relief and block-wise contingency plans to be developed for implementation in near future. Participatory irrigation management through Water Users Association and information technology are useful tools in transmitting technological know hows, meteorological information to the end users. Capacity building of the stakeholders, linkages with government and non-government organisation, public-private partnership will strengthen dissemination of technologies to the beneficiaries and collecting feedback too. Water being a state subject requires high level coordination from different departments and neighbouring states for its efficient management. Lastly to arrest the excess water, to prevent overexploitation, to control pollution and to recycle wastewater, policies are to be framed and implemented. This will lead towards sustainable development of surface water, groundwater and wastewater in the country in terms of quantity and quality.

References

AICRP on GWU. 2010, 2011, 2012, 2013. Annual Report 2009-10, 2010-11, 2011-12, 2012-13. AICRP on Groundwater Utilization. Directorate of Water Management, Bhubaneswar, Odisha 751023. Website: www.dwm.res.in

AICRP on WM. 2010, 2011, 2012, 2013. Annual Report 2009-10, 2010-11, 2011-12, 2012-13. AICRP on Water Management. Directorate of Water Management, Bhubaneswar, Odisha 751023. Website: www.dwm.res.in

Asian Development Bank (ADB). 2009. Water Resources Development in India: Critical Issues and Strategic Options. New Delhi: Asian Development Bank. Website: http://www.adb.org/Documents/Assessments/Water/IN D/Water-Assessment.pdf.

Central Water Commission (CWC), Ministry of Water Resources, Government of India. 2010. Water and Related Statistics. Water Planning and Projects Wing. Website http://www.indiaenvironmentportal.org.in/files/water%20and%20related%20statistics.pdf

Narsimhan, T.N. 2008. A note on India's water budget and evapotranspiration. Journal of Earth System Science. Vol. 117. No. 3. pp. 237- 240.

National Water Mission (NWM). 2009. Ministry of Water Resources, Government of India. website: http://india.gov.in/allimpfrms/alldocs/15658.pdf.

Planning Commission. 2011. Faster, Sustainable and more Inclusive Growth: An approach paper to 12th Five Year Plan (2012-2017). Planning Commission, Government of India. Website www.planningcommission.nic.in.

18

Wastewater Management and Climate Smart Agriculture

Mausumi Raychaudhuri and S. Raychaudhuri

Abstract

Water is one of the world's most valuable resources, yet it is under constant threat due to climate change and resulting drought, explosive population growth, and waste. One of the most promising efforts to stem the global water crisis is industrial and municipal water reclamation and reuse. Water reuse allows communities to become less dependent on groundwater and surface water sources and can decrease the diversion of water from sensitive ecosystems. Additionally, water reuse may reduce the nutrient loads from wastewater discharges into waterways, thereby reducing and preventing pollution. This 'new' water source may also be used to replenish overdrawn water sources and rejuvenate or re-establish those previously destroyed. The present paper give an insight into the appropriate technology for treatment of wastewater and its reuse in agriculture for sustaining crop productivity in the arena of climatic change.

Key words: *waste water, reuse, treatment, agriculture, productivity,*

Introduction

Wastewater use in agriculture is growing to meet the fresh water scarcity and huge demand for food production to feed the growing population of the country. It is also a readily chosen option by the periurban farmers to combat climatic variability to provide life saving irrigations whenever there is moisture stress due to deficit and scanty rainfall during the cropping season. Climate smart agriculture ensures sustainability in terms of productivity and income, enhances resilience to climate change and variability, reduces agriculture's contribution to climate change by reducing green house gas emissions and increasing carbon storage on farm lands. Wastewater irrigation certainly acts as an additional water resource to adapt climatic variability and improves livelihood of the farmers but the undecomposed organic matter contained in it may contribute towards green house gas emissions and carbon sequestration both. With the increase in frequency of drought situation, frequency of wastewater use in agriculture also increases which is of concern. The surface water bodies are also being polluted day by day owing to discharge of untreated domestic wastewater in them and becoming poor quality water. The water quality monitoring results obtained during 1995 to 2011 by Central Pollution Control Board (CPCB) indicate that the organic and bacterial contamination are continued to be critical in water bodies. Besides, hardness, alkalinity and heavy metal content especially lead (Pb), cadmium (Cd), chromium (Cr) and nickel (Ni) also increased beyond the permissible limits. With changes in the rainfall pattern receiving water bodies do not get adequate water for dilution. Therefore, the oxygen demand, bacterial pollution and heavy metal contamination is increasing day by day posing threat to environment and mankind.

Water availability

India with geographical area of 329 M ha covers only 2.42 per cent of total earth contains 4 per cent of the total available fresh water to support 17 per cent (Census, 2011) of world's population. It receives about 4000 BCM of rain water annually but often marred by spatio-temporal variations. The water resources potential of the country which occurs as natural runoff in the rivers is estimated at about 1869 BCM. Considering both uneven distribution of resource over space and time about 1121.3 BCM of the total potential can be put to beneficial use, 690 BCM through surface water resources and 431.3 BCM by groundwater. The total annual fresh water requirement in India by 2025 and 2050 is estimated to be 1093 and 1447 BCM respectively where domestic requirement will be 30 to 40 per cent more than present requirement with concomitant increase in domestic wastewater generation to the tune of 58 and 82 BCM (Table 1). By 2050 there is a rapid decline in future availability of so-far-unused water resources (Fig. 1). Under such circumstances untreated wastewater becomes an alternative source for irrigation and food production. The urban conglomerations which are hub of multifarious activities will produce huge quantity of wastewater with channelised discharges from domestic, service, industrial sectors (mainly small and medium sector units in urban centres) etc will be a huge resources to address water scarcity issues if managed scientifically otherwise may have serious environmental repercussions.

Wastewater production and treatment

The estimated sewage generation from Class I cities and Class II towns together in India is 38254 mld , out of which only 11787 mld is being treated with a capacity gap of 26467 mld which needs urgent attention of all concerned (CUPS /70/2009-10). The total wastewater constitutes about 63 per cent from domestic discharges and 37 per cent from industrial discharges. The treatment capacity available for domestic wastewater is only 26 per cent as against 59 per cent for industrial wastewater. About 74 per cent of generated domestic wastewater is disposed of untreated in the water bodies directly contaminating fresh water. A good number of sewage treatment plants under various river action plans have also been created over the past two decades in India. Numerous technologies like activated sludge process, trickling filter, waste stabilization ponds, UASB and other new technologies have been applied. The huge installation (approximately 80 120 lakh per MLD of average pollution load), maintenance and operational costs (approximately 172 lakhs per annum

Table 1: Sector-wise water requirement in India

Annual Requirement of Water in India (in BCM)			
	Standing Sub-Committee of MoWR		
Water uses	**2010**	**2025**	**2050**
Domestic	56	73	102
Irrigation	688	910	1072
Industry	12	23	63
Energy	5	15	130
Others	52	72	80
Total	**813**	**1093**	**1447**

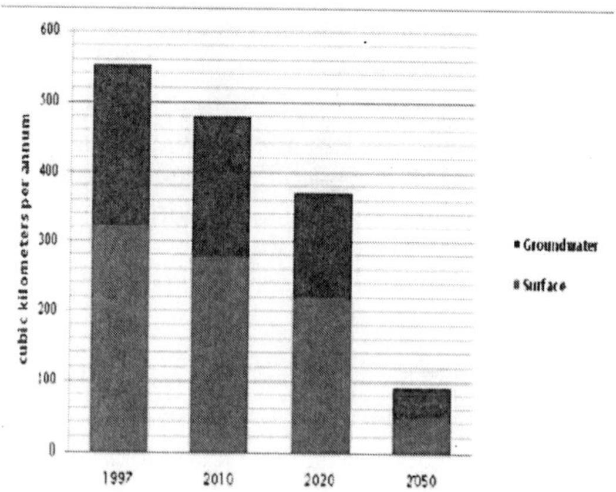

Fig. 1. Availability of unused water resources in India.

Source : World Bank, 2005.

per MLD excluding the cost of electricity) of conventional biological treatment (Kodarkar and Joshi, 2009) coupled with land scarcity and technological difficulty are the limitations and makes it unrealistic proposition and yet unsustainable, to treat all generated wastewater (sewage) in any growing city on sustainable basis.

Wastewater use in agriculture (Indian Scenario)

In India, wastewater is generally used for irrigating crops in periurban areas and as aquaculture stabilisation ponds. Strauss and Blumenthal (1990) estimated the area under wastewater irrigation to be over 73000 ha although no official estimates are available but actual figures seems to be much higher. Wastewater is used for irrigation along rivers which flow through rapidly growing cities as Delhi, Kolkata, Coimbatore, Hyderabad, Indore, Kanpur, Patna, Vaḍodara, Varanasi, Dharwad, Bhubaneswar etc. Along the rivers, the water is diverted via anicuts (weirs) to canals and often to tanks and then channeled to the fields for irrigation. If such uses were included, a much higher figure than 73000 ha would be obtained as in Musi river, Hyderabad alone approximately 40500 ha is irrigated with wastewater (van der Hoek, 2004). Kolkata (formerly Calcutta) has a long history of using wastewater stabilisation tanks for aquaculture. An estimated 2.4 t/ha (Gopal *et al.,* 1991) of fish is produced annually in Kolkata from about 3200 ha of ponds with inflow of about 3 m^3/sec. Table 2 shows the area under wastewater use reported upon. Another estimation says that wastewater contributes to irrigating 600,000 hectares in India (http://www.iwmi.cgiar.org /assessment/files_new/ research_projects/ Urban%20Wastewater- Full_Report.pdf).

A variety of food and non-food crops are irrigated with urban wastewater water supporting the livelihoods of the households of millions of farmers in India. In a research project funded by the German Federal Ministry of Economic Cooperation and Development (BMZ) and coordinated IWMI, risks and benefits of wastewater irrigation in six villages along the Musi River found that contrary to previous suggestions (Clemett and Ensink, 2006), biodiversity under wastewater irrigation did not decline. A great variety of vegetables was produced with Musi water (Jacobi *et al.,* 2009). Van der Hoek (2004) reported approximately 40500 ha is irrigated with wastewater in Hyderabad along Musi river. Both direct and lift irrigation from canals using pumps exists. Vegetables and paragrass are cultivated mostly in the periurban zones whereas paddy is the major crop in the rural zone. The rural areas downstream of Vadodara in Gujarat, India, present an interesting case where wastewater supports annual agricultural production worth Rs 266 million (US$5.5 million). Both food crops and cash crops are irrigated by domestic wastewater and industrial effluent (Bhamoriya, 2004). A large area in and around Bhubaneswar city are irrigated with Gangua nala which receives municipal sewage of the city. A number of

Table 2: Area under wastewater use in India

City	Direct	Indirect use (ha)	Vegetables use (ha)	Rice	Other cereals	Fodder	Fruit trees	Cotton	Fish	References
Ahmedabad	890			“	“	“				Juwarkar *et al,* 1988
Amritsar	1214				“					Juwarkar *et al,* 1988
Bhilai	607		“	“	“					Juwarkar *et al,* 1988
Bikaner	40		“		“					Juwa\rkar *et al,* 1988
Bhubaneswar		2000	“	“						Pers. Comn.
Calcutta		12900	“	“					“	Mara and Carncross, 1989, Edwards. 2001
Delhi	1214		“		“					Juwarkar *et al,* 1988, Farooqui, 2002
Gwalior	202		“	“	“					Juwarkar *et al,* 1988
Hubli-Dharwad		“			“	“	“			

(*Contd...*)

Hyderabad	110	40500	“	“		“	“	“	Buechler and Devi, 2002
Jamshedpur	113				“	“			IWMI (Bhamoriya, 2002)
Kanpur	1300		“	“	“				Strauss and Blumerthal, 1990
Lucknow	150		“	“					Juwarkar *et al,* 1988
Madras	133					“			Juwarkar *et al,* 1988
Madurai	77					“			Juwarkar *et al,* 1988
Nagpur	1500		“	“	“	“			Juwarkar *et al,* 1988
Trivandrum	37					“			Juwarkar *et al,* 1988
Vadodara		14567			“				IWMI (Bhamoriya, 2002)

Source : http://old.cseindia.org/programme/health/pdf/conf2006/water2_senguptaAK.pdf

crops viz. paddy, vegetables and sugarcane are grown using Gangua water. A study in DWM on impact of long-term urban wastewater irrigation on soil did not indicate any soil degradation (Raychaudhuri *et al,* 2014). Weckenbrock *et al,* (2009) concluded contrary to common perception, wastewater can be regarded as a valuable resource. If properly managed, wastewater irrigation can make an important contribution to providing fresh, affordable food to rapidly growing populations of cities in water scarce regions.

Impact of Wastewater Use

Wastewater use in agriculture especially sewage water use has been beneficial to the farming community in comparison to fresh water use due to higher crop productivity resulting from high nutrient value of sewage and consequent lower use of fertiliser (Minhas and Samra, 2004). A comparative study on wastewater and fresh water irrigation in five cities by IWMI showed in most cases sewage-fed farming is more profitable (Table 3) (Amerasinghe *et al,* 2013).

Table 3. Comparative incomes from freshwater and wastewater irrigation

City	Crop/ Produce	Variable	With Freshwater	With Wastewater
Delhi	Okra	Net income (INR/month)	7,200	15,700
Kanpur	Buffalo	Net profit/buffalo/day (INR)	28.75	67.15
Kanpur	Rose	Net income (INR/ha)	72,319	65,201
Kanpur	Fodder	Net income (INR/ha)	15,370	2,296
Kanpur	Rice	Net income (INR/ha)	4,455	10,621
Kanpur	Wheat	Net income (INR/ha)	8,259	9,213

In many places, the practice of using sewage for agriculture has been going on for decades and has acquired some degree of formalisation. Several municipalities in Gujarat collect charges for farmers for using wastewater often at the same rates as that applicable to lifting water from notified rivers. For example, the Municipal Corporation of Bhavnagar charges Rs 750/ha and the Municipal Corporation of Rajkot charges Rs 2,500-3,000/ha (Palrecha *et al,* 2012). In addition, in Rajkot city, the arrangement forbids the farmers from withdrawing water from Lalpari Lake that supplies the city in exchange for access to the wastewater. Farmers' cooperatives also take part in auctions for wastewater; examples can be found in the Kutch district of Gujarat (Palrecha *et al,* 2012).

Raychaudhuri *et al,* 2014 reported overall increase in fertility status of soils in terms of major nutrients (N, P and K) along with increase in organic matter when Ganguanala wastewater irrigation was practised in vegetables at Bhubaneswar. They reported increase in Fe, Mn, Cr and Pb concentration by 13, 94, 72 and 21 per cent respectively in surface soils. They reported transfer of heavy metals from soil to plant for different vegetables were in the order, Amaranths Red = Tomato> Amaranths green > Water melon > Malabar Climbing > Bitter gourd > Ladies Finger > Ridge gourd. This shows that consumption of leafy vegetables grown with wastewater irrigation is risky due to higher accumulation of metals in leaves.

Studies conducted (Raychaudhuri *et al,* 2013) in the Patna bye-pass area where farmers grow a variety of vegetables and field crops viz., red spinach, cowpea, okra, radish, bittergourd, sugarbeet, cabbage, turnip, sponge gourd and pumpkin by lifting sewage water through pumps and open wells. Over the last 50 years the area receives a continuous disposal of sewage sludge from the city. Soils, plants and groundwater samples were collected from sixteen sampling sites and analysed to estimate the heavy metal content. The soils nearest to the discharge point of sewage treatment plant contained appreciably higher amount than other sampling sites away from the discharge point. They also reported that leafy vegetables contain more amounts of heavy metals which are of concern.

It has been reported by Anita Singh *et al,* 2010 that in certain areas of Varanasi city, wastewater from Dinapur sewage treatment plant is used for irrigating vegetable plots. They revealed that wastewater used for irrigation had the highest concentration of Zn followed by Pb, Cr, Ni, Cu and Cd. Continuous application of wastewater for more than 20 years has led to accumulation of heavy metals in the soil. Consequently, concentrations of Cd, Pb and Ni have crossed the safe limits for human consumption in all the vegetables. Per cent contribution of fruit vegetables to daily human intake for Cu, Ni, Pb and Cr was higher than that of leafy vegetables, while the reverse was true for Cd and Zn. Target hazard quotient showed health risk to the local population associated with Cd, Pb and Ni contamination of vegetables.

In another study of heavy metal contamination in Musi River and its environs at Hyderabad, Sridhara Chary *et al,* 2008 assessed heavy metals in soils, forage, grass, milk from cattle, leafy and non-leafy vegetables. They found high content of Pb, Zn, Cr and Ni as compared to permissible limits and also reported high hazard quotient (HQ) for Zn, Cr and Pb particularly in leafy vegetables like spinach and amaranthus.

Industrial effluents those are relatively less toxic such as food/beverage sector, for example, breweries, sugar mills and fruit processing units are being used in agriculture. The SABMiller brewery in Andhra Pradesh has been distributing treated wastewater free of charge to local farmers; Ugar Sugar Works in Karnataka has been selling treated effluent from its sugar mill at Rs 3,000/ha to local farmers. However, since industrial wastewater is typically more toxic, agricultural reuse efforts are likely to be concentrated mainly on municipal sewage for the foreseeable future.

Management Strategies

Assessment of wastewater resources

Generation of wastewater resources is increasing day by day and its lion's share is disposed in the environment without any treatment posing threat to ecology and mankind. Further when this untreated water is used for irrigation the soil, crop, the producers and consumers are also affected. No proper data is available citing the actual area of cultivation under wastewater. Using modern tools like remote sensing and GIS it is necessary to assess and map the wastewater irrigated areas putting stress on periurban areas where actually it is being practised. The spatio-temporal variations in wastewater characteristics offer opportunities for decentralised treatment, segregation of different qualities of water to reduce treatment costs. This will help the researchers, policy-makers and planners in finding out the causative factor of pollution, developing decentralised treatment facilities to reduce contaminants, deriving policies for proper use of wastewater based on quality and also to monitor and study the impact of wastewater on natural resources and human health.

Conjunctive use

The higher content of salinity and heavy metals in marginal water can be reduced or diluted through conjunctive use with fresh water and in this way a considerable amount of fresh water can be saved. An experiment in peri-urban Bhubaneswar area showed that conjunctive use of wastewater with other source may save up to 50 per cent of fresh water. The heavy metal concentrations were within permissible in edible parts of vegetables grown with conjunctive irrigation with river water and were lower than with only wastewater irrigation. (Raychaudhuri *et al,* 2014, and Raychaudhuri *et al,* 2013). In another experiment conducted at Bundi district, Udaipur revealed that 33 per cent of fresh water can be saved using poor quality saline groundwater for irrigation in wheat without any yield loss. Conjunctive use of canal water and marginally saline groundwater for wheat cultivation under calcareous soil of Bundi district of Udaipur resulted in of saving of good quality water without any adverse effects (AICRP GWU, 2011).

Water Quality Index (WQI)

To prevent the negative impact of poor quality water/wastewater a measuring tool is required to assess the suitability for different uses which was addressed through development of WQI. It is a single score derived by considering different important parameters of water quality. It is an integration of the individual effect of all the parameters in right proportion in deciding the quality of water. The index was validated using Central Groundwater Board data and found trustworthy and easy to compute. Its applicability will help in decision-making in reusing poor quality water safely for irrigation and also help in finding the causative factor responsible for contaminating the water. This index will be of immense use to the researcher, policy- maker, users and other stakeholders in near future (Raychaudhuri *et al,* 2014).

Decision support system

Wastewater use in agriculture is spontaneous and indispensable. Development of a computer based decision support system for usage of irrigation water according to its quality could be a base for reduction the impact of low quality water used in irrigated agriculture on food and environment. Under Indo-Bulgaria bilateral Project a methodology has been developed and its application to assess suitability of water quality for irrigation will helping minimising the risk of use of wastewater in agriculture. Creating a computer based decision support system, the methodology will be used by local agriculture extension services, unions of water users, administrations of irrigation systems enterprises, as well as by farmers (Kalcheva et al 2010)

Small scale filter

Farmers are more inclined towards using urban wastewater as they found it as a cheap nutritional resource unaware of the poor water quality that contaminates soil, crop and groundwater. ICAR-Institute of Water Management (IIWM, Formerly Directorate of Water Management) has initiated a project to devise a on-line filter that reduces the undesirable/unwanted/toxic contaminants present from the wastewater (domestic and municipal) is a boon to the small scale holders. Laboratory studies conducted revealed that combination of gravel, quartz sand, and jute fibre (in order of flow) was most satisfactory in significantly reducing suspended particles as well as microbial load including total coliform and E. coli. To reduce heavy metal contamination from aqueous solution modified charcoal was used. Initial results have shown continuous availability of 1 lps clean water with much reduced heavy metal contamination (DWM, 2014). There is a scope of further research and modification of the device for increased discharge and more reduction in the contaminants.

Conclusion

Wastewater is a reliable source to adapt climatic variability if managed scientifically. Wastewater generation will increase vehemently and large centralised treatment systems appear to be unsustainable and require huge expenses due to requirement of large tracts of land, high energy and skilled labour for maintenance. Thus, an inclusive management system needs to be adopted for urban wastewater where intervention should start from the point of generation followed by exploring reuse options with in the urban area and safe use in agriculture in the downstream. This could save fresh water for high value potable or non-potable use. Suitable protocols need to be developed for safe use in a multi barrier approach like, proper crop selection, appropriate irrigation methods, use of ameliorative agents and low cost treatments.

References

AICRP on GWU. 2011. Annual Report 2010-11. AICRP on Groundwater Utilization. Direcorate of Water Management, Bhubaneswar, Odisha 751023. Website: www.dwm.res.in

Amerasinghe, P., Bhardwaj, R.M., Scott, C., Jella, K., and Marshall, F. 2013. *Urban Wastewater and Agricultural Reuse Challenges in India*. International Water Management Institute (IWMI) Research Report 147.

Anita Singh, Rajesh Kumar Sharma, Madhoolika Agrawal, Fiona M. Marshall, 2010. Health risk assessment of heavy metals via dietary intake of foodstuffs from the wastewater irrigated site of a dry tropical area of India. Food and Chemical Toxicology 48 (2010) 611619.

Bhamoriya, V. 2004. Wastewater Irrigation in Vadodara, Gujarat, India: Economic Catalyst for Marginalised Communities In : Wastewater Use in Irrigated Agriculture Coordinating the Livelihood and Environmental Realities Eds C.A. Scott N.I. Faruqui and L. Raschid-Sally, CAB *International in association with IWMI and IDRC*, pp. 127-134.

Bhamoriya, V. 2002. Wastewater and Welfare: Pump Irrigation Economy of Peri-urban Vadodara. Annual Partners' Meet 2002, *IWMI Tata Water Policy Program*, Anand, Gujarat, India. Census 2011. http://www.censusindia.gov.in

Clemett, A. and Ensink J. 2006. Farmer driven wastewater A case study from Faisalabad, Pakistan treatment: http//wedc.lboro.ac.uk/conferences/pdfs/32/ Clemett.pdf

CUPS 70/2009-10. Control of Urban Pollution Series: Status of water supply, wastewater generation and treatment in class-I cities and class- II towns of India. CPCB, New Delhi, p. 89.

DWM 2014. Annual Report 2013-14. Directorate of Water Management (ICAR) Bhubaneswar, India. 96p. Website: www.dwm.res.in

Gopal, B., Bandhopadhyay, S., Sah, M., and Chatterjee, K. (1991), *Land Application of Municipal Sewage for Resource Recovery: the Indian Experience*. National Institute of Ecology, New Delhi, India.

Jacobi, J, Drescher, A, Amerasinghe, P. 2009. Agricultural Biodiversity: Strengthening Livelihoods in Periurban Hyderabad, India. *Urban Agriculture Magazine* 22 : 45 - 47.

Juwarkar, A. S. 1988. Jal Vigyan Samiksha, 3(1) : 82-96.

Juwarkar, A. S. 1991. Proc. Intntl. Conf. On Environment, NEERI, Nagpur.

Kalcheva, Spaska, Dimitrova, Ivanka, Kumar Ashwani, Raychaudhuri, Mausumi and Dimitrov, Peter. 2010. Model and Software Development for Applicability of Irrigation Water Depending on Its Quality. BALWOIS. Proceedings electronic. 25-29 May 2010, Ohrid, Republic of Macedonia, pp. 1-5.

Kodarkar, Mohan and Joshi, Sandip 2009. IBM Impact Story Ecological restoration of highly polluted stretch of Ahar river, Udaipur abs ecological improvement of Udaisagar lake, Rajasthan, India.

Minhas, P.S., and Samra, J.S. 2004. *Wastewater Use in Peri-urban Agriculture: Impacts and Opportunities.* Bulletin No. 2, Central Soil Salinity Research Institute, Karnal, India.

Palrecha, A., Kapoor, D., and Malladi, T. 2012. Wastewater irrigation in Gujarat: An exploratory study. IWMI-TATA Water Policy Program. Available at: www.iwmi.cgiar.org/iwmi-tata/PDFs/2012_Highlight-30.pdf

Raychaudhuri, Mausumi, Singh, A. K., Jain, S. K. Chandra, Ravish and Kumar, Ashwani. 2013. Web site https://indiawaterweek.water.tallyfox.com/documents/heavy-metal-accumulation-crops-irrigated-poor-quality-water

Raychaudhuri, Mausumi, Raychaudhuri, S., Jena, S. K., Kumar Ashwani and Srivastava, R. C. 2014. WQI to monitor water quality for irrigation and potable use. DWM Bulletin No. 71, Directorate of Water Management, Bhubaneswar, Odisha, 43p.

Raychaudhuri, Sachidulal, Raychaudhuri, Mausumi, Rautaray, Sachin Kanta and Kumar Ashwani, 2014, DWM Bulletin No. 64, Directorate of Water Management, Bhubaneswar, Odisha, p. 31.

Raychaudhuri, S., Raychaudhuri, Mausumi, Rautaray, S. K. and Kumar, Ashwani. 2013. Conjunctive use of urban wastewater with other water source for vegetable production. India Water Week, 2013.(e publication -*issuu.com/indiawaterweek.in/docs/s-13_dr_sachidulal_raychaudhuriý*)

Sachidulal Raychaudhuri, Mausumi Raychaudhuri, Sachin Kanta Rautaray, Ashwani Kumar. 2014. Impact of urban wastewater irrigation on soil and crop. DWM Bulletin No. 64. Directorate of Water Management, Bhubaneswar, Odisha.

Sridhara Chary, N., C.T. Kamala and R.D. Samuel Suman. 2008. Assessing risk of heavy metals from consuming food grown on sewage irrigated soils and food chain transfer. *Ecotoxicology and Environmental Safety* 69: 513-524.

Strauss, M. and Blumenthal, U. 1990. Use of human waste in agriculture and aquaculture. IRCWD Report No. 08/90, Duebenforf, Switzerland

Van der Hoek, W. 2004. A Framework for a Global Assessment of the Extent of Wastewater Irrigation: In: The Need for a Common Wastewater Typology, Scott, C. A., Farunqui, N. I., Raschid-Sally, L. Wastewater Use in Irrigated Agriculture, Confronting the Livelihood and Environmental Realities, 11-24, CAB International, International Water Management Institute, International Development Research Centre, Trowbridge.

Weckenbrock, P., Dresher, A. Amarasinghe, P., Simmons, R. W. and Jacobi, J. 2009. Lower than expected risk of wastewater irrigated agriculture along the Musi river, India. 2nd German Indian Conference on Research for Sustainability, Bonn.

World Bank, 2005. *India's Water Economy: Bracing for a Turbulent Future*. Report No. 34750-IN. World Bank Agriculture and Rural Development Unit, South Asia Region.

Website:

http://www.iwmi.cgiar.org /assessment/files_new/research_projects/ Urban%20 Wastewater- Full_Report.pdf

19

Runoff Yield Modelling Using Artificial Neural Networks for a Small Agricultural Watershed of Dhar, Madhya Pradesh

Satish K. Sharma, M.K. Hardaha and Vijay K.S ingh

Abstract

Runoff is an important variable in hydrological cycle making proper estimation of runoff water yield which further needful in crop management decisions and planning. Also watershed development programme requires estimation of runoff right from watershed prioritisation to design of soil and water conservation structures. Runoff models are the key components of the watershed modelling in spatial and temporal basis. A number of parametric models have been developed to predict runoff and soil loss and these models are based on soil, land use, land cover, and land forms, climatic and other topographic features of watershed. Accurate estimation of runoff yield is required for proper watershed management and other developmental works. An attractive feature of ANNs is their ability to extract the relationship between input and outputs of a process without the physics being explicitly provided to them. These properties suggest that ANNs may be well suited for modelling for estimation of runoff as well as soil loss in any watershed. Artificial neural network (ANN) models are developed to predict runoff for a small agricultural watershed in well manner. Simulated runoff was compared with observed values further evaluation and validation criterions were used for selecting the best performing models for its prediction. The study area is a small agricultural watershed viz. Kalidevi watershed, where this study was carried out comes under the Bagh block of the Kukshi tehsil of district Dhar of Madhya

Pradesh has been chosen for developing artificial neural network model for estimation of runoff. Daily rainfall, API and days since 1st June data were taken as input variables of period 2003-2005 during rainy seasons have been used for the analysis and development of model for runoff. For runoff modelling, it was found that an ANN model with three input variables with 25 neurons single hidden layer and one output and learned with Levenverg-Marquardt (LM) algorithm performed best.

Introduction

Runoff is one of the major processes of the hydrological cycle. It is also necessary component for any water balance assessment for different watershed planning design operation and management studies including hydrology, irrigation water management flood forecasting for this modelling runoff is required. Runoff is perhaps most difficult to estimate owing to complex interaction between the components of the soil water plant atmosphere system. Hence, it is an important variable in making the runoff water yield and crop management decision. Also estimation of the runoff yield from the soil surface is important to understand the degree of soil degradation which restricts the land for potential agricultural use. Watershed development programme requires estimation of runoff right from watershed prioritisation to design of soil and water conservation structures also. Runoff models are the key components of the watershed modelling in spatial and temporal basis. A number of parametric models have been developed to predict runoff and soil loss and these models are based on soil, land use, land cover, and land forms, climatic and other topographic features of watershed. Furthermore such models proves themselves highly cost as well as time effective to predict runoff and soil loss from a particular area in particular time period used in combination. Artificial Neural Networks are highly adaptive to non-parametric data distribution and while other statistical methodologies require a set of assumptions to be fulfilled, the former make no prior hypotheses about the relationship between the variables. Neural networks are also less sensitive to error term, assumptions and they can tolerate noise, chaotic components and heavy tails better than most of the other methods.

An attractive feature of ANNs is their ability to extract the relationship between input and outputs of a process without the physics being explicitly provided to them. These properties suggest that ANNs may be well suited for modelling for estimation of runoff as well as soil loss in any watershed. Artificial neural network (ANN) models are developed to predict runoff and soil loss for a small agricultural watershed. Simulated runoff and soil loss were compared with observed values further evaluation and validation criterions were used for selecting the best performing models for prediction.

Study Area

The study area Kalidevi watershed where proposed study was carried out comes under the Bagh block of the Kukshi tehsil of District Dhar (MP). The Dhar district is having a total geographical area of about 0.82 million hectares. The Dhar district of Madhya Pradesh is located between Malwa Plateau and Nimar valley centrally divided by Vindhyan range of hills and covers two different Agro-climatic zones viz. zone-XI Nimar Valley and zone-XII Jhabua Hills (Mehta, 2004).The district's northern region comes in the catchments of Mahi and Chambal rivers and southern region is in Narmada River catchments through Bagh and Man tributaries of the river. The study area Kalidevi Watershed lies between 74° 50′ to 74° 55′E longitude and 22° 30′ to 22° 35′ N latitude and having total geographical area of 3135.12 hectares. The watershed is located in Agro-climatic zone-XII Jhabua Hills of Madhya Pradesh. The topography of the watershed is undulating with 1/6th of its total area as wasteland or forest and only 40 per cent land is under cultivation and remaining part have very high altitude (190 m to 800 m MSL). Maximum part of the study area has shallow soil depth (Anonymous, 2006).

Material and Method

Climate

The climate plays an important role in soil development and watershed management. The intensity of weathering depends on climate. Dhar district has been classified as transitional ecosystem with annual rainfall of 1085 mm and of moist semi-arid and dry sub-humid climate type with index (-) 41.9. The mean annual temperature varies from 18 to 22°C with annual mean of 24.5° C. The region experiences hot summer and mild winter with mean summer temperature of 29.5°C and mean winter temperature of 18.7 °C.

Topography and Cropping Pattern

Maximum part of the study area was having of shallow soil depth (25-50 cm), some area was under slightly deep soil category (50-75 cm) and also little part has deep soil (>100 cm).Various types of crop cover (land use) were reported about the watershed for the year 2003 to 2005 Report about the study area was dominated mainly by Soybean+ Maize crop and larger area of the watershed was found under mixed cropping system. Open forest, dense forest, water body and wasteland were also reported in during 2003 to 2005 years.

Data collection

Daily rainfall, temperature, runoff and soil loss data for the period 2003-2005 during rainy seasons were recorded by the project Land Use Planning and Management of Natural Resources of Dhar District through Remote

Sensing, College of Agriculture, Indore (M.P.) have been used, Verma *et.al* (2006) for the analysis and development of model for runoff.

Development of ANN based runoff model

To develop ANN based model for runoff estimation; one input layer, one hidden layer and one output layer has been considered to design an ANN network. The input variables are rainfall, average temperature, Antecedent Precipitation Index (API- last five days rainfall amount), and days since 1st June. All models for runoff estimation use rainfall as input variable. Temperature of the day effect the evaporation rate, thus affect the runoff. Initially the ANN architecture consisted of all four input variables, 45 numbers of neurons and training algorithm as Levenberg-Marquardt (Coulibaly, *et al.* 2000). Sixty per cent of total length of data was used for training of the model and rest 40 percent for validation. The "tansig" and "purelin" activation function were used as the transfer function at hidden and output layers respectively Fausett, L. (1994), Haykin, S. (1994) and Haykin, S. (1999).

ANN architecture

The input dataset for runoff modelling includes daily rainfall, average temperature of the day, Antecedent Precipitation Index (API) and days since 1st June for 276 days of record for the rainy season of year 2003, 2004 and 2005. The ANN models have been developed in Neural Network Module of MATLAB (6.5 version, 2006). Model performance has been evaluated in terms of R, RMSE and MAE. The basic ANN architecture was optimized in terms of training algorithm, number of neurons in the hidden layer, input variables and length of data set for training of the model, for runoff estimation. Twelve algorithms for training the neural network have been evaluated. Various combinations of input variables have been studied to improve the model efficiency. Model performance has been evaluated with neurons varied from 10 to 90 in the hidden layer.

Normalisation methods

Normalisation of input and target data was done by two methods namely, Minimum-Maximum and Mean-Standard Deviation. By using function mapminmax to scale inputs and targets, so that they fall in the range [-1,1]. Minimum value of data vector is assigned -1 and maximum values are assigned 1. Another approach for scaling network inputs and targets is to normalise the mean and standard deviation of the training set. The function mapstd normalises the inputs and targets so that they will have zero mean and unity standard deviation. These two methods have been used for training the network with algorithm- trainlm and varying numbers of neuron. The method of normalisation that performed better during training and validation was selected for further improvement of network.

Training algorithms

During the training of the network, the weight and biases of the network are iteratively adjusted to minimise the network performance function. The basic back propagation algorithm used here are, in which the weights are moved in the direction of the negative gradient. Back propagation training algorithms have been used evaluate the performance of ANN. Total 12 no. of training algorithms were tried in ANN runoff modelling viz. trainlm, traingdtraingdm, traingda, traingdx, trainrp, traincgf, traincgp, traincgb, trainscg, trainbfg,andtrainoss.

ANN models with combinations of input variables

Four input variables namely, rainfall, average temperature, API, days since 1st June and one output variable- runoff depth used in the model development. It was assumed that all four variables have significant effect on performance of ANN model. Various models developed with combinations of input variables are listed as M1,M2,M3,M4,M12,M13,M14,M123, M124,M134 and M1234.

Table 1: ANN models with combinations of input variables.

Model	*Output*	*Input*			
	Y	*X1*	*X2*	*X3*	*X4*
	Runoff, mm	*Rainfall, mm*	*Av. Temp. °C*	*API*	*Days since 1st June*
M1	√	√			
M2	√		√		
M3	√			√	
M4	√				√
M12	√	√	√		
M13	√	√		√	
M14	√	√			√
M123	√	√	√	√	
M124	√	√	√		√
M134	√	√		√	√
M1234	√	√	√	√	√

Number of neurons

ANN models have been developed and evaluated for their performance had 45 numbers of neurons. Model with more number of neurons takes more time for training. Also very high number of neurons does not help in generalisation of the model. Hence by trial and error, optimum number of neurons was selected by evaluating the model performance with varying number of neurons i.e. 25, 35, 45, 55, and 65. Number of neurons that performed better was considered for further improvement of ANN model.

Length of record for training and validation

The datasets included the daily record of four input variables for the total duration of 276 days during monsoon periods of year 2003, 2004 and 2005. M50 considered 50 per cent of total data set for training and 50 per cent for validation, similarly M60 considered 60 per cent of total data set for training and 40 per cent for validation and so on for M70 and M80.

Results and Discussion

Comparative performance of the two best ANN based models among other, in Table 2 for small Agricultural watershed are discussed as under:

Model M-134- Daily rainfall, API and days since the 1st June are the input variables for this model. The model performed very well as depicted in Fig. 2. Mean squared error obtained during the training is 0.00022747. Relationship between estimated runoff by the model and observed runoff shows that most of the data points are on or around the 1:1 line showing good agreement between them. The correlation coefficient between estimated and targeted runoff is 0.975 during training and is 0.973 during model validation. Model performed well in training as well validation. RMSE is 1.56 and 3.48 during training and validation; whereas MAE is 0.88 and 1.13 during training and validation, respectively.

Model M-1234

The model M-1234 includes all four input variables and has been already discussed in the section 4.2.1.1. Performance of the model has been shown in Fig. 1. Model M-1234, in which all four variables have been considered as input, performed well with R at training and validation as 0.941 and 0.987 respectively. Model M-134 resulted into R as 0.975 and 0.973, at training and validation respectively. During training performance of model M-134 is better than model M-1234, but during validation model M-1234 performed better than model M134. The same is the case with RMSE and MAE (Table- 2). It becomes difficult to decide which model is superior. Taking average of parameter values during training and validation, R for model M-134 is 0.974 and for M-1234 is

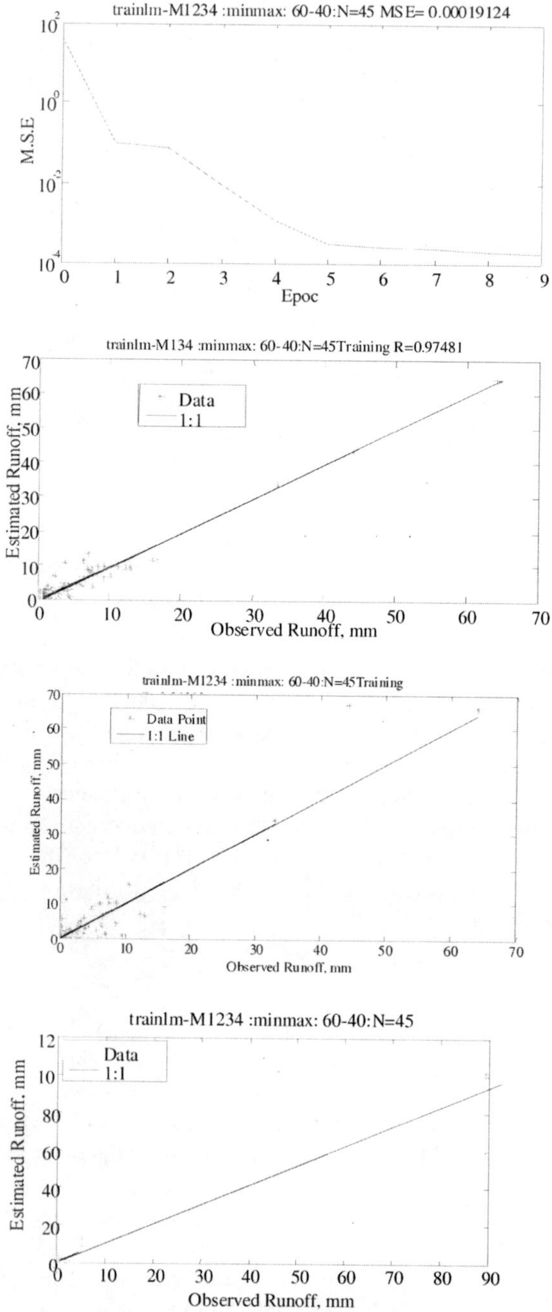

Fig. 1. Performance of ANN model M-1234 for runoff estimation.

Table 2 : Performance of ANN models with different input variables for runoff estimation (trainlm, N = 45, minmax, 60-40)

Model	*R(Trg) R (Val)*		*MSE(Trg) MSE (Val)*		*RMSE(trg) RMSE (Val)*		*MAE(trg) MAE (Val)*	
	Trg	*Val*	*Trg*	*Val*	*Trg*	*Val*	*Trg*	*Val*
M1	0.934	0.991	0.0026	0.0157	2.98	2.17	1.32	1.04
M2	0.448	0.025	0.0186	0.0511	6.21	10.23	2.52	2.78
M3	0.386	0.038	0.0147	0.0594	6.38	11.01	2.56	3.46
M4	0.353	-0.036	0.0100	0.0942	6.85	13.97	2.92	5.04
M12	0.946	0.731	0.0024	0.0237	2.23	6.92	1.31	2.42
M13	0.953	0.929	0.0009	0.0081	2.10	4.36	1.17	1.11
M14	0.603	0.017	0.0016	0.0979	9.09	15.90	5.61	5.25
M123	0.961	0.872	0.0009	0.0152	2.19	5.55	1.75	2.35
M124	0.908	0.078	0.0009	0.1348	3.13	10.62	1.56	2.47
M134	0.975	0.973	0.0002	0.0100	1.56	3.48	0.88	1.13
M1234	0.941	0.987	0.0002	0.0086	2.79	3.28	1.24	1.05

(Trg = Training; Val = Validation)

0.964. Average RMSE and MAE for model M-134 is 2.520 and 1.005; and for M-1234 is 3.035 and 1.145, respectively. Hence it can be inferred that performance of the model M-134 is the best. Model M-1234 stands second and model M-1 stands third with average R = 0.9625, RMSE = 2.575 and MAE = 1.180. Performance of models, in which rainfall is not considered as input variable, has been observed poor. It reveals that ANN model for daily runoff estimation must include daily rainfall as input variable. Model M-134 has been further refined in terms of number of neurons in the hidden layer of artificial neural network model.

Conclusion

The investigation on ANN based modeling for estimation of runoff is undertaken to develop an ANN model for estimation of runoff from the Kalidevi watershed situated at Kukshi tehsil of Dhar district of MP. For this small agricultural watershed an Artificial Neural Network (ANN)based model for estimation of runoff has been developed. One input layer, one hidden layer and one output layer has been considered in the ANN architecture. Two methods (MinMax and Std) of normalization of input vector and target vector has been evaluated. MinMax method normalizes the data in the range of -1to1. Std method normalizes the data such that mean is zero and standard deviation is unity. For ANN based modeling, 'MinMax' method of normalization is found better than 'Std' method. Levenverg-Marquardt training algorithm of back propagation performed better than other twelve algorithms evaluated. With 45

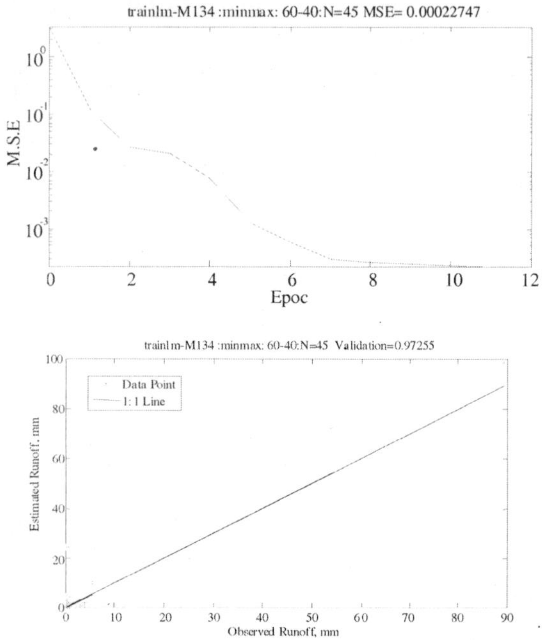

Fig. 2. Performance of ANN model M-134 for runoff estimation.

neurons in the hidden layer, three input variables (rainfall, API and days since 1st June), ANN model for runoff resulted in to R as 0.975 and 0.973 during training and validation. Model with all four input variables resulted R as 0.897 and 0.947 during training and validation respectively. With 25 neurons in the hidden layer, three input variables (rainfall, API and days since 1st June), ANN model for runoff resulted in to R as 0.982 and 0.980 during training and validation respectively. While with 45 neurons in the hidden the model resulted R as 0.974 and 0.972 during training and validation respectively. ANN based model M-134 for runoff estimation has been developed. The model has 25 neurons in a single hidden layer, three input variables, and Levenberg Marquardt training algorithm. The ANN model for runoff has correlation coefficient of 0.973.

References

Anonymous (2006). Land Use Planning and Management of Natural Resources of Dhar District through Remote Sensing. *An annual report of AICRP project, College of Agriculture Indore MP-submitted to ISRO Bangalore.*

Coulibaly, P., F. Anctil and B. Bobee (2000). Previous hydrologique par researc de neurones artifacts: et at de l'art, Reveue Canadienne de genie Civil, Vol. 26(3), pp. 293-304.

Fausett, L. (1994). Fundamentals of neural networks. Prentice Hall. Englewood Cliffs, New Jersey.

Haykin, S. (1994). Neural networks a comprehensive foundation. IEEE press.

Haykin, S. (1999). Neural Networks, Prentice Hall, NJ, 1999, 2nd edn.

Verma, S. K, J. Singh, S.C.Tiwari and D.H. Ranade (2006). Runoff estimation by using remote sensing and GIS from watershed. *J. Soil Wat. Conserv*. 5(3): 49-5.

20

Rainfall-Runoff Modelling and Determination of Threshold Rainfall for Runoff Generation

N.H.Patel and T.M.V. Suryanarayana

Abstract

A runoff model is a mathematical model describing the rainfall-runoff relations of a rainfall catchment area, drainage basin or watershed. More precisely, it produces the surface runoff hydrograph as a response to a rainfall hydrograph as input. In other words, the model calculates the conversion of rainfall into runoff. The aim of this paper is to develop rainfall-runoff model and also check the correlation between rainfall and runoff. The study area considered is Devgadhbaria of Panam catchment area, Gujarat, India. The data considered are for the months of June, July, August, September and October for 10 years, i.e. 1996 to 2005. The whole length of data is divided into two parts for development of model and for validation of the model. In 70:30 data set where 70 per cent data are used for development of the model and 30 per cent data are used for validation process and similarly for 60:40 and 80:20 data sets. The model is developed using curve expert. Three linear models i.e., Model 1, Model 2 and Model 3 have been developed considering data set of 70:30, data set of 60:40 and data set of 80:20 respectively. Also, non-linear models i.e., Model 4, Model 5 and Model 6 have been developed considering the same data sets as described above. The best model from linear models and also from non-linear models has been analysed, which are based on various performance indices. The auto- and cross-correlation analysis carried out reveals that the rainfall and runoff are highly correlated to each other. There is indistinct seasonal pattern between rainfall and runoff. The best linear model is Model 3 and best non-linear model is model 5, and Model 5 is better than Model 3, which may be used in the area considered under study for estimation of future runoff for any given amount of rainfall. The threshold rainfall have been determined for the area considered under study, which is 12 mm. So it gives a fair idea of the minimum amount of rainfall required for runoff generation in Devgadhbaria of Panam catchment area.

Introduction

A runoff model is a mathematical model describing the rainfall-runoff relations of a rainfall catchment area, drainage basin or watershed. In other words, the model calculates the conversion of rainfall into runoff. Rainfall-Runoff is very complex phenomena and its often required attention for modelling to estimate the runoff.

Almost everyone has a definition of the word model. Models are simplified systems that are used to represent real life systems and may be substitutes of the real systems for certain purposes. The hydraulic engineer builds a model physical model to answer questions about the prototype structure. They are material or physical models. The hydrologic models are formalised ideas or concepts. Conceptual models are of many kinds and are used in many disciplines. Hydrologic models are best defined rigorously in relation to the concept of a system. Hydrologic modelling can be viewed as a means of getting useful information about watersheds. The model needs to develop, to know how much water is available, when and where it is available.

Rainfall-runoff models have a wide application. Some of the models are used for hydrological simulations, some for runoff estimation of ungauged basins, for prediction of catchment response to changed conditions as well as for water quality investigations and measurements of meteorological impacts. At present the watershed management gains the attention as well.

In statistics, regression analysis is a statistical process for estimating the relationships among variables. It includes many techniques for modelling and analysing several variables, when the focus is on the relationship between a dependent variable and one or more independent variables. More specifically, regression analysis helps one understand how the typical value of the dependent variable changes when any one of the independent variables is varied, while the other independent variables are held fixed. The term "regression" was coined by Francis Galton in the nineteenth century to describe a biological phenomenon. For Galton, regression had only this biological meaning, but his work was later extended by Udny Yule and Karl Pearson to a more general statistical context. Regression analysis is a statistical tool for the investigation of relationships between variables.

There are several methods for extending stream flow data records, ranging from simple rainfall-runoff correlation, to the complex analysis involving the physics of catchment dynamics and its response to rainfall inputs. The model is based on an extension of rainfall-runoff correlation techniques and is particularly convenient because for most catchments in England and Wales, there are considerably longer rainfall records, often much earlier than 1920, which could form the basis for extending short stream flow data records. However, before using such a model for record extension, it is important to ensure that the model performs adequately during calibration and verification, i.e. its reconstructed data record must reproduce adequately the relevant characteristics of the historic record. Panchal *et al.* (2014) developed rainfall-runoff modelling using Adaptive Neuro-Fuzzy Inference System (ANFIS). Different combinations of rainfall were considered as the inputs to the model, and runoff was considered as the output. Džubáková (2010) classified rainfall-runoff models, their applications, strengths, weaknesses, and possible future development. Although rainfall-runoff modelling has already achieved important advancement in the past hundred years, there is still wide room for its further development. Mainly, the introduction of computing technologies offered, and is still offering, a lot of new possibilities. Nawaz and Adeloye (1999) presented Results of further evaluation tests of a monthly rainfall-runoff model used for extending streamflow data records in England and Wales. Hundecha *et al.* (2001) described that Rainfall-runoff models are used to describe the hydrological behaviour of a river catchment. Many different models exist to simulate the physical processes of the relationship between precipitation and runoff. Some of them are based on simple and easy-to-handle concepts, others on highly sophisticated physical and mathematical approaches that require extreme effort in data input and handling. Recently, mathematical methods using linguistic variables, rather than conventional numerical variables applied extensively in other disciplines, are encroaching in hydrological studies. Panchal *et al.* (2014) developed rainfall runoff modelling using fuzzy logic.

The determination of flood discharges that fits to the physical condition of the watershed requires an effort in developing flood discharges estimation model that meets the physical condition of the watershed. The research uses two models in estimating the flood in any watershed (Anwar, 2011).

Study Area

The Panam dam site situated near village Kel-Dezar on the border of Lunawada and Santrampur taluka of Panchmahal district in Gujarat. It is about 20 km from Shehera on Godhra-Lunawada state highway No. 5. The scheme envisages the construction of masonry dam across river Panam and Ogee shaped gated spillway with ten radial gates.

The location of Devgadhbaria is at 22.7022° N, 73.9150° E. Panam River is the major tributary of Mahi River. There are 14 raingauge stations in the whole catchment area. Among that 14 raingauge stations, Devgadhbaria raingauge station is selected for the study.

Data Collection

The daily rainfall and runoff data is to be collected from the State Water Data Centre, Gandhinagar from the year 1996 to 2005 of the months June, July, August, September and October.

Methodology

Regression models provide the scientist with a powerful tool, allowing predictions about past, present, or future events to be made with information about past or present events. Regression analysis is widely used for prediction and forecasting. Regression analysis is also used to understand which among the independent variables are related to the dependent variable, and to explore the forms of these relationships.

The root-mean-square error (RMSE) is a frequently used measure of the differences between value predicted by a model or an estimator and the values actually observed. RMSE is a good measure of accuracy, but only to compare forecasting errors of different models for a particular variable and not between variables.

In statistics, the coefficient of determination (r^2) is a number that indicates how well data fit a statistical model. Its main purpose is either the prediction of future outcomes or the testing of hypotheses, on the basis of other related information. It provides a measure of how well-observed outcomes are replicated by the model, as the proportion of total variation of outcomes explained by the model.

Before modelling by the use of any data, there is need to check how the data are correlated to each other. So, rainfall auto correlation, runoff auto correlation and rainfall-runoff cross-correlation have been done using the data collected for the study area.

To carry out regression analysis, the form of the function f must be specified. $Y=f(X)$, Where, Y is the dependent variable i.e., runoff for this study in mm and X is the independent variable i.e., rainfall for this study in mm.

The whole length of data is divided into two parts for development of model and for validation of the model.

The three data sets have been made i.e., 70:30, 60:40 and 80:20. In 70:30 data set where 70 per cent data are used for development of the model and 30 per cent data are used for validation process and similarly for 60:40 and 80:20 data sets. The model is developed using curve expert.

Three linear models i.e., Model 1, Model 2 and Model 3 have been developed considering data set of 70:30, data set of 60:40 and data set of 80:20 respectively. Also, non-linear models i.e., Model 4, Model 5 and Model 6 have been developed considering the same data sets as described above.

The best model from linear models and also from non-linear models has been analysed, which is based RMSE and co-efficient of determination (r^2). So here Model 1, Model 2 and Model 3, linear model have been developed for estimation of future runoff and also Model 4, Model 5 and Model 6, non-linear model have been developed for estimation of future runoff of the study area considered.

The best model has been sorted out on the basis of RMSE and co-efficient of determination (r^2).

It should be noted that a certain amount of rainfall is always required before any runoff occurs. By putting different values of rainfall in the best model developed, the threshold rainfall has been obtained where the value of runoff exceeds zero.

Results and Analysis

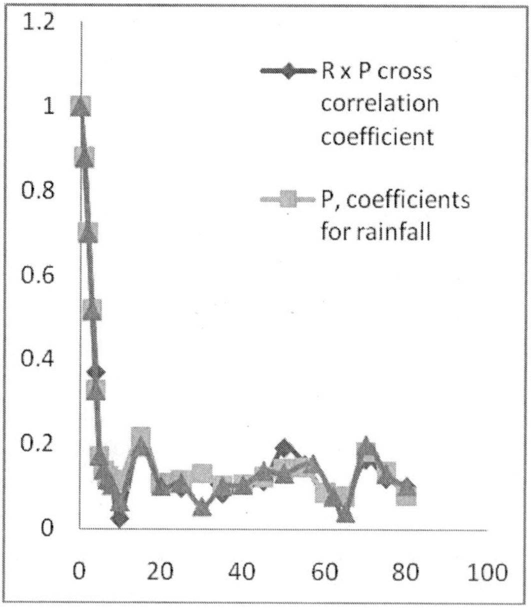

Fig.e 1. Auto and Cross-correlation coefficients for Devgadhbaria catchment.

Figure 1 shows, the auto and cross-correlations for daily rainfall and runoff for the study area considered. Figure 1 shows that rainfall and runoff have higher serial correlation with each other. Rainfall and runoff show indistinct seasonal pattern.

The developed regression models give estimation of the runoff of the study area considered.

(1) *Linear Models*

Model 1, Model 2 and Model 3 are the linear models. The obtained Root Mean Square Error (RMSE) by development of the model and obtained RMSE by validation of the model and also co-efficient of determination (r^2) are as below:

From Table 1, the obtained RMSE of the Model 3 is less than other two models and co-efficient of determination (r^2) of Model 3 is nearer to one than other two models. So, Model 3 is best linear fit for the area considered under study.

Table 1: RMSE and r^2 of Developed Linear Models

Model	r^2		*RMSE (mm)*
Model 1	0.9815	By development By validation	6.47 6.91
Model 2	0.9915	By development By validation	6.47 6.81
Model 3	0.9819	By development By validation	5.14 5.48

The best linear model 3 is shown as below:

Model 3: $y = a + bx$

Where,

$a = -1.13E+01$

$b = 8.39E-01$

(2) *Non-Linear Models*

Model 4, Model 5 and Model 6 are the non-linear models. The obtained Root Mean Square Error (RMSE) by development of the model and obtained RMSE by validation of the model and also co-efficient of determination (r^2) are as below:

Table 2: RMSE and r^2 of Developed Non-linear Models

Model	r^2		*RMSE (mm)*
Model 4	0.9926	By development By validation	3.90 4.88
Model 5	0.9927	By development By validation	3.91 4.59
Model 6	0.9926	By development By validation	4.02 4.87

From Table 2, the obtained RMSE of the Model 5 is less than other two models and co-efficient of determination (r^2) of Model 5 is more nearer to one than other two models. So, Model 5 is best non-linear fit for the area considered under study.

The best non-linear Model 5 is shown as below:

Model 5: $y = a + bx + cx^2 + dx^3 + ex^4$

Where,

a = -2.22E+00

b = 5.75E−02

c = 1.19E−02

d = -5.86E−05

e = 9.07E−08

Non-linear relationship of rainfall and runoff is more suitable than linear relationship. The values of co-efficient of determination of Model 3 and Model 5 are 0.9919 and 0.9927 respectively. So Model 5 is the best fit for the area considered under study.

It should be noted that a certain amount of rainfall is always required before any runoff occurs. This amount, usually referred to as threshold rainfall, represents the initial losses due to interception and depression storage as well as to meet the initially high infiltration losses. The threshold rainfall depends on the physical characteristics of the area and varies from catchment to catchment. The threshold value of rainfall, by the best model of the study, is 12 mm.

Conclusions

The auto- and cross-correlation analysis carried out reveals that the rainfall and runoff are highly correlated to each other. There is indistinct seasonal pattern between rainfall and runoff. The best model among all the models is selected on the basis of RMSE and co-efficient of determination (r^2). So, among linear models, Model 3 is the best linear model with RMSE of 5.14 mm by development and 5.48 mm by validation of Model 3 and also r^2 value is 0.9919. And also among non-linear models, Model 5 is the best non-linear model with RMSE of 3.91 mm by development and 4.59 mm by validation of Model 5 and also r^2 value is 0.9927. Therefore, non-linear Model 5 is the best suited model for area considered. This model may be used in the Devgadhbaria of the Panam catchment area for estimation of future runoff for any given amount of rainfall. The threshold rainfall by the best model is obtained to be 12 mm, which is applicable to area considered for the study.

References

Anwar, M.R., (2011). "The Rainfall-runoff model using of the watershed physical characteristic approach", International Journal of Civil & Environmental Engineering, Vol. 11, pp. 71-75.

Džubáková Katarína, (2010). "Rainfall-Runoff modeling : Its development, Classification and possible applications", ACTA Geographica universitatis comenianae, Vol. 54, No. 2, pp. 173-181.

Hundecha, Y., Bardossy, A., Theisen, H., (2001). "Development of a fuzzy logic-based rainfall-runoff model", Hydrological Sciences-Journal~des Sciences Hydrologiques, 46(3), pp. 363-376.

Nawaz, N.R., Adeloye, A.J., (1999). "Evaluation of monthly runoff estimated by a rainfall-runoff regression model for reservoir yield assessment", Hydrological Sciences—Journal—des Sciences Hydrologiques, 44, pp. 113-134.

Panchal, R., Suryanarayana, T.M.V., Parekh, F.P., (2014). "Adaptive Neuro-Fuzzy Inference System For Rainfall-Runoff Modeling", Int. Journal of Engineering Research and Applications, Vol. 4, 6, pp. 202-206.

Panchal, R., Suryanarayana, T.M.V., Parekh, F.P., (2014). "Rainfall-Runoff Modeling: A Fuzzy Logic Approach", International Journal for Scientific Research & Development, Vol. 2, 05, pp. 81-83.

21

Estimation of Groundwater Recharge through Water Harvesting Structures in Meghal River Basin using Remote Sensing and GIS

N. K. Gontia and P. P. Rathod

Abstract

Meghal is one of the river basins of Saurashtra region of Gujarat falling in the district of Junagadh and being an agricultural area, the major demand of water is for agriculture. Implementation of various water harvesting and groundwater recharge schemes in Meghal river basin has improved the groundwater table and thus quality of groundwater and by that decreased the rate of seawater intrusion in the region. The identification of location of the water harvesting structures and estimating their storage capacity in Meghal river basin became necessary to assess its water resources. Remote sensing has advantage of spatial, spectral and temporal variability of covering large and inaccessible areas within short time and has become a very handy tool in accessing and monitoring the water resources. The satellite images of the IRS P6 LISS III of October 14, 2005 and IRS P6 LISS III of February 1, 2008, SOI toposheet and personal visits of the study site were used to generate land use/land cover maps, to locate as well as quantify number of water harvesting structures and their storage capacity. In the Meghal river basin during rabi *season about 25,978 ha area is under various crops. About 900 water harvesting structures were identified through remote sensing images. In the month of November, these structures have water spread area of 711 ha with storage capacity of 1,559 ha m. The water harvested by these structures can irrigate about 3400 ha area of* rabi *crops. Besides irrigation, these structures also recharge groundwater, which enhances the groundwater quantity and quality in the region.*

Keywords: *Water harvesting structure, Groundwater recharge, Remote sensing and GIS.*

Introduction

The artificial groundwater recharge aims at augmentation of groundwater reservoir by modifying natural movement of surface water utilising suitable civil construction techniques. Artificial groundwater recharge structures include percolation tank, check dam, gabion structure etc. The efficiency and feasibility of these structures is more in hard rock formation where rocks are highly fractured and weathered. While, check dams are constructed across small streams having gentle slope and are feasible both in hard rock as well as alluvial formation. Remote sensing and GIS are playing a rapidly increasing role in the field of hydrology and water resources development. Remote sensing provides multi-spectral, multi-temporal and multi-sensor data of the earth's surface (Choudhury, 1999; Gontia, 2006). One of the greatest advantages of using remote sensing data for hydrological investigations and monitoring is its ability to generate information in spatial and temporal domain, which is very crucial for successful analysis, prediction and validation (Saraf, 1999; Gontia and Tiwari, 2005; Gontia and Tiwari, 2008; Gontia and Tiwari, 2010; Gontia and Tiwari, 2011).

However, the use of remote sensing technology involves large amount of spatial data management and requires an efficient system to handle such data. The GIS technology provides suitable alternatives for efficient management of large and complex databases. Information from satellites is becoming more and more important for environmental research. An important part of this information concerns water—an element most essential for man, its phases and peculiarities.

The study focuses on development of remote sensing and GIS-based analysis and methodology in artificial groundwater recharge studies. In order to demonstrate the integrated remote sensing and GIS-based methodology, an area which forms Meghal river basin situated in the state of Saurashtra region (Gujarat) has been taken.

Materials and Methods

Study Area

The study area is Meghal river basin 5G1D2, Dist. Junagadh, Saurashtra region of Gujarat. It is located between 20° 58' N to 21° 17' N latitude and 70° 13' E to 70° 32' E longitude. The study area falls under Survey of India toposheet number 41K/08 and the IRS LISS III images of 91/57 (Path/Row). It comprises of 541 square km of land with total length of river is about 72 km. The climate of area can be classified as sub-tropical and semi-arid. January is the coldest month with mean monthly temperature varying from 14°C to 20°C and maximum monthly temperature varies between 30°C to 44°C in the month of May. The location map of the study area is shown in Fig. 1.

Satellite Data and Software Used

The LISS III (Linear Imaging Self Scanning Sensor) scene of Indian Remote Sensing Satellite P6 (IRS P6) of date of pass October 14, 2005 and February 1, 2008 was used for generation of landuse and landcover of the area. The PCI Geomatica V10.1 and ArcGIS V9.2 software were used for all remote sensing and GIS applications.

Estimation of the storage capacity of water harvesting structures using remote sensing and GIS

The water harvesting structures existing in watershed were identified using satellite images of Cartosat digital data and snapshots from Google Earth Pro. Then areas under submergence in these structures were identified. Taking the average depth of storage from ground truth information of the study area the volume of water stored in the structures was estimated. From this volume of impounding evaporation losses in study area deducted. From local information and personal survey the duration for which the structures were remain filled was determined. Normally two or three fillings are expected in a year (Sivanappan, 1997). Based on this total groundwater recharge from these structures was estimated. Volumes of water impounded in water harvesting structures were estimated using following formula;

$$V_s = C * A_s * H \quad \ldots(1)$$

where, V_s is volume of impounding of water in structures (ha m), A_s is water spread area in ha, H is average depth of storage, and C is storage coefficient.

Results and Discussion

The landuse/landcover map of study area were prepared for *kharif* season (October, 2005) and *rabi* season (February, 2008) and presented in Fig. 2 and Fig. 3 and listed in Table 1 and 2 respectively. In *kharif* and *rabi* season groundnut and wheat are the major crop grown in area is about 11982 and 9084 ha respectively.

Identification of water harvesting structures and estimation of their storage capacity using remote sensing and GIS

The Meghal river basin area water resources map, the toposheets, LISS III image of Oct. 14, 2005 and the land use/land cover map of *kharif* season of Meghal river basin generated using same image were used for identifying the water harvesting structures and their water spread area and storage capacity.

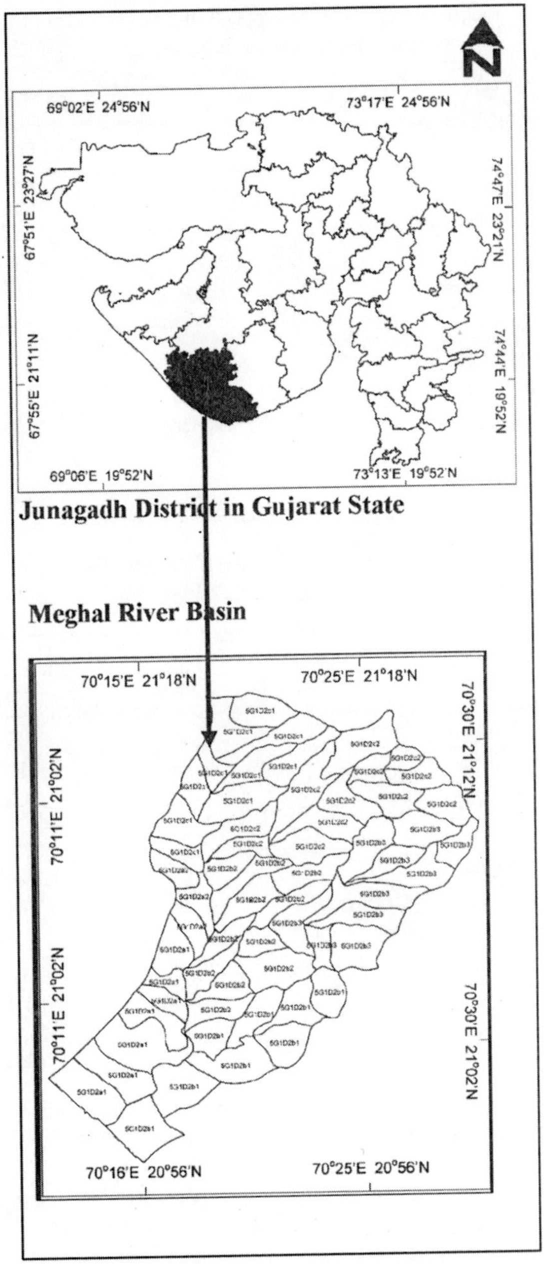

Fig.1. Location map of study area (Meghal River Basin).

Table 1: Land use/ Land cover classification of Meghal river basin for Oct. 2005

Sr. No.	*Name of class*	*Area*	
		Ha	*Per cent*
1.	Water bodies	662	1.22
2.	Horticultural crops	2664	4.92
3.	Orchards	4209	7.77
4.	Other crops	9370	17.30
5.	Groundnut crops	11982	22.12
6.	Grass land/ Open Forest	12683	23.42
7.	Fallow land/ Degraded forest	7225	13.34
8.	Establishment/ Scrub land	4657	8.60
9.	Waste land	706	1.30
	Total	**54158**	**100**

Table 2: Land use/ Land cover classification of Meghal river basin for Feb. 2008

Sr. No.	*Name of class*	*Area*	
		Ha	*Per cent*
1.	Water bodies	384	0.71
2.	Horticulture crops	7993	14.76
3.	Wheat crop	9084	16.77
4.	Other *rabi* crops	8901	16.44
5.	Grass land	10013	18.49
6.	Fallow land/ Open forest	7801	14.40
7.	Pasture/ Establishment	6220	11.48
8.	Degraded forest/ Scrub land	2832	5.23
9.	Waste land	931	1.72
	Total	**54158**	**100**

The water bodies in Meghal river basin area were identified based on their reflectance signatures obtained in FCC image of Oct. 14, 2005 and verified with ground survey visits, SOI toposheets and LU/LC map of *kharif* season. The identified water harvesting structures were digitized from FCC as a topological polygon vector layer using GIS to identify their location, perimeter and water spread area. The water spread area and perimeters of these water harvesting structures were generated through attribute table of pond vectors. Water harvesting structures that have water spread area greater than 1000 m^2 (0.1 ha) were considered for irrigation purpose. The water harvesting structures identified in Meghal river basin is shown in Fig. 4.

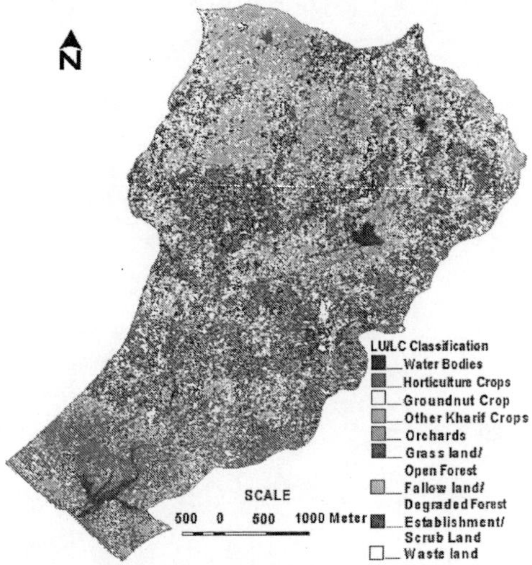

Fig.2. Landuse/landcover classification map of Meghal river basin for the* kharif *season 2005.

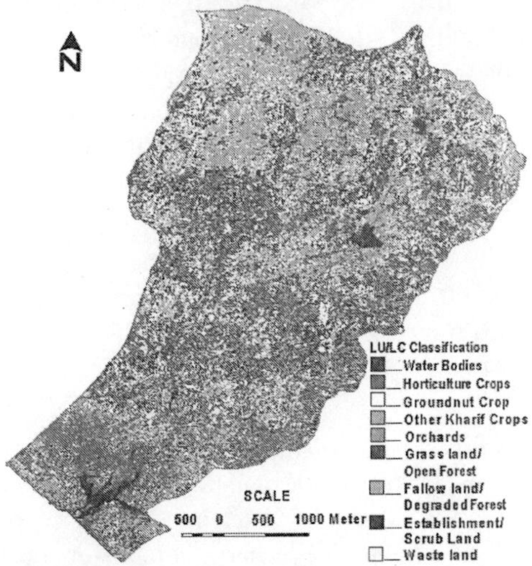

Fig. 3. Landuse/landcover classification map of Meghal river basin for the* rabi *season 2007-08.

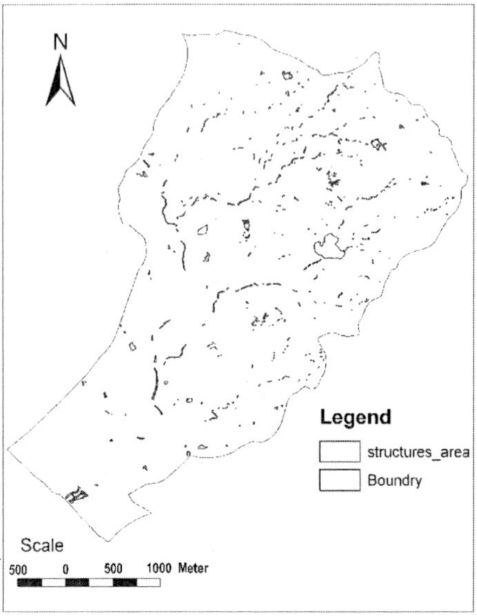

Fig. 4. Location of water harvesting structures in Meghal river basin.

The water harvesting structures identified using remote sensing images were grouped in nine ranges of water spread areas as less than 0.1, 0.1 to 0.5 ha, 0.5 to 1.0 ha, 1.0 to 3.0 ha, 3.0 to 5.0, 5.0 to 10.0 ha, 10.0 to 20.0 ha, 20.0 to 50.0 ha and greater than 50 ha. Based on the ground truth the average depths were considered for these groups of water harvesting structures as 0.5, 1.0, 1.5, 2.0, 2.5, 3.0, 3.5, 4.0 and 4.5 m respectively. The actual area of submergence of each pond was computed from the attribute table of the water harvesting structures vector and summed for that group to generate the total area of submergence.

Inventory and quantification of water harvesting structures and area under irrigation in *rabi* crops estimated using remote sensing and GIS is presented in Table 3. The number of water harvesting structures identified under nine groups were 453, 311, 53, 52, 17, 5, 4, 4 and 1 respectively with storage capacity of 9.7, 48.9, 36.5, 122.6, 113.4 77.3, 133.7 302.3 and 714.8 respectively.

The evaporation loss in the study area for the month of July to December was found 145.2 mm. Using volume of impounding after subtracting evaporation and other losses from volume of water impounded the recharge volume of water comes as 1456.1 ha m. The recharge volume for area less than 20 ha and greater than 20 ha were estimated assuming recharge as three times and two times that of volume of impounding respectively. Total volume of recharge from structures was estimated as 3400 ha m in the study area. On an average considering the seasonal water requirement of *rabi* crop in the Meghal river

Table 3: Estimated total recharge capacity of the water harvesting structures

Group		*No. of structure*	*Avg. depth (m)*	*Avg. water spread area (ha)*	*Volume (ha m)*	*Volume of water - (Evaporation losses) (ha m)*	*Total recharge (ha m)*	*Area under irrigation (ha)*
1		2	3	4	5 = (col.3* col.4* 0.7)	6 = (col.5 - 0.145* col.4)	7 =	8 = (col. 7/ 0.50)
I	<0.1 ha	453	0.50	27.8	9.7	5.7	17.1	34.3
II	0.1-0.5 ha	311	1.00	69.8	48.9	38.8	116.3	232.6
III	0.5 - 1 ha	53	1.50	34.8	36.5	31.5	94.4	188.9
IV	1 - 3 ha	52	2.00	87.6	122.6	109.9	329.7	659.4
V	3 - 5 ha	17	2.50	64.8	113.4	104.0	312.0	624.0
VI	5 - 10 ha	5	3.00	36.8	77.3	72.0	216.0	431.9
VII	10 - 20 ha	4	3.50	54.6	133.7	125.8	377.5	754.9
VIII	20 - 50 ha	4	4.00	108.0	302.3	286.6	573.3	1146.5
IX	>50 ha	1	4.50	226.9	714.8	681.9	1363.7	2727.4
Total		**900**		**711.1**	**1559.2**	**1456.1**	**3400.0**	**6780.0**

basin as 50 cm, the overall 6780 ha crop area can be irrigated from 900 identified potential water harvesting structures. The land use/land cover classification generated using remote sensing image shows that about 25,978 ha area is under various crops in the *rabi* season in Meghal river basin. The area irrigated from these water harvesting structures is about 26 per cent of the total *rabi* crops grown in the river basin area.

Conclusions

Groundwater is a precious resource of finite extent. Over the years, increasing population, urbanisation and expansion in agriculture has lead to the unscientific exploitation of groundwater creating a water stress condition. This alarming situation calls for a cost and time-effective technique for proper evaluation of groundwater resources and management planning. Groundwater development programme needs a large volume of data from various sources. As demonstrated successfully in this study that integrated remote sensing and GIS can provide the appropriate platform for convergent analysis of large volume of multi-disciplinary data and decision-making for groundwater studies. The following conclusions are drawn from the above study:

1. In the present study an integrated remote sensing and GIS-based methodology has been developed and demonstrated for evaluation of groundwater resources.
2. The present study has demonstrated that the recharge sites situated on a gentle slope and lower order streams are likely to provide artificial recharge to a larger area.
3. The area irrigated from these water harvesting structures is about 26 per cent of the total *rabi* crops grown in Meghal river basin.

References

Gontia, N. K. and Tiwari, K. N., 2005. Identification of ponds and estimation of their storage capacity in Tarafeni South Main Canal (TSMC) command using remote sensing and GIS. National conference on watershed management for sustainable production, livelihood and environmental security, May 19-21, 2005. p. 32.

Gontia, N. K., 2006. Crop Water Stress Indices and Evapotranspiration Estimation for Irrigation Scheduling and Yield Modelling of Wheat Crop Using Remote Sensing and GIS. Unpublished Ph. D. thesis submitted to the Indian Institute of Technology, Kharagpur.

Gontia, N.K. and Tiwari, K.N., 2008. Development of crop water stress index of wheat crop for scheduling irrigation using infrared thermometry. Agricultural Water Management. 95 (10): 1144-1152.

Gontia, N.K. and Tiwari, K.N., 2010. Estimation of spatially and temporally distributed crop coefficient and evapotranspiration of wheat (Triticum aestivum) in an irrigation command using remote sensing and GIS. Water Resources Management, Springer Publication, 24 (7): 1399-1414.

Gontia, N.K. and Tiwari, K.N. 2011. Yield Estimation Model and Water Productivity of Wheat Crop (Triticum aestivum) in an Irrigation Command Using Remote Sensing and GIS, Journal of the Indian Society of Remote Sensing, 39(1): 2737.

Saraf, A. K. 1999. A report on Landuse Modelling in GIS for Bankura District, Projet sponsored by DST, NRDMS division, Govt. of India.

Sivanappan, R. K., 1997. Technologies for water harvesting and soil moisture conservation in small watersheds for small scale irrigations. Proceedings of the seminar on Irrigations Technology Transfer in support of Food Security, FAO, 14-17, April 1997.

22

Cropping Pattern Optimisation for Surface and Drip Irrigated Canal Command Area

R.M. Satasiya and R. Subbaiah

Abstract

Judicious utilisation of land water resources in irrigation command can be attained through optimised cropping pattern, sound irrigation planning and crop scheduling. An attempt was made to optimise the cropping pattern for Uben irrigation scheme by planning conjunctive use models for surface and drip irrigated command using linear programming approach. Rainfall runoff in the catchment are exponentially related to each other. Per hectare annual cost and benefit of drip irrigation system was Rs 28135.85 and Rs 41206.28. Existing annual return (Rs 94311572.00) of the command was increased by 87.95 per cent and 103.37 per cent revised cropping pattern with surface irrigation and drip irrigation suggested through the linear programming models. Benefit cost ratio of 1.46 was attained for canal command under drip irrigation system. There is plenty of the scope to adopt the system for saving precious water resources and attaining more productivity per drop of water.

Introduction

The vulnerability of the agricultural sector to climate changes has driven to set up programmes and policies heading towards maximising the utilisation of its limited resources, mainly, the irrigation water supply and arable land to increase the crop productivity per drop of water and crop revenue in irrigation commands. Various modelling approaches have been applied to optimise the cropping pattern worldwide including the linear and non-linear optimisation models (Haouari and Azaiez, 2001; Sahoo *et al,* 2001; Singh *et al.*, 2001; Cheng *et al.*, 2009; Montazar and Rahimikob 2008; Kaur *et al.*, 2010); chance-constrained linear programming models (Sethi *et al.*, 2006), the interactive fuzzy multi-objective optimisation approach (Zhou *et al.*, 2007), the goal programme approach (Vivekanadan *et al.*, 2009), the multi-objective fractional goal programming approach (Fasakhodi *et al.*, 2010), and the genetic algorithm model (Karamouz *et al.*, 2010). Most of these models failed to explain the additional crop revenue that can be accrued and amount of water resources saving due to adoption of trickle irrigation in irrigated canal command areas. It is also perceived that adoption of micro irrigation (MI) system leads to increase in yield (25-30 per cent); real water saving over surface irrigation (30 to 50 per cent); and expansion in area under irrigation (20 per cent), all resulting in social benefits (INCID, 1994; Kumar *et al.*, 2008). Keeping this in view an effort is made in the present study to compare the performance of surface irrigated canal commands and drip irrigated commands on crop revenue and water saving.

Study area

The Uben command area lies between 21° 37′ N latitude and 70° 38′ E longitudes with an altitude of 65 m above mean sea level. Uben irrigation scheme has a cultivable command area of 2500 ha and is spread over three villages of Junagadh taluka of Junagadh district and five villages of Jetpurtaluka of Rajkot district.The climate of the area is subtropical and semi-arid with an average pan evaporation of 900 mm per year. The average annual rainfall is 696.00 mm mostly received during *Kharif* season. Major *kharif* crops are groundnut, cotton and greengram while *Rabi* crops are wheat, onion. In summer, seasaṃe and groundnut are sown.The soils of the study area are observed to be sandy clay loam to silt loam in nature.

Model formulation

Two decision models were developed to allocate optimal land area among different crops for attaining maximum net benefit from the command area using linear programming. In the first model, the irrigation to the crops were given through surface irrigation system and in the second model, the irrigation to the crops is met through drip irrigation systems. The model assume that: (a) spatial and temporal variation on soil physical and chemical properties in the command are ignored; (b) each unit of land receive the same management practices for a particular crop; (c) date of sowing and duration of crop remain invariant. (d) rainfall is uniformly distributed spatially over the study area.

Surface irrigated command area model

The objective function is to maximise the total annual net benefits from irrigation water. The benefits can be calculated as:

$$\text{Max} \sum_{j=1}^{n} A_j \, NR_j - (CCW * ACW + CGW * AGW)$$

where A_j= cropping areas under different crops, NRj = Net return of crop *j* excluding water charges, Rs/ha, *CCW*= Cost of canal water, *ACW*=Annual canal water utilised by different crops grown in the command area, ham, *CGW*=Cost of ground water, Rs/ham, *AGW*=Annual ground water utilised by different crops in the command area, ham.

Constraints

1. Total area under different crops during any weeks should be less than or equal to the total cultivable command area (CCA) of the irrigation project. Land availability constraints for the different weeks are presented in the Table 1.
2. The continuity ofreservoir storage in any week should not exceed the live storage capacity of reservoir, which is fixed as 1560 ham.
3. Storage of water at the beginning of week (S_t) should not exceed the volume of water left over from the previous week (S_{t-1}) plus inflow into the reservoir from the previous week (I_{t-1}) minus water transferred to the command area to meet the weekly water requirement of different crops (CW_{t-1}) and water losses during the same week (seepage and evaporation).
4. Volume of canal water to be transferred should not exceed the weekly water transferring capacity of the main canal (C_c). Discharge capacity of main canal is 1.5 cumec, so weekly volume of water transfer to the command area by the canal is 90.72 ham.
5. Weekly ground water exploitation should not exceed weekly ground water availability and total ground water utilisation in a year should not exceed annual ground water availability.
6. Total irrigation requirements for all crops during any week in the command area should be met from the water releases from canal (CW_t) and water diverted from the aquifer (GW_t) in the command during the same time period.
7. A minimum per cent of CCA is to be demarcated towards food crops like groundnut (10 per cent), green gram(1 per cent), wheat (2 per cent) and onion (1 per cent) to cater the local demands even though the benefit from these crops may not be attractive as compared to commercial crops like cotton.

Drip irrigated command area lp model

The above model can be used for drip irrigated command area also. In the above objective function, per hectare net return of the crops (NR_j) will be increased according to the increase in the yield of the crop under drip irrigation system and other parameters remain unchanged. The irrigation requirement constraint, weekly net irrigation requirement of the different crops (IR_{jt}) will be reduced according to the water saving of the crop under drip irrigation system and other parameters remain unchanged.

Table 1: Land availability constraint

Std. week No.	*Name of crop*	*Land availability constraint*
25 to 45	Groundnut, Green gram and Cotton	$A1+A2+A3d \leq 2500$
46 to 51	Cotton, Wheat and Onion	$A3+A4+A5d \leq 2500$
52 and 1 to 5	Wheat and Onion	$A4+A5d \leq 2500$
6 to 21	Wheat, Onion, Summer groundnut and Summer sesame	$A4+A5+A6+A7d \leq 2500$

Estimation of model parameters

The data on meteorological parameters were collected from JAU observatory, Junagadh. Daily reservoir level, cropping pattern, canal network, numbers of wells in the command and annual water table fluctuation were collected from the Irrigation Division, Junagadh. Pumping tests were performed at four different sites on in the Uben command area for the estimation of ground water quantity. Total numbers of bores/wells in the command area were 922. Average discharge capacity of pumping unit on the well was found as 6 lps based on the survey conducted in the command area. Average pumping hours (hr/day) available during the month of October to December, January to April and May to June was observed to be 6, 4 and 2 respectively. Analysis of pumping test data was done by Papadopulos and Cooper (1967) method. Storage co-efficient was found in the range of 0.000001 to 0.04448 with average value as 0.01675. From the field survey, annual water table fluctuation was found as 15 m. Annual ground water availability was estimated by multiplying the annual water table fluctuation, storage co-efficient and gross command area. Annual ground water storage was estimated as 628.125 ham. The weekly ground water extractable was calculated by multiplying the number of wells in the command area. On this basis weekly ground water extractable were calculated as 83.6438 ham, 55.7625 ham and 27.8812 ham during the standard week No. 40 to 52, 1 to 17 and 18 to 39 respectively. Total annual ground water quantity available was estimated to be 487.5 ham.

Weekly reservoir level data was used to calculate runoff entered into reservoir during the respective week. Rainfall-runoff relationship was established using least square method. Power function $Y = 0.01X^{1.1338}$ (Y = Runoff from the catchment area, ham and X = Rainfall, mm) had given higher value of R^2 (0.7695). Weekly live storage available into the reservoir was calculated by considering inflow as water yield from the catchment area and outflows as irrigation demand, spill over, seepage and evaporation losses from the reservoir. The water lost by evaporation and seepage was calculated as 6.21 per cent of storage available into the reservoir. Weekly net irrigation requirement was obtained using FAO CROPWAT model. The net return was obtained by subtracting the cost of cultivation from the gross income.The existing net annual returns from the command area are estimated to be Rs 94311572.00 based on the data on area under different crops in the command, productivity of the crops and sale price of crop product at the prevailing prices of2013.

Results and Discussion

Surface irrigated command area model

The model was solved using linear programming with seven crop activities. The optimal cropping pattern for surface irrigated command is presented in Table 2. The net annual return from the surface irrigated command was arrived to be as Rs 117133500 which was 8.21 per cent higher than the existing net annual return of the project. Maximum area allocation was given to green gram (62 per cent) followed by groundnut (10 per cent). Area under cotton (A3), wheat (A4) and onion (A5) was found as 700.02 ha, 1271.17 ha and 25.00 ha respectively. Season wise area allocation was 2500.00 ha and 1296.17 ha during the *Kharif* and *Rabi* seasons respectively. Area allocation during summer was found nil. Cropping intensity was found as 151.85 per cent. Area under green gram was increased due to its less water requirement. Cotton area was increased due to its profitability. During *Rabi* season, per cent fallow land area was decreased from 37.7 per cent to 20.15 per cent of CCA in surface irrigated command. Wheat area was increased by 4.6 per cent than the existing cropping pattern. Area under onion was decreased by 5 per cent than existing cropping pattern. Reduction in onion area may be due to its higher water requirement. During summer season, area allocation under the crop was nil due to lack of water availability.

Water utilisation pattern from reservoir and ground in surface irrigated command area

Weekly live storage available in the reservoir (S_t), water transfer from the reservoir (CW_t) and ground water utilised (GW_t) are presented in the Fig. 1. Maximum weekly live storage available in the reservoir (875.337 ham), water transfer from the reservoir (71.764 ham) and ground water utilised (61.253 ham) was found during the standard week number 33, 37 and 51 respectively. Water transfer from the reservoir starts from the 25th standard week and ends in the 49th standard week. Ground water was utilised during 49th standard week to 10th standard week. Volume of canal water and ground water utilised was found as 755.563 ham and 628.125 ham respectively.

Table :2. Optimal cropping pattern for surface and drip irrigated command area

Sr. No.	Crop name and notation	Area under different crops, ha		
		Existing	Surface irrigated command	Drip irrigated command
Kharif season				
1	Groundnut (A1)	2011.12	250.00	1762.43
2	Green gram (A2)	235.06	1549.98	25.00
Two seasonal crop				
3	Cotton (A3)	253.82	700.02	712.57
Total		**2500.00**	**2500.00**	**2500.00**
Fallow land in Kharif		Nil	Nil	Nil
Rabi season				
4	Wheat (A4)	1154.30	1271.17	50.00
5	Onion (A5)	149.00	25.00	1672.75
Total		**1303.30**	**1296.17**	**1722.75**
Fallow land in Rabi		942.88 (37.7 %)	503.81(20.15 %)	64.68(2.5 %)
Summer season				
6	Summer groundnut (A6)	48.10	0	0
7	Summer sesame (A7)	37.50	0	0
Total		**85.60**	**0**	**0**
Fallow land in Summer		2414.40(96.5 %)	2500(100%)	2500(100 %)
Net return, Rs.		**108249677.44**	**117133500.00**	**220149200**
Cropping intensity, %		155.56	151.85	168.91
Per cent increase of annual net return as compare to existing net return of the project			8.21	103.37
Per cent increase of annual net return as compare to surface irrigated command				87.95

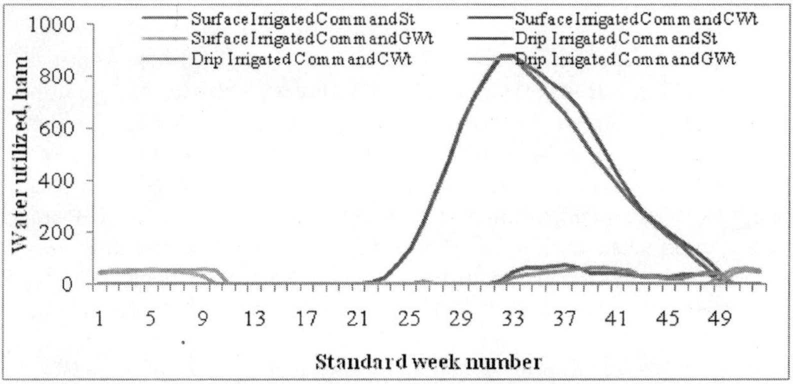

Fig. 1. Weekly live storage available in the reservoir (St), water transfer from the reservoir (Wt) and ground water utilised (Gt) for surface and drip irrigated command.

Drip irrigated command area model

Area allocation for drip irrigated command area model

The optimal cropping pattern for surface irrigated command suggested by model is presented in the Table 2. The net annual return from the drip irrigated command was arrived to be as Rs 220149200 which is 103.37 per cent higher than the existing net annual return of the project and 87.95 per cent higher than the net annual return in surface irrigated command. Season wise area allocation was calculated as 2500.00 ha and 1722.75 ha during the Kharif and Rabi seasons respectively. Cropping intensity was increased by 13 per cent and 17 per cent than existing and surface irrigated command. During Kharif season, maximum area was covered by groundnut followed by cotton and green gram. Area under groundnut was decreased by 248.69 ha and increased by 1512.43 ha than existing cropping pattern and surface irrigated command area respectively. Area under cotton was increased to 17.8 per cent and 0.5 per cent than existing and surface irrigated command. This may be due to improved performance under drip irrigation system and increased profitability of cotton. Area under green gram reduced drastically to 25 ha compared to existing and surface irrigated cropping pattern. During Rabi season fallow land was decreased to 2.5 per cent of CCA in drip irrigated command from 37.7 per cent as in existing cropping pattern and 20.15 per cent in surface irrigated command. Area under wheat in drip irrigated command was reduced to 44.2 per cent and 48.8 per cent than existing and surface irrigated command. Onion became the major crop in Rabi season due to water saving and increased crop yield under drip irrigation system. During summer season, area allocation under the crop was nil due to insufficient water resources.

Efficient reservoir and ground water utilisation for drip irrigated command area

Weekly storage available in the reservoir (S_t), water transfer from the reservoir (CW_t) and ground water utilised (GW_t) are presented in Fig.1 for drip command area. Maximum weekly storage in the reservoir (876.461 ham), water transfer from reservoir (61.617 ham) and ground water utilised (55.566 ham) was found during the standard week number 33, 39 and 9 standard weeks respectively. Water transfer from the reservoir was initiated from 25th standard week and terminated during 49th standard week. Water demand after 49th standard week is met from the ground water supply. Volume of canal water and ground water utilised was found as 714.684 ham and 628.125 ham respectively.

Economic analysis of adopting drip irrigation system in canal commands

If drip irrigation system is adopted in canal-based irrigation system then need for auxiliary storage structures arises to assure water supply from drip irrigation system. Number and capacity of the intermediate storage structures was decided on the basis of maximum weekly water transfer from the reservoir and irrigation requirement of crops in the command area. Cost towards drip irrigation system and intermediate storage structures are to be accounted for adopting drip irrigation system in canal command areas. Costs were calculated by considering depreciation value, interest on original cost of the project and maintenance cost. Life span of 25 years and 10 years were assumed for intermediate storage structure and drip irrigation system respectively. one per cent for service reservoir and two per cent for drip irrigation system were considered for maintenance cost. Nine per cent of initial invest for intermediate storage structures and drip irrigation system was taken for the calculation of interest amount. Payback period for canal based drip irrigation system was calculated by dividing the total initial investment towards drip irrigation system and intermediate storage structure with net annual benefit of the project due to drip irrigation system. The economic viability of drip irrigation system, benefit cost ratio was estimated using the procedure suggested by Srivastava et. al., 2010. The net annual return under drip irrigated command enhanced to per cent and per cent than existing and surface irrigated command areas (Table 1). The benefit per year in drip irrigated command enhanced to per cent and per cent than existing and surface irrigated command areas.

Capacity and cost of auxiliary storage structures

For drip irrigated command, maximum weekly water transfer from the reservoir was 61.617 ham during the 39th standard week. Volume of water storage per day in the intermediate storage structure was 88024 m^3. To accommodate that much volume of water, one hundred intermediate storage structures across the command has to be constructed each of 880.24 m^3 capacity to cater irrigation needs of the crops in the command area. For 880.24 m^3 capacity, dimension of intermediate storage structure (length x width x depth) was kept as 15.8 m x 15.8 m x 3.53 m. The cost of one intermediate storage structure was calculated as Rs 1269127.99 accounting dimensions, quantity, and rate.

Drip system costing

Area under each intermediate storage structure was 25 ha out of which 70.5 per cent, 1 per cent and 28.5 per cent are allocated under groundnut, green gram and cotton during Kharif and66.9 per cent and 2 per cent under onion and wheat during Rabi season. Design of drip irrigation was done with 0.76 m lateral spacing and 0.4 m dripper spacing for groundnut and green gram whereas for cotton this was 1.52 m and 0.4 m. Dripper discharge of 2 lph was used. The turbidity of the canal water is a major impediment in using it through drip irrigation. To assess the threshold value of the maximum permissible turbidity, a laboratory experiment was conducted to evaluate the effect of turbidity on dripper discharge. The water of different turbidity was used and its effect on discharge of 2l ph pressure compensating drippers with time was recorded. To evaluate the filtration efficiency of different components of the system, the turbidity of the water was measured in canal, in the adjunct reservoir, in catch well and after filtration. Although it was tried to take samples after each filter, but sampling could not be done during running of the system and therefore the sample was taken from the water emitting out of the dripper. To assess the efficiency of the pressurised irrigation system, the irrigation efficiency of this system was evaluated in comparison to surface irrigation system in another part of command of the same outlet with similar cropping system. The parameters evaluated were conveyance efficiency, application efficiency, and uniformity of the pressurised system. The standard procedure for estimating these parameters were adopted (Michael, 1978). For measurement of application efficiency, initial and final moisture content in the root zone was monitored by gravimetric method during the evaluation time.

Two 10 hp pumping units were provided for groundnut (17.62 ha), green gram (0.25 ha) and cotton (7.13 ha). Cost of drip irrigation system including pumping unit for groundnut, green gram and cotton crops was calculated as Rs 3049221.00, Rs 39296.92 and Rs 788976.48 respectively. The same drip irrigation system can be used for onion and wheat crop in *rabi* season. So, the cost of drip irrigation system for *rabi* crops was not taken. Total cost of drip irrigation system including pumping unit was Rs 3877494.40 for 25 ha area.

Annual cost, benefit and payback period by adopting drip irrigation system

The cost benefit analysis of adopting drip irrigation in the command area was evaluated by considering the cost of drip irrigation system and intermediate storage structures (Table 4). Per hectare net annual cost and benefit of drip irrigation system was Rs 28135.85 and Rs 41206.28 and the benefit cost ratio was 1.46 which indicates the economic viability of adopting drip irrigation system in canal command areas.

Table 3: Annual cost of the system under drip irrigation system and benefit cost analysis

S. No.	*Item*	*Cost in Rs*
1.	Investment on intermediate storage structure for 25 hectare area	1,269,127.99
2.	Investment on drip irrigation system including pumping system for 25 hectare area	3,877,494.40
	Sub total	**5,146,622.39**
Annual cost of intermediate storage structure for 25 hectare area		
1.	Depreciation on intermediate storage structure for 25 hectare area (assuming 25 years life and 50 per cent junk value)	25,382.56
2.	Maintenance @ 1 per cent	12,691.28
3.	Interest on investment @ 9 per cent	57,110.76
	Sub total (A)	**95,184.60**

Annual cost of drip irrigation system for 25 hectare area		
1.	Depreciation of drip irrigation system for 25 hectare area assuming 10 years life and 10 per cent junk value	348,974.50
2.	Maintenance of the system @ 2 per cent	77,549.89
3.	Interest on investment @ 9 per cent	174,487.25
4.	Electricity charges @ Rs. 30 per hp per month (two pumping unit each of 10 hp)	7,200.00
	Sub total (B)	**608,211.63**
Total annual cost of the system for 25 hectare area (A + B)		**703,396.23**
	Annual cost per hectare	28,135.85
	Annual benefit per hectare	41206.28
	Benefit cost ratio of the system	**1.46**

Payback period was calculated by dividing the total initial investment towards drip irrigation system and intermediate storage structure with net annual benefit of the project due to adoption of the drip irrigation system. Cost of one intermediate storage structure and cost of drip irrigation system (25 ha) were Rs 1269127.99 and Rs 3877494.40 respectively. Total initial investment for 25 ha area was Rs 5146622.39. Net annual benefit of drip irrigated command was Rs 103015700 from 2500 ha command area. Net annual benefit for 25 ha was Rs 1030157. Payback period was is within five years duration.

Conclusion

It can be concluded that the canal-based pressurised irrigation system is a feasible option in flow-based minor irrigation system and it increases the irrigation efficiency very significantly and reduce the stress on the existing water resources. The problem of turbidity of canal water can be managed by making it to pass the adjunct reservoir, catchwell and a three-stage filtration process.The amount of fallow land can be drastically reduced in *kharif* and *rabi* seasons by adopting micro irrigation. Nearly 100 auxiliary storage structures are suggested across the command area to cater the needs of crops. The drip system should be designed for 25 ha in Uben command area to get a benefit cost ratio of 1.46.

References

Allen, R.G., L.S. Pereira, D. Raes and M. Smith (1998). Crop evapotranspiration. FAO Irrigation and Drainage paper No. 56, Rome, pp. 281.

Anonymous-b (1992). Plan for operation and maintenance, Uben Irrigation Scheme, Water and Land Management Institute, Government of Gujarat, Narmada and Water Resource Department, Gandhinagar.

Haouari, M. and M.N. Azaiez (2001). Optimal cropping patterns under water deficits, European Journal of Operational Research, Vol. 130(1) : 133-146.

Fasakhodi, A., Seyed, H. Nour and Amini M. (2010). Water resources sustainability and optimal cropping pattern in farming systems; a multi-objective fractional goal programming approach. Water Resource Manage, Vol. 24: 4639 4657.

INCID (1994). Drip Irrigation in India, Indian National Committee on Irrigation and Drainage, New Delhi.

Karamouz, M., B. Zahraie, R. Kerachian and A. Eslami (2010). Crop pattern and conjunctive use management: A case study. Irrigation and Drainage, Vol. 59: 161173.

Kaur, B., R.S. Sidhu, V. Kamal (2010). Optimal crop plans for sustainable water use in Punjab. Agricultural Economics Research Review, 23: 273284.

Kumar, M. D., S. Rajvanshi and S. K. Dash (2008). Social costs and benefit of micro irrigation system adoption in canal commands: a study from IGNP command area of Bikaner in Rajasthan. International Water Management Institute, Conference Papers number h041874. 67-82.

Michael, A.M. (2008). Irrigation - Theory and Practices, Vikas Publishing House Private Limited, New Delhi.

Montazar, A. and A. Rahimikob (2008).Optimal water productivity of irrigation networks in arid and semi-arid regions, Irrigation and Drainage, Vol. 57(4) : 411-423.

Papadopulos, I.S. and H.H. Cooper (1967). Drawdown in a well of large diameter, Water Resource Research, Vol. 3(1) : 241-244.

Satasiya, R.M. (2011). Uben command area planning for efficient water management. Ph.D. thesis submitted to Junagadh Agricultural University, Junagadh.

Sethi, L.N., S.N. Panda and M.K. Nayak (2006). Optimal crop planning and water resources allocation in a coastal groundwater basin, Orissa, India, Agricultural Water Management, Vol. 83(3) : 209-220.

Singh, D.K., C.S. Jaiswal, K.S. Reddy, R.M. Singh, and D.M. Bhandarkar (2001). Optimal cropping pattern in a canal command area. Agricultural Water Management, Vol. 50(1) : 1-8.

Srivastava, R.C., S. Mohanty, R.B. Singandhuppe, R.K. Mohanty, M.S. Behera, L.I.P. Ray and D. Sahoo (2010). Feasibility Evaluation of Pressurized Irrigation in Canal Commands, Water Resource Management, Vol. 24(12) : 3017-032.

Zhou, H., P. Hui and Z. Chi (2007). An interactive fuzzy multi-objective optimization approach for crop planning and water resources allocation. In: Bio-Inspired Computational Intelligence and Applications, Lecture Notes in Computer Science, 4688: 335-346, Springer, Berlin-Heidelberg.

Vivekanandan, N., K. Viswanathan and S. Gupta (2009). Optimization of cropping pattern using goal programming approach. OPSEARCHRCHRCH, Vol. 46: 259-274.

23

Development of Land Capability Classification for Soil and Water Conservation Planning and Management using Remote Sensing and GIS

H. V. Parmar, R. Subbaiah, G. V. Prajapati, P. A. Damor and A. B. Kheni

Abstract

The land capability is governed by the different land attributes such as the types of soil, its depth and texture, underlying geology, topography, hydrology, etc. To get the maximum benefit out of the land, proper use is inevitable. The common way of determination of land quality from land characteristic is mainly by assessing and grouping the land types in orders and classes according to their aptitude. The attempt was made to apply Remote Sensing and GIS for the development of land capability classification of the Ozat catchment situated in Junagadh district, Gujarat state, India. The thematic layers were generate for slope, soil texture, soil depth, drainage, soil erosion and land use/land cover using Remote Sensing and GIS. The supervised classification technique in Geomatica V10.0 Software was used to assess Land use/ land cover conditions using IRS P6, LISS III images. The intersect overlay technique was used to overlay the different thematic layers to generate land capability classification.The analysis reveals that Class I, II, IV, V and VI were present in the study area. Class I which was much suitable for agriculture crop production accounts 69.21 per cent. Soils in this class require careful soil management, including conservation practices and measures, to prevent deterioration or to improve soil fertility. The Class-IV which was cultivated land but with restrictions with limited plant choice and requires careful management accounts 9.31 per cent. This land was recommended for limited crop use, pasture, woodland, range or wildlife with the adoption of engineering for the soil and water conservation. Class VI was most susceptible to land degradation which accounts for 16.54 per cent. This class was identified as most suitable for deciduous (Moist/Dry) forest, scrub forest and other forest plantation to conserve the soil and water using vegetative measures.

Key words: *Remote Sensing, GIS, Land Capability Classification, Land Use.*

Introduction

The problem of land and water degradation is on the rise due to the over exploitation and mismanagement of these resources. The both resources are limited and with increasing population, the demands for become more competitive. The land store water and nutrients required for plants and other living organisms. The use of land is not only determined by the user but also by its land capability. The land capability is governed by the different land attributes such as the types of soil, its depth and texture, underlying geology, topography, hydrology, etc. These attributes limit the extents of land available for various purposes. To get the maximum benefit out of the land, proper use is inevitable. Sustainable development occurs only when management goals and actions are ecologically viable, economically feasible and socially desirable.

The basic principle is to use the land according to its capability and deal with its problems according to the needs of the people. Therefore, the knowledge of land capability classification is a pre-requisite for determining the capability of land planning and implementation of soil and water conservation programmes. United States Department of Agriculture has provided specific guidelines for Land Capability Classification (LCC). United States Department of Agriculture (USDA)'s LCC system includes eight classes of land designated by Roman numerals from I to VIII. Remote Sensing has shown great potential in Land capability mapping and natural resources planning and management due to its advantages over traditional procedures in terms of cost and time effectiveness in the availability of information over larger areas. The surface reflectance spectra over a wide range of objects and conditions should be identified and interpreted into meaningful outputs prior to decision-making and applications. Geographical information system (GIS) can be defined as a system used to store, retrieve, manipulate, analyse and output geographically referenced data or geo-spatial data, in order to support decision-making for planning and management of land use, natural resources, environment, transportation, urban facilities and other administrative records.Teotia *et al.*, (1993) utilised data Landsat TM to provide land use information for the semi-arid areas of the region of Brazil. They derived maps for the objectives of regional planning under this investigation Land use/Land cover map, Soil Associations map, and Land Capability Units map. Ali and Kotb (2010) used the spatial analysis technique to assess the soil capability in the Nile Delta, Egypt using satellite data (Landsat ETM and SRTM images) with GIS. Teotia and Djail (2010) used remotely sensed data and GIS technology for resources management in Brazil. By re-coding the land use land cover classes, three maps, such as land use/land cover, major soil associations and land capability classification maps were prepared. The results were found very beneficial for land development, natural resources management, land evaluation, soil conservation and land reforms programmes of the Federal Government of Brazil and for the development of semi-arid region of northeastern. Panhalkar (2011) used Geographic information systems and remote sensing for development Land capability classification for integrated watershed development in southern Maharashtra. The Class II, III, IV and VI were present in the study region. Maryati (2012) evaluated land capability of reclamation area in Indonesia using climate properties, slope, erosion risk, soil properties, and natural hazard risk. The area categorised be cultivated which was varying from class II, III and IV based on land characteristic and limitation factor. Mahmut *et al.* (2010) updated the soil map in Turkey by employing GIS and remote sensing techniques. Land Capability Classification I, II and III cover 87.3 per cent of the total geographical area whereas soils of VI and VII class only 12.7 per cent.

Watershed management is an endowed approach to mitigate the gap between demand and supply of water and other natural resources, particularly in the fragile arid and semi-arid tropics. Aher *et al.* (2014) used watershed management information system (WATMIS) toolkit for integrated watershed planning and management of its natural resources using multiple technologies like GIS, Remote Sensing (RS), Global Positioning System (GPS), hydrological modelling and soft computing tools. Ma (2004) applied the GIS in management of upper and lower parts of a watershed equally for sustenance of environment. Watershed Management has assumed urgency for planned development of land and water resources and to arrest land degradation process to preserve environment and ecological balance. Decision support to such management planning requires scientific knowledge of resources information, expected runoff and sediment yield, priority classification of watersheds for conservation planning, monitoring of watershed for environmental impact assessment and technologies of GIS for data base creation, scenario development and appropriate decision-making. Remote sensing technique is ideally suited to evolve such a management strategy (Chakraborti, 1993). The study was undertaken to identify the different classes as per land capability classification for soil and water conservation planning and management using remote sensing and GIS.

Materials and Methods

Study Area

The study area lies between latitude 21°32′ 35.31"N to 21°12′20.22"N and longitude 70° 39′15.71" E to 70°46′35.38"E. The study area falls in the catchment of Ozat river of Junagadh District of Gujarat state. The major crops of the study area are Groundnut, Cotton, Mango, Wheat, etc. Total catchment study area of Ozat catchment is 1409.16 km². The location of the study area is shown in Fig. 1.

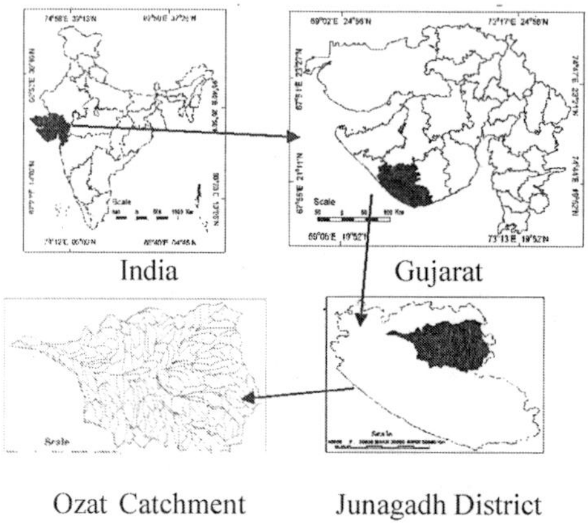

Fig. 1: Location Map of Study Area Ozat catchment.

Climate of Study Area

The climate of study area is subtropical and semi-arid type which receives rainfall South-West monsoon with an average of 929.81 mm. It is received from the mid of the June to the mid of the October month. The average annual pan evaporation of 5.6 mm/day. January is the coldest month with mean monthly temperature varying from 7°C to15 °C and maximum monthly temperature is recorded in the month of May varying between 29.50 °C to 39.40 °C. The relative humidity is varying between 45 per cent and 89 per cent. The wind speed is varying between 2 to 9.70 km/h.

Data Collection and Software Used

The satellite images of IRS P6, LISS- III, Soil maps, etc. were collected from Bhaskaracharya Institute for Space Application and Geoinformatics (BISAG), Gandhinagar. Geomatica V10.0 was used for different spatial operations, image correction, image processing, intersect overlay analysis, etc. for the preparation of different thematic maps and land capability classification.

Use of Remote Sensing and Geographical Information System (GIS)

The satellite data collected from BISAG, Gandhinagar were utilised for the remote sensing and GIS application for the study. Geometica V 10.0 was used for different remote sensing and GIS operations.

Visual interpretation: The image of the catchment was interpreted using fundamental picture elements interpreted in the satellite imageries. Visual interpretation was done on the basis of the image characteristics like tone/ colour, texture, pattern, shape, size, location, shadow as well as the associated elements.

Geo-Referencing: The digitised topographic map of the catchment was registered over the satellite image in Geomatica V 10.0 software. Due care was taken to register it as accurate as possible so as to get the exact features in ground survey.

Soil Characteristics/ Thematic Maps

The spatial information in the form of maps required for the preparation of land capability classification. The soil characteristics maps like soil texture, soil erosion, soil depth and slope maps were prepared using Geomantice V 10.0.

Land Use/ Land Cover Map

Land use/land cover map was prepared from satellite image of IRS P6, LISS III digital data of Ozat catchment. Based on visual image interpretation of satellite imagery, the land use/land cover map was generated. Area statistics and spatial distribution of different land use/land cover pattern were also determined through the help of software. Saadat *et al.* (2011) proposed a new protocol for LULC classification using Landsat ETM+ images. The summer Landsat ETM+ image yields the best results with an overall accuracy of 95 per cent, while a spring image yields the poorest accuracy (82 per cent). Zhixin *et al.* (2012) pro-posed a new four-component algorithm for land use and land cover (LULC).

Land Capability Classification

Land capability classification refers to a systematic arrangement of different kinds of land according to those properties that determine the ability of the land. Two broadest, classification of land cultivable lands and uncultivable lands. The cultivable land is further classified into four classes I to IV and the uncultivable land is also classified into four classes V to VIII.

Class I Land (Green colour): Class I land is the best type of land available for agricultural purposes. *Class II Land (Yellow colour)*: The class II land is a reasonably good land that can be readily cultivated after adopting certain improved practices. *Class III Land (Red colour):* The land under this class is considered moderately good land that could be cultivated on a regular basis under a good crop rotation with regular intensive remedial measures. *Class IV Land (Blue colour):* Soil under this land class is fit for cultivation but is restricted by different factors. Hence it requires careful management. *Class V Land (Dark green):* This is the land in the second group which includes all lands that are not suitable for cultivation. *Class VI Land (Orange):* This is another class of land that is not suitable for cultivation but only for pastures, wildlife and forest cover. *Class VII Land (Brown colour):* It is subject to severe restrictions or severe hazards for its use as a grazing and/ or forest land. *Class VIII Land (Purple colour):* This is the most unfavourable type of land. It is not suitable either for cultivation, grazing or forestry.

Preparation of Land Capability Classification Maps

The generalised methodology for LCC using GIS is depicted in the form of a flow chart (Fig. 2). The geospatial techniques help in generation of a reliable spatial and non-spatial information database. The land capability classification has been carried out by applying parameters like soil depth, soil texture, land use /land cover and slope of land. The information related with soil depth and soil texture about study region has been used from BISAG Gandhinagar and thematic maps of the study region were created. These parameters were integrated in GIS environment by using intersect overlay technique. Moreover the surface and overlay analysis capabilities in GIS can effectively facilitate in handling vast amount of spatial information. The Intersect tool calculates the geometric intersection of input feature classes. The features or portion of features that are common to (intersect) all inputs were written to the output feature class as final LCC map.

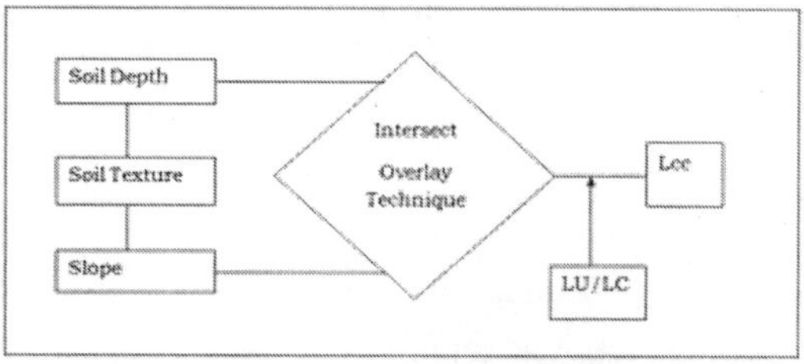

Fig. 2. Flow chart for land capability classification in GIS.

Ground truth verification and error estimation: The ground truth verification for the different themes have been carried out. The primary and secondary data were collected and information on the thematic map were verified under the ground truth verification. The accuracy was also checked for the land use land cover map.

Results and Discussion

Use of Remote Sensing and GIS

Soil Characteristics Maps

The thematic maps of different soil characteristics were prepared using information provided by BISAG Gandhinagar. The separate maps were prepared for each soil characteristic.

Soil Texture Map

The soil texture map of the study area is shown in Fig. 3. The maximum area as 1027.50 km^2 (72.91 per cent) was under fine soil texture in the Ozat catchment. Only 3.73 per cent of the total area of the catchment was under clayey soil texture.

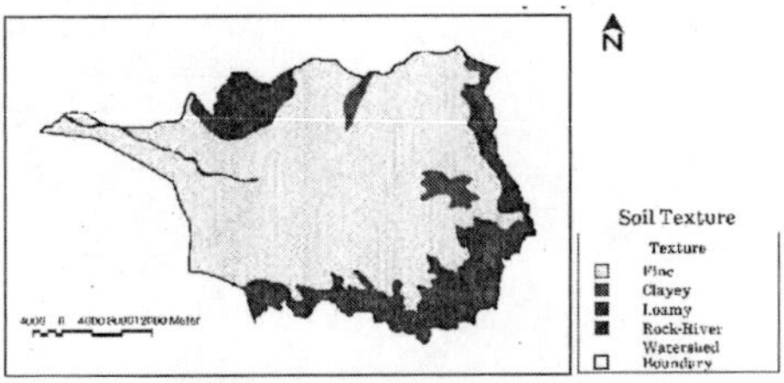

Fig. 3. Soil texture map of Ozat catchment.

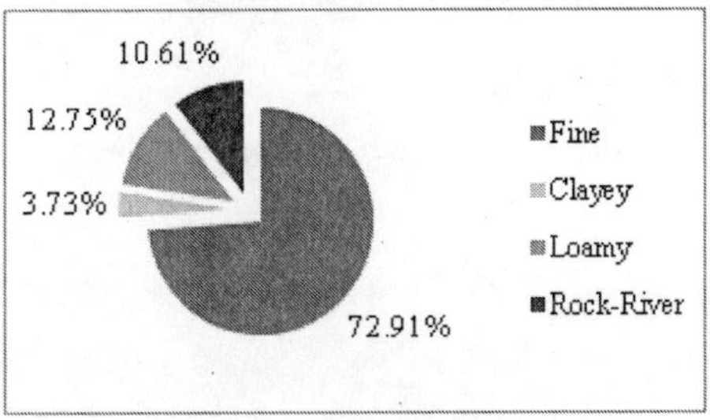

Fig. 4. Per cent area under different Soil texture.

Soil Depth Map

The soil depth map of the study area is shown in Fig. 5, in which area under moderate deep (75-100 cm) was 229.67 km^2 (16.30 per cent), under shallow (25-50 cm) was 51.72 km^2 (3.67 per cent), under very shallow (10-25 cm) was 175.85 km^2 (12.48 per cent), under river was 9.23 km^2 (0.66 per cent) and under rock was 140.34 km^2 (9.96 per cent). The maximum area as 802.35 km^2 (56.93 per cent) was under moderate shallow (50-75 cm) soil depth.

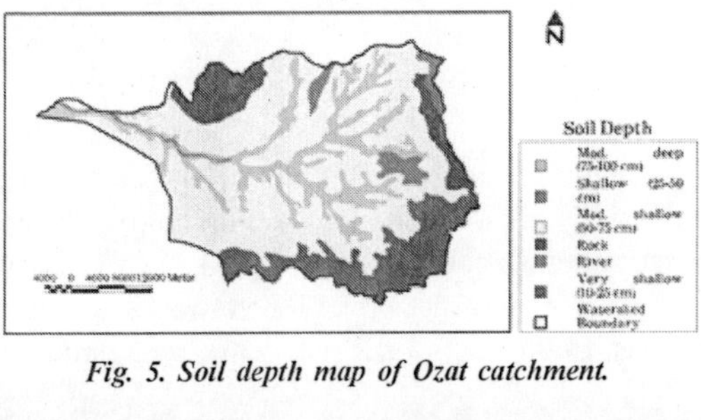

Fig. 5. Soil depth map of Ozat catchment.

Soil Erosion Map

The soil erosion map of the study area is shown in Fig. 6. There were three groups of soil erosion were indemnified in the study area, *i.e.* slight erosion, severe erosion and moderate erosion.The maximum area as 833.23 km^2 (59.15 per cent) was under moderate soil erosion in the Ozat catchment. Out of three classes, minimum area was under severe soil erosion as 194.20 km^2.

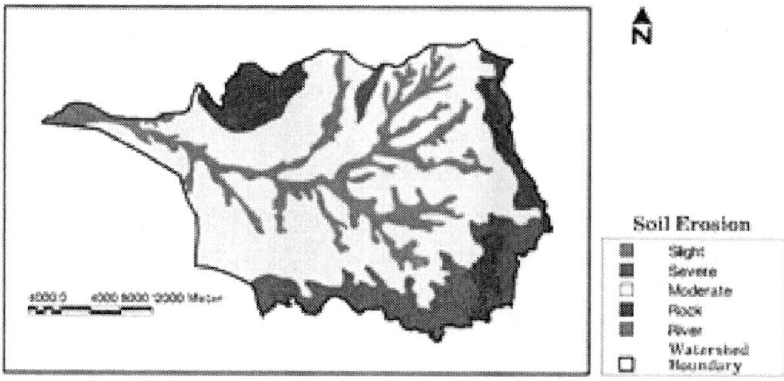

Fig. 6. Soil erosion map of Ozat catchment.

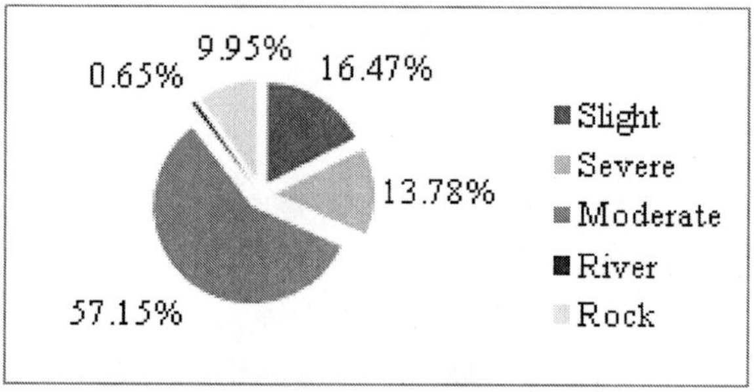

Fig. 7. The per cent area under soil erosion.

Slope (per cent) Map

The land slope map of Ozat catchment is shown in Fig 8. Out of total area 72.17 per cent (1017.05 km^2) area falls under very gently sloping (1-3 per cent) area.

Land Use / Land Cover Map

As per the land use/land cover, total six major group of land use were identified in the study area. In which area under agriculture was 1004.79 km^2 (71.30 per cent), under wastelands was 115.26 km^2 (8.18 per cent), under built-up was 16.15 km^2 (1.15 per cent), under forest was 225.29 km^2 (15.99 per cent), under others was 19.94 km^2 (1.42 per cent), and under water bodies was

27.73 km^2 (1.96 per cent). The maximum71.30 per cent area was under agriculture (Fig. 9).

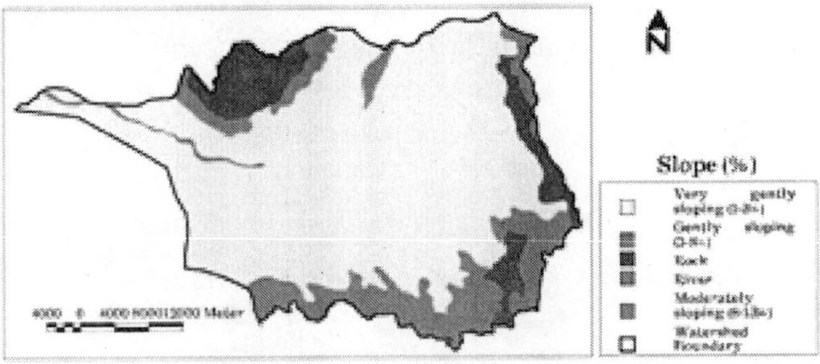

Fig.8. Slope (per cent) map of Ozat catchment.

Table: 1. Area under different slope class (per cent)

Sr. No.	*Particular*	*Area, km^2*	*Per cent (%)*
1.	Very gently sloping (1-3%)	1017.05	72.17
2.	Gently sloping (3-8%)	164.82	11.7
3.	Moderately sloping (8-15%)	77.72	5.52
4.	River	9.23	0.66
5.	Rock	140.34	9.95
	Total	**1409.16**	**100.00**

Table: 2. Area under different land use

Sr. No.	*Particular*	*Area, km^2*	*Per cent (%)*
1.	Agriculture	1004.79	71.30
2.	Wastelands	115.26	8.18
3.	Built-up	16.15	1.15
4.	Forest	225.29	15.99
5.	others	19.94	1.42
6.	Water bodies	27.73	1.96
	Total	**1409.16**	**100.00**

Overlay Map of Soil Texture, Soil Depth and Slope Map

The soil depth map, soil texture map and slope map of the study area were overlay by intersect overlay technique. The single map was generated by overlaying these three layers as shown in Fig 10. The attribute of all three layers were also overlayed and total nine group were generated. As per the overlay analysis maximum area as 749.18 km^2 (53.16 per cent) of the total area was under fine texture soil, moderately shallow soil and gently sloping area. The minimum area was under clayey texture, shallow depth (25-50 cm) and gently sloping (3-8 per cent) land as 28.24 km^2 (2.00 per cent).

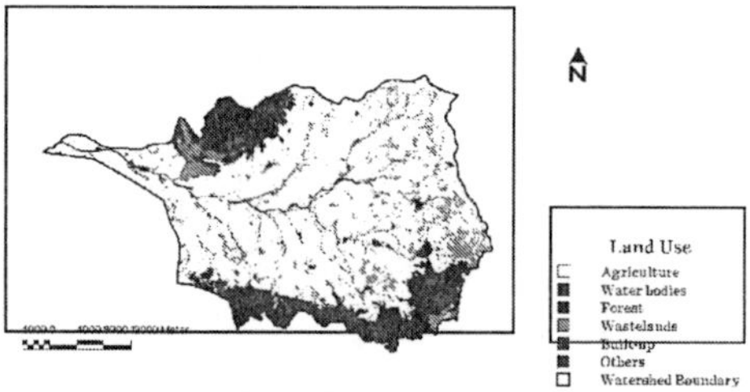

Fig. 9. Land usemap of Ozat catchment

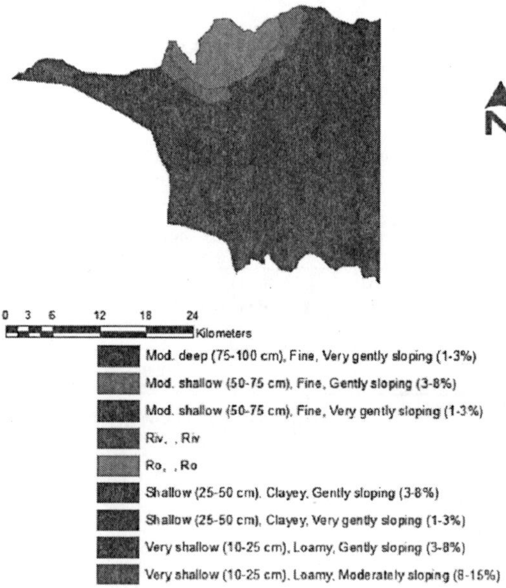

Fig. 10. Overlay map of soil texture, soil depth and slope map

Land Capability Classification Map

The overlay map prepared from the texture map, soil depth and slope map using intersect overlay technique and the land use/land cover map were used to prepare the land capability classification map. The Land capability classification of Ozat catchment is shown in Fig 11.

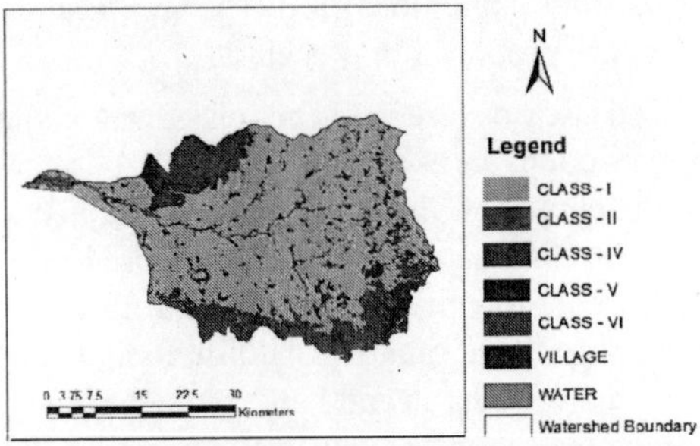

Fig.12. The per cent area under LCC of Ozat catchment.

The soil parameters, slope parameters and land use/land cover information were used to identify the land capability classes. Total five classes were identified in the study area as per land capability classification. The area of different class and percentage are given in Table 3. The maximum area as 957.34 km^2 (69.21 per cent) was under class I, most suitable for cultivation in the study area.

Table 3: Area under different land capability classes.

Sr. No.	*Particular*	*Area, km^2*	*Percent (%)*
1.	Class - I	975.34	69.21
2.	Class - II	7.67	0.55
3.	Class - IV	131.16	9.31
4.	Class - V	14.42	1.02
5.	Class - VI	233.12	16.54
6.	Village	20.52	1.46
7.	River	26.93	1.91
	Total	**1409.16**	**100.00**

Class I: Class I cover 975.34 km^2 area (69.21 per cent) in Ozat catchment, this class was identified as most suitable for agricultural crop production (Kharif + Rabi (Double cropped). Soils in this class require careful soil management, including conservation practices, to prevent deterioration or to improve air and water relations when the soils are cultivated. The limitations were few and the practices are easy to apply for this class.

Class II: The total area covered by this class was about 7.67 km^2 (0.54 per cent). Soils in class II have more restrictions than those in class I and when used for cultivated crops, the conservation practices are usually more difficult to apply and to maintain. Limitations of soils in class II restrict the amount of clean cultivation; timing of planting, tillage, choice of crops and harvesting. The soils used for plantations crop. The area treatment for the soil and water conservation are proposed in this class.

Class IV: The total area covered by this class was about 131.16 km² (9.31 per cent).The restrictions in used for soils in class IV are greater than those in class III and the choice of plants was more limited. When these soils were cultivated, more careful management was required and conservation practices were more difficult to apply and maintain. Soils in class IV may be used for crops, pasture, woodland, range or wildlife food and cover. The soil used in this class with Land with Scrub and Land without Scrub. Many sloping soils in class IV was suited to occasional but not regular cultivation. The engineering measures are proposed for soil and water conservation in this area.

Class V: The total area covered by this class was about 14.42 km² (1.02 per cent). This class soil was not suitable for cultivation. This class was identified as uncultivated waste land suitable for Prosophis, etc.

Class VI: This class ranks second in the study region as it coved area about 233.12 km² (16.54 per cent). This class was identified as most suitable for Deciduous (Moist/Dry), Scrub Forest, Forest Plantation. The vegetative measures for the soil and water conservation proposed in this area.

Village, River: The total area covered under village settlement was about 20.52 km² (1.46 per cent) and total area covered by river was 26.93 km² (1.91 per cent).

Summary and Conclusions

Land is the most valuable natural resource, which needs to be harnessed according to its potential. Land is a limited resource and with increasing population, the demands for land become more competitive. Land capability is the basis of sustainable watershed development. The basic principle of soil and water conservation is to use the land according to its capability and treat the land according to its needs. The following conclusion were drawn from the present study:

1. The area under moderate soil erosion condition was maximum as 883.23 km² (59.12 per cent) of the total area.
2. As per the land use/land cover area was divided into six major groups and maximum area as 1004.79 km² (71.30 per cent) was under agriculture.
3. From the overlay analysis of the soil texture, soil depth and slope layers of the study area, maximum 749.18 km² (53.16 per cent) area was under fine texture soil, moderately shallow depth soil (55-75 cm) and very gently sloping (1-3 per cent) land.
4. As per the land capability classification, total five classes (Class- I, II, IV, V and VI) were identified in the Ozat catchment.

5. The maximum area falls in Class-I as 975.34 km^2 (69.21 per cent), *i.e.* most suitable for agriculture crop production.
6. The area covered under Class-II was about 7.67 km^2 (0.55 per cent), restrict the amount of clean cultivation; timing of planting, tillage, choice of crops and harvesting.
7. The area under Class-IV was 131.16 km^2 (9.31 per cent), *i.e.* cultivated land but with restrictions, in use with limited plant choice and requires careful management.
8. The total area covered under Class-V was about 14.42 km^2 (1.02 per cent), the uncultivated waste land suitable for prosophis, etc.
9. The total area under Class- VI was 233.12 km^2 (16.54 per cent), uncultivated land and area under forest and plantation.

Acknowledgement

The acknowledgement to the BISAG-Gandhinagar for providing the necessary satellite images for the research work.

References

Aher, P. D., Adinarayana, J., Gorantiwar, S. D. and Sawant, S. A., 2014. Information System for Integrated Watershed Management Using Remote Sensing and GIS. Remote Sensing Applications in Environmental Research, Society of Earth Scientists Series, 2014. DOI: 10.1007/978-3-319-05906-8_2, 2014, pp. 17-34.

Ali, R. R. and Kotb, M. M., 2010. "Use of Satellite Data and GIS for Soil Mapping and Capability Assessment", Nature and Science. Vol. 8(8): 104-115.

Chakraborti, A. K., 1993. "Strategies for Watershed Management planning using remote sensing technique", Journal of the Indian Society of Remote Sensing. Vol. 21 (2): 87-97.

Ma, T., 2004. "GIS Application in Watershed Management", Nature and Science, Vol. 2 (2): 1-7.

Mahmut, D., Oztekin, M. E., Akca, E. and Senol, S., 2010. "Updating soil characteristics of Cukurova region (Southern Turkey) using geographical information systems and ilsen software", Indian Journal of Agricultural Research, Vol. 44(4) : 316-320.

Maryati, S., 2012. "Land Capability Evaluation of Reclamation Area in Indonesia Coal Mining Using LCLP Software", International Symposium on Earth Science and Technology, CINEST 2012. Procedia Earth and Planetary Science, Vol. 6 : 465-473.

Panhalkar, S., 2011. "Land capability classification for integrated watershed development by applying remote sensing and GIS Techniques", ARPN Journal of Agricultural and Biological Science. Vol. 6 (4) : 46-55.

Saadat, H., Adamowski J., Bonnell R., Sharifi F., Mohammad N. and Ale-Ebrahim S., 2011. "Land use and land cover classification over a large area in Iran based on single date analysis of satellite imagery", ISPRS Journal of Photogrammetry and Remote Sensing. Vol. 66(5) : 608-619.

Teotia, H. S. and Djail, S., 2010. "Remote sensing and GIS based land cover, soil and land capability information for resources management in semi-arid region of Paraiba, Brazil", ASPRS 2010 Annual Conference San Diego, California. April (2010) : 26-30.

Teotia, H.S., Ulbricht, K. A. and Civco, D. L., 1993. "Application of spot to GIS/LIS in land capability evaluation for regional planning of Jaico semi-arid region of Piaui, Brazil", VII-simposio brasileiro de sensoriamento remoto curitiba - pr, Brazil. May (1993) : 114-154.

Zhixin, Q., Gar-On Yeh, A., Xia, L. and Zheng L., 2012. "A novel algorithm for land use and land cover classification using RADARSAT-2 polarimetric SAR data", Remote Sensing of Environment. Vol.118 : 21-39.

24

Remote Sensing Derived Vegetation Indices to Assess the Crop Water Requirement in Ozat-II Canal Command

H. V. Parmar and N. K. Gontia

Abstract

The canal command areas suffer from either excessive or inadequate water supply resulting in wide gap between irrigation demand and supply. For effective water management, the irrigation schemes need up-to-date real time information on the irrigated area and the crops being grown within the command area of each canal, also data on the amount of water being used by each crop. This is not easy to obtain as many irrigation schemes are vast and cover hundreds and sometimes thousands of square kilometers and have vast numbers of farms; their scale severely limiting the effectiveness of data collection. Satellite images provide a scientific basis for identification of crops and when used along with other related data of soil and meteorological information, it provides clear information for land use and crop identification. Estimation of crop evapotranspiration from climatic parameters provides objective criteria for irrigation management. The real-time crop coefficient (K_c) generated from remote sensing responds to actual crop conditions in the field and captures the variability between fields. The remotely-sensed vegetation indices can be derived to predict crop coefficients at field and regional scales. Therefore, the possibility of directly estimating the crop coefficient from satellite reflectance of a crop for the estimation of evapotranspiration was investigated in the Ozat-II canal command of Junagadh District of Gujarat State, India. Satellite images of Landsat-8 OLI and TIRS (March, April, May-2014) were used to calculate the NDVI, SAVI and establish a *general relationship between indices and the crop coefficients (K_c). The remote sensing based K_c estimation follow the similar pattern of the seasonal variation of crop fraction, NDVI and SAVI. The daily crop evapotranspiration (ET_c) ranging from 4.836 to 7.321 mm/day based on K_{cNDVI}. The ET_c was ranging from 3.213 to 8.204 mm/day based on K_{cFAO} for different growth stages of summer groundnut for the Ozat-II canal command. The correlation coefficient between ET_c estimated with satellite data and meteorological data was 0.7362.*

Key words: *Remote Sensing, Vegetation Indices, crop coefficient, canal command.*

Introduction

Water is the most valuable, renewable and inexhaustible resource of nature. Irrigation is mainly dependent on various sources, including the availability of canal water and ground water. Water use efficiencies are comparatively less in canal command areas than command areas that depend on groundwater. In India, most of the prominent canal command areas suffer from either excessive or inadequate water supply resulting in wide gap between irrigation demand and supply. Another significant problem that is expected in the future is the increasing need for alternative demands for water supply due to urbanisation and industrialisation. These demands create more pressure on water resources and in turn on irrigation sector. Irrigation in the future will certainly face the challenge of maximising efficiency. Hence, in order to enhance the irrigation efficiency, estimation of irrigation demand is really important coupled with efficient management of water in the canal command area before releasing the water to the crops. The irrigation water requirement of a command area is dependent on the nature of the crop and the nature of the soil where the crop is grown. This shows the need to obtain up-to-date and real time information pertaining to the irrigated area, the type of soil and the type of crops being grown within the command area of each canal.

The evapotranspiration estimation is crucial for water requirement, water use efficiency and resulting crop productivity in canal command. For effective water management, the irrigation schemes need up-to-date real time information on the irrigated area and the crops being grown within the command area of each canal, also data on the amount of water being used by each crop. This is not easy to obtain as many irrigation schemes are vast and cover hundreds and sometimes thousands of square kilometers and have vast numbers of farms; their scale severely limiting the effectiveness of data collection. Satellite images provide a powerful tool for the identification of crops and when used in conjunction with Geographical Information Systems (GIS) have proved to be effective for land use and crop identification.

Estimation of crop evapotranspiration from climatic parameters provides objective criteria for irrigation management. The crop evapotranspiration represents the crop water demand and governed by weather and crop conditions. The value of crop coefficient taken from literature may provide a practical guide for scheduling irrigation, but considerable error in estimating crop water requirement can occur due to their empirical nature (Jagtap and Jones, 1989). The real-time crop coefficient generated from remote sensing responds to actual crop conditions in the field and captures the variability between fields. The Food and Agriculture Organisation (FAO) Penman Monteith empirical calculation is an accepted method for calculating crop water requirements and has been tried and tested worldwide (Penman, 1948. Monteith 1965; Allen *et al.* 1998; Gontia and Tiwari, 2004; Gontia and Tiwari, 2010; Papadavid *et al.*, 2011). This empirical calculation uses standard meteorological data to estimate the evapotranspiration (ET) of a reference crop, which is in turn modified by a crop factor to estimate the ET of a particular crop. The latter is affected by the local climatic conditions, crop characteristics, length of growing season, soil moisture and the time of planting (Doorenbos and Pruitt, 1977). Gontia and Tiwari (2004) estimated the evapotranspiration of wheat crop using spectral vegetation indices generated by spectral data using NDVI. Singh and Irmak (2009) used surface energy balance algorithm for land model for estimating the spatial distribution of ET_c for major agronomic crops. The Kc values were estimated based on ET_c and ET_r (i.e., $K_c = ET_c/ET_r$). A regression model was developed to establish the relationship between NDVI and the ET_r -based crop coefficients (K_{cr}) for corn, soybeans, sorghum, and alfalfa. Farg *et al.* (2012) estimated crop coefficient (K_e) and crop evapotranspiration (ET_c) using SPOT-4 satellite data integrated with the meteorological data and FAO-56. The reference evapotranspiration (ET_0) were estimated using FAO Penman-Monteith and tabled single crop coefficient values were adjusted to real values. SPOT-4 images geometrically and radio-metrically corrected were used to drive the vegetation indices (NDVI and SAVI). Kamble *et al.* (2013) estimated crop coefficient (K_c)-based crop evapotranspiration for irrigation water management. Similarities between the crop coefficient curve and a satellite-derived vegetation index showed potential for modelling a crop coefficient as a function of the vegetation index. Knudby (2004) developed a semi-deterministic model that allows assessment of groundnut yields from NOAA AVHRR data for the Peanut Basin in Senegal. Linear regression, using yield as the dependent variable and integrated NDVI during the reproductive period of the groundnuts, minus integrated NDVI during a period before the growing season as the independent variable. Bandyopadhyay *et al.* (2005) used the weekly and seasonal field water balance components for the estimation of actual evapotranspiration (ET_a) including the capillary contribution into root zone. Mean crop coefficients were determined for each 7-day period of growth and were related to leaf area index and growing degree-days.

Average seasonal values of ET_a of peanut were 434, 391 and 356 mm for the three treatments, respectively, for 115 days of growth. The maximum average K_c of 1.19 occurred about nine weeks after sowing relative to grass reference ET (ET_0).

The objective of the study was to investigate the applicability of time-series Landsat NDVI, SAVI data to the development of spatially representation of K_c. To establish the relationship between the crop coefficients and remote sensing-based vegetation indices for the estimation of crop evapotranspiration/ water requirement in Ozat-II canal command.

Materials and Methods

Study Area

The study area comprises the canal command area of Ozat-II comprises of a composite dam across river Ozat near Badalpur, Junagadh district, Gujarat, India. The location of the command area lies between latitude 21°12'46"N to 21°33'04"N and longitude 70°25'07"E to 70°53'24"E. The gross and live storage capacity of the reservoir is 36.20 Mcum and 27.71 Mcum respectively. The total catchment of the dam is 475.24 sq. km. The gross command area of the project is 6150 hectare and culturable command area is 4918 hectare. The canal system comprises of 20.60 km long main canal. Two farms in the canal command with summer groundnut crops were also selected for the study. The location of Farm-1: Latitude: 21°23'15.31" N, Longitude: 70°33'51.17" E and the location of farm-2: Latitude: 20°22'44.79" N and Longitude: 70°33'12.25" E. The location of the study area is shown in Fig. 1.

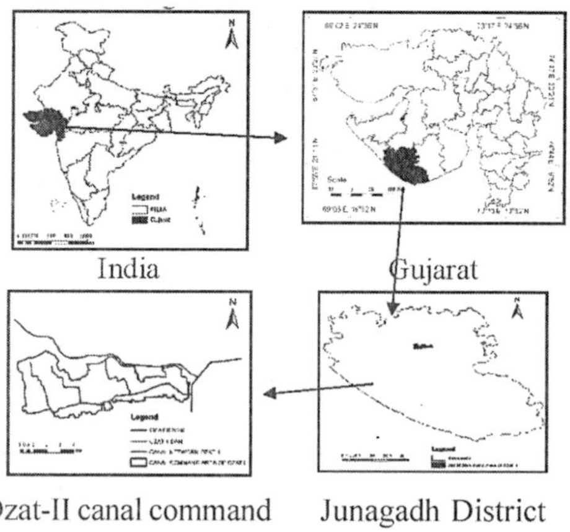

Fig. 1: Location Map of Study Area Ozat-II Canal command.

Climate of Study Area

The climate of study area is subtropical and semi-arid type which receives rainfall South-West monsoon with an average of 929.81 mm. It is received from the mid of the June to the mid of the October month. The average annual pan evaporation of 5.6 mm/day. January is the coldest month with mean monthly temperature varying from 7°C to 15 °C and maximum monthly temperature is recorded in the month of May varying between 29.50 °C to 39.40 °C. The relative humidity is varying between 45 per cent and 89 per cent. The wind speed is varying between 2 to 9.70 km/h.

Data Collection and Software Used

The satellite images of Landsat-8 Operational Land Imager (OLI) and Thermal Infrared Sensor (TIRS) images consist of nine spectral bands from 1 to 11 were downloaded from www.earthexplorer.usgs.gov/ The Approximate scene size is 170 km north-south by 183 km east-west. The images having 16-bit: radiometric resolution, GeoTIFF: format, UTM: Map projection and WGS 84: Datum were used in the study. Total 6 Landsat-8 images of date 09 March, 25 March, 10 April, 26 April, 12 May and 28 May, 2014 representing different growth stage were used in this study. The climatological daily data like, maximum and minimum temperature, sunshine hours, wind speed, maximum and minimum relative humidity, etc were collected from Agrometeorology Cell, Junagadh Agricultural University (JAU), Junagadh. The information of Ozat-II canal command were collected from Junagadh Irrigation Project Division, Junagadh. Geomatica V10.0 was used for different spatial operations, image correction, image processing, composition of different bands, mosaicking of different scene, clipping of area of interest.

Use of Remote Sensing and Geographical Information System (GIS)

The downloaded satellite data of Landsat-8 were used in Geometica V 10.0 software for different remote sensing and GIS applications. The bends of Landsat-8 images were composited into a single file. The canal command were demarcated and the satellite images were clipped using command area boundary layer. The primary and secondary data were collected and information on the thematic map were verified under the ground truth verification.

Vegetation Indices (VI)

The vegetation indices include NDVI (Rouse *et al.,* 1974), SAVI (Huete, 1988) etc., has been tested to predict crop coefficients at field and regional scales (Choudhury *et al.,* 1994; Ray and Dadhwal, 2001; Gontia and Tiwari, 2004; Knudby, 2004; Farg *et al.,* 2012).

Normalised Difference Vegetation Index (NDVI)

The Normalized Difference Vegetation Index (NDVI) is a measure of the amount and vigour of vegetation at the surface. Green leaves have a reflectance of 20 per cent or less in the 0.5 to 0.7 range (green to red) and about 60 per cent in the 0.7 to 1.3 ìm range (near infrared). The value is then normalised to -1 $\leq$ NDVI $\leq$ 1 to partially account for differences in illumination and surface slope.

$$NDVI = \frac{(NIR - R)}{(NIR + R)} \quad \text{...(1)}$$

The Landsat-8 NDVI product is generated at 30-meter spatial resolution on a Universal Transverse Mercator (UTM) mapping grid set to World Geodetic System 1984 (WGS-84). NDIV in Landsat-specific terms is,

$$NDVI = \frac{NIR(Band\,5) - R(Band\,4)}{NIR(Band\,5) + R(Band\,4)} \quad \text{...(2)}$$

Where,

NIR = Near infrared band

R = Red band

Soil Adjusted Vegetation Index (SAVI)

SAVI is the Soil Adjusted Vegetation Index, which was introduced by Huete (1988). This index attempts to be a hybrid between the ratio-based indices and the perpendicular indices.

$$SAVI = \frac{(NIR - R)(1 + L)}{(NIR + R + L)} \quad \text{...(3)}$$

The Landsat-8 SAVI product is generated at 30-meter spatial resolution on a UTM set to WGS-84. SAVI is calculated with a soil brightness correction factor (L) defined as 0.5 to accommodate most land cover types.

$$SAVI = \frac{[NIR(Band\,5) - R(Band\,4)] * [1 + 0.5]}{[NIR(Band\,5) + R(Band\,4) + 0.5]} \quad \text{...(4)}$$

Estimation of Crop Coefficient

The FAO-56 Table 12 (Allen *et al.*, 1998), contains the typical values for crop coefficients of various crops for initial, mid season and late season of crop growth stages. The crop coefficients and vegetation indices are affected by leaf area index and fractional ground cover and can be correlated using regression equation. The advantage of remote sensing derived crop coefficient over traditional crop coefficient is that it represents a real-time crop coefficient that responds to actual crop conditions in the field and captures the variability between fields. The empirical equation used for calculation of basal crop coefficient K_{cb} (K_{cSAVI}) is taken from Gonzalez-Dugo and Meteos, 2008 as below:

$$K_{cb} = K_{cbmax} \frac{(SAVI - SAVI_{min})}{(SAVI_{max} - SAVI_{min})} \qquad ...(5)$$

Where,

K_{cb} = K_{cSAVI} = Basal crop coefficient

K_{cbmax} = Basal crop coefficient at effective full growth cover, 1.30 (Hunsaker, 2005)

$SAVI_{max}$ = Maximum value of SAVI associated with dense vegetation

$SAVI_{min}$ = Minimum value of SAVI associated with bare soil.

Crop coefficient can also estimate using remote sensing (NDVI) based ET_c and ET_o as given below:

$$K_{cNDVI} = ET_c / ET_o \qquad ...(6)$$

The crop evapotranspiration (ET_c) is determined using FAO Penman-Monteith with canopy parameters estimated from satellite imagery. The following equation is used for estimation of ET_c based on Satellite data and meteorological data.

$$ET_c = \frac{1}{\lambda}\left(\frac{\Delta(Rn - 0.4e - 0.5LAI) + \frac{Cp\rho(e - e)U}{124}}{\Delta + \gamma(1 + \frac{u}{0.6LAI})}\right) \qquad ...(7)$$

ET_c = ET_{cNDIV} = Crop Evaporation, mm/day,

λ = Latent Heat of vaporisation of water, MJ/Kg

Δ = Slope of vapour pressure curve, kPa/°K

Rn = Net radiation, MJ/m^2 day

LAI = Leaf Area Index,

C_p = Specific Heat, MJ/Kg °K

ρa = Specific Density, Kg/m^3

es - ea = Vapour pressure deficit, kPa

u = Wind Speed, m/sec

γ = Phychromatric constant, kPa/°K

For calculation of LAI following two equations are used for growth and decline stages:

$$LAI_{Growth} = 0.038 * e^{(4.836*NDVI)} \qquad ...(8)$$

$$LAI_{Decline} = 1.444*e^{(1.24*NDVI)} \quad ...(9)$$

Estimating of Reference Crop Evapotranspiration (ET_0)

The FAO Penman-Monteith method is selected as the method by which the evapotranspiration of the reference surface (ET_0) can be unambiguously determined, and this method provides consistent value of ET_0. The reference evapotranspiration (ET_0) can be estimated using FAO Penman-Monteith method using equation,

$$ET_0 = \frac{0.408\Delta(R_n - G) + \frac{\gamma(900)}{T+273} u_2 (e_s - e_a)}{\Delta + \gamma(1 + 0.34 \ u_2)} \quad ...(10)$$

where,

ET_0 : reference evapotranspiration (mm/d)

R_n : net radiation at crop surface (Mj/m^2 d),

G : soil heat flux density (Mj/m^2 d),

T : mean daily temperature at 2 m height (°C),

e_s : saturation vapor pressure at T_c (kPa),

e_a : actual vapour pressure (kPa),

$e_s - e_a$: saturation vapour pressure deficit (kPa),

Δ : slope of the e_s, temperature relationship (kPa/°C),

γ : psychrometric constant (kPa/ °C) and

u_2 : wind speed at 2 m height (m/s)

Crop Evapotranspiration (ET_c): The effects of various weather conditions on evaporation are incorporated into crop ET_0; the effects of characteristics that distinguish the cropped surface from the reference surface are integrated into the crop coefficient K_c. By multiplying ET_0 with the crop coefficient, ET_c is determined as

$$ET_c = K_c \times ET_o \quad ...(11)$$

Results and Discussion

Use of Remote Sensing and GIS

Vegetation Indices (VI)

The vegetation indices maps of NDVI (Fig. 2) and SAVI (Fig. 3) were prepared using Landsat-8 images for all five dates. The value of NDVI ranges from 0.10434 for farm-1 for date 25/3/2014 to maximum as 0.59435 for Command area for date 26/04/2014. The SAVI value ranges from 0.15651 for farm-1, Dt. 25/3/2014 to 1.78305 of command area Dt. 26/04/2014.

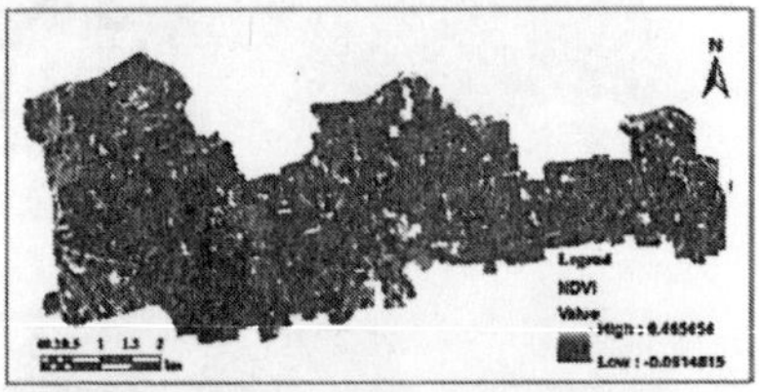

NDVI map Dt. 25/03/2014

NDVI map Dt. 10/04/2014

NDVI map Dt. 26/04/2014

NDVI map Dt. 12/05/2014

Fig. 2. Normalised Difference Vegetation Index (NDVI) maps for March, April, May, 2014.

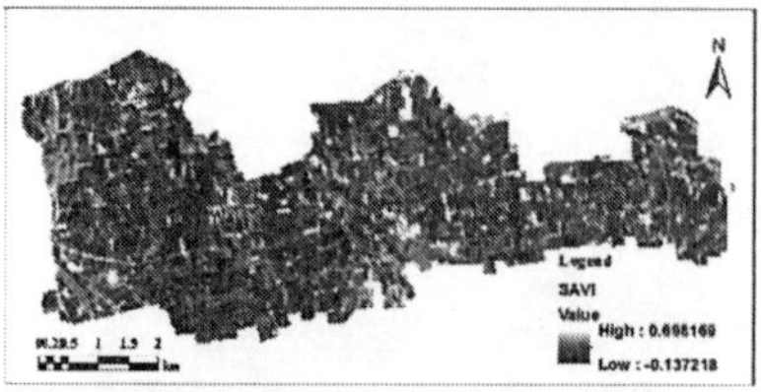

SAVI map Dt. 25/03/2014

SAVI map Dt. 10/04/2014

SAVI map Dt. 26/04/2014

SAVI map Dt. 12/05/2014

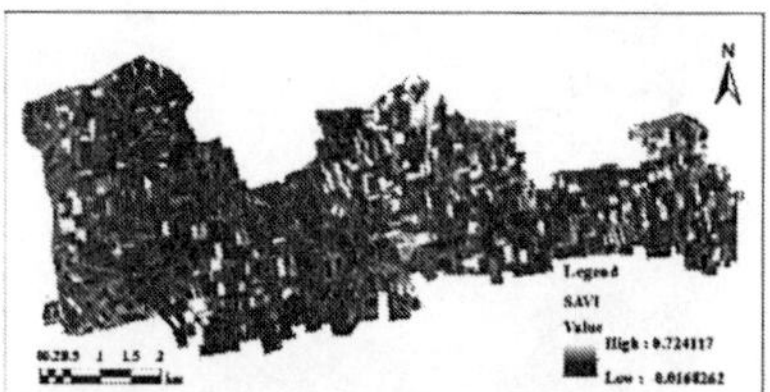

SAVI map Dt. 28/05/2014

Fig. 3. Soil-Adjusted Vegetation Index (SAVI) maps for March, April, May, 2014.

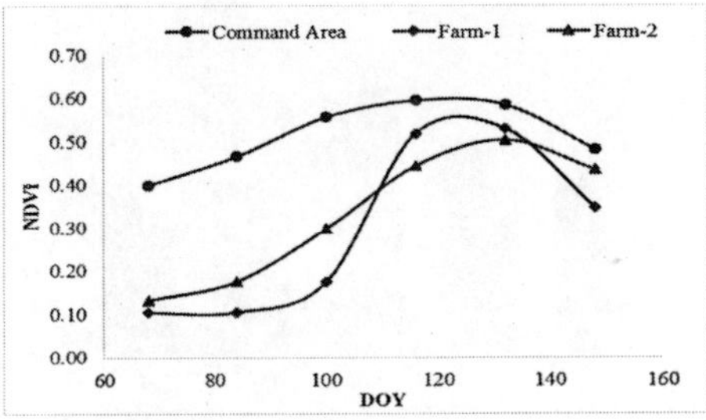

Fig. 4. Seasonal progression of NDVI.

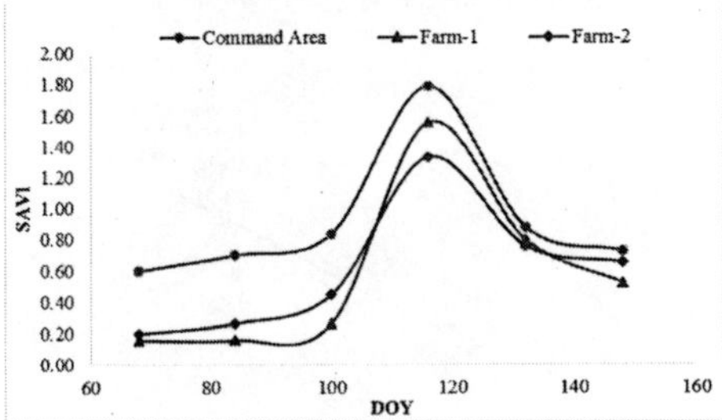

Fig. 5. Seasonal progression of SAVI.

Estimation of Crop Evapotranspiration (ET_c)

The reference evapotranspiration (ET_0) values were computed using the FAO-56 Penman-Monteith Equation-10. The computed ET_0 values were cumulated for daily basis and multiplied with the corresponding crop coefficient values (K_{cNDVI}, K_{cFAO}) and corresponding actual crop evapotranspiration (ET_c) were estimated. The crop coefficients were estimated using Eq. 5 and Eq. 6. The maximum crop coefficient for summer groundnut crop, K_{cSAVI} was 1.15, K_{cNDVI} was 1.037 and K_{cFAO} was 1.1. The daily ET_c based on K_{cNDVI} were estimated as 5.055, 6.643, 7.321, 5.298 and 4.836 mm/day and based on K_{cFAO} were estimated as 3.213, 5.125, 8.204, 5.451 mm/day for 25 March, 10 April, 26 April, 12 May and 28 May, 2014 (Day of year- DOY as 84, 100, 116, 132 and 148) respectively (Fig. 6). The correlation coefficient between ET_c estimated with satellite data and meteorological data was 0.7362 (Fig.7).

The relationships between NDVI and SAVI with indices based estimated K_c (Fig. 8 and Fig. 9) were also developed. It may help in estimation of crop coefficients for different growth stages by knowing the values of NDVI or SAVI for the estimation of crop water requirements using a time series of images.

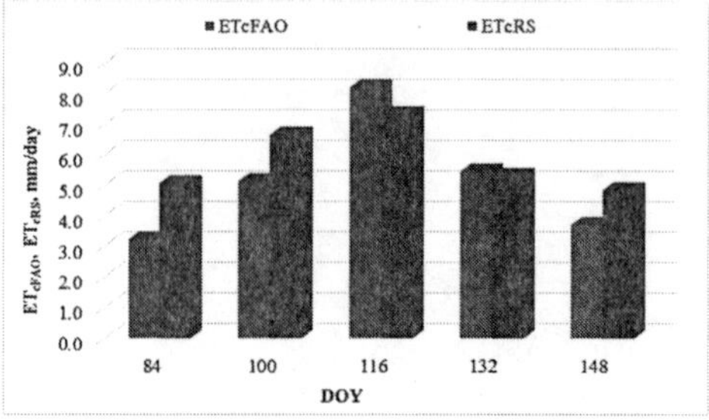

Fig. 6. Comparison of ET_c estimated with satellite data and meteorological data.

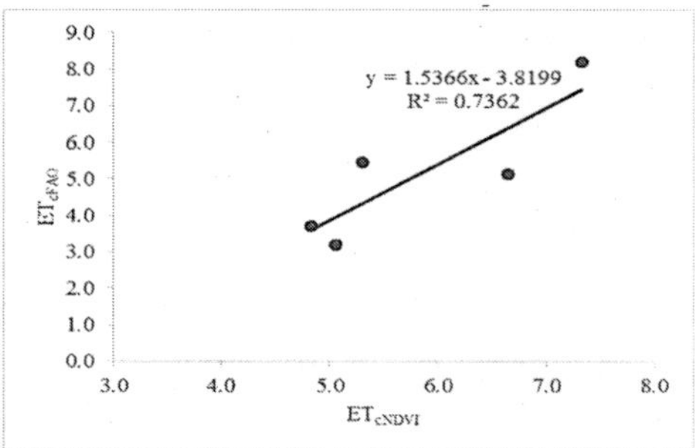

Fig. 7. Relationship between ET_c estimated with satellite data and meteorological data.

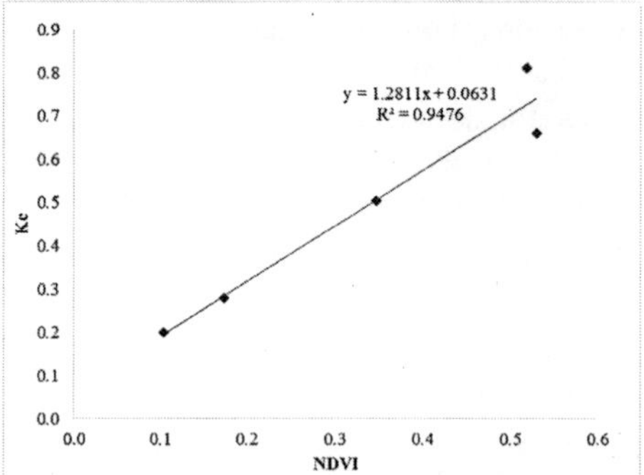

Fig. 8. Relationship between Landsat-8 NDVI and K_{cNDVI} for different growth stages.

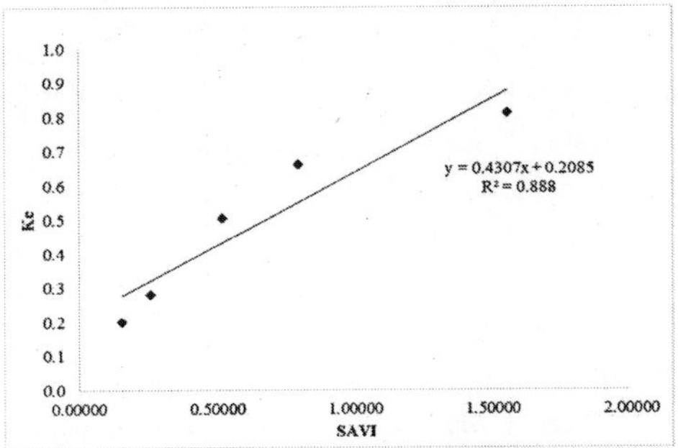

Fig. 9. Relationship between Landsat-8 SAVI and K_{cNDVI} for different growth stages.

Summary and Conclusions

Crop evapotranspiration under standard conditions (ET_c) is the evapotranspiration from the disease-free, well-fertilised crops, grown in the large fields, under optimum soil water conditions and achieving full production under the given climatic conditions. The FAO Penman-Monteith method adapted to satellite data was used to estimate the actual crop evapotranspiration for summer groundnut crop using satellite data with ground meteorological data. The crop coefficient was estimated using the NDVI, SAVI. The relationships between Landsat-8 NDVI and SAVI with K_{cNDVI} and K_{cSAVI} for different growth stages were developed to predict the value of K_c. The maximum crop coefficient for summer groundnut crop, K_{cSAVI} was 1.15, K_{cNDVI} was 1.037 and K_{cFAO} was 1.1. The remote sensing based K_c estimation follow the similar pattern of the seasonal variation of crop fraction, NDVI and SAVI. The daily crop evapotranspiration (ET_c) ranging from 4.836 to 7.321 mm/day based on K_{cNDVI}. The ET_c was ranging from 3.213 to 8.204 mm/day based on K_{cFAO} for different growth stages of summer groundnut for the Ozat-II canal command. The correlation coefficient between ET_c estimated with satellite data and meteorological data was 0.7362.

Acknowledgement

The acknowledgement to NASA for free Landsat data, Agrometeorology Cell, JAU, Junagadh, for climatic data and Junagadh Irrigation Project Division, Junagadh for data and information of Ozat-II dams for the research work.

References

Allen, R.G., Pereira, L.S., Raes, D. and Smith, M., 1998. Crop evapotranspiration, guidelines for computing crop water requirements, Irrig. and Drain., Paper No. 56. FAO, Rome, Italy, 300 pp.

Bandyopadhyay, P. K., Mallick, S. and Rana, S. K., 2005. Water balance and crop coefficients of summer-grown peanut (*Arachis hypogaea* L.) in a humid tropical region of India. Irrigation Science, 23: 161169.

Choudhury, B.J., Ahmed, N.U., Idso, S.B., Reginato, R.J., Daughtry, C.S.T., 1994. Relations between evaporation coefficients and vegetation indices studied by model simulations. Remote Sens. Environ. 50, 117.

Doorenbos, J. and Pruitt, W.O., 1977. Guidelines for predicting crop water requirements. FAO Irri. and Drain.Paper No. 24, FAO, Rome, Italy.

Farg, E., Arafat, S.M., Abd El-Wahed, M.S. and EL-Gindy, A.M., 2012. Estimation of Evapotranspiration ET_c and Crop Coefficient K_c of Wheat, in south Nile Delta of Egypt Using integrated FAO-56 approach and remote sensing data. The Egyptian Journal of Remote Sensing and Space Sciences, 15: 83-89.

Gontia, N. K. and Tiwari, K. N., 2004. Crop Evapotranspiration estimation using temporally distributed normalized difference vegetation index. Paper presented in Int. Conference on Emerging Technologies in Agricultural and Food Engineering (etae-2004) held at IIT Kharagpur, India, Dec. 14-17, 2004.

Gontia, N.K. and Tiwari, K.N. 2010. Estimation of spatially and temporally distributed crop coefficient and evapotranspiration of wheat (Triticum aestivum) in an irrigation command using remote sensing and GIS by Water Resources Management, 24 (7): 1399-1414.

Gonzalez-Dugo, M. P. and Mateos, L. 2008. Spectral vegetation indices for benchmarking water productivity of irrigated cotton and sugarbeet crops. Agr. Water Manage., 99: 48-58.

Huete, A.R., 1988. A soil-adjusted vegetation index (SAVI). Remote Sens. Environ. 25: 295-309.

Hunsaker, D. J., Printer, P. J., and Kimball, B. A. 2005. Wheat basal crop coefficients determined by normalised difference vegetation index. Irrig. Sci., 24: 1-14.

Jagtap, S.S. and Jones, J.W., 1989. Sensitivity of crop coefficients to climate and irrigation management. Irrig. Science, 10:231-244.

Kamble, B., Kilic, A. and Hubbard, K., 2013. Estimating Crop Coefficients using Remote Sensing-Based Vegetation Index. Remote Sensing, 5(4): 1588-1602.

Knudby, A., 2004. An AVHRR-based model of groundnut yields in the Peanut Basin of Senegal, International Journal of Remote Sensing, 25(16):3161-3175.

Monteith, J. L., 1965. Evaporation and environment, Symp. Soc. Experimental Biology, 19: 205-234.

Papadavid, G., Hadjimitsis, D., Michaelides, S. and Nisantzi, A., 2011. Crop evapotranspiration estimation using remote sensing and the existing network of meteorological stations in Cyprus. Adv. Geosci., 30, 3944.

Penman, H.L., 1948. Natural evaporation from open water, bare soil and grass. Proc. Roy. Soc. of London, 193:120-146.

Ray, S.S., Dadhwal, V.K., 2001. Estimation of crop evapotranspiration of irrigation command area using remote sensing and GIS. Agric. Water Manage. 49: 239-249.

Rouse, J.W., Haas, R.H., Schell, J.A., Deering, D.W., Harlan, J.C., 1974. Monitoring the vernal advancement and retrogradation of natural vegetation. NASA/GSFC, Type III, Final Report, Greenbelt, MD, pp. 1371.

Singh, R. K. and Irmak, A., 2009. Estimation of Crop Coefficients using Satellite Remote Sensing. J. Irrig. Drain Eng., 135(5): 597-608.

25

Groundwater Recharge Potential of Harvested Roof Water in Junagadh Region

P. B. Vekariya, H. D. Rank, J. H. Kelaiya, D. M. Paradava
R. J. Patel, H. V. Parmar and H. H. Mashru

Abstract

Rain water is free from organic matter, soft in nature and free from biologically impurities. In urban areas, huge volume of rain water can be collected from large number of roof of buildings. This water can be stored in storage tanks or it can be directly diverted for recharging the aquifers through existing GW tapping arrangements and thereafter can be utilised. But groundwater recharge in combination of buffer storage tank may prove better groundwater recharge solution. The roof water recharge system was installed at Department of Soil and Water Engineering, College of Agricultural Engineering and Technology, Junagadh Agricultural University, Junagadh. 58 years of rainfall data of Junagadh station were considered in this study. 110 mm PVC pipe along with required fittings and water meter were used for installing the system. Nearby tube wells and existing sump were also used in the experiment for water level observations and collecting excess water during storm. The quantity of available roof water under five days maximums rainfall were estimated. It was found that at 10 years return period the quantity of roof water varies from 0.22 to 0.42 cu.m per sq.m against one to five days maximum rainfall respectively. Cost of roof water harvesting tank for storing total volume of roof water runoff per sq. m. for one to five days maximum rainfall varies from Rs 50 to 94 respectively. The average annual roof water runoff was estimated 0.62 cu.m per sq.m. At full head condition hydraulic conductivity was found 13.32 sq.m per hour. At 10 years return period annual roof water runoff estimated 0.95 cu.m per sq.m. Buffer storage sump of 0.207 cu.mper sq. m of roof is required for safe operation of system in combination of groundwater recharge through tube well with buffer sump system.

Key Words: *rain water, roof water harvesting, groundwater recharge.*

Introduction

Artificial groundwater recharge is a process by which the groundwater reservoir is augmented at a rate exceeding the augmentation rate under natural conditions. In some parts of India, due to over-exploitation of groundwater, decline in groundwater levels resulting in shortage of supply of water and intrusion of saline water in coastal areas have been observed. In such areas, there is need for artificial recharge of groundwater by augmenting the natural infiltration of precipitation or surface-water into underground formations by methods such as water spreading, recharge through pits, shafts, wells etc. The choice of a particular method is governed by local topographical, geological and soil conditions.

The direct recharge from rainfall in hard rock area is approximately 10 per cent only. Geological formation of Saurashtra region is mainly made of basaltic hard rocks formation. Rainfall occurrence is mostly in storm. Mostly storms are of short duration, therefore some part is stored in top soil, small part infiltrated as recharge and most of water goes into sea as run off. Shallow unconfined and fractured zones are recharged during monsoon at slow rate but deep confined and semi-confined aquifers not getting sufficient recharge water.

Groundwater recharge through roof water harvesting has a great scope; the urbanisation has increased large roof areas. The roof water almost goes out if not properly guided in to the ground. Runoff coefficient in case of urban area is reported very high upto 0.95 because of concrete roof and most of parts like paved roads and other paved utilities. Very high quality water can be collected and supply to the ground which enhance ground water storage and improves groundwater quality at very reasonable cost.

The roof water harvesting system was designed and installed by AICRP on Irrigation Water Management project at southern part of Department of Soil and Water Engineering, College of Agricultural Engineering and Technology, Junagadh Agricultural University, Junagadh. Nearby tube wells were used for recharging as well as observation wells. Junagadh is located at the foothills of mount Giranar. Geographically it is situated at 21° 31'N latitude, 70° 31' E longitudes and 85 km away from Arabian Sea. Junagadh has a tropical climate and receives rainfall during South-West monsoon (June-November). The details of material used and methodology adopted in the experiment are given as per following details.

Objective: To estimate roof water harvesting potential for Junagadh region. To determine groundwater recharge potential through harvested roof water for Junagadh region. To determine the buffer storage sump in combination with tube well recharging and cost for Junagadh region.

Materials and methods

Rainfall analysis: Roof water harvesting systems collects rainwater from the roof catchments during rainy days. Therefore, the amount and distribution of rainfall are major factors influencing the systems. Starting from 1957 to 2014 of 58 years rainfall data were collected from the observatory of JAU Junagadh and rainfall data were analysed. The data were processed to determine one to five days maximum rainfall by adding consecutive neighbouring daily rainfall values and obtain year-wise one to five days maximum values of rainfall and each maximum values of rainfall for 10 years recurrence interval were determined to evaluate roof water runoff for maximum five days storm for Junagadh region. Recurrence interval is the average number of years during which a storm of given magnitude (maximum depth or intensity) may be exceed to occur once. Frequency F is the percentage of years during which a storm of given magnitude may be equaled or exceed, Weibull equations were used for analysis.

$$T = \frac{n+1}{m} \qquad \ldots(1.1)$$

The frequency F (expressed as per cent of time) of that storm magnitude (having recurrence interval T) is given by

$$F = \frac{1}{T} \times 100\% \qquad \ldots(1.2)$$

The probability that a T-year storm (and frequency $F = 1/T \times 100\%$) may not occur in any series of N years is

$$P_{(N,0)} = (1-F)^N \qquad \ldots(1.3)$$

And that it may occur is

$$P_{EX} = 1-(1-F)^N \qquad \ldots(1.4)$$

Where, P_{EX}=probability of occurrence of a T-year storm in N-years.

Hydraulic design of roof water harvesting system

Available head:

$$h_{av} = H \frac{flv^2}{2gd} \qquad \ldots(1.5)$$

Where, h_{av} = Available head, m, H = Total head, 4.7 m, f = Friction factor for pvc pipe, 0.01, l = Total length of pipe, 98 m, v = Flow velocity, 3 m/s (assumed), g = gravity constant, 9.81 m/s^2, d = diameter of pipe = 0.11 m. So, Head loss = 4.08 m and available head, h_{av} = 0.62 m, Head loss due to rapid sand filter assumed = 20 cm , Net available head = 0.62–0.2 = 0.42 m

Expected maximum system velocity:

$$V = \sqrt{2gh} = 2.84 \text{ m/s} \qquad \text{...(1.6)}$$

Flow through pipe, Q = A x V = 0.027 m^3/s

Where, A is cross-sectional area of pipe = 0.0094 m^2

Peak runoff rate of roof water: The following model developed by AICRP groundwater utilisation Junagadh for rainfall intensity-duration-frequency relationship was used for calculation of rainfall intensity.

$$I = \frac{68.86T^{0.265}}{(t+0.86)^{1.022}} \qquad \text{...(1.7)}$$

Where, I = rainfall intensity (mm/hr); T = return period (10 years); t = duration (hours) = 20 min (CGWB);

So, Rainfall intensity, I =106.58 mm/hr.

Peak runoff rate: $$Q = \frac{CIA}{360} \qquad \text{(1.8)}$$

Where, C= Runoff coefficient, I = Rainfall intensity, mm/hr., A = Roof catchment area, ha

For estimationof runoff from storm event, c value for roof top was suggested from 0.85 to 0.95 (Aditya Kumar Patra and Sneha Gautam, 2011, CPWD, 2002). So considering safer side runoff coefficient C was 0.95 accepted, Roof area 767 sq.m,. Therefore peak roof water runoff, Q = 0.021 m^3/s. So, the design of roof water collecting pipe system is safe.

Experimental layout: The Experimental lay out of roof water harvesting system is shown in Fig. 1

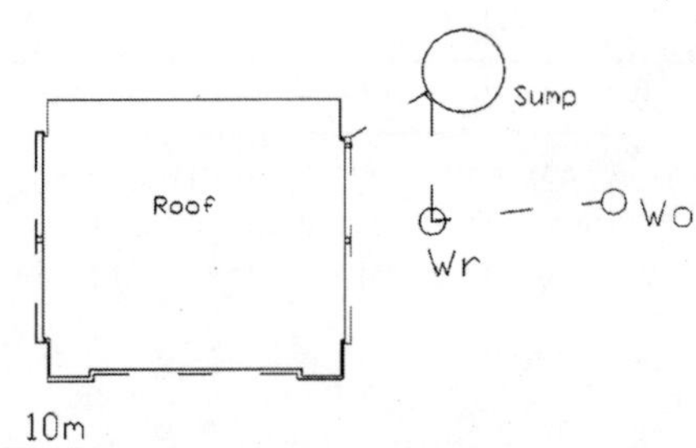

Fig. 1. layout of experiment.

Auto distribution System: The roof water harvesting system was made of 110 mm pvc pipe. The pipe is fitted to connect all six outlets of roof area at inlet point of collection pipe, disposal end was fixed tightly on tube well keeping nipple above sump level open to atmosphere and bypass was connected to sump which facilities to divert automatic excess water to sump while runoff from roof exceeds the intake capacity of tube well. So observation of recharge and storage in sump can be recorded. The details of roof water harvesting system and tube wells are shown in Table 1 and 2.

Filter: The major problem in the roof water harvesting from the RCC surface is leaves and plastic thin sheets deposit on roof, which flow with roof water runoff and may clogged the roof water conveyance system. So before water entry in the roof water conveyance pipe gravel materials as rapid sand filter was used to prevent entry of such debris materials.

Table 1: Details of roof water harvesting system.

Name of material	*Size (mm)*	*Quantity*	*Rate (Rs)*	*Cost (Rs)*
Pipe	110	112 m	140/m	15680
Elbow	110	26 No.	60 /No	1560
GI clamps	110	68 No.	8 /No	544
Coupler	110	16 No.	35 /No	560
Tee	110	9 No.	30 /No	270
Labour cost	-	-	-	7000
Total Cost	-	-	-	25614
Water meter	40	1 No.	2000	

Table 2: Details of tube wells.

Wells	*Diameter, cm*	*Depth, m*
W_r		12
W_o	15	7.2

Where, W_r = Recharging tube well, W_o = Observation tube wells

Tube well Intake capacity: During storm period when surrounding area is also recharging by rain water at the same time the observations for intake flow rate and water levels of recharge tube wells and surrounding observation well were recorded.

The hydraulic gradient was estimated at the time of storm. The hydraulic conductivity was determined as per Darcy's law.

$$Q = kia \qquad \ldots(1.9)$$

Where, k = hydraulic conductivity, m^2 /hr, i = hydraulic gradient, a = Cylindrical area of tube well, (4.2 m^2, effective depth of well ,12 3 casing pipe = 9 m)

Sump size and cost: The roof water harvesting volume against one to five days maximum rainfall of 10 years return period was determined. Plots of roof water runoff against days of maximum rainfall of five days storm were prepared. The cost of roof water harvesting of maximum five days rainfall in sump was estimated for Junagadh.

The cost of system was determined considering yearly tube well recharge and buffer storage requirement for excess roof water runoff. Also sump size and cost was estimated by subtracting the tube well recharge capacity from possible maximum roof water runoff of five days maximum rainfall at highest intensity in combination of recharge and buffer sump system were determined for the Junagadh.

Result and Discussion

Potential of roof water harvesting: The rainfall of the location play important role which is main source of roof water flow. Daily rainfall data of 58 years were collected from JAU observatory. Storm of one to five days maximum values of rainfall for 10 years return period were determined as per shown in Table 3 and corresponding volume of roof water runoff were determine and presented in Fig. 2. It shows that after four days the curve becomes just horizontal, it means the increment of roof water runoff between four and five consecutive days is only 2.8 cu.m. five days maximum rainfall for 10 years return period was 439.4 mm. Estimated cost of harvesting storm of one to five days consecutive maximum rainfall of 10 years return period storm in sump varies from Rs 50 to 94 per sq.m of roof respectively, also estimated corresponding roof water runoff varies from 0.22 to 0.42 cu.m per sq.m of roof respectively as per shown in Table 3.

Table 3: Roof water harvesting potential for five days maximum rainfall of Junagadh.

Days maximum for 10 years return period	*Maximum Rainfall (mm)*	*Area (m^2)*	*Runoff Volume (m^3)*	*Runoff Volume (m^3) per sq.m*	*Sump Cost per cu.m, Rs.*	*Total cost of sump Rs.*	*Cost of sump per sq.m*
1.	233.2		169.9	0.22		38227	50
2.	324.9		236.7	0.31		53257	69
3.	392.6		286.1	0.37		64372	84
4.	435.6	767	317.4	0.41	225	71415	93
5.	439.4		320.2	0.42		72038	94

Run off coefficient 0.95 for concrete roof, cost of sump 225 per m^3 (CGWB 2007, Chatrabhuji and Barad 2013).

Tube well intake and hydraulic conductivity: During the storm surrounding area was also recharging ground water at the same time roof water was recharging tube well, intake capacity of recharge well was observed at full head as per Table 4.The 0.094 hydraulic gradient was determined during storm time when surrounding area also supplying water in to the ground and at full recharge head condition hydraulic conductivity was determined 13.32 m per hour.

Table 4: Observations of water meter during full intake of recharge tube well.

Observation	*Time taken* Δ *T, min, sec*	*Roof water runoff recharged* Δ *v, liter*	*Recharge Rate (lpm)*	*Average recharge Rate ((lps)*
1.	11	1000	90.9	1.46
2.	12	1000	83.3	
3.	11, 8	1000	89.8	

Buffer storage in combination with recharging tube well during storm: The capacity of buffer storage sump for collection of excess roof water then real time tube well intake capacity of one to five days maximum rainfall of 10 years return period was determined as per following Table 5. The capacity of storage sump in combination with recharge well was evaluated by allowing roof water runoff of one day maximum rainfall in to tube well and excess then intake volume was estimated as required storage capacity of buffer sump, same work done up to five days maximum rainfall. If excess is positive then buffer storage sump is required else not. Only in case of one day maximum rainfall buffer storage tank of 43.76 m^3 is required, so, 0.057 cu.m per sq. m. of roof area of buffer storage tank

Table 5: Buffer storage in combination with recharge tube well for five consecutive days maximum storm.

Days Maximum maximum for 10 years return period	*Area Rainfall (mm)*	*Runoff (m^2)*	*Runoff water runoff (m^3)*	*Tube Intake (m^3)*	*Excess roof water runoff (m^3)*	*Require-ment Buffer Storage tank (m^3)*	*Require-ment of buffer storage, (m^3/sq.m)*
1.	233.2		169.9	126.14	43.76	Yes	0.057
2.	324.9		236.7	252.28	-15.58	No	-
3.	392.6		286.0	378.42	-92.42	No	-
4.	435.6	767	317.4	504.56	-187.16	No	-
5.	439.4		320.2	630.7	-310.5	No	-

Run off coefficient = 0.95 for concrete roof, cost of tank 225 per m^3 (CGWB 2007, Chatrabhuji and Barad 2013).

is required. In remaining cases intake is more than roof water for given duration. But rates of runoff water and tube well intake is not matching always therefore temporary roof pounding may occurs in case system without buffer storage tank so buffer storage tank is advisable.

The annual average, minimum, maximum and 10 years return period of rainfall of Junagadh were determined and presented in Table 6 and their corresponding roof water runoff, groundwater recharge and required size of buffer storage sump were determined. Also 0.6, 0.1, 1.86 and 0.95 cu.m per sq. m roof water runoff for annual average, lowest, highest and 10 years return period of rainfall were determined respectively. The rainy days of average, lowest, highest and 10 years return period of rainfall were recorded 33, 13, 52 and 41 respectively. Tube well intake of annual average, lowest, highest and 10 years return period of rainfall years was determined considering their respective rainy days. Rainy days of respective rainfall years were consider as duration of tube well intake and roof water runoff was compared with intake of tube well. Table 8 shows yearly intakes for all the cases were found higher as compare to yearly roof water runoff. So no buffer sump required.

Table 6: Annual rainfall of Junagadh

Rainfall, mm				
Maximum	*Average*	*Minimum*	*Rainfall of 10 years return period*	*Number of years*
2793.6	936.65	145.6	1427	58

In previous both the cases, case of storm and case of annual events, it cannot be possible that always roof water runoff matches with intake capacity of tube well, so decision on buffer storage requirement may analysed on the basis of safest side. But it is concluded from both the cases that tube well intake is more as compare to roof water runoff.

For the determination of safest buffer storage sump in combination with groundwater recharge through tube well were determined considering one to five days maximum rainfall, suppose occurs at highest rainfall intensity of 10 years return period and time of concentration equal to 20 min. It means it generates peak rate of runoff for all five days rainfall duration. Also day-wise rainfall duration was distributed by subtracting duration of one days maximum rainfall from duration of two days maximum rainfall and corresponding roof water runoff, same work done up to five days consecutive maximum rainfall. Also day wise duration of rainfall occurrence was used to calculate the tube well recharge capacity. When tube well intake for duration equal to one day maximum rainfall at highest rainfall intensity of 10 years return period was subtracted from roof water runoff of two days maximum rainfall, it was found

158.42 cu.m and per sq.m of roof it was 0.207 cu.m and for remaining days it was 0.081, 0.060, 0.038 and 0.003 cu. m for two, three, four and five days maximum rainfall respectively. Cost of system per sq. m of roof was found Rs 79.87, 51.67, 46.89, 41.97 and 34.16 for one to five days maximum rainfall respectively. It is found total intake during maximum five days rainfall is higher as compare to roof water runoff but referring Table 9. 0.207 cu.m of buffer storage sump is recommended for safest combination of buffer storage and with groundwater recharge system for Junagadh region.

The observations of groundwater recharge and roof water runoff collected in the sump were recorded as per shown in Table 7. The excess roof water runoff then tube well recharge naturally allowed to collect in to sump for whole monsoon season, it was found that out of total runoff 0.87 cu.m per square meter of roof, 0.26 cu.m of runoff per square meter of roof was contributed as ground water recharge and 0.61 cu.m of runoff per square meter of roof was collected in to sump. Cost of system was determined Rs 172 per square meter of roof.

Table 7: Roof water runoff coefficient of RCC roof for Junagadh region.

Sr. No.	*Annual Rainfall in 2014, mm*	*Available roof water run off, m³*	*Roof water run off contributed to tube recharge*	*Roof water run off contributed into sump, m³*	*Roof water run off harvested m³*	*Annual roof water run off coefficient*	*Total Cost of system without recharge Rs*	*Total Cost of system with recharge Rs*
	1	2	3	4	5	6		
1	1234	946	196	469	665	0.703(~0.7)	176139	132039
Analysis per square meter of roof area								
2			0.26	0.61	0.87		230	172

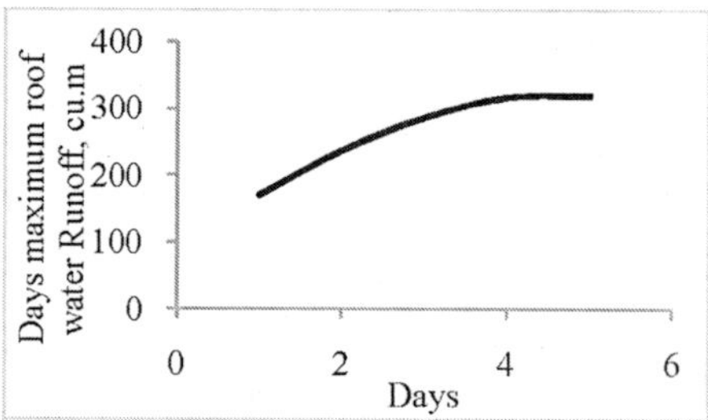

Fig. 2. Maximum roof water runoff against days.

Conclusions

1. Five days maximum rainfall of Junagadh for 10 years return period was 439.4 mm. Estimated cost of harvesting maximum storm of one to

Table 8: Annual roof water potential, tube well intake and annual buffer storage requirement for Junagadh region.

Rainfall Events	*Rainfall (mm)*	*Annual runoff fraction*	*Area of roof, sq.m*	*Annual roof water runoff (m^3)*	*Annual roof water runoff (m^3)/ sq.m*	*Tube well intake lps*	*Rainy days*	*Annual intake cu.m*	*Intake (m^3/ sq.m)*	*Buffer storage requirement per sq.m*
Annual average	936.65	0.7	767	478	0.62	1.46	33	4163	5	-4.78
Lowest	145.6			74	0.10		13	1640	2	-2.04
Highest	2793.6			1425	1.86		52	6559	9	-6.60
Annual rainfall of 10 years return period	1427			728	0.95		41	5172	7	-5.75
Run off coefficient = 0.95 for concrete roof, cost of tank 225 per m^3 (CGWB 2007, Chatrabhuji and Barad, 2013)										

Table 9: Tube well intake, buffer storage and cost of system under 5 days maximum rainfall storm for Junagadh region

Days maximum for 10 years return period	*Maximum Rainfall (mm)*	*roof water runoff(m^3)*	*Peak run off rate (m^3/hr)*	*Duration of storm, hr.*	*Tube well intake (lps)*	*Tube well intake, m^3*	*Requirement of buffer storage, cu.m*	*Distribution of duration of storm day wise, hr.*	*Tube well intake for distributed day wiseduration, cu.m*	*Buffer storage day wisecu.m*	*Tube Intake (cu.m)*	*Buffer storage cu.m per sq.m of roof*	*Cost of sump per sq.m of roof*	*Cost of system per sq.m of roof*	*Total Cost per sq.m of roof*
1	2	3	4	5	6	7	8	9	10	12	14	15	16	17	18
1	233.2	169.9	77.66	2.19	1.46	11.50	158.42	2.19	11.50	158.42	126.14	0.207	46.47	33.4	79.87
2	324.9	236.7		3.05		16.02	220.72	0.86	4.52	62.30	252.29	0.081	18.27	33.4	51.67
3	392.6	286.1		3.68		19.36	266.71	0.64	3.34	45.99	378.43	0.060	13.49	33.4	46.89
4	435.6	317.4		4.09		21.48	295.92	0.40	2.12	29.21	504.58	0.038	8.57	33.4	41.97
5	439.4	320.2		4.12		21.67	298.50	0.04	0.19	2.58	630.72	0.003	0.76	33.4	34.16
Roof catchment area = 767 sq.m and run off coefficient = 0.95, For 10 years return period															

five days consecutive rainfall in sump varies from Rs 50 to 94 per sq.m. of roof for 10 years return period respectively. Also estimated runoff of storm of maximum one to five consecutive days rainfall varies from 0.22 to 0.42 cu.m. per sq.m of roof for 10 years return period respectively for Junagadh region.

2. The hydraulic conductivity was found 13.32 sq.m. per hour at full recharge and at full head condition for shallow aquifer of Junagadh region.
3. The capacity of buffer storage tank for collection of excess roof water then real time tube well intake capacity of one to five days maximum rainfall of 10 years return period was determined, only in case of one day maximum rainfall buffer storage sump of 43.76 m^3 is required, so, 0.057 cu.m per sq. m. of roof area of buffer storage tank is required. In remaining days intake is more than roof water for given time duration.
4. The roof water runoff for annual average, lowest, highest and 10 years return period of rainfall was determined 0.6, 0.1, 1.86 and 0.95 cu.m per sq. m roof respectively. The rainy days of average, lowest, highest and 10 years return period of rainfall were recorded 33, 13, 52 and 41 respectively. Tube well intakes for annual average, lowest, highest and 10 years return period of rainfall years were determined considering intake duration equal to their respective rainy days and roof water runoff was compared with intake of tube well yearly intakes for all the cases were found higher as compare to yearly roof water runoff. So no buffer sump was recommended. But roof water runoff and tube well intake is not always matching therefore temporary roof pounding may occurs in case system without buffer storage tank so buffer storage tank is advisable.
5. For the determination of safest buffer storage sump in combination of groundwater recharge through tube well were determined considering maximum one to five days rainfall, suppose occurs at highest rainfall intensity of 10 years return period and time of concentration equal to 20 min, it means it generates peak rate of runoff for all five days rainfall duration. Day-wise duration of rainfall occurrence was used to calculate the tube well recharge capacity. When Tube well intake for duration equal to one day maximum rainfall at highest rainfall intensity of 10 years return period was subtracted from roof water runoff of one day maximum rainfall, it was found 158.42 cu.m. and it means 0.207 cu.m. per sq.m. of roof and for remaining days it was 0.081, 0.060, 0.038 and 0.003 cu. m. per sq.m. for 2, 3, 4 and 5 days maximum rainfall respectively. Cost of system per sq. m of roof was found Rs 79.87, 51.67, 46.89, 41.97 and 34.16 for one to five days maximum rainfall respectively. It is found total intake during maximum five days rainfall is higher as compare to

roof water runoff. However 0.207 cu.m. of buffer storage sump is recommended for safest combination of buffer storage sump with groundwater recharge.

6. The observations of groundwater recharge and roof water runoff collected in the sump were recorded. The excess roof water runoff then tube well recharge naturally allowed to collect in to sump for whole monsoon season. It was found that out of total runoff 0.87 cu.m. per square meter of roof, 0.26 cu.m. of runoff per square meter of roof was contributed as ground water recharge and 0.61 cu.m. of runoff per square meter of roof was collected in to sump. Cost of system was determined Rs 172 per square meter.
7. The annual roof water runoff coefficient for RCC roof for Junagadh region was found 0.7

References

Anonymous, 2002. "Rainwater harvesting and conservation Manual", Govt. of India, Central public works department, N. Delhi.

Anonymous, 2010. "A Guide on Artificial Recharge to Groundwater", Central Ground Water Board, Govt. of India, New Delhi.

Chatrabhuji, A.V. and D. G.Barad. 2013. "Study on domestic roof water harvesting for Junagadh city", Unpublished B.Tech thesis CAET, JAU, Junagadh 362001.

Patra, A.K. and S. Gautam. 2011. " A Pilot Scheme for Rooftop Rainwater Harvesting at Centre of Mining Environment, Dhanbad", International Journal of Environmental Sciences, Vol. 1-7, pp. 1542-48.

Ragunath, H.M.,1967. "Hydrology-Principles, Analysis and Design", New Age International Private Limited, New Delhi.

26

Climate Change Assessment for Ozat Basin of Saurashtra Region of Gujarat state, India

D. M. Paradava, R. J. Patel, H. V. Parmar, H. H. Mashru, P. B. Vekariya, H. D. Rank and J. H. Kelaiya

Abstract

The realisations of the importance of warming impacts on agriculture and spatial variability of warming trend on planet earth had led the present study on climate change assessment in Ozat basin (3176.24 km^2) having four watersheds. The downscaled global climate model simulations (CMIP3) data of total 10 weather grid points falling in basin were used for the analysis. The time series of weekly, monthly, seasonal and annual average of daily maximum, minimum and mean temperature along with solar radiation (1979-2013) for each of four watersheds and basin were analysed separately using Mann-Kendall, Sen's slope and best fit trend line methods. The seasonal and annual series of daily minimum and mean temperature were found significantly increasing in all four watershed and basin also. The seasonal average of maximum temperature was significantly increasing during winter (0.3° C/ decade) and annual (0.1 °C/decade). However, the seasonal average of daily minimum and mean temperature were found increased significantly at 0.3, 0.3, 0.2 and 0.3 °C/decade and 0.3, 0.2, 0.1 and 0.2 °C/decade respectively during winter, summer, monsoon and annual. The solar radiation was found increasing significantly during monsoon but insignificantly during rest of seasons.

Key words: *climate change, temperature, Mann-Kendall, Sen's slope, warming trend.*

Introduction

The earth's climate has remained dynamic throughout 4.5 billion years history of the earth, showing changes through a natural cycle periodically.

These climate changes in the past geological time had profound influence on sea level, rainfall patterns and temperature related weathering processes. Most of the evidences of past climate change, however, are circumstantial. The combination of a unique level of temperature increase in the late 20th century and improved constraints on the role of natural variability provides further evidence that the greenhouse effect has already established itself above the level of natural variability of the last 1000 years and is greater than the best estimate of global temperature change for the last interglacial period. The global warming potential of carbon dioxide was estimated by the Swedish Chemist Svante Arrhenius and subsequently many researchers concluded that human consumption of fossil fuels was leading to significant increase in atmospheric-CO_2 and global average temperature (Callender, 1938). Huang *et al.* (2000) have reported 20th century to be the warmest of the past five centuries and later Huang (2004) reported 20th century warming is a continuation to a long-term warming started before the onset of industrialisation. The Fourth Assessment Report of IPCC (2007a) has also mentioned that the change in earth's climate has been in an unprecedented manner in the past 40,000 years, but greatly accelerated during the last century, due to rapid industrialisation and indiscriminate destruction of natural environment.

Climate change and agriculture are interrelated processes both of which take place on global scale. Global warming is projected to have significant impacts on conditions affecting agriculture, including temperature, CO_2, glacier runoff, precipitation and the interaction of these elements. These condition in term determine the carrying capacities of different ecosystems to produce enough biomass including food for human population and domesticated animals. The overall effect of climate change on agriculture will depend on the balance of these effects. Assessment of the effects of global climate change on agriculture might help to properly anticipate and adapt farming to maximise agricultural production. At the same time, agriculture has been shown to produce significant effects on climate change, primarily through the emission of GHGs as about 12.5 per cent of total emission is from agricultural products globally. Land use change such as deforestation and desertification, together with the use of fossil fuels are the major anthropogenic sources of carbon dioxide. The various assessment reports brought out since 1990s by the IPCC have reconfirmed that the global atmospheric concentrations of carbon dioxide (CO_2), methane (CH_4) and nitrous oxide (N_2O) have increased precedent. The increase in greenhouse gases (GHGs) was 70 per cent between 1970 and 2004 and the IPCC has shown through a series of observations and modelling studies that these increases in GHGs have resulted in warming of the climate system by 0.74°C during 1906-2008 (Aggarwal, 2008). The IPCC (2007a) has projected that the temperature increase by the end of this century is expected to be in the range of 1.8-4.0°C and it is also likely that future tropical cyclones will become more intense with larger peak wind speeds and heavier precipitation. Increase in the amount of precipitation are likely in high latitudes while decreases are expected in most sub-tropical land regions, continuing observed patterns in recent trends. The projected sea level rise by the end of this century is likely to be 0.18- 0.59 m (Aggarwal, 2008). The IPCC (2007b) has projected 0.50-1.20 °C rise in temperature by 2020, 0.88-3.16 °C by 2050 and 1.56-5.44 °C by 2080 depending upon the scenario of future development in south Asia including India.

Climate change impacts on agriculture are being witnessed all over the world, but countries like India are more vulnerable in view of the huge population dependent on agriculture, excessive pressure on natural resources and poor coping mechanisms. The warming trend in India over the past 100 years has indicated on increase of 0.60°C. The projected impacts are likely to further aggravate field fluctuations of many crops thus impacting food security. There are already evidences of negative impacts on yield of wheat and paddy in parts of India due to increased temperature, water stress and reduction in number of rainy days. Significant negative impacts have been projected with medium-term (2010-2039) climate change, e.g. yield reduction by 4.5 to 9 per cent, depending on the magnitude and distribution of warming. Since agriculture makes up roughly 15 per cent of India's GDP, 4.5 to 9.0 per cent negative impact on production implies cost of climate change to be roughly at 1.5 per cent of GDP per year. Enhancing agricultural productivity, therefore, is critical for ensuring food and nutritional security for all, particularly the resource poor small and marginal farmers who would be affected most. In the absence of planned adaptation, the consequences of long-term climate change could be severe on the livelihood security of the poor. Keeping in views of importance of warming impacts on agriculture and spatial variability of warming trend on planet earth, the present effort aims to study climate change assessment for Ozat basin of Saurashtra region, Gujarat state, India.

Methodology

The downscaled global climate model simulations (CMIP3) data (1979-2013) of daily maximum and minimum temperature (°C) and solar radiation (MJ/sq.m) for total 10 weather grid points falling in basin were used for the analysis. The weekly, monthly, seasonal and annual average weather parameters were estimated taking average of weather parameter of days of the respective periods.

Climate change analysis

The climate change was assessed by the trend analysis of the different time series data of maximum, minimum, mean temperature and solar radiation using the standard method as described by Gilbert (1987) and Kendall and Gibbons (1990) along with best fit trend analysis.

Result and Discussion

The trend analysis of weekly, monthly, seasonal and annual average of weather parameters like daily maximum, minimum, mean temperature and solar radiation for Ozat basin is given in Tables 1 and 2 and Figs 1 and 2.

Temperature Trend

Watershed 5G1C2

Weekly average T_{max}, T_{min} and T_{mean}

The Mann-Kendall (Z) and Sen's slope statistics as well as best fit trend lines of weekly average T_{max}, T_{min} and T_{mean} for each time series of 52 standard weeks showed that the significant increasing trend was found during 2, 8, 9, 11, 12, 34, 43-45, 47-49 and 51 for T_{max}. The T_{min} was in increasing trend in most of the standard weeks, and T_{mean} were found increasing significantly during standard weeks of 5, 6, 11, 13, 14, 31, 33, 35 and 37 at 10 per cent significant level, 2, 3, 7, 8, 18, 29, 30, 34, 36, 42, 44, 46, 47, 49, and 51 at five per cent significant level, 12, 43 and 48 at one per cent significant level, and 45 at 0.1 per cent significant level respectively. However, the significant decreasing trend was found for T_{max} only during standard weeks 20, 23, 24, 28 and 38. While T_{min} and T_{mean} did not showed significant decreasing trend during any standard week. The non-significant increasing trend was found during 1, 3-7, 10, 13-15, 17, 18, 27, 29-33, 35, 36, 42, 46, 50 and 52 standard week for T_{max}, 1, 4, 24, 28, 39, 50 and 52 standard week for T_{min} and 1, 4, 15-17, 19-22, 25-27, 32, 38, 40, 41, 50 and 52 standard weeks for T_{mean} while the non-significant decreasing trend for T_{max} was found during 16, 19, 21, 22, 25, 26, 37 and 39-41 standard week and 23, 24, 28, and 39 standard weeks for T_{mean} respectively.

Monthly average T_{max}, T_{min} and T_{mean}

Out of twelve months in a year, the monthly average maximum temperature was found increased during four month, i.e. January (10 per cent level), March (5 per cent level), November (1 per cent level) and December (5 per cent level). The monthly average minimum temperature increased during all months and the mean temperature increased during nine months (Jan to March, May, July, Aug, October to Dec). The monthly average of maximum temperature was found decreased during one month (June at 5 per cent level). The insignificant trend could be found during seven months(Feb, April, May, July-Oct.) for monthly average T_{max} and during three months (April, June, Sept) for T_{mean}.

Seasonal average T_{max}, T_{min} and T_{mean}

The result showed that insignificant decreasing tend in seasonal average maximum temperature was found during summer and monsoon season. It was increasing significantly during winter season. The seasonal average daily minimum temperature was found increased significantly during winter and summer (0.1 per cent level) and monsoon (1 per cent) seasons. The seasonal average of daily mean temperature was found increased during all three seasons.

Annual Average T_{max}, T_{min} and T_{mean}

It was found that the annual average of daily minimum, mean and maximum temperature was found significant increasing at 0.1 per cent, 0.1 per cent and 5 per cent significant level respectively.

Watershed 5G1C3

Weekly average T_{max}, T_{min} and T_{mean}

Time series of 52 standard weeks showed that the significant increasing trend was found during 2, 3, 7-9, 11-14, 18, 44, 45 and 47-51 for T_{max}. The weekly average T_{min} was found in increasing trend during 2, 3, 5, 9, 12-14, 16, 18, 20-23, 25, 30, 32, 33, 35-37, 41, 42, 45, 48 and 49th standard weeks. The T_{mean} were found increasing significantly during standard weeks of 3, 5, 6, 11, 13, 42, 43, 46 and 47 at 10 per cent significant level, 2, 7-9, 14, 17, 44, 48, 49 and 51 at five per cent significant level, 12 and 18 at one per cent significant level, and 45 at 0.1 per cent significant level respectively. However, the significant decreasing trend was found for T_{max} during standard weeks 23, 24, 28, 38 and 39. The T_{mean} was found deceasing significantly during 23, 24 and 39 standard weeks. The significant decreasing trend was not found during any standard week for T_{min}. The non-significant increasing trend was found during 4, 5, 6, 10, 15, 17, 29, 30, 42, 43, 46 and 52 standard week in T_{max}, during 1, 4, 6-8, 10, 11, 15, 17, 19, 24, 26, 27, 29, 31, 34, 38-40, 43, 44, 46, 47 and 50-52 standard week in T_{min} and during 4, 10, 15, 16, 19-22, 25, 27, 29, 30, 33-36, 41, 50 and 52 standard weeks in T_{mean}. The non-significant decreasing trend was found during 1, 16, 19-22, 25-27, 31-33, 35-37, 40 and 41 standard week in T_{max}, during 28th standard week in T_{min} and during 1, 26, 31, 32, 37, 38 and 40 standard weeks in T_{mean}.

Monthly average T_{max}, T_{min} and T_{mean}

Out of twelve month in a year, the monthly average maximum temperature was found increased during five months *i.e.* Jan (10 per cent level), Feb (10 per cent level), March (1 per cent level), Nov (5 per cent level) and Dec (5 per cent level). The monthly average minimum temperature increased during ten months *i.e.* Feb (10 per cent level), March (1 per cent level), April (5 per cent level), May (1 per cent level), June (5 per cent level), Aug (5 per cent level), Sept (1 per cent level), Oct (5 per cent level), Nov (5 per cent level) and Dec (5 per cent level). The monthly average mean temperature was found increased during ten month (Feb to June and Sept to Dec). However, no significant trend could be found during six months(April, May, July-Oct.) for monthly average T_{max}, during two month (Jan and July) for monthly average T_{min} and T_{mean}.

Seasonal average T_{max}, T_{min} and T_{mean}

The result showed that no significant decreasing tend in seasonal average maximum temperature was found during summer and monsoon season while, it was found increasing significantly trend (5 per cent level) during winter season. The seasonal average minimum and mean temperature was found increased highly significant (0.1 per cent level) during summer season while increased significant (1 per cent level) during winter and monsoon season of the year.

Annual Average T_{max}, T_{min} and T_{mean}

The annual average of minimum and mean temperature was found significantly increasing (0.1 per cent level) while annual average of maximum temperature was increasing insignificantly.

Watershed 5G1C4

Weekly average T_{max}, T_{min} and T_{mean}

The results showed that the significant increasing trend was found for T_{max} during standard weeks 2, 7-14, 17, 18 and 44-51. The T_{min} was found in increasing trend during 2, 9, 10, 12, 13, 16, 20-23, 25, 26, 30, 32-37, 42, 45, 48 and 49th standard weeks. The T_{mean} were found increasing significantly during standard weeks of 2, 3, 7-14, 16-18, 21, 34, 44-49 and 51 respectively. However, the significant decreasing trend was found for T_{max} during standard weeks 23, 24, 28, 38 and 39. The significant decreasing trend was found during 23, 28 and 39 standard weeks for T_{mean}. The significant decreasing trend were not found during any standard weeks for T_{min}. The non-significant increasing trend was found during 1, 3, 3-6, 15, 16, 29-31, 34-36, 42, 43 and 52 standard week for T_{max}, 2, 9, 10, 12, 13, 16, 20-23, 25 ,26, 30, 32-37, 42, 45, 48 and 49 standard week of T_{min} and 1, 4-6, 15, 19, 20, 22, 25-27, 29-33, 35-37, 40, 42, 43, 50 and 52 standard weeks in T_{mean} while the non-significant decreasing trend of T_{max} was found during 19-22, 25-27, 32, 33, 37, 40 and 41 standard week, 28 and 39th standard week of T_{min} and 24, 38 and 41th standard weeks of T_{mean} respectively.

Monthly average T_{max}, T_{min} and T_{mean}

Out of twelve months in a year, the monthly average maximum, minimum and mean temperature was found increased during five months *i.e.*

Jan (5 per cent level), Feb (5 per cent level), March (1 per cent level), Nov (1 per cent level) and Dec (5 per cent level), during nine months *i.e.* March (1 per cent level), April (10 per cent level), May (1 per cent level), June (5 per cent level), Aug (5 per cent level), Sept (5 per cent level), Oct (10 per cent level), Nov (10 per cent level) and Dec (5 per cent level), and during eight month (Jan to May, Aug, Nov and Dec). However, no significant trend could be found during six months (April, May, July-Oct.) for monthly average T_{max}, during three months (Jan, Feb and July) for monthly average T_{min} and during four month (June, July, Sept and Oct.) for monthly average T_{mean}.

Seasonal average T_{max}, T_{min} and T_{mean}

The results showed that insignificant trend was found in seasonal average maximum and mean temperature during monsoon season while, increasing significantly during winter and summer season. The seasonal average minimum temperature was found increased significantly (one per cent level) during all seasons.

Annual Average T_{max}, T_{min} and T_{mean}

The annual average of daily minimum, maximum and mean temperature was found increasing significantly.

Watershed 5G1C5

Weekly average T_{max}, T_{min} and T_{mean}

The trend analysis results of weekly average T_{max}, T_{min} and T_{mean} for each time series of 52 standard weeks showed that the significant increasing trend was found during standard weeks 2, 3, 7-14, 17, 18, 44 and 46-51 for T_{max}. The T_{min} was found in increasing trend during standard weeks 9, 12, 16, 20-23, 25, 26, 30, 32, 33, 35-37, 45, 48 and 49th standard weeks, and T_{mean} were found increasing significantly during standard weeks of2, 6-9, 11, 12, 14, 17, 18, 44-49 and 51 respectively. However, the significant decreasing trend was found for T_{max} only during standard weeks 22-24, 28, 38, 39 and 41 and for T_{mean} during 23, 24, 28, 38 and 39 standard weeks. However, no significant decreasing trend were found in T_{min} during any standard week. The non-significant decreasing trend was found during 4, 19-21, 25-27, 31-33, 35-37 and 40 standard week for T_{max}, during 1, 15, 27, 28 and 39th standard week for T_{min} and during 1, 4, 19, 25-27, 31, 32, 40 and 41th standard weeks for T_{mean} respectively. The non-significant increasing trend was found during 1, 5, 6, 15, 16, 29, 30, 34, 42, 43, 45 and 52 standard week for T_{max}, during 2-8, 10, 11, 13, 14, 17-19, 24, 29, 31, 34, 38, 40-44, 46, 47 and 50-52 standard week for T_{min} and during 3, 5, 10, 13, 15, 16, 20-22, 29, 30, 33-37, 42, 43, 50 and 52 standard weeks for T_{mean} respectively.

Monthly average T_{max}, T_{min} and T_{mean}

Out of twelve months in a year, the monthly average daily maximum was found increased during five months *i.e.* Jan (5 per cent level), Feb (10 per cent level), March (1 per cent level), Nov (5 per cent level) and Dec (5 per cent level) while minimum and mean temperature was found increased during seven months *i.e.* (March, May, June, Aug, Sept, Nov and Dec). The monthly average maximum temperature was found decreasing significantly during two months i.e. June and Sept(5% and 10% level). However, insignificant trend could be found during five months(April, May, July, Aug and Oct.) in monthly average T_{max}, and during five month(Jan, Feb, April, July and Oct.) in monthly average T_{min} and T_{mean}.

Seasonal average T_{max}, T_{min} and T_{mean}

The result showed that insignificant increasing and decreasing trend was found in seasonal average maximum temperature respectively during summer and monsoon season. However, it was found increasing significantly during winter season. The seasonal average minimum and mean temperature was found increased significantly during all three season of the year.

Annual Average T_{max}, T_{min} and T_{mean}

The significant increasing trend was found for annual average minimum and mean temperature and insignificant increasing trend for annual average maximum temperature.

Ozat Basin

Weekly average T_{max}, T_{min} and T_{mean}

The results showed that the significant decreasing trend was found during standard weeks 23, 24, 28, 38 and 39 for T_{max}, and during 23, 24, 28 and 39 standard weeks for T_{mean}. The significant decreasing trend was not found in T_{min} during any standard week. However, the significant increasing trend was found in T_{max} during standard weeks 2, 3, 7-9, 11-14, 18 and 44-51 respectively. T_{min} was found in increasing trend during standard weeks 2, 9, 10, 12-14, 16, 18-23, 25, 26, 30, 32-37, 42, 45, 48 and 49th standard weeks. The T_{mean} were found increasing significantly during standard weeks of 2, 3, 7-9, 11, 12, 14, 17, 18, 34, 42-49 and 51 respectively. The insignificant increasing trend was found during 1, 4-6, 10, 15-17, 27, 29, 30, 34, 36, 42, 43 and 52 standard week in T_{max}, during 3-8, 11, 15, 17, 24, 27, 29, 31, 38-41, 43,44, 46, 47 and 50-52 standard week in T_{min} and during 1, 5, 6, 10, 13, 15, 16, 19-22, 25-27, 29-33, 35-37, 50 and 52 standard weeks in T_{mean}. The insignificant decreasing trend was found in T_{max} during 19-22, 25, 26, 31-33, 35, 37, 40 and 41 standard week, during 1 and 28th standard week in T_{min} and during 4, 26, 38, 40 and 41th standard weeks in T_{mean} respectively.

Monthly average T_{max}, T_{min} and T_{mean}

Out of twelve month in a year, the monthly average of daily maximum, minimum and mean temperature was found increased during five month i.e. Jan (5 per cent level), Feb (10 per cent level), March (1 per cent level), Nov (5 per cent level), Dec (5 per cent level), during ten months i.e. Feb (10 per cent), March (1 per cent level), April (5 per cent), May (1 per cent level), June (5 per cent level), Aug (1 per cent level), Sept (1 per cent level), Oct (5 per cent level), Nov (5 per cent level), Dec (5 per cent level) during seven month (Jan, Feb, March, May, Aug, Nov and Dec) respectively. The monthly average maximum temperature was found decreased significantly during one month i.e. June (5 per cent level). However, insignificant trend could be detected during six months (April, May, July- Oct.) in monthly average T_{max}, during two month (Jan, July) in monthly average T_{min} and during five months (April, June, July, Sept. and Oct.) in monthly average T_{mean}.

Seasonal average T_{max}, T_{min} and T_{mean}

The result showed that insignificant increasing and decreasing trend was found in seasonal average maximum temperature during summer and monsoon season respectively, while increasing significantly during winter season. The seasonal average of daily minimum and mean temperature was found increased significantly during all three seasons.

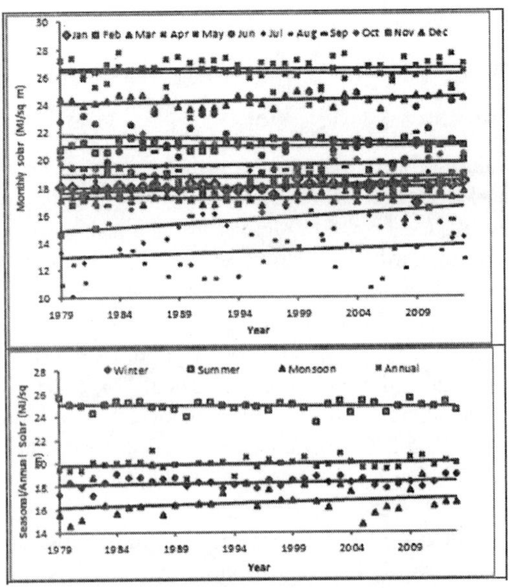

Fig. 1. Trend of seasonal/annual average of daily T_{max} (°C), T_{min} (°C) and T_{mean} (°C) in Ozat river basin.

Annual Average T_{max}, T_{min} and T_{mean}

The annual average of daily minimum, maximum and mean temperature was found increasing significantly(0.1%, 10% and 1% level) respectively in the Ozat river basin.

Solar radiation (MJ/sq.m.)

Watershed 5G1C2

Weekly average solar radiation (MJ/sq.m.). Out of 52 weeks of the year, the insignificant increasing and decreasing trend of weekly average solar radiation was found during twenty four and seventeen weeks respectively. However, the significant increasing and decreasing trend of weekly average solar radiation was found during six and five weeks out of 52 weeks of the year respectively.

Monthly average solar radiation (MJ/sq.m.)

The result showed that the monthly average solar radiation was found significantly increasing during one month (March) only. The insignificant trend was found for the rest of the months.

Seasonal/annual average solar radiation (MJ/sq.m.)

The insignificant increasing trend during winter and monsoon season while insignificant decreasing trend during summer season was found. The annual average solar radiation was found increasing insignificantly.

Watershed 5G1C3

Weekly average solar radiation (MJ/sq.m.)

The significant increasing and decreasing trend in weekly average solar radiation was found during three and three weeks out of 52 weeks of the year respectively. However, the insignificant increasing and decreasing trend of weekly average solar radiation was found during thirty one and fifteen weeks.

Monthly average solar radiation (MJ/sq.m.)

Out of 12 months in a year, the insignificant trend could be found during nine months except month of Jan, March and April. It was found significantly increasing during two months (Jan and March) and decreasing during April month only.

Seasonal/annual average solar radiation (MJ/sq.m.)

Out of three seasons of the year, the insignificant increasing trend during winter season and annual and insignificant decreasing trend during summer season was found. However, the significant increasing trend was found during monsoon season.

Watershed 5G1C4

Weekly average solar radiation (MJ/sq.m.)

Out of 52 weeks of the year, the significant increasing and decreasing trend of weekly average solar radiation was found during five and four weeks of the year respectively. However, the insignificant increasing and decreasing trend of weekly average solar radiation was found during thirty and thirteen weeks respectively.

Monthly average solar radiation (MJ/sq.m.)

It was found that the monthly average of solar radiation increased significantly during two months(Jan and March). The insignificant trend was found during rest of 10 months.

Seasonal/annual average solar radiation (MJ/sq.m.)

Out of three seasons of the year, the insignificant increasing and decreasing trend in seasonal average solar radiation was found during winter and summer season respectively. However, the significant increasing trend was found during monsoon season.

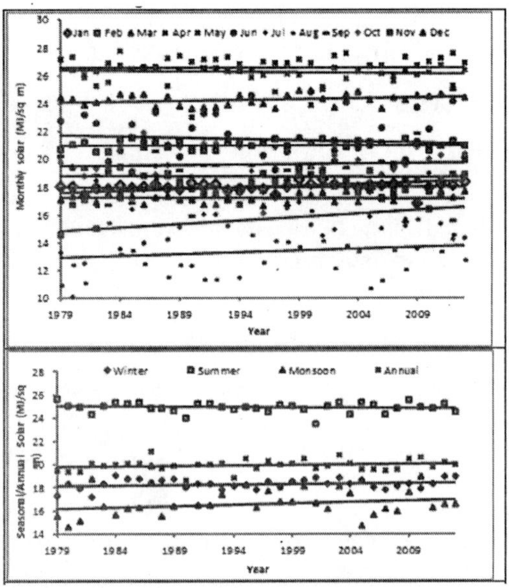

Fig. 2. Trend of monthly and seasonal/annual average of daily solar radiation (MJ/m²) for Ozat river basin.

Watershed 5G1C5

Weekly average solar radiation (MJ/sq.m.)

The significant increasing and decreasing trend of weekly average solar radiation was found during four and two weeks out of 52 weeks of the year respectively. However, the insignificant increasing and decreasing trend of weekly average solar radiation was found during twenty nine and seventeen weeks respectively.

Monthly average solar radiation (MJ/sq.m.)

The monthly average of solar radiation was found significantly increasing during two months (Jan and March) only. Out of 12 months in a year, the insignificant trend could be found during ten months.

Seasonal/annual average solar radiation (MJ/sq.m.)

Out of three seasons of the year, the insignificant increasing and decreasing trend in seasonal average solar radiation was found during winter and summer season respectively. However, the significant increasing trend was found during monsoon season. The annual average solar radiation was found increasing insignificantly.

Ozat basin

Weekly average solar radiation (MJ/sq.m.)

The insignificant increasing trend of weekly average solar radiation was found during twenty seven weeks while insignificant decreasing trend during sixteen weeks of the year respectively. Out of 52 weeks of the year, the significant increasing and decreasing trend was found during six and three weeks respectively.

Monthly average solar radiation (MJ/sq.m.)

Out of 12 months in a year, the insignificant trend in monthly average solar radiation could be found during ten months. The monthly average solar radiation was found significantly increasing during two month i.e. Jan and March.

Seasonal/annual average solar radiation (MJ/sq.m.)

Out of three seasons of the year, the insignificant increasing and decreasing trend in seasonal average solar radiation was found during winter and summer season respectively. However, the significant increasing trend was detected during monsoon season. The annual average solar radiation was found insignificantly increasing for the entire river basin.

Conclusions

The warming trend was detected in Ozat basin. However, the increasing trend of solar radiation was noticed only during monsoon season. The cereal crops like wheat may be replaced by some oilseed or pulse crop due to warming trend.

Acknowledgements

The authors are grateful to the Directorate of Water Management, ICAR, Bhubaneswar for the financial supports under NICRA project.

References

Aggarwal, P.K., 2008. "Global climate change and Indian agriculture: impacts, adaptation and mitigation Ind", J. Agril. Sci. 78(11): 911-919.

Callender, G.S., 1938. "The artificial production of carbon dioxide and its influence on temperature" Quart. J. Royal Meteorol. Soc. 64:223-240.

Gilbert, R.O., (1987). "Statistical methods for environmental pollution monitoring", Van Nostrand Reinhold, NY.

Huang, S., H.N. Pollack and P.Y. Shen. 2000. "Temperature trends over past five centuries reconstructed from borehole temperatures," Nature, 403: 756-758.

Huang, S., 2004. "Merging information from different resources for new insights into climate change in the past and future" Geophy. Res. 31: L13205.

IPCC, 2007a."Summary for policymakers Climate Change: The Physical Science Basis", Contribution of Working Group I to the Fourth Assessment Report of the IPPC, Cambridge University Press, Cambridge.

IPCC, 2007b."Climate Change: Climate Change Impacts, Adaptation and Vulnerability, Summary for policymakers", Inter-Governmental Panel on Climate Change.

Kendall, M. and J. D. Gibbons.1990. "Rank Correlation Methods", 5th Eds. Oxford University Press, NY. pp. 272.

Table 1: The trend parameters of weekly, monthly, seasonal and annual average of weather parameters like daily T_{max}, T_{min}, T_{mean}(°C) and solar radiation (MJ/m^2) for Ozat river basin

S. No.	*Parameters*	*Period*	*Increasing trend*		*Decreasing trend*	
			Significant (No.)	*Insignificant (No.)*	*Significant (No.)*	*Insignificant (No.)*
1.	Daily maximum temperature-T_{max} (°C)	Week	18	16	5	13
		Month	5	2	1	4
2.	Daily minimum temperature-T_{min} (°C)	Week	25	25	-	2
		Month	10	2	-	-
3.	Daily mean temperature-T_{mean} (°C)	Week	20	23	4	5
		Month	7	4	-	1
4.	Daily mean solar Radiation (MJ/sq m)	Week	6	27	3	16
		Month	2	7	-	3

Table 2: Trend analysis of Seasonal and Annual average daily T_{max}, T_{min}, T_{mean} (°C) and solar radiation (MJ/m^2) for Ozat river basin

Season	*Mann-Kendall (Z)*	*Conf. Level (per cent)*	*Sen's Slope*	*Lower and Upper limit of Sen's slope (1 per cent)*		*Lower and Upper limit of Sen's slope (5 per cent)*		*Slope of best fit trend*	*R^2*
Maximum Temperature (°C)									
Winter	1.97**	97.58	0.032	-0.015	0.074	0.000	0.065	0.03	0.16
Summer	1.01NS	84.34	0.011	-0.017	0.042	-0.010	0.034	0.02	0.08
Monsoon	-0.53NS	70.04	-0.008	-0.062	0.037	-0.044	0.028	-0.01	0.01
Annual	1.43*	92.43	0.013	-0.014	0.042	-0.006	0.034	0.01	0.06
Minimum Temperature (°C)									
Winter	2.77***	99.72	0.033	0.002	0.066	0.012	0.060	0.03	0.21
Summer	3.25****	99.94	0.033	0.008	0.056	0.014	0.051	0.03	0.23
Monsoon	2.51***	99.40	0.021	-0.001	0.040	0.006	0.035	0.02	0.19
Annual	3.31****	99.95	0.027	0.007	0.051	0.014	0.045	0.03	0.32
Mean Temperature (°C)									
Winter	2.60***	99.53	0.032	0.000	0.063	0.008	0.055	0.03	0.24
Summer	2.37***	99.12	0.024	-0.001	0.049	0.003	0.043	0.02	0.19
Monsoon	1.41*	92.02	0.012	-0.016	0.031	-0.007	0.026	0.01	0.03
Annual	2.80***	99.74	0.024	0.002	0.040	0.008	0.036	0.02	0.25
Solar radiation (MJ/sq.m)									
Winter	0.75NS	77.42	0.006	-0.016	0.030	-0.011	0.024	0.01	0.05
Summer	-0.75NS	77.42	-0.008	-0.025	0.014	-0.021	0.009	-0.01	0.02
Monsoon	1.55*	93.92	0.032	-0.028	0.097	-0.011	0.075	0.02	0.03
Annual	0.95NS	82.93	0.009	-0.012	0.034	-0.007	0.026	0.01	0.04

**** = Significant at 0.1% level, *** = Significant at 1% level, ** = Significant at 5% level, * = Significant at 10% level

27

Estimating the Water Balance Component in Canal Command Area of Jhansi Minor (Left Bank Canal of Bargi)

Sumeet Meshram, Shri Kant, Arun Jhajhariya and Navneet Singh Parmar

Abstract

The study present analysis of water balance components of Jhansi minor district of Madhya Pradesh.Water balance techniques have been extensively used to make quantitative estimates of water resources and the impact of man's activities on the hydrologic cycle. On the basis of the water balance approach, it is possible to make a quantitative evaluation of water resources and its dynamic behaviour under the influence of man's activities. In this paper, an attempt has been made to describe the methodologies to understand and evaluate the various recharge and discharge components of ground water balance equation and to establish the recharge coefficient with a view to work out the ground water potential of an area. The study observed that in year 2008 monsoon and non-monsoon period inflow in the form of rainfall is 287.04 ha-m and for the year 2009 monsoon and non monsoon period inflow in the form of rainfall is 293.28 ha-m.

Introduction

Water is essential for survival of living body. With continuous increase in population, the demand on land and water resources has been increasing for enhancing agricultural production. Ahmad *et.al.* (2010) found out a new technique to estimate net groundwater use across large irrigated areas by combining remote sensing and water balance approaches, Rechana Doab,

Pakistan Contribution of groundwater is so significant that more than 70 per cent of population uses groundwater for its domestic needs and also more than half of irrigation needs are met from this source. Maraux *et al.* (1998) compared two functional models, FAO (Doorenbos and Pruitt, 1977) and Ritchie (1985) models, with a mechanistic model (Maraus and Lafolie, 1998) to simulate the soil water balance of maize and sorghum grown in sequence in Nicaragua. Comparison of the three models was first made according to a deterministic approach with parameters obtained from the same database. They then considered a stochastic approach, for which 800 hydraulic characteristics of the soil were generated, according to the spatial variability observed at the field scale and to the scaling theory applied to similar porous media. A distribution of the stochastic parameters used in the three models was thus derived. Results showed that the order of magnitude of the evapo-transpiration was similar for the three models. Adding a capillary rise mechanism in the functional models improved moderately the soil water balance. ET and drainage showed moderate sensitivity to spatial variability in soil hydraulic properties, whereas final water storage showed a greater sensitivity.Water balance techniques have been extensively used to make quantitative estimates of water resources and the impact of man's activities on the hydrologic cycle. On the basis of the water balance approach, it is possible to make a quantitative evaluation of water resources and its dynamic behaviour under the influence of man's activities. The water balance equation incorporates contributions from both precipitation and evapotranspiration, but because evapotranspiration is linearly related to temperature fluctuations in groundwater levels are often strongly correlated with regional temperature and precipitation data. S. Irmark (2008) was carried out study on Evapotranspiration he found that efficient use of water resources in semi-arid and arid agro ecosystem of the world has become increasingly important because of rapid depletion of water resources, drought condition and degradation of ground and surface water quality in many regions. He carried out these studies to define and discuss several ET terms including potential, reference and actual ET. Introduction of canal irrigation in the state of Madhya Pradesh has also caused deteriorating situations in the command areas due to rise in the water table, specifically in Tawa command and Chambal commands. An assessment of groundwater potential for irrigation that for optimal utilisation of water resources in irrigated command area, assessment of surface and groundwater resources is essential activity. Mane *et al.* (2008). The remedial measures of combating water logging could not restore the area during last ten years. In view of this studies were undertaken at Rani Avanti Bai Sagar Irrigation Project (Bargi Project) for planned groundwater use as a precautionary measure to avoid water logging problem. Canal irrigation projects in the country have brought the mixed fortunes of increased cropping intensity which enhanced crop productions but simultaneously water logging and salinity experienced in almost every canal irrigation project. Canal irrigation project have really increased the total productivity of the food grain within few years. This has played an important role to change the national economy and to elevate the socio-economic conditions of the farmers. The continuous seepage in the canal distribution system cause rise in water table. It affects the water economy of the distribution system. It is reported that about six million ha have so far been affected due to seepage from irrigation canals.

At the same time, canal irrigation released bad effects namely, water logging, soil salinity and alkalinity in the command area which are further responsible for water born diseases, infestation of weeds. All these problems ultimately results in degradation of productive agricultural lands into barren land which leads to unemployment and migration of people from the villages. The general equation of the soil-water budget is derived by considering the mechanisms by which water can enter, exit, or be stored in a predefined region of the vadose zone (Stephens, 1995). Rise in water table was due to seepage from canal and over irrigation. Seepage from canal was estimated using different formulae Shukla (1987). The seepage loss through unlined channels in heavy black soil was 2.83 cumec per M sq.m wetted area. The loss was reduced to 2.2 cumec per M sq.m wetted area after compaction of the channel bed Patel (1988). The seepage losses in earthen channels to the tune of 154.4 $l/m^2/day$ and 138.4 $l/m^2/day$ with soil bulk densities of 1.62 gm/cm^3 respectively Patel *et al.* (1977). The seepage loss in observed in unlined channels as 8.90 lit/m^2/day in the sandy clay loam soil Agrawal and Khan (1983).

Material and Methods

Parameter Estimation

The water balance has been used for computing seasonal and geographic patterns of irrigation demand, prediction of stream flow. Although the prediction may be approximate, they are sufficiently accurate to indicate whether a scheme is hydrologically sound or not. Water balance equation for the specific condition of the area 208 ha is represented as:

$$NR = NI\ NO$$

$$NI = R * 208 + WC + WI$$

$$NO = ETC + EU + GWD + SR \pm \text{ÄSMS} \pm Qn$$

$$NR = R * 208 + WC + WI - ETC - EU - GWD - SR \pm \text{ÄSMS} \pm Qn$$

where,

NR = Net recharge or GWS i.e. change in groundwater storage (ha-m).

NI = Net inflow into the area (ha-m).

NO = Net outflow from the area (ha-m).

R = Precipitation (m).

WC = Seepage from the minors (ha-m).

WI = Irrigation from the minors (ha-m).

ETC = Evapotranspiration from cultivated area (ha-m).

EU = Evaporation from uncultivated area (ha-m).

GWD = Ground Water Draft (ha-m)

SR = Surface Runoff (ha-m).

Δ SMS = Change in soil moisture storage (ha-m).

Qn = Net subsurface flow (ha-m).

Qn = Qi Qo

Qi = subsurface inflow into the study area (ha-m).

Qo = subsurface outflow from the study area (ha-m).

The components of the above water balance equation were estimated and are presented in the following paragraphs.

Table 1: Standard meteorological weeks for sowing and harvesting of different crops

Crops	*Sowing week*	*Harvesting week*
Paddy	28	46
Soya bean	26	44
Wheat	46	11
Gram	43	09
Lentil	41	07
Peas, Vegetables and other	41	02

Inflow components (NI)

It includes rainfall, seepage from the canal and irrigation from the minor i.e.

$$NI = R*208 + WC + WI$$

Methods of Seepage Measurement:

The direct methods of seepage measurement from the direct approach are as follows:

(i) Inflow outflow method
(ii) Ponding method
(iii) Current meter method

Out of these three methods, current meter method is used to determine the seepage rate in this study. Discharge from selected points of the minor was calculated through current meter method for seepage analysis in the minor.

Irrigation from the minor (WI)

For determining the quantity of irrigation water duration, depth of flow, discharge of minor at outlet to the fields was considered as volume of water delivered for irrigation. Since the water courses are yet to be constructed by total volume of water delivered at the head of the minor minus seepage losses at the minor.

Outflow Components (NO)

Evapotranspiration from cultivated land, evaporation from uncultivated land, ground water draft, surface runoff, change in soil moisture storage and subsurface inflow and subsurface outflow are considered as outflow components which are discussed below.

Evapotranspiration from cultivated land (ETC)

Evapotranspiration from cropped land (ETC) was computed considering the sowing and harvesting weeks of the crops. The sowing and harvesting weeks (Table 1) of the crops dominating the area are as follows:

Evapotranspiration from cropped land during different years are presented in the Table 1 by using equation:

$ETC = PET * K_c * Area$

PET = Potential Evapotranspiration

K_c = crop coefficient

Ground water draft (GWD)

Ground water draft has been estimated by calculating the total discharge through tube wells. Annual ground water drafts were taken as 5.50 ha-m during non-monsoon period and 0.50 ha-m for monsoon period (Table 2).

Surface runoff (SR)

Runoff coefficient for paddy and haveli system were fixed based on amount and distribution of rainfall for the duration under study and on the basis of percolation rates of soil of the project area. Surface runoff was computed as

SR = Rainfall * Area * coefficient of runoff.

Table 2: Water Balance for Monsoon and non-monsoon period

S. No.	Particulars	Monsoon period		Nonmonsoon period	
		2008	2009	2008	2009
1.	Rainfall (mm)	1.380	1.410	0.018	0.135
2.	No. of days	123	123	242	242
3.	Runoff coefficient for paddy field	0.32	0.30	0	0
4.	Runoff coefficient for other field	0.48	0.43	0	0
5.	Surface runoff (ha-m)	137.7	127.5	0	0
6.	Evapotranspiration from cultivated land (ha-m)	63.4	73.9	56.2	58.7
7.	Evaporation from uncultivated land (ha-m)	0	0	0	0
8.	Seepage from minor (ha-m)	0	0	36.9	38
9.	Irrigation from minor (ha-m)	0	0	40	41.2
10.	Ground water draft (ha-m)	0.50	0.50	5.50	5.50
11.	Change in soil moisture storage (ha-m)	39.93	39.93	39.9	39.9
12.	Subsurface flow (ha-m)	0	0	0	0

Non-Monsoon Period

Subsurface inflow and outflow (Qn)

Darcy's equation was used to calculate the volume of subsurface inflow and outflow.

Qn = Qi Qo

Qi = K * Ii * Li * D * T

Qo = K * Io * Lo * D * T

Qi = Subsurface inflow (ha-m)

Qo = Subsurface outflow (ha-m)

K = hydraulic conductivity (m/day)

Ii and Io = Inflow and Outflow gradients in (m/m)

Li and Lo = Length of Inflow and Outflow boundaries (m)

D = Effective depth of aquifer in (m)

T = Duration in days for which calculation is made.

Net Recharge (NR)

The different components as calculated above were used for estimating the net recharge. Net recharge during monsoon and non-monsoon period for the year 2008 and 2009 was found to be 33.07 ha-m1, 40.54 ha-m and 43.73 ha-m, 135.81 ha-m respectively (Table 1).

Results and Discussion

Water Balance Component

Table 2 presents the computation of water balance components for monsoon and non-monsoon period of Jhansi minor. Result shows the average data of nine different water balance parameters i.e. the rainfall, seepage from minor, irrigation from the minor, evapotranspiration from cultivated land, evaporation from uncultivated land, surface runoff, ground water draft, change in soil moisture storage and net subsurface flow.

The results for the year 2008 for monsoon period showed that the rainfall (Fig. 1) was 1380 mm, seepage (Fig. 3) from minor zero, irrigation (Fig. 2 and 4) from the minor zero, evapotranspiration (Fig. 5) from cultivated land 63.4 ha-m, evaporation from uncultivated land 0 ha-m, surface runoff (Fig. 6)137.77 ha-m, ground water draft (Fig. 6) 0.50 ha-m, change in soil moisture storage 39.93 ha-m and net subsurface flow were taken as 0 ha-m respectively. During non monsoon period it was estimated as 18.8 mm, 36.94 ha-m, 40 ha-m, 56.22 ha-m, 0 ha-m, 0 ha-m, 5.50 ha-m, 39.93 ha-m and 0 ha-m respectively (Table 1).

In year 2009 it was estimated for monsoon period that rainfall (Fig. 3) 1410 mm, seepage (Fig. 3) from minor 0 ha-m, irrigation (Fig. 4) from the minor 0 ha-m, evapotranspiration (Fig. 5) from cultivated land 73.98 ha-m, evaporation from uncultivated land 0 ha-m, surface runoff (Fig. 6) 127.57 ha-m, ground water draft (Fig. 7) 0.50 ha-m, change in soil moisture storage 39.93 ha-m and net subsurface flow were taken as 0 ha-m respectively. During non-monsoon period it was estimated as 135.4 mm, 38 ha-m, 41.21 ha-m, 58.72 ha-m, 0 ha-m, 0 ha-m, 5.50 ha-m, 39.93 ha-m and 0 ha-m respectively.

In the year 2008 it was estimated that the average rainfall (Fig. 1) during monsoon and non-monsoon period was 1380 mm and 18.8 mm respectively. It may be due to the precipitation occurrence in monsoon period. Result shows (Table 2) the runoff coefficient for haveli and paddy field during monsoon and non-monsoon period was worked out to be 0.32 and 0 respectively while runoff coefficient for other field was 0.48 and 0 respectively. The Table 1 shows the average surface runoff (Fig. 6) for monsoon and non monsoon period in year 2008 was estimated as137.77 ha-m and 0 ha-m respectively. Evapotranspiration (Fig. 5) from cultivated land during monsoon and non-monsoon period was observed 63.4 ha-m and 56.22 ha-m respectively, while evaporation from uncultivated land was observed 0 ha-m for both seasons. The Table 2 shows the seepage (Fig. 3) from minor was 0 ha-m and 36.94 ha-m during monsoon and non-monsoon period respectively, while irrigation from minor was estimated as 0 ha-m and 40 ha-m during monsoon and non-monsoon period respectively. In order to quantify the ground water draft Table 1 shows the results for monsoon and non-monsoon period and it was observed as 0.50 ha-m and 5.50 ha-m respectively. While the change in soil moisture storage for monsoon and non-monsoon period was 39.93 and 39.93 ha-m respectively (Table 2). During the computation of water balance component, it was assumed that net subsurface flow was 0 ha-m for both seasons.

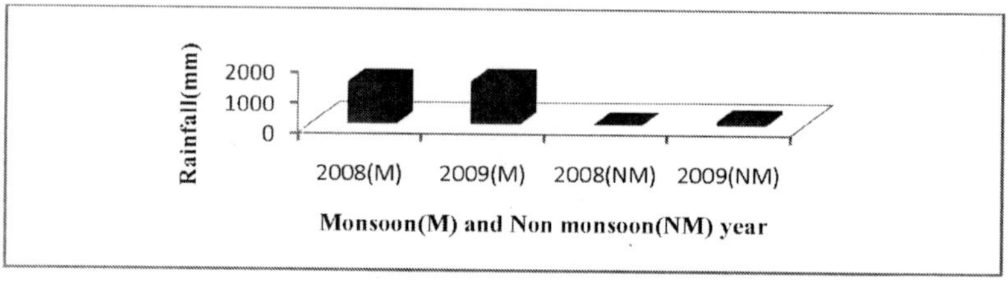

Fig. 1. Rainfall during Monsoon and Non-Monsoon Period.

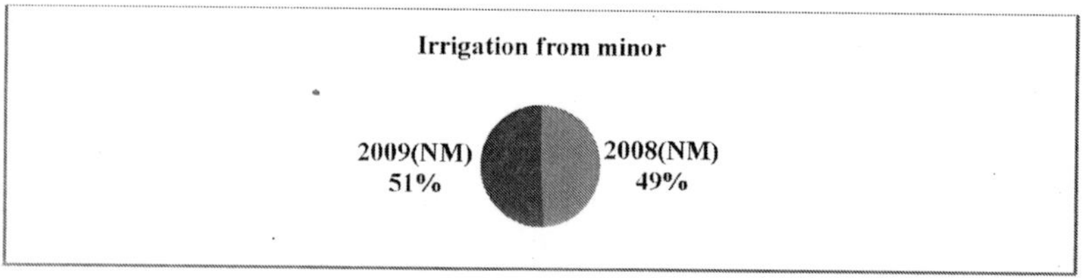

Fig. 2. Available irrigation during Monsoon and Non-Monsoon Period from minor.

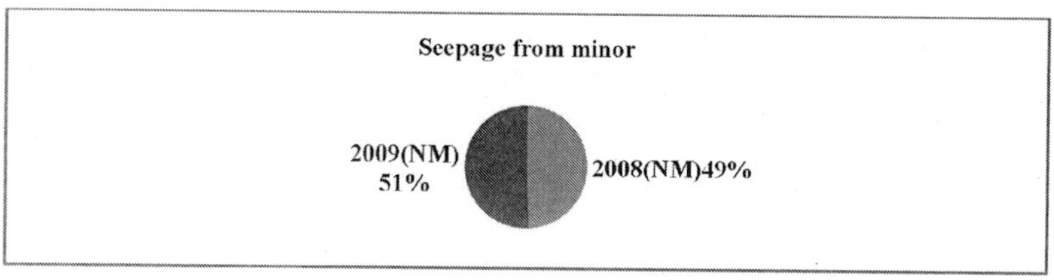

Fig. 3. Seepage loss in Monsoon and Non-Monsoon Period from minor.

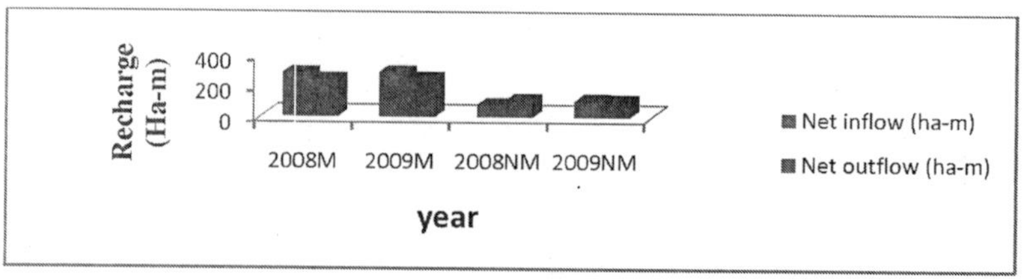

Fig. 4. Net Inflow and Net outflow.

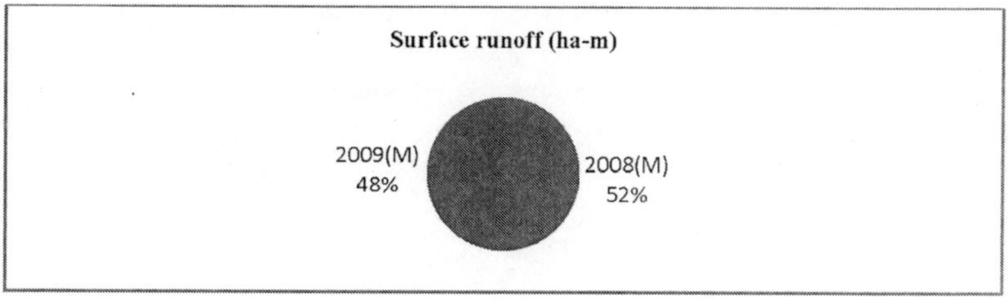

Fig. 5. Surface Runoff during Monsoon and Non-Monsoon Period.

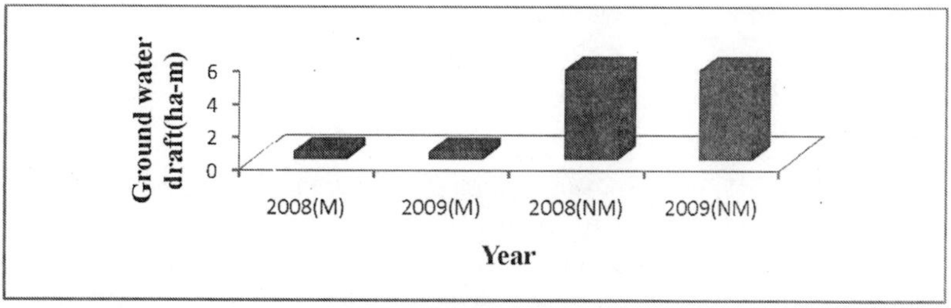

Fig. 6. Performance of Ground Water Draft in Monsoon and Non-Monsoon Period.

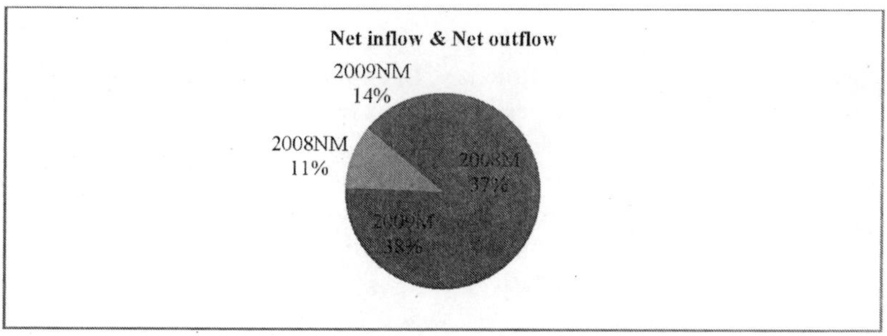

Fig. 7. Performance of Net Inflow (NI) and Net Outflow (NO) in Monsoon and Non-Monsoon Period

In order to compute the net inflow (NI), all the preliminary calculations were made. For the year of 2008 net inflow was estimated as 287.04 ha-m and 80.85 ha-m respectively during monsoon and non-monsoon period. During monsoon and non-monsoon period net outflow components were estimated as 241.60 ha-m and 108.89 ha-m respectively (Fig. 8). The ground water storage was computed as 45.44 ha-m and -28.03 ha-m respectively.

The Table 2 shows the results of seepage from minor which was 0 ha-m and 39 ha-m during monsoon and non-monsoon period respectively, while irrigation from minor were estimated 0 ha-m and 41.21 ha-m during monsoon and non-monsoon period respectively. In order to quantify the ground water draft as depicted in Table 2 for monsoon and non-monsoon period were estimated 0.50 ha-m and 5.50 ha-m respectively.While change in soil moisture storage were 39.93 and 39.93 ha-m respectively. During the computation of water balance component. In order to compute the net inflow (NI), all the preliminary calculations were made. For the year 2009 net inflow (Fig. 8) were estimated as 293 ha-m and 107.37 ha-m respectively during monsoon and non-monsoon period.

During monsoon and non-monsoon period net outflow components were estimated 249.55 ha-m and104.15 ha-m respectively. After deducting the inflow to the net outflow (Fig. 8) ground water storage were computed as 51 ha-m and 3.22 ha-m respectively.

Conclusion

The water available through minor supply into the command area 41.21 ha-m and from tube well 5.50 ha-m. For the year 2008 monsoon and non-monsoon period inflow in the form of rainfall is 287.04 ha-m, 3.91 ha-m, seepage from minor is 0 and 36.94 ha-m and irrigation from minor is 0 and 40 ha-m respectively. And the outflow for monsoon and non-monsoon period in the form of evapotranspiration from cultivated land is 63.4, 56.22 ha-m and evaporation from uncultivated land is 0, 0 ha-m, surface runoff is 137.77, 0 ha-m, ground water draft 0.50, 5.50 ha-m, change in soil moisture storage is 39.93, 39.93 ha-m and subsurface flow is 0, 0 ha-m respectively. For the year 2009 monsoon and non-monsoon period inflow in the form of rainfall is 293.28 ha-m, 28.16 ha-m, seepage from minor is 0 and 38 ha-m and irrigation from minor is 0 and 41.21 ha-m respectively. And the outflow for monsoon and non-monsoon period in the form of evapotranspiration from cultivated land is 73.98, 58.72 ha-m and evaporation from uncultivated land is 0, 0 ha-m, surface runoff is 127.57, 0 ha-m, ground water draft 0.50, 5.50 ha-m, change in soil moisture storage is 39.93, 39.93 ha-m and subsurface flow is 0, 0 ha-m.

References

Agrawal, R., S.K. Sodhi and M.P. Kaushal (1999). Seasonal estimation of groundwater potential of South-West Punjab. Deptt. of Soil and Water Engg. PAU Ludhiana (Punjab).

Ahmad Mobin-Ud Din, Wim G.M. Bastiaanssen and Reinder A. Feddes (2010), Vol. 13, pp. 653-664.

Doorenbos, J. and Pruit, W. O. (1977). Crop water requirements Irrigation and Drainage paper no. 24, Food and Agriculture Organisation of United Nation, Rome, Italy.

Irmark, S., (2008). Evapotranspiration, Encyclopedia of ecology, p. 1432-1438.

Jadhav, P.W., G.U.Satpute, Gitte N. P. and S.P. Divekar (2003). Water balance model for estimation of ground water recharge in a watershed. *Journal of Soil and Crops* (2003), 13:2, 253-256.

Kaledhonkar, J.M. (2007). Regional scale modelling for conjunctive water use planning. Central Soil Salinity Research Institute, Karnal-132 001 Haryana.

28

Prioritisation of Watershed Development by Estimating Soil Loss using Remote Sensing and GIS

R. J. Patel, H. V. Parmar, H. H. Mashru, P. B. Vekariya, H. D. Rank, J. H. Kelaiya and D. M. Paradava

Abstract

Adoption of soil conservation measures is necessary for the optimal utilisation of natural resources and to increase the productivity on the sustainable basis. Remote Sensing and GIS techniques for generation of development plans for the watershed area in consonance with the production potential and limitations of terrain resources and for assessing the impact of the measures undertaken before their implementation in the field. The successful and fruitful outcome of the adoptions of soil conservation can only be harnessed if they planned and executed on watershed basis. The sound planning of watershed development and management required detailed surveys on topography, land use, erosion hazard and drainage network. The survey can be done either by manually or by using remote sensing and GIS. The manual survey is very cumbersome, time consuming and costly. Remote Sensing and GIS technique can be effectively used for prioritisation of catchments and this helps in the way the catchments can be taken for treatment for conservation measures. Therefore, the study was mainly aimed to plan and propose watershed development and management strategies for the sub-watershed (5G1C5e). The sub-watershed of 16940 ha comprising of 23 micro watersheds. The thematic maps were prepared using the Remote Sensing images, soil maps and reports prepared by NBSS and LUP. All these maps were digitised in the ARCVIEW module of ARC/INFO GIS software. In the study, soil loss was predicted employing USLE method and an integrated analysis in GIS was carried out for soil conservation and for delineation suitable conservation unit. Satellites and other collateral data were used to identify the problems and potential in the watersheds and measures for soil conservation. The estimated soil losses were ranged from 2.9903 to 4.9870 tons/ha/year. The minimum and maximum EHI obtained in the study area were 0.8614 and 0.9131 respectively.

Introduction

Watershed is an area that catches the rainwater from precipitation and then it drained by a river and its tributaries. Watershed area is the area starting from highest point of the area (ridge line) to the outlet of the *nalah* or the natural stream. Soil erosion is one of the most serious problems in recent times which affects both cultivable and forest lands. It creates serious problems in agriculture and water resources management by removing the top fertile soil and its subsequent deposition in reservoirs and lakes. Soil and Water Conservation measures have been long practiced to protect the productive lands. Remote Sensing and GIS techniques have been used recently to arrive at cost-effective plans for conservation and development measures for watersheds. However, there is a necessity to assess the impact of various potential measures on runoff yield before implementation of the plans. This study was aimed at assessing the impact of these measures on runoff from a region. The objectives of this study were to prioritise watersheds on the basis of runoff yields, to generate development plans suitable to the terrain and production potential of local resources, and to evaluate the hydrologic impact of plans/land use.

These primary phases were integrated in the GIS environment for generating the secondary maps. The erosion maps showing the intensity of the erosion was ranging from 0.1 to 3.8 kg/m^2 (Morgan and Finney, 1984). Estimation of soil erosion through an Integrated Land Information System (ILIS) for evaluating potential soil erosion having several information layers on the basis of the USLE model and preliminary results were evaluated by comparison with ground data and were considered promising (Conese *et al.*, 1993). An attempt was made to apply the Universal Soil Loss Equation (USLE) to eroded soils following an analysis of soil erosion by water. Due to a lack of data, rainfall erosivity could only be roughly estimated. Soil erodibility was low, and was considered to be a secondary cause of soil erosion. Variations in vegetation cover were an important factor of soil erosion. It was suggested that care needs to be taken because rough estimations were made for several factors, and that the USLE, being a multiplicative model, magnifies the errors of the estimates (Nyssen, 1997). Process-based model was applied by Yadav *et al.* (2001) to quantify soil erosion and to prioritise the watershed on the bases of soil erosion. Topographical maps were used to prepare base map and contour map. Satellite data of IRS 1C was used for the preparation of land use/

land cover map. The estimation of runoff by SCS curve number and soil loss estimation using modified USLE, by the application of GIS software ARCVIE, was done for the Ozat river basin and suggested an action plan to minimise the soil loss. Soil loss can be minimised with increase in vegetative cover (Patel, 2012).

Das *et al.* (2005) estimated annual soil loss from a catchment in Assam and compared with that of base year using Remote sensing and satellite data. Chandramohan and Durbude (2002) simulated sediment yield from a catchment using GIS for lumped and distributed models using USLE model (Wischmaier and Smith, 1978). Mongkolsawat *et al.* (1994) carried out a study to establish spatial information of soil erosion using the Universal Soil Loss Equation in GIS environment. The results indicated that low erosion classes are located in areas of land that is used for field where slope gradient is very low. High erosion classes concentrated in areas where land is used for field crops with no conservation practice. Hesadi *et al.* (2003) used some factors that include soil characteristics, climatic parameters, runoff, topography, landuse to estimate erosion based on capability of RS and GIS in a watershed in Iran and compared results with traditional studies. Significant difference between traditional and new methods was observed. Martin *et al.* (2007) obtained soil data from soil resource inventory and land and climate were derived from the remote sensing satellite data (Landsat TM, bands 1 to 7) and integrated the data in GIS environment to obtain the soil erosion loss using USLE model for the watershed area. The priorities of different sub-watershed areas for soil conservation measures were identified. Land productivity index was also used as a measure for land evaluation. Sutaria *et al.* (2005) predicted some parameters of the Universal Soil Loss Equation like R, K and C factors for prevailing conditions of Middle Gujarat Agro Climatic Zone. The EI was found as a good index for the prediction of soil losses.

Methodology

The part of Jhanjheari sub-watershed, Ozat watershed forms the part of the Bhesan and Visavadar talukas of Junagadh district was selected as a study area. The selected sub-watershed comprising of 16940 ha has 23 micro watersheds having erratic rainfall pattern with an average annual rainfall of 824.7 mm (1959-2000). The study area falls in Survey of India (SOI) Topographical Map No. 41-K/10, 41-K/11, 41-K/14 and 41-K/15. The study area has predominantly got medium black soils, formed out of alluvium. These are blackish in colour, easy to work upon and suitable to dry climates. The soil depth in study area is shallow to moderately deep i.e. 25-100 cm. Geologically, the study area comprises of basaltic rocks of the Daccan Trap. A unique feature of the Deccan Trap in this area is the occurrence of a variety of igneous rocks intruding the lava flows. The various intrusive bodies are supposed to belong to a slightly later age than the main bedded flows but are part and parcel of the same igneous episode.

Data Collection

Satellite data: The FCC of IRS-1C PAN+LISS-III merged geocoded products on 1:25,000 scale were used for classification of various themes. The IRS-1C PAN+LISS III (merged data) image of the study area is shown in Fig. 1.

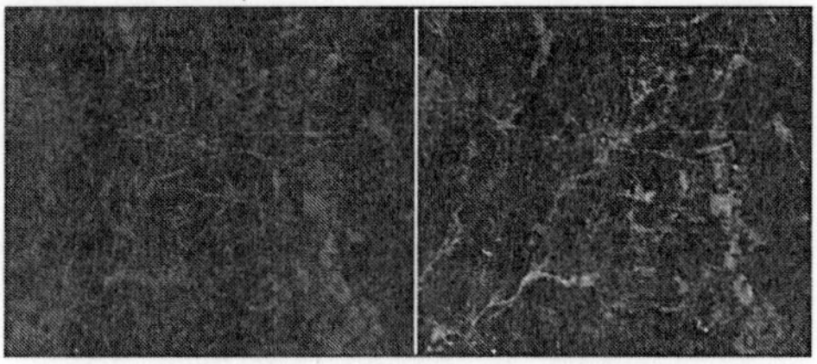

Fig.1. Images of the Study Area.

Table 1: Details of IRS-1C PAN+LISS-III Merged Geocoded Products

No.	*Name of the Satellite*	*Sensor*	*Scale*
1.	IRS-1C	PAN+LISS-III	1:25,000
2.	IRS-1C	PAN+LISS-III	1:25,000

Soil toposheet: Study area (sub-watershed) has been delineated from Survey of India (SOI) topographic sheet. The sheet number 41-K/10, 41-K/11, 41-K/14 and 41-K/15 of 1:50,000 scale were used in the study.

Soil Map and report: Soil map and report prepared by National Bureau of Soil Survey and Land Use Planning was used.

Rainfall data: Daily hourly rainfall data of the rain gauge stations at head-quarters of Bhesan, Junagadh, Munjiyasar (near to Bagsara) and Visavadar for 10 years in mm were collected.

Generation of thematic map

All these maps are digitised in the ARCVIEW module of ARC/INFO Geographic Information System (GIS).

Study area preparation: The study area is delineated using watershed concept. This map was then superimposed on the satellite data products and matched with the features present on the satellite imagery for the interpretation purpose. The study area map is shown in Fig. 2.

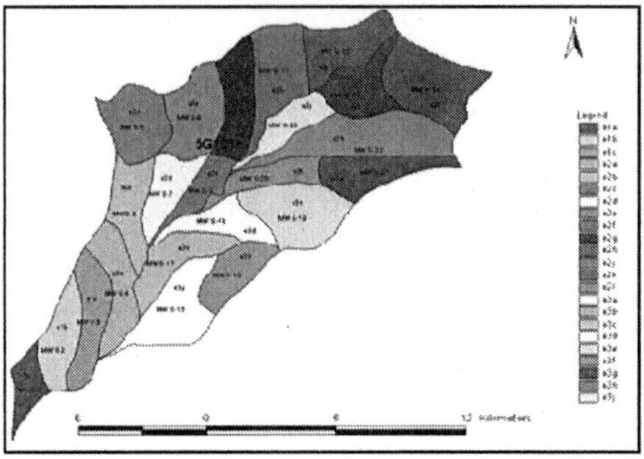

Fig.2. Study Area Map.

Land use/land cover map: The interpretation and correlation of imagery with objects involved the comparison of spectral response of each type of objects with tinge characteristics. The interpreted details are checked on ground to verify the interpretation and doubtful areas. Based on the ground verification, boundaries of the different land use/land cover units were finalised.

Hydrogeomorphological map: Using standard visual interpretation techniques 8 hydrogeomorphological units were mapped.

Estimation of Soil Loss

The micro-watersheds were prioritised based on the estimated quantum of soil loss.

Universal Soil Loss Equation

Universal Soil Loss Equation (USLE), developed by Wischmeier and Smith (1978) was used to study the degradation of land, in order to prioritise micro-watersheds and adopt suitable soil conservation measures.

$$A = R \times K \times L \times s \times C \times p$$

Where, A is the average annual soil loss in tonnes/ha/year, R is the rainfall erosivity factor, K is the soil erodibility factor, L is the slope length factor, s is the slope steepness factor, C is the cover and management factor and p = the support practice factor.

Rainfall Erosivity Factor (*R*)

The rainfall erosivity factor (R) measures the erosive force of rainfall and runoff. The R factor, sometimes called the rainfall erosion index, takes into account the erosive effects of storms.

$$R = \left[\frac{1}{100} (210.3 + 89 \log li) hi \right] I_{\max} 30$$

where, li is the intensity and hi is the amount of the rain during a period of rain, $I_{max} 30$ is the 30-minute maximum rainfall intensity of a storm.

Soil Erodibility Factor (*K*)

The soil erodibility factor (*K*) indicates the inherent erodibility of a soil. The soil erodibility or *K* factor normally varies from near zero to about 0.6. *K* was calculated using the data on soil properties.

$$K = 1.2917\left[2 \times 10^{-4} \times M^{1.14}(12-a) + 3.25(b-2) + 2.5(c-3)\right] \times \frac{1}{100}$$

where, *M* is (%silt + very fine sand) (100% clay), *a* is per cent organic matter, *b* is soil structure code used in soil classification, and *c* is permeability class.

Topographic Factor (*LS*)

The topographic factor (*LS*) reflects the influence of length and steepness of slope. The *LS* was calculated using the empirical equation from slope length and slope per cent,

$$LS = \left(\frac{L}{22.13}\right)^{m} \left(0.065 + 0.045s + 0.0065s^{2}\right)$$

where, '*m*' is an exponent varying from 0.2 to 0.6 depending on the percent slope.

Cover and Management Factor (*C*)

The cover and management factor (*C*) indicates the influence of cropping systems and management variables on soil loss. '*C*' factor for various cropping and management practices for the study area are given in Table 2.

Support Practice Factor (*p*)

It is the ratio of soil loss with given support practice to the corresponding loss when crop culture is up and down the slope.

Table 2: Values of *C*-Factors under Different Landuse/Land cover Classes

No.	*Level I*	*Level II*	*C*
1.	Agriculture	Double Crop	0.40
		Single Crop	0.50
2.	Waste Land	Land With Scrub	0.80
		Land Without Scrub	0.95
3.	Habitation	Urban/Rural	0.20
4.	Water Bodies	River	0.00
		Reservoir/Stream	0.00
5.	Others	Prosophis	0.10

Soil Loss Tolerance (*T*) Values

It is the threshold soil loss value. Any increase in soil loss above this limit would result in the reduction of crop productivity on sustainable basis. Table 3 shows the soil sub-groups, their association and the assigned tolerance values.

Table 3: Showing Soil Loss Tolerance Values for Different Soil Types Present in the Study Area

No.	*Sub-Soil Group*	*T (tonnes/ha/year)*
1.	Lithic Ustorthents	2.2
2.	Lithic Ustochrepts	4.5
3.	Vertic Ustochrepts	6.7

Erosion Hazard Index (*EHI*)

It is the ratio of soil loss to the tolerance limit. *EHI* was estimated for micro-watersheds by considering area-weighted values.

Results and Discussion

The study was carried out to suggest the soil conservation measures for the purpose of conserving soil.

Generation of Thematic Map

Drainage, micro-watershed and surface water body map: Micro-watershed boundaries are delineated using drainage and surface water body map.

23 micro-watersheds are delineated ranging from 366.62 to 1332.51 ha. Details of micro-watershed code and their areas are given in Table 4. Drainage map with the details of water bodies is shown in Fig. 3.

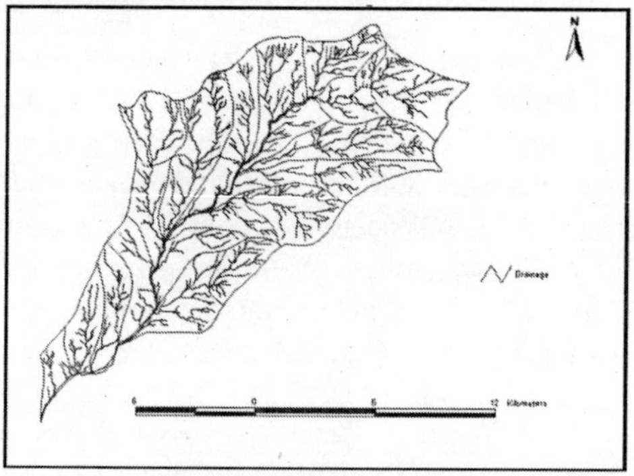

Fig. 3. Drainage Map with Water Bodies.

Table 4: Details of MWS and their Extent Lying under the Study Area

Sr. No.	*MWS Code*	*MWS*	*Area of MWS (Hectares)*	*percentage of Extent*
1.	e1a	MWS-1	366.62	2.16
2.	e1b	MWS-2	635.07	3.75
3.	e1c	MWS-3	719.25	4.25
4.	e2a	MWS-4	733.26	4.33
5.	e2b	MWS-5	657.75	3.88
6.	e2c	MWS-6	824.35	4.87
7.	e2d	MWS-7	572.22	3.38
8.	e2e	MWS-8	881.91	5.21
9.	e2f	MWS-9	399.49	2.36
10.	e2g	MWS-10	735.22	4.34
11.	e2h	MWS-11	970.40	5.73
12.	e2j	MWS-12	776.10	4.58
13.	e2k	MWS-13	674.50	3.98
14.	e2l	MWS-14	1158.24	6.84
15.	e3a	MWS-15	953.33	5.63
16.	e3b	MWS-16	580.31	3.43
17.	e3c	MWS-17	687.90	4.06
18.	e3d	MWS-18	547.01	3.23
19.	e3e	MWS-19	1046.61	6.18

(Contd...)

20.	e3f	MWS-20	481.28	2.84
21.	e3g	MWS-21	721.42	4.26
22.	e3h	MWS-22	1332.51	7.87
23.	e3j	MWS-23	485.37	2.87

Land use/land cover map: It was found that river Ozat is flowing in the middle portion of the watershed. About 11.37 per cent areas are occupied by double crop which is largely lying in areas where irrigation facilities are available. All most portions (i.e. 78.46 per cent) of the watershed are occupied by single cropped area. Wastelands occupied about 7.61 per cent areas and are found scattered all over the watershed. Water bodies scattered in the upper portion are found to be small. The land use/land cover map is shown in Fig. 4. Statistics of areal distribution of different land use/land cover categories is shown in Table 5.

Table 5: Areal Extent of Land use/Land cover Condition Delineated from Remotely Sensed Data

No.	*Level-I*	*Level-II*	*Area (ha)*	*percentage of Extent*
1.	Agriculture	Double Crop	1925.75	11.37
		Single Crop	13291.48	78.46
2.	Waste Land	Land With Scrub	1205.55	7.12
		Land Without Scrub	83.03	0.49
3.	Habitation	Urban/Rural	204.97	1.21
4.	Water Bodies	River	224.20	1.32
		Reservoir/Stream	1.11	0.01
5.	Others	Prosophis	4.01	0.02

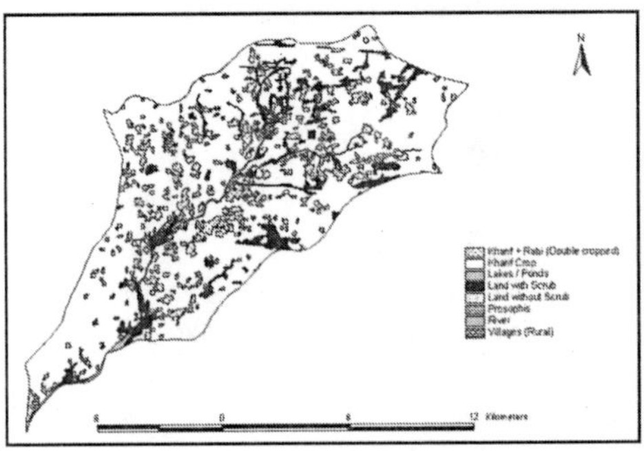

Fig. 4. Land cover/land use Map.

Slope map: All the study area is found to be normally sloping *i.e.* 0-1 per cent.

Soil map: The southeast and middle portion of north is largely occupied by Lithic Ustorthents having area of 5127.01 ha. Lithic Ustochrepts is the soil, which occupy very small area about 1116.06 ha in the east and west portion of watershed. These are very shallow-to-shallow in depth. The soil sub-group Vertic Ustochrepts occupies big area of 10697.04 ha in the middle portion, east and west portion of the watershed. The soil map is shown in Fig. 5. Table 6 gives the areal extent of various soil sub-groups in the study area.

Table 6: Areal extent of soil sub-groups

No.	*Soil category*	*Soil sub-group*	*Area (ha)*	*percentage of Extent*
1.	Soils of pediment-undulating soils of pediment	Lithic Ustorthents	5127.01	30.26
2.	On gently sloping surface soils of pediment	Lithic Ustochrepts	1116.06	6.59
3.	Plain with narrow valleys	Vertic Ustochrepts	10697.04	63.15

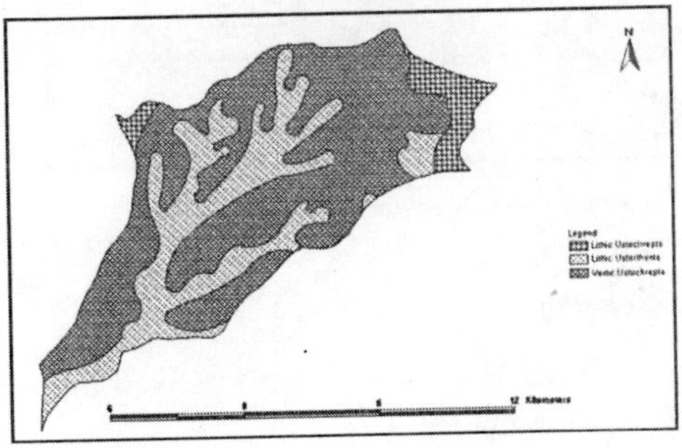

Fig. 5. Soil Map.

Hydrogeomorphological map: The geomorphic unit, buried pediplain (shallow) is found largely through all over the sub-watershed which occupied 13686.18 ha area. Pediment occupied the area about 1583.85 ha, which is scattered all over. About 684.64 ha area is occupied by buried pediplain (medium) and it is found in the middle portion of south in large extent. Only 24.46 ha area is occupied by Flood plain. The hydrogeomorphological map is shown in Fig. 6. Areal statistics of various hydrogeomorphic units is given in Table 7.

Table 7: Details of Hydrogeomorphic Units and their Extent in the Watershed

No.	*Hydrogeomorphic units*	*Area (ha)*	*percentage of Extent*
1.	Built-up	46.37	0.27
2.	Flood Plain	24.46	0.14
3.	Flood Plain-Eroded	683.39	4.03
4.	Linear Ridge/ Dyke	57.01	0.34
5.	Moderately weathered/ moderately buried Pediplain	684.64	4.04
6.	Pediment/ Valley Floor	1583.85	9.35
7.	River	174.23	1.03
8.	Shallow weathered/ shallow buried Pediplain	13686.18	80.79

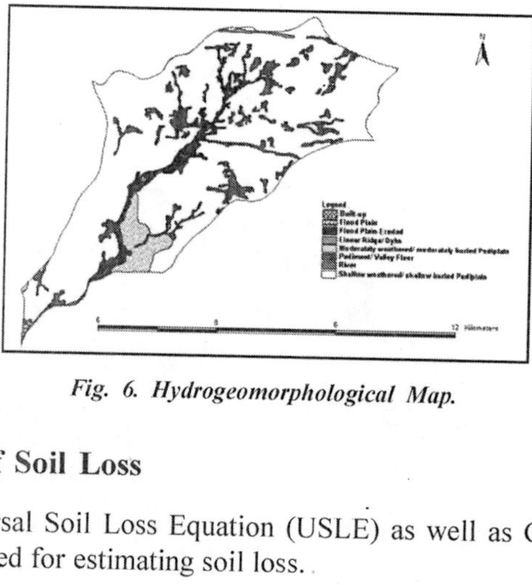

Fig. 6. Hydrogeomorphological Map.

Estimation of Soil Loss

The Universal Soil Loss Equation (USLE) as well as GIS application software was used for estimating soil loss.

Estimating Rainfall Erosivity Factor *(R)*: The *R*-factor comprises a measure of the erosive force of precipitation i.e. the rainfall erosivity index (EI_{30}). Its calculation involves the computation of the kinetic energy of a rainfall event (K.E.) and the maximum 30-min rainfall intensity I_{30}. In the calculation of kinetic energy, rainfall data at 30 minute interval is required, but rainfall data at 30 minute intervals is not available these are interpolated from hourly rainfall data. The following equation is used for calculation of *R*-factor,

$$R = EI_{30}$$

where, $EI_{30} = 1/100$ (Total K.E. of the storm × $Max.\ I_{30}$), K.E. = $(210.3 + 89 \log li)$, Total K.E. = K.E.$x\ (hi)$

For example, for a storm occurred on 8-8-1994 between 18.30 and 20.30 hours, *(EI_{30})* is estimated as given in following Table 8.

Table 8: Tabulation for calculating EI_{30}

Time	*Rainfall (cm)*	*Intensity (cm/h)*	*K.E. (t/h)*	*Total K.E. (t/h)*
18:30-19:00	0.05	0.1	121.3	6.065
19:00-19:30	0.05	0.1	121.3	6.065
19:30-20:00	0.45	0.9	206.23	92.802
20:00-20:30	0.45	0.9	206.23	92.802
	Total *K.E.* of storm event			197.734

Maximum 30-min intensity,

Max. $I_{30} = 0.9$ cm/h

$EI_{30} = 1/100\,(197.734 \times O.9)$

$EI_{30} = 1.779$ t/ h

$R = 1.779$ t/h

details of rainfall erosivity factor values are given in Table 9. Mean value for all the 10 years are found. Each of the mean annual rainfall erosivity factor (*R*) is assigned to the corresponding Thiessen polygon of the weighted area basis.

Table 9: Details of Rainfall Erosivity Values.

No.	*Year*	*Rainfall Erosivity Factor of different rain gauge stations*			
		Bhesan	*Bagsara*	*Junagadh*	*Visavadar*
1.	1995	128.28	70.63	301.30	243.68
2.	1996	214.18	685.84	279.34	182.59
3.	1997	104.20	219.67	125.30	146.26
4.	1998	186.10	393.68	125.08	160.10
5.	1999	24.88	212.62	42.99	204.05
6.	2000	30.72	176.82	61.52	161.77
7.	2001	195.72	136.70	99.61	195.72
8.	2002	266.81	457.44	251.73	266.81
9.	2003	339.31	251.66	275.50	509.29
10.	2004	204.64	93.11	118.87	299.50
Mean		**169.48**	**269.82**	**168.12**	**236.98**

Estimation of Soil Erodibility Factor *(K)* : It is calculated on the basis of texture, percentage organic content, structure and permeability. Soil erodibility factor '*K*' for all the 23 micro-watershed is found. Soil properties for all three-soil sub-groups are given in the Table 11. The *K*-factor obtained for each soil sub-group is given in the Table 10 and weighted mean K-factor for all the 23 micro-watersheds is given in Table 12 which is vary from 0.2075 to 0.3386.

Table 10: *K*-factors Assigned to Different Soil Sub-Groups

No.	*Sub-Soil Group*	*Association*	*K-Factor*
1.	Lithic Ustorthents	Lithic Ustochrepts	0.2635
2.	Lithic Ustochrepts	Lithic Ustochrepts	0.1261
3.	Vertic Ustochrepts	Lithic Ustochrepts	0.3386

Preparation of Slope Length Factor Layer *(L)*: Slope length is defined as the distance from the point of origin of over land flow to the brink of storage of water. Lengths of slopes are derived using drainage map.

Preparation of Slope Steepness Factor Layer *(s)*: The slope steepness factor layer is prepared using the slope map. Average value of the slope category, viz. 0-1 per cent, obtained for the study area is considered to be 0.5.

Estimating Topographic Factor *(LS)*: Instead of considering *L* and *s* factors separately, these are combined to generate a new layer called topographic factor *(LS)* layer. Mean slope-length/topographic factors for all the 23 micro-watersheds are given in Table 12 which is varying between 0.1299 and 0.1561.

Table 11: Physical and Chemical Properties of Soil Sub-Group

Soil Properties	*Soil Sub-Group*		
	Lithic Ustorthents	*Lithic Ustochrepts*	*Vertic Ustochrepts*
Organic Content (per cent)	0.58	0.25	0.86
Organic Matter (per cent)	2.5	1.5	1.5
Very Fine Sand (per cent)	7	4	25
Silt (per cent)	40	25	20
Clay (per cent)	50	58	35
Soil Structure Code (*b*)	4	2	4
Profile Permeability Class (*c*)	3	4	3

Estimating Cover and Management Factor *(C)*: The '*C*' factor combines the effects on erosion of cover and management. These effects are significantly interrelated and cannot be treated separately. Weighted mean c-factors assigned to for all micro-watersheds are given in Table 12 which is varying from 0.4629 to 0.5383.

Table 12: Mean Values of Different Factors and Estimated Total Soil Loss Values for Different Micro-Watersheds.

Micro-Watershed Code	*K-Factors*	*LS-Factors*	*C-Factors*	*Soil Loss (tonnes/ha/year)*
MWS-1	0.2952	0.1382	0.4629	3.9866
MWS-2	0.3293	0.1385	0.5090	4.9006
MWS-3	0.3204	0.1454	0.4841	4.7608
MWS-4	0.2958	0.1550	0.4985	4.8248
MWS-5	0.3094	0.1419	0.4860	4.5043
MWS-6	0.2075	0.1427	0.4784	2.9903
MWS-7	0.2948	0.1394	0.4861	4.2170
MWS-8	0.3005	0.1353	0.4990	4.2828
MWS-9	0.2725	0.1329	0.4955	3.7881
MWS-10	0.3155	0.1415	0.5036	4.7460
MWS-11	0.3032	0.1299	0.5240	4.3567
MWS-12	0.3386	0.1300	0.5133	4.7697
MWS-13	0.3354	0.1354	0.5202	4.9870
MWS-14	0.2076	0.1418	0.5281	3.2818
MWS-15	0.3001	0.1347	0.5125	4.3734
MWS-16	0.3113	0.1329	0.5291	4.6209
MWS-17	0.3386	0.1452	0.4746	4.9257
MWS-18	0.3309	0.1353	0.5010	4.7350
MWS-19	0.3242	0.1372	0.5187	4.8705
MWS-20	0.3188	0.1561	0.4743	4.9827
MWS-21	0.3033	0.1428	0.5383	4.9217
MWS-22	0.2953	0.1334	0.5012	4.1679
MWS-23	0.2963	0.1339	0.4930	4.1290

Estimating Support Practice Factor *(p)*: Since no soil conservation measures were in practice in the study period, the *p*-factor is not taken into consideration by keeping its value one for the pre-conservation scenario (i.e. for existing land use conditions).

Estimating Soil Loss: The soil loss is estimated by multiplying the *R, K, LS, C* and *p* factors.

$A = R \times K \times LS \times C \times p$

Where, A = soil loss in tonnes/ha/year

Soil loss is determined on micro-watershed basis and these values are shown in Table 12. The estimated soil losses were ranged from 2.9903 to 4.9870 tonnes/ha/year.

Soil Loss Tolerance *(T)* Values: If any increase in soil loss above Soil loss tolerance *(T)* value, there will be reduction in crop productivity on sustainable basis. Table 13 shows micro-watershed-wise weighted mean soil loss tolerance values.

Table 13: Showing Soil Loss Tolerance Values for Different Micro-Watershed.

Micro-Watershed Code	*EHI*	*Soil Loss Tolerance Values (tonnes/ha/year)*
MWS-1	0.8960	4.1013
MWS-2	0.9059	6.1419
MWS-3	0.9077	5.6105
MWS-4	0.9064	4.1328
MWS-5	0.9026	4.9477
MWS-6	0.8621	4.0321
MWS-7	0.8966	4.0761
MWS-8	0.8956	4.4161
MWS-9	0.8843	2.7391
MWS-10	0.9041	5.3162
MWS-11	0.8926	4.5805
MWS-12	0.9028	6.7000
MWS-13	0.9055	6.5094
MWS-14	0.8614	5.3433
MWS-15	0.8951	4.3916
MWS-16	0.8973	5.0651
MWS-17	0.9121	6.6986
MWS-18	0.9043	6.2390
MWS-19	0.9038	5.8378
MWS-20	0.9131	5.5147
MWS-21	0.9014	5.6627
MWS-22	0.8927	5.2441
MWS-23	0.8933	4.1677

Erosion Hazard Index *(EHI)*: It is the ratio of soil loss to the tolerance limit. The minimum and maximum EHI obtained in the study area are 0.8614 and 0.9131 respectively. EHI was also estimated for micro-watersheds by considering area-weighted values. Table 13 shows the EHI for different micro-watersheds.

Soil Conservation Units

The degree of erosion and site conditions is considered to evaluate and identify conservation units. The site characteristics constitutes slope, soil depth, soil texture and land cover. Table 14 shows site characteristics and suggested soil conservation measures in identified units for the study area.

Table 14. Site characteristics and suggested soil conservation measures in identified units

No.	*Potential Erosion*	*Soil Depth and Texture*	*Land Use/ Land cover*	*Suggested Control Measures*
1.	5-10	Shallow/ moderately Shallow Clay loam	Waste Land with/without Scrub, Single cropped area	Afforestation Horticulture practices
2.	2-5	Moderately deep Clay	Waste lands Single crops	Contour strip cropping, inter cropping, vegetative bunding

Conclusion

The study demonstrates the feasibility of using integrated remote sensing and GIS for development of a watershed and for assessment of its hydrologic response, to various land use and management changes. The spatial analysis of thematic information, which can be derived from remote sensing helps in the assessment of development plans before they are implemented in the field. This approach is thus an effective tool for selection of the best management plan to be implemented. The remotely sensed data has been found to be highly valuable in the delineation of land cover with greater precision of type and extent and to evaluate the appropriate annual cover factors. Implementation of Universal Soil Loss Equation, using integration procedures of GIS enabled the prediction of soil loss rates and in the identification of units for suitable protection measures. The mean soil loss rate is estimated to be 4.44 tonnes/ha/year. Suitable agronomic and mechanical measures are suggested for soil conservation in each unit. The estimated soil losses and EHI were ranged from 2.9903 to 4.9870 tonnes/ha/year and 0.8614 to 0.9131 respectively.

References

Chandramohan, T. and D. Durbude. 2002. "Estimation of Soil Erosion Potential using Universal Soil Loss Equation", Journal of the Indian Society of Remote Sensing. 30(4): 131-142.

Conese, C., T. Filippis, F. Maselli and P. Winkler. 1993. "Estimation of Soil Erosion through an Integrated Land Information System", Remote sensing for monitoring the changing environment of Europe: Proceedings of the 12th Symposium, Hungary 8-11 September, pp. 235-238.

Hesadi, H., K. Jalili and M. Hadidi. 2003. "Applying Remote Sensing and GIS for Soil Erosion and Sediment Estimation by MPSIAC Model - A case study of Kenesht watershed", http://www.gisdevelopment.net/application/natural_hazards/landslides/ma06_175pf.htm.

Martin, D. and S.K. Saha. 2007. "Integrated Approach of using Remote Sensing and GIS to Study Watershed Prioritisation and Productivity", Journal of the Indian Society of Remote Sensing, 35(1), 21-30.

Mongkolsawat, C., P. Thirangoon and S. Sriwongsa. 1994. "Soil Erosion Mapping with Universal Soil Loss equation and GIS", http://www.aars.org/acrs/proceeding/ACRS1994/Papers/DIS94- 1.htm.

Morgan and Finney. 1984. "Predicting Erosion Losses from Crop Land east of Rocky Mountains", USDA Science and Education Administration Handbook No. 374.

Nyssen, J. 1997. "Soil erosion in the Tigray Highlands (Ethiopia). II. Soil loss Estimation", Geo Eco Trop. 21(1/4): 27-49.

Patel, R. 2012. "Watershed Management Planning Using Remote Sensing and GIS", A book published in LAP-Lambert Academic Publishing, Germany.

Sutaria, G.S., Sharma, G.R. and Limbasiya, B.B. 2005. "Prediction of some parameters of the Universal Soil Loss Equation", Proceeding National Conference on Sustainable Management of Water Resources, held during 28-29 January, 2005, at CAET, JAU, Junagadh, pp. 250.

Wischmeier, W.H. and D.D. Smith. 1978. "Predicting Rainfall Erosion Losses - A Guide to Conservation Planning", Agriculture Hand Book No. 537, United State Department of Agriculture, USA.

Yadav, P., K. Durga Rao, and K. K. Das. 2001. "Soil Erosion Modeling Through Morgan Model, using Remote Sensing and GIS Technique", National Symposium on Advances in Remote Sensing Technology with Special Emphasis on High Resolution Imagery, December 11th 13th 2001, ISRS, Ahmedabad.

29

Impact of Climate Change on Hydraulic Properties, Profile Water Storage Capacity, Reserve Water and Water Balance in the Coastal Soils of Southern Saurashtra Region of Gujarat

S. G. Savalia, J. D. Gundaliya, S. P. Kachhadia and H.P. Patel

Abstract

A study was undertaken on "Impact of climate change on hydraulic properties, profile water storage capacity, reserve water and water balance in the Coastal Soils of Southern Saurashtra region of Gujarat". Result revealed that the weighted mean of saturated hydraulic conductivity and available water content of the coastal soils varied from 0.02 to 0.22 cm hr^{-1} and 0.22 to 0.32 m^3 m^{-3}, respectively. The infiltration rate ranged from 0.54 to 1.23 cm hr^{-1}. The total water storage capacity of the profiles varied from 0.255 to 1.539 m ha^{-1}. The soils belongs to Fluventic Calciustepts of piedmont plain area recorded the highest total profile water storage capacity (1.539 m ha^{-1}) followed by Vertic Haplustepts (0.537 m ha^{-1}) in lower piedmont. The PAWC / WSC ratio varied from 0.377 to 0.498 of which P_4 belong to Fluventic Calciustepts in piedmont plain area has the highest ratio i.e. 0.498 indicating the soils of this area have 50 per cent plant available water capacity. The reserve water and water balance varied from 0.156 to 1.021 m ha^{-1} (mean of 0.461 m ha^{-1}) and -0.945 to + 0.615 m ha^{-1} (+ 0.085 m ha^{-1}), respectively. The total profile water storage capacity and reserve water increases with decrease of an elevation. The negative water balance observed -0.945 m ha^{-1} in the profile belong to Fluventic Calciustepts in piedmont plain area indicating lead to ground water depletion every year and ultimately resulted in the ingress of seawater.

Introduction

The local effects of climate change on soil moisture, however, will vary not only with the degree of climate change but also with soil characteristics. Climate change also may affect soil characteristics, perhaps through changes in water logging or cracking, which in turn may affect hydraulic properties viz. soil moisture storage properties. Infiltration capacity and water-holding capacity of many soils are influenced by the frequency and intensity of freezing (Kumar, 2012). Knowledge of hydraulic properties of a soil is essential for scheduling agricultural operations and preparation of any management strategies in water conservation, irrigation scheduling, drainage and solute migration and management of irrigated agriculture. The hydrologic properties of soils are the basic information required in planning of an effective system of soil and water management particularly under semi-arid regions in respect to climate change. Information on the hydraulic properties of soil may help in formulating improved water management strategies and contingency crop planning for irrigated as well as unirrigated areas for improving the prospects of yield enhancement and sustainable agriculture under climate change.

Methods

The study area of different landforms of Hiran toposequence of Southern Saurashtra in Gujarat- lies between 20°10' to 21°05' N latitude and 70°25' to 70°35' E longitude. It belongs to semi-arid type of climate with a mean annual rainfall of 732 mm.

IRS 1A LISS II FCC imagery on 1:50,000 scale in conjunction with Survey of India topographical (SOI) map referred above on 1:50,000 scale were used to identify various land forms. The four typical pedons representing different landforms units in Hiran toposequence viz., Lithic Ustorthents (P_1) from hill slope, Lithic Ustorthents (P_2) from upper piedmont, Vertic Haplustepts (P_3) from lower piedmont and Fluventic Calciustepts (P_4) from piedmont plain were studied. Soil samples (horizon-wise) from each pedon were processes (< 2 mm) and were analysed for hydraulic characteristics by following usual procedures. Some hydraulic parameters were computed as per Table 1.

Result and Discussion

Hydrological Characteristics

Pedon 1: Lithic Ustorthents {Sasan (Gir)} from Hill Slope

A close look at Table 2 shows that the maximum water holding capacity varied from 35.88 to 50.18 per cent (weighted mean of 46.33 per cent). The saturated hydraulic conductivity decreased with increasing depth. This might be due to the dominant proportion of smectite clay at surface as compared to

Table 1: Computation of hydraulic parameters.

No.	*Hydraulic Parameters*	*Computation*	*Reference*
1.	PAWC (cm ha^{-1}/horizon)	AWC (m^3m^{-3}) x Depth of horizon (cm)	Sharma (2000)
2.	Total PAWC of Profile (m ha^{-1})	Sum of PAWC (cm ha^{-1}) of each horizon ÷ 100	Sharma (2000)
3.	MWHC v/v (%)	MWHC (%) x B.D. (Mg m^{-3})	DACS (2001)
4.	Moisture retention (cm ha^{-1})	MWHC v/v (%), 100 x Depth of horizon (cm)	DACS (2001)
5.	Total water storage capacity (WSC) of profile (m ha^{-1}).	Sum of moisture retention (cm ha^{-1}) of each horizons of profile ÷ 100	DACS (2001)
6.	Reserve water of the profile (m ha^{-1}).	Total WSC-(m ha^{-1})_ Total PAWC-(m ha^{-1})	—
7.	Water balance (Runoff/ percolation loss) of the profile (m ha^{-1}).	Annual rainfall-(m ha^{-1})_ Total WSC of profile (m ha^{-1})	—

sub-surface horizon and have much more micro pores causing poor water transmission properties. The infiltration rate was 123 cm hr^{-1} indicating moderately slow infiltration rate. The moisture retention at field capacity (0.03 MPa) and wilting point (1.5 MPa) varied from 0.37 to 0.46 m^3m^{-3} (weighted mean of 0.42 m^3m^{-3}) and 0.17 to 0.21 m^3m^{-3} (weighted mean of 0.19 m^3m^{-3}), respectively (Table 2). The variation was restricted to low tension (0.03 and 0.05 MPa) as the structure and pore geometry of the soils play an import role in retention, whereas at higher tension (0.10 to 1.5 MPa) there was a little variations possibly due to moisture retention on surface of the clays. Higher field capacity values associated with clay showed a corresponding increase in water retention at 1.5 MPa. These findings are in vicinity of those reported by Tiwary *et al.* (1989) and Parmasivam (1992). The amount of available watent content varied between 0.18 to 0.25 m^3m^{-3} (weighted mean of 0.23 m^3m^{-3}). The moisture held between the field capacity (0.033 MPa) and permanent wilting point (PWP) (1.5 MPa) may not be actually available to crops though it is considered available. Hence, plant available water capacity (PAWC) was calculated to draw an inference for water retention behaviour of the soil. The maximum water holding capacity (v/v), plant available water capacity and moisture retention varied from 47.36 to 64.23 per cent, 1.44 to 4.50 cm ha^{-1} and 3.79 to 11.56 cm ha^{-1}, respectively. Suitable management practices can be adopted to minimise the risk of crop failure in the rainfed agriculture with the knowledge of water storage capacity of soil in addition to rainfall characteristics. The total plant available water capacity (Total PAWC) of the profile was recorded 0.099 m ha^{-1} out of 0.255 m ha^{-1} of total water storage capacity of pedon P_1. The lower PAWC is due to shallow depth in hill slope area as well as loss of water due to runoff. The PAWC/WSC ratio was found 0.388 indicating the soils belong to Lithic Ustorthents have 39.0 per cent plant available water capacity.

Pedon 2: Lithic Ustorthents (Borvav) from Upper Piedmont

The data of Table 2 show that maximum water holding capacity ranged from 34.50 to 48.83 (weighted mean of 42.89) per cent. The saturated hydraulic conductivity varied from 0.08 to 0.18 cm hr^{-1} with a weighted mean of 0.14 cm hr^{-1} indicating the soils are slow in hydraulic conductivity. The rate of hydraulic conductivity decreased with increasing depth. The reduction in hydraulic conductivity with depth may be due to the soils on the surface is better aggregated (B.D. 1.35) as compared to the soils in the lower horizons (B.D. > 1.4). These findings are corroborated by those reported by Paramsivam (1992). The infiltration rate was 0.78 cm hr^{-1} indicating the moderately slow infiltration rate. This may be due to the dominant proportion of smectite clay and have much more micro pores causing poor water transmission properties. The moisture held at 0.03 MPa and 1.5 MPa varied between 0.41 to 0.44 m^{3}m^{-3} (weighted mean of 0.43 m^{3}m^{-3}) and 0.19 to 0.22 m^{3}m^{-3} (weighted mean of 0.21 m^{3}m^{-3}), respectively (Table-2). Most of the water (50 to 60 per cent) retained by the soil is depleted up to 0.50 MPa tension. The similar results were reported by Kaushal *et al.* (1986). Above 0.50 MPa tension the water released with each increment of tension was very slow and in narrow range at different depth. The low content of water was found at 1.5 MPa. A close look at Table-2 shows that available water content (AWC) and volumetric moisture percentage of MWHC (v/v) were found between 0.20 and 0.23 m^{3}m^{-3} (weighted mean of 0.22 m^{3}m^{-3}) and 48.30 and 65.92 per cent (weighted mean of 57.0 per cent), respectively. The plant available water capacity and moisture retention ranged form 2.30 to 4.40 and 6.59 to 11.83 cm ha^{-1}, respectively. The vertical distribution of plant available water capacity with depth showed a similar pattern as with the moisture retention. The total PAWC of the profile recorded 0.097 m ha^{-1} (38.0 per cent) out of 0.257 m ha^{-1} of total water storage capacity of pedon P_2 (Table-2). This low profile water storage capacity which might be due to the shallow depth. The PAWC/WSC ratio was recorded 0.377 indicating the soils of Lithic Ustortents have 38.00 per cent plant available water capacity (Sharma, 2000).

Pedon 3: Vertic Haplustepts (Maljinjava) from the Lower Piedmont

The maximum water holding capacity (MWHC) varied from 40.00 to 56.06 per cent (weighted mean of 47.89 per cent). The MWHC was found to be

Table 2: Hydraulic characteristic of the representative pedons of Hiran toposequence

Horizon	Depth (cm)	MWHC (%)	Sat. H.C	IR	Moisture retention ($m^3 m^{-3}$) at different tensions (MPa)						AWC (m^3m^{-3})	MWHC (v/v %)	PAWC	Moisture retention	Total* PAWC	Total WSC	PAWC/ WSC
			(cm hr^{-1})		0.03	0.05	0.10	0.50	1.0	1.5			(cm ha^{-1})		(m ha^{-1})		
P_1: Hill Slope, Lithic Ustorthents, Sasan (Gir), Tal. Talala, Dist. Junagadh, MSL : 162 m																	
A_1	0-17	47.18	0.38	1.23	0.40	0.24	0.22	0.19	0.18	0.17	0.23	59.45	3.91	10.11	0.099	0.255	0.388
A_{12}	17-35	50.18	0.13	-	0.46	0.37	0.31	0.25	0.23	0.21	0.25	64.23	4.50	11.56	-	-	-
AC	35-43	35.88	0.09	-	0.37	0.30	0.23	0.21	0.20	0.19	0.18	47.36	1.44	3.79	-	-	-
Weighted mean	-	46.33	0.22	-	0.42	0.31	0.26	0.22	0.20	0.19	0.23	59.20	-	-	-	-	-
P_2: Upper Piedmont, Lithic Ustorthents , Borvav (Gir), Tal. Talala, Dist. Junagadh, MSL : 125 m																	
Ap	0-20	46.21	0.18	0.78	0.44	0.31	0.28	0.26	0.24	0.22	0.22	59.14	4.40	11.83	0.097	0.257	0.377
A_{12}	20-30	48.83	0.13	-	0.42	0.34	0.27	0.23	0.21	0.19	0.23	65.92	2.30	6.59	-	-	-
AC	30-45	34.50	0.08	-	0.41	0.32	0.27	0.24	0.23	0.21	0.20	48.30	3.00	7.25	-	-	-
Weighted mean	-	42.89	0.14	-	0.43	0.32	0.27	0.25	0.23	0.21	0.22	57.03	-	-	-	-	-
P_3: Lower Piedmont, Vertic Haplustepts ,Maljinjava, Tal. Talala, Dist. Junagadh, MSL : 70 m																	
Ap	0-20	56.06	0.03	0.63	0.46	0.38	0.32	0.30	0.20	0.19	0.27	74.56	5.40	14.91	0.210	0.537	0.391
A_{12}	20-40	53.80	0.05	-	0.47	0.38	0.32	0.29	0.21	0.20	0.27	74.78	5.40	14.96	-	-	-
B_{21}	40-60	41.70	0.07	-	0.49	0.40	0.35	0.30	0.25	0.23	0.26	60.00	5.20	12.01	-	-	-
IIC_1	60-80	40.00	0.16	-	0.47	0.36	0.32	0.28	0.26	0.22	0.25	59.20	5.00	11.83	-	-	-
Weighted mean	-	47.89	0.08	-	0.47	0.38	0.33	0.29	0.23	0.21	0.26	61.15	-	-	-	-	-
P_4 : Piedmont Plain, Fluventic Calciustepts , Kajali, Tal. Veraval, Dist. Junagadh, MSL : 11 m																	
Ap	0-50	43.80	0.02	0.54	0.50	0.38	0.32	0.29	0.23	0.21	0.29	57.82	14.50	28.91	0.767	1.539	0.498
B_{21}	50-120	45.20	0.03	-	0.54	0.41	0.34	0.30	0.24	0.22	0.32	64.18	22.40	44.93	-	-	-
IIC_1	120-150	44.70	0.02	-	0.55	0.40	0.35	0.30	0.23	0.22	0.33	64.82	9.90	19.45	-	-	-
IIC_2	150-200	48.20	0.02	-	0.59	0.45	0.37	0.33	0.27	0.24	0.35	71.52	17.50	35.76	-	-	-
$IIIC_3$	200-240	40.10	0.02	-	0.54	0.45	0.37	0.33	0.25	0.23	0.31	62.16	12.40	24.86	-	-	-
Weighted mean	-	44.58	0.02	-	0.54	0.42	0.35	0.31	0.24	0.22	0.32	64.13	-	-	-	-	-

*Assuming the water loss through evaporation, transpiration, etc. considered negligible.

decrease with depth in the profile, which may be attributed to the decrease level of clay content (Sharma, 1995). The saturated hydraulic conductivity varies from 0.03 to 0.16 cm hr^{-1} (weighted mean of 0.08 cm hr^{-1}) indicating very low hydraulic conductivity. It might be due to the increase in bulk density (1.33 to 1.48 Mg m^{-3}) with increasing the soil depth. The infiltration rate was recorded 0.63 cm hr^{-1} indicating moderately slow infiltration rate. It might be due to the soils having high clay content (35 to 43.50 per cent). The moisture content at 0.03 and 1.5 MPa ranged from 0.46 to 0.49 m^3 m^{-3} (weighted mean of 0.47 m^3m^{-3}) and 0.19 to 0.23 m^3 m^{-3} (weighted mean of 0.21 m^3m^{-3}), respectively. The moisture retention was highest in B_{21} horizon up to 0.10 MPa tension might be due to high $CaCO_3$ content in this horizon. Similar results were also evidenced by Prasad *et al.* (1998). The results show that most of water retained by the soil is depleted up to 0.5 MPa tension. The available water capacity and MWHC (v/v) ranged from 0.25 to 0.27 m^3 m^{-3} (weighted mean of 0.26 m^3 m^{-3}) and 59.20 to 74.78 per cent (weighted mean of 61.15 per cent), respectively. The plant available water capacity and moisture retention ranged from 5.00 to 5.40 cm ha^{-1} and 11.83 to 14.96 cm ha^{-1}, respectively. The total PAWC of the profile was observed 0.210 m ha^{-1} out of 0.537 m ha^{-1} of total water storage capacity of the pedon P_3. The similar results were also reported by Paramsivam (1992). The fraction of PAWC/WSC was identified 0.391 indicating the soils of Vertic Haplustepts have 39.0 per cent plant available water capacity.

Pedon 4: Fluventic Calciustepts (Kajali) from the Piedmont Plain

The maximum water holding capacity lie from 40.10 to 48.20 per cent (weighted mean of 44.6 per cent). The saturated hydraulic conductivity varied from 0.02 to 0.03 cm ha^{-1} (weighted mean of 0.02 cm hr^{-1}) indicating very slow hydraulic conductivity. This might be due to the high exchangeable sodium which induces high dispersion of the clay in these soils and which leads to clogging of most of the micro pores, thus causing compaction in the subsurface horizon (Paramsivam, 1992). The infiltration rate recorded 0.54 cm hr^{-1} indicating that the soils of this pedon were moderately slow in infiltration rate. This might be due to the surface sealing of pores. The non-capillary pores may become capillary in size and smaller capillary pores might be sealed which may affect the rate of movement of water and decrease the infiltration (Ghildyal and Tripathi, 1987). The amount of water held at 0.03 and 1.5 MPa varied from 0.50 to 0.59 m^3m^{-3} (weighted mean of 0.54 m^3m^{-3}) and 0.21 to 0.24 m^3m^{-3} (weighted mean of 0.22 m^3m^{-3}), respectively. The available water capacity and MWHC (v/v) vary from 0.29 to 0.35 (weighted mean of 0.32) m^3 m^{-3} and 57.82 to 71.52 per cent, respectively. The PAWC and moisture retention ranged from 9.90 to 22.40 and 19.45 to 44.93 cm ha^{-1}, respectively. The total PAWC of the profile was 0.767 m ha^{-1} out of 1.539 m ha^{-1} of total water storage capacity of pedon indicating very high water storage capacity of the profile (Table-2). The very high water storage capacity of this pedon might be due to very deep soils depth (2 m), smectite type of clay and high clay coupled with higher exchangeable sodium. The results are also corroborated by the finding of Paramsivam (1992). The wide PAWC/WSC ratio of 0.498 indicating the soils belong to Fluventic Calciustepts has 50 per cent plant available water capacity.

In general, the soils belongs to Fluventic Calciustepts recorded the highest total profile water storage capacity (1.539 m ha^{-1}) followed by Vertic Haplustepts (0.537 m ha^{-1}) and Lithic Ustorthents {0.257 m ha^{-1} (P_2) and 0.255 m ha^{-1} (P_1) which have more soil depth in lower elevated topography. The lowest water retention observed in Lithic Ustorthents which are dominated by sandy materials in higher elevated topography.

Impact of climate change on Profile water storage capacity, Reserve water and water balance in coastal soils

The data on total water storage capacity, reserve water and water balance based on annual rainfall, total WSC and total PAWC of profiles in the different land slopes of coastal soils of Southern Saurashtra are presented in Table 3.

Table 3: Reserve water and water balance based on annual rainfall, total WSC and PAWC in the different land slopes of Hiran toposequence

Pedon	*Soil depth (m)*	*Average annual rainfall (m ha^{-1})*	*Total WSC (m ha^{-1})*	*Total PAWC (m ha^{-1})*	*Reserve water (m ha^{-1})*	*Water balance (m ha^{-1})*
P_1 : Hill Slope, Lithic Ustorthents,	0.43	0.870	0.255	0.099	0.156	+0.615
P_2 : Upper Piedmont, Lithic Ustorthents	0.45	0.870	0.257	0.097	0.160	+0.613
P_3 : Lower Piedmont, Vertic Haplustepts	0.80	0.594	0.537	0.210	0.505	+0.057
P_4 : Piedmont Plain, Fluventic Calciustepts	2.40	0.594	1.539	0.767	1.021	-0.945
Mean	**1.02**	**0.732**	**0.647**	**0.293**	**0.461**	**+0.085**

The average rainfall varied from 0.594 to 0.870 m ha^{-1} (mean of 0.732 m ha^{-1}). The total water storage capacity ranged from 0.255 to 1.539 m ha^{-1} (mean of 0.647 m ha^{-1}) which was increases with decrease of an elevation. The total PAWC varied from 0.097 to 0.767 m ha^{-1} with the mean of 0.293 m ha^{-1}. The reserve water and water balance ranged from 0.156 to 1.021 (mean of 0.461 m ha^{-1}) and 0.945 to +0.615 m ha^{-1} (mean of 0.085 m ha^{-1}). The negative water balance observed -0.945 m ha^{-1} in the soils belongs to Fluventic Calciustepts in piedmont plain area indicating lead to ground water depletion year after year resulted in the ingress of seawater in to the ground water. This is the main reason for the poor quality of saline water observed in the piedmont plain area. With the use of poor quality water, the soils also become saline and/or sodic due to secondary salinisation in this area. In addition to this, the salinity problem also alarming in depression area as well as the tidal water inundated and mixed with fresh resorvior water thus the salinity spread over the piedmont plain area due to climate change (Kumar, 2012). As a remedial measure, the use of soil amendments and drainage provision should be done accordingly. The salt tolerant crops and varieties like pearl millet, sorghum, cotton Var. dhumad and paddy Kalarada can be grown in coastal plain area.

Conclusion

It may conclude that the saturated hydraulic conductivity was recorded slowly to very slow. The infiltration rate was observed moderately slow. The PAWC/WSC ratio was between 0.377 and 0.498 of which P_4 belong to Fluventic Calciustepts in piedmont plain area has the highest i.e. 0.498 indicating the soils of this area have 50 per cent plant available water capacity. The reserve water and water balance varied from 0.156 to 1.021 m ha^{-1} and -0.945 to +0.615 m ha^{-1} with the mean of 0.461 and +0.085 m ha^{-1}, respectively. The negative water balance observed -0.945 m ha^{-1} in soils belong to Fluventic Calciustepts in piedmont plain area indicating lead to ground water depletion every year may be due to climate change and ultimately resulted in the ingress of seawater. The soils belongs to Fluventic Calciustepts recorded the highest total profile water storage capacity (1.539 m ha^{-1}) followed by Vertic Haplustepts (0.537m ha^{-1}) and Lithic Ustorthents {0.257 m ha^{-1} (P_2) and 0.255 m ha^{-1}(P_1). The lowest water retention observed in Lithic Ustorthents in higher elevated topography.

References

DACS, 2001. Annual AGRESCO Report. Deptt. of Agril. Chem. and Soil Sci., GAU, Junagadh.

Ghildyal, B.P. and Tripathi, R.P., 1987. Soil Physics, Wiley Easter Ltd., New Delhi.

Kaushal, G.S., Tembhase, A. and Sinha, S.B., 1986. J. Indian Soc. Soil Sci., Vol. 34, pp. 329-333.

Kumar, C.P., 2012. "Climate change and its impact on Groundwater Resources", International Journal of Engineering and Science ISSN: 2278-4721, Vol. 1(5), pp. 43-60.

Paramasivam, P., 1992. "Pedology and soil resources management of lower Bhavani Project Command area of Tamilnadu", Ph.D. Thesis, TNAU, Coimbatore.

Prasad, P., Subbaiah, R.K., Satyanarayana, V. and Shrinivas, R.C., 1998. J. Indian Soc. Soi Sci., Vol. 46, pp. 171.

Sharma, S.S., 2000. "Characterization and classification of salt affected soils for land use planning in sub-humid southern plains of Rajasthan", Ph.D. Thesis, R.A.U., Bikaner Campus, Udaipur.

Sharma, R.K., 1995. Characterization and classification of soils in a trensect of Haldighati region of Rajasthan. M.Sc. (Agri.) Thesis, RAU, Bikaner, Udaipur.

Tiwary, J.R., Jha, P.P and Mishra, B.B., 1989. J. Indian Soc. Soil Sci., Vol. 37, pp. 518-523.

30

Analysis of Climate Change Impact on Rainfall for Gujarat State

N. S. Vithlani, H. D. Rank, J. K. Christian and H. J. Vala

Abstract

Climate change is expected to affect agriculture, following food security and farmers' livelihood. Though several international level interventions have taken a step forward, there is a need to carry out disaggregated analysis at the regional level, particularly, within the state in order to fine-tune the hot spot areas that need immediate interventions. Keeping this in view, and the fact that there exists a dearth of systematic literature with reference to climate change in Gujarat, the present study aims to assess the impacts of climate change particularly with reference to rainfall parameters notwithstanding the relationship between climate change and the vulnerability of people living in different districts of Gujarat. The evidences have shown the impacts of climate change on rainfall and its pattern in the globe which was location specific. The historical records of rainfall at different station of Gujarat and trend was analysed using best fit trend line. The each time series data of monsoon rainfall and its 3, 5 and 10 years moving average were analysed. The slope of the best fit trend line was taken as slope of the change per year. Analysing the data statistically and the trend graphs showed that climate change impacts on rainfall was not found much in Gujarat state as out of 204 talukas. The rainfall was found increasing, decreasing and no trend in 16, 14 and 174 talukas. However, some indications of the decreasing trend in Saurashtra region, increasing trend in North Gujarat and no trend in Kutch, Central Gujarat and South Gujarat could be found.
Keywords: *Climate change, rainfall variability/deviations, adaptation.*

Introduction

Climate fluctuations and human activities have a profound impact on various elements of the hydrologic cycle (Ohmura and Wild, 2002; Barnett *et al.*, 2008; Cong *et al.*, 2009). Scientific evidence shows that climate change has begun to manifest itself, globally, in the form of increased downpours and storms, rising temperature and sea level, retreating glaciers, etc. Climate change refers to the variation in the earth's global climate or in regional climates over time. It describes changes in the variability or average state of the atmosphere over time scales ranging from decades to millions of years. Initially, these changes were looked upon as cyclical changes but evidences pointed to what seemed like a trend, which could not be explained away as normal cyclical climatic changes. Considering, the space-time variations of climatic parameters over a region, trend analysis for change detection is one of the reliable methods. Rain variability in space and time is one of the most relevant characteristics of Mediterranean climate that is associated with economic, social and ecological implications. Extreme rain events have significant environmental consequences that cause considerable damages in urban as well as in rural areas.

Description of study area

Gujarat is a state with semi-arid zones and encompasses the longest coastline. It is located between latitude 23°13'27.83"N to 21°31'34.74"N and longitude 68°35'03.51"E to 74°20'12.60"E. These characteristic features make the state relatively more vulnerable to climate change.The dynamic patterns in climate started being observed since the mid-20th century and Gujarat is no exception to this. Agricultural growth, food security, human health, biodiversity, and overall development could be in jeopardy if the issue is not addressed upon in a timely and systematic manner. Western India is expected to receive higher than normal rainfall as temperatures continue to rise. This change in the quantum of rainfall and its time of occurrence will adversely affect agriculture in the state. Coastal areas are likely to be the worst hit as the agricultural land is susceptible to inundation and salinity. Extreme weather conditions in the form of droughts floods and cyclones wipe out standing crops and the farmers are left impoverished and helpless.

This study is aimed to characterise the long-term movement or trend of rainfall pattern in the Gujarat state considering 19 districts.

Data Sources

Annual Rainfall were obtained from the Gujarat Water Resources Development Corporation (GWRDC), Gandhinagar, IMD, Ahmedabad, and Flood Control Cell, Gandhinagar.

The location details of the different rain gauge station of the Gujarat state are presented in appendix-B. The ARC/INFO 8.1 GIS software developed by ESRI, USA was used for locating the different rain gauge station in the map.

The map were prepared using the GIS.One of the difficulties encounter in the interpretation of environmental field data is the quantification of trends (e.g. calculation of slope) and demonstration that this estimation of trend is statistically different from zero. The focus here is on one non-parametric method used in determining the presence of slope and is known simply as Sen's Nonparametric Estimator of Slope.

Material and Methods

The non-parametric Mann Kendall (MK) test (Mann, 1945; Kendall, 1975) was applied to detect trends and identify change points of precipitation in the Gujarat state.

Mann-Kendall Analysis

The Mann-Kendall test is a non-parametric test for identifying trend sin time series data. The test compares the relative magnitude so sample data rather than the data values themselves (Gilbert,1987). One benefit of this test is that the data need not conform to any particular distribution. Moreover, data reported as non-detects can be included by assigning them a common value that is smaller than the smallest measured value in the data set.The procedure that will be described in the subsequent paragraphs assumes that here exist only one data value per time period. The data values are evaluated as an ordered time series. Each data value was compared to all subsequent data values. The initial value of the Mann-Kendall statistic, S, was assumed to be 0 (*e.g.*, no trend). If a data value from a later time period was higher than a data value from an early time period, S was incremented by 1. On the other hand if the data value from a later time period was lower than a data value, S was decremented by 1. The net result of all such increments and decrements gave the final value of S. Mann-Kendall statistic (S) was calculated by following equation,

$$S = \sum_{j=k}^{n-1} \sum_{j=k+1}^{n} \sin(x_j - x_k) \quad \text{...(1)}$$

Sin (x_j x_k) = 1, 0 and -1, if $x_j > x_k$, $x_j = x_k$ and $x_j < x_k$ respectively. *n* is the number of data points in time series.

A very high positive value of S was an indicator of an increasing trend and a very low negative value indicated a decreasing trend. However, it was necessary to compute the probability associated with S and the sample size, n, to statistically quantify the significance of the trend. The variance of S, i.e. VAR(S) for each data series was estimated by the following equation.

$$\text{VAR(S)} = \frac{1}{18}[(n(n-1)(2n+5) -$$

$$\sum_{p=1}^{m}(t_p(t_p-1)(2t_p+5)] \quad \text{...(2)}$$

Where n was the number of data points, m was the number of tied groups (a tied group was a set of sample data having the same value), and t_p was the number of data points in the p group. The normalised test statistic Z was computed as follows.

$$Z = \{[S-1]/[VAR(S)]^{1/2}\} \quad \text{if } S>0$$

$$=0 \quad \text{if } S=0$$

$$=\{[S+1]/[VAR(S)]^{1/2}\} \quad \text{if } S<0$$

If $Z_{cal} > 0$ and $Z_{cal} > Z_{tab,}$ where, $Z_{tab} = 3.090, 2.326, 1.645, 0.282$, the trend was considered as increasing and if $Z_{cal} < 0$ and $-Z_{cal} > Z_{tab}$ the trend was considered as decreasing at 0.1 per cent, 1 per cent, 0.5 and 10 per cent respectively.

Sen's slope method

One of the difficulties encounter in the interpretation of environmental field data is the quantification of trends (e.g. calculation of slope) and demonstration that this estimation of trend is statistically different from zero. The focus here is on one non-parametric method used in determining the presence of slope and is known simply as Sen's Nonparametric Estimator of Slope. Both the methodology for Sen's method is discussed.

Several tests are available for the detection and/or quantification of trends. The first step in analysing any data set, however, is to graph the data, usually as a function of space or location. A graphical representation of data facilitates observation of general trends and cycles which may assist in the selection of an appropriate statistical test.

Sen's method for the estimation of slope requires a time series of equally spaced data. Sen's method proceeds by calculating the slope as a change in measurement per change in time, as shown here in Equation (3). The true slope of an existing trend (as change per year) was determined using the Sen's nonparametric method. The Sen's method could be used in cases where the trend could be assumed to be linear.

$$f(t) = m\,t + c \qquad \ldots(3)$$

where m is the slope and c is a constant. The slope m_i within between two values of pair of all data value was estimated using the following equation (4).

$$m_i = (x_j - x_k)/(j-k) \qquad \ldots(4)$$

where,

$k = 1, 2, 3 \ldots (n-1)$.

$j = k+1 = 2, 3 \ldots n$.

$i = 1$ to N $[N = n(n-1)/2]$

The Sen's estimator of slope was estimated using the following expression (5)

$$m = m_{(N+1)/2}, \quad \text{if } N \text{ is odd}$$
$$m = \tfrac{1}{2}\left[m_{N/2} + m_{(N/2)+1}\right] \quad \text{if } N \text{ is even.} \qquad \ldots(5)$$

If multiple data measurements are collected at a given time, two options exist if multiple measurements are recorded for a given time step. The first option is to simply combine the measurements for a given time step into a single measurement of central tendency (e.g. mean, median) and proceed as above. The second option is to calculate a slope for each individual measurement. Note that the slope between measurements collected at the same time is not calculated.

Best fit trend line analysis

The each time series data of weather parameter and its 3, 5, and 10 years moving average were analysed for the best fit trend line. The slope of the best fit trend line was taken as slope of the change per year.

Results and Discussion

This chapter includes climatic variability's and climate change impacts on rainfall for the different talukas of the various districts of Gujarat State.

Rajkot district

The Mann-Kendall and Sen's slope statistics show that the rainfall is in no trend at 60.71 per cent confidence level. Also its three and five years moving average showed no trend. If the same trend of climate change will be continued, the expected increasing rainfall will be 0.13 mm/year. Out of 14 talukas, the rainfall is increasing in Jamkandorana, Lodhika and Wankaner while no trend in rest of the talukas.

Amreli district

The rainfall is in no trend at 86.78 per cent confidence level. However its three years moving average showed significant decreasing trend. The five years moving average showed no trend at 78.30 per cent confidence level. If the same trend of climate change will be continued the expected increasing rainfall will be 5.27 mm/year.

Surendranagar district

The analysis showed that the rainfall is in increasing trend at 91.37 per cent confidence level. Also its three and five years moving average showed significant increasing trend. If the same trend of climate change will be continued, the expected decreasing rainfall will be 35.78 mm/year. Out of 14 talukas, the rainfall is decreasing in Halvad and Surendranagar city, no trend in rest of the talukas.

Surat district

The rainfall is in no trend at 77.56 per cent confidence level. Also its three and five years moving average showed highly significant increasing trend. If the same trend of climate change will be continued the expected increasing rainfall will be 6.51 mm/year. The rainfall is increasing in Nizar and Mandavi, no trend in rest of the talukas.

Sabarkantha district

The results showed that the rainfall is in increasing trend at 87.56 per cent confidence level. Also its three and five years moving average showed highly significant increasing trend. If the same trend of climate change will be continued the expected increasing rainfall will be 2.98 mm/year. The rainfall is increasing in Vijaynagar, Himmatnagar and Sabarkantha district, no trend in rest of the talukas.

Vadodara district

The Mann Kendall and Sen's slope statistics show that the rainfall is in no trend at 67.53 per cent confidence level. Also its three years moving average showed no trend. If the same trend of climate change will be continued the expected increasing rainfall will be 0.79 mm/year. The rainfall is decreasing in Padara, no trend in rest of the talukas.

Valsad district

The Mann Kendall and Sen's slope statistics show that the rainfall is in no trend at 61.91 per cent confidence level. Also its three and five years moving average showed no trend. If the same trend of climate change will be continued the expected increasing rainfall will be 1.59 mm/year. The rainfall is decreasing in Gandevi, no trend in rest of the talukas.

Jamnagar district

The analysis showed that the rainfall is in no trend at 61.91 per cent confidence level. Also its three years moving average showed no trend. The five years moving average showed significant increasing trend at 94.59 per cent confidence level. If the same trend of climate change will be continued the expected decreasing rainfall will be -2.45 mm/year. The rainfall is decreasing in Jamnagar, no trend in rest of the talukas.

Dang district

The analysis showed that the rainfall is. in no trend at 52.11 per cent confidence level. However its three and five years moving average showed highly significant increasing trend. If the same trend of climate change will be continued the expected increasing in rainfall will be 11.19 mm/year. The rainfall is stable in Dang district.

Ahmadabad district

The Mann-Kendall and Sen's slope statistics show that the rainfall is in no trend at 89.7 per cent confidence level. Also its three and five years moving average showed no trend. If the same trend of climate change will be continued the expected increasing rainfall will be 1.26 mm/year. The rainfall is increasing in Dahegam and Sanand while no trend in rest of the talukas.

Gandhinagar district

The results showed that the rainfall is in no trend at 71.05 per cent confidence level. However its three and five years moving average showed highly significant increasing trend. If the same trend of climate change will be continued the expected increasing rainfall will be 4.19 mm/year.

Banaskantha district

The Mann-Kendall and Sen's slope statistics show that the rainfall is in highly significant increasing trend at 97.26 per cent confidence level. Also its three years moving average showed highly significant increasing trend. The five years moving average showed no trend at 99.53 per cent confidence level. If the same trend of climate change will be continued the expected increasing rainfall will be 9.87 mm/year. The rainfall is increasing Disa, Dhanera, Palanpur, Radhanpur, Vadgam, Vav, Santalpur and Banaskantha district, no trend in rest of the talukas.

Bhavnagar district

The Mann-Kendall and Sen's slope statistics show that the rainfall is in no trend at 71.94 per cent confidence level. Also its three and five years moving average showed no trend. If the same trend of climate change will be continued the expected increasing rainfall will be 33.09 mm/year. The rainfall is stable in all talukas.

Junagadh district

The Mann-Kendall and Sen's slope statistics show that the rainfall is in decreasing trend at 70.20 per cent confidence level. However its three and five years moving average showed no trend. If the same trend of climate change will be continued the expected decreasing rainfall will be 5.43 mm/year. The rainfall is decreasing in Keshod, Bhesan, Maliya, Mangrol, Ranavav and Junagadh district and no stable in rest of the talukas.

Kheda district

The Mann-Kendall and Sen's slope statistics show that the rainfall is in no trend at 85.56 per cent confidence level. However its three and five years moving average showed highly significant increasing trend. If the same trend of climate change will be continued the expected increasing rainfall will be 24.15 mm/year. The rainfall is stable in all talukas.

Kutch district

The Mann-Kendall and Sen's slope statistics show that the rainfall is in no trend at 64.76 per cent confidence level. However its three and five years moving average showed highly significant increasing trend. If the same trend of climate change will be continued the expected decreasing rainfall will be 1.18 mm/year. The rainfall is stable in all talukas.

Bharuch district

The Mann-Kendall and Sen's slope statistics show that the rainfall is in no trend at 67.53 per cent confidence level. Also its three and five years moving average showed no trend. If the same trend of climate change will be continued the expected increasing rainfall will be 6.70 mm/year. The rainfall is stable in all talukas.

Mehsana district

The Mann-Kendall and Sen's slope statistics show that the rainfall is in no trend at 77.56 per cent confidence level. However its three and five years moving average showed highly significant increasing trend. If the same trend of climate change will be continued the expected increasing rainfall will be 22.83 mm/year. The rainfall is increasing in Patan, no trend in rest of the talukas.

Panchmahal district

The Mann Kendall and Sen's slope statistics show that the rainfall is in no trend at 81.83 per cent confidence level. However its three and five years moving average showed highly significant increasing trend. If the same trend of climate change will be continued the expected increasing rainfall will be 2.02 mm/year. The rainfall is increasing in Limkheda and Santrampur, no trend in rest of the talukas.

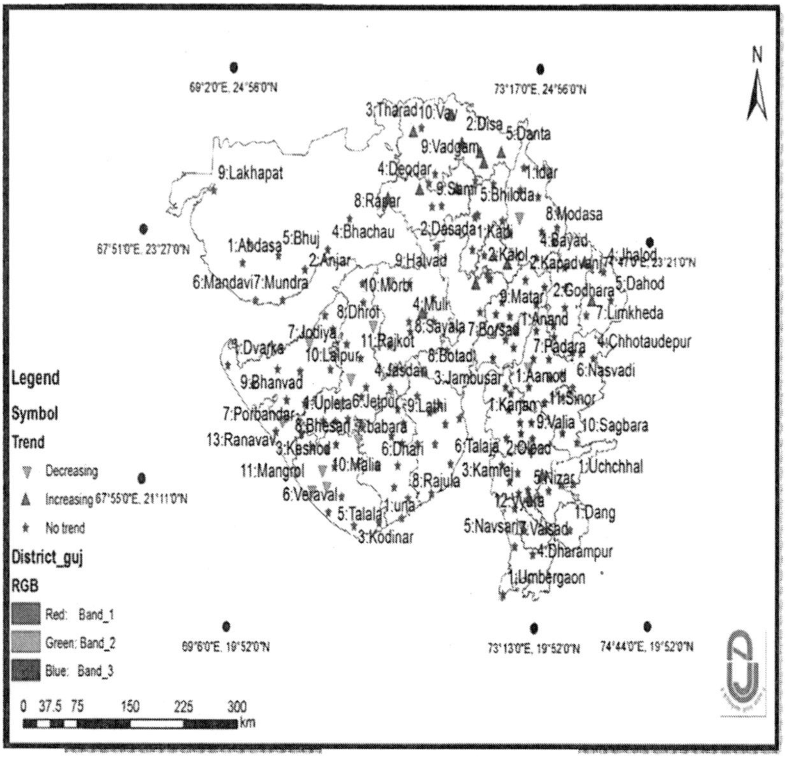

Fig. 1. Climate change impacts on rainfall in Gujarat state.

Adaptation options for improving rural livelihoods in Gujarat

International communities have tried to respond to these changing phenomena though establishing high level Intergovernmental Panel on Climate Change (IPCC). Adaptation is the response in natural or human systems to actual or expected climatic stimuli and their effects, which moderates harm or exploits beneficial opportunities. It is the ability of a system to adjust to climate change (including climate variability and extremes) to moderate potential damages, to take advantage of opportunities, or to cope with the consequences.

Since the agricultural sector was found to have the greatest bearing towards the overall vulnerability of different districts to climate change in, there is a need to shift focus towards investments in adaptation research capacity: particularly, in the development of climate-proof crops (drought resistant and heat tolerant varieties) as well as redeploying the existing improved crop varieties that can cope with wide range of climatic conditions.

The percentage area under forests must also be increased as less than 10 per cent of Gujarat's land is designated forest area, and much of that is seriously degraded. The villages of Gujarat continue to depend on firewood as a major source of energy. About 65 per cent of all energy in villages is generated through wood, a report by the State Forest Department on social forestry in Gujarat, has revealed. Thus, there is a call for more effective afforestation programmes and changes in policies *e.g.* resource conservation technologies and resource use efficiency, use of non-renewable energy sources and bio-degradable materials, pricing of resources and credit for transition to adaptation technologies. New strategies must be built around 'green' agricultural technologies, such as adaptive plant breeding, pest forecasting and rainwater harvesting and fertiliser microdosing.

Following the agricultural indicators, the occupational indicators were found to be the second largest contributors towards overall vulnerability in the selected districts of Gujarat.

The migration problem in some of the districts like Surendranagar was noticed to be high because of the water scarcity in the region. New industries hardly come to this district. In the whole of Gujarat, Surendranagar district gets lowest amount of new industrial investments. Thus, the policymakers must focus on generating better employment opportunities for the people in this region.

Apparently, the Northwest Arid Zone was found to be the most vulnerable zone among all the agro-climatic zones due to extreme deviations in the rainfall pattern. State-sponsored interventions in Kutch have not only failed to mitigate water scarcity but have exacerbated problems in some areas. The drought severity is the symptomatic manifestation of decades of dryland blindness. But strategies such as rainwater harvesting, livestock development and techniques to enhance dryland agriculture can help overcome many of these constraints.

Thus, for a state like Gujarat, what is required is a development strategy that integrates climate change policies with sustainable development strategies.

Conclusion

Gujarat state has the fastest growing economy in the country. It is rapidly expanding its production and consumption activities. Thus, the state not only contributes to climate change but is equally vulnerable to its impacts. The result showed that climate change impacts on rainfall was not found much in Gujarat state as out of 204 talukas, the rainfall was found increasing, decreasing and no trend in 16, 14 and 174 talukas. However, some cases of the decreasing trend in Saurashtra region, increasing trend in North Gujarat and no trend in Kutch, Central Gujarat and South Gujarat could be found. There is a pressing need to balance this development by simultaneously acting upon climate change and other issues which are putting tremendous pressure on the environment's carrying capacity. The evidences are overwhelming; it is only the time to act! What is most pleasing is that Gujarat has already begun to tackle this menace by opting for a more sustainble and holistic development for itself.

References

Amer, Z. S. and K. A. Emad. (2001). An early warning system for wheat production in low rainfall areas of Jordan. Journal of Arid Environments, **49**(3): 631-642.

Archer, D. R. and Fowler, H.J. (2008). Using meteorological data to forecast seasonal runoff on the river Jhelum, Pakistan. Journal of Hydrology, **361**(1-2):10-23.

Babu, P.G. (1998). Some issues in climate change decision-making. International Journal of Environment and Pollution, **9**(2/3): 179-185.

Basishtha, A., Goel, N. K., Arya, D. S. and Gangwar, S. K. (2007). Spatial patterns of trends in Indian sub-divisional rainfall. Jalvigyan Sameeksha, 22. 47-57.

31

Quality of Irrigation Water in Coastal Bhavnagar District of Gujarat

K. B. Polara and S. G. Rajput

Abstract

An investigation was undertaken to study the quality of irrigation waters by analysing 220 samples from eleven taluka of coastal Bhavnagar district. The results showed that the overall pH values of irrigation water ranged from 6.95 to 8.78 with mean value of 7.72 indicating waters of the region were alkaline in reaction. Whereas the overall EC values ranged from 0.38 to 18.2 with mean value of 2.77 dSm^{-1}. In water samples, among the cations, Na^+ (mean 20.93 me l^{-1}) was dominant and among the anions Cl^- (mean 20.5 me l^{-1}) was dominant. The waters are mostly Na-Ca-Mg type with dominance of chloride followed by bicarbonates and sulphate. On the basis of EC, 55.9 and 36.8 per cent samples were found in C_3 and C_4 classes of EC, respectively. Whereas on the basis of SAR, 49.5, 34.1, 14.1 and 2.3 per cent samples were found under S_1, S_2, S_3 and S_4 classes of SAR, respectively. Overall, on the basis of EC and SAR of irrigation water samples 41.4, 14.5, 20.0, and 14.4 per cent samples were categorised as C_3S_1, C_3S_2, C_4S_2, and C_4S_3 classes, respectively. Highly significant positive relations were observed between EC with SSP and SAR. The pH was highly significantly and positively correlated with RSC. The SSP was highly significantly and positively correlated with SAR.

***Key words:** Quality of irrigation water, EC, pH, SAR, RSC, SSP.*

Introduction

Quality of irrigation water is one of the main factor to be understood in irrigated agriculture. "Injudicious irrigations" even with good quality waters, turn many a good soils into saline or alkali conditions. In majority of arid, semi-arid and coastal regions, groundwater is either the major or the only source of irrigation to supplement the scanty rainfall. The majority of ground waters in these regions are of poor quality due to excessive salt concentration and high values of SAR and RSC. Scarcity of good quality of waters forces the farmers to use available saline or sodic water for irrigation (Girdhar, 1996). Such waters are being used in Bhavnagar district of Gujarat, knowing full well that the quality and yield of the crops are continuously decreased. Therefore, the present investigation was undertaken to study the quality of irrigation water in coastal Bhavnagar district of Gujarat.

Materials and Methods

Two hundred twenty underground well/tube well water samples were collected from the eleven *talukas* of coastal Bhavnagar district viz., Mahuva, Talaja, Palitana, Shihor, Umarala, Gariyadhar, Gadhada, Botad, Vallabhipur, Bhavnagar, and Ghogha. These water samples were analysed for the parameters viz., electrical conductivity, pH, calcium, magnesium, sodium, potassium, chloride, carbonate, bicarbonate and sulphate content as per the standard methods outlined by Richards (1954).The water qualities were determined as per USSL classification.

Results and Discussion

The data presented in Table 1 showed that the overall pH values of irrigation water ranged from 6.95 to 8.78 with mean value of 7.72 indicating waters of the district were alkaline in reaction. The highest mean pH value of 8.07 was recorded in Vallabhipur taluka,whereas the lowest mean pH value of 7.44 was recorded in Talaja *taluka*. The overall EC values ranged from 0.38 to 18.2 with mean value of 2.77 dSm^{-1} (Table-1).The highest (6.59 dSm^{-1}) and lowest (1.12 dSm^{-1}) EC mean value were recorded with Bhavnagar and Gariadhar *taluka*, respectively. The higher mean value of EC (6.59 dSm^{-1}) is an indicative of potential development of saline soils in Bhavnagar district. The variation in pH and EC might be due to hydrological influence of seawater. Well and tube well waters were mostly alkaline in reaction and saline in nature. This finding is in conformity with the findings of Verma *et al.* (2003) and Polara and Kabaria (2005).

Among the cations, overall higher proportion of Na^+ (20.93 me l^{-1}) was observed, which was followed by Ca^{++} (4.45 me l^{-1}), Mg^{++} (3.64 me l^{-1}) and K^+ (0.11 me l^{-1}). The presence of large proportion of Na^+ in most of the area under investigation is indicative of a potential danger for the alkalinity hazards (Table-2). The overall concentration of Ca^{++}, Mg^{++}, Na^+ and K^+ varied from 0.83 to 31.10, 0.39 to 26.20, 1.46 to 139.5 and 0.00 to 1.37 me l^{-1}, respectively. The highest mean value for Na^+ (48.1 me l^{-1}) was found in Bhavnagar *taluka* followed by Vallabhipur (39.8 me l^{-1}) and Shihor (23.5 me l^{-1}) and the lowest mean value for Na^+ (7.59 me l^{-1}) was reported in Gariyadhar *taluka* (Table 2). In case of anions, the highest overall mean value of 20.47 me l^{-1} was recorded for Cl^- and it was followed by HCO_3^- (6.16 me l^{-1}), SO_4–(1.34 me l^{-1}) and CO_3–(0.27 me l^{-1}). The overall range values for CO_3–, HCO_3^-, Cl^- and SO_4– were 0.0 to 4.10, 1.90 to 38.40, 1.90 to 135.44 and 0.00 to 18.23 me l^{-1}, respectively (Table 2). The highest mean value of Cl^- (51.63 me l^{-1}) and SO_4– (3.95 me l^{-1}) was observed in Bhavnagar and Vallabhipur talukas, respectively, whereas CO_3–(0.74 me l^{-1}) and HCO_3^- (11.32 me l^{-1}) in Vallabhipur taluka. The higher sodium and chloride concentration of ground water is a definite proof of seawater intrusion. This is in close agreement with finding of Maliwal and Timbadia (2000) and Dhanushkodi and Kannathasan (2009).

The SAR value is an important for studying the alkali hazards in irrigation water. The overall mean value of SAR was 9.80 and it was varied from 1.32 to 26.3. The lowest (1.32) and highest (26.3) SAR values were reported in samples collected from Palitana and Botad talukas, respectively (Table 1). The highest (15.3) and lowest (5.22) mean SAR value were registered in water samples collected from Bhavnagar and Gariadhar talukas, respectively. All talukas, except Mahuva, Shihor, Gadhada, Vallabhipur and Bhavnagar talukas had mean SAR value less than 10. This finding is in conformity with those of Nilsood *et al.,* (1998) and Maliwal and Timbadia (2000).

The RSC values ranged from 0.0 to 4.17 with mean value of 0.39 me l^{-1}. The highest mean RSC value (0.75 me l^{-1}) was recorded in Vallabhipur taluka, whereas the lowest (0.18 mel^{-1}) in Talaja talukas (Table 1). For all talukas, the mean RSC values were lower than permissible limit of 2.50 mel^{-1}. Overall 87.7, 5.5 and 6.8 per cent samples fall under safe, marginal and unsafe classes of RSC, respectively. Most of (87.7 per cent) well/tube well water samples had RSC value less than 1.25 and only 6.8 per cent samples had RSC value greater than 2.5 mel^{-1} The overall mean value of SSP was 68.7, which varied from 30.3 to 90.5 (Table 1). The highest mean value (72.7) was recorded in Vallabhipur and Bhavnagar talukas whereas, the lowest mean value (60.7) was recorded in Gariyadhar taluka. For all the talukas, mean SSP values were greater than 60. Overall, 25.5 and 74.5 per cent samples fall under safe and unsafe classes of SSP, respectively.

The percentage distribution of water samples into different quality indices viz., EC and SAR showed that overall 0, 7.3, 55.9 and 36.8 per cent samples were found under C_1, C_2, C_3 and C_4 classes of EC, respectively (Table 3). More than half (55.9 per cent) of water samples had EC value in range of 0.75 to 2.25 dSm^{-1}(C_3 class). In Botad taluka, 80 per cent water samples fall under C_3 class of EC. The Bhavnagar, Vallabhipur and Shihor taluka registered more than 60 per cent samples under C_4 class. So, the salinity hazard of irrigation water is the cause of the development of secondary salinisation in the soils of Bhavnagar district.

Table1: Taluka-wise Range and Mean Values of EC, PH and Different Indices of Irrigation Water Samples of Coastal Bhavnagar District

Name of Taluka	*EC (dS m^{-1})*	*pH*	*RSC(me l^{-1})*	*SSP*	*SAR*
Mahuva	0.81-7.19 (2.35)	7.13-8.69 (7.65)	0.00-2.64 (0.34)	34.6-87.7 (68.3)	1.73-18.6 (10.2)
Talaja	0.74-5.91 (1.90)	7.13-8.51 (7.44)	0.00-3.66 (0.18)	39.4-83.4 (67.4)	2.07-21.5 (8.39)
Palitana	0.59-5.84 (1.74)	7.19-8.51 (7.67)	0.00-2.18 (0.21)	30.3-90.5 (65.2)	1.32-19.5 (7.50)
Gariyadhar	0.63-2.37 (1.12)	7.01-8.51 (7.52)	0-1.920.0 (0.26)	36.1-83.3 (60.7)	1.83-10.2 (5.22)
Shihor	0.95-8.53 (3.01)	7.38-8.74 (7.95)	0.00-4.04 (0.65)	52.1-83.2 (70.2)	4.26-19.7 (11.0)
Umarala	0.72-6.52 (2.42)	6.97-8.45 (7.60)	0.00-2.65 (0.36)	40.7-79.7 (67.3)	2.03-18.6 (9.02)
Gadhada	0.38-4.64 (2.39)	6.95-8.33 (7.27)	0.00-1.10 (0.07)	44.6-81.9 (70.8)	1.58-18.3 (10.1)
Botad	0.85-8.51 (2.10)	7.36-8.45 (7.75)	0.00-2.76 (0.43)	49.1-83.6 (70.8)	3.84-26.3 (9.53)
Vallabhipur	0.45-18.2 (5.21)	7.03-8.78 (8.07)	0.00-4.17 (0.75)	56.6-87.6 (72.7)	2.82-26.1 (13.6)
Bhavnagar	1.36-16.2 (6.59)	7.12-8.71 (7.94)	0.00-3.04 (0.52)	60.2-79.9 (72.6)	6.41-26.1 (15.3)
Ghogha	0.68-3.78 (1.65)	7.36-8.59 (8.01)	0.00-3.03 (0.52)	55.8-85.6 (70.4)	3.17-18.1 (7.93)
Overall	0.38-18.20 (2.77)	6.95-8.78 (7.72)	0.00-4.17 (0.39)	30.3-90.5 (68.7)	1.32-26.3 (9.80)

The overall 49.6, 34.1, 14.1 and 2.3 per cent samples were found under S_1, S_2, S_3 and S_4 classes of SAR, respectively (Table 3). Based on EC and SAR of irrigation water samples, 41.4, 20.0, 14.5 and 14.1 per cent were found under C_3S_1, C_4S_2, C_3S_2, and C_4S_3 , respectively. Similar finding were also reported by Polara and Kabaria (2005) and Maliwal and Timbadia (2000) for Amreli district of Gujarat and Verma *et al.* (2003) for Churu district of Rajasthan.

The correlation among the different properties of irrigation waters (EC_{iw}, pH_{iw}, RSC_{iw}, SSP_{iw} and SAR_{iw}) were worked out.The results showed that the highly significant positive relations were observed between EC_{iw} with SSP_{iw} (0.2874**) and SAR_{iw} (0.8089**). The pH_{iw} was highly significantly and positively correlated with RSC_{iw} (0.6818**) and poor correlation with SSP (0.1739*). The RSC_{iw} was highly significantly and positively correlated with SSP_{iw} (0.2738**) and SSP with SAR_{iw} (0.7252**). The almost similar relationship was also obtained by Paliwal and Maliwal (1971).

Table 2: Taluka-wise Range and Mean Values of Cations and Anions of Well/ Tube Well Water Samples of Coastal Bhavnagar District

Name of Taluka	Cations in me l^{-1}				Anions in me l^{-1}			
	Ca^{++}	Mg^{++}	Na^{+}	K^{+}	CO_3^{-}	HCO_3^{-}	Cl^{-}	SO_4^{-}
Mahuva	2.02-14.52 (4.20)	0.97-10.52 (3.04)	2.82-65.63 (19.07)	0.02-0.97 (0.21)	0.00-1.20 (0.14)	2.10-15.24 (5.98)	6.07-70.03 (18.45)	0.00-4.25 (0.88)
Talaja	0.97-11.83 (3.10)	1.35-8.58 (3.11)	3.28-46.26 (14.01)	0.02-0.59 (0.11)	0.00-1.80 (0.09)	2.10-9.35 (4.60)	4.59-48.91 (13.82)	0.00-5.58 (0.90)
Palitana	1.04-6.03 (2.89)	0.47-5.36 (2.55)	1.96-46.53 (13.03)	0.02-0.20 (0.08)	0.00-1.20 (0.09)	2.08-9.60 (4.87)	2.64-46.36 (12.06)	0.00-5.04 (0.73)
Gariyadhar	1.06-4.74 (2.46)	0.70-4.72 (1.99)	2.41-17.70 (7.59)	0.01-0.14 (0.05)	0.00-1.19 (0.14)	1.90-7.20 (3.98)	2.11-17.42 (7.02)	0.00-2.76 (0.14)
Shihor	2.03-12.34 (4.87)	1.54-9.65 (3.77)	6.16-65.40 (23.56)	0.01-0.11 (0.05)	0.00-4.10 (0.57)	3.51-19.04 (7.76)	5.28-55.19 (21.90)	0.00-4.89 (1.20)
Umarala	1.55-8.32 (3.83)	1.25-6.90 (3.22)	3.00-51.32 (17.78)	0.02-0.42 (0.09)	0.00-0.80 (0.09)	3.30-11.32 (5.87)	3.80-50.06 (18.06)	0.00-6.57 (0.91)
Gadhada	1.23-6.93 (3.42)	0.49-5.88 (3.22)	1.46-36.72 (18.61)	0.02-0.46 (0.10)	0.00-0.50 (0.03)	1.90-6.30 (4.44)	1.90-37.39 (19.43)	0.00-6.87 (1.10)
Botad	1.40-8.32 (3.30)	0.83-7.56 (2.40)	5.75-74.00 (16.82)	0.00-0.34 (0.07)	0.00-1.10 (0.22)	2.25-14.60 (5.08)	5.21-64.78 (15.10)	0.00-6.48 (1.01)
Vallabhipur	0.83-31.10 (8.59)	1.06-26.20 (6.96)	2.94-139.5 (39.82)	0.02-1.37 (0.15)	0.00-3.18 (0.74)	2.00-38.40 (11.32)	2.53-135.4 (37.23)	0.00-18.23 (3.95)
Bhavnagar	2.01-29.35 (9.90)	1.26-22.65 (7.87)	10.57-116.2 (48.06)	0.02-1.09 (0.28)	0.00-3.10 (0.45)	3.30-25.48 (9.72)	7.60-125.1 (51.63)	0.00-10.38 (3.71)
Ghogha	1.03-4.02 (2.41)	0.39-4.53 (1.94)	3.93-32.73 (11.88)	0.01-0.24 (0.06)	0.00-1.90 (0.42)	2.09-7.01 (4.12)	3.66-29.6 (10.43)	0.00-2.96 (0.22)
Overall	0.83-31.10 (4.45)	0.39-26.20 (3.64)	1.46-139.5 (20.93)	0.00-1.37 (0.11)	0.00-4.10 (0.27)	1.90-38.40 (6.16)	1.90-135.4 (20.47)	0.00-18.23 (1.34)

Table 3: Percentage Distribution of Well/Tube Well Water Samples into Different EC and SAR Classes

Name of Taluka	EC (dSm^{-1})				SAR Classes				Major irrigation water quality class											
									C_2				C_3				C_4			
	C_1	C_2	C_3	C_4	S_1	S_2	S_3	S_4	S_1	S_2	S_3	S_4	S_1	S_2	S_3	S_4	S_1	S_2	S_3	S_4
Mahuva	0	0	55	45	50	25	25	0	0	0	0	0	50	5	0	0	0	20	25	0
Talaja	0	5	65	30	65	20	15	0	5	0	0	0	60	5	0	0	0	15	15	0
Palitana	0	10	75	15	65	30	5	0	10	0	0	0	55	20	0	0	0	10	5	0
Gariyadhar	0	35	65	0	85	15	0	0	35	0	0	0	50	15	0	0	0	0	0	0
Shihor	0	0	35	65	30	50	20	0	0	0	0	0	30	5	0	0	0	45	20	0
Umarala	0	10	50	40	50	40	10	0	5	5	0	0	45	5	0	0	0	30	10	0
Gadhada	0	5	50	45	45	50	5	0	5	0	0	0	35	15	0	0	5	35	5	0
Botad	0	0	80	20	50	35	10	5	0	0	0	0	45	35	0	0	0	5	10	5
Vallabhipur	0	5	35	60	20	45	30	5	5	0	0	0	15	20	0	0	0	25	30	5
Bhavnagar	0	0	35	65	20	35	30	15	0	0	0	0	20	15	0	0	0	25	30	10
Ghogha	0	10	70	20	65	30	5	0	10	0	0	0	50	20	0	0	5	5	5	0
Overall	0 (0)	7.3 (16)	55.9 (123)	36.8 (81)	49.5 (109)	34.1 (75)	14.1 (31)	2.3 (5)	6.5 (15)	0.5 (5)	0 (0)	0 (0)	41.4 (91)	14.5 (32)	0 (0)	0 (0)	1.0 (2)	20.0 (44)	14.1 (31)	1.8 (4)

Conclusion

From the above results it can be concluded that the well/tube well water of Bhavnagar district are alkaline in reaction and saline in nature and poor in quality. Almost half of the waters samples of well/tube well from cultivator's fields were saline (EC 0.75 dSm^{-1} and above).This is indicative of the potential development of saline soils in the Bhavnagar district.

References

Dhanushkodi,V. and M. Kannathasan (2009). Quality of irrigation water and their effect on soil properties in coastal region of Ramanathapuram district in Tamil Nadu. Journal of the Indian Society of Coastal Agricultural Research, 27(1): 26-29.

Girdhar, I.K. (1991). Effect of sea water irrigation on the establishment of mesquite (*Prosopis juliflora*) in coastal saline soils of Gujarat. Journal of the Indian Society of Coastal Agricultural Research, 9(1-2): 373-379.

Maliwal, G.L. and N.K.Timbadia (2000). Quality of underground irrigation water and their effect on soil properties of coastal area of Amreli district of Gujarat. Journal of the Indian Society of Coastal Agricultural Research,18:30-35

Paliwal, K.V. and G. L. Maliwal (1971). Some relationships between constituents of irrigation water and properties of irrigated soils of western Rajasthan. Journal of the Indian Society of Soil Science, 19: 299-303.

Polara, J. V. and B. D. Kabaria (2005). Quality of underground irrigation water in Coastal Amreli district of Gujarat. Paper presented in 8th National Seminar on Strategies for Improving Farming and Ecology Security of Coastal Region during 21-24 December, 2005.

Richards, L.A. (1954). *Diagnosis and Improvement of Saline and Alkali Soils*. Hand Book No. 60, Oxford and IBH pub. Co., Calcutta-16.

Verma, B.L.; Sharma, Y. and Sighania, R.A. (2003). Quality of under-ground irrigation waters of Churu district in Rajasthan. Journal of the Indian Society of Soil Science, 51(2): 214-216.

32

Influence of Climate Change on Different Land Forms of Coastal Soils in Meghal Irrigation Command of Gujarat

H. P. Patel, S. G. Savalia and S.T. Shirgire

Abstract

An attempt has been made to quantify the hydraulic properties and water balance impacts on climate change in the coastal soils of Meghal Irrigation Command area of Southern Saurashtra region of Gujarat. The weighted mean of saturated hydraulic conductivity and available water content varied from 0.04 to 0.21 cm hr^{-1} and 0.21 to 0.29 m^3 m^{-3}, respectively. The total water storage capacity of the profiles varied from 0.241 to 0.647 m ha^{-1}. The soils belongs to Fluventic Haplustepts of Coastal plain area recorded the highest total profile water storage capacity (0.351 m ha^{-1}) followed by Typic Ustothrents (0.096 m ha^{-1}) in Undulating Upper Pediment. The PAWC / WSC ratio varied from 0.398 to 0.542 of which P5 belong to Fluventic Haplustepts in coastal plain area has the highest ratio i.e. 0.542 indicating the soils of this area have 50 per cent plant available water capacity. The reserve water and water balance varied from 0.145 to 0.377 m ha^{-1} (mean of 0.249 m ha^{-1}) and -0.053 to + 0.629 m ha^{-1} (+ 0.350 m ha^{-1}), respectively. The total profile water storage capacity increases with decrease of an elevation. The negative water balance observed -0.053 m ha^{-1} in the profile belong to Fluventic Haplustepts in Coastal plain area indicating lead to groundwater depletion every year and ultimately resulted in the ingress of seawater.

Key words: *Profile water storage, Hydraulic characteristics, and Calcareous basaltic soils.*

Introduction

The soil moisture and water profile plays an emerging role on plant growth which directly affects the productivity of crop. The catchment is the area which catches rainfall and drains it's a single outlet. The structure would be provided at the outlet and falling it through the command area. The local effects of climate change on soil moisture, however, will vary not only with the degree of climate change but also with soil characteristics. Climate change also may affect soil characteristics, perhaps through changes in water logging or cracking, which in turn may affect hydraulic properties *viz.* soil moisture storage properties. Infiltration capacity and water-holding capacity of many soils are influenced by the frequency and intensity of freezing (Kumar, 2012). Knowledge of hydraulic properties of a soil is essential for scheduling agricultural operations and preparation of any management strategies in water conservation, irrigation scheduling, drainage and solute migration and management of irrigated agriculture. The hydrologic properties of soils are the basic information required in planning of an effective system of soil and water management particularly under semi-arid regions. Information on the hydraulic properties of soil may help in formulating improved water management strategies and contingency crop planning for irrigated as well as unirrigated areas for improving the prospects of yield enhancement and sustainable agriculture under climate change.

Material and Methods

The study area (Meghal Irrigation Command area) lies between 21°14' to 21°01' N latitude and 70°24' to 71°16' E longitude encompassing parts of Maliya and Veraval taluka of Junagadh district of Southern Saurashtra.

During the traverse, IRS 1A LISS II FCC imagery on 1:50,000 Scale in conjunction with Survey of India topographical (SOI) map having 1:50,000 scale were used to identify various landforms units of the soils of Meghal Irrigation Command area of Southern Saurashtra region of Gujarat namely, undulating upper pediment (LS-1), upper pediment (LS-2), lower pediment (LS-3), alluvial plain (LS-4) and coastal plain (LS-5) in Southern Saurashtra. The five representative pedons in the land slopes of the study area were selected according to heterogeneity and change in morphological characteristics.

Result and Discussion

Hydrological Characteristics

Pedon 1 (Dedekiyal) from the Undulating Upper Pediment: A close look at Table 1 shows that the maximum water holding capacity varied from 34.96 to 45.49 per cent with maximum of 45.49 per cent in A1 horizon due to high clay content (29.55 per cent) in this horizon (Table 1). The saturated hydraulic conductivity of the pedon ranged from 0.15 to 0.30 cm hr^{-1} decreased with increasing depth indicating the slow rate of water movement. This might be due to the dominant proportion of smectite clay at surface as compared to sub surface horizon and have much more micro pores causing poor water transmission properties (Savalia, 2005).

Computation of water storage capacity and water balance of the profile:

S. No.	*Hydraulic Parameters*	*Computation*	*Reference*
1.	PAWC (cm ha^{-1}/horizon)	AWC (m^3m^{-3}) x Depth of horizon (cm)	Sharma (2000) and Savalia (2005)
2.	Total PAWC of Profile (m ha^{-1})	Sum of PAWC (cm ha^{-1}) of each horizon ÷ 100	Sharma (2000) and Savalia (2005)
3.	MWHC v/v (%)	MWHC (%) x B.D. (Mg m^{-3})	DACS (2001) and Savalia (2005)
4.	Moisture retention (cm ha^{-1})	MWHC v/v (%) ÷ 100 x Depth of horizon (cm)	DACS (2001) and Savalia (2005)
5.	Total water storage profile capacity (m ha^{-1}).	Sum of moisture retention (cm ha^{-1}) of each horizons of profile ÷ 100	DACS (2001) and Savalia (2005)
6.	Reserve water of	Total WSC-(m ha^{-1}) - the profile (m ha^{-1}). Total PAWC- (m ha^{-1})	Savalia (2005)
7.	Water balance (Runoff/ percolation loss) of the profile (m ha^{-1}).	Annual rainfall - (m ha^{-1})_Total WSC of profile (m ha^{-1})	Savalia (2005)

The moisture retention at field capacity (0.03 MPa) and wilting point (1.5 MPa) varied from 0.42 to 0.45 and 0.21 to 0.23 m^3 m^{-3}, respectively (Table 1). Higher field capacity values associated with clay showed a corresponding increase in water retention at 1.5 MPa. These findings are in vicinity of those reported by Parmasivam (1992) and Savalia, (2005). The amount of available water content varied between 0.20 to 0.23 m^3 m^{-3}.The maximum water holding capacity (v/v), plant available water capacity and moisture retention varied from 48.59 to 60.05 per cent, 2.00 to 4.40 cm ha^{-1} and 5.46 to 10.20 cm ha^{-1}, respectively. The MWHC (v/v), PAWC and moisture retention values were recorded higher in subsurface as compared to surface horizon (Savalia, 2005). The total plant available water capacity (Total PAWC) of the profile was recorded 0.096 m ha^{-1} (40 per cent) out of 0.241 m ha^{-1} of total water storage capacity of pedon P1 (Table 1) indicating very little plant available water capacity and low water storage capacity. This is due to shallow depth in undulating upper pediment area as well as loss of water due to runoff. The PAWC/WSC ratio was found 0.398 indicating 40 per cent plant available water capacity of the soils of undulating upper pediment of Meghal Irrigation Command area. This fact also corroborated by the findings of Savalia (2005).

Podon 2 (Gadhali) from the Upper Pediment: The data of Table 1 showed that maximum water holding capacity ranged from 45.68 to 50.42 per cent. The saturated hydraulic conductivity varied from 0.09 to 0.14 cm hr^{-1} with a weighted mean of 0.11 cm hr^{-1} indicating the soils are very slow in hydraulic conductivity. The hydraulic conductivity increased with increasing depth in the solumn of profile. The reduction in hydraulic conductivity with depth irrespective clay content may be attributed to the soils on the surface are better aggregated (B.D. 1.29) as compared to the soils in the lower horizons (B.D. > 1.37). These findings are corroborated by those reported by Paramsivam (1992) and Savalia (2005).

The moisture held at 0.03 MPa and 1.5 MPa varied between 0.46 to 0.49 and 0.24 to 0.26 $m^3 m^{-3}$, respectively (Table 1). The moisture retention characteristic curve of pedon P2 (weighted mean) are 0.47 to 0.25 $m^3 m^{-3}$, respectively. It appears that most of the water (50 to 60 per cent) retained by the soil is depleted upto 0.50 MPa tension. The similar results were reported by Kaushal *et al.* (1986) and Savalia (2005). Above 0.50 MPa tension the water released with each increment of tension was very slow and in narrow range at different depth. The low content of water was found at 1.5 MPa.

A close look at Table 1 shows that available water content (AWC) and volumetric moisture percentage of MWHC (v/v) were ranged from 0.22 to 0.23 $m^3 m^{-3}$ and 58.03 to 65.55 per cent, respectively. The available water content and MWHC (v/v) observed higher in subsurface horizon as compared to surface horizon. The plant available water capacity and moisture retention ranged form 1.60 to 3.30 and 4.30 to 9.04 cm ha^{-1}, respectively. The total PAWC of the profile recorded 0.101 m ha^{-1} (37.0 per cent) out of 0.274 m ha^{-1} of total water storage capacity of pedon P2 (Table 1) indicating very low PAWC, which might be due to the shallow depth of soil (45 cm) in upper pediment area. These findings are parallel to those of Singh (1999) and Walia *et al.* (1999) who reported that water storage capacity of soil influenced by soil depth. The PAWC/WSC ratio was recorded 0.367 indicating that 37 per cent plant available water capacity out of total water storage capacity of the soils of upper pediment area of Meghal Irrigation Command area.

Pedon 3 (Ghumti) from the Lower Pediment: An appraisal of data on hydrological characteristics are depicted in Table 1 unveils that the maximum water holding capacity (MWHC) ranged between 37.27 to 45.52 per cent which was maximum at the surface (45.52 per cent) in Ap horizon. The saturated hydraulic conductivity gradually increased from 0.05 cm hr^{-1} in the surface horizon to 0.20 cm hr^{-1} in the lower most horizon with the weighted mean of 0.12 cm hr^{-1} (Table 1) indicating very low hydraulic conductivity. It might be due to the increase in bulk density (1.31 to 1.48 Mg m^{-3}) with increasing the soil depth. Similar result was also reported by Savalia (2005). The moisture content at 0.03 and 1.5 MPa ranged from 0.41 to 0.48 and 0.18 to 0.23 $m^3 m^{-3}$, respectively (Table 1). The available water capacity and MWHC (v/v) ranged from 0.20 to 0.26 $m^3 m^{-3}$ and 51.06 to 59.74 per cent, respectively (Table 1). The plant available water capacity and moisture retention ranged from 3.20 to 5.80 and 8.60 to 14.27 cm ha^{-1}, respectively. The vertical distribution of plant available water capacity with depth showed a similar trend as with the available water content, MWHC (v/v) and moisture retention. The total PAWC of the profile was observed 0.226 m ha-1 out of 0.553 m ha^{-1} (41 per cent) of total water storage capacity of the pedon P3 (Table 1). Savalia (2005) reported 0.210 m ha-1 out of 0.537 m ha-1 (39 per cent) of total water storage capacity of the soils of lower Pediment area of Southern Saurashtra. The fraction of PAWC/WSC was identified 0.408 indicating that 41 per cent PAWC out of total water storage capacity of the studied soils (Table 1).

Pedon 4 (Sukhpur) from the Alluvial plain: A perusal of data on hydrological characteristics are given in Table 1 stipulate that the maximum water holding capacity lie from 26.16 to 37.00 per cent. The saturated hydraulic conductivity of this pedon varied from 0.01 to 0.08 cm hr^{-1} indicating very slow hydraulic conductivity. This might be due to the high exchangable sodium which induces high dispersion of the clay in these soils and leads to clogging of most of the micropores, thus causing compaction in the subsurface horizon (Paramsivam, 1992 and Savalia, 2005).

The amount of water held at 0.03 and 1.5 MPa varied from 0.39 to 0.46 and 0.15 to 0.22 $m^3 m^{-3}$, respectively. The available water content ranged from 0.21 to 0.27 $m^3 m^{-3}$. The available water capacity and MWHC (v/v) vary from 0.21 to 0.27 $m^3 m^{-3}$ and 40.29 to 52.65 per cent, respectively. The available water content and MWHC (v/v) increased with depth in the solum of the pedon. The PAWC and moisture retention ranged from 2.50 to 9.50 and 5.20 to 18.43 cm ha^{-1}, respectively. The total PAWC of the profile was 0.299 m ha^{-1} (about 49 per cent) out of 0.606 m ha^{-1} of total water storage capacity of pedon (Table 1) indicating high plant available water capacity and water storage capacity of the profile. This might be due to smectite type of clay and high clay coupled with higher exchangeable sodium. Thus, it also corroborated by the finding of Paramsivam (1992) and Savalia (2005). The wide PAWC/WSC ratio of 0.492 indicating the 49 per cent plant available water capacity out of total water storage capacity of the soils of alluvial plain area of Meghal Irrigation Command area. This may be due to very deep soil (125 cm) which have more storage capacity.

Pedon 5 (Simar) from the Coastal plain: The data concerning on hydrological characteristics are furnished in Table 1 showed that the water holding capacity varied from 27.00 to 45.36 per cent. The saturated hydraulic conductivity of the pedon was observed 0.01 cm hr^{-1} to 0.10 cm hr^{-1} indicating very slow hydraulic conductivity. This might due to dominance of smectite type of clay, high ESP (10.32 to 11.16) which decompose the organic matter and dispersed the clay resulting in completely clogging of pores which stop the water transmission in these soils. These findings are in agreement with those of Khan and Afzal (1993) and Savalia (2005).

The moisture held at 0.03 and 1.5 MPa varied between 0.44 to 0.51 and 0.09 to 0.22 m^3 m^{-3}, respectively (Table 1). The AWC and MWHC (v/v) ranged from 0.24 to 0.38 m^3 m^{-3} and 38.07 to 67.13 per cent, respectively (Table 1). The vertical distribution of MWHC (v/v) with depth was not in systematic manner. The PAWC and moisture retention varied from 3.75 to 11.40 and 7.83 to 20.14 cm ha^{-1}, respectively (Table 1). The vertical distribution of PAWC with depth was found irregular. The total PAWC of the profile was observed 0.351 m ha^{-1} (54 per cent) out of 0.647 m ha^{-1} of total water storage capacity having the PAWC/WSC ratio of 0.542 indicating 54 per cent plant available water capacity out of total water storage capacity of the pedon of studied area and have very high water storage capacity which might be due to enough soil depth (120 cm), high ESP and other cations in coastal plain area.

Overall Hydrological Characteristics of Southern Saurashtra

In general, the maximum water holding capacity of the soils of Meghal Irrigation Command area ranged from 33.73 to 47.41 per cent with the overall mean of 39.87 per cent. The saturated hydraulic conductivity ranged from 0.04 to 0.21 cm ha^{-1} with the overall mean of 0.10 cm ha^{-1} indicating very slow saturated hydraulic conductivity in soils of Meghal Irrigation Command area which may be due to high clay content, dominant expandable smectite clay minerals, high bulk density in lower pediment to coastal plain (1.40 to 1.51 Mg m^{-3}), high ESP and SAR in alluvial plain and coastal plain areas of Meghal Irrigation Command area.

The moisture held at 0.03 and 1.5 MPa ranging from 0.43 to 0.47 and 0.18 to 0.25 m^3 m^{-3} with the overall mean of 0.45 and 0.21 m^3 m^{-3}, respectively. The AWC and MWHC (v/v) ranged between 0.21 to 0.29 m^3 m^{-3} and 48.51 to 60.98 per cent with the overall mean of 0.24 m3 m-3 and 54.43 per cent, respectively. The total PAWC and water storage capacity of the soils associated in different land slopes ranged from 0.096 to 0.351 and 0.241 to 1.647 m ha^{-1} with the mean of 0.215 and 0.464 m ha^{-1}, respectively. The PAWC/WSC ratio ranged from 0.367 to 0.542 with the mean of 0.441 indicating the 44 per cent water is available for plant out of 0.647 m ha^{-1} of total water capacity of the soils of different land slopes of Meghal Irrigation Command area of Southern Saurashtra. (Table 1)

Table 1: Hydrological characteristics of the representative pedons of Meghal Irrigation Command area

Horizon	Depth (cm)	MWHC (%)	Sat. H.C	Moisture retention ($m^3 m^{-3}$) at different tensions (MPa)						AWC ($m^3 m^{-3}$)	MWHC (v/v %)	PAWC	Moisture retention	Total* PAWC	Total WSC	PAWC/ WSC
			($cm\ hr^{-1}$)	0.03	0.05	0.10	0.50	1.0	1.5			($cm\ ha^{-1}$)		($m\ ha^{-1}$)		
LS1: P1: Undulating Upper Pediment, Dedeldyal, Tal. Maliya, Dist. Junagadh, MSL : 115 m (Typic Ustothrents)																
A_1	0-10	42.62	0.30	0.43	0.33	0.29	0.26	0.24	0.23	0.20	54.55	2.0	5.46	0.096	0.241	0.398
A_2	10-24	45.49	0.25	0.45	0.34	0.33	0.31	0.26	0.22	0.23	60.05	3.2	8.41	-	-	-
AC	24-45	34.96	0.15	0.42	0.27	0.25	0.23	0.22	0.21	0.21	48.59	4.4	10.20	-	-	-
Weighted Mean	-	39.94	0.21	0.43	0.31	0.28	0.26	0.24	0.22	0.21	53.48	-	-	0.096	0.241	0.398
LS2: P2: Upper Pediment, Gadhali, Tal. Maliya, Dist. Junagadh, MSL : 90 m (Typic Ustothrents)																
Ap	0-13	46.80	0.10	0.46	0.34	0.30	0.28	0.25	0.24	0.22	58.03	2.9	7.54	0.101	0.274	0.367
A_2	13-20	47.97	0.12	0.49	0.36	0.34	0.33	0.32	0.26	0.23	61.40	1.6	4.30	-	-	-
AC	20-30	50.42	0.14	0.48	0.37	0.35	0.32	0.29	0.25	0.23	65.55	2.3	6.55	-	-	-
Ck	30-45	45.68	0.09	0.47	0.33	0.30	0.29	0.26	0.25	0.22	60.30	3.3	9.04	-	-	-
Weighted Mean	-	47.41	0.11	0.47	0.35	0.31	0.30	0.28	0.25	0.22	60.98	-	-	0.101	0.274	0.367
LS3 : P3: Lower Pediment, Ghumti, Tal. Maliya, Dist. Junagadh, MSL : 45 m (Vertic Haplustepts)																
Ap	0-15	45.52	0.05	0.48	0.33	0.29	0.27	0.24	0.22	0.26	57.36	3.9	8.60	0.226	0.553	0.408
Bw_1	15-40	40.73	0.10	0.41	0.28	0.25	0.24	0.21	0.18	0.23	52.95	5.8	13.24	-	-	-
BC_1	40-55	44.25	0.08	0.42	0.32	0.28	0.27	0.25	0.21	0.21	59.74	3.2	8.96	-	-	-
BC_2	55-75	37.27	0.12	0.43	0.34	0.31	0.30	0.28	0.23	0.20	51.06	4.0	10.21	-	-	-
C	75-100	39.92	0.20	0.46	0.35	0.33	0.31	0.28	0.23	0.23	57.09	5.8	14.27	-	-	-
Weighted Mean	-	41.08	0.12	0.44	0.32	0.29	0.28	0.25	0.21	0.23	55.28	-	-	0.226	0.553	0.408

(Cond...)

LS4 : P4: Alluvial Plain, Sukhpur, Tal. Maliya, Dist. Junagadh, MSL : 15 m (Calcic Haplustepts)																
Ap	0-20	37.00	0.08	0.43	0.32	0.28	0.26	0.24	0.21	0.22	47.36	4.4	9.47	0.299	0.606	0.492
Bk_1	20-40	36.80	0.04	0.45	0.32	0.28	0.28	0.25	0.22	0.23	52.26	4.6	10.45	-	-	-
Bk_2	40-75	36.31	0.02	0.46	0.31	0.30	0.25	0.22	0.19	0.27	52.65	9.5	18.43	-	-	-
Bk_3	75-90	31.60	0.01	0.42	0.29	0.26	0.29	0.25	0.21	0.21	46.77	3.2	7.02	-	-	-
Bk_4	90-100	34.19	0.01	0.40	0.27	0.25	0.23	0.19	0.15	0.25	51.97	2.5	5.20	-	-	-
Ck	100-125	26.16	0.05	0.39	0.28	0.27	0.24	0.21	0.16	0.23	40.29	5.8	10.07	-	-	-
Weighted	-	33.73	0.04	0.43	0.30	0.28	0.26	0.23	0.19	0.24	48.51	-	-	0.299	0.606	0.492
Mean																
LS5: P5: Coastal Plain, Simar, Tal. Veraval, Dist. Junagadh, MSL : 5 m (Fluventic Haplustepts)																
Ap	0-15	37.83	0.02	0.48	0.36	0.31	0.28	0.25	0.21	0.27	52.21	4.1	7.83	0.351	0.647	0.542
B_{21}	15-45	27.00	0.04	0.44	0.32	0.29	0.25	0.23	0.20	0.24	38.07	7.2	11.42	-	-	-
B_{22}	45-75	40.18	0.10	0.47	0.34	0.29	0.25	0.22	0.09	0.38	58.26	11.4	17.48	-	-	-
B_{23}	75-105	45.36	0.02	0.51	0.37	0.33	0.31	0.27	0.22	0.29	67.13	8.7	20.14	-	-	-
BC	105-120	37.16	0.01	0.46	0.37	0.32	0.28	0.24	0.21	0.25	55.74	3.75	8.36	-	-	-
Weighted Mean	-	37.17	0.04	0.47	0.35	0.31	0.27	0.24	0.18	0.29	53.88	-	-	0.351	0.647	0.542
Overall Weighted Mean		39.87	0.10	0.45	0.32	0.29	0.27	0.25	0.21	0.24	54.43	-	-	0.215	0.464	0.441

Profile water storage capacity, Reserve water and water balance

Evaluation of Soil and Water Resources

The data on reserve water and water balance based on annual rainfall, total WSC and total PAWC of profiles in the different land slopes of Meghal Irrigation Command area are presented in Table 2.

Undulating Upper Pediment: The minimum reserve water balance were observed 0.145 m ha^{-1} (Table 2). The water balance was observed 0.629 m ha^{-1}, indicating that if intensity of rainfall is greater than infiltration rate in undulating upper pediment area, the positive water balance (0.629 m ha^{-1}, Table 4.25) observed might be due to runoff, which can be utilize as surface reservoir. If intensity of rainfall is less than infiltration rate then it percolates into the underground water in the studied area. Similar results were also reported by Savalia (2005) in the soils of Southern Saurashtra region.

Upper Pediment: The reserve water and positive water balance were 0.173 and 0.596 m ha-1, respectively (Table 2) indicating that the positive water balance (0.596 m ha^{-1}) as observed in upper pediment area might be percolated but when intensity of rainfall is less than infiltration rate then it could be percolate into ground water. Soil and water conservation measures should be followed along with cover crops like groundnut, black gram and bean across the slopes.

Lower Pediment: The reserve water and water balance were 0.327 and 0.317 m ha^{-1} (Table 2) indicating that the positive water balance observed in pedon P3 in lower pediment area might be percolated when intensity of rainfall is less than infiltration rate and this percolating water can be utilise for ground water recharge. The soil and water conservation measures should be followed along with cover crops like groundnut, black gram and bean across the slope. However, the lower and upper pediment areas have the intermediate hydrological problems hence management practices should be done accordingly.

Alluvial Plain: The reserve water and water balance were 0.377 and 0.264 m ha^{-1}, respectively (Table 2) indicating that the percolation may be comparatively delay and less as soil depth is more and in association with very slow permeability, therefore the runoff will take place even moderate intensity of rainfall and water logged condition take place during the monsoon. As a corrective measure for this, the drainage provision should be done. An establishment of drainage network with the sowing of high water requirement crops like sugarcane and banana should be adopted.

Coastal plain: The reserve water and water balance were observed 0.296 and 0.053 m ha^{-1}, respectively (Table 2). Due to the negative water balance (-0.053 m ha^{-1}) in coastal area lead to underground water depletion year after

Table 2: Reserve water and water balance based on annual rainfall, total WSC and PAWC in the different land slopes of Meghal Irrigation Command area of Southern Saurashtra

Pedon / Toposequence	*Soil depth (m)*	*Average annual rainfall (m ha^{-1})*	*Total WSC (m ha^{-1})*	*Total PAWC (m ha^{-1})*	*Reserve water (m ha^{-1})*	*Water balance (m ha^{-1})*
LS-1 Undulating Upper Pediment, MSL : 115 m						
P_1	0.45	0.870	0.241	0.096	0.145	+0.629
LS-2 Upper Pediment, MSL : 90 m						
P_2	0.45	0.870	0.274	0.101	0.173	+0.596
LS-3 Lower Pediment, MSL : 45 m						
P_3	1.00	0.870	0.553	0.226	0.327	+0.317
LS-4 Alluvial Plain, MSL : 15 m						
P_4	1.25	0.870	0.606	0.229	0.377	+0.264
LS-5 Coastal Plain, MSL : 5 m						
P_5	1.20	0.594	0.647	0.351	0.296	-0.053
Overall mean	**0.87**	**0.814**	**0.464**	**0.215**	**0.249**	**+0.350**

year resulted in the ingress of seawater in to the ground water. This is the main reason for the poor quality of saline water observed in the coastal area Savalia, (2005). With the use of poor quality water, the soils also become saline and/or sodic due to secondary salinisation in this area. It is inferred from the data (Table 2) that the problem of coastal plain was more serious as compared to alluvial plain. In addition to this, the salinity problem also alarming in depression area as well as the tidal water inundated and mixed with fresh reservoir water thus the salinity spread over the coastal area. As a remedial measure, the use of soil amendments and drainage provision should be done accordingly. The salt tolerant crops and varieties like pearl millet, sorghum, cotton Var. Dhumad and paddy kalarada can be grown in coastal plain area.

In general, to reserve water and water balance of different land slopes of Meghal Irrigation Command area varied from 0.145 to 0.377 m ha^{-1} and -0.053 to +0.629 m ha^{-1} with the mean of 0.249 and +0.350 m ha^{-1}, respectively. The negative water balance in coastal plain (-0.053 m ha^{-1}) lead to ground water depletion every year and ultimately resulted in the ingress of sea water.

Conclusion

It may conclude that the saturated hydraulic conductivity was recorded slow to very slow. The infiltration rate was observed moderately slow. The PAWC/WSC ratio was between 0.367 and 0.542 of which P5 belong to Fluventic Haplustepts in coastal plain area has the highest i.e. 0.542 indicating the soils of this area have 50 per cent plant available water capacity. The reserve water and water balance varied from 0.145 to 0.377 m ha^{-1} and -0.053 to +0.629 m ha^{-1} with the mean of 0.249 and +0.350 m ha^{-1}, respectively. The negative water balance observed -0.053 m ha^{-1} in soils belong to Fluventic Haplustepts in coastal plain area indicating lead to ground water depletion every year and ultimately resulted in the ingress of seawater. The soils belongs to Fluventic Haplustepts recorded the highest total profile water storage capacity (0.647 m ha^{-1}) followed by Typic Ustothrents (0.241 m ha^{-1}). The lowest water retention observed in Typic Ustothrents in higher elevated topography.

References

DACS, (2001). Annual AGRESCO Report. Deptt. of Agril. Chem. & Soil Sci., GAU, Junagadh.

Kaushal, G. S., Tembhase, A. and Sinha, S. B. (1986). Morphology and taxonomy of black soils under Bargi irrigation project in Madhya Pradesh. J. Indian Soc. Soil Sci., 34: 329-333.

Kumar, C.P. (2012). Climate change and its impact on Groundwater Resources, International Journal of Engineering and Science ISSN: 2278-4721, Vol. 1, Issue 5, pp. 43-60.

Khan, G. S. and Afzal, T. (1993). Hydraulic characteristics of some important soil series of Pakistan. Soils Fertil., 56: 320-326.

Khan, Z. H., Mazumdar, A. R., Hussein, M. S. and Mohiuddin, A. S. M. (1997). Total and DTPA extractable Fe, Mn, Cu and Zn content in some soils of Bangladesh. J. Indian Soc. Soil Sci., 45: 485-489.

Paramsivam, P. C. (1992). Ph.D. Thesis, Tamilnadu Agricultural University, Coimbatore, India.

Savalia, S. G. (2005). Characterisation, classification and evaluation of soil and water resources across the toposequences of Southern Saurashtra. Ph.D Thesis submitted to J.A.U., Junagadh.

Sharma, S. S. (2000). Characterisation and classification of salt affected soils for land use planning in sub-humid southern plains of Rajasthan. Ph.D. Thesis, R.A.U., Bikaner Campus, Udaipur.

Singh, S.K. (1999). Studies on variability in soil properties as related to genesis in Vertisols of Rajasthan. Ph.D. Thesis, R.A.U., Bikaner Campus, Udaipur.

Walia, C. S., Rao, V. S. and Balade, S. V. (1999). Waters retention characteristics of some sedentary and alluvial soils of Bundelkhand region. Agropedology, 9: 105-113.

33

Recent Environmental Archaeological Investigation in Bhima Basin: A case study from Man, Sina and Bhima River from parts of Solapur District of Maharashtra

S. D. Kshirsagar, P. D. Sabale, S. V. Joshi and Goyal Pankaj

Abstract

A detail environmental archaeological investigation was carried out in the parts of Man, Sina and main channel, in Bhima river basin at the border of Maharashtra and Karnataka state. Since 2003 to till date, the continuous efforts were taken and this studies revealed invention of number of important Prehistoric period archaeological sites of Paleolithic and Mesolithic culture. Example such as site from the Sina River basin Khadki, Borgaon, Khairav and Nilaj, and Mudhavi, Ozewadi from Man River basin while Ambe and Bramhapuri from Bhima River basin. From the stratigraphical section of habitation deposit of archaeological mound, the faunal remains were recovered. To understand the details of the paleo environment, apart from the other type of studies, Fluorine dating method is applied to understand the period of deposition. The result of this technique indicates Middle to Late Holocene climate in which the fossil is preserved in the archaeological sediment section. The present paper is deals with the multidimensional studies of the archaeo-sediment which required for the reconstruction of the paleo environment of the area.

Study Area

Solapur district is south part of the Maharashtra state (17° 07" to 18° 03" N and 74° 36" to 76° 26" E) (Fig. I). Physiographically, it forms part of

Maharashtra plateau and Middle reaches of Bhima Basin. The major drainage is formed by south-east flowing Bhima River and its tributaries such as Man, Sina, Bori and Bor rivers. Man and Bor rivers flow from south-west to north-east and join Bhima River at right bank (Fig. II). The drainage is dendritic controlled locally by joints and fractures. The area is entirely covered by basaltic lava flows overlain by sporadic duricrust, soil and alluvium at places. The soil is of varying thickness, reddish brown to black in colour and loamy in nature, which popularly known as black cotton soil. Alluvial spread is along the banks of Bhima and Sina River. It is formed of consolidated to semi-consolidated, poorly sorted, arenaceous to rudaceous sediments. The grains are sub-angular to sub-rounded and the cemented with calcareous material. The alluvium attains a maximum thickness of 15 m. *Kankar* development on the soil is sporadic and may attain a thickness of 1 to 2 m. The basalt flows are all Thoelelitic. Both pahoehoe and Aa Aa flows with flows showing mixed characters of both types. The pahoehoe flows are less in number. They pipe amygdules squeeze-up and spherical vesicles. The red bole bed marks the top of most of the basalt flows. The bole bed varies in thickness from few cm to 2 m.

The climate of this district is characterised by general dryness during major part of the year with temperature ranging from 13°C to 40°C with an annual average rainfall of 750 mm.

The lands are fertile and well irrigated and are highly suitable for Jowar, Bajara, Maize, Wheat, Groundnut, Sugarcane and Pulses are the main crops (Fig. III).

Sites of Bhima, Man and Sina River basin Ambe

This site is located on the right bank of Bhima River, 16 km east from Pandharpur Tehsil of Solapur district. Animal Teeth fossil is recovered from the river bed loose sandy pebbly gravel with cultural material of Mesolithic period Microlithic tools such as blade, flake, core, point etc. made on chalcedony, agate and chart raw material (Fig. 1, Plate 1, 2).

Begampur

This site is locate on the left bank of Bhima river, 15 km north-east from Mangalwedha Tehsil of Solapur district. Fossil of Turtle Carpacae and animal fossils are recovered from the river bed sandy pebbly gravel (Fig. 2).

Borgaon

This site is located on the left bank of Sina River, 12 km east of Karmala Tehsil of Solapur district. Animal fossil recovered from the loose sandy pebbly gravel with the cultural material of Mesolithic period microlithic tools. The litho sections are exposed on locality 1 and 2. The locality 1 is 30-35 ft. sandy pebbly loose gravel is exposed on the left bank of Sina River, and the locality 2 right bank of Sina river 25-30 cm semi-compact sandy pebbly gravel is rest on the basalt bed rock, 1-1.5 m. loose silt clay deposit on the gravel (Plate 3, 4, 5).

Khadaki

This site is located on the right bank of the Sina River, 18 km north of Karmala Tehsil of Solapur district. Animal Teeth fossil is recovered from the river bed loose sandy pebbly gravel (Plate 6).

Khairav

This site is located on the left bank of Sina River, north-east of Madha Tehsil of Solapur district. Animal Teeth fossil recovered from compact sandy river conglomerate gravel (Fig Plate 7, 8).

Mudavi

This site is located on the right bank of Man River, 16 km north from Mangalwedha Tehsil of Solapur district. Animal Teeth fossil are recovered from Dug-well section of bridge foundation from compact sandy, pebbly, boulder river conglomerate gravel associated with cultural material of Lower Paleolithic tools of Achuelian type such as large flakes (Fig. 3, Plate 9).

Nilaj

This site is located on the right bank of Sina River, 14 km north-east of Karmala Tehsil of Solapur district. Animal Teeth and bone fossil are recovered from loose sandy pebbly gravel of river bed (Fig. 4, Plate 9, I, II).

Relative Dating of Bones Found in Bhima, Man and Sina River Basin, District Solapur

Providing chronological framework to archaeological findings has one of the primary goals of interdisciplinary science of archaeological chemistry. Bones is an excellent material for dating by using a method of fluorine dating, which has become a standard procedure in archaeological chemistry today (Joshi 2006-07). The examination of fluorine content in bones was initially undertaken by Morichine over two centuries ago in England. Towards the end of 19th century Carnot in France also examined the method of fluorine dating. (Middleton, 1844). In India, collective efforts of the (Late) R.V. Joshi, V.D. Gogte, and A.A. Kshirasagar have given a new dimension to the subject of relative dating. The hypothesis is that there is a marked difference in the fluorine contents of bones from prehistoric to historical period, implying that the amount of fluorine in bones goes on increasing with antiquity. During the exploration around the Bhima, Man and Sina River. Dr Shivaji Dadaso Kshirsagar has found fossilised and semi-fossilised bones. It was not possible to securely date of these sites in the absence of datable material and statigraphy. Bones from these sites were analysed to put them into proper chronological framework by using relative dating technique by fluorine method.

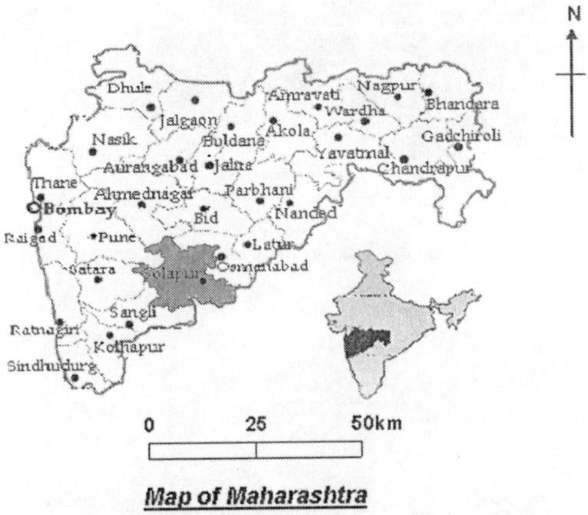

Fig. I. Location Map of Study Area.

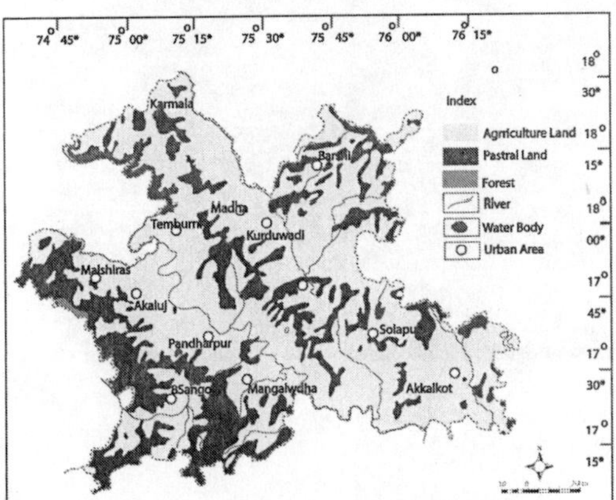

Fig. II. Land Use Map of Study Area.

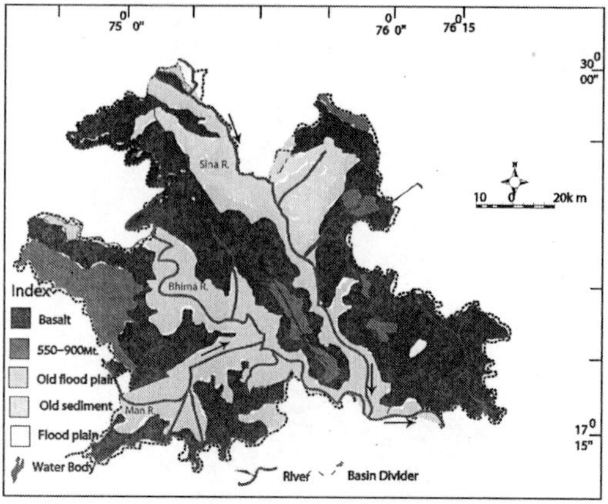

Fig. III. Geomorphology Map of Study Area.

Plate I. Location of Mudavi Site.

Conclusion

Fluorine and Phosphate analysis of these samples give the ratio ($100F/P_2O_5$) in between the range of 0.050 to 1.0 (Table 1 and 2). Ratio ranges from 0.050 to 1.0 which gives the dates at the border of Mid to Late Holocene period. Dr Anupama Kshirsagar has proposed a chart correlating the fluorine values in bones with dates obtained by other dating method; it becomes clear that the bones in this study fall in the relative time scale covering a small time span of late to Mid Holocene period. Analysis indicates that these bones are semi-fossilised. Amount of fluorine accumulation in the bone is very less.

Plate II. Location of Mudavi Site

Table 1: Phosphate and Fluorine Values of Bones from Bhim, Man and Sina River Site

Sr. No.	*Description*	*Percantage of Phosphorus*	*Percantage P_2O_5*	*Percantage Fluorine*	*Ratio*
1	Begampur, Bhim	13.37	30.62	0.281	0.91
2	Borgaon 2, Sina	6.62	15.17	0.031	0.28
3	Borgaon 1, Sina	10	22.9	0.044	0.17
4	Ambe Bhima	40.875	11.16	0.17	0.15
5	Khadaki, Sina	10	22.9	0.021	0.091
6	Mangalwed-ha, Bhima	10.75	24.61	0.014	0.058

Acknowledgment

All authority of Deccan College Post-Graduate and Research Institute, Pune, and Indian Council of Historical Research, New Delhi.

Table 2: Fauna from Bhim, Man and Sina River Sites

Sr. No.	*Site Name*	*Fauna*
1.	Begampur, Bhima	Turtle Carapacae
2.	Bramhapuri, Bhima	Domestic Pig (Mandible with P.3) *Sus Domesticus*
3.	Ozewadi, Man	Cattle, Molar. *Boss Indicaus*
4.	Borgaon 2, Sina	Cattle, Redius- Distel end. *Boss Indicaus*
5.	Borgaon 1, Sina	Cattle, Humerus- Distel end, *Bos Indicaus*
6.	Nilaj 1, Sina	Large Size Mammal Long Bone
7.	Nilaj 2, Sina	Cattle, Teeth, *Boss Indicaus*
8.	Nilaj 3, Sina	Cattle, Mollar- Upper Jaw, *Boss Indicaus*
9.	Borgaon 2, Sina	Cattle, Mollar- Lower Jow, *Boss Indicus*
10.	Khairav, Sina	Cattle, Mollar- Lower Jaw, *Boss Indicus*
11.	Ambe 1, Bhima	Cattle, Mandible with Mollar 1, 2. Right Side Lower Jow, *Boss Indicaus*
12.	Khadaki Sina	Cattle, Teeth with Mollar 1, 2. Left Side Lower Jow, *Boss Indicaus*

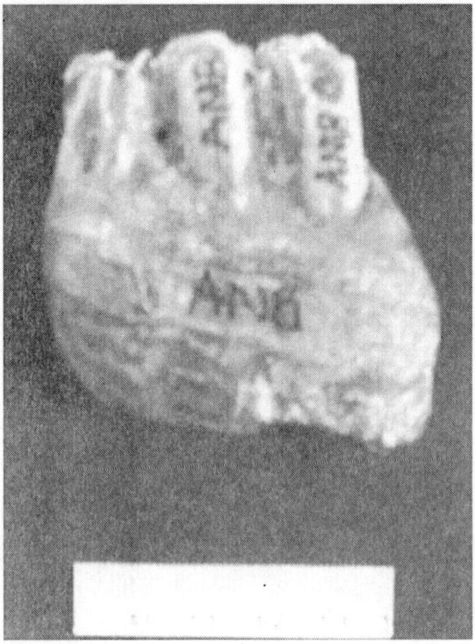

Plate 1. Ambe Animal Teeth Fossil.

Plate 2. Location of Ambe Site.

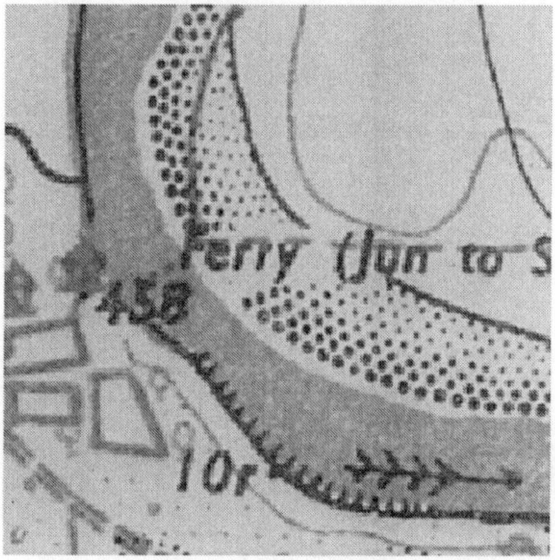

Fig. 1. Location Map of Ambe site.

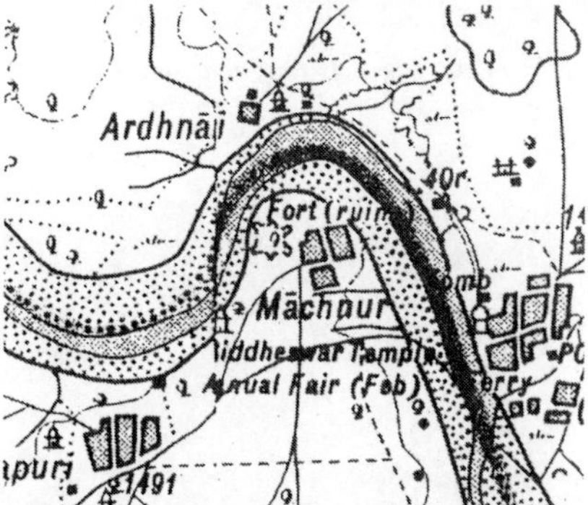

Fig. 2. Location Map of Begampur Site.

Plate 3. Borgaon Animal Bone Fossil.

Plate 4. Borgaon Section.

Plate 5. Location of Borgaon Site.

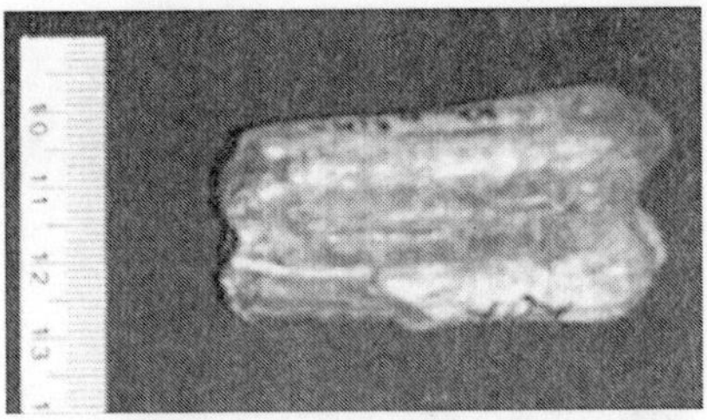

Plate 6. Khadaki Teeth Fossil.

Plate 7. Khairav Animal Teeth Fossil.

Plate 8. Khairav Litho section.

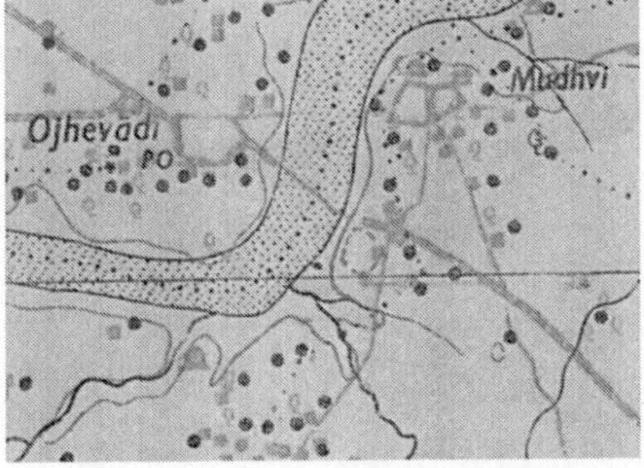

Fig. 3. Location Map of Mudavi Site.

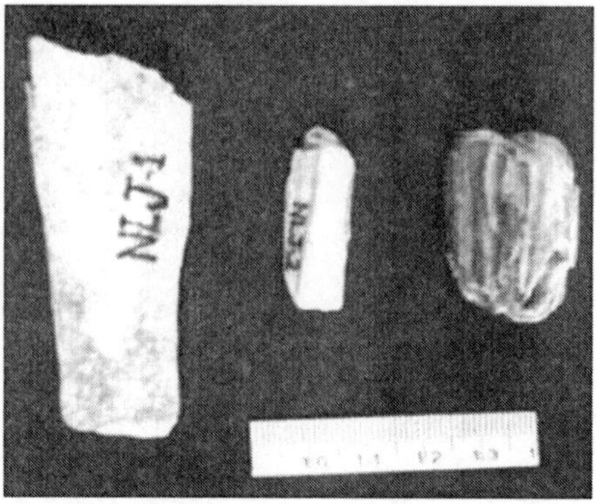

Plate 9. Nilaj Animal Teeth and Bone.

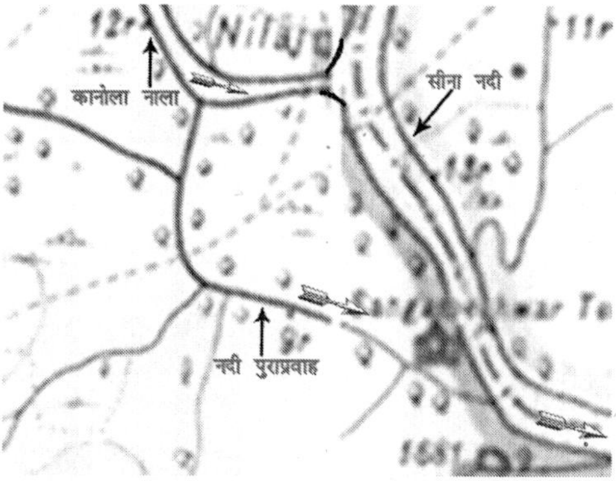

Fig. 4. Location of Map of Nilaj Site.

References

District Resource Map of Solapur, 2001. Geological Survey of India.

District Planning Map of Solapur, 1994. Survey of India.

Hampe Santosh S. and S.D. Kshirsagar. 2012. Effect of Flood Disaster Activity on the Temple Architect: A Case Study of Mahadev Temple at Mallewadi Village. In Proceeding of International Conference on Art and Architecture in the Indian Subcontinent.(Multi Disciplinary) Organised by D.B.F. Dayanand College of Arts and Science, Solapur. Held on 17th to 19th Feb 2012 at Solapur. pp. 103-106, ISBN No. 978-93-81948-08-8.

Joshi, S.V. 2006-07. Fluorine dating of the Pleistocene vertebrate fossils from Siwaliks, Himachal Pradesh, *Bulletin of the Deccan College Post Graduate and Research Institute, Pune,* Vol. 66-67, pp. 285-289.

Kshirsagar, A. A. and V. D. Gogte, 1990. Fluorine Determination in fossil Bones with Ion Selective Electrode. *Bulletin of the Deccan College Post Graduate and Research Institute, Pune,* Vol. 49, pp 213-216.

Kshirsagar, S.D., 2010. Geoarchaeology of Bhima Basin in Solapur District, Maharashtra. Unpublished Ph.D. Thesis of Deccan College Post-Graduate and Research Institute, Pune 411006.

Khirsagar S.D. 2012. Palaeochannel Studies of Sina River Basin by Google Earth (Remote Sensing) Technique. Paper presented in National Symposium on "Research Advances in Environmental Sciences: The Indian Perspective" (RAESIP-2012) Organised by School of Earth Sciences, Swami Ramanand Teerth Marathwada University, Nanded. On 16th and 17th March, 2012.

Middleton, J.J., 1844. On fluorine in Bones, its source and its application to the determination of geological age of fossil bones, *Proc. Geol. Sci. of London* 4, part 2, 99: 431.

Quadrangle Map no. 470, 2001. Geological Survey of India.

Sushama, G. Deo, Shivaji D. Kshirsagar, S. N. Rajguru, Santosh Hampe, P.P. Jogalekar. 2005. Geoarchaeological Observation of Man Basin in Solapur District, Maharashtra. Puratattva, Vol. 35, pp. 111-117. ISSN. No. 0970-2105

Sabale, P.D. and S.D. Kshirsagar. 2010. Recent Advances in Gearchaeology of Sina River Basin in Solapur District, Maharashtra. Paper presented in National Conference on Recent Advances in Indian Archaeology During 3rd and 4th December 2010, Organised by the Department of Archaeology and Ancient History, Faculty of Arts, The Maharaja Sayajirao University of Baroda.

Sabale, P.D. and S.D. Kshirsagar. 2010. Palaeochannel Studies of the Man River Basin by Remote Sensing Technique. Annual Convention of ISRS and National Symposium on GIS and Remote Sensing in Infrastructure Development, held on 1-3rd December 2010 at Sinhgad Technical Education Society, Lonavala, Pune.

34

Watershed Characteristics and Prioritisation of Shetrunji River Basin using GIS and Remote Sensing Techniques

Wandre S.S. and H. D. Rank

Abstract

The study area i.e. Shetrunji basin falling in the district of Bhavanagar, Amreli and Junagadh, is a major one among 71 river basins of Saurashtra region of the Gujarat state, India. In the present paper, an attempt has been made to study the morphometric characteristics and prioritisation of watersheds of Shetrunji river basin. So the entire study area has been further divided into 17 sub-watersheds and has been taken up for prioritisation based on morphometric analysis using GIS and remote sensing techniques. According to severity ranking of each watershed the compound parameters were estimated by averaging the priorities. The final priority was allotted to the watersheds according to the values of compound parameters. The value of compound parameter for all 17 watersheds varied from 6 to 11.50. The watershed 5G2B5d (compound parameter value 6) receives the highest priority value and so it becomes potential and prioritised candidate for applying soil conservation measures, in contrast to the other watersheds with low priorities are subjected to lower degree of erosion. Such kind of watershed prioritisation becomes helpful to overall watershed management planning. The morphometric properties determined for this basin as a whole and for each watershed will be useful for the sound planning of water harvesting and groundwater recharge projects on watershed base.

Introduction

Land and water resources are limited and their wide utilisation is imperative, especially for countries like India, where population pressure is increasingly continuous. Morphometric analysis requires measurement of linear features, gradient of channel network and contributory ground slopes of the drainage basin (Nautiyal, 1994). Morphometric analysis is the measurement of the three dimensional geometry of land forms and has traditionally been applied to watershed, drainages, hill slopes, and other group of terrain features (Barber, 2005). It is a significant tool for prioritisation of sub-watersheds even without considering the soil map (Biswas *et al.,* 1999). The success of watershed development programme revolves around the conservation of soil and water resources in that watershed and hence, it is imperative to prioritise the watersheds on the basis of conservation and developmental needs. River basins comprise a distinct morphologic region and have special relevance to drainage pattern and geomorphology (Doornkamp *et al.*, 1971; Strahler, 1957).

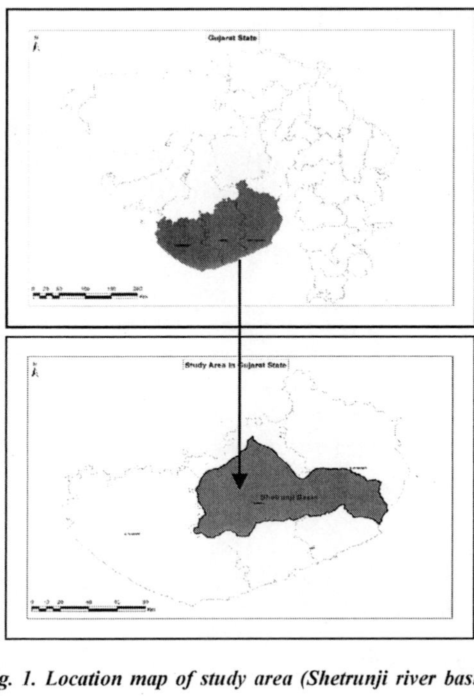

Fig. 1. Location map of study area (Shetrunji river basin).

Study area

The study area is Shetrunji river basin in Saurashtra region of Gujarat. It is located between 21° 00' to 21° 47' North latitude and 70° 50' to 72° 10' East longitude. The climate of the project area can be classified as tropical and sub-tropical. January is the coldest month with mean monthly temperature varying from 14.5°C to 20°C and maximum monthly temperature varies between 30°C to 44°C in the month of May. The river Shetrunji originates at Chchai hills in Gir forest of Junagadh district at 380 m.s.l. and flows towards east direction till it confluence with Gulf of Khambhat near Santrampur port. Its length is 227 km having 5646 km^2 catchment area with average annual rainfall of 604.52 mm on varied slope of basin (1:1000 to 1:5000).

Methodology

The remote sensing and GIS software used for the study is PCI Geomatica V10.1 and ArcGIS-ArcMap 9.2. The satellite images of IRS P6 LISS III captured in Oct 2005 and Feb 2008 having resolution of 23.5 m x 23.5 m and images from Google Earth Pro of study area were used. The order was given to each stream by following Strahler (1964) stream ordering technique. The linear parameters, areal and relief parameters were determined using GIS. Prioritisation rating of all the seventeen watersheds of Shetrunji river basin is carried out by calculating the compound parameter values.

Result and Discussion

The basin is divided into 17 watersheds with codes viz. 5G2B2a, 5G2B2b, 5G2B2c, 5G2B3a, 5G2B3b, 5G2B3c, 5G2B4a, 5G2B4b, 5G2B4c, 5G2B5a, 5G2B5b, 5G2B5c, 5G2B5d, 5G2B6a, 5G2B6b, 5G2B6c and 5G2B6d.

Linear aspects of the Watershed

Linear aspects of the 17 watersheds, related to the channel patterns of drainage network where in the topological characteristics of the stream segments in terms of open links of the stream network system are analysed. The results have been tabulated in the Table 1 as a whole and sub-watersheds. The present study has adopted the widely used method of Strahler. The study area is a seventh order drainage basin covering an area of 5646.54 km^2. Lower stream lengths are likely to have lower runoff (Chitra *et al,* 2011).

Bifurcation ratio

The analysis of bifurcation value shows that the basin and its watersheds possesses well developed drainage network as the bifurcation ratio ranges between 2.8 to 4.7 i.e. low value.

Stream length ratio

Stream length ratio may be defined as the ratio of the mean length of the one order to the next lower order of the stream segment (Horton, 1945).

The value of stream length ratio ranges widely between 1.35 to 176 which shows the early stage of maturity of the watershed.

Length of overland flow

A larger value of length of overland flow indicates longer flow path and thus, gentler slopes (Sethupathi *et al,* 2011). The Length of overland flow for basin 0.3132 km and for watersheds ranges from 0.2026-0.4419 km. The watersheds 5G2B4b, 5G2B4c, 5G2B5d, 5G2B6c and 5G2B6d are having lower values of length of overland flow comes under the influence of high structural disturbance, low permeability, and steep to very steep slopes and high surface runoff. Other remaining watersheds having length of overland flow greater than 0.25 are under very less structural disturbance, less runoff conditions and having higher overland flow. For basin it is greater than 0.25 it comes under very less structural disturbance, less runoff conditions and having higher overland flow (Mishra *et al,* 2011).

Aerial aspects of the Watershed

The parameters which are governed by the area of the drainage basin are classed as areal aspects of the basin. Results have been given in Table 1.

Drainage density

The drainage density for watersheds varies from 0.1314 to 3.0857. The watersheds 5G2B4a, 5G2B4b, 5G2B5a, 5G2B5d, 5G2B6c and 5G2B6d show high drainage density (greater than 2 km/km^2) due to the presence of impermeable sub-surface material, sparse vegetation and high relief. Whereas remaining watersheds fall under low drainage density indicate the region has highly permeable subsoil and dense vegetation cover (Mishra *et al,* 2011).

Stream frequency

The stream frequency for watersheds varies from 0.9393 to 3.50931 and for basin 1.4671 it is low due to permeable rocks the surface runoff is low and infiltration capacity is high within in the study area. The stream frequency for all 17 watersheds of the study area show positive correlation with the drainage density which indicate that the stream population increases with the increase of drainage density (Chitra *et al,* 2011).

Elongation Ratio

The elongation ratio of the basin watersheds is varies from 0.45 to 0.85. The watersheds 5G2B2b, 5G2B2c, 5G2B4a, 5G2B4b, 5G2B4c, 5G2B5b, 5G2B5c, 5G2B6a and 5G2B2d are elongated in nature. While remaining are less elongated in nature. For the basin it is 0.6206 indicating less elongated in nature (Nautiyal, 1994).

Form Factor

The form factor for basin is 0.3023 and for basin watersheds varying from 0.16 to 0.56. The watershed 5G2B5a, 5G2B5d and 5G2B6c are circular in shape showing less side flow for shorter duration and high main flow for longer duration (Chitra *et al,* 2011).

Circularity ratio

A dimensionless circularity ratio (R_e) is the ratio of basin area to the area of circle having the same perimeter as the basin (Strahler, 1964; Miller, 1953). The circularity ratio for watersheds varies from 0.37 to 0.77 and for basin it is 0.3853 indicating elongated in shape, low discharge of runoff and highly permeability of the subsoil condition.

Compactness coefficient

The compactness coefficient for watersheds ranges from 1.14 to 1.79 and for basin is 1.6106. They have elongated shape so they have enough time for discharge (Hadely *et al.*, 1961).

Drainage texture

The texture ratio for watersheds varies from 3.3 to 10.7. For watershed 5G2B4a, 5G2B4b 5G2B4c and 5G2B6c it is greater than eight indicating very fine texture *i.e* higher runoff potential while 5G2B5d, 5G2B5c, 5G2B6a and 5G2B6d is moderate in nature. The rest watersheds are coarser in nature i.e. having less runoff potential. For basin it is 19.3095 showing very fine nature (Pal *et. al.*, 2012).

Relief aspects of the Watershed

The relief aspects of drainage basin are also important in water resources studies. The character of the distribution of slope, angles sampled over the whole basin depends on the height distribution within it.

Relief

The relief of basin is 0.605 km. The study area is of high relief region as it is greater than 0.3 km. The high relief value indicates low gravity of water flow as well as infiltration and high runoff conditions (Nongkynri *et al,* 2011). The relief for watersheds varies from 0.095 to 0.445 km. The watersheds 5G2B5b and

5G2B6d are of low relief region, 5G2B2c, 5G2B3a, 5G2B5d and 5G2B6a are of high relief region and remaining are of moderate relief region.

Relief ratio

The relief ratio for watersheds varies from 0.00273 to 0.019 and for basin it is 0.004427. It was noticed that the higher values of relief ratio indicated steep slope and high relief (5G2B6d watershed), while the lower values in case of watershed 5G2B5a indicated the presence of basement rocks that are exposed in the form of small ridges and mounds with lower degree of slope (GSI, 1981).

Relative relief

The relative relief for watersheds varies from 0.0915 to 0.5399 and for basin it is obtained as 0.1410. The watersheds having higher relative relief have higher runoff potential than others. Therefore, the watershed 5G2B5a and 5G2B5d are having the lowest and highest runoff potential.

Channel slope

For watersheds it varies between 0.001549 km/km to 0.01224 km/km and for basin it is 0.002820 km/km. The higher channel slopes in 5G2B6d watershed indicated less time of concentration i.e. peak flow occurs in short time while lower slope in 5G2B5a watershed indicated less peaked flow for longer duration. Therefore, while constructing the water harvesting structures on channel of watershed 5G2B6d, the outlet should be designed of higher discharge capacity and the rest components like headwall, sidewall and wing wall should also be of higher height for the designed storage capacity (Suresh, 2002). The drop structures in series in the channels of this watershed are recommended.

Ground slope

For watersheds it is obtained as 0.00307 km/km to 0.0192 km/km and for basin 0.004427 km/km. The higher ground slopes in case of 5G2B6d lying in upper reach of the basin indicates lower time of concentration of overland flow. Also, the possibilities of soil erosion will be higher in this watershed.

Watershed prioritisation

The watershed prioritisation is done on the basis of linear parameters. The higher the values of these parameters, higher will be the degree of hazardous. The shape parameters like elongation ratio, circularity ratio, form factor and compactness coefficient varying inversely with the same (Mishra *et al.*, 2010). The prioritisation ranking was given based on the degree of hazardous and the final prioritization was done on the basis of compound parameter obtained from averaging the values of the rankings of the linear and shape parameters allotted to the watersheds. The final prioritized map of the study area and prioritization ranks of watersheds is shown in figure 2.

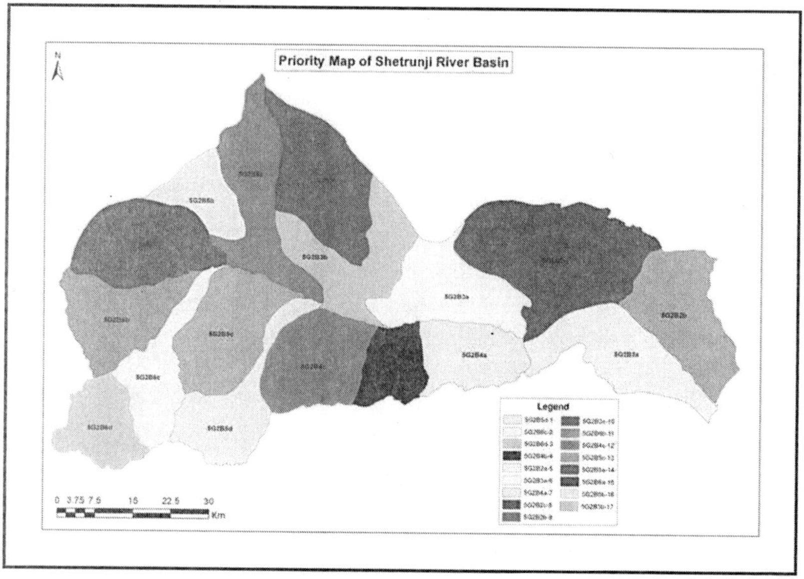

Figure 2: Final priority map of Shetrunji river basin

Conclusions

Usually, morphometric analysis of drainage system is prerequisite to any hydrological study. The present study demonstrates the usefulness of GIS for morphometric analysis and prioritization of the watersheds of Shetrunji river basin, Gujarat. The watershed 5G2B5d possess highest priority 1 which indicate greater degree of erosion and it becomes potential candidate for applying soil conservation measures, while the other watersheds with lower priorities are subjected to lower degree of erosion. Thus the morphometric properties determined for this basin as whole and for each watershed will be useful for the sound planning of water harvesting and groundwater Recharge projects on watershed base

Table 1: Morphometric parameters of watersheds of Shetrunji River basin

Aspects	Watersheds																
	5G2B2a	5G2B2b	5G2B2c	5G2B3a	5G2B3b	5G2B3c	5G2B4a	5G2B4b	5G2B4c	5G2B5a	5G2B5b	5G2B5c	5G2B5d	5G2B6a	5G2B6b	5G2B6c	5G2B6d
Linear aspects of watersheds of Shetrunji River basin																	
Length of overland flow, km	0.4778	0.438	0.3872	0.3844	0.4202	0.4419	0.1921	0.162	0.1919	0.4115	0.3798	0.3127	0.2026	0.41	0.379	0.2494	0.2116
Bifurcation ratio	4.1092	4.0736	3.1635	3.1647	2.864	3.3716	3.071	2.8815	2.6554	1.8461	3.94	3.185	3.7709	·3.0528	3.3262	3.3778	3.6403
Stream length ratio	1.3586	2.0111	1.7125	175.54	1.5789	1.8697	2.0303	1.5036	1.6702	2.2711	0.5643	20.052	2.1779	1.249	1.7262	2.0648	1.767
Aerial aspects of watersheds of Shetrunji river basin																	
Drainage density (km/km^2)	1.2262	1.1416	1.2915	1.3007	1.1899	1.1314	2.6025	3.0857	2.605	1.215	1.3164	1.5991	2.4678	1.2195	1.3191	2.0047	2.3628
Stream frequency (1/km^2)	0.9522	0.9393	1.1905	1.1605	0.9777	1.0463	2.5792	3.5093	2.1403	1.1738	1.266	1.1643	2.6968	1.1895	0.9564	2.1275	3.2235
Elongation ratio	0.606	0.7362	0.8031	0.6505	0.6314	0.5702	0.7456	0.7043	0.7681	0.4672	0.7748	0.7646	0.4882	0.7119	0.6956	0.4503	0.8467
Circularity ratio	0.3709	0.5981	0.5744	0.4242	0.2807	0.5839	0.6002	0.6727	0.7752	0.3047	0.5669	0.6475	0.3125	0.7575	0.5097	0.3087	0.5582
Form factor	0.2883	0.4254	0.5063	0.3219	0.3129	0.2553	0.4365	0.3894	0.4631	0.1714	0.4712	0.4589	0.1871	0.3978	0.3798	0.1592	0.5628
Compactness coefficient	1.6416	1.2927	1.3191	1.5349	1.887	1.3083	1.2905	1.219	1.1354	1.8112	1.3279	1.2425	1.7885	1.1487	1.4003	1.7994	1.3382
Drainage texture (1/km)	3.1191	4.0018	6.5082	4.0333	2.7056	4.6002	8.8469	10.716	9.7871	3.5797	3.5212	5.092	7.2247	5.6945	3.3952	5.3236	9.9494
Relief aspects of watersheds of Shetrunji river basin																	
Relief km	0.34	0.2	0.315	0.445	0.13	0.11	0.22	0.175	0.14	0.115	0.075	0.15	0.315	0.095	0.16	0.275	0.375
Relative relief km/km	0.3065	0.2236	0.2635	0.4325	0.105	0.1163	0.3065	0.3069	0.1889	0.0915	0.1217	0.1768	0.2925	0.1197	0.1829	0.2701	0.5399
Relief ratio	0.0096	0.00668	0.00876	0.01356	0.00393	0.00273	0.00926	0.00828	0.00517	0.00243	0.00393	0.00528	0.00802	0.00307	0.0056	0.00687	0.0192
Channel slope km/km	0.0061	0.00426	0.00558	0.00864	0.00250	0.00174	0.0059	0.00527	0.00330	0.00155	0.00251	0.00336	0.00511	0.00196	0.00356	0.00438	0.01`224
Ground slope, km/km	0.0096	0.00668	0.00876	0.01356	0.00393	0.00273	0.00926	0.00828	0.00517	0.00243	0.00393	0.00528	0.00802	0.00307	0.0056	0.00687	0.0192

Table 1(continue): Morphometric parameters of watersheds of Shetrunji River basin

Aspects	Order	5G2B2a	5G2B2b	5G2B2c	5G2B3a	5G2B3b	5G2B3c	5G2B4a	5G2B4b	5G2B4c	5G2B5a	5G2B5b	5G2B5c	5G2B5d	5G2B6a	5G2B6b	5G2B6c	5G2B6d
No of Streams	1	251	274	580	312	249	335	474	448	534	18	172	328	608	341	222	410	530
	2	68	62	148	78	63	76	113	113	136	346	36	77	133	81	56	99	122
	3	19	16	35	17	13	16	31	30	39	78	6	20	27	21	14	24	28
	4	6	4	10	5	5	5	12	13	12	2	2	4	7	4	3	6	8
	5	1	1	3	1	2	2	3	4	3	3	1	1	2	2	1	1	2
	6	0	0	1	1	2	1	1	2	1	2	0	1	1	2	1	1	1
	7	1	1	1	1	1	0	1	1	0	1	0	1	0	1	0	1	0
Stream length, km	1	227.4	207.5	478.8	231.7	209.8	273.8	316	275.7	477.76	277.81	49.48	318.76	374	293.11	226.75	275.3	267.57
	2	94.3	120.2	166.7	135.6	96.83	96.54	124.1	98.02	185.59	82.21	140.3	140.28	165.4	91	85.7	116.7	112.46
	3	80.46	53.98	118.4	54.28	31.6	41.99	103.2	59.05	115.56	43.63	13.65	77.56	89.07	47.55	61.91	57.59	64.15
	4	21.12	28.9	43.12	14.19	26.76	37.88	64.73	68.68	59.61	26.79	13.05	29.1	31.12	12.82	17.52	18.15	38.93
	5	0.068	11.58	20.24	8.62	13.85	9.56	8.34	28.2	39.89	12.12	9.18	19.39	29.44	9.05	16.68	6.69	23.23
	6	20	0	11.4	0.02	17.4	10.6	12.1	5.11	4.01	9.2	0	0.073	22.92	8.01	1.09	33.79	0.15
	7	22.19	12.97	15.37	20.83	11.45	0	12.12	2.46	0	14.02	0	8.16	0	1.84	0	2.5	0

Table 2: Final priority of watersheds based on compound parameter

Watershed	Linear parameters								Shape parameters				C_p	Final Priority
	R_b	F_s	L_g	T	D_d	H	R_{hp}	R_h	R_e	R_f	R_c	C_c		
5G2B2a	2	16	6	16	11	3	3	6	5	5	5	13	7.58	5
5G2B2b	1	17	2	12	16	8	9	9	11	11	12	6	9.50	9
5G2B2c	11	8	7	6	10	5	5	3	16	16	10	8	8.75	8
5G2B3a	10	12	8	11	14	1	2	2	7	7	6	12	7.67	6
5G2B3b	17	14	3	17	15	12	14	16	6	6	1	17	11.50	17
5G2B3c	7	13	1	10	17	15	16	15	4	4	11	7	10.00	10
5G2B4a	13	4	15	4	3	7	4	5	12	12	13	5	8.08	7
5G2B4b	14	1	17	1	1	9	6	4	9	9	15	3	7.42	4
5G2B4c	15	5	16	3	2	13	12	10	14	14	17	1	10.17	12
5G2B5a	16	10	4	13	13	14	17	17	2	2	2	16	10.50	14
5G2B5b	3	7	9	14	9	17	13	13	15	15	9	9	11.08	16
5G2B5c	9	11	11	9	7	11	11	12	13	13	14	4	10.42	13
5G2B5d	4	3	14	5	4	4	7	7	3	3	4	14	6.00	1
5G2B6a	12	9	5	7	12	16	15	14	10	10	16	2	10.67	15
5G2B6b	8	15	10	15	8	10	10	11	8	8	7	11	10.08	11
5G2B6c	6	6	12	8	6	6	8	8	1	1	3	15	6.67	2
5G2B6d	5	2	13	2	5	2	1	1	17	17	8	10	6.92	3

$\mathbf{R_b}$-Bifurcation ratio, $\mathbf{F_s}$- Stream frequency, $\mathbf{L_g}$-Length of overland flow, **T**-Drainage Texture, $\mathbf{D_d}$-Drainage Density, **H**-Relief, $\mathbf{R_{hp}}$-Relief ratio, $\mathbf{R_h}$-Relative relief, $\mathbf{R_e}$-Elongation ratio, $\mathbf{R_f}$-Form factor, **Rc**-Circularity ratio, $\mathbf{C_c}$-Compactness Coefficient $\mathbf{C_p}$,-Compound parameter

References

Babar, M., 2011. "Hydrogeomorphological analysis for watershed development in Jintur Tahsil, Parbhani district, Maharashtra", J. of Indian Streams Research., Vol. 6, pp. 275-298.

Biswas, S., Sudhakar, S., and V.R. Desai, 1999. "Prioritization of sub-watersheds based on morphometric analysis of drainage basin: a remote sensing and GIS approach", J. of the Indian Society of Remote Sensing., Vol. 27, pp.155-166.

Chopra, R., Dhiman R., and P.K. Sharma, 2005. "Morphometric analysis of sub-watersheds in Gurdaspur district, Punjab using remote sensing and GIS techniques", J. of the Indian Society of Remote Sensing, Vol. 33, pp. 531-539.

Chitra, C., Alaguraja, P., Ganeshkumari, K., Yuvaraj, D. and Manivel, M. 2011. "Watershed characteristics of Kundah sub basin using Remote Sensing and GIS techniques", International J. of Geomatics and Geosciences.,Vol. 2, pp. 311-335.

Doornkamp, J. C. and C.A.M. King, 1971. "Numerical analysis in Geomorphology: An Introduction". St. Martin's Press, New York, p. 372.

GSI, 1981. "Geological and Mineralogical Map of Karnataka and Goa", Geological Survey of India.

Hadely, R.F. and S.A. Schumm, 1961. "Sediment sources and drainage basin characteristics in upper Cheyenne River basin". United States Geological Survey water-supply paper, Vol.1531-B, pp. 137-196.

Horton, R.E., 1932. "Drainage basin characteristics". Trans. Amer. Geophysics Union,Vol.13, pp. 350-361.

Pal, B., Samanta S., and D. K. Pal, 2012. "Morphometric and hydrological analysis and mapping for Watut watershed using Remote Sensing and GIS techniques", International J. of Advances in Engineering and Technology, Vol. 2, pp. 357-368.

Miller, V.C. 1953. "A quantitative geomorphic study of drainage basin characteristics in the Clinch Mountain area, Varginia and Tennessee", Project NR 389042, Tech. Rept. 3, Columbia University, Department of Geology, ONR, Geography Branch, New York.

Mishra, A., Dubey D.P., and R. N. Tiwari, (2011). "Morphometric Analysis of Tons basin, Rewa District, Madhya Pradesh, based on watershed approach", J. of Earth Science India, Vol. 4, pp.171-180.

Motsara, M.R., 1998. "Need for the Right Mix". The Hindu Survey of Indian Agricultural 1999, Kasture and sons, National Press, Chennai: pp.169-174.

Nautiyal, M. D., 1994. "Morphometric analysis of a drainage basin, district Dehradun, U. P.". J. of Indian Soc. Remote Sensing, Vol. 22, pp. 251-261.

Shreedevi, P.D., Owais, S., Khan, H.H. and S. Ahmed, 2009. "Morphometry analysis of a watershed of south India using SRTM and GIS", J. of Geological Society of India., Vol.73, pp. 543-552.

Strahler, A.N., 1964. "Quantitative geomorphology of drainage basins and channel networks". In: V.T. Chow (Ed.), Handbook of Applied Hydrology. McGraw-Hill, New York: Vol.12, pp. 4.39-4.76.

Suresh, R., 2002. "Soil and water conservation engineering", Standard Publishers Distributors., Delhi: pp.793-812.

Vijith, H. and R. Satheesh, 2007. "GIS based morphometric analysis of two major upland sub-watersheds of Meenachil River in Kerala", J. of Indian Soc. Rein. Sens., Vol. 34, pp.181-185.

35

Quality of Irrigation Water in Junagadh District of Gujarat

N. B. Babaria, K. B. Polara and K. P. Sojitra

Abstract

An investigation was undertaken to study the quality of irrigation waters by analysing 280 samples from fourteen taluka *of coastal Junagadh district. The results showed that the overall pH values of irrigation water ranged from 6.75 to 8.86 with mean value of 7.71 indicating waters of the district was alkaline in reaction. Whereas the overall EC values ranged from 0.38 to 16.2 with mean value of 3.10 dSm^{-1}. In water samples, among the cations, Na^+ (mean 22.28 me l^{-1}) was dominant and among the anions Cl^- (mean 22.16 me l^{-1}) was dominant. The waters are mostly Na-Ca-Mg type with dominance of chloride followed by bicarbonates and sulphate. On the basis of EC 54.3 and 38.9 per cent samples were found in C_3 and C_4 classes, respectively. Whereas on the basis of SAR. 46.0, 30.0, 18.6 and 5.4 per cent samples were found under S_1, S_2, S_3 and S_4 classes, respectively. Overall, on the basis of EC and SAR of irrigation water samples 40.0, 17.9, 16.4, and 11.8 per cent samples were categorised as C_3S_1, C_4S_2, C_4S_3 and C_3S_2 classes, respectively.*

***(Key words:** Quality of irrigation water, EC, pH, SAR, RSC, SSP)*

Introduction

Quality of irrigation water is one of the main factors to be understood in irrigated agriculture. In majority of arid, semi-arid and coastal regions, groundwater is either the major or the only source of irrigation to supplement the scanty rainfall. The majority of ground waters in these regions are of poor quality due to excessive salt concentration and high values of SAR and RSC. Scarcity of good quality of waters forces the farmers to use available saline or sodic water for irrigation.Therefore, the present investigation was undertaken to study the quality of irrigation water in coastal Junagadh district of Gujarat.

Materials and Methods

Two hundred eighty underground well/tube well water samples were collected from the eleven *talukas* of coastal Junagadh district (Junagadh, Bhesan, Visavadar, Mendarda, Vanthali, Manavadar, Seshod, Mangrol, Maliya, Talala, Veraval, Sutrapada, Kodinar, Una). These water samples were analysed for the parameters viz., electrical conductivity, pH, calcium, magnesium, sodium, potassium, chloride, carbonate, bicarbonate and sulphate content as per the standard methods outlined by Richards (1954). The water qualities were determined as per USSL classification.

Results and Discussion

The results presented in Table 1 showed that the overall pH values of irrigation water ranged from 6.95 to 8.78 with mean value of 7.72 indicating waters of the district were alkaline in reaction. The highest (8.11) mean value of pH was recorded in Talala *taluka*, whereas the lowest (7.46) mean value of pH was recorded in Maliya *taluka*. The overall EC values ranged from 0.38 to 16.2 with mean value of 3.10 dSm^{-1}.The highest (4.88 dSm^{-1}) and lowest (1.57 dSm^{-1}) EC mean value were recorded with Talala and Sutrapada *talukas,* respectively. The higher value of EC (4.88 dSm^{-1}) is an indicative of potential development of saline soils in Junagadh district. The variation in pH and EC might be due to hydrological influence of seawater. This finding is in conformity with the findings of Verma et al., (2003) and Polara and Kabaria (2005).

The overall mean value of SAR was 8.80 and it was varied from 1.23-28.9. The lowest (5.15) and highest (12.9) mean SAR values were registered in samples collected from Vanthali and Talala *talukas*, respectively (Table 1). All the *talukas*, except Mendarda, Seshod, Mangrol, Talala, and Una *talukas* had mean SAR value less than 10. The RSC values ranged from 0.0 to 8.80 with mean value of 0.54 me l^{-1}. Overall 87.7, 5.5 and 6.8 per cent samples fall under safe, marginal and unsafe classes of RSC, respectively. Most of (85.0 per cent) well/tube well water samples had RSC value less than 1.25 and only 8.2 per cent samples had RSC value greater than 2.5 mel^{-1}. This finding is in conformity with those of Maliwal and Timbadia (2000). The overall mean value of SSP was 64.3, which varied from 33.3 to 87.1 (Table 1). The highest mean SSP value (71.8) was recorded in Talala and Una *taluka*, respectively. Whereas, the lowest mean value (57.4) was recorded in Vanthali *talukas*.

The relative proportion of cations showed that overall the highest proportion of Na^{+} (22.28 me l^{-1}) was observed, which was followed by Ca^{++} (5.55 me l^{-1}), Mg^{++} (4.70 me l^{-1}) and K^{+} (0.11 me l^{-1}). The presence of large proportion of Na^{+} in most of the area under investigation is indicative of a potential danger for the alkalinity hazards (Table-2). The overall concentration of Ca^{++}, Mg^{++}, Na^{+} and K^{+} were varied from 0.86 to 35.1, 0.40 to 30.6, 1.34 to

Table 1: Taluka-wise Range and Mean Values of EC, pH and Indices of Irrigation Water

Name of talukas	*EC (dSm^{-1})*	*pH*	*RSC*	*SSP*	*SAR*
Junagadh	0.65-4.42 (2.23)	1.02-8.52 (7.71)	0.0-2.47 (0.41)	34.4-80.0 (62.7)	1.53-15.3 (7.51)
Bhesan	0.38-6.05 (2.17)	6.98-8.45 (7.47)	0.0-3.30 (0.17)	34.9-79.3 (63.9)	1.23-13.8 (7.58)
Visavadar	1.33-11.4 (2.70)	7.03-8.71 (7.72)	0.0-5.70 (0.32)	46.1-81.0 (64.1)	3.46-17.5 (7.91)
Mendarda	0.90-12.8 (4.16)	7.48-8.57 (7.82)	0.0-3.60 (0.45)	48.4-80.4 (67.3)	4.14-28.9 (11.1)
Vanthali	0.65-4.26 (1.57)	6.97-8.51 (7.71)	0.0-3.30 (0.55)	34.4-72.3 (57.4)	1.53-11.1 (5.15)
Manavadar	0.38-6.52 (2.23)	6.75-8.15 (7.49)	0.0-0.0 (0.0)	35.0-78.5 (62.8)	1.23-15.7 (7.58)
Keshod	0.45-16.2 (4.17)	7.03-8.78 (7.92)	0.0-7.40 (0.94)	48.4-77.0 (66.6)	2.54-23.8 (10.1)
Mangrol	0.38-16.1 (4.60)	6.75-8.57 (7.56)	0.0-4.50 (0.28)	35.0-79.7 (64.3)	1.23-22.9 (10.8)
Maliya	0.99-6.52 (2.69)	6.75-8.15 (7.46)	0.0-0.0 (0.0)	39.4-75.2 (62.1)	2.49-15.7 (7.92)
Talala	0.45-16.1 (4.88)	7.03-8.86 (8.11)	0.0-5.50 (1.26)	53.6-87.1 (71.7)	2.54-23.9 (12.9)
Veraval	0.90-12.8 (3.72)	7.17-8.71 (7.85)	0.0-5.22 (0.95)	47.2-79.5 (65.3)	2.85-23.3 (9.61)
Sutrapada	0.59-4.26 (1.57)	7.02-8.51 (7.66)	0.0-3.60 (0.34)	33.3-80.8 (57.9)	1.53-14.1 (5.76)
Kodinar	0.38-6.05 (2.28)	6.99-8.74 (7.59)	0.0-8.80 (0.89)	35.0-80.0 (69.4)	1.23-15.3 (7.83)
Una	0.90-16.2 (4.40)	7.12-8.71 (7.87)	0.0-6.70 (1.06)	46.1-87.1 (71.8)	3.46 -27.2 (12.7)
Overall	0.38-16.2 (3.10)	6.75-8.86 (7.71)	0.0-8.80 (0.54)	33.3-87.1 (64.3)	1.23-28.9 (8.88)

120.2 and 0.00 to 1.95 me l^{-1}, respectively (Table 2).The relative proportion of anions showed that the highest overall mean value of 22.16 mel^{-1} was recorded for Cl^- and it was followed by HCO_3^- (6.57 mel^{-1}), SO_4–(1.65 mel^{-1}) and CO_3–(0.32 mel^{-1}), respectively (Table 2). The overall concentration of CO_3^-, HCO^3,–Cl^- and SO4– were ranged from 0.0-4.80,1.90-29.5,1.80-130.7 and 0.0-12.58 mel^{-1}, respectively (Table 1). The higher sodium and chloride concentration of ground water is a definite proof of seawater intrusion. This is in close agreement with finding of Maliwal and Timbadia (2000).

The percentage distribution of water samples into different quality indices viz., EC and SAR showed that overall 0, 7.3, 55.9 and 36.8 per cent samples were found under C_1, C_2, C_3 and C_4 classes of EC, respectively (Table 3). About 70 per cent water samples of Vishavadar talukas fell under C_3 classes of EC. The Junagadh, Mangrol,Talala and Una taluka registered more

Table 2: Taluka-wise Range and Mean Values of Cations of Well/ Tube Well Water Samples of Coastal Junagadh District

Name of Talukas	Water Soluble Cations (me L^{-1})				Water Soluble Anions (me L^{-1})			
	Ca^{++}	Mg^{++}	Na^{+}	K^{+}	CO_3^{-}	HCO_3^{-}	Cl^{-}	SO_4^{-}
Junagadh	1.10-8.80 (4.11)	1.10-7.90 (3.65)	2.21-33.39 (15.59)	0.01-1.95 (0.18)	0.0-1.72 (0.25)	2.40-12.80 (6.47)	4.0-32.50 (15.19)	0.0-3.79 (0.76)
Bhesan	1.40-15.0 (4.09)	0.40-10.0 (3.45)	1.34-40.43 (14.82)	0.02-0.46 (0.10)	0.0-0.80 (0.07)	1.90-9.70 (4.96)	1.80-51.21 (16.73)	0.0-5.47 (0.43)
Visavadar	2.10-17.10 (4.75)	1.10-16.4 (3.98)	6.97-64.2 (16.92)	0.03-0.50 (0.12)	0.0-3.10 (0.29)	2.70-18.10 (5.82)	7.20-54.94 (15.56)	0.0-8.51 (1.24)
Mendarda	1.1-20.6 (6.8)	0.4-20.8 (6.32)	6.01-109.54 (31.63)	0.02-0.07 (0.14)	0.0-1.5 (0.28)	2.5-16.32 (6.86)	3.6-108.8 (30.93)	0.0-12.58 (2.90)
Vanthali	1.10-8.80 (3.34)	1.10-7.90 (2.91)	2.21-30.67 (9.52)	0.01-0.14 (0.05)	0.0-1.72 (0.32)	1.90-9.40 (4.93)	3.6-32.5 (10.07)	0.0-3.78 (0.19)
Manavadar	1.40-11.20 (4.05)	0.40-8.20 (3.36)	1.34-48.92 (15.10)	0.02-0.28 (0.07)	0.0-0.0 (0.0)	1.9-7.9 (4.83)	1.8-55.5 (16.3)	0.0-6.87 (1.33)
Keshod	0.86-35.10 (7.99)	1.12-23.60 (6.00)	2.70-119.25 (29.90)	0.02-1.04 (0.16)	0.0-4.8 (0.69)	2.0-25.48 (9.01)	2.4-130.76 (28.97)	0.0-10.38 (2.28)
Mangrol	1.40-35.10 (8.49)	0.40-30.60 (6.84)	1.34-110.25 (34.10)	0.02-1.04 (0.12)	0.0-2.3 (0.19)	1.9-25.48 (6.15)	1.8-130.76 (36.45)	0.0-12.58 (3.47)
Maliya	1.40-11.20 (5.03)	1.10-10.0 (4.27)	4.75-48.92 (18.10)	0.02-0.28 (0.08)	0.0-0.0 (0.0)	2.05-9.70 (5.28)	6.0-55.5 (20.14)	0.0-6.87 (1.85)
Talala	0.86-35.10 (8.41)	0.60-23.60 (6.57)	2.70-108.26 (37.77)	0.0-1.04 (0.14)	0.0-4.12 (1.16)	2.0-29.5 (10.72)	2.4-130.76 (34.13)	0.0-11.09 (3.04)
Veraval	1.0-26.40 (6.69)	1.0-20.8 (6.09)	4.28-100.55 (27.07)	0.02-0.22 (0.08)	0.0-2.0 (0.34)	2.50-10.9 (6.55)	3.6-108.8 (28.18)	0.0-10.20 (1.80)
Sutrapada	1.30-8.80 (3.35)	1.30-7.90 (2.64)	2.21-30.67 (10.27)	0.01-0.18 (0.06)	0.0-1.72 (0.23)	2.4-9.86 (4.98)	2.0-32.5 (9.95)	0.0-3.79 (0.53)

(Contd...)

Name of Talukas	Water Soluble Cations (me L^{-1})				Water Soluble Anions (me L^{-1})			
	Ca^{++}	Mg^{++}	Na^{+}	K^{+}	CO_3^{-}	HCO_3^{-}	Cl^{-}	SO_4^{-}
Kodinar	1.40-13.50 (4.24)	0.40-10.0 (3.65)	1.34-40.43 (15.91)	0.02-0.42 (0.07)	0.0-3.7 (0.23)	1.9-12.8 (6.38)	1.8-47.21 (15.74)	0.0-5.47 (0.66)
Una	1.30-30.10 (6.41)	0.40-23.60 (6.10)	6.01-120.25 (34.18)	0.01-1.04 (0.17)	0.0-3.10 (0.52)	2.5-25.48 (9.11)	3.6-130.76 (31.89)	0.0-10.38 (2.57)
Overall	0.86-35.10 (5.55)	0.40-30.60 (4.70)	1.34-120.25 (22.28)	0.0-1.95 (0.11)	0.0-4.8 (0.32)	1.9-29.5 (6.57)	1.8-130.76 (22.16)	0.0-12.58 (1.65)

Note: Values in parenthesis are mean values.

Table 3: Percentage Distribution of Water Samples into Different EC And SAR Classes (USDA, Richard, 1954)

Name of Taluka	EC				SAR				Irrigation water quality classes											
									C_2				C_3				C_4			
	C_1	C_2	C_3	C_4	S_1	S_2	S_3	S_4	S_1	S_2	S_3	S_4	S_1	S_2	S_3	S_4	S_1	S_2	S_3	S_4
Junagadh	0	15	35	50	50	30	20	0	15	0	0	0	35	0	0	0	0	30	20	0
Bhesan	0	5	60	35	55	35	10	0	5	0	0	0	50	10	0	0	0	25	10	0
Visavadar	0	0	70	30	45	35	15	5	0	0	0	0	45	25	0	0	0	10	5	15
Mendarda	0	0	55	45	40	30	15	15	0	0	0	0	40	15	0	0	0	15	15	15
Vanthali	0	25	55	20	80	20	0	0	25	0	0	0	55	0	0	0	0	20	0	0
Manavadar	0	10	60	30	50	45	5	0	5	5	0	0	45	15	0	0	0	20	10	0
Keshod	0	5	55	40	30	30	35	5	5	0	0	0	25	20	10	0	0	10	25	5
Mangrol	0	5	45	50	40	25	20	15	5	0	0	0	35	5	5	0	0	20	15	15
Maliya	0	5	55	40	45	40	15	0	0	5	0	0	45	10	0	0	0	25	15	0
Talala	0	5	45	50	25	10	50	15	5	0	0	0	20	10	15	0	0	0	35	15
Veraval	0	0	60	40	40	40	10	10	0	0	0	0	40	20	0	0	0	20	10	10
Sutrapada	0	15	65	20	70	25	5	0	15	0	0	0	55	10	0	0	0	15	5	0
Kodinar	0	5	50	45	45	30	25	0	5	0	0	0	40	10	0	0	0	20	25	0
Una	0	0	50	50	30	25	35	10	0	0	0	0	30	15	5	0	0	10	30	10
Overall	0	6.79 (19)	54.29 (152)	38.93 (109)	46.07 (129)	30 (84)	18.57 (52)	5.36 (15)	6.07 (17)	1.07 (3)	0	0	40 (112)	11.79 (33)	2.5 (7)	0	0	17.86 (50)	16.43 (46)	5.36 (15)

Note: Values in parenthesis are mean values.

than 50 per cent samples fell under C_4 classes of EC. So, the salinity hazard of irrigation water is the cause of the development of secondary salinisation in the soils of Junagadh district. The overall 46.0, 30.0, 18.6 and 5.4 per cent samples were found under S_1, S_2, S_3 and S_4 classes of SAR, respectively (Table 3). Based on EC and SAR of irrigation, water samples, 40.0, 17.9, 16.4, and 11.8 per cent were found under C_3S_1, C_4S_2, C_4S_3, and C_3S_2, respectively. Similar finding were also reported by Polara and Kabaria (2005) and Maliwal and Timbadia (2000) for Amreli district of Gujarat and Verama *et al.*, (2003) for Churu district of Rajasthan.

Conclusion

From the above results it can be concluded that the well/tube well water of Junagadh district are alkaline in reaction and saline in nature and poor in quality. Almost half of the waters samples of well/tube well from cultivator's fields were saline (EC 0.75 dSm^{-1} and above).This is indicative of the potential development of saline soils in the Bhavnagar district. Therefore, proper management practices viz., use of amendments, conjunctive use of saline well /tube well waters by growing low water requirement crops and salt tolerant varieties.

References

Maliwal, G.L. and N.K. Timbadia (2000). Quality of underground irrigation water and their effect on soil properties of coastal area of Amreli district of Gujarat. Journal of the Indian Society of Coastal Agricultural Research, 18:30-35.

Paliwal, K.V. and G.L. Maliwal (1971).Some relationships between constituents of irrigation water and properties of irrigated soils of western Rajasthan. Journal of the Indian Society of. Soil Science, 19: 299-303.

Polara, J.V. and B.D. Kabaria (2005). Quality of underground irrigation water in Coastal Amreli district of Gujarat. Proceeding of 8th National Seminar on Strategies for Improving Farming and Ecology Security of Coastal Region during 21-24 December, 2005.

Richards, L.A. (1954). "Diagnosis and Improvement of Saline and Alkali Soils". Hand Book No. 60, Oxford and IBH pub. Co., Calcutta-16.

Verma, B.L.; Y. Sharma, and R.A. Sighania, R.A. (2003). Quality of under ground irrigation waters of Charu district in Rajasthan. Journal of the Indian Society of Soil Science, 51(2): 214-216.

36

Porous Pavement: A new way of recharging ground water

Neelima Palagani, Alka Singh, Tatte Sumathi and M Sree Devi

Abstract

Porous pavement is one of the unique and effective methods to mitigate the important environmental issues and support green and sustainable growth. It plays a vital role in rain water harvesting, a natural way of renewing natural resources. In contrast to the traditional pavements, porous paving allows the rain water to percolate through it, where the water is naturally filtered and make free from the pollutants. At the same time a porous pavement can reduce the storm water runoff and recharge the groundwater, so the water table can be raised. It reduces the erosion and siltation of surface water bodies. The heavy metals that fall on them are captured, preventing them from washing downstream and accumulating inadvertently in the environment. Naturally occurring micro-organisms in the void spaces digest car oils, leaving little but carbon dioxide and water. Adaptation of porous pavement particularly in urban areas can have multipurpose benefits on ecological balance as it can resist traffic above the ground and effectively harvesting the water below the ground. By this way porous pavement offers a wide range of advantages in view of maintaining water quality, raising the water table, balancing the water cycle, reduces the storm water runoff and erosion. Considering the advantages of porous pavement it seems to be the right pave to adapt it for sustainable water use management in the climate smart agriculture.

Introduction

As urbanisation increases in India and many more parts on the globe the problem of water logging and requirement of drainage is also increasing. This is partly due to impervious nature of the impervious and concrete pavements. Pervious pavement which has an open cell helps significantly to provide high permeability due to its interconnected pores. Pervious pavement (also called porous pavement or permeable pavement) is a special type of pavement with a high porosity that allows water from precipitation and other sources to pass directly through, thereby reducing the runoff from a site and allowing groundwater recharge. India is facing a typical problem of groundwater table falling at a fast rate due to reduced recharge of rainwater into subsoil and unplanned water withdrawal for agriculture and industry by pumping. NFC if adopted for construction of pavements, platform/walkways, parking lots designed for lighter load. It can become a component of rainwater harvesting schemes being prepared by Government of India on a priority basis. It also reduces the runoff from the pavement areas hence reduce load on drainage systems. In the current situation larger amount of rainwater falling on impervious surfaces such as parking lots, driveways, sidewalks, and streets rather than soaking into the soil is subjected to run off. This creates an imbalance in the natural ecosystem and leads to problems *viz.*, erosion, floods, ground water level depletion and pollution of rivers, as rainwater rushing across pavement surfaces picks up everything from oil and grease spills to de-icing salts and chemical fertilisers. In this contest porous paving offers a great solution. Porous pavement is a permeable pavement surface with a stone reservoir underneath. The reservoir temporarily stores surface runoff before infiltrating it into the subsoil. Runoff is thereby infiltrated directly into the soil and receives some water quality treatment. Porous pavement often appears the same as traditional asphalt or concrete but is manufactured without "fine" materials, and instead incorporates void spaces that allow for infiltration.

By stopping stormwater from pooling and flowing away, porous paving can help recharge underlying aquifers and reduces peak flows and flooding. That means that streams flow more consistently and at cooler temperatures, contributing to healthy ecosystems. Stormwater pollutants are broken down in the soil instead of being carried to surface waters. Below is a graphic that illustrates the relationship between surface flow, groundwater flow and aquifers.

Need of Choosing Permeable Pavement

- To prevent the soil erosion
- Better storm water management
- Increases the cropping area
- Helps in recharging the groundwater which will

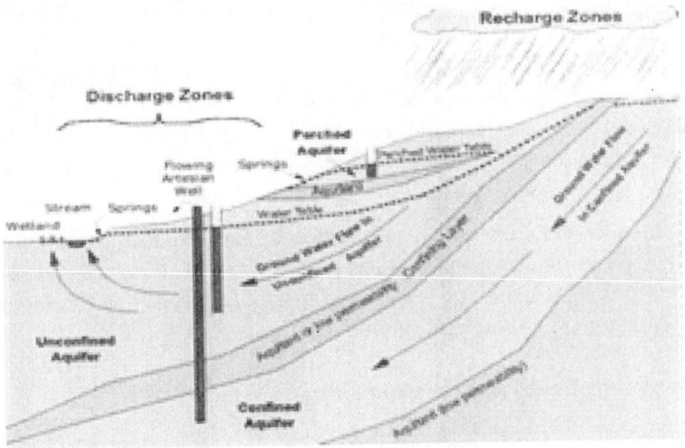

Illustration: http://imnh.isu.edu/digitalatlas/hvdr/concepts/gwater/imgs/6comp.ipg

- Benefit plants and gardening
- Enhance the underground water table
- Contribute to the betterment of the environment by
  - reducing the storm water contaminants in streams and rivers
  - reducing stormwater runoff volume
  - balancing the eco system

Benefits of Porous Pavement

Environmental Benefits

Porous pavement efficiently reduced stormwater runo, minimises the expensive storm water management practices. Recharge groundwater, efficient land use by reducing the need for retention of ponds. Porous pavement in the gardens plays a vital role in sustaining the flora and offers best gardening management by providing water for longer periods. Porous paving also mitigates the level of pollutants in the runoff water. Keeps the pavement surfaces dry even in wet situations. It offers better penetration of air and water in to the soil so that the growth of vegetation will be enhanced. It collects the irrigation water and it can be recycled. Allows a project to claim LEED® points. (Leadership in Energy and Environmental Design is a rating system developed by the U.S. Green Building Council to evaluate the environmental performance of a building.)

EconomicBenefits- efficient management reort made in preventing runo during heavy rainfall is prevented. Reduces cost to maintain large detention ponds. A parking lot properly constructed from pervious concrete has a life span ten times as long as an asphalt lot, thereby providing excellent long-term benefits. It is true that the initial costs for pervious pavement may be slightly higher due to the preparation of the sub-base, but those who look long-term will realise the economic benefits (Neetu *et al.*, 2013). Increase the land utilisation value as there is no need to purchase additional land for large ponds for retention and any other filtering systems.

Easy installation. Less money will be needed for labour, construction and maintenance of ponds, drainage systems and other storm water management systems. Easy maintenance that consists primarily of prevention of clogging through pressurised washing and power vacuuming.

Pockets of Application of Porous Pavement

- Pervious concrete as a road pavement
- Pathways and sidewalks
- Residential roads and colonies
- Public and private building compounds
- Gardens and parks
- Parking places
- Slope stabilisation
- Hydraulic structures
- Playing grounds and tennis courts
- Rural roads

Materials used for permeable paving?

1. Porous Asphalt

Porous asphalt is the same as normal asphalt except it is manufactured with the fine materials omitted, leaving open spaces that allows water to filter through to a recharge or drainage system.

Porous Asphalt's primary advantages are:

1. Quantity and flood control
2. Better water quality treatment
3. Recharges the groundwater to underlying aquifers
4. Reduction of stormwater infrastructure (Piping, Catch-basins, Retention ponds, etc.)
5. Allows for reduced salt and sand usage due to low/no blackice development
6. Reduced spray from travelling vehicles, reduced roadway noise
7. Extended pavement life due to well drained base

Typical Porous Asphalt cross-section

4" of Porous asphalt
4" thickness of ¾" crushed stone
8-12" thickness of open graded reservoir subbase
4" thickness of ¾"> crushed stone for frost protection
Soil permeability > 0.5 in/hr

Advantages

Contaminants on the surface of pavement tend to become part of runo but with porous asphalt, they flow down to the soil, where beneficial bacteria and other natural processes cleanse them. Open-graded asphalt roads have shown lower night time surface temperatures as compared to impervious pavements.

Benefits include noise reduction, a decrease in splash and spray kicked up by vehicles in heavy downpours (Shivam Gupta).

2. Pervious Concrete

Pervious concrete is used for a number of applications, but its primary use is in road pavement such as in rural and urban areas. Pervious concrete is a zero-slump, open-graded material consisting of cement, coarse aggregate, admixtures and water. Pervious concrete contains little or no fine aggregates such as sand, it is sometimes referred to as "no-fines" concrete.

Pervious concrete pavement is a unique and effective means to achieve important environmental issues and support green, sustainable growth. By capturing storm water and allowing it to seep into the ground, porous concrete is instrumental in recharging the groundwater and reducing storm water runoff.

Properties of pervious concrete (Neetu *et al.*, 2013)

Density and porosity

The density of pervious concrete depends on the properties and proportions of the materials used, and on the compaction procedures used in placement. In place densities on the order of 1600 kg/m^3 to 2000 kg/m^3 are common, which is in the upper range of lightweight concretes. The porosity of pervious concrete largely depends on aggregate size.

Permeability

The flow rate through pervious concrete depends on the materials and placing operations. Typical flow rates for water through pervious concrete are

288 in./hr, 120 L /m^2/min, or 0.2 cm/s to 770 in. /hr, 320 L /m 2/min, or 0.54 cm/s, with rates up to 1650 in. /hr, 700 /m 2/min,1.2 cm/s and higher having been measured in the laboratory.

Compressive strength

Pervious concrete mixtures can develop compressive strengths in the range 3.5 MPa to 28 MPa, which is suitable for a wide range of applications. Typical values are about 17 MPa. As with any concrete, the properties and combinations of specific materials, as well as placement techniques and environmental conditions, will dictate the actual in-place strength.

Flexural strength

Flexural strength in pervious concretes generally ranges between about 1 MPa and 3.8 MPa. Many factors influence the flexural strength, particularly degree of compaction, porosity, and the aggregate: cement (A/C) ratio. However, the typical application constructed with pervious concrete does not require the measurement of flexural strength for design.

Shrinkage Drying

Shrinkage of pervious concrete develops sooner, but is much less than conventional concrete. Specific values will depend on the mixtures and materials used, the material's low paste and mortar content is a possible explanation. Roughly 50 per cent to 80 per cent of shrinkage occurs in the first 10 days, compared to 20 per cent to 30 per cent in the same period for conventional concrete. Because of this lower shrinkage and the surface texture, many pervious concretes are made without control joints and allowed to crack randomly.

Benefits of Pervious concrete

1. It reduces the storm water runoff. 2. Eliminates the need for detention ponds and other costly storm water management practices. 3. Mitigates surface runoff. 4. Replenishes the aquifers and water table. 5. Allows more efficient land development. 6. Prevents water from entering into the stream and also prevents it from being polluted.

Applications of Pervious concrete

Pervious pavement for parking lots. Rigid drainage layers under exterior mall areas. Greenhouse floors to keep the floor free of standing water. Structural wall applications where lightweight or better thermal insulation characteristics, or both, are required. Pavements, walls, and floors where better acoustic absorption characteristics are desired. Base course for city streets, county roads, driveways, and airports. Surface course for parking lots, tennis courts, zoo areas, and animal barns and stalls. Bridge embankments. Swimming pool decks. Beach structures and seawalls.

3. Concrete/brick Pervious Pavers

Precast concrete or brick manufactured in many sizes and shapes are laid with a drainage base and permeable joint material, allowing water to slowly seep into the ground.

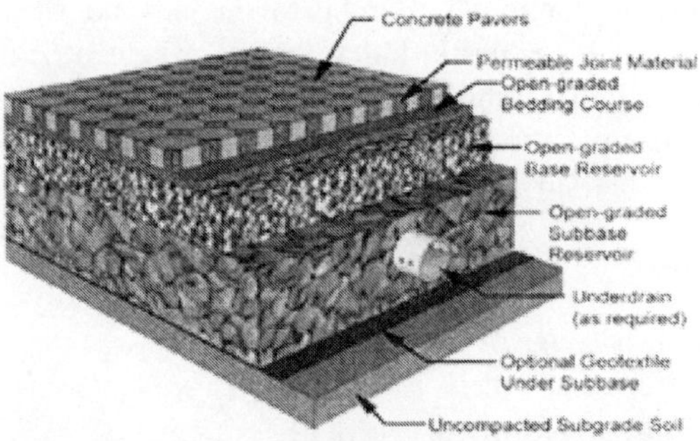

There are different types of brick pavers which include:

Open-Celled pavers

Open-celled pavers are made by installing a plastic or concrete grid over a bed of drainage material and soil. Then the voids are seeded with grass or turf plugs are embedded. Alternatively the voids may be filled with aggregate. They must be constructed with drainage.

Spot clogging may be fixed by drilling 0.5" holes through the pavement layer every few feet.

Concrete/brick pervious pavers

Periodically add joint material (sand) to replace material that has been transported.

Open-Celled Pavers with aggregate fill

Refill displaced gravel when necessary. Plastic cells may need to be replaced periodically.

Open-Celled Pavers with grass fill

Conclusion

Porous pavement is a cost-effective and environmentally friendly solution to support sustainable construction and friendly environment. Its ability to capture storm water and recharge groundwater while reducing storm water runoff enables. Porous pavement to play a vital role. Porous pavements a modern sustainable option with very high efficiency. Porous pavements an ideal solution to control soil erosion, storm water, recharging of groundwater, and flood control at downstream and sustainable land management. As it is a low cost construction, if it gets utilised in Indian context then it proves to be very beneficial to ameliorate environmental issues and water logging problems which are the major issues in India. Porous pavement is the modern way to build the green India. If this kind of porous paving will get implemented in all the pockets of application then India become self-sufficient in water utilisation and creates an aura in better flora and fauna development.

References

http://www.concrete.org

http://imnh.isu.edu/digitalatlas/hvdr/concepts/gwater/imgs/6comp.ipg

http://www.concretenetwork.com/pervious

Neetu, B. Y., Jayesh, A. S. and Rushabh, A. S. (2013). Pervious Concrete: Solution for Low Cost Construction. International Journal of Innovative Science and Modern Engineering. 1(10)-38-41.

Shivam Gupta. (2010). Design of Porous Pavements Termpaper - CE682: Infrastructure and Transportation Planning. 1-13.

List of Contributors

Authors	Page No.
A. B. Kheni, College of Agricultural Engineering and Technology, JAU, Junagadh, Gujarat, India	253
A.N. Kunapara, Senior Engineer, Netafim Irrigation India Pvt. Ltd, Vadodara, (India).	158
Alka Singh, Department of Floriculture and Landscape Architecture, ACHF, NAU, Navsari	407
Arun Jhajharia, Department of Farm Engineering, Institute of Agricultural Sciences, Banaras Hindu University, Varanasi 221005 Uttar Pradesh	97, 309
Bhavin Ram, Research Associate, Polytechnic in Agril. Engg., Anand Agricultural University, Dahod-389151 (India).	158
Chaitali V. Thakare, Ex - M. Tech. Student, Department of Soil and Water Conservation Engg., Dr. PDKV, Akola	180
D.M. Paradava, Research Associate,College of Agricultural Engineering and Technology, Junagadh Agricultural University, Junagadh-362 001	70, 148, 282, 294, 320
F.G. Sayyad, Assistant Professor, Polytechnic in Agril.Engg., Anand Agricultural University, Dahod- 389151 (India).	158
Falguni. P. Parekh, Offg. Director & Associate Professor, Water Resources Engineering and Management Institute, Faculty of Technology Engineering, The M.S. University of Baroda, Samiala 391 410, Gujarat, India.	109
G. Pranuthi, Research Scholar, Department of Water Resources Development and Management, Indian Institute of Technology, Roorkee-247667, India.	9
G.L. Chunale, Assistant Professor of Agril. Engineering, College of Agriculture, Karad, District Satara (M.S.).	34
G.U. Satpute, Associate Professor, Department of Soil and Water Conservation Engineering, Dr. PDKV., Akola.	34, 108
G.V. Prajapati, Centre of Excellence on Soil and Water Management, R.T.T.C, Junagadh Agricultural University, Junagadh-362 001 (Gujarat), India	139, 253
Goyal Pankaj, Deccan College Post-Graduate and Research Institute, (Deemed University) Pune- 411006.	376
H. J. Vala, Department of Soil and Water Engineering, College of Agricultural Engineering and Technology, Junagadh Agricultural University, Junagadh – 362 001	346
H. P. Patell, Assistant Professor, College of Agriculture, NAU, Waghai.	364

Authors	Page No.
Mausumi Raychaudhuri, Indian Institute of Water Management, Bhubaneswar, Odisha-751023	193, 202
N. K. Gontia, College of Agricultural Engineering and Technology, Junagadh Agricultural University, Junagadh-362001.	267
N. K. Gontia, Principal and Dean, College of Agricultural Engineering and Technology Junagadh Agricultural University, Junagdh-362001	232
N. S. Vithlani, Department of Soil and Water Engineering, College of Agricultural Engineering and Technology, Junagadh Agricultural University, Junagadh – 362 001	346
N.H. Patel, Temporary Teaching Assistant, Water Resources Engineering and Management Institute, Faculty of Technology Engineering, The M.S. University of Baroda, Samiala 391 410, Gujarat, India.	224
N.J. Rankja, Department of Agricultural Statistics, College of Agriculture, Junagadh Agricultural University, Junagadh.	172
Navneet Singh Parmar, Departments of Soil and Water Conservation Engineering, JNKVV Jabalpur-482004 MP, India	309
Neelima Palagani, Department of Floriculture and Landscape Architecture, ACHF, NAU, Navsari	407
P. D. Sabale, Deccan College Post-Graduate and Research Institute, (Deemed University) Pune- 411006.	376
P. P. Rathod, Principal and Dean, College of Agricultural Engineering and Technology Junagadh Agricultural University, Junagdh-362001	232
P.A. Damor, Technical Assistant College of Agricultural Engineering and Technology, Junagadh Agricultural University, Junagadh	139, 253
P.B. Maraviya, Department of Agricultural Statistics, College of Agriculture, Junagadh Agricultural University, Junagadh.	172
P.B. Vekariya, Assistant Professor, ICAR Network Project on National Initiatives on Climate Resilient Agriculture (NICRA), AICRP on Irrigation Water Management.	70, 148, 282, 294, 320
P.M. Kyada, Subject Matter Specialist, Krishi Vigyan Kendra, Lokbharti Gramvidhyapith, Sanosara, Bhavnagar	59
R. Subbaiah, College of Agricultural Engineering and Technology Junagadh Agricultural University, Junagadh-362001	242
R. Subbaiah, Research Scientist (Agril. Engg), Centre of Excellence on Soil and Water Management, R.T.T.C, Junagadh Agricultural University, Junagadh-362 001 (Gujarat), India	45, 119, 129, 139, 253
R.A. Kachhadiya, Subject Matter Specialist, Krishi Vigyan Kendra, Lokbharti Gramvidhyapith, Sanosara, Bhavnagar	59
R.C. Purohit, Professor, Maharana Pratap University of Agriculture Technology, Udaipur-323001 Rajasthan, India.	84
R.J. Patel, Assistant Professor,College of Agricultural Engineering and Technology, Junagadh Agricultural University, Junagadh-362 001	70, 148, 282, 294, 320

Water Management and Climate Smart Agriculture

Proceedings of National Seminar on
Water Management and Climate Smart Agriculture

(13th – 14th February, 2015)

Vol – III

Adaption of Climatic Resilient Water Management and Agriculture

R. Subbaiah

G. V. Prajapati

Centre of Excellence on Soil & Water Management
Research, Testing & Training Centre
Junagadh Agricultural University, Junagadh, Gujarat, India

GYAN

ISBN: 978-81-212-1317-2

Price: ₹ 4500 (set of 3 Vols)
₹ 1700 (Vol–III)

First Published, 2015

Published by

Gyan Publishing House
5, Ansari Road, Daryaganj,
New Delhi-110002
Phones: 9811692060
E-mail: books@gyanbooks.com

Printed at: G. Print Process, Delhi

Cataloging in Publication Data—DK
Courtesy: D.K. Agencies (P) Ltd. <docinfo@dkagencies.com>

National Seminar on Water Management and Climate Smart Agriculture (2015 : Junagadh Agricultural University)
Proceedings of National Seminar on Water Management and Climate Smart Agriculture (13th-14th February, 2015) / R. Subbaiah, G.V. Prajapati.
3 volumes cm
National seminar organized by Junagadh Agricultural University.
Includes bibliographical references.
Contents: vol. I. Vulnerability of agriculture to climatic aberrations — vol. II. Climate change impact on water cycle — vol. III. Adaption of climatic resilient water management and agriculture.
ISBN 9788121213172

1. Water-supply, Agricultural—India—Management—Congresses. 2. Climatic changes—India—Congresses. 3. Crops and climate—India—Congresses. 4. Agriculture—Environmental aspects—India—Congresses. 5. Agriculture—Economic aspects—India—Congresses. I. Subbaiah, R., editor. II. Prajapati, G. V., editor. III. Junagadh Agricultural University.organizer. IV. Title.

S494.5.W3N38 2015 DDC 333.9130954 23

Contents

1

The Gujarat Way of Adaption and Mitigation Strategies of Climatic Change on Agricultural Productivity through Micro Irrigation Systems

R.K. Sugoor and S. Raman

Abstract

Agriculture in developing countries must undergo a significant transformation in order to meet the related challenges of achieving food security and responding to climate change. Projections based on population growth and food consumption patterns indicate that agricultural production will need to increase by at least 70 per cent to meet demands by 2050. Most estimates also indicate that climatechange is likely to reduce agricultural productivity, production stability and incomes in some areas that already have high levels of food insecurity. Developing climate-smart agriculture like adoption of Micro Irrigation, is thus crucial to achieving future food security and climate change goals. Climate-Smart Agriculture seeks to increase productivity in an environmentally and socially sustainable way, strengthen farmers' resilience to climate change, and reduce agriculture's contribution to climate change by reducing greenhouse gas emissions and increasing carbon storageon farmland. Climate-smart agriculture includes proven practical techniques - such as mulching, intercropping, conservation agriculture, crop rotation, integrated crop-livestock management,agro forestry, improved grazing, and improved water management. This paper examines the water management in the form of adoption of Micro Irrigation System in agriculture production, a crucial factor that requires immediate attention to maintain sustainable growth in agriculture.The Government of Gujarat formed a special purpose vehicle called GGRC to implement

Micro Irrigation Scheme in uniform mode throughout the State. The practices and procedure followed by GGRC led to great success in bringing more area under assured irrigation system thereby increasing growth in agriculture as well as income of farmers. It is all about what GGRC done is getting existing technologies off the shelf and put into the hands of farmers by promoting new technologies such as Micro Irrigation coupled with other agro input services to meet the demands of the changing climate.

Introduction

Over the past six decades world agriculture has become considerably more efficient. Improvements in production systems and crop and livestock breeding programmes have resulted in a doubling of food production while increasing the amount of agricultural land by just 10 per cent. However, climate change is expected to exacerbate the existing challenges faced by agriculture. The purpose of this paper is to highlight how one can address the issue of climatic change by adopting smart agriculture technology like Micro Irrigation System coupled with other scientific agro inputs to ensure the food security by maintaining a sustainable growth in agriculture production.

Estimates show that world population will grow from the current 6.7 billion to nine billion by 2050 with most of the increase occurring in South Asia and sub-Saharan Africa. Taking into account the changes in the composition and level of consumption associated with growing household incomes, FAO estimates that feeding the world population will require a 70 per cent increase in total agricultural production (Bruinsma, 2009).

At the same time, climate change threatens production's stability and productivity. In many areas of the world where agricultural productivity is already low and the means of coping with adverse events are limited, climate change is expected to reduce productivity to even lower levels and make production more erratic IPCC 2007).

Long-term changes in the patterns of temperature and precipitation, that are part of climate change, are expected to shift production seasons, pest and disease patterns, and modify the set of feasible crops affecting production, prices, incomes and ultimately, livelihoods and lives. Preserving and enhancing food security requires agricultural production systems to change in the direction of higher productivity and also, essentially, lower output variability in the face of climate risk and risks of an agro-ecological and socio-economic nature. In order to stabilise output and income, production systems must become more resilient, i.e. more capable of performing well in the face of disruptive events. More productive and resilient agriculture requires transformations in the management of natural resources (e.g. land, water, soil nutrients, and genetic resources) and higher efficiency in the use of these resources and inputs for production. Transitioning to such systems could also generate significant mitigation benefits by increasing carbon sinks, as well as reducing emissions per unit of agricultural product.

Transforming small holder systems is not only important for food security but also for poverty reduction. Countries like India, increasing productivity to achieve food security is clearly a priority, which is projected to entail a significant increase in emissions from the agricultural sector in developing countries (IPCC, 2007). Achieving the needed levels of growth, but on a lower emissions trajectory will require a concerted effort to maximise synergies and minimise tradeoffs between productivity and mitigation. Ensuring that institutions and incentives are in place to achieve climate-smart transitions, as well as adequate financial resources, is thus essential to meeting these challenges. To mitigate the effect of climate change on agriculture production system, the Government of Gujarat decided in the year 2005 for investing in Micro Irrigation Scheme (a crucial factor in agriculture production) in a larger scale by forming a special purpose vehicle called Gujarat Green Revolution Company Limited. In this context mitigation finance can play a key function in leveraging other investments to support activities that generate synergies.

Climate Change and Water Resources

The IPCC has projected a rise in temperature of 0.5°C-1.2°C by 2020, 0.88°C-3.16°C by 2050 and 1.56°C-5.44°C, depending on the scenario of future development (IPCC, 2007b). Further there is immense danger on the quality of groundwater especially for the coastal aquifers of Gujarat and Tamil Nadu. In short the climate change will have pronounced effect on the quantity and quality of ground water which is the major source of irrigation in the country. From the climate change viewpoint, India's groundwater hotspots are concentrated in arid and semi-arid areas of western and peninsular India, especially in the states of Punjab, Haryana, Rajasthan, Maharashtra, Karnataka, Gujarat, Telengana, Andhra Pradesh, and Tamil Nadu.

Mitigation and adaptation are the two ways of copping up with the climate change. Through mitigation we aim at reducing the green house gases and by way of adaptation we try to adjust our agriculture with the climatic changes. Adaptation will be necessary to address impacts resulting from the warming which is already unavoidable due to past emissions. Proper irrigation technology can address both the issues, though the mitigation to a lesser extent. Inspite of the fact that in India traditionally used surface storage and gravity flow to water crops is still the major method of irrigation, during the last 40 years; there is a significant shift towards groundwater irrigation through millions of shallow and deep tube wells (Fig. 1).

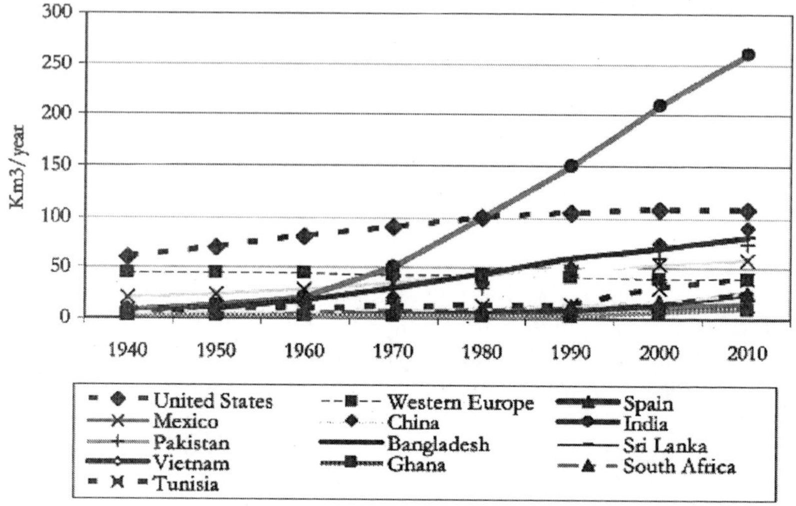

Fig. 1: Growth of Groundwater Use in Different Countries (Shah et al, 2007).

In fact, as per a recent report there are around 17 million tube wells in the country of which 11.7 million are in the Western and Southern India and about five million in Eastern India. Groundwater pumping with electricity and/or diesel accounts for an estimated 16–25 million MT of carbon emissions, which contributes to four to six per cent of India's total carbon emission. Between the two, the carbon emission is more through electrical pumping (Fig. 2) than diesel (Nelson and Robertson, 2008).

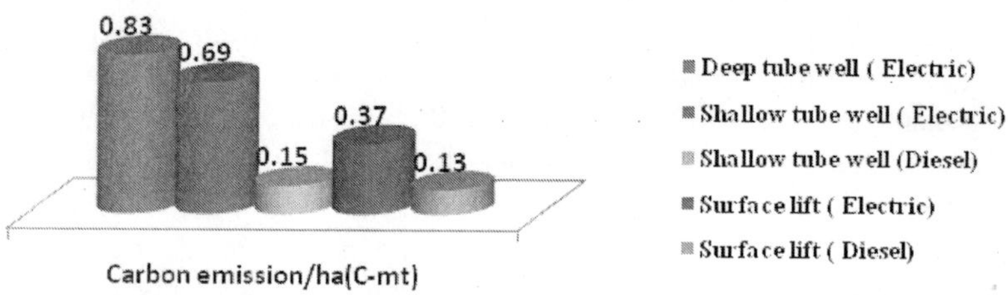

Fig. 2: Estimated Carbon Emission through different Water Lifting Devices.

With decline in water level there is shift towards electrical pumping from diesel. Shukla *et al.* (2003) reported that for every meter decline in pumping water levels increases GHG emissions by 4.37 per cent in Haryana and 6 per cent in Andhra Pradesh. In fact, while 80 per cent of the tube wells in the western and southern India are operated through electricity and 20 per cent through diesel it is reverse in eastern India.

Our present problem in groundwater irrigated areas is arresting the groundwater decline and grows crops with less water. In other words, there is an imminent need to increase the water use efficiency in groundwater irrigated agriculture. Adoption of Climate Smart Agriculture can address these issues. Through Climate Smart Agriculture we aim at Sustainable increase in the productivity, Adaptation and Building Resilience to Climate Change and Reduce and/or Remove Green House Gas Emissions.

The micro irrigation, due to favourable soil and water conditions during crop growth results in sustainable increase in productivity, due to less water requirement will be able to adapt to lesser groundwater available conditions and can arrest the decline groundwater table. Further, use of micro irrigation when adopted in a contiguous large area coupled with fertigation can, in long run, reduce the green house gas emissions. Thus, micro irrigation adoption can achieve the goal of food security through climate- smart agriculture. Realising the importance of micro irrigation under Indian grounds water utilisation scenario and in the wake of climate change conditions, the central and state governments are in the process of popularising the micro irrigation in this country. Gujarat state is implementing this project of popularising micro irrigation in a unique way establishing a special purpose vehicle namely, the Gujarat Green Revolution Company Ltd. exclusively for this purpose.

Micro Irrigation Potentiality in the Country

The potentiality assessment for micro irrigation in each State was made based on source-wise irrigated area in different States and the suitability of different crops for micro irrigation. It has been assessed that there is potentiality of bringing around 45million hectares under micro irrigation in the country. Out of this 45 million about 30 million hectares are suitable for sprinkler irrigation for crops like cereals, pulses and oilseeds in addition to fodder. This is followed by drip with a potentiality of around 12 million hectares and the major crops suitable for this system are cotton, sugarcane, fruits and vegetables, spices and condiments and some pulse crop like red gram etc. In addition to the above two, there is potentiality for bringing an area of about 2.8 million hectares under mini sprinkler for crops like potato, onion, garlic, groundnut, short stature vegetable crops like cabbage, cauliflower etc. Based on a rough estimate of multiple usage of a system for different crops in a field, the system potentiality of micro irrigation business can be estimated as 30 million hectares.

Micro Irrigation Adoption in the Country

As per the NCPAH data, as on March, 2013, the total area under micro irrigation in the country has been reported to be 6.6 million hectares.Out of 6.6 million hectares, sprinkler's contribution is 3.75 million hactares and that of drip is 2.85 million hectares. Though Rajasthan leads with 1.25 million, it is predominated by sprinkler with coverage of 1.16 million hectare leaving only 84,000 hectares under drip (Fig. 3).

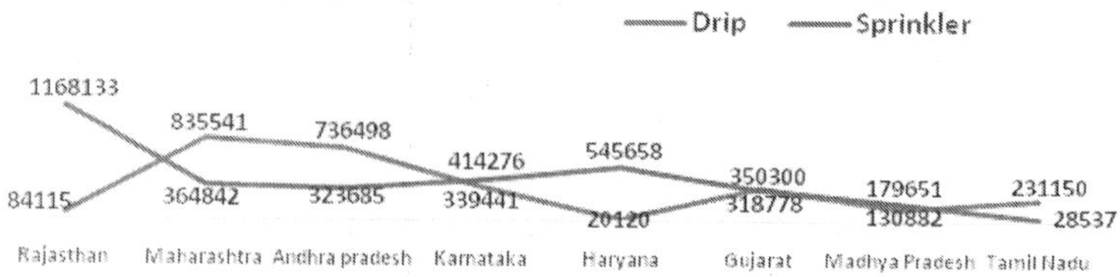

Fig. 3: Drip and Sprinkler system distribution in different States.

On the other hand, in State of Maharashtra, which also has 1.2 million hectares under micro irrigation, the drip : sprinkler is almost 2.3:1. Gujarat occupies the third position in drip adoption and fifth position in overall micro irrigation adoption with an almost equal contribution from sprinkler and drip.

Government's Efforts in Popularising Micro Irrigation

Apart from the research and development activities through the agricultural universities and ICAR institutes, the central and state governments assist the farmers financially.The financial assistance by the central government started from 1990s and the percentage assistance varied in different years. Since last decade it was 50 per cent. But many States included their additional contributions and increased the subsidy to 60/75/90 per cent. In Tamil Nadu, it was as high as 100 per cent for tribal farmers with limitation on hectare. Except in Gujarat where hectarage limit is not there while in all other States the subsidy is limited to five hectares. Recently, previous National Mission on Micro Irrigation was integrated into "On Farm Water Management" under National Mission on Sustainable Agriculture. The subsidy levels have been modified at the central level, but most of the State governments provide the same level of subsidy as existed before.

Gujarat Model – The way of Implementing MI scheme in Gujarat

The central government scheme is being operated by different agencies in the States. Most of them are through the departments of agriculture / horticulture/soil conservation. While, in Andhra Pradesh and Gujarat separate special purpose vehicles were formed for implementation in the form of APMIP in Andhra Pradesh and Gujarat Green Revolution Company Limited (GGRC) in Gujarat.

Though subsidising micro irrigation scheme by different states and central government had been in practice since 1990s, only during last six years the adoption rate is picking up in many States. The Government of Gujarat has brought a major shift in the country in the implementation of micro irrigation scheme through a major policy change, by establishing a special purpose vehicle called GGRC which is later popularly known as Gujarat model. This move of Gujarat government was considered as one of the innovative change brought in the implementation of MIS scheme which call the attention by other States to adopt with suitable modifications. The main reason for the success of Gujarat model in the State are transparency adopted by GGRC in implementing scheme with the use of information technology right from registration to subsidy disbursement to final payment by single nodal agency. The practices adopted by the GGRC were declared as one of the best in the country by the NABARD Consultancy Services in its evaluation report submitted to the Government of India during the year 2009. The salient feature of Gujarat model and result obtained by the State are discussed below.

Implementation Modality for Micro Irrigation Scheme under Gujarat Model

A unique Government Resolution (GR) different from other State was issued by Government of Gujarat in 2005, wherein any farmer can adopt MIS system for any area under any crop as per his choice of GGRC registered MIS supplier. The other features of GR are that farmer who adopts micro irrigation system can get agriculture electricity connection on overriding priority. The model adopted by GGRC is based on business approach keeping farmer interest at the centre (Fig. 4).

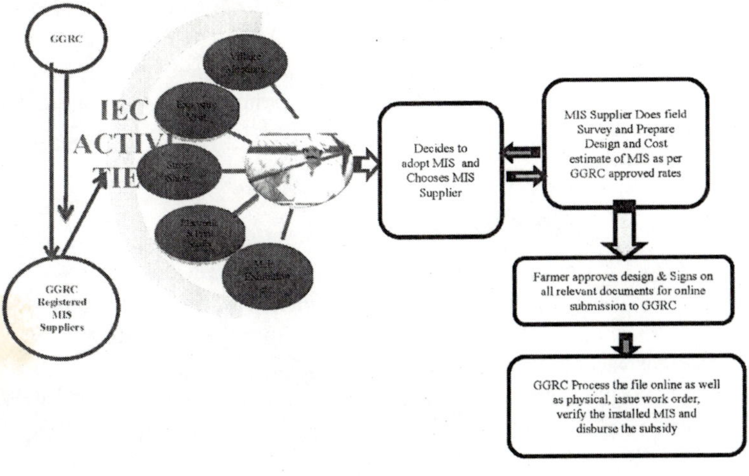

Fig. 4: Implementation modality adopted by GGRC under Gujarat Model.

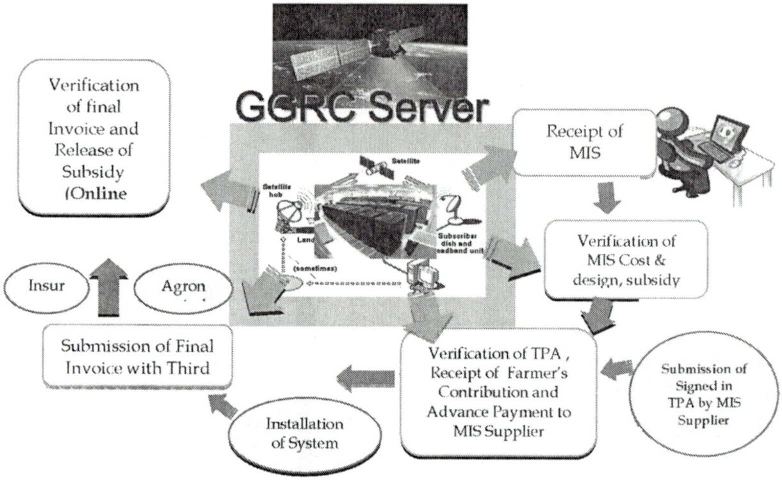

Fig. 5: The MIS Application processing procedure adopted by GGRC under Gujarat

The GGRC developed and established its own state-of-the-art software module to process the application received from the farmer. The procedure adopted for the processing of the MIS application has been shown in Fig. 5.

Unique Features of the Gujarat Model

The success for the Gujarat model adopted by GGRC is attributed to its unique feature inbuilt in the implementation modality. The main features of this model are (1) Simple and Flexible in its Approach, (2) Transparency in its Approach, (3) Effective Quality Monitoring and Assurance System, (4) Use of IT-based application to monitor the installed MIS, (5) Basket of Services, (6) Implementation cost for the government is nil. Above all implementation is based on the concept of PPP Mode and overall improvement in the quality of the service.

The Best Practices Followed by GGRC as Documented by NABARD

The practices and approach followed by GGRC was evaluated by NABARD Consultancy Services (NABCONS) appointed by Govt. of India during the year 2009 and identified GGRC practices mainly (1) Self Sustaining Set-Up, (2) Effective use of Information Technology, (3) Effective Checks and Balances, (4) Innovative package and (5) Objective and Consultative Unit-Cost Revision Methodology as best practices in the country.

Achievement under Gujarat Model

Before establishment of GGRC (1991-2005), the average coverage per annum under micro irrigation was 15000 hectare. After implementation of Gujarat model (2005- December-2014), the total area brought under MIS is 10.35 lakh hectares benefitting more than 6.40 lakh farmers (Fig. 6) with an average coverage per annum approximately one lakh hectares. In fact, since last couple of years the installations are more than two lakh annually.

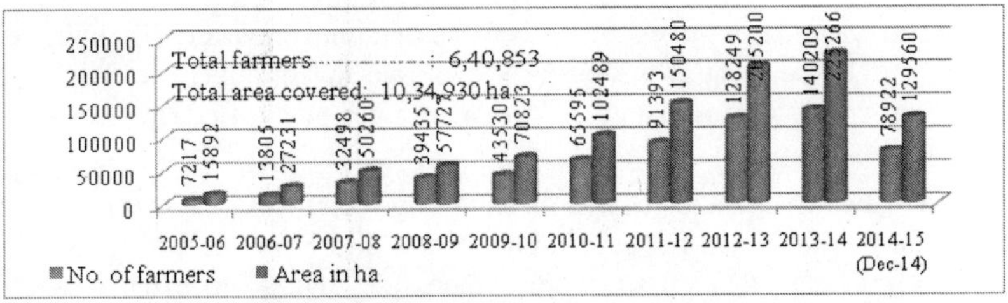

***Fig. 6:** Physical achievement in terms of area and No. of beneficiaries covered under the Micro Irrigation Scheme from 2005-06 to 2014-15 (up to Dec-14) in Gujarat State.*

Impact of Micro Irrigation Adoption in Gujarat

Since last nine years, Gujarat has been emerging as one of the major micro irrigation adopting states in the country. Among the five leading micro irrigation systems adopting states, the increase has been observed to be maximum in Gujarat (1844 per cent) followed by Karnataka (1313 per cent). In fact, since last couple of years the installations are more than two lakhs annually. This impact is being concurrently monitored by Agricultural Finance Corporation every year. The observations for the year 2012 are discussed below (Anon, 2014).

Impact on Crop Productivity

The increase in the productivity varies from 300 kg/ha in sprinkler irrigated *kharif* groundnut to 25 tonnes/ha in the drip irrigated sugarcane (Table 1). The impact of drip irrigation on the productivities in various crops indicates that the mean productivities of the crops varied between 25 and 30 per cent.

Impact on Water Savings

The estimated water saved due to drip varied from 200 mm/ha in orchard crops to 1162 mm/ ha in banana (Fig. 3). Due to mini-sprinkler the water saved ranged from 1709 mm/ ha in groundnut to 180 mm/ha in potato.While the sprinkler in wheat is estimated to save 170 mm of water per hectare in *kharif* groundnut the estimation was to the tune of 90 mm/ha.

The water savings through drip, varied from 27 per cent in potato to 49 per cent in vegetable crops. In the mini sprinkler systems it was 21 per cent and in the sprinkler system it varied from 20 per cent in *kharif* groundnut to 28 per cent in wheat (Fig. 4).

But these percen tages are considerably leser than the potential of around 60 per cent.

Percent Savings in Fertilisers

The scientific study conducted by the agricultural university also indicated that 40 to 60 per cent of the fertiliser could be saved through fertigation in different crops. The estimated crop-wise nitrogen savings is depicted below (Fig. 5)

But the survey indicated that on an average there was a savings of only22 per cent in nitrogen,15 per cent in phosphorus and 11 per cent in potash.The survey indicated that on an average there was reduction in the usage of pesticides which got reduced from 3.6 liters under traditional method to 2.6 liters under drip. Thus, there was 27 per cent reduction in the usage of pesticides due to drip. Among the different crops the estimated reduction in the pesticide use ranged from 24 per cent in castor to 42 per cent in vegetables. Similar to fertigation, no reduction in the pesticide usage has been reported by any of the farmers adopting sprinkler system.

Impact on Energy Savings

Since, under micro irrigation, there will be less pumping time needed for irrigating a unit area of the crop, there will be reduction in the energy requirement. Based on the above, the energy savings by different crops is estimated as under (Fig. 6). The per hectare estimated savings were more than 1000 kWhr/ha in bananas and sugarcane.It was minimum and around 200 under the mini sprinkler system. Based on the total Crop wise installations during the year under report, the total savings in energy was estimated to the tune of23.1 MKWH

Payback Periods

The crop-wise payback period was found to range from 1.3 crop seasons in banana with drip to 5.1 in groundnut with mini sprinkler. When the subsidy factor is considered the payback period is reduced by about 50 per cent. Cotton is the major crop under drip and many farmers use the same system for other crops like groundnut/watermelon. Similarly, potato, groundnut/ watermelon is a common sequence followed with micro irrigation. Under these multiple use of the systems in a sequence the payback period will work out to be half or lesser than that. Thus, in a cropping sequence the payback period for cotton and potato may become less than one year with subsidy and less than two years without subsidy. Some farmers use the same system for three crops like groundnut-potato-groundnut/watermelon. In such cases the payback period will be only a few months.

Micro Irrigation for Saline Water Use

Climate change may deteriorate the groundwater quality specially by increasing salinity and drip irrigation can play an important role in the management of saline water use. Experiments have proven that with the use of drip saline water can used without affecting the crop and soil productivity.

Conclusion

Micro irrigation can play a big role on the impact of climate change on water resources in the country and help in smart agriculture. The micro irrigation, due to favourable soil and water conditions during crop growth results in sustainable increase in productivity, due to less water requirement

Table 1: Estimated yield increase (t/ha) in some crops due to micro irrigation in Gujarat

Crop	*Estimated yield increase (t/ha)*	*Crop*	*Estimated yield increase (t/ha)*
Banana	10.5	Potato (MS)	3.1
Castor	0.86	Sugarcane	25
Cotton	0.74	Vegetables	3
Groundnut (D)	0.6	Others (Drip)	2
Groundnut (MS)	0.39	Groundnut (S)	0.3
Orchard	2	Wheat (S)	0.5
Potato (D)	4.3	Others	0.5

D = Drip, MS= Mini sprinkler, S=Sprinkler

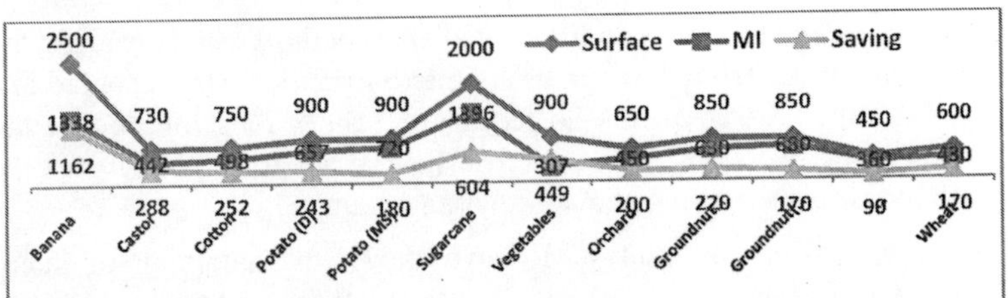

D-Drip, MS- Mini sprinkler, S-Sprinkler.

Fig. 3: Estimated Water Use and Savings (mm/ha).

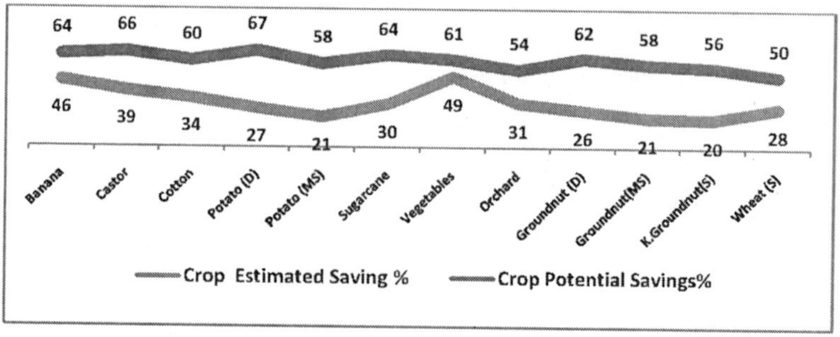

Fig. 4: Estimated Water savings in different Crops (per cent).

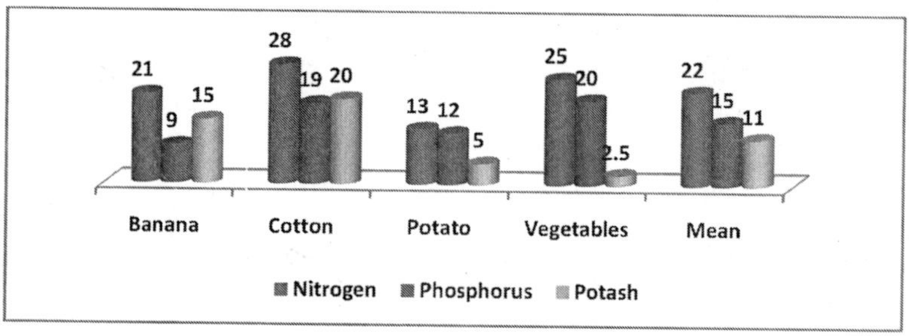

Fig. 5: Estimated Savings in Major Nutrients (per cent).

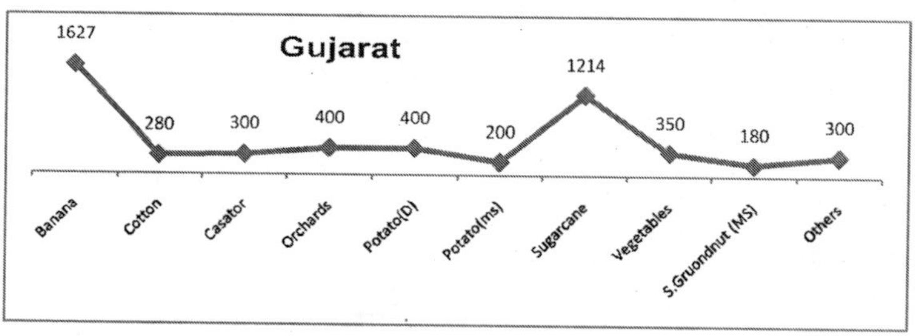

Fig. 6: Cropwise Estimated Energy Savings (kWhr/ha).

will be able to adapt to low groundwater available conditions and can arrest the decline groundwater table. Further, use of micro irrigation when adopted in a contiguous large area coupled with fertigation can, in long run, reduce the green house gas emissions. Thus, micro irrigation adoption can achieve the goal of food security through climate- smart agriculture.

Field data from different states have shown that adoption of micro irrigation is techno economically viable at farmers' field conditions, especially under the anticipated climate change scenario, by way of improving water, fertilizer and energy use efficiencies with simultaneous increase in the productivity levels. Further, micro irrigation is a good option for efficient use of saline water to improve the crop productivity without affecting the soil health.

References

Anonymous (2001). AGRESO Report, Danti, SWMRU, GAU, Navsari.

Anonymous (2010). International Commission on Irrigation and Drainage Report.

Anonymous (2012). National Committee on Use of Plastics in Horticulture.

Anonymous (2014). Concurrent Monitoring Report of AFC for the year 2011- 12, GGRC, Vadodara.

Chandrakanth, M.G. Thamanadevi, G.L, Nagraj, N, Priyanka, C.N. Mmantha, M. Srikanthamurthy.

P.S. Prakashkumar M, Nagaraju, YU, Sanjeev, G.Ravi Hosur, Umesh Pradhani, Basavaraj Biradar and Ranganath Mangalvedkar (2012). Micro Irrigation : Economics and Outreach in Karnataka, in Micro Irrigation Economics and Outreach, Pub. Macmillan Publishers.

Dhaka, J.M. (2007). Economic Evaluation of Sprinkler Irrigation in Jaipur Region of Rajasthan State, RWSP project report, submitted to Directorate of Horticulture, Jaipur.

IPCC (2007). Climate Change 2007: The Physical Science Basis. Contribution of Working Group I to the Fourth Assessment Report of the Intergovernmental Panel on Climate Change, pp. 996.

Krish Iyengar and Naresh Modi (2013) NCPAH. Personal communication.

Narayanmoorthy, A. (2012). Drip method of irrigation in Maharashtra : Status, Economics and Outreach in Micro Irrigation Economics and Outreach, Pub. Macmillan. pp. 120-139.

Nelson, G. C. and Robertson R. 2008 cited as Personal communication by Shah Tushar (2009).

Raman, S. (2012). Potential of Micro Irrigation in India : A State-wise Assessment, in Micro Irrigation Economics and Outreach, Pub. Macmillan. pp. 11–26.

Shah *et al*. (2007). Cross reference Shah Tushar 2009.

Shah Tushar (2009). Environ.Res.Lett. 4 IWMI, Colombo.

Shukla *et al*. (2003). Cross reference Shah Tushar 2009.

2

Evaluation of Drip Fertigation in Aerobic Rice

V. Praveen Rao, V. Ramulu, M. Uma Devi, K. Avil Kumar, R. Raghavaiah and K. Radhika

Abstract

A field study was conducted to evaluate the potential for improving water productivity and yield of rice by adopting variable surface drip irrigation schedules and N-fertigation levels in aerobic rice. The experiment was conducted for two years during kharif, *2012 and 2013 in split plot design with three replications under surface drip irrigation at college farm, College of Agriculture, Rajendranagar, Hyderabad. The main treatments consisting of three drip irrigation schedules (at 100 per cent, 150 per cent and 200 per cent pan evaporation replenishment) and the sub-treatments comprises of 4 N fertigation levels (0,60,120,180 kg N/ha). Irrigation equivalent to 150 per cent pan evaporation replenishment produced significantly higher grain yield over 100 per cent pan evaporation replenishment. Further, it was noticed that application of water equivalent to 200 per cent pan evaporation replenishment decreased the grain yield but the reduction was not statistically significant as compared to 150 per cent pan evaporation replenishment. Each higher level of nitrogen significantly enhanced the grain yield of rice up to 120 kg N/ha. Application of N at 180 kg N/ha did not prove to be advantageous. The seasonal water use in drip irrigated aerobic rice ranged from 729 to 979 mm and 617 to 1108 mm under different drip irrigation schedules when compared to 1650 and 1681 mm observed under conventional flood irrigated rice in* kharif, *2012 and 2013, respectively.*

Introduction

Food security depends on the ability to increase agricultural production with decreasing availability of water to grow crops. Rice an important staple food crop in the world and most widely grown under irrigation with seasonal water needs ranging between 1650 to 3000 mm depending upon soil and climatic conditions (Tuong and Bouman, 2003; Lampayan and Bouman, 2005). To produce one kg of rough (unmilled) rice on an average about 2500 mm of water is needed (Bouman, 2009). In Asia, more than 80 per cent of the developed freshwater resources are used for irrigation purposes; about half of which is used for rice production alone (Dawe *et al.*, 1998). Until recently, this amount of water has been taken for granted, but now the global "water crisis" threatens the sustainability of irrigated rice production as a fundamentally different approach to reduce water inputs in rice is to grow the crop like an irrigated-dry crop, such as result of decreasing water quality (chemical pollution, salinisation), decreasing water resources (e.g., falling groundwater tables, silting of reservoirs), and increased competition from other sectors such as urban and industrial users (Belder *et al.*, 2004; Bouman, 2007). On the other hand, to keep pace with the population growth and income induced demand for food, it is estimated that rice production has to be increased by 56 per cent over the next 30 years (IRRI, 1997). For food security and livelihood of rice producers and consumers, it is essential to "produce more rice with less water" (Guerra *et al.,* 1998; Bouman, 2007). Reducing water input in rice production can have high societal and environmental impact, if the water saved can be diverted to areas where competition is high for water. Therefore, farmers and researchers are looking for ways to decrease water use in rice production, nitrogen (257 kg/ha), medium in available P_2O_5 (45 kg/ha) and K_2O (312 kg/ha).The average soil bulk density and infiltration rate wheat or corn using modern irrigation systems. Unlike lowland rice, drip irrigated aerobic rice crop is grown in non-puddled, non-saturated aerobic soil with an optimally balanced soil water-air conditions in the rhizosphere. Irrigation scheduling in ET-based (when rainfall is insufficient) with water applied equivalent to replenish the soil water content in the root zone back to the field capacity during each irrigation event after it has reached a certain lower permissible threshold. Thus, the field water supply matches the evapotranspiration requirements of the rice crop. Nitrogen is an important nutrient element in the soil for rice crop but unlike low land puddle soil, the nitrogen availability in aerobic conditions is low. There are reports that aerobic rice responds to nitrogen application up to 100 (Danial and Wahab, 1994), 150 (Reddy *et al.*,1993 and Maheswari *et al.*, 2007, Lateefpasha, 2004),180 kg/ha (Ghobrial, 1983). Zhang *et al.* (2009) reported that there was no interaction effect between N rates and water management. Limited research has been done on optimising nutrient management to produce high yield of aerobic rice. Hence, a field study was conducted to evaluate the variable surface drip irrigation schedules along with the N-fertigation dose for aerobic rice in the semi-arid region of India.

Materials and Methods

A field experiment was conducted for two seasons during *kharif,* 2012 and 2013 at Water Technology Centre, College Farm, Prof. Jayashankar Telangana State Agricultural University (Formerly part of Acharya N.G. Ranga Agricultural University), Rajendranagar, Hyderabad). Geographically the farm is located at 78°24'43" E–Longitude and 17°19'16" N–Latitude with an altitude of 542.6 m above mean sea level. The soil was sandy clay loam in texture, alkaline in reaction (pH 8.0) and non-saline with C1 class (EC = 0.7 dS/m). The soil was low in organic carbon (0.15) and available during the last ploughing. Likewise the recommended K_2O @ 40 kg/ha was fertigated uniformly to all plots along with N as per the treatments at weekly intervals using water soluble fertilisers viz., urea and sulphate of potash. The fertigation commenced from 15 DAS and continued up to 90 DAS. Iron deficiency was noticed and was corrected by foliar application (three times at weekly intervals) of ferrous sulphate @ 2%. Further, a transplanted conventional flood irrigated control rice plot (non-replicated) with same variety was also maintained to compare the water input use. Irrigation water was scheduled to each treatmental plot at a predetermined pan evaporation replenishment factor after adjusting the effective precipitation received and the amount of water discharged through drip was measured through watermeter attached at drip irrigation system control unit. The water productivity (kg/ m^3 of water) was calculated by following equation

$$WP = Y/ WA_{(IR+ER)}$$

Where Y= Grain yield (kg/ ha)

WA(total water used) = (IR- irrigation +ER-effective rainfall)

The data collected in the present experiment on rice yield was analysed statistically and wherever the treatment differences were found significant (F test), the critical difference was calculated at five per cent probability.

Results and Discussion

Grain yield of rice was significantly influenced by drip irrigation levels and N-fertigation doses (Table-1) in both the years. Irrigation equivalent to 150% pan evaporation replenishment ($I_{1.5}$) produced significantly higher grain yield over 100 per cent pan evaporation replenishment ($I_{1.0}$). Further, it was noticed that application of water equivalent to 200 per cent pan evaporation replenishment ($I_{2.0}$) decreased the grain yield but the reduction was not statistically significant as compared to 150 per cent pan was 1.48 g/cm^3 and 24 mm/hour, respectively. Likewise the average volumetric soil water content at field capacity (0.33 bars) and permanent wilting point (15 bars) was 0.284 cm^3/cm^3 and 0.161 cm^3/cm^3, respectively. Whereas the available soil moisture estimated was 110.8 mm/m depth of soil. The drip system was installed as per the statistical design in the field. The drip lateral used was

Table-1: Grain yield of aerobic rice as influenced by drip irrigation schedules and N-fertigation levels.

Treatments	*Grain yield (kg/ha)*	
	2012	*2013*
$I_{1.0}$(100% Epan)	2991	2903
$I_{1.5}$(150% Epan)	3978	3907
$I_{2.0}$(200% Epan)	3720	3876
SEm ±	109	157
C.D. (P=0.05)	429	616
N_0 (No Nitrogen)	2052	3238
N_1(60 kg/ha Nitrogen)	3702	3377
N_2(120 kg/ha Nitrogen)	4360	3911
N_3(180 kg/ha Nitrogen)	4137	3722
SEm ±	84	151
C.D. (P=0.05)	250	449
Interaction (I x N)	**	NS

integral pressure compensated dripper line spaced at 0.8 m with each dripper line feeding four rice crop rows (Fig.1). The emitter spacing was 0.4 m with an emitter flow rate of 2 iph. The resultant application rate was 6.25 mm/hr.

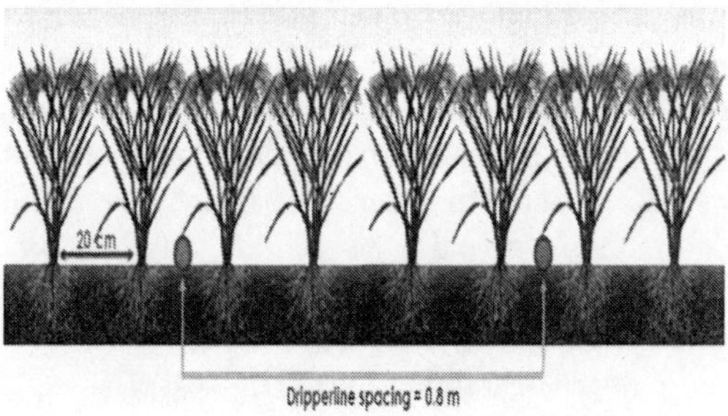

Fig. 1. Dripper line location in drip irrigated aerobic rice.

The rice experiment was laid out in split plot design with three replications. There were three treatments consisted of drip irrigation schedules *viz.,* Irrigation at evaporation (from USWB Class A Pan evaporimeter) replenishment of 100 per cent, 150 per cent and 200 per cent in main-plots. While four nitrogen (N) fertigation levels *viz.,* 0 kg N/ha (control), 60 kg N/ha, 120 kg N/ha and 180 kg N/ha were allocated in the sub-plots. A medium duration (125-130 days) rice variety MTU – 1010 was sown in both the years using a seed rate of 80 kg/ha in solid rows at 20 cm apart.The recommended P_2O_5 @ 60 kg/ha in the form of Single Super Phosphate along with Zinc sulphate @50 kg/ha was applied uniformly to all the plots as basal dose evaporation replenishment ($I_{1.5}$). Whereas, each higher level of nitrogen fertigation significantly enhanced the grain yield of rice up to 120 kg N/ha. Application of Nat 180 kg N/ha did not prove to be advantageous and it decreased the yield marginally in comparison to 120 kg/ N /ha but the decrease was not statically significant.

Table 2: Grain yield (kg/ha)of aerobic rice as influenced by drip irrigation schedules and N-fertigation levels during kharif, 2012.

N-fertigation levels	*Drip irrigation schedules*		
	$I_{1.0}$(100% Epan)	*$I_{1.5}$(150% Epan)*	*$I_{2.0}$(200% Epan)*
N_0(No Nitrogen)	1960	2156	2042
N_1(60 kg/ha Nitrogen)	2979	3980	4148
N_2(120 kg/ha Nitrogen)	3690	4842	4550
N_3(180 kg/ha Nitrogen)	3337	4933	4142
I x N Interaction effect		SEm ±	C.D. (P=0.05)
N at same level of irrigation		146	434
Irrigation at same or different level of N		167	565

The results of experimental study revealed that drip irrigated aerobic rice crop water use (ETc) ranged from 729 to 979 mm and 617 to 1108 mm under different irrigation schedules in 2012 and 2013, respectively (Table-3).The water productivity (weight of produce per unit volume of water used) which is an indicator used to compare different water saving systems/technologies. Irrigation equivalent to 150 per cent pan evaporation replenishment had higher water productivity as compared to either $I_{1.0}$ or $I_{2.0}$ irrigation treatments (Fig. 2) in both the years of study. Typical water productivity (WP) values for flood

Table 3: Water requirement of aerobic rice by evapotranspiration based drip irrigation at Hyderabad.

	*Total Water used (mm)**		*Water Prod. (kg/m³)*	
	2012	*2013*	*2012*	*2013*
$I_{1.0}$-Epan	729	617	0.41	0.47
$I_{1.5}$-Epan	848	800	0.46	0.48
$I_{2.0}$- Epan	979	1108	0.38	0.34
Con.flood-ing	1650	1681	-	-

irrigated rice crop were reported to range from 0.2 to 0.4 (Bouman and Tuong, 2001) but in the present study, the water productivity of rice was enhanced (0.34 – 0.48 kg/m^3) over conventional flood irrigated rice. Thus, shifting from conventional flooded systems to aerobic rice raised under surface drip irrigation in non-puddled, non-flooded fields reduced the water requirement by 150 per cent pan evaporation replenishment ($I_{1.5}$) and nitrogen fertigation at 120 kg N/haproduced significantly higher rice grain yield. Further, by adoption of surface drip irrigation scheduling based on evaporation replenishment factor in aerobic rice realised higher water productivity with considerable water saving over conventional flood irrigated rice.

References

Belder, P., Bouman, B.A.M., Cabangon, R., Lu, G., Quilang, E.J.P., Li, Y., Spiertz, J.H.J. and T.P. Tuong (2004). Effect of water saving irrigation on rice yieldand water use in typical lowland conditions in Asia. Agricultural Water Management,65: 193 – 210.

Bouman, B.A.M. (2007). A conceptual framework forthe improvement of crop water productivity at ifferent spatial scales. Agricultural Systems, 93: 43 – 60.

Bouman, B.A.M. (2009). How much water does rice use. Rice Today, (Jan – Mar) 2009: 28 – 29.

Bouman, B.A.M. and T.P.Tuong. (2001). Field water management to save water and increase its productivity in irrigated rice. Agricultural Water Management, 49(1): 11-30. Daniel, K.V. and K.Wahab.(1994). Levels andtime of nitrogen in semi dry rice. Madras Agricultural Journal, 81(6): 357-358.

Dawe, D., Barker, R., and D. Seckler. (1998). Water supply and research for food security in Asia. In "Proceedings of theWorkshop on Increasing Water Productivity and Efficiency in Rice-Based Systems," July 1998, International.

Rice Research Institute, Los Banos, Philippines.

Ghobrial, G.I. (1983). Response of irrigated dry seeded rice to nitrogen level, inter row spacing and seeding rate in a semi-arid environ. IRRI News letter (8):4.

Guerra, I.C., Bhuiyan, S.I., Tuong, T.P. and R.Barker. (1998). Producing more rice with less water from irrigated systems. SWIM Paper 5, IWMI/IRRI, Columbo, Sri lanka, p. 24.

Lampayan, R. M. and B.A.M. Bouman. (2005). Management strategies for saving water and increase its productivity in lowland rice-based ecosystems. In: proceedings of the First Asia-Europe Workshop on Sustainable Resource Management and Policy Options for Rice Ecosystems *SUMAPOL*), 11–14 May 2005, Hangzhou, Zhejiang Province, P.R. China. On CDROM, Altera, Wageningen, Netherlands.

Lateefpasha, M.D. (2004). Performance of dryseeded irrigated rice under different seed densities and N levels. Un published M.Sc.(Ag.) thesis submitted to ANGRAU, Hyderabad.

Maheswari, J., Maragatham, N. and G.J. Martin. (2007). Relativelysimple irrigation scheduling and N application enhances the productivity of aerobic rice (*Oryza sativa*.L.). American Journal of Plant Physiology, 2(4):261-268.

Medley, J.C. and L.T. Wilson.(2008). The use of subsurface drip irrigation in rice. Paper presented at 32nd Rice Technical Working Group Meeting, Westin San Diego 400 West Broadway San Diego, CA February 18-21, 2008.

Reddy, M.D., Kumar, S.S.,Vinod, S and V.N. Reddy. (1993). Management of direct seeded irrigated rice for north Telangana. Indian Farming, April: 3-5.

Tuong, T. P. and B.A.M. Bouman. (2003). Rice production in water-scarce environments. In: "Water Productivity in Agriculture: Limits and Opportunities for Improvement" (J. W. Kijne, R. Barker, and D. Molden, eds.). pp. 53–67, CAB International, Wallingford, UK. Zhang, L., S. Lin, S., Bouman, B. A. M., Xue,C., Wei, F., Tao, H., Yang, X,Wang, H., Zhao, D., and K.Dittert. (2009). Response of aerobic rice growth and grain yield to N fertilizer at two contrasting sites near Beijing, China. Field Crop Research, 114:45-53.

3

Enhancing Productivity of Coriander (*Coriandrum sativum* L.) through Micro-irrigation under Climate Change Condition

B.K. Sagarka, R.K. Mathukia, M.A. Shekh and R.L. Davaria

Abstract

A field experiment was conducted during 2008 to 2010 at Junagadh to evaluate methods (drip, mini sprinkler and flood irrigation) and levels of irrigation (0.4, 0.6 and 0.8 PEF) in coriander. The results revealed that mini sprinkler and drip irrigation significantly improved growth and yield attributes viz., *plant height, branches/ plant, umbels/plant and test weight and thereby increased seed and stalk yields along with higher net returns, B:C and WUE over flood irrigation. Similarly, irrigation at 0.8 PEF enhanced growth and yield parameters* viz. *plant height, branches/plant, umbels/plant and test weight and ultimately increased seed and stalk yield with higher net returns and B:C ratio over irrigation at 0.6 and 0.4 PEF.*

Introduction

Coriander (*Coriandrum sativum* L.), commonly known as "Dhania" or "Dhana", is a well known spice crop for its uses as medicine, oil, perfumery and culinary purpose. It is consumed in large quantities and earns a large sum of foreign exchange every year through export. It is probably the first spice to be used by man as a common flavouring substance.

India is a largest producer of coriander with a production of almost 90 per cent of the total world's supply. In Saurashtra region, irrigation facilities are very meagre. Major source of irrigation is lift irrigation through wells due to non-availability of major irrigation projects. Coriander is remunerative and short duration crop cultivated in *rabi* season. The low productivity of coriander in India can be ascribed to lack of improved agronomic practices. Among these, water management is the most important one. Hence, the present experiment was undertaken to evaluate methods and levels of irrigation for coriander.

Materials and Methods

A field experiment was conducted at Department of Agronomy, College of Agriculture, Junagadh Agricultural University, Junagadh during 2008 to 2010. The soil of the experimental plot was clayey in texture and slightly alkaline in reaction (pH 7.9 and EC 0.32 dS/m) as well as low in available nitrogen (226 kg/ha), available phosphorus (24.5 kg/ha) and medium in available potash (233 kg/ha). The experiment comprising of three methods of irrigation (M_1- Drip irrigation, M_2- Mini sprinkler irrigation and M_3- Flood irrigation) and three levels of irrigation (I_1- 0.4, 0.6 and 0.8 PEF or IW/CPE) were laid out in split plot design with four replications. The coriander variety 'Gujarat Coriander 1' was sown at 30-60-30 cm paired rows. The crop was fertilised with 20-10-0 kg N-P_2O_5-K_2O/ha as basal. The crop was raised as per the recommended package of practices. In drip system, laterals were laid out at 90 cm spacing (in between each paired row) and drippers (4 lph capacity) at 60 cm spacing. While in mini sprinkler system, laterals were laid out at 1.80 m spacing and mini sprinklers (35 lph capacity) at 2.50 m spacing. Drip and mini sprinkler system was operated on alternate day at 1.2 kg/cm^2 pressure on the basis of pan evaporation fraction (PEF) of 0.4, 0.6 and 0.8. Flood irrigation each of 5 cm depth was scheduled on the basis of IW/CPE ratio of 0.4, 0.6 and 0.8.

Results and Discussion

Growth and yield attributes

Data presented in Table-1 indicated that different methods of irrigation significantly influenced growth and yield attributes of coriander. Significantly the highest plant height (50.3 cm), branches/plant (4.09), umbels/plant (6.83) and test weight (22.34 g) were recorded with mini sprinkler irrigation (M_2), however it remained statistically at par with drip irrigation (M_1) in respect of branches/plant. Conversely, flood irrigation (M_3) registered significantly the lowest values of these growth and yield contributing parameters.

Levels of irrigation did cause their significant impact on growth and yield attributes (Table-1). Irrigation at 0.8 PEF (I_3) significantly enhanced plant height (49.9 cm), branches/plant (4.03), umbels/plant (6.82) and test weight (22.44 g) over 0.6 (I_2) and 0.4 PEF (I_1).

Crop yield

A perusal of data furnished in Table-2 revealed that seed yield and stalk yield of coriander were significantly influenced by various irrigation methods.

Mini sprinkler irrigation (M_2) gave significantly the highest seed yield (1205 kg/ha) and stalk yield (2075 kg/ha). The drip irrigation (M_1) stood next to mini sprinkler in this respect. On the other hand, significantly the lowest seed and stalk yields were observed under flood irrigation (M_3). Pooled over three years, mini sprinkler (M_2) and drip irrigation (M_1) increased seed yield to the tune of 70.7 and 42.9 per cent and stalk yield to the tune of 36.8 and 13.6 per cent, respectively over flood irrigation (M_3). The results corroborate the findings of Rajaraman *et al.* (2011).

Table 1: Effect of methods and levels of irrigation on growth and yield attributes of coriander (Pooled over three years)

Treatment	*Plant height (cm)*	*Branches/ plant*	*Umbels/ plant*	*1000-seed weight (g)*
Methods of irrigation				
M_1- Drip	47.2	3.72	6.55	21.46
M_2- Mini sprinkler	50.3	4.09	6.83	22.34
M_3- Flood	46.2	3.39	6.39	20.97
S.Em.±	0.6	0.1.1	0.07	0.20
C.D. at 5%	2.3	0.43	0.27	0.80
C.V.%	8.06	15.10	5.38	4.92
Levels of irrigation				
I_1- 0.4 PEF	45.8	3.48	6.33	20.58
I_2- 0.6 PEF	48.1	3.69	6.63	21.73
I_3- 0.8 PEF	49.9	4.03	6.82	22.44
S.Em.±	0.4	0.10	0.06	0.17
C.D. at 5%	1.6	0.26	0.16	0.50
C.V.%	5.44	13.80	4.65	4.19
M x I	NS	NS	NS	NS

Seed and stalk yields significantly influenced by levels of irrigation (Table-2). Each increase in irrigation level significantly increased seed and stalk yields. Irrigation at 0.8 PEF (I_3) recorded significantly the highest seed yield (1075 kg/ha) and stalk yield (1964 kg/ha). Obviously, irrigation at 0.4 PEF (I_1) produced significantly the lowest seed and stalk yields. Datta *et al.* (2006) also obtained similar results.

Economics and WUE

Irrigation through mini sprinkler (M_2) recorded the highest gross returns (Rs 38225/ha), net returns (Rs 24697/ha), B:C (2.83) and WUE (4.82 kg/ha-mm), followed by drip irrigation (M_1) having gross returns of Rs 31994/ha, net returns of 18527/ha and WUE of 4.04 kg/ha-mm.

Table 2: Effect of methods and levels of irrigation on yield of coriander and economics (Pooled over three years)

Treatment	*Seed yield (kg/ha)*	*Stalk yield (kg/ha*	*Gross returns (Rs/ha)*	*Net returns (Rs/ha)*	*B:C*	*WUE (kg/ha-mm)*
Methods of irrigation						
M_1- Drip	1009	1724	31994	18527	2.38	4.04
M_2- Mini sprinkler	1205	2075	38225	24697	2.83	4.82
M_3- Flood	706	1517	22697	13810	2.55	2.82
S.Em.±	19	27				
C.D. at 5%	57	81				
C.V.%	11.78	9.23				
Levels of irrigation						
I_1- 0.4 PEF	871	1544	27674	20404	3.81	4.36
I_2- 0.6 PEF	974	1806	31026	23756	4.27	3.90
I_3- 0.8 PEF	1075	1964	34214	26944	4.71	3.58
S.Em.±	17	31				
C.D. at 5%	49	89				
C.V.%	10.55	10.58				
M x I	NS	NS				

Among the levels of irrigation, 0.8 PEF (I_3) accrued maximum gross returns (Rs 34214/ha), net returns (Rs 26944/ha) and B:C (4.71). However, the WUE decreased with each increase in irrigation level. Analogous results were reported by Lakpale *et al.* (2007).

References

Datta, S., R. Chatterjee and A.C. Sinha (2006). Effect of irrigation level on growth, yield and evapotranspiration in coriander. Indian Journal of Horticulture, 63(4): 428-432.

Lakpale, R., G.K. Shrivastava and R.S. Tripathi (2007). Effect of irrigation schedule on growth, yield and economics of spice crops. Indian Journal of Agricultural Sciences, 77(3): 170-173.

Rajaraman, G., P. Paramaguru, P. Aruna and I.P. Sudagar (2010). Effect of fertigation on growth and physiology in coriander. Asian Journal of Horticulture, 5(1): 85-88.

4

Cotton Yield Response to Moisture Regimes and Nomograph for Optimal Land and Water Resources Planning

B. B. Limbasiya, H. D. Rank, P. B. Vekariya, V. K. Sardhara

Abstract

The field experiment having total 32 treatments, replicating four times was laid out adopting split plot design at university farm, Junagadh, Gujarat, India during 2003-04 and 2004-05. Among 32 treatments, 30 treatments were combination of 2 irrigation methods, three water application levels and five growth periods and two were controls, i.e. one under each irrigation method. The observed data of cotton yield and evapotranspiration for different irrigation scheduling levels during various growth stages of cotton under surface and drip methods were used to develop the mathematical models and nomographs. The yield production models relating water consumption or irrigation scheduling levels were found quadratic. The Jenson multiplicative model was found best fit among all the dated yield production functions. The developed dated yield production models and irrigation scheduling level-water consumption relationships were used to develop the nomograph separately for surface and drip irrigation methods. The developed nomographs were used for cotton crop to forecast the maximised net return, optimal acreage with stage-wise water allocation and irrigation scheduling levels during different growth periods. The developed nomograph was found the versatile tool for land and water resources planning for cotton crop.

Key words : *Growth periods/stages, Scheduling level, Water consumption, cotton yield, model.*

Introduction

The cotton is highly remunerative crop because of higher yield due to favourable soil and climatic conditions in the Saurashtra Region of Gujarat,

India. However, water scarcity is the main constraint in its adoption for large scale. Increasing demand for water, declining water tables and increasing energy cost for pumping emphasise the need for conserving water in irrigated agriculture. Irrigation for maximum yield is no longer feasible in water scared areas. Deficit irrigation through efficient irrigation method is one of the major ways to increase the cotton acreage and water use efficiency. The irrigation water management planning requires information on stage-wise optimal water allocation and irrigation schedule levels during different growth stages. The documented empirical evidence is not yet available for the soils and agro climatic conditions of Saurashtra Region of the Gujarat state. Thus, the lack of information on these important aspects provided the necessary impetus to initiate the present study for H-8 variety of cotton (*Gossypium hirsutum* L.) crop successfully and extensively adopted in the Saurashtra region. Keeping this in mind, the present effort was made (i) to develop the yield production functions and (ii) to develop the nomograph for surface and drip irrigation methods that can be understood and used by farmers.

The seasonal yield is expressed as a function of stage-wise inputs of transpiration, evapotranspiration, applied water or irrigation schedule level (Rank, 1991). Moore (1961) was one of the firsts to give the concept of the dated yield production function assuming that the soil moisture stress in each cycle has an equal effect upon plant growth. In contrast to Moore, Flinn and Musgrave (1967) assumed that the potential growth function of a plant is of sigmoidal shape with different rate of growth in each cycle. The differential effects of water stress on yields of many crops, depending on growth stages have been demonstrated by Hall and Butcher (1968) and Jensen (1968). The dated yield production functions can be either additive or multiplicative. There are disagreements on whether the water stresses occurring during different growth stages of crop have either additive or multiplicative effects on yield. The models of Stewart *et al.* (1977), Hiler and Clark (1971), Sudar *et al.* (1981), Rao *et al.* (1988), Shalhevet and Fowden (1993), and Subbaiah and Rao (1992) are of former type and Hall and Butcher (1968), Jensen (1968), Hiller *et al.*, (1973), Hanks (1974), Minhas *et al.* (1974), Hill *et al.* (1979) and Kumar and Khepar (1980) are of later type. The detailed reviews and discussion on these different types of models have been done by Rank (2005).

Methodology

Field experiment

The selected hybrid cotton variety (H-8) had total growing periods of 210 days from sowing. The cotton was sown on 1st of July during both the years. In the experiment, 3 - factorial split plot design was adopted. The main factor, sub factor and subsub factor were respectively the irrigation methods (2 no.), water application levels (3 no.) and growth periods (5 nos). Total 32 treatments were evaluated replicating four times. Among these 30 treatments were consisted of different combinations two irrigation methods, three water application levels and five growth periods. One control under each irrigation method was also studied, under which moisture stress-free condition was maintained through out the growing period. The irrigation methods were border (B) and drip (D). Three water application levels were L_1 (80 per cent depletion of ASM for border and IW/CPE = 0.4 for drip), L_2 (60 per cent depletion of ASM for border and IW/CPE = 0.6 for drip) and L_3 (40 per cent depletion of ASM for border and IW/CPE = 0.8 for drip). The growth stages were classified as (a) The vegetative stage (S_1) from 46 days after sowing until 20 per cent of flowering (90 days after sowing), (b) The flowering stage (S_2) from 20 per cent until 80 per cent flowering (135 days after sowing), (c) The boll development stage (S_3) from 80 per cent flowering until opening of 20 per cent bolls (180 days after sowing), and (d) Boll maturity stage (S_4) from opening of 20 per cent to 95 per cent bolls (210 days after sowing). The whole season was considered as 46 to 210 DAS. To keep the moisture stress-free conditions, the irrigations were scheduled at level L(water application scheduled at IW/CPE = 1.0 in case of drip irrigation and at 20 per cent depletion of ASM in case of surface irrigation). The irrigation was given using the considered irrigation scheduling level during the considered growth stage only and before as well as after that stage, the cotton was irrigated under stress free conditions. The treatment plot (sub sub plot) was having three rows at a distance of 1.22 m. The length of the row was 5.4 m having 12 plants at 0.45 m distance in a row. The net plot size of treatment was 3.66 m x 5.4 m. The common irrigations, if needed after rain, were applied as per requirement to keep the plant stress-free during the establishment stage (0 to 45 days after sowing). The water meters were used to apply the metered quantity of irrigation water to the treatments. The irrigation in different treatments under drip was applied daily having depth equal to pan evaporation of last day multiplies by respective IW/CPE value. The irrigation to surface treatments was applied at moisture depletion level- L_j having depth equals to 90 to 95 per cent of calculated depth (d). [$d = ASM \times Z \times L_j$; where ASM, Z and L_j are the available soil moisture (mm/mm), root zone depth (mm) and depletion level of ASM respectively (fraction)]. The ASM is the difference between field capacity and permanent wilting point. The actual evapotranspiration from each treatment plot was estimated using the water balance approach.

Model development

Modeling water consumption and irrigation scheduling level

The models relating water consumption and irrigation scheduling level were developed separately for individual growth stage-i (Eq. 1). The models for the different stages like vegetative (46-90 DAS), flowering (91-135 DAS), boll development (136-180 DAS) and boll maturity (181-210 DAS) were developed by regression analysis. This model is useful to predict the water consumption for given irrigation scheduling level. Also, if the water availability/allocation for the particular stage is known, the irrigations schedule can be predicted for that stage using this model.

$$ET_{ai} = f(L_{ij}); \text{for all stage } i = 1, 2, 3, \textit{and } 4 \quad \ldots(1)$$

where, ET_{ai} is the actual evapotranspiration(mm) during the stage-i, L_{ij} is the irrigation level-j during the growth stage-i. The model combining all growth stages was also developed relating seasonal water consumption and irrigation scheduling levels-L_{ij}(Eq. 2). The seasonal water consumption can be predicted for a set of given irrigation scheduling level for different stages.

$$ET_a = f(L_{ij}) \quad \ldots(2)$$

where, ET_a is the seasonal actual evapotranspiration (mm/season).

Modelling seasonal cotton yield and water consumption

The model relating yield and water consumption is necessary to convert the effect of irrigation management practices into monetory terms. The model establishing the relationships between seasonal cotton yield and seasonal water consumption is known as seasonal yield production model. If it is expressing the relationship between seasonal cotton yield and stage-wise water consumptions, it is known as dated yield production model (Eq. 3).

The experimental methodology to obtain this type of model is now well established and documented. The additive as well as multiplicative models were tested to fit the observed data of seasonal cotton yield and stage-wise water consumptions and the best fit model was proposed based on the goodness of fit. The seed cotton yield and evapotranspiration during different growth period-i observed in respective control plot of each irrigation method were taken as maximum yield (Y_m) and maximum evapotranspiration (ET_{mi}) respectively while developing the dated model for that method (Prieto and Angueira, 1999). And, and data sets of Y and ET so observed under and scheduling levels were taken as the actual yield (Y_a) and actual evapotranspiration (ET_{ai}). The cotton was irrigated under water stress-free conditions during the rest of growing periods other than stressed stage-i. Thus, three sets of data of (ET_{ai}, Y_a) and one set of (ET_{mi}, Y_m) were used.

$$\frac{Y}{Y_m} = f\left(\frac{ET_{ai}}{ET_{mi}}\right) \quad \ldots(3)$$

where, Y_a = observed cotton yield (kg/ha) under moisture stress environment; Y_m= observed maximum crop yield (kg/ha) under moisture stress-free environment maintained throughout the growing season; ET_{ai} = actual evapotranspiration (mm) observed under moisture stress environment during i[th] stage; ET_{mi} = maximum evapotranspiration (mm) observed under moisture stress-free environment during i[th] stage.

Nomographs development

The nomograph is a graphical relationship of multiple variables on two-dimensional plane. The nomographs were developed to predict the optimal irrigation schedules under varied water availability and land area. The following relationships (Eqs. 1 and 4) were used to develop the nomographs.

NLP Model

$$Max\ NR = A\{[P\ Y] - [CCC + FCI + VCC + VCI]\} \qquad ...(4)$$

Subject to set of land and water resources constraints.

Where, Max NR = Maximised net return ; A =Area to be brought under cotton cultivation (ha); P = Price of the seed cotton yield (Rs/kg); $Y = Y_m \times f(ET_{ai} / ET_{mi})$, a dated yield production model; Y_m = Maximised yield under moisture-stress free environment; ET_{ai} = Optimal water allocation (mm) to each stage-i; CCC = common cost of cultivation (Rs/ha/ season); FCI= Fixed cost of irrigation (Rs/ha/ season); VCC= Variable cost of cultivation (Rs/ha/ season) = $C_h \times Y$; C_h = is the cost of harvesting cotton yield (Rs/kg); VCI= Variable cost of irrigation (Rs/ha/ season)= $C_{iw} \times S(ET_{ai})$; C_{iw} = is the cost of irrigation water (Rs/ha.mm). The above nonlinear mathematical model (Eq. 4) was furnished using the observed data. The upper and lower limits of water allocation to each stage were imposed. The upper and lower limits were based on the observations during the experimentations. The model was used to find the optimal water allocation for each stage under varied water availability and area, which could maximise the net return.

The maximised net return, optimal seasonal water depth and optimal stage-wise water consumption obtained by NLP model (Eq. 4) and relationships (Eq. 1) between water consumption (ET_{ai}) and irrigation scheduling level (L_j) developed individually for the stages i = 1, 2, 3, and 4 were used to develop the nomograph by setting the space and scale of the various axes. The developed nomographs for surface and drip irrigation are depicted in Fig. 1 and Fig. 2 respectively.

Results and Discussion

The average of the observed seasonal cotton yield and actual evapotranspiration during different growth stages of cotton during the experimental year 2003-04 and 2004-05 were as given in Table 1.

Model development

The data presented in Table 1 were used to develop the models and nomographs as described in following subheads.

Table 1: Stage wise water consumption and seasonal cotton yield under different treatments (average of 2003-04 & 2004-05)

Drip Treatments	*Water consumption during stage*					*Cotton Yield*	*Surface Treatments*	*Water consumption during stage*					*Cotton Yield*
	S_0 (mm)	S_1 (mm)	S_2 (mm)	S_3 (mm)	S_4 (mm)	(kg/ha)		S_0 (mm)	S_1 (mm)	S_2 (mm)	S_3 (mm)	S_4 (mm)	(kg/ha)
DL_1S	74	64	86	79	50	1356	BL_1S	74	62	118	98	41	887
DL_2S	74	96	130	118	75	1826	BL_2S	74	115	217	174	62	1319
DL_3S	74	128	173	157	100	2047	BL_3S	74	140	256	194	71	1459
DLS	74	160	216	197	125	2183	BLS	74	156	284	219	79	1564
DL_1S_1	74	64	216	197	125	1985	BL_1S_1	74	61	242	197	75	1253
DL_2S_1	74	96	216	197	125	2100	BL_2S_1	74	115	256	208	76	1405
DL_3S_1	74	128	216	197	125	2136	BL_3S_1	74	140	270	212	78	1480
DL_1S_2	74	160	86	197	125	2015	BL_1S_2	74	156	127	197	75	1265
DL_2S_2	74	160	130	197	125	2109	BL_2S_2	74	156	232	208	76	1449
DL_3S_2	74	160	173	197	125	2150	BL_3S_2	74	156	264	212	78	1510
DL_1S_3	74	160	216	79	125	2091	BL_1S_3	74	156	284	110	75	1447
DL_2S_3	74	160	216	118	125	2147	BL_2S_3	74	156	284	191	76	1582
DL_3S_3	74	160	216	157	125	2174	BL_3S_3	74	156	284	204	78	1541
DL_1S_4	74	160	216	197	50	2305	BL_1S_4	74	156	284	219	47	1646
DL_2S_4	74	160	216	197	75	2265	BL_2S_4	74	156	284	219	68	1613
DL_3S_4	74	160	216	197	100	2231	BL_3S_4	74	156	284	219	74	1549
ET_o	133	142	204	176	94	-	ET_o	133	142	204	176	94	-

Modelling water consumption with irrigation scheduling level.

(a) Individual stage

When the available water supply is not adequate to maintain or fulfill the crop water requirement, it is not possible to irrigate the crop adequately. This results in declining of the crop yield. The yield reduction is not only the function of degree of water stress but also the growth stage during which it occurs. The water management strategies under scarcity can be planned for different growth stages if such information is available in mathematical forms.

The model describing the relationships between water consumption and scheduling level individually for each of the four stages and whole season (46-210 DAS) were developed in different mathematical frame. However, the following form was identified as the best fit based on the goodness of fit.

$$ET_{ai} = a + b\,(L_i)\ \textit{for all stages} \qquad \ldots(5)$$

Where ET_{ai} is the water consumption(mm) during stage-i if irrigation is scheduled at level-L_j. The a, and b are empirical constants. The developed models are presented in Table 2. Using these models, the water consumption can be determined for particular irrigation scheduling level during particular growth stage. It can be seen in Table 2 that for all the cases, the relationships between water consumption and irrigation scheduling level are linear. Also, it can be seen that the values of empirical coefficient 'b' for drip irrigation is positive and for surface irrigation it is negative. The reason is that for the model of drip irrigation, the L_i is the IW/CPE level, while it is the level of depletion (fraction) of ASM at which the irrigation is scheduled. The water consumption will increase with increase in IW/CPE due to higher moisture regimes while it will decrease with increase in level of depletion of ASM due to increase in moisture stress level.

Table 2: Developed models of individual growth periods/stages relating water consumption and irrigation scheduling level

Growth period/ stage	*Empirical coefficient*	*Model : $ET_{ai} = a + b\,L_j$ (Eq. 4)*					
		2003-04		*2004-05*		*Pooled*	
		Drip	*Surface*	*Drip*	*Surface*	*Drip*	*Surface*
S_1 (46-90 DAS)	a	638.13	933.25	584.02	879.81	611.08	906.53
	b	140.8	-306.2	179.6	-326.5	160.2	-316.3
	R^2	1.00	0.97	1.00	0.97	1.00	0.97
S_2 (91-135 DAS)	a	548.23	996.31	562.12	907.82	555.18	952.06
	b	230.7	-424.2	201.5	-353.8	216.1	-389.0
	R^2	1.00	0.91	1.00	0.91	1.00	0.91
S_3 (136-180 DAS)	a	569.27	944.94	580.42	864.92	574.85	904.94
	b	208.9	-265.91	183.20	-221.9	196.05	-243.9
	R^2	1.00	0.83	1.00	0.84	1.00	0.83
S_4 (181-210 DAS)	a	654.23	868.74	638.62	804.29	646.43	836.51
	b	124.70	-70.68	125.00	-66.90	124.85	-68.79
	R^2	1.00	0.91	1.00	0.91	1.00	0.91

(b) Seasonal

The data of seasonal water consumption observed for the various irrigations scheduling levels during different growth stages were used. Among the various forms tried, the following was found best one based on the R^2.

$$Et_a = a + b\,L_{1j} + c\,L_{2j} + d\,L_{aj} + eL_{4j} + fL_{1j}^{\,2} + g\,L_{2j}^{\,2} + h\,L_{aj}^{\,2} + i\,L_{4j} \quad ...(6)$$

where, ET_a is the water consumption (mm); L_{1j}, L_{2j}, L_{3j}, L_{4j} is the irrigation scheduling level during the stages S_1, S_2, S_3, and S_4 respectively. The developed model for surface and drip irrigation is given in Table 3. The R^2 varying from 0.98 to one indicates that the developed model can be used accurately for predicting the water consumption for the given set of scheduling levels. If the water allocation /availability is known, the irrigation scheduling plan for all the growth stages can be made.

Table 3: Developed models combining all growth period/stages relating water consumption and irrigation scheduling level

Empirical coefficient	*Model : $ET_a = a + b\,L_{1j} + c\,L_{2j} + d\,L_{3j} + e\,L_{4j} + f\,L_{1j}^2 + g\,L_{2j}^2 + h\,L_{3j}^2 + i\,L_{4j}^2$ (Eq. 5)*					
	2003-04		*2004-05*		*Pooled*	
	Drip	*Surface*	*Drip*	*Surface*	*Drip*	*Surface*
a	72.49	736.38	74.22	685.46	73.44	710.89
b	140.94	51.20	179.76	74.87	160.22	63.06
c	230.84	329.69	201.66	279.02	216.12	304.39
d	210.87	310.15	183.36	260.98	196.99	285.59
e	124.84	117.06	125.16	106.06	124.87	111.59
f	-0.12	-302.05	-0.13	-339.49	-0.02	-320.80
g	-0.12	-643.18	-0.13	-539.91	-0.02	-591.57
h	-0.70	-494.94	-0.13	-414.71	-0.31	-454.85
i	-0.12	-157.29	-0.13	-145.04	-0.02	-151.19
R^2	1.00	0.99	1.00	0.99	1.00	0.98

Modelling cotton yield and water consumption

The dated yield production models are most suitable for irrigation water planning purposes as they incorporate effects of both timing and quantities of water applications on crop yield. The extraplotability to other site specific conditions are more. The empirical dated yield production functions were derived for drip and surface irrigation separately using the data observed in field experiment. Among the various additive as well as multiplicative dated yield models tried, the following (Eq. 6) was found the best fit to the experimental data.

$$\frac{Y}{Y_m} = \left(\frac{ET_{a1}}{ET_{m1}}\right)^{\lambda_1} x \left(\frac{ET_{a2}}{ET_{m2}}\right)^{\lambda_2} x \left(\frac{ET_{a3}}{ET_{m3}}\right)^{\lambda_3} x \left(\frac{ET_{a4}}{ET_{m4}}\right)^{\lambda_4} \quad ...(7)$$

where, the λ_1, λ_2, λ_3 and λ_4 are crop sensitivity factors for stage S_1, S_2, S_3, and S_4 respectively. The crop sensitivity factor (λ_i) for each growing periods were computed through regression analysis. The set of λ_i (here i= 1, 2, 3, 4) were determined separately for each method.

The derived crop sensitivity factors for the developed models were as presented in Table 4. The R^2 for the developed yield production model was found as 0.93, 0.97 and 0.97 under surface and 0.88, 0.94 and 0.98 under drip respectively for the year 2003, 2004 and pooled results. The R^2 values indicated that the developed dated yield production model could explain well the yield-ET behaviour.

Table 4 : The dated yield production function for surface and drip irrigation obtained under different stressed conditions for the year 2003, 2004 and pooled

Sensitivity Factor	*SURFACE*			*DRIP*		
	2003	*2004*	*POOLED*	*2003*	*2004*	*POOLED*
	$Y/Y_m = (ET_{a1}/ET_{m1})^{\lambda_1} \times (ET_{a2}/ET_{m2})^{\lambda_2} \times (ET_{a3}/ET_{m3})^{\lambda_3} \times (ET_{a4}/ET_{m4})^{\lambda_4}$					
λ_1	0.193	0.191	0.192	0.085	0.106	0.096
λ_2	0.281	0.255	0.267	0.048	0.109	0.082
λ_3	0.112	0.121	0.117	0.015	0.064	0.042
λ_4	-0.111	-0.097	-0.104	-0.095	-0.038	-0.064
R^2	0.93	0.97	0.97	0.88	0.94	0.98

The crop sensitivity factor for the first, second, third and fourth stages of cotton were found as 0.096, 0.082, 0.042 and –0.064 under drip and 0.192, 0.267, 0.117 and -0.104 under surface respectively. The highest sensitivity factor was found for first stage under drip and second stage under surface irrigation. The data depicted in Table 4 showed that the sensitivity factor (λ) under the surface irrigation was higher than that of drip if any one individual stage was stressed. This means that under surface irrigation, the cotton was more sensitive to moisture stress than the drip irrigation. The negative value of λ during the fourth stage showed that the certain water stress was desirable during fourth stage to increase the yield. Therefore, the deficit water application should be adopted during last stage for saving scarce water with benefit of yield increase. In this region, water availability is the main limiting factor during latter growth of cotton. Therefore, this finding is a promising result for the cotton growers in Saurashtra Region having water scarcity.

Developed nomographs

The nomographs as depicted in Figs. 1 and 2 were developed separately for drip and surface irrigation methods respectively. The developed nomographs are useful to forecast the maximised net return, the optimal area to be brought under cotton cultivation, optimal seasonal irrigation depth, stage-wise optimal water allocation and optimal irrigation scheduling during different stages for varied availability of water and land resources. The nomograph was applied for the case of farmer having land and water resources of 4 ha and 180 ha.cm respectively. Using the respective nomograph of surface and drip irrigation, the results obtained for the given availability of water and land are as presented in Table 5. It can be seen in Table 5 that the maximised net return of Rs 41000 can be obtained by cotton cultivation in 3 ha under surface irrigation. However, using the same water resources, the net return can be increased to the tune of Rs 85000 by bringing cotton under 4 ha if drip system is adopted.

Table 5: Land and irrigation water planning for drip and surface irrigation methods using nomographs.

Particulars	*Irrigation method*							
Irrigation methods	*Drip*				*Surface*			
Water availability, (ha.mm)	1800				1800			
Land availability, (ha)	4				4			
Optimal seasonal water allocation, (mm)	450				600			
Optimal area allocation, (ha)	4				3			
Maximised net return	85000				41000			
optimal water allocation (mm)	During growth periods				During growth periods			
	S_1	S_2	S_3	S_4	S_1	S_2	S_3	S_4
	160	153	87	50	158	280	114	48
Stage-wise optimal irrigation scheduling level	IW/CPE during growth period				Depletion (fraction) of ASM during growth period			
	S_1	S_2	S_3	S_4	S_1	S_2	S_3	S_4
	1.0	0.72	0.44	0.40	0.20	0.25	0.75	0.73

The drip irrigation helped to bring additional 33 per cent area under cotton with assured irrigation from the same water availability level. The drip irrigation favours to irrigate entire 4 ha, while under surface, only 3 ha can be brought under irrigation. The net return per hectare is also high under drip (21250 Rs/ha) than the surface (13667 Rs/ha) irrigation.

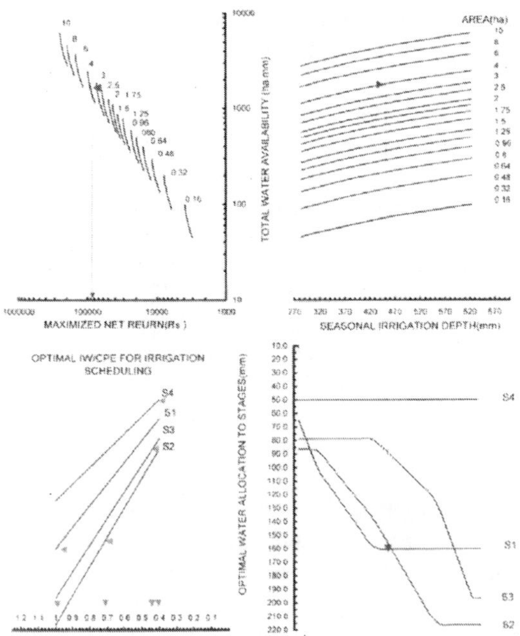

Fig. 1. Nomograph for optimal land and water resources planning for cotton crop under drip irrigation.

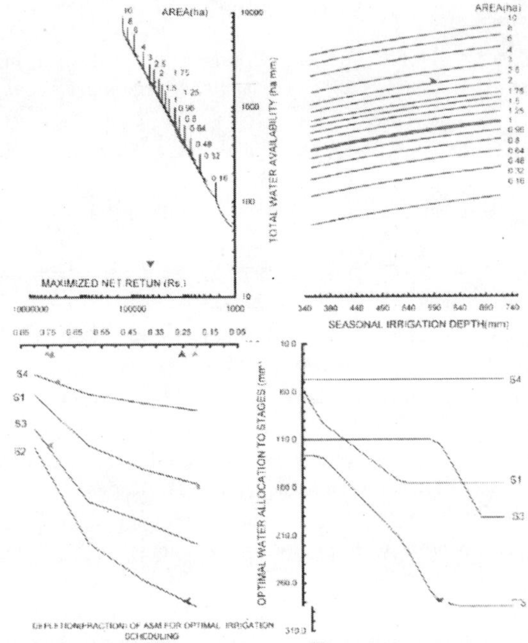

Fig. 2. Nomograph for optimal land and water resources planning for cotton crop under border irrigation.

The nomograph allocates optimally the seasonal water depth of 450 mm and 600 mm respectively under drip and surface. It also distribute seasonal irrigation depth optimally to first, second, third and fourth stage as 160 mm, 153 mm, 87 mm and 50 mm under drip and 158 mm, 280 mm, 114 mm and 48 mm under surface method respectively. The nomographs also suggest that the optimal allocated stage-wise water should be scheduled at IW/CPE of 1.0, 0.72, 0.44 and 0.4 under drip and at depletion of 0.2, 0.25, 0.75 and 0.73 of ASM during the first, second, third and fourth stage respectively.

Conclusions

The following conclusions could be drawn from the present study.

- The relationship between irrigation scheduling level and water consumption is linear for all individual stages and quadratic for combined growth stages.
- The Jensens multiplicative dated yield production model is the best fit to describe the cotton yield response to stage wise-water inputs.
- The developed nomogaphs can be used to forecast the maximized return, optimal area and seasonal water depth, stage wise water depth and irrigation scheduling level during different growth stages.

References

Flinn, J.C. and Musgrave, W.F. (1967). Development and analysis of inputs-output relations for irrigation water. *Australian Journal of Agricultural Economics*, 11: 1-19.

Hall, W.A. and Butcher, W.S. (1968). Optimal timing of irrigation. J. Irrig. Drain. 94:267-275.

Hanks, R.J. (1974). Model for predicting plant yield as influenced by water use. *Agron. J.*, 66(5): 660-665.

Hill, R.W., Johnson, D.R. and Ryan, K.M.(1979). A Model for Predicting Soybean Yields from Climatic Data. *Agron. J.*, 71: 251-256.

Hiller, E.A. and Clarke, R.N. (1971). Stress day index to characterise effects of water stress on crop yields. *Trans. ASAE*,14(4): 757-761.

Hiller, E.A. and Howell, T.A. (1973). Grain sorghum response to trickle and subsurface irrigation. *Trans. ASAE*, 16(4):792-893.

Jensen, M.E. (1968). Water consumptions by agricultural plants. In: Koslowski T.T. (Ed.), Water deficits in plant growth. Vol. 2, pp. 1-22. Academic Press, New York.

Kumar, R. and Khepar, S.D. (1980). Decision models for optimal cropping patterns in irrigations based on crop water production functions. *Agric. Water Manage.*, 3(1):65-76.

Minhas, B.S., Parikh, K.S. and Srinivasan, T.N. (1974). Towards the structure of a production function for wheat yield with dated inputs of irrigation water. *Water Resour. Res.*,10(3): 383-393.

Moore, C.V. (1961). A general analytical framework for estimating the production function of crops using irrigation water. *Journal of Farm Economics*, 43:876-888.

Prieto, D. and Angueira. (1999). Water stress effects on different growing stages of cotton and its influencing yield reductions. In: Kirda, C., Moutonnet, P., Hera, C., and Nielsen, D. R. (Eds.), *Crop yield response to deficit irrigation,* Kluwer Academic Publishers, Dordrecht, The Netherlands.

Rank, H.D. (1991). Nonlinear programming approach to agricultural resources management. Unpub. M. Tech. Thesis, sub. to IIT, Kharagpur, India.

Rank, H.D. (2005). Effect of irrigation systems and moisture regimes on physiological response function of cotton. Unpub. Ph. D. thesis, sub. to Junagadh Agricultural university, Junagadh, India.

Rao, N.H., Sarma, P.B.S. and Chander, S. (1988). Irrigation scheduling under a limited water supply. *Agric. Water Manage.*, 15(2) : 165-175.

Shalhevet, J. and Fowden, L. (1993). Plants under salt and water stress. In: Mansfield, T. and Stoddart, J. (Eds.) *Plant adaptation to environmental stress*, pp. 133-154.

Stewart, J.I., Cuenca, R.H., Pruit, W.O., Hagan, R.M. and Tosso, J. (1977). Determination and utilisation of water production functions for principal California crops. *W-67 Calif. Contributing proj. Rep.* Univ. of California, Davis.

Subbaiah, R. and Rao, K.A. (1992). Water stress response function for wheat. J. of Indian Water Resources Society. Vol.12(3&4) : 168-171.

Sudar, R.A., Saxton, K.E. and Spomer, R.G.(1981). A predictive model of water stress in corn and soybean. *Trans. ASAE*, 21(3): 97-102.

5

Micro-irrigation and Water Stress in Summer Groundnut (*Arachis hypogaea* L.) under Climate Change Condition

D. M. Panara, R.K. Mathukia, B.K. Sagarka and K.C. Patel

Abstract

*A field experiment was conducted during the summer seasons of 2009 to 2011 at Junagadh (Gujarat, India) to study the effect of different methods of irrigation (drip, micro sprinkler and surface) and water stress (up to 20, 30 and 40 DAS) on groundnut (*Arachis hypogaea *L.). The results indicate that drip irrigation method significantly enhanced growth, yield and quality attributes viz., plant height, branches/plant, LAI, dry matter/plant, mature pods/ plant, pod weight/plant, test weight and shelling, which in turn resulted in higher pod and haulm yield as well as WUE over surface irrigation method. Moisture stress up to 40 DAS decreased plant height and dry matter/plant, while increased branches/plant and LAI over stress up to 20 DAS. However, yield and quality parameters as well as pod and haulm yields were remained statistically equivalent under various stress periods. The WUE was increased with each increment in stress period. It is suggested that drip irrigation method and water stress up to 30 DAS can be adopted for efficient utilisation of irrigation water and to enhance the yield of groundnut in summer season.*

Introduction

Groundnut, the king of oilseeds grown in India, accounts for 29 per cent of total area and 36 per cent of total production. It is the main oilseed crop of Gujarat and Saurashtra region of the state is considered as the groundnut oil bowl of the country. Besides predominantly grown in rainy season, it can be grown successfully in summer season also, but lack of sufficient irrigation facility limits its area. This situation calls for efficient utilisation of irrigation water, which can be achieved through adoption of drip or sprinkler method of irrigation. With saving of irrigation water, transient water stress in early vegetative phase regulate growth, shorten flowering period and diverts photosynthates to reproductive sinks and thereby increase pod yield and water use efficiency (Ravindra *et al.,* 1989). With these points in view, the present experiment was undertaken to evaluate the effect of method of irrigation and water stress in early vegetative phase on growth, yield and quality of groundnut in summer season.

Materials and Methods

A field investigation was carried out during summer seasons of 2009 to 2011 at Department of Agronomy, Junagadh Agricultural University, Junagadh. The soil was clayey in texture and slightly alkaline in reaction (pH 7.9 and EC 0.30 dS/m) with available N 235 kg/ha, available P_2O_5 26 kg/ha and available K_2O 276 kg/ha. Field capacity and permanent wilting point were 28.4 and 12.8 per cent, respectively, whereas bulk density was 1.40 g/cm^3 with 43.2 per cent porosity. There were three main plots assigned to methods of irrigation viz., drip irrigation, micro sprinkler irrigation and surface irrigation and three sub-plots allocated to stress periods viz., water stress up to 20, 30 and 40 days after sowing (DAS). The experiment was laid out in split plot design with four replications. The crop was fertilised with 25 kg N and 50 kg P_2O_5/ha. The groundnut variety 'GG 2' was sown at 30 cm row spacing using seed rate of 125 kg/ha. A common surface irrigation was given to all the plots immediately after sowing for uniform germination. The laterals of drip system were laid in alternate inter-rows (60 cm) and drippers of 4 LPH were spaced at 60 cm. Micro sprinklers were laid at 2.50 m × 2.50 m spacing. Both these systems were operated on alternate days at pan evaporation fraction (PEF) of 1.0 with a pressure of 1.20 kg/cm^2. In surface method, irrigations each of 50 mm depth were applied at 1.0 IW/CPE. Water stress was imposed by curtailing irrigation after the first irrigation at sowing. Leaf area was measured by leaf area meter (Licor, 3000) at 60 DAS and leaf area index (LAI) was worked out accordingly. The oil content of kernel was determined using Nuclear Magnetic Resonance Spectro Photometer (Oxford, 4000).

Results and Discussion

Method of irrigation

Different methods of irrigation could impose their significant influence on growth, yield and quality attributes (Table 1). Drip irrigation method recorded significantly the highest values of growth and yield attributes viz., plant height, number of branches/plant, LAI at 60 DAS, dry matter/plant, number of mature pods/plant, pod weight/plant, test weight and shelling. Drip irrigation also registered significantly the highest oil content though remained at par with micro sprinkler method. Micro sprinkler recorded significantly the lowest values of growth parameters viz., plant height, branches/plant, LAI and dry matter/plant, while surface irrigation method registered significantly the lowest values of yield attributes viz., mature pods/plant, pod weight/plant, test weight and shelling. Pod and haulm yield were significantly affected due to irrigation methods in all the three years and pooled results (Table 2). The drip irrigation method produced significantly the highest pod yield during 2010, 2011, 2012 and pooled results. Similarly, drip irrigation method recorded significantly the highest haulm yield in individual and pooled results, except in 2000 wherein it was at par with micro sprinkler method. On an average, drip irrigation resulted in 48 and 27 per cent higher pod and haulm yield with WUE of 2.253 kg/ha-mm over surface irrigation method. Increased yield and WUE with drip irrigation were also reported earlier by Patil *et al.* (2009), Wane *et al.* (2009) and Sripunitha *et al.* (2011).

Table 1: Effect of different methods of irrigation and water stress on growth, yield and quality characters of groundnut (Pooled over three years)

Treatments	*Plant height (cm)*	*Branches /plant*	*LAI at 60 DAS*	*Dry matter /plant (g)*	*Mature pods/ plant*	*Pod weight/ plant (g)*	*100-seed weight (g)*	*Shelling (per cent)*	*Oil (per cent)*
Irrigation method									
Drip	20.92	5.45	3.40	24.12	15.66	13.25	36.54	73.55	49.19
Micro sprinkler	15.94	4.10	3.15	20.63	14.13	11.61	34.78	72.42	48.55
Surface	19.24	4.98	3.25	22.62	13.30	10.76	33.40	70.75	47.86
CD (P=0.05)	1.10	0.29	0.10	1.09	0.75	1.31	1.34	1.08	0.93
Water stress									
20 DAS	19.35	4.58	3.22	22.87	14.38	11.73	34.73	71.93	48.33
30 DAS	19.00	4.94	3.27	22.65	14.75	12.08	35.12	72.47	48.57
40 DAS	17.75	5.01	3.31	21.86	13.97	11.81	34.87	72.32	48.70
CD (P=0.05)	0.78	0.25	0.05	0.71	NS	NS	NS	NS	NS

Water stress

Different periods of water stress exhibited their significant consequence on growth characters, while yield and attributes were remained almost equal under varying stress periods (Table 1). Water stress up to 20 DAS significantly excelled plant height and dry matter/plant over stress up to 40 DAS. Whereas, maintenance of stress up to 40 DAS, being at par with stress up to 30 DAS, resulted in higher number of branches/plant and LAI at 60 DAS over stress up

Table 2: Effect of different methods of irrigation and water stress on yield of groundnut, WUE and net returns

Treatments	*Pod yield (q/ha)*				*Haulm yield (q/ha)*				*WUE (kg/ha-mm)*
	2009	*2010*	*2011*	*Pooled*	*2009*	*2010*	*2011*	*Pooled*	
Irrigation method									
Drip	14.31	23.71	16.89	18.30	30.13	55.50	42.11	42.58	2.253
Micro sprinkler	12.93	17.39	11.35	13.89	27.81	36.67	28.94	31.14	1.697
Surface	8.47	17.62	11.03	12.37	21.64	45.28	33.33	33.42	1.517
CD (P=0.05)	1.17	1.62	1.19	1.42	2.90	1.82	1.38	2.52	
Water stress									
20 DAS	11.50	19.82	12.88	14.73	28.46	45.50	35.00	36.32	1.623
30 DAS	12.49	19.97	13.49	15.31	26.45	47.50	33.39	35.78	1.857
40 DAS	11.72	18.93	12.90	14.52	24.67	44.44	36.00	35.04	1.987
CD (P=0.05)	NS	NS	NS	NS	2.39	NS	1.79	NS	

to 20 DAS. Various stress periods did not influence pod yield during individual and pooled results (Table 2). Whereas, haulm yield was significantly influenced by water stress in 2010 and 2012, but not in 2011 and pooled results. In the year 2010, water stress up to 20 DAS recorded significantly the highest haulm yield but held at par with stress up to 30 DAS. While in 2012, significantly the highest haulm yield was registered under stress up to 40 DAS, which was at par with stress up to 20 DAS. The highest WUE of 1.987 kg/ha-mm was realised with stress up to 40 DAS, followed by stress up to 30 DAS (1.857 kg/ha-mm). Nautiyal *et al.* (1999; 2002) also reported analogous results.

References

Nautiyal, P.C., V. Ravindra and Y.C. Joshi (2002). Dry matter partitioning and water use efficiency under water-deficit during various growth stages in groundnut. Indian Journal of Plant Physiology, 7(2): 135-139.

Nautiyal, P.C., V. Ravindra, P.V. Zala and Y.C. Joshi (1999). Enhancement of yield in groundnut following the imposition of transient soil-moisture-deficit stress during the vegetative phase. Experimental Agriculture, 35(3): 371-385.

Patil, S.T., M.S. Mane and B.L. Ayare (2009). Effect of irrigation methods on yield and water requirement of summer groundnut. Journal of Soils and Crops, 19(1):147-152.

Ravindra, V., P.C. Nautiyal, V.G. Koradia and Y.C. Joshi (1989). Manipulation of water stress to increase productivity in groundnut. Groundnut News, 2: 1:5.

Sripunitha, A., K. Sivasubramaniam, S. Manikandan, K. Selvarani and K.K. Shyla (2011). Sub surface drip irrigation studies on seed and field quality of groundnut cv. VRI 2. Legume Research, 34(4):311-313.

Wane, S., H. Bagde and R.C. Bhuyar (2009). Effect of different micro-irrigation methods on growth, yield and quality of summer groundnut in Vidharbha. Green Farming, 2(6):351-353.

6

Micro-Irrigation, Fertigation and Weed Management in Summer Groundnut under Climate Change Condition

M. A. Shekh, B. K. Sagarka, R. K. Mathukia and R. L. Davaria

Abstract

A field investigation was carried out at Junagadh during the summer seasons of 2011 and 2012 to evaluate the comparative efficacy of irrigation methods, fertigation and weed management in groundnut. The results revealed that irrigation-fertigation through sub-surface drip system was proved to be superior in respect of pod yield, weed biomass, WUE and economics over surface irrigation. Among weed management practices, hand weedings at 30 and 60 DAS and Oxyfluorfen 0.24 kg/ha as pre-emergence followed by hand weeding at 30 DAS were found effective in controlling weeds along with higher pod yield, WUE and net returns.

Introduction

The South Saurashtra region, an oil bowl of the country, is the most potential pocket for cultivation of summer groundnut too. But scarcity of irrigation water especially during summer months restricts its area. Adoption of micro-irrigation methods under such a situation not only offers a quite good water economy but also provides high yield and fertiliser use efficiency as well as lessens the weed infestation (Bhakare and Chavan, 2013; Pawar *et al.*, 2013; Suresh *et al.*, 2013). Keeping these points in view, the present experiment was carried out to evaluate the efficacy of different methods of irrigation, fertigation and weed management in summer groundnut.

Materials and Methods

A field experiment was conducted during the summer seasons of 2011 and 2012 at Instructional Farm, Department of Agronomy, Junagadh Agricultural

University, Junagadh. The soil was clayey with pH 8.1 and EC 0.30 dS/m having 0.73 per cent organic carbon, 17 kg/ha available P and 240 kg/ha available K. The treatments comprised four methods of irrigation (**I_1**- Surface drip + fertigation, **I_2**- Sub-surface drip + fertigation, **I_3**- Micro-sprinkler + fertilisers applied to soil and **I_4**- Surface irrigation + fertilizers applied to soil) as main plots and three weed management practices (**W_1**- Oxyfluorfen 0.24 kg a.i./ ha as pre-emergence, **W_2**- Oxyfluorfen 0.24 kg a.i./ha as pre-emergence + 1 HW at 30 DAS and **W_3**- 2 HW at 30 and 60 DAS) as sub-plots were arranged in split plot design and replicated four times. The test variety 'GG 2' was sown at 30 cm row spacing in first week of February and harvested in last week of May during both the years. The crop was fertilised with 25-50 kg N-P_2O_5/ha through urea and diammonium phosphate. The laterals of drip system were laid out at 60 cm spacing (in alternate rows) and drippers of 4 LPH capacities were placed at 45 cm distance on each lateral. In case of sub-surface drip, the laterals were buried 10 cm below the soil surface. The micro-sprinklers of 35 LPH capacities were laid out at 2.5 m × 2.5 m spacing. Irrigations through drip and micro-sprinkler systems were scheduled at 0.8 CPF, while surface irrigation was scheduled at 1.0 IW:CPE with a water depth of 50 mm. The herbicide oxyfluorfen was sprayed to soil surface using 500 litres of water on the next day of sowing.

Results and Discussion

Irrigation-fertigation

An appraisal of data bestowed in Table 1 expounds that different treatments of irrigation-fertigation exerted their significant influence on pod yield of groundnut during both the years as well as in pooled results. The treatment comprising sub-surface drip + fertigation recorded significantly the highest pod yield in individual and pooled results, while the haulm yield remained unaffected under different treatments. The sub-surface drip + fertigation also registered significantly the lowest dry weight of weeds and maximum water use efficiency (WUE). The maximum gross returns of Rs 39740 was realised with sub-surface drip + fertigation, while maximum net returns were accrued under micro-sprinkler method closely followed by sub-surface drip + fertigation. Patil *et al.* (2009), Wane et al. (2009) and Sripunitha *et al.* (2011) also reported similar results.

Weed management

A perusal of data presented in Table 1 indicates that pod yield of groundnut significantly influenced by different weed management practices. Two hand weedings at 30 and 60 DAS registered significantly the highest pod yield during both the years and in pooled results. This treatment also resulted in significantly the lowest dry weight of weeds and maximum gross and net returns. The next best treatment was Oxyfluorfen 0.24 kg/ha as pre-emergence followed by hand weeding at 30 DAS. Similar results were also obtained earlier (Bhagat *et al.*, 2002; Solanki *et al.*, 2005; Sonwalkar and Londhe, 2011).

Table 1: Effect of different treatments on yield of groundnut, weed biomass, WUE and economics

Treatments	*Pod yield (q/ha)*			*Haulm yield (q/ha)*	*Weed dry weight (q/ha)*	*WUE (kg/ha-mm)*	*Gross returns (Rs/ha)*	*Cost of cultivation (Rs/ha)*	*Net returns (Rs/ha)*
	2011	*2012*	*Pooled*						
Irrigation-fertigation									
Surface drip + fertigation	21.79	15.93	18.86	32.74	20.73	3.56	34838	18247	16591
Sub-surface drip + fertigation	25.58	18.30	21.94	34.15	19.54	4.14	39740	18747	20993
Micro-sprinkler	23.39	17.14	20.27	35.69	23.83	3.82	37536	15527	22009
Surface irrigation	18.87	15.19	17.03	34.68	25.39	2.62	32481	11847	20634
CD (P=0.05)	2.40	1.64	1.35	NS	4.26	-	-	-	-
Weed management									
Oxyfluorfen	20.70	15.25	17.98	33.99	30.38	3.21	33761	15740	18021
Oxyfluorfen + HW	22.73	15.88	19.31	35.13	24.55	3.45	35984	16080	19904
2 HW	23.80	18.79	21.30	33.83	11.45	3.80	38709	16455	22254
CD (P=0.05)	1.68	1.23	1.01	NS	5.36	-	-	-	-

On the basis of two years' field study, it seems quite logical to conclude that irrigating and fertigating summer groundnut through sub-surface drip system along with manual weeding at 30 and 60 DAS or Oxyfluorfen 0.24 kg/ha as pre-emergence followed by hand weeding at 30 DAS was found to be advantageous in respect of yield, weed control, WUE and economics.

References

Bhagat, P.S., H.N. Sethi, A.P. Karunakar and D.J. Jiotode (2002). Efficacy of herbicides for weed control in groundnut (*Arachis hypogaea* L.). Research on Crops, 3(3): 542-545.

Bhakare, B.D. and D.N. Chavan (2013). Yield, cost economics and water saving in summer groundnut as influenced by various fertilizer recommendations and irrigation methods. Journal of Agriculture Research and Technology, 38(3): 451-456.

Patil, S.T., M.S. Mane and B.L. Ayare (2009). Effect of irrigation methods on yield and water requirement of summer groundnut. Journal of Soils and Crops, 19(1): 147-152.

Pawar, D.D., S.K. Dingre and D.M. Nanaware (2013). Yield and quality of summer groundnut under different irrigation scheduling through microsprinkler in clay loam soils of western Maharashtra. Journal of Agriculture Research and Technology, 38(1): 102-106.

Solanki, R.M., V.B. Bhalu, K.V. Jadav and G.R. Kelaiya (2005). Studies on integrated weed management in irrigated groundnut. Indian Journal of Weed Science, 37(1/2):119-120.

Sonwalkar, S.N. and T.B. Londhe (2011). Effect of various methods of weed control and planting layouts on weed intensity, weed control efficiency, weed index and yield of groundnut. Journal of Maharashtra Agricultural Universities, 36(1):142-144.

Sripunitha, A., K. Sivasubramaniam, S. Manikandan, K. Selvarani and K.K. Shyla (2011). Sub surface drip irrigation studies on seed and field quality of groundnut cv. VRI 2. Legume Research, 34(4): 311-313.

Suresh, K., D. Balaguravaiah, V. Ramulu and C.H.S. Rao (2013). Comparative efficiency of sprinkler irrigation over check basin irrigation in groundnut at different irrigation schedules. International Journal of Plant, Animal and Environmental Sciences, 3(2): 9-13.

Wane, S., H. Bagde and R.C. Bhuyar (2009). Effect of different micro-irrigation methods on growth, yield and quality of summer groundnut in Vidharbha. Green Farming, 2(6): 351-353.

7

Optimisation of Irrigation and Fertilizer for Sweet Corn (*Zea mays* L. var. *saccharata* Sturt) under Climate Change Conditions

B.S. Gohil, R.K.Mathukia, P.R. Mathukia and S.K. Chhodavadia

Abstract

A field experiment was conducted during rabi *seasons 2010 and 2011 on clayey soil of Junagadh (Gujarat) to study the response of sweet corn (*Zea mays *L.var.* saccharata *Sturt) to irrigation (0.6, 0.8 and 1.0 IW:CPE) and fertilizer (control, 90-45, 120-60 and 150-75 kg N-P_2O_5/ha). The results revealed that irrigation at 0.8 IW:CPE being at par with 1.0 IW:CPE enhanced growth and yield attributes and thereby green cob and fodder yield along with higher net returns and B:C ratio over irrigation at 0.6 IW:CPE. Application of 120-60 kg N-P_2O_5/ha being at par with 150-75 kg N-P_2O_5/ha improved growth and yield attributes and ultimately higher green cob and fodder yields with higher net returns and B:C ratio over control and 90-45 kg N-P_2O_5/ha. Therefore, irrigation at 0.8 IW:CPE and fertilizer dose of 120-60 kg N-P_2O_5/ha could be applied for higher yield and economical realisation from* rabi *sweetcorn along with appreciable saving of water and fertiliser under climate change conditions.*

Introduction

Speciality corns *viz*., sweet corn, pop corn, baby corn, high-oil corn *etc.* assume tremendous market potential not only in India but also in the international market. These specialty corns with their high market value are perfectly suitable to *peri*-urban agriculture. Thus, they promise higher income to maize growers. Out of the various speciality corns, sweet corn

(*Zea mays* L.var. *saccharata* Sturt) has a big market potential. It is a hybridised variety of maize specifically bred to increase the sugar content. Sustainability of scientific sweet corn cultivation practices must be ensured to attain the goal of agricultural sustainability. In order to achieve higher cob yields, irrigation and fertilisers are the most important factors. Thus, there is need to work out an optimum irrigation scheduling in relation to other agronomic factors. Judicious use of fertilizers is a key to bumper maize production as they alone contribute 40-60 per cent of the crop yield (Dayanand, 1998). Maize is an exhaustive crop and requires high quantities of nitrogen and phosphorus. Low soil fertility is one of the bottlenecks to sustain agricultural production and productivity in India. Anthropogenic factors such as inappropriate land use systems, mono-cropping, nutrient mining and inadequate supply of nutrients are aggravated the situation.To minimise input costs and environmental damage, farmers have to resort to producing corn with less irrigation and fertiliser in the future. Considering the present trends of global climate change, atmospheric CO_2 and temperature levels are likely to increase in future, which will affect yields, water and fertiliser requirements of the crops in a given region.The agronomic requirements like irrigation and fertiliser for maize crop has been worked out, but the recommended irrigation and fertiliser dose for normal maize may not be applicable to sweet corn. In India much work has not been done so far for the sweet corn (Kumar, 2009). Therefore, the present experiment is conducted to work out irrigation, nitrogen and phosphorus requirement of sweet corn.

Materials and Methods

A field experiment was conducted during *rabi* 2010 and 2011 at Department of Agronomy, College of Agriculture, Junagadh Agricultural University, Junagadh (Gujarat). The experimental soil was clayey in texture and slightly alkaline in reaction with pH 7.8 and EC 0.33 dS/m. It was low in available nitrogen (227 kg/ha), available phosphorus (22.5 kg/ha) and medium in available potash (289 kg/ha). The experiment comprised twelve treatment combinations consisting three levels of irrigation (0.6, 0.8 and 1.0 IW:CPE) and four fertiliser levels (Control, 90-45, 120-60 and 150-75 kg N-P_2O_5/ha). These treatments were replicated four times in a split plot design. The sweet corn hybrid 'Sugar 75' was sown during second week of November and harvested during third week of March in 2010 and 2011. The gross plot size was 5.0 m × 3.6 m and net plot size was 4.0 m x 2.4 m with 60 cm x 15 cm spacing. Flood irrigation each of 5 cm depth was scheduled on the basis of IW:CPE of 0.6, 0.8 and 1.0. The entire dose of phosphorus and half dose of nitrogen were applied as basal application in form of urea and DAP at just before sowing in the furrows. Remaining half dose of nitrogen was top dressed as urea in two equal splits at 30 and 45 DAS.

Results and Discussion

Growth and yield attributes

The data (Table 1) indicated that irrigation did exert significant influence on growth attributes of sweet corn viz., plant height, leaf area index (LAI) and dry matter per plant. However, number of leaves per plant and stem diameter remained unaffected under various levels of irrigation. Irrigating the crop at 1.0 IW:CPE recorded significantly the highest plant height (173 cm) and LAI (3.68) and dry matter per plant (145 g), which was statistically at par with 0.8 IW:CPE. The significant reduction in plant growth with decreased frequency of irrigation seems to be the resultant of water stress, which might have reduced the availability and uptake of water and nutrients and inhibited photosynthesis and cell division. The results are in close accordance with finding of Golparvar and Karimi (2011) and Thorat *et al.* (2011). The irrigation at 1.0 IW:CPE, being at par with 0.8 IW:CPE, recorded significantly the highest number of cobs/plant (1.41), cob length (17.2 cm), cob girth (15.0 cm), fresh weight of cob (129 g), number of kernels/cob (273) and fresh weight of 100-kernels (24.9 g). The enhanced yield components under 0.8 and 1.0 IW:CPE might be due to significant improvement in overall growth of the crop expressed in terms of plant height, leaf area index and dry matter accumulation by virtue of increased photosynthetic efficiency. The present findings are in close agreement with the results obtained by Patil *et al.* (2012) and Painyuli e*t al.* (2013).

The data furnished in Table 1 revealed that significantly the highest plant height (180 cm), number of leaves per plant (14.9), stem diameter (2.17 cm), LAI (3.69) and dry matter per plant (148 g) were recorded with application of 150-75 kg N-P_2O_5/ha, which remained at par with that of 120-60 kg N-P_2O_5/ha. Profound influence of N and P on crop growth seem to be due to congenial nutritional environment of plant system on account of their greater availability from soil media, which might have resulted in greater synthesis of amino acids, proteins and growth promoting substances, which seems to have enhanced the meristematic activity and increased cell division and their elongation. Further application of 150-75 or 120-60 kg N-P_2O_5/ha might have increased interception, absorption and utilisation of radiant energy which in turn increased photosynthesis and thereby plant height, stem diameter and finally accumulation of dry matter. The enhanced growth with fertilisers was also reported by Massey and Gaur (2006) and Khazaei *et al.* (2010). Application of 150-75 kg N-P_2O_5/ha, being at par with that of 120-60 kg N-P_2O_5/ha, significantly enhanced yield attributes viz., number of cobs/plant (1.47), cob length (17.4 cm), cob girth (15.2 cm), fresh weight of cob (129 g), number of kernels/cob (276) and fresh weight of 100-kernels (25.1 g) over application of 90-30 kg N-P_2O_5/ha and control. Fertility levels of 150-75 and 120-60 kg N-P_2O_5/ha did cause about significant improvement in overall growth of the crop by virtue of increased photosynthetic efficiency. Thus, greater availability of photosynthates and nutrients to develop reproductive structures seems to have resulted in increased productive plants, cob girth, cob length and cob weight with these fertility levels. The present findings are within the close vicinity of those reported by Massey and Guar (2006) and Khazaei *et al.* (2010).

Table 1. Effect of irrigation and fertiliser on growth and yield attributes of sweet corn (Pooled over two years)

Treatments	*Plant height (cm)*	*Number of leaves/ plant*	*Stem diameter (cm)*	*Leaf area index*	*Dry matter per plant (g)*	*Number of cobs/ plant*	*Cob length (cm)*	*Cob girth (cm)*	*Fresh weight of cob (g)*	*Number of kernels/ cob*	*100-kernels fresh weight (g)*
Irrigation (IW:CPE)											
0.6	160	12.4	1.89	3.39	126	1.14	15.4	13.8	115	246	21.5
0.8	168	12.8	2.09	3.61	139	1.32	16.7	14.9	124	263	23.3
1.0	173	13.1	2.17	3.68	145	1.41	17.2	15.0	129	273	24.9
SEm±	2	0.3	0.04	0.05	3	0.04	0.3	0.3	3	5	0.6
CD ($P=0.05$)	8	NS	NS	0.18	9	0.13	1.1	1.1	9	19	2.0
Fertilizer (kg N-P_2O_5/ha)											
Control	150	10.4	1.85	3.37	122	1.08	15.1	13.6	114	241	21.2
90-45	162	11.1	2.02	3.54	134	1.21	16.2	14.4	119	254	22.3
120-60	176	14.7	2.15	3.62	143	1.41	17.0	15.1	128	271	24.4
150-75	180	14.9	2.17	3.69	148	1.47	17.4	15.2	129	276	25.1
SEm±	2	0.2	0.03	0.04	2	0.04	0.3	0.3	2	4	0.5
CD ($P=0.05$)	6	0.6	0.09	0.11	7	0.11	0.8	0.8	7	12	1.4

Selling price: Green cob: Rs 10.00/kg, Green fodder: Rs 1.00/kg.

Nutrient uptake

Irrigation at 1.0 IW:CPE registered significantly the highest uptake of N (116 and 124 kg/ha) and P (53 and 39 kg/ha) during 2010 and 2011, however it was at par with 0.8 IW:CPE (Table 2).The significant improvement in uptake of nutrients might be attributed to their respective higher concentration in cob and fodder associated with higher yields. This might also be attributed to better availability of nutrients in the soil under non-competitive environment with these irrigation levels. The results are in close agreement with the finding of Patil *et al.* (2012).

Application of150-75 kg N-P_2O_5/ha recorded significantly the highest uptake of N (142 and 143 kg/ha) and P (62 and 47 kg/ha) in 2010 and 2011, however it remained at par with that of 120-60 kg N-P_2O_5/ha (Table 2). The increased uptake of N and P might be attributed to their respective higher concentration in cob and fodder with higher cob and fodder yields. This might also be attributed to better availability of nutrients in the soil under these fertility levels. The results are in close vicinity with the report of Kumar (2009).

Crop yield

An appraisal of data (Table 2) showed that different irrigation schedules imparted their significant influence on green cob yield. Significantly the

Table 2. Effect of irrigation and fertiliser on nutrient uptake, yield and economics of sweet corn

Treatments	N uptake (kg/ha)		P uptake (kg/ha)		Green cob yield (t/ha)		Green fodder yield (t/ha)		Net return (Rs/ha)		B:C	
	2010	2011	2010	2011	2010	2011	2010	2011	2010	2011	2010	2011
Irrigation (IW:CPE)												
0.6	101	106	41	28	7.11	6.82	31.87	30.18	59,424	54,834	2.36	2.26
0.8	111	118	49	35	7.78	7.44	35.95	33.97	69,204	63,824	2.55	2.43
1.0	116	124	53	39	7.84	7.51	36.13	34.11	68,984	63,664	2.51	2.40
SEm±	3	3	2	1	0.15	0.14	1.25	1.17				
CD (P=0.05)	10	9	6	5	0.51	0.49	4.25	3.98				
Fertilizer (kg N-P_2O_5/ha)												
Control	53	63	26	16	6.75	6.49	31.45	29.50	62,110	57,560	2.69	2.56
90-45	108	119	44	28	7.52	7.20	33.43	31.78	68,459	63,609	2.70	2.58
120-60	134	139	59	45	8.01	7.67	36.66	35.44	75,645	71,025	2.84	2.73
150-75	142	143	62	47	8.02	7.69	37.15	35.89	75,291	70,731	2.79	2.68
SEm ±	3	3	2	1	0.14	0.13	1.08	1.09				
CD (P=0.05)	9	8	5	4	0.4	0.39	3.14	3.16				

Selling price: Green cob: Rs 10.00/kg, Green fodder: Rs 1.00/kg

highest green cob yield (7.84 and 7.51 t/ha) and green fodder yield (36.13 and 34.11 t/ha) were recorded with 1.0 IW:CPEduring 2010 and 2011, and it was found at par with irrigation at 0.8 IW:CPE. Averaged over two years, irrigation at 0.8 and 1.0 IW:CPE increased green cob yield by 9.3 and 10.2 per cent and green fodder yield by 12.7 and 13.2 per cent over irrigation scheduled at 0.6 IW:CPE, respectively. Improved individual plant performance under 0.8 and 1.0 IW:CPE ultimately reflected in higher cob and fodder yields over 0.6 IW:CPE. The present finding is in close agreement with the results obtained by Patil *et al.* (2012) and Painyuli *et al.* (2013).

Fertiliser dose of 150-75 kg N-P_2O_5/ha recorded significantly the highest green cob yield (8.02 and7.69 t/ha) and green fodder yield (37.15 and 35.89 t/ha), however remained at par with 120-60 kg N-P_2O_5/ha during both the years (Table 2). On an average of two years, application of 120-60 and 150-75 kg N-P_2O_5/ha increased green cob yield by 18.4 and 18.7 per cent and green fodder yield by 18.3 and 19.9 per cent, respectively over control. Significant increase in green cob and fodder yields under these fertility levels appears on account of their influence on growth and development of the crop. The present finding is in close agreement with the results obtained by Massey and Gaur (2006) and Sahoo and Mahapatra (2007).

Interaction between irrigation and fertiliser levels was found to be non-significant in respect of growth, yield attributes, nutrients uptake and yield.

Economics

Irrigating the crop at 0.8 IW:CPE gave maximum net returns of Rs 69,204 and 63,824/ha with B:C ratio of2.55 and 2.43, followed by 1.0 IW:CPE, which recorded net returns of Rs 68,984 and 63,664/ha and B:C ratio of2.51 and 2.40 during 2010 and 2011, respectively. Patil *et al.* (2011) also reported similar results.

Fertilising the crop with 120-60 kg N-P_2O_5/ha accrued maximum net returns of Rs. 75,645 and 71,025/ha with B:C ratio of2.84 and 2.73, closely followed by fertiliser dose of 150-75 kg N-P_2O_5/ha having net returns of Rs. 75,291 and 70,731/ha with B:C ratio of2.79 and 2.68 during 2010 and 2011, respectively. Thorat *et al.* (2011) and Kumar (2009) also reported alike results.

References

Dayanand. (1998). Principles governing maize cultivation during rainy season. Indian Farming, 48(1):84-87.

Golparvar, A.R. and M. Karimi (2011). Effects of drought stress on seed yield and yield components of sweet corn (*Zeamays* L.) cultivars. Research on Crops, 12(2): 346-351.

Khazaei, F., M.A. Alikhani, I. Yari and A.Khandan (2010). Study the correlation, regression and path coefficient analysis in sweet corn (*Zea mays* var. *saccharata*) under different levels of plant density and nitrogen rate. Journal of Agricultural and Biological Science, 5(6): 212-216.

Kumar, A. (2009). Production potential and nitrogen-use efficiency of sweet corn (*Zea mays*) as influenced by different planting densities and nitrogen levels. Indian Journal of Agricultural Sciences, 79(5): 351-355.

Massey, J.X. and B.L. Gaur (2006). Effect of plant population and fertility levels on growth and NPK uptake by sweet corn (*Zea mays* L.) cultivars. Annals Agricultural Research, New Series, 27(4): 365-368.

Painyuli, A., M.S. Pal, A. Bhatnagar and A.S. Bisht (2013). Effect of planting techniques and irrigation scheduling on productivity and water use efficiency of sweet corn (*Zeamayssaccharata*). Indian Journal of Agronomy, 58(3): 344-348.

Patil, S.A., U.V. Mahadkar, S.P. Gosavi and P.G. Parab (2011). Cost economics of sweet corn (*Zea mays saccharata*) as influenced by irrigation and fertigation levels. Journal of Soils and Crops, 21(2): 209-212.

Patil, S.A., U.V. Mahadka, N.C. Mohiteand A.H. Karpe (2012). Effect of irrigation and fertigation levels on yield, quality and nutrient uptake of *rabi* sweet corn (*Zea mays saccharata*). Journal of Soils and Crops, 22(1): 100-104.

Sahoo, S.C. and P.K. Mahapatra (2007). Yield and economics of sweet corn (*Zea mays*) as affected by plant population and fertility levels. Indian Journal of Agronomy, 52(3): 239-242.

Thorat, T.N., R.T. Thokal, U.V. Mahadkar and D.J. Dabke (2011). Effect of irrigation regimes and integrated nutrient management on yield of sweet corn (*Zea mays saccharata*). Journal of Maharashtra Agricultural Universities, 36(1): 18-21.

8

Response of Summer Pearl millet (*Pennisetum glaucum* L.) to Irrigation, Mulches and Antitranspirant under South Saurashtra Conditions

S. P. Kachhadiya, B. B. Ramani and P. M. Kyada

Abstract

A field experiment was conducted during summer seasons of 2007 and 2008 at Instructional Farm, College of Agriculture, Junagadh Agriculture University, Junagadh comprising three irrigation schedules, three levels of mulch and two levels of antitranspirant under South Saurashtra agroclimatic conditions. The results revealed that CU, WUE and moisture extraction increased with increase in IW: CPE ratio from 0.6 to 1.0. Similarly application either of wheat cut straw or groundnut shell mulch @ 5 t ha^{-1} and spraying six per cent kaolin at 20 and 50 DAS recorded significantly higher CU, WUE and moisture extraction over their Control. The positive effect of the efficient utilisation of irrigation water significantly improved yield parameters and thereby grain yield of summer pearl millet when the crop was irrigated at 1.0 IW: CPE ratio using groundnut shell much @ 5 t ha^{-1} at 35 DAS and spraying six per cent kaolin at 20 and 50 DAS.

Introduction

Pearlmillet (*Pennisetum glaucum* L.) is one of the major cereal crop grown in the arid and semi-arid regions of India. Summer cultivation particularly in the irrigated areas of Gujarat state has got importance because of the assurance of targeted crop yield. In Gujarat state during summer season, it is grown in 3.30 lakh hectares with production of 8.11 lakh tonnes and productivity of 2458 kg ha^{-1} (Anonymous, 2013). Irrigation water in summer pearlmillet is one of the major inputs of crop production and efficient utilisation of this scarce resource is need of this time. The information regarding water requirement and irrigation scheduling as well as the use of mulches and antitranspirant for summer pearlmillet crop is lacking for the medium black soils of South Saurastra agroclimatic zone, therefore, the present experiment was conducted.

Methods

A field experiment was conducted during summer seasons of 2007 and 2008 at Instructional Farm, College of Agriculture, Junagadh Agriculture University, Junagadh. The experiment comprised three irrigation schedules, three levels of mulch and two levels of antitranspirant (Table 1) in split plot design with four replications. The irrigation treatments were imposed after a common irrigation of 50 mm depth. The mulches were applied at 35 days after sowing of the crop in the respective treatments, leaving the pearlmillet rows open. Water spray as a control treatment and six per cent kaolin were sprayed on the crop plants at 20 and 50 days after sowing of the crop during both the years. The kaolin @ 600 l ha^{-1} solution was applied with a hand operated knapsack sprayer fitted with a fan type nozzle in the morning time under dry and clear sun conditions. The data were recorded for consumptive use (CU), water use efficiency (WUE) as well as yield and yield parameters of pearlmillet.

Table 1. Consumptive use (CU), water use efficiency (WUE) and moisture extraction as influenced by levels of irrigation, mulches and antitranspirant (mean of two years)

Treatment	*Mean CU (mm)*	*Mean WUE (kg ha-mm^{-1})*	*Percentage of moisture extracted - Soil depth (cm)*		
			0-15	*15-30*	*30-45*
Irrigation levels					
0.6 IW : CPE ratio	297.6	14.18	25.39	18.69	18.35
0.8 IW : CPE ratio	398.4	12.07	28.50	20.06	16.41
1.0 IW : CPE ratio	500.3	9.98	32.21	21.42	18.77
Mulches					
Control (no mulch)	281.4	15.47	28.20	19.37	18.03
Wheat cut straw @ 5 t ha^{-1}	306.1	15.56	28.61	19.82	18.08
Groundnut shell @ 5 t ha^{-1}	311.4	15.93	28.80	20.08	18.16
Antitranspirant					
Control (Water spray)	286.3	15.60	28.25	19.62	18.00
6% Kaolin spray at 20 and 50 DAS	291.0	16.90	28.35	19.80	18.03

Result and Discussion

Consumptive Use of Water

Mean consumptive use of water (Table 1) was increased from 297.6 to 500.3 mm with increase in quantity of irrigation water (increasing IW: CPE ratio from 0.6 to 1.0). Among different mulches, the highest mean consumptive use of water (311.4 mm) was recorded under groundnut shell mulch @ 5 t ha^{-1} while Control (no mulch) recorded the lowest consumptive use of water (281.4 mm). Application of mulches might have provided the looseness in the upper soils of the root zone in pearlmillet crop ultimately providing more water to the root system resulting into higher consumption of soil water. The higher mean consumptive use of water (291.0 mm) was recorded under six per cent kaolin spray over Control (286.3 mm). This might be due to increase in growth and yield attribu-tes by using six per cent kaolin spray as compared to control treatment. The results are in confirmation with the results of Lal *et al.* (1992) and Kaushik and Lal (1996).

Water Use Efficiency

The mean water use efficiency (kg ha-mm^{-1}) was remarkably influenced due to application of different mulches as well as antitranspirant. The data indicated that the highest mean water use efficiency (15.93 kg ha- mm^{-1}) was observed under groundnut shell mulch @ 5 t ha^{-1} followed by wheat cut straw mulch @ 5 t ha^{-1}. While the lowest mean water use efficiency (15.47 ha-mm^{-1}) was recorded under Control (no mulch). The higher mean water use efficiency (16.90 kg ha mm^{-1}) was observed under six per cent kaolin spray than Control which recorded the lowest mean water use efficiency (15.60 kg ha- mm^{-1}). The increase in water use efficiency might be due to higher grain yield which was obtained under these treatments (Lal *et al.* 1992).

Moisture Extraction

The maximum per cent of moisture was extracted from 0-15 cm soil depth (Table 2), followed by 15-30 cm soil depth for all the treatments of irrigation schedules. Thus, the moisture extraction was increased with the increase in quantity of irrigation water from upper layers of soil (0-15 and 15-30 cm soil depth) as compared to deeper layers of soil (30-45 cm soil depth). Similarly application of mulches either of wheat cut straw or groundnut shell mulch @ 5 t ha^{-1} as well as antitranspirant six per cent Kaolin spray at 20 and 50 DAS observed maximum per cent of moisture extraction as compared to Control from 0-15 cm soil depth followed by 15-30 cm soil depth.

Yield and Yield parameters

Being at par with each other irrigation levels of 0.8 and 1.0 IW:CPE ratio recorded significantly higher values of yield parameters viz., leaf area index at

Table 2. Yield and yield parameters of summer pearlmillet as influenced by levels of irrigation, mulches and antitranspirant (pooled over two years)

Treatment	*Leaf area index at harvest*	*Effective tillers*	*Length of earhead (cm)*	*Girth of earhead (cm)*	*Grain Yield (kg ha^{-1})*
Irrigation levels					
0.6 IW : CPE ratio	4.85	3.89	18.36	9.17	4229
0.8 IW : CPE ratio	5.32	4.20	20.76	9.60	4810
1.0 IW : CPE ratio	5.41	4.32	21.44	10.02	5044
LSD(P=0.05)	0.18	0.14	2.25	0.28	320
Mulches					
Control (no mulch)	4.96	3.98	19.22	9.28	4356
Wheat cut straw @ 5 t ha^{-1}	5.28	4.19	20.29	9.69	4763
Groundnut shell @ 5 t ha^{-1}	5.33	4.24	21.05	9.83	4963
LSD(P=0.05)	0.16	0.14	0.66	0.26	279
Antitranspirant					
Control (Water spray)	5.09	4.03	19.78	9.36	4466
Six per cent Kaolin spray at 20 and 50 DAS	5.29	4.25	20.59	9.83	4922
LSD(P=0.05)	0.11	NS	0.41	0.18	212

harvest, effective tillers and length and girth of earhead ultimately resulted in significantly higher grain yields (4810 and 5044 kg ha^{-1}, respectively) over 0.6 IW:CPE ratio (Table 2). The increase in grain yield was to the tune of 13.74 and 19.27 per cent respectively. In case of mulches, application of either wheat cut straw or groundnut shell mulch @ 5 t ha^{-1} observed significantly higher values of said yield parameters and produced significantly higher grain yields (4763 and 4963 kg ha^{-1}, respectively) over Control. The magnitude of increase in grain yield was 9.34 and 13.93 per cent respectively over Control. Spraying of antitranspirant six per cent Kaolin at 20 and 50 DAS recorded significantly higher yield parameters except effective tillers over Control and produced significantly higher grain yield (4922 kg ha^{-1}) over Control treatment.

Conclusion

To obtain maximum grain yield of summer pearlmillet with efficient use of irrigation water, the crop should irrigate at 1.0 IW: CPE ratio using groundnut shell much @ 5 t ha^{-1} at 35 DAS and spraying six per cent kaolin at 20 and 50 DAS under South Saurashtra agroclimatic conditions.

References

Anonymous, 2013. Crop-wise area, production and yield of food and nonfood crops in Gujarat State for the year 2009-10 to 2011-12. Directorate of Agriculture, Gujarat State, Gandhinagar.

Kaushik, S.K. and K. Lal 1996. Effect of cultural and moisture conservation practices on pearl millet (*Pennisetum glaucum*). Annals of Agriculture Research, 17(4): 450-452.

Lal, B, S. K. Kaushik and R.C. Gautam 1992. Response of rainfed pearlmillet (*Pennisetum glaucum*) to nitrogen fertilisation and moisture conservation. Indian Journal of Agricultural Science, 62(7):476-478.

9

Influence of the Eco-Friendly Treatment 'Soil Solarisation' Period on Raising Nursery of Fruits and Vegetables in Saurashtra Region

I.U. Dhruj and P. M. Chauhan

Abstract

In this study, the effect soil solarisation period i.e. 15 and 30 days on soil profile temperature, weed and pathogen control and, increasing in growth response (IGR) of tomato, chillies, brinjal and custard apple nursery plants and soil nutrients status was studied for Junagadh (21.53°N, 70.53°E) climatic conditions and performance was compared with non-treated soil. Efforts have also been made to study the economics of the 15 days soil solarisation treatment for raising nursery. The results revealed that 15 days soil solarisation treatment under Junagadh climatic conditions is good enough to serve the purpose. The 15 days soil solarisation treatment effectively improved the plant growth characters and transplantable seedlings of crops, efficiently managed the seedling blight disease in custard apple and reduced weeds population drastically. The microflora were significantly reduced after treatment though at the time of up-rooting. The uptake of macro and micro - nutrients was significantly improved in the solarised soil.

Introduction

Soil solarisation is a method of heating of soil by covering it with 25 to 100 micron thick transparent polyethylene (PE) sheets during the hottest part of the year. The land to be solarised is thoroughly cultivated and then levelled to minimise such profusion. The major benefit of soil solarisation is reduction of soil borne pathogens; however there are many other possible additional beneficial effects that can result in an increased growth response and yield of plants because direction and rate of most of the physical, chemical and biological processes in soil are influenced by soil temperature. Physical processes such as evaporation, aeration and water movement are very much influenced by soil temperature and has direct effect on biological process like seed germination, seedling emergence, root development nutrient water uptake and the microbial activity. Besides, the above potential advantages, soil solarisation is a non-chemical method hence not hazardous to the user and does not involve substances toxic either to consumer or host plant or other organisms. In the right instances, it is less expansive than other methods and easy to teach to less trained farmers and it can be applied in large areas, by machine or manually in small plots.

In most of the literature minimum 30 days of soil solarisation is advocated for controlling soil born diseases, pests, weeds and increasing the yield of field crops. In recent studies, Yaduraju and Ahuja (1990) reported that grass population was substantially reduced (67 per cent) by polyethylene (100 micron) mulching of wet soil for 30 days but not for 15 days under Delhi's climatic conditions. Kumar *et al.* (1993) reported that pre-sowing soil solarisation significantly reduced the weed population and increased growth of soybean and yield up to 78 per cent. Mansoori and Jaliani (1996) concluded that in hot regions, soil borne diseases of watermelon can be controlled by applying solarisation treatment to the soil for 30 days.

In this study, efforts have been made to study the impact of reduction in treatment period for local climatic condition i.e. from 30 days to 15 days for raising fruits and vegetable nursery crops which are normally affected by various biotic factors like seed born fungi, soil born pathogen, weeds, hard coat and seed size resulting poor germination of seeds and improper growth of plants which might be overcome by soil solarisation. The efforts have also been made to study the economics of the treatment.

Materials and Methods

Experimental set-up

The experiments were conducted at APA nursery, College of Agril. Engineering and Technology, Junagadh Agricultural University, Junagadh (21.53°N, 70.53°E), India. Soil and farmyard manure (FYM) were sieved through a wire mesh to reject pebbles and debris. Three beds of size 2.5 m x 10 m were prepared with boundary of stones. The beds were filled up to 20 cm height with clay loam soil and FYM mixture (3:1 soil and FYM) and moistened with help of shower and kept for overnight. Next day early morning beds were covered with clear transparent polyethylene sheets of 25 microns thickness and all the edges were sealed in soil and compacted to prevent heat and soil moisture loss. One plot was kept uncovered to be considered as non-treated soil while out of two remaining plots, one was covered for 15 days and another was for 30 days during the month of May.

Weather data and soil parameters

The monthly average diurnal variation in ambient temperature (T_a) and solar flux (I) for soil solarisation duration were measured using copper constant (K-type) thermocouple wire and pyranometer sensors attached with DT-600 Data-Taker data logger. The average of these data are presented in Fig.1. The soil samples from three locations of each experimental plot were drawn just before covering and after removal of the PE sheet by using the core-cutter sampler for obtaining moisture content, density and porosity.

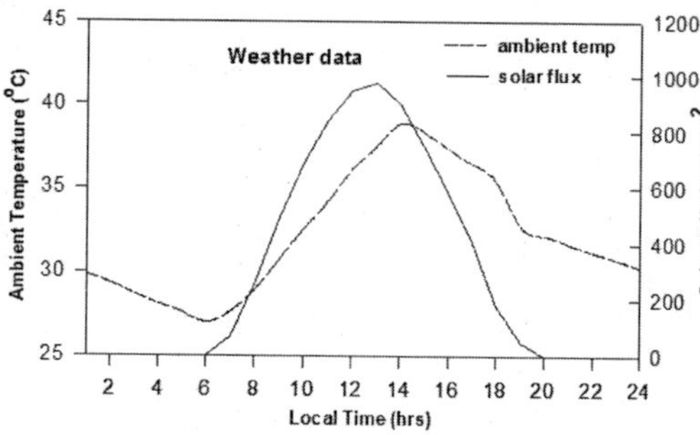

Fig.1. Diurnal variation in ambient temperature and solar flux for May month.

Plant growth response

The experiment was conducted by using complete randomised design-factorial with two replications. The three levels of solarisation i.e. without solarisation (non-treated), 15 days solarisation and 30 days solarisation and three vegetable crops: tomato (Var. Junagadh Rubi), chillies (Var. Resham patta) and brinjal (Var. Guj. Brinjal long-1) and, one fruit crop i.e custard apple were used in the experiment. The sowing of vegetable nursery was done on 5th July and custard apple seeds on 11th September. The beds were divided into 90 x 90 cm small beds. Three vegetable crops in two replications were sown keeping 10 cm distance between two rows and custard apple seed seeds were sown in rows keeping 10 cm and 5 cm inter and intra row spacing, respectively in each treatment. The growth parameters i.e. plant height, root length, green and dry weight were recorded from each treatment by randomly selecting 10 plants for vegetable crops and eight plants for custard apple. The observations were taken at 15 and 30 days after germination for vegetable crops and after two months of sowing for custard apple. The data obtained during the investigation were subjected to statistical analysis using complete randomised block design.

Weeds

The weed count, green and dry weight of weeds were recorded from entire bed (90 x 90 cm^2) and subjected to statistical analysis. The weeds species were also identified.

Soil micro flora

The soil samples were randomly collected at appropriate stages from eight different places of each bed to a depth of 0 -15 cm mixed thoroughly and filled in polyethylene bags. The soil samples were kept at 5°C until processed. Fifty grams of soil sample from each treatment was processed using serial dilution and plate count method for the study. An aliquot of 0.1 ml of the desired end dilution from each treatment was inoculated and spread aseptically on sterilised pre-cooled 90 mm petridishes of potato dextrose agar (PDA) for total fungal count, nutrient agar for total bacterial count, yeast mannitol agar W/ congored for *Rhizobium spp.* and *Azotobacter* agar(Mannitol) for *Azotobactor* spp. The ready-made media and sterilised gamma irradiated disposable petridishes of HiMedia Laboratories Limited, Mumbai were used. The petridishes were incubated at 27°C to 30°C for fungal and bacterial counts. Colonies per petridishes were counted at regular interval and the data were statistically analysed. The media and their compositions are given in Appendix.

Macro and micronutrients uptake by custard apple

Plant samples were collected at the time of uprooting the saplings of custard apple, air dried and used for further chemical analysis. The uptake of micronutrients was determined by using methanol di-acid extract and assessed by Atomic Absorption Spectrophotometer Model AA 2:3:13. The weighted quantity of plant sample was digested on hot plate with di-acid mixture of HNO_3 and $HClO_4$ (3:1) as per method described by Johnson and Ulrich (1969). The phosphorus and potash uptake were analysed by using the method of Jackson (1974). The nitrogen uptake was determined according to Kanwar and Chopra (1976).

Results and Discussion

Soil profile temperature characteristics and properties

The average moisture content (wb) of soil samples obtained from non-treated, 15 days and 30 days solarised experimental plots before covering and after removal of sheet was found as 26.55, 24.11, 24.04 and 19.22, 21.38, 19.85 per cent respectively. The bulk density of soil of the above plots before covering and after removal of sheet was found as 1456, 1450,1367 kg/m^3 and 1347, 1406, 1310 kg/m^3 respectively. The particle densities of the above plots were found to be 2608, 2572 and 2551 kg/m^3 respectively and porosity was computed as 44.15, 43.78, 46.36 and 48.25, 45.31, 48.60 per cent, respectively before and after soil solarisation. Fig. 2 shows the 15 days average hourly variation in soil profile temperatures at 5.0, 10.0 and 15.0 cm depths. At above depths, the average minimum temperatures of uncovered and covered plots were found to be 31.73, 33.71, 34.38°C and 33.63, 35.54 and 38.22°C respectively whereas the maximum temperatures at corresponding depths and plots were found to be 47.05, 41.35, 38.50°C and 56.53, 50.81 and 46.84°C respectively. Thus, due to soil solarisation treatment, increment in minimum and maximum soil temperatures were observed to be 1.9, 1.83, 3.84°C and 9.48, 9.46, 8.34°C at 5, 10 and 15 cm depths respectively.

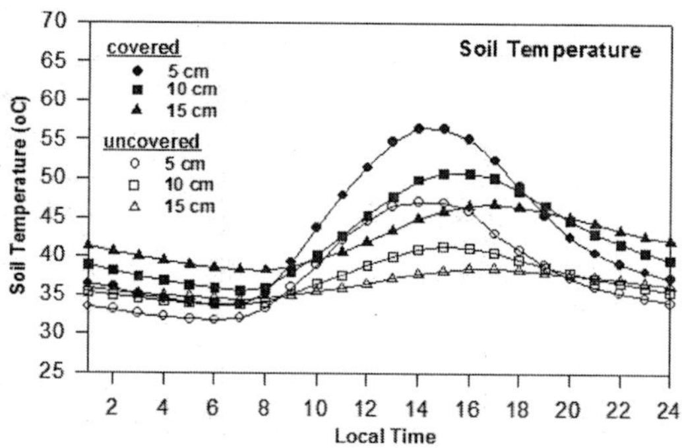

Fig. 2. Diurnal variation in soil profile temperature.

Effect of Soil Solarisation on Nursery Raising

1. Plant growth response

From Table-1 it can be observed that 30 and 15 days soil solarisation significantly increased the growth of vegetable crops. The maximum response was observed in tomato followed by chillies and brinjal for shoot length, fresh and dry weight of plant. Whereas, the maximum root length was recorded in brinjal crop that was at par with chillies. The soil solarisation of 30 and 15 days effectively improved all the plant growth characters significantly. The interaction effects were statistically significant except, root length x soil solarisation and dry weight x period. The interaction of crop x solarisation for shoot, fresh and dry weight showed that the soil solarisation treatment was effective. However, 15 and 30 days soil solarisation treatment was at par for brinjal and chillies, but they differed in case of tomato.

Table 1: Effect of soil solarisation period and crops

Source	*Mean*			
	Length (cm/ plant)		*Weight (gram/plant)*	
	Shoot	*Root*	*Fresh*	*Dry*
Crops (C):				
Tomato	26.69	04.55	04.85	00.33
Brinjal	13.60	05.78	02.42	00.24
Chillies	16.65	05.38	02.58	00.27
S. Em. ±	00.49	00.15	00.18	00.02
C.D.(5%)	01.46	00.44	00.53	00.05
Solarization (S):				
Control	14.77	03.98	01.87	00.14
15 days	19.84	05.30	03.34	00.29
30 days	22.33	06.44	04.64	00.42
S. Em. ±	00.49	00.15	00.18	00.02
C.D.(5%)	01.46	00:44	00.53	00.05
Period (P) :				
15 days	08.02	03.29	00.93	00.06
30 days	29.94	07.18	05.63	00.50
S. Em. ±	00.40	00.12	00.14	00.01
C.D.(5%)	01.19	00.36	00.43	00.04
Interactions :				
1. CXS S. Em. ±	00.85	00.26	00.30	00.03
C.D.(5%)	02.52	NS	00.91	00.09
2. CXP S. Em. ±	00.69	00.21	00.25	00.03
C.D.(5%)	02.06	00.63	00.74	NS
3. SXP S. Em. ±	00.69	00.21	00.25	00.03
C.D.(5%)	02.06	00.63	00.74	00.07
1. CXSXP S.Em. ±	01.20	00.37	00.43	00.04
C.D.(5%)	NS	NS	01.29	NS
C.V.%	8.94	09.90	18.67	21.94

The results on effect of solarisation on transplantable seedlings (Table 2) indicated that the main effects of solarisation on crops as well as their interaction were significant. The maximum transplantable seedlings were recorded in brinjal (305.47) that was at par with tomato. The solarisation treatment significantly increased the number of transplantable seedlings. However, duration (days) of solarisation was non-significant. The crop x solarisation interaction revealed that solarisation treatment significantly increased the transplantable seedlings as compared to control, but 15 and 30 days solarisation interactions were at par for all crops.

Table 2: Effect of solarisation on number of transplantable seedlings

Crops	Solarisation			Mean
	Control	15 days	30 days	
Tomato	223.88	304.88	318.88	280.89
Brinjal	167.63	365.63	382.05	305.47
Chillies	147.38	200.25	207.00	184.86
Mean	179.63	290.25	302.64	-
Effect	Crop (C)	Solarization (S)	C X S	-
S.Em. ±	11.61	11.61	20.07	-
C.D. 5%	33.62	33.62	58.23	-
C.V.%	15.62			

From Table-3 it can be observed that 30 and 15 days soil solarisation significantly improved all growth parameters except root length of custard apple. The shoot length, green and dry weights of plants were highest in 30 days soil solarisation. The per cent of plant emergence at the initial stage was non-significant. However, the plant stand at harvest was significantly highest in 15 days soil solarisation.

2. Root rot disease of custard apple

The plant mortality during experimentation had occurred due to seedling blight disease caused by soil inhabitant fungus *Rhizoctonia solani.* The minimum plant mortality was recorded in the 15 days soil solarisation treatment followed by 30 days soil solarisation. However, the highest mortality was observed in control plot. Number of seedlings per square meter was 127 in 15 day's soil solarisation, which was 47.67 per cent higher than the control.

3. Weed count and flora

The weed count (Table 4) indicated that the solarisation drastically reduced the weed population, green and dry weight as compared to control irrespective of crops. The effect of crop and interaction of crop x solarisation was found non-significant for all three characters under study, except dry weight where crop effect was found to be significant. It clearly indicated that solarisation is effective irrespective of crop and duration of solarisation. The duration of solarisation (30 and 15 days) was also remained statistically at par for all the parameters of weeds studied.

The weed flora in control plots was identified species-wise. It was observed that the total 21 weed species belonging to 13 families were present in the control plots, whereas only monocot weeds such as *Cynodon dactylon* and *Cyperus rotundus* were present in the solarised plots. The monocot weeds contributed the 56.40 per cent of total weed flora. The major monocot weeds were *Cynodon dactylon* (21.67 per cent), *Cyperus rotundus* (10.00 per cent) and *Commelina duffusa* (03.04 per cent). The *Alternanthera sessilis* was the dominant dicot (31.94 per cent) weed followed by *Tridex procumbens*, *Phyllanthus fraternus* and *Digera muricata*. Other dicot weeds were less than one per cent.

Table 3: Effect of soil solarisation on plant growth parameters and plant stand in custard apple

Treatment	Growth parameters				Plant emergence per cent	Plant stand at harvest per cent	Plant mortality per cent	Sapling /m^2	Per cent increase over control
	Length(cm)		Weight (g)						
	shoot	Root	Green	Dry					
Control	18.63	23.35	03.91	01.83	52.70 (63.13)	60.53 (76.92)	30.18 (25.27)	86	-
15 days	22.69	24.25	09.90	04.53	58.57 (72.15)	88.19 (99.03)	01.81 (00.96)	127	47.67
30 days	24.94	25.75	11.87	05.32	56.60 (68.75)	84.20 (96.96)	05.79 (03.03)	119	38.37
S.Em.±	00.54	00.74	00.63	00.30	02.54	02.09	02.09	-	-
C.D.at 5%	01.57	NS	01.84	00.88	NS	06.14	06.14	-	
C.V.%	06.85	08.60	20.64	21.61	12.85	07.60	13.92	-	-

Table 4: Effect of soil solarisation on weeds

Name of crop	Weed count			
	Control	15 days	30 days	Mean
Tomato	11.53 (132.94)	02.03 (04.12)	01.73 (02.99)	05.10 (26.01)*
Brinjal	10.76 (115.78)	01.87 (03.50)	02.12 (04.49)	04.92 (24.21)
Chillies	12.04 (144.96)	02.09 (04.37)	02.45 (06.00)	05.53 (30.58)
Mean	11.44 (130.87)	02.00 (04.00)	02.10 (04.41)	-
	Crop	Solarization	C x S	-
S.Em.±	0.219	0.219	0.379	-
C.D. 5%	NS	0.700	NS	-
C.V.%	10.35			

**values in parenthesis are retransformed value.*

4. Soil micro flora

Effect of soil solarisation on soil microflora *viz.*, total fungi, total bacteria, *Rhizobium* spp., and *Azotobacter* spp. was studied and results are presented in Table 5. The soil microfloral population was counted before solarisation, after solarisation, 15 days after sowing the nursery and at uprooting stage of

Table 5: Effect of soil solarization on soil microflora papulation per gram sail

Stage of sampling	*Total Fungi* $(x10^4)$			*Total Bacteria (xl06)*			*Rhizobium spp.* $(x10^4)$			*Azotobacter spp.* $(x10^4)$		
	Control	*15 days*	*30 days*	*Control*	*15 days*	*30 days*	*Control*	*15 days*	*30 days*	*Control*	*15 days*	*30 days*
Before soil solarizatian	4.32 (18.66)*	4.42 -19.66	4.3 -18.66	11.5 (132.33)*	11.37 -129.33	11.14 -124.33	7.43 (55.33)*	7.25 -52.66	7.18 -51.66	6.05 (36.66)*	5.91 -35	5.82 -34
After soil solarizatian	6.24 -39	1.72 -3	1.52 -2.33	21.94 -483.33	5.79 -33.66	4.65 -21.66	14.21 -202.33	3.73 -14	2.3 -9	11.59 -134.33	3.05 -9.33	2.52 -6.33
15 days after planting	8.41 -71	6.66 -44.33	6.31 -40	29.79 -896.66	23.28 -800	24.6 -606.66	19.73 -356.66	17.87 -320	15.75 -250	16.11 -260	147 -216.66	13.29 -176.66
At up rooting of nursery	10.34 -107	9.7 -94.33	9.47 -90	30.87 -956.66	31.34 -983.33	28.26 -800	20.53 -430	19.82 -396.66	19.3 -373.33	16.53 -273.33	15.18 -250	15.49 -240
Treatment average	7.33 -58.92	5.62 40.33	5.4 -50.33	23.53 -617.25	19.19 486.88	17.16 -388.16	15.49 -261.08	12.17 -195.83	11.31 -171	12.57 -176.08	9.87 -127.75	9.28 -114.25
	Stage	**Treatment**	**Stage x treatment**	**Stage**	**Treatment**	**Stage x treatment**	**Stage**	**Treatment**	**Stage x treatment**	**Stage**	**Treatment**	**Stage x treatment**
S.Em.±	0.16	0.14	0.27	0.52	0.45	0.9	0.4	0.35	0.7	0.15	0.13	0.26
C.D. 5%	0.46	0.4	0.8	1.52	1.32	2.64	1.18	1.02	2.04	0.43	0.37	0.75
C.V.%		7.72			7.85			9.3			4.19	

seedling. The Table shows that the total fungi, total bacteria, Rhizobium spp. and Azotobacter spp counts were more or less similar in all treatments before solarisation. However, microflora were significantly reduced when solarisation was done. At the time of uprooting, the microflora population increased in all the treatments. The results clearly indicated that soil microfloral population differences were non-significant between control and 15 days soil solarisation at the time of uprooting the nursery. The Table 5 shows that in case of 30 days soil solarisation treatment, the fungi ($x10^4$) and total bacteria counts ($x10^6$) before solarisation were 4.30 and 11.14 respectively, which reduced to lowest level (1.52 and 4.65). Though, after 15 days of planting it again increased to 6.31 and 24.60 respectively which further increased to the maximum level (9.47 and 28.26) at up-rooting stage. The Table shows that the both counts were found at par with 15 days treatment at all stages. The maximum fungi and bacteria counts 10.34 and 31.34 noticed in control and 15 days soil solarisation treatment respectively at up-rooting stage were found at par with the remaining treatments. The lowest average count ($x10^4$) *of Rhizobium* spp. (11.31) was observed in 30 days solarisation followed by 15 days solarisation. These treatments were significantly differed from control. Before solarisation, the *Rhizobium* spp. counts statistically non-significant. After solarisation, the counts were significantly lower in solarisation treatments than the control. The minimum count (2.99) was noticed in 30 days solarisation. However, both the solarisation treatments were found at par. At 15 days after sowing of nursery, the *Rhizobium* spp count was minimum in 30 days solarisation treatment, which differed significantly from rest of the treatments. The interaction of treatment x stage was significant at soil solarisation stage and non-significant at the time of uprooting the nursery (Table 5). The highest *Azotobacter spp* count (15.94) was observed at the time of uprooting the nursery. The treatment average count indicated that *Azotobacter spp.* counts were significantly lower in the solarisation treatment as compared to non - solarisation (control). After solarisation stage, the minimum count 2.52 was noticed in the 30 days solarisation and it was significantly differed from rest of the treatments. At the time of uprooting the nursery, the maximum count 16.53 was found in control followed by the 15 days solarisation (15.81). However, the differences between these treatments were non-significant.

5 Uptake of macro and micro nutrients

The results on effect of soil solarisation on the uptake of macro (NPK) and micronutrients (Fe, Mn, Zn and Cu) by custard apple are presented in Table 7. It indicates that macro nutrients NPK uptake was significantly improved in solarisation treatments. Among the macro-nutrients, the highest uptake 11.25 (kg/ha) was noticed in case of nitrogen. The interaction of treatment x macronutrient was significant. The nitrogen uptake was significantly highest (14.70 kg/ha) in 30 days soil solarisation treatment. More or less similar trend was noticed in case of phosphorus uptake. However, both solarisation treatments remained at par. The potash uptake was higher in case of 15 days soil solarisation, which was at par with 30 days soil solarisation treatment. It was observed that the interaction of treatment x micronutrient in soil solarisation was significantly improved the micronutrients as compared to non-solarisation treatment. The Fe was significantly highest (71.47 g/ha) in 15 days soil solarisation, while Zn was highest in 30 days soil solarisation treatment. The Mn and Cu uptake was higher in 30 days soil solarisation treatment. However, they were at par with 15 days soil solarisation.

Conclusion

The results revealed that under Junagadh (21.53°N, 70.53°E) climatic conditions, the 15 days soil solarisation treatment during hottest period of the season is good enough to control weeds and pathogens and improved growth characters and transplantable seedlings of vegetables and custard apple.

Acknowledgement

The authors are grateful to the Indian Council of Agricultural Research (ICAR), New Delhi for providing financial support to this study.

References

Jackson, M.L., 1974. "Soil Chemical Analysis". Prentice Hall India Private Ltd, New Delhi.

Johnson, C. M. and Ulrich, P., 1969. "Analytical Method for Use in Plant Analysis". California Agr. Exp. St. Bull. pp. 766.

Kanwar, J.S. and Chopra, S.L.,1976. "Analytical Agricultural Chemistry". Kalyani Publishers, Ludhiana - New Delhi.

Kirkham Don and Powers W.L., 1972. "Advanced Soil Physics". Willey Intersciences. New york, 534.

Kumar B., Yaduraju N.T., Ahuja K.N. and Prasad D.1993. "Effect of soil solarisation on weeds and nematodes under tropical Indian conditions". Weed Research, 33(5): 423-429.

Mansoori B. and Jaliani N.K. 1996. "Control of soil borne pathogens of watermelon by solar heating". Department of plant and disease research, Iran.

Yaduraju, N.T. and Ahuja, K.N. 1990. "Weed control through soil solarisation". Indian J. Agron. 35(4):440-442.

10

Utilisation of Seaweed Extract Fertiliser in Climate Smart Agriculture

Arti Joshi, A.J. Bhatt, A.Y. Desai, D.T. Vaghela, and S.I. Yusufzai

Abstract

Seaweed is a macroscopic, marine algae inhabiting near the seabed. The term includes some members of the red, brown and green algae. There is a wide range of useful and commercially important products made from these different types of seaweeds. Many such commercially important seaweed species are available on Gujarat coast. Seaweed extract is one such important product currently used in plant cultivation, with a greater scope for its wide use in different parts of the country. The main source of seaweed extract is different species of brown marine algae. Nowadays, seaweeds and products obtained from marine algae constitute the subject of interest in agriculture with emphasis on its application in Climate Smart Agriculture. A wide range of beneficial effects have been reported from the use of liquid seaweed extracts, including increased crop yields, resistance of plants to frost, increased uptake of inorganic constituents from the soil, more resistance to stress conditions, and reductions in storage losses of fruit. This paper deals with scopes and current status of utilisation of seaweed extract fertiliser in agriculture. This may replace the chemical fertiliser to save our climate and atmosphere.

Introduction

Seaweed is a macroscopic, marine algae inhabiting near the seabed. There is a wide range of useful and commercially important products made from these different types of seaweeds. Many such commercially important seaweed species are available on Gujarat coast. There are about 10,000 kinds of seaweeds in the world. They fall into three main groups: green, brown and red. Seaweed or marine algae are the oldest living organisms on the earth that are rather wrongly considered as "sea plant". Seaweed is the group of plants that live either in marine or brackish aquatic environment. Like the terrestrial plants, seaweed also contains photosynthetic pigments that help them to trap solar energy to produce carbohydrate, with the absorbed nutrients that are present in the seawater. Seaweed does not contain true leaf, stem or roots so is called as thallus.

One big difference between the groups is the pigments that they have in their cells.

- Green seaweeds have mainly chlorophyll. This green pigment captures sunlight for photosynthesis (a process that uses sunlight to produce food).
- Brown and red seaweeds also contain chlorophyll, but their cells also have other pigments that mask the green.

Red and brown are almost exclusively marine, whilst green algae are also common in freshwater (rivers and lakes), and even in terrestrial (rocks, walls, houses, and tree bark in damp places) situations. Many of these algae are very ancient organisms, and although lumped together as "algae", are not really very closely related, having representatives in five or six kingdoms of organisms. The brown colour of these results from the dominance of the xanthophyll pigment fucoxanthin. Green algae may be unicellular (one cell), multicellular (many cells), colonial (living as a loose aggregation of cells). Most green are aquatic and are found commonly in freshwater (mainly charophytes) and marine habitats. The red algae *Kappaphycus* and *Betaphycus* are now the most import sources of carrageenan, a commonly used ingredient in food, particularly yoghurts, chocolate milk and prepared puddings.

Seaweeds grow abundantly along the Tamil Nadu and Gujarat coasts and around Lakshadweep and Andaman and Nicobar islands. There are also rich seaweed beds around Mumbai, Ratnagiri, Goa, Karwar, Varkala, Vizhinjam and Pulicat in Tamil Nadu and Chilka in Orissa. Out of approximately 845 species of marine algae found in both intertidal and deep water zones of the Indian coast, nearly 60 species are commercially important.

On the west coast, especially in the state of Gujarat, abundant seaweed resources are present on the intertidal and subtidal zones. The greatest variety of red seaweed found subtropical and tropical water, while brown seaweed are common in cooler and temperate water. These resources have great potential for the development of seaweed-based industries in India. Brown seaweed plays an important role in pharmaceutical and cosmetic industries as emulsifiers and gel formers. They are also used as thickener in pharmaceutical and chemical industries, such as toothpaste, shampoo, a myriad of other products, in rubber products, paper, paints, and cosmetics.

Seaweed extract is used as a foliar spray, application to soil and for soaking of seed before sowing. It enhances the germination of seeds, increase uptake of nutrient and give resistance to frost and fungal disease. These fertilisers are used to supply important nutrients to soil in order to promote plant growth. The fertilisers are safe to use, accurate in composition and are offered at industry leading price. These extracts are excellent in quality and useful in increasing yield and prolonging shelf life of crops. This is non-toxic and widely used by farmers, agricultural units and fruit and vegetable growers.

Preparation of seaweed liquid extracts (SLE)

Dried seaweeds were hand crushed and powdered with coffee-grinder. Algae were heated with sterile distilled water in a ratio 1:100 (w/v) at 60°C for 45 min. Then the extracts were filtered through a filter paper and stored at 4°C for further experimental studies. The filtrates were 10-2 g of dried seaweeds per milliliter (gSW mL^{-1}) extracts. Different concentrations of SLE were prepared by diluting of these extracts with distilled water.

What use are they?

Industrial utilisation is at present largely confined to extraction for phycocolloids and, to a much lesser extent, certain fine biochemicals. Fermentation and pyrolysis are not being carried out on an industrial scale at present but are possible options for the future, particularly as conventional fossil food run out. Seaweeds are being used in cosmetics, as fertilisers. They have the potential to be used as a source of long- and short-chain chemicals with medicinal and industrial uses. Marine algae may also be used as energy-collectors and potentially useful substances may be extracted by fermentation and pyrolysis. Seaweed extracts appear in the oddest of places: you have probably eaten some sort of seaweed extract in the last 24 hrs as many foods contain seaweed polysaccharides such as agars, carrageenans and alginates.

Seaweed liquid fertiliser

Liquid Seaweed Extracts: Fertilisers uses of seaweeds date back at least to the 19th century. Today with the rising popularity of organic farming, there has been some revival of the industry, but not yet on a large scale. The growth area in sea weed fertilisers is in the production of liquid seaweed extracts. They are especially effective in the growing of vegetables and some fruits. A wide range of beneficial effects have been reported from the use of liquid seaweed extracts including increased crop yields, resistance of plants to frost, increased uptake of inorganic constituents from the soil, more resistance to stress conditions, and reductions in storage losses of fruit.

Studies of seaweed extracts have shown that although in some bioassay systems, for example the radish leaf expansion bioassay, high levels of cytokinin activity are recorded, in others, for example, the Amaranthus seedling assay, low levels are found. Glycine betaine, one of the structurally simple betaines, was first extracted from sugar beet and was found to have chlorophyll-retention properties. In growth tests it was found to have an activity similar to that of cytokinins in several other growth tests and so it was considered that some of the cytokinin-like activity in sugar beet extracts was due to glycine betaine. It has been shown also to be a major osmoticum (controlling water movement in and out of plant cells) in certain higher plant families adapted to either salt or water stress and it has been suggested that other betaines and tertiary sulphonium compounds have a similar function in other species. Also it has been claimed that glycine betaine has a role in frost resistance.

Betaines have been recorded for most of the species of marine algae used in the manufacture of seaweed extracts. Ascophyllumnodosum yields c-aminobutyric acid betaine, d-aminovaleric acid betaine and laminine whilst Laminaria species have a range of betaines including glycine betaine. Commercial seaweed extracts have been examined for their betaine content and the compounds detected were those reported for the algal species used in the manufacture of the extracts. Because of the reported effects from the application of commercial seaweed extracts and the known properties of compounds such as glycine betaine, the circumstantial evidence for at least part of the activity of the seaweed extracts being due to compounds of this type is strong. Moreover, some of the discrepancies in the results obtained for the cytokinin contents when the extracts are bioassayed using different procedures may be explained by the presence of betaines in the extracts.

Fertilisers and soil conditioners

There is a long history of coastal people using seaweeds, especially the large brown seaweeds, to fertilise nearby land. Generally drift seaweed or beach-washed seaweed is collected, although in Scotland farmers sometimes cut *Ascophyllum* exposed at low tide.

Seaweed meal is dried, milled seaweed, and again it is usually based on the brown seaweeds because they are the most readily available in large quantities. Species of *Ascophyllum*, *Ecklonia* and *Fucus* are the common ones. They are sold as soil additives and function as both fertiliser and soil conditioner. They have a suitable content of nitrogen and potassium, but are much lower in phosphorus than traditional animal manures and the typical N:P:K ratios in chemical fertilisers. The large amounts of insoluble carbohydrates in brown seaweeds act as soil conditioners (improve aeration and soil structure, especially in clay soils) and have good moisture retention properties. Their effectiveness as fertilisers is also sometimes attributed to the trace elements they contain, but the actual contribution they make is very small compared to normal plant requirements.

Blunden (1991) describes an interesting application of *Ascophyllum* as a soil conditioner in controlling losses of top soil. Like all brown seaweeds, *Ascophyllum* contains alginate, a carbohydrate composed of long chains. Seaweed extracts have given positive results in many applications. There are probably other applications where they have not made significant improvements, but these receive less, if any, publicity. However, there is no doubt that seaweed extracts are now widely accepted in the horticultural industry. When applied to fruit, vegetable and flower crops, some improvements have included higher yields, increased uptake of soil nutrients, increased resistance to some pests such as red spider mite and aphids, improved seed germination, and more resistance to frost.

Table 1: Marine algae along the Indian coasts

Algae	*Order**	*Family**	*Genus**	*Species variation**	*Total**
Chlorophyta	7	19	43	179+ 37	216
Phaeophyta	6	13	37	159 + 32	191
Rhodophyta	16	36	136	406 + 28	434
Xanthophyta	1	1	1	3 + 1	4
Total	**30**	**69**	**217**	**747 + 98**	**845**

State	*Chlorophyta**	*Phaeophyta**	*Rhodophyta**	*Total**
Gujarat	64	70	113	247
Maharashtra	52	47	99	198
Goa	22	29	32	83
Karnataka	13	20	23	56
Kerala	23	12	39	74
Lakshadweep	40	24	58	122
Total	**214**	**202**	**364**	**780**

(*Source*: Oza and Zaidi, 2001) (* In number)

Seaweed extracts in plant cultivation

Some research show that the application of seaweed extracts in plant cultivation exhibits positive effect on cultivated plants. Algae extracts improve plant resistance to frost and drought and increase crop yields. Plants sprayed with the use of seaweed extracts are also characterised by higher resistance to pests and pathogens and more efficient consumption of nutrients from soil. Seaweed extracts contribute to the recovery of damages caused by insects and bacterial or fungal diseases. Formulations basing on algal extracts are rich in phytohormones (gibberelins, auxins, cytikinins), amino acids and fatty acids which are responsible for plant growth, development and resistance to pathogens. Seaweed extracts can be delivered to plants in many ways. Soaking of seeds in algal extracts is one of the methods. This way of treatment can affect seeds germination. In addition to application methods, also the concentration of seaweed extract, species of algae, plant variety affects the efficiency of seaweed extracts as plant biostimulants. Both, micro- and macroalgae extracts have been used to increase crop yields and food production in various regions of the world. It is due to beneficial effect of seaweed extracts on the soil which is dependent on the type of crop and environmental conditions.

Chemical constituents of seaweed extract

It has been already established that the beneficial effect is due to the presence of trace elements and in particular the growth hormone like substances such as cytokinin in seaweed. The chemical constituent for different seaweed extract have been reported by various workers. Similarly the presence of natural cytokinin in seaweed extract.

References

Blunden, G. 1991. *Agricultural uses of seaweeds and seaweed extracts*. pp. 65-81. *In* Guiry and Blunden, 1991, q.v.

Indergaard, M. (1983). The aquatic resource. I. The wild marine plants: a global bioresource. In Biomass utilisation. (Cote, W. A., editors), 137-168. Plenum Publishing Corporation.

Stein, J.R. and Borden, C.A. (1984). Causative and beneficial algae in human disease conditions: a review. Phycologia, 23: 485-501.

Tseng, C.K. (1981a). Marine phycoculture in China. Proceedings of the International Seaweed Symposium, 10: 124-152.

Tseng, C.K. (1981b). Commercial cultivation. In : *The Biology of Seaweeds*. (Lobban, C. S. and Wynne, M. J., editors), 680-725. Blackwell Scientific Publications, Oxford.

Oza, Rohit M. and Zaidi, S.H. (2001). A revised checklist of Indian marine algae. Central Salt and Marine Chemical Research Institute, Bhavnagar Publication. pp. 341.

11

Cotton and Climate Change: Impacts and Options to Mitigate and Adaption

G. K. Kataria, M. G. Valu and M. D. Khanpara

Abstract

Climate change is effect on cotton production as a result of higher concentrations of CO_2 and increases in temperature. Cotton has certain resilience to high temperatures and drought due to its vertical taproot. The crop is, however, sensitive to water availability, particularly at the height of flowering and boll formation. Rising temperatures favour plant development, unless day temperatures exceed 32°C. Increases in atmospheric CO_2 will also favour plant development. In order to produce cotton lint, 5019 m_3/tonne of blue and 15198 m^3/tonne green water were required, making Indian cotton one of the most water-intensive cotton of the country. In turn increased pests, water stress, diseases, and weather extremes will pose adaptation challenges. Limited increases in temperatures could favour cotton plant growth and lengthen the cotton growing season. It has not been established that 41.8°C is the upper limit, particularly in India, has shown that heat stress is a big constraint to increasing yields.

Introduction

Cotton is a perennial with an indeterminate growth habit. Climate change impacts on cotton growth and development that influence yield and fibre quality will most likely be a result of the net effects of increases in CO_2 concentration, reduced water availability and increased atmospheric evaporative demand as a result of lower rainfall and relative humidity and increases in temperature (Ton, 2011). Cotton is grown on every continent except Antarctica, and in over 60 countries in the world. In many countries, cotton is one of the primary economic bases which provide employment and income for millions of people involved in its production, processing, and marketing (Chaudhry *et al.*, 2003). Mauney (1986) stated that all processes leading to square, blossom and boll initiation, and maturation are temperature-dependent.

Impact of climatic changes on cotton:

Cotton plants respond to changing environments. The response depends on the stage of development of the plant. Key stages in cotton plant development are: (a) conditions at the time of planting; (b) plant development in early season; (c) flowering, (d) boll formation and (e) conditions towards the end of the season. Indian cotton production has a very high evaporative (800-1000mm). Rain comes during a short period of time (but with a large quantity) which results in low yields. Calculations were made to estimate the water footprint of cotton by (Chapagain *et al.*, 2006). They also assessed the requirement of effective rainfall (green water) and irrigation water (blue water) used (m^3/ha) during the entire period of the plant's growth to the corresponding crop yield (ton/ha). The result showed that in order to produced cotton lint 5019 m^3/ton of blue and 15198 m^3/ton green water were required, making Indian cotton one of the most water-intensive cotton in the world (Freeland *et al.*, 2007 and Chapagain *et al.*, 2006), using approximately 15 per cent of the water resources of the country (ICAC. 2007).

Effect of high temperatures on leaf Photosynthesis

Climate change may increase the frequency of these high temperatures. Cotton plants maintain optimum growing temperatures by opening stomates in the leaves, allowing water to pass out and evaporate, thus cooling leaves (transpiration). Excessively high temperatures (greater than 35°C) during the day can reduce photosynthesis while warm nights (above 25°C) mean that leaf temperature remains high, and respiration remains high, consuming stored assimilates. Maintenance respirationapproximately doubles for every 10°C rise in temperature (Fig. 2).

Effects of climate on CO_2 level:

Higher CO_2 levels in the immediate surroundings of the cotton plant will increase photosynthetic activity. Cotton will grow more vigorously as the amount of CO_2 in the air increases. Leaves will likely be larger, thereby giving plants a greater photosynthetic surface area, which subsequently facilitates growth. With more atmospheric CO_2, greater numbers of branches and fruiting sites will likely develop, and this in turn, should ultimately provide for higher lint yields (ICAC, 2007). According to Reddy *et al.* (1998), at temperatures greater than 30° C most of the fruit was aborted regardless of CO_2 concentration (Fig. 3).

Effects of high temperature on boll retention

Boll retention is more sensitive to high temperatures can produce bud shedding, which is the most common reason for loss of fruit forms (Reddy *et al.*, 1999). They also observed that temperature regimes alter boll development: boll size and the maturation period both decreased as the temperature increased. It determined that boll growth decreases significantly and fruit is shed three to five days after blossom in temperatures over 32°C (ICAC, 2007). Thus, the upper limit of cotton for blossom and fruit period is 32°C(Table 1). However, referring to the monthly average maximum temperature and stresses that were cotton production is currently viable also in hotter environments. Cotton is successfully grown at 28.2°C in China and 37.6°C in India, 36.8° C in Pakistan and 41.8°C in Sudan. It has not been established that 41.8°C is the upper limit, but experience in many countries, particularly in India, Pakistan and Syria, has shown that heat stress is a big constraint to increasing yields. These countries successfully developed heat tolerant varieties during the 1970s and 1980s (ICAC, 2009).

Table 1: Optimum Climate Needs for Cotton Crop

Growth Stage	*Average Daily Temperature Celsius**	*Daily Crop Water Use (mm)**
Planting (Soil)	18° Minimum	>0
Planting (Air)	>21°	
Vegetative Growth	21°-27°	1-2
1st Square	-	2-4
Reproductive Growth	27°-32°	3-8
Peak Bloom	-	8
1st Open Boll	-	8-4
Maturation	21°-32°	4

Effect of high temperature on cotton quality

Quality is much less sensitive to high temperatures than yield. In general, cotton growing in a hot climate will have a higher micronaire due to the thicker rings of cellulose that are deposited daily in the fiber. Cotton fiber needs a minimum of 40 to 50 days to mature regardless of temperature. The boll period is less accelerated by hot weather than other developmental stages, such as seedling growth. Under hot weather especially if the plant is adequately watered, each daily ring will be thicker resulting in thick fibers and increased micronaire (Bange, M. (2007). Additionally, hot temperatures often cut short the boll-setting period, decreasing late setbolls that tend to have lower micronaire. As photosynthesis increases with temperature in the absence of water stress, more resources are available to mature the fibres thus increasing micronaire (Fig. 4).

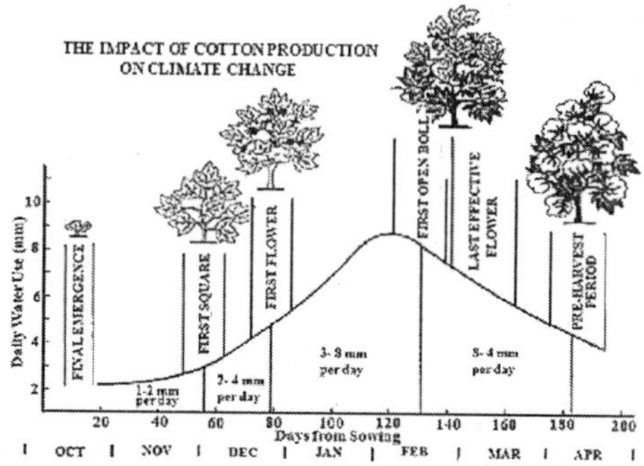

Fig. 1.Different growth stages for cotton and their demand for water. (Source: Freeland et al, 2007)

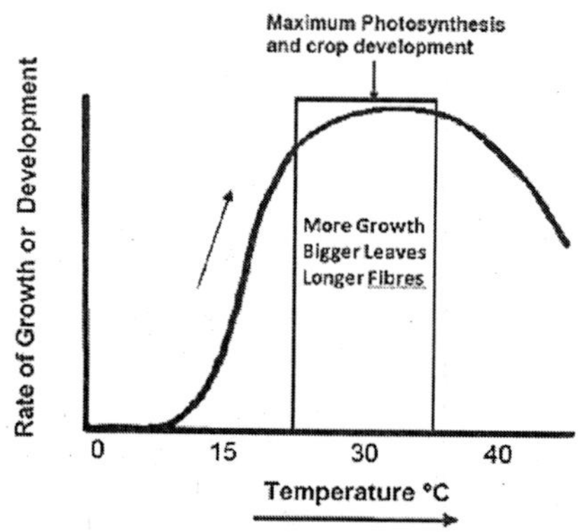

Fig. 2: Effect of high temperatures on leaf Photosynthesis.

Effect of temperature on the seed cotton yield:

When hot temperatures occur prior to bloom or after boll set, yield is often increased. Hot temperatures pre-bloom speed the arrival of the bloom period and occur at a time when water use is low and the root system is still expanding into fresh soil moisture. Hot temperatures after boll set hasten the maturation and opening of the crop. High night temperatures are detrimental to young boll set and boll size regardless of the moisture status, because the plant does not cool itself at night. Minimum night temperatures in the 80's decrease yield due to the high respiration and reduced supply of carbohydrates, resulting in the same "adjustment of bollload".

Effect of temperature on the agronomy of cotton

Cotton needs favourable growing conditions with respect to temperature, sunshine and soil moisture. Cotton requires a total of 105 to 125 days of sufficient soil moisture to grow. In tropical regions, 2 to 4 mm of water are needed daily at the beginning and the end of the growing period, while at the height of flowering 5 to 7 mm are required daily according to climatic zone. Thus, 500 to 700 mm of water is sufficient for the crop to develop fully (Fig. 1). Cotton is resilient to sub-optimal growing conditions. Cotton responds to loss of vegetation or fruiting parts (buds, flowers, bolls) through so-called 'compensatory growth'. If a flower bud, flower or boll is shed, the cotton plant quickly tries to compensate that loss through the production of more flower buds or even retaining buds that would otherwise have been shed (Chaudhry and Guitchounts, 2003).

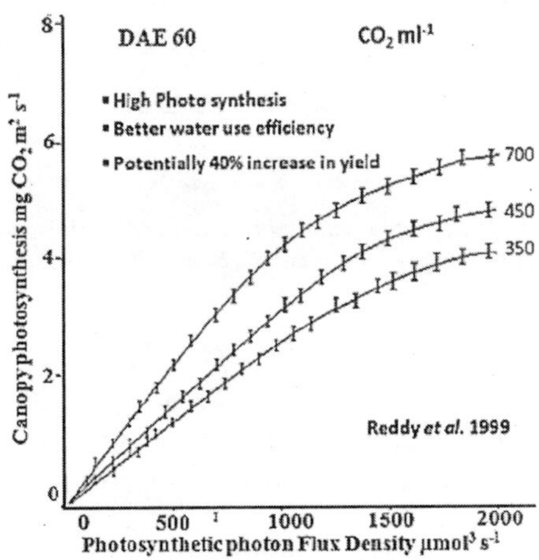

Fig. 3: Effects of photosynthetic photon flux density on canopy photosynthesis.

Effect of temperature on attack of insect pest

Insects are a recognised threat to cotton production throughout the world. Most insects can adapt their body temperature to the temperature of the environment. The effect of global warming on living organisms is slow enough for cotton insects to adjust to rising temperatures and other changes accruing from global warming. Thus, the insects currently plaguing cotton are expected to continue to be pests and possibly thrive in new environmental conditions.

In light of increased temperatures, drought and humidity, pest control will become more critical in achieving optimal growth and yield. Wu *et al.* (2007) report hat genetically-modified Bacillus thuringiensis (Bt) cotton shows less Bt toxin after exposure to elevated CO_2, which might affect plant-bollworm interactions. Karl *et al.* (2009) state that higher temperatures that reduce the effectiveness of certain classes of pesticides (pyrethroidsandspinosad).

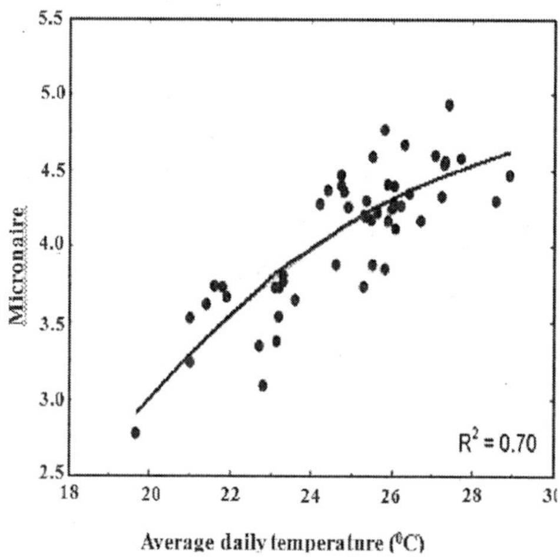

Fig. 4: Impact of average minimum temperature on fibre micronaire during boll filling (Source: Bange M., 2007).

Effect of temperature on attack of diseases :

Higher CO_2 levels will increase the severity of diseases, induce fungal growth and spore formation, and will destroy more plant tissue. In general, the disease problem will become more important (ICAC, 2007).

Effect of temperature on water availability

Plants need adequate water to grow and to maintain their temperature within an optimal range. Without water for cooling, plants may suffer heat stress. The amount and timing of water availability during the growing season, through precipitation or irrigation, are critical for cotton. If water supply variability increases, it will affect plant growth and cause reduced yields (Karl *et al.*, 2009). Cotton surface that is dedicated to irrigation is already high; about 53 per cent of the total area (Soth, 1999). However, yields for irrigated cotton are much higher (3,000–4,000 kg of seed cotton/ha) than in rain-fed cotton (1,000–2,000 kg of seed cotton/ha). Therefore, no less than 73 per cent of all cotton fibre worldwide has actually been grown under some conditions of irrigation (full or supplementary irrigation). The water footprint is defined as the total volume of freshwater that is used to produce a unit of a good or service. Cotton's share of the global agricultural water footprint is estimated at three per cent. This is proportionate to cotton's global land use footprint of 2.5 per cent (Cotton Incorporated, 2009) but will of course be very pronounced in large irrigated production areas.

Overall impact of climate changes in India:

India is the second largest producer of cotton worldwide. Cotton is grown in India in three distinct zones: Central, South and North. Temperatures are expected to increase all over India. Rainfall intensity during monsoons may become an increasing problem. Higher temperatures in already hot areas may hinder cotton development and fruit formation. Rain-fed cotton production may suffer from higher climate variability leading to periods of drought or flooding (India, 2004). Irrigated cotton, particularly in northern India, may suffer from lower water availability due to the upstream reduction of snow and ice from Himalayan and Tibetan Plateau glaciers and snowfields (UNFCCC, 2008).

Options to Adaptation Against Climate Change

As climate change alters the economics of production, rural cotton farming communities will have to formulate different adaptation strategies including planting different crops and seeking alternative non-farm income streams. At the production level, the cotton plant's genetic make up allows it to make limited adjustments to changes in climatic conditions (ICAC, 2007). The following options can be distinguished to adapt to climate change:

(1) Stop any unnecessary loss of nutrients for the farming system, preventing soil erosion and abandoning the burning of cotton crop residues where still applied.

(2) Favour a cropland design that has plant diversity and that favours soil fertility management; for example, through the inclusion of cover crops or perennials.

(3) Adjust sowing dates to offset moisture stress during the warm period, to prevent pest outbreaks and to make best use of the length of the growing season.

(4) Minimise the period that land lays bare, in order to slow down loss of organicmatter and soil humidity and soil erosion in general.

(5) Minimise soil tillage in order to prevent loss of soil organic matter – a natural source of soil fertility and storing water for plant uptake.

(6) Breed cotton varieties that are more resistant to heat stress, drought spells, weeds, pests and diseases, etc.

(7) Optimise the use of sustainable, natural fertilising sources in cotton production, including nitrogen-fixing crop rotations, compost and composted manure.

(8) Optimise the efficiency of additional fertiliser use where required, because of its costs and carbon fuel footprint. Synthetic fertiliser use is particularly high in irrigation.

(9) Optimise the water use efficiency in the production of irrigated cotton, because of the irrigation water's costs and carbon fuel footprint.

(10) Optimise the use of industrial preparations such as pesticides, herbicides and defoliants because of their costs and carbon fuel footprint.

References

Bange, M. (2007).Effects of climate changes on cotton growth and development. In:*The Australian Cotton Grower* (June-July 2007), pp. 41-45.

Chaudhry, M. Rafiq and Andrei Guitchounts (2003). Cotton Facts, International Cotton Advisory Committee, Technical Paper No. 25 of the Common Fund for Commodities. The U N, New York.

Chaudhry, R. and A. Guitchounts (2003). Cotton Facts International Cotton Advisory Committee (ICAC).Washington, D.C., USA.

Chapagain, K.A. Hoekstra, Y.A. Savenije, G.H.H. and Gautam, R., (2006). The water footprint of cotton consumption: An assessment of the impact of worldwide consumption of cotton producing countries. Ecologic. *Economics*. 60:186-203.

Cotton Incorporated (2009). Summary of Life Cycle Inventory Data for Cotton (Field to Bale–version 1.1–2 July 2009). Cotton Incorporated, USA.

Freeland, T.B. Jr, S.M. Martin, W.M. Ebelhar and W.R. Meredith (2004). Yield, quality and economic impacts of the 2002 harvest season rainfall in the Mississippi Delta. In: Proc. Beltwide Cotton Conf., San Antonio,Texas, 5–9 Jan. 2004. India (2004). Ministry of Environment and Forests, India.

ICAC (2007). Global warming and cotton production–Part 1. *In: ICAC Recorder*, **25(4):**12-16. International Cotton Advisory Committee.

ICAC (2009). Global warming and cotton production–Part 2. *In: ICAC Recorder*, **27(1):** 9–13. International Cotton Advisory Committee.

Karl, T.R.; Melillo, J.M. and Peterson,T.C. (2009).Global climate change impacts in the United States. Cambridge Uni. Press, USA.

Mauney, J.R. (1986).Vegetative growth and development of fruiting sites. *In: Cotton Physiology* (J.R. Mauney and J. McD. Stewart, eds). Memphis, Cotton Foundation.

Reddy, K.R., H.F. Hodges and J.M. McKinion, (1998). Photosynthesis and environmental factors. In: Proc. Beltwide Cotton Conf., San Diego, California, 5–9 January 1998. Vol. **2**. Memphis, *Natl. Cotton Counc. Am.*

Reddy, K.R., G.H. Davidonis, A.S. Johnson and B.T. Vinyard, (1999). Temperature regime and carbon dioxide enrichment alter cotton boll development and fiber properties. *Agron. J.*,**91:**851–858.

Soth. J. (1999). The impacts of cotton on fresh water resources and ecosystems. A preliminary synthesis. WWF.

Ton, P. (2011). Cotton and Climate Change. Impacts and options to mitigate and adapt. International Trade Centre (ITC) UNCTAD/WTO, Switzerland.

UNFCCC (2008). Climate change: Impacts, vulnerabilities and adaptation in developing countries. United Nations Framework Convention on Climate Change.

Wu G., F.J. Chen, F. Geand Y.C. Sun (2007). Effects of elevated carbon dioxide on the growth and foliar chemistry of transgenic Bt cotton. *In: Journal of Integrative Plant Biology* **49(9):** 1361-1369.

12

Effect of Irrigation Schedule and Nitrogen Management in Bt. Cotton (*Gossypium* Spp.)

J.M. Modhvadia, M.D. Khanpara, R.M. Solanki and B.S. Gohil

Abstract

A field experiment was conducted at Cotton Research Station, Junagadh Agricultural University, Junagadh (Gujarat) during kharif *season of 2009-10, 2010-11 and 2011-12 to study the effect of irrigation schedule and nitrogen management on Bt. Cotton on medium black clayey soil. The experiment was laid out in Split Plot Design. The treatments comprised viz., three irrigation levels based on IW: CPE ratios, I_1-0.8, I_2-1.0 and I_3-1.2 were assigned to main plot and four levels of nitrogen i.e. N_1-160, N_2-200, N_3-240 and N_4-280 kg ha^{-1} in sub-plots which replicated four times. The results revealed that significantly higher seed cotton (3652 kg ha^{-1}) and stalk (5057 kg ha^{-1}) yields as well as plant height (cm), number of monopodial and sympodial branches per plant, number of bolls/plant, boll weight (g) and lint yield (kg ha^{-1}) were recorded with crop was irrigated at an IW: CPE ratios of 1.2. Application of nitrogen 280 kg ha^{-1} produced significantly higher seed cotton yield (3592 kg ha^{-1}) and stalk yield (4776 kg ha^{-1}), growth and yield attributes.*

Introduction

Cotton is the most important non-food cash crop in Gujarat, covering more than 27 lakh hectares area with production of about 110 lakh bales of lint. Area under cotton in Saurashtra is about 18 lakh hectares and contributes around 65 per cent of total production in the state. Bt cotton varieties are high yielding, fertiliser and irrigation responsive and hence farmers are attempting to grow more and more cotton in this region. Water and nitrogen are the key inputs for improving cotton productivity, which must be used in most efficient manner to sustain the cotton productivity at higher level.

Nitrogen play very important role in crop productivity and is an important determinant of the growth and yield of irrigated cotton (Gadhiya *et al.,* 2009). However, the productivity of cotton in India (555 kg lint/ha) is below the world average (790 kg lint/ha). Since, both irrigation and nitrogen are costly inputs, efficient utilisation of these resources through optimum synergistic combination is essential for higher productivity and input-use efficiency of Bt cotton. Hence, the study was taken up to find out optimum water nitrogen requirement for Bt cotton.

Materials and Methods

A field experiment was conducted at Cotton Research Station, Junagadh Agricultural University, Junagadh (Gujarat) during *kharif* season of 2009-10, 2010-11 and 2011-12 to study the effect of irrigation schedule and nitrogen management in Bt. cotton. The soil of the experiment site was clayey in texture and has organic carbon 3.9 g kg^{-1}, nitrogen 212 kg ha^{-1}, available phosphorus 25.5 kg ha^{-1}, and available potassium 297 kg ha^{-1} with $pH_{2.5}$ 7.6. The experiment was laid out in SPD. The treatments *viz.*, three irrigation levels based of IW: CPE ratios viz., I_1-0.8, I_2-1.0 and I_3-1.2 IW: CPE ratios were assigned to main plot and four levels of nitrogen (i.e. N_1-160, N_2-200, N_3-240 and N_4-280 kg ha^{-1}) in sub-plots. Nitrogen as per treatments was applied in four split, each of 25 per cent at basal and remaining at 30, 60 and 90 days after sowing (DAS). A common irrigation was given immediately after sowing the crop for satisfactory seed germination and proper establishment of crop. Remaining irrigations each of 50 mm depth were given as per treatments. The quantity of irrigation water was measured with 7.5 cm throat Parshall Flume installed under free flow condition in the water channel. Cotton variety NHH-44 Bt. was sown at spacing of 120 x 45 cm at 29th June 2009, 09th June 2010 and 11th July 2011, respectively and harvested at 03rd January 2010, 22nd January 2011 and 06th January 2012, respectively. The experimental data recorded for growth parameters, yield attributes, yield, and economics were statistically analysed for level of significance and pooled of the season.

Results and Discussion

Effect of irrigation on growth yield attributes and yields

The results showed the significant effect of irrigation levels on seed cotton and stalk yields. Significantly maximum seed cotton yield of 4612, 3952, 2393 and 3652 kg/ha was recorded when crop was irrigated at 1.2 IW/CPE ratio during 2009-10, 2010-11, 2011-12 and in pooled results, respectively. Significantly lower seed cotton yield recorded when crop was irrigated at 0.8 IW/CPE ratios during individual's years and in pooled. The per cent increased due to 1.2 IW/CPE ratios over 0.8 IW/CPE ratio was, 19.14, 23.73, 54.59 and

27.20 per cent corresponding. Similarly, stalk yield was also significantly influenced by irrigation levels and significantly higher yield of 7480, 5017, 2674 and 5057 kg/ha was produced when crop was irrigated at1.2 IW/CPE ratio during 2009-10, 2010-11, 2011-12 and in pooled data, respectively (Table 1). The increase in stalk yield due to 1.2 IW/CPE ratios over 0.8 IW/CPE ratios was, 28.59, 24.89, 33.83 and 28.22 per cent correspondingly. Irrigation levels had significant effect on growth and yield attributing characters. Significantly, higher plant height (128.59 cm), numbers of monopodial and sympodial branches per plant (2.53 and 18.30), number of bolls per plant (47.76), boll weight (3.59) and lint yield (1281 kg/ha) were recorded when crop received irrigation at 1.2 IW/CPE ratio on pooled data basis (Table 2). Similar results were obtained by Adarsha *et al.* (2004) and Sankarnarayan, *et al.*(2004).

Effect of nitrogen levels on yields, growth and yield attributes

Seed cotton and stalk yields were significantly influenced by nitrogen management practice during individual years and in pooled results. Fertilising the crop with 280 kg N/ha produced significantly maximum seed cotton yield of 4597, 3917, 2261 and 3592 kg/ha during 2009-10, 2010-11, 2011-12 and on pooled data basis. The respective increase in seed cotton yield with fertilised 280 kg N/ha over lower levels of nitrogen was to tune of 18.97, 19.90, 27.67, and 21.02 per cent (Table 1). Similar trend was also observed in stalk yield, wherein application of 280 kg N/ha produced significantly maximum stalk yield with fertilised 280 kg N/ha during individual years and in pooled results and remained statistically at

Table 1: Effect of different treatments on seed cotton yield of *Bt.* cotton.

Treatments	*Seed cotton yields (kg/ha)*				*Stalk yields (kg/ha)*			
	2009-10	*2010-11*	*2011-12*	*pooled*	*2009-10*	*2010-11*	*2011-12*	*pooled*
Irrigation levels (IW/CPE)								
I_1:0.8	3871	3194	1548	2871	5817	4017	1998	3944
I_2:1.0	4122	3537	2051	3237	6875	4537	2382	4598
I_3:1.2	4612	3952	2393	3652	7480	5017	2674	5057
S.Em.±	128	105	89	62	300.97	179.86	112.15	122.71
C.D. at 5 %	444	363	308	186	1041.51	622.42	388.12	364.59
Nitrogen (kg/ha)								
N_1:160	3864	3267	1771	2968	6308	4327	2101	4245
N_2:200	4024	3400	1910	3112	6707	4503	2246	4485
N_3:240	4322	3659	2046	3342	6822	4526	2529	4626
N_4:280	4597	3917	2261	3592	7059	4729	2530	4776
S.Em.±	108	99	71	54	150.15	72.65	78.74	61.48
C.D. at 5 %	314	289	208	153	435.76	210.80	228.51	173.28

par with fertilised 240 kg N/ha in 2011-12 individual year. Nitrogen management treatments also significantly affect the growth and yield attributing parameters. Significantly higher plant height (128.53 cm), numbers of monopodial and sympodial branches per plant (2.60 and 18.73), number of bolls per plant (48.18), boll weight (3.53) and lint yield (1256 kg/ha) were recorded when crop was fertilised with application of nitrogen 280 kg/ha (Table 2). Similar positive response on growth and yield attributes was observed by Basavanneppa (2005) and Meena *et al.* (2007).

Interaction effect of irrigation and nitrogen. The interaction effect found non-significant for yields, growth and yield attribute characters.

Effect of different treatments on N content and uptake by seed cotton and stalk

Effect of irrigation: Significantly the higher N content and uptake in seed cotton (1.88 per cent and 68.64 kg/ha) and stalk (0.93 per cent and 47.02 kg/ha) is found with irrigated at 1.2 IW/CPE ratios over lower level (Table 2).

Effect of nitrogen levels: Significantly the higher N content and uptake in seed cotton (1.90 per cent and 68.59 kg/ha) and stalk (0.92 per cent and 43.62 kg/ha) were observed with application of nitrogen 280 kg/ha and on stalk remained statistically at par with fertilized 240 kg N/ha in pooled data (Table 2).

Interaction effect: The interaction effect was not found significant on nitrogen uptake on seed cotton and stalk.These findings are in close agreement with Kote *et al.* (2005). Gadhiya *et al.*(2009) indicated increase in Bt cotton yield by increasing nitrogen levels from 160 to 240 kg/ha under irrigated condition.

Table 2. Effect of different treatments on growth and yield attributing characters (Average of three years)

Treatment	*Plant height at harvest (cm)*	*Number of mono-podia branches / plant*	*Number of sympodia branches / plant*	*Number of bolls / plant*	*Boll weight (g)*	*Lint yield (kg/ha)*	*N content (per cent)*		*N uptake (kg/ha)*	
							Seed cotton	*Cotton stalk*	*Seed cotton*	*Cotton stalk*
Irrigation levels (IW/CPE)										
I_1:0.8	117.68	1.99	16.12	40.69	3.16	1014	1.82	0.82	52.65	32.23
I_2:1.0	123.58	2.44	17.35	44.61	3.30	1131	1.85	0.85	60.28	38.58
I_3:1.2	128.59	2.53	18.30	47.76	3.59	1281	1.88	0.93	68.64	47.02
S.Em.±	1.44	0.07	0.19	0.92	0.05	23.31	0.012	0.014	1.38	1.32
C.D. at 5 %	4.28	0.20	0.55	2.72	0.16	69.26	0.036	0.043	4.11	3.92
Nitrogen (kg/ha)										
N_1:160	117.40	2.07	15.81	42.01	3.16	1048	1.77	0.78	53.05	33.54
N_2:200	122.60	2.30	16.67	42.23	3.27	1095	1.85	0.86	57.87	38.04
N_3:240	124.59	2.30	17.83	45.00	3.45	1168	1.88	0.91	62.59	41.90
N_4:280	128.53	2.60	18.73	48.18	3.53	1256	1.90	0.92	68.59	43.62
S.Em.±	1.09	0.07	0.25	0.74	0.05	21.52	0.01	0.014	1.12	0.81
C.D. at 5 %	3.07	0.20	0.70	20.7	0.14	60.65	0.03	0.039	3.16	2.27

Effect of different treatments on total nitrogen uptake (kg/ha) by crop

The data showed that different total nitrogen uptake by Bt. cotton crop was significantly affected due to irrigation and nitrogen levels. Significantly higher N total uptake (115.66 kg/ha) was observed with irrigated 1.2 IW/CPE ratio and higher total N uptake (112.20 kg/ha) was observed with 280kgN/ha.

Water Use efficiency and water saving

Among the irrigation ratios, the $I_{0.8}$ and $I_{1.0}$ recorded higher WUE (2.73 kg/ha-mm) and also saved 22.22 and 11.11 per cent water, respectively over $I_{1.2}$ (Table 3)

Table 3. Water use efficiency (kg/ha-mm) as influenced by different irrigation levels (Average three years)

Irrigation levels (IW/CPE)	*Number of irrigation*	*Quantity of water (mm)*	*WUE (kg/ha-mm)*	*parcentage Saving of water over I_3*
I_1:0.8	7	350	8.20	22.22
I_2:1.0	8	400	8.09	11.11
I_3:1.2	9	450	8.12	-

Economics

The average economics of three years of different treatments is presented in Table. It can be seen that the highest net realisation of 112669 Rs/ha and B:C ratio of 2.49 was achieved by adopting irrigated 1.2 IW/CPE ratio followed by 1.0 IW/CPE ratio. Among the different nitrogen management treatments, application of 280 kg N/ha recorded the highest net realisation 110178 Rs/ha and B:C ratio of 2.44, respectively (Table 4).

Table 4. Economics of different treatments in *Bt.* cotton (average of three years)

Treatment	*Yield (kg/ha)*		*Gross Realisation (Rs/ha)*	*Total cost of cultivation Rs/ha*	*Net Realisation (Rs/ha)*	*B:C ratio*
	Seed cotton	*Stalk*				
Irrigation levels (IW/CPE)						
I_1:0.8	2871	3944	124314	41265	82869	2.01
I_2:1.0	3237	4598	139975	43166	96809	2.24
I_3:1.2	3652	5057	157907	45238	112669	2.49
Nitrogen (kg/ha)						
N_1:160	2967	4255	128304	41510	86836	2.09
N_2:200	3111	4484	134532	42493	92083	2.17
N_3:240	3341	4625	144460	43776	100727	2.30
N_4:280	3591	4775	155220	45115	110178	2.44

Conclusion

From the above results it can be concluded that the application of nitrogen 240 kg ha^{-1} and crop was irrigated with IW: CPE ratios of 1.2 gave higher yields of Bt. cotton in medium black soils of Saurashtra region of

Gujarat. The farmers of South Saurashtra Agro Climatic Zone growing cotton Bt. Hybrid (NHH 44Bt) in *Kharif* season are advised to apply irrigation of 50 mm depth at an interval of 8 to 10 day after cessation of rainfall (IW/CPE ratio of 1.2) and fertilise crop with 280 kg N/ha (four equal splits i.e. basal, 30, 60 and 90 DAS) for getting higher seed cotton yield and net realisation.

References

Adarsha, T. S., Patil, B. C., Vanaja, M. and Srinivas, K. 2004. Nitrogen and phosphorus, management of cotton under different moisture regimes. Paper presented in International Symposium Strategies for Sustainable Cotton Production. *Int. Symp. Strat. Sust. Cotton Prod., A Global Vision,* Univ. Agric. Sci., Dharwad, November, p. 23-25.

Basavanneppa, M. A., 2005. Productivity and profitability of cotton influenced by fertiliser management under Tungabhadra project area of *Karnataka. J. Cotton Res. Dev.,* 19:66-68.

Gadhiya, S.S., Patel B.B., Jadav, N.J., Patel, M.V. and Patel, V.R. (2009). Effect of different level of nitrogen, phosphorus and potassium on growth, yield and quality of Bt. cotton. *An Asian J. Soil Science*, 4(1): 37-42.

Kote, J. M., Giri, A. N. and Kausale, S. P, 2005. Nutrient concentrationand uptake of different cotton (*Gossypium hirsutum* L.) genotypes as influenced by intercrops and fertiliser level underrainfed conditions. *J. Cotton Res. Dev.,* 19:188-190.

Meena, R. L., Babu, V. R. and Nath, A, 2007. Effect of fertilizer management on cotton under saline soils of Gujarat. Bhartiya Krishi Annusandhan Patrika, 22:206-210.

Sankarnarayana, K., Nalayini, P. and and Praharaj, C. S. 2004. Agronomic requirements of Bt cotton hybrid in relation to plant densities and fertilizer requirements. *Int. Symp. Strat.Sust. Cotton Prod., A Global Vision,* Univ. Agric. Sci., Dharwad, November, p. 149.

13

Effect of Saline Water on Tomato (*Lycopersiconesculentum*) Yield and Water Use under Drip Irrigation in Coastal Belt Area of Saurashtra

H.Y. Maheta, G.V. Parajapati, Kalpesh Kumar and C.R. Bharodia

Abstract

Saline water has been included as an important substitutable resource for fresh water in agricultural irrigation in many fresh water scarce regions. In order to make good use of saline water for agricultural irrigation in coastal belt area of Saurashtra, a field experiment was carried out to study the possibility of using saline water for supplement irrigation of tomato. The average electrical conductivity of irrigation water (EC_{iw}) was 2.0, 4.0, 6.0 and 8.0 mmhos/cm throughout tomato-growing season and saline water was applied after tomato seedling stage. The experimental results revealed that tomato fruit number per plant and yield decreased by 10.49 per cent in 2.0 mmhos/cm of EC_{iw}. The maximum yield loss was around 66.29 per cent for EC_{iw} of 8.0mmhos/cm. The highest benefit cost ratio of 2.88 attained for EC_{iw} of 2 mmhos/cm followed by 2.64 for ECiw of 4.0 mmhos/cm. The minimum benefit cost ratio of 1.08 found for ECiw of 8.0 mmhos/cm. So in coastal belt area of Saurashtra no enough fresh water for irrigation, saline water up to 4.0 mmhos/cm can be used to irrigate field culture tomato at the expense of some yield loss.

Keywords: *Saline water, Drip irrigation, yield.*

Introduction

The availability of fresh water for agricultural use is declining in many areas of the world due to the increasing water needs of industries and municipalities. Thus, agriculture faces challenges of using low quality wastewater and saline water for crop production. Many studies indicate that these water resources traditionally classified as unsuitable for irrigation can be used successfully to grow crops without long-term hazardous consequences to crops and soils if proper management strategies are established. These strategies include adopting advanced irrigation technology, selecting appropriately salt-tolerant crops, leaching salts below the crop root zone (Rhoades *et al.,* 1992; Oster, 1994; Shalhevet, 1994).

Saline water up to 11 dS/m has been used successfully for commercial irrigation of a number of crops globally (Hoffman *et al*., 1990; Rhoades *et al*., 1992). However, in order to assure maximum yields from crops irrigated with saline water, it is necessary to develop special management procedures (Pasternak and De Malach, 1994). Presently, drip irrigation is widely regarded as the most promising irrigation system to use with saline water (e.g. Meiri *et al*., 1992; Suarez, 1992). Several factors contribute to the good results obtained with saline water when using drip irrigation (Dasberg and Or, 1999): (i) less water use (high application efficiency) results in less salt deposited on the field, (ii) avoidance of leaf burn, (iii) high frequency drip irrigation prevents the soil from drying out between irrigation events, thereby avoiding peaks in salt concentration and concomitant high osmotic potentials and (iv) salts are continuously leached out from the wetted section and accumulate at the wetting front away from the active root zone.

Saline water has been included as an important substitutable resource for fresh water in agricultural irrigation in many fresh water scarce regions. In order to make good use of saline water for agricultural irrigation in coastal belt area of Saurashtra the experiment was conducted with the objectives to study the responses of open cultivated tomato to saline water irrigation, and explore the possibility of using saline water for supplement irrigation of tomato in Saurashtra; and to optimise saline water irrigation management strategies to maintain tomato productivity and the salt balance in the soil.

2. Materials and Methods

2.1. *Experimental site*

A field experiment was conducted at Fruit Research Station, Mangrol, district Junagadh of South Saurashtra agro climatic zone. The climate of the area is sub-tropical and semi-arid type with an average annual rainfall of 669 mm and average pan evaporation of 900 mm. May is the hottest month with mean weekly pan evaporation of 108 mm and mean monthly temperature varying between 35°C to 45°C.

2.2. *Experimental design*

The experiment consisted of four saline water treatments with an EC_{iw} of 2.0, 4.0, 6.0 and 8.0 mmhos/cm. Artificial saline water were produced by adding fresh water in the existing saline water (2–10 dS/m) resource in coastal belt area of Saurashtra. All of the treatments were replicated five times following a randomised block design. The size of each plot for one replication was 3.6 m x 4.2 m and number of plant in each row was 30.

2.3. *Irrigation*

During tomato-growing periods irrigation was applied at alternate days with IWCPE ratio of 0.8 using drip irrigation system, except at seeding stage. In order to ensure plants grow normally, fresh water was applied immediately after seeding and during the tomato seedling stage (0–35 days after seeding). The treatment was initiated at the end of seedling stage.

2.4 *Observation and equipments*

2.4.1. *Weather data*

Meteorological measurements, such as precipitation, air temperature, relative humidity, solar radiation, and wind speed were recorded in the experimental station.

2.4.2. *Soil salinity*

Soil samples for each treatment were obtained from soil cores extracted between rows with an auger (0.025 m in diameter and 0.1 m high) at the end of the experiment. The distances to drip from the sample locations were 0, 5, 10, 15 and 20 cm, and sample depths were 0, 0–15,15–30 and 30 cm depth. All soil samples were air-dried and sieved through a 2 mm sieve. The EC_e values were based on soil: water 1:5 extracts (EC1:5, on a volume basis), and were determined using a conductivity meter.

2.4.3. *Yield*

The produce was harvested in manually at two to four days interval and separately weighted for each plot. The crop was harvested in eight phases and weights of the produce were recorded separately for each treatment. The fruit number and the total fruit mass per plot were measured at each time.

2.4.4. *Statistical analysis*

Analysis of variance (ANOVA) was conducted and significance of differences among treatment was tested using the least significant difference (LSD). Differences were declared significant at $P < 0.05$.

2.4.5. *Techno-economic feasibility*

The techno-economic feasibility was determined by comparing total seasonal input with seasonal output. Drip installation cost was determined considering two season utility of system and life of drip considered 10 years with 10 per cent rate of interest. The present worth cost of drip system was calculated using method given by Goodman, 1984 and per ha benefit cost ratio was determined for each treatment under tomato crop.

3. Results and discussion

3.1. *EC_e in the soil profile*

Figs. 1 to 4 demonstrates the distributions of EC_e in the vertical transect perpendicular to the drip lateral for each treatment. It was observed from the figures that distribution of salt under point source of irrigation is less at all profiles and depths within the moisture zone because of high metric potential and it increase with increase in the lateral distance from the point source. It was observed that as the salinity of irrigation water is increased, the salt concentration throughout the soil also increased, the same result was also reported by Seifert *et.al.* (1975). It was also revealed from the observation that if the drip irrigation system use as source of irrigation application, less area is salinised, but it is more at surface between the emitters than in lower depth.

Fig. 1. Salt Distribution for ECIW 2 mmhos/cm.

3.2. *Tomato yield*

Fruit number per plant and tomato yield kg per ha under different treatments are presented in Fig 5. It is evident from the result that more yields were obtained in the case of crop with use of saline water of 2 mmhos/cm. Increase in the soil salinity resulted in reducing water uptake which was directly related to fruit yield reduction, this due to the osmotic effects of salt on

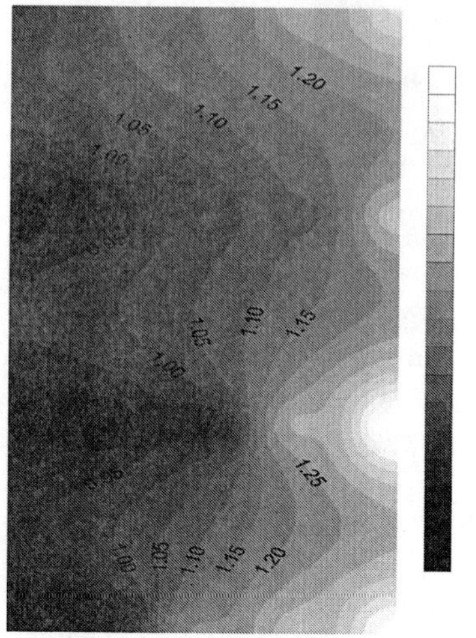

Fig. 2. Salt Distribution for EC_{iw} 4 mmhos/cm.

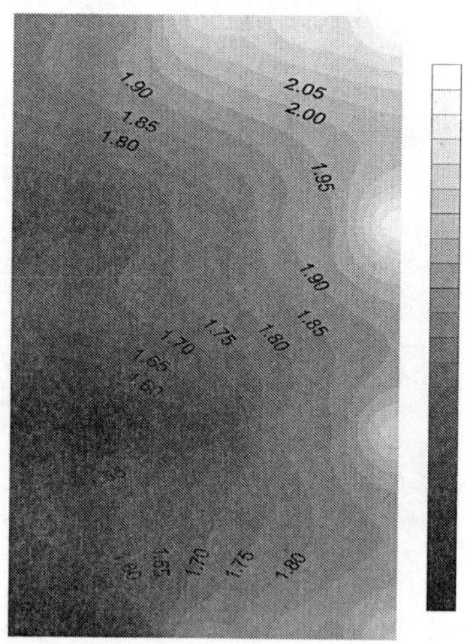

Fig. 3. Salt Distribution for EC_{iw} 6 mmhos/cm.

plant roots and causing low water availability, toxic accumulation of salinity ions on plants tissues and nutritional imbalance due to excess of salinity ions on plant tissues. The experimental results revealed that tomato fruit number per plant and yield decreased by 10.49 per cent in 2.0 mm hos/cm of EC_{iw}. The maximum yield loss was around 66.29 per cent for EC_{iw} of 8.0 mmhos/cm. The reduction in yield was linear with increase in the salinity of water applied. Statistical analysis showed a significant ($P < 0.05$) effect of EC_{iw} on fruit number and yield in and both of them decreased as EC_{iw} increased. In the experiment, the major contribution to total yield reduction under saline condition was fruit number. This is in accordance with the results of Twersky *et al.* (1979).

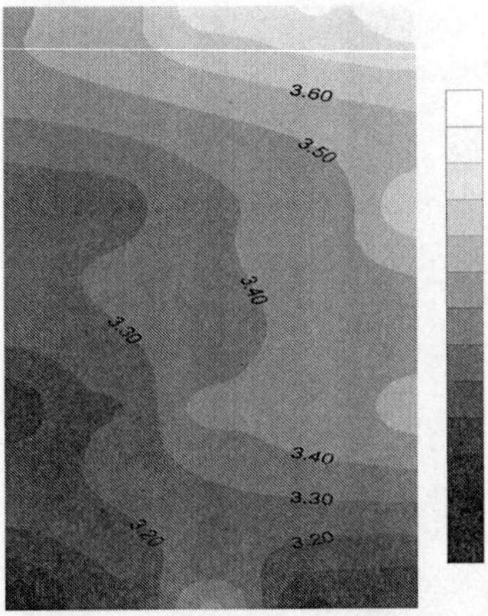

Fig. 4. Salt Distribution for EC_{iw} 8 mmhos/cm.

3.3. *Techno-economic feasibility*

The benefit cost ratio for all treatments were determined using method given by Goodman, 1984. The highest B/C ratio was obtained in the case

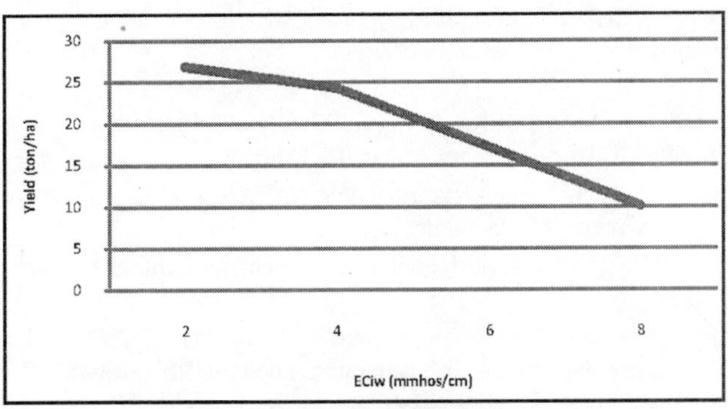

Fig. 5. Effect of EC_{iw} on Tomato Yield.

of EC_{iw} of 2 mmhos/cm followed by 2.64 for EC_{iw} of 4.0 mmhos/cm. Engineering point of view one can have precise system when adapted with few percent losses i.e. treatment of EC_{iw} of 4.0 mmhos/cm otherwise treatment with EC_{iw} of 2 mmhos/cm is found best in all treatments.

4. Summary and conclusions

A field experiment was carried out to study the possibility of using saline water for supplement irrigation of tomato. The average electrical conductivity of irrigation water (EC_{iw}) was 2.0, 4.0, 6.0 and 8.0 mmhos/cm throughout tomato-growing season and saline water was applied after tomato seedling stage. The experimental results revealed that tomato fruit number per plant and yield decreased by 10.49 per cent in 2.0 mmahos/cm of EC_{iw}. The maximum yield loss was around 66.29 per cent for EC_{iw} of 8.0 mmhos/cm. The highest benefit cost ratio of 2.88 attained for EC_{iw} of 2 mmhos/cm followed by 2.64 for EC_{iw} of 4.0 mmhos/cm. The minimum benefit cost ratio of 1.08 found for EC_{iw} of 8.0 mmhos/cm.

References

Dasberg, S., Or, D., 1999. Drip Irrigation. Springer-Verlag, Berlin, Germany, 162 pp.

Goodman, A.S., 1984. Principal of water resource planning and practices. Prentice Hall, Inc. England cliffs, New Jersey.

Hoffman, G.J., Rhoades, J.D., Letey, J., Sheng, F., 1990. Salinity management. In: Hoffman, G.J., Meiri, A., Frenkel, H., Mantell, A., 1992. Cotton response to water and salinity under sprinkler and drip irrigation. Agron. J. 84:44–50.

Oster, J.D., 1994. Irrigation with poor quality water. Agric. Water Manage. 25:271–297.

Pasternak, D., De Malach, Y., 1994. Crop irrigation with saline water. In: Pessarakli, M. (Ed.), Handbook of Plant and Crop Stress. University of Arizona, Marcel Dekker Inc., New York, U.S., pp. 599–622.

Rhoades, J.D., Kandiah, A., Mashali, A.M., 1992. The use of saline waters for crop production. In: FAO Irrigation and Drainage Paper 48, FAO, United Nations, Rome.

Seifert, W.J, Hller, E.A. and Howell, T.A., 1975. Trickle irrigation with water of different salinity levels. Trans. of ASAE. 18(1): 89-94.

Shalhevet, J., 1994. Using water of marginal quality for crop production: major issues. Agric. Water Manage. 25: 233–269.

Suarez, D., 1992. Perspective on irrigation management and salinity. Outlook Agr. 21 (4): 287–291.

Twersky, M.D., Pasernak, I., Borovic, I., and Malach, Y.D.E., 1979. Irrigation of processing tomatoes with brackish water. Hassadeh, 58(9): 1889-1898.

14

Yield and Quality of Late Sown Chickpea (*Cicer arietinum* L.) as Influenced by Irrigation Scheduling and Sulphur Levels

D. V. Srinivasulu, R. M. Solanki and P. J. Prajapati

Abstract

A field experiment was carried out during rabi *2010-11 to study the growth, yield and quality of chickpea as influenced by irrigation and sulphur levels. Irrigation and sulphur have shown significant influence on growth, yield, quality, moisture extraction pattern and water use efficiency. Among four irrigation schedules, irrigation scheduled at 0.9 IW/CPE ratio recorded higher values for all the growth parameters at various stages, yield attributes, grain and stover yield, quality parameters as well as net return and BCR while, 0.7 IW/CPE ratio remained on par. Amount of moisture extracted from surface layers was more irrespective of irrigation treatment. Depletion of soil moisture increased and water use efficiency decreased with increasing frequency of irrigation. Application of 40 kg S ha^{-1} recorded higher grain yield, protein content and protein yield, net return and BCR and remained on par with 20 kg S ha^{-1}.*

Key words: *Chickpea, Irrigation, IW/CPE ratio, Sulphur, Water use efficiency, BCR.*

Introduction

Chickpea (*Cicer arietinum* L.) is the most important pulse crop of India accounting 34.6 per cent area and 48.4 per cent production of total pulses with a productivity of 841 kg ha^{-1}. Gujarat occupied 2.46 per cent of chickpea area and 2.80 per cent of production of the country, respectively with an average productivity of 977 kg ha^{-1} (Singh, 2010). Since many years farmers were following the same irrigation schedule irrespective of the varieties cultivated without knowing it's feasibility under today's climatic conditions. Hence, today's limited water resources along with changing cropping patterns calls urgent need for application of water at an appropriate intervals for ensuring better water use efficiency. Inspite of this, recent studies on soil fertility across the country shown that long-term application of N, P and K fertilisers alone reported sulphur deficiency in most of the states including Saurashtra region of Gujarat and further, sulphur was known to increase yield and quality in chickpea. Precise information regarding appropriate irrigation interval and suitable sulphur dose for chickpea crop in recent years is very limited in Saurashtra region thus, the present investigation was carried out.

Materials and Methods

The field experiment on yield and quality of chickpea as influenced by irrigation and sulphur levels was conducted at the Instructional Farm, Department of Agronomy, College of Agriculture, Junagadh Agricultural University, Junagadh during *rabi* 2010-11. The soil was clayey in texture, slightly alkaline (pH-7.9) in reaction, rich in organic carbon with 0.76 per cent, low in available nitrogen (178.75 kg ha^{-1}), K_2O (112.90 kg ha^{-1}) and sulphur (8.15 ppm) and medium in available P_2O_5 (38.40 kg ha^{-1}). The bulk density of the soil is 1.36 Mg m^{-3} with field capacity 28.4 per cent and permanent wilting point 12.8 per cent. Further, no rainfall was received throughout the crop period.

The experiment was laid out in split plot design comprising four levels of irrigation based on IW/CPE ratios [I_1= 0.5, I_2= 0.7, I_3= 0.9 and I_4= farmer's practice (1st irrigation immediately after sowing, 2nd irrigation at 10-12 DAS and rest of three at an interval of 18-20 days)] as main plot and three levels of sulphur (S_1= 0, S_2= 20 and S_3= 40 kg S ha^{-1}) as sub-plot treatments replicating thrice in 36 plots each of size 5.0 m × 3.6 m. Sowing of chickpea (JG-16) using 60 kg seed ha^{-1} was done at a spacing of 45 cm × 10 cm. One intercultivation followed by a hand weeding was done at 40 DAS. Immediately after sowing and at 12 DAS light irrigations of 50 mm depth were given for proper germination and ensuring better establishment of the crop irrespective of cumulative pan evaporation readings. Afterwards irrigation of 50 mm depth was provided as per treatments based on cumulative pan evaporation readings. Amount of water to be irrigated was measured using parshall flume of 7.5 mm throat placed at the head irrigation channel. After initial two common irrigations, a total of three (41, 62 and 82 DAS), four (33, 50, 65 and 79 DAS), five (29, 44, 54, 66 and 77 DAS) and three (29, 47 and 68 DAS) irrigations were received by I_1, I_2, I_3 and I_4 treatments, respectively. The quantity of water received by I_1, I_2, I_3 and I_4 treatments is 250 mm, 300 mm, 350 mm and 250 mm, respectively. Sulphur was applied in soil as per treatments 10 days prior to sowing in elemental form.

Recommended dose of both nitrogen (25 kg ha^{-1}) and phosphorus (50 kg ha^{-1}) was supplied through Urea and DAP, respectively at the time of sowing. Observations on growth parameters, yield attributes and yield and quality parameters as well as moisture studies were recorded.

Results and discussion

Growth parameters: Giving five irrigations to chickpea (after initial two common irrigations) at an IW/CPE ratio of 0.9 resulted in significantly higher plant height, plant spread and dry matter accumulation at 60 DAS and at harvest, branches per plant, nodules and nodule dry weight per plant and was at par with 0.7 IW/CPE ratio (Table 1). This may be due to the adequate availability of moisture at all critical stages of growth and development contributing to luxurious uptake of nutrients, favourable physiological processes and active cell division. The results obtained by Dixit *et al.* (1993a) are in corroborative with the above results. Further, irrigation at 0.9 IW/CPE ratio delayed the flowering and maturity due to prolonged vegetative growth under high moisture availability compared to farmer's practice which resulted in early maturity due to moisture deficit coupled with high temperatures.

Application of 20 kg S ha^{-1} recorded significantly higher plant height at harvest, plant spread and dry matter accumulation at 60 DAS being at par with 40 kg S ha^{-1}. But plant height at 60 DAS, plant spread and dry matter accumulation at harvest, number of nodules and nodule dry weight per plant were significantly higher with the application of 40 kg S ha^{-1} which was at par with 20 kg S ha^{-1} (Table 1). Increase in growth parameters with increased levels of sulphur may be due to it's higher availability and uptake as well as it's active involvement in synthesis of amino acids, regulation of various metabolic and enzymatic processes along with enhanced nitrogen fixation and biomass accumulation. Singh *et al.* (2004) reported the similar results.

Yield attributes: Scheduling irrigation to chickpea at 0.9 IW/CPE ratio recorded significantly maximum number of pods per plant, seeds per pod, test weight, grain and stover yield per plant followed by 0.7 IW/CPE ratio (Table 1 and 2). The harvest index was not influenced significantly by various irrigation levels but decreased with increasing frequency of irrigation. The irrigation intervals at 0.9 IW/CPE ratio were nearly coincided with that of farmer's practice and further provided two more irrigations one at peak vegetative stage and another at maturity stage. Thus, it may provided higher vegetative growth and synthesis of more photosynthates coupled with better translocation and partitioning from source to sink resulting in more number of well filled pods with more number of seeds. This finally resulted in higher grain and stover yield per plant as well as test weight. These results are in close conformity with Reddy and Ahlawat (1998). Application of sulphur @ 40 kg ha^{-1} resulted significantly higher number of pods per plant followed by 20 kg S ha^{-1}.

Significantly maximum values in test weight, grain and stover yield per plant were recorded with 20 kg S ha^{-1} followed by 40 kg S ha^{-1} (Table 1 and 2). Influence of sulphur on yield attributes was further more pronounced as it plays vital role in improving the vegetative structure for nutrient absorption and provides strong sink strength through the development of reproductive structures as well as production of assimilates to fill economically important sink (Ashok Kumar *et al.*, 2006).

Yield: The extent of increase in grain and stover yields of chickpea at 0.9 IW/CPE ratio was to the tune of 16.88 and 30.68 per cent over farmer's practice, respectively. And it remained at par with 0.7 IW/CPE ratio. The probable reason for the potential increase in grain and stover yields with increasing frequency of irrigation may be contribution of growth and yield attributes to the yields. These results are in complete agreement with those obtained by Chandrasekhar and Saraf (2005).

Sulphur @ 40 kg ha^{-1} recorded significantly higher grain yield and was at par with 20 kg S ha^{-1}. Whereas, maximum stover yield was obtained with 20 kg S ha^{-1} and was at par with 40 kg S ha^{-1}. This potential increase of grain and stover yields with the application of sulphur might be due to it's key role in photosynthetic rate and translocation as well as transformation of assimilates into yields. Shrikrishna *et al.* (2004) reported significant increase in grain and stover yields with increasing levels of sulphur in chickpea.

But interaction of irrigation scheduled at 0.7 IW/CPE ratio with application of 20 kg S ha^{-1} (I_2S_2) recorded 32 per cent higher grain yield over farmer's practice (I_4S_1) followed by I_2S_3 and I_3S_3.

Quality characters: Increasing frequency of irrigation significantly increased the protein content and protein yield and was recorded higher at 0.9 IW/CPE ratio. It was due to increased availability and uptake of nutrients under high moisture conditions which boosted the synthesis of amino acids and proteins (Dixit *et al.*, 1993a and Reddy and Ahlawat, 1998).

Increased application of sulphur significantly increased the protein content and protein yield in chickpea and was recorded maximum when fertilised with 40 kg S ha^{-1}. This was due to the synthesis of more sulphur containing amino acids. Narendra Kumar *et al.* (2003) also reported increase in protein content with the application of sulphur in chickpea.

Moisture studies: Higher amount of moisture was extracted from surface layers irrespective of irrigation schedules. About 60-64 per cent of moisture was extracted from 0-30 cm soil depth and around 90-95 per cent moisture was extracted from 0-60 cm depth. With increasing frequency of irrigation from 0.5 to 0.9 IW/CPE ratio the per cent moisture extracted was also increased. It was further revealed that moisture extracted from deeper layers was higher in lower IW/CPE ratios. Dixit *et al.* (1993b) also reported the same trend in

Table 1. Effect of irrigation and sulphur levels on growth and yield parameters of chickpea

Treatments	Plant height (cm) at		Plant spread (cm)		Branches /plant	Dry matter accumulation (g)		Days to 50 per cent flowering	Days to maturity	Nodules/ plant	Nodule dry weight/ plant (g)	No. of pods/ plant	No. of seeds/pod
	60 DAS	Harvest	60 DAS	Harvest		60 DAS	Harvest						
Irrigation: (IW/CPE ratios)													
I_1: 0.5	29.6	33.5	26.5	18.8	6.4	3.7	14.8	52.8	86.4	20.8	0.197	53.1	1.27
I_2: 0.7	32.3	38.2	30.3	22.8	7.1	5.3	19.7	55.4	93.3	26.6	0.303	58.4	1.25
I_3: 0.9	34.0	42.2	32.1	23.4	7.9	5.7	20.9	57.4	95.9	31.1	0.342	61.0	1.37
I_4: Farmer's practice	31.5	34.7	29.8	18.5	7.4	4.2	16.3	50.9	85.8	22.9	0.214	54.9	1.19
S.Em.±	0.84	1.02	1.05	0.94	0.28	0.21	0.88	0.15	0.23	1.23	0.015	1.34	0.03
C.D. at 5%	2.91	3.53	3.62	3.24	0.97	0.73	3.06	0.51	0.81	4.25	0.053	4.63	0.10
Sulphur levels (kg ha^{-1})													
S_1: 0	30.9	35.7	28.6	19.63	6.6	4.2	16.7	54.2	90.3	22.6	0.244	52.5	1.26
S_2: 20	32.1	38.3	30.3	21.45	7.4	5.0	18.2	54.0	90.5	25.2	0.263	58.0	1.29
S_3: 40	32.5	37.5	30.1	21.52	7.7	5.0	18.8	54.3	90.3	28.3	0.286	60.1	1.25
S.Em.±	0.50	0.61	0.36	0.42	0.24	0.16	0.51	0.25	0.26	0.91	0.009	0.93	0.02
C.D. at 5%	NS	1.84	1.09	1.25	0.71	0.50	1.52	NS	NS	2.74	0.028	2.79	NS
Interaction (I X S)													
S.Em.±	1.00	1.23	0.73	0.84	0.48	0.34	1.02	0.51	0.53	1.83	0.019	1.86	0.04
C.D. at 5%	NS	NS	NS	NS	NS	NS	NS	NS	NS	5.48	NS	5.57	NS

Table 1. Effect of irrigation and sulphur levels on yield parameters and yield, quality, water use and production economics of chickpea

Treatments	Yield parameters			Yield		Harvest index (per cent)	Quality of grain		Water use		Production economics		
	Test weight (g)	Grain yield/ plant (g)	Stover yield/ plant (g)	Grain yield/ ha (kg)	Stover yield/ha (kg)		Protein content (per cent)	Protein yield (kg/ha)	CUW* (mm)	WUE** (kg ha^{-1} mm^{-1})	Gross reali- sation (₹/ha)	Net Reali- sation (₹/ha)	BCR
Irrigation: (IW/CPE ratios)													
I_1: 0.5	17.05	9.2	13.8	1744	2503	41.28	18.80	326	211	6.98	40241	14483	1.56
I_2: 0.7	17.74	11.8	17.5	2199	3472	38.69	20.57	455	248	7.33	50866	24826	1.95
I_3: 0.9	19.03	13.2	18.7	2243	3791	37.27	22.10	496	282	6.41	51984	25661	1.97
I_4: Farmer's practice	16.45	10.4	15.4	1919	2901	40.02	20.58	396	224	7.67	44338	18580	1.72
S.Em.±	0.28	0.47	0.86	103.2	209.5	1.24	0.28	20.9	6.3	0.41			
C.D. at 5%	0.95	1.64	2.97	357.2	724.9	NS	0.98	72.5	21.7	NS			
Sulphur levels (kg ha^{-1})													
S_1: 0	16.62	9.4	15.6	1919	2965	39.65	19.46	374	238	6.75	44364	19349	1.77
S_2: 20	18.65	12.6	17.1	2035	3291	38.28	20.92	429	239	7.10	47104	21134	1.81
S_3: 40	17.44	11.4	16.4	2124	3245	40.00	21.16	451	246	7.44	49088	22165	1.82
S.Em.±	0.18	0.32	0.31	46	71	1.0	0.23	11	6.76	0.16			
C.D. at 5%	0.53	0.95	0.94	137	212	NS	0.70	32	NS	0.49			
Interaction (I X S)													
S.Em.±	0.35	0.63	0.62	91.51	141.74	1.62	0.47	21.67	13.52	0.32			
C.D. at 5%	1.1	1.9	NS	274.4	NS	NS	1.4	64.9	NS	0.97			

Table 3. Depth-wise moisture extraction pattern (per cent) by chickpea as influenced by irrigation and sulphur levels

Treatments	*Soil depth (cm)*				
	0-15	*15-30*	*30-45*	*45-60*	*60-75*
Irrigation: (IW/CPE ratios)					
I_1: 0.5	31.83	29.00 **(60.83)**	16.73 **(77.56)**	12.46 **(90.02)**	9.98 **(100.00)**
I_2: 0.7	32.63	29.55 **(62.18)**	18.74 **(80.92)**	13.47 **(94.39)**	5.61 **(100.00)**
I_3: 0.9	34.01	30.55 **(64.56)**	21.20 **(85.76)**	11.51 **(97.27)**	2.73 **(100.00)**
I_4: Farmer's practice	33.78	29.12 **(62.90)**	17.88 **(80.78)**	10.67 **(91.45)**	8.55 **(100.00)**
Sulphur levels (kg ha^{-1})					
S_1: 0	33.64	30.83 **(64.47)**	18.59 **(83.06)**	10.07 **(93.13)**	6.87 **(100.00)**
S_2: 20	31.97	29.27 **(61.24)**	18.86 **(80.10)**	13.13 **(93.23)**	6.77 **(100.00)**
S_3: 40	33.57	28.56 **(62.13)**	18.47 **(80.60)**	12.88 **(93.48)**	6.52 **(100.00)**

Table 4. Interaction effect of irrigation and sulphur levels on number of nodules, yield, quality of grain and production economics of chickpea

Treatment	*No. of nodules/plant*	*Grain Yield (kg ha^{-1})*	*Protein content (per cent)*	*Protein yield (kg ha^{-1})*	*Gross realisation (Rs ha^{-1})*	*Total expenditure (Rs ha^{-1})*	*Net realisation (Rs ha^{-1})*	*BCR*
I_1 S_1	18.9	1860	17.7	328	42775	24804	17972	1.72
I_1 S_2	19.8	1636	18.5	300	37861	25758	12103	1.47
I_1 S_3	23.9	1736	20.2	351	40075	26711	13364	1.50
I_2 S_1	22.7	1914	18.8	359	44364	25086	19278	1.77
I_2 S_2	23.4	2353	22.2	523	54414	26041	28374	2.09
I_2 S_3	33.6	2330	20.7	484	53826	26993	26832	1.99
I_3 S_1	27.0	2122	21.6	457	49126	25369	23757	1.94
I_3 S_2	35.0	2276	22.0	500	52777	26323	26454	2.00
I_3 S_3	31.5	2330	22.8	531	54035	27276	26760	1.98
I_4 S_1	21.8	1782	19.8	352	41221	24804	16418	1.66
I_4 S_2	22.5	1875	21.0	395	43373	25758	17614	1.68
I_4 S_3	24.2	2099	20.9	440	48401	26711	21690	1.81
C.D. at 5%	5.5	274.4	1.4	64.9	-	-	-	-

moisture extraction in chickpea (Table 3). The present study further revealed that scheduling irrigation from 0.5 to 0.9 IW/CPE ratio increased total consumptive use of water and decreased water use efficiency (Table 2). This was due to more consumption of water due to higher vegetative growth and comparatively lower grain yield per unit quantity of water at higher IW/CPE ratios. Same trend in CUW and WUE were observed by Arya *et al.* (2005).

It was also revealed that application of 40 kg S ha^{-1} markedly increased the WUE in chickpea. Significant interaction between irrigation and sulphur was observed in number of nodules, pods and grain yield per plant, test weight, grain yield per ha, protein content and protein yield. Significant interaction between irrigation and sulphur was also reported by Patel and Patel (2005) in chickpea.

Economics: Irrigating chickpea at 0.9 IW/CPE ratio and fertilising with 40 kg S ha^{-1} recorded maximum net returns and BCR individually (Table 2). But interaction of irrigation scheduled at 0.7 IW/CPE ratio with 20 kg S ha^{-1} recorded higher net returns and BCR (Table 4). Arya *et al.* (2005) and Pramanik *et al.*, (2009) also recorded higher net returns and higher BCR at higher IW/CPE ratios.

Conclusion

From the present investigation it was revealed that irrigating chickpea (cv JG-16) at an IW/CPE ratio of 0.7 along with the application of 20 kg S ha^{-1} including recommended dose of fertilisers recorded higher yield, net realisation and higher BCR under clayey soils of South Saurashtra agro-climatic zone.

References

Arya, K. L., Lalit Kumar, Singh, K. K., and Kushwaha, B. L. (2005). Effect of fertilizers and tillage management in rice (*Oryza sativa*)- chickpea (*Cicer arietinum* L.) cropping system under varying irrigation schedules. *Indian J. of Agron.* **50**(4): 256-259.

Ashok Kumar, Shambhu Prasad and Kumar, S. B. (2006). Effect of boron and sulphur on performance of chickpea (*Cicer arietinum* L.). *Indian J. of Agron.* **51**(1): 57-59.

Chandrasekhar, K. and Saraf, C. S. (2005). Influence of irrigation and fertilizer levels on growth and yield of late sown chickpea (*Cicer arietinum* L.). *The Andhra Agricl. J.* **52** (3 and 4): 322-325.

Dixit, J. P., Dubey, O. P. and Soni, N. P. (1993a). Effect of sowing date and irrigation on yield and nutrient uptake by chickpea (*Cicer arietinum* L.) cultivers under Tawa Command area. *Indian J. of Agron.* **38**(2): 227-231.

Dixit, J. P., Soni, N. K. and Namdeo, K. N. (1993b). Moisture use pattern and yield of chickpea (*Cicer arietinum*) in relation to planting date, variety and irrigation. *Indian J. Agron.* **38**(4): 573-577.

Narendra Kumar, S. S., Khangarot and Raj Pal Meena. (2003). Effect of sulphur and plant growth regulators on yield and quality parameters of chickpea (*Cicer arietinum* L.). *Annals of Agricl. Res.* New series **24**: 434-436.

Patel, R. A. and Patel, R. H. (2005). Response of chickpea (*Cicer arietinum*) to irrigation, FYM and sulphur on a sandy clay loam soil. *International Chickpea and Pigeonpea Newsletter* (**12**): 22-24.

Pramanik, S. C., Singh, N. B. and Singh, K. K. (2009). Yield, economics and water use efficiency of chickpea (*Cicer arietinum* L.) under various irrigation regimes on raised bed planting system. *Indian J. of Agron.* **54**(3): 315-318.

Reddy, N. R. N. and Ahlawat, I. P. S. (1998). Response of chickpea (*Cicer arietinum*) genotypes to irrigation and fertilizers under late sown conditions. *Indian J. Agron.* **43**(1): 95-101.

Shrikrishna, Sharma, A. P. and Chandra Bhushan. (2004). Nitrogen and sulphur nutrition of chickpea (*Cicer arietinum*) grown under semi arid conditions of central Uttar Pradesh. *Legume Res.* **27**(2): 146-148.

Singh, N. P. 2010. Project Co-ordinators Report 2009-10, Annual group meet Aug. 29-31, 2010. AICRP on chickpea, IIPR, Kanpur, pp. 30-32.

Singh, S., Saini, S. S. and Singh, B. P. (2004). Effect of irrigation, sulphur and seed inoculation on growth, yield and sulphur uptake of chickpea (*Cicer arietinum*) under late sown conditions. *Indian J. Agron.* **49**(1): 57-59.

15

Effect of Irrigation Practices on the Quality of Cotton Yield

P.A. Pandya, H.D. Rank, R.J. Patel, P.B. Vekariya, J.H. Kelaiya and D.M. Paradava

Abstract

The cotton response to irrigation practices was evaluated through field experiment for the two consecutive years (2003-04 and 2004-05) adopting 3-factorial split plot with four replications. The main factor was irrigation method (2 no.) viz. drip and border irrigation, sub-factor was water application level (3 no.) viz. 80 per cent, 60 per cent, 40 per cent depletion of ASM for border and IW/CPE of =0.4, 0.6 and 0.8 for drip irrigation. Five growth periods viz. vegetative, flowering, boll development, boll maturity and whole season were used as sub sub-factor. The interaction effects of all the factors on cotton yield were found to be significant for both the years and pooled results. The highest yield was obtained as 2305 kg/ha under pooled results with drip irrigation at IW/CPE of 0.4 during boll maturity stage. It was found that for the highest yields, the cotton should be irrigated at IW/CPE of 1.0 during first three stages and at IW/CPE of 0.4 during last stages by drip irrigation. The cotton yield quality was evaluated based on parameters like 2.5 per cent length of fiber (mm), fiber fineness (micronaire value), bundle strength of fiber (g/tex), uniformity ratio (per cent) and maturity ratio, the cotton seed quality like seed index (g), oil content and seed cotton quality by lint percentage. The cotton yield quality parameters were improved under drip over border method. The highest fiber length, oil content and seed index was found as 26.9 mm, 21.14 per cent and 6.31 g respectively under drip with 0.8 IW/CPE. All these qualities were increased with increased in moisture regimes. The lint percentage decreased with increase in IW/CPE and highest was found as 39.61 per cent with 0.4 IW/CPE drip. All these parameters were improved under drip over border method irrigation. The highest bundle strength of 22.6 (g/tex) was found under border irrigation with 40 per cent depletion of ASM during the boll development stage.

Introduction

Many efforts were made to know the effects of irrigation methods and water application levels on cotton yield. (Constable and Hodgson, 1990; Smith *et al.*, 1991; Berami and Ofen, 1993; Cetin *et al.,* 1994; Nawar *et al.*, 1995; Cruz and Agulto, 1996; Maas *et al.*, 1996; Dippenaar *et al.*, 1997; Cruz and Agulto, 1998; Feng *et al.*, 1998; Nikolov, 1998; Rosario *et al.* , 2001; Cetin and Bilgel , 2002;) It was clear from the results presented by them that drip irrigation, with few exceptions, increased the seed cotton yield in comparison to other methods of irrigation and cotton yield was increased with water application up to a certain level i.e. the optimum water application level, and then after the yield was reduced.

However, the quality of cotton yield is also very important as the market prices are governed by it. Lots of efforts were made to assess the effects of irrigation practices on the quality of the yield. The fiber quality through genetic character may be modified to some extent by environmental factors such as water, nutrients and climatic conditions. As the availability of the soil water decreased, the fiber fineness improved whereas fiber length decreased as reported by Singh and Bhan (1993) from Kanpur. Shette and Shilke (1990) in respect of seed index, Li and Xu (1991) in respect of oil content in seeds observed higher value under higher moisture than lower moisture regimes. On the contrary, Nawar *et al.* (1995) found that drip irrigation increased fiber maturity by 13.5 per cent compared with the traditional system. Tank (1995) observed that there were neither adverse nor favourable effects on moisture and nitrogen stresses on the fiber properties viz., fines, strength and 2.5 per cent span length. The seed oil content was adversely influenced due to moisture and nutrient stresses. Xiao *et al.* (2000) conducted an experiment in which the water was applied at 0.85, 0.70, 0.55 or 0.40 ET. The small but non-significant difference in lint percentage, fiber uniformity and micronaire were observed.

The past results of research in the world reviewed above showed that the cotton yield quality response to irrigation practices varies from location and cultivars. The scarcity of water in this region has forced the farmers to adopt deficit irrigation. Also it led the farmers to adopt the drip irrigation, which is the most water use efficient method. Therefore, the present efforts were made to know the yield quality response to irrigation methods drip and border and water application levels.

Materials and Methods

The cotton response to irrigation practices was evaluated through two years field experiments at Instruction Farm of the College of Agricultural

Engineering & Technology, JAU, Junagadh. The 3-factorial split plot design was adopted. The irrigation method was border and drip. Three water application levels were L_1 (80 per cent depletion of ASM for border, IW/CPE = 0.4 for drip), L_2 (60 per cent depletion of ASM for border, IW/CPE = 0.6 for drip) and L_3 (40 per cent depletion of ASM for border, IW/CPE=0.8 for drip). The growing periods were vegetative (S_1: 45-90 DAS), flowering (S_2: 91-135 DAS), boll development (S_3: 136-180 DAS), boll maturity (S_4: 181-210 DAS) and whole season (S: 45-210 DAS). The moisture stress free condition was maintained during the days other than the days of particular growth period for water stress by irrigating at IW/CPE = 1.0 under drip and at 20 per cent depletion of ASM under border. One control under each irrigation method was also studied, under which the moisture stress-free condition was maintained throughout the growing period. Under drip irrigation, all the treatments were irrigated daily having the depth equal to respective IW/CPE multiplies pan evaporation of last day.

The treatment plot (sub sub-plot) was having three rows at a distance of 1.22 m. The length of the row was 5.4 m having 12 plants at 0.45 m distance in a row. The net plot size of treatment was 3.66 m x 5.4 m. The water meters were used to apply the metered quantity of irrigation water to the treatments. Under the drip irrigation treatments, the lateral of 16 mm diameter and 5.4 m length with 4 lph discharge capacity drippers at 0.9 m spacing was used. The lateral to lateral distance was 1.22 m. The water meter was fitted to middle lateral line in each treatment plot. The normal operating pressure of 1 kgf/cm^2 was maintained. The seed rate, fertiliser dose, plant and row spacing were adopted as recommend by the Agronomy Department of the university. The interculturing, hand weeding and plant protection measures were adopted as per the requirements. The interaction effects of irrigation methods, water application levels and stages/growth periods of water application levels on cotton yield were evaluated for both the years and pooled results.

The samples from each replication of the treatments were taken for the quality analysis. The samples collected from each of four replications were mixed homogeneously for each treatment. The homogeneously mixed samples of each treatment collected during both years were again mixed to get a representative sample of the treatment. These reprehensive samples were ginned at Ginning Lab of the Cotton Research Station, Junagadh Agricultural University, Junagadh, to separate the seeds and lint from the seed cotton. The 100 gm lint samples were sent to Central Institute for Research in cotton technology, Regional station, Surat for the lint quality analysis. The quality of the lint yield was evaluated by assessing the quality parameters like 2.5 per cent length of fiber, fiber fineness, bundle strength, uniformity ratio and maturity ratio. The seed quality was evaluated by assessing the seed index and oil content. The quality of seed cotton yield was evaluated by examining the lint percentage.

(a) 2.5 per cent length of fiber (mm)

The average of individual length of fibers comprising of a representative sample of cotton lint, usually expressed in millimeters, was determined using digital fibrograph machine.

(b) Fiber fineness micronaire value (mv)

The direct measurement of fiber fineness was cumbersome and hence indirect method was adopted. The micronaire instrument was used to find coarseness of fiber. The micronaire value above 3.5 mv indicates coarseness of fiber, while value below 3.5 mv indicates fiber fineness.

(c) Bundle strength (g/tex)

The fiber strength was measured as the strength of a bundle of fibers. The Pressley strength tester was used for determining the bundle strength.

(d) Fiber uniformity ratio (per cent)

It is a ratio of mean length of fiber to mean length of half of the upper length of fiber. This ratio indicates the variation in the length of fiber. It was measured by digital fibrograph.

(f) Maturity ratio

It represents the percentage of matured, half matured and immature fibers. The fiber maturity ratio was determined by micronaire instrument with 3/8" spacer and in the usual manner without spacer. The fiber maturity ratio was worked out by following formula :

$$MR = 0.1579(B) + 0.4670 \quad \ldots(1)$$

where, MR is the maturity ratio; B is the difference between reading with and without spacer.

(e) Seed index (g)

The seed index was obtained weighing 100 seeds on electronic balance of 0.01 g LC.

(f) Oil content (per cent)

The oil content of seed was determined directly by Nuclear Magnetic Resonance (NMR) analyser.

(g) Lint percentage

The lint percentage or ginning percentage is the ratio of weight of lint to that of seed cotton expressed as percentage. The lint percentage was worked out by using the following formula as suggested by Sikka and Joshi (1960).

$$\text{Lint (\%)} = 100 \times (\text{weight of lint (kg ha}^{-1}) / (\text{weight of seed cotton (kg ha}^{-1}) \quad \ldots(2)$$

Results and Discussions

1. Quantitative Parameters

The interaction effects of irrigation methods, water application levels and stages/growth periods of water application levels on cotton yield were found to be significant for both the years and pooled results as presented in Table 1. It was observed that during both the years and pooled; the highest yield was obtained under the treatment combination of DL_1S_4 (under drip irrigation, water application level of IW/CPE of 0.4 during stage-4). The highest yield was 2075 kg/ha, 2535 kg/ha and 2305 kg/ha respectively in the year 2003-04, 2004-05 and pooled results, which were 1.07, 1.04 and 1.06 times the yield obtained in control under drip irrigation. It was 1.44, 1.51 and 1.47 times the yield obtained in the control under border irrigation during the year 2003-04, 2004-05 and for pooled data. It was found that for the highest yields, the cotton should be irrigated at IW/CPE of 1.0 during first three stages and at IW/CPE of 0.4 during last stages by drip irrigation. The lowest yield were obtained when continuous water application level-L_1 (irrigation at 80 per cent depletion of ASM) were applied through border method for entire growing period. It was found as 864 kg/ha, 910 kg/ha and 887 kg/ha respectively during the year 2003-04, 2004-05 and for pooled data. The lowest yield were 0.45, 0.37 and 0.41 times the yield under control of drip and 0.60, 0.54 and 0.57 times the yield obtained under control of border irrigation respectively during the year-2003-04, 2004-05 and pooled data.

It was observed that the yield (1356 kg/ha) obtained under drip when irrigation applied at IW/CPE of 0.4 throughout the whole season was statistically at par with the yield (1319 kg/ha) obtained under border when irrigation applied at 60 per cent depletion of ASM throughout the whole season.

2. Qualitative Parameters

The results of effect of irrigation methods, water application levels and stages/growth periods of water application levels on cotton quality under different treatments were found as presented in Table 2.

(a) 2.5 per cent length of fiber

The fiber length is the most important quality of lint yield as the market price largely depends on it. The results showed that highest fiber length of 26.9 mm could be attained when the cotton was irrigated by drip continuously at IW/CPE of 0.8. The lowest of 23.9 mm was found when cotton was irrigated by border at 80 per cent depletion of ASM continuously throughout the season.

Table 1: Interaction effect of method of irrigation, water application levels and considered stages/growth periods for water application levels on cotton yield

Year	*Method*	*Water Level*	*Stages /growth periods of water application(stress) level*					*Con-trol*	*SEm*	*CD*
			S_1	S_2	S_3	S_4	S			
2003-04	Drip	L_1	1757	1833	1862	2075	1234	1933	12.3	34.8
		L_2	1891	1898	1975	2061	1625			
		L_3	1931	1957	1978	2052	1801			
	Border	L_1	1159	1158	1336	1518	864	1443		
		L_2	1282	1329	1498	1498	1213			
		L_3	1357	1398	1400	1465	1349			
2004-05	Drip	L_1	2213	2196	2319	2535	1479	2432	12.0	33.9
		L_2	2310	2321	2319	2469	2027			
		L_3	2342	2343	2370	2411	2293			
	Border	L_1	1347	1372	1558	1773	910	1684		
		L_2	1529	1569	1666	1727	1425			
		L_3	1602	1622	1682	1632	1569			
Pooled	Drip	L_1	1985	2015	2091	2305	1356	2183	15.8	49.3
		L_2	2100	2109	2147	2265	1826			
		L_3	2136	2150	2174	2231	2047			
	Border	L_1	1253	1265	1447	1646	887	1564		
		L_2	1405	1449	1582	1613	1319			
		L_3	1480	1510	1541	1549	1459			

Control: (DC: control under drip and BC : control under border irrigation)
1-Drip irrigation: Continues water application level of IW/CPE of 1.0 (moisture stress free conditions) throughout the season
2-Border irrigation: Continuous water application at 20 per cent depletion of ASM (moisture stress free conditions) throughout the season.

However, the highest fiber length of 21.2 mm at IW/CPE of 0.6 was found by Sagaraka (1998) while neither adverse nor favourable effects of moisture regimes was found by respectively, in which, the stress free conditions was maintained for entire season. This showed that the moisture stress-free environment could not achieve highest fiber length. As can be seen in Table-1 that the fiber length could be found increased with increasing the water application during all the stages except stage two under border irrigation. The fiber length was mainly influenced by water application/stress level particularly when the same level was maintained throughout the entire season. However, when the different water stress/application level maintained during individual stage only, the effects could not be found much. Also it can be seen that the fiber length was found more under drip irrigation than under border irrigation. The reason may be frequent water applications under drip. The fiber length could be improved by increasing water availability as reported by Singh and Bhan (1993).

(b) Fiber fineness

It is the second most important quality of lint after fiber length. The fiber fineness is Tank (1995) for the same cotton (H-8) under drip. The fiber length of 25.7 mm and 26.7 mm was found in the control of border and drip irrigation

Table 2: Cotton Yield Quality Parameters under Treatment Combinations

Quality parameter : 2.5 per cent length of fiber(mm)							
Irrigation method	***Stress level***	***Stage of water stress level***					***Control-C***
		S_1	S_2	S_3	S_4	***Whole season (S)***	
Drip	L_1	25.9	25.8	24.3	24.7	24	26.7
	L_2	26.4	25.8	24.4	24.8	24.3	
	L_3	26.5	25.8	24.8	24.9	26.9	
Border	L_1	25.3	25	24.3	24.4	23.9	25.7
	L_2	25.4	25.10	24.3	24.5	24.1	
	L_3	25.5	25.2	24.8	24.8	25.7	
Quality parameter: Fiber fineness micronaire value							
Drip	L_1	3.4	3.7	4.3	3.7	3.6	4.3
	L_2	3.6	3.6	3.5	3.4	3.5	
	L_3	4.0	3.6	3.6	3.8	3.6	
Border	L_1	3.2	3.4	3.5	3.7	3.7	3.6
	L_2	3.7	4.0	3.4	3.7	3.8	
	L_3	3.5	4.1	3.8	3.8	3.8	
Quality parameter : Bundle strength of fiber(g/tex)							
Drip	L_1	21	20.4	18.6	18.10	19.8	20.3
	L_2	18.8	19.4	19.5	19.20	19	
	L_3	19.9	19.7	19.5	20.1	22.10	
Border	L_1	20.7	19.4	19.3	19.8	20.4	20.0
	L_2	20.6	19.8	19.6	18.6	19.0	
	L_3	20.4	20.1	22.6	18.6	18.9	
Quality parameter: uniformity ratio (per cent)							
Drip	L_1	48	47	50	47	50	48
	L_2	50	46	49	48	47	
	L_3	50	47	46	46	46	
Border	L_1	48	49	49	46	50	48
	L_2	51	49	49	47	48	
	L_3	47	48	49	49	48	
Quality parameter : Maturity ratio							
Drip	L_1	0.80	0.81	0.82	0.79	0.80	0.83
	L_2	0.80	0.80	0.80	0.79	0.80	
	L_3	0.82	0.80	0.80	0.81	0.81	
Border	L_1	0.80	0.79	0.80	0.81	0.81	0.81
	L_2	0.81	0.82	0.79	0.80	0.80	
	L_3	0.80	0.82	0.82	0.80	0.81	
Quality parameter: Seed index (g)							
Drip	L_1	5.84	5.83	5.75	5.70	5.35	6.31
	L_2	5.84	5.84	5.82	5.81	5.60	
	L_3	6.29	6.24	6.19	6.18	6.16	
Border	L_1	5.81	5.82	5.72	5.69	5.29	6.25
	L_2	5.82	5.83	5.81	5.80	5.45	
	L_3	6.24	6.20	6.18	6.16	6.15	
Quality parameter: Oil content (per cent)							
Drip	L_1	19.10	19.05	18.83	18.80	18.28	21.11
	L_2	21.04	20.56	20.35	20.12	19.50	
	L_3	21.11	21.01	20.70	20.65	21.14	
Border	L_1	18.81	18.70	18.41	18.52	18.25	20.91
	L_2	19.85	19.81	19.80	19.79	19.21	
	L_3	20.49	20.49	20.51	20.61	20.93	
Quality parameter : Lint percentage							
Drip	L_1	38.46	38.47	39.05	39.45	39.61	38.42
	L_2	38.23	38.25	38.85	39.15	39.04	
	L_3	38.01	38.15	38.35	38.82	38.51	
Border	L_1	38.44	38.43	38.95	39.44	39.59	38.40
	L_2	38.23	38.15	38.84	39.15	39.00	
	L_3	38.10	38.15	38.35	38.80	38.45	

measured in terms of micronaire value, which indicates its coarseness. The lesser the values of micronaire, higher will be the fineness of the fiber. The highest fiber coarseness (micronaire value) of 4.3 mv was found under control of drip irrigation. When the cotton was irrigated continuously for whole season maintaining any same water application/stress level, the fiber coarseness was found more under border than the drip irrigation.

However, except this, no other trend could be identified from the results data of fiber coarseness. Constable and Hodgson (1990) also noted that the drip irrigation reduced the fiber fineness. The data showed that no remarkable differences in the micronaire values were found among treatments. The results are comparable to reported by Tank (1995), Luz *et al.* (1997), Reddy and Thimmegowda (1998), Enciso *et al.* (2000) and Xiao *et al.* (2000). However, as reported by Singh and Bhan (1993), the fiber finess could be improved by increasing the moisture stress.

(c) Bundle strength of fiber (g/tex)

The results showed that the highest bundle strength (g/tex) of 22.6 was found under border irrigation when the cotton was irrigated at 40 per cent depletion of ASM during the boll development stage. The second highest of 22.1 was recorded under drip when it was irrigated continuously for the 0.8 IW/CPE throughout the season. The effect of increased water application level during boll development stage was in favor of increasing the fiber strength. However, the increased water applications for rest of the stages could not give any clear trend of fiber strength. The fiber strength under control (no water stress for entire season) of border and drip irrigation was found as 20.3 and 20.0 respectively. The effects of irrigation method on bundle strength could not be found. However, Sagarka (1998) reported that the lint produced under alternate furrow irrigation gave higher bundle strength than the drip irrigation. Also the significant difference in fiber strength of lint produced under drip and furrow irrigated cotton was noticed by Razaz *et al.* (1997). The irrigation practices could not influence significantly the fiber strength as reported by Constable and Hodgson (1990), Luz *et al.*, 1997; Reddy and Thimmegowda (1998) and Enciso *et al.* (2000).

(d) Uniformity ratio (per cent)

The highest uniformity ratio of 51 per cent under BL_2S_1 treatment and lowest of 46 per cent under DL_2S_2, DL_3S_3, DL_3S_4, DL_3S and BL_1S_4 was found. The lint produced in the control under both methods gave same uniformity ratio of 48 per cent. However, the treatments results could not give any clear trend of influence on the uniformity ratio of the lint yield. The literature could not be found on results having effects of irrigation practices on the uniformity ratio (per cent) of lint.

(e) Maturity ratio

It could be seen that the maturity values varied from 0.79 to 0.83 among the treatments. Most of the treatments gave maturity ratio of 0.80. The control under border produced the lint having maturity ratio of 0.81. The highest of 0.83 were found in the control of drip irrigation. The effects of irrigation method could not give any indication. However, Nawar *et al.* (1995) reported that drip irrigation increased fiber maturity by 13.5 per cent compared with the traditional system in the Nile Valley. No other experimental evidence on this line could be found in literature.

(f) Seed index (g)

The highest seed index of 6.31 g and lowest of 5.29 g was found respectively when the cotton crop irrigated stress-free for entire season under drip (control) and at 80 per cent depletion of ASM under border irrigation throughout the season. The effects of water application level during the boll development and maturity stage could influence more than the earlier stages. The seed index was found increased with increasing the water application level for all the stages. The drip irrigation could improve the seed index over border irrigation. The highest seed index under both methods was found under the respective control i.e. water stress-free environment throughout the season. The results are comparable with reported by Sagaraka (1998), Shette and Shilke (1990).

(g) Oil content (per cent)

The oil content under drip and border irrigation was found as 21.11 per cent and 20.91 per cent respectively in control under each method. The highest oil content of 21.14 per cent under drip and 20.93 per cent under border was found when the same water application level-L_3 was maintained throughout the season. The drip could improve the oil content at all application levels. The increased water application levels could help to increase the oil content under both irrigation methods. The similar results were found by Li and Xu (1991), Tank (1995), Sagarka (1998), Anac *et al.* (1999).

(h) Lint percen1 tage

The lint percentage under control of drip and border irrigation was found as 38.42 and 38.40 respectively. The highest lint percentage under drip and border irrigation was found as 39.61 per cent and 39.59 per cent respectively when the cotton irrigated at water application level L_2 for entire season. The lint percentage under drip irrigation was improved over border irrigation. However, as reported by Reddy and Thimmegowda (1998), lint percentage was higher with furrow irrigation than drip. The increase in water application decreased the lint percentage, this showed that the deficit irrigation can increase the lint percentage. However, the differences among the treatments were not much. Xiao *et al.* (2000) also reported that under different water applications, small but non-significant difference was observed in lint percentage. As reported by Anac *et al.* (1999), the sever water deficit substantially reduce the oil percentages in the seed and more fiber materials with 20 per cent lower oil and protein content compared to a adequately irrigated crop. Therefore, the increase in the lint percentage under deficit irrigation was on the cost of reduced oil and protein content.

Conclusions

- The cotton yield quality parameters like fiber length, lint percentage, seed index and oil content can be improved adopting drip replacing border method.
- All these qualities were increased with increased in moisture regimes except lint percentage. The lint percentage was decreased with increase in water applications.
- The effects of irrigation practices on lint quality like uniformity ratio and maturity coefficient could not be found much.

References

Anac, MS; Ul, AM; Tuzel, IH; Anac, D; Okur, B; Hakerlerler, H, 1999. Optimum irrigation scheduling for cotton under deficit irrigation conditions. In: C Kirda, P Moutonnet, C Hera, and D R Nielsen, (Eds.) Crop yield response to deficit irrigation, Dordrecht, The Netherlands, Kluwer Academic Publishers.

Anonymous 1992. Evaluation of drip over surface method of irrigation for cotton (Achhalia) : Salient feature of plasticulture experiments' results 1991-1992. Published by Water Management Project, GAU, Navsari, p. 29.

Berami, A; Ofen, A, 1993. Irrigation Engineering : Sprinkler, Trickle and Subsurface Irrigation - Principal, Design and Agricultural practices, AGRIPRO, Kfargalim, 30865, Israel, pp. 165-166.

Cetin, O; Bilgel, L, 2002. Effects of different irrigation methods on shedding and yield of cotton. Agric. Water Manage., 54: 1-15.

Cetin, O; Ozyurt, E; Sener, S, 1994. The effects of different irrigation methods on the yield and water use efficiency of cotton in Harran plain. 17th ICID European Regional Conference on Irrig. and Drain., Bulgaria, 16-22 May, Vol.1: yield response to water as affected by irrigation schedule, 241-246.

Constable, GA; Hodgson, AS 1990. A comparison of drip and furrow irrigated cotton on a cracking clay soil. 3. Yield and quality of four cultivars. Irrig. Sci., 11(3):149-153.

Cruz, RS; Agulto, IC, 1996. Economics of a designed trickle irrigation system using a twin wall emitter tubing. Cotton Research Journal, 9(1-2): 52-58.

Cruz, RS; Agulto, IC, 1998. Seed cotton yield increase and water saving with a designed trickle irrigation system. Philippine Journal of Crop Science, 23(1):24.

Dippenaar, MC; Barnard, C; Pretorius, MM, 1997. Yield and gross margin of cotton under drip and sprinkler irrigation. Applied-Plant-Science, 11(1): 7-12.

Enciso, MJ; Unruh, BL; Henggeler, JC; Multer, W L, 2002. Effect of row pattern and spacing on water use efficiency for subsurface drip irrigated cotton. Trans. ASAE, 45(5): 1397-1403.

Feng, SY; Huang, GH; Wang, F X; Wang, GX;, Wu, HQ; Liu, ZG, 1998. Effects of water-fertilizer coupling on cotton yield under drip irrigation. Journal of China Agricultural University, 3(6):59-62.

Li, ZD; Xu, X, 1991. Effect of soil moisture on seed development. Field Crop Ab., 44(10):7524.

Luz, MJ; Bezerra, JRC; Barreto, AN; Santos, JW, 1997. Effect of water stress on cotton yield and fiber quality. Revista de Oleaginosas e Fibrosas., 1(1) : 125-133.

Maas, SJ; De TWR; McLaughlin, JR; Thullen, RJ; Ayars, JH; Camp, CR, 1996. Water relations of furrow and drip irrigated cotton. In : Sadler, EJ; Yoder, R E. (Ed.). Evapotranspiration and irrigation scheduling. Proc. of the Inter. Conf., San Antonio, Texas, USA, 838-844. ASAE, St Joseph; USA.

Mateos, L; Berengena, J; Orgaz, F; Diz, J; Fereres, E, 1992. A comparison between drip and furrow irrigation in cotton at two levels of water supply, Irrigation and Drainage Abstracts, pp. 125.

Nawar, MT; Zaher AM; El-Sahhar K; Abdel-Rahim SA, 1995. Effect of salinity on botanical characters and fiber maturity of three Egyptian cotton cultivars. Proc. 1995 Beltwide Cotton Conferences. 568-575, San Antonio, TX.

Nikolov, G, 1998. Effect of drip irrigation in cotton. Soil Science Agrochemistry and Ecology, 33(5) : 35-36.

Razaz, MM; Makram, EA; Mousa, IA, 1997. Evaluation of Egyptian cotton cultivars under drip irrigation system in sandy soils of east delta. Arab Universities Journal of Agricultural Sciences, 5(2): 227-235.

Reddy, DVS; Thimmegowda, S, 1998. Effect of methods and levels of irrigation on the fiber quality of Jayalaxmi hybrid cotton. Current Research. UAS, Bangalore, 27(2):32-33.

Rosario, GA; Cruz, RS; Ganotisi, ND; Penaranda, HD; Cosico, VB; Rosario, MR Orpia, ED, 2001. Drip fertigation for cotton production. PCARRD-Highlights-2000(Philippines). P:16-18.

Sagarka, BK, 1998. Effect of varying moisture regimes and nitrogen levels on growth yield and quality of drip irrigated rabi-cotton. Ph.D. thesis. Sub. GAU Jaunagdh.

Shette, CTG; Shelk, VB, 1990. Response of differential irrigation and fertilizer levels to cotton genotypes. J. Maharashtra Agric. Univ., 15(1):110-112.

Sikka, SM; Joshi, AB, 1960. Cotton in India. A monograph. Vol. I. ICCC, Bombay, 145-146.

Singh, RP Bhan, S, 1993. Yield quality and economics of summer cotton (Gossypium species) as influenced by frequency of irrigation and moisture conservation practices. Indian J. Agron., 38(3): 439-442.

Smith, RB; Oster, JD; Phene, C, 1991. Subsurface drip produced highest net return in west lands area study. Calif. Agric., 45(2): 8-10.

Tank, BA, 1995. Effect of varying moisture regimes and nitrogen levels on growth yield and quality of drip irrigated rabi-cotton. Ph.D. thesis, GAU Navsari.

Xiao, JF; Liu, ZG; Yu, XG; Zhang, JY; Duan, AW, 2000. Effects of different water application on lint yield and fiber quality of cotton under drip irrigation. Acta Gossypii Sinica, 12(4): 194-197.

16

Effect of Drip Irrigation and Nitrogen on Growth, Yield and WUE of Summer Sesame

R.M. Solanki, B.K. Sagarka, B.M. Dabhi and M.A. Shaikh

Abstract

A field experiment was carried out during summer season of 2010, 2011 and 2012 at the Instructional Farm, Department of Agronomy, College of Agriculture, Junagadh Agricultural University, Junagadh, Gujarat. The twelve treatment combinations consisted of four irrigation levels through drip viz., *I_1- 0.6, I_2-0.8, I_3- 1.0 pan evaporation fraction (PEF) and I_4- surface irrigation of 50 mm depth at 1.0 IW/CPE ratio were allotted to main plot and three nitrogen levels* viz., *N_1- 30, N_2-40 and N_3-50 kg N ha^{-1} were assigned to sub-plot and laid out in split plot design with three replication. Sesame variety Guj. Sesame-2 was sown in 30-60-30 cm paired row. Irrigations and nitrogen were applied as per treatments. Results revealed that significantly highest plant height, number of branches and capsules per plant, seed (1413 kg ha^{-1}) and stalk (2043 kg ha^{-1}) yields, respectively were recorded when crop was irrigated at 1.0 PEF through drip. Maximum water use efficiency of 2.12 kg $ha^{-1}mm^{-1}$ was obtained when crop was irrigated through drip at 0.6 PEF. Highest gross and net return as well as benefit cost ratio of ₹ 71080 and 48457 ha^{-1} and 3.14, respectively were obtained with the application of irrigation through drip at 1.0 PEF. Application of N @ 50 kg ha^{-1} recorded significantly maximum plant height, number of branches, capsules per plant, seed (1232 kg ha^{-1}) and stalk (1898 kg ha^{-1}) yields of sesame. Application of nitrogen was found to be beneficial for improving WUE. Crop fertilised with 50 kg N ha^{-1} noted higher WUE of 2.02 kg $ha^{-1}mm^{-1}$ followed by application of 40 kg N ha^{-1}. Application of nitrogen @ 50 kg ha^{-1} gave maximum gross (62075 ha^{-1}) and net (39434 ha^{-1}) realisation along with B:C ratio of 2.74.*

Introduction

Sesame (*Sesamuum indicum*, L.) is one of the most ancient oilseed crops of India. The crop is cultivated almost throughout India for its high quality oil and it has tremendous potential for export. It ranks third in terms of area and production in the country. Sesame occupies a position of prominence in the oilseed scenario of Gujarat state. It ranks second in area and production next to groundnut in the state. Sesame traditionally grown as rainy season crop in the state but with enhanced irrigation facilities and availability of high yielding varieties makes its cultivation feasible in the summer season, yield performance of sesame in *kharif* season in general is low, it has shown a better yield potential when introduced in summer particularly in areas like Saurashtra region of Gujarat. Among various factors responsible for the low productivity levels of sesame, proper scheduling of irrigation and nitrogen fertilisation is the most important one.

Irrigation water is key input in crop production. Full benefit of crop production technologies such as high yielding varieties, fertiliser use, multiple cropping and plant protection measures can be derived only when adequate supply of water is assured. Enhanced irrigation facilities and availability of high yielding varieties sesame make its cultivation feasible in summer season, particularly in areas where this crop is yet to find a place in the existing cropping system.

It is well established fact that there is a positive correlation between fertiliser and productivity. At least 40-50 per cent increase in agricultural production in recent years can be credited to fertiliser alone (Tandon, 1994). Therefore, supply of balanced nutrients is considered as one of the basic needs to achieve the potential yield. Fertilisers even though comparatively a costly input of production are essential for securing higher yields. The prudent use of fertilisers with appropriate method and time of application are the prime importance in securing higher and economic yield. Among the various nutrients, nitrogen is one of the decisive as well as expensive inputs and the chief promoter of plant growth. As a constituent of protoplasm, it is, intimately, involved in the process of photosynthesis and ultimately, in the dry matter production. Nitrogen also, plays as important role in synthesis of chlorophyll and amino acid, which contributes to the building units of protein and, thus, the growth of plants. Insufficient nitrogen may reduce yield drastically and deteriorates the quality of crop production. Information on the response of drip irrigation and nitrogen requirement for summer sesame is not well documented, especially for summer crop in Saurashtra region of Gujarat. Therefore, the present investigation was undertaken to study the response of summer sesame to drip irrigation and nitrogen on growth and yield components, water use efficiency and yield.

Materials and Methods

A field experiment was carried out during summer season of 2010, 2011 and 2012 at the Instructional Farm, Department of Agronomy, College of Agriculture, Junagadh Agricultural University, Junagadh, Gujarat. The soil was medium black clayey, with 7.9 pH having, 212.0 kg ha^{-1} available nitrogen, 19.5 kg ha^{-1} available phosphorus and 337.0 kg ha^{-1} available potash. The farm is located at about 21.5° N and 70.5° E longitude with an average altitude of 60 m above mean sea level under semi-arid tropic having a little extreme of weather condition. The investigation included 12 treatments which were the combinations of four irrigation levels through drip *viz.*, 0.6 (I_1), 0.8 (I_2), 1.0 (I_3) alternate day evaporation pan fraction (PEF) and surface irrigation of 50 mm depth at 1.0 IW/CPE ratio (I_4) devoted to main plot and three nitrogen levels *viz.*, 30 (N_1), 40 (N_2) and 50 (N_3) kg N ha^{-1} were assigned to sub-plots and laid out in split design with three replications. Sesame variety 'Gujarat Sesame-2' was sown at 30-60-30 cm paired row. Phosphorus @ 50 kg ha^{-1} was applied as basal to all the plots in the form of single super phosphate, 50 per cent nitrogen was applied as basal and remaining 50 per cent as split at 30 DAS as per treatment was applied in the form of ammonium sulphate and urea, respectively. The drip irrigation system was laid out at lateral spacing of 90 cm with dripper spacing of 60 cm and operated at pressure of 1.2 kg cm^{-2} with 4 LPH discharge rate. The scheduling of irrigation was done with 0.6, 0.8 and 1.0 PEF at alternate day irrigations were applied through drip and in surface method, 50 mm depth irrigations were applied at 1.0 IW/CPE ratios. The data on growth and yield attributes and yields were recorded at maturity. Water use efficiency, gross and net realisation was worked out by using current market prices of produce and inputs used. Other package of practices was followed as per recommendations made for the crop.

Results and Discussion

Growth and yield attributes

Significantly highest plant height, number of branches and capsules per plant were recorded when crop was irrigated at 1.0 PEF through drip (I_3) and remained at par with crop received irrigation at 0.8 PEF through drip (I_2). Whereas, minimum growth and yield attributing characters were noted when irrigation was applied as per 0.6 PEF (I_1) through drip (Table 2). Higher values of growth and yield attributes under higher irrigation level might be due sufficient moisture available to the crop during entire crop period and hence availability and uptake of nutrients are more than stress condition, which ultimately resulted improvement in growth and yield components. Similar results were also reported by Ashok Kumar and Prasad (1993), Chandrakar *et al.* (1994), Dutta *et al.* (2000) and Saren *et al.* (2005).

The perusal of the data (Table 2) indicated that varying levels of nitrogen had significant influence on growth and yield attributes of sesame on pooled data basis. Application of N @ 50 kg ha^{-1} recorded maximum plant height, number of branches and capsules per plant. Results confirms the findings of Ashok Kumar *et al.* (1996), Chandrakar *et al.* (1994), Daury and Mandal (2006) and Shrivastava and Tripathi (1992).

Table 1: Effect of irrigation and nitrogen on seed and stalk yields of summer sesame

Treatment	*Seed yield (kg ha^{-1})*				*Stalk yield (kg ha^{-1})*			
	2010	*2011*	*2012*	*Pooled*	*2010*	*2011*	*2012*	*Pooled*
Irrigation								
I_1 : 0.6 PEF	1043	957	922	974	1586	1620	1385	1530
I_2 : 0.8 PEF	1207	1122	1074	1134	1823	1845	1488	1718
I_3 : 1.0 PEF	1417	1517	1304	1413	2220	2100	1810	2043
I_4 : 1.0 W:CPE	1277	1399	1157	1278	2103	2148	1652	1968
C.D. at 5%	126.6	157.9	142.1	74.8	225.8	182.5	186.5	104.3
Nitrogen levels (kg ha^{-1})								
N_1 : 30	1192	1203	1034	1143	1850	1853	1476	1726
N_2 : 40	1277	1262	1132	1224	1925	1933	1605	1821
N_3 : 50	1238	1281	1176	1232	2024	2000	1670	1898
C.D. at 5%	NS	54.7	104.2	47.0	NS	78.5	131.3	80.4

Table 2: Effect of irrigation and nitrogen levels on growth characters, yield attributes, WUE and economics of summer sesame

Treatment	*Plant height (cm)*	*No. of branches /plant*	*No. of capsules /plant*	*Gross Return (Rs ha^{-1})*	*Net return (Rs ha^{-1})*	*B:C ratio*	*Water use efficiency (kg ha^{-1} mm^{-1})*
Irrigation							
I_1 : 0.6 PEF	57.5	2.5	36.5	49083	26460	2.17	2.12
I_2 : 0.8 PEF	59.7	2.7	39.5	57130	34507	2.52	1.96
I_3 : 1.0 PEF	64.7	3.3	43.7	71080	48457	3.14	2.02
I_4 : 1.0 W:CPE	62.8	3.1	42.9	64392	42188	2.90	1.83
C.D. at 5%	1.8	0.4	1.9				
Nitrogen levels (kg ha^{-1})							
N_1 : 30	59.5	2.6	38.4	57581	35186	2.57	1.88
N_2 : 40	61.1	2.8	40.5	61655	39137	2.73	2.01
N_3 : 50	62.8	3.2	42.9	62075	39434	2.74	2.02
C.D. at 5%	1.3	0.4	1.3				

Seed and Stalk Yields

Seed and stalk yields of sesame progressively increased with increase in levels of drip irrigation from I_1 to I_3 during individual years and on pooled basis, except during 2011 irrigating the crop at an IW/CPE ratio of 1.0 through surface method produced significantly maximum stalk yield. Significantly the highest sesame seed and stalk yields of 1413 and 2043 kg ha^{-1}, respectively on pooled basis were obtained when crop was irrigated through drip at 1.0 PEF (I_3) and it was at par with irrigating the crop through surface method at an IW/CPE ratio of 1.0 (I_4). The per cent increase in seed and stalk yields under I_3 was to the tune of 45.1 and 33.5 per cent, accordingly over I_1 i.e. irrigating the sesame through drip at 0.6 PEF. The increase in seed and stalk yields might be due to the effect of timely and frequent irrigations with sufficient quantity of irrigation water under drip irrigation under I_3 provided constant wet root zone, increased the nutrient availability, enhance the growth and yield attributes and ultimately higher seed and stalk yields. These results are collaborated with that of Ashok Kumar *et al.* (1996), Chandrakar *et al.* (1994), Dutta *et al.* (2000), Saren *et al.* (2005) and Shrivastava and Tripathi (1992).

Application of nitrogen to summer sesame found significant during 2011, 2012 and in pooled results. Significantly maximum seed (1232 kg ha^{-1}) and stalk (1898 kg ha^{-}) yields of sesame on pooled basis were produced when sesame crop was fertilised with 50 kg N ha^{-1} (N_3) which was closely followed by application of nitrogen @ 40 kg ha^{-1} (N_2). Application of nitrogen @ 50 kg ha^{-1} increased the seed and stalk yields to the extent of 7.8 and 10.0 per cent, respectively over 30 kg ha^{-1} on pooled basis (Table 1). Increase in seed and stalk yields with higher dose of nitrogen might be due to enhanced plant height, number of branches and capsules per plant. Similar results of higher seed and stalk yields of sesame due to higher N application were explained by Ashok Kumar (1993), Ashok Kumar *et al.* (1996), Chandrakar *et al.* (1994), Duary and Mandal (2006), Dutta *et al.* (2000), Saren *et al.* (2005) and Shrivastava and Tripathi (1992).

Water Use Efficiency

Maximum water use efficiency of 2.12 kg ha^{-1} mm^{-1} (Table 2) was obtained when crop was irrigated through drip at 0.6 PEF (I_1) followed by drip irrigation at 1.0 PEF (2.02 kg ha^{-1} mm^{-1}) . Ashok Kumar *et al.* (1996), Chandrakar *et al.* (1994) and Dutta *et al.* (2000) reported similar results.

Application of nitrogen was found to be beneficial for improving WUE. Crop fertilised with 50 kg N ha^{-1} (N_3) noted higher WUE of 2.02 kg ha^{-1} mm^{-1} followed by application of 40 kg N ha^{-1} (N_2). Ashok Kumar *et al.* (1996) and Chandrakar *et al.* (1994) also reported similar trend.

Economics

Highest gross (Rs 71080 ha^{-1}) and net (Rs 48457 ha^{-1}) realisation as well as benefit cost ratio of 3.14 were obtained with the application of irrigation through drip at 1.0 PEF followed by irrigating the crop through surface method at 50 mm depth of irrigation at an IW/CPE ratio of 1.0 (Table 2). These results confirm the findings of Ashok Kumar and Prasad (1993).

Application of nitrogen @ 50 kg ha^{-1} gave maximum gross (Rs 62075 ha^{-1}) and net (Rs 39434 ha^{-1}) realisation along with B:C ratio of 2.74 which was closely followed by application of nitrogen @ 40 kg ha^{-1}. Similar results were also reported by Duary and Mandal (2006) and Ashok Kumar and Prasad (1993).

Conclusion

Thus, from the present study it could be concluded that summer sesame sown by paired row at 30-60-30 cm spacing and irrigated through drip at 1.0 PEF along with the application of nitrogen @ 50 kg ha^{-1} significantly improved growth and yield attributes; seed and stalk yields as well as gross and net return under medium black clayey soils of Saurashtra region of Gujarat.

References

Ashok Kumar and Prasad, T.N. (1993). Response of summer sesame (*Sesamum indicum* L.) to irrigation and nitrogen in calcareous soil. Indian Journal of Agronomy. 38 (1): 145-147.

Ashok Kumar; Prasad, T.N. and Prasad, U.K. (1996). Effect of irrigation and nitrogen on growth, yield, oil content, nitrogen uptake and water use of summer sesame (*Sesamum indicum* L.). Indian Journal of Agronomy. 41 (1) : 111-115.

Chandrakar, B.L.; Sekhar, N.; Tuteja, S.S. and Tripathi, R.S. (1994). Effect of irrigation and nitrogen on growth and yield of summer sesame (*Sesamum indicum* L.). Indian Journal of Agronomy. 39 (4) : 701-702.

Duary, B. and Mandal, S. (2006). Response of summer sesame, *Sesamum indicum* L. to varying levels of nitrogen and sulphur under irrigated condition. Journal of Oilseeds Research. 23 (1) : 109-112.

Dutta, I. D.; Jana, P. K.; Bandyopadhyay, P. and Maity, D. (2000). Response of summer sesame (*Sesamum indicum* L.) to irrigation. Indian Journal of Agronomy. 45 (3) : 613-616.

Saren, B. K.; Nandi, P. and Tudu, S. 2005. Effect of irrigation and sulphur on yield attributes and yield, oil content and oil yield and consumptive use efficiency of summer sesame, *Sesamum indicum L.* Journal of Oilseeds Research. 22 (2) : 383-384.

Shrivastava, G.K. and Tripathi, R.S. (1992). Effect of irrigation, mulch and nitrogen levels on growth and yield of summer sesame (*Sesamum indicum L.*). Indian Journal of Agronomy. 37 (3) : 602-604.

Tandon, H. L. S. (1994). "Fertilizer management in Dryland Agriculture" FDCO, New Delhi.

17

Performance Evaluation of Horizontal Sand Filter of Drip Irrigation

H.N. Dalsania, H.D. Rank, Dezy Sahu, R.J. Patel, P.B. Vekariya, S.B. Dugad

Abstract

The study was carried out to generate basic information such as pressure drop across the filter and filtration efficiency at different times for clean water and water with suspended load concentration levels of 50,100, and 300 mg/l and also for three different sizes of silica sand as 0.7 to 1.2 mm, 1.0 to 1.4 mm and 1.4 to 2.0 mm. Filtration efficiency and pressure drop both decreased as the sand size increased from finer to coarser. The maximum filtration efficiency was achieved as 70.84 per cent for sand size of 0.7 to 1.2 mm and suspended load concentration 50 mg/l. Hence, sand size of 0.7 to 1.2 mm was found suitable to use in sand filters to achieve higher filtration. Filtration efficiency decreased with increase in suspended load concentration. The pressure drop increased with time as the suspended load concentration increased from 50 mg/l to 300 mg/l. The developed nomogram was found beneficial for deciding backwash frequency for the horizontal model of sand filter.
Keywords: *Sand Filter, Filtration Efficiency, Pressure drop, Silica Sand, Suspended load Concentration, Nomogram.*

Introduction

Drip irrigation also known as trickle irrigation or micro irrigation is an efficient method of application of water at a low rate over a long period of time at frequent intervals with low pressure delivery systems. The water supplied from different sources contains dissolved and suspended impurities which can lead to physical, chemical, biological restrictions in the emitter, resulting in uneven application rate of water. Uniformity is greatly reduced even when one to five per cent of the emitters are clogged (Nakayama and Bucks, 1981).

Due to clogging of emitters the maintenance cost of system is increased and life of system is decreased. Capra and Scicolone (2004) observed that the performance of drip irrigation systems using waste water is mainly limited by emitter clogging, and this discourages farmers from introducing it. Clogging will adversely affect the rate of water application and uniformity of water distribution and thus increasing the operation cost.

Removal of particles causing clogging is essential for the satisfactory performance of a drip irrigation system. Nakayama *et al.* (1978) reported that adequate water filtration is primary requirement for reliable emitter operation. Padmakumari and Sivanappan (1980) concluded that the different chemicals are useful only to improve relative discharge rates. Prevention is best solution to reduce clogging of emitters. Filtration of water is absolutely essential. The filters commonly used in drip irrigation system are sand or gravel filter, screen filter, hydro cyclone filter or centrifugal separator and disc filters. Sand media filters consist of layered beds of graded sand and gravel placed in one or more cylindrical tanks. They may be vertical or horizontal types. Sand media filter is used widely in drip irrigation where water contains organic impurities such as algae, fungi etc. or inorganic suspended impurities such as sand, silt or clay. When too much foreign material has accumulated within the sand bed, it becomes difficult to force the water though the filter, and the head loss across the filter is increased. This increased head loss indicates that the filter needs to be cleaned by backwashing. It must be noted that poor filtration not only ruins the trickle irrigation systems performance but also reduces the crop yield and quality regardless of the best cultural practices. Hence, it is essential to select the appropriate filter and know its performance characteristic. Gideon *et al.* (1982) found out removal efficiency or the filtration efficiency by using the formula given by equation. The quality of effluent produced by a media filter depends upon the flow rate though the filter, and on the type of sand used. In general, the lower the flow rate and the finer the sand, the better the filtration will be. However, a lower flow rate means more filters, and hence a higher cost and finer sand may result in a greater head loss across the filter and a more frequent backwashing requirement. Rodgers *et al.* (2004) reported that accumulation of biomass and deposition of suspended solids at the surface of a sand filter can lead to clogging of the filter media. Thus, the cost of a media filter station must be weighed against the required effluent quality, and a tradeoff must be made. Converse and Tyler (2004) observed that clogging of filter media is less of a concerned when using coarse sand. Siag (1997) concluded that the average particle size of sand and depth of filter should be 0.85 mm and 60 cm to get greater life of filter. Wadate and Khanal (1997) found that there was a poor relationship between filtration efficiency and sediment load concentration showing that the filtration efficiency decreased with sediment load concentration. Khaire *et al.* (2008) found that pressure drop across the sand media filter increases with increase in flow rate of clean water and water with sediment load of different concentrations. The pressure drop across the sand media filter was increased with increase in sediment load concentration levels. Clean water pressure drop nearly tends to constant as time increases. Therefore, the research study was attempted to suggest the operational strategies for the sand filter for the best filtration efficiency with least pressure drop and backwash interval of sand media filter in drip irrigation system with the following objectives (a) to assess the performance of commonly used horizontal type model of sand filter; and (b) to prepare a nomogram for sand filter of given flow rate to determine the backwash interval for different levels of suspended load concentration.

Materials and Methods

The experiments were conducted at Research and Development Laboratory, Jain Irrigation Systems Ltd., Jalgaon (M.S.), during the year 2008-09. The test assembly was specially fabricated and installed for the study. The entire unit consisted of external water source (bore well), three pumps, one mild steel tank, one screen filter of 200 mesh, two screen filters of 120 mesh, horizontal model of sand filter (50 m^3/h capacity), two digital pressure gauges, one flow meter, one graduated cylinder of 220 l and pipe network. Water from the external source was filled into the mild steel tank of size 698 × 125 × 125 cm pumped to the system using centrifugal pump of 20 H.P. with bypass arrangement. Suspended load concentrations were prepared with special set up and injected to the inlet water simultaneously (Figs. 1 and 2). Two ports for collecting sample of inlet and outlet water with sediment load concentration levels of 50, 100 and 300 mg/l were provided. The water from filter was taken into further in mild steel tank with screen filter connected for turbidity free water to the steel tank for circular flow of water for the test.

The physical properties of the filtering media like porosity, density, particle size distributions were determined. Also sand was tested for impurity, specific gravity, chemical resistance and loss on ignition. The water used for the sand filter testing was tested in laboratory for its qualities. The quality parameters like pH, EC, TDS and hardness were found as 7.7, 1.832 mmhos/cm, 1005 ppm and 618 ppm respectively. The turbidity meter was calibrated for four ranges with 0-2 NTU, 2-20 NTU, 20-200 NTU and 200-2000 NTU. A calibration graph was drawn and the best fit equation was developed. Fine soil of size 105-180 micron was used to prepare suspended loads. Total 12 tests were conducted with two replications. The testing for each replication of all of the 24 tests was continued till the pressure drop though the filter was just exceeded 0.5 kg/cm^2 which is the critical limit recommended for the sand filter. The filtration efficiency was found out by using the formula suggested by Gideon (1982).

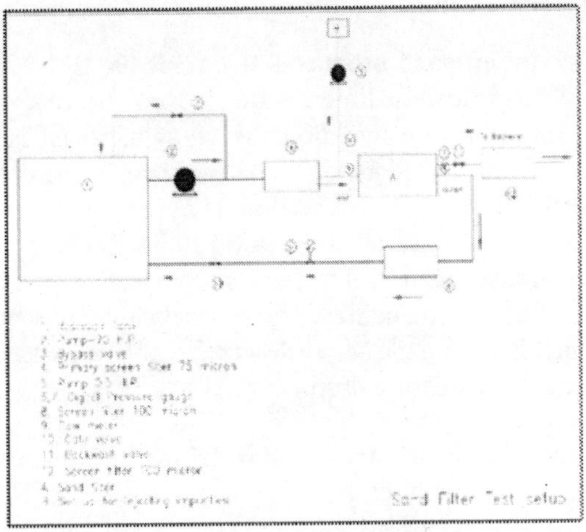

Fig. 1. Sand filter test set up.

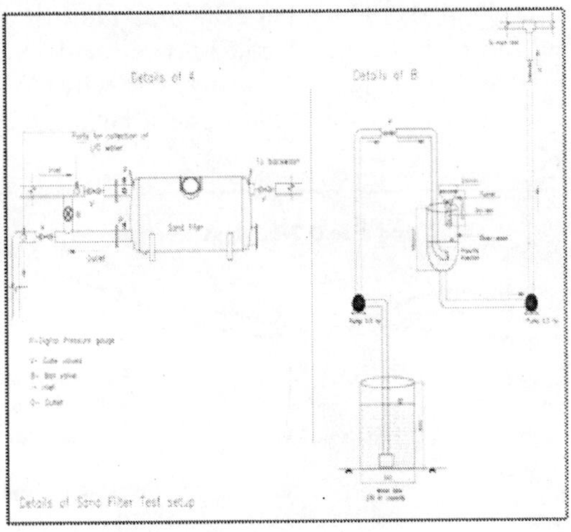

Fig. 2. Details of sand filter test set up.

$$F_e = 100 \times \left(1 - \frac{S_o}{S_i}\right) \quad \ldots(1)$$

where, F_e = Filtration Efficiency, (per cent). S_o = Suspended load concentration at outlet, (NTU), S_i = Suspended load concentration at inlet, (NTU). The data thus obtained was analysed to know the relationship between pressure drop across the filter and time for clean water or water with sediment load concentration, relationship between filtration efficiency and time and in developing nomogram.

Results and Discussion

The quality of effluent produced by a media filter depends upon concentration of suspended sediment load and on the type of sand used. Keeping these points into consideration, this study has been carried out. The results obtained are presented below. The physical properties of Filtering Media are presented in Table 1. The test sand has 0.144 per cent acid soluble matter an the loss on ignition was found as 0.1395 per cent for the three sand size groups used in investigation. The results of all tests showed that the sand used in the filters was of good quality. The vertical model of sand filter for the above filtering media having the above described properties was tested for the filtration efficiency and pressure drop.

Effect of suspended load concentration of water on the filtration efficiency for different sizes of sand media

Figs. 3, 4 and 5 shows the filtration efficiency of Horizontal sand filter for sand size A (0.7-1.2 mm), B (1.0 – 1.4 mm) and C (1.4- 2.0 mm) respectively with suspended loads of 50 mg/l, 100 mg/l and 300 mg/l. In Fig. 3, initially, the filtration efficiency was 70.84 per cent, 65.63 per cent and 60.15 per cent for suspended loads 50 mg/l, 100 mg/l and 300 mg/l respectively. Then it started decreasing for loads of 50 mg/l and 100 mg/l because initially the wide range of

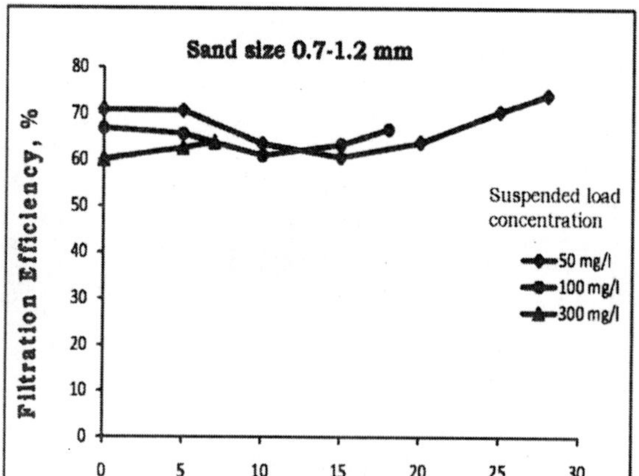

Fig. 3. Filtration Efficiency-Time relationship for sand size A (0.7-1.2 mm).

pore spaces of media was available which allowed the suspended particles to escape though the outlet of the sand filter causing decrease in filtration efficiency. After some time when the suspended particles have been trapped on the sand bed and the pore spaces got clogged due to which very less contamination were escaped though the media so the filtration efficiency increased. It was found as 74.03 per cent, 66.62 per cent and 63.84 per cent for 50 mg/l, 100 mg/l and 300 mg/l respectively. For suspended load of 300 mg/l, the filtration efficiency increased from the start to the end of the test. The reason for this is that the suspended load was so high that the media bed got clogged within 7 min only. Also from the Fig. 3, it was observed that filtration efficiency was more at lower suspended load concentration (50 mg/l) and less with higher suspended load concentration (300 mg/l).

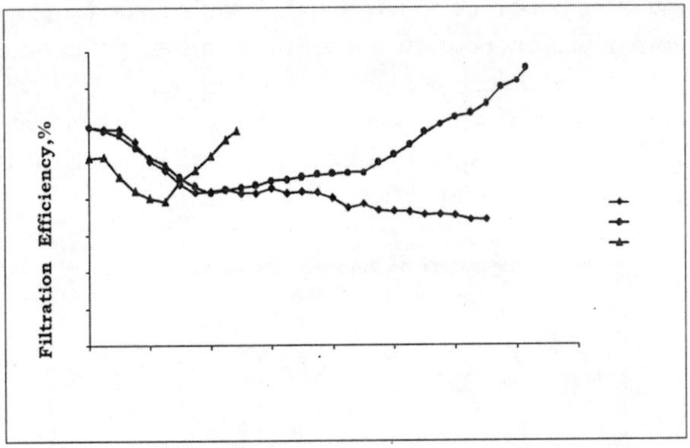

Fig. 4. Filtration Efficiency-Time relationship for sand size B (1.0-1.4 mm).

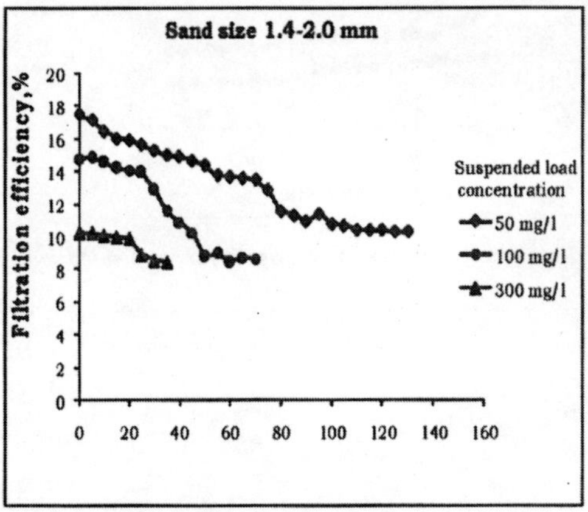

Fig. 5. Filtration Efficiency-Time relationship for sand size C (1.4 -2.0 mm).

Effect of sand size of filter media on the filtration efficiency for different suspended load concentration of the water

Figs. 6, 7 and 8 shows the filtration efficiency-time relationship for horizontal sand filter for suspended load of 50 mg/l, 100 mg/l and 300 mg/l respectively with sand sizes A, B and C.

For sand size A (Fig. 6), initially, the filtration efficiency decreased because more pore spaces were available but after 15 min, there was increase in filtration efficiency till the end of test as the pores started to be clogged due to trapping of suspended particles over sand bed. In case of sands B and C, filtration efficiency decreased with time and after a course of time it tended to be constant. Hence, filtration efficiency was more for finer sand (A) and vice versa. The Fig. 7 shows the same trend of filtration efficiency for suspended load of 100 mg/l as in Fig. 6 for the sand A. For sand B, initially, filtration efficiency decreased but over a course of time it started increasing. For sand C, the filter showed very poor performance in terms of filtration efficiency. Efficiency with this sand decreased from 14.80 per cent to 8.64 per cent. The efficiency was too low with this sand than the former sands because of its coarser size. About 90 per cent of suspended particles escaped though the media thus gives lower filtration efficiency.

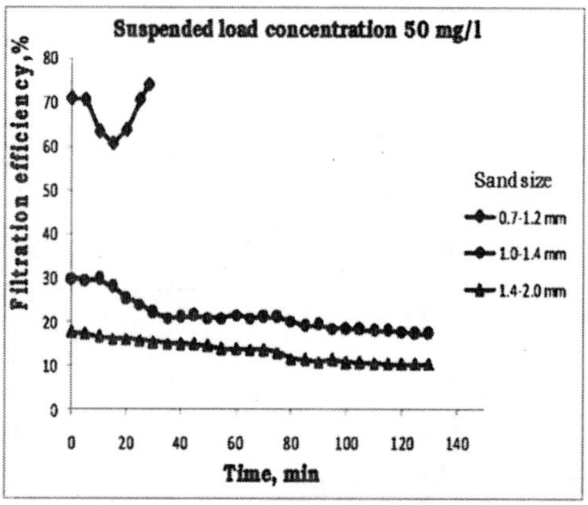

Fig. 6. Filtration Efficiency-Time relationship for suspended load of 50 mg/l.

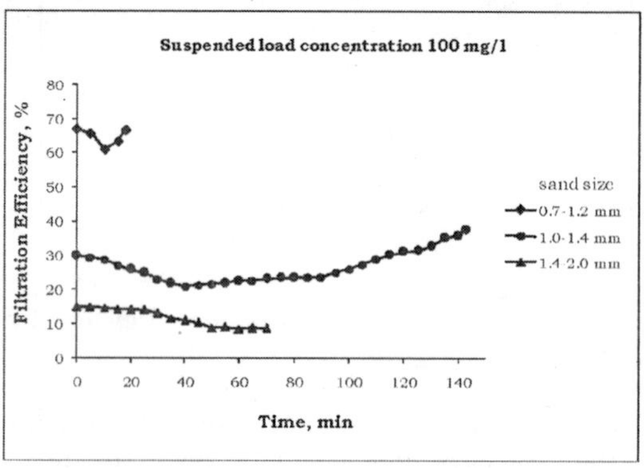

Fig. 7. Filtration Efficiency-Time relationship for suspended load of 100 mg/l.

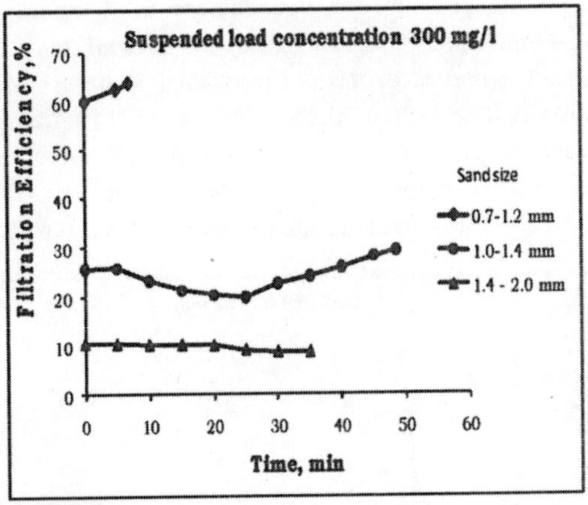

Fig. 8. Filtration Efficiency-Time relationship for suspended load of 300 mg/l.

Sand of size A was finest among the three sands, so it provided highest filtration efficiency (Fig. 8). For this sand, filtration efficiency increased with time for suspended load of 300 mg/l. As this was heavy load, so the pores got clogged within 7 min. Finally the filtration efficiency was recorded as 63.84 per cent. Filtration efficiency for sand B initially decreased for 25 min but after that it started increasing. Filtration efficiency increases because of clogging of pores of media. Finally, the efficiency was recorded as 29.25 per cent. For the coarsest sand C, filtration efficiency decreased because only a small quantity of suspended particles could be trapped over the sand bed.

Effects of suspended load concentration of water on the pressure drop for different sizes of sand media

Figs. 9, 10 and 11 shows pressure drop with time for horizontal sand filter for sand size A, B and C respectively with clean water and water with suspended loads of 50 mg/l, 100 mg/l and 300 mg/l. For clean water it was nearly constant for one hour for all sand sizes. For water with suspended loads pressure drop increases with time. In Fig. 9, at beginning it was 0.15 kg/cm^2, 0.17 kg/cm^2 and 0.15 kg/cm^2 for 50 mg/l, 100 mg/l and 300 mg/l respectively. Pressure drop reaches to 0.55 kg/cm^2 after 28 min for 50 mg/l, after 18 min for 100 mg/l and after 7 min for 300 mg/l. As the suspended load concentration increased from 50 mg/l to 300 mg/l, the time to reach pressure drop of 0.5 kg/cm^2 decreased. This was because of clogging of media due to suspended particles. Hence, pressure drop increased at faster rate on increasing suspended load concentration.

Fig. 10 shows pressure drop was nearly constant for suspended load of 50 mg/l. Pressure drop increased for both the concentrations of 100 mg/l and 300 mg/l but it increased at faster rate for 300 mg/l than 100 mg/l. As 300 mg/l was the heavier load, pressure drop reached to 0.54 kg/cm² in 47 min and 0.52 kg/cm² in 145 min for 100 mg/l. Fig. 11 shows for all suspended loads, there was no significant pressure drop, as this was the coarsest sand among the three. It was initially 0.15 kg/cm², 0.15 kg/cm² and 0.14 kg/cm² for loads of 50 mg/l, 100 mg/l and 300 mg/l respectively. It was to 0.16 kg/cm² in 130 min for 50 mg/l, 0.16 kg/cm² in 70 min for 100 mg/l and 0.15 kg/cm² in 35 min for 300 mg/l. Pressure drop for the three loads tends to be constant after some time.

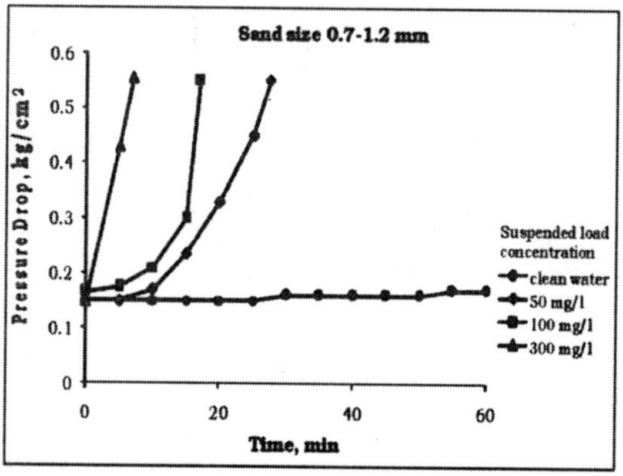

Fig. 9. Pressure drop-Time relationship for sand size A.

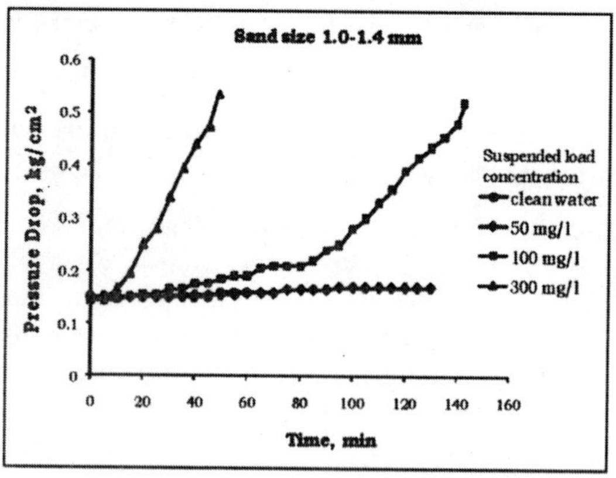

Fig. 10. Pressure drop -Time relationship for sand size B.

Effect of sand size of filter media on the pressure drop for different suspended load concentration of water

The Figs. 12, 13 and 14 shows the Pressure drop-time relationship the sand filter for suspended loads of 50 mg/l, 100 mg/l and 300 mg/l respectively with sand sizes A, B and C. It was initially 0.15 kg/cm² for the three sand sizes.

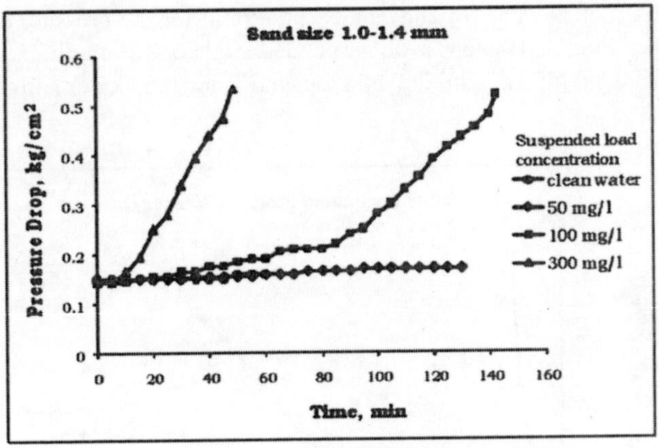

Fig. 11. Pressure Drop-Time relationship for sand size C.

In Fig. 12, the pressure drop increased with time for sand A but it remains constant for the sands B and C. As these two were coarser sands than the former, so maximum suspended particles passed though it and thus, there was no clogging of media. This resulted in no pressure drop. Hence, pressure drop was more with sand A as the suspended particles trapped there. But because of clogging of media due to suspended particles, this sand gave higher filtration efficiency.

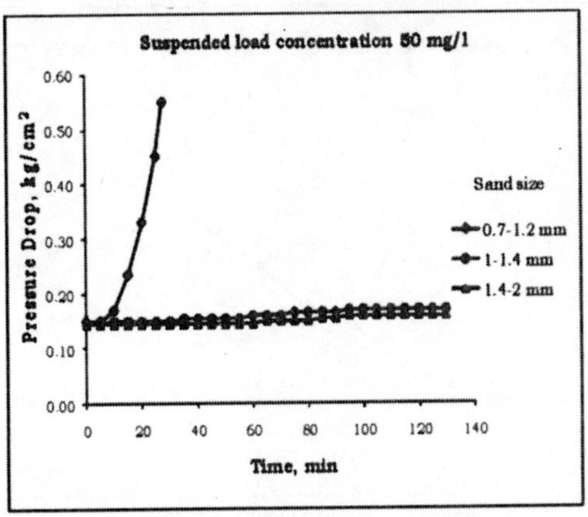

Fig.12. Pressure Drop-Time relationship for suspended load of 50 mg/l.

Fig. 13 Pressure drop increased with time for sand A and B but it remained constant for the sand C. Hence, for suspended load of 100 mg/l, there was significant pressure drop for sands A and B because the suspended particles were trapped on these sand beds. This was also helpful in achieving higher filtration efficiency.Fig. 14 shows the same trend for the pressure drop as in Fig. 13 for sand A, B and C with suspended loads of 100 mg/l. The pressure drop reached to 0.56 kg/cm^2 in 7 min for sand A and 0.54 kg/cm^2 in 49 min for sand B.

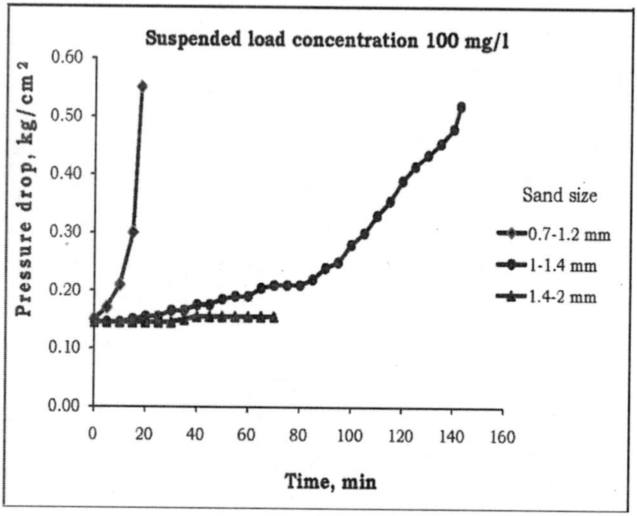

Fig. 13. Pressure Drop-Time relationship forsuspended load of 100 mg/l.

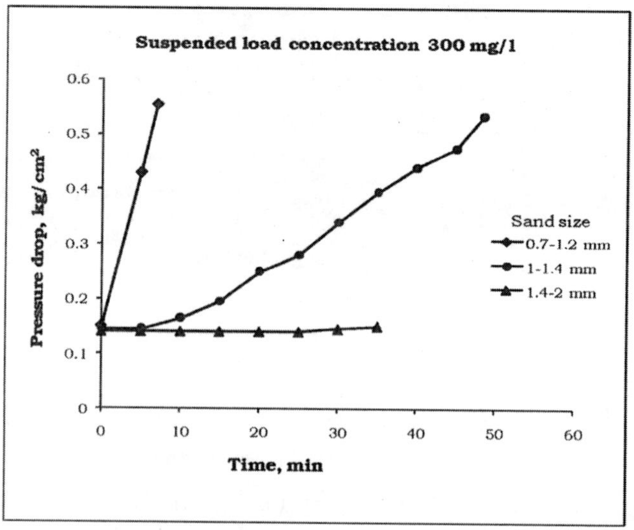

Fig. 14. Pressure Drop-Time relationship for suspended load of 300 mg/l.

Nomogram

Nomogram was prepared for Horizontal filter (Clean Master) for sand size A which was found best. The data observed on backwash interval for the suspended load concentration of 50, 100 and 300 mg/l was used to develop the nomograms. The nomogram can be helpful to set the backwash interval for the given suspended load concentration. It will be very beneficial for fully automatic backwash systems to set the backwash time. Fig. 15 shows nomogram for suspended load concentrations and backwash interval for the sand filter with sand size A mm. It was clear that if clean water was used or suspended load concentration was below 50 mg/l then the backwash interval could be one to two hours. As the impurity concentration was increased, the backwash interval decreased. For 50 mg/l, it was 30 min and for 100 mg/l it was 18 min. At any suspended load concentration from 0-600 mg/l, we can find the backwash interval with the help of this nomogram. For example, if suspended load concentration is 20 mg/l then backwash interval will be about 70 min.

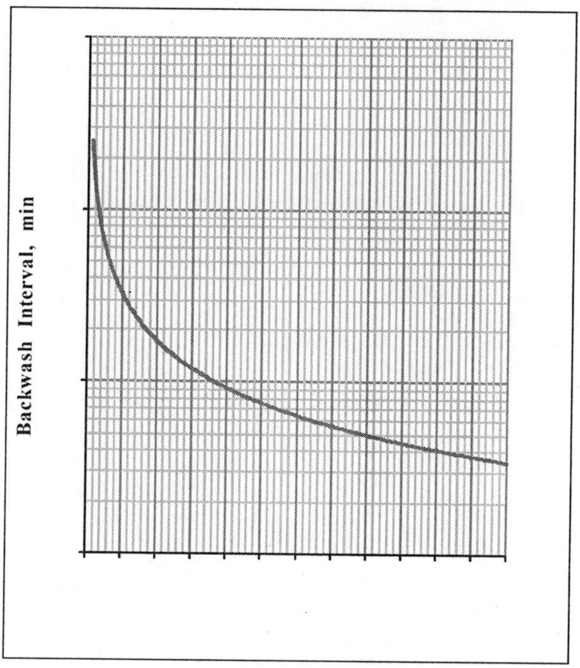

Fig. 15. Nomogram for suspended load concentration and backwash interval for horizontal sand filter.

Conclusion

Based on the data collected and its analysis, the following conclusions could be drawn from the present investigation.

- The pressure drop, across the sand media filter for clean water, up to flow rate of 50 m3/h for horizontal model can be within the permissible limit of2.0 m.
- The filtration efficiency can be increased by decreasing the sand size from coarser to finer.
- There was a poor relationship between filtration efficiency and sediment load concentration showing that the filtration efficiency decreased with sediment load concentration.
- The pressure drop decreased as the sand size changed from finer to coarser for the sand filter. The pressure drop was constant for one hour for all three sand sizes when clean water was passed.
- There was a direct relationship between pressure drop and filtration efficiency. The higher the pressure drop more would be the filtration efficiency.
- In all, it is suggested to adopt horizontal sand filter for lesser pressure drop using filtering media having sand size of 0.7-1.2 mm (16-25 mesh).
- The developed nomogram seems the best and handiest tool for the decision supports to optimise the backwash interval of the sand filter.

References

Capra, A. and B. Scicolone, 2004. "Emitter and filter tests for wastewater reuse by drip irrigation", Agric. Water Manage,(2):135-149.

Converse and Tyler, 2004. "Fabrication and evaluation of sand media filters used intrickle irrigation", Proceeding of Third Irrigation Seminar, Trans. ASAE, 40 (4):850 - 865.

Gideon, O., B. A. Jiftah, and D. Yoel, 1982."Effluent in trickle irrigation of cotton in arid zones", J. Irrig. and Drain. Div., ASCE, 108(IR2): 115 – 125.

Khaire, S. K., D. K. Khemnar and K. C. Sanap, 2008. "Hydraulic performance of sand media filters in drip irrigation system", Unpublished B.Tech (Agril. Engg.) thesis submitted to M.P.K.V. Rahuri (M.S.).

Nakayama, F. S., R. G. Gilbert, and D. A. Bucks, 1978. "Water treatments in trickle irrigation systems", J. Irrig. and Drain. Div., ASCE, 104(IR1): 23-34.

Nakayama, F.S. and D.A. Bucks, 1981."Emitter clogging effects on trickle uniformity", Trans. ASAE, 24(1): 77-78.

Padmakumari, O. and, R.K. Sivanappan 1980. "Drip irrigation", College of Agricultural Engineering, Tamilnadu AgriculturalUniversity, Coimbatore, pp. 332-353.

Rodgers, M., J. Mulqueen, and M. G. Healy, 2004. "Surface clogging in an intermittent stratified sand filter", Soil Sci. Soc. Am. J., 68 (6): 1827-1832.

Siag, M. 1997. "Studies on efficacy of sand filter for artificial groundwater recharge", M.Tech. (Agril. Engg.) thesis submitted to P.A.U., Ludhiana.

Wadate, A.P. and P. Khanal, 1997. "Comparative performance of disc and screen filter used in micro irrigationsystem", Unpublished B. Tech. (Agril.Engg.) thesis submitted to Dr. Annasaheb Shinde College of Agricultural Engineering, M.P.K.V., Rahuri, and Maharashtra, India.

18

Optimisation of Irrigation and Fertilisation for Summer Sesame (*Sesamum indicum* L.) under Influence of Changing Climate

R.B. Thanki**, R.M. Solanki, J.M. Modhawadia and B.S. Gohil

Abstract

*A field experiment was conducted on optimisation of irrigation and fertilisation for sesame during summer season of 2007 and 2008 on a clayey soil at Cotton Research Centre, Khapat Farm, Junagadh Agricultural University, Porbandar (Gujarat) to find out the combined effect of irrigation at critical growth stages and fertility levels on yield, yield attributes, quality and water use efficiency of sesame (*Sesamum indicum *L.). There were total twenty four treatment combinations consisted of four levels of irrigation assigned to main plot and six levels of fertility in sub-plot laid out in split plot design with four replications. The results revealed that irrigating the crop at branching, flowering, capsule and seed development stages along with the application of 75.0:50.0:50.0 kg* $N:P_2O_5:K_2O$ ha^{-1} *were found significant for number of effective branches per plant, length of capsule, No. of capsules per plant, No. of seeds per capsule, seed yield per plant, test weight, protein and oil content, oil yield, NPK uptake and seed yield and stover yields of sesame along with appreciable saving of water and fertiliser under climate change conditions. Whereas, water use efficiency was maximum when the crop was irrigated at flowering and capsule development stages only.*

Introduction

Sesame (*Sesamum indicum* L.) is an important oilseeds crop which has been cultivated since ancient times. As sesame seeds contain 50-60 per cent oil and 20-30 per cent protein, it is a significant crop of agrarian-industrial importance which occupies 19.0 lakh ha area with annual production of 8.10 lakh tonnes having productivity of 426 kg ha^{-1} in India and in Gujarat area of sesame is 2.47 lakh ha with annual production of 1.17 lakh tonnes having productivity of 474 kg ha^{-1} (DOD, 2014). However, with increase in availability of irrigation facility in the state, area under summer sesame cultivation is increasing day by day due to more productivity, higher market price and net monetary returns. Summer crop is generally grown after harvest of wheat in Saurashtra region of Gujarat state. To achieve higher productivity potential, irrigation scheduling and balanced fertilisation are the key factors affecting the yield. Irrigation water is a key and costly input in crop production and plays a crucial role in improving the seed and oil yields of sesame. Narang and Gill (1998) reported that seed yield of summer sesame increased with increase in number of irrigations. Keeping in view, fast ever diminishing water resources and increasing competition from and with in agriculture for water, its economical and efficient utilisation becomes quite imperative. Under limited water supply, higher seed yield can be obtained by scheduling irrigation at critical growth stages of crop growth (Choradia and Gaur, 1986). It is well established fact that there is a positive correlation between nutrient application and productivity. Therefore, supply of balanced nutrients is considered as one of the basic needs to achieve the potential yield. Very little work has so far been carried out on irrigation scheduling and nutritional requirement of summer sesame, present investigation was undertaken to study the effect of irrigation scheduling at critical growth stages and fertility levels on yield, quality and water use efficiency of summer sesame under medium black clayey soil of South Saurashtra region.

Materials and Methods

A field experiment was conducted during summer season of 2007 and 2008 at Cotton Research Centre, Khapat Farm, Junagadh Agricultural University, Porbandar (Gujarat). The soil was clayey in texture with pH: 7.7, EC: 0.74 dSm^{-1}, low in available nitrogen (270 kg ha^{-1}) medium in available phosphorus (38.8 kg P_2O_5 ha^{-1}) and high in available potassium (278 kg K_2O ha^{-1}). The experiment was laid out in a split plot design with four replications. All possible 24 treatment combinations consisting of four irrigation schedules at critical growth stages *viz.*, I_1- irrigation at branching + flowering + capsule development + seed development stage, I_2- irrigation at branching + capsule development + seed development stage, I_3- irrigation at branching + flowering + seed development stage and I_4- irrigation at flowering + capsule development stage as main plot treatments, six fertility levels *viz.*, F_1- control, F_2- 25.0:12.5:00.0 kg $N:P_2O_5:K_2O ha^{-1}$, F_3- 50.0:25.0:00.0 kg $N:P_2O_5:K_2O$ ha^{-1}, F_4- 50.0:50.0:25.0 kg $N:P_2O_5:K_2O$ ha^{-1}, F_5-75.0:50.0:00.0 kg $N:P_2O_5:K_2O$ ha^{-1} and F_6- 75.0:50.0:50.0 kg $N:P_2O_5:K_2O$ ha^{-1} as sub-plot treatment. Half dose of nitrogen and full dose of phosphorus and potassium was applied at the time of sowing in the form of urea, single super phosphate and muriate of potash as per treatments and remaining half dose of nitrogen was given at branching stage in the form of urea. A common irrigation was given immediately after sowing the crop for satisfactory seed germination and proper establishment of crop. Remaining irrigations each of 50 mm depth were given as per treatments. The quantity of irrigation water was measured with 7.5 cm throat Parshall Flume installed under free flow condition in the water channel. Sesame cv. 'GT-2' was sown at 30 cm spacing using seed rate of 2.5 kg/ha on 15th and 10th February of 2007 and 2008 and harvested on 19th and 15th March of 2007 and 2008, respectively. The experimental data recorded for growth parameters, yield attributes, yield, quality parameters and economics were statistically analysed for level of significance and pooled analysis for both the season.

Results and Discussion

Effect of irrigation on yield attributes, yield, nutrient uptake and quality parameters

Two years pooled results furnished in showed that significantly maximum yield attributes, seed (971 kg ha^{-1}) and stover yields (1325 kg ha^{-1}) were recorded when the crop was irrigated at branching + flowering + capsule development + seed development stages (I_1). The mean seed yield increased under I_1 over I_2, I_3 and I_4 were to the tune of 27.7, 45.1 and 47.6 per cent, respectively. The per cent increased in stover yield in I_1 over I_2, I_3 and I_4 were to the tune of 6.8, 9.6 and 19.0 per cent, accordingly. Increase in yield under I_1 might be due to the fact that crop might have received right quantity of irrigation water at right crop growth stage and hence, soil moisture content remained optimum in the surrounding root zone area which finally resulted in improvement crop growth and yield. The findings are in close vicinity with the results obtained by Patra (2001), Chang *et al*, (2005). Application of higher dose of 75.0-50.0-50.0 Kg NPK ha^{-1} to summer sesame resulted in significantly higher NPK uptake of 65.38, 13.22 and 19.03 Kg NPK ha^{-1}, respectively. The increase in uptake of nutrients by sesame crop appears due to the cumulative effect of increased seed and stover yields. The application of N, P and K might have increased their concentration in soil solution which increased their availability and uptake by plant. The results are substantial the findings of and Patra (2001). Different irrigation schedules significantly affect the protein and oil content as well as oil yield. Significantly higher protein and oil content of 27.57 and 52.62 per cent as well as oil yield of 511 kg ha^{-1} were observed under I_1. It might be due to fact that protein content in seed was directly reflected to nitrogen content and uptake by seed. The increased uptake of nitrogen and water which in turn enhanced assimilation of amino acid leading to increased synthesis of protein. Mondal *et al.* (1997) found that oil and protein yields were increased as irrigation frequency increased.

Effect of fertility level on yield attributes, yield, nutrient uptake and quality parameters

Application of 75.0:50.0:50.0 kg $N:P_2O_5:K_2O$ ha^{-1} (F_6) recorded significantly higher yield attributes, seed and stover yields of 836 and 1287 kg ha^{-1}, respectively and remained at par with F_5. The per cent increase in seed and stover yields under F_6 over control (F_1) were to the tune of 17.6 and 14.7 per cent, respectively (Table 1). Increase in seed and stover yields with the higher level of NPK might be due to the fact that balanced nutrient supply increases the absorptive power of soil for cation and anion, created a situation favourable for higher uptake of NPK by plant. These absorbed ions are released slowly for the entire growth period resulted in better nutrient availability at active growth of the crop. The finding confirms the results obtained by Patra (2001). Higher seed and stover yields under 75.0:50.0:50.0 kg $N:P_2O_5:K_2O$ ha^{-1} were also probably a consequence of greater amount of nutrient uptake by the seed and stover. The increase in yield with increased fertiliser levels is in close conformity with the findings of Kathiresan (2002) and Thanunathan *et al.* (2006). Application of higher dose of 75.0-50.0-50.0 kg NPK ha^{-1} resulted in significantly higher NPK uptake of 65.38, 13.22 and 19.03 kg NPK ha^{-1}, respectively. The increase in uptake of nutrients by sesame crop appears due to the cumulative effect of increased seed and stover yields. The application of N, P and K might have increased their concentration in soil solution which increased their availability and uptake by plant. Protein and oil content as well as oil yield were significantly increased with 75.0:50.0:50.0 kg $N:P_2O_5:K_2O$ ha^{-1} (F_6) over rest of the treatments on pooled data basis (Table 2). Higher content of protein under higher level of fertility might be due to higher content and uptake of nitrogen by seed under this treatment. The results confirm the findings of Mankar *et al.* (1995) and Patra (2001).

Interaction effect of irrigation and fertility levels

Pooled data of two years (Table 3) revealed the presence of significant interaction between irrigation and fertility levels (IxF). Significantly higher seed and stover yields, protein content and oil yield were recorded when crop was irrigated at branching, flowering, capsule development and seed development stages along with the application of 75.0:50.0:50.0 kg $N:P_2O_5:K_2O$ ha^{-1} (I_1F_6). Treatment combination I_1F_6 recorded significantly higher seed yield than rest of the combinations, but it remained at par with I_1F_4 combinations. These might be due to application of sufficient amount of water along with optimum amount of nutrients might have increased physiological processes like cell division and cell expansion which improves growth and yield attributes and ultimately higher yields. These results are in close vicinity of the results obtained by Khade *et al.* (1996) and Ravinder *et al.* (1996).

Table 1. Effect of irrigation schedules and fertility levels on growth, yield attributes and seed and stover yields of summer sesame (pooled data of two years)

Treatment	*Plant Height at Harvest (cm)*	*No. of branches/ plant*	*No. of Capsules/ plant*	*Length of capsule*	*Seeds/ Capsule*	*1000-Seed weight (g)*	*Seed weight/ plant (g)*	*Seed yield Kg/ha*	*Stover yield Kg/ha*
Irrigation Level									
I_1	73.19	3.50	48.79	2.03	59.90	3.12	4.06	971	1325
I_2	71.16	3.15	46.34	2.02	58.71	3.09	3.74	760	1240
I_3	65.42	2.90	43.61	1.86	55.78	2.95	3.56	669	1209
I_4	64.98	2.67	42.34	1.84	53.88	2.93	3.37	658	1113
S.Em. ±	0.83	0.06	0.43	0.02	0.72	0.033	0.06	13	19
C.D. at 5 per cent	2.46	0.19	1.26	0.07	2.15	0.10	0.16	39	57
Fertility level									
F_1	66.43	2.72	41.80	1.85	53.45	2.93	3.22	711	1122
F_2	66.53	2.94	43.99	1.89	55.11	2.97	3.37	730	1200
F_3	68.05	3.04	44.40	1.93	56.53	3.00	3.59	746	1224
F_4	69.54	3.05	45.87	1.96	57.33	3.03	3.73	776	1229
F_5	70.52	3.27	47.30	1.98	59.93	3.10	4.03	786	1268
F_6	71.07	3.31	48.24	2.00	60.04	3.11	4.15	836	1287
S.Em. ±	0.87	0.06	0.44	0.02	0.67	0.03	0.05	12	17
C.D. at 5 per cent	2.43	0.16	1.24	0.06	1.88	0.07	0.14	35	46
I X F Interaction	NS	S	S	S	S	S	S	S	S

Table 2: Effect of irrigation schedules and fertility levels on total NPK uptake, protein and oil content, oil yield, water use efficiency and economics of summer sesame (pooled data of two years)

Treatment	*Total NPK uptake (kg/ha)*			*Protein content*	*Oil content (per cent)*	*Oil yield (kg/ha)*	*WUE (kg/ha-mm)*	*Net returns (Rs/ha)*	*B:C ratio*
	N	*P*	*K*						
Irrigation Level									
I_1	71.63	14.16	20.20	27.57	52.62	511	3.24	45594	3.62
I_2	58.63	11.54	17.50	27.07	51.98	395	3.04	32463	2.91
I_3	53.33	10.14	15.39	27.01	51.45	345	2.67	26446	2.56
I_4	50.15	9.27	14.03	26.80	50.73	334	3.29	26172	2.59
S.Em. ±	0.92	0.17	0.32	0.07	0.58	9	0.05	853	0.05
C.D. at 5 per cent	2.72	0.52	0.95	0.22	NS	26	0.15	2534	0.15
Fertility level									
F_1	50.84	9.42	14.93	26.21	49.83	353	2.86	31195	3.07
F_2	55.33	10.45	15.92	26.85	50.65	372	2.93	31435	2.96
F_3	57.74	10.90	16.25	27.10	51.70	386	2.99	31534	2.86
F_4	59.25	11.57	17.22	27.09	51.95	405	3.09	33295	2.94
F_5	62.07	12.11	17.34	27.58	52.45	413	3.16	32825	2.79
F_6	65.38	13.22	19.03	27.85	53.58	448	3.34	35728	2.91
S.Em. ±	0.73	0.15	0.23	0.07	0.58	8	0.05	799	0.05
C.D. at 5 per cent	2.06	0.42	0.65	0.20	1.63	21	0.14	2238	0.13
I X F Interaction	S	NS	NS	S	NS	S	-	-	-

Table 3. Interaction effect of irrigation and fertility level on seed and stover yields; protein content, oil yield and yield attributes in summer sesame (pooled data of two years)

Irrigation Level / *Fertility level*	*Seed yield (kg ha^{-1})*				*Stover yield (kg ha^{-1})*				*No. of capsules/ plant*				*Length of capsule (cm)*			
	I_1	I_2	I_3	I_4	I_1	I_2	I_3	I_4	I_1	I_2	I_3	I_4	I_1	I_2	I_3	I_4
F_1	914	682	597	653	1208	1219	1094	1021	43.7	42.7	40.6	40.2	1.95	1.93	1.82	1.72
F_2	879	713	692	636	1289	1240	1163	1107	46.6	44.7	41.9	42.7	2.00	1.94	1.85	1.77
F_3	928	740	683	632	1363	1232	1221	1081	47.9	45.7	43.8	40.1	1.97	1.98	1.94	1.84
F_4	1046	768	671	621	1310	1219	1231	1157	49.2	46.3	43.4	44.6	1.95	1.97	1.95	1.98
F_5	972	793	663	717	1426	1260	1274	1114	51.4	47.0	45.5	44.9	2.11	2.12	1.82	1.88
F_6	1086	865	706	689	1351	1271	1274	1197	53.9	51.2	46.4	41.4	2.19	2.18	1.81	1.84
C.D. at 5 per cent	69				80				2.48				0.11			
	Protein content (per cent)				*Oil yield (kg ha^{-1})*				*No. of seeds/capsule*				*Seed weight (g/plant)*			
	I_1	I_2	I_3	I_4	I_1	I_2	I_3	I_4	I_1	I_2	I_3	I_4	I_1	I_2	I_3	I_4
F_1	26.63	26.13	26.23	25.87	450	342	301	321	57.15	55.30	52.58	48.78	3.51	3.34	3.10	2.92
F_2	27.28	26.46	27.15	26.53	451	374	351	311	60.13	56.43	53.83	50.08	3.69	3.35	3.31	3.12
F_3	27.96	27.14	26.21	27.08	496	376	357	315	60.15	59.83	54.35	51.80	3.97	3.38	3.55	3.47
F_4	27.57	27.15	27.00	26.65	567	399	340	313	61.10	59.08	51.73	57.43	4.19	3.87	3.62	3.25
F_5	28.07	27.72	27.50	27.02	510	424	343	375	59.53	58.13	61.73	59.35	4.15	3.94	4.09	3.93
F_6	27.91	27.82	27.99	27.68	594	454	375	367	61.33	63.50	60.50	54.85	4.85	4.57	3.68	3.51
C.D. at 5 per cent	0.40				38				3.76				0.27			

Water use efficiency

Significantly higher WUE of 3.29 kg $ha^{-1}mm^{-1}$ was recorded under irrigation at flowering and capsule development stages (I_4). The lowest WUE was observed under higher irrigation level i.e. I_1 - irrigation at branching, flowering, capsule and seed development stages (Table 2). The transpiration, evaporation and excessive vegetative growth and development increased seed and stover yields might be due to higher water use and ultimately lower WUE. The results completely collaborate with the findings of Mitra and Pal (1999). In case of fertility levels, higher WUE of 3.34 kg $ha^{-1}mm^{-1}$ was found when crop was fertilised with the 75.0:50.0:50.0 kg N:P_2O_5:K_2O ha^{-1} (F_6) which was remained at par with the fertility level F_5.

Correlation studies

The relationship was studied between seed yield and growth, yield and quality parameters as well as nutrient contents and uptake. The data on intercept (a), slope (b), correlation co-efficient (r) and coefficient of determination (R^2) are furnished in Table 4. It was observed that plant height (cm) at 60 DAS, number of effective branches per plant, number of capsules per plant, length of capsule (cm), number of seeds per capsule, seed yield (g/plant), stover yield (kg/plot), 1000 seed weight (g), oil content (per cent) in seed, oil yield (kg ha^{-1}), protein content (per cent) in seed, N, P and K content (per cent) and uptake (kg ha^{-1}), and WUE (kg $ha^{-1}mm^{-1}$) showed highly significant positive correlation with seed yield of sesame. Among the growth parameters, correlation with seed yield was the highest in case of number of effective branches per plant at harvest (0.5769, 0.7258 and 0.6640 in pooled results; among quality parameters it was recorded highest with oil yield (0.9901, 0.9458 and 0.9291) in the year and pooled results, while from nutrient content and uptake, N uptake showed the highest correlation (0.9437, 0.9361 and 0.9444) in and pooled results. From the above results, it is apparent that about 89.18 per cent variation in seed yield could be explained by variation in N uptake followed by variation in oil yield (86.32 per cent) and P uptake (83.96 per cent).

Economics

The economics of different treatments showed that irrigation at branching, flowering, capsule development and seed development (I_1) proved superior and recorded maximum net return in 2007, 2008 and pooled of/48189, 42999 and 45594/ha, respectively (Table 2).This treatment also recorded maximum B:C ratio of 3.84, 3.40 and 3.62 in years 2007, 2008 and in pooled data, respectively. The higher net return might be due to higher seed and stover yields. Fertilising the crop with 75-50-50 kg NPK/ha (F_6) recorded the 12.97, 16.25 and 14.52 per cent higher net return with maximum B:C ratio of 3.25, 2.89

Table 4. Correlation coefficient (r), coefficient of determination (R^2) and regression equation of seed yield (dependent variable) with individual growth and yield attributes and nutrient uptake (Pooled)

Character	*Intercept (a)*	*Slope (b)*	*Correlation coefficient (r)*	*$100r^2$*
Plant height (cm) at 60 DAS	-0.0697	0.0110	0.5391**	29.06
No. of effective branches/plant	0.1399	0.1794	0.6640**	44.10
No. of capsules/plant	-0.1531	0.0186	0.6424**	41.26
Length of capsules (cm)	0.0664	0.3208	0.4865**	23.67
No. of seeds/capsule	0.0121	0.0118	0.4647**	21.60
Seed yield (g/plant)	0.1763	0.1390	0.6289**	39.55
Stover yield (kg/plot)	0.0561	0.5747	0.5801**	33.65
1000-seed weight (g)	0.4341	0.0840	0.1543**	2.38
Oil content (per cent) in seed	0.6696	0.0004	0.0105^{NS}	0.01
Oil yield (kg/plot)	0.0966	1.7140	0.9291**	86.32
N content (per cent) in seed	0.2625	0.0981	0.1148^{NS}	1.32
Protein content (per cent) in seed	0.2613	0.0157	0.1152^{NS}	1.33
N content (per cent) in stover	-0.3047	0.4824	0.5385**	29.00
P_2O_5 content (per cent) in seed	0.0781	0.7356	0.3979**	15.83
P_2O_5 content (per cent) in stover	-0.1974	2.2305	0.6548**	42.88
K_2O content (per cent) in seed	-0.0493	0.9463	0.5196**	27.00
K_2O content (per cent) in stover	-0.1562	0.9636	0.4988**	24.89
N uptake	-0.0487	0.0140	0.9444**	89.18
P uptake	0.0997	0.0580	0.9163**	83.96
K uptake	0.0898	0.0396	0.8433**	71.12
Water use efficiency(kg/ha-mm)	0.0296	0.2152	0.6797**	46.20

and 3.07 in the years 2007, 2008 and in pooled data, respectively. The higher net return in treatment F_6 might be due to higher seed and stover yields. The results are in agreement with those of Dutta *et al.* (2000).

References

Chang, A. H., Sheikh, S. A., Jamro, G. H., Jamro, G. M. and Memon, R. B. 2005. Growth and yield response of sesame (*Sesamum indicum* L.) to different NP combinations. *Indian Journal of Plant Science*, 4(1): 32-37.

Choradia, R. K. and Gaur, B. L. 1986. Effect of irrigation and mulch on seed yield of safflower. *Indian Journal of Agronomy*, 31(3): 298-299.

DOD, 2014. Status paper on oilseed crops. http://www.dancnet.nic.in.

Dutta, D, Jana, P K, Bandyopadhyay, P. and Maity, D. 2000. Response of summer sesame (*Sesamum indicum* L.) to irrigation. *Indian Journal of Agronomy*, 45(3): 613-616.

Kathiresan, G. 2002. Response of sesame (*Sesamum indicum*) genotypes to levels of nutrients and spacing under different seasons. *IndianJournal of Agronomy*, 47(4): 537-540.

Khade, V. N., Jadhav, S. N. and Khanvilkar, S. A. 1996. Studies on scheduling of irrigation and phosphate fertilisation to sesamum. *Journal Maharashtra Agricultural University*, 21(3): 410-411.

Mankar, D. D., Satao, R. N., Solanke, V. M. and Ingole, P. G. 1995. Effect of nitrogen and phosphorus on quality, uptake and yield of *sesame. PKV Research Journal*, 19(1): 69-70.

Mitra, S. and Pal, A. K. 1999. Water use and productivity of summer sesame as influenced by irrigation and nitrogen. *Journal of Indian Society of Soil Science*, 47(3): 400-404.

Mondal, D. K., Sounda, G., Panda, P. K., Ghosh, P., Maitra, S. and Roy, D. K. 1997. Effect of different irrigation levels and nitrogen doses on growth and yield of sesame (*Sesamum indicum* L.). *Indian Agriculturist*, 41(1): 15-21.

Narang, R. S. and Gill, M. S. 1998. Irrigation scheduling for higher WUE in oilseeds and pulses. *Fertilizer News*, 43 (3):57-67.

Patra, A. K. 2001. Yield and quality of sesame (*Sesamum indicum* L.) as influenced by N and P during post-rainy season. *Annals of Agricultural Research*, 22(2): 249-252.

Ravinder, N., Satyanarayana, V., Rao, V. P., Latchanna, A. and Varaprasad, P. V. 1996. Influence of irrigation and fertilisation on seed yield, nutrient uptake and fertilizer use efficiencies in summer sesame (*Sesamum indicum* L.). *J. Oilseeds Res.*, 13(2): 173-177.

Thanunathan, K., Thiruppathi, M., Leopold, M. and Imayavaramban, V. 2006. Nutrient management for sesame under water constraint situation. *Research on Crops*, 7(2):426-428.

19

Effect of Drip Irrigation and Plant Geometry in Summer Groundnut (*Arachis hypogaea* L.)

P. M. Vaghasia, V. B. Bhalu and, V. H. Kachhadiya

Abstract

The field experiments were conducted for consecutive three years to assess the crop performance and economics of drip irrigation along with three levels drip system and four levels of plant geometry with surface irrigation as control for groundnut (GG-6) during summer season of year 2009-10 to 2011-12 at the Main Oilseeds Research station, Junagadh Agricultural University, Junagadh.The results revealed that alternate day drip irrigation at 100 per cent of cumulative pan evaporation (CPE) recorded highest pod yield (2623 kg ha^{-1}), haulm yield (4725 kg ha^{-1}), oil yield (915 kg ha^{-1}) and yield attributing characters with 3.89 kg ha^{-1} mm^{-1}water use efficiency. Among the drip system highest net return (60024 ha^{-1}) and B:C ratio (2.38) obtained by drip irrigation at 100 per cent CPE and magnitude of increase in pod yield over border strip irrigation at 1.0 IW/ CPE ratio was to the tune of 18.0 per cent. Border strip irrigation recorded lower pod yield (2221 kg ha^{-1}), haulm yield (3841 kg ha^{-1}) and oil yield (723 kg ha^{-1}) with net return (57275 ha^{-1}) and B: Cratio (2.91).The drip irrigation at 60 per cent CPE found better one having the highest water use efficiency of 5.01 kg ha^{-1} mm^{-1}. Also, in case of deficit water conditions (60 per cent of CPE) the drip was not that profitable. Significantly maximum pod yield of 2670 kg ha^{-1} was recorded by the conventional sowing (row to row: 30 cm) with one lateral to each row. One lateral between two rows in conventional system layout found more economical (net return, 61401 ha^{-1}) and cheaper in terms of capital investment (B:C ratio, 2.55).

Introduction

Groundnut (*Arachis hypogaea* L.) is cultivated predominantly in the tropics and subtropics. In Gujarat, it is grown on 16.86 lakh hectares area with production of 27.17 lakh tonnes and 1662 kg ha^{-1} productivity (Anonymous, 2012). The wide difference in productivity of groundnut in our country as compare to that in USA (3540 kg ha^{-1}) and China (2664 kg ha^{-1}) indicate that productivity of groundnut can be increased to a great extent by providing better crop husbandry practices and well planned irrigation management practices. During the summer season, there is a higher yield potential of groundnut in Saurashtra and Kutch region of Gujarat State (Patel *et al.*, 1990). However, the biggest constraint is the limited water resources. One of the critical challenges to water resources management is to shift from the extensive supply oriented approach to the one focusing upon deficit applications; there are several ways of increasing efficiency in irrigation. One way is changing from surface to pressurised methods of irrigation and second is to apply deficit water. The extent to second is to apply deficit irrigation is to increase the water use efficiency of a crop by eliminating irrigations that have little impact of yield (Rank, 2007). The resulting yield reduction may be small compared with the benefits gained through diverting the saved water to irrigate additional area or other crops for which water would normally be insufficient under traditional practices. Before implementing a deficit irrigation programme, it is necessary to know crop yield responses to water applications. The drip irrigation has the potential to provide consistently high yields with non-uniform precipitation while conserving soil, water, and energy. Drip irrigation systems have the capability of frequently supplying water to the root zone while reducing the risk of cyclic water stress that is typical of other irrigation systems or non-irrigated production. Root deterioration can be minimised when using drip irrigation systems (Radin *et al.*, 1989). Planting geometry with lateral arrangement respect to crop rows to find out the possible reduction of lateral requirement. The soil wetting patterns under different plant geometry are controlled by irrigation levels. Hence, it is essential to use limited water for irrigation as a judiciously as possible. Thus, micro irrigation especially drip under planting geometry with lateral arrangement has been tried for summer groundnut.

Materials and Methods

The experiment was carried out at Main Oilseed Research Station, Junagadh Agricultural University, Junagadh, Gujarat, India to assess the summer groundnut crop performance under drip irrigation system at various water application levels with system lay out during summer season for consecutive three years (2009-10 to 2011-12). The soil of the experimental plot was clayey in texture and slightly alkaline in reaction. The soil has an organic carbon content of 0.57 per cent and was medium in available nitrogen and potash and sulphur, low in phosphorus. The moisture content of the experimental plot at field capacity and permanent wilting point were 29.4 and 13.8 per cent, respectively, while the bulk density was 1.44 g cm^{-3}. The experiment was laid out in split plot design with four replications. Four levels of drip irrigation (alternate day irrigation at 100 per cent (I_1), 80 per cent (I_2) and 60 per cent (I_3) of average daily pan evaporation) were taken in main plot, while four planting geometryviz., conventional sowing (row to row distance 30 cm) with one lateral to each row (G_1), one lateral between two rows (G_2), paired-rows (15/30 cm apart) with one median lateral to each pair (G_3) and one lateral between two pairs (G_4). Surface irrigation at 1.0 IW/CPE (IW=50 mm) as control (I_4) was compared with each drip levels by employing two sample "t" test. Drip system operated at 1.2 kg/cm^2 pressure and dripper were placed at 45 cm apart in lateral. Two common surface irrigation were given just after sowing and eight days after sowing (DAS) to all the treatments for uniform germination of crop. The groundnut variety GG-6 was sown at a seed rate 125 kg/ha and fertiliser 25:50:00 (N:P:K) kg/ha as per agronomic recommendations for this region.

In drip system, the irrigation was done once in two days by calculating two days cumulative pan evaporation. By multiplying the two days cumulative pan evaporation with plot area and the treatment percentage, the depth of irrigation water and volume of water in litres were calculated. With the dripper discharge rate and number of drippers the operating time was calculated and the irrigation was done for the calculated duration for each set of treatments with the help of flow control valve fitted in each lateral. The water meters were used to measure the volume of water applications.The control plot was irrigated through the sub-main end cap by following IW/CPE=1.0 by closing all the lateral flow control valves.

The cost of cultivation excluding cost of irrigation (C_c) included cost of various inputs like cost towards land preparation, seeds, seed treatment, fertiliser, sowing, agro chemicals, weeding, inter culturing, harvesting, threshing, cleaning and packing etc. The cost of seeds, fertiliser and agro chemicals were taken following the recommended package of agronomic practices. The cost of irrigation (C_i) includes the cost of labour, electricity and maintenance required for the irrigation application. The fixed cost (C_t) included the cost of pumping/delivery and irrigation system. It was assumed that the 7.5 HP pumping system can serve six ha area (three ha/season, two seasons/year) for 15 years as per USDA (1959). Also, it was assumed that the irrigation system could be useful for two seasons per year. The following expressions were used for assessing the economics.

$$C_t = C_c + C_i \text{ in which, } C_i = C_v + C_f$$

$$C_t = \frac{P1 \times i(1+i)^m}{2\times 3\{(1+i)^m \quad 1\}} + \frac{P2 \times i \times (1+i)^m}{2\{(1+i)^n - 1\}}$$

where,

C_t = Total cost of cultivation (kg/season/ha); C_c = Cost of cultivation excluding cost of irrigation(kg/ha/season); C_i = Cost of irrigation (kg/ha/season); C_f = Fixed cost of the irrigation system (kg/ha/season); C_v = Variable cost of irrigation water applications (kg/ha/season); P_1 = Cost of pumping and delivery system, kg; m =Life of the pumping system, years; P_2 = Cost of irrigation system, kg/ha; n = Life of the irrigation system, years; i = Prevailing rate of interest.

Results and Discussion

(A) Effect of drip Irrigation

(a) Pod and haulm yields

The experiment was carried out from summer 2010 to 2012. Three years pooled data on pod and haulm yields are presented in Table 1. The results indicated that alternate day drip irrigation at 100 per cent of cumulative pan evaporation (I_1) gave significantly the highest pod yield (2623 kg ha^{-1}) and haulm yield (4725 kg ha^{-1}) as compared to rest of treatments. The lowest pod yield (2231 kg $ha^{-1)}$ and haulm yield (4004 kg ha-1) was produced by drip irrigation at 60 per cent of cumulative pan evaporation (I_3). The magnitude of per cent pod yield reduction under I_2 and I_3 treatments of irrigation was to tune of 12.65 per cent and 14.94 per cent in pooled results. Similar findings were also reported by Patel *et al.* (1990), Ghetia *et al.* (1998) and Manavadaria *et al.* (2004).

(b) Yield attributes and quality characters

Irrigation treatments significantly influenced pods per plant, shelling per cent, test weight, oil per cent and oil yield (Table 1). The alternate day drip irrigation at 100 per cent of cumulative pan evaporation (I_1) recorded significantly highest pods per plant (19.40) shelling per cent (73.06 pre cent), 100 kernel higher (35.03 g), oil per cent (47.67 per cent) and oil yield (915 kg ha^{-1}) than rest of treatments. Similar results were obtained by Patel *et al.* (1990), Ghetia *et al.* (1998), Manavadaria*et al.* (2004) and Rathod and Trivedi (2011).

(B) Border strip irrigation method v/s drip levels

Significantly higher pod yield (2623 kg ha-1) of summer groundnut was recorded under alternate day drip irrigation at 100 per cent of cumulative pan evaporation (I_1) than border strip irrigation (I_4) at 1.0 IW/ CPE ratio (2221 kg ha^{-1}). The magnitude of increase in pod yield in drip irrigation at I_1 treatments over border strip irrigation at 1.0 IW/ CPE ratio (I_4) was to the tune of 18.0 per cent. Whereas maximum haulm yield, pods per plant, oil per cent and oil yield was also observed in I_1 treatment followed by I_2, I_3 and border strip irrigation treatments (I_4).This may be due to the fact that drip irrigation minimised the conventional losses that usually occur in surface method. These finding are in agreement with those of Sigte *et al.* (1997), Manavadaria *et al.* (2004) and Rathod and Trivedi (2011).

Table 1. Effects of Different Treatments on Yield Attributes, Yield and Water Use Efficiency of Groundnut (Pooled Data of Three Years)

Treatments	*No. of pods/ plant*	*Shel-ling (per cent)*	*100 kernel weight (g)*	*Pod yield (kg ha^{-1})*	*Haulm yield (kg ha^{-1})*	*Oil (per cent)*	*Oil yield (kgha^{-1})*	*Total water applied (mm)*	*Water use efficiency (kg-ha/mm)*
Main plot: Drip irrigation									
I_1: Alternate day drip irrigation at 100 per cent of CPE	19.40	73.06	35.03	2623	4725	47.67	915	675	3.89
I_2: Alternate day drip irrigation at 80 per cent of CPE	18.71	71.84	34.28	2291	4186	47.15	778	570	4.02
I_3: Alternate day drip irrigation at 60 per cent of CPE	16.42	71.40	33.89	2231	4004	46.80	747	445	5.01
CD (P=0.05)	0.70	0.94	0.54	292	503	0.36	111	-	-
Border strip irrigation v/s drip levels									
I_4: Border strip irrigation at 1.0 IW/ CPE ratio irrigation water depth (5 cm)	16.34	71.08	34.09	2221	3841	46.98	723	700	3.17
CD (P=0.05)	0.65	NS	NS	312	566	0.31	131	-	-
Sub plot: Plant geometry									
G_1: Conventional sowing (row to row distance 30 cm) with one lateral to each row	19.10	72.52	34.80	2670	4839	47.59	924	567	4.71
G_2: Conventional sowing (row to row distance 30 cm) with one lateral between two rows	18.61	72.49	34.42	2564	4560	47.10	858	567	4.52
G_3: Paired-rows (15/30 cm apart) with one median lateral to each pair	18.22	72.03	34.49	2270	4024	47.40	777	567	4.00
G_4: Paired-rows (15/30 cm apart) with one lateral between two pairs	16.44	71.35	33.90	2073	3799	46.74	695	567	3.66
CD (P = 0.05)	0.58	NS	NS	124	290	0.33	45	-	-
Main x Sub : CD (P = 0.05)	NS	NS	NS	NS	NS	NS	NS	-	-

(C) Effect of plant geometry

(a) Pod and haulm yield

The pod yield variation due to plant geometry was observed significant (Table 1). Significantly maximum pod yield of 2670 kg ha^{-1} was recorded by the 30 cm row spacing with one lateral to each row (G_1) which remained at par with 30 cm row spacing with one lateral between two rows (G_2). The results showed that 17.62 per cent and 28.80 per cent pod yield increased due to one lateral to each row (G_1) than paired rows with median one lateral to each pair (G_3) and one lateral between two pairs (G_4), while 12.95 per cent and 23.68 per cent increased pod yield in respect of one lateral between two rows (G_2). In case of haulm yield, G_1 treatment was found superior by producing significantly higher haulm yield of 4839 kg ha^{-1} and it was found at par with G_2 treatment. Lower pod yield in one lateral between two pairs (G_4) treatment might be due to wider lateral spacing and unequal distribution of water hence moisture stress condition was experienced by the outer rows of the crop. The results were in agreement with the finding of Lamm *et al.* (1992), Singte *et al.* (1997) and Manavadariya *et al.* (2004).

(b) Yield attributes and quality characters

Shelling per cent and 100 kernel weight were found non-significant due to plant geometry treatments. Significantly highest pods per plant (19.10), oil per cent (47.59 per cent) and oil yield (924 kg ha^{-1}) were recorded under conventional sowing (row to row : 30 cm) with one lateral to each row (G_1), which remained at par with one lateral between two rows (G_2) in respect of pods per plant and oil per cent.

Interaction between drip irrigation and plant geometry (I x G) remained unaffected with regards to yield attributes and yield.

(D) Water use efficiency and water saving

The highest water use efficiency of 5.01 kg ha^{-1} mm^{-1}, was obtained under alternate day drip irrigation at 60 per cent of cumulative pan evaporation (I_3) treatment (Table 1) followed by drip irrigation at 80 per cent of cumulative pan evaporation (4.02 kg ha^{-1} mm^{-1}, I_2), drip irrigation at 100 per cent of cumulative pan evaporation (3.89 kg ha^{-1} mm^{-1}, I_1) and border strip irrigation (3.17 kgha^{-1}mm^{-1}, I_4). In respect of plant geometry, maximum WUE was recorded by 30 cm row to row distance sowing with one lateral to each row (4.71 kg ha^{-1} mm^{-1},G_1). Dutta and Mondal (2006) also reported increase in moisture supply leading to increase in water use and reduction in water use efficiency of groundnut.

Maximum amount of water (700 mm) was applied in border strip irrigation (I_4) which was closely followed by drip irrigation I_1, I_2 and I_3. The average

Table 2: Effects of Different Treatments on Economics of Groundnut (Pooled Data of Three Years)

Treatments	*Gross realization (kg /ha)*	*Total cost of cultivation (kg /ha)*	*Net realization (kg /ha)*	*B:C ratio*
Main plot: Irrigation				
I_1:Alternate day drip irrigation at 100 per cent of CPE	103618	43593	60024	2.38
I_2: Alternate day drip irrigation at 80 per cent of CPE	90650	42993	47657	2.11
1_3: Alternate day drip irrigation at 60 per cent of CPE	88095	42393	45702	2.08
Border strip irrigation v/s drip levels				
I_4: Border strip irrigation at 1.0 IW/ CPE ratio irrigation water depth (5 cm)	87338	30063	57275	2.91
Sub plot: Plant Geometry				
G_1: Conventional sowing (row to row distance 30 cm) with one lateral to each row	105548	50784	54764	2.08
G_2: Conventional sowing (row to row distance 30 cm) with one lateral between two rows	101140	39739	61401	2.55
G_3: Paired-rows (15/30 cm apart) with one median lateral to each pair	89510	45985	43525	1.95
G_4: Paired-rows (15/30 cm apart) with one lateral between two pairs	82053	37384	44668	2.19

per cent of saving of water against border strip irrigation in I_1, I_2, and I_3 drip system were 3.57 per cent, 18.57 per cent and 35.0 per cent, respectively.

(D) Economics

The average of three years of different treatments is presented in Table 2. The results showed that irrigation cost the alternate day drip irrigation at 100 per cent of cumulative pan evaporation (I_1) gave maximum net return (60024 kg ha^{-1}) and B: C ratio (2.38). Among the different plant geometry treatments conventional sowing (row to row: 30 cm) with one lateral between two rows (G_2) recorded maximum net return (61401 kg ha^{-1}) with 2.55 B: C ratio. While, border strip method of irrigation recorded net return of 57275 kg ha^{-1} and B : C ratio of 2.91. Higher net return recorded under treatments conventional sowing (row to row: 30 cm) with one lateral between two rows winch might be due to high yield and wider lateral spacing. Manavadariya *et al.* (2004) and Rathod and Trivedi (2011) also observed similar results.

Conclusions

Under limited water resources optimum summer groundnut production along with water saving can be achieved by irrigating the crop through drip irrigation system at 100 per cent of CPE (lateral spacing 30 cm, dripper spacing 45 cm, dripper discharge 2 lph, operating time one hour and five minutes with pressure 1.2 kg/cm^2) as compared to border strip irrigation. Sowing the crop with 30 cm spacing with one lateral to between two rows resulted in maximum pod yield and net realisation.

References

Anonymous, (2012). Annual Report (*Kharif*). "Annual meeting of groundnut Researchers", Directorate of Groundnut Research, Govt. of India.

Dutta, D. and Mondal, S. S. (2006). Response of summer groundnut to moisture stress, organic manure and fertilizer with and without gypsum under lateritic soil of West Bengal. Indian Journal of Agronomy, 51(2):145-148.

Ghetia, N. R.; Khistaria, M. K.; Khanpara, V. D. and Malavia, D. D. (1998). Drip irrigation in summer groundnut. GAU Res. J. 23: 103-105.

Lamm, F. R.; Stone, L.R. and Monges, H. L. (1992). Optimum lateral spacing for drip irrigation corn. Paper American Society of Agricultural Engineers. No. 92-2575, p.16.

Manavadaria, S. M.; Modhavadia, M. M. and Khanpara, V.D. (2004). Efficient water management through drip irrigation in summer groundnut. *In:* National Symposium on Enhancing productivity of groundnut for sustaining food and nutrition security, Junagadh. Oct. 11-13, 2004, pp. 99-100.

Patel, B. G; Khanpara, V. D.; Malavia, D. D.; Maravia, R. B. and Kaneria, B. B. (1995). Economics feasibility of drip irrigation in garlic. Indian Journal. Agronomy, 40:143-145.

Radin, J. W., J. R. Mauney, and P. C. Kerridge. 1989. Water uptake by cotton roots during fruit filling in relation to irrigation frequency. Crop Science, 29: 1000 -1005.

Rank, H. D. (2007). Summer groundnut crop performance and economics under micro sprinkler irrigation at various water application levels. Legume Research, 30 (4):261-265.

Rathod, A .B. and Trivedi, S. A. (2011). Summer groundnut crop performance and economics under drip irrigation at various water application levels. *In*: National Conference on recent trends in engineering and technology. 13-14 May 2011, pp. 1-2.

Single, M. B.;Yamger, V. T.; Kathmalc, D. K. and Gaikwad, D. T. (1997). Growth, productivity and water use of turmeric (*curcuma longa*) under drip irrigation. Indian Journal Agronomy, 42:547-549.

USDA, (1959). SCS National Engg. Handbook, Washington, D.C., pp. 8-1 to 8.15.

20

Growth and Yield of Fenugreek (*Trigonella foenum-graecum* L.) as Influenced by Irrigation Levels and Sowing Methods

P. J. Prajapati, R. M. Solanki, J. B. Vasave and D. V. Srinivasulu

Abstract

An experiment was conducted during rabi *season of 2011-12 at Instructional Farm, Department of Agronomy, Junagadh Agriculture University, Junagadh (Gujarat) to study the growth and yield of fenugreek (*trigonella foenum-graecum L.*) as influenced by irrigation levels and sowing methods. Experiment was laid out in split plot design with four irrigation levels based on IW/CPE ratio were assigned to main plot and two sowing method in sub-plot with four replications. Result revealed that almost all the growth and yield attributes, grain and straw yields were significantly higher when crop was irrigated at an IW/CPE ratio of 1.0 followed by 0.8 IW/CPE ratios. Numerically higher gross and net realisation as well as B:C ratio were also observed under IW/CPE ratio of 1.0 followed by 0.8 IW/CPE ratio. Sowing the crop at 30 cm row spacing recorded higher values of almost all the growth and yield attributes, yields, gross and net realisation as well as BCR. More than 65 per cent moisture was extracted from upper 30 cm soil depth irrespective of IW/CPE ratio and method of sowing. Water use efficiency observed maximum when crop was irrigated at 0.4 IW/CPE ratio followed by 0.6 and 0.8 IW/CPE ratio.*

***Key words**: Fenugreek, Irrigation, Spacing, Water use efficiency, Economics.*

Introduction

Fenugreek (*Trigonella foenum-graecum* L.) commonly known as *methi* is an important seed spices grown in arid and semi arid region of India during *rabi* season and cultivated in an area of 82,900 hectares with production of 1.25 lakh tonnes of seed in the country (Anonymous, 2012). In India, it is mainly grown in Rajasthan, Gujarat, Madhya Pradesh, Maharashtra, Haryana, Punjab and Karnataka. Water is a major factor for every metabolic activity of plant and limited quantity of water is available for irrigation, which calls for scheduling of irrigation to improve water productivity of fenugreek. Recently irrigation is being scheduled on the basis of climatological approach which now consider as most scientific, since it indicate all whether parameters in a given climate plant continuum giving them natural weighted (Pariher *et. al,* 1976). Maintenance of optimum plant population is essential for interception of solar radiation without exerting competition for nutrient and water in plants. Precise information regarding appropriate scheduling of irrigation based on IW/CPE ratio and method of sowing for fenugreek is meagre. Keeping in view the above considerations this research programme was planned to study the judicious use of irrigation water which may provide information not only on water use by crop but also when to irrigate and to find suitable method of sowing for fenugreek. Hence there is an urgent need to generate precise information on irrigation requirement and method of sowing.

Materials and Methods

A field experiment was conducted at Instructional Farm, College of Agriculture, JAU, Junagadh (Gujarat) during *rabi* season of 2011-12. The soil of the experimental field was clayey in texture having field capacity moisture of 28.88 per cent, PWP was 12.86 per cent and soil moisture before sowing was 22.15 per cent, pH 7.9 Ec 0.33 dS m^{-1}. The soil in experimental site was medium in available nitrogen (267.0 kg ha^{-1}), P_2O_5 (38.3 kg ha^{-1}), K_2O (232.0 kg ha^{-1}) and rich in organic carbon (0.83 per cent). The experiment was laid out in split plot design with four replication. Keeping four levels of irrigation [0.4, 0.6, 0.8 and 1.0 irrigation water (IW) cumulative pan evaporation (CPE) ratio] in main plot and two sowing method [M_1= line sowing at 30 cm and M_2= broadcasting] in sub-plots. The fenugreek (GM-2) was sown in first week of November-2011 as per treatments using a seed rate of 15 kg ha^{-1}. A full dose of 20 kg nitrogen and 40 kg phosphorus per hectare was drilled manually through DAP and urea as a basal at sowing. Two common irrigations each of 50 mm depth were applied, first immediately after sowing and second at 5 DAS for better germination and establishment of seedlings. Rest of the irrigations were applied as per treatments. A total of 1400 mm water was measured by parshall flume, installed in the field channel and total of 5, 6, 8 and 9 irrigations of 5.0 cm each at IW/CPE ratio of 0.4, 0.6, 0.8 and 1.0 respectively were prepared cumulative pan evaporation was taken as the sum of the daily pan evaporation from USWB class-A evaporimeter. The soil samples collected from five different depths *viz.,* 0-15, 15-30, 30-45, 45-60 and 60-75 cm were taken before irrigation and 48 hours after irrigation and cumulative pan evaporation (CPE) values were calculated from daily pan evaporation measured with the help of USWB Class-A open pan evaporimeter installed at the meteorological observatory, Instructional Farm, College of Agriculture, Junagadh Agricultural University, Junagadh which was in the proximity of the experimental plot. The samples were weighed and oven dried for a day at 105°C. Other packages of practises were followed as per recommendation made for the crop in the region. The crop was harvested on 5th March, 2012.

Results and discussion

Effect of irrigation

Significantly higher plant height, dry matter accumulation, number of branches per plant, pods per plant, pod length, number of seeds per pod, test weight, seed and straw yields were recorded at IW/CPE ratio of 1.0 than rest of the treatments but it remained at par with IW/CPE ratio of 0.8 (Table-1). While in case of water use efficiency it was maximum (6.46 kg ha^{-1} mm at 0.4 IW/CPE ratio (Table-2). Significantly higher seed and straw yields (2031 and 3551 kg ha^{-1}), were produced at higher IW/CPE ratio 1.0 it might be due to frequent irrigation at shorter interval under this treatment which maintain of optimum moisture level in the soil during the entire growth period resulted in better translocation of photosynthates from source to sink which in turn resulted higher yield attributes and yield of fenugreek. These findings are in close agreement with studies of Dutta and Chatterjee (2006).

Table 2. Effect of irrigation levels and sowing method on water use, WUE and economics of fenugreek

Treatments	*Total no. of irrigation applied*	*Total quantity of irrigation water applied (mm)*	*WUE (kg/ha^{-1} mm)*	*Realisation (Rs ha^{-1})*		*BCR*
				Gross	*Net*	
Irrigation (IW/CPE ratio)						
I_1: 0.4	5	250	6.46	42571	16337	1.62
I_2: 0.6	6	300	5.80	45816	18972	1.71
I_3: 0.8	8	400	4.95	52274	24210	1.86
I_4: 1.0	9	450	4.51	53566	24892	1.87
Sowing method						
M1: 30 Line sowing	-	-	5.85	52564	25980	1.98
M2: Broadcasting	-	-	5.01	44563	16239	1.57

Effect of sowing method

The higher values of growth and yield attributed; seed and straw yields were observed when crop was sown at 30 cm row spacing (Table 1). Significantly the maximum seed yield (1996 kg ha^{-1}) was produced under line sown crop at

Table 1. Effect of irrigation levels and sowing method on growth, yield attributes and yields of fenugreek

Treatments	*Plant height (cm)*	*Branches/plant*	*Dry matter (g)*	*Pods/plant*	*Pod length (cm)*	*Seeds/pod*	*Test weight (g)*	*Protein yield (kg ha^{-1})*	*Yield (kg ha^{-1})*	
									Seed	*Straw*
Irrigation (IW/CPE ratio)										
I_1: 0.4	42.6	5.9	14.1	18.1	6.2	10.9	12.3	351	1615	2777
I_2: 0.6	45.1	6.7	15.7	19.9	7.0	11.2	12.8	385	1740	2892
I_3: 0.8	50.5	7.2	21.6	23.1	7.8	12.8	15.2	451	1981	3517
I_4: 1.0	52.9	7.8	22.9	24.9	8.4	13.4	15.3	477	2031	3551
S. Em ±	1.54	0.22	0.84	0.66	0.20	0.23	0.20	20.0	83.0	174.5
C.D. at 5 per cent	4.95	0.71	2.6	2.11	0.66	0.76	0.65	64.0	265.7	558.2
C.V. per cent	9.1	9.1	12.8	8.7	8.0	5.6	4.1	13.6	12.7	15.5
Sowing method										
M1: 30 Line sowing	50.1	7.2	19.5	22.6	7.8	12.4	14.1	457	1996	3331
M2: Broadcasting	45.5	6.6	17.6	20.4	6.9	11.8	13.7	375	1688	3037
S. Em ±	0.82	0.13	0.45	0.30	0.11	0.14	0.09	10.2	42.2	51.7
C.D. at 5 per cent	2.54	0.41	1.3	0.93	0.35	0.44	0.28	31.5	130.1	159.6
C.V. per cent	6.9	7.8	9.8	5.6	6.3	4.7	2.6	9.8	9.1	6.5
Interaction (I x M)	NS	Sig.	NS	NS	NS	Sig.	Sig.	Sig.	Sig.	Sig.

30 cm row spacing which was 18.2 per cent higher than broadcast method of sowing. The probable reason for higher yield under line sowing might be due to intra-space competition and competitive interactions among the plant can be avoided by providing suitable stand geometry by which all the strata of leaves can get copious sunlight and carry out effective photosynthesis and consequently higher dry matter and yield (Duncan, 1977). Yadav *et al.* (2000) also reported higher seed yield of fenugreek under 30 cm row spacing.

Interaction effect of irrigation and sowing method

Data of one year revealed presence of significant interaction of irrigation (I) and sowing method (M). Significantly for the higher no. of branches per plant, seeds per pod, test weight, protein yield as well as seed and straw yields were recorded when crop was irrigated at 1.0 IW/CPE ratio along with 30 cm row spacing (Table-3). Treatment combination I_4M_1 recorded significantly higher seed yield than rest of the combinations, but it remained at par with I_3M_1 combinations (irrigation at 0.8 IW/CPE ratio along with 30cm row spacing). Irrigating the crop at an IW/CPE ratio 1.0 and sowing the crop at 30 cm row spacing significantly increased yields which might be due to the contribution of more pods, pod length and seeds per pod as evidenced by test weight and grain yield per plant resulting from more nutrition to plants. These findings are corroborating with the reports of Davut Karaaslan (2007) in sesame.

Table 3. Interaction effects of irrigation and sowing method on growth, yield attributes, seed and straw yield

	Branches/plant				*Seeds/pod*				*Test weight (g)*			
	I1	*I2*	*I3*	*I4*	*I1*	*I2*	*I3*	*I4*	*I1*	*I2*	*I3*	*I4*
M1	6.1	6.5	7.6	8.4	11.1	11.0	13.4	14.0	12.4	12.8	15.4	15.9
M2	5.7	6.9	6.7	7.1	10.7	11.4	12.2	12.7	12.3	12.7	15.1	14.7
C.D. at 5%	0.83				0.88				0.57			
	Protein yield (kg ha^{-1})				*Seed yield (k gha^{-1})*				*Straw yield (kg ha^{-1})*			
	I1	*I2*	*I3*	*I4*	*I1*	*I2*	*I3*	*I4*	*I1*	*I2*	*I3*	*I4*
M1	376	394	493	565	1729	1792	2108	2354	2913	2933	3583	3896
M2	326	376	410	389	1500	1688	1854	1708	2642	2850	3450	3206
C.D. at 5%	63.0				260.3				329.2			

Irrigation level- I_1=0.4 IW/CPE ratio, I_2=0.6 IW/CPE ratio, I_3=0.8 IW/CPE ratio, I_4= 1.0 IW/CPE ratio;

Sowing method- M_1= 30 cm line sowing, M_2=Broadcasting

Moisture extraction pattern

The maximum amount of moisture was extracted from 0-15 cm soil depth irrespective of irrigation and it was decreased with soil depth (Fig. 1). About

60.91 per cent soil moisture was extracted from the top 30 cm of soil depth under 0.4 IW/CPE ratio but in case of 1.0 IW/CPE ratio more than 65.57 per cent soil moisture was extracted at the same depth. This indicated that irrigation levels had promi-nent effect on soil moisture extraction pattern.

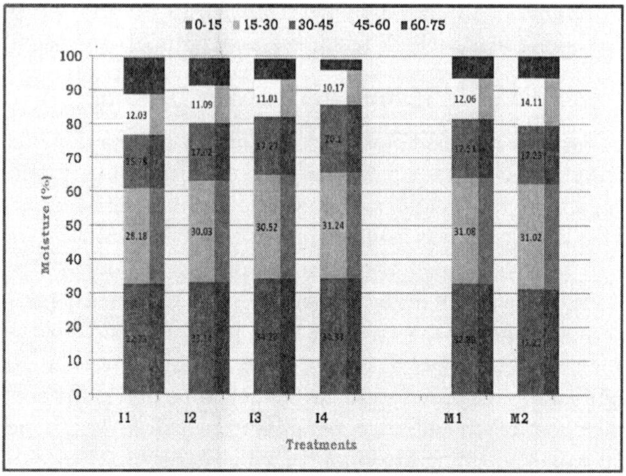

Fig. 1. Moisture extraction pattern of fenugreek under different IW/CPE ratio and sowing method.

Soil moisture extraction increased from 15 cm soil depth increased with increasing irrigation level. The higher amount of moisture extraction from surface layer under different irrigation levels might be due to increased surface evaporation, shallow root density and higher water uptake due to availability of irrigation water. These findings are also corroborate with the findings of Dutta and Chatarjee (2006). In lower level of irrigation treatments, moisture extraction was higher from deeper (30-60 cm) soil layer. In this condition moisture stress compelled the roots to go into deeper soil layer for utilisation of water for their proper growth and development. In case of sowing method maximum moisture extraction of 63.97 per cent was noted in line sowing followed by broadcast method of sowing with 62.24 per cent, respectively.

Economics

The results presented in Table 2 on economics of different irrigation levels indicated that an application of irrigation water at 1.0 IW/CPE ratio recorded highest realisation (gross Rs 53566 and net Rs 24892 as well as B: C ratio 1.87 per hectare) due to more yield followed by 0.8 IW/CPE ratio. In case of sowing method highest gross and net realisation as well as B: C ratio recorded under 30 cm lines sowing Rs 52564, Rs 25980 as well as B : C ratio of 1.98 per hectare, respectively.

Conclusion

The study thus indicated that application of irrigation at IW/CPE ratio 1.0 with 30 cm row spacing is better for getting higher yield of fenugreek and water use efficiency.

References

Anonymous, 2012. Department of Agriculture and Cooperation, Ministry of Agriculture, Government of India. *http:// agricoop.nic.in/agristatistics. htm.*

Davut, Karaaslan; Erkan, Boydak; Sinan, Gercek and Mehmet, Simsek. 2007. Influence of Irrigation Intervals and Row Spacing on Some Yield Components of Sesame Grown in Harran Region. Asian Journal of Plant Sciences, 6(4): 623-627.

Duncan, W. G. 1977. Cultural manipulation for higher yields in "Physiological Aspects of Crop Yield" Am. Soc. Agron. and Crop Sci., America (USA). pp. 327-339.

Dutta, S. and Chatarjee, R. 2006. Effect of irrigation regimes on moisture extraction pattern, evaporation and yield of fenugreek *(Trigonella foenum-graecum* L). Journal of Spices and Aromatic Crops, 15(2):125–129.

Parihar, S. S.; Khera, K. L.; Sandhu, K. S. and Sandgu B. S. 1976. Comparision of irrigation schedule based on pan evaporation and growth stages in wheat. Agronomy Journal, 68:650-653.

Yadav, J.S.; Jagdev, S.; Virender, K.; Yadav, B.D.; Singh, J. and Kumar, V. 2000. Effect of sowing time, spacing and seed rate on seed yield of fenugreek (*Trigonella foenum-graecum* L.) on light textured soil. Haryana Agri. Univ. Res. 30(3-4): 107-111.

21

Hydraulic Performance of Drip Emitters under Field Condition

*F.K. Savani**, J.M. Hemani, H.H. Mashru, K. C. Patel and G. V. Prajapati*

Abstract

Gujarat, which is one of the water starving States, receives an annual rainfall of 625 mm only. In Gujarat, the total irrigated area is 3.7 M ha. against the total cultivated area of 6.5 M ha. Therefore, it is necessary to economise the use of water in agriculture to bring more area under irrigation, reduce the cost of irrigation per hectare and increase the productivity per unit area and unit quantum of water. Irrigation performance assessment is of high priority in irrigation project. A field study was undertaken to evaluate the performance of on-line non pressure compensating emitters of 2, 4-8 lph discharge ratings and inline non-pressure compensating emitter of 2 lph capacity. The system was tested for its uniformity coefficient, emission uniformity, manufacturing coefficient of variation and head-discharge relationship. The study revealed that for better discharge of on-line emitters of 2, 4 and 8 lph capacity, optimal pressure of 0.7, 0.7 and 1 kg/cm^2 is required respectively to achieve uniformity coefficient more than 90 per cent. In 2, 4 and 8 lph capacity emitters, emission was 89.50, 91.12 and 86.64 per cent respectively. For inline emitter optimal pressure 0.6 kg/cm^2 is required to achieve uniformity coefficient of 95.09 per cent and emission was 88.51 per cent respectively.

Keywords: *Discharge, pressure, uniformity coefficient, emission uniformity.*

Introduction

Drip irrigation is referred to as a regulated and slow application of water for longer period through evenly spaced emitters at frequent intervals of time near the root zone of the plants. Drip system covered the area is about 3.5 lakh in India. Drip irrigation is gaining popularity and momentum over the past few years as it claims lot of advantages over conventional surface irrigation methods, the superiority of drip irrigation system lies in saving of water and increasing the water use efficiency. Drip system permits the controlling of discharge and flexibility in time of water application. In drip system, 30-70 per cent water is saved without significantly affecting the crop yield (Satpute and Pandey, 1989 and Pandey *et. al.,* 2003). Drip irrigation systems are mostly used for irrigating orchards, vegetables, spices and cash crops like sugarcane and cotton (Kumar, 2001). Emitters are used for precise delivery of water in drip irrigation system. The uniformity of water application from a micro irrigation system is affected both by water pressure distribution in the pipe network and the hydraulic properties of the emitters used. The emitter hydraulic properties depend on emitter design, water quality, water temperature and other factors on emitter flow rate. The capacity of the emitters available in the market varies from 2 to 16 lph. There are two types of emitters: pressure and non-pressure compensating. The former show no variation in discharge due to the corresponding change in the pressure head but in the latter the discharge changes with pressure. Little scientific information is available on the flow characteristics of different emitters under operating pressure. Keeping this in view a field test was done to evaluate the hydraulics of on-line, non-pressure compensating emitters of different discharges ratings and inline non-pressure compensating emitters.

Materials and Methods

The study was conducted in a Instructional farm, Centre of Excellence on Soil and Water Management, Research, Testing and Training Centre, Junagadh Agricultural University, Junagadh. The drip irrigation system was installed in the field. The field lies at Longitude 70° 27' E and latitude 21° 30' N. The system was tested for its uniformity coefficient, emission uniformity and manufacturing coefficient of variation. The water supply for the system was drawn from an open well located in the nearby area. A filtration unit was used to remove the fine particles from water. The system were operated under different pressure i.e. 0.20, 0.30, 0. 40, 0.50, 0.60, 0.70, 0.80, 0.90 and 1.0 kg / sq.cm. for one hours and discharge through each drippers were collected in catch cans and measured it.

Uniformity Coefficient

Uniformity coefficients of emitters were tested using the Christiansen's formula (1942). It gives the information that how efficiently water is distributed in the field.

$$C_u = \left(1 - \left(\frac{\frac{1}{n}\sum_{i=1}^{n} | q_i - q_{ave} |}{q_{ave}}\right)\right)100 \qquad ...(1)$$

where,

C_u = Uniformity coefficient (per cent)

q_i = Emitter discharge

n = Number of observed discharge

q_{ave} = Average of discharge values

Emission Uniformity

Another measure of emitter uniformity (EU) is typically used to evaluate manufacturing quality of emitters. The EU is the ratio between the average discharge in the quarter receiving less water and average discharge at the system level. It is used to describe the predicted emitter flow variation along a lateral line and can be assumed as synonymous to that of distribution uniformity (DU).The following formula was used to calculated Emission Uniformity (Keller and Karmeli, 1974)

$$EU = 100\left(1 - 1.27 C_v n^{-\frac{1}{2}}\right)\left(\frac{q_{min}}{q_{ave}}\right) \qquad ...(2)$$

where,

EU = Emission Uniformity

C_v = Manufacturer's coefficient of variation

n = Number of emitter per plant

q_{min} = Minimum emitter discharge rate

q_{ave} = Average of discharge rate

Head – Discharge relationship

The head discharge relationship for emitters was expressed by the formula (Karmeli, 1977 and Wu and Gitlin, 1977).

$$Q = K_d H^x \qquad ...(3)$$

where,

Q = Discharge rate of dripper (lph)

K_d = Discharge coefficient

H = Pressure head

x = Dripper flow exponent

Results and Discussion

Pressure Discharge Relationship: Emitter discharge was measured at different operating pressure. Four replications for the 2 lph emitters were taken. Data revealed that for a average discharge of 2 lph, the emitter discharge varied from 1.28 to 2.43 lph. The closest to the average discharge i.e. 2.01 lph was obtained at 0.70 kg/cm^2 and the discharge variation was only 0.01. Similar was the case with the emitters of the other discharge ratings as shown in Table 3 above. The variation in the discharge of the emitters was within the acceptable limit for 2, 4 and 8 lph capacity emitter (Wu and Gitlin,1981). For the better performance, emitters of 2, 4 and 8 lph capacity require 0.70, 0.70 and 0.80 kg/cm^2 optimum pressure, respectively. Similarly the inline emitter of 2 lph was operated at different pressures and the corresponding discharge was observed. Four replications for the 2 lph inline emitter were taken. Data revealed that for a average discharge of 2 lph, the emitter discharge varied from 1.05 to 2.44 lph. The closest to the average discharge i.e. 1.98 lph was obtained at 0.60 kg/cm^2 and the discharge variation was only 0.02. Thus, the optimum pressure at which the inline dripper gives the rated discharge was 0.60 kg/cm^2.

Uniformity Coefficient: Table 3 shows the value of Uniformity coefficient of 2 lph online dripper for the four replications. The average value of Cu ranges from 93.36 per cent to 95.80 per cent. As per ASAE (1989) recommendation, it was concluded that the emitters were of good quality. The maximum uniformity coefficient is 95.80 per cent at 90 kPa pressure in 2 lph capacity emitter and for 2 lph inline emitter, the maximum uniformity coefficient is 95.50 per cent.

Emission Uniformity: The maximum emission uniformity is 90.03 per cent at 90 kPa pressure for 2 lph online emitter. For 4 and 8 lph online emitters, the emission uniformity are 95.14 per cent and 89.98 per cent respectively at 100 kPa. For 2 lph inline emitter the emission uniformity is 89.28 per cent at 100 kPa.

Table 1: Average Discharge of Online Drippers

Pressure (kg/cm^2)	q_{ave}		
	2 lph	*4 lph*	*8 lph*
0.20	1.28	2.63	7.25
0.30	1.45	3.25	7.35
0.40	1.56	3.46	7.50
0.60	1.78	3.71	7.71
0.70	2.01	4.00	7.85
0.80	2.16	4.07	8.06
0.90	2.30	4.12	8.03
1.00	2.43	4.18	8.00

Table 2: Average Discharge of 2 lph Inline Drippers

Pressure (kg/cm²)	q_{ave}
0.20	1.05
0.30	1.21
0.40	1.36
0.60	1.98
0.70	2.10
0.80	2.19
0.90	2.31
1.00	2.44

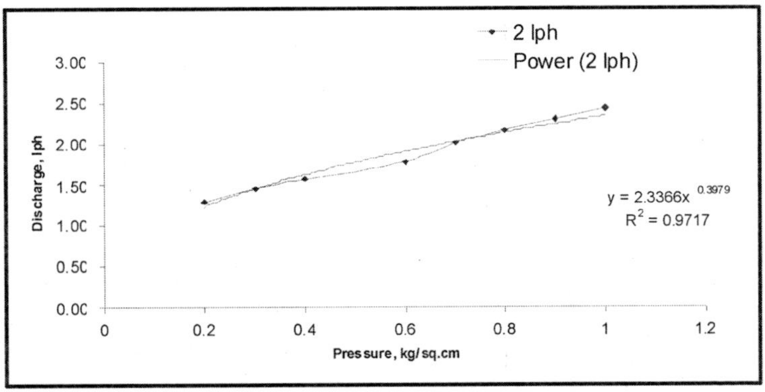

Fig. 1: Pressure Discharge Relationship of 2 lph Online Drippers.

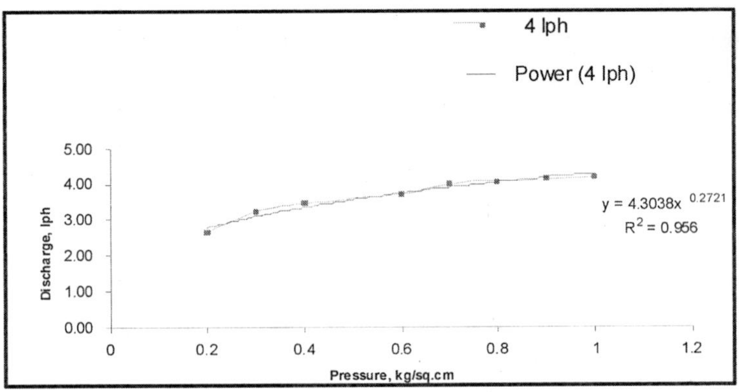

Fig. 2: Pressure Discharge Relationship of 4 lph Online Drippers.

Summary and Conclusion

Study revealed that for better discharges of online emitters of 2, 4 and 8 lph rated discharge, operating pressure of 70, 70 and 100 kPa rated discharge and inline emitter of 2 lph rated discharge operating pressure of 70 kPa is required, respectively to achieve uniformity coefficient of more than 80 per cent. Emission uniformity were 88.11 per cent, 88.33 per cent and 82.88 per cent for 2, 4 and 8 lph online emitters and 86.46 per cent for 2 lph inline-drip emitter respectively. According to ASAE (1989) standards, the emitters were of good quality.

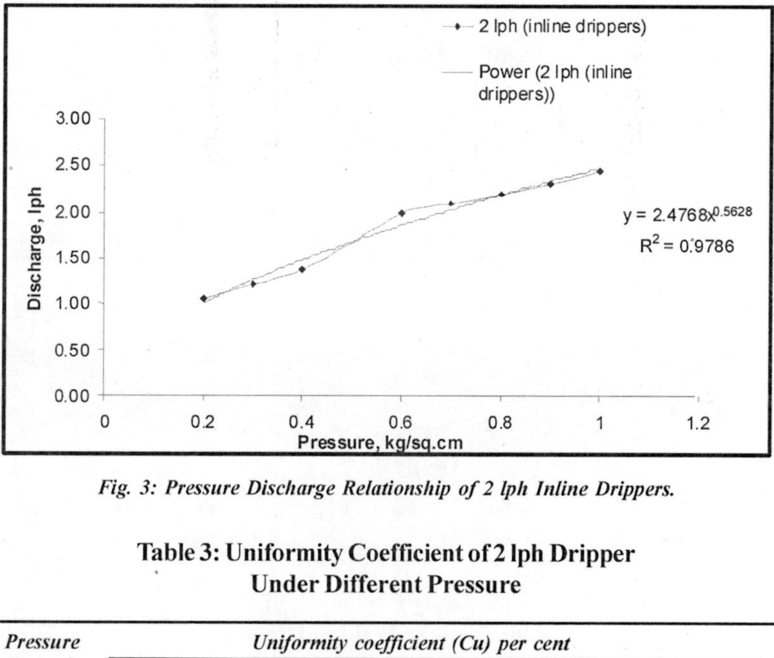

Fig. 3: Pressure Discharge Relationship of 2 lph Inline Drippers.

Table 3: Uniformity Coefficient of 2 lph Dripper Under Different Pressure

Pressure (kg/cm²)	*Uniformity coefficient (Cu) per cent*						
	R1	*R2*	*R3*	*R4*	*Min.*	*Max.*	*Average*
0.20	92.92	94.25	94.04	92.23	92.23	94.25	93.36
0.30	94.52	94.92	95.03	93.53	93.53	95.03	94.50
0.40	94.57	94.77	95.45	93.98	93.98	95.45	94.69
0.60	93.35	95.55	95.82	94.07	93.35	95.82	94.69
0.70	95.23	94.98	95.54	95.32	94.98	95.54	95.27
0.80	95.50	95.84	95.54	95.97	95.50	95.97	95.71
0.90	95.83	95.76	95.97	95.65	95.65	95.97	95.80
1.00	95.53	94.98	95.78	95.60	94.98	95.78	95.48

Table 4: Uniformity Coefficient of 4 lph Dripper Under Different Pressure

Pressure (kg/cm²)	*Uniformity coefficient (Cu) per cent*						
	R1	*R2*	*R3*	*R4*	*Min.*	*Max.*	*Average*
0.20	95.65	95.35	95.03	95.30	95.03	95.65	95.34
0.30	95.95	95.94	95.62	94.93	94.93	95.95	95.61
0.40	96.62	95.72	95.64	95.76	95.64	96.62	95.94
0.60	96.65	95.87	94.97	95.53	94.97	96.65	95.76
0.70	95.83	96.08	95.95	95.58	95.58	96.08	95.86
0.80	95.94	95.91	95.87	95.46	95.46	95.94	95.79
0.90	95.93	95.82	95.71	95.92	95.71	95.93	95.84
1.00	95.68	95.62	95.51	95.93	95.51	95.93	95.68

Table 5: Uniformity Coefficient of 8 lph Dripper Under Different Pressure

Pressure (kg/cm²)	*Uniformity coefficient (Cu) per cent*						
	R1	*R2*	*R3*	*R4*	*Min.*	*Max.*	*Average*
0.20	95.52	93.75	94.44	93.32	93.32	95.52	94.26
0.30	93.80	95.95	94.64	94.22	93.80	95.95	94.65
0.40	93.22	93.61	94.78	95.19	93.22	95.19	94.20
0.60	94.91	93.54	95.50	95.35	93.54	95.50	94.82
0.70	95.53	95.52	93.90	95.23	93.90	95.53	95.05
0.80	95.76	95.67	95.87	95.95	95.67	95.95	95.81
0.90	95.71	94.90	95.50	95.60	94.90	95.71	95.43
1.00	95.42	95.73	95.79	95.71	95.42	95.79	95.66

Table 6: Uniformity Coefficient of 2 lph Inline Dripper Under Different Pressure

Pressure (kg/cm²)	*Uniformity coefficient (Cu) per cent*						
	R1	*R2*	*R3*	*R4*	*Min.*	*Max.*	*Average*
0.20	93.10	92.69	93.16	92.21	92.21	93.16	92.79
0.30	93.96	94.34	93.30	90.79	90.79	94.34	93.10
0.40	92.83	93.27	91.62	92.80	91.62	93.27	92.63
0.60	95.75	95.28	95.00	94.34	94.34	95.75	95.09
0.70	93.80	95.68	95.51	94.37	93.80	95.68	94.84
0.80	93.17	94.13	95.67	95.31	93.17	95.67	94.57
0.90	94.60	93.53	95.94	95.73	93.53	95.94	94.95
1.00	96.80	94.31	95.24	95.63	94.31	96.80	95.50

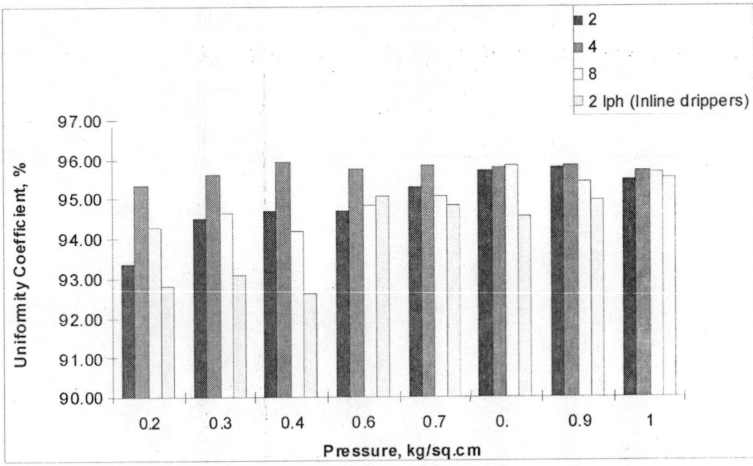

Fig.4: Uniformity Coefficient of 2, 4, 8 lph Online Dripper and 2 lph Inline Dripper.

Table 7: Emission Uniformity of 2, 4 and 8 Lph Online Dripper Under Different Pressure

Pressure (kg/cm²)	*Emission uniformity (per cent)*		
	2 lph	*4 lph*	*8 lph*
0.20	84.14	89.25	77.96
0.30	87.33	80.13	79.63
0.40	88.00	82.44	78.65
0.60	87.61	84.69	79.01
0.70	89.50	91.12	83.23
0.80	89.36	91.24	86.64
0.90	90.03	92.60	87.96
1.00	88.93	95.14	89.98

Table 8: Emission Uniformity of 2 lph Inline Dripper Under Different Pressure

Pressure (kg/cm²)	*Emission Uniformity (per cent)*
0.20	83.32
0.30	84.51
0.40	83.41
0.60	88.51

(Contd.)

0.70	87.23
0.80	87.64
0.90	87.78
1.00	89.28

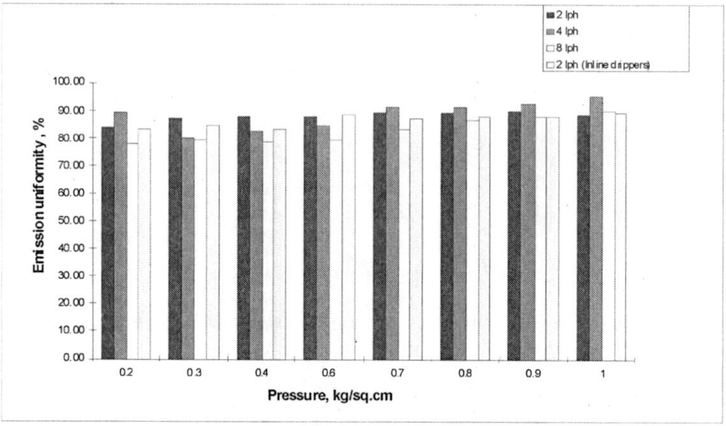

Fig.5: Emission Uniformity of 2, 4, 8 lph Online Drippers and 2 lph Inline Drippers.

Reference

ASAE, 1989. American Society of Agricultural Engineers. "Design, installation and performance of trickle systems".

Karmeli, D. 1975. "Trickle irrigation design", Ed and publ. by Rain Bird Sprinkler Manufacturing Crop. Glendora, CA91740.

Keller , J. and D. Karmeli. 1974. Trickle irrigation design parameters. Transactions of the ASAE, Vol. 17, pp. 678-684.

Kumar, A. 2001. "State of art and research and development needs in micro irrigation".

Satpute, G.V. and Pandey, N.N. 1988. Effect of drip layout and planting geometry of tomato and cost of drip system. Proc Agri. Engg. Conference, Bangkok, Thailand; AIT(1992), pp. 773-781.

Wu, I.P. and H.M. Gitlin, 1981. Preliminary concept of a drip irrigation network design. Trans. ASAE, Vol. 24(2), pp. 330-337.

22

Effect of Sea Water Irrigation and Sowing Time on Seed Yield and Component of Salicornia (*S. brachiata* Roxb.)

Bhoomika B. Patel, J. M. Patel and B. K. Patel

Abstract

Climate change enhances the intrusion / inundation of sea water in coastal areas under South Gujarat conditions. Under such condition seawater agriculture is the strategy to link ecology and livelihood. Production alternatives include development of salt-tolerant plant that already exists in salt flat areas. Salicornia *is a genus of succulent, halophyte (salt tolerant) plants that grow in salt marshes, on beaches locally known as* 'Merchar'. *Because of excessive exploitation of* Salicornia *by the local it is at the zenith of extinction. A field experiment was conducted during* kharif-rabi *season of 2012-13 on coastal salt affected soils of Coastal Soil Salinity Research Station, NAU, Danti-Umbharat in split plot design with four replications. The three dates of sowing* viz., *D_1: 1st week of June, D_2: 3rd week of June and D_3: 1st week of July were allotted to main plot and three levels of irrigation* i.e., *I_1: 0.6 IW/CPE, I_2: 0.8 IW/CPE and 1.0 IW/CPE in sub-plot. On overall basis, almost all the growth parameters were found to increase with the delay in sowing* i.e., *3rd week of June or 1st week of July. This was also true for the fresh biomass and seed yield also. In case of sea water irrigation regimes, irrigation at 0.8 IW/CPE ratio out yielded rest of the irrigation regimes. The interaction effect was also significant on fresh biomass yield of* salicornia. *Treatment combinations involving 0.8 IW/CPE ratio with sowing date D_3 produced significantly higher fresh biomass (34.24 t/ha) as compared to early sowing (23.78 and 31.12 t/ha). In the case of seed yield late sowing and frequent irrigation* *had an edge over early sowing with lower levels of irrigation. For getting higher biomass production and seed yield of* salicornia *under South Gujarat conditions, it is essential to sow the crop during either 3rd week of June or 1st week of July and irrigating should be scheduled at 0.8 IW/CPE ratio.*

Key word: *Salicornia, seawater, irrigation, date of sowing.*

Introduction

Land is a limiting entity and the arable land productivity is not sufficient to withstand the growing population demand. We need production alternatives to sustain the demand. Production alternatives include development of salt-tolerant plant that already exists in salt flat areas (Ungar, 1987). There is abundance of exotic species of plants that can be domesticated in stress areas. One of the exotic species that can articulate in saline ingresses salt marsh is *salicornia*. An obligate halophyte *salicornia*, is inhabited to salt flats marshes. *Salicornia* is a genus of succulent, halophyte (salt tolerant) plants that grow in salt marshes, on beach sand among mangroves. *Salicornia* species are native to North America, Europe, South Africa and South Asia. Common names for the genus include glasswort, pickleweed and marsh samphire; these common names are also used for species not in *salicornia*. The *salicornia* species are small, usually less than 30 cm tall, succulent herbs with a jointed horizontal main stem and erect lateral branches. The leaves are small and scale-like and as such the plant may appear leafless. *Salicornia* species can generally tolerate immersion in salt water. *Salicornia* is efficient halophytes to be cultivated in saline and sodic soils of the coastal semi-arid regions of Gujarat. *Salicornia* (*marcher*) naturally growing in coastal areas of South Gujarat. Agronomical practices *viz.*, method of sowing and land preparation for this crop was standardised but information about date of sowing is not available. Gujarat having the longest sea coast in India needs imperative concern. The information regarding sowing time of halophyte production with sea water will be a rejuvenation of naturally occurring edible halophyte in its most suitable habitat as well as it can be a most socio-economically viable option for the farmers of the coastal areas.

Field experiment was conducted at the Coastal Soil Salinity Research Station, Danti-Umbharat of Navsari in the South of Gujarat near the Arabian Sea during 2011-12. Geographically, Danti- Umbharat is situated at 20°83' N latitude and 72°50' E longitudes with an elevation of 2.5 meters above sea level on the western coastal belt of India. Genesis of the Danti soil is Onjal-1 and Dandi series are sub-grouped as *Typic Halaquepts* while Att series is *Fluventic Halaquepts*. Seawater from the Arabian Sea was applied as irrigation water source. The seeds were supplied from Coastal Soil Salinity Research Station, NAU, Danti-Umbharat, Navsari. Healthy seeds of *Salicornia brachiata* Roxb, obtained from the previous year cultivation was selected. The recommended dose 250:75:50 kg NPKha^{-1} was applied. N was applied in three splits *i.e.*, 50 per cent as basal, 25 per cent at 60 days after sowing and 25 per cent at 120 days after sowing. The seed rate was 2.5 kg ha^{-1}. A field experiment with nine treatment combinations comprising of three levels each of date of sowing as main plot treatment (D_{-1}: 1st week of June, D_2: 3rd week of June and D_3: 1st week of July) and surface irrigation using sea water as sub-plot treatment (I_1: 0.6 IW/CPE, I_2: 0.8 IW/CPE and I_3: 1.IW/CPE) was conducted in split plot design with four replications during *kharif-rabi* season of 2011-12. The biometric observations, yield, oil content etc., recorded/determined during the course of investigation were analysed statistically by standard statistical method. The plant heights, plant canopy, number of main branches per plant, number of spike per branch, spike length and segment per spike were recorded at the time of fresh biomass harvesting stage are presented in Table 1.

Table 1: Effect of date of sowing and irrigation level on different growth parameters of *salicornia*

Treatment	*Plant height (cm)*	*Plant canopy (cm^2)*	*Number of main branch per plant*	*Spike per branch*	*Spike length*	*Segment per spike*
			Date of sowing			
D_1: 1st week of June	30.77	72	20.98	382	8.28	10.03
D_2: 3rd week of June	32.22	112	25.23	478	8.39	14.92
D_3: 1st week of July	31.54	103	23.95	462	8.37	12.53
SEm±	1.000	3.2	0.58	7.50	0.86	0.20
CD at 5per cent	NS	11	2.00	26	NS	0.68
CV per cent	10.99	11.72	8.56	5.89	-35.65	5.43
			Irrigation levels			
I_1 : 0.6 IW/CPE	27.55	79	19.49	392	6.39	8.75
I_2 : 0.8 IW/CPE	34.06	92	24.47	458	8.67	14.35
I_3 :1.0IW/CPE	32.92	116	26.20	472	9.99	14.38
SEm±	0.787	4.1	0.72	19.30	0.29	0.42
CD at 5 per cent	2.34	12	2.13	57	0.87	1.24
CV per cent	8.65	14.88	10.62	15.16	12.18	11.53
Significant interaction	-	D x I	-	-	-	-

All growth parameters were significantly affected due to main plot treatments as well as sub-plot treatments, in case of plant height and spike length were not significantly affected sowing dates (D). The interaction effect of D x I was significantly only on plant canopy. In case of different date of sowings, D_2 (3rd week of June) recorded significantly higher values of all the growth parameters than D_1 (1st week of June) and D_2(1st week of July). Among all the irrigation levels, 0.8 IW/CPE (I_2) recorded significantly taller plant, wider canopy, more number of main branches per plant, number of spike per branch, spike length and segment per spike than 1.0 IW/CPE (I_3) and 0.6 IW/CPE (I_1) irrigation levels.The combination D_2I_3 (of crop sown on 3rd June and irrigated at 1.0 IW/CPE) recorded significantly higher values of plant canopy (125 cm^2) than rest of the combinations (Table 2). In all the cases, D_2 *i.e.,* sowing of *salicornia* during 3rd week of June showed superiority over D_1 *i.e.,* 1st week of June. However in most of the cases, D_2 was found to be at par with D_3 (1st week of July). This implies that optimum sowing time for *salicornia* crop in coastal areas at South Gujarat is between 3rd week of June to 1st week of July. In other words, this period is matching with regular period of onset of monsoon of South Gujarat region. With respect to effect of irrigation levels on growth parameters, irrigation level I_3 *i.e.,* scheduling irrigation at 1.0 IW/CPE recorded significantly higher values of all the growth parameters as compared to I_1 (0.6 IW/CPE). Only exception to this trend was plant height where I_2 was better than I_1. Though, I-$_3$ showed superiority over I_1,yet it was at par with I_2 in all the parameters except plant canopy where in I_3 was found superior over I_2.

Table 2: D x I effect on plant canopy (cm^2) of *salicornia*

D x I	*I1*	*I2*	*I3*	*D-Mean*
D1	45	65	108	72
D2	93	119	125	112
D3	99	93	115	103
I-Mean	79	92	116	
SEm±	7.10	CD at 5 per cent	21	

Yield

The results pertaining to the fresh biomass yield, dry biomass yield seed yield and oil content in seed of *salicornia* presented in Table 3 indicated that individual effect of date of sowing (D), level of irrigation (I) significantly influenced all the yield attributes of *salicornia.* Among the dates of sowing, D_3: 1st week of July recorded significantly higher fresh biomass yield and seed yield than D_1:1st week of June and D_2: 3rd week of June. In case of dry biomass yield and oil content D_2: 3rd week of June recorded higher yield than D_1:1st week of June and D_3: 1st week of July. Among the irrigation levels, 0.8 IW/CPE (I_2) was recorded higher fresh biomass yield, dry biomass yield, seed yield and oil content, but it was statistically at par with treatment 1.0 IW/CPE (I_3). The interaction effect of date of sowing and irrigation level was also found to be significant on fresh biomass yield and oil content in seed. The combination D_2I_2 recorded significantly higher fresh biomass yield (38.05 t ha^{-1}) and oil content (27.55 per cent) over rest of the possible combinations. In case of fresh biomass yield, it remained statistically at par with the treatment combination

Table 3: Effect of different treatments on fresh and dry biomass yield, seedyield and oil content

Treatment	*Fresh biomass yield (t ha^{-1})*	*Dry biomass yield (t ha^{-1})*	*Seed yield (t ha^{-1})*	*Oil content (per cent)*
Date of sowing				
D_1: 1st week of June	23.78	5.74	634	25.16
D_2: 3rd week of June	31.12	7.38	821	26.41
D_3: 1st week of July	34.24	7.10	839	25.58
SEm±	1.24	0.30	21.4	0.079
CD at 5 per cent	4.28	1.05	74	0.27
CV per cent	14.41	15.61	9.67	1.07
Irrigation levels				
I_1 : 0.6 IW/CPE	22.15	5.63	613	24.71
I_2 : 0.8 IW/CPE	34.92	7.38	865	26.26
I_3 :1.0IW/CPE	32.08	7.20	816	26.19
SEm±	1.03	0.36	22.6	0.07
CD at 5 per cent	3.07	1.06	67	0.20
CV per cent	12.05	18.34	10.25	0.92
D x I				
$D_1 I_1$	16.28	4.58	520	24.71
$D_1 I_2$	29.47	6.19	706	24.56
$D_1 I_3$	25.59	6.44	675	26.21
$D_2 I_1$	19.66	6.67	606	25.13
$D_2 I_2$	38.05	7.81	955	27.55
$D_2 I_3$	35.66	7.65	902	26.56
$D_3 I_1$	30.49	5.65	712	24.29
$D_3 I_2$	37.24	8.15	935	26.66
$D_3 I_3$	35.00	7.51	871	25.80
SEm±	1.79	0.62	39.2	0.12
CD at 5 per cent	5.32	NS	NS	0.35

D_3I_2, D_2I_3 and D_3I_3. Similarly, beneficial effects of D_2I_2 were also observed on growth parameters. This might have ultimately resulted in the higher fresh biomass yield of *salicornia*. Contrary to this, the dry yield of *salicornia* was affected significantly only due to main effect of D and I and not by their interaction. In dry biomass yield, I_2 and I_3 as well as D_2 and D_3 were at par with other but conspicuously more than I_1 and D_1. Though not significant, but here also D_2I_2 maintained their superiority over rest of the combinations. The results suggest that only individual effect of D and I was found to significant on seed yield of *salicornia*. Here also, I_2/I_3 and D_2/D_3 maintained their superiority over I_1 and D_1. The reasons for recording higher values of growth parameter and biomass as well as seed yields of *salicornia* with D_2 and I_2 are as follow. All the above favourable conditions ultimately culminated into significantly higher growth and biomass as well as seed yields of *salicornia* in delayed sowing and frequently irrigated crop as compared to that of early sowing and less frequently irrigated *salicornia* crop. The growth and yield advantages under frequent irrigation/lower salt concentration in soil during initial growth stages by sea water dilution was also reported by Joshi and Iyenger (1982) and Keiffer *et al.* (1994).The yield advantage in delayed sowing was also reported by Sumrah *et al.* (2003) comparing different species of *salicornia* in their study where in they compared day light period under growth chamber condition. Similarly, Ventura *et al.* (2011) also reported early maturity induced due to exposure of plant to longer day length during initial growth stage of *salicornia* grown under south Israel condition. In the present study, while harvesting for fresh biomass (just before flower initiation) plants were kept for seed production. From the seeds oil content was determined. Significant differences with respect to oil content in seeds due to D, I and D x I effects were observed. Sowing of *salicornia* during 3rd week of June seems to be optimum time of sowing for getting higher oil content in seed. Similarly, higher level of irrigation resulted into more oil content in *salicornia* seed than less irrigation. Almost similar trend was observed for oil yield per hectare. Irrespective of treatments the mean oil content in *salicornia* seed is 25.72 per cent which is comparable with most of the oil seed crops grown in normal soil with irrigation facility and that too in salt affected land with sea water irrigation. Grattan *et al.*(2003) reported *salicornia*, which is one of the most salt tolerant of the vascular plants, is sold as a salad supplement in Europe and produces oil in its seed that is equal in quantity and quality to soybean oil. Further, Tikhomirova *et al.* (2008) determined biochemical composition of the *Salicornia europaea* oil and found that plant lipids are characterised by a high unsaturation degree mainly due to alpha linolenic and linoleic acids. *Salicornia bigelovii* oil is similar to safflower oil in fatty acid composition (Glenn *et al.*, 1991). This suggests that in coastal areas of South Gujarat *salicornia* has an excellent scope for salad as well as edible oil purposes.

Conclusion

Salicornia seed sowing during 3rd week of June or during 1st week of July and irrigated either at 0.8 IW/CPE ratio or at 1.0 IW/CPW recorded significantly higher fresh biomass yield than rest of the combinations. The crop sown during this period had 30 and 24 per cent more biomass production than two weeks earlier sown crop. Almost similar effect of date of sowing and irrigation were also observed on dry biomass yield and seed yield of *salicornia.* Sowing of *salicornia* during 3rd week of June and irrigation scheduled at 0.8 IW/CPE resulted into more oil content in *salicornia* seed (27.55 per cent). Based on the results of present study it is concluded that for getting higher yield of *salicornia* and ultimately the net returns from the extremely saline soil and water conditions, sowing of the *salicornia* during either 3rd week of June or 1st week of July and irrigating the crop at 0.8 IW/CPE ratio (8-10 irrigation) is essential.

References

Glenn, E. P., O'Leary, J. W.,Watsin, M. C.,Thompson, T. L. and Kuehl, R. O. (1991). *Salicornia bigelovii* Torr. An Oilseed Halophyte for Sea water Irrigation. *Science,* 4997 (251): 1065-1067.

Grattan, S. R. and Oster, J. D. (2003).Use and reuse of saline-sodic waters for irrigation of crops. *Journal of Crop Production,* 1-2(7): 131-162.

Joshi, A. J. and Iyengar, E. R. R. (1982). Effect of salinity on the germination of *Salicornia brachiata* Roxb. *Plant Physiology,* 25: 65-70.

Keiffer, C. H., Mccartfiy, B. C. and Ungar, I. A. (1994). Effect of Salinity and Waterlogging on Growth and Survival of *Salicornia europaea* L., an Inland Halophyte. *Ohio J. Sci.* 94 (3): 70-73.

Sumrah, M. A., Bakhsh, A. and Ahmad, S. (2003). Effect of sowing time on growth behavior of beet root in sub-mountainous climate condition. *Asian Journal of Plant Science,* 2(3): 354-357.

Tikhomirova, N. A., Ushakova, S. A., Tikhomirova, A. A., Kalacheva, G. S. and Gros, J. B. (2008). *Salicornia europaea* L. plants as possible constituent of Bio-regenerative Life Support Systems Phototrophic Link. *Journal of Siberian Federal University Biology,* 2 (1):118-125.

Ungar, I. A. (1987). Population characteristics, growth and survival of the halophyte *Salicornia europaea. Ecology.* 68 (3): 569-575.

Ventura Y., Wuddineh W. A., and Sagi M. (2011). Effect of day length on flowering and yield production of Salicornia and Salicornia species. *Scientia Horticulturae,* 130: 510-51.

23

Effect of Nozzle Size and Operating Pressure on the Performance of Raingun Irrigation System

H.R. Vadar, H. D. Rank, S.V. Gajera, G.J. Vaghela, H.V. Parmar, H.H. Mashru and P.A. Damor

Abstract

The experiment was conducted to evaluate the performance parameters like pressure-discharge-nozzle size relationships, uniformity co-efficient, and pattern efficiency for all the four nozzle sizes and different operating pressure. The catch cans were arranged in radial direction considering raingun as a centre. The pattern efficiency and uniformity co-efficient can be greatly affected by the operating pressure. The model for relationship among shift distance (S, in m)-pressure (in (kg/sq.cm) and nozzle size (in mm) could be developed. The developed nomograms for nozzle size-pressure-discharge relationship and nozzle size-pressure-shift distance relationship can be used to design the operational parameters of the raingun irrigation system.

Key words: *Raingun, irrigation, uniformity, distribution, pattern efficiency, shift distance.*

Introduction

The raingun irrigation is one of the efficient MIS. After realisation of importance and necessity of MIS by Government of Gujarat, it has established the "Gujarat Green Revolution Company Limited" from April 1, 2005. The company has resolved that any farmer would be given 50 per cent subsidy with 45 per cent loan without acreage limitation. After installation, the company would provide warranty and agro engineering consultancy for five years. Therefore, a comprehensive strategy is needed for the conservation and development of national water resources. In the course of action we needed to deal with various factors like availability, quality, location, distribution and variation in the occurrence of water, climatic conditions and nature of soil, competing demand and socio-economic conditions. In dealing with each of these, it is required to make every effort for the best use of water in order to increase the level of agricultural production per unit volume of water. Application of excess water to crop, either due to rainfall or due to surface irrigation resulted in reduction of yields, rise in the ground water table and water logging. Rise of ground water level in irrigated areas brings up harmful salts to the surface layers and such soils may go out of cultivation. The adaptation of efficient irrigation methods is therefore important in view of increasing irrigation as well as water use efficiency. High application efficiencies can only be obtained by pressurised irrigation methods such as sprinkler, raingun and drip irrigation system. Besides this, these methods can be adopted for almost all crops under variables topographic conditions and on different type of soil. Even one conditions out of several, make these system better choices than surface irrigation methods. The surface irrigation methods like border, check basin, furrow etc. have poor water application efficiency. There is a loss of huge quantity of water applied by surface irrigation methods through deep percolation and evaporation from wet soil. Also surface methods of irrigation are labour consuming and are requiring land grading and land levelling which adds cost inputs.

Raingun is a powerful mega sprinkler that throws a large amount of water (up to 500 liters per minute) to a good distance (radius of 30 meter and even more) as artificial rain. It offers a number of benefits to the farmer. It reduces water consumption by 30 per cent as compared to flood irrigation in achieving the same yield. As a result of the reduced water consumption with the raingun irrigation system, large savings can be accrued. Irrigation time comes down (50 per cent time is saved), and power consumption comes down. Also, raingun irrigation is less labour intensive than flood irrigation. It increases crop yield by 10 per cent, as farmers have experienced. Fertilizers can also be applied with the raingun irrigation system, reducing consumption of fertilisers. Irrigation with the raingun washes away pests like aphids, white flies etc. The raingun irrigation systems supports the highly recommended practice of trash mulching in sugarcane, which is a process of converting trash into nutrient for the crop. However, in surface irrigation method the farmers do not make ready use of this available nutrient and resort to clearing it or setting it on fire to manage the huge quantity of trash. The raingun irrigation system gives farmers the ability to practice trash mulching successfully. Raingun irrigation is a versatile means of applying water to the surface of any crop or soil. Raingun irrigation system has several advantages such as (a) It may be adopted on porous soil for good distribution which is not possible by surface methods. (b) It is better option for land having steep slop and erodable soils. (c) It is good for shallow soils, the topography of which prevents proper levelling. (d) Raingun have been adopted for many agricultural and non-agricultural uses. Some of these include water distribution for compaction. (e) It can also be used for fire and frost protection and (f) It can also be possible to apply slurry by this methods.

The above facts show that raingun is a great hope for irrigation water management particularly in water scared area. However, there is a dire need to have its technical evaluation for the decision supports in its adoptions and operations. Therefore, keeping all things in mind, the present study was planned for the objective to determine the optimal shift distance of raingun for various nozzle sizes and available operating pressure based on the uniformity coefficient, water application efficiency, water distribution efficiency, pattern efficiency for various nozzle size and operating pressure.

Materials and Methods

The experiment was conducted at Research-cum-Demonstration Farm of Centre of Excellence on Soil and Water Management, JAU, Junagadh.

Materials

The raingun with all accessories were used for the experiment. The 10,12,14 and 18 mm nozzle sizes at various operating pressure were tested for the experiment. The dug well (Khodiyar well) of 4.25 m diameter having total depth of 23 m was used as water source. The water level from the ground surface was varied from 6 m to 14 m during the season. The electric submersible motor of 5 HP x 1 stage having outlet size of 65 mm was used for pumping water from the well. The delivery line was of HDPE pipe having size of 75 mm x 6 kg/ sq.cm. The bypass assembly of 75 mm size was fitted on main line. The pressure gauge having range of 0-7 kg/cm^2 and LC of 0.1 kg/cm^2 was used and it was fitted at the bottom of the raingun stand. The PVC pipe line of 75 mm diameter was used as main line. The flexible hose pipe of 63 mm was used to convey the water from main line to raingun. The hose pipe was connected to raingun using the coupling assembly of the raingun. The catch cans having diameter of 8 cm and 4.97 cm height were used for collecting the water sprayed by raingun at particular operating pressure. The collected water in the can was measured volumetrically by measuring transparent cylinder of 250 ml and then converted it in to depth of water using can area. The barrel of 225 litres capacity was used for determining the discharge of the raingun at various operating pressure. The digital thermo-anemometer was used to measure the instantaneous wind speed and air temperature. The digital clock feature available in the mobile was used for measuring the durations of the particular test.

Methodology

A raingun is a high volume sprinkler that operates at high pressure. It requires high energy than conventional sprinkler and is less susceptible to clogging than drip emitter and sprinkler nozzle. It has much larger area of coverage than drip emitter and sprinkler. It has not been as popular as sprinkler and drip system due to its recent introduction into the market.

Pressure-discharge relationship

Total three nozzles (10 mm, 12 mm and 14 mm) of raingun were tested for measuring their discharge at varying pressure. The barrel of 225 litres was used. The time required to fill that barrel by keeping constant pressure of 0.5, 1.0, 1.5, 2.0, 2.5 and 3 kg/cm^2 separately was noted. The input pressure to raingun was measured by pressure gauge connected at inlet of raingun. Pressure of 0.5 kg/cm^2, 1 kg/cm^2, 1.5 kg/cm^2, 2 kg/cm^2, 2.5 kg/cm^2 and 3 kg/cm^2 were maintained constant during the test and discharge was measured. Raingun flow rate expressed as a function of pressure for the selected nozzle was established (Eq. 1a). The general mathematical relationship for the raingun nozzle as a function of pressure and nozzle size was established by testing the various forms of the mathematical models (Eq.1b).

$$Q = f(P) \quad \ldots(1a)$$

$$Q = f(P, D) \quad \ldots(1b)$$

where, Q = raingun discharge (klph); P = operating pressure (kg/cm^2);

Raingun performance efficiency

The four sectors of 90° were considered. Among these four lines, two were parallel to wind direction and two were perpendicular to wind direction. The required catch cans were placed on each of these four radii. The first catch can was placed one meter from the center of the raingun and thereafter at two meter intervals. Each catch can was set in an upright position with its top parallel to the ground. Surrounding vegetation was removed from all the sides of the catch cans if found necessary. The volume of water in each catch can was measured with the help of a graduated measuring cylinder.

The irrigation system was operated for one hour periods at 1, 2 and 3 kg/cm^2 pressure. The amount of water collected in each catch can was measured. The data of depth of water in catch canes were collected separately at each pressure. The following performance parameters were found using the observation data at various pressure inputs.

Co-efficient of uniformity

The uniformity co-efficient of the raingun system was estimated at prevailing wind speed using irrigation depth of water applications at different locations within the wetted perimeter of raingun. The following equation was used for calculating uniformity co-efficient of raingun (Eq. 2)

$$C_u = 100\ [1-(dev_{mean}/d_{mean})] \qquad ...(2)$$

where, C_u = Uniformity co-efficient, per cent; d_{mean} =Average value of all observed depth of irrigation by raingun, mm; dev_{mean} = Average of all numerical deviations of individual observed irrigation depths from the average application depth, mm.

Distribution efficiency/pattern efficiency

It was calculated after obtaining the depth of water at each location on four radii originating from raingun center. The minimum depth was calculated considering average of the lowest one-fourth of the depth collected in catch cans set in particular test.

$$Ed = 100\ d_{1/4mean}/d_{mean} \qquad (3)$$

where; E_d=Distribution efficiency/pattern efficiency (per cent); $d_{1/4mean}$ = 1/4th minimum water depth (mm) i.e. average of the lowest one-fourth of depths of water collected in the catch cans for particular set.

Application rate

Application rates using different raingun with different discharge capacity at various operating pressure were measured and application rate was taken as ratio of raingun discharge (lph) to Spraying area of raingun (m^2).

Shift distance of raingun

The percentage of average application rate at various distance from the center of the raingun on four radial directions were calculated for each of the four raingun nozzles at each of operating pressure of 1, 2 and 3 kg/cm^2. The shift distance was taken as the twice of the distance at which the application rate was 50 per cent of the average application rate. The mathematical relationships for the shift distance as a function of the nozzle size and the operating pressure were established (Eq. four). The various forms of the model were tested to fit the observed data of the shift distance of the raingun at various operating pressure for the different nozzle sizes.

$$S = f(P, D) \qquad ...(4)$$

where, S = Shift distance of the raingun (m), P = Operating pressure of the raingun (kg/sq.cm), D = Diameter of the raingun nozzle (mm), out of radial arrangement of catch cans by considering Raingun as centre.

Results and Discussions

The wind velocity and air temperature during experimental test were 6.84 km per hour (1.9 m/s) and 36.3°C. Under this experimental set up, each of the following four nozzles of raingun was tested. The discharge for various nozzle sizes and operating pressure were measured and the mathematical relationships were developed.

Evaluation of raingun performance

Pressure discharge relationship

The observed data of discharge of different nozzle sizes at various operating pressure were used to develop the models. The various forms of mathematical model were tested for fitting the observed data. However, the power form of the mathematical model was found best fit for the pressure-discharge relationships (Table 1).

The exponent of power function for the pressure was increased with nozzle size which indicated that the sensitivity of the nozzle to pressure for the discharge was increased with increase in nozzle size. The results is supported by Yasin *et al.* (1999). Also the exponent of power function for the pressure was less than 0.5, which indicated that the flow through the nozzle was not orifice flow.

Table 1: Best fit model for the pressure-discharge-nozzle size relationships

S.No.	*Nozzle size (mm)*	*Best fit Model*	R^2
1.	10	$Q = 5842\ P^{0.315}$	0.996
2.	12	$Q = 7249\ P^{0.372}$	0.975
3.	14	$Q = 9293P^{0.369}$	0.961
4.	Any size D(mm)	$Q = 175.1\ D^{1.5051}\ P^{0.3657}$	0.983

where, Q = discharge of the raingun (lph), D = diameter of nozzle (mm), P = pressure (kg/ cm^2).

The model for the pressure-discharge-nozzle size (Table 1) showed that the exponent of the power form for the nozzle size was less than two which indicated that the flow through the raingun nozzle is not an orifice flow. Using the developed model of raingun discharge-pressure-nozzle size relationships, the nomogram was developed (Fig. 1) which can be very useful to decide the nozzle size for available pressure and discharge of the pumping system. It showed that the discharge of the raingun increase with pressure and nozzle size. The results is supported by Chennakesavulu, B. (2014).

Average application rate

The average rate at which irrigation water could be applied for various operating pressure and different nozzle sizes were calculated using observed data of discharge and radius of throw. It could be seen that the average application rate decreased with increase in pressure but increased with increase nozzle sizes. Similar results was also reported by Latif *et al.* (1999) and Yasin *et al.* (2000).

Co-efficient of uniformity

The co-efficient of uniformity was determined using observed data obtained through the experiment. The results obtained through analysis are given in Fig. 2 for 10 mm, 12 mm, 14 mm and 18 mm nozzle sizes. The Fig. 2 shows the graphical relationship between Co-efficient of uniformity and pressure. It can be seen that no definite relationship among the Co-efficient of uniformity, nozzle size and operating pressure existed.

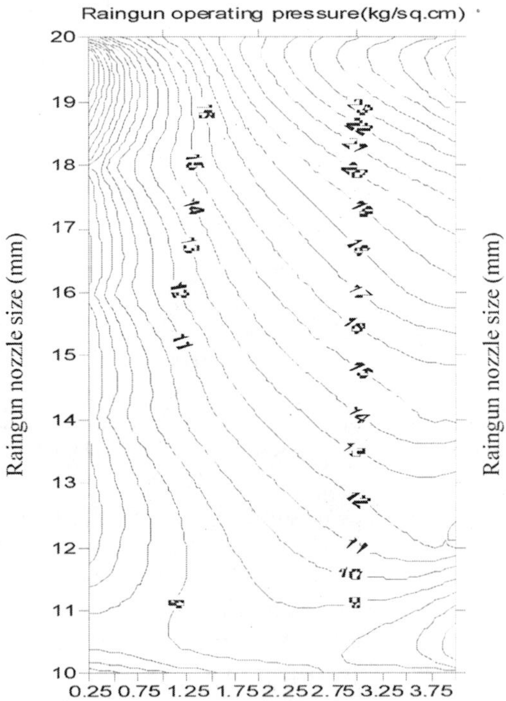

Fig. 1. Nomogram of raingun for discharge (KLPH) at various operating pressure (kg/sqm) for different nozzle sizes.

The maximum co-efficient of uniformity of 83 per cent was obtained for 1 kg/sq.cm and 10 mm nozzle size while that was minimum of 52 per cent for 1 kg/sq/cm pressure and 18 mm nozzle size. The similar results also found by Latif *et al.* (1999).

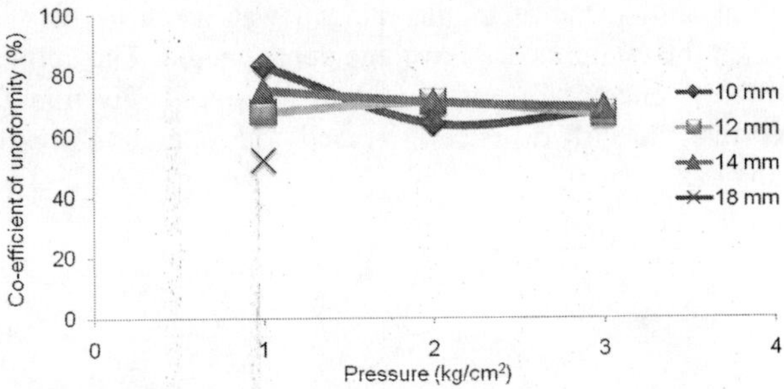

Fig. 2. Coefficient of uniformity (per cent) obtained at different operating pressure for various nozzle sizes.

Pattern efficiency

The graphical relationship between pattern efficiency and pressure are plotted in Fig. 3 for 10, 12, 14 and 18 mm nozzle sizes. It can be seen that no definite relationship among the pattern efficiency, nozzle size and operating pressure existed. The maximum pattern efficiency of 71 per cent was obtained for 1 kg/sq.cm and 12 mm nozzle size while that was minimum of 33 per cent for 3 kg/sq/cm pressure and 10 mm nozzle size. The results is similar to reported by Richards P.J. and Weatherhead E.K. (1990) and Joshi and Monpara (2009).

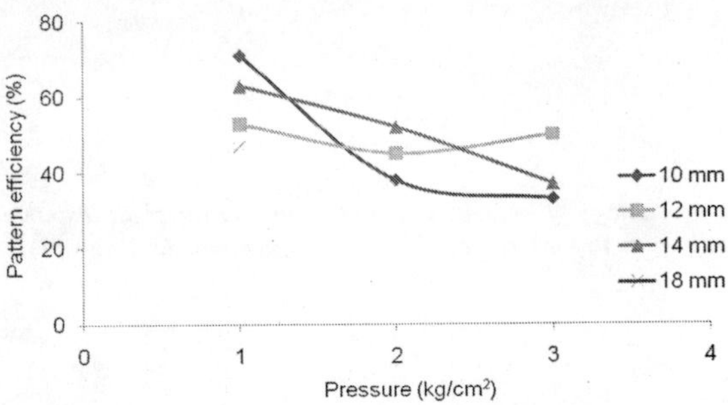

Fig. 3: Pattern efficiency (per cent) obtained at different operating pressure for various nozzle sizes.

Shift distance

The observed data of depth of water caught at different distance from the raingun in all four radial directions were used for calculating the percentage of average application rate for different nozzle sizes and different operating pressure. The percentage of average depth at different operating pressure for 14 mm nozzle sizes and 3 kgf/sq.m are shown in Fig. 4.

The optimal shift distance of the raingun was taken as twice of the distance at which the 50 per cent of average depth caught. The optimal shift distance at different operating pressure and nozzle sizes is given in Table 2. Fig. 5 showed that the shift distance increased with operating pressure but decreased with nozzle sizes.

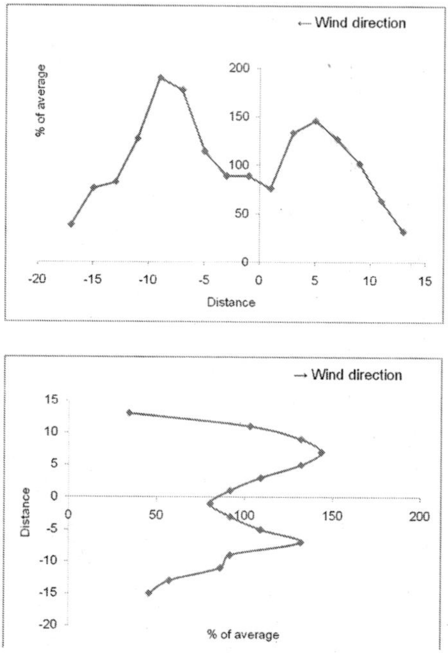

Fig. 4: Percentage of average irrigation depth at different distance from rain gun in four radial direction (14 mm nozzle 3 kg/sq.m).

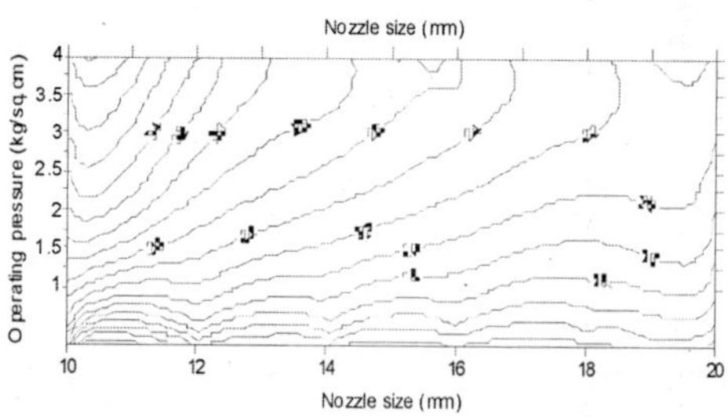

Fig. 5 : Nomogram of iso-shift distance of raingun for various nozzle size and operating pressure.

Table 2: Shift distance at different operating pressure for different nozzle Sizes

Nozzle size	*Shift distance (m)*		
	1 kg/cm²	*2 kg/cm²*	*3 kg/cm²*
10	24	32	34
12	23	29	31
14	22	25	28
18	10	Not tested	Not tested

The best fit general formula for shift distance (m) for different nozzle size (mm) and different operating pressure (kg/cm^2) was found as below.

$$\mathbf{S = 174.9798\ D^{-0.8401}\ P^{0.3244}}$$

where, S = Shift distance of the raingun (m); D = Diameter of the nozzle (mm); P = Operating pressure (kg/sq.cm)

Conclusions

- The pressure-discharge relationships can as $Q = 175.1\ D^{1.5051}\ P^{0.3657}$ can be used for the different nozzle sizes. Where; Q = raingun discharge (lph); P = operating pressure (kg/cm^2); D= nozzle inside diameter (mm).
- The shift distance for 10 mm nozzle size should be taken as 24 m, 32 m, and 34 m when the water supply is available at pressure of 1, 2, and 3 kg/cm^2 respectively.
- The shift distance for 12 mm nozzle size should be taken as 23 m, 29 m, and 31 m when the water supply is available at pressure of 1, 2, and 3 kg/cm^2 respectively.
- The shift distance for 14 mm nozzle size should be taken as 22 m, 25 m, and 28 m when the water supply is available at pressure of 1, 2, and 3 kg/cm^2 respectively.
- The shift distance for 18 mm nozzle size was found as 10 m when the water supply was available at pressure of 1 kg/cm^2 respectively.
- It was found that the shift distance of the rain gun can be increased by increasing the operating pressure and decreasing the nozzle size.
- The relationship among shift distance-pressure-nozzle size as $S = 174.9798\ D^{-0.8401}\ P^{0.3244}$ can be used. Where, S = Shift distance of the raingun (m); D = Diameter of the nozzle (mm); P = Operating pressure (kg/sq.cm).

- The developed nomograms for nozzle size-pressure-discharge relationship and nozzle size-pressure- shift distance relationship can be used to design the operational parameters of the raingun irrigation system.

References

Chennakesavulu, B. 2014. "Performance evaluation of raingun sprinkler irrigation system with different operating pressures", International Journal of Food, Agriculture and Veterinary Sciences, Vol. 4(1), pp.127-129.

Joshi, S. D. and P. L. Monpara. 2009. "Performance evaluation of raingun irrigation system". B Tech.(Agril. Engg.) thesis sub. JAU, Junagadh.

Latif, M.; M. S. Ahmad, and. S. Mahmood. 1999. "Evaluation locally manufactured raingun sprinkler system", Journal of Drainage and Water Management (Pakistan). Vol. 3(1-2), pp. 34-41.

Richards, P. J. and E. K. Weatherhead. 1990. "Prediction of Raingun Application Patterns in Windy Conditions", Journal of Agricultural Engineering Research, Vol. 54(4), pp. 281-291.

Yasin, M.; Badruddin, S. Ahmad and P. M. Moshabbir. 2000. "Hydraulics performance of the raingun (PY1-30) with 6 mm diameter nozzle", Pakistan Journal of Scientific and Industrial Research (Pakistan), Vol. 43(6), pp. 351-356.

Yasin, M.; A. G. Mangrio; M. Aslam and M. J. Siddiqui. 1999. "Pressure discharge relations of the raingun (PY1-30) sprinkler irrigation system", Sarhad-Journal-of-Agriculture, Vol.15(2), pp. 107-113.

24

Co-ordinated Influence of Irrigation Systems, Regimes and Mulching on Bt. Cotton

G. V. Prajapati and R. Subbaiah

Abstract

Lower unit cost of production and increase in yield of Bt. Cotton with its special characteristics like; very low naps, good luster, more wax on fibers surface, very low dye absorption appeased farmers to adopt on large scale cultivation. Adverse environmental conditions like high evaporative conditions, scarcity of groundwater, and deficient rainfall condition caused poor yields of cotton necessitated the farmers to adopt drip irrigation with mulch. Lack of knowledge on combined impact of drip irrigation regimes and mulch on water saving and its economy constrained mass adoptability of this technique. Keeping this in view, an experiment was conducted at Junagadh Agricultural University, Junagadh for two consecutive years (2012-13 and 2013-14) to assess the combined impact of irrigation regimes (0.6 ET_c, 0.8 ET_c, 1.0 ET_c), mulch material (silver black plastic mulch, biodegradable mulch), and irrigation systems (drip and furrow) on yield of Bt. Cotton (Hybrid-6 BG-II). Split plot design with three replications was adopted. Drip without mulch outperformed over furrow irrigation system (4.82 kg $ha^{-1}mm^{-1}$). Seed cotton yield of drip with silver black plastic mulch was 2.1 folds over furrow method, 1.40 folds over drip irrigated cotton and 1.15 folds over biodegradable mulch at 0.8ET_c. Benefit cost ratio was more in silver black plastic mulch (2.64) at 0.8 ET_c compared to biodegradable plastic mulch (1.19), drip without mulch (2.38) and furrow (1.60) grown cotton.

Key words: *Cotton, irrigation regimes, mulching, WUE, furrow.*

Introduction

Water demand has significantly increased over the last decades while available water resources are becoming increasingly scarce. This is mainly due to the combined effect of climate change, persistent drought and the increase of water demands related to increase in irrigated surfaces, urbanisation and tourism recreational projects. In this context, improvement of water management in agriculture, which is the biggest water consumer, is necessary to enhance agricultural productivity in order to meet food demands of the growing population. Cotton is the major crop of Gujarat occupying an area of 35 lakh ha. However, the cotton productivity is still below the potential yield because of high evaporative conditions, scarce groundwater availability, deficient rainfall, and poor water management practices. Sound irrigation scheduling help farmers to increase crop yield and save water in irrigated areas. Drip irrigation reduce input cost, increase yield, give more water productivity than surface irrigation and reduce the risk of yield reduction due to inter-irrigation dry spells (Sivanappan, 2004). Managing congenial soil moisture in the crop root zone and preventing losses from soil is very much essential for attaining more productivity per drop of water. Mulching reduced water loss, regulate soil temperature, control weeds and increase water use efficiency in cotton (Ravi and Lourduraj, 1996; Naeini and Cook, 2000; Murugan and Gopinath 2001; Jordan, 2004; Zhang *et al.*, 2005). Relatively few studies were conducted to analyse the conjunctive performance of drip irrigation with silver black plastic mulch and biodegradable plastic mulch and comparison among irrigation systems on productivity of Bt. Cotton. Hence, present investigation was aimed at studying the techno-economic feasibility of drip irrigation with various irrigation regimes and mulch material on productivity of Bt. Cotton.

Materials and Methods

A field experiment was conducted at Junagadh Agricultural University for two consecutive years (2012-13 and 2013-14) to determine the impact of different drip irrigation levels (0.6ET_c; 0.8ET_c; 1.0ET_c) with mulch (Silver black plastic mulch; Biodegradable plastic mulch) and two irrigation systems without mulch (Drip and Furrow) on productivity of Bt. Cotton (Hybrid Cotton-6). The treatment table is presented in Table 1. Split plot design was adopted with twelve treatment combinations. Each treatment was replicated thrice. CROPWAT Model was used to estimate effective rainfall, irrigation water requirements, as it provide consistent estimates in synchronisation with actual crop water use data (Allen *et al.*, 1998). Irrigation was given at every alternative day in drip treatments whereas furrow method received 12 irrigations as per the conventional approach. Hybrid Bt.Cotton-6 was sown at 45 cm x 120 cm spacing. Standard agronomical practices recommended by University were followed. Morphological parameters and yield attributes at the end of experiment were monitored.

Results and Discussion

Yield attributes

Incorporation of mulch with drip enhanced seed cotton yield significantly over no mulch. An increase of 28.74 per cent over no mulch was observed. This

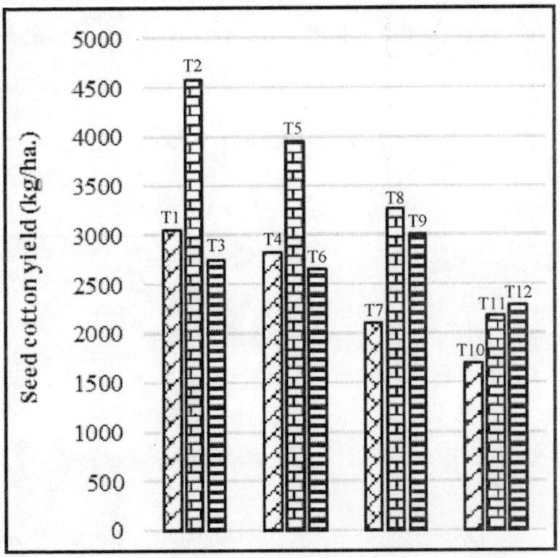

Fig. 1: Response of Seed Cotton yield of different treatment at Etc.

could be attributed to more conversion and consequently higher availability of moisture, which might have helped in improving the growth and yield components. Soil temperature regulation by mulch could be another mechanism causing yield increase. The combined effect of mulch, irrigation regimes and irrigation systems on cotton productivity was depicted in Fig. 1. The results revealed that T_2 recorded significantly higher seed cotton yield (4583.77 kg ha^{-1}) followed by T_5 (3956.03 kg ha^{-1}), T_8 (3266.4 kg ha^{-1}), T_{11} (2180.57 kg ha^{-1}) (Table 2). Significant difference among all the treatments was observed.

Morphological parameters

Positive impact was observed in mulched plots on the morphological attributes compared with no mulch. Mulched plot stored the soil moisture effectively, thereby enhancing the vegetative growth in terms of plant height; sympodial branches; and other attributes. Highest average height (1.93 m) cotton plant was attained at 0.8ET$_c$, followed by 1.0ET$_c$ (1.44 m) and 0.6ET$_c$ (1.41 m) for drip irrigated silver black plastic mulch which is superior to all other treatments at all irrigation regimes. Significant difference was found among all the treatments. More sympodial branches (23.60 and 20.65) were observed in the drip irrigated mulched plots (silver black plastic and biodegradable plastic) than the furrow method (9.8) at 0.8ET_c. The enhanced sympodial branches might be due to better partitioning of assimilates towards reproductive parts as the source was not limiting under mulching. Better root growth and assimilation of photosynthates under mulching caused higher partitioning of assimilates towards reproductive structure as could be seen from the production of more bolls (66.46 bolls/plant) as against 29.4 bolls/plant under the furrow method. Treatment T_1 was at par with T_9. Treatment T_7 was at par with T_{11}. This shows that drip irrigation was performing better than furrow irrigation system without mulch.

Table 1. Treatments details

Treatment No.	*Irrigation System*	*Mulch material*	*Irrigation regime*
T_1	Drip	Silver plastic	0.6
T_2	Drip	Silver plastic	0.8
T_3	Drip	Silver plastic	1.0
T_4	Drip	Biodegradable plastic	0.6
T_5	Drip	Biodegradable plastic	0.8
T_6	Drip	Biodegradable plastic	1.0
T_7	Drip	No mulch	0.6
T_8	Drip	No mulch	0.8
T_9	Drip	No mulch	1.0
T_{10}	Furrow	No mulch	0.6
T_{11}	Furrow	No mulch	0.8
T_{12}	Furrow	No mulch	1.0

Irrigation water

Using CROPWAT model, ET_c, effective rainfall and irrigation requirement for both the years were determined. The total and effective rainfall was 428 mm, and 183.4 mm whereas, actual water used by the plant was 560.7 mm and actual irrigation requirement was 396.0 mm during the year 2012-13 Fig. 2. During the year 2013-14, the total and effective rainfall received was 1500.9 mm, and 548.2 mm whereas, actual water used by the plant was 569.4 mm and actual irrigation requirement was 327.6 mm Fig. 3.

All the systems were scheduled at same irrigation regimes. This will yield irrigation water delivered for drip and mulched plots irrigation systems were 225.96 mm, 301.28 mm, and 376.61 mm, whereas for furrow it was 338.95 mm, 451.93 mm and 564.92 mm at 0.6ET_c, 0.8ET_c and 1.0ET_c respectively. Results reveal that irrigation amount increase when scheduling was done at higher ET_c. Relationship between irrigation water applied and yield of cotton was depicted in Fig. 4. Treatments possessed quadratic relationship between irrigation water and seed cotton yield for all the treatments. Maximum yield of 4583.77 kg ha.$^{-1}$ was obtained with 301.29 mm of water for treatment T_2 whereas, lowest yield (2180.6 kg ha.$^{-1}$) was recorded for T_{11} and 451.93 mm water application.

Water Use Efficiency

The mean water use efficiency (WUE) for different treatments were shown in Fig 6. The mean WUE values ranged from 15.21 to 4.05. Highest mean WUE (15.21 kg ha^{-1}mm^{-1}) was found for treatment T_2 followed by T_5 (13.13 kg ha^{-1} mm^{-1}), T_8 (10.84 kg ha^{-1}mm^{-1}) and T_{10} (5.02 kg ha^{-1}mm^{-1}). Mulching recorded higher WUE over no mulch. High WUE with mulch was due to reduction in ET, increase in ASM and higher yield. Due to high water holding capacity of mulch, available soil moisture status has been maintained more in surface layers.

Table 2. Morphological and Yield parameters of Bt. Cotton

Treatment	*Seed cotton yield (kg ha^{-1})*		*No. of Bolls/ plant*		*No. of Sympodial/plant*		*Mena Plant height (m)*	
T_1	3052.3		44.2		15.95		1.41	
T_2	4583.7		66.4		23.60		1.93	
T_3	2749.8		39.8		19.52		1.44	
T_4	2826.1		39.5		12.18		1.31	
T_5	3956.1		55.3		20.65		1.44	
T_6	2653.7		37.8		17.28		1.35	
T_7	2110.1		29.5		9.32		1.06	
T_8	3266.3		45.7		16.38		1.33	
T_9	3011.8		42.3		14.45		1.19	
T_{10}	1698.5		22.9		9.08		0.88	
T_{11}	2180.5		29.4		9.77		1.03	
T_{12}	2290.3		30.9		10.45		1.08	
	SEM	**CD at 5%**	**SEM**	**CD at 5%**	**SEM**	**CD at 5%**	**SEM**	**CD at 5%**
I	55.17	179.9	0.76	2.48	0.47	1.54	0.05	0.19
M	50.52	145.02	0.73	2.09	0.44	1.26	0.04	0.13
I at M	87.50	251.18	1.26	3.62	0.76	2.18	0.08	0.22

Economics

Economics of cotton cultivation influenced by mulching, irrigation regimes and types of irrigation considering the yield and biomass production is

Table 3: Cotton seed yield and biomass of different treatments.

Irrigation level	Silver black plastic mulch	Bio- degradable plastic mulch	Drip Irrigation	Furrow Irrigation
Mean seed cotton yield (kg/ha)				
$ET_c = 0.6$	3052.30	2826.13	2110.12	1698.52
$ET_c = 0.8$	4583.77	3956.03	3266.35	2180.57
$ET_c = 1.0$	2749.83	2653.67	3011.82	2290.34
Mean biomass (T/ha)				
$ET_c = 0.6$	5.19	4.54	3.06	3.01
$ET_c = 0.8$	6.95	5.93	4.17	3.95
$ET_c = 1.0$	6.22	5.74	4.91	4.00

Table 4: Economics of cotton cultivation as influenced by mulching and drip irrigation.

	Particulars	Silver black plastic mulch	Biodegradable plastic mulch	Drip Irrigation	Furrow Irrigation
1.	**COST**				
	(A) Fixed Cost				
	1) Drip cost	42500	42500	42500	53500
	2) Other costs	9883	9883	9883	0
	Total of (A)	52383	52383	52383	53500
	(B) Variable cost = Irrigation + mulching				
	I_1 – 0.6 ETc	24360	39560	7560	6720
	I_2 – 0.8 ETc	26880	42080	10080	8960
	I_3 – 1.0 ETc	29400	44600	12600	11200
	Total Cost (Fixed + Variable cost)				
	I_1 – 0.6 ETc	76743	91943	59943	60220
	I_2 – 0.8 ETc	79263	94463	62463	62460
	I_3 – 1.0 ETc	81783	96983	64983	64700
2.	**INCOME**				
	(A) Cotton Seed Yield				
	I_1 – 0.6 ETc	137354	127176	94955	76433
	I_2 – 0.8 ETc	206270	178021	146986	98126
	I_3 – 1.0 ETc	123743	119415	135532	103065
	(B) Biomass				
	I_1 – 0.6 ETc	2074	1815	1222	1204
	I_2 – 0.8 ETc	2778	2371	1667	1580
	I_3 – 1.0 ETc	2489	2296	1963	1600
3.	**GROSS INCOME**				
	I_1 – 0.6 ETc	139428	128991	96178	77637
	I_2 – 0.8 ETc	209048	180392	148653	99706
	I_3 – 1.0 ETc	126232	121711	137495	104665
4.	**BENEFIT-COST RATIO**				
	I_1 – 0.6 ETc	1.82	1.40	1.60	1.29
	I_2 – 0.8 ETc	2.64	1.91	2.38	1.60
	I_3 – 1.0 ETc	1.54	1.25	2.12	1.62

Note: Water cost @ Rs. 80 per hour, Silver black plastic mulch required 160kg @ Rs. 105/kg, Biodegradable mulch required 160 kg @Rs. 200/kg. Average Price of Seed Cotton Rs.45/kg, and Cotton Stalk Rs. 0.40/kg

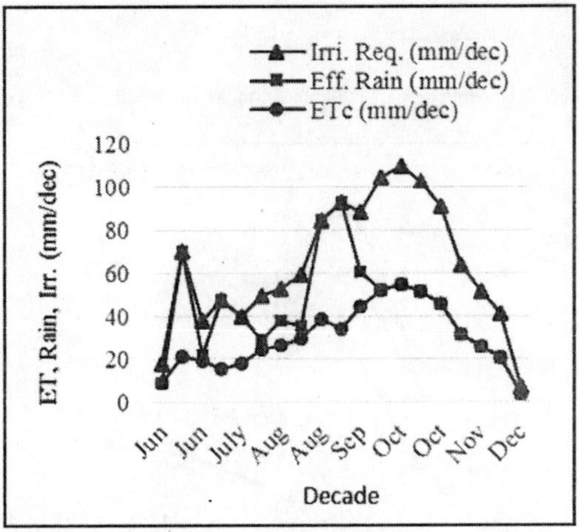

Fig. 2: ETc. Effective rainfall and Irrigation requirement of Bt. Cotton for the year 2012-13.

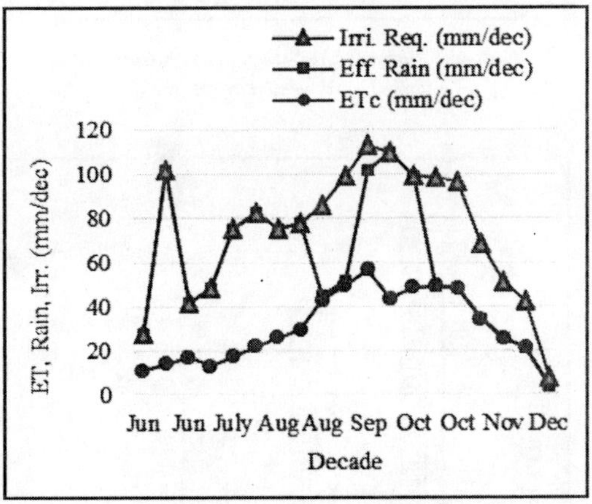

Fig. 3: ET_c. Effective rainfall and Irrigation requirement of Bt. Cotton for the year 2012-13.

presented in Table 3 and 4. The net benefit cost ratio for drip with silver plastic mulch, biodegradable plastic mulch, drip without mulch and furrow irrigation was found to be 2.64, 1.38, 2.38 and 1.60 respectively. System cost can be recovered within a year. High initial investment in mulching can overcome by availing the benefit of Govt. subsidy as well as availing the agricultural loan from the bank at low interest rates. Even if you count the interest also the payback period of the system remains invariant because of the additional revenue from saved irrigation water and towards weeding cost.

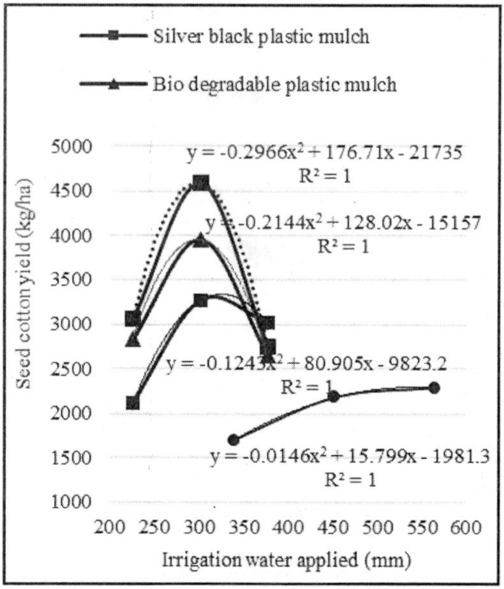

Fig. 4. Relationship between irrigation water applied and seed cotton yield.

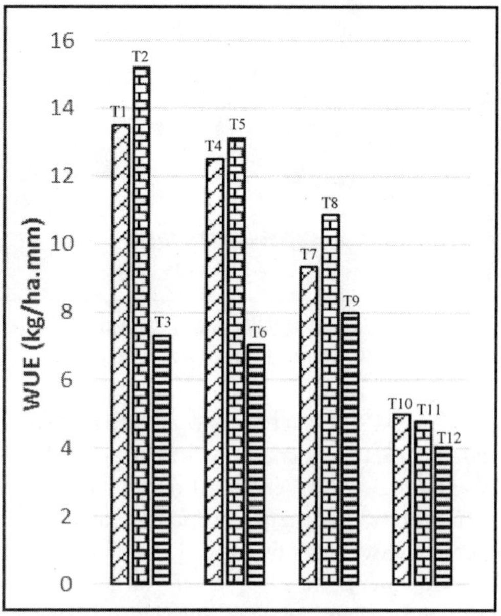

Fig. 5. Irrigation water use efficiency of different treatments at ET_c.

Conclusion

Drip irrigation coupled with silver black plastic mulch scheduled at $0.8ET_c$ resulted in better performance of morphological variables, yield attributes, and WUE than biodegradable plastic mulch, drip without mulch and furrow system. Drip irrigation with mulch scheduled at $0.6ET_c$ was at par with drip irrigation without much at $1.0ET_c$. Drip irrigation without mulch scheduled at $0.6ET_c$ was at par with the furrow irrigation scheduled at $0.8ET_c$. Drip irrigated silver plastic mulch enhanced the yield by 13.69 per cent than biodegradable plastic mulch, 28.74 per cent than drip irrigation without mulch and 52 per cent than furrow irrigation at $0.8\ ET_c$. Mulch to drip irrigation saved the irrigation water by 34 per cent over furrow irrigation. Benefit cost ratio for drip with silver plastic mulch, biodegradable plastic mulch, drip without mulch and furrow irrigation system was 2.64, 1.38, 2.38 and 1.60 respectively.

References

Allen, R.G., Pererira, L.S., Raes, D. and Smith, M. 1998. Crop evapotranspiration. Guidelines for computing crop water requirements. *FAO irrigation and drainage paper No. 56.* FAO, Rome.

Jordan, C.F. 2004. Organic farming and agroforestry: Alley cropping for mulch production for organic farms of southeastern United States. *Agroforestry Systems.* 61/62:79–90.

Muruga0,n M., Gopinath, G. 2001. Effect of organic and inorganic mulches on growth and flowering of crossandra (Crossandra undulaefolia Salisb) cv. "Saundarya". *Research on Crops*. 2: 346-350.

Naeini, S.A.R.M. and Cook, H.F. 2000. Influence of municipal compost on temperature, water, nutrient status and the yield of maize in a temperate soil. Soil Use and Management 16, 215–221.

Ravi, V. and Lourduraj, A. C. 1996. Comparative performance of plastic mulching on soil moisture content, soil temperature and yield of rainfed cotton. *Madras Agricultural Journal* 83: 709-711.

Sivanappan, R. K. 2004. Irrigation and rainwater management for improving water use efficiency in cotton. In: Proceedings of the International Symposium on Strategies for Sustainable cotton production, A Global Vision. 23-25, November, 1998, Dharward, Karnataka, India.

Zhang, D.Q., Liao, Y.C. and Jia, Z. K. 2005. Research advances and prospects of film mulching in arid and semi-arid areas. *Agric. Res. Arid Areas*. 23(1): 208-213.

25

Mitigation Strategies for Green House Gas Emission through Agriculture

G. V. Prajapati, H. Y. Maheta, H. V. Parmar, V. K. Sardhara and P. A. Damor

Abstract

Agricultural lands occupy 40-50 per cent of the earth's land surface. Agriculture accounted for 10-12 per cent of total global anthropogenic emissions of greenhouse gases. Globally, agricultural CH_4 and N_2O emissions have increased by nearly 17 per cent from 1990 to 2005, an average annual emission increase of about 60 $MtCO_2$-eq/yr. Despite large annual exchanges of CO_2 between the atmosphere and agricultural lands, the net flux is small. Many agricultural practices can potentially mitigate greenhouse gas emissions, the most prominent of which are improved cropland and grazing land management and restoration of degraded lands and cultivated organic soils as well as significant mitigation potential is also provided by water and rice management, set-aside, agroforestry, land use change manure and livestock management.

Key words: *Agriculture, Climate Change, Greenhouse gas.*

Introduction

Agriculture accounts for roughly 14 per cent of global GHGs or about 6.8 Gt of CO_2 equivalent per year (IPCC, 2007). GHG emissions from land-use change, including deforestation in tropical areas, are around 17 per cent of total GHG emissions. In most countries they are associated with agricultural activities and exceed emissions from all other agricultural sources. About 74 per cent of total agricultural emissions originate in developing countries. The technical mitigation potential of agriculture has been calculated as 5.5-6 Gt of CO_2 e per year by 2030 (IPPC, 2007). This potential is extremely large, especially relative to emissions from the sector. About 89 per cent of this potential could be achieved through soil carbon sequestration. Mitigation of

CH_4 can provide nine per cent (through improvements in rice management and livestock/manure management) and mitigation of N_2O can provide 2 per cent (primarily through cropland management). The majority of the potential (70 per cent) can be realised in developing countries (Smith *et al.* 2007). Livestock manure, nitrogenous fertilisers and irrigated paddy are said to be responsible for producing most agricultural nitrous oxide and methane emissions. Of these emissions, methane accounts for 3.3 Gt equivalents and nitrous oxide for 2.8 Gt CO_2 equivalents annually, while net emissions of CO_2, at only 0.04 Gt CO_2 equivalents per year, are small. While, in India agriculture sector contributes 28 per cent of the total GHG emissions from India (NATCOM, 2004). In a more recent estimate Bhatia *et al.*, 2004 estimated for the base year 1994-95, that the CH_4 and N_2O emissions from Indian agricultural fields were 2.9 Tg (61 Tg CO_2 equivalent) and 0.08 Tg (39 Tg CO_2 equivalent), respectively.

Although challenges remain, agriculture has much to offer in helping to reduce net GHG emissions to the atmosphere, while at the same time improving the environment and the sustainability of the agricultural sector. Further research and development will result in improved assessments of GHG contributions from agriculture, increases in agriculture's contribution to renewable energy, better ways to manage lands, and design of more efficient policies. Government policy plays an important role in making best management practices and biofuel production economically attractive, and farmers will adopt best management practices for GHG reduction only if they seem profitable.

Role of Agriculture in mitigating climate change

Several farming practices and technologies can reduce greenhouse gas emissions and prevent climate change by enhancing carbon storage in soils; preserving existing soil carbon; and reducing carbon dioxide, methane and nitrous oxide emissions.

Cropland Management

Numerous field experiments show that soil carbon stocks tend to attain levels that are roughly proportional to the annual rate of carbon added. A number of management practices are available to increase cropland soil carbon inputs. Increasing soil carbon inputs by increasing the productivity of crops is largely in line with farmer's management goals of achieving high productivity. Carbon inputs to soil can also be increased by using crop rotations with high residue yields, by reducing or eliminating the fallow period between successive crops in annual crop rotations, and by making efficient use of fertiliser and manure. On annual croplands, soil carbon losses can be reduced by decreasing the frequency and intensity of soil tillage, in particular through conversion to no-till practices.

(1) Use of high-residue crops and grasses

Annual crops that produce large amounts of residues such as corn and sorghum typically result in higher soil carbon levels than many other crops. Hay and pasture lands also tend to have high carbon inputs because perennial grasses allocate a large portion of their total carbon assimilation to root growth. For example, long-term experiments at two sites in Ohio (Dick *et al.* 1998) show about 10 t ha^{-1}. More soil carbon after 30 years under continuous corn crops or with corn-oat-hay rotations than for corn soybean rotations, equal to an average gain of 0.3 t ha^{-1} yr^{-1}. Conversion from continuous cereal cropping to cereal-hay rotations was estimated to increase soil carbon by about one per cent per year, or about 0.5 t ha^{-1} yr^{-1} (Smith *et al.,* 1997a).

(2) Reduction of fallow periods between crops

In semi-arid regions like the Great Plains, summer fallow was developed as a way of storing soil moisture to improve yields and reduce crop failure. However, summer fallow practices caused high rates of soil carbon loss and soil degradation in large areas. More recently, new cropping systems that combine winter wheat with summer season crops in rotation using no-till practices have proved successful in both improving soil moisture and increasing soil carbon (Peterson *et al.* 1998). In more humid regions, where fields may be left fallow in winter, it is often feasible to grow winter cover crops, usually legumes or annual grasses, and thus maintain vegetation year round. Cover crops serve several functions, including taking up excess soil nutrients to reduce leaching or other losses to the environment, fixing atmospheric nitrogen, and controlling weeds; but they also serve to augment the input of plant residues, thereby increasing soil carbon content.

(3) Efficient use of manures, nitrogen fertilisers, and irrigation

As a general rule, promoting the efficient use of inputs such as fertiliser, manure, and irrigation will yield the best results for GHG mitigation (Paustian *et al.* 2000). If high rates of crop production are achieved primarily through increased nitrogen fertilisation and irrigation, increases in other GHG emissions, particularly nitrous oxide, can offset part or all of the gains in soil carbon. Tailoring fertiliser and manure applications to satisfy crop nitrogen demands, so that less nitrogen is left behind in the soil, can reduce nitrous oxide emissions while building soil carbon stocks. Efficient use of irrigation water will similarly reduce nitrogen losses, including nitrous oxide emissions, and minimise CO_2 emissions from energy used for pumping while maintaining high yields and crop-residue production.

Conservation Agriculture (CA)

CA aims to pursue more sustainable agriculture and rural development through the application of three major principles: minimal or no soil disturbance, permanent soil cover an crop rotations.

(i) Minimum/No Soil Disturbance

Conservation tillage is any tillage and planting system in which 30 per cent or more of the crop residue remains on the soil surface after planting to reduce soil erosion by water. Where soil erosion by wind is the primary concern, conservation tillage is any system that maintains at least 1,000 kg ha^{-1} of flat, small-grain residue equivalent on the surface throughout the critical wind erosion period. Conservation tillage can include specific tillage types such as no-till, ridge-till, mulch-till, zone-till, and strip-till systems that meet the residue requirements. These tillage systems are chosen by farmers to address soil type, crop grown, machinery available, and local practice. Although these systems were originally developed to address water quality, soil erosion, and agricultural sustainability problems, they also lead to higher soil carbon and increased fuel efficiency.

(ii) Permanent Soil Cover

Permanent soil cover needs to be integrated into farming systems to obtain additional benefits. Crop residues will not be burnt since they are made part of the permanent soil cover, and air pollution will thus be reduced where burning is stopped. Residues from previously planted crops, other cover crops, and green manure cover crops are utilised for permanent or semi - permanent organic soil cover. The dead residue biomass of the cover crops functions as mulch, protecting the soil physically from sun, rain and wind. Soil mulch reduces water evaporation, conserves moisture, and helps moderate soil temperature, making conditions more hospitable for below ground biota. Mineralisation and nutrient losses are reduced, and more satisfactory levels of organic soil matter are built up and maintained.

(iii) Crop Rotations

The use of crop rotation will help to control pests, diseases, weeds and other biotic factors. Well balanced crop rotations can neutralise many of the possibly negative aspects of minimum/no-tillage, such as pest build-up, as they increase the diversity of favourable insects and organisms that can help maintain checks on the spread and impact of pests and diseases.

Benefits of Conservation Agriculture

CA can bring socio-economic benefits to farmers and ecological/ environmental benefits to climate change mitigation. Conventional tillage is a fossil energy intensive process. CA reduces the energy consumption of farming operations and increases energy productivity; this is the yield output per energy input in the range of 15 per cent -50 per cent and 25 per cent -100 per cent, respectively. It also helps to reduced soil erosion, increased organic matter and improved soil fertility/health, recharge of the aquifers through improved water infiltration, decreased soil compaction, reduced air pollution through reduced release of carbon gases and nitrous oxides, increased carbon sequestration, more micro biota and improved biodiversity. The adoption of CA techniques has seen a steady increase worldwide since the 1990s.

Grazing Lands Management

Grazing lands occupy 1,900 to 4,400 Mha, depending on definitions recent global estimates have placed the area of grazing lands at 3,200 Mha (FAO, 1999). Grasslands contain 10-30 per cent of the world's soil carbon including approximately 200-420 Gt organic C and 470-550 Gt of carbonate C to a depth of 1 m. Management activities can significantly affect carbon storage by reducing carbon loss in degradation processes or increasing carbon inputs and residence times.

Permanent grasslands used as pastures, rangelands, and hayfields can maintain large soil carbon stocks due to several characteristics. Perennial grasses allocate a high proportion of photo synthetically fixed carbon below ground, maintain plant cover year round, and promote formation of stable soil aggregates. Improved land conditions where overgrazing is addressed reduces erosion; reduces methane emissions by reducing animal numbers and improving intake quality; and is likely to reduce nitrous oxide emissions as a result of lower levels of excreted nitrogen and probably greater uptake by vegetation.

Agro-forestry

Agroforestry encompass a wide variety of practices, including crop fallow rotations, complex agro-forests, simple agro-forests, silvo-pastoral systems, and urban agro-forestry. Agro-forestry practices in the temperate regions include planting trees at wide spacing in combination with pastures or crops; this practice results in increasing carbon density. Examples in Canada include the Permanent Cover Programme under the Prairie Farm Rehabilitation Administration. Agroforestry in North America was reviewed by Lassoie and Buck (1999). Agroforestry systems can be superior to other land uses at the global, regional, watershed, and farm scales because they optimise tradeoffs between increased food production, poverty alleviation, and environmental conservation (Izac and Sanchez, 2000). They can also be inferior to other land uses, particularly when the technology is inappropriate or the accompanying policies are not enabling (Sanchez, 1995). Analysis of tradeoffs between private farmer benefits and global environmental benefits provide a solid basis for partitioning benefits arising from global environmental conventions and protocols.

Forest Management

Forest management is the application of biological, physical, quantitative, managerial, social, and policy principles to the regeneration, tending, utilisation, and conservation of forests to meet specified goals and objectives while maintaining forest productivity. Management intensity spans the range from wilderness set-asides to short-rotation woody cropping systems. Forest management encompasses the full cycle of regeneration, tending, protection, harvest, utilisation, and access.

A vast number and variety of activities exist and are being developed by researchers to manage forests. The end use of wood products is important for two reasons. First, wood products provide a stored carbon stock that depends on the life span of the product. Wood in use stocks are growing larger in many countries and management choices can contribute to further growth in these stocks. Second, greater utilisation of wood allows reduced use of fossil fuel by utilising the wood either directly for energy production or to replace energy intensive products such as steel, aluminium, plaster board, and bricks. There are many possible forest management activities, the following are examples that are likely to alter carbon stocks like; Forest regeneration, Forest fertilisation, Pest management, Forest fire management, Harvest quantity and timing, Low impact harvesting, Reducing forest degradation.

Wetlands Management

Wetlands are defined as areas of land that are inundated for at least part of the year, leading to physico-chemical and biological conditions characteristic of shallowly flooded systems (IPCC, 1996b). Anaerobic conditions associated with inundation slow decomposition rates and allow accumulation of large stores of carbon over long time scales, even in systems of relatively low productivity. Although wetlands occupy only four to six per cent of the Earth's land area (0.53-0.57 Gha) (Matthews and Fung, 1987; Aselmann and Crutzen, 1989), they store an estimated 20-25 per cent of the world's soil carbon (350-535 Gt C) (Gorham, 1995). Most of the wetland area and associated carbon storage is in peatlands in temperate and boreal regions; roughly 10-30 per cent is in the tropics.

Decomposition under anaerobic conditions produces methane greenhouse gas. Wetlands are the largest natural source of methane to the atmosphere, emitting roughly 0.11 Gt CH_4 yr^{-1} of the total of 0.50-0.54 Gt CH_4 yr^{-1} (Fung *et al.*, 1991). Using a Global Warming Potential (GWP) of 21 for CH_4, emissions of 1.7 g CH_4 $m^{-2}yr^{-1}$ will offset the CO_2 sink equivalent to a 0.1 Mg C $ha^{-1}yr^{-1}$ accumulation of organic matter. The range of CH_4 emissions from freshwater wetlands ranges from 7 to 40 g CH_4 $m^{-2}yr^{-1}$; carbon accumulation rates range from small losses up to 0.35 t C $ha^{-1}yr^{-1}$ storage. Most freshwater wetlands therefore are small net GHG sources to the atmosphere. Two exceptions are forested upland peats, which may actually consume small amounts of methane (Moosavi and Crill, 1997) and coastal wetlands, which do not produce significant amounts of methane (e.g., Magenheimer *et al.*, 1996). Wetlands appear to be relatively small sources of N_2O to the atmosphere, except when they are converted for agricultural use. Increased decomposition rates in warmer temperatures if associated with drier conditions may lead to large carbon losses to the atmosphere, particularly from northern peatlands (Gorham, 1995); warmer temperatures may also lead to enhanced CH_4 emissions. Changes in regional hydrology caused by precipitation changes may cause loss or new growth of wetlands locally. Wetlands management takes several forms: conversion to agriculture, drainage for forestry or agriculture, conversion for urban/industrial land uses, creation through construction of dams (energy uses), direct harvesting, and wetlands reconstruction.

Reducing Livestock-related methane and itrous oxide emissions

Livestock-related emissions from digestive processes and animal wastes account for 26 per cent of total agricultural emissions. Although enteric (digestive tract) emissions are more significant (70 per cent of agricultural CH_4 emissions), emissions from livestock wastes have a greater potential for mitigation. Improving manure handling facilities, for example, by covering animal-waste lagoons and capturing and burning the CH_4, can reduce emissions while providing renewable energy and income. Capture and combustion of CH_4 from animal wastes also reduces other environmental problems, including odours and nitrate pollution. Overall the best option for reducing digestive process emissions is to increase the efficiency of livestock production.

Manure storage and management

Manure management in the United States currently accounts for 25 per cent of agricultural CH_4 and six per cent of agricultural N_2O emissions. In addition to GHG production, problems associated with odour and nutrient pollution from animal wastes are widespread. Hence, improvements in manure handling that address both GHG reductions and odour and nutrient problems are of great interest. Manure produced by livestock can emit N_2O and/or CH_4 during storage and following application to soil. The type and rate of these emissions are highly dependent on the storage conditions and characteristics of the manure. In general, storage under anaerobic conditions will produce CH_4 while N_2O emissions will be suppressed. Conversely, piled storage and composting of manure will promote largely aerobic decomposition, suppressing CH_4 emissions but promoting N_2O emissions. Both types of handling and storage methods are widely used in the United States. Opportunities for mitigating N_2O emissions from stockpiled or composted manure are relatively limited. One way to reduce N_2O emissions is by increasing the ratio of bedding material to manure. This immobilises more nitrogen by converting it into organic compounds. Perhaps the most effective measure for reducing manure-related N_2O emissions from stockpiled or composted manure is to apply the manure at rates based on crop needs, thus maximising plant uptake of manure derived nitrogen, and minimising the amount of residual nitrogen left in the soil.

Enteric fermentation

Methane is produced in the digestive tract of animals, particularly in ruminants such as cows, sheep, goats, and camels. This source of CH_4 emissions is termed enteric fermentation. In the United States these emissions amount to about 5.4 MMT of CH_4 yr^{-1} (30.7 MMT carbon-equivalent), which represents about 70 per cent of agricultural CH_4 emissions and 20 per cent of total agricultural GHG emissions on a carbon-equivalent basis. Because CH_4 emissions from enteric fermentation are influenced by the feed quality and digestive efficiency of the animals, improving these will reduce CH_4 emissions. One area where substantial improvements are possible is in improving forage quality for grazing animals on smaller livestock operations through better pasture management (DeRamus *et al.* 2003). Various feed additives such as edible vegetable oils and certain antibiotics can also be used to inhibit the rumen bacteria that produce CH_4 (Teather and Forster, 1998). For a given animal type and food quality, CH_4 production will be roughly proportional to food intake. Thus, increasing the amount of product (meat, milk) per unit of food consumed will effectively reduce CH_4 emissions per unit of product. Ways to increase the production efficiency of individual animals include improved animal genetics (breeding) and animal health.

Summary and Conclusions

Farmers' decisions about whether to adopt new management practices and whether to grow energy crops will ultimately determine the level of success of any agricultural sector GHG mitigation strategy. Farmers' decisions are motivated first and foremost by what they perceive to be most profitable. Thus, mitigation practices must be economically attractive to farmers. If farmers can be persuaded to adopt desired practices, the impacts on GHG emissions could be significant. It is technically feasible that 70 to 220 MMT of carbon could be added to U.S. agricultural soils annually over two to three decades. This would remove 260 to 810 MMT of carbon dioxide (CO_2) from the atmosphere annually, offsetting 4 to 11 per cent of current U.S. GHG emissions. Economic potential to store carbon varies substantially by region, and current studies suggest that at prices of \$50 t^{-1} of carbon (\$13 t^{-1} CO_2), soil carbon increases would be limited to 70 MMT yr^{-1}. If an aggressive research and development programme succeeds in substantially improving per acre yields of energy crops and reducing costs of conversion technologies, biomass from agricultural sources could supply up to 19 per cent of total current U.S. energy consumption. This would yield GHG savings on the order of 180 to 470 MMT of carbon, which is equivalent to reducing CO_2 emissions by 670 to 1,710 MMT CO_2 yr^{-1} (by substituting for fossil fuels) or 9 to 24 per cent of total U.S. year-2004 GHG emissions.

Thus, agricultural mitigation of GHGs includes utilisation of agriculture's sink capacity, reduction of agricultural emissions. Utilisation of agriculture's sink capacity is primarily accomplished through increasing soil carbon stocks. Soil carbon increases, which are typically in the 0.1 to 1 tonnes per hectare per year range, could be achieved through adoption of practices such as:

- Reducing the frequency and intensity of soil tillage,
- Including more hay crops in annual rotations,
- Production of high-residue-yielding crops and reduced fallow periods,
- Improved pasture and rangeland management, and
- Conservation set-asides and restoration of degraded lands.

  Since nitrogen fertiliser use is an important component of modern, high-yield agriculture, more efficient use of nitrogen inputs is the key to reduction of nitrous oxide emissions through:

- Use of soil testing to determine fertiliser requirements;
- Better timing and placement of fertiliser; and
- Use of nitrification inhibitors and controlled-release fertiliser.

References

Agrawal, P.K. 2002. Climate Change and Agriculture: Information Needs and Research Priorities, Presentation at FICCI's International Conference on Capacity Building for Climate Change, October 21, 2002.

Bhatia, A., H. Parikh and P.K. Agarwal, 2004. Inventory of methane and nitrous oxide emissions from agricultural soils of India and their global warming potential. Current Science, 87: 317-324.

Dick, W.A., R.L. Blevins, W.W. Frye, S.E. Peters, D.R. Christensen, F.J. Pierce, and M.L. Vitosh. 1998. Impacts of agricultural management practices on C sequestration in forest-derived soils of the eastern corn belt.

IPCC. 2007. Climate Change 2007: Mitigation. Contribution of Working Group III to the Fourth Assessment Report of the Intergovernmental Panel on Climate Change [B. Metz, O.R. Davidson, P.R. Bosch, R. Dave, L.A. Meyer (eds)]. Chapter 8 Agriculture, pp. 498-540. Cambridge University Press.

Lal, R., L.M. Kimble, R.F. Follett, and C.V. Cole. 1998. The Potential of U.S. Cropland to Sequester Carbon and Mitigate the Greenhouse Effect. Chelsea, MI: Ann Arbor Press.

Lal, R., R.F. Follett, and J.M. Kimble. 2003. Achieving soil carbon sequestration in the United States: A challenge to policy makers. Soil Science 168: 827–845.

Larson, W.E. 1979. Crop residues: energy production or erosion control? Journal of Soil and Water Conservation 34(2): 74–76.

Mall, R.K. *et al.,* 2006. Impact of Climate Change on Indian Agriculture: A Review, Climate Change 78:445-478.

Ministry of Earth Science Annual Report 2006-2007.

NATCOM, 2004. India's initial National communication to the United Nations framework convention on climate change. National communication project, Ministry of Environment and Forest, Govt. of India.

Pathak, H. *et al* (2005), "Greenhouse Gas Emissions from Indian Rice Fields: Calibration and Upscaling Using the DNDC Model", Biogeoscience, 2: 113-123.

Paustian, K., E.T. Elliott, J. Six, and H.W. Hunt. 2000. Management options for reducing CO_2 emissions from agricultural soils. Biogeochemistry 48: 147–163.

Peterson, G.A., A.D. Halvorson, J.L. Havlin, O.R. Jones, D.J. Lyon, and D.L. Tanaka. 1998. Reduced tillage and increasing cropping intensity in the Great Plains conserves soil carbon. Soil Tillage Research 47: 207–218.

Smith, P., D. Martino, Z. Cai, D. Gwary, H.H. Janzen, P. Kumar, B.A. McCarl, S.M. Ogle, F. O.Mara, C. Rice, R.J. Scholes, O. Sirotenko, M. Howden, T. McAllister, G. Pan, V. Romanenkov, U.A. Schneider, and S. Towprayoon. 2007. Policy and technological constraints to implementation of greenhouse gas mitigation options in agriculture. Agriculture, Ecosystems and Environment 118:6-28.

Smith, P., D.S. Powlson, M.J. Glendining, and J.U. Smith. 1997a. Potential for carbon sequestration in European soils: preliminary estimates for five scenarios using results from long-term experiments. Global Change Biology 3:67-79.

26

Mitigation of Climate Change through Organic Agriculture

N. K. Dhamsania, G.V. Prajapati, K. C. Patel, H.Y. Maheta, H. V. Parmar and A.J. Bhatt

Abstract

Organic farming system has proven its advantages in terms of resource efficiency, ecosystem functioning, soil fertility conservation and economic impact by refraining the use of chemosynthetic fertilisers, pesticides and pharmaceuticals in a wide range of experiments. Although organic farming is the oldest form of agriculture; few attentions have been paid to develop this approach. In fact, while organic farming has promoted as environmental friendly approach and developed during the last few years in most developed countries, there are few emphasising on developing countries such as India. Traditional agriculture in India is a kind of small scale and low cost non-certified organic, because almost all practices and processes in these agro-ecosystems are compatible with organic agriculture and holds the key to rural transformation and sustainable livelihood in India. Furthermore, organic agriculture is reported to be climate change resilience farming systems and have a competency to diminish Greenhouse Gas Emission mainly due to the less use of fossil fuel and chemical fertilisers, thereby acting as a good option for adaptation to climate change. Further research needed on the contribution of organic agriculture for climate change and yields.

Key words: *Organic agriculture, Climate Change, Greenhouse gas.*

Introduction

In the world total 50 per cent of farmers in the developing country - rural population were smallholders (farming three hectares or less of crop land) and they are generally much poorer than the rest of the population in these countries. Studies from developing countries show that farmers who adopt organic agriculture earn higher income and achieve better living standards. As a result, new opportunities for global partnerships that aim to improve livelihood through safe-food production in developing countries are rapidly emerging. In India more than 60 per cent of arable land is under traditional agriculture, where no synthetic inputs are being used, while only 0.3 per cent of agricultural land coverage is under organic agriculture (Organic India Onesource, 2009). Increase in Organic agricultural land will enables Indian farmers to achieve household food security and gain modest incomes, while regenerating the land, enhancing biodiversity and supplying quality food to local communities. By working with natural processes and making use of locally available assets, poor smallholder farmers can build up the fertility, productivity and makes soil less susceptible to erosion of their farms and avoid dependence on expensive external inputs likes fertilisers, pesticide, use of fossil fuel etc. in agriculture which contributes 10-12 per cent of total green house gas emission in the world. As an engine for sustainable development, organic agriculture has a role to play in the areas to improve livelihood and mitigate climate change.

Organic agriculture holds the key to rural transformation and sustainable development in India. There is an organic revolution brewing throughout the country. The urge to preserve biodiversity and improve farm production without the use of chemical inputs has encouraged farmers to become less dependent on unsustainable inputs. This approach is contributing towards rural livelihood opportunities.

Agriculture: Source and Victim of Climate Change

The current change in global climate is a phenomenon that is largely due to the burning of fossil energy (coal, oil, natural gas) and to the mineralisation of organic matter as a result of land use. These processes have been caused by mankind's exploitation of fossil resources, clearing of natural vegetation and use of these soils for arable cropping. These activities have primarily led to a measurable increase in the carbon dioxide, methane and nitrous oxide content of the atmosphere, an increase which results in global warming. The global warming potential of methane and nitrous oxide is twenty and three hundred times greater than that of CO_2 (International Trade Centre, 2007).

As per the Intergovernmental Panel on Climate Change (IPPC), the annual amount of greenhouse gases emitted by the agricultural sector is estimated at between 5.1 and 6.1 gigatonnes CO_2 equivalents[2] in 2005 (Barker *et. al.*, 2007). This represents approximately 10-12 per cent of total greenhouse gas emissions.

Livestock manure, nitrogenous fertilisers and irrigated paddy are said to be responsible for producing most agricultural nitrous oxide and methane emissions. Of these emissions, methane accounts for 3.3 Gt equivalents and nitrous oxide for 2.8 Gt CO_2 equivalents annually, while net emissions of CO_2, at only 0.04 Gt CO_2 equivalents per year, are small (Smith *et. al.,* 2007). While, in India agriculture sector contributes 28 per cent of the total GHG emissions from India (NATCOM, 2004). In a more recent estimate Bhatia *et al.* (2004) estimated for the base year 1994-95, that the CH_4 and N_2O emissions from Indian agricultural fields were 2.9 Tg (61 Tg CO_2 equivalent) and 0.08 Tg (39 Tg CO_2 equivalent), respectively.

Mitigation of Climate Change through Organic Agriculture

Organic agriculture not only enables ecosystems to better adjust to the effects of climate change but also offers potential to reduce the emissions of agricultural greenhouse gases by complements emission reduction efforts with its major sequestration potential, which is based on the intensive humus production (requiring CO_2) of the fertile soils. In comparison to conventional agriculture, organic agriculture also directly contributes to emission reductions as it emits less N_2O from nitrogen application (due to lower nitrogen input), less N_2O and CH_4 from biomass waste burning (as burning is avoided), and requires less energy, mainly due to zero chemical fertiliser use.

Reduction of nitrous oxide

The global warming potential of conventional agriculture is strongly affected by the use of synthetic nitrogen fertilisers and by high nitrogen concentrations in soils. One per cent of global fossil energy consumption is used for chemical nitrogen fertiliser production, emitting about 0.23 Gt CO_2eq. Organic agriculture avoids these emissions, as no chemical nitrogen fertilisers are used. Furthermore, due to lower utilisation of nitrogen input in organic agriculture than in conventional agriculture leads to a potential emission reduction of 1.2-1.6 Gt CO_2eq (Muller *et. al.,* 2009).

In organic agriculture, nitrogen input stems from application of manure and compost is fixed from the air by leguminous plants. Badgley *et.al.* (2007) calculated the potential nitrogen production by leguminous plants via intercropping and off-season cropping to be 154 million tonnes, a potential which exceeds the nitrogen production from fossil fuel by far and which is not fully exploited by conventional farming techniques.

Reduction of Methane

Methane accounts for about 14 per cent of the greenhouse gas emissions (Barker *et. al.,* 2007). Two thirds of this is of anthropogenic origin and mainly from agriculture. Methane emissions stem to a large extent from enteric fermentation and manure management and in consequence are directly proportional to livestock numbers. Avoidance of methane emissions of anthropogenic origin and especially of agricultural origin is of particular importance for mitigation. Organic agriculture has an important, though not always superior, impact on reduction as livestock numbers are limited in organic farms (Weiske *et. al.*, 2006; Kotschi and Muller-Samann, 2004). The data available on methane emissions from livestock is limited, especially with respect to the reduction of GHG emissions from ruminants and manure heaps.

Methane emissions from organic rice production and ruminant production do not differ substantially from those of conventional production. Better rice production practices in organic and conventional agriculture, such as avoiding continuous flooding or choosing low methane emitting varieties (Smith and Conen, 2004) could enhance reduction of methane emissions.

Kotschi and Muller-Samann (2004) suggested that composting and biogas production reduced the methane as well as longevity of animals on organic farms contributes to reduction of methane emission. In organic agriculture, biomass is not burned which leads to reduce the CH_4 and N_2O emissions by about 0.6-0.7 Gt CO_2eq in comparison to conventional agriculture (Smith *et. al.*, 2007).

Carbon sequestration

By sequestering carbon dioxide in the soil, agriculture may contribute to the carbon cycle in a positive way. Whether the balance is positive or negative depends to a large extent on the cutoff points for the comparison (for instance if the fertiliser industry is included or not) as well as the management and farming practices applied.

Soil carbon sequestration is enhanced through agricultural management practices such as increased application of organic manures, use of intercrops, green manures, higher shares of perennial grasslands and trees or hedges etc., which promote greater soil organic matter (and thus soil organic carbon) content and improve soil structure (Niggli *et. al.*, 2008; IFOAM, 2006). Increasing soil organic carbon in agricultural systems has also been pointed out as an important mitigation option by IPCC (2007b). Very rough estimates for the global mitigation potential of organic agriculture amount to 3.5–4.8 Gt CO_2 from carbon sequestration (around 55–80 per cent of total global greenhouse gas emissions from agriculture) and a reduction of N_2O by two-thirds (Niggli *et. al.*, 2008). For sound estimates, however, more information on the mitigation potential of organic agriculture duly differentiated according to climatic zones, local climatic conditions, soil characteristics, variations in crops and cultivation practices, etc. More research is also needed on how organic agriculture performs regarding emissions per unit product.

Concluding Remarks

Organic agriculture as a strategy holds the key to rural transformation and sustainable development in India. The urge to preserve biodiversity and improve farm production without the use of chemical inputs has encouraged farmers to become less dependent on unsustainable inputs. In India, agriculture farming looks a lot like organic agriculture due to no synthetic inputs are being used and benefits have been demonstrated by diversified and integrated organic farm systems that are economically viable, environmentally sound, the flexibility to cope with adverse effects of climate change. This leads to higher economic and ecological stability through optimised ecological balance and risk-spreading. Further research on this, however, still needed. Organic agriculture is labour intensive so it improve the employment and small farmers can earn higher income and achieve better living. But the major constraints restricted the small farmers to fetch higher profit are acknowledgement of the competence of non-international certification bodies, domestic market and training and education of organic agriculture. Organic agriculture, as a mitigation strategy enables ecosystems to better adjust to the effects of climate change and offers potential to reduce the emissions of agricultural greenhouse gases (CO_2, CH_4 and N_2O) with its major sequestration potential, less use of fossil fuel and chemical fertilisers. More research on emissions per unit product in organic agriculture is still needed.

References

APEDA, 2007. Present status of organic agriculture in India. Available at: <http://www.apeda.com/organic/PresentStatus.htm>

Badgley, C., J. Moghtader, E. Quintero, E. Zakem, M. Chappell, K. Avilez-Vazquez, A. Samulon, and I. Perfecto, 2007. Organic agriculture and the global food supply. Renewable Agriculture and Food Systems, 22(2): 86–108.

Barker, T., I. Bashmakov, L. Bernstein, J. E. Bogner, P. R. Bosch, R. Dave, O. R. Davidson, B. S. Fisher, S. Gupta, K. Halsnaes, G.J. Heij, S. Kahn Ribeiro, S. Kobayashi, M. D. Levine, D. L. Martino, O. Masera, B. Metz, L. A. Meyer, G. J. Nabuurs, A. Najam, N. Nakicenovic, H. H. Rogner, J. Roy, J. Sathaye, R. Schock, P. Shukla, R. E. H. Sims, P. Smith, D. A. Tirpak, D. Urge-Vorsatz, D. Zhou, 2007. Technical Summary. In: Climate Change 2007: Mitigation. Contribution of working group III to the fourth assessment report of the Intergovernmental Panel on Climate Change [B. Metz, O. R. Davidson, P. R. Bosch, R. Dave, L. A. Meyer (eds)], Cambridge University Press, Cambridge, United Kingdom and New York, NY, USA. Available at: <http://www.nl/ipcc/pages_media/Fardocs/final_pdfs_ar4/TS.pdf>

Baron, R., and J. Ellis, 2006. Sectoral crediting mechanisms for greenhouse gas mitigation: institutional and operational issues. Organisation for Economic Cooperation and Development and International Energy Agency, COM/ENV/

EPOC/IEA/SLT(2006)4. Paris: OECD and IEA. Available at <http://www.iea.org/ textbase/papers/2006/greenhouse.pdf.>

Bhatia, A., H. Parikh, and P. K. Aggarwal, 2004. Inventory of methane and nitrous oxide emissions from agricultural soils of India and their global warming potential. Current Science, 87 : 317-324.

Hinostroza, M.L., C-C. Cheng, X. Zhu, J. Fenhann, C. Figueres, F. Avendano, C. Figueres and F. Avendano, 2007. "Potentials and barriers for end-use energy efficiency under programmatic CDM." CD4CDM Working Paper Series 3. Roskilde, Denmark: UNEP Risoe Center. Available at: <http://cd4cdm.org / Publications/pCDM&EE.pdf.>

IFOAM (International Federation for Organic Agriculture Movements), 2006. Mitigation and adaptation to climate change: the role of organic agriculture. IFOAM side event to the 24th session of the subsidiary bodies, Bonn, Germany, May 24, 2006.

International Trade Centre, 2007. Organic farming and climate change. Available at : <*http://*www.ifoam.org/...organic/.../FiBLStudy Organic_Farming_and_*Climate_ Change.pdf*>

IPCC, 2007b. Summary for policy makers. In IPCC Fourth Assessment Report, "Working Group III Report: Mitigation of Climate Change, Available at : < http://www.ipcc.ch/pdf/assessment-report/ar4/wg3/ar4-wg3-spm.pdf and http:/ /www.ipcc.ch/ipccreports/ar4-wg3.htm.>

Kotschi, J. and K. Muller-Samann, 2004. The role of organic agriculture in mitigating climate change. International Federation of Organic Agriculture Movements (IFOAM), Bonn.

Lampkin, N., C. Foster, S. Padel, and P. Midmore, 1999. The policy and regulatory environment for organic farming in Europe. In Organic Farming in Europe: Economics and Policy, Vol. 2, Stuttgart- Hohenheim University.

Muller, A., J.S. Davis, P. R. Happerly, and U. Niggli, 2009. Reducing global warming: the potential of organic agriculture - a policy brief. Available at : <http://*mailman.anu.edu.au/pipermail/hepr-vn/.../attachment-0001.pdf*>

NATCOM, 2004. India's initial National communication to the United Nations framework convention on climate change. National communication project, ministry of environment and forest, Govt. of India.

Niggli, U., A. Fliessbach, P. Hepperly, and N. Scialabba, 2008. Low greenhouse gas agriculture: mitigation and adaptation potential of sustainable farming systems. Rome: FAO.

Organic India Onesource, 2009. Organic agricultural scenario in India. Available at : <http://www.organic-india-onesource.50megs.com/about.html>

Smith, K. A., and F. Conen, 2004. Impacts of land management on fluxes of trace greenhouse gases. Soil Use and Management, 20: 255-263.

Smith, P., D. Martino, Z. Cai, D. Gwary, H. Janzen, P. Kumar, B. McCarl, S. Ogle, F. O'Mara, C. Rice, B. Scholes, and O. Sirotenko, 2007. Agriculture in climate change, 2007: mitigation. Contribution of Working Group III to the Fourth Assessment Report of the Intergovernmental Panel on Climate Change [B. Metz, O.R. Davidson, P.R. Bosch, R. Dave, L.A. Meyer (eds)], Cambridge University Press,

Cambridge, UK, and New York, NY, USA. Available at : <http://www.ipcc.ch/pdf/assessment- report/ar4/wg3/ar4-wg3-chapter8.pdf>

Weiske, A., A. Vabitsch, J.E. Olesen, K. Schelde, J. Michel, R. Friedrich, and M. Kaltschmitt, 2006. Mitigation of green house gas emission in European conventional and organic dairy farming. Agriculture, Ecosystems and Environment, 112: 221-232.

27

Domestic Rooftop Water Harvesting for Junagadh City

H.H. Mashru, P.B. Vekariya, H.D. Rank, J.H. Kelaiya, D.M. Pardava, R.J. Patel and H.V. Parmar

Abstract

Water is as important for survival of human being as much as food, air etc, but hardly any attention is paid for its economical use and conservation of this precious resource. Due to indiscriminate pumping of ground water, the water table is going down abnormally and if the problem is not given a serious look, then the future generations may have to face severe crisis of water. Rains are the main source of water and if rain water is harvested, the scarcity of water can be eliminated altogether. This is an ideal solution of water problem where there is inadequate groundwater supply quantitatively and qualitatively and surface sources are either lacking or insignificant. Rain water is bacteriologically pure, free from organic matter and soft in nature. In urban areas, rain water available from rooftop of buildings, paved and unpaved areas goes waste. This water can be stored in tank and can be used directly and also indirectly by diverting to recharge the aquifers through existing GW tapping arrangements and thereafter can be utilised gainfully at the time of need. The study is based on the analysis of survey record of around 50 houses of different rooftop areas of Junagadh city. The estimation of the appropriate size of the water tanks and their costs required to fulfill the annual drinking water demands through Domestic Rooftop Water Harvesting (DRWH) from rooftop of different areas are done. A mathematical equation expressing the relationship between the required size of water tank and different rooftop areas is developed. The DRWH systems for all houses are designed considering the existing rain water outlets and cost estimation for each individual house is done. A cost model expressing the relationship between rooftop area and cost of DRWH system is developed.

Keyword: *Rooftop water harvesting, Domestic water demand.*

Introduction

Water is the most common or major substance on the earth, covering more than 70 per cent of its surface. Out of the total volume of water available on the surface of the earth, only two per cent is fresh water. The fresh water is used for the purpose of human use, industries and agriculture. In India, the water availability per capita is declining. The per capita availability of water at the national level is reduced from about 5,177 m^3 in the year 1951 to the present level of 1,869 m^3. The prominent reasons behind these are the increasing demand for water due to the increasing population and extensive use of water by agricultural sector, which continues to be the single largest consumer of water (CGWB, 2000 and IE-NLC, 2006). In fact, India is blessed with adequate rainfall as a whole, yet there are large swathes of dry and drought prone area. In many places the quality of groundwater is not good. In such places rainwater harvesting may provide lifeline for survival.

Need for rainwater harvesting

Due to over population and higher usage levels of water in urban areas, water supply agencies are unable to cope up demand from available surface sources especially during summer seasons. Mudrakartha and Chopade (2002). This has led to digging of individual tube wells by house owners. Even water supply agencies have resorted to groundwater sources by digging tube-wells in order to augment the water supply. The replenishment of ground water is drastically reduced due to paving of open areas. Indiscriminate exploitation of ground water results in lowering of ground water table (GWT) rendering many bore-wells dry, which has led to drilling of bore wells of greater depth. This further lowers the water table such frequent fluctuations in GWT results in presence of higher concentration of salts in ground water. In coastal areas, over exploitation of ground water results in seawater intrusion thereby rendering fresh ground water bodies' saline LRHF (2001).

Rooftop water harvesting

Rooftop water harvesting is a process of collecting of runoff during rains from impermeable surfaces on houses or close to houses, its storage in water proof vessels and its subsequent use for the inhabitants of the houses. The use may be temporary (within a day or so following a rainstorm), seasonal (throughout the rainy season) or permanent (throughout the year) except in years of exceptionally low rainfall. The rain water from the roof may also be used for recharging the ground water through nearby water sources such as open dug wells or bore wells, Kumar (2004). It may be achieved in the case of storing the harvested water from roof by diverting the excess water for ground water recharge and in absence of storing vessel by diverting all the water for ground water recharge. Rooftop rainwater harvesting for household purpose only represent a small part of the total water balances. In areas with significant variations in the annual rainfall pattern, the matching of water supply and water demand may be difficult. However, in terms of economic and human welfare it has a crucial role to play. Rainwater in many cases is the easiest to access, most reliable, and least polluted source. It can be collected and controlled by the individual household or community as it is not open to abuse by other users.Water scarcity is a characteristic of north-western states of India, such as Gujarat. Over time, the continuous increase of the population as well as the financial, administrative and technical deficiencies of the new supply system has lead to the deterioration of the water service in the city areas. Patra, and Sneha (2011). In the meantime, the water demand has considerably increased due to the improvement of standards of living. This has resulted in an increasing pressure on underground water resources, which has lead to an alarming depletion of aquifers. Hence, keeping in view all the problem, Domestic Rainwater harvesting system can be considered as a best solution for fighting against scarcity of water. Moreover, owing to its simple technique, ease of construction and installation and low cost of investment, this technique again suits for implementation in residential area. Rainwater harvesting can meet potable and non-potable water demands and also control flooding and can easily meet the potable water demand and also able to save money which is being spends for procuring potable-water. Jethoo and Poonia (2006) In this way potable water can be conserve and harvested rainwater plays major part in conserving it. Keeping this in view, the present study was undertaken to determine the optimum size of water storage tank for different rooftop area and cost model of domestic rooftop water harvesting system.

Study area

Junagadh is located at the foothills of mount Girnar. Geographically it is situated at 21°31' N latitude, 70°31' E longitudes at an altitude of 61 m above mean sea level, 85 km. away from Arabian Sea at Somnath. Junagadh has a tropical climate and receives rainfall during South-west monsoon (June-September). Monthly rainfall records from year 1971 - 2011 were collected from Agro-meteorological observatory of Junagadh Agricultural University, Junagadh to analyse the average monthly rainfall amount available. Survey was carried near to Motibagh campus area of Junagadh city. Around 50 houses having different rooftop area are surveyed. The DRWH potential for different sizes of rooftop is estimated. The annual rainfall quantum in Junagadh City is less and hence the gross demand of a family can not be fulfilled by DRWH and at the same time number of days of rainfall is another constraint. The estimation of the size of the water tanks to fulfill drinking and cooking water demand @ 8.00 LPCD and their costs required to fulfill the annual drinking water demand through DRWH from rooftop of different areas are done. A general mathematical equation expressing the relationship between the size and cost of water tank and rooftop areas of different houses is developed.

The cost estimations of proposed DRWH system for all houses are done and a cost model expressing the relationship between rooftop area and cost of DRWH system is developed.

Tank size and GW recharge from rooftop area

The estimations of tank sizes to fulfill the drinking and cooking water demands @ 8.00 LPCD are done. The sizes of water tanks and the overflow from tanks for GW recharge are estimated. The month-wise status of quantity of water available in the water tanks of different capacities are estimated. A calculation for 50 m^2 rooftop area is shown in Table 1.

Percentage of water demand fulfilled by DRWH from different rooftop area

As the non-monsoon rainfall is almost negligible in Junagadh city, hence it is not feasible to fulfill the gross water demand throughout the year from DRWH. Even during monsoon period this demand cannot met from rooftop area less than 50 sq. m. except for the month of July. It is evident from the Table 2 that the gross water demand for four monsoon months can be met from the roof top area 100 sq.m and above. Quantity of water available at the end of each month from DRWH systems are shown in Table 3.

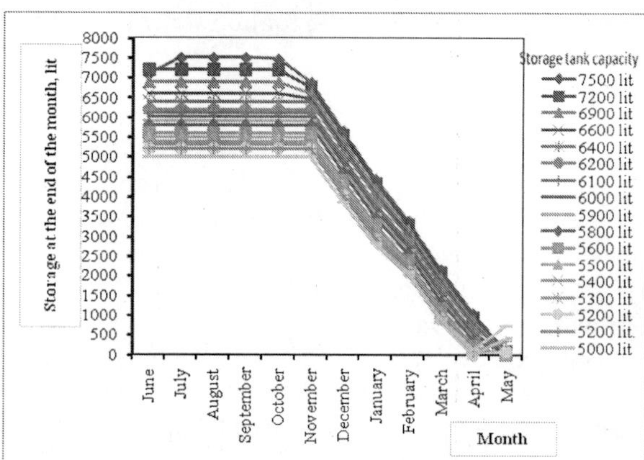

Fig. 1: Quantity of water available at the end of month in the storage tank, lit.

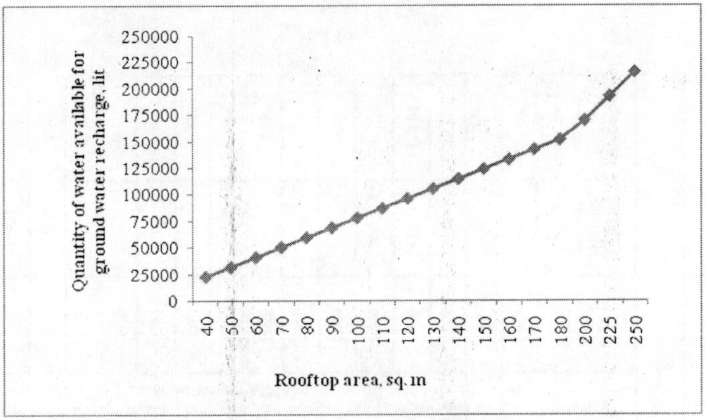

Fig. 2: Quantity of water available for ground water recharge, lit.

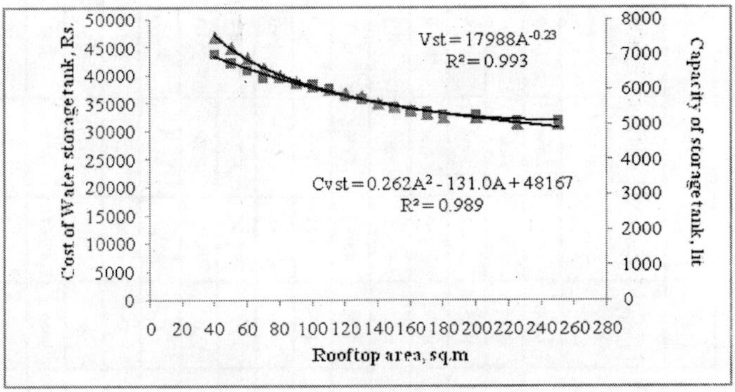

Fig. 3: Cost of water storage tank and capacity of water storage tank vs. rooftop area.

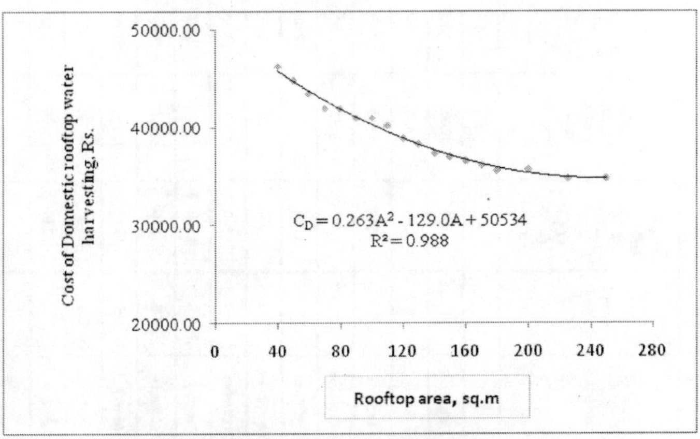

Fig. 4: Cost model of DRWH system.

Table 1: Estimation of water tank's capacity to fulfill the drinking and cooking water demand by DRWH from rooftop area 50 m²

Month	*Average Rainfall (mm)*	*Demand Drinking and cooking @ 8 LPCD (for 5 person)*	*Cumulative demand (Drinking and Cooking)*	*Demand (Overall) @ 80 LPCD*	*Cumulative demand (Overall @ 80 LPCD*	*Rooftop Area = 50 sq. mt.; R.O. Coefficient = 0.95*							
						DRWH Volume	*Cumul-ative DRWH*	*DRWH Volume - Demand*	*CDRWH – Cumu. Demand*	*Remarks*	*End of month storage*	*DRWH available for GW Storage*	*Percentage of Total demand can met from DRWH*
						Litres	*Litres*	*Litres*	*Litres*		*Litres*	*Litres*	
1	**3**	**4**	**5**	**6**	**7**	**8**	**9**	**10 = 8 - 4**	**11 = 9 - 5**	**12**	**13**	**14**	**15**
June	218.0	1200	1200	12000	12000	10354	10354	9154	9154	**Drinking and cooking water demand can be met from DRWH. The size of the storage tank = 7200 liters. The Surplus water for ground water recharge = 31384 liters.**	**7200**	1954	86.29
July	341.7	1240	2440	12400	24400	16230	26584	14990	24144		**7200**	14990	130.89
August	212.6	1240	3680	12400	36800	10097	36681	8857	33001		**7200**	8857	81.42
September	135.5	1200	4880	12000	48800	6436	43117	5236	38237		**7200**	5236	53.63
October	31.7	1240	6120	12400	61200	1507	44624	267	38504		**7200**	267	12.15
November	16.0	1200	7320	12000	73200	760	45383	-440	38063		6760	0	6.33
December	0.4	1240	8560	12400	85600	20	45404	-1220	36844		5540	0	0.16
January	0.8	1240	9800	12400	98000	39	45443	-1201	35643		4339	0	0.32
February	1.6	1120	10920	11200	109200	77	45520	-1043	34600		3296	0	0.69
March	0.3	1240	12160	12400	121600	13	45534	-1227	33374		2070	0	0.11
April	1.8	1200	13360	12000	133600	84	45617	-1116	32257		953	0	0.70
May	7.7	1240	14600	12400	146000	366	45984	-874	31384		80	0	2.96
								-7120	**31384**			**31304**	

Table 2. Percentage of total water demand fulfilled by DRWH from different rooftop area

Month	Rooftop area, sq.mt.																	
	40	50	60	70	80	90	100	110	120	130	140	150	160	170	180	200	225	250
	% of total demand fulfilled from DR WH																	
Jan.	0.25	0.32	0.38	0.44	0.51	0.57	0.63	0.70	0.76	0.82	0.89	0.95	1.01	1.08	1.14	1.27	1.43	1.58
Feb.	0.55	0.69	0.83	0.97	1.11	1.24	1.38	1.52	1.66	1.80	1.93	2.07	2.21	2.35	2.49	2.76	3.11	3.45
March	0.09	0.11	0.13	0.15	0.17	0.20	0.22	0.24	0.26	0.28	0.30	0.33	0.35	0.37	0.39	0.43	0.49	0.54
April	0.56	0.70	0.84	0.97	1.11	1.25	1.39	1.53	1.67	1.81	1.95	2.09	2.23	2.37	2.51	2.78	3.13	3.48
May	2.36	2.96	3.55	4.14	4.73	5.32	5.91	6.50	7.09	7.68	8.27	8.87	9.46	10.05	10.64	11.82	13.30	14.78
June	69.03	86.29	103.5	120.8	138.0	155.3	172.5	189.8	207.0	224.3	241.6	258.8	276.1	293.3	310.6	345.1	388.2	431.4
July	104.7	130.8	157.0	183.2	209.4	235.6	261.7	287.9	314.1	340.3	366.4	392.6	418.8	445.0	471.2	523.5	588.9	654.4
Augu.	65.14	81.42	97.71	113.9	130.2	146.5	162.8	179.1	195.4	211.7	227.9	244.2	260.5	276.8	293.1	325.6	366.4	407.1
Sept.	42.91	53.63	64.36	75.08	85.81	96.54	107.2	117.9	128.7	139.4	150.1	160.8	171.6	182.3	193.0	214.5	241.3	268.1
Oct.	9.72	12.15	14.58	17.01	19.44	21.87	24.30	26.74	29.17	31.60	34.03	36.46	38.89	41.32	43.75	48.61	54.69	60.76
Nove.	5.06	6.33	7.60	8.86	10.13	11.39	12.66	13.92	15.19	16.46	17.72	18.99	20.25	21.52	22.79	25.32	28.48	31.65
Dec.	0.13	0.16	0.20	0.23	0.26	0.29	0.33	0.36	0.39	0.43	0.46	0.49	0.52	0.56	0.59	0.66	0.74	0.82

Table 3. Quantity of water available at end of month from DRWH from different rooftop area and storage tank capacity

Month	Rooftop area, sq.m																	
	40	50	60	70	80	90	100	110	120	130	140	150	160	170	180	200	225	250
	Storage tank capacity, lit																	
	7500	7200	6900	6600	6400	6200	6100	6000	5900	5800	5600	5500	5400	5300	5200	5200	5000	5000
June	7084	7200	6900	6600	6400	6200	6100	6000	5900	5800	5600	5500	5400	5300	5200	5200	5000	5000
July	7500	7200	6900	6600	6400	6200	6100	6000	5900	5800	5600	5500	5400	5300	5200	5200	5000	5000
August	7500	7200	6900	6600	6400	6200	6100	6000	5900	5800	5600	5500	5400	5300	5200	5200	5000	5000
September	7500	7200	6900	6600	6400	6200	6100	6000	5900	5800	5600	5500	5400	5300	5200	5200	5000	5000
October	7466	7200	6900	6600	6400	6200	6100	6000	5900	5800	5600	5500	5400	5300	5200	5200	5000	5000
November	6873	6760	6611	6463	6400	6200	6100	6000	5900	5800	5600	5500	5400	5300	5200	5200	5000	5000
December	5649	5540	5396	5252	5192	4997	4901	4805	4709	4613	4417	4321	4225	4129	4033	4041	3851	3862
January	4441	4339	4203	4067	4015	3827	3739	3651	3563	3475	3287	3199	3111	3023	2934	2958	2788	2818
February	3383	3296	3176	3055	3019	2847	2774	2701	2629	2556	2384	2311	2238	2166	2093	2148	2016	2085
March	2153	2070	1952	1834	1801	1631	1561	1491	1421	1351	1181	1111	1041	971	901	962	837	912
April	1020	953	852	751	734	581	528	475	421	368	215	162	109	55	2	96	13	130
May	73	80	52	24	81	1	21	41	61	81	1	21	41	61	81	322	422	722

Size of water storage tank and cost of water storage tank :

Figure 3 shows the cost involved for construction of DRWH water storage tank for different rooftop area. This figure also shows the size of DRWH storage tank required for different rooftop area. Cost involved for construction of water storage tank to fulfill the drinking and cooking water demand through DRWH are estimated as per Schedule of Rate, Executive Engineer (R & B), Junagadh 2011-12. Following mathematical model is developed to relate the cost involved for construction of DRWH storage tank to rooftop area

$$C_{wst} = 0.262\,A^2 - 131.0A + 48167 \qquad ...(1)$$

where C_{wst} = Cost of construction of water storage tank, Rs., A = Area of rooftop, sq. mt

The mathematical model expressing the relationship between the required size of water storage tank and rooftop area is

$$V_{st} = 17988\,A^{-0.23} \qquad ...(2)$$

where V_{st} = Volume of water storage tank, lit

A = Area of roof top, sq. mt.

Cost model of DRWH system

A survey of around 50 individual houses in Junagadh city was conducted; most of houses are duplex type. The DRWH is planned for these houses considering the existing roof's drainage system. The minimum size of the conveyance pipe is taken as 63 mm. The quantity of pipe and other relevant items measured as on actual up to the boundary limit of plot of the house, in such a way that the harvested water can be used for purpose of ground water recharging or and for direct use or for both. The plot of cost of DRWH versus the area of rooftop of surveyed houses is shown in Fig. 4. The following mathematical model for finding out the cost of DRWH system for different rooftop area required to fulfill the drinking and cooking water demand is developed.

$$C_D = 0.263\,A^2 - 129.0\,A + 50534 \qquad ...(3)$$

where C_D = Cost of DRWH system in Rs., A = Rooftop area, sq.m

References

GWB. 2000. "Guide on Artificial Recharge to Ground Water prepared by Central Ground Water Board", Ministry of Water Resources, Government of India, New Delhi, May.

Kumar, D. 2004. "Roof Water Harvesting for Domestic Water Security: Who Gains and Who Loses?" International Water Management Institute, Gujarat, India. Water International, Vol. 29, No. 1, pp. 65-76.

IE-NLC. 2006. "Rain Water Harvesting and Water Management" Proceedings of 22nd National Convention of Environmental Engineers, 11-12 November 2006, Organized by Institution of Engineers (India) Nagpur Local Centre, November.

Jethoo and M.P. Poonia. 2006. "Traditional Domestic Rain Water Harvesting System Suitable for Rajasthan State and Their Sizing- A Review", Jaipur (India). International Journal of Modern Engineering Research (IJMER). Vol.1, Issue 2, pp. 715-718.

LRHF. 2001. "A Report Domestic Roof Water Harvesting and Water Security in the Humid Tropics" prepared by Lanka Rainwater Harvesting Forum under Domestic Roof Water Harvesting in the Humid Tropic, June.

Mudrakartha and Chopade. 2002. "Demand Supply gap through Rain Water Harvesting- A Case Study" International Symposium on Artificial Recharge (ISAR-4), Adelaide, Australia.

Patra, A.K. and G. Sneha. 2011. "A Pilot Scheme for Rooftop Rainwater Harvesting at Centre of Mining Environment, Dhanbad." International Journal of Environmental Sciences, Vol. 1, No. 7, pp. 1542-1548.

28

Water Use Efficiency and Productivity of *Bt.* Cotton as Influenced by Precision Application of Irrigation and Fertilisers Through Drip System under South Gujarat Condition

V. P. Usadadia, J. G. Patel, C. M. Sutaria, K. B. Sankat and C. K. Patel

Abstract

A field experiment was conducted at Main Cotton Research Station, Surat during Kharif *2008-09 to 2010-11. The soils of experimental site is deep black, low in organic carbon, medium in available P and high in available K with 7.9 pH. The experiment consisted of sixteen treatments in which, four levels of irrigation scheduling (0.6, 0.8, 1.0 Etc through drip irrigation and surface irrigation at 0.8 IW/CPC) and four levels of nitrogen (75, 100 and 125 per cent RDN in six splits with 15 days interval through drip and 100 per cent RDN in three splits with 30 days interval as soil application) and treatments evaluated under split plot design with three replications. RDN for* Bt *cotton was 240 kg/ha. Growth, yield and yield attributes were significantly highest recorded with irrigation scheduling at 0.8 Etc treatment, but which was at par with surface irrigation in case of seed cotton yield. As regarding WUE and WP, irrigation scheduling at 0.8 Etc was best followed by surface irrigation. Levels of nitrogen had no significant effect on yield and yield attributes, yet 125 per cent RDN through drip in six equal splits yielded maximum seed cotton compared to conventional fertilizers applied in soil. The maximum NUE was noted under 75 per cent RDN applied through drip system. Irrigation scheduling at 0.8 Etc recorded the maximum net return of Rs 151126/ ha with BCR of 7.44. In case of fertiliser levels, 125 per cent RDN recorded the highest net return of Rs 132818/ha with BCR 6.08, whereas the highest BCR (6.33) recorded under 75 per cent RDN applied through drip system in six equal splits. The treatment combination irrigation scheduling at 0.8 Etc through drip system along with 75 per cent RDN applied in six equal splits at 15 days interval produced the highest seed cotton yield and net return.*

Introduction

Cotton is an important commercial crop and it plays an important role in state economy. Although *Bt* cotton is cultivated in about 83 per cent of the cotton growing area, the agronomic recommendations developed for conventional hybrids are being advocated to *Bt* hybrids also. However, considering the higher yield potential, altered morphoframe, shorter duration and more synchronous boll setting, more so in the form of the first flush, it would be appropriate to develop site-specific nutrient and water management technologies to harness the higher yield potential, if any of *Bt* cotton. Application of irrigation based on crop evapotranspiration (Etc.) and fertigation are the new approaches for water and nutrient management. Management of water and nutrients play a key role in breaking the undesired tempo for enhancing the productivity.

The maximum water use efficiency (WUE) under drip irrigation, due to precise application of water (Manjunath *et al.*, 2010). Fertiliser is also applied through drip system with irrigation water NUE reach as high as 90 per cent of applied nutrients. Hence, an experiment was planned to find out suitable irrigation schedule and fertiliser requirement for *Bt* cotton under South Gujarat conditions.

Materials and methods

A field experiment was conducted at Main Cotton Research Station, NAU, Surat during *Kharif* 2008-09 to 2010-11. The soils of experimental site is deep black, low in organic carbon, medium in available P and high in available K with 7.9 pH. The experiment consisted of sixteen treatments in which, four levels of irrigation scheduling (0.6, 0.8, 1.0 Etc through drip irrigation and surface irrigation at 0.8 IW/CPC) and four levels of nitrogen (75, 100 and 125 per cent RDN in six splits with 15 days interval through drip and 100 per cent RDN in three splits with 30 days interval as soil application) and treatments evaluated under split plot design with three replications. For *Bt* cotton 240 kg RDN/ha. Cotton variety "RCH-2" was sown on last week of June at 120 x 45 cm during all the three years of experimentation. The on-line drippers of 4 LPH discharge capacity were operated at a pressure of 1 kg/cm^2. In drip system, water was applied on every alternate day while in surface method 60 mm depth of irrigation water was applied at 0.8 IW/CPE ratio.

Results and discussion

Irrigation scheduling

The data presented in Table 1 showed that plant height and number of sympodia per plant were significantly influenced due to irrigation scheduling. Significantly the highest plant height (81.4 cm) and sympodia per plant (16.2) were noticed with I_2 (irrigation scheduling at 0.8 Etc) treatment. Irrigation scheduling significantly increased number of bolls/plant, seed cotton/plant and seed cotton yield. Irrigation scheduling at 0.8 Etc recorded maximum number of bolls/plant (41.7) and seed cotton/plant (164.6 g), which was at par with irrigation given in surface irrigation treatment, while significantly the highest seed cotton yield was produced under the same treatment of irrigation, whereas boll weight remained unaffected due to irrigation scheduling. As regarding WUE and WP, irrigation scheduling at 0.8 Etc was best followed by irrigation given by surface method. This is in confirmation with the findings of Ramamurthy *et al.* (2009) that can increase the cotton yield in addition to water saving. The minimum water use efficiency of 62.54 kg/ha- mm was obtained in conventional method of irrigation *i.e.* surface method (Table 3).

Levels of nitrogen

Levels of nitrogen had no significant effect on yield and yield attributes, yet 125 per cent RDN through drip in six equal splits yielded maximum seed cotton (3312 kg/ha) compared to fertilisers applied in soil. However, plant height was significantly influenced, the highest plant height (79.1 cm) was observed under F_2 (100 per cent RDN *i.e.* 240 kg N/ha through drip in six equal splits with 15 days interval) treatment, whereas, number of sympodial branches per plant did not showed significant results. The maximum NUE was noted under 75 per cent RDN applied through drip irrigation treatment. These results have close conformity to those reported by Zhenan *et al.* (2007) and Pawar *et al.* (2013) that fertiliser applied through drip help in better nutrient uptake and development of cotton plant and ultimately increased the production.

Quality parameters were not significantly affected by irrigation scheduling, nitrogen levels as well as their interaction except, LUR (Table 2).

Economics

Irrigation scheduling at 0.8 Etc recorded the maximum net return of Rs. 151126/ ha with BCR of 7.44 (Fig. 1). In case of fertiliser levels, 125 % RDN recorded the highest net return of Rs 132818/ha with BCR 6.08, whereas the highest BCR (6.33) recorded under F_1 (75 per cent RDN applied through drip system in six equal splits). The treatment combination I_2F_1 (Irrigation scheduling at 0.8 Etc through drip system along with 75 per cent RDN applied in six splits at 15 days interval) recorded the highest seed cotton yield and net return.

Table. 1 : Growth and Yield Attributes as Influenced by Different Irrigation Scheduling and Nitrogen Levels in *Bt.* Cotton (Pooled of three years)

Treatments	Plant Height (cm)	Sympodia /plant	No. of bolls/plant	Boll weight (g)	Seed cotton /plant (g)	Seed cotton yield (kg/ha)
Irrigation schedule (I)						
I_1- 0.6 Etc	70.9	15.2	33.4	3.88	132.0	2837
I_2- 0.8 Etc	78.5	16.2	41.7	3.95	164.6	3637
I_3- 1.0 Etc	81.4	15.8	35.8	3.97	138.5	3017
I_4- Surface irrigation	79.5	16.0	38.5	3.98	150.8	3266
S. Em ±	0.75	0.17	1.29	0.04	4.02	98.50
C.D. at 5%	2.20	0.50	4.47	NS	13.93	341
Nitrogen levels (F)						
F_1: 75 % RDN in 6 splits with 15 days interval through drip	75.9	15.9	37.5	3.91	145.9	3146
F_2: 100 % RDN in 6 splits with 15 days interval through drip	79.1	15.9	37.2	3.98	144.9	3172
F_3: 125 % RDN in 6 splits with 15 days interval through drip	78.3	16.1	37.8	3.95	150.6	3312
F_4:100 % RDN in 3 splits with 30 days interval as soil application	77.2	15.4	36.9	3.93	144.4	3127
S. Em ±	0.70	0.19	0.81	0.04	2.80	60.84
C.D. at 5%	1.96	NS	NS	NS	NS	NS

Table 2: Fiber Quality of *Bt.* Cotton as Affected by Irrigation Scheduling and Nitrogen Levels

Treatments	G.P. (per cent)	2.5 per cent span length (mm)	LUR	Micronire value (mv)	Fibre strength (g/tax)	Maturity ratio	Elongation	SFI
Irrigation schedule (I)								
I_1	29.85	30.3	47.08	4.42	20.56	0.83	5.89	8.13
I_2	31.03	30.2	46.50	4.41	19.94	0.82	5.87	8.84
I_3	30.33	30.3	47.75	4.33	20.40	0.82	5.88	8.18
I_4	30.90	30.2	47.41	4.50	20.64	0.83	6.01	7.61
S. Em ±	0.266	0.23	0.327	0.057	0.220	0.001	0.019	0.406
C.D. @ 5%	NS	NS	NS	NS	NS	NS	6.69	NS
Nitrogen levels (F)								
F_1	30.66	30.3	46.33	4.34	20.14	0.82	5.90	8.61
F_2	30.90	30.1	47.50	4.45	20.14	0.82	5.90	8.10
F_3	30.42	30.1	46.91	4.50	20.60	0.83	5.91	8.28
F_4	30.12	30.7	48.00	4.39	20.65	0.83	5.95	7.78
S. Em ±	0.324	0.25	0.389	0.045	0.181	0.001	0.029	0.280
C.D. @ 5%	NS	NS	1.13	NS	NS	NS	NS	NS

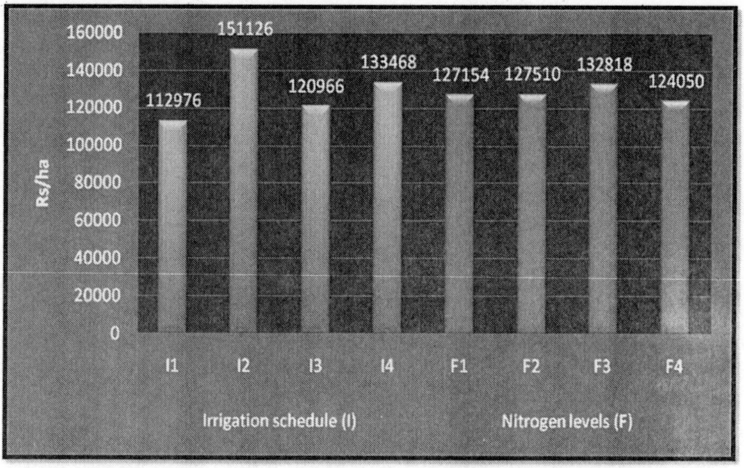

Fig.1: Economics as Influenced by Different Treatments of Irrigation Scheduling and Nitrogen Levels.

Conclusion

It could be concluded that *Bt* cotton produced higher seed cotton yield and net profit, cultivated under South Gujarat conditions and applied irrigation scheduling at 0.8 Etc along with 75 per cent RDN *i.e.* 180 kg N/ha in six equal

Table 3: Nutrient Use Efficiency, Water Use Efficiency and Water productivity in *Bt.* Cotton as Influenced by Different treatments

Treatments	*Nutrient use efficiency (kg/kg)*	*Water use efficiency (kg/ha mm)*	*Water productivity (Rs/M³)*
Irrigation schedule (I)			
I_1:0.6 ETc	--	76.46	1.92
I_2: 0.8 ETc	--	97.69	2.35
I_3: 1.0 ETc	--	65.39	1.63
I_4:Surface irrigation	--	62.54	1.62
Nitrogen levels (F)			
F_1: 75 % RDN in 6 splits with 15 days interval through drip	17.48	--	--
F_2: 100 % RDN in 6 splits with 15 days interval through drip	13.22	--	--
F_3: 125 % RDN in 6 splits with 15 days interval through drip	11.04	--	--
F_4:100 % RDN in 3 splits with 30 days interval as soil application	08.69	--	--

splits at 15 days interval through drip system for saving 20 per cent water and 25 per cent nitrogen as compared to conventional cultivation of *Bt* cotton.

References

Manjunath, V.; Hebbara, M. and Ravishankar, G. (2010). Response of cotton (*G. hirsutum*) to single and paired row method of sowing under drip and furrow irrigation in saline Vertisols. *Indian Journal of Agricultural Sciences* 80(1): 38-41.

Pawar, D. D.; Dingre, S. K.; Bhakre, B. D. and Surve, U. S. (2013). Nutrient and water use by *Bt* cotton (*G. hirsutum*) under drip fertigation. *Indian Journal of Agronomy* 58(2): 237-242.

Ramamurthy, V.; Patil, N. G.; Venugopalan, M. V. and Challa, O. (2009). Effect of drip irrigation on productivity of water use efficiency of hybrid cotton (*G. hirsutum*) in Typic Haplusters. *Indian Journal of Agricultural Sciences* 79(2): 118-21.

Zhenan, H.; Pinfang, L.; Baoguo, L.; Jiang, G. and Yanna, W. (2007). Effect of fertigation schedule on uptake and N use efficiency in cotton. *Plant soil* 290: 115-126.

29

Rainwater Harvesting Methods in Rural and Urban Areas in India

Shri Kant, Sumeet Meshram and Arun Jhajharia

Abstract

Water management is very critical for the growth and development of any economy, more so in a developing countries. However, resource is now under stress, because of excessive groundwater abstraction in the course of socioeconomic development and meeting increasing needs of growing population. The most part of the earth surface i.e. about 71 per cent is covered by water. Out of total volume of water available on the surface of the earth 97 per cent is saline water, two per cent water is in the form of ice and glaciers and only one per cent is fresh and potable water. The average annual rainfall in India is about 1190 mm. This is considerable variation in rain both temporarily and spatially. Regional varieties are also extreme, ranging from a low value of 100 mm in Western Rajasthan to over 11,000 mm in Meghalaya in North-Eastern India. Most rain-falls in the monsoon season (June-September), necessitating the creation of large storages for maximum utilisation of the surface runoff. It is necessary to opt for rainwater harvesting measures for fulfillment of water requirement.

Keyword: *In-situ rainwater harvesting, Runoff harvesting and Sub-surface flow harvesting.*

Introduction

According to FAO, the population of the least developed countries in the world is still predominantly rural: nearly 70 per cent reside in the countryside (FAOSTAT, 2012). Despite massive progress in reducing poverty in some parts of the world, over the past two decades notably in East Asia. The majority of the rural poor affected by food insecurity can be found in semi-humid and semi-arid areas. These areas are especially dependent on rainfed agriculture, and global increases in food prices can exacerbate food insecurity. But these challenges simultaneously provide opportunities. With the rise of market prices and increased knowledge about productive sustainable land and water management systems, these areas have the potential to become at least self-sufficient, or even net exporters of food (Tiffen, Mortimore and Gichuki, 1994). Rainfed agriculture is practiced in almost all the agro-ecological/hydro-climatic zones of the world. Yields can be high in temperate regions, with relatively reliable rainfall and productive soils; and also in tropical regions, particularly in the sub-humid and humid zones. But in drylands, which cover approximately 40 per cent of the global land area, yields of the major crops tend to be relatively low; between a quarter and half of their potential (Rockstrom *et al.*, 2007; Wani *et al.*, 2009; Scheierling *et al.*, 2013). Challenges within rainfed farming are many in arid, semi arid, sub-humid and even in humid regions. Water for production continues to be a key constraint to agriculture, due to highly variable rainfall, long dry seasons, and recurrent droughts, as well as floods. If rainfall is less than crop water requirements, then clearly actual yields will be less than the potential; moreover the impact of variable rainfall is strongly affected by the nature of the soil and the stage of the growing period (Critchley and Scheierling, 2012).

Beside the challenges of coping with water scarcity and stress due to climatic variability, land degradation resulting from soil erosion by wind and water, and poor management of soil fertility contributes to low rainwater use efficiency. Poor land and water management practices are major causes of low crop productivity. Water in an agricultural production system can be lost due to evaporation from the soil surface, surface runoff and through deep percolation/drainage, which sometimes can be later recovered for irrigation elsewhere. These rainfall losses, however, can be transformed into productive "green water" meaning soil water directly used by transpiration for plant growth. Then losses become advantages: runoff feeds water harvesting systems that store water directly in the soil profile. Losses can also be turned into useful "blue water" i.e. water collected in water bodies and thus made available for irrigation. An extra 10-25 per cent of water runoff harvested and made available during critical periods of plant growth can double or triple yields (Liniger *et al.*, 2011). Low cost farm ponds are a better option for collecting rainwater excess during monsoon periods for utilisation for irrigation in North Western Himalayan Region, India. The most efficient and cheapest way of conserving rainwater at the agricultural farm was found to be in-situ runoff management, which also reduces soil losses and increases the opportunity time for ground water recharging. The earthen embankment for rainwater harvesting has cost benefit ratio of 1.38:1. In addition, good results of harvesting and storage are being achieved in ferro-cement water storage structures of different dimensions of 3 to 5 m deep and 1 to 3 m in diameter (Rohitashw Kumar *et al.*, 2011). India is one of the developing countries.

Due to faster industrialisation and urbanisation and increase in population water demand is increasing day by day. Rainfall in India is highly irregular. Most of it is concentrated during a few months of the year and maximum amount flows away resulting in poor recharge of ground water. There is significant spatial imbalance in water resource available and water demand. Therefore, it is becoming necessary to bring water from distant places increasing the costof conveyance. It is also a common observation that underground water table is depleting due to uncontrolled extraction of water. Conventional sources like open well, bore well and piped water supplies have failed due to depleting water tables, poor water quality and high cost involved in operation and maintenance. Every year a great amount of wateis being lost that falls on terraces, all of which find sits way to the storm water drains. Rain water harvesting can play an important role for solving the water problems.

Purpose and Scope of Rain Water Harvesting

Water is the most essential natural resource for life. It is likely to become a critical scarce resource in many regions of the world in the coming decades. Although water is the most abundant substance on earth, it is not equally distributed. Variations in latitude, rainfall patterns, topography, etc. affect its availability. The aim of Water harvesting is to collect runoff or groundwater from areas of surplus or where it is not used, store it and make it available, where and when there is water shortage. This results in an increase in water availability by either (a) impeding and trapping surface runoff, and (b) maximising water runoff storage or (c) trapping and harvesting sub-surface water. Water harvesting makes more water available for domestic, livestock and agricultural use by buffering and bridging drought spells and dry seasons through storage. There are various system of water harvesting depending upon the source of water supply and places as classified below (Fig. 3).

1. In-situ Rainwater harvesting

(i) Bunding: Bunding is one of the mechanical measures, used for retaining the water creating obstruction thus to control erosion. Bunds are simply embankment like structure constructed across the land slope. They reduce run off and impound water for longer time to help infiltration into the soil. Their construction and design will depend upon rainfall, soil type, slope, and such other factors. These measures are costly but if properly maintained will improve the land over a long period of time (Fig. 1).

(ii) Terracing: A terrace is an embankment of ridge of earth constructed across the slope to control runoff and to minimise soil erosion. A terrace reduces the length of the hill side slope, thereby reducing sheet and rill erosion and prevents formation of gullies (Fig. 2).

Fig. 1. Bunding.

Fig. 2. Terracing.

(iii) Vegetative contour barriers. Vegetative barriers (also referred to as grass hedges) are narrow, parallel strips of stiff, erect, dense grass planted close to the contour. These barriers cross concentrated flow areas at convenient angles for farming. This practice differs from other conservation buffers because vegetative barriers are managed in such a way that any soil berms that develop are not smoothed out during maintenance operations. It Control sheet and rill erosion, trap sediment, and facilitate benching of sloping cropland (Fig. 4).

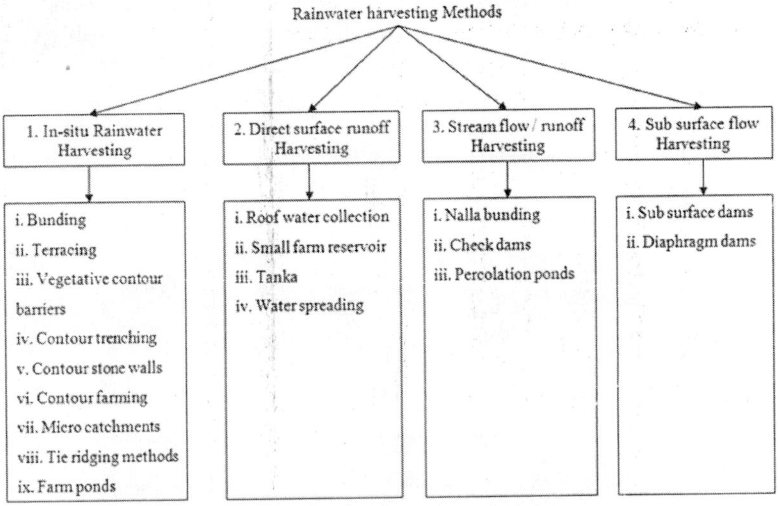

Fig. 3. Rainwater harvesting methods.

Fig. 4. Vegetative contour barriers.

(iv) Contour trenching: It consist of making a series of deep pit (*i.e.* 2 two feet. wide and one foot deep) or trenches across the slope at convenient distance. The soil excavated from the trenches is deposited on the lower edge of the trenches where forest trees are planted.

(v) Contour stone walls: A contour stone wall is a traditional method of soil and water conservation, which was improved with the help of projects and innovative farmers. The stone walls are positioned along the contours. There are different ways of placing the stones, depending on the slope and the amount of water. Compared to earth walls, the stone walls are permeable. The flow speed of the run-off water is slowed down and the water can spread over the whole field (Fig. 5).

Fig. 5. Contour stone walls.

(vi) Contour farming: The practice of tilling sloped land along lines of consistent elevation in order to conserve rainwater and to reduce soil losses from surface erosion. These objectives are achieved by means of furrows, crop rows, and wheel tracks across slopes, all of which act as reservoirs to catch and retain rainwater, thus permitting increased infiltration and more uniform distribution of the water (Fig. 6).

Fig. 6. Contour farming.

(vii) Micro catchments. This system collects surface runoff, as sheet flow over a short distance, from a small catchments area (around 1000 sq. m). Runoff water is usually applied to agriculture area where it is stored in root zone that can be used directly by the plants and also can be stored in small reservoir around the plant for later use (Fig. 7).

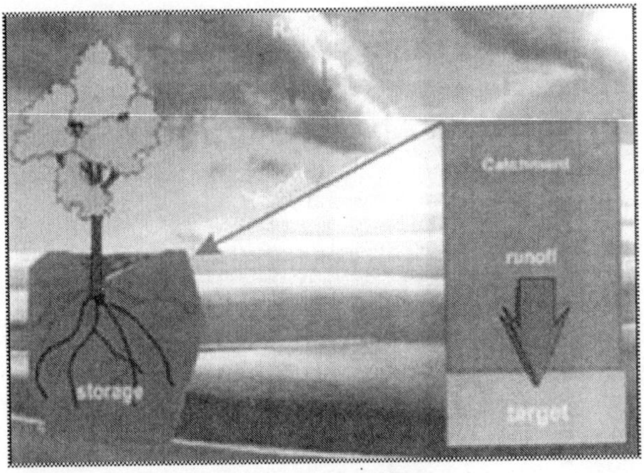

Fig. 7. Micro catchments.

(viii) Tie ridging methods: To offset agricultural losses related to low rainfalls and variations in rainfall related to these vagaries of nature, researchers investigated tillage methods that conserve soil moisture. The major objective in moisture conservation tillage methods is to conserve moisture in the soil in order to increase germination and yields. Runoff is greatly reduced and infiltration rates are increased. Moisture conservation tillage technologies include tied-ridging, mulching, contour ploughing and minimum tillage.

(ix) Farm ponds: A farm pond is a large hole dug out in the earth, usually square or rectangular in shape, which harvests rainwater and stores it for future use. It has an inlet to regulate inflow and an outlet to discharge excess water. The pond is surrounded by a small bund, which prevents erosion on the banks of the pond. The size and depth depend on the amount of land available, the type of soil, the farmer's water requirements, the cost of excavation, and the possible uses of the excavated earth. Water from the farm pond is conveyed to the fields manually, by pumping, or by both methods.

1. Direct surface runoff harvesting

(i) Roof water collection: Rooftops systems collect and store rainwater from roof of the houses or large buildings, greenhouses, courtyards, and similar impermeable surfaces, including roads. Modern roofing materials and gutters, for example, allow the collection of clean water suitable for drinking and other domestic uses, especially in rural areas without tap water. Such systems provide a low cost water supply for humans and animals in remote areas. Although mainly used for domestic purposes, this technique also has agricultural uses (Fig. 8).

Fig. 8. Roof water collection.

(ii) Small farm Reservoir: These reservoirs may range in capacity between 1000 to 500,000 cu.m. The most important feature is to have a spillway with sufficient capacity to allow for the excessive peak flows that may pass through the wadi (Fig. 9).

Fig. 9. Small farm Reservoir.

(iii) Tanka: Tanka is a traditional rainwater harvesting technique, common to the Thar desert region of Rajasthan, India. It is meant to provide drinking water for a single or a small group of families and is an important element of water security in these arid regions.

(iv) Water spreading: In this technique, part of the wadi flow, is forced to leave its natural course and is then conveyed to nearby areas for the crops. The flow is directed by levee, which should run a little off the contour, away from the wadi path. One important point to consider is that the slope of the conveyance canal should allow a flow velocity sufficient to prevent the accumulation of sediments near the structure, otherwise these will block the flow and entail high maintenance costs (Fig. 10).

Fig. 10. Water spreading.

3. Stream flow/runoff harvesting

(i) Nalla bunding: Nalla bunding is comprehensive activity. Nalla bunds are embankments constructed across nalla for checking velocity of runoff, increasing water percolation and improving soil moisture regime. In fact, nalla bunds and percolation tanks both terms are synonymous but still they are being called differently at different places (Fig. 11).

Fig. 11. Nalla bunding.

(ii) Check dams: The structures which are to constructed to check the velocity of flowing water in the gully is known as check dams. They reduce the degree of the gully bed by constructed a series of checks at regular interval, across the width, in the gully. These check dams divide the longitudinal slope of gully bed into a series of steps with low series and long flat treads. Classified as below (Fig.12).

(iii) Percolation ponds: Percolation Ponds are a pretty great way to refill the dent in the Bay Area that was left a hundred years ago. The ground sunk so they are filling it up slowly. Percolation ponds are man-made ponds in certain areas. People are filling the up ponds with reservoir water and then the water slowly are pumped down to fill up the area underneath the ground. When needed the people in charge can pump up the water for daily usage for farmers and others (Fig. 13).

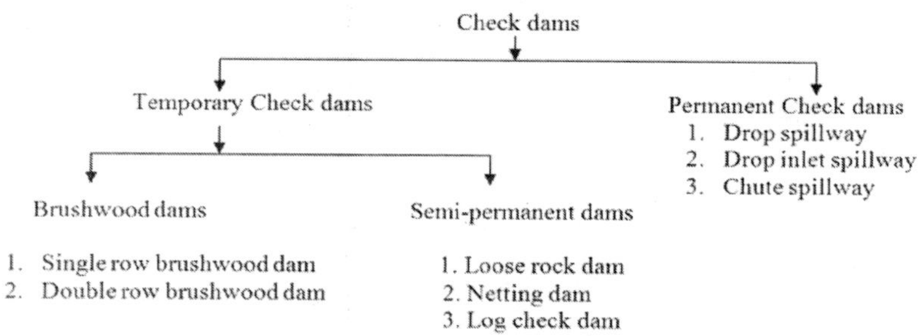

Fig.12. Classification of check dams.

Fig. 13. Percolation ponds.

4. Subsurface flow harvesting

(i) Subsurface dams: Subsurface dams built of soil across river beds have proved to be the most reliable and inexpensive water source in arid and semi-arid areas. Subsurface dams prevent floodwater, which has filled the spaces between the sand particles in riverbeds, from being drained downstream thereby drying up the sand in riverbeds. Subsurface dams let rainwater pass without reducing the flow to people living downstream. A dyke is a hard layer under the sand that already acts as a dam. Subsurface dams can be constructed in riverbeds with no dykes, but they might produce less water (Fig. 14).

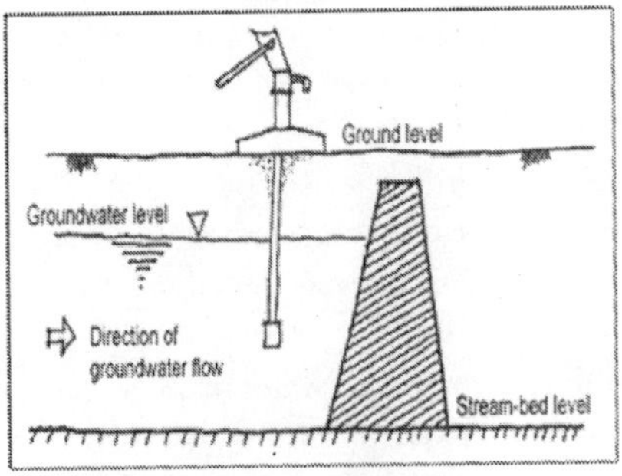

Fig. 14. Subsurface dams.

(ii) Diaphragm dams: This type of dam is constructed with pervious materials, with a thin impervious diaphragm in the central part to prevent seepage of water. The thin impervious diaphragm may be made of impervious clayey soil, cement concrete or masonry or any impervious material. The diaphragm can be constructed in the central portion or on the upstream face of the dam. The main difference in zoned and diaphragm type of dams depend on the thickness of the impervious core or diaphragm. The thickness of the diaphragm is not more than 10 m (Fig. 15).

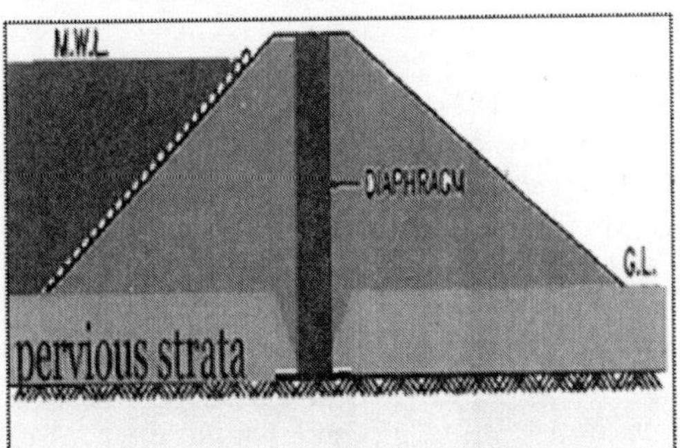

Fig. 15. Diaphragm dams.

Conclusions

1. The erratic and uneven distribution of rainfall both spatially and temporally, necessitates rainwater harvesting to increase and sustain the agricultural productivity.
2. It is very important to make water everybody's business. It means a role for everybody with respect to water. Every household and community has to become involved in the provision of water and in the protection of water resources.
3. Water conservation and water harvesting will go a long way towards reducing misuse of government funds. It will also develop the ownership (own water supply systems), they will also take good care of them. In this way it is possible to solve water problems facing the county in the 21st century.
4. Rainwater harvesting should be made mandatory. Sequential water use (reuse, recovery and recycling of waste waters) should be planned wherever possible so that the load on fresh water can be reduced.
5. For agriculture, an integrated water management practice consisting of three main components rain water harvesting, water-saving micro-irrigation, and highly efficient crop production should be adopted. Conservation of water should be taken as a way of life and widely adopted.

References

FAO. 2012. Discovery based learning in land and water management, a practical guide for farmer field schools (16 modules). Module 10: managing rainwater; Module 11: water for crops; Module 12: harvesting water for people and for livestock. Food and Agriculture Organisation (FAO), Rome, Italy.

Tiffen, M., Mortimore, M. and F. Gichuki. 1994. More people, less erosion, environmental recovery in Kenya. John Wiley & Sons, Chichester, UK.

Rockström, J., Hatibu, N., Oweis, T.Y., Wani, S., Barron, J., Bruggeman, A., Farahani, J., Karlberg, L. and Z. Qiang. 2007. Managing Water in Rainfed Agriculture. In Molden, D. (ed). Water for Food, Water for Life: Comprehensive Assessment of Water Management in Agriculture. Earthscan and International Water Management Institute (IWMI), London and Colombo.

Wani, S.P., Sreedevi, T.K., Rockström, J. and Y.S. Ramakrishna. 2009. Rainfed Agriculture: Past Trends and Future Prospects. In Wani, S.P., Rockström, J. and T. Oweis (eds). Rainfed Agriculture: Unlocking the Potential. Comprehensive Assessment of Water Management in Agriculture 7

Scheierling, S.M., Critchley, W. R. S., Wunder, S. and J.W. Hansen. 2013. Improving Water Management in Rainfed Agriculture: Issues and options in water-constrained production systems. Water Paper, Water Anchor, The World Bank.

Critchley, W. and S. Scheierling. 2012. Water Harvesting for Crop Production in Sub-Saharan Africa: Challenges, Concepts and Practices. In Critchley, W. and J. Gowing (eds). Water Harvesting in Sub-Saharan Africa. Earthscan.

Liniger, H., Mekdaschi Studer, R., Hauert, C. and M. Gurtner. 2011. Sustainable Land Management in Practice: Guidelines and Best Practices for Sub-Saharan Africa. TerrAfrica, World Overview of Conservation Approaches and Technologies (WOCAT) and Food and Agriculture Organisation of the United Nations (FAO).

Rohitashw, Kumar, Thaman, S., Agrawal, G. and Sharma, Poonam. 2011. Rain Water Harvesting and Ground Water Recharging in North Western Himalayan Region for Sustainable Agricultural Productivity, Universal Journal of Environmental Research and Technology, Volume 1, Issue 4: 539-544.

30

Agronomic Management of Grain Legumes for Higher Productivity under Water-Limited Environments: A Review

Amit M. Pujar, Fakeerappa Arabhanvi and R. H. Patil

Abstract

Grain legumes (pulses) are important contributors to world food supplies and nutrition, especially in the developing world. They are also major factor in enhancing sustainability through intensification and diversification of agricultural production systems. The current total annual production of grain legumes in India is about 17.08 m.t. In terms of utilisation, about two-third of world grain legumes are consumed as human food and quarter ends up as animal feed. The average yield of grain legumes varies from 780 to 870 kg per ha depending upon environmental conditions. The poor productivity of pulses is mainly due to reason that these crops are generally grown under rainfed and dryland condition. These crops usually exposed to biotic and abiotic stresses at some time during their cycle. Among the stresses, moisture stress is most important and limits the productivity of grain legumes. Since they are grown mostly as a rainfed and post-rainy season, water stress during vegetative and/or reproductive growth stages is one of the most limiting factors for the pulses. It affects nodulation, biological nitrogen fixation and yield. The crop performance depends on meagre amount of rainfall received and quantity of moisture stored in the soil. Selection of drought tolerant genotypes and short duration variety are important management strategies to realise higher yield and water use efficiency and seed soaking with 2 per cent KH_2PO_4 or 2 per cent $CaCl_2$ and growth regulator have been found to increase the drought resistance in pigeonpea. Intercropping of grain legumes with pearlmillet or sorghum increases the moisture *use efficiency, minimises the risk of crop failure and maintains high B:C ratio. Wherever feasible, providing one irrigation at branching, or at pre- flowering and pod filling stage increases the grain yield as well as water use efficiency. Different soil and moisture conservation practices like opening of furrows between two rows, ridges and furrow, tied ridges, mulching and deep tillage increases the soil moisture retention and extended availability due to reduced evaporation and increases the yield.*

Key words: *Abiotic stress, Grain legumes, Genotypes, Pulses, Water-limited environment.*

Introduction

Water is the most important factor that greatly influences agricultural production as it affects the food, fodder and feed security and the sustainability of production system across the world. Crop production under rainfed conditions dominates global agriculture as more than 80 per cent of world's total cultivated land cannot be irrigated (FAO, 2011) and depends only on precipitation. In India close to two-third of total arable land is under drylands. Therefore, amount of rainfall received in a particular locality greatly determines the productivity, especially in drylands where rainfall is not only scanty but also erratic in nature. Water limited environments are arid, semi-arid and sub-humid regions, and are at times collectively called as drylands. Water deficiency leading to mild and sever moisture stress conditions cause reduction in the yield. However, such conditions occur repeatedly and extend for long periods may lead to drought.

Grain legume crops are by nature known to be drought tolerant, hence historically cultivated in water limited environments. For instance in India, close to 90 per cent of the total legume / pulse crops are grown under water limited environments, more so in drylands, thus, the productivity is very low. The average yield of grain legumes varies from 780-870 kg/ha depending upon environmental condition and these values drop further down for India. In terms of consumption pattern, about two-third of global grain legumes are consumed as human food and the rest fed to animals. For a country like India where vast majority solely depend on legumes / pulses for their daily dietary protein requirement it is important, but over the time the per capita consumption of pulses in India has declined from 69 g/day (1960-61) to 36 g/day (2007-08), whereas WHO for India recommends minimum of 80 g/day.

Apart from contributing to the global food supplies and nutrition, especially in the developing world, grain legumes play an important role in enhancing sustainability through intensification and diversification of agricultural production system (Byerlee and White, 2000). The current total world production of grain legumes is about 61.34 m t per annum (FAO, 2011-12). The poor productivity of grain legumes is mainly due to the fact that these crops (Table 1) are predominantly grown under water limited environments, often under dryland condition. Hence, these crops usually face soil water deficits at some time for a short or extended period during their life cycle.

Table 1: Some of commonly used grain legumes in India (Source: Anonymous, 2012)

Sl. No.	*Pulses*	*Botanical name*	*Common name*
1.	Pigeonpea (redgram)	*Cajanuscajan*	Arhar and tur
2.	Blackgram (urdbean)	*Vignamungo*	Urd, kalai
3.	Greengram (mungbean)	*Vignaradiata*	Mung, moong
4.	Chickpea (bengalgram)	*Cicerarietinum*	Chana, gram
5.	Cowpea	*Vignaungiculata*	Lobia, chowli,
6.	Soybean	*Glycine max*	Soy, bhat

Moisture stress affects nodulation, biological nitrogen fixation, total biomass accumulation and final grain yield (Turner *et al*, 1986). This situation may get worst in drylands with changing climate as the latest climate projections indicate increased unpredictability of rainfall distribution pattern over both time and space, more often than not, leading to occurrence of drought like situations. Therefore, water conservation and its judicious use become even more important as part of climate smart agriculture to maintain and improve grain legume yields in years to come for national food and nutritional security.

Adoption of good management and appropriate water conservation practices and its subsequent efficient use improving agricultural crop production are possible under dry land (Wang et al., 2004; Li et al., 2001). This paper reviews some of this work under the following headings and puts things in perspective.

Agronomic management strategies to minimise drought effects in grain legumes

Soil management techniques

Tillage during off season or pre-rainy season, helps with rainwater intake by breaking the hard soil and keeping its surface under rugged condition to make soil surface more permeable to rain water, especially suited for high intense pre-monsoon rains. Summer deep ploughing improves soil porosity, water intake rate and reduces the runoff, whereas shallow tillage and repeated harrowing during monsoon conserve the stored soil water through decreased evaporation by reduced soil capillaries, that results in reduced evaporation and acts as soil mulch (Mishra *et al.*, 2005). This allows water to seep into deeper soil layers and keeps the soil wet for longer time. Off season tillage also controls weed and destroys, egg, cocoons and larvae of some pests by exposing them to hot sun of summer. Soil moisture content in tilled field is found to be higher (13 per cent) compared to untilled land (seven per cent) (Anonymous, 2011).Tillage with mould board plough leading to increased yield in soybean by 19 per cent compared with no-tillage has been reported elsewhere (Edwards et al., 1988). It was due to availability of more moisture during the sowing of the crop. In addition to improved soil moisture regimes, tillage would affect other soil physical and chemical properties for better. (Bauer and Kucera, 1978). However, many tillage experiments showed inconstancies in their effect on crop yield and are due to the complexity of the changes in soil properties caused by tillage, which needs further systematic study in years to come.

Crop management techniques

Selection of genotype

Most of the pulses are grown under marginally low soil moisture conditions. These crops invariably suffer from moisture stress to varying degrees, particularly during pre- and post-flowering stage, and also at pod filling period. This influences various physiological processes associated with crop growth and development, as a result, the final grain yield. Selection of varieties adapted to such environments becomes very critical. For instance, short duration varieties which could complete their life cycle relatively early and escape from post-flowering or terminal drought, thus permits the crop to maintain favourable water balance during critical phase of crop growth. In this regard, incorporation of early vigour trait by integrating several physiological factors like early germination, more water uptake with faster root and shoot development and leaf expansion followed by deep rooting pattern needs to be targeted in crop improvement programme (Biradar *et al.*, 2007). For example, chickpea genotype ICC 13124 was found to yield better under moisture stress environments because of higher drought tolerant efficiency and higher harvest index (Yadav *et. al.,* 1996; Parameshwarappa and Salimath, 2008).

Intercropping cereals with legumes

It refers to growing more than one crop, preferably cereal and legume crop on same land in definite row proportion and pattern. It provides insurance against total crop failure in drought prone dryland area (e.g., sorghum + pigeonpea, pearl millet + cowpea) apart from providing nutritional food security. The main advantage of intercropping is the more efficient and staggered utilisation of the available resources, especially limited soil moisture, resulting in increased productivity compared with each sole crop of the mixture. Intercropping when mixed with grain legumes along with cereals can conserve soil water by providing shade by the fast spreading legumes, reduce wind speed by taller cereal crop, increase infiltration with mulch layers and improve soil structure and fertility with biological N fixation by the legume crop. Legume crops such as cowpea, mungbean, soybean and groundnuts are known to fix and accumulate significant amount of nitrogen (80 to 350 kg nitrogen/ha).

For instance, intercropping maize with cowpea has been reported to increase total light interception, reduce water evaporation, and thus improve conservation of the soil moisture compared with maize alone (Hamid *et al.,* 2014). Growing pearl millet + pigeonpea (2:1) along with an application of PMA (320 ppm) and 50 kg N + 17.2 kg P_2O_5 /ha has been found to improve each component's crop yield and the system productivity with higher Water Use Efficiency (Ansari and Rana, 2012). Similar synergies between cereal + legume crops leading to improved WUE and productivity have been reported elsewhere before (Hulugalle and Lal, 1986; Garba and Renard, 1991). Intercropping with legumes also control soil erosion as legumes act as best cover crops (Kariaga, 2004; Seran and Brintha, 2010), whereas taller cereal crops act as wind barrier (Reddy and Reddi, 2007), thus acting in complementarily to each other. Sorghum + cowpea intercropping reducing runoff by 20 to 55 per cent compared with sole sorghum or cowpea monoculture has been reported (Zougmore *et al.*, 2000). Pearlmillet + pigeonpea in 2:1 (70/30 cm) row ratio resulting in better land utilisation, higher yield as well as profitability and WUE than their sole crop been reported.

Surface mulching to conserve soil moisture

Surface mulching either by timely or repeated inter-cultivation or by covering the soil surface with plant residue conserves moisture to benefit the crop as it reduces water evaporation from the soil, contains run-off and weed population (Tiwari *et al.,* 1991). Use of crop residues as mulch adds organic matter to the soil and improves soil quality and fertility. Organic mulching in pigeonpea enhanced yield attributing characters like number of flowers/plant (41 per cent), number of pods/plant (34 per cent), test weight(15 per cent), number of seeds/pod and seed yield (48 per cent) over control plot with no mulch, and was attributed to maintenance of higher soil moisture in root zone at all stages of crop growth which resulted in better nutrient uptake and yield (Jawahar *et al.,* 2012). Abubakkar *et al.* (2004) also recorded similar observations in summer green gram. Application of straw mulch at 5 t/ha and application of Butachlore at 1.25 kg/ha (PE) + one hand weeding (at 35 DAS) recorded higher seeds/pods, seed yield, WUE and lower weed dry matter at harvest over control (Sabyasachi Mitra and Bhattacharya, 2005). Subba Reddy *et al.* (2006) proposed that establishment of soil mulch through additional intercultural practices at early and mid-stress periods would increase crop yield to the extent of 12 to 19 per cent as compared to no soil mulch. Kumar and Rana (2007) found that creating dust mulch +FYM @ 5 t /ha + kaolin, and dust mulch + FYM @5 t/ha caused significant improvement in growth attributes of pigeonpea at all the growth stages of crop over dust mulch. This could be attributed to an extended period of moisture availability, improved soil physical properties, hydraulic conductivity as well as availability of NPK due to FYM and mulching. Similar results were also reported by Jat and Gautam (2000). Similar effect was also found in mungbean under dust mulch +FYM @ 5 t/ha + kaolin 6 per cent spray. This was attributed to reduction in crop evapotranspiration, suppression of weeds and conservation of soil moisture, which ultimately led to extended period of water availability (Kumar and Rana, 2007). Most of the research works elsewhere on mulching practices have been found to increase in the yield of pulses anywhere between 30 to 40 per cent compared with no mulching, which proves beyond doubt that this technology could easily be popularised as climate smart agricultural practices to be adopted in dryland regions to cope with climate change.

Soil and rain water conservation technique

Conservation and judicious use of natural resources like soil and water play an important role in sustainable crop production under rainfed / dryland farming in semi - arid tropics (Reddy *et al.*, 2009). Every drop of rain water makes a difference during drought and effective conservation followed by

Table 2: Land treatments for in situ rainwater conservation suiting diverse soil and rainfall environment (Source: Anonymous, 2011)

Sl. No.	Area	Rainfall	System of land configuration
1.	*Alfisols*	Low	Dead furrows at 2-3 cm intervals
2.	*Alfisols* (semi-arid)	Medium	Sowing across the slope and ridging the land between row with later inter-cultivation practice
3.	*Vertisols* (Semi-arid)	Low	Contour farming
4.	*Vertisols* (Semi-arid)	Medium	Broad bed and furrow with a bed of 100 cm and 50 cm of shallow furrow having a depth of 15 cm
5.	*Vertisols* (sub-humid)	High	Graded furrows at a grade of 0.2 to 0.3 per cent and with a cross-section of 0.1 to 0.15 cm and spacing of 8-10 m interval

(Very low : <500 mm, Low: 500-750 mm, Medium : 750-1000 mm and high : >1000 mm)

efficient utilisation of rainwater is the key to mitigate drought. Mechanical barriers across the direction of flow of water to retard or retain the runoff for conservation of soil and water could be followed. For example, contour bunds are recommended on land with 0.5 – 6 per cent slope and receiving annual rainfall of 750 – 1000 mm. Graded bunds are recommended on lands with upto 10 per cent slope and annual rainfall exceeding 800 mm. Bench terraces are followed in hilly areas with slope of upto 33 per cent. Conservation bench terraces or Zing terraces are designed to control runoff and soil erosion, and stabilise yields of rainfed crops by soil moisture storage and inter-plot water harvesting. Bunding or terracing treatments help to intercept runoff, reduce its soil cutting velocity increase infiltration opportunity time and disposal of surplus runoff. Today the concept of 'catch the drop where it falls' is being popularised and is also referred as *in-situ* water management. Using broad bed and furrows, compartment bund and inter-row water harvesting and reutilisation of stored water at critical stages techniques for improved productivity have been recorded elsewhere (Kantwa *et al*., 2005; Singh *et al*., 2006).

In chickpea field bunding + green manure and field bunding + FYM @ 10 t/ha accounted for maximum net return (Meena *et al.,* 2006) and field bunding was found to increase critical consumptive use (CU) of water over no bunding leading to increased WUE. Ridges and furrow + mulching practice in cowpea recorded higher nodules/plant and nodule dry weight at 25 and 50 DAS, higher root length and grain yield, and was on par with compartment bunding + mulching (Ramesh and Devasenapathy, 2007). Makkhan *et al.* (2003) also reported higher growth and yield parameters of pearlmillet grown under ridge and furrow moisture conservation techniques. Tumbare and Bhoite (2003) also observed enhanced growth and yield parameters in crops grown under moisture conservation techniques. Ramesh and Devaseenapathy (2007) in an another study found that sowing of cowpea on either ridges and furrows or tied ridges along with mulching by locally available crop residues at the rate of 2.5 t/ha could be viable technique for sustainable productivity of cowpea under rainfed *alfisols* conditions.

Application of 100 per cent RDF + FYM at 5 t/ha + opening of shallow furrows between two rows, seed soaking with two per cent KH_2PO_4 or two per cent $CaCl_2$ and growth regulator recorded significantly higher plant height, leaf area, no. of pods/plant, seed yield of pigeonpea and B:C ratio (Sharma *et al.,* 2010) suggesting additive effect of more than one technique followed at a time. Selviraj (2002) reported that following compartmental bunding (CB) and broad bed furrow (BBF) helped conserve 17.3 per cent and 11.4 per cent higher moisture content, respectively over traditional flatbed in sorghum + pigeon pea intercropping system, and these strictures have been effective in conserving sufficient soil moisture to support crop growth even during drought situations. However, the performance of CB and BBF varied with rainfall pattern, as former one was found to be better under low rainfall year and the latter on during high rainfall year. Similar effects of modified land configuration (Desai *et al.*, 2000; Kantwa *et al.*, 2005) have also been reported elsewhere

Conclusions

Selection of drought tolerant genotypes coupled with adoption of effective soil and water conservation techniques are important strategies to realise higher yield and WUE in drylands. Intercropping pearl-millet or sorghum with grain legumes increases the moisture use efficiency, minimises the risk of crop failure and maintains soil fertility. Providing irrigation wherever feasible using stored rainwater at branching, pre-flowering and pod filling stage increases grain yield and B:C ratio of grain legumes. Different soil and moisture conservation practices like opening of furrows between two rows, ridges and furrow, tied ridges, mulching and deep tillage not only increases the soil moisture available to crop, but also extends its period during crop growing cycle thus increases the yield.

References

Abubakkar, S., B. Protit and A. Zaman. 2004. Effect of irrigation date of sowing and mulching on yield attributes, yield and B:C ratio of summer greengram. Indian Agriculturist, Vol. 48(3 and 4), pp.145–148.

Anonymous, 2011. Sustainable development of dryland agriculture in India.

Anonymous, 2012. Text book on Agronomy of field crops, Kalyani Publishers, India, pp.78-95.

Ansari, M.A. and K.S. Rana. 2012. Effect of transpiration suppressants and nutrients on productivity and moisture-use efficiency of pearlmillet (*Pennisetum glaucum*)–pigeonpea (*Cajanuscajan*) intercropping system under rainfed conditions. Indian Journal of Agricultural Sciences, Vol. 82(8), pp. 676–680.

Bauer, A. and H.L. Kucera. 1978. Effect of tillage on soil physical properties and on annually cropped spring wheat yields. Agricultural experimentation station. North Dakota, state university. Bulletin, p. 506.

Biradar, K.S., P.M. Salimath and R.L. Ravikumar. 2007. Association of Early Vigour with Drought Tolerance in Greengram, Karnataka Journal of Agricultural Sciences, Vol. 20(3), pp. 10-12.

Byerlee, D.E. and R. White. 2000. Agricultural system intensification and diversification through food legumes: technological and policy option. Linking research and marketing opportunities for pulses in the 21st century, Kluwer, Academic publishers, The Netherland, pp. 31-47.

Desai, N.C., R.B. Ardesna and C.G. Intwala. 2000. Increasing pigeonpea productivity by providing land configuration on vertisol. *International Chickpea and Pigeonpea, Newsletter,* Vol. 7, pp. 66-67.

Edwards, J. H., Thurlo and J. T. Eason. 1988. Influence of tillage and crop rotation on yields of corn, soybean and wheat. Agronomy Journal, Vol. 80, pp. 76-80.

FAO, 2011.The State of Food Insecurity in the World, 2001. FAO, Rome.

Garba, M. and C. Renard. 1991. Biomass production, yields and water use efficiency in some pearl millet/legume cropping systems at Sadore, Niger. Proceedings of the Niamey Workshop, February 1991. IAHS Publ., No. 199, pp. 431-441.

Hamid, R.M., R.V. Mohamad and K. Rigi. 2014. Effect of intercropping on resource use, weed management and forage quality. Internationl Journal of Plant, Animal and Environmental Sciences. Vol. 4(2), pp. 706-713.

Hulugalle, N. R. and R., Lal. 1986. Soil Water Balance in Intercropped Maize and Cowpea Grown in a Typical Hydromorphic soil in Western Nigeria. Agronomy Journal, Vol. 77, pp. 86- 90.

Jat, M.L. and R.C. Gautam. 2000. Effect of summer ploughing and *in situ* moisture conservation practices on growth and yield of rainfed pearlmillet (*Pennisetum glaucum*). Indian Journal of Agronomy, Vol. 45 (4), pp. 722-727.

Jawahar, S., A. Veeramani and P. Prema. 2012. Weed dynamics, yield and economics of pigeonpea as influenced by growth promoters and mulching. Indian Journal of Weed Science, Vol. 44(3), pp. 186–190.

Kantwa, S.R., I.P.S. Ashlawat, and B. Gangaih. 2005. Effect of land configuration, post-monsoon irrigation and phosphorus on performance of sole and intercropped pigeonpea (*Cajanus cajan)*. Indian Journal of Agronomy,Vol. 50 (4), pp. 278-280.

Kariaga, B.M. 2004. Intercropping of Maize with Cowpeas and Beans for Soil and Water Management in Western Kenya. Proceedings of the 13th International Soil Conservation Organisation Conference, July 2004, Conserving Soil and Water for Society, Brisbane. pp. 1- 5.

Kumar, A. and K.S. Rana. 2007. Effect of cropping systems, moisture conservation practices and fertility levels on growth and yield of pigeonpea and mungbean in intercropping system. Annals of Agriculture Research, Vol. 28(1), pp. 63-67.

Li, X.Y., J.D. Gong, Q.Z. Gao and F.R. Li. 2001. Incorporation of ridge and furrow method of rainfall harvesting with mulching for crop production under semi-arid conditions. Agric. Water Management, Vol. 50, pp. 173-183.

Makkhan L., K.K. Bora, I. Singh, R.Verma and P.C.Yadav. 2003. Productivity of pearl-millet as influenced by in situ moisture conservation practices and integrated nutrient management in arid region of Rajasthan. Current Agriculture, Vol. 27. pp. 77-80.

Meena, L.R., R.K. Singh and R.C. Gautam. 2006. Effect of moisture conservation practices, phosphorus levels and bacterial inoculation on growth, yield and economics of chickpea (*Cicer arietinum* L.). Legume Res., Vol. 29 (1). pp. 68 – 72.

Mishra, J.S., V.P. Singh and N.T. Yaduraju. 2005. Effect of tillage practices and herbicides on weed dynamics and yield of wheat (*Triticum aestivum*) under transplanted rice (*Oryza sativa*) – wheat system in *vertisols*. Indian Journal of Agronomy, Vol. 50(2), pp. 106-109.

Parameshwarappa, S.G. and P.M. Salimath. 2008. Field screening of chickpea genotypes to drought resistance. Karnataka Journal of Agricultural Science, Vol. 21(1), pp. 113-114.

Ramesh and P. Devasenapathy. 2007. Physical indicators of sustainability in rainfed cowpea (*Vignaunguiculata* L.) as influenced by in situ rainwater harvesting. Legume Research, Vol. 30 (4), pp. 256 – 260.

Reddy, T.Y., G.H.S. Reddi. 2007. Principles of Agronomy, Kalyani Publishers, India, pp. 468-489.

Selviraj, R. and T.N. Balsubramaniam. 2002. Effect of Moisture Conservation Practices on Water Use, Micro-Climate and Productivity of Intercropping in Dryland Vertisols. Journal of Agricultural Physics. Vol. 1, pp. 112-116.

Sabyasachi Mitra and B.K., Bhatacharaya. 2005. Water use and productivity of greengram as influenced by spacing, mulching and weed control under rainfed situation. Journal of Agricultural Sciences, Vol. 75 (1), pp. 54-56.

Seran, T.H., I. Brintha. 2010. Review on Maize Based Intercropping. Journal of Agronomy,Vol. 9 (3), pp. 135–145.

Sharma, A., P.S. Rathod and C. Chavan. 2010. Response of pigeonpea to drought management practices under rainfed conditions. Karnataka Journal of Agricultural Sciences, Vol. 23(5), pp. 693 -700.

Singh, R.S., L.K. Idani and R.K. Rai. 2006. Grain yield, water use efficiency, economics and soil moisture extraction pattern of summer greengram (*Vignaradiata var. Radiata*) as influenced by planting and irrigation methods; irrigation schedules and vam-inoculation. Annals of Agricultural Research,Vol. 16(3), pp. 289–291.

Subba, R., G. V. Maruthi and M. SreeRekha. 2006. Drought management options for rainfed pigeonpea in alfisols, Indian Journal of Dryland Agricultural Research and Development, Vol. 21, pp. 7-11.

Tiwari, A. N., K. Singh, S.K. Singh and U.S. Tiwari. 1983. Crop weed competition in pigeonpea intercropped with blackgram. Indian Journal of Agonomy. Vol. 34(1), pp. 140-142.

Tumbare, A.D. and S.U. Bhoite. 2003. Effect of moisture conservation techniques on growth and yield of pearlmillet-gram sequence in watershed. Indian Journal of Dryland Agricultural Research and Development, Vol. 15, pp. 94-95.

Turner, N.C., G.C. Wrighte, K.H.M. Sidhique. 1986. Adaptation of grain Legumes to water limited environments. Advances in Agronomy, Vol. 71, pp. 193-230.

Wang, F.H., X.Q. Wang and K. Sayre. 2004. Conventional, flood irrigated, flat planting with furrow irrigated, raised bed planting for winter wheat in China. Field Crop Research, Vol. 87, pp. 35-42.

Yadav, V.K., N. Yadav and R.D. Singh. 1996. Metabolic changes and their impact on yield in chickpea under water stress. Plant Physiology of Biochemistry,Vol. 23, pp. 49-52.

Zougmore, R., F. N. K. Ouattara, S. Guillobez. 2000. Sorghum-cowpea intercropping: An effective technique against runoff and soil erosion in the Sahel (Saria, Burkina Faso). Arid Land Research and Management,. Vol.14, pp. 329-34.

31

Soil Moisture Conservation Practices for Enhanced Crop Productivity in Drylands: A Review

Amit M. Pujar, Fakeerappa Arabhanvi and R. H. Patil

Abstract

Of the 142 m ha of net cultivated land in India, irrigated area constitutes only 30.5 per cent whereas 69.5 per cent is under dryland and rainfed area where, more often than not, crops are exposed to low and erratic rainfall thus experience moisture stress resulting in lower yields. Therefore, only about 44 per cent of the total food grain production comes from rainfed/dryland areas. Current national annual food grain basket is around 260 m t (2012-13) which needs to go by additional 100 m t to feed its estimated population of 1.5 billion by 2030 (FAO, 2012-13). This target cannot be realised only from irrigated areas the maximum achievable gross irrigation potential in India is only 113 m ha. Main reasons for declined productivity in dryland areas include soil erosion, low organic matter content, low moisture retention capacity, and low soil fertility status. Hence, there is a need to adopt the suitable soil moisture conservation practices viz., *compartment bunding, ridges and furrows, mulching practices, application of compost and/or farm yard manure application and use of antitranspirants which helps to reduce the crop water demand, Thus, enhance water use efficiency and sustainable production in dryland situation. For instance, application of 50 per cent RDF + microbial consortia + FYM @ 2.5 t/ha increases rainwater use efficiency and resulting in higher yields on a sustainable basis. Different land configurations (compartment bunding and ridges and furrows) with* in-situ *moisture conservation practices helps to reduce the bulk density, increases infiltration rate and porosity, improves root growth thereby increasing grain yield of rabi sorghum. Spreading of* Glyricidia *leaves and green weed biomass between crop rows conserved more soil moisture.* In-situ *moisture conservation practices* viz., *ridges and furrows along with mulches (dried leaves of sapota, guava and mango) have also been found beneficially influence growth and yield of chilli under dryland. Application of soil mulch + FYM @ 5 t/ha + Kaolin 6 per cent spray was found to be the best moisture conservation practice by recording the higher values of pigeon pea- equivalent yield (pigeonpea + green gram) and water use efficiency.*

Keywords: *Soil moisture conservation, Drylands, Water use efficiency, Crop productivity.*

Introduction

Drylands in India face twin problems of poor soil fertility and inadequate soil moisture for successful crop production resulting in very low crop productivity. Being at the mercy of highly erratic and unpredictable monsoon make the dryland farming a gamble in nature, and the situation is projected to become even more worst under future climates, hence making our effort to improve dryland productivity much more difficult task, thus calls for immediate action. Dryland agriculture in India occupies 65 per cent of the total arable land and contributes little over 40 per cent to the national total food grains production. Some 60 per cent of coarse grain and close to 90 per cent of pulses and legumes production comes from dryland areas, thus underlining its importance if India were to achieve nutritional food security. With the fast increasing demand for food, oilseeds and pulses by the growing human population the need of the hour is to adopt climate smart agricultural practices to improve and sustain dryland productivity.

In fact, severe soil moisture deficiency and land degradation is threatening the scope of dryland agriculture (Patil *et al*., 1994; Meerasab and Guled, 2013). Day by day the area under cultivation is contracting to meet the developmental need of nation. Apart from low and erratic rainfall pattern, the major reasons for low and declining productivity in dryland areas include soil erosion, low organic matter content, land degradation, low moisture retention capacity and low soil fertility status. These problems unfortunately are interrelated and act additively on the productivity of land. Hence, there is a need to adopt the suitable soil and moisture conservation practices.

Land configuration for improved *in-situ* soil moisture conservation

Soil moisture being the most critical factor for growth and development in drylands it needs to be conserved to the best advantage of crop plants.

Table 1: Effect of land configuration on grain yield, consumptive use and WUE in sorghum (Source: Patil and Sheelavantar, 2004)

Treatments	*Grain yield (kg/ha)*	*Consumptive use (mm)*	*WUE kg/ha-cm)*
Flat bed	1276	272	46.79
Ridge and furrows (60 cm)	1603	275	58.36
Compartment bunding (3 x 3 m)	1567	273	57.22
S.Em +	38	-	-
C.D. @ 5 per cent	125	-	-

In-situ soil moisture conservation practices *viz.,* compartment bunding (CB), tied ridges and furrows, mulching practices, application of compost and/or farm yard manure are reported to help hold and conserve the rainfall in soil profile, reduce runoff by improved water percolation, provide more opportunity time for ponded water to infiltrate in the soil profile leading to reduced runoff (Patil *et al.* (1994). This further helps in extended availability of stored rain water to the crop during various stages of crop growth and especially during dry-spells resulting in better crop growth and higher yields (Mahale *et al.* 1998). Benefits of CB (Surkod and Itnal, 1998; Selvaraju *et al.* 1999) and tied ridges and furrows (Selvaraju *et al.* 1999) on improved soil moisture content, in comparison with flatbed method, have been demonstrated beyond doubt long time ago. Further, such *in-situ* practices also enhanced Water Use Efficiency (WUE) and sustainability of crop production in dryland agriculture (Selvaraju *et al.*, 1999). Patil and Sheelavantar (2004) reported that land configuration of compartmental bunds, ridges and furrows practices resulted in higher yield and water use efficiency of sorghum in the *vertisols* of Bijapur, Karnataka significantly over flatbed due to increased availability of moisture and nutrients in the soil profile. Work of various scientists referred above and elsewhere clearly suggests that adoption of appropriate soil moisture practices play significant role to enhance water use efficiency and crop productivity under water limited dryland environments.

Allolli *et al.* (2008) reported that *in-situ* moisture conservation practice *viz.,* ridges and furrows along with mulch, resulted in beneficial effect on clusterbean for getting higher growth and yields. Similarly, Kiran and Lingaraju (2005) observed that *in-situ* moisture conservation practices *viz.*, ridges and furrows, and compartment bunding beneficial in conserving higher soil moisture and resulting in higher grain yield of sorghum (16 q ha^{-1} and 15 q ha^{-1}, respectively) compared with flatbed sowing method (12 q ha^{-1}). Beneficial effect of compartment bunding and broad bed furrows on improved soil moisture to an extent of 17 per cent and 11 per cent, respectively, over traditional flatbed system was also observed elsewhere (Selvaraju and

Balasubramanian, 2001). Meerasab and Guled (2013) observed that border strip method of *in-situ* moisture conservation practice recorded higher grain yield, WUE and sorghum equivalent yield (sorghum + chickpea) under dryland tracts of Bijapur. In another study, soil moisture conservation practices coupled with mulching resulted in higher soil moisture content than farmer's practice (Ramesh and Devasenapathy, 2007). Broad bed and furrow planting increasing pigeonpea equivalent yield (pigeonpea + *urd* bean), consumptive water use and WUE over flat sown was also observed by Kantwa *et al.* (2005). *In-situ* moisture conservation practices *viz.,* ridge-furrow and bed-furrows at 90 cm spacing using tractor drawn furrower tool increased the sorghum grain yield on average by 32.5 per cent and 22 per cent during 2006-08 respectively over no furrows (control) as noted by Reddy *et al.* (2009). *In-situ* soil moisture conservation through 2:1 or 3:1 skip row method of planting along with the application of either organic manures or organic manures + inorganic fertilisers to preceding crop (sesame) and no organic manures/inorganic fertilisers to succeeding crop (sorghum) produced higher yields of both crops in sesame-sorghum sequence. Sastry (2002) reported that CB, contour farming and BBF system conserve the rain water *in-situ* and reduce soil erosion. Ramesh and Rathika (2009) tested different land configurations *viz.,* compartmental bunding, ridges and furrow and tide ridging which helped conserve more rain water and improved soil moisture ultimately increased yield attributes and yield of many field crops under rainfed condition in *alfisols* ecosystem. Similar effect of ridge and furrow sowing on the yield pearlmillet-Indian mustard sequence over flatbed method of were observed by Parihar *et al.* (2009). The positive effects of soil moisture conservation practices viz., ridges and furrows; in enhancing the plant height and yield attributes of sorghum, cowpea, bengalgram and sunflower have also been reported by Somasundaram *et al.* (2000). The above referred studies clearly suggest that soil moisture conservation based land configurations as mentioned above play significant role in enhancing WUE and crop productivity under water limited environment.

Cropping systems for soil moisture conservation

Intercropping is a practice to have an opportunity to diversify cropping system to make multiple use of available natural resources and utilise them more effectively while providing insurance against failure of one crop under dryland situations. Any factor that increases yield in drylands will result in increased WUE. Similarly, any factor reducing evapotranspiration with no serious effect on yield will also increase WUE (Eastin and Sullivan, 1984). Higher WUE has been reported for maize + soybean (2:1) and maize + mungbean (2:1) (De and Singh, 1981), pearlmillet + greengram (2:1) and pearlmillet + cowpea (3:1) (Goswami *et al.* 2002) intercrops in comparison with their respective mono-crops. Tetarwal and Rana (2006), and Kumar and Rana

(2007) tried different crop row proportions and crop combinations and found pearlmillet + mothbean (2:1) and pigeonpea + green gram (2:1) resulting in higher WUE over sole crop. The main advantage of intercropping is the more efficient utilisation of the available resources and the increased productivity compared with each crop grown in isolation. Further, intercropping conserves soil water by providing shade, reducing wind speed and increasing infiltration with mulch layers and improves soil structure (Mobasser *et al.,* 2014).

Therefore, above studies underline the importance of optimising and adopting cereal + legume based cropping systems as climate smart agricultural practices not only to diversify cropping system for sustainability, but also to enhance conservation and efficient use of scarce resources (soil moisture) to adapt to future projected climates in water limited environments.

Crop and input management practices for soil moisture conservation

Better crop and input management practices do help enhance soil moisture conservation, thus improve crop productivity. For example, incorporation of *Leucaena leaves* and FYM (@2.5 t ha^{-1}) in comparison with application of vermicompost (@1 t ha^{-1}) recorded slightly higher soil moisture in top 60 cm soil profile from sowing till harvest (Patil and Sheelavantar (2004). Tetarwal and Rana (2006) reported that application of FYM @ 5 t/ha + dust mulch + straw mulch recorded significantly higher pearl millet equivalent yield, nutrient uptake, WUE and economics. In a study carried out by Pandey *et al.* (1988)

Table 2: Soil physical properties, water use efficiency, and pigeonpea equivalent yield, as influenced by set-furrow method of cultivation in pigeonpea + pearl millet and pigeonpea + sesame intercropping systems in shallow soil under dryland conditions (Source: Arjun Sharma and Guled, 2011)

Treatments	*P (per cent)*	*MWHC (per cent)*	*IR (cm/hr)*	*WUE (Kg/ha-mm)*	*PEY (q/ha)*
Sole pigeonpea	52	48.6	0.69	2.38	13.41
Sole pearlmillet	51	48.4	0.65	1.64	7.14
PP+PM (1:2)-Flat bed	52	48.34	0.85	2.19	14.08
PP+PM (1:1)-Flat bed	52	46.94	0.73	2.04	13.12
PP+PM (2:4) Flatbed (paired row)	52	46.61	0.7	1.78	11.44
PP+PM (1:1) Set-furrow	52	48.44	1.03	2.45	15.55
PP+PM (1:2) Set-furrow	53	48.82	1.1	2.62	16.76
PP+PM (1:2) Set-furrow + vermicompost @ 2.5 t ha^{-1}	53	50.36	1.26	2.68	17.12
S.Em. ±	53	48.82	1.1	2.62	16.76
C.D. at 5 per cent	1.53	1.71	0.28	0.16	1.71

Note: P- porosity, IR- infiltration rate, PEY- pigeonpea equivalent yield, WEU- water use efficiency.

pre-sowing seed treatment with KNO_3 and Kaolin spray on pearl millet greatly increased the grain yield and WUE compared with the untreated control. Jat and Gautam (2001) reported that straw mulch and straw + Kaolin spray pearl millet was superior to all other treatments in terms of yield, consumptive use and WUE of rainfed pearlmillet. Anti-transpirants, wherever feasible could be used as they reduce the extent of water loss through transpiration, thus extending availability period of limited soil moisture during the crop growing cycle (Rana, 2007). Among the organic inputs application of vermicompost @ 2.5 t/ha in set-furrows significantly increased the water stable aggregates, per cent pore space, maximum water holding capacity and infiltration rate as compared to flatbed system (Arjun Sharma and Guled, 2011). The positive effects of FYM and mulch on crop yield of maize and wheat, and soil properties were also reported by Kumar et al. (2009).

Optimising plant population/row orientation for improved WUE

Rathore *et al.* (2008) reported sowing pearlmillet in 45 cm x 12 cm resulted in higher WUE owing to larger crop canopy cover leading to proper utilisation of soil moisture as well as nutrients under optimum population in dryland area. Similarly, Pandey *et al.* (1988) observed 2,00,000 plants/ha as optimum plant density for pearl millet as it recorded higher consumptive use, rate of moisture use and WUE as compared to a plant density of 1,00,000 plants/ha. This again was attributed to faster canopy coverage for efficient use of soil moisture.

Table 3: Effect of tillage practices on yield of rainfed maize (Source: Sanjay Arora and Rajan Bhatt, 2005)

Treatments	Grain yield (kg/ha)	Straw yield (kg/ha)
Shallow tillage	2272	4647
Deep tillage	2315	5422
Ridge and furrows	2493	6247
Farmers practices	2254	4268
Lsd (p=0.05)	88	190

Conclusions

This review clearly suggests that although crop production in dryland is highly risky and productivity is very low adoption of suitable soil and water conservation structures, legume based cropping systems and optimisation of crop and input management practices help improve WUE, build soil fertility towards sustainability and climate smart agriculture practices to cope with future projected climate change.

References

Allolli, T.B., U.K. Hulihalli and S.I. Athani. 2008. "Influence of *in-situ* moisture conservation practices on the performance of dryland cluster bean", Karnataka Journal of Agricultural Sciences. Vol. 22, pp. 250-252.

Allolli, T.B., U.K. Hulihalli and S.I. Athani. 2008. "Influence of *in-situ* moisture conservation practices on the performance of dryland Chilli", Karnataka Journal of Agricultural Sciences. Vol. 21, pp. 253-255.

Anil Kumar, B. C. Sharma, Kahan Bassi and Mahender Singh. 2009. "Effect of tillage and mulching on *in-situ* moisture conservation under rainfed conditions in maize-wheat system", Haryana Journal of Agronomy. Vol. 25, pp. 54-58.

Arjun Sharma and M.B. Guled. 2011. "Effect of set-furrow cultivation in pigeonpea + pearlmillet and pigeonpea + sesame intercropping systems in shallow black soil under rainfed conditions", Karnataka Journal of Agricultural Sciences. Vol. 24, pp. 643-650.

Arjun Sharma, Pandit S. Rathod and Mohan Chavan. 2010. "Response of pigeonpea (*Cajanus cajana*) to drought management practices under rainfed conditions", Karnataka Journal of Agricultural Sciences. Vol. 23, pp. 693-700.

De, R. and S. Singh. 1981. "Management practices for intercropping systems, In proceedings of International Workshop on Intercropping", ICRISAT, Hyderabad, India, pp. 17-21.

Eastin, J.D. and C.Y. Sullivan. 1984. "Environmental stress influences on plant persistence, physiological and production physiological basis of crop growth and development". American Society of Agronomy. Madison, Wisconsin, pp. 201-236.

Goswami, V.K., S.K. Kaushik and R.C. Gautam. 2002. "Effect of intercropping and weed control on nutrient uptake and water use efficiency of pearlmillet (*Pennisetum glaucum*) under rainfed conditions". Indian Journal of Agronomy. Vol. 47, pp. 504-508.

Mobasser, H. R., Mohammad Reza Vazirimehr, Khashayar Rigi. "Effect of intercropping on resources use, weed management and forage quality".

Jat, M. L. and R.C. Gautam. 2001. "Productivity and water use of rainfed pearl millet (*Pennisetum glaucum*) as influenced by summer ploughing and *in-situ* moisture conservation practices under semi-arid conditions of northwest India". Indian Journal of Agronomy. Vol. 46, pp. 266-272.

Kantwa, S.R., I.P.S. Ahlawat and B. Gangaiah. 2005. "Effect of land configuration, post-monsoon irrigation and phosphorus on performance of sole and intercropped pigeonpea (*Cajanus cajan*)", Indian Journal of Agronomy. Vol. 50, pp. 278-290.

Kiran, J.A. and B.S. Lingaraju. 2005. "Effect of *In situ* moisture conservation practices and nitrogen levels on growth and yield of rabi sorghum in vertisols under rainfed condition" Karnataka Journal of Agricultural Sciences.Vol.18, pp. 294-296.

Kumar, Ajay and K.S. Rana. 2007. "Performance of pigeonpea (*Cajanus cajan*) + greengram (*Phaseolus radiates*) intercropping system as influenced by moisture conservation practice and fertility level under rainfed conditions", Indian Journal of Agronomy, Vol. 52, pp. 31-35.

Mahale, D.M., P.P. Patil and, S.M. Pokharkar. 1998. "Effect of tied ridging on soil moisture and yield of pearlmillet", Indian Journal of Dryland Agricultural. Research and Development. Vol. 13, pp.11–13.

Meerasab, H.J. and M.B. Guled. 2013. "Effect of sara method of in situ moisture conservation practices and intercropping system on yield of rabi sorghum and chickpea", Karnataka Journal of Agricultural Sciences. Vol. 26, pp. 356-359.

Pandey, S.K., S.K. Kaushik and R.C. Gautam.1988. "Response of rainfed pearlmillet (*Pennisetum glaucum*) to plant density and moisture conservation". Indian Journal of Agricultural Sciences. Vol. 58, pp. 517-20.

Parihar, C. M., K. S. Rana and M.D. Parihar. 2009. "Crop productivity, quality and nutrient uptake of pearl millet (*Pennisetum glaucum*) Indian mustard (*Brassica juncea*) cropping system as influenced by land configuration and direct and residual effect of nutrient management". Indian Journal of Agricultural Sciences. Vol. 79, pp. 927–930.

Patil, S.L. and M.N. Sheelavantar. 2004. "Effect of cultural practices on soil properties, moisture conservation and grain yield of winter sorghum (*Sorghum bicolar* L. Moench) in semi-arid tropics of India" Agricultural Water Management. pp. 49–67.

Patil, S.N., M.S. Khakare and R.S. Raut. 1994. "Effect of *in-situ* moisture conservation measure on yield, water consumption, evapotanspiration and water use efficiency of grain sorghum under rainfed condition", PKV Research Journal. Vol. 18, pp.170-172.

Ramesh, T. and P. Devasenapathy. 2006. "Physical indicators of sustainability in rainfed cowpea (*Vigna unguiculata* walp) as influenced by *in-situ* rainwater harvesting", Legume Research. Vol. 30, pp. 256 – 260.

Ramesh, T. and S. Rathika. 2009. "Land configuration techniques for rain fed alfisols ecosystem- review". Green Farming. Vol. 2, pp. 879-881.

Rathore, B.S., V.S. Rana and R.K. Nanwal. 2008. "Effect of plant density and fertilizer levels on growth and yield of pearl millet (*Pennisetum glaucum*) hybrids under limited irrigation conditions in semi-arid environment". Indian Journal of Agricultural Sciences. Vol.78, pp. 667-670.

Reddy B.S. V. Maruthi, R.V. Adake and U.K. Madal. 2009. "Effect of different land configuration practices on productivity of sorghum – pigeonpea intercropping system in shallow alfisols" Indian Journal of Dryland Agricultural Research and Development. Vol. 24, pp. 57-62.

Sanjay Arora and Rajan Bhatt. "Impact of improved soil and in-situ water conservation practices on productivity in rainfed foothill region of North-west India".

Sastry, A.P., 2002. "Evolution of land treatment for *in-situ* moisture conservation in maize crop on medium deep soil". Crop Research Hisar. Vol. 23, pp. 423-427.

Selvaraju, R. and T.N. Balasubramanian. 2001. "Effect of moisture Conservation practices on water use, micro climate and productivity of intercropping in dryland vertisols" Journal of Agricultural Physics. Vol. 23, pp. 112-116.

Selvarajua, R., P. Subbiana, A. Balasubramaniana and R. Lal. 2005. "Land configuration and soil nutrient management options for sustainable crop production on alfisols and vertisols of southern peninsular India" Soil & Tillage Research. Vol. 52, pp. 203-216.

Somasundaram, E., Jauhar Ali, A., Manoharan, M.L. and Arokiaraj, A. 2000. "Response of crops to different land management practices under sodic soil conditions". Indian Journal of Agronomy, pp. 45-50.

Tetarwal, J.P. and K.S. Rana. 2006. "Impact of cropping system, fertility level and moisture-conservation practice on productivity, nutrient uptake, water use and profitability of pearlmillet (*Pennisetum glaucum*) under rainfed conditions", pp. 23-27.

32

Green Roof: The Sustainable Recharge

Maguluri Sree Devi, S. L. Chawla, Neelima Palagani and Tatte Sumathi

Abstract

Roof garden is a modern approach in urban and peri urban landscaping, which has various advantages in maintaining the ecological balance and can ameliorate the adverse effects caused by climate change. One of the advantages is storm water management regarding the water quality. Extensive green roofs are the single most effective solution to storm water management. Vegetation layers are installed over impervious roof surfaces and do not take up additional land. They do not compromise contemporary architectural design —suitable for almost every building's roof surface thus retaining storm water for small storm events and benefiting Sustainable urban drainage system (SUDS) strategy. Urban areas are dominated by hard, non-porous surfaces that contribute to heavy runoff, which can overburden existing storm-water management facilities and cause combined sewage overflow into lakes and rivers. Unlike conventional storm-water management techniques that include storage reservoirs and ponds, constructed wetlands, and sand filters, that are surface-area intensive technologies and may be difficult to implement in dense urban centres, green roofs are ideal for urban storm-water management because they make use of existing roof space and prevent runoff before it leaves the lot. Green roofs store water during rainfall events, delaying runoff until after peak rainfall and returning precipitation to the atmosphere through evapotranspiration. These also show feasible solution in contrast to the costly sewer systems built to cope with the recurring problem. Green roofs also reduce the chance of localised flooding, and relieve pressure on local drainage systems. Considering the many advantages roof gardens have to offer, implementation of roof gardens in our country can help in better water recycling and management.

Introduction

When it rains on forested and open, undisturbed land, water goes through its natural cycle. About 30 per cent of the water reaches shallow aquifers that feed plants, another 30 per cent percolates and nourishes deeper aquifers, and approximately 40 per cent is almost immediately returned into the atmosphere through plant evaporation and transpiration. There is rarely any surface runoff. In metropolitan areas with buildings and streets comprising 75 to 100 per cent impervious surface cover, rainwater is distributed much differently. Only five per cent infiltrates to shallow and deep groundwater aquifers and 15 per cent evaporates into the air through vegetation. A staggering 75 per cent of the rainwater becomes surface runoff. Storm water that does not soak into the ground becomes surface runoff, which either flows into surface waterways or is channelled into storm sewers. Stormwater is of concern for two main issues - one related to the volume and timing of runoff water (flood control and water supplies) and the other related to potential contaminants that the water is carrying- pollutants including sediment, heavy metals, petroleum-based hydrocarbons, nutrients, pesticides, chlorides, bacteria and oxygen-demanding organic matter. (water pollution). As impervious surfaces (parking lots, roads, buildings) do not allow rain to infiltrate into the ground, more runoff is generated than in the unpaved surfaces.Stormwater runoff quality is a particular concern for developments that drain directly to the river system. Impacts include sediment contamination, toxicity to aquatic life, bacterial contamination, and excessive phosphorus and other nutrients in the water body (causing unwanted algae growth in rivers and lakes). This additional runoff can erode water courses (streams and rivers) as well as cause flooding when the storm water collection system is overwhelmed by the additional flow. Because the water is flushed out of the watershed during the storm event, little infiltrates the soil, replenishes groundwater, or supplies stream base flow in dry weather. Often coping with storm water runoff and its after effects involve huge amount of capital for engineering and implementation of drainage systems as they generally do not go hand in hand with the existing city plans and building layouts.

A better way to address storm water management in view of integrated approach is the use of roof gardens. A "green roof" is a permanent rooftop planting system containing live plants in a lightweight engineered soil medium. These systems are designed to retain precipitation where the water is taken up by plants and transpired into the air. As a result, much less water runs off the roof compared to conventional rooftops. Studies show that green roofs, dependent upon their substrate depth, can retain 60-100 per cent of incoming rainfall.

Green roofs can also mitigate storm-water runoff from building surfaces by collecting and retaining precipitation, thereby reducing the volume of flow into storm-water infrastructure and urban waterways. Green roofs offer sustainable, cost-effective and eco-friendly solutions for harvesting and reusing storm water for both a potential alternative water supply for non-drinking uses and a means to further reduce storm water pollution in our waterways.

Unlike conventional storm-water management techniques that include storage reservoirs and ponds, constructed wetlands, and sand filters, that are surface-area intensive technologies and may be difficult to implement in dense urban centers, green roofs are ideal for urban storm-water management because they make use of existing roof space and prevent runoff before it leaves the lot. This makes green roofs a sustainable answer and an eco-friendly solution to the water challenges faced today in view of Water Sustainable Urban Design (WSUD). Green roofs store water during rainfall events, delaying runoff until after peak rainfall and returning precipitation to the atmosphere through evapotranspiration.

There are two distinctly different types of green roofs: intensive and extensive. Intensive green roofs require a minimum of one foot of soil depth to create a more traditional roof garden, with large trees, shrubs and other manicured landscapes. In contrast, extensive green roofs range from as little as one to five inches in soil depth. Extensive green roofs are primarily built for their environmental benefits. Extensive green roofs are the single most effective solution to storm water management. Vegetation layers are installed over impervious roof surfaces and do not take up additional land. They do not compromise contemporary architectural design —suitable for almost every building's roof surface thus retaining storm water for small storm events and benefiting Sustainable urban drainage system (SUDS) strategy.

Green roofs have many benefits to the building, both inside and out, as well as many environmental benefits like reduced air pollution, reduced heat island effect, better management of biodiversity (Kim, 2004; Brenneisen, 2005) and controlled water pollution.

Runoff Management

Green roofs can be an important element in an integrated water sensitive design and planning approach, as the roof is often the first point of contact in the storm water chain. By intercepting the rain runoff at the source, the green roof eliminates the potential multiplying effect further downstream of the runoff chain. Vegetation assists the management of runoff by reproducing many of the hydrological processes normally associated with the natural environment. Many elements contribute to this process. Storage detention or retention rates for green roofs depend on many variable factors. Studies have found considerable seasonal variations for rainwater retention between summer and winter due to the greater amount of water being returned to the atmosphere insummer through evaporation and transpiration. Retention rates in summer can be between 70-100 per cent, but in winter may be 40-50 per cent. In a more recent study (Hutchinson *et al.,* 2003) involving stormwater monitoring of two eco-roofs in Portland, Oregon, it was found that the summer retention rate can be up to 100 per cent, and down to 69 per cent at other times of the year for a 100-150 millimetre thick green roof. These studies also showed that for small to moderate storms, the green roofs virtually retained all the rain that fell on them.

Water Quality

Green roofs not only retain rainwater, but also moderate the temperature of the water and act as a natural filter for any runoff. This occurs through infiltration and the bioretention process in the substrate. A recent innovative development in green roof design is to install constructed wetlands as an integral part of the green roof system. Taking the known benefits of constructed wetlands and adding this element to the runoff management role of green roofs is an important WSUD technique. A demonstration case is the John Deere Works in Mannhein, Germany where it was decided to begin treating the wastewater from the company's manufacturing and assembly operations with a constructed wetland on the factory roof. The wetland includes a combination of sedges, rushes and irises grown hydroponically in 50 millimetres of water. It breaks down carbon and nitrogen compounds present in the wastewater while sequestering phosphates and heavy metals (Earth Pledge, 2005)

Pollutant removal

(Pollutant removal efficiency using green roofs as compared to water runoff from impervious surfaces)

Total Suspended Solids -80 per cent

Nutrients - Total Phosphorus / Total Nitrogen removal- 50 per cent

Metals - Cadmium, Copper, Lead, and Zinc removal- 40 per cent

Future research directions

Much green-roof plant selection depends on lists from German research. Further research is needed to identify suitable plant species for living roofs in many other climatic regions. Furthermore, most green-roof plant combinations are selected for full sun exposure, using species that originate in permanently open (non-forested) habitats such as rock outcrops (e.g., most *Sedum*), cliffs, dunes, and heathland (Lundholm, 2006). New selections are being investigated to identify plants suitable for shaded roof conditions. Researchers are also investigating plants that provide other services, such as removing contaminants from storm water and providing resources (e.g., pollen) for native insects and other animals.

Conclusion

The recent drought and concerns about climate change have all highlighted the need tomanage our water resources more sustainably. Expanding the use of stormwater runoff to add to our water supply and reduce water pollution are important objectives for the Indian Government. Stormwater is now recognised as a valuable resource, rather than a nuisance to be disposed of quickly, especially in large urban centres.

Over recent years, utilisation of roof gardens as a component of stormwater harvesting and reuse have emerged as a new field of sustainable water management. Harvesting and reusing stormwater by means of roof gardens in combination with other components offer both a potential alternative water supply for non-drinking uses and a means to further reduce stormwaterpollution in our waterways. Landscaping of precisely engineered green roofs thereby harvesting stormwatercomplements other approaches to sustainable urban water management, including rainwater tanks, greywater systems, effluent reuse and demand management.

References

Earth Pledge (2005). Green Roofs: Ecological Design & Construction, Schiffer Publishing, USA.

Hutchinson, D., Abrams, P., Retzlaff, R., Liptan, T. (2003). Stormwater Monitoring for Ecoroofs in Portland, Oregon, USA. Greening Rooftops for Sustainable Communities, Chicago.

Kim, K., (2004) - The application of the biosphere reserve concept to urban areas: the case of green rooftops for habitat network in Seoul. Annals of the New York Academy of Sciences 1023, pp. 187–214.

Lundholm JT. 2006.Green roofs and facades: A habitat template approach.

Urban Habitats 4: 87–101.

33

Rain Water Harvesting- Adapting to climate change

Tatte Sumathi; Dr S.L. Chawla and Neelima Palagani

Abstract

Water is predicted to be the primary medium through which early climate change impacts will felt by the people, ecosystems and economics. Extreme climate events like drought,flood, cyclone and stormy rainfall are expected to leave an impact on human society. They are also expected to generate widespread response to adapt and mitigate the sufferings associated with these extremes. However, climate and culture can interact in numerous ways. People may resort to modify dwelling environments by adapting new strategies to optimise the utility of available water by harvesting rain water. Rainwater harvesting has significant potential to provide environmental and economic benefits by reducing storm water runoff and pollution, reduces erosion in urban environment, conserving potable water and provides inexpensive supply of water. Rainwater harvesting, collecting rainwater from impervious surfaces and storing it for later use was a technique that has been used for millennia. It has not been widely employed in industrialised societies that rely primarily on centralised water distribution systems, but with limited water resources and storm water pollution recognised as serious problems and the emergence of green building, the role that rainwater harvesting can play for water supply is being reassessed. Among the various alternative technologies to augment freshwater resources, rainwater harvesting and utilisation is a decentralised, environmentally sound solution, which can avoid many environmental problems often caused in conventional large-scale projects using centralised approaches.

Introduction

Climate change has been one of the most studied environmental concerns over the past two decades. Today there is no doubt that the environmental health of the planet is in a critical state. The main issues affecting water resource management are (i) The natural conditions, low rainfall with high evaporation rates, which together create low availability of runoff; (ii) Population growth and economic development lead to greater water demand and increased pollution of available resources; (iii) The policies pertaining to management of water resources determine the approach taken by relevant authorities to managing the resource and directly impacts other driving forces and pressures.

A comprehensive knowledge of climate fluctuations and corresponding adaptation by human society is crucial for our progress towards sustainability. The archaeological and historical records of rainwater harvesting systems and water management worldwide to understand the climate–rainwater harvesting. Since we expect that adaptation through rainwater collection may be particularly effective in tropical monsoon regions, where the seasonal cycle in rainfall is large.

The most vital resource, water, is becoming inaccessible for many people and it is obvious that old patterns of its management must be changed. At the same time, many centuries ago, mankind's wisdom has foreseen how to go through critical situations. When dramatic climate change had transformed flourishing landscapes into deserts; man has started to collect the precious raindrops. Rain is the ultimate source of fresh water. With the ground area around houses and buildings being cemented, particularly in cities and towns, rainwater, which runs off from terraces and roofs, was draining into low-lying areas and not percolating into the soil. Thereby, precious rainwater is squandered, as it is drained into the sea eventually. Rain water harvesting is a system by which, the rainwater that collects on the roofs and the area around the buildings is directed into open wells through a filter tank or into a percolation chamber, built specifically for this purpose.

Importance of rain water harvesting

Major parts of our country have been facing continuous failure of monsoon and consequent deficit of rainfall over the last few years. Also, due to ever increasing population of India, the use of ground water has increased drastically leading to constant depletion of ground water level causing the wells and tubewells to dry up. At some places, excessive heat waves during summer createa situation similar to drought. Rain water harvesting is increases the infiltration rate of soil, ground water recharge and improves the ground water storage, reduce the soil erosion. It fulfilled the fresh water demand to meet the drinking water needs of the people in the country besides irrigation and domestic needs. Harvested water can also used after kitchen garden and other home needs.

Methods to harvest rain water

Rainwater can be harvested in a variety of ways:

- Directly from roof tops and stored in tanks.
- Monsoon run off and water in swollen streams during the Monsoon and storing it in underground tanks.
- Water from flooded rivers can be stored in small ponds.

Rainwater harvesting is principally the management of these two partitioning points in the water flow. At the local scale, such as a farmer's field, the flows partition the incoming rainfall at the soil surface, either infiltrating the water into the soil or diverting the water as surface runoff. Within the soil, the second partitioning point is at the plant roots where water is either taken up by the vegetation, or contributes to the recharge of shallow or deep groundwater. Depending on the soil surface conditions, infiltration can range from 100 per cent ina well managed agricultural soil, to 100 per cent runoff from a paved road or rooftop. The second partitioning point can be managed, indirectly, through planting different plant species, improving crop uptake capacity, modifying plant root depths, and altering soil management practices in agriculture, through enhancing/depleting soil health, including soil organic matter content, etc.

In domestic rooftop rain water harvesting systems, rainwater from the house roof is collected in a storage vessel or tank to use during the periods of scarcity. Usually these systems are designed to support the drinking and cooking needs of the family at the doorstep. Such a system usually comprises a roof, a storage tank and guttering to transport the water from the roof to the storage tank. In addition, a first flush system to divert the dirty water which contains roof debris collected on the roof during non-rainy periods and a filter unit to remove debris and contaminants before water enters the storage tank are also provided.

Roof Catchment

The roof of the house is used as the catchment for collecting the rainwater. Roofs are made of corrugated iron sheet, asbestos sheet, tiles or concrete can be utilised as such for harvesting the rainwater. But thatched roofs are not suitable as it gives some colour to water and also the water carries pieces of roof material (such as palm leaves).

Gutters

Gutters are channels fixed to the edges of roof all around to collect and transport the rainwater from the roof to the storage tank. Gutters can be prepared in semi-circular and rectangular shapes. Locally available material such as plain galvanised iron sheet can be easily folded to required shapes to prepare semi-circular and rectangular gutters. Semi-circular gutters of PVC material can be readily prepared by cutting the PVC pipes into two equal semi-circular channels. Bamboo poles can also be used.

Downpipe

Down pipe is the pipe, which carries the rainwater from the gutters to the storage tank. It is joined with the gutters at one end, and the other end is connected to the filter unit of the storage tank. PVC or GI pipes of diameter 50 mm to 75 mm (two inches to three inches) are commonly used for down-pipe.

First Flush Pipe

Debris, dirt and dust collect on the roofs during non-rainy periods. When the first rains arrive, this unwanted material will be washed into the storage tank by roof water. This causes contamination of water collected in the storage tank thereby rendering it unfit for drinking and cooking purposes. Therefore, a first flush system is incorporated to dispose off the water from 'first rain' so that it does not enter in the tank. There are two such simple systems. One is based on a simple manually operated arrangement, whereby, the down pipe is moved away from the tank inlet and replaced again once the first flush water has been disposed. In another simple and semi-automatic system, a separate vertical pipe is fixed to the down pipe with a valve provided below the "T" junction.

After the first rain is washed out through first flush pipe, the valve is closed to allow the water to enter the down pipe and reach the storage tank.

Filter Unit

The filter unit is a container or chamber filled with filter media such as coarse sand, charcoal, coconut fiber, pebbles and gravels to remove the debris and dirt from water that enters the tank. The container is provided with a perforated bottom to allow the passage of water. The filter unit is placed over the storage tank. Commonly used filters are of two types. One is a ferro-cement filter unit, which is comparatively heavy and the other is made of either aluminium or plastic bucket. The latter is readily available in market and has the advantage of case in removing, cleaning and replacing. Another simple way of filtering the debris and dust particles that came from the roof along with rainwater is to use a fine cloth as filter media. The cloth, in two or three layers, can be tied to the top of a bucket or vessel with perforations at the bottom.

Storage Tank

Storage tank is used to store the water that is collected from the Rooftops. Common vessels used for small scale water storage are plastic bowls, buckets, jerry cans, clay or ceramic jars, cement jars, old oil drums etc. For storing larger quantities of water the system are required a bigger tank with sufficient strength and durability.

In rural areas, houses have usually inclined roof made up of asbestos sheets, bricks or cemented sheets. Thus, water from the roof comes to the edges easily and is collected using channels fitted and is brought to storage tank through PVC pipes. The storage tank is made up of ferro-cement. Cloth is the material used as filter to clean out the water from the roof. These methods used to store water either for agricultural purpose or for domestic use.

Rock catchments

Just as the roofs of buildings can be used for the collection of rainwater, rock outcrops can also be used as collecting surfaces. Indeed, if access to the catchment area by animals, children, etc can be prevented, a protected catchment can collect water of high quality as long as its surfaces are well flushed and cleaned before storage takes place.

Sometimes, it proves difficult to prevent the collected water from being polluted. If so, it is sensible to use this water for purposes that do not require a potable water supply, such as house cleaning, laundry, horticulture, etc, and reserve for drinking water, cooking and personal hygiene the better quality water that has been collected from a clean roof.

Conclusion

Climate change was affect rainfall; increase the temperature, evaporation, which will put increasing pressures on our ecosystems services. At the same time, rapid growth of population affect our ecosystems as we increase our demands for services, including reliable and clean water. Rainwater harvesting will continue to be an adaptation strategy for people living with high rainfall variability, both for domestic supply and to enhance crop, livestock and other forms of agriculture. Access and right to land can be a first step to rainwater harvesting interventions. Special measures should be in place so rainwater harvesting interventions also benefit land-poor and landless in a community. Establish enabling policies and cost –sharing strategies, (including subsides) to be provided together with technical know-how and capacity building.

References

In Barron, J. (Ed), (2009). Rainwater harvesting: a lifeline for human well-being. A report prepared for UNEP by Stockholm 17 Environment Institute York, UK/ Stockholm Resilience Centre, Stockholm, Sweden. ISBN: 978-92-807-3019-7.

Mukheibir, P., (2008). Water Resources Management Strategies for Adaptation to Climate-Induced Impacts in South Africa. Water Resources Management 22:1259–1276.

Mwenge Kahinda, J., Boroto, R.J. and Taigbenu, A. E. 2005. Developing an Integrated water resources management and rainwater harvesting systems in South Africa. Paper presented at the 12th SANCIAHS Symposium. September 5-7, 2005, Johannesburg, RSA, (CD format).

Pandey, D.N., Gupta, A.K. and Anderson, D.M., 2003. Rainwater harvesting as an adaptation to climate change. Current Science 85 (1), 46-59.

Salas, J.C., Konig, K.W. and Lo, A., (2009). Rainwater harvesting providing adaptation opportunities to climate change.

34

Effect of Irrigation Regimes on Plant Growth Parameters and Yield of Wheat (Triticum *Aestivum* L.)

P. S. Jayswal, G. R. Sharma, K. N. Sondarva and M. S. Dudhat

Abstract

In view of the limited water resources in Saurashtra, the effect of magnitude of water deficit on crop growth and yield is of major importance. Keeping this in mind, the study was conducted to find out the effect of irrigation regimes on crop yield and water use efficiency. Five treatments were adopted i.e. 0.6 ET_c, 0.8 ET_c, 1.0 ET_c, 1.2 ET_c and control. ET_c was calculated using daily weather data for wheat crop growth period. Irrigation was applied at 10 to 12 days interval as per the treatments. In control treatment crop 60 mm fixed depth of irrigation was applied. The significantly higher grain yield (2808.65 kg/ha) and dry matter (3990.76 kg/ha) were attained for control crop. The grain water use efficiency and dry matter water use efficiency were found insignificant.

Introduction

Water is a crucial input for agriculture, and major resource constraint that limits economic development and food grain production in India. Water is necessary for industrial production, energy generation and navigation. Its availability is becoming scarce and costly.

Planning of the allocation and use of irrigation water needs proper predictive estimation of production output status associated with given input levels. A basic crop production function attempts to estimate the contribution of water, fertiliser and chemical amendments etc. to crop production for the purpose of understanding the nature of crop response to water. Even the water is considered the only controllable input, the crop-water-yield relationship is quite complex. The food production has to be increased through maximisation of crops yield per unit of available water for which information on yield response to water for various crops grown under different environmental condition is imperative.

In view of the limited water resources in Saurashtra, the effect of magnitude of water deficit on crop growth and yield is of major importance in scheduling available but limited water supply. Therefore, the efforts are made to find out the effect of irrigation strategies on yield of wheat, considering water use simulation approach during the investigation to study the effect of irrigation regimes on plant growth parameters and yield of wheat.

Materials and methods

The experiment was conducted at Instructional Farm of Soil and Water Engineering Department, College of Agricultural Engineering and Technology, Junagadh Agricultural University, Junagadh undertaken during the year 2009-10 in Rabi season. Geographically Junagadh is situated at 21.5°N latitude and 70.58°E longitude with an altitude of 60 meter above mean sea level on the western side of the foothills of Girnar. The area is situated in South Saurashtra Agroclimatic Zone of Gujarat State.

The study area is having typically subtropical and semi-arid climate, characterised by fairly cold and dry winter, hot and dry summer and warm and moderately humid during monsoon. Partial failure of monsoon once in three to four years is common in this region. Winter sets in the month of November and continues till the end of February. January is the coldest month of winter. Summer commences in the second fortnight of February and ends in the middle of June. April and May are the hottest months of summer.

The daily weather data was collected from JAU observatory. The climatic variation at the experimental site during the experiment period (Nov. 2009 to Mar. 2010), the minimum pan evaporation, temperature and relative humidity were observed as 2.4 mm, 9.5°C, 5 per cent respectively while maximum these were observed as 8.5 mm, 40.2°C, 94 per cent respectively.

The randomised block design (RBD) was adopted. The following treatments were taken:

1. T_1 : 0.6 ET_c
2. T_2 : 0.8 ET_c
3. T_3 : 1.0 ET_c
4. T_4 : 1.2 ET_c
5. T_5 : control (60 mm depth of water)

All these treatments were replicated six times. The treatment plot was having 12 rows having spacing of 22.5 cm. The gross plot size of treatment was 6.00 m x 3.60 m. The net plot size of treatment was 4.00 m x 2.70 m. The water meters were used to apply the calculated quantity of irrigation water to the treatments. Water meters and valves were connected with flexible 63 mm hosepipes to regulate the water flow.

Crop Evapotranspiration

$$ET_C = K_C \times ET_0 \quad ...(1)$$

where, K_C is Crop coefficient and ET_0 is Reference evapotranspiration (mm/day)

$$ET_0 = \frac{0.408\Delta(R_n - G) + \gamma\frac{900}{T+273}u_2(e_s - e_a)}{\Delta + \gamma(1+0.34u_2)} \quad ...(2)$$

where, R_n is net radiation at the crop surface (MJ/m²/day), G is soil heat flux density (MJ/m²/day), T is mean daily air temperature at 2 m height (°C), u_2 is wind speed at 2 m height (m/s), e_s is saturation vapour pressure (kPa), e_a is actual vapour pressure (kPa), e_s-e_a is saturation vapour pressure deficit (kPa), is slope of vapour pressure curve (kPa/°C), and a is psychrometric constant (kPa/°C).

Crop Coefficient

Empirically determined crop coefficient (K_c) can be used to relate ET_0 to maximum crop evapotranspiration (ET_c) when water supply meets the water requirements of the crop. Crop evapotranspiration is also known as maximum crop evapotranspiration (ET_m). The values of K_c for crop given by Doorenbos and Kassam (1979) are used for calculated the crop evapotranspiration.

Water-Use Efficiency

water use efficiency was calculated by the following relationship:

$$WUE_g = \frac{\text{Grain yield}}{\text{Total amount of water applied}} \quad ...(3)$$

$$WUE_d = \frac{\text{Dry matter production}}{\text{Total amount of eater applied}} \quad ...(4)$$

Crop Growth Rate

The values of CGR was calculated for the stage between 60 and 90 DAS with formula (Cheema *et al.*, 1991)

$$\text{CGR (g/m}^2\text{/day)} = \frac{W_2 - W_1}{t_2 - t_1} \qquad \ldots(5)$$

where, W_1 and W_2 are weight of dry matter of plant (g/m²) at first and second stages and t_1 and t_2 are time in days of first and second stages.

Results and discussion

Yield Response

The significantly higher yield was obtained (2808.65 kg/ha) in T_5 (control) as shown in Table 1. By increasing amount of irrigation / levels of irrigation, the grain yield of wheat was not increasing significantly. Among other treatments, T_4 (1.2 ET_c) was found superior but it was at par with T_3 (1.0 ET_c). T_1 (0.6 ET_c) and T_2 (0.8 ET_c) was found at par with each other. Also T_1 (0.6 ET_c), T_2 (0.8 ET_c) and T_3 (1.0 ET_c) were found at par with each other. The results of the present study were in good agreement with findings of Ejaz *et al.* (2008) and Mahdi *et al.* (1997). They reported that as amount of water applied reduced, the grain yield also reduced.

By increasing irrigation amount there was no significant increase in grain yield. Therefore, farmers who have scarcity of water in their area should apply irrigation as per treatment T_1 (0.6 ET_c) to save water which should be used to increase area under irrigation.

Table 1: Grain Yield (kg/ha) and Dry Matter Production (kg/ha) Under Different Irrigation Regimes

S.No.	*Treatment*	*Grain yield (kg/ha)*	*Dry matter production (kg/ha)*
1.	0.6 ET_c (T_1)	1475.32	2203.72
2.	0.8 ET_c (T_2)	1527.78	2226.09
3.	1.0 ET_c (T_3)	1765.44	2513.13
4.	1.2 ET_c (T_4)	2117.29	2998.47
5.	Control (T_5)	2808.65	3990.76
6.	SEm+	179.0291	191.0286
7.	CD at 5%	528.15	563.54
8.	CV%	22.62	16.79

Dry matter production was significantly affected by irrigation regimes. Significantly higher dry matter was obtained in T_5 (control) (3990.76 kg/ha) and it was lowest in T_1 (0.6 ET_c) (2203.72 kg/ha) (Table 1). By increasing the amount of irrigation/irrigation levels) i.e. T_1 (0.6 ET_c) to T_4 (1.2 ET_c), the dry matter production of wheat was not increased significantly. T_1 (0.6 ET_c), T_2 (0.8 ET_c) and T_3 (1.0 ET_c) were found at par with each other. Similar results were obtained by Mahdi *et al.* (1997) and Pandey *et al.* (2001) that, as amount of irrigation water increased, dry matter production also increased.

Response of Plant Growth Parameters to Irrigation Regimes

Plant height

As the irrigation levels/amount of irrigation applied increased the height of plants also increased. The plant height was found significantly higher in T_5 (control) as compared to other treatments (Table 2). Treatment T_3 (1.0 ET_c), T_4 (1.2 ET_c) and T_5 (control) were found at par with each other. T_1 (0.6 ET_c) and T_2 (0.8 ET_c) were found at par with each other.

Similar results were obtained by, Mirbahar *et al.* (2009) and Ali *et al.* (2010). Richards *et al.* (2001) and Ghodsi (2004) reported that one of the major effect of water stress was to decrease plant height, which also caused a reduction in dry matter accumulation and subsequently plant production.

Number of tillers per plant

Number of tillers per plant was found significantly affected by irrigation regimes. As amount of irrigation increased, number of tillers also increased. It was found significantly higher in T_5 (control) treatment (Table 2). Among the irrigation levels, T_1 (0.6 ET_c) and T_2 (0.8 ET_c), T_3 (1.0 ET_c) and T_4 (1.2 ET_c) were found at par. T_4 (1.2 ET_c) was also found at par with T_5 (control).

Similar results were found by Mesbah (2009) that number of tillers increased significantly with increasing irrigation regimes. The highest values number of tillers was 5.91 for I_3 (1850 m^3) than the 4.17 for I_2 (1600 m^3).

Table 2: Effect of Irrigation Regimes on Plant Height, Number of Tillers per Plant and Number of Spikelet per Spike

S.No.	*Treatment*	*Plant height (cm)*	*No. of tillers per plant*	*No. of spikelet per spike*
1.	0.6 ET_c (T_1)	56.66	2.33	12.50
2.	0.8 ET_c (T_2)	57.33	2.67	12.08
3.	1.0 ET_c (T_3)	62.78	4.17	13.33
4.	1.2 ET_c (T_4)	64.15	4.67	14.33
5.	Control (T_5)	66.56	5.17	15
6.	SEm+	1.65	0.2392	0.5307
7.	CD at 5%	4.876	0.7057	1.5656
8.	CV%	6.583	15.42	9.67

Number of spikelet per spike

Number of spikelet per spike was found significantly higher for T_5 (control) (Table 2). As the irrigation levels increased the spikelet also increased. Treatment T_4 (1.2 ET_c) was found at par with T_3 (1.0 ET_c) as well as with T_5 (control). Also T_1 (0.6 ET_c), T_2 (0.8 ET_c) and T_3 (1.0 ET_c) were found at par with each other.

Results obtained in this study were in good agreement with findings of Qadir *et al.* (1999) that water stress reduced the number of spikelet per spike in wheat crop.

Crop growth rate

Crop growth rate was found significantly higher in T_5 (control) (Table 3) as compare to other treatments. It was observed that crop growth rate increase with increase in irrigation levels. T_1 (0.6 ET_c) and T_2 (0.8 ET_c) were found at par with each other.

Table 3: Effect of Irrigation Regimes on Crop Growth Rate

S.No.	*Treatment*	*CGR (g/m²/day)*
1.	0.6 ET_c (T_1)	4.95
2.	0.8 ET_c (T_2)	5.29
3.	1.0 ET_c (T_3)	7.64
4.	1.2 ET_c (T_4)	9.3
5.	Control (T_5)	12.34
6.	SEm+	0.4746
7.	CD at 5%	1.40
8.	CV%	14.71

Water Use Efficiency

The results of this study were in close agreement with the results obtained by Singh *et al.* (2007), having water use efficiency 6.93 kg/ha-mm for 50 per cent PET crop was more in comparison to 4.64 kg/ha-mm 100 per cent PET and 6.61 kg/ha-mm control. Similar results reported by Abbate *et al.* 2004 and Tavakkali and Oweis (2004).

At higher irrigation regimes, the grain and dry matter water use efficiencies were obtained less as compare to limited irrigation. Similar results were reported by Li *et al.*, 1999, and Howell *et al.*, 1998 which supported the results of this study.

Table 4: Effect of irrigation regimes on water use efficiency of wheat crop

S.No.	*Treatment*	*Grain WUE (kg/ha-mm)*	*Dry matter WUE (kg/ha-mm)*
1.	0.6 ET_{-c} (T_1)	6.67	9.96
2.	0.8 ET_c (T_2)	5.55	8.09
3.	1.0 ET_c (T_3)	5.36	7.64
4.	1.2 ET_c (T_4)	5.53	7.83
5.	Control (T_5)	5.85	8.31

Wheat yield production function

The wheat yield response function take into account consumptive water use and yield. The observed grain yield, dry matter production and consumptive water use are shown in Fig. 1. The developed wheat yield response functions were as presented in Table 5.

Table 5: Wheat Yield Response Function for Grain and Dry Matter Production

Yield	*Wheat Yield Response Function*	*R^2 value*
Grain	Y = 5.349 W + 133.2	0.948
Dry matter	Y = 7.188 W + 360.1	0.925

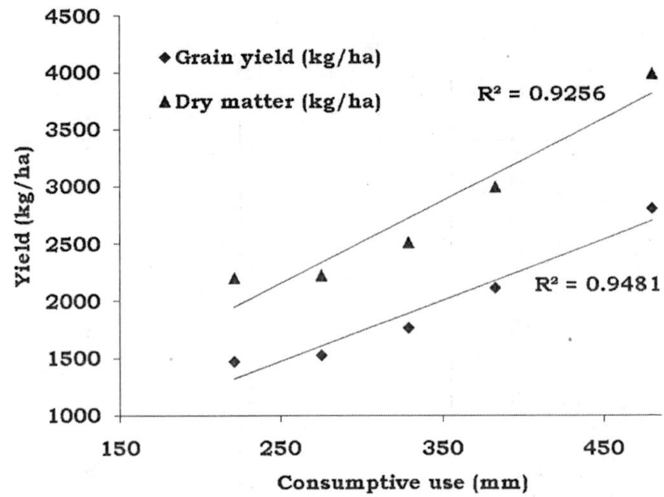

Fig. 1: Relationship between observed grain yield and water consumption.

The grain yield and dry matter production both show increasing trend to irrigation depth. The results revealed that the yield of wheat was not increasing significantly with the increase of irrigation levels. Also T_1 (0.6 ET_c), T_2 (0.8 ET_c) and T_3 (1.0 ET_c) were at par with each other.

Conclusion

The major conclusions which were drawn from the study are

- The significantly higher grain yield and dry matter production (2809 kg/ha and 3991 kg/ha) were attained when 60 mm irrigation was applied.
- By increasing the irrigation levels/amount of irrigation, the grain yield and dry matter production was not increased significantly.
- The higher values of grain and dry matter water use efficiencies were observed.
- At higher levels, the grain and dry matter water use efficiencies were observed less as compared to low irrigation levels.

References

Abbate, P.E., J.L. Dardanelli, M.G. Cantarero, M. Maturano, R.J.M. Melchiori and E.E. Suero, 2004. Climatic and Water Availability Effects on Water Use Efficiency in Wheat, Crop Science, Vol. 44, pp. 474-483.

Ali, A.M., N.B. Amru and S.B. Syed, 2010. The Performance of Durum and Bread Wheat Genotypes Associated with Yield and Yield Component under Different Water Deficit Conditions. Australian J. of Basic and Applied Sci., Vol. 4 (1), pp. 106-113.

Cheema, S.S., B.K. Dhaliwal and T.S.Sahota, 1991. "Agronomy, Theory and Digest", Kalyani Publishers, New Delhi.

Doorenbos, Kassam, 1979. Yield Response to Water. Irrigation and Drainage paper No. 33, FAO, Rome. p.193.

Ejaz, A.W., R. Ahmad, A. Ali and S. Ullah, 2008. Irrigation and Nitrogen Effects on Grain Development and Yield in Wheat (*Triticum aestivum* L.). Pak. J. Bot., Vol. 39(5), pp. 1663-1672.

Ghodsi, M., 2004. Ecophysiological Aspects of Water Deficit on Growth and Development of Wheat Cultivars. PhD thesis, University of Tehran, Iran.

Howell,T.A., J.A. Howell, A.D. Tolk, Schneider and S.R. Evett, 1998. Evapotranspiration, Yield and Water Use Efficiency of Corn Hybrids Differing in Maturity. Agron. J., Vol. 90, pp. 3–9.

Li, F.M., A.H. Guo and H.Wei, 1999. Effects of Clear Plastic Film Mulch on Yield of Spring Wheat. Field Crops Res., Vol. 63, pp. 79–86.

Mahdi, M.A., A. Berrada and M. Stack, 1997. Evaluation of Irrigation Scheduling Program and Spring Wheat Yield Response in Southwestern Colorado. Agril. Water Mgmt., Vol. 34 (2), pp. 137-148.

Mesbah, E.A.E., 2009. Effect of Irrigation Regimes and Foliar Spraying of Potassium on Yield, Yield Components and Water Use Efficiency of Wheat (*Triticum aestivum* L.) in Sandy Soils. World J. of Agril. Sciences, Vol. 5(6), pp. 662-669.

Mirbahar, A.A., G.S. Markhand, A.R. Mahar, S.A. Abro and N.A. Kanhar, 2009. Effect of Water Stress on Yield and Yield components of Wheat (*Triticum aestivum* L.) Varieties, Pak. J. Bot., Vol. 41(3), pp. 1303-1310.

Pandey, R.K., J.W. Maranville and A. Admou, 2001. Tropical Wheat Response to Irrigation and Nitrogen in a Sahelian Environment. I. Grain Yield, Yield Components and Water Use Efficiency, Euro. J. of Agron, Vol. 15(2), pp. 93-105.

Richards,R.A., A.G. Condo and G.J. Rbetzke, 2001. Trait to Improve Yield in Dry Environments In: Reynold, M.P., Oritz - Monasterio, J. I. and McNab, A. (Eds) Application physiology in wheat breading. Mexico, D.F, CIMMYT, pp. 88-100.

Singh, R.S., A.S. Rao, and Laxminarayan, 2007. Water and Heat-Use Efficiency of Mustard (Brassica juncea L. Czern. & Coss) and Its Yield Response to Evapotranspiration Rates under Arid Condition. J. of Agromet., Vol. 9(2), pp. 236-241.

Qadir, G., M. Saeed, and M.A. Cheema, 1999. Effect of Water Stress on Growth and Yield Performance of Four Wheat Cultivars, Pak. J. of Biological Sci., Vol. 2(1), pp. 236-239.

Tavakkoli, A.R. and T.Y. Oweis, 2004. The Role of Supplemental Irrigation and Nitrogen in Producing Bread Wheat in the Highlands of Iran. Agril. Water Mgmt., Vol. 65(13), pp. 225-236.

35

Differential Water Production Function for Wheat

V. K. Sardhara, R. Subbaiah, G. V. Prajapati and H. D. Rank

Abstract

The experimental investigations were carried at Junagadh to determine the differential response of moisture regimes on the crop yield. The optimum moisture values for each physiological stage are determined by analysing the experimental investigations. Milk ripe stage was found to be the most sensitive to moisture stress followed by dough ripe, flowering, late tillering and crown root initiation stages. Maximum and minimum grain yields of 4180 and 3024 kg ha^{-1} were observed when the wheat crop was subjected to 20 per cent and 80 per cent moisture stress respectively at milk ripe stage. Maximum grain yield can be attained by maintaining the moisture stress levels at 40, 40, 20, 20 and 20 per cent respectively during crown root initiation, late tillering, flowering, milk ripe, and dough ripe stages. The water stress response function for different physiological crop stages with respect to degree of moisture stress have been fitted with quadratic equation. The stress sensitivity factor for crown root initiation, late tillering, flowering, milk ripe and dough ripe stages for wheat cultivar was found to be 0.147, 0.178, 0.234, 1.19 and 0.583 respectively.

Key words: *Differential response of moisture regimes, water stress response function.*

Introduction

Experiments have indicated that crops respond differently to water allocations at different times of their following seasons. The response is more pronounced for different physiological growth stages of the crops. Also some growth stages are more sensitive than others to water deficits. This is because the yield of a crop is a function of total dry matter produced and its partitioning between the vegetative and reproductive organs. This is governed by the various plant parts that are produced during the growth (easy tillers, ears, flowers, seeds etc.). Water stress at some of these stages influences the relationships among the yield producing components and with dry matter and consequently the yield (Sinha *et al.*, 1985). The relationships are complex characteristics of 'developmental plasticity' in exhibited by most non-forage crops (Turner and Bruch, 1983). For this reason, irrigation system management in practice, when faced with water shortages has tended to be reactive in nature and almost entirely based on water supply considerations alone. If the empirical research findings from irrigation experiments were generalised so that they may be applicable across locations and seasons, it would be possible to balance water demands and supplies to a higher degree than currently feasible. This implies that available information from empirical experiments has to be carefully analysed so as to abstract generalised mathematical relationships from them between crop water use and yield. This is feasible if the entire soil water balance is considered in the analysis. Finally, the relationships have to be combined with the physical limitations of the water systems to determine the optimal irrigation strategy. Therefore, keeping the above facts into consideration an experiment was undertaken for wheat cultivar to attain optimum yield under deficient water supply conditions

Material and Methods

A field experiment was conducted to determine the impact of soil moisture depletion levels on each physiological stages of wheat cultivar. In the present study five physiological stages were considered. They include crown root initiation stage (S_1), late tillering stage (S_2), flowering stage (S_3), milk ripe stage (S_4) and dough ripe stage (S_5). In each physiological stage four levels of soil moisture depletion were considered. They include 20 per cent (T_1), 40 per cent (T_2), 60 per cent (T_3) and 80 per cent (T_4) of the available water. The experiment was conducted on split plot design with three replications. The size of the plot was 2.0 m × 1.0 m. A spacing of 1.0 m was kept between main plots as well as between sub-plots and to avoid the interference of the moisture content. A spacing of 2.0 m was maintained between replications. An exact quantity of precalculated irrigation water was applied to the plots using flexible PVC pipe network. Wheat (GW- 496 variety) sowing was done on consecutively for two years. Fertiliser dosages as well as intercultural operations as advocated in standard agronomical practices were followed for growing the crop. The daily soil moisture content from various plots was monitored using gravimetric method. The maximum yield (Y_m) and the yield sensitivity factor was estimated using the agro-ecological zone method (Doorenbos and Kassam, 1979) and the procedure outlined by FAO 33.

Results and discussion

Variation of Grain Yield

The mean grain yield obtained in different treatments are presented in Fig. 1. Maximum grain yield of 4180 kg ha^{-1} was observed when the crop was subjected to 20 per cent moisture stress during milk ripe stage. Lowest grain yield value of 3024 kg ha^{-1} was observed when the milk ripe stage was subjected to 80 per cent moisture stress level. The control treatment gave a yield value of 4177 kg ha^{-1} (99.93 per cent of Y_m). The yields obtained in various treatments are expressed in terms of relative yield decrease (Y_{rd}) = [1 - (actual yield/maximum yield)]. A minimum Y_{rd} of 0.0012 was observed when crown root initiation stage was subjected to 20 per cent moisture stress. The Y^{rd} was increasing with increasing moisture stress. The yield obtained at 80 per cent moisture stress during crown root initiation, late tillering and flowering were found to be higher than 60 per cent stress during milk ripe stage. At higher levels of moisture stress milk ripe stage in wheat cultivar was found to be sensitive to moisture stress than dough ripe stage and the converse is true for lower levels of moisture stress. Therefore minimum Y_{rd} was observed at 20 per cent moisture stress during milk ripe stage. A marginal difference between Yrd values were observed between 60 per cent and 80 per cent moisture stress values during crown root initiation stage. Yield variation between optimum and maximum moisture stresses during milk ripe stage was found to be higher than the difference of yield values obtained at all the stages of wheat cultivar. Milk ripe and dough ripe stages were found to be most sensitive to moisture stress. A drastic reduction in the yield values was observed with increasing moisture stress levels particularly for milk ripe and dough ripe stages. The crown rooting initiation and late tillering stage were found to be resistance to moisture stress than milk ripe and dough ripe stage and is at par with flowering stage. The yield difference between 20 per cent and 40 per cent moisture stress was found to be negligible during crown root initiation and late tillering stages. Higher yield was observed at 60 per cent moisture stress than 80 per cent moisture stress during all the stages.

Dry Matter Yield

Mean dry matter yield obtained at each stage for different moisture levels are presented in Fig. 1. Control treatment gave a dry matter of 4255 kg/ha. Maximum and minimum dry matter yields were observed when the dough ripe stage is subjected to 20 per cent and 80 per cent moisture stress respectively. Dry matter yield was found to be decreasing with increasing moisture stress from optimum level at all physiological stages of the wheat. Considering WUE as a criteria the optimum dry matter yields of 4220, 3897, 4217, 4185 and 4192 kg ha^{-1} were observed respectively during crown root initiation stage at

40 per cent moisture stress, late tillering stage at 80 per cent moisture stress, flowering stage at 40 per cent stress, milk ripe stage at 40 per cent stress and dough ripe stage at 40 per cent stress. Marked difference between dry matter yields at 20 per cent and 60 per cent moisture stresses during crown root initiation stage as well as 40 per cent and 60 per cent moisture stresses during late tillering and flowering stage were not observed.

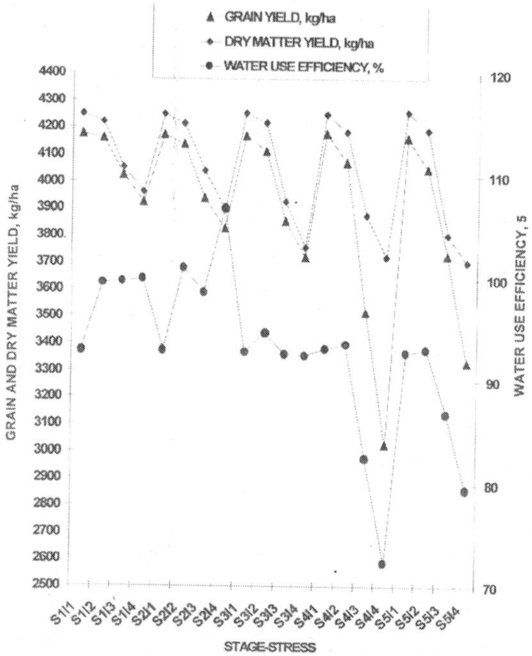

Fig. 1: Effects of Physiological Stressing on Grain.

Signification between Stages

Mean grain yield was obtained for each physiological stage of the wheat cultivar by taking the arithmetic mean of the yield values obtained at various moisture stresses to know the significance among the stages. The average grain yield was found to be 4070, 4018, 3964, 3817 and 3698 kg ha^{-1} respectively for the crown root initiation stage, late tillering stage, flowering stage, dough ripe stage and milk ripe stage. All the stages are responding differentially with each other. The crop maturity stage is insensitive to moisture stress as compared to other stages. Milk ripe stage was most sensitive followed by dough ripe stage, flowering stage, late tillering stage, and crown root initiation stage. Similar analysis on dry matter yield is presented in Table-1. Average dry matter yield obtained were found to be 4120, 4100, 4038, 4007 and 3989 kg ha^{-1} respectively during crown root initiation, late tillering, flowering, milk ripe and dough ripe stages. The dry matter yield is significantly differing with moisture stress levels. The result suggests that dough ripe and milk ripe stage are most sensitive to moisture stress followed by flowering, late tillering, and crown root initiation stages.

Significance between Water Stresses (Sub factor)

Significance between moisture stresses (Table- 1) indicated that the mean grain yield obtained at 20 per cent, 40 per cent, 60 per cent and 80 per cent moisture stress treatments was found to be respectively 4172, 4106, 3813 and 3580 kg ha^{-1}. Marked differences were not observed between 20 per cent and

Table 1: Mean Yield Variation of Wheat Crop

Main Plot	*Mean yield of grain, (kg/ha)*	*Mean yield of dry matter (kg/ha)*	*Sub Plot*	*Mean yield of grain, (kg/ha)*	*Mean yield of dry matter (kg/ha)*
S_1	4070.5	4120.4	I_1	4172	4251
S_2	4018.5	4100.4	I_2	4106	4206
S_3	3964.6	4038.1	I_3	3813	3939
S_4	3698.6	4007.5	I_4	3580	3807
S_5	3817.5	3989.1			
Sem.	126.50	182.00	Sem.	4054.50	1351.00
CD at 1 %	5.095	4.412	CD at 1 %	9.679	5.054
Cv %	0.10	0.1174	Cv %	0.292	0.1642
Interaction (I x S)					
Sem.	1116427.00	171872.00			
CD at 1 %	20.778	16.047			

Table 2: Mean Values of Grain and Dry Matter Yields (Kg Ha^{-1})

Stage	*Yield*	*a*	*B*	*C*	*Rc₂*	*F-test*
S_1	GY	4184	101	-547	0.92	17.9
	DMY	4300	-120	-397	0.88	11.9
	IR	47.49	-14.38	2.94	0.98	29.8
S_2	GY	4229	-126	-494	0.89	13.4
	DMY	4267	85	-697	0.95	26.8
	IR	47.19	-12.87	-1.25	0.95	25.3
S_3	GY	4268	-320	-478	0.88	12.1
	DMY	4313	-39	-850	0.88	11.6
	IR	46.59	-8.39	0.50	0.99	92.5
S_4	GY	4426	-890	-716	0.83	8.5
	DMY	4370	-387	-562	0.87	11.4
	IR	46.42	-8.48	3.50	0.99	85.2
S_5	GY	4171	351	-1762	0.99	251.4
	DMY	4462	-820	-209	0.76	5.6
	IR	46.11	-6.95	2.22	0.97	49.5

40 per cent moisture stressing values. The results also revealed that 80 per cent moisture stress was significantly different than others. Mean dry matter yield obtained at 20 per cent, 40 per cent, 60 per cent and 80 per cent moisture stress levels was 4251, 4206, 3939 and 3807 respectively. The significance of combinations (Tables 1 and 2) for grain as well dry matter is shown in Table 3. Combinations which do not differ significantly have been underlined by bar.

Table 3: The Significances of Combinations of Grain and Dry Matter Yields (Kg Ha^{-1})

Particular	*Grain yield*					*Dry Matter Yield*				
Combinations	S_4I_1	S_1I_1	S_2I_1	S_3I_1	S_5I_1	S_5I_1	S_3I_1	S_4I_1	S_1I_1	S_2I_1
Mean yield (kgha^{-1})	4180	4175	4172	4168	4164	4259	4252	4250	4249	4247
Combinations	S_1I_2	S_2I_2	S_3I_2	S_4I_2	S_5I_2	S_1I_2	S_3I_2	S_2I_2	S_5I_2	S_4I_2
Mean yield (kgha^{-1})	4160	4136	4112	4072	4048	4220	4217	4215	4192	4185
Combinations	S_1I_3	S_2I_3	S_1I_4	S_3I_3	S_2I_4	S_1I_3	S_2I_3	S_1I_4	S_3I_3	S_2I_4
Mean yield (kgha^{-1})	4025	3940	3922	3855	3825	4052	4041	3960	3927	3897
Combinations	S_5I_3	S_3I_4	S_4I_3	S_5I_4	S_4I_4	S_4I_3	S_5I_3	S_3I_4	S_4I_4	S_5I_4
Mean yield (kgha^{-1})	3727	3722	3717	3330	3024	3875	3802	3756	3720	3702

Physiological Water Stress Response and Combined Response Function

The water stress response function for different physiological crop stages with respect to the degree of moisture stress have been fitted with second order polynomial equation which can be shown as:

$$Y = a + bX + cX^2 \quad \ldots(1)$$

where, Y = Grain yield or dry matter yield (kg ha^{-1}), X = Degree of moisture stress (function), a, b and c = empirical coefficients; The value of empirical

Table 4: Values of Empirical Coefficients in Different Water Strees Response Function

Stage	*Yield*	*a*	*B*	*C*	*Rc_2*	*F-test*
S_1	GY	4184	101	-547	0.92	17.9
	DMY	4300	-120	-397	0.88	11.9
	IR	47.49	-1438	2.94	0.98	29.8
S_2	GY	4229	-126	-494	0.89	13.4
	DMY	4267	85	-697	0.95	26.8
	IR	47.19	-12.87	-1.25	0.95	25.3
S_3	GY	4268	-320	-478	0.88	12.1
	DMY	4313	-39	-850	0.88	11.6
	IR	46.59	-8.39	0.50	0.99	92.5
S_4	GY	4426	-890	-716	0.83	8.5
	DMY	4370	-387	-562	0.87	11.4
	IR	46.42	-8.48	3.50	0.99	85.2
S_5	GY	4171	351	-1762	0.99	251.4
	DMY	4462	-820	-209	0.76	5.6
	IR	46.11	-6.95	2.22	0.97	49.5

coefficients for grain yield, dry matter yield and irrigation requirements are presented in Table 4. The table also gives values of statistical parameters related with each physiological water stress response function. Using these empirical coefficients the grain yield, dry matter yield and irrigation requirements of wheat cultivar can be determined for particular moisture stress levels at different physiological crop growth stages.

The yield response due to subjecting the crop to varied moisture stress levels at two or more stages cannot be predicted through these individual physiological response functions. In such circumstances the yield response can be predicted by a combined water stress response function and is expressed mathematically for grain yield, dry matter yield and irrigation requirements as:

Grain yield,

$$Y = 4333 + 75X_1 - 135X_2 - 327X_3 + 562X_4 + 227X_5 - 527X_1^2 - 487X_2^2 - 472X_3^2 - 2542X_4^2 - 1662X_5^2 \quad \ldots(2)$$

Dry matter yield,

$$Y = 4634 - 199X_1 - 23X_2 - 23X_3 - 382X_4 - 654X_5 - 333X_1^2 - 610X_2^2 - 862X_3^2 - 567X_4^2 - 343X_5^2 \quad \ldots(3)$$

Irrigation requirements,

$$IR = -21.3 - 14.7X_1 - 7.31X_2 - 7.95X_3 - 6.79X_4 + 12.17X_5 + 0.22X_1^2 - 0.37X_2^2 + 3.08X_3^2 + 2.09X_4^2 + 55.73X_5^2 \quad \ldots(4)$$

where, X_1, X_2, X_3, X_4 and X_5 = degree of water stress for physiological stage S_1, S_2, S_3, S_4 and S_5 respectively. The correlation coefficient and F value for eqs. 2, 3 and 4 were found to be 0.97, 0.90, 0.98 and 71.2, 17.4, 41.75 respectively.

Water Use Efficiency (WUE)

WUE is the ratio of grain yield and irrigation water applied. Maximum WUE of 106.8 kg ha^{-1} cm^{-1} was obtained during late tillering stage by applying 80 per cent water stress, while a minimum value of 72.3 kg ha^{-1} cm^{-1} was observed during milk ripe stage at 80 per cent water stress. During milk ripe stage WUE decreased with increasing moisture stress attained lowest value at 80 per cent stress. Same trend was prevailed for all other stages except crown root initiation stage. WUE decreases with increase in moisture stress. The optimum moisture stress levels for attaining high WUE are found to be 80, 80, 40, 40 and 40 per cent respectively during S_1, S_2, S_3, S_4, and S_5 stages. The controlled treatment yielded a WUE value of 93.1 kg ha^{-1} cm^{-1}.

Estimation of Crop sensitivity factor

Agro-ecological zone method described in FAO 33 is adopted for estimating maximum yields of wheat crop based on climatological parameters and crop parameters.

The maximum yield was calculated for the past 24 years. The probability analysis was done to obtain the maximum yield that can be accrued at five years return period. This was found to be 4185 kg/ha (Fig. 3). The response of yield to water supply is quantified through the yield sensitivity factor λ which relates relative yield (Y/Y_m) to relative evapotranspiration (ET_a/ET_m). The λ

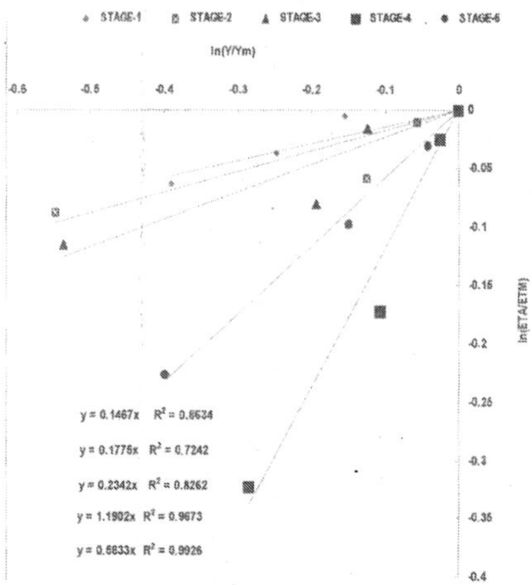

Fig. 2: Crop Sensitivity Factor for Various.

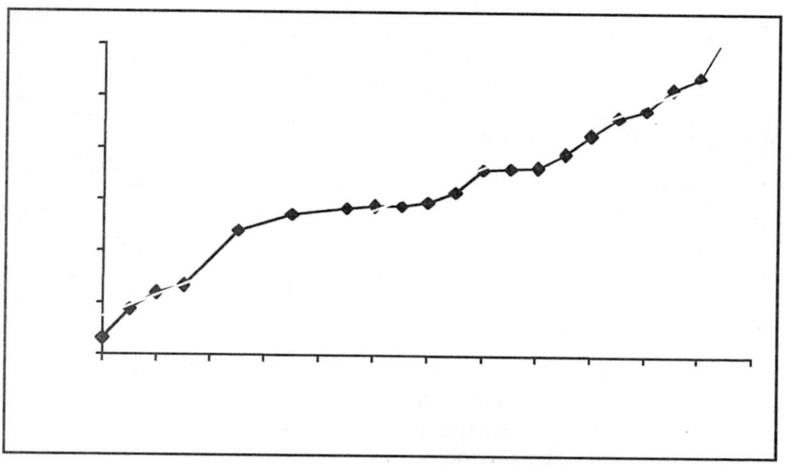

Fig. 3: Estimation of Maximum Yield of Wheat Cultivar.

values for most of the crops are derived on the assumption that the relationship between relative yields (Y/Y_m) and relative evapotranspiration (ET_a/ET_m) is linear and are valid for water deficits up to 50 per cent. In the present study the values of relative yield and the values of relative evapotranspiration were plotted on a log-log paper. The slope of the linear fit equation gives the estimate of λ. Adopting the above procedure the values of stress sensitivity factor was determined for various physiological stages. The values are found to be 0.147, 0.1778, 0.234, 1.19 and 0.583 for crown root initiation, late tillering, flowering, milk ripe and dough ripe stages.

Conclusions

1. Milk ripe stage was found to be the most sensitive to moisture stress followed by dough ripe, flowering, late tillering and crown root initiation stages.
2. Maximum and minimum grain yields of 4180 and 3024 kg ha^{-1} were observed when the wheat crop was subjected to 20 per cent and 80 per cent moisture stress respectively at milk ripe stage.
3. Maximum and minimum dry matter yields of 4259 and 3702 kg ha^{-1} were observed when the wheat crop was subjected to 20 per cent and 80 per cent stress respectively at dough ripe stage.
4. Maximum grain yield can be attained by maintaining the moisture stress levels at 40, 40, 20, 20 and 20 per cent respectively during crown root initiation, late tillering, flowering, milk ripe, and dough ripe stages.
5. Maximum dry matter yield can be attained by maintaining the moisture stress levels at 40, 40, 40, 20 and 20 per cent respectively during crown root initiation, late tillering, flowering, milk ripe, and dough ripe stages.
6. Physiological water stress response function with respect to degree of moisture stress have been fitted with quadratic equation.
7. Maximum yield obtainable at five years return period in the study area for wheat cultivar was found to be 4185 kg ha^{-1}.
8. Stress sensitivity factor for crown root initiation, late tillering, flowering, milk ripe and dough ripe stages for wheat cultivar was found to be 0.147, 0.178, 0.234, 1.19 and 0.583 respectively.

References

Doorenbos, J. and Kassam, A. H. (1977). Yield Response to Water. Irrigation and Drainage Paper No. 33, FAO, Rome, Italy.

Singh, P. (1985). Simulation Model and Optimisation Technique for Irrigation Scheduling of Wheat under Limited Water. Ph.D. Thesis, Technical University, West Berlin.

Singh, K. P. and Kumar V. (1979). Influences of Irrigation on the Leaf Water Potential and Yield of Wheat and Barley at Two Date of Sowing. Field Crops Research, 2(2): 117-124.

Sinha, S. K., Agarwal, P. K. and Khanna Chopra, R. (1985). Irrigation in India: A Physiological and Phenological Approach to Water Management in Grain Crops. In D. Hillel (ed): Advances in Irrigation Vol. 3, Academic Press, New York.

36

Seasonal Variations in Biomass and Intertidal Distribution of Few Species of Phaeophyceae Seaweed at Narara Coast- Marine National Park, Gujarat

A.J. Bhatt, N.H. Joshi and A.Y. Desai

Abstract

Three species of Phaeophyceae group were studied for their seasonal variations in biomass and intertidal distribution at Narara coast. Sargassum johnstonii, *was dominant at all the tidal levels, viz. Upper, Middle and Lower littoral. The annual mean biomass distribution was higher at lower littoral, the seasonal variations for the species showed dominance of* Sargassum Johnstonii, *in winter whereas,* Cystoseira indica *and* Iyengaria stellata *showed higher biomass in summer. Basic physico-chemical parameters (temperature, pH, salinity, DO, Nitrate and phosphate) of seawater were correlated positively with the biomass of* Sargassum Johnstonii. *Random survey by belt transacts method showed dominance of Phaeophyceae species.*

Introduction

The distribution of marine macro algae is controlled by several factors. The ecology of beach is influenced by two specific environmental requirements. These are the presence of light; sufficient to drive photosynthesis and another important common requirement is to have a firm substrate of attachment. The seaweeds are most commonly found in littoral zone, on rocks and on sandy shore. The commonly occurring marine macro algae as individual species was not studied much for their seasonal distribution in relation to environmental factors at Marine National Park area of Gulf of Kutch. With this as an objective, the commonly occurring species of Phaeophyceae were studied along the coast of Narara. The Phaeophyceae group algae of the present study are important in research and their utilisation in algin industry (Lubchenco and Gaines, 1981).

Materials and methods

The present investigation has been conducted at Narara-Marine National Park, Near Jamnagar (22° 15.51' N, 69° 30' 53" E) (Fig.1). Narara has a flat rock intertidal belt, provides with tide pools, gullies and crevice. Intertidal belt has stretched vertically about 2000 m and on average, gets exposed at zero chart datum. The sandy beach gradually slopes into the rocky intertidal belt from landward side. In the present study, the area along the inter-tidal at 0.5 m above zero of chart datum has been considered as the lower littoral. The area around 1.5 m above zero of chart datum was considered as the middle littoral zone, while area around 20. M was considered as the upper littoral zone. However, in nature, there is no such demarcation since each part of the inter-tidal zone merge gradually with the other and there was no clear cut zonation of different algal species along the inter-tidal (Dadolahi *et al.*, 2012; John et al., 2012).

Quantitative survey

Algal vegetation was surveyed for its quantitative phyto-sociological attributes by monthly sampling survey during low water of the spring tides. Belt transects of one miter width were run across the shore covering all the tidal levels i.e. upper, middle and lower parts of the inter-tidal zones. Along each transect, structure and composition of the seaweed vegetation was

Fig. 1. Map showing the study area- Narara, Marine National Part, Jamnagar.

studied in samples (quadrate of 1 m^{-2}) at regular intervals. In all, minimum of 20 sample plots were studied each month covering the upper, middle and lower littoral zone.

Density

The number of individuals of each species falling in a sampling plot is abundance of the species, from the pool replicate values of the abundance from all the sample plots survey- mean density of a species was calculated.

Biomass

The entire biomass falling within each sample plot was harvested right from the base. The collected biomass from each sample plot was sorted out species-wise. The biomass fraction, thus sorted species-wise were collected into separate polythene bags and brought to the laboratory. The fractions of biomass were oven-dried at 70 °C to constant weight and thus dry weight for each species from each sample were recorded. The mean value was calculated from the pooled replicates for each species and expressed as gm^{-2}.

Physico-chemical parameter of seawater

The basic seawater parameters such as temperature, salinity, pH, dissolved oxygen, nitrite and phosphate were measured for the month of September-2013 to March- 2014 three species of Phaeophyceae were collected during low tide using the belt transect and quadrate from the Narara coast. The temperature, salinity, pH, dissolved oxygen, nitrite and phosphate were measured following the standard methods.

Results and Discussion

Variations in the distribution of all three species occurred at the every belt transect. All the three species was found growing throughout the study period at middle littoral. The entire intertidal area at Narara was thoroughly surveyed first to get an idea of the characteristics of the coast and to make qualitative assessment of the seaweed flora inhibiting there throughout the period of study.

Density of Phaeophyceae

Variations in the distribution of all three species occurred at the every belt transect. All the three species was found growing throughout the study period at middle littoral. While moderate density of *Sargassum Johnstonii* was observed during the months of November to January for Upper littoral, September to January in Middle and lower littoral zones. There was no presence of *Iyengaria Stellata* during September and October in Upper littoral and February and March in lower littoral zone. On the other hand Middle littoral shows the higher to moderate density of all the three species studied.

Mean Biomass of the Phaeophyceae species

After recording the density, the entire biomass falling within each sample plot was harvested right from the base. The collected biomass from each sample plot was sorted out species-wise. The biomass fraction, thus, sorted were collected into separate polythene bags and bought to the laboratory. Fresh weight was recorded after blotting. The fractions of biomass were oven-dried at 70 °C to a constant weight and thus dry weight was determined. The oven-dried weight for each species from each sample were recorded. The mean value was calculated from pooled replicates for each species and expressed as gm^{-2} (Table 2) (Fig. 1, 2 and 3).

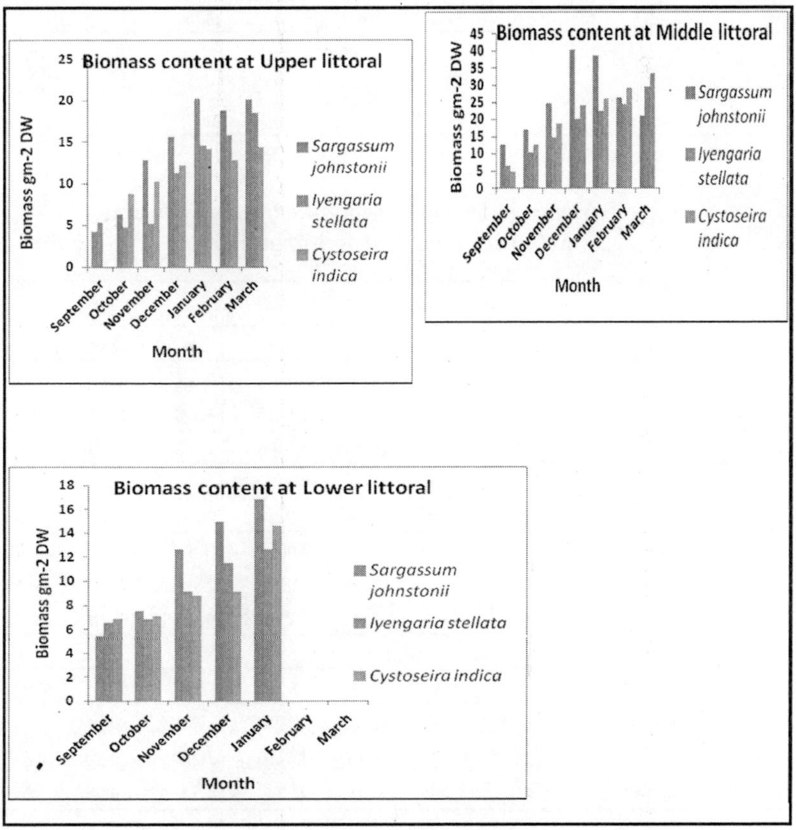

Fig.1, 2 and 3. Biomass content at upper, middle and lower littoral.

Table 1: Density of Phaeophyceae group of species at Narara Intertidal area

	Sep	Octo	Nov	Dec	Jan	Feb	March
Upper Littoral							
Sargassum Johnstonii	+	+	++	++	++	+	+
Iyengaria stellata	+	0	+	++	++	+	+
Cystoseira indica	0	0	+	+	+	+	+
Middle Littoral							
Sargassum johnstonii	+++	+++	+++	+++	+++	+	+
Iyengaria stellata	+++	++	++	++	++	+++	+++
Cystoseira indica	+++	+++	+++	++	++	+++	+++
Lower Littoral							
Sargassum johnstonii	+++	+++	+++	+++	+++	0	0
Iyengaria stellata	+++	++	++	++	++	0	0
Cystoseira indica	+++	+++	+++	++	++	0	0

* 0 = No species available, + Less density, ++ Moderate density, +++ High density.

Table 2: Mean biomass of Phaeophyceae group of species at Narara Intertidal area (g m^{-2} Dw)

	Sep	Octo	Nov	Dec	Jan	Feb	March
Upper littoral							
Sargassum johnstonii	4.20	6.30	12.80	15.60	20.24	18.80	20.12
Iyengaria stellata	5.36	4.80	5.12	11.25	14.60	15.80	18.50
Cystoseira indica	0	8.80	10.20	12.16	14.12	12.80	14.40
Middle littoral							
Sargassum johnstonii	12.60	16.90	24.80	40.50	38.80	26.50	21.12
Iyengaria stellata	6.55	10.40	14.60	20.20	22.40	24.40	29.50
Cystoseira indica	4.80	12.80	18.80	24.15	26.12	29.40	33.50
Lower littoral							
Sargassum johnstonii	5.40	7.50	12.65	14.90	16.80	0	0
Iyengaria stellata	6.50	6.80	9.12	11.50	12.60	0	0
Cystoseira indica	6.90	7.12	8.80	9.12	14.60	0	0

Upper littoral

Species biomass recorded from upper littoral were very scanty as compared to middle. *Sargassum Johnstonii* exhibited maximum biomass 20.24 and 20.12 gm^{-2} DW in the month of January and March respectively. For Iyengaria Stellata it was 18.50 gm^{-2} DW in the month of March, while lowest (4.80 gm^{-2} DW) in the month of October. *Cystoseira indica* was not available in the month of September, while highest biomass 14.50 gm^{-2} DW was observed in the month of March.

Middle littoral

Species biomass recorded from middle littoral were very luxuriant as compared to upper and lower littoral. *Sargassum Johnstonii* exhibited maximum biomass 40.50 and 38.80 gm^{-2} DW in the month of December and January respectively. For Iyengaria Stellata it was 29.50 gm^{-2} DW in the month of March, while lowest (6.55 gm^{-2} DW) in the month of September.

Cystoseira indica showed maximum biomass in the month of March (33.50 gm^{-2} DW) while lowest (4.80 gm^{-2} DW) in the month of September.

Lower littoral

Biomass in lower littoral was dominated by *Sargassum* in the month of January with 16.80 gm^{-2} DW, followed by *Cystosira indica* (14.60 gm^{-2} DW) and *Iyengaria Stellata* (12.60 gm^{-2} DW). However Cystosira and Iyengaria was not at all available in the month of February.

Physico-chemical Parameters

Physico-chemical characters viz. temperature, pH, DO, Salinity, Nitrate nitrogen and Phosphate were measured once in a month (Table 3). The temperature, pH and DO was in the range of standard figures of the open seawater. In general when DO, Nitrate Nitrogen, salinity and Phosphate increase in the seawater, growth of the algae also increases (Joshi and Murthy, 2004). The same trend was also observed in this study Sarojini, 2013.

Table 3. Physico-chemical characters of seawater

Parameters	*Sep*	*Oct*	*Nov*	*Dec*	*Jan*	*Feb*	*March*
Temperature (°C)	29	30	30	26	27	28	31
pH	8	8.2	8.2	8.1	8.3	8.2	8.3
Dissolved Oxygen (μgl^{-1})	4.5	4.6	5	5.5	5.8	4.8	4.5
Salinity (ppt)	33.4	35.2	34.5	34.5	34.2	36.4	36.7
Nitrate Nitrogen (μgl^{-1})	1.65	1.23	2.23	2.49	3.1	2.24	2.45
Phosphate (μgl^{-1})	0.23	0.57	1.52	2.32	1.58	1.75	2.12

References

Dadolahi - Sohrab, A., M. Garavand - Karimi, H. Riahi and H. Pashazanoosi 2012. Seasonal variation in biomass and species composition of seaweeds along northern coasts of Persian Gulf (Bushehr Province). *J. Earth Syst. Sci.,* 121 :241-250.

John Peter Paul, J. and D. Patrie Raja 2012. Seasonal variability of some *Enteromorpha* species in Kanyakumari region, the southern coastal region of Tamil Nadu. *Asian J. Biol. Life Sci.,* 1 :147 - 149.

Joshi, N.H., Murthy, M.S. 2004. Ecological studies on Sargassum species at Port Okha (Gujarat). *Seaweed Res. Utiln*, 26 Special Issue: 63-71.

Sarojini, Y., P. Santa Rao, B. Sujatha and K. Lakshminarayana 2013 a. Distribution and diversity of marine macro algae in relation to environmental factors at Visakhapatnam coast. *Seaweed Res. Utiln*, 35: 55-64.

37

Role of Agricultural Machinery in Soil Moisture Conservation in Mitigation of Climatic Change

Pramod Mohnot

Abstract

India is an agricultural-based country mainly dependent on rainfall, having only about one-third area under irrigation. The population is increasing at an alarming rate and to meet the ever-increasing demand of food and fiber of our country it is necessary to bring more and more area under cultivation with the available irrigation facilities. This situation has become more difficult under climate change condition. It is well known fact that during rainfall a considerable amount of water goes to sea due to high rate of runoff with soil erosion. It is very well necessary to check the runoff and erosion and to conserve the soil moisture. On the other hand to upgrade the living standard of Indian farmers and to check the migration to urban areas especially under critical situation of climatic change an attempt has been made to provide information of using suitable agricultural machinery and tillage practices for soil moisture conservation to mitigate the effect of climate change. Right from seed bed preparation to harvesting of crop several agricultural equipments are being used to provide favourable environment for the plant growth. The selection of suitable agricultural machinery/ equipment may result in saving of water and precious inputs in the form of seeds, fertiliser, insecticides, pesticides, irrigation and above all the human labour. Some of the machinery/ equipments like subsoiler, rotavator, leveller, mould board plough, disc plough, para plough, chisel plough, disc harrow, cultivator, blade harrow, rollers, bed and furrow former, raised bed planter, weeder, earthening up and mulching equipment, etc. can be used to increase soil aeration, reduce runoff and soil erosion, conserve the soil moisture, provide favourable environment for plant growth and ultimately increase quality and quantity of produce.

Introduction

The technological improvements in Indian agriculture since mid-sixties have brought about revolutionary increase in agricultural production. Interestingly, the growth rate of food grain production particularly in case of wheat and rice was much higher than the growth rate of population. The country was facing acute food shortages till eighties has now become not only self-sufficient but also a net exporter of food grains. This has been made possible due to evolution of high yielding crop varieties, increased use of chemical fertilisers, development of irrigation facilities, plant protection and above all the use of improved agricultural machinery to raise their use efficiencies though mechanisation. Normally, there are good chances to reduce the cost of production if farm operations are mechanised as it saves labour, both human and bullock. In the absence of suitable machinery, the ever-increasing wage rate of human labour and cost of upkeep of draught animals could have increased the cost of production much higher. Further, large scale production means less per unit cost on the farms. Moreover, it reduces the weather risk and risk of non-availability of labour and thus wastage is minimised. Timely harvesting, threshing, grading and marketing is also made possible by quick mechanical machinery. Further, the area under fodder and feed for draught animals could be reduced due to decline in their use. The land, thus, released can be brought under commercial crops.

The use of farm mechanisation enlarges the employment opportunities both on farms and in non-farm sectors through increase in area under plough, multiple cropping, development of agro-industries and related services. On the other hand, displacement of human labour does take place and demand for semi-skilled labour in place of unskilled labour is increased. Also, the drudgery for human labour is reduced and unhygienic operations such as handling of farm yard manure can be done with machinery. The traditional farm tools and implement mainly relied on use of animate power. Improved farm tools, implements and machinery, which use both animate and mechanical power were devised from time to time. The average size of farm holding being small, animate power is widely used in many parts of the country. Mechanical power is making its impact in Indian agriculture with steady increase in land and labour productivity.

Seedbed preparation

The optimum seedbed preparation for raising crops, involves the following unit operations:

(i) Loosening of soil

This is done to achieve a desired granular soil structure for a seedbed and to allow rapid infiltration and good retention of moisture, to provide adequate air exchange capacity within the soil and to minimise resistance to root penetrationa and shoot growth. Local plough and blade harrow (Bakhar) are traditional implements used for loosening of soil. These are simplest tools designed to break the topsoil and multi-passes are carried out to prepare seedbed. Mould board plough, disc plough, soil stirring plough, ridger plough, tool frames/carriers with mould board plough or tillage sweeps, etc. are improved implements designed for breaking soil. Ploughs are used to break soil and invert furrow slice to control weeds, etc.

(ii) Clod size reduction

Clod breaking operation is required to produce a granular soil structure in the final seedbed. Tine cultivator and disc harrow are used for breaking of clods. Generally these are operated after one pass of mould board plough or ridger plough. Direct harrowing or cultivator operation is also performed when the fields are clean and free from plant residues of previous crop. Clod crushers is very effective for clod crushing under favourable soil moisture conditions but their effect is confined to soil surface only. Power driven implements like rotavators disintegrate the clods over a wide range of soil moisture and provide uniform and fine size clods or aggregates in seedbed.

(iii) Clod sorting

Operation of tools with narrow tines such as comb harrow and spike tooth harrow, in loosened soil, produces a sorting effect, bringing larger clods and aggregates on surface. The sorting effect increases with increasing forward inclination of tines and share width and decreasing speed and soil moisture. Large size clods on the surface are recommended because of their stability under rainfall, which helps in reducing soil erosion.

(iv) Compaction and consolidation

Wide, backward inclined implements compact soil as well as break clods in top surface of soil. Direct compaction at seed depth can best be achieved using narrow press wheels/discs. Planking is widely used to compact the soil at the surface.

(v) Smoothening

Smoothening of seedbed is required for proper operation of sowing machines, better distribution of irrigation water and quick disposal of excess rainwater. Smoothening can be best achieved by using wide backward inclined blades, such as levelling boards, floats and patela harrow with closely spaced shallow working narrow tines. Wooden plank, patela harrow, are recommended for smoothening operation.

Deep tillage

- Tillage beyond 30 cm. depth can be termed as deep tillage which is normally performed after two to three years to break the hard pan below the top soil to augment the rainfall or runoff water, thus reducing the soil erosion.
- Deep tillage of the compacted plots removes the compaction effect, improves yield in all treatment, gives the higher values for grain yield and yield components, increases in depth of penetration of the crop root and reduces the erosion (Hamdeh, 2003).
- The heavy traffic in the farm of various agricultural operations carried out using tractor with plough, harrow and other equipment creates a hard pan below the ploughed soil, like a plaster.
- This hard layer becomes the harder for the movement of water and root growth and checks the aeration. This results low infiltration rate causing water logging or runoff.

Subsoiling

- Subsoiler is very useful for deep tillage. It breaks the hard pan without disturbing the topsoil. Continuous slits made in the field increases infiltration rate and allow the runoff water to enter into the ground.
- The subsoiler does not disturb the topsoil, the moisture available in the topsoil will not evaporate or conserve moisture. It is necessary to do subsoiling after everytwo to three years.
- Conventionally single bottom subsoiler is being used for more deep tillage. For medium deep tillage operation, a two-bottom subsoiler operated by tractor is very much suitable.

It is used to break hardpan of the soil, loosening of the soil and helps the water to seep into the soil for improving drainage. A mole ball can be attached to create a small tunnel in the soil, which serves as drainage channel for water. The sub-soiler is operated against the natural slope of land so that run off water is stored in furrows for ground water recharging. When soil surface moisture is reduced the moisture is made available through capillary rise in the furrow.

By using two bottom subsoiler mounted on heavy duty cultivator frame tine, cost and energy can be saved when root zone and soil depth is less.

Some of the very useful and improved agricultural machinery is discussed here especially for moisture conservation.

Tractor drawn rotavator

It consists of a steel frame, a rotary shaft on which blades are mounted, power transmission system and gearbox. The blades are of L-type, made from medium carbon steel or alloy steel, hardened and tempered to suitable hardness. The PTO of tractor drives the rotavator. Rotary motion of the PTO is transmitted to the shaft carrying the blades through gear box and transmission system. A good seedbed and pulverisation of the soil is achieved in a single pass of the rotavator. Under hard, dry, grassy, full of stubbles, clods and soil with plant residues, it is very difficult to operate any other machine for land preparation or require three to four operations. This rotavator is useful in digging soil, breaking clods, crushing the stubbles and bury them in soil which improve aeration and water holding capacity. It can replace three or four operations of plough, cultivator and harrow for seed bed preparation.

Crop residue spreader and cover

Crop residue of previous crop grown can be cut by this machine and spread in the field as soil cover to maintain soil moisture.

Tractor drawn para plough

For mulch tillage and moisture conservation under dry farming condition the para plough is a primary tillage implement used for deep ploughing without inversion. This enhances infiltration but reduces hazard of erosion. The implement consists of boxed frame structure with standard three point hitch and two bottoms one right hand and one left hand arranged facing outside. Field trails with equipment showed that ploughing with para plough reduced the bulk density and increase the hydraulic conductivity.

Tractor/bullock drawn mould board plough

Cutting, inverting and pulverisation of soil is with burring of weeds and unwanted plants. Rectangular slice of soil is cut with MB plough.

Improved plough

The plough consists of mild steel sheet body to which a replaceable share is attached. The body also serve the purpose of frame and the other component like handle, tie bar and hitch assembly is also attached to it. The shape of the share is similar to flat iron section with a tang and is held on the body with fasteners. The share is usually fabricated from used leaf spring steel; tip is forged to a sharp edge and hardened. The share can be replaced after it has worn out and the cutting edge can be sharpened upon becoming dull.

Tractor/bullock drawn disc harrow

The harrow consists of a frame and shafts on which discs are mounted. The shafts are held in bearings for smooth rotation of discs. The shafts are called gangs on which discs are arranged and equally spaced with the help of spools. Single gang and double gang disc harrows are equally popular and operated using tractor or a pair of animals. The discs are fabricated from carbon steels and the cutting edges are hardened.

Tractor drawn spiked clod crusher

It is used as a combination tillage tool with tractor drawn harrow or cultivator. The clod crusher consists of a mild steel sheet drum and pegs are welded on its surface, a rectangular frame made from mild steel angle section, hitch frame and a shaft for carrying the drum. The shaft is mounted on bearing pedestals. The soil clods are pierced and broken by the pegs of the crusher.

Bullock drawn land leveler

The bullock drawn land leveler is fabricated from steel sheets and two handles are fixed to it with the help of fasteners. Mild steel round bar is bent to a suitable shape with holes at its extremities for fixing to the scraper. This steel round has a semi-circular loop to which the rope from the yoke of the animal can be attached for towing. One pair of animals is required to operate the implement. The levelers are of different design and shape, but the scoop type and terrace blade type are most common. The leveler is used for filling minor depression and levelling minor irregularities in the field. Scoop type has got a bucket and is operated by a person walking behind it. The terrace blade type has got sitting arrangement is provided with pneumatic wheels. Initial ploughing and loosening of the soil is necessary before the leveler could be put to use.

Tractor drawn land leveler

It is a tractor-mounted implement controlled by tractor hydraulics and three-point linkage. It consists of hitch system, replaceable cutting blade with sharp edge, and a curved plate with side wings, which form a bucket. The blade is made from medium carbon steel or low alloy steel, hardened and tempered to suitable hardness. During operation, the blade digs into the soil and extra soil is collected in the bucket, which is released in the depressions of the field. The angle and pitch of leveler is adjustable. The leveler can also angled left or right, or reversed for back filling.

Tractor drawn laser land leveler

A rotating laser light source (like a miniature lighthouse) is located somewhere in the field. As the laser rotates rapidly, a virtual "plane" of light is produced in the field. A "receiver" is mounted on the leveling equipment and connected hydraulically to the actual earthmoving blade. When activated, the receiver (and thus, the blade) will "lock on" to the laser source.

If the earthmover has to climb over a high spot in the field, the blade will dig the soil and drag the soil mass. If the earthmover goes over a low spot, the blade will dump that soil into the low spot. If the rotating source is tilted according to the prescribed grade, a grade can be installed in the field. Laser land-leveling in itself can improve irrigation efficiency by reducing high spots in a field.

Bullock drawn multipurpose pipe frame implement

Previously wooden implements were used for almost all the operations performed by bullocks. The initial cost and repair of such equipments are high. Pipe frame implement can replace all the local as well as improved implements. It consists of main pipe frame, beam, prongs, tynes and soil working tools like blades, cultivator shovels, plough bottoms, ridger and bund former.

Bullock/tractor drawn cultivator

The cultivators are available in fixed or expandable frame designs. The unit consists of reversible tines, frame, handle. The working width of the cultivator can be adjusted by varying the spacing between the tines or expanding the frame. For operation the implement is attached to

The tines dig into the soil and cut a small furrow slice. During weeding operation, the tines also uproot weeds and create soil mulch. Seeding attachment can also be provided on the frame, increasing the versatility of the unit. Cultivators are used for seedbed preparation both in dry and wet soils. It is also used for interculture purpose by adjusting the tynes in wider row crops. It is also used for puddling purposes.

Tractor drawn spring tyne cultivator

Cultivator consists of a frame, tynes with reversible shovels, land wheel, hitch system and heavy-duty springs. The function of springs is to save the cultivator tynes from breaking when some hard object comes in contact with the shovel or under the tyne. The shovels are made of heat-treated steel for longer life. The implement is mounted type and is controlled by the hydraulic system of the tractor. Sometimes V-shape or duck foot type cultivators can be used for interculturing operation.

Bar point cultivator

The cultivator consists of rectangular frame made of mild steel angle or channel section. Rigid tynes of heavy flat section are mounted on the frame with the help of clamps. Bar point shares are joined to the end of tynes. Bar points shares are made of medium carbon or low alloy steel, hardened and tempered to suitable hardness. Bar points can be replaced on wearing or becoming dull. In place of bar points, duck foot shovels can be attached to the cultivator. The cultivator is mounted type and is operated by tractor. The tynes of cultivator can be adjusted vertically by moving in the clamp.

Power tiller operated slasher

It cuts the weeds, grass, crop residues and spread over the soil surface providing mulch, reduces evaporation, run-off as well as erosion and improves soil moisture.

Tipping type bullock cart for manure spreading

To improve the fertility and microbial activities of soil for proper aeration and soil moisture conservation with organic manure. It is useful for spreading FYM directly in to furrows.

Tractor drawn trencher

It digs soil to form rectangular trench of 30 x 30 cm for placing pipes for irrigation and drainage purpose.

Ditcher

It consists of two curved wings with cutting blades, front cutting point, tie bars for adjusting wingspan, and hitch assembly with three point linkages. The cutting blades and cutting point are made of medium carbon or alloy steel, hardened and sharpened. The ditcher is operated by tractor and controlled by hydraulic system. The ditcher penetrates in the soil due to its own weight and suction of the cutting point. Upon drawing the ditcher in the field, it opens the soil in the shape of ditch with either 'V' bottom or flat bottom. The wings enable the ditcher to slice and roll the tough sod, brush and root sets. The depth and width of the ditch is adjusted from the operator's seat. The front cutting point and wings cutting edges are replaceable. Sometimes, the rotary trencher can be used. In this cutter excavates soil, which is uniformly distributed to one side. The deflection of the soil can be adjusted by the deflector.

Bullock drawn ridger

The ridger plough consists of single share and double mould board attached to a short (steel) or long beam (wooden). Two handles are provided to balance the ridger during operation. During operation the share cuts furrow slice and pulverises the soil, which moves on to the mould board and is inverted. During the onward movement, a furrow is formed and ridges get formed in its return pass.

Tractor drawn bed furrow former

The tractor drawn bed-furrow former consists of mild steel angle iron frame; three-point linkage, lifting pin, furrow former, bed former and stiffeners. The bed and furrow formers are made of mild steel sheet and bent in required shape. The implement is operated in the tilled soil. Forms alternate beds and channels in ploughed fields. The beds are suitable for planting crops like maize, cotton etc. The bed and furrow system is ideally suitable for efficient irrigation and save water.

Tractor drawn raised bed planter

Improves irrigation efficiency and provide drainage.

Bullock drawn bund former

The implement consists of two blades, flat iron frame bent at an angle a handle attached to the frame with tie bars and wooden beam. The operator's handle is made of wood for providing better grip and convenience in operation and attached to the frame with the help of suitable brackets. The frame is bent at an angle and has holes for adjusting the space between the blades. The profile of blades is made to a shape so that bund formed is trapezoidal and remains stable. The blades are attached to the frame with fasteners. For operation, a pair of bullock pulls the implement, the blades gather the loose soil and accumulate it in the form of bund. Bund formers are used to make borders or check basins for proper distribution of irrigation water.

Hand operated general purpose tools

Hand operated tools like Sickle, khurpi, spade, pick axe, etc. are being used for several operations of hoeing, weeding, soil mulching for conservation of soil moisture.

Hand operated improved dry land weeder

For weeding in row crops for removing shallow rooted weeds. It has been designed ergonomically for easy operation. Useful in dry land and garden land crops and is ideal at a soil moisture content of eight to ten per cent. It creates shallow soil mulch for moisture conservation.

Bullock drawn blade harrow/weeder

The implement consists of a blade, blade holding prongs, wooden or iron body and a beam. The blade is attached to the bakher frame prongs with the help of rings, which are hammered into place. It cuts a thin slice of soil and removes the weeds etc and requires a pair of bullocks to pull the implement. The V-shape of the blade has a self-cleaning advantage, which prevents adherence of weeds and its roots. Mainly twin blade harrow is used to cover two rows at a time.

Power tiller operated duck foot type cultivator for interculturing

It is operated by power tiller and it can also be used for weeding and making shallow soil mulch.

Bullock drawn hoe-cum-fertilizer drill

Place the second dose of fertiliser automatically in moist soil near root zone along with interculturing operation increasing fertilizer use efficiency.

Mini tractor drawn weeder

It is very difficult to do interculturing in narrow spaced row crops with the help of conventional large tractor as it damage the plants. Mini tractor drawn weeder is suitable even at later stage of crop growth.

Tractor drawn weeding-cum earthening up equipment

For weeding and intercultural operations in between row crops. Mechanical weed control kills the weed between the rows of crop and also keeps the soil surface loose. An intercultivator-cum-earthing up equipment fitted to a standard tractor drawn ridger. Three number of sweep type blades are affixed to the ridger frame for accomplishing the weeding operation in between standing rows of crops. Three ridger bottom which are fitted behind the sweep blade, work on the loosened soil mass and aid in earthing up by forming ridges and furrows.

Tractor drawn rotary weeder

In case of hard soil when simple blade harrow cannot dig the soil and uproot the weeds between wider row spaced crops, this rotary weeder is suitable.

38

Efficient Water Management for Higher Productivity of Rice: A Review

Manjula Udikeri, Keshavamurthy G. M.,
Shashidhar, G. B. and R.H. Patil

Abstract

Rice crop under paddy fields consumes more water than most of the arable crops and is due to both avoidable and unavoidable losses. Thus, water requirement of rice can be lowered greatly by reducing water losses due to seepage, percolation and evaporation from the soil. Promising technologies include maintaining saturated soil culture and intermittent irrigation during the growing period. However, these technologies still use prolonged periods of flooding, so water losses remain unacceptably high. To achieve efficient water management in rice cultivation new technologies required to be adopted. Modified SRI method of cultivation loss promising as it significantly improves dry matter per hill, grain yield, straw yield and WUE as compared to normal method of paddy cultivation. Similarly, alternate wetting and drying method of irrigation in paddy field help significantly increase grain yield and irrigation WUE compared with continuous flooding of paddy field. Irrigation scheduling at seven days after disappearance of ponded water has also been found to significantly increase total dry matter, grain yield, straw yield, and field WUE as compared with continuous submergence of water in paddy field. Further, latest concept of growing of aerobic rice by keeping the soil only at saturated condition (no standing of water) has also been reported to significantly increase grain yield, nutrient uptake and WUE compared with other irrigation schedules. If IW/CPE ratio was followed a 1.2 ratio has been found to significantly increase nutrient uptake, grain yield, straw yield, net returns and B:C ratio compared with other irrigation schedules. Alternatively, irrigating paddy once in 14 days up to tillering, twice a week until flowering and once in 10 days until maturity has also been found to significantly increase grain yield while improving water productivity. Locally suitable methods out of these new methods need to be popularised to save avoidable water losses and improve water productivity for sustained rice productivity under future climate of water limited environment.

Key words: *Rice productivity, irrigation levels, irrigation scheduling, water regimes and WUE.*

Introduction

Water is not only valuable but also increasingly becoming scarce natural resource in many regions of the world. With the projected increase in Indian population by 2050 the current average per capita availability of water, which is around 1,600 m^3/capita/year, is expected to come down to 1,000 m^3/capita/year (Narula and Lall, 2010). Similarly, agriculture's share of freshwater will be reduced by 8-10 per cent by 2025 because of increasing competition from the urban and industrial sectors (Mahajan *et al.,* 2009). Thus, water for rice cultivation which is staple food of Asians will become increasingly scarce. In the north-western part of the Indo-Gangetic Plains (NW-IGP) increasing extraction of groundwater for rice cultivation has led to decline in water table by 0.1-1.0 m each year, resulting in further water scarcity while increasing costs of pumping water (Rodell *et al*., 2009). To fulfill the increased rice demand with declining water resources, it is imperative that rice productivity needs to be improved with less water. A growing scarcity of fresh water will pose problems not only for rice production but to improve yields to meet expected demands in coming years, especially under future warmer climates. The successful transition from traditional lowland cultivation (flooded) to water saving methods should be adopted and the concept of keeping the soil "wet" but not flooded or saturated (Ramana Murthy and Reddy, 2013) needs to be popularised. In that context, this review paper tries to identify advanced or efficient water management techniques in rice cultivation.

Methods of establishment

Geethalakshmi *et al.* (2010) studied different methods of establishment *viz*., transplanted rice (conventional), direct sown rice (wet seeded), alternate wetting and drying (AWD) method (once in three days), system of rice cultivation (SRI) and aerobic rice cultivation using rice hybrid CORH-3. SRI produced higher grain yield (6014 kg/ha) (Table 1), followed by transplanted rice. However, aerobic rice cultivation registered the lowest grain yield of 3582 kg/ha. With respect to water productivity, SRI method of rice cultivation recorded the highest water productivity followed by AWD and aerobic rice cultivation. The conventional rice cultivation and direct sown rice as expected produced the lowest water productivity.

Table 1: Water usage, grain yield and water productivity in rice as influenced by different system of rice cultivation (Source: Geethalakshmi *et al.*, 2010)

System of rice cultivation	*No. of irrigations*	*Total water used (m^3)*	*Grain yield (kg/ha)*	*Water productivity (kg/m)*
Transplanted	32	16120	5732	0.36
SRI	29	14085	6014	0.43
AWD	24	13636	5376	0.39
Wet seeded rice	37	15763	5175	0.33
Aerobic rice	26	9687	3582	0.37

SRI = System of rice intensification and AWD = Alternate wetting and drying.

Manjunatha *et al.* (2010) observed that the method of planting has significant influence on the grain yield of rice as they reported significantly higher grain yield (6342 kg/ha) in modified SRI method, that was 24.23 per cent higher over the normal method. The better performance of rice under modified SRI was attributed to favourable micro climatic conditions created for crop growth resulting in more tillers, grains per panicle and longer panicles. Further, significantly lower sterility was recorded in modified SRI method which also added to grain yield. Straw yield also followed same trend as that of grain yield. The water requirement was reduced by 47 per cent over normal method. The water use efficiency (WUE) in modified SRI method was the highest (74.7 kg/ha-cm) followed by recommended SRI method (73.1 kg/ha-cm) and both were significantly better over normal SRI method (40.9 kg/ha-cm).

Rashid *et al.* (2009) in Bangladesh tried different establishment methods in both *boro* and *aman* rice and reported that the grain yield response to the crop establishment method was not significant in *aman* but in *boro* (Table 2). In *aman*, CT-DrumR yielded 6.0 t/ha, which was at par with CT-TPR1 transplanted on the day of seeding of CT-DrumR (6.1 t/ha) and CT-TPR2 was transplanted 30 days after seeding of CT-DrumR (6.1 t/ha). In *boro*, the yield of CT-TPR2 was similar to that of CT-DrumR but the earlier transplanted rice (CT-TPR1) gave a lower yield than CT-DrumR and CT-TPR2 which may have been affected by low temperature in the vegetative and reproductive phases. The higher straw yield under CT-DrumR than with CTTPR1 and CT-TPR2 was mainly due to more tillers per unit area. Higher straw yield was produced in wet seeded rice than in transplanted rice. Drum-seeded rice and CTTPR2 produced similar straw yield, which was higher than that of CT-TPR1 in *boro* due to higher plant height and tiller density. Total water input was similar in both CT-DrumR and CT-TPR1 but was significantly higher than in CT-TPR2. The WUE for grain production was higher in CT-TPR in both the seasons.

Table 2: Establishment methods as influenced by water used, WUE, grain yield and straw yield in wet drum seeded and transplanted rice in *aman* and *boro* (Source: Rashid *et al.*, 2009)

Establishment methods	*No. of irrigations*	*Irrigation water applied (mm)*	*Rain fall (mm)*	*Total water per crop (mm)*	*WUE (Kg/ha-mm)*	*Grain yield (t/ha)*	*Straw yield (t/ha)*
			Aman (2006)				
CT-Drum R	6	209b	1339	1548a	3.9b	6.0a	7.4a
CT- TPR1	5	222ab	1323	1545a	3.9b	6.1a	6.6b
CT- TPR2	5	238a	849	1087b	5.6a	6.1a	6.3b
			Boro (2006-07)				
CT-Drum R	21	818c	181	999b	6.4a	6.4a	6.8a
CT- TPR1	17	928a	142	1070a	5.6c	6.0b	6.1b
CT- TPR2	15	873b	202	1075a	6.1b	6.5a	6.7a

Yushi *et al*. (2013) reported that the alternative wetting and drying (AWD) produced marginally higher yields (5.7 per cent in 2010 and 6.6 per cent in 2011, respectively) than continuous flooding (CF). The water use efficiency (WUE) including both irrigation and total water use efficiencies (IWUE and TWUE) were improved significantly. AWD method enhanced root growth, facilitated remobilisation of photosynthates to the grains, accelerated grain filling and improved grain yield. Compared with CF, IWUE and TWUE in AWD were increased by 82 per cent and 29 per cent in 2010 and 49 per cent and 20 per cent in 2011 (Table 3).

Table 3: Grain yield, IWUE, TWUE, Total above ground N uptake (TANU), Agronomic use efficiency (AEN) and N recovery efficiency (REN) as affected by different water management practices 2010-2011 (Source: Yushi *et al.*, 2013)

Treatments	*Gr. yield (kg/ha)*	*IWUE (Kg/m^3)*	*TWUE (Kg/m^3)*	*TANU (Kg/ha)*	*AEN (Kg/kg)*	*REN (per cent)*
Continuous flooding	6927^b	1.67^b	0.64^b	170.2^a	14.0^b	47.5^b
Alternate wetting and drying	7388^a	2.48^a	0.77^a	179.2^a	15.3^a	50.7^a

Sun *et al*. (2012) using differential responses to physiological and biochemical changes of rice showed that higher activities of ammonia assimilation enzymes (glutamine synthetase, glutamate synthase, and glutamate dehydrogenase) which are the main enzymes involved in plant N metabolism were obtained under intermittent irrigation than flooded irrigation. Therefore, rice root system was promoted and well developed, plant N metabolism was strengthened and the leaf photosynthetic rate was further improved, which produced more assimilates for the above-ground parts resulting in high yield. This study further demonstrates that AWD would not only reduce water consumption but also increase root and panicle biomass to maintain high grain yield.

Feng *et al.* (2007) reported that AWD reduced irrigation water and total water consumption by around 37 per cent and 22 per cent, respectively. In other study Belder *et al.*, (2004) also found the same effect of AWD and reported a saving of about 6–14 per cent and 15–18 per cent, respectively. Yao *et al.* (2012) showed that AWD saved 24 per cent and 38 per cent irrigation water in 2009 and 2010, respectively. Belder *et al.* (2007) in another study summarised data for Asia and reported that on an average AWD decreased total water input by 15–30 per cent with yields comparable to rice in CF, which is a significant saving and this saved water could be used for expanding irrigated area.

Adoption of AWD meant significant reduction in not only irrigation water and total water consumptions, but also in irrigation frequency which meant that labour force (by 38 per cent), fossil fuel energy used to pump water and water resources were economised (Rejesus *et al.*, 2011).

Irrigation scheduling

Mallareddy *et al.*, (2013) tested different irrigation scheduling / methods only to find that soil saturation maintained throughout the crop growing cycle recorded significantly higher grain and straw yield. Pasha *et al.* (2013) found that irrigation at seven days interval during vegetative stage and four days interval during reproductive stage recorded higher grain yield compare with other treatments due to improved nutrients uptake resulting in more number of tillers. The water productivity was higher in irrigation at seven days interval during vegetative stage and in seven days interval during reproductive stage.

Dhar *et al.* (2008) reported that the maximum grain yield of 5289 kg/ha was harvested when the crop was irrigated seven days after disappearance of water (seven DADW) which was significantly superior to other treatments except continuous submergence (4957 kg/ha) and alternate irrigation (4934 kg/ha) for which grain yield was comparable (Table 4). In fact, they reported reduction in yield by 6.3 per cent under continuously submerged, 6.7 per cent when the crop was irrigated alternately, 14.8 per cent when irrigated three DADW, 7.3 per cent 5 DADW and 20.3 per cent after nine DADW. This could be due to aerobic soil condition resulting from no or less standing water in the field between two irrigations. This may have facilitated roots respiration through aeranchyma in the root cells.

Murthy and Reddy (2013) in their study showed significant increase in grain yield of rice with increase in IW/CPE irrigation schedule from 0.8 to 1.2. The improvement in yield was due to better availability of moisture, which lead to efficient physiological activity. High level of dry matter production and efficient translocation of photosynthates from source to sink may have been responsible for increased level of yield structure (Table 5). It is further reported that rice plants when grown under saturated condition develop more plant stature, leaf area, root volume, productive tillers, resulting in higher yields. The highest gross returns, net returns and benefit cost ratio were realised with the irrigations scheduled at IW/CPE ratio of 1.2, and were significantly higher than other irrigation schedules tested.

Table 4: Effect of irrigation regimes on grain yield, straw yield and field water use efficiency of rice (Source: Dhar *et al.*, 2008)

Treatments	*Total dry matter (kg/ha)*	*Grain yield (kg/ha)*	*Straw yield (kg/ha)*	*Amount of water supplied (ha-mm)*	*Field WUE (kg/ha-mm)*
Continuous submergence	14869	4957	9912	1012	2.94
Alternate irrigation	14804	4934	9870	920	3.10
3 DADW	13649	4507	9142	870	2.92
5 DADW	14714	4899	9719	630	3.77
7 DADW	15855	5289	10470	450	4.72
9 DADW	13385	4215	9170	390	3.98
LSD (P = 0.05)	473	251	379	-	-

Table 5: Nutrient uptake (kg/ha), grain yield (kg/ha), straw yield (kg/ha) and net returns (Rs/ha) of rice under aerobic culture as influenced by irrigation schedule (Source: Murthy and Reddy, 2013)

Treatments	Nutrient uptake			Grain yield	Straw yield	Net returns	B:C
	N	P	K				
Irrigation schedules							
IW/CPE of 0.8	25.79	20.57	136.88	4095	6768	32560	3.56
IW/CPE of 1.0	27.10	20.67	137.86	4230	6923	34005	3.67
IW/CPE of 1.2	30.93	26.62	144.20	4702	7627	39048	4.05
S.Em±	0.72	0.71	0.61	110	156	1191	0.09
C.D at 5%	2.82	2.79	2.38	432	612	4676	0.37

Shekara *et al.* (2011) in a multi-year study found that irrigation scheduling at IW/CPE ratio of 2.5 recorded significantly higher grain yield (6.41 t/ha) but was found at par with irrigation scheduling at IW/CPE ratio of 2.0 (6.22 t/ha).

This could be attributed to increase in yield components and growth characters at various phenological stages with adequate moisture availability throughout the crop growth period at IW/CPE ratio of 2.0 and 2.5. Maheshwari *et al.* (2008) reported that the IW/CPE ratio 1.2 significantly recorded higher grain yield compare to other treatments due to higher yield parameters. The water productivity was significantly higher under IW/CPE ratio 0.8 due to less number of irrigations.

Pandey *et al.* (2010) reported that the water and irrigation requirement were the highest under irrigation schedule of continuous submergence (CS) followed by irrigation given under one and three day after disappearance (DAD) of ponded water. The irrigation at three DAD of ponded water gave the highest water use efficiency (WUE) (88.51 kg $ha^{-1}days^{-1}$) but grain yield was reduced substantially as compared to both the irrigation schedules. The irrigation schedule of one DAD of ponded water although gave WUE less than three DAD, the grain yield was similar to that of CS. As compared to CS, reduction in irrigation requirement at irrigation schedule of one DAD of ponded water was to the extent of 507 mm and 700 mm, respectively during first and second year.

Irrigation levels

Gill (2008) found that irrigation at 2 days interval enabled the crop to produce significantly more effective tillers (371.4/m^2), 179.4 grains/panicle, 1000 grains weight (22.1 g) and grain yield (7.23 t/ha), than under three days interval. Cyclic submergence of 0-5 cm water give higher grain yield (5.86 t/ha), which was at par with UAS Bangalore package (5.41 t/ha). Significantly higher WUE (62 cm/m^3) recorded in treatment with field capacity saturation (Ganesh, 2001). Soriano (2007) in a study looked at rice grain yield and water productivity with respect to irrigation input plus rainfall. Total rainfall depth during the growing season (January-June, 2007) was 425 mm. Water productivity decreased as total water input increased. Highest water productivity occurred in every 14 days until tillering, weekly until flowering, twice a week from one week before to one week after flowering, once every 10 days after flowering until maturity. W_3 (0.48 kg grain/m^3 water input), and the lowest was found in twice a week irrigation, W1 with a water productivity of 0.29 kg grain/m^3 water. In this experiment, average grain yield under W_3 was lower by 36 per cent but lower water inputs by 54 per cent with a corresponding higher water productivity of 31% as compared with the average grain yield in irrigated rice of 5 t/ha with water input of 15,000 m^3/ha.

Farmers notmally use higher irrigation level than continuous submergence, which is not only expensive but may have environmental consequences besides lowering paddy yield in the long run. There is significant saving of water, land, and environment in addition to higher crop productivity on sustainable basis if efficient use of irrigation water was coupled with an optimum dose of nutrients (Yang and Zhang, 2010). Amiri *et al* (2009) suggested five day interval irrigation instead of submerged irrigation.

Thus, in view of the above discussion, it can be concluded that submerged irrigation is neither productive nor economical. Irrigation levels are negatively correlated with WP, indicating that WP decreases with increase in irrigation level (Tao *et al,* 2006).

Conclusion

Water saving irrigation methods *viz*., modified SRI, maintaining saturation and aerobic methods are better options for increasing in WUE in water limited / scarce environments for rice cultivation. Maintenance of saturation throughout crop period with proper drainage gives better productivity of rice. Further, higher rice grain yield and WUE is also obtained in AWD method of cultivation. Therefore, among these available better technologies farmers could adopt the one that suits to their local condition in order to improve rice productivity on sustainable basis and make rice cultivation in Asia and India a climate smart production system to cope with water limited future climates.

References

Amiri, E, A. Ebrahim, K, Morteza, H. R. Bozorgi, S. M. Sadeghi, M. Rezaei, 2009. "Response of rice varieties to water limit conditions in north Iran". *World Applied Science Journal* 6(9): 1190–1192.

Belder, P., B.A.M. Bouman, R.J. Cabangon, G. Lu, E.J.P. Quilang, Y., Li, J.H.J. Spiertz, T.P. Tuong, 2004. "Effect of water-saving irrigation on rice yield and water use in typical lowland conditions in Asia". *Agriculture Water Manage*. 65, 193–210.

Belder, P., B.A.M. Bouman, J.H.J. Spiertz, 2007. "Exploring options for water savings in lowland rice using a modelling approach". *Agriculture System*. 92, 91–114.

Dhar, R., N. K. Gupta, and A. Samanta, 2008. "Effect of irrigation scheduling on the performance of kharif rice grown under different establishment methods". *Journal of research, SKUAST-J,* 7(2):277-280.

Feng, L., B.A.M. Bouman, T.P. Tuong, R.J., Y. Cabangon, G. Li, Y. Lu, Feng, 2007. "Exploring options to grow rice using less water in northern China using a modelling approach". I. Field experiments and model evaluation. *Agriculture Water Manage*. 88, 1–13.

Ganesh, H. V., 2001. "Effect of different moisture regimes on the grain and straw yield of paddy". *Karnataka Journal of Agricultural Sciences*, 14 (2): 277-280.

Geethalakshmi, V. A. Lakshmanan, and T. Ramesh, 2010. "Assessment of water requirement for different system of rice cultivation". *Field Crops Research,* 114: 274–281.

Gill, M. S., 2008. "Productivity of direct seeded rice (*Oryza sativa)* under varying irrigation levels". *Indian Journal of Agricultural Sciences,* 78(9): 766-770.

Mahajan, G., B.S. Chauhan, and D.E. Johnson. 2009. "Weed management in aerobic rice in northwestern Indo-Gangetic Plains". *Journal of Crop Improvement,* 23:366–382.

Maheswari, J., N. Margatham, G. J. Martin, 2008. "Relatively simple irrigation scheduling and N application enhances the productivity of aerobic rice (*Oryza sativa L.)*". *American Journal of Plant Physiology*. 2, 261–268.

Mallareddy, M., B. Padmaja, G. Veeranna, and D. Vishnu Vardhan Reddy, 2013. "Response of aerobic rice to irrigation scheduling and nitrogen doses under drip irrigation". *Journal of ANGRAU*, 41(2):144-148.

Manjunath, B. N., R. Basavarajappa, and B. T. Pujari, 2010. "Effect of age of seedling on growth, yield and water requirement by different system of rice intensification". *Karnataka Journal of Agricultural Sciences*, 23 (2): 231-234.

Murthy, K. V. R., and D. S. Reddy, 2013, "Effect of irrigation and weed management practices on nutrient uptake and economics of production of aerobic rice". ISOR, Journal of Agriculture and Veterinary Sciences. 3 (1): 15-21.

Narula, K. K., and U. Lall, 2010. "Challenge in securing India's water future". *Journal of Crop Improvement,* 24:85–91.

Pandey, N., A. K. Verma, and R.S. Tripathi, 2010. "Response of hybrid rice to scheduling of nitrogen and irrigation during dry season". *Indian Journal of Agricultural Sciences,* 47(1):34-37.

Pasha, L., M. D. Reddy, M. G. Reddy, and M. Uma Devi, 2013. "Influence of irrigation schedule, weed management and nitrogen levels on grain yield, nutrient uptake and water productivity of aerobic rice". *Indian Journal of Agricultural Research,* 47(1):26-34.

Ramana Murthy, K.V. D.S. and Reddy, 2013. "Effect of irrigation and weed management practices on Nutrient uptake and Economics of production of Aerobic rice". *IOSR Journal of Agriculture and Veterinary Science (IOSR-JAVS),* Volume 3, Issue 1 (May. - Jun. 2013), pp. 15-21.

Rashid, H. M., M. M. Alam, A. H. M. Khan, and J. K. Ladha, 2009. "Productivity and resource use of direct-(drum)-seeded and transplanted rice in puddled soils in rice–rice and rice–wheat ecosystems". *Field Crops Research,* 113: 274–281.

Rejesus, R.M., F.G. Palis, D.G.P. Rodriguez, Lampayan, R.M., Bouman, B.A.M., 2011. Impact of the alternate wetting and drying (AWD) water-saving irrigation technique: evidence from rice producers in the Philippines. *Food Policy* 36, 280–288.

Rodell, M., I. Velicigna, and J.S. Famiglietti. 2009. Satellite-based estimates of groundwater depletion in India. *Nature* 460:999–1,002.

Shekar, B. G., Bandi, A. G., Shreedhara, D., Sharanappa and Krishnamurthy, N., 2011. Effect of irrigation schedules on growth and yield of aerobic rice under varied levels of farm yard manure. *Oryza*, 48(4):324-328.

Soriano, J. B., 2007. Water regimes and nutrient requirements for aerobic rice production system. *International rice research institute, Los banos, Philippines*, pp. 270.

Sun, Y., Ma, J., Sun, Y., Xu, H., Yang, Z., Liu, S., Jia, X., Zheng, H., 2012. The effects of different water and nitrogen managements on yield and nitrogen use efficiency in hybrid rice of China. *Field Crops Research*. 127, 85–98.

Tao, H, Brueck, H, Dittert, K, Kreye, C, Lin, S, Sattelemacher, B. 2006. "Growth and yield formation for rice (*Oryza stiva* L.) in the water-saving ground cover rice production system (GCRPS)". *Field Crop Research*, **95**: 1–12.

Yang, J.C., and J. H. Zhang, 2010. "Crop management techniques to enhance harvest index in rice". *Journal of Experimental Botany*, **61** (12): 3177–3189.

Yao, F., J. Huang, K. Cui, L. Nie, J. Xiang, X. Liu, W. Wu, M. Chen, S. Peng, 2012. "Agronomic performance of high-yielding rice variety grown under alternate wetting and drying irrigation". *Field Crops Research*, 126, 16–22.

Yushi, Y., L. Xinqiang, Yingxu Chena, B. Jin, G. Jiatao, and R. Guob, 2013. "Alternate wetting and drying irrigation and controlled-release nitrogen fertilizer in late-season rice". *Field Crop Research*, 144: 212–224.

39

Conservation Agriculture Practices to Improve Water Use Efficiency in Rainfed Crops: A Review

Keshavamurthy G.M. Manjula Udikeri, R. Basavarajappa and R. H. Patil

Abstract

Conservation agriculture (CA) is seen as climate smart agriculture practice to conserve natural resources, build resilience to changing climate and enhance productivity to achieve the goals of sustainable agriculture. CA practices help hold and store more plant available water than the conventional tillage practices. The high soil moisture content under conservation tillage is due to both improved soil structure and decrease in the evaporation loss under continuous crop residue mulch cover. Increase in the available water content under conservation tillage, particularly in the surface horizon, increases the consumptive use of water by crops and hence improves the water use efficiency (WUE). Reduced and zero tillage practices have been reported to enhance infiltration rate, soil moisture content and WUE as compared to conventional tillage in maize. In wheat, minimum tillage with maize stover mulch @ 5 t/ha at 25 DAS of wheat has shown significant increase in water use, grain yield and WUE over other treatments. Maize + soybean (2:2) recorded significantly higher maize equivalent yield, WUE and economic returns over sole maize. Application of FYM @ 5 t/ha + incorporation of mustard straw @ 5 t/ha recorded significantly higher seed yield, higher organic carbon, WUE and net returns as compared to control. However, it was found on par with application of FYM @ 5 t/ha + incorporation of taramira *straw @ 5 t/ha. These studies clearly show the potential of conservation agriculture practices to enhance WUE and rainfed crop yields in water limited environments.*

Key words: *Conservation agriculture, tillage, cropping systems, residue management.*

Introduction

Rainfed agriculture is mainly practiced in arid, semi-arid and sub-humid dry regions in India. With about 68 per cent of total population living in villages (Kumar *et al.*, 2009), these regions are home to little over 80 per cent of rural poor (Rao *et al.*, 2005). In rainfed regions, the annual rainfall is much lower than the evapotranspiration demand particularly in arid and dry semi-arid zones. Coarse cereals (85 per cent), pulses (83 per cent), oilseeds (70 per cent), cotton (65 per cent) and upland rice (45 per cent) are the predominant crops grown in rainfed areas in India (Anon., 2009).

Rainfed agriculture has traditionally been considered as a gamble with monsoon, but the situation might turn to even worst under future climates where dry and hot regions get drier and hotter with global warming. In order to feed 1.3 billion populations expected by 2020 India will have to produce 5-6 m.t. of additional food grains each year (Anon., 2009). Although rainfed / dryland agriculture occupies 67 per cent of net sown area, it contributes just over 40 per cent of total food grain production and supports only 40 per cent of the total population. This is mainly because of the low productivity in comparison to irrigated regions. However, even after realisation of full irrigation potential in the country we would still continue to have at least half the net sown area under rainfed, especially under drylands.

These regions suffer from a number of biophysical and socio-economic constraints which affect productivity of crops and livestock. These include low and erratic rainfall, land degradation and low fertility, among others. On top of it the changing climate leading to hotter and drier climate coupled with changed patterns in rainfall are threatening the sustainability of agriculture practices in rainfed / dryland regions. In this context conservation agriculture concept is being proposed to address current rainfed / dryland problems while adapting and building resilience to global warming.

Conservation agriculture–way forward: The term CA refers to the system of raising crops without tilling the soil while retaining crop residues on the soil surface. The key elements which characterise CA include (i) minimum soil disturbance by adopting no or minimum tillage for field operations, (ii) crop residues management on the field surface, and (iii) adoption of diversified legume-based cropping systems to realise maximum benefits from inputs and minimise adverse environmental impacts (Avtar Singh *et al.*, 2013).

Conservation tillage (CT)

There are many practices in CT concept primarily aimed at reducing soil disturbance, and conserving and managing crop residue to reduce soil and rain water erosion. These include no-till, reduced-till, ridge-till, mulch-till or strip-till (Veer Singh *et al.*, 2011). These practices reduce labour, save time and fuel, and in the long-run improve farm profitability, machinery wear and its traction on the field. Further, increase organic matter and soil biological diversity, trap rain water more effectively to improve water availability and water use efficiency (WUE). At farm or landscape level these practices reduce soil erosion, improve wildlife habitat, water and air quality and sequester more carbon to mitigate climate change (Dev Narayan and Lal, 2009). CT practices *viz.*, reduced and zero tillage are reported to enhance infiltration rate (6.2 and 5.9 cm/hr), soil moisture content (34 and 36 cm at 60 cm depth) and WUE (9.3 and 9.9 kg/ha/mm), respectively, as compared with conventional tillage in maize, and has been attributed to improved soil structure resulting in more water held for longer period (Avtar Singh *et al.*, 2013). In another study fresh raised bed method of maize sowing recorded significantly lower bulk density (1.3 Mg/m^3), higher water productivity (3.2 kg/M^3), grain yield (3.4 t/ha) and net returns (Rs 18600/ha) (Table 1) over conventional method and was found at par with zero tillage and permanent raised bed method (Veer Singh *et al.*, 2011).

Table 1: Effect of tillage methods on maize productivity (Source: Veer Singh *et al.*, 2011)

Tillage practices	*Organic carbon (g/kg)*	*Bulk density (Mg/M^{-3})*	*Water productivity (kg/m^{-3})*	*Grain yield (t/ha)*	*Net Income (Rs/ha)*	*Energy Use Efficiency*
Zero tillage	5.8	1.34	2.17	2.82	13400	6.75
Conventional tillage	5.2	1.48	1.68	2.91	12300	5.21
Fresh raised bed	5.6	1.28	3.15	3.35	18600	5.71
Permanent raised bed	5.9	1.38	3.01	3.08	16900	7.29
S. Em ±	0.10	0.02	0.12	0.12	0.65	0.21
CD (P = 0.05)	0.40	0.07	0.41	0.40	2.89	0.97

Minimum tillage coupled with sunnhemp residue incorporation (@ 4 t/ha) has also helped increase soybean yield (1351 kg/ha), WUE (2.9 kg/ha-mm) and net returns (Rs 9541/ha) as compared with conventional tillage (Anon., 2009). Dev Narayan and Lal (2009) reported that reduced tillage (*i.e.*, ploughing to a depth of 7 cm and harrowing) with *in-situ* surface mulching recorded significantly higher sorghum grain yield (3.9 t/ha), stover yield (11.3 t/ha), water stable aggregates (48.1), infiltration rate (5.6 cm/hr) and lower bulk density (1.4 Mg/m^3) as compared with traditional tillage practices. This is attributed to improved soil structure influenced by reduced tillage while mulching reduces the evaporation losses of water. Sharma *et al.* (2013) in a maize-wheat sequence showed that minimum tillage soon after maize harvest and using maize stover as mulch (@ 5 t/ha) at 25 DAS after wheat was sown improved water use (353.4 mm), grain yield (2537 kg/ha) and WUE (7.2 kg/ha-mm) over other treatments.

Hari Ram *et al.* (2013) while working on soybean - wheat sequence tried permanent raised beds + straw mulch @ 5 t/ha in soybean and permanent raised beds in wheat only to find significantly lower soil temperature (33.5 °C), higher nodule weight (174.1 mg/plant), biomass (5.3 t/ha) and WUE (23.5 kg/ha-cm) over conventional tillage in soybean, and similar beneficial effects were also observed in the following wheat crop. However, it was on par with permanent raised beds in soybean and wheat without mulching suggesting both individual and synergistic effect of reduced tillage with mulch. Alexandru *et al.* (2010) carried out a study in Romania on tillage in soybean and maize, and found strip tillage and no tillage significantly increasing water use (332 and 325 mm), yield (2098 and 1941 kg/ha) and WUE (6.3 kg/ha-mm) as compared with other tillage practices.

Crop Residues (CR) Management

Most crop residues, especially legumes, in general have higher nutrient content than FYM and its use along with chemical fertilisers help not only to sustain the productivity, but reduce the input cost on chemical fertilisers and long-term impact on the environment. Increased use of inorganic fertilisers alone in unbalanced way leads to decline in soil quality and multiple nutrient deficiencies (Regar *et al.*, 2010). Efficient CR management practices play an important role in building soil productivity as well as in increasing the use efficiency of chemical fertiliser. Among different mulches, gliricidia leaves @ 2 t/ha in soybean crop was found to significantly increase seed yield (1869 kg/ha), WUE (4.0 kg/ha-mm), gross return (Rs 31904/ha) and net return (Rs 19404/ha) (Table 2) as compared with no mulching (Anon., 2009). Application of FYM @ 5 t/ha + incorporation of mustard straw @ 5 t/ha recorded significantly higher seed yield (1444 kg/ha), higher organic carbon (0.7 per cent), WUE (5.9 kg/ha-mm), net returns (Rs 15398/ha) and B:C (2.6) as compared to control. However, it was on par with application of FYM @ 5 t/ha + incorporation of *taramira* straw @ 5 t/ha (Regar *et al.*, 2010).

This might be due to increase in organic matter and physico-chemical properties of soil due to addition of FYM and crop residues. This further enhanced water holding capacity of the soil. Bharath Bhushan Vashisht *et al.* (2013) reported that use of *subabul* twigs and leaves as mulch improved grain yield in maize and wheat (40.6 q/ha and 27.6 q/ha) over no mulch. Sugarcane trash when used as mulch not only improved crop yields but improved WUE while reducing soil loss. This might be due to sugar cane trashes decompose at slower rate as compared to other mulches and it was helped to more water to infiltrate and lower evaporation losses.

Table 2: Influence of soil and bio mulches on productivity of *kharif* soybean (Source: Anon, 2009; Indore)

Treatment	*Yield (kg/ha)*	*WUE (Kg/ ha/mm)*	*Gross return (Rs/ha)*	*Net return (Rs/ha)*
No mulching	1177	2.53	18839	9832
Gliricidia mulch @ 2t/ha	1804	3.88	30864	18364
Straw mulch @ t /ha	1869	4.02	31904	19404
Weed biomass @ 5 t/ha	1842	3.97	30972	18972
Soil mulch	1621	3.49	67336	15436
Polythene mulch	1956	4.21	33796	20796
CD (p = 0.05)	479	-	-	-

Singh *et al*. (2012) tested different crop residues for incorporation. Among them mungbean residue incorporation with irrigation recorded significantly lower bulk density, particle density, pore space and WHC over control. Rajendra Reddy *et al*. (2002) revealed that wheat straw (@ 5 t/ha) + minimum tillage after harvest of blackgram (with or without its residues) crop recorded significantly lower BD, hydraulic conductivity, infiltration rate, water stable aggregates and moisture retention over deep tillage by iron plough without residue. Nalatwadmath *et al*. (2006) stated that among different crop residue management practices, sorghum + *dolichos* incorporated into the soil at 45 days after sowing recorded significantly lower pH (8.5), higher OC (4.0 g/kg), available N (202 kg/ha), soil aggregate size (696 microns), lower runoff loss (99 mm), soil loss (3141 kg/ha), total soil moisture in 60 cm depth (10.7 cm) as compared to control (8.8, 3.7 g/kg, 165 N kg/ha, 582 microns, 124 mm, 4424 kg/ ha and 7.2 cm) respectively.

Sharma *et al*. (2009) reported that 6 t/ha of *leucaena* air dried leaves' incorporation recorded lower BD (1.4 g/cm^{-3}), higher water stable aggregates (34.1 per cent), available water capacity (12.8 per cent), infiltration rate (0.7 cm/hr^{3}), OC (0.72 per cent) and soil pH (6.4) as compared to lower levels and control. With respect to days after incorporation, 30 and 45 days after incorporation found to be better.

Crop Diversification

Improved cropping systems

Cropping sequence, intercropping, alley cropping, double cropping and alternate land use systems etc., mixing cereals with legume / oil seed and fibre crops prevent soil depletion, maintain soil fertility, reduce soil erosion, increase soil moisture holding capacity, minimise pest load and reduce reliance on chemicals. Ram *et al.* (2008) reported that, rice-fenugreek-okra cropping system has recorded significantly higher rice equivalent yield (25.7 t/ha), WUE (33 kg/ha-mm), net return (Rs 96286 ha^{-1}) and B:C ratio (2.8) and it was on par with rice-chickpea-sesame cropping system.

Among different legume based cropping systems, rice-mustard-green gram system has recorded significantly higher REY (12.8 t/ha), WUE (15.2 kg/ha-mm), net return (Rs 48415 /ha) and B:C (2.2) over rice–wheat–fallow system. Narendra Kumar *et al.* (2013) reported that normal planting of pigeonpea + inclusion of one row of black gram recorded significantly higher PEY (25.8 q/ha), LER (1.70), gross return (Rs 120050/ha), net return (Rs 99396/ha), B:C (5.80) and WUE (4.55 kg/ha-mm) as compared with sole pigeonpea, but was found at par with paired row planting of pigeonpea + black gram in two rows' proportion.

Similarly, Ansari and Rana (2012) revealed that paired row sowing of pearl millet + one row of pigeonpea recorded significantly higher pearl millet equivalent yield (4.8 t/ha), LER (1.6), consumptive use (443.6 mm), moisture use rate (3.0 mm/day), WUE (10.6 kg/ha-mm), net returns (Rs. 38130/ha) and B:C (3.0) as compared to sole pearlmillet (Table 3).

Table 3: Effect of cropping systems on yield, moisture use indices and economics of pearl millet + pigeonpea (Ansari and Rana, 2012)

Treatment	*PEY (t/ha)*	*LER*	*Consumptive Use (mm)*	*WUE (kg/ha-mm)*	*Net Return (Rs/ha)*
C_1	3.34	1.00	367.17	8.95	25030
C_2	4.20	1.00	340.77	4.65	25280
C_3	4.82	1.35	443.59	10.57	38130
S. Em ±	0.17	0.10	3.82	0.15	-
CD at 5%	0.54	0.32	11.41	0.44	-

C_1-Sole pearlmillet (PM), C_2 – Sole pigeonpea (PP) and C_3 – Paired row of PM + one row of PP.

Conclusions

Conservation tillage coupled with crop residues mulching or incorporation is helpful for enhancing rain water conservation as well as its retention and utilisation for achieving higher yield and WUE. Improved cropping systems – crop rotation, intercropping, alley cropping, alternative land use systems etc., improves the soil fertility and productivity in rainfed situations. Zero/Minimum tillage practices with crop residue retention can be adopted in rainfed agriculture for improving crop productivity, water use efficiency, and soil fertility.

Conservation agriculture practices store more plant available water than the conventional inversion tillage practices. The high soil moisture content under conservation tillage is due to both improved soil structure and decrease in the evaporation loss under continuous crop residue mulch cover. Increase in the available water content under conservation tillage, particularly in the surface horizon, increases the consumptive use of water by crops and hence improves the water use efficiency (WUE).

References

Alexandru, I. C., V. Z. Gina, and C. Nicolae, 2010. "Tillage system effects on water use and grain yield of winter wheat, maize and soybean in rotation". www.incda-fundulea.ro.

Anonymous, 2009. "All India Coordinated Research Project for Dryland Agriculture, Central Research Institute for Dryland Agriculture". Annual Report, Hyderabad.

Ansari, M. A. and K. S. Rana, 2012. "Effect of transpiration suppressants and nutrients on productivity and moisture use efficiency of pearl millet – pigeonpea intercropping system under rainfed conditions". *Indian Journal of Agricultural Sciences*, 82 (8): 676-680.

Avtar Singh, J. K., Kang, Maninder Kaur and Ashu Goyal, 2010. "Irrigation scheduling in zero-till and bed planted wheat (*Triticum aestivum*)". *Indian Journal of Soil Conservation,* **38** (3): 194-198.

Bharat BhushanVashisht, B. S. Sidhu, Satvinder Singh and Nilesh Biwalkar, 2013. "Effect of different mulches on soil erosion and carry-over of residual soil moisture for sowing of *rabi* crop in maize – wheat cropping sequence in rainfed Shivaliks of Punjab". *Indian Journal of Soil Conservation*, 41 (2): 136-140.

Dev Narayan and B. Lal, 2009. "Rainwater conservation and yield of sorghum as influenced by tillage and cover management practices in red soils". *Indian Journal of Agronomy*, **54** (4): 438-443.

Hari Ram, Yadvinder Singh, K. S. Saini, D. S. Kler, and J. Timsina, 2013. "Tillage and planting methods effects on yield, water use efficiency and profitability of soybean–wheat system on a loamy sand soil". *Experimental Agriculture*, 35(2): 1-19.

Kumar, P., P. K. Joshi, P.S. Birthal, 2009. "Demand projections for Foodgrains in India". *Agricultural Economics Research Review,* 22(2): 237-243.

Nalatwadmath, S. K., S. L. Patil, R. N. Adhikari, and S. Mana Mohan, 2006. "Effect of crop residue management on soil erosion, moisture conservation, soil properties and sorghum yield on *Vertisols* under dryland conditions of semi-arid tropics in India". *Indian Journal of Dryland Agricultural Research & Development*, 21 (2):99-104.

Narendra Kumawat, P. S. Rajendra, and K. Rakesh, 2013. "Productivity, economics and water use efficiency of rainfed pigeonpea + black gram intercropping as influenced by integrated nutrient management". *Indian Journal of Soil Conservation*, 41 (2): 170-176.

Ram, A. J., R. A. Dungrani, M.K. Arvadia, and K.L. Sahrawat, 2008. "Diversification of rice (*Oryza sativa* L.) based cropping systems for higher productivity, resource-use efficiency and economic returns in south Gujarat of India". *Indian Journal of Agricultural Sciences,* 84 (7): 887-891.

Rao, K.P.C., M.C.S. Bantilan, K. Singh, S. Subrahmanyam, P. P. Deshingkar, Rao, Parthasarathy and B. Shiferaw, 2005. "Overcoming Poverty in Rural India: Focus on Rainfed Semi-Arid Tropics". Patancheru 502 324, Andhra Pradesh, India: International Crops Research Institute for the Semi-Arid Tropics, p. 96.

Rajendra Reddy, R. G., G. U. Malewar, and B. G. Karle, 2002. "Effect of crop residue incorporation and tillage operations on soil properties of *vertisols* under rainfed agriculture". *Indian Journal of Dryland Agricultural Research & Development*, 17(1):55-58.

Regar, P.L., S.S. Rao, and S.P. Vyas, 2010. "Crop residue management for sustainable production of Indian mustard (*Brassica juncea*) in arid and semi-arid region". *Indian Journal of Agronomy*, 37 (2): 118-122.

Sharma, N. K., S. C. Mohan, and P. C. Tyagi, 2009. "Effect of *Leucaena* mulch incorporation on soil properties under maize-wheat crop rotation". *Indian Journal of Soil Conservation*, 37 (2): 123-125.

Sharma, N.K., B.N. Ghosh, O.P.S. Khola, and R.K. Dubey, 2013. "Residue and tillage management for soil moisture conservation in post-maize harvesting period under rainfed conditions of North-West Himalayas". *Indian Journal of Soil Conservation*, 41 (2): 170-176.

Singh, R. P., P.N. Yadav, S. K. Uttam, S. C. Katiyar, and A. K. Tripathi, 2012. "Effect of moisture conservation and nutrient management on growth, yield use efficiency of sorghum (*Sorghum bicolor*) under rainfed conditions". *Current Advances in Agricultural Sciences,* 4 (1): 37-40.

Veer Singh, Shri Ram, B. Amit, and S. S. Uma, 2011. "Effect of tillage methods on soil properties and productivity of quality protein maize (*Zea mays*) – wheat (*Triticum aestivum*) system". *Indian Journal of Agronomy,* 56 (2): 83-87.

40

Waste Water for Irrigation as a Climate Smart Agricultural Strategy to Improve Crop Productivity– A Review

Hanamant M. Halli, S. S. Angadi and R. H. Patil

Abstract

Water is a fast dwindling resource on earth but needed for the overall development of human society including agriculture. With global warming and projected changes in temperature and rainfall patterns the situation may further deteriorate. Therefore, as a part of climate smart agriculture practices, use of wastewater holds a great potential, which is yet to be fully tapped and utilised in Indian agriculture. India produces 22,900 million litres of wastewater each day, but only one-fourth is treated before disposed. The total wastewater generated by the major cities alone is around 16,662 million litres per day, which with treatment could well be used in agriculture as a supplementary source of irrigation. Current and future fresh water demand in agriculture could be met by enhancing water use efficiency and by exploring multiple waste water sources: effluent from distillery units, city sewage, Vinasse, crude oil refinery and tannery industries. Spentwash is one such source and its application in optimised dose has positive impact on soil properties, crop qualities and yield. It contributes substantial amount of N, P, K, Ca, S and micronutrients to the crops. Conjunctive use of spentwash with other amendments in crops like sugarcane, rice, wheat, maize etc., in different locations has been found to increase crop yield over use of only fresh water. It can substitute at least 50 per cent N, 70 per cent P and 100 per cent K to sugarcane and 62 per cent P and 100 per cent K to wheatcrop. Thus, waste water/low quality water is emerging as potential source for improving the crop productivity and sustainability.

Key words: *Waste water, fresh water, spentwash, conjunctive use, sugarcane, cereals.*

Introduction

Water is an important scarce resource for the overall growth and development of human society including agriculture and protection of this dwindling natural resource efficiently has become the prime importance. India is located in a sub-tropical region and most of it falls under semi-arid tropics. Therefore, limitation of water resources is one of the major factors in the agricultural development. India supports more than 16 per cent of the world's population with only four per cent of the world's fresh water resources (Singh, 2003). Although agriculture sector in this country has been a major user of water, share of water allocated to irrigation is likely to be decreased by 10–15 per cent in next two decades (CWC, 2000).

Today, Indian agriculture is facing severe scarcity of fresh water due to climate uncertainties and population pressure. Per capita average annual freshwater availability has been declining since 1951 from 5177 m^3 in 2001 and 1588 m^3 in 2010 hence, the gap between the supply and demand for water is widening and is reaching alarming levels in many parts of the country and the world (Milano *et al.,* 2012). Considering the urgency to improve the water productivity, prevent pollution of our water bodies and preserve precious water resources exploiting and managing alternate water resources assumes high priority. In this respect wastewater has been an untapped resource in India till now. Out of 22,900 Mld (Million litre per day) of wastewater generated in the country, only about 5900 Mld (26 per cent) is treated before letting out, the rest i.e., 17000 Mld is disposed of untreated. The level of treatment available in major cities with existing treatment plant varies from 2.5 per cent to 89 per cent of the sewage generated (Bhardwaj, 2005).

What is waste water?

Wastewater in general comprises liquid waste generated by households, industry and commercial / industrial sources, as a result of daily usage, production, and consumption activities. In many regions of the world, particularly in water-scarce urban and periurban areas, and where demand for water is very high, wastewater is being used for agricultural purposes in one or the other way.

Types of wastewater

As the wastewater used in agriculture for irrigation has different sources and covers wastewater of different qualities, ranging from raw to diluted, and generated by various urban and industrial activities (Raschid-Sally and

Jayakody, 2008). Urban wastewater is usually a combination of one or more of the following: (i) domestic effluent consisting of black water (excreta, urine and associated sludge) and greywater (kitchen and bathroom wastewater), (ii) effluent from commercial establishments and institutions, including hospitals, (iii) industrial effluent, and (iv) storm water and other urban run-off. These could be used directly untreated or after dilution or treatment, but it carries potential risks in addition to benefits. Wastewater has the potential to cause health impacts like diseases because it contains bacteria, viruses and parasites. Further, heavy metals if present in wastewater, especially industrial and oil refinery waste can be very dangerous to human health (Hussain *et al.*, 2001). Leaching of nutrients and excess salts present in wastewater has the potential to affect the quality of both surface and ground water resources when used untreated and in excess amount, thus requires treatment / dilution before use and in optimised levels. However, in this review detail on treatment and its process is not included, but rather more on how different sources of wastewater are being used in agriculture for irrigating crops and its impacts on both crop and soil.

1. Distillery Spentwash

Distilleries are important agro-based industries in India linked to sugar industry. These produce ethyl alcohol from molasses for potable and industrial uses, and generate large volumes of foul smelling dark coloured waste water known as spentwash (SW). For every litre of alcohol produced, 12-15 litres of spentwash are produced. Some 40 billion litres of spentwash are generated every year from 285 distilleries in the country (Chonker *et al.*, 2000). The spentwash due to its heavy organic load and high level of suspended and soluble solids, salts etc. is a concern for environmental if disposed untreated, thus is environmentally unacceptable. In fact, according to Central Pollution Control Board (CPCB), distilleries are classified as one of the most polluting industries (Chandraju *et al.*, 2011).

Kumar and Chopra (2011) analysed spentwash from Doon distillery, Dehradun and found it acidic in nature (pH 5.2) with high BOD (3,266 mg/L), COD (8,653 mg/L), alkalinity (765 mg/L), hardness (2,139.0 mg/L), Ca^{2+} (1,855 mg/L), Fe^{2+}(31 mg/L), total available N (573 mg/L) and SO_4(1,246 mg/L). Radha and Srivastava (2012) studied the nutrient composition of crude and digested SW, and reported higher concentrations of essential nutrients (P, S, Fe, Mn, Zn and Cu) and heavy metals (Cd, Cr, Ni and Pb) in crude SW as compared to the digested SW.

Chandraju *et al.* (2011) used distillery SW diluted with fresh water in different proportion on cotton and groundnut. They and found that with increase in dilution germination increased and 100 per cent germination was found at 1:3 ratio of spentwash:normal water irrigation whereas, only

95 per cent germination was recorded with normal water irrigations. Chandraju *et al.* (2012) in another study also found the same effect on sprouting and growth of *Anthurium* plant and concluded that the diluted spentwash can be used for irrigation purpose without any adverse effect on the crop. In another study, Doddamani *et al.* (2011) also reported significant increase in plant height, leaf number, LAI, chlorophyll content, and yield parameters and grain yield of maize. Sharma (2014) used SW as a source of nutrient to substitute chemical fertiliser in sugarcane, and his work revealed that combined use of 25% NPK + 75% SW gave the highest mean cane yield (82 t/ha) apart from significant improvement in sugar yield (9.75 q/ha) over 100 per cent NPK through chemical fertilisers alone. These studies and by others clearly suggest that SW, after optimum dilution, could be used as a source of irrigation and nutrient in agriculture and realise higher and better quality crop yield (Bhaskar *et al.*, 2003).

However, its use on agricultural fields may have adverse impact on soil and ground water sources, which is an environmental concern. Shenbagavalli *et al.* (2011) reported that open wells near the SW applied fields in Namakkal district contained large amount of salts, particularly K and Cl. This suggests that the water was contaminated due to the application of distillery SW and more than 50 per cent of the samples collected from these open wells found unsuitable for irrigation purpose due to salinity hazards. In another study Selvamurugan *et al.* (2013) showed that use of bio-methanated distillery SW increased both pH and EC of the soil, but the salt content was within the critical limit of 1.00 dS/m.

Farid *et al.* (2012) observed that both raw and treated SW had high load of chemical and organic pollutants. Therefore, raw SW was diluted to 50 per cent and 75 per cent of water and its use decreased bulk density and improved infiltration rate. Similarly, Rubina (2013) used diluted SW for longer period and reported improvement in the physico-chemical properties of the soil. Tripathi *et al.* (2011) investigated the effect of application of various dosages of distillery SW on agricultural field and showed that diluted application substantially increased the enzyme activities. Similarly, Rubina (2013) also reported that the soil dehydrogenase and phosphatase enzyme activities increased when used with dilution on the long-run. These studies again show that with optimum dilution SW could be safely used as a source of irrigation and nutrients with positive effect on both crops and soil.

One long term study with use of SW on sugarcane by Shivraj (2014) clearly showed that significant improvement in physical properties of soil *viz.*, lower bulk density, better aggregate stability, improved WHC and infiltration rate, and lower erosion index even after 15 to 20 years of use. The reduction in bulk density was due to addition of high organic matter by SW. However, beyond 20 years there was slight reduction in these parameters which might be due to clogging of pores with high particulate organic carbon. Application of SW at the beginning has also found to stimulate microbial activity and secretion of microbial polysaccharides, which helped in stabilisation of soil aggregates. Increase in infiltration rate in spentwash treated plots was ascribed to higher organic matter content and changed distribution of pore sizes of the soil. An increase in maximum water holding capacity was due to increased number of small pores caused by better and fine grade aggregation.

However to avoid long-term negative effects of SW and its heavy organic load on soil properties combined approach of aerobic and anaerobic methods are found to be more cost effective and efficient in removal of pollutants. Therefore, optimised dose (after dilution or treatment) of SW application has positive impact on soil properties, crop qualities and yield improvement. It contributes substantial amount of N, P, K, Ca, S and micronutrients to the crops. Conjunctive use of SW with normal water in optimum proportion in crops like sugarcane, rice, wheat, maize etc., in different locations has been found to increase crop yield over use of only normal water while substituting chemical fertiliser input to a varying degree depending on the crop and soil.

2. Sewage effluent

The total wastewater generated by the major cities in India is >16,500 Mld approximately 81 per cent of the total water supplied. It is further projected that, with the current way of expansion of cities across India, the wastewater from urban centres may cross 120,000 Mld by 2051 and that from rural India may go up to 50,000 Mld (Bhardwaj, 2005). Therefore, we need to have technology and practices in place to use this huge source of wastewater productively and safely in the long run. Studies have shown city sewage could be used for irrigation after dilution in various proportions and has been found to improve soil organic matter and fertility status (Yadav *et al.*, 2002). Further it helped in build-up of total N, P and available K in surface 15-30 cm soil. However, indiscriminate use may cause environmental problems. For instance, traces of NO_3- N (upto 2.8 mg/l), available P (58 kg/ha), Pb (upto 0.35 mg/l) and Mn (upto 0.23 mg/l) were observed in well waters near the disposal point of city sewage indicating initiation of ground water contamination, although the heavy metals content in crops sampled from the area were below the permissible critical levels.

Though many similar studies elsewhere have confirmed that the domestic sewage can be effectively used as a resource of irrigation only after dilution, but there is a need for continuous monitoring of the concentrations of potentially toxic elements in soil, plants and ground water (Yadav *et al.*, 2002). Use sewage effluent leads to a greater saving in N and K fertilisers (Martinez *et al.*, 2013). Other studies elsewhere also strongly suggest that sewage effluent after advanced could be safely used, in terms of both environmental sustainability and public health safety, for vegetable crop production, concurrently promoting water use efficiency in dry areas (Anastasis Christou *et al.*, 2014; Dhir *et al.*, 2014).

Long-term studies on the long-term effect of sewage irrigation on heavy metal content in soils, large number of crops and groundwater under Keshopur Effluent Irrigation Scheme (KEIS) of Delhi, India (Rattan *et al.*, 2005) has shown that sewage effluent contained higher amount of P, K, S, Zn, Cu, Fe, Mn and Ni compared to groundwater. While there was no significant variation in Pb and Cd concentrations in these two sources water and levels of metal content were within the permissible limits for its use as irrigation water. However, sewage irrigation for 20 years and more resulted into significant build-up of Zn, Cu, Fe, Ni, and Pb in sewage-irrigated soils over adjacent tube-well water irrigated soils. Risk assessment in respect of metal contents in some vegetable crops grown on these sewage-irrigated soils indicated that these vegetables can be consumed safely by human (Rattan *et al.*, 2005). Long term studies to see the effect on soil microbes have also shown that repeated application of treated sewage wastewater had no signicant effect on the soil microbial population and possibly the bacteria might have adapted to the changed soil environment by developing various levels of metal resistance (Hayat *et al.*, 2001).

3. Vinasse effluent

Vinasse is a waste material from distillery industries which has the potential to cause major environment problems across the world. It is also referred to as stillage or molasses SW in Asian countries (Sajbrt *et al.*, 2010 and Muhammad *et al.*, 2012). Irrigation at (300 m^3/ha) as 1/5th dilution of Vinasse about two irrigation cycles + 60:25:80 kg N/ha in Cuba increased the sugarcane yield up to 64. 4 per cent (Armengol *et al.*, 2003). Application of Vinasse @ 100 m^3/ha just prior to planting in sugarcane increased the yield up to 9.18 per cent in Mauritius (Soobadar and Kwong, 2013). Application of Post Methonated SW up to 3 months of planting + 100 per cent NPK in sugarcane increased the yield up to 63.3 per cent in India (Janaki and Velu, 2010). Green Leaf Manure (6.5 t ha^{-1}) + Post Methonated SW at 125 m^3/ha) increased rice yield up to 102.8 per cent in India (Das *et al.*, 2010). Application of molasses spentwash at 5 cm to soyabean and 2 cm to wheat increased the wheat yield up to 94.7 per cent in India (Biswas *et al.*, 2009). Rock phosphate and Spentwash (1:20) in fodder sorghum increased the yield up to 94.7 per cent in India (Kumari and Pogat, 2012).

4. Crude oil refinery waste

For processing the oil, refineries use a large amount of water and generate about 0.6 barrels of wastewater per barrel of oil processed. This water contains waste oil and other impurities from refining processes. An estimate suggests that crude oil refinery waste comes to around 1.7 billion gallons of wastewater which must be treated before disposed to the environment (Rajesh *et al*., 2009). Its uncontrolled disposal may result in the degradation of environment and depreciation of aesthetic and natural habitat quality.

Rajesh *et al*., (2009) asessed the effect of refinery effluent at different dilution levels on soil physico-chemical and biological properties using pot culture experiments. They observed significant increase in pH, EC, Ca, Mg, alkalinity, Cl, SO_4, Fe, P, Na, K, heavy metals, soil dehydrogenase activity and organic content of soil at all intervals of time except bulk density and specific gravity. Therefore, only treated oil refinery effluent and after dilution could be used as an alternate source for irrigating crops (Rajesh *et al*., 2009).

5. Tannery effluents

Tanning refers to the process by which collagen fibers in a hide react with a chemical agent (tannin, alum or other chemicals). However, the term leather tanning also commonly refers to the entire leather-making process. Hides and skins have the ability to absorb tannic acid and other chemical substances that prevent them from decaying, make them resistant to wetting, and keep them supple and durable. The flesh side of the hide or skin is much thicker and softer. The three types of hides and skins most often used in leather manufacture are from cattle, sheep, and pigs. Out of 1000 kg of raw hide, nearly 850 kg is generated as solid wastes in leather processing. Only 150 kg of the raw material is converted into leather. A typical tannery generate huge amount of waste:Fleshing: 56-60 per cent, Chrome shaving, chrome splits and buffing dust: 35-40 per cent, Skin trimming: 5-7 per cent Hair: 2-5 per cent (Salman Zafa, 2014).

Currently, about 6.5 million tonnes of wet salted hides and skins are processed worldwide annually. About 3.5 million tonnes of various chemicals are used for leather processing. A considerable part of this amount is discharged into the effluent. At an average consumption of 45-50 m^3 of the waste liquor and 800 kg of solid wastes per tonne of raw hide are discharged by the leather industry. The composition of liquid and solid wastes generated in tanneries worldwide are: COD, 1470; BOD, 619; suspended solids, 920; chromium, 30; sulphur, 60 and solid wastes (trimmings, fleshing, shaving and buffing dust), 3000 thousand tonnes (Kanagaraj *et al*., 2006).

A study by Singh *et al.* (2001) has shown that disposal of tannery effluent into city sewage lowered the pH but increased EC of the sewage drain water. Further, K, Ca, Na, Zn, Fe, Mn, B, Al, Cr, Hg, Ni, Cr and Pb content were increased with the addition of tannery effluents. Some of these elements are dangerous to both crop and soil, thus tannery effluent can be used as source of nutrients or irrigation water only after effective pre-treatment before its disposal into the sewage drain (Singh *et al.*, 2001).

Conclusion

With growing population, economy and expanding industrialisation demand for fresh water would go up on one hand and wastewater generation would increase on the other. Therefore, we need to explore and popularise the safe use of these multiple wastewater sources in agriculture which is faced with scarcity of water. This would not only reduce demand on fresh water, but also improves water use efficiency while reducing chemical fertiliser input in agriculture to a great extent. Further this would also contain or limit environmental pollution. Most of the studies reviewed in this paper clearly show the potential benefits of using multiple sources of wastewater on crop yields by supplementing water and nutrients provided the wastewater is adequately treated or diluted before application.

Table 1. Effect of different periods of spentwash application on sugarcane yield and soil physical properties. (Source: Shivraj, 2014)

Treatment	*No. of malleable Canes (000/ha)*	*Single cane weight (kg/cane)*	*Bulk density (Mg/m)*	*MWHC (per cent)*
Control	111.80	1.11	1.43	67.48
Spent wash application for 5 to 10 years	114.12	1.18	1.42	70.84
Spent wash application for 10 to 15 years	105.50	1.03	1.33	71.98
Spent wash application for 15 to 20 years	93.04	0.90	1.25	73.34
S.Em.±	1.53	0.03	1.23	74.46
CD (P = 0.05)	4.73	0.07	0.02	0.44

References

Anastasis Christou, Grivas Maratheftis, Elena Eliadou, Costas Michael, Evroula Hapeshi, Despo Fatta-Kassinos, 2014. Impact assessment of the reuse of two discrete treated waste waters for the Irrigation of tomato crop on the soil geochemical properties, fruit safety and crop productivity. Agriculture ecosystems and environment, 192:105-114.

Armengol, J. E., Lonrenzo, Noemi, F., 2003. Use of vinasse dilutions in water as an alternative for improving chemical properties of sugarcane planted vertisols. Cltivostropicales. 24(3):73-76.

Baskar, M., Kayalvizhi, C., Subash, M., Chandra Bose, 2003. Eco-Friendly Utilisation of Distillery Effluent in Agriculture-A Review, Agricultural Reviews: 24 (1), p. 16.

Bhardwaj, Rajendra M., 2005. Status of Wastewater Generation and Treatment in India, Paper presented at Intersecretariat Working Group on Environment Statistics (IWG–Env) Joint Work Session on Water Statistics, Vienna, 20–22.

Biswas, A. K., Mohanty, M., Hati, K. M., Misra , A. K. 2009. Distillary effluents effect on soil organic carbon and aggregate stability of a versitol in India. *Soil. Till. Res.*104: 241-246.

Chandraju, S., Thejovathi, C. and Chidankumar, C. S., 2012. Impact of distillery spentwash irrigation on sprouting and growth of anthurium flowering plant. *Res. J. Pharm., Boil. Chem. Sci.*, **3**(4): 629- 635.

Chonker, P. K., Datta, S. P., Joshi, H. C. and Pathak, H., 2000. Impact of industrial effluents on soil health and agriculture-Indian experience. Part I: Distillery and paper mill effluents. *J. Scientific Ind. Res.*, **59**: 350–361.

CWC, 2000. Water and related statistics. Information Systems Organisation, Water Planning and Organisation Wing, Central Water Commission, RK Puram, New Delhi.

Das, M., Chakrabarty, H., Hati, K. M., Mishra, A. K., 2009. Utilisation of distillery waste water for improving production in under productive paddy grown area in India, *J. Sci. Ind. Res.* **69**(7):560-6563.

Dhir, B. and Srivastava, S., 2014. Effect of short term irrigation with treated and untreatedwastewater on growth and physio-biochemical parameters of wheat, *Indian Journal of Plant Physiology,* **19**(3), pp. 257-262.

Doddamani, M. B., Gali, S. K., Patagar, R. G and Mukharib, D. S., 2011. Effect of long-term distillery spentwash application on crop growth, yield and soil potassium status in vertisols. In: *Nation. Symp. Potassium nutrition in enhancing yield and quality of crops*, Univ. Agric. Sci., Dharwad, pp. 116-117.

Farid, A., Ajay, K., Awasthi, R. and Bhawana, S., 2012. Physico-chemical characterisation of distillery effluent and its dilution effect at different levels. *Arch. Appl. Sci. Res.*, **4**(4): 1705-1715.

Hayat, S., Iqbal Ahmad, Azam, Z. M., Ahmad, A., Inam, A., Samiullah, 2009. Effect of long-term application of oil refinery wastewater on soil health with special reference to microbiological characteristics, Bioresource Technology. 84 (2002):159–163.

Hussain, I., Raschid, L., Hanjra, M.A., Marikar, F and Van der Hoek, W., 2001. A framework for analyzing socioeconomic, health and environmental impacts of wastewater use in agriculture in developing countries: Working Paper 26. Colombo, Sri Lanka: International Water Management Institute (IWMI).

Janki, D., Velu, V., 2010. Effect of distillery spentwash on yield and quality of sugarcane var COC 86032, *Madras Agric*. J. **97**(10-12):560-563.

Jimenez, B, Asano, T., 2008. Water reuse: an international survey of current practice, issues and needs. Scientic and Technical Report No. 20. International Water Association, The Hague.

Kanagaraj, J., Velappan, K. C., Chandra Babu and Sadulla, S., 2006. Solid wastes generation in the leather industry and its utilisation for cleaner environment- A review. *Journal of Scientific & Industrial Research.* 6, pp. 541-548.

Kumar, V. and Chopra, A. K., 2011. Impact on physico-chemical characteristics of soil after irrigation with distillery effluent. *Arch. Appl. Sci. Res.*, **3**(4): 63-77.

Kumari, K., Phogat, V. K., 2012. Solubilisation of Low- grade Rock Phosphate by Spentwash and its effect on Germination, Yield and P uptake by sorghum as Fodder Crop. J. Indian Soc. Soil Sci. **60**(3): 244-248.

Madhu Thawani, Priti Garg, 2013. Waste water management - A study, Dynamics of Public Administration, Vol. 30, No. 2, pp. 137-146.

Mallikarjun, T. M., 2010. Effect of spentwash and tillage on growth, yield and juice quality of sweet sorghum (*Sorghum bicolour* (L.) Moench). Department of Agronomy, College of Agriculture, Dharwad. University of Agril. Sci., Dharwad-580005. Karnataka, India.

Martinez, S., Segura, M. L., Suay, R., Moreno, J., 2013. Reuse of tertiary municipal wastewater effluent for irrigation of *Cucumismelo* L. *Irrig. Sci.* **3**:661–672.

Milano, M., Ruelland, D., Fernandez, S., Dezetter, A., Fabre, J., Servat, E., 2012. Facing climatic and anthropogenic changes in the Mediterranean basin: what will be the medium-term impact on water stress, C. R. Geosci. 344, 432–440.

Muhammad, T. M., Muhammad, Y. K., Mushtaq, A. B., Taj, M. J., 2012. Effect of spentwash of ethanol industry on groundwater: A case study of Rahumyar Khan district, Pak. *J. Envion. Sci. Water Resour.* **4**:85-94.

Nhapi, I., Gijzen, H.J., 2004. Wastewater management in Zimbabwe in the context of total water withdrawal for agriculture. Available from http://www.fao.org/ nr/water/aquastat/globalmaps.

Radha, J. and Srivastava, S., 2012. Nutrient composition of spentwash and its impact on sugarcane growth and biochemical attributes. *Physiol. Mol. Biol. Plants*, **18**(1):95–99.

Rao, KVGK, Kumbhare, P. S. Kamra, S. K., 1988. Studies on performance of subsurface drainage system – a case study of Sampla. Annual Report, Central Soil Salinity Research Institute, Karnal, pp. 64 – 65.

Raschid-Sally, Liqa and Priyantha Jayakody, 2008. Drivers and characteristics of wastewater agriculture in developing countries: Results from a global assessment. Colombo, Sri Lanka: International Water Management Institute. 35p. (IWMI Research Report 127).

Rattan, R. K., Datta, S. P., Chhonkar, P. K., Suribabu, K., Singh, A. K., 2005. Long-term impact of irrigation with sewage effluents on heavy metal content in soils, crops and groundwater—a case study. Agriculture, Ecosystems and Environment 109: 10–322.

Rhoades, J. D., Bingham, T. P., Letey, J., Hoffman, G. J., Dedrick, A. R., Pinter, P.J., Replogle, J. A., 1989. Use of saline drainage water for irrigation: Imperial Valley study. Agric. Water Mgmt. **16**:25 – 36.

Rijsberman, F., 2004. Sanitation and access to clean water. In: Lomborg B (ed) Global crises, global solutions. Cambridge University Press, London.

Rubina, 2013. Response of maize to long-term biomethanated spentwash application under vertisols of Northern Transition Zone of Karnataka. *M.Sc. (Agri.) Thesis*, Univ. Agric. Sci., Dharwad, Karnataka (India).

Sajbrt, V., Rosol, M, Dip, 2010. A comparison of distillery stillage disposal methods, Acta Poletechnica. **50**:2.

Salman Zafa, 2014. Wastes Generation in Tanneries, *Environment, Industrial Waste Management*. p. 1-3.

Selvamurugan, M., Ramkumar, V. R., Doraisamy, P. and Maheswari, M., 2013. Effect of biomethanated distillery spentwash and bio-compost application on soil quality and crop productivity. *Asian. J. Sci. Tech.*, **10**: 124-129.

Sharma, A., 2014. Effect of spentwash and chemical fertilizer on yield and nutrient uptake by sugarcane. *Annals of Plant and Soil Res.*, **16**(1): 32-34.

Sharma, D. P, Tyagi, N. K., 2004. On-farm management of saline drainage water in arid and semi-arid regions. Irrig. Drain **53**:87 – 103.

Sharma, D. P., Singh, K. N. and Kumbhare, P. S., 2005. Response of sunflower to conjunctive use of saline drainage water and non-saline canal water irrigation. Archives of Agronomy and Soil Science; **51**(1): 91 – 100.

Shenbagavalli, S., Mahimairaja, S. and Kalaiselvi, P., 2011. Impact of biomethanated distillery spentwash application on soil and water quality: A field appraisal. *Intl. J. Envi. Sci.*, **1**(7): 1753-1759.

Shivaraj, S. Kambale, 2014, Effect of distillery spentwash use on growth and yield of crops and soil properties. *M.Sc. (Agri.) Thesis*, Univ. Agric. Sci., Dharwad, Karnataka (India).

Shuval, H. I., Adin, A., Fattal, B., Rawitz, E. and Yekutiel, P., 1986. Wastewater irrigation in developing countries: health effects and technical solutions. Technical Paper No. 51. The World Bank, Washington, D.C.

Singh, A. P., Brar, M. S., Arora, C. L., 2001, Effect of Tannery Complex Effluents on the Composition of Raw Sewage Water for Irrigation of Crops. Journal of Research. **38:**3-4.

Singh, A. K., 2003. Water resources and their availability. In: Souvenir, National Symposium on Emerging Trends in Agricultural Physics, Indian Society of Agrophysics, New Delhi, 22-24, pp. 18–29.

Soobadar, A., Rene N, Keekwong, K. F., 2012. Impact of high rates of vinasse on some pertinent soil characteristics and on sugarcane yield in Mauritius. J. Sustain. Agric. **36**(1):36-53.

Tripathi, D. M., Tripathi, S. and Tripathi, B. D., 2011. Implications of secondary treated distillery effluent irrigation on soil cellulose and urease activities. *J. Environ. Prot.*, **2**:655-661. .

World Health Organisation, 2006. WHO Guidelines for the Safe Use of waste water Excreta and Grey water.

World Water Assessment Programme (WWAP) (2012). The United Nations World Water sustainability. Water Policy :501–517.

Yadav, R. K., Goyal, B., Sharma, R. K., Dubey, S. K. and Minhas, P. S., 2002. Post-irrigation impact of domestic sewage effluent on composition of soils, crops and ground water—A case study. Environment International, **28**:481– 486.

Yagoubi, M., Foutlane, A., Bourchich, J., Jellal, Wittland, C. and Yachioui, M. El. 2000. Study on the performance of the wastewater L. stabilisation pond of Boujaad, Morocco, Aqua (Oxford). **49(4):**203-209.

41

Farmers and Agriculture: A Key for Climate Change Adaptation

K. C. Patel, H. Y. Maheta and G. V. Prajapati

Abstract

Agricultural activity has always included adaptation to a number of diverse stresses and opportunities–elements that continue influencing farmers and agri-food sector. Farmers around the world already bear the consequences of climate change – some crops will not be resistant to the changes brought about by climate change; and harvests will wane. Climate and weather conditions are a good example of factors that require on-going adaptation. With climate change they take on even more significance. Based on data from agriculture producers, this article identifies several climate and weather risk adaptation strategies currently in use and notes their close links to sustainable agriculture practices. The article concludes that the mutually supportive relationship between farmers, agriculture and climate change adaptation could be used to justify more government support for sustainable agriculture policies and programmes.

Key words: *Organic agriculture, Rural Livelihood, Climate Change.*

Introduction

In the world total 50 per cent of farmers in the developing country - rural population were smallholders (farming three hectares or less of crop land) and they are generally much poorer than the rest of the population in these countries. Studies from developing countries show that farmers who adopt organic agriculture earn higher income and achieve better living standards. As a result, new opportunities for global partnerships that aim to improve livelihood through safe-food production in developing countries are rapidly emerging. In India more than 60 per cent of arable land is under traditional agriculture, where no synthetic inputs are being used, while only 0.3 per cent of agricultural land coverage is under organic agriculture (Organic India onesource, 2009). Increase in organic agricultural land will enables Indian farmers to achieve household food security and gain modest incomes, while regenerating the land, enhancing biodiversity and supplying quality food to local communities. By working with natural processes and making use of locally available assets, poor smallholder farmers can build up the fertility, productivity and makes soil less susceptible to erosion of their farms and avoid dependence on expensive external inputs likes fertilisers, pesticide, use of fossil fuel etc. in agriculture which contributes 10-12 per cent of total green house gas emission in the world. As an engine for sustainable development, organic agriculture has a role to play in the areas to improve livelihood and mitigate climate change.

Organic agriculture in India

Organic farming can be seen as an approach to agriculture where the aim is to create integrated, humane, environmentally and economically sustainable agricultural production systems. The term 'organic' is best thought of as referring not to the type of inputs used, but to the concept of the farm as an organism, in which all the components, the soil minerals, organic matter, microorganisms, insects, plants, animals and humans, interact to create a coherent, self-regulating and stable whole. Reliance on external inputs, whether chemical or organic, is reduced as far as possible (Lampkin *et.al.,* 1999). Organic farming is a method of crop and livestock production that involves much more than choosing not to use certain pesticides, fertilizers, genetically modified organisms, antibiotics and growth hormones that are not permitted by organic standards. In developing countries, certified organic agricultural production is limited. However, significant shares of agricultural land are under traditional or alternative production methods. Such areas could be converted to certified agriculture relatively easily, provided that certification costs can be kept low (Vossenaar and Jha, 2007). In India traditional agriculture farming was the main structure of farming communities for centuries looks a lot like organic agriculture in many ways because more than 60 per cent of India's arable land is under traditional agriculture, where no synthetic inputs are being used. Although, all has so far not defined as organic products but the products grown under such systems mean they are genuine organic products. There are many areas in India where crop residues, manures, legumes and neem are being used to grow their crops. They rely on crop rotation and inter-planting, to do their job. These farmers practice, farming through organic inputs.

At present 3.4 lakh ha (1.14 lakh farmers) certified organic agricultural land in the country and only 31 products and 19456 M.T. of quantity is exported from India (APEDA, 2007). Despite the promising potential and opportunities for Indian organic producers and exporters, the country faces a number of challenges. The major challenge is the acknowledgement of the competence of non-international certification bodies. Countries have a diverse range of approaches to organic agriculture owing to cultural differences, practical innovations and different approaches in advocating standards of production and certification on the basis of ecological principles. This has led to limited recognition of each other's standards, resulting in de facto trade barriers. To overcome these barriers and to promote environment friendly organic agriculture benefiting farmers and consumers across the globe, countries may need to come together to bring about some harmonisation even in diverse situations, to the extent feasible. India would like to support initiatives in this direction. Perhaps, one way could be to set up an inter-governmental forum to take this forward. In any case, since the agro climatic conditions in most of the Asian countries is quite similar, it would be worth considering working on uniform standards within the region with the view to facilitating trade in organic products.

Studies from developing countries in Asia show that small farmers who adopt organic agriculture earn higher income and achieve better living standards. In addition, organic food trade is now the most rapidly growing food sector worldwide due to mounting concerns surrounding food safety and quality. As a result, new opportunities for global partnerships that aim to reduce poverty through safe-food production in developing countries are rapidly emerging. As an engine for sustainable development, organic agriculture has a role to play in the areas of poverty alleviation, human development, gender equality and environmental protection (Vossenaar and Jha, 2007).

Organic operation is labour intensive; therefore organic agriculture can improve the employment especially for woman compared with the conventional production system. In Karnataka, India, for example, the demand for female labour for crops such as tea and spices has increased by percent. Organic agriculture will enables Indian smallholders to achieve household food security and gain modest incomes while regenerating the land, enhancing biodiversity and supplying quality food to local communities. Benefits have been demonstrated by diversified and integrated organic farm systems that are economically viable, environmentally sound, the flexibility to cope with adverse effects of climate change and variability, such as changed rainfall patterns and supportive of cultural traditions. This leads to higher economic and ecological stability through optimised ecological balance and risk-spreading. These environmental, economic and social objectives are essential to rural development in India. Study carried out on organic agriculture and rural livelihood in Karnataka, India reported that the conversion from conventional to organic agriculture had improved their livelihoods in a range of ways. They pointed out that over the long term the conversion had improved their net-farm incomes, reduced the risk of pesticide poisonings, lead to more self-sufficiency, improved food safety and reduced vulnerability and improved the access to networks supporting knowledge exchange and political participation. However, risk and uncertainty related to the conversion period, such as temporarily declining yields and the lack of experiences and information, were mentioned as major constraints preventing in particular asset-poor households from adopting organic farming (Lukas and Cahn, 2008).

Doubts have been frequently expressed about the capacity of organic agriculture to produce as much as conventional agriculture. Recent research has however shown that organic agriculture, particularly in developing countries and arid regions, can have considerably higher yields than conventional agriculture (Badgley *et. al.,* 2007; Pretty *et. al.,* 2006), and it is acknowledged as being able to contribute to food security (FAO, 2007). In particular, organic agriculture performs better than conventional agriculture under water scarcity (Badgley *et. al.,* 2007).

Organic agriculture as a strategy does not hinge on large additional financing for the organic agriculture farming system itself. (Additional costs come from extension services, the general provision of information, and, if certified, certification costs). However, it is crucial to have access to international markets and to develop local markets for the products. In the transition phase to organic agriculture, additional financing for the farms may be necessary: training and extension services need to be provided and lower yields for the two to three years of the transition period may necessitate some additional support. It is sensible, then, to emphasise knowledge transfer and infrastructure building (including access to markets, etc.), rather than direct monetary transfers only, although such may be necessary in certain cases. The economic viability of organic farming is also likely to increase with increasing energy prices, due to low-risk farming strategy with reduced input costs and, therefore, lower risks with partial or total crop failure due to extreme weather events or changed conditions in the wake of climate change and variability (El-Hage Scialabba and Hattam, 2002; Eyhorn, 2007). As such, it is a viable alternative for poor farmers (which makes conventional farming more expensive, due to the energy costs for production of fertilisers and pesticides) and with decreasing levels of subsidies for conventional agriculture. Several options to meet the financial requirements exist in principle. Examples are governmental support and research programmes for agriculture, microfinance strategies, biodiversity conservation initiatives (Carroll *et. al.,* 2007), and, for Organic agriculture as a mitigation strategy, the programmatic or sectoral Clean Development Mechanism (Hinostroza *et. al.,* 2007; Sutter, 2007; or Baron and Ellis, 2006).

Organic agriculture holds the key to rural transformation and sustainable development in India. There is an organic revolution brewing throughout the country. The urge to preserve biodiversity and improve farm production without the use of chemical inputs has encouraged farmers to become less dependent on unsustainable inputs. This approach is contributing towards rural livelihood opportunities.

Agriculture: Source and victim of climate change

The current change in global climate is a phenomenon that is largely due to the burning of fossil energy (coal, oil, natural gas) and to the mineralisation of organic matter as a result of land use. These processes have been caused by mankind's exploitation of fossil resources, clearing of natural vegetation and use of these soils for arable cropping. These activities have primarily led to a measurable increase in the carbon dioxide, methane and nitrous oxide content of the atmosphere, an increase which results in global warming. The global warming potential of methane and nitrous oxide is twenty and three hundred times greater that of CO_2 (International Trade Centre, 2007).

As per the Intergovernmental Panel on Climate Change (IPPC), the annual amount of greenhouse gases emitted by the agricultural sector is estimated at between 5.1 and 6.1 gigatonnes CO_2 equivalents[2] in 2005 (Barker *et. al.*, 2007). This represents approximately 10-12 per cent of total greenhouse gas emissions.

Livestock manure, nitrogenous fertilisers and irrigated paddy are said to be responsible for producing most agricultural nitrous oxide and methane emissions. Of these emissions, methane accounts for 3.3 Gt equivalents and nitrous oxide for 2.8 Gt CO_2 equivalents annually, while net emissions of CO_2, at only 0.04 Gt CO_2 equivalents per year, are small (Smith *et. al.,* 2007). While, in India agriculture sector contributes 28 per cent of the total GHG emissions from India (NATCOM, 2004). In a more recent estimate Bhatia *et al.* (2004) estimated for the base year 1994-95, that the CH_4 and N_2O emissions from Indian agricultural fields were 2.9 Tg (61 Tg CO_2 equivalent) and 0.08 Tg (39 Tg CO_2 equivalent), respectively.

Mitigation of climate change through organic agriculture

Organic agriculture not only enables ecosystems to better adjust to the effects of climate change but also offers potential to reduce the emissions of agricultural greenhouse gases by complements emission reduction efforts with its major sequestration potential, which is based on the intensive humus production (requiring CO_2) of the fertile soils. In comparison to conventional agriculture, organic agriculture also directly contributes to emission reductions as it emits less N_2O from nitrogen application (due to lower nitrogen input), less N_2O and CH_4 from biomass waste burning (as burning is avoided), and requires less energy, mainly due to zero chemical fertiliser use.

Reduction of nitrous oxide

The global warming potential of conventional agriculture is strongly affected by the use of synthetic nitrogen fertilisers and by high nitrogen concentrations in soils. one per cent of global fossil energy consumption is used for chemical nitrogen fertiliser production, emitting about 0.23 Gt CO_2eq.

Organic agriculture avoids these emissions, as no chemical nitrogen fertilisers are used. Furthermore, due to lower utilisation of nitrogen input in organic agriculture than in conventional agriculture leads to a potential emission reduction of 1.2-1.6 Gt CO_2eq (Muller *et. al.*, 2009).

In organic agriculture, nitrogen input stems from application of manure and compost is fixed from the air by leguminous plants. Badgley *et.al.* (2007) calculated the potential nitrogen production by leguminous plants via intercropping and off-season cropping to be 154 million tonnes, a potential which exceeds the nitrogen production from fossil fuel by far and which is not fully exploited by conventional farming techniques.

Reduction of Methane

Methane accounts for about 14 per cent of the greenhouse gas emissions (Barker *et. al.*, 2007). Two-thirds of this is of anthropogenic origin and mainly from agriculture. Methane emissions stem to a large extent from enteric fermentation and manure management and in consequence are directly proportional to livestock numbers. Avoidance of methane emissions of anthropogenic origin and especially of agricultural origin is of particular importance for mitigation. Organic agriculture has an important, though not always superior, impact on reduction as livestock numbers are limited in organic farms (Weiske *et. al.*, 2006; Kotschi and Muller-Samann, 2004). The data available on methane emissions from livestock is limited, especially with respect to the reduction of GHG emissions from ruminants and manure heaps.

Methane emissions from organic rice production and ruminant production do not differ substantially from those of conventional production. Better rice production practices in organic and conventional agriculture, such as avoiding continuous flooding or choosing low methane emitting varieties (Smith and Conen, 2004) could enhance reduction of methane emissions.

Kotschi and Muller-Samann (2004) suggested that composting and biogas production reduced the methane as well as longevity of animals on organic farms contributes to reduction of methane emission. In organic agriculture, biomass is not burned which leads to reduce the CH_4 and N_2O emissions by about 0.6-0.7 Gt CO_2eq in comparison to conventional agriculture (Smith *et. al.*, 2007).

Carbon sequestration

By sequestering carbon dioxide in the soil, agriculture may contribute to the carbon cycle in a positive way. Whether the balance is positive or negative depends to a large extent on the cutoff points for the comparison (for instance if the fertiliser industry is included or not) as well as the management and farming practices applied.

Soil carbon sequestration is enhanced through agricultural management practices such as increased application of organic manures, use of intercrops, green manures, higher shares of perennial grasslands and trees or hedges etc., which promote greater soil organic matter (and thus soil organic carbon) content and improve soil structure (Niggli *et. al.,* 2008; IFOAM, 2006). Increasing soil organic carbon in agricultural systems has also been pointed out as an important mitigation option by IPCC (2007b). Very rough estimates for the global mitigation potential of organic agriculture amount to 3.5–4.8 Gt CO_2 from carbon sequestration (around 55–80 per cent of total global greenhouse gas emissions from agriculture) and a reduction of N_2O by two-thirds (Niggli *et. al.*, 2008). For sound estimates, however, more information on the mitigation potential of organic agriculture duly differentiated according to climatic zones, local climatic conditions, soil characteristics, variations in crops and cultivation practices, etc. More research is also needed on how organic agriculture performs regarding emissions per unit product.

Concluding Remarks

Organic agriculture as a strategy holds the key to rural transformation and sustainable development in India. The urge to preserve biodiversity and improve farm production without the use of chemical inputs has encouraged farmers to become less dependent on unsustainable inputs. In India agriculture farming looks a lot like organic agriculture due to no synthetic inputs are being used and benefits have been demonstrated by diversified and integrated organic farm systems that are economically viable, environmentally sound, the flexibility to cope with adverse effects of climate change. This leads to higher economic and ecological stability through optimised ecological balance and risk-spreading. Further research on this, however, still needed. Organic agriculture is labour intensive so it improve the employment and small farmers can earn higher income and achieve better living. But the major constraints restricted the small farmers to fetch higher profit are acknowledgement of the competence of non-international certification bodies, domestic market and training and education of organic agriculture. Organic agriculture, as a mitigation strategy enables ecosystems to better adjust to the effects of climate change and offers potential to reduce the emissions of agricultural greenhouse gases (CO_2, CH_4 and N_2O) with its major sequestration potential, less use of fossil fuel and chemical fertilisers. More research on emissions per unit product in organic agriculture is still needed.

References

APEDA, 2007. Present status of organic agriculture in India. Available at <http://www.apeda.com/organic/PresentStatus.htm>

Badgley, C., J. Moghtader, E. Quintero, E. Zakem, M. Chappell, K. Avilez-Vazquez, A. Samulon, and I. Perfecto, 2007. Organic agriculture and the global food supply. Renewable Agriculture and Food Systems, 22(2): 86–108.

Barker, T., I. Bashmakov, L. Bernstein, J. E. Bogner, P. R. Bosch, R. Dave, O. R. Davidson, B. S. Fisher, S. Gupta, K. Halsnaes, G.J. Heij, S. Kahn Ribeiro, S. Kobayashi, M. D. Levine, D. L. Martino, O. Masera, B. Metz, L. A. Meyer, G. J. Nabuurs, A. Najam, N. Nakicenovic, H. H. Rogner, J. Roy, J. Sathaye, R. Schock, P. Shukla, R. E. H. Sims, P. Smith, D. A. Tirpak, D. Urge-Vorsatz, D. Zhou, 2007. Technical Summary. In: Climate Change 2007: Mitigation. Contribution of working group III to the fourth assessment report of the Intergovernmental Panel on Climate Change [B. Metz, O. R. Davidson, P. R. Bosch, R. Dave, L. A. Meyer (eds)], Cambridge University Press, Cambridge, United Kingdom and New York, NY, USA. Available at <http://www.nl/ipcc/pages_media/Fardocs/final_pdfs_ar4/TS.pdf>

Baron, R., and J. Ellis, 2006. Sectoral crediting mechanisms for greenhouse gas mitigation: institutional and operational issues. Organisation for Economic Cooperation and Development and International Energy Agency, COM/ENV/EPOC/IEA/SLT(2006)4. Paris: OECD and IEA. Available at <http://www.iea.org/textbase/papers/2006/greenhouse.pdf.>

Bhatia, A., H. Parikh, and P. K. Aggarwal, 2004. Inventory of methane and nitrous oxide emissions from agricultural soils of India and their global warming potential. Current Science, 87 : 317-324.

Carroll, N., J. Fox, and R. Bayon, 2007. Conservation and biodiversity banking: a guide to setting up and running biodiversity credit trading systems. London: Earthscan.

El-Hage Scialabba, N., and C. Hattam, 2002 "Organic agriculture, environment and food security." Environment and Natural Resources Service, Sustainable Development Department, Food and Agriculture Organisation of the United Nations (FAO). Available at <http://www.fao.org/docrep/005 /y4137e /y4137e00.htm>

Eyhorn, F. 2007. Organic farming for sustainable livelihoods in developing countries: the case of cotton in India. PhD diss. Department of Philosophy and Science, University of Bonn. Available at <http://www.zb.unibe.ch/download /eldiss/06eyhorn_f.pdf.>

FAO (Food and Agriculture Organisation), 2007. International conference on organic agriculture and food security, Rome, May 3–5, 2007. Available at <http://www.fao.org/organicag/ofs/index_en.htm.>

Hinostroza, M.L., C-C. Cheng, X. Zhu, J. Fenhann, C. Figueres, F. Avendano, C. Figueres and F. Avendano, 2007. "Potentials and barriers for end-use energy efficiency under programmatic CDM." CD4CDM Working Paper Series 3. Roskilde, Denmark: UNEP Risoe Center. Available at <http://cd4cdm.org / Publications/pCDM&EE.pdf.>

IFOAM (International Federation for Organic Agriculture Movements), 2006. Mitigation and adaptation to climate change: the role of organic agriculture. IFOAM side event to the 24th session of the subsidiary bodies, Bonn, Germany, May 24, 2006.

International Trade Centre, 2007. Organic farming and climate change. Available at <http://www.ifoam.org/...organic/.../FiBLStudyOrganic_ Farming_and_ *Climate_ Change.pdf*>

IPCC, 2007b. Summary for policy makers. In IPCC Fourth Assessment Report, "Working Group III Report: Mitigation of Climate Change". Available at < http://www.ipcc.ch/pdf/assessment-report/ar4/wg3/ar4-wg3-spm.pdf and http://www.ipcc.ch/ipccreports/ar4-wg3.htm.>

Kotschi, J. and K. Muller-Samann, 2004. The role of organic agriculture in mitigating climate change. International Federation of Organic Agriculture Movements (IFOAM), Bonn.

Lampkin, N., C. Foster, S. Padel, and P. Midmore, 1999. The policy and regulatory environment for organic farming in Europe. In Organic Farming in Europe: Economics and Policy, Vol. 2, Stuttgart- Hohenheim University.

Lukas, M., and M. Cahn, 2008. Organic agriculture and rural livelihoods in Karnataka, India. 16th IFOAM Organic World Congress, Modena, Italy, June 16-20, 2008. Available at < http://orgprints.org/view/projects/conference.html>

Muller, A., J.S. Davis, P. R. Happerly, and U. Niggli, 2009. Reducing global warming : the potential of organic agriculture - a policy brief. Available at <http://*mailman.anu.edu.au/pipermail/hepr-vn/.../attachment-0001.pdf*>

NATCOM, 2004. India's initial National communication to the United Nations framework convention on climate change. National communication project, ministry of environment and forest, Govt. of India.

Niggli, U., A. Fliessbach, P. Hepperly, and N. Scialabba, 2008. Low greenhouse gas agriculture: mitigation and adaptation potential of sustainable farming systems. Rome: FAO.

Organic India onesource, 2009. Organic agricultural scenario in India. Available at <http://www.organic-india-onesource.50megs.com/about.html>

Pretty, J.N., A.D. Noble, D. Bossio, J. Dixon, R.E. Hine, F. Penning de Vries, and J. Morison. 2006. Resource-conserving agriculture increases yields in developing countries. Environmental Science and Technology 40(4): 1114–19.

Smith, K. A., and F. Conen, 2004. Impacts of land management on fluxes of trace greenhouse gases. Soil Use and Management, 20: 255-263.

Smith, P., D. Martino, Z. Cai, D. Gwary, H. Janzen, P. Kumar, B. McCarl, S. Ogle, F. O'Mara, C. Rice, B. Scholes, and O. Sirotenko, 2007. Agriculture in climate change, 2007: mitigation. Contribution of Working Group III to the Fourth Assessment Report of the Intergovernmental Panel on Climate Change [B. Metz, O.R. Davidson, P.R. Bosch, R. Dave, L.A. Meyer (eds)], Cambridge University Press, Cambridge, UK, and New York, NY, USA. Available at <http://www.ipcc.ch/pdf/assessment- report/ar4/wg3/ar4-wg3-chapter8.pdf>

Sutter, C., 2007. Will It Fly?. Carbon Finance (July 2007): 16. Available at <http://www.carbonfinanceonline>

Vossenaar, R., and V. Jha, 2007. Trading opportunities for organic food products from developing countries. United Nations.

Weiske, A., A. Vabitsch, J.E. Olesen, K. Schelde, J. Michel, R. Friedrich, and M. Kaltschmitt, 2006. Mitigation of green house gas emission in European conventional and organic dairy farming. Agriculture, Ecosystems and Environment, 112: 221-232.

42

Mitigating Climate Change Through Off-Seasonally Raising Seedling of Papaya (*Carica Papaya* L.) In Protected Environment

Joshi Viral P. and P. M. Chauhan

Abstract

An experimental study on off-season raising seedling of papaya (Carica papaya L.) in protected environment was conducted. Major difficulty of raising papaya seedling is mortality of seedling, among abiotic factors of environment like unfavourable microclimate; especially temperature restricts raising off-season fruit nursery under open field condition in arid region. The Low Tunnel Greenhouse (S_1) and Black Shade net house (S_2) were taken as different structures. Papaya Seedling are fetching very high price during off-season availability. The daily environmental data were recorded and computed mean monthly average for the analysis. The response of papaya seedling was evaluated in terms of germination percentage, seedling height, collar diameter, no. of leaves, diameter of shoot, fresh and dry weight of above ground underground biomass, sturdiness, root shoot ratio, vigour index length, vigour index mass in root media. The temperature during the entire study period was found lower in black shade net structure compared to ambient. The minimum time taken four days of germination, germination percentage, and seedling height, collar and shoot diameter was found under the treatment S_2 with Variety i.e. Madhubindu and Soil Media (S_2VM). The highest germination percentage was observed for S_2VM, higher number of leaves, sturdiness, root-shoot ratio, vigour index length and vigour index mass was found under treatment S_2VM.

Key words:*Black Shade net house, Greenhouse, Environmental parameters, Plant parameters.*

Introduction

Agriculture is the riskiest profession in the world, since natural factors like temperature, precipitation, hail and thunder storms influence crop yield and thereby the economic fate of the farmer. Indian agriculture has long been described as a gamble in monsoon. Major difficulty in open field condition is of a high temperature and low relative humidity, in such condition, if crop is protected from this unfavourable condition, more yields can be obtained.

The media should be reach enough to sustain seedling for about a year (Srivastava *et. al.* 1998). A good potting medium is characterised by light weight, friable, easy blendable, excellent water capacity, porosity, drainage, low bulk density, free from fungal spores and insects and low inherent fertility etc (Chakrabarti *et. al.* 1998).

Major difficulty of raising papaya seedling is mortality of seedling, which is a very serious problem. Therefore, greenhouse or shade net house is used to raise the papaya seedling. It is difficult to maintain the temperature particularly when outside temperature is low or high. Hence greenhouse and shade net house are supposed to be a better option for region like Saurashtra. Generally black shade net of 75 per cent is used for raising of Fruit Seedling. Light band is the important factor affecting the quality of the seedling. Therefore, in this study different structures have been used to find the effect of protected structure on microclimate and crop parameters. Microclimate, seedling quality and number of healthy seedlings per bag also affect the cost of cultivation hence, economics is also worked out.

Among abiotic factors of environment like unfavourable microclimate, especially temperature restricts raising off-season nursery under open field condition. Mostly the farmers are raising papaya seedling during their normal growing season by sowing of seeds and when such seedling are sold for marketing, the markets are flooded with these seedlings and sometime growers are not getting back their cost of production. But the same seedling are fetching very high price during off-season availability.

Considering the popularity of an area under papaya and its potential for capturing markets, but very few study has done on raising papaya seedling in protected environment and it is priority to take up research work on this important fruit crop in Saurashtra region.

Materials and Methods

Study Area

The experiment conducted at Instructional Farm of Department of Soil and

Water Engineering, CAET, JAU, Junagadh (21.5°N, 70.1°E) in Saurashtra region of Gujarat.

Details of Structures

The one single span low tunnel naturally ventilated greenhouse of (6.08 m x 30.4 m x 2.4 m) structure was used for the experiment purpose. Structure was made of 1.0" diameter G.I pipe and 200 micron UVS diffused light polyethylene sheet of good quality to be fixed appropriately over all sides of the greenhouse. The Black shade net house was constructed by using 1.0" diameter G.I pipe with 75 per cent black shade net covering all sides of the structure. The care was taken that shadow of one structure do not fall on other structure as well as wind do not interrupted by one structure for other. The construction detail of greenhouse and shade net structure is shown in Table 3.1, and Fig. 3.1.

Table: 1. Details of different structures

Details	*Low tunnel Greenhouse*	*Black shade Net house*
Type	Single Span low tunnel	Black shade net 75% shading
Floor Area	6.08 m x 30.4 m	3 m x 3 m
Ridge height	2.4 m	2.4 m
Ventilation	Natural ventilation and fogging	Natural ventilation
Orientation	North-South	North-South
Cooling mechanism	Natural ventilation and fogging	Natural ventilation
Covering	200 micron UVS diffused light polyethylene sheet of good quality to be fixed appropriately over all sides of the greenhouse.	

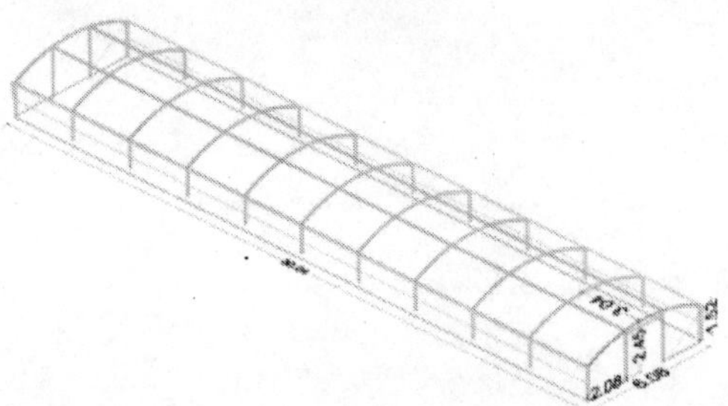

Fig. 1. Constructional details and isometric view of greenhouse.

Fig. 2. Low tunnel Greenhouse.

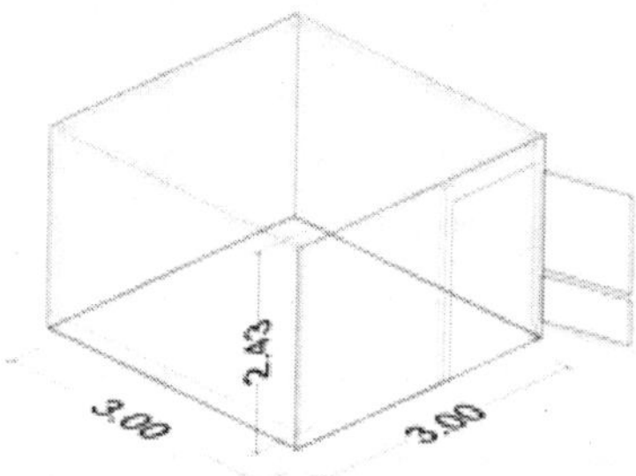

Fig. 3. Constructional details and isometric view of black shed net house.

Fig. 4. Black shed net house.

Experimental Details

Parameter considered

(A) Structure (S) with two levels viz.

(1) Low tunnel greenhouse (S1)

(2) Black shade net house (S2)

(B) Variety: Madhubindu

(C) Media: Soil (Soil + Sand + FYM)

Observations Recorded

Environmental observations

The environmental parameters like temperature, relative humidity, solar radiation, were measured inside and outside of greenhouse, net house daily. These data were used to calculate the monthly average data, which were used for observing variation in microclimate during the day.

Temperature

For the measurement of temperature digital thermometer was used. The observations were recorded daily at two hours intervals from 8:00, to 18:00 hrs.

Relative humidity

For the measurement of relative humidity, digital thermo-hygrometer was used. The observations were recorded daily at two hours intervals from 8:00 to 18:00 hrs.

Solar radiation

Solar radiation was measured using digital Solarimeter. The observations were recorded daily at two hours intervals from 8:00 to 18:00 hrs.

Morphological observations

Germination percentage

Total number of germination seeds under each treatment will count daily, right from the first emergence of the seed up to the period of completion of seed germination. The percentage of seed germination was calculated by No. of seed germinated / Total no. of seed taken for germination multiply by 100.

Seedling height

Height from all seedlings from each structure was measured in centimeter at 15, 30, 45 and 60 days after sowing (DAS) and mean height of seedling was calculated.

Collar diameter (cm)

The diameter was measured of the same seedling, which was used for measuring the seedling height in each structure. It was measured at the bottom portion of the stem with the help of vernier calipers at 15, 30, 45 and 60 days after sowing (DAS) and average girth was calculated.

Number of leaves

Number of leaves in the entire seedling in each structure was counted at 15, 30, 45 and 60 days after sowing (DAS) and mean number of leaves per seedling was calculated.

Diameter of shoot (cm)

The shoot diameter was measured of the same seedling, which was used for measuring the seedling height. It was measured at the top portion of the seedling with the help of measuring tape at 15, 30, 45 and 60 days after sowing (DAS) and average calculated.

Biomass Production

Fresh and dry weight of above ground biomass

The fresh weight of above ground biomass was recorded of the same seedling that was used for measuring the seedling height in each treatment and weights of samples biomass were calculated. Samples will oven dried at 70 °C till constant weight and dry weight of above ground biomass of seedling was recorded.

Fresh and dry weight of below ground biomass

The fresh and dry weight of below ground biomass was recorded of the same seedling that was used for measuring the seedling height in each treatment. Samples will oven dried at 70 °C till constant weight and dry weight of above ground biomass of seedling was recorded.

Quality parameters

Sturdiness

The sturdiness of plant was measured by using height to diameter ratio of seedling (Chauhan and Sharma, 1997).

Root shoot ratio

The root to shoot ratio of plant was measured by use of biomass of roots and biomass of shoots of seedling (Chauhan and Sharma, 1997).

Vigour index length (Vigour index I)

Vigour index was calculated as per procedure prescribed by Abdul-Baki and Anderson (1973), as follow:

Vigour index-I = Germination Per cent × Seedling length (cm)

Vigour index mass (Vigour index II)

Vigour index in terms of mass was determined by multiplication of germination percentage with seedling dry weight on the day of final count as per procedure prescribed by Abdul-Baki and Anderson (1973), as follow:

Vigour index- II = Germination Per cent × Seedling Dry Weight (g).

Results and Discussion

Net House Microclimate Temperature

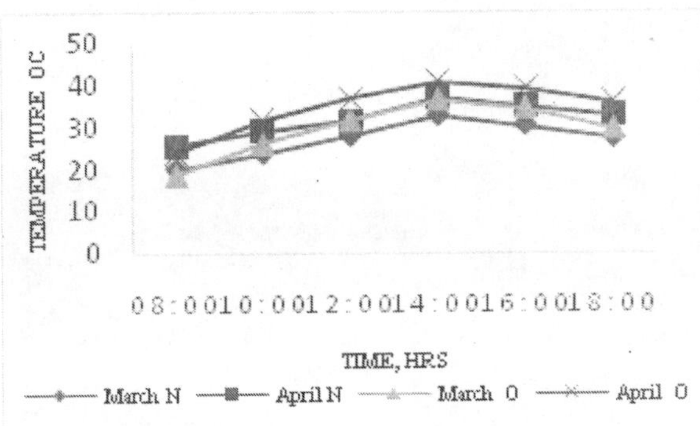

Fig. 1. Diurnal variation of temperature in shade net and open field.

The diurnal variations in monthly average temperature for March to April months for Black shed net house and open field are depicted in Fig. 1.

Relative humidity

The diurnal variations in monthly average RH for March to April months for Black shed net house and open field are depicted in Fig. 2.

Solar radiation

Due to 70 per cent black shed net house, the variations in monthly average solar radiation w/m^2 for March to April months for Black shed net house and open field are depicted in Fig. 3.

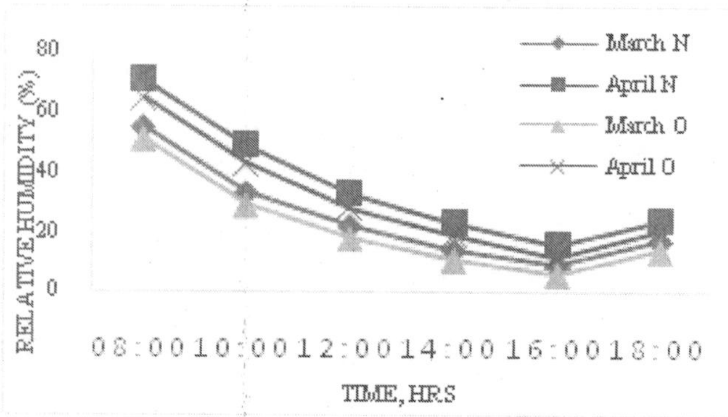

Fig. 2. Diurnal variation of RH in shade net and open field.

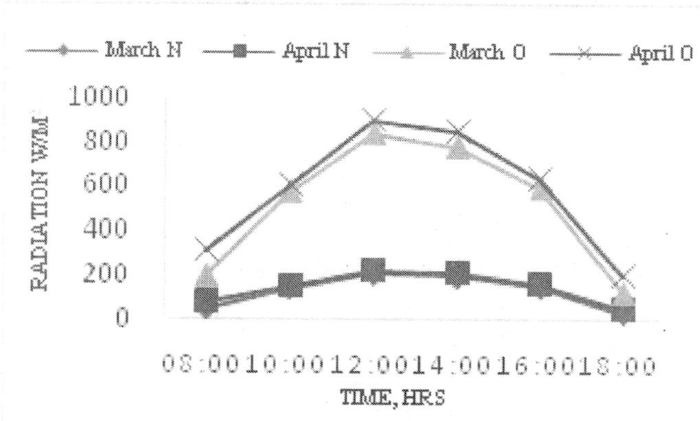

Fig. 3. Variation of solar radiation in shade net and open field.

Greenhouse Microclimate

Temperature

The diurnal variations in monthly average temperature for March to April months for greenhouse and open field are depicted in Fig. 4.

Relative humidity

The diurnal variations in monthly average RH for March to April months for greenhouse and open field are depicted in Fig. 5.

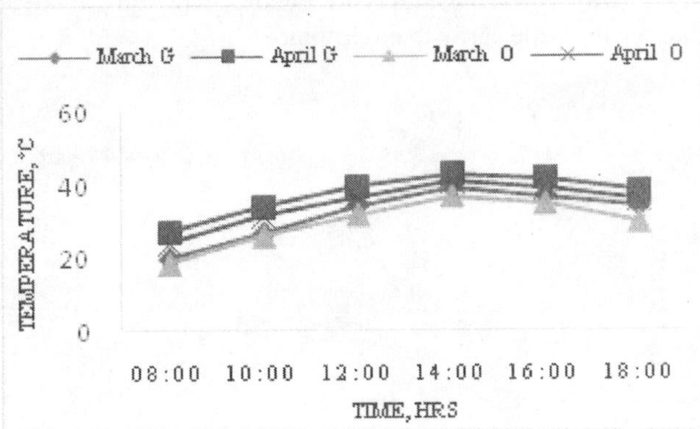

Fig. 4. Diurnal variation of temperature in greenhouse and open field.

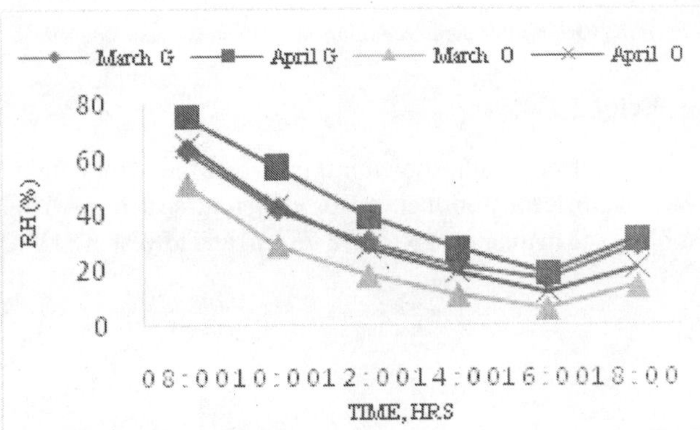

Fig. 5. Diurnal variation of RH in greenhouse and open field.

Solar radiation

Due to plastic covering of greenhouse, the variations in monthly average solar radiation for March to April months for greenhouse and open field are depicted in Fig. 6. From figure, it is revealed that as glazing material almost 50 per cent less radiation reach in greenhouse crop as compare to open field.

Morphological observations

Germination percentage

Germination percentage of papaya seedling was significantly influenced by different structures and media. The observation found that more germination was found in Treatment S1 (Greenhouse) (96.50) followed by S2 (Black shade

Net house) (95.38). The soil media is provided high mixture of organic matter, responsible for the better growth environment.

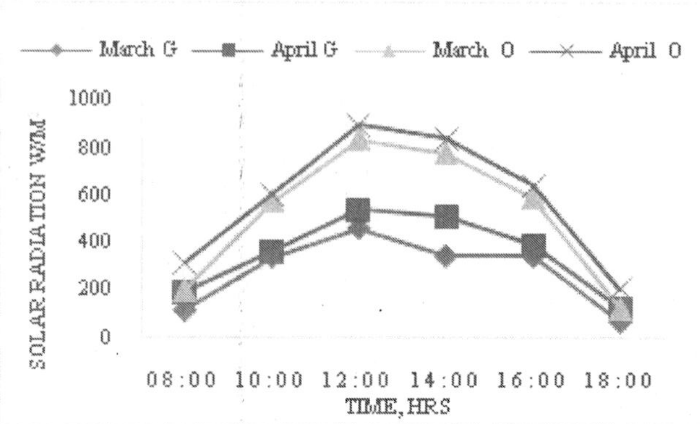

Fig. 6. Variation of solar radiation in greenhouse and open field.

Seedling height

It is apparent from graph shown in Fig. 7 that the plant height of papaya seedling was significantly influenced by different structures. Highest plant height was observed in treatment S2V (38.75 cm) and lowest in S1V (29.56 cm).

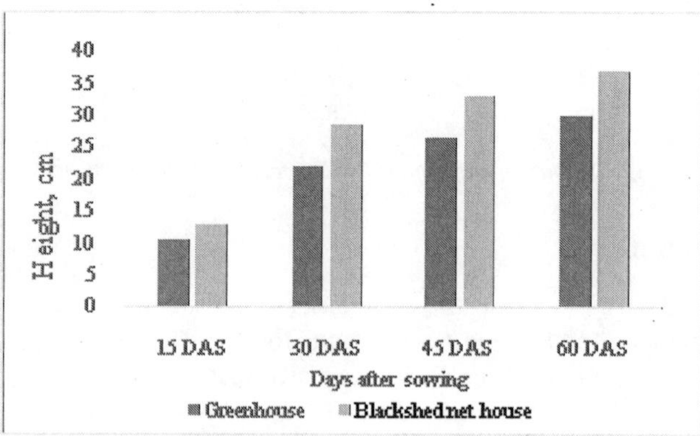

Fig. 7. Variation of seedling height in different structures.

Collar diameter (cm)

A perusal of data presented Fig. 8 indicated that the collar diameter of papaya seedling was significantly influenced by different structures. However, due to favourable environmental conditions highest collar diameter was observed in treatment S2 (6.06 cm) and lowest in S_1 (5.73 cm).

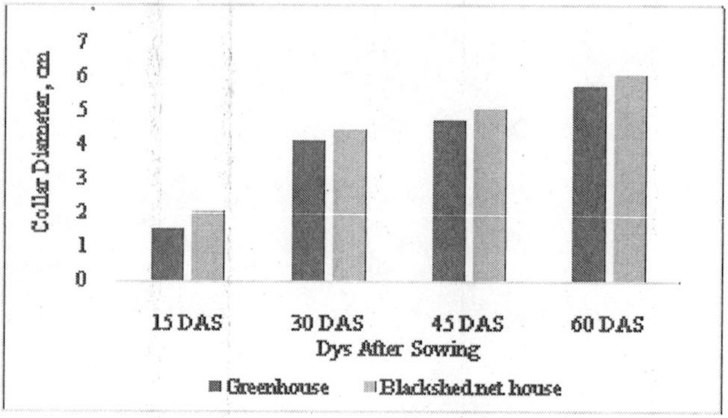

Fig. 8. Variation of collar diameter in different structures.

Number of leaves

It is apparent from graph shown in Fig. 9 that the number of leaves of papaya seedling was significantly influenced by different structures. Treatment S2 (Black shade Net house) gave significantly higher number of leaves compared to Treatment S1 (Greenhouse).

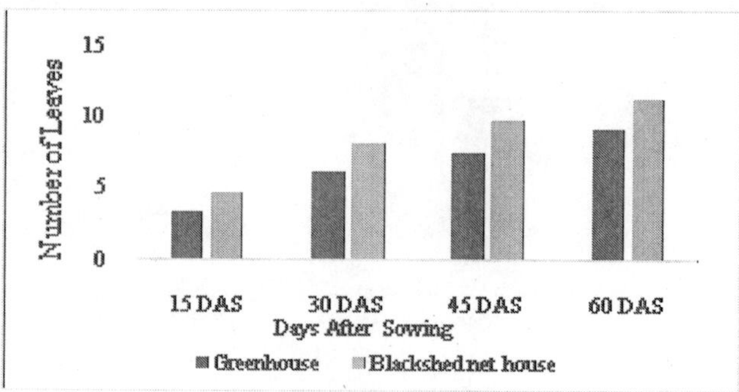

Fig. 9. Number of leaves of seedling in different structures.

Diameter of shoot (cm)

A perusal of data presented in Fig. 10 indicated that the shoot diameter of papaya seedling was significantly influenced by different structures and media. It is obvious that Black shade net house provides favourable condition for growth of seedling with media having mixture of soil, sand and FYM.

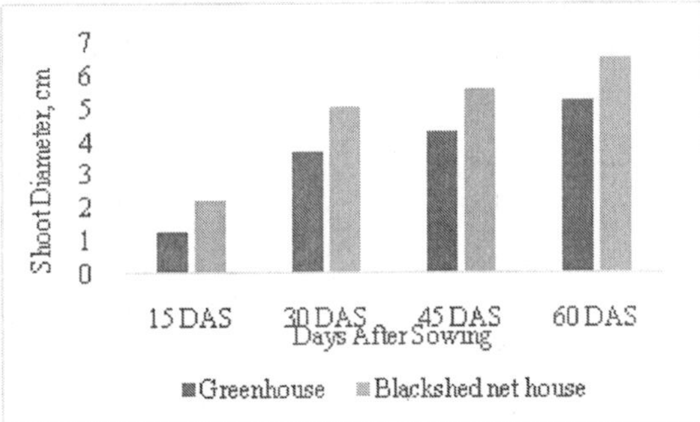

Fig. 10. Shoot diameter of seedling in different structures.

Biomass Production

Fresh and dry weight of above ground biomass

A perusal of data presented in Fig. 11 indicated that the fresh and dry weight of above ground biomass of papaya seedling was significantly influenced by different structures, and media. Treatment S2 (Black shade Net house) gave significantly higher fresh and dry weight of above ground biomass (21.89 g), and (4.16 g) followed by treatment S1 (Greenhouse)

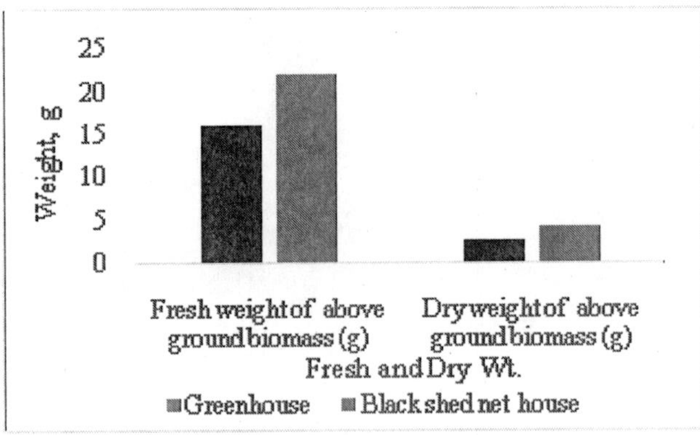

Fig. 11. Fresh and dry weight of above ground biomass.

Fresh and dry weight of below ground biomass

As data presented in Fig. 12 revealed that the fresh and dry weight of below ground biomass of papaya seedling was significantly influenced by different structures, and media. Treatment S2 (Black shade Net house) gave significantly higher fresh and dry weight of below ground biomass (1.56 g), and (0.86 g) followed by treatment S1 (Greenhouse)

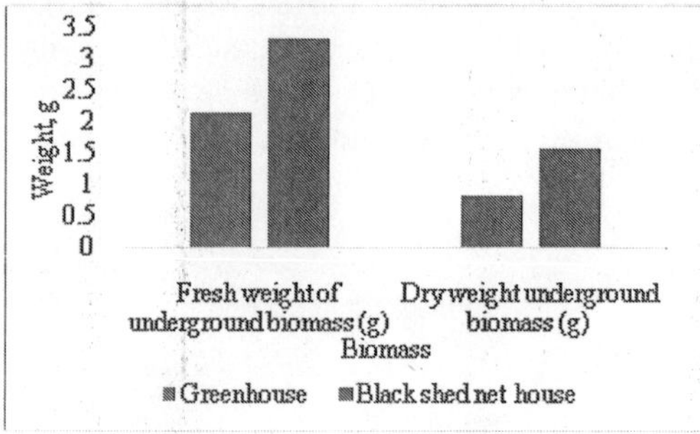

Fig. 12. Fresh and dry weight of below ground biomass.

Quality parameters

Sturdiness

It reflects the stocky or spindly nature of the seedlings. Although a good indicator of the ability to withstand physical damage in all stock-types. From Fig. 13 it revealed that Black shed net house is best in this.

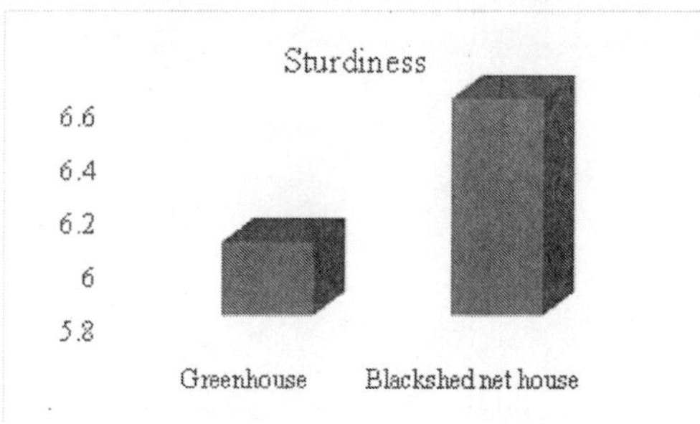

Fig. 13. Sturdiness of seedling in different structure.

Root shoot ratio

It is an indication of healthier plant. Here Treatment S2 indicates higher root-shoot ratio as compared to treatment S1. This result are in agreement with the findings of Rathore *et. al.* (2004) that more sturdiness and root-shoot ratio was noted in Casuarina equisetifolia seedling with higher amount of soil and FYM in the media mixture.

Vigour index length (Vigour index I)

It is a growth indication of healthier plant. Treatment S2 (4777.7) indicates higher root-shoot ratio as compared to treatment S1 (3934.6). These proprieties were very significantly with the effect of treatments, evident from present investigation. It is obvious that Black shade net house provides favourable condition for growth of seedling with media having mixture of soil, sand and FYM.

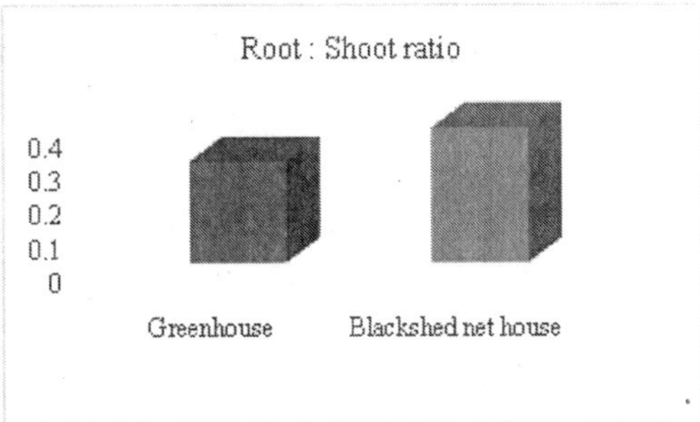

Fig. 14. Root-shoot ratio of seedling in different structures.

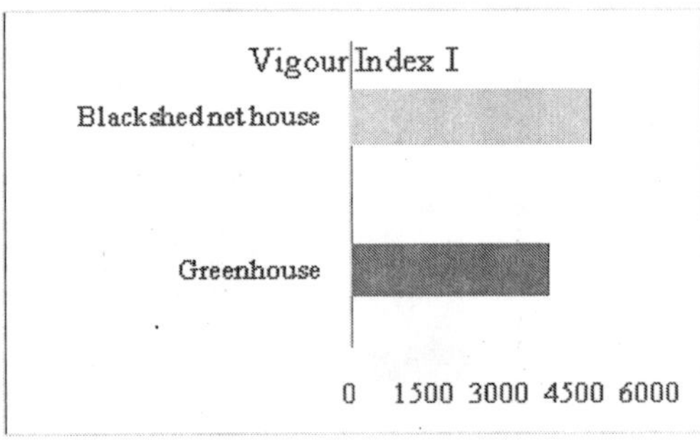

Fig. 15. Vigour index length of seedling for different structures.

Vigour index mass (Vigour index II)

It is also a growth indication of healthier plant. As discuss in material and methods it is measured accordingly. Treatment S2 (439.36) indicates higher root-shoot ratio as compared to treatment S1 (336.16). These proprieties were very significantly with the effect of treatments, evident from present investigation. It is obvious that Black shade net house provides favourable condition for growth of seedling with media having mixture of soil, sand and FYM.

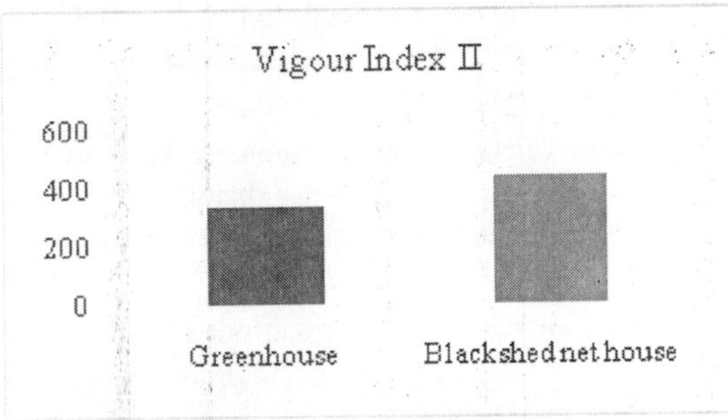

Fig. 16. Vigour index mass of seedling for different structures.

Summary and Conclusions

Protected cultivation is the concept of growing potential crops in the modified natural environment for ensuring optimum growth of the crop plant without any or lest stress, hence offers great scope to harness this potential to growing the high value by achieving independence of climate and weather means some level of control over plant microclimate to alleviate one or more of abiotic stress for optimum plant growth this can be achieved by poly house/ net house. It is important to produce quality seedling by inducing morpho-physiological changes in plant for making them competent enough to bear the shock in the field after planting and enhancing their productivity through protected cultivation than those under open field conditions, quality of seedling is also superior, higher input use efficiencies are achieved. The result of this study reveals that

1. The highest inside air temperature was found in greenhouse structure as compared to shade net house and control condition house during March to April month. Among all treatment combination mean monthly Relative Humidity was found lower for treatment ambient and higher in Black shed net house.
2. The maximum germination was recorded under black shade net house S2 than greenhouse S1. There was significant effect of structure and media on favourable condition for germination.
3. The maximum seedling height was recorded under black shade net house S2 followed by greenhouse S1. There was significant effect of structure and media on seedling height at different growth stage.
4. The higher collar diameter was recorded under black shade net house S2 than greenhouse S1.
5. The maximum number of leaves was recorded under black shade net house S2 than greenhouse S1.

6. The higher shoot diameter was recorded under black shade net house S2 than greenhouse S1. Effect of microclimate effect on biomass. Also media is helping as organic matter it consist.
7. The higher fresh weight of above ground, underground was recorded higher under black shade net S2 house than greenhouse S1.The media is reach enough to sustain seedling for about a year. (Srivastava *et. al.* 1998).
8. The sturdiness, root: shoot ratio, vigour index I and vigour index II was recorded superior under treatment S2 than treatment S1.

References

Anonymous, 2010 a. "National Horticulture Board, Sector 18, Gurgaon (Haryana)", http//www.nhb.gov.in

Anonymous, 2010 b. "The Reports from Director of Horticulture, Gujarat State, Ganghinagar", http://agri.gujarat.gov.in

Abdul-Baki, A.A., and Anderson, J.D. 1973. "Vigour determination in Soybean by multiple criteria", Crop Science, 13: 630-637.

Behera, P. C., Rao, K. A., and Mitra, B. N., 1990. "Design, development and management of plastic house for nursery raising", Proceedings of XI International Congress of use of plastic in agriculture P: E 91, 26th Feb-2nd March 1990 at New Delhi, India.

Bisla, S. S., Singhrot, R. S. and Chauhan, S. S. 1984. "Effect of growing media on seed germination and growth of Ber", Haryana Journal Horticulture Science", 13 (3-4): 118-122.

Chauhan, S. K. and Sharma, R., 1997. "Seedling quality evaluation: Important in afforestation", Advances in Forestry Res. in India. XVII :59-27.

Kasetsart, U., 1981. "Effects of growing media on germination and vigor of papaya (*Carica papaya*) seedlings", Bangkok (Thailand). 14 leaves.

Kasetsart, U., 1987. "Studies on nitrogen, phosphorus and potassium deficiency symptom of papaya seedling", Bangkok (Thailand). 1987. 54 leaves.

Moller, M., and Assouline, S., 2007. "Effects of a shading screen on microclimate and crop water requirements", Journal of Irrigation science, 25:171–181.

Rajamanickam, C., Balasubramanyan, S. and Natarajan, S., 2010. studies on nursery management in papaya (*carica papaya* l.) var. co2. Acta Hort. (ISHS) 851:307-312.

Rathore, T. S., Annaourna, D., Joshi, G., and Srivastava, A., 2004. "Studies on potting mixture and size of containers on the quality of seedling production in Casurina equisetifolia frost" Indian Forester, 130 (3): 323-332.

Shrivastava, R., Nanhorya, R. and Upadhyay, J. K., 1998. Selection of proper potting mixture for root trainer of eucalyptus hybris", Indian Forester, 124 (7): 503-510.

43

Mitigating Climate Change in Drylands through Advanced Water Conservation Options

S. Somasundaram and K. Ramaswamy

Abstract

In changing climate scenario, sustainable dryland management depends on identifying best management options for effective use of limited rainwater. Field experiments were conducted to identify technologies with the combination of minimum tillage, crop residue and supplementary irrigation through harvested rainwater to mitigate climate change at Agricultural Engineering College and Research Institute, Kumulur during 2012-2014. The test crop was groundnut in 2012-13 and blackgram in 2013-14. The harvested rainwater in the farm pond was 842 m^3 and 1350 m^3 with a rainfall of 299 mm and 336 mm during 2012-13 and 2013-14 respectively. The water used in control without irrigation, one, two and three supplementary irrigation treatments was 240, 260, 280 and 300 mm in groundnut and 269, 299, 319 and 399 mm in blackgram. Minimum tillage + crop residue @ 5 t ha^{-1} + three supplementary irrigation through raingun sprinklers registered higher groundnut pod yield (1325 kg ha^{-1}) and blackgram grain yield (986 kg ha^{-1}) and was comparable with the combined effect of minimum tillage + crop residue @ 5 t ha^{-1} + two supplementary irrigation through raingun sprinklers in both crops. The above management strategies may be effectively used for mitigating climate change and increasing productivity in drylands of Tamil Nadu, India.

Introduction

Indian agriculture is highly vulnerable to climate change and notably drylands are at risk. Climate change will significantly affect the drylands in ways that without mitigation will ultimately reduce yield of crops. The increase in resilience to climate change is being achieved through identifying technologies which facilitate water conservation and efficient use of limited rainwater in drylands. This identifying process is very difficult in water scared and drought prone dryland regions of Tamil Nadu, India where the rainfall is low and erratic.

Practicing conservation agriculture (combining minimum tillage and crop residue) and use of harvested rainwater through micro-irrigation systems may be superior options to mitigate climate change and increase yield in drylands. Conservation agriculture can be used across different agroeco-systems, to mitigate and adapt to climate change (Silici, 2010; Delgado, 2010). Harvesting rainwater in farm ponds and efficient use of harvested rain water for supplementary irrigation for increasing yield was proved (Sandhu and Sandhu,1995). Now raingun sprinklers are gaining momentum among farmers and success of the system in providing supplementary irrigation may increase the option of irrigating more area with less water. But limited attempt have been carried out in identifying the best combination of the above technologies in dryland tracts of Tamil Nadu. Hence with the above background, the present investigation was carried out to identify suitable minimum tillage + crop residue and supplementary irrigation options for mitigating climate change.

Materials and Methods

Two field experiments were conducted at Agricultural Engineering College and Research Institute Farm, Kumulur during North East monsoon season of 2012-2014 under dryland situation. A total quantity of 299 mm and 336 mm of rainfall was received in 21 rainy days during 2012-13 and 2013-14 cropping season. Effective rainfall was determined by the water balance method suggested by Dastane (1974). The soil of the experimental field was sandy loam with pH 7.8, organic carbon 0.53 per cent, available N (186 kg ha^{-1}), available phosphorus (18 kg ha^{-1}), available potassium (215 kg ha^{-1}), EC 0.56 dSm^{-1}, WHC 57.1 per cent and IR 13.2 cm hr^{-1}. Rain waterwas harvested in the farm pond of capacity 1350 m^3 with well connected water ways. The harvested water was used for supplementary irrigation using raingun sprinklers. The raingun sprinkler system with a discharge rate of 8-10 lps, operating pressure of 8 bars and wetting diameter of 20-25 meters was designed as per the treatment structure. The raingun was attached to a 5 HP diesel engine from the farm pond with harvested rainwater for providing supplementary irrigation. The experiment was laid out in strip plot design with four main plots viz., M_1 – Conventional tillage, M_2 – Minimum tillage without crop residue, M_3 – Minimum tillage + crop residue @ 2.5 t ha^{-1} and M_4 – Minimum tillage + crop residue @ 5 t ha^{-1} and four sub-plots viz., S_1- one supplementary irrigation through raingun sprinklers, S_2- two supplementary irrigation through raingun sprinklers, S_3- three supplementary irrigation through raingun sprinklers, S_4- No supplementary irrigation. The treatments were replicated thrice. The data on supplementary irrigation, water used and yield (statistically) were analysed and presented.

Results and Discussion

Supplementary irrigation and water used

The data on effective rainfall, supplementary irrigation depth and water used was given in Table 1. The rainfall during the cropping season was 299 and 336 mm and effective rainfall was 240 and 269 mm during 2012-13 and 2013-14. The water harvested in the farm pond was 842 m^3 and 1350 m^3 during 2012-13 and 2013-14. The irrigation depth was 20, 40 and 60 mm for one, two and three supplementary irrigation in groundnut during 2012-13 and 30,50 and 70 mm for one, two and three supplementary irrigation in blackgram during 2013-14. The water used in control without irrigation, one, two and three supplementary irrigation treatments was 240, 260, 280 and 300 mm in groundnut and 269, 299, 319 and 399 mm in blackgram. The water used was higher in three supplementary irrigation in both groundnut (300 mm) and blackgram (339 mm). The water used in three supplementary irrigated plots was 6.6, 13.3 and 20 per cent higher in groundnut and 5.8, 11.7 and 20.6 per cent higher in blackgram respectively compared to two, one and no supplementary irrigation plots. Results were in same line as reported by Byzedi (2011).

Table 1: Effect of treatments on supplementary irrigation depth and water used

Treatment	*Effective Rainfall (mm)*		*Supplementary irrigation depth (mm)*		*Water used (mm)*	
	GN (2012-13)	*BG (2013-14)*	*GN (2012-13)*	*BG (2013-14)*	*GN (2012-13)*	*BG (2013-14)*
No SI	240	269	0	0	240	269
SI – one time	240	269	20	30	260	299
SI – two times	240	269	40	50	280	319
SI – three times	240	269	60	70	300	339

SI – supplementary irrigation, GN – Groundnut, BG – Blackgram

Yield

During both the years of experimentation, minimum tillage + crop residue and supplementary irrigation had significant influence on pod and grain yield of groundnut and blackgram and was presented in Table 2.

Minimum tillage + crop residue @ 5 t ha^{-1} produced higher yield (1127 kg ha^{-1} in groundnut and 838 kg ha^{-1} in blackgram) and followed by minimum tillage + crop residue @ 2.5 t ha^{-1} (966 kg ha^{-1} in groundnut and 755 kg ha^{-1}

blackgram). Significantly lower yield was recorded by conventional tillage. Minimum tillage + crop residue @ 5 t ha^{-1} out yielded conventional tillage practiced by farmers in this region by 26 per cent in groundnut and 33 per cent in blackgram.

Table 2: Effect of tillage + crop residue and supplementary irrigation on yield of groundnut pod and green gram grain (kg ha^{-1})

Treat ments	*Groundnut pod yield (2012-13)*					*Blackgram grain yield (2013-14)*				
	Tillage + Crop Residue					*Tillage + Crop Residue*				
	M1	*M2*	*M3*	*M4*	*Mean*	*M1*	*M2*	*M3*	*M4*	*Mean*
Supp. Irrg										
S1	803	760	984	1026	893	595	612	786	796	697
S2	969	870	1040	1280	1040	631	737	838	967	793
S3	1082	1085	1152	1325	1161	719	798	870	986	843
S4	492	651	686	878	677	292	398	526	604	455
Mean	836	841	966	1127		559	636	755	838	
	M	S	M at S	S at M		M	S	M at S	S at M	
SEd	61	43	37	77		31.0	22.8	19.2	54.5	
CD (5%)	150	90	78	159		75.8	47.0	42.5	94.0	

M_1 – Conventional tillage, M_2 – Minimum tillage without crop residue, M_3 – Minimum tillage + crop residue @ 2.5 t ha^{-1}, M_4 – Minimum tillage + crop residue @ 5 t ha^{-1}, S_1- one supplementary irrigation through raingun sprinklers, S_2- two supplementary irrigation through raingun sprinklers, S_3- three supplementary irrigation through raingun sprinklers, S_4- No supplementary irrigation.

Three supplementary irrigation through raingun sprinklers (1161 kg ha^{-1} in groundnut and 843 kg ha^{-1} in blackgram) enhanced the yield compared to two supplementary irrigation through raingun sprinklers (1040 kg ha^{-1} in groundnut and 793 kg ha^{-1} in blackgram). Giving three supplementary irrigation through raingun sprinklers from harvested rain water in the farm pond boosted the yield up to 41 per cent in groundnut and 46 per cent in blackgram, compared to crops grown without supplementary irrigation.

Interaction effect was significant during both the years. Minimum tillage + crop residue @ 5 t ha^{-1} + three supplementary irrigation through raingun sprinklers registered higher groundnut pod (1325 kg ha^{-1}) and blackgram grain (986 kg ha^{-1}) and was comparable with the combined effect of minimum tillage + crop residue @ 5 t ha^{-1} + two supplementary irrigation through raingun sprinklers (1040 kg ha^{-1} in groundnut and 793 kg ha^{-1} in blackgram). Combined effect of conventional tillage and no supplementary irrigation drastically reduced the yield. Minimium tillage + crop residue with supplementary irrigation increases the soil moisture content and reduces the stress during critical period and thereby resulted in higher yield. The above explanation was supported by Krishna *et al.* (1987) and Oweisand Hachum (2004).

Conclusion

The analysis of the results identifies which combination of technologies has proved to be better suited to mitigate climate change. Conservation agriculture and supplementary irrigation is a highly efficient practice with great potential for increasing agricultural production and improving livelihoods in the dry lands. Conservation agricultural practices especially minimum tillage + crop residue @ 5 t ha^{-1} combined with supplementary irrigation to a amount of 60 mm through micro-irrigation methods was more promising to be adopted in dryland tracts of Tamil Nadu, India.

References

Byzedi, M., B. Mehdinejadian, H. Zolnorian, M. Siosemarde. 2011. "Studying the Effects of Supplementary Irrigation (Sprinkler Systems) to Dry Farming Wheat Cultivars", World Academy of Science, Engineering and Technology, Vol. 55, pp. 551 – 553.

Dastane, N.S., 1974. "Effective rainfall", Irrigation Drainage paper, 25 FAO, Rome, p. 69.

Delgado, J.A., 2010. "Crop residue is a key for sustaining maximum food production and for conservation of our biosphere", Journal of Soil and Water Conservation, Vol. 65(5), pp. 111A-116A.

Krishna, J. H., F. G. Arkin, and J. R. Martin. 1987. "Runoff impoundment for supplemental irrigation in Texas." Water Resources. Bulletin, Vol. 23(6), pp. 1057-1061.

Oweis, T., and A. Hachum. 2004. "Water harvesting supplementary irrigation for improved water productivity of dry farming systems in West Asia and North Africa", In: New Direction for a Diverse Planet. Proceedings of the 4th International Crop Science Congress, Brisbane Australia, 26 September to 1 October 2004.

Sandhu, K.S., and A.S. Sandhu. 1995. "Response of dryland wheat supplemental irrigation and rate and method of N application", Nutrient Cycling in Agroecosystems, Vol. 45(2), pp. 135-142.

Silici, L., 2010. "Conservation Agriculture and Sustainable Crop Intensification in Lesotho", Rome, Italy: Food and Agriculture Organisation of the United Nations (FAO).

44

Climate Smart Technology through Sustainable/Conservation Agriculture

Revathi, P, Madu Bindu, G. S., Rajeswari, B, and Parag, P

Abstract

Climate change particularly, decreased rainfall and caused extreme weather patterns which disrupted food production and agriculture development, creating economic volatility and food insecurity. It has great impact in areas where hunger and malnutrition are already persistent problems, particularly in Africa and South Asia. Climate smart technology is to focus from production oriented to profit oriented sustainable farming. It is possible only by controlling emission of green house gases. Carbon sequestration is a long run (50 to 100 years) process but with the help of conservation agriculture, carbon storage has occurred within two to three years. Soil organic carbon is fundamental in the development of soil quality and sustainable food production system. Conservation agriculture has significant contribution to mitigate climate change by the use of agricultural machinery for reduction in the emission of green house gases (GHG) as well as carbon sequestration. Conservation agriculture through machinery/precision farming/ weather forecast/protected cultivation will lead to sustainable farming and will thrust to profit oriented farming in future. The end result is reduction in the emission of greenhouse gases per tonne of farm produce, sequestration of carbon dioxide in the soil and increased crop resilience to heat, drought and floods. Collaborative research and using innovative precision data technology are critical steps to make agriculture more 'climate smart.'

Introduction

From current date towards 2050, the world's population will increase by one-third. Most of these additional two billion people will live in developing countries. At the same time, more people will be living in cities. If current income and consumption growth trends continue, FAO estimates that agricultural production will have to increase by 60 per cent by 2050 to satisfy the expected demands for food and feed. Agriculture must therefore transform itself if it is to feed a growing global population and provide the basis for economic growth and poverty reduction. Climate change will make this task more difficult under a business-as-usual scenario, due to adverse impacts on agriculture, requiring spiralling adaptation and related costs. To achieve food security and agricultural development goals, adaptation to climate change and lower emission intensities per output will be necessary. This transformation must be accomplished without depletion of the natural resource base. Climate change is already having an impact on agriculture and food security as a result of increased prevalence of extreme events and increased unpredictability of weather patterns. This can lead to reductions in production and lower incomes in vulnerable areas. These changes can also affect global food prices. Developing countries and small holder farmers and pastoralists in particular are being especially hard hit by these changes. Many of these small-scale producers are already coping with a degraded natural resource base. They often lack knowledge about potential options for adapting their production systems and have limited assets and risk-taking capacity to access and use technologies and financial services. Enhancing food security while contributing to mitigate climate change and preserving the natural resource base and vital ecosystem services requires the transition to agricultural production systems that are more productive, use inputs more efficiently, have less variability and greater stability in their outputs, and are more resilient to risks, shocks and long-term climate variability.

A proposed means to achieve this is increased adoption of a 'climate-smart agriculture' (CSA) approach (FAO, 2010). The agricultural technologies and practices that constitute a CSA approach are, in most cases, not new, and largely coincide with those of sustainable agriculture and sustainable intensification. It resembles understanding of adoption costs and benefits in various agro-ecological and socio-economic contexts.

Few of the farmers were successfully using weather bulletins of the meteorological department to make their agriculture climate smart but it is made perfect only through conservation agriculture. It is associated with crop production that strives to achieve acceptable profits together with high and sustained production levels without degrading the environment. Conservation agriculture machinery/ precision farming/ protected cultivation/ sustainable agriculture/ farming system-cropping systems all will lead to sustainable farming and will thrust to profit oriented farming in future.

Conservation agriculture (CA) is a system of integrated management of soil, water and biological resources combined with external inputs with a view to conserve natural resource. It is associated with crop production that strives to achieve acceptable profits together with high and sustained production levels, without degrading the environment. Interventions such as soil tillage are reduced to an absolute minimum and the use of external inputs such as agro-chemicals and nutrients of mineral or organic origin are applied at an optimum level without interfering with soil biological processes. Conservation tillage, as compared to conventional practices, has shown better performance in terms of increased benefit cost ratio (2.0-2.5) and water operational energy (5-26 per cent). Zero tillage improves soil drainage by subverting the sponge effect created by normal cultivation. The effect of direct seeding/planting is advantageous in terms of improved trafficability, decreased soil compaction in the long run and reduced soil erosion. It also leads to decreased water evaporation and increased availability of water.

Climate smart with conservation agriculture composed of three main pillars: sustainably increasing agricultural productivity and incomes; adapting and building resilience to climate change and reducing and/or removing greenhouse gases emissions relative to conventional practices (FAO, 2013). These three components jointly aim to maintain a permanent or semi-permanent organic covering on the soil, either with dead mulch or the crop itself. This covering protects the soil from erosion and provides a more amenable environment for soil biota (Gowing and Palmer, 2008; Knowler and Bradshaw, 2007). In contrast, mechanical tilling degrades soil microstructure, buries organic material and disturbs natural biological processes. CA's avoidance of mechanical tilling helps maintain soil porosity, allowing water to infiltrate the soil, reach crop roots and slow erosion processes. CA does not avoid the use of synthetic chemicals. For instance, weed control previously achieved through tilling may be replaced by increased herbicide use, at least initially. However, a long-term reduction in input requirements is often stated as a benefit of CA (Knowler and Bradshaw, 2007; Hobbs, 2007).

Sustainably increasing agricultural productivity

There are a number of meta-studies which attempt to quantify the average impacts of CA based on disparate uses. These assessments pertain both to environmental impacts and to a lesser extent, yield impacts. CA improves yields most directly by improving soil moisture, especially for the lowest productivity systems. Lal (2009) concluded that mulching and no-till clearly improved soil health, sometimes improved yields (depending on conditions) and usually improved profits (due to lower inputs). Farooq *et al.* (2011) in 25 long-term CA trials (mainly from North America, Australia and Europe) found that crop yields showed a slight increase, which grew over time relative to conventional tillage. The CA advantage is most pronounced in dry conditions. Pretty *et al.* (2006) gathered evidence on the effect of CA from 286 developing country case studies. The selected cases were examples where 'best practice' sustainable agriculture interventions had occurred. For interventions related to small holder CA, average yield improvement was over 100 per cent. Branca *et al.* (2011) undertook a comprehensive, empirical meta-analysis of 217 individual studies on CA globally. Their empirical analysis showed that improved agronomic practices such as cover crops, crop rotations (especially with legumes) and improved varieties has increased cereal productivity by 116 per cent on an average across the studies consulted. Similarly, reduced tillage and crop residue management caused an increase of 106 and 25 per cent, respectively and agro-forestry techniques caused an increase of 69 per cent. Tillage management and agro-forestry were found to be particularly beneficial in dry agricultural areas.

CA is primarily a water harvesting strategy. Consequently the technique has the greatest benefits during poor seasons and thus represents a useful climate change adaptation strategy. The disadvantage of this is an elevated danger of water-logging in wet seasons. CA is thus best suited to dry areas and allows for dry season land preparation, allowing for crop sowing immediately after the first rains arrive. However, dry season land preparation is not always possible.

There is also a strong evidence of benefit in annual food crop yields from agro-forestry in Malawi, both as a substitute and a complement for inorganic fertilisers. Agro-forestry involves the planting of selected tree and shrub species either alongside, or sequentially (during fallow) to an annual food crop. The trees help maintain soil cover, improve nutrient levels, increase soil organic matter (via the provision of mulch), improve water filtration, and provide a secondary source of food, fodder, fiber and fuel. There is abundant evidence that such practices can deliver dramatic increases in crop yields. For instance, intercropping *Glyricidia sepium* led to an improvement in maize yields up to an average of 345 per cent, based on trials at five different sites in Malawi (David *et al*). Other species have shown similarly dramatic benefits. Published yield improvements from Malawi field trials of intercropping with *Tephrosia vogelii, Sesbania sesban* and *Cajanus cajan* ranged between 55 and 255 per cent over mono-cropped unfertilised maize. Under a future situation of higher fertiliser prices – a distinct possibility given high global energy prices – the profitability of agro-forestry (either unfertilised or partially fertilised) relative to mono-cropped maize improves considerably.

Though there is general agreement that CA can improve yields under at least some circumstances, a debate continues over how extensive these circumstances are in practice. Nkala *et al.* (2011), Giller *et al.* (2009), Gowing and Palmer (2008), Knowler and Bradshaw (2007) and Lal *et al.* (2004) outlined a number of reasons why CA may not be suitable in particular contexts. For instance, crop residues are often used as animal feed. The benefits of mulching with crop residues may not be worth the trade-off of reduced livestock numbers. Similarly, there may be a trade-off between labour saved on tillage and labour spent on increased weeding, in the absence of herbicides.

Some authors found that costs were higher under the CA system, primarily due to increased labour requirements (David *et al*). CA requires labour-intensive digging of planting basins although this labour requirement diminishes in subsequent seasons. It also requires increased weed control, either by herbicide application or hand weeding. Both reduce the profit advantage of CA relative to conventional tillage. Although several studies on CA effects in Zambia are positive, more recent work has raised questions about its widespread viability (Arslan *et al*. 2013). Thus, there is a need for further research to identify in what contexts CA is best promoted.

Adapting and building resilience to climate change

Despite the benefits of CA outlined above, and its promotion over the last fifteen years, adoption is relatively limited. The circumstances in which CA is beneficial are related to the time after adoption. Erenstein (2002) concluded that in the short term, yields can rise, fall or remain unchanged. Over time the productive benefits of improved soil chemistry, biology and structure are more likely to deliver increased yields, however, discrepancy between short-term and long-term yields can hinder adoption (Nkala *et al*. 2011; Giller *et al*, 2009; Fowler and Rockstrom, 2001).

The available literature highlights some consistent determinants of adoption: (1) households with a larger pool of labour or larger land holdings are more likely to adopt; (2) Agro-forestry that provides an additional marketable product (e.g. nuts or fruit from fertiliser trees) or can be planted directly from seed is more likely to be adopted; and (3) a poorly functioning fertiliser tree seed market is a serious constraint as they are often exposed to bush fires and livestock grazing especially in the absence of perennial private rights on the land. Of the competing approaches, *C. cajan* (pigeon pea) was preferred and most often adopted, especially by female farmers.

Adoption of agro-forestry in Malawi (and across southern Africa) has been slow (David *et al*) despite the profit and food security advantages. It is argued by a number of authors that agro-forestry projects will be slower to become self-sustaining and self-diffusing than earlier 'Green Revolution' advances. Agro-forestry uptake is particularly complex due to the multiple components and multiple years through which testing and adaptation takes place.

Overall, adoption is based less on a desire for long-term soil regeneration and more on short-term alternative food or fuel wood production. High labour requirements (even in densely populated areas), lack of access to seed markets (for both purchasing and selling of seed), and lack of access to improved legume genotypes are constraints to adoption. Research focused on the multi-functionality of agro-forestry products, as well as ways to facilitate seed markets may help identify the ways forward.

Mitigation of green house gas emissions

The third key component of the Climate Smart Agriculture approach is the mitigation of green house gas emissions. Improved agricultural practices, such as agroforestry and CA, can deliver significant carbon mitigation and reduced greenhouse gas emissions (Bianca, *et al.* 2011).

There are two ways of assessing and reporting the sequestration potential of agricultural systems. Some studies report a *rate* (per year) of carbon sequestration. This rate is likely to vary depending on the quality of the season and the maturity of the system, and only applies while the system has not yet reached its carbon equilibrium. Others report the total soil organic carbon quantity, which represents the total amount of carbon that could be sequestered if an improved system is maintained.

There is considerable evidence that agro-forestry leads to large carbon sequestration benefits, globally, and specifically under the conditions found in Malawi and Zambia. Estimates are highly variable, reflecting different soil and climatic conditions, tree species, tree densities and plot maturity. The evidence for carbon sequestration under reduced tillage is more mixed. Although a number of authors have found positive evidence from various locations, some argue that the benefits are insignificant. The literature consulted for this review suggested that CA systems have the potential to sequester between 0.2 – 0.7 tonnes C ha^{-1} yr^{-1}. Agro-forestry systems have a higher potential of between two to five tonnes C ha^{-1} yr^{-1}. Short-term rates (such as during establishment) can be higher.

Kaonga and Coleman (2008) studied carbon sequestration in experimental fallows at Msekera, eastern Zambia. In a comparison with maize monoculture after *ten years,* they found that soil carbon stocks were more than six tonnes C ha^{-1} greater in coppicing fallows, and over three tonnes C ha^{-1} greater in non-coppicing fallows, using a variety of agro-forestry tree species. These higher carbon stocks were achieved by accumulation of an extra 0.7 - 1.4 tonnes C ha^{-1} yr^{-1} relative to unfertilised monocropped maize. C_4 plants like sugarcane and maize sequestrate more carbon inside the soils due to rubisco enzyme. In the future through genetic modification C_3 rice plant could be converted to C_4 rice plant to overcome abiotic stresses. The C_4 G.M rice or maize type of plants may capture more rubisco to sequestrate more carbon inside the soil. Aerobic rice also help in mitigating the emission of green house gases.

Based on this, some authors have argued that there is the potential for carbon offset payments, which could pay for investments in improved farming, sequester emissions and simultaneously improve food security. For such a possibility to occur, however, precise estimates of carbon sequestration quantities under different farming regimes are required.

Further, Ngwira *et al.* (2012) conducted soil carbon measurements in farmers' fields in Kasungu and Mzuzu districts in Africa and found that organic carbon levels were 41 percent higher under zero till, in comparison to conventional till maize, over a four year period. Over a five year period, organic carbon levels were 75 per cent higher. Precision farming, by application at right time, right place and right method of avoids; protected cultivation (cultivating crops in green houses) by reflecting U.V radiation and farming system/ cropping system/ organic farming by adding residues to the soil were found eco-friendly, contributing to reduce green house gases and enhance natural ecological processes which contribute to climate resilience and reduced vulnerability to yield, reducing disease/ pest problem.

CA for sustainability

CA provides a truly sustainable production system, not only conserving but also enhancing availability of natural resources and increasing the variety of soil biota, flora and fauna in agricultural production systems without sacrificing yields. When a tillage -based production system is to be transformed into a CA-based system, it involves a shift in the prevailing on–farm mechanical technologies, some of which would remain but with only marginal changes in farm power requirements and land use for sustainable intensification.

Further research needs

Opportunities for further research are discussed briefly under the following three categories: (1) research on outstanding biophysical questions, (2) research on outstanding socio-economic questions, and (3) research on the institutional dimensions of CSA promotion. There is an emphasis on the second and third categories due to the greater lack of understanding in these areas.

Biophysical research

The scientific understanding of CA would benefit from further research into its yield impacts and carbon sequestration potential. Both topics are the source of some debate. While a number of studies in Zambia have assessed the yield impact of CA, these have tended to use small case studies. Geographically broad, time series analysis of CA performance under farmer-managed conditions and climate change indicators would be valuable. Further research that disentangles the separate yield impacts of different components of CA would help refine the technique.

Socio-economic research

There remains a need to better understand the farm level economic implications of both agro-forestry and CA. The summary of results from this review suggests that there is some disconnect between gross profit margins results and adoption rates. Both technologies have been promoted for some time, often with incentives (e.g. subsidised inputs), yet uptake remains limited despite the often positive benefits purported. Dis-adoption rates of CA in Zambia are reported to be high (Baudron, 2007). Potential reasons include constraints to adoption that are poorly understood, slow dissemination of CA knowledge, or smaller benefits than field trials suggest in the absence of incentives provided by CA programmes. Understanding the reasons for this will be useful for future CA policy development.

A better understanding of this could help improve the effectiveness of CA and agro-forestry promotion efforts. Fuel wood availability and extra harvests of nuts and fruits from agro-forestry, for instance, may be important for farm level decision making. It is suggested by a number of researchers that adoption is considerably influenced by consideration of such auxiliary benefits.

Conclusions from previous studies suggest that research on agro-forestry and CA adoption should concentrate on long-term analyses of specific practices in the field in addition to the fundamental determinants (such as farmer characteristics) that drive adoption. Such research can better inform policy that needs to target different farming approaches under different agro-ecological zones.

Institutional context

There is a need to identify how CA and agro-forestry can complement, or in some cases, substitute for inorganic fertiliser input, and how policies regarding these approaches can achieve synergies. There are a number of reasons why CSA alternatives have an important role to play alongside fertiliser policies. For instance, the best yield, soil health and drought resistance occurs when both inorganic fertiliser and the alternative techniques are combined. However, there is little existing research on how to improve incentives for their parallel use. Research focusing on the interaction between fertiliser subsidies and alternative agricultural practices would thus be valuable. Research focusing on how to improve the efficiency of fertiliser use alongside its combination with these practices is also necessary to decrease the budgetary burden of fertiliser programmes in these countries and to potentially provide mitigation side-benefits.

There is also a need to consider the cultural context in which agroforestry and CA are promoted. For instance, there may be gender equality implications for different types of agricultural policies. CA is suggested to increase the burden of labour on women due to an increase in weeding responsibilities, while decreasing the burden of labour on men due to a reduction in tillage responsibilities.

For the sake of clarity and better communication, various technologies and interventions were classified into five groups: water-smart, energy-smart, nutrient-smart, crop-smart, and weather-smart interventions (Aggarwal *et al.*, 2004). The specifications of the technology basket are provided in Table 1.

Table 1. Selected technology option for choice of the experiment

Type of technology	*Definition*
1. Water-smart technologies	**Interventions that reduce water requirements to produce the same or a higher level of yield.**
Rainwater management	In situ rainwater storage in rice paddies with 20–25 centimetres (cm) bunds. This technique is for rice only.
Laser land levelling	Levelling of land with a laser leveller.
System of rice intensification	7- to 10-day-old seedlings are transplanted at 20 cm spacing with 1–2 seedlings per hill.
Furrow-irrigated raised bed	Growing crops on ridges or beds. Irrigation is applied through furrows separating the beds.
2. Energy-smart technologies	**Technologies that help reduce energy consumption during land preparation without affecting yield levels. These also help reduce water requirements for crops.**
Direct-seeded rice	Dry seeds are sown either by broadcasting or drilling in line.
Zero tillage / minimum tillage	The crop is seeded through a seeder in an untilled field, and the crop residue is incorporated into the soil. At present, this technique is limited to wheat only.
3. Nutrient-smart technologies	**Technologies that save/supplement/avoid chemical fertilizer use for crops and enrich carbon in the soil.**
Green manure	Cultivation of legumes in a cropping system. This practice improves nitrogen economy and soil health/quality.
Integrated nutrient management	Integrated use of organic and chemical fertilizers to partially (25 per cent to 50 per cent) reduce NPK (nitrogen, phosphorus, and potassium) requirements without affecting productivity and improve soil health.
Leaf colour chart	Standardised colour charts are used to identify nutrient deficiency to estimate fertiliser doses in different field locations.

4. Weather-smart instruments	**Interventions that provide services related to financial security and weather advisories to farmers.**
Crop insurance	Crop-specific insurance to compensate income loss due to vagaries of weather.
Weather advisories	Information and communication technology–based forecasting about the weather.
5. Introduction of stress tolerant crops and diversification	**Tolerant crops withstand biotic and abiotic stresses and crop diversification reduces water demands and helps in harnessing nutrients from different soil layers.**
Drought-tolerant variety	Seed variety that is tolerant to drought or relatively dry weather conditions.
Crop diversification (maize-wheat cropping)	Rice is replaced by maize on part of the land to economize on water use.

Final Remarks

Improving food security in the face of rapidly growing populations and climate change impacts, is a daunting challenge. However, considerable recent success of conservation agriculture programmes and agro-forestry programmes are cause for optimism David *et al.* This review documents considerable, positive evidence for the benefits of both. It also identifies the substantial need for further research, particularly with regard to the socio-economic impacts of these approaches.

The main impacts of climate change on agriculture are summarised as well as the contribution of agriculture to global greenhouse gas emissions. The principles guide necessary changes of systems: more efficiency in the use of resources, to increase production while reducing emission intensity of the food produced and consumed and more resilience, to get prepared to variability and change. Certain issues are to be addressed to implement climate-smart agriculture and progress towards efficient and resilient food systems. It requires comprehensive policies at every level, adequate institutions and proper governance to make the necessary choices. It also requires new financing to address the needs in terms of investments and research and to enable the farmers to overcome barriers to adoption of new practices including up-front costs and income foregone during the transition period. The last section articulates the concept of CSA closely linked issues of sustainable intensification, green growth and sustainable development.

References

Food and Agriculture Organisation of the United Nations, Climate Smart Agriculture Sourcebook. This is the major source document for climate-smart agriculture.

Garima Tanga, Barun Deb Pal, Pramod K. Agarwal and N.K.Tyagi et al.

Farmers preference for climate –smart Agriculture, IFPRI Discussion paper 01337 April 2014.

Hand book of Agricultural Engineering, Indian Council of Agricultural Research, New Delhi, ISBN: 978-81-7164-134-5

FAO (2010). '"Climate-Smart" Agriculture: Policies, Practices and Financing for Food Security, Adaptation and Mitigation', Food and Agriculture Organisation of the United Nations, Rome.

FAO, 2013. 'Sourcebook on Climate Smart Agriculture, Forestry and Fisheries', Food and Agriculture Organisation of the United Nations (FAO), Rome, Italy. http://www.fao.org/climatechange/37491-0c425f2caa2f5e6f3b91 62d39c8507fa3.pdf

Farooq, M, Flower, K, Jabran, K, Wahid, A and Siddique, K (2011). 'Crop yield and weed management in rainfed conservation agriculture', *Soil and Tillage Research,* 117: 172-83.

Gowing, J and Palmer, M (2008). 'Sustainable agricultural development in sub-Saharan Africa: The case for a paradigm shift in land husbandry', *Soil Use and Management,* 24(1): 92-9.

Hobbs, P (2007). 'Conservation agriculture: What is it and why is it important for future sustainable food production?', *Journal of Agricultural Science,* 145(2): 127-37.

Knowler, D and Bradshaw, B (2007). 'Farmers' adoption of conservation agriculture: A review and synthesis of recent research', *Food Policy,* 32(1): 25-48.

Lal, R (2009). 'Soils and food sufficiency. A review', *Agronomy for Sustainable Development,* 29(1): 113-33.

David, *et al* (2012). Climate-Smart Agriculture? A review of current practice of agroforestry and conservation agriculture in Malawi and Zambia (pdf)

Arslan, A, McCarthy, N, Lipper, L, Asfaw, S and Cattaneo,.A. (2013 – forthcoming in *Agriculture, Ecosystems and Environment*) 'Adoption and intensity of adoption of conservation farming practices in Zambia,' ESA Working Paper 13-01, Rome, FAO. (Also published as IAPRI Working Paper No. 71).

Ngwira, A, Aune, J. and Mkwinda, S (2012). 'On-farm evaluation of yield and economic benefit of short term maize legume intercropping systems under conservation agriculture in Malawi', *Field Crops Research,* 132: 149-157.

Branca, G, McCarthy, N, Lipper, L and Jolejole, M (2011). 'Climate-smart agriculture: A synthesis of empirical evidence of food security and mitigation benefits of from improved cropland management', Food and Agriculture Organisation of the United Nations (FAO), Rome.

Giller, K, Witter, E, Corbeels, M and Tittonell, P (2009). 'Conservation agriculture and smallholder farming in Africa: The heretics' view', *Field Crops Research,* 114: 23–34.

Lal, R, Griffin, M, Apt, J, Lave, L and Morgan, M (2004). 'Managing Soil Carbon', *Science,* 304(5669): 393.

Nkala, P, Mango, N, Corbeels, M, Veldwisch, G and Huising, J (2011). 'The conundrum of conservation agriculture and livelihoods in Southern Africa', *African Journal of Agricultural Research,* 6(24): 5520-8.

Pretty, J, Noble, A, Bossio, D, Dixon, J, Hine, R, De Vries, F and Morison, J (2006). 'Resource-conserving agriculture increases yields in developing countries', *Environmental Science and Technology*, 40(4): 1114-9.

Kaonga, M and Coleman, K (2008). 'Modelling soil organic carbon turnover in improved fallows in eastern Zambia using the RothC-26.3 model', *Forest Ecology and Management*, 256(5): 1160-6.

Erenstein, O (2002). 'Crop residue mulching in tropical and semi-tropical countries: An evaluation of residue availability and other technological implications', *Soil and Tillage Research*, 67(2): 115-33.

By C.D. Singh (2014). Training programme on Application of sensors, wireless sensor network and instrumentation in precision agriculture.

45

Catalysed Role of Value Addition and Processing Industries in the Arena of Climate Change

S. P. Cholera, V. K. Chandegara, J. N. Nandasana, K. C. Patel, M. T. Kumpavat and M. H. Jethva

Abstract

This paper discusses the effects of value addition and food processing industries on climate changes. The various food processing operations, viz., drying, refrigeration, transportation, storage, etc., are the main sources of greenhouse gas emissions (GHG) in the environment. Meat and dairy production has a huge impact on climate change and it is estimated to be responsible for 18 per cent of global emissions. The processing of meat and dairy is also responsible for the majority of agricultural methane emissions. In addition to this, food and drink manufacturer also credited large amounts of green house gases in the atmosphere. Changing in food habits of consumers and lack of government policies also increases the effects of green house gases. The emissions in turn accelerate climate change, causing a pernicious feedback loop. The paper gives particular attention to consumer, processor, industrialist, entrepreneurs, and government to adapt new advance techniques to minimise the changes in climate.

Introduction

Agriculture and the food system are likely to be substantially affected by both climate change and greenhouse gas emission. Agricultural practices are major anthropogenic contributors to climate change. According to the World Bank (2008), they account for 30 per cent of greenhouse emissions. Furthermore, the agricultural sector is also very susceptible to climate change, which threatens food security globally, especially with the projected population increase. Various agencies report yield declines as one of the main foreseen consequences, which will result in decreased foodstuffs availability and incremental prices (World Bank, 2010). These trends are especially significant for basic grains. Hence, their consumption as well as of animal protein is thought to decline. The decreased consumption is projected to reduce caloric intake and thus increased childhood malnutrition by 2050 (Nelson *et al.*, 2009).

The last two decades marked the emergence, rapid proliferation, growth and development of food processing industries in India. This is due to increasing demand for processed foods particularly in urban areas. The raw materials for the processed food industries are mainly agricultural, from where finished products such as beverages, edible oils, sugar and other sweeteners, drinks (both alcoholic and non-alcoholic), fish and meat products emerge (Ojo, 1998).

It is becoming increasingly clear that the food system has a major impact on climate change. This is not surprising considering how complicated it is to get food onto our plates. Fields are sprayed with fertiliser, animals have to be fed and food is flown across the world to be processed in factories. The combined effects of all the stages of the food system mean that food and drink (and tobacco) are the single largest source of greenhouse gas emissions from consumption (Francis, 2004).

Food security and food availability is a consequence not only of food production, but of food processing. Food processing and distribution is a major source on employment and income, especially in developing economies. According to the World Bank (2008), agribusinesses account for approximately the 30 per cent of the GDP in transforming and urbanised countries. In low income countries, industries and services linked to agriculture amount to 50 per cent of the total manufacturing value added (World Bank, 2008; UNIDO, 2009). The transformation, storage and transport of raw agricultural products involve activities that cut across the agricultural, manufacturing and transport sectors.

India's productivity of rice (3590 kg/ha) is lower than China (6686 kg/ha), Bangladesh (4219 kg) and Myanmar (4081 kg). Whereas India's productivity of wheat (1661 kg) is lower than China (4838 kg) in 2011. China performed better in the productivity of coarse grains (5470 kg) and pulses (1533 kg) when compared to India which stood at 1591 kg and 699 kg respectively. India's per hectare production of pulses is the lowest when compared to its six neighbours - Bangladesh, Bhutan, China, Myanmar, Nepal and Sri Lanka. Besides all our efforts to enhance agricultural productivity through a huge network of institutions, we are still lagging behind due to numerous problems that are inherent in our system. At present, our aim should be to enhance the agricultural productivity without causing much damage to the natural resources and production environment. This calls for a stringent action to evolve varieties and technologies/innovations which can enhance the production/ productivity of agricultural crops under changing climatic scenarios. This is possible through some innovative techniques like inducing C4ness in rice, developing multiple resistant cultivars through MAGIC (multiparent advanced generation intercross), development of super hybrids, application of GIS/ Remote sensing technologies, nanotechnology, crop modelling etc. All the new generation technologies require policy support for their development as well as adoption for harnessing their full potential benefits. At the same time, the institutions involved in research, extension activities, development departments, banking, planning and execution should all come together and work in coherence to make this a successful venture (Shetty *et al.*, 2013).

Today, food industries in the country are so many that they could be sub-divided into thirteen (13) categories. These are flour and grain; soft drinks and carbonated water; breweries; starch and miscellaneous food products; meat, poultry and fish; tea, coffee and other beverages; fruit juices; animal feed; sugar; distilleries and blending of spirits; cocoa, chocolates and sugar confectioneries; agricultural and food chemicals and industrial packaging. Food processing projects involve the processing and packaging of meat products, fish and shell fish, dairy products, fruits and vegetables, grains and beverages production. It includes refinement, preservation, and improvement of product; storage, handling, packaging and canning. The processing may involve receiving and storing raw or partially processed plant, animal or other food materials, processing the raw materials into finished products, and packaging and storing the finished products (Ojo, 1998).

The food processing and distribution system is not vulnerable to climate in the same way that production agriculture is, but there could be some significant impacts, particularly in the sectors that are more energy intensive. Agricultural processing operations may experience two kinds of climate change impacts: those experienced directly and those related to climate change impacts on agricultural production (SVG, 2014).

Impact of Meat and Dairy Production

Meat and dairy production has a huge impact on climate change. It is estimated to be responsible for 18 per cent of global emissions (FAO, 2006). There are three main reasons for this:

(1) It is a major driver of deforestation to create pasture and to grow feed;

(2) Huge amounts of fertiliser is used to grow the animal feed;

(3) The animals themselves release high levels of greenhouse gases by enteric fermentation i.e. belching and passing wind.

With increased prosperity, people are consuming more meat and dairy products. Globally, meat production is set to more than double from 2001 to 2050, predominantly in the developing world as their diets become more like our own. One of the effects of this will be that more forest will be cleared to grow feed and provide pasture – activities already thought to be responsible for around seven per cent of global greenhouse gas emissions (FAO, 2006).

The production of meat and dairy is also responsible for the majority of agricultural methane emissions (BFCC, 2007). And as meat consumption increases methane emissions from livestock are predicted to rise by up to 60 per cent by 2030 (FAO, 2006). Delgado *et al.* 1999 reported that animals themselves release most of the methane by enteric fermentation during digestion, and the rest comes from the manure they produce. There are indications that as agricultural methods become more intensive, such as moving cows from fields to sheds, methane emissions from livestock usually increase too.

Livestock also accounts for over half of man-made nitrous oxide emissions (FAO, 2006). Intensification of livestock production is partly responsible for this. One of the major trends in livestock intensification is the use of animal feed instead of allowing animals to graze (Delgado *et al.* 1999). Nitrous oxide is released by the fertiliser used to grow the feed, by the manure and urine produced by the animals, and from the storage of manure in intensive rearing systems. Intensive farming techniques increase emissions from all of these sources (FAO, 2006).

Impact of Food Industries

If we look beyond greenhouse gas emissions from agriculture and deforestation, the food system continues to produce greenhouse gases during processing and manufacturing. Most of these emissions come from energy use. Using energy causes the release of the greenhouse gas carbon dioxide. This gas is emitted when fossil fuels are burnt to produce energy. The chart below shows how much each stage of food production contributes to the total energy used – and therefore CO_2 emitted by the food industry. This includes the energy used in our homes but not in transportation (DEFRA, 2007).

Changes in the way we cook and preserve our food can have a significant impact on emissions as a large proportion of emissions are from energy we use in our homes when cooking and refrigerating food. But the food and drink manufacturing, food retail and catering sectors are currently responsible for approximately four per cent of the UK's annual greenhouse gas emissions, so it is vital that industry plays its part too (DEFRA, 2007).

(a) Food and Drink Manufacturing

Food and drink manufacturers can help curb carbon dioxide emissions by reducing the amount of energy they use. So far food manufacturers are having mixed success at doing so. They have managed substantial cuts in energy use by improving efficiency, although some of their progress has been cancelled out by the drive to increase productivity (DEFRA, 2006). But they will still have to increase their efforts if they are to reach their carbon dioxide emission reduction targets currently set at 20 per cent from 1990 levels by 2020 by the Government's Food Industry Sustainability Strategy (BFCC, 2007).

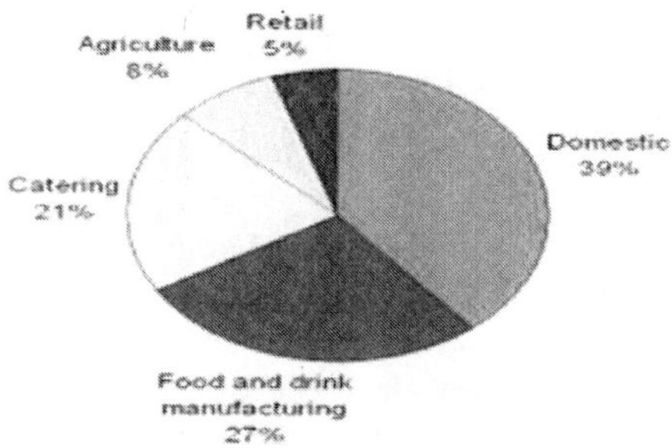

Chart-1The proportion of primary energy used during the different stages offood production in the UK. (Source – DEFRA 2006)

To achieve this it has been suggested that more research is needed to determine which areas of manufacturing are most wasteful, although some sectors such as meat and dairy processing have already been identified as using high amounts of energy (DEFRA, 2006). Some processed food such as ready-meals are also particularly damaging because they have to be cooked and cooled more than once, therefore adding to both the manufacturing and domestic emissions shown in Chart 1.

(b) Food Refrigeration

Garnett (2007) reported that the refrigeration has been identified as an area where dramatic emission cuts could be made relatively easily, by using and maintaining energy-efficient equipment correctly. This would be an important step to make because food refrigeration is estimated to account for up to 3.5 per cent of the UK's greenhouse gas emissions. One of the reasons for this being so high is the shift in our food culture and tastes. For example, increased consumption of processed foods and chilled drinks has increased refrigeration needs. So too has our desire to eat large amounts of meat and dairy – the food products that require the most refrigeration throughout their lifecycle, and therefore generate the most carbon dioxide emissions through energy use.

Similar to the issue of needing more refrigerated transportation, ambient temperature food storage may need to be supplemented with increased refrigerated storage as ambient temperatures rise or more intense heat waves occur. Depending on their locations, storage facilities may also need to adapt to climate change impacts such as flooding or droughts (SVG, 2014).

(c) Transportation

Food miles are now talked about by everyone from shoppers to retailers and the media. Crops and animals are shipped, driven and flown to factories, food is transported to shops and then we drive to the supermarket to buy it – but what impact does this actually have? This is a tricky question to answer because of all the different stages of transportation. The range of journeys and vehicles involved mean that estimates of the impact are extremely varied: estimates of emissions range from 1.8 per cent of the UK's carbon dioxide (DEFRA, 2005), to at least 3.5 per cent of our total greenhouse gases (Garnett, 2000). Although the current impact of food transport on climate change is low in comparison to other parts of the food system, the system is currently growing in a way that will increase food miles. It is, therefore, important that we curb this trend.

Extreme weather events and sea level rise may affect transportation infrastructure, which will affect delivery of produce and livestock to processing facilities as well as shipping of goods from the facilities. An increase in average temperatures also poses challenges in the way goods are shipped, as warmer temperatures may require a shift from ambient transport to refrigerated transport (SVG, 2014).

What is clear is that the food industry makes up a significant proportion of the UK's total transport emissions, as show by the following figures:

- Food transport accounts for one quarter of all heavy-goods vehicle miles in the UK (DEFRA, 2005).
- Around one in ten car journeys are for food shopping (NTS, 2007).
- Even though most of us could walk to the shops to buy basics like bread and milk, many of us don't. The food miles 'wasted' on unnecessary car journeys are thought to be the equivalent of over half a million transatlantic flights a year (NTS, 2007).
- The average number of miles that our food travels has doubled in 30 years (DEFRA, 2005).
- Air freight is the fastest growing mode of food transport – it accounts for 11 per cent of the food industry's transport emissions despite only carrying one per cent of the food and making up just 0.1 per cent of the food miles (DEFRA, 2005). The rise in air freight is a symptom of increasing globalisation of agriculture, promoted by the world trade

system – large-scale agricultural business gain strength, but local British farmers lose out.

The Government has identified some of the causes for the massive increase in food miles (DEFRA, 2005). First, food and feed is being imported and exported all over the world much more than ever before. The fact that fewer, larger companies are supplying is also contributing to the problem.

Farmers and the environment can benefit from relying on local and regional markets rather than on global markets where rules are not always fair and parameters not under their control. To ensure this we need fairer trade rules and national and international food policies that benefit developing countries and small farmers. But we can all play our part by supporting our local shops, eating seasonal produce and buying local produce whenever possible (BFCC, 2007).

The food system has a major impact on climate change at every stage of production, and this must be tackled. A lot of the solutions already exist – good livestock management techniques, organic farming and local food economies. But to reach our Government's goal of 80 per cent reductions in Greenhouse gases by 2050 further changes are needed. Relying on technological solutions alone is not enough because this often means that damaging activities are allowed to continue. Instead we need to confront the root causes of the food system's huge climate impact (BFCC, 2007).

As with transportation scheduling, processing scheduling may be affected by climate change impacts experienced by producers. Livestock and crop harvest times and yields differ depending on ambient temperatures as well as extreme temperatures and extreme weather events (SVG, 2014).

Food safety is likely to be impacted by climate change through several mechanisms. Food borne pathogens, such as cholera and mycotoxins, are likely to expand their geographic range, and outbreaks are often associated with extreme weather events. In the meat slaughter sector, higher temperatures would increase the costs of refrigeration, packaging, handling, and storage of perishable meats that are vulnerable to pathogens such as E-coli (SVG, 2014).

Processing facilities looking to expand due to product success or benefits reaped from new crops or greater yields may experience barriers. For example, those requiring water for their activities may be prone to water quality and quantity issues resulting from flooding, drought or salinisation (SVG, 2014).

Role of Industrial Biotechnology Techniques in Processing Industries

Industrial biotechnology, also known as white biotechnology, uses enzymes and micro-organisms to make biobased products in sectors as diverse as chemicals, food and feed, healthcare, detergents, paper and pulp, textiles and energy. Agricultural products, biomass and organic waste, including food processing waste and effluents (also referred to as renewable raw materials) are transformed into other substances, in the same way as crude oil is used as a feedstock in the production of chemicals. Existing energy infrastructure and production processes are largely based on fossil fuels, which result in high levels of GHG emissions and affects the climatic conditions. In a bio-based economy society is no longer wholly dependent on fossil fuels and industrial raw materials (OECD, 2011).

Industrial biotechnology techniques were first introduced within traditional industries, such as the food industry, where enzymes and yeasts are used in many food processing techniques, resulting in an increased efficiency and reduced environmental and GHG impact. The experiences developed in the food industry have found application in a number of traditional industries, such as the pulp and paper, leather and textile sectors. Enzymes and microorganisms can carry out processes that traditional manufacturing systems would perform with much higher quantities of energy, using aggressive chemicals, and producing significant amounts of potentially hazardous waste products. In addition, the biotechnological alternatives can deliver productivity improvements and can help to reduce the impact of GHG emissions (OECD, 2011).

Enzymes have been used in food manufacturing for hundreds of years, mainly based on fermentation by micro-organisms. The last ten years in particular have seen an increase in new enzyme applications in food. Before that, the dominant new enzyme innovations were aimed at the production of high fructose syrups from corn starch. Today, novel enzyme applications are also being implemented in baking, fruit and vegetable processing, brewing, wine-making, processing of vegetable oils, cheese manufacturing, and meat- and fish-processing. Many enzymes applications in the food industry are advantageous due to their impact on processing conditions in food manufacturing plants, where enzyme use may result in savings of energy and chemicals (OECD, 2011).

Recently a special amylase has been commercialised that diminishes the crystallisation of starch, reducing waste of bread by allowing bread to stay fresh and moist longer. This effect has provided industrial bakeries with new opportunities for changing their production and delivery setup in order to produce at larger, centralised bakeries and make fewer deliveries to retailers. The industry can save both money and energy, while less waste bread also means more efficient use of agricultural raw materials. There are significant environmental gains due to better utilisation of agricultural raw materials. A reduction of GHG emissions of up to 54 t per million loaves of bread sold can be obtained. The major contribution to the reduction of CO_2 emissions is also, in this case, the agricultural load. Some 65 per cent of the reduction stems from savings in the production of wheat, including agricultural emissions, production of fertilizer and traction. Some 15 per cent of the reduction comes from savings in energy consumption during milling and baking and 15 per cent comes from reduced transportation (OECD, 2011).

Provisioning services cover the products of renewable biotic resources including foods, fibers, fuels, water, bio-chemicals, medicines, pharmaceuticals, as well as the genetic material of interest to the CBD. The production, processing and consumption of these things all have consequences both for the net emission of greenhouse gases and for the capacity of the system to accommodate the effects of climate change (OECD, 2011).

In principle, evaluation of the biological causes of climate change requires estimation of the multiple ways in which the production, processing and consumption of foods fuels and fibers are associated with climate drivers—emissions of GHGs. Combustion of fossil fuels is the dominant source of CO_2, but agriculture is a major source of CH_4 and N_2O (OECD, 2011).

The purpose of this study is to fill the gap about the impacts of climate change on the food industry. Specifically, the objective of this study is to understand the decline of crop yield as a result of climate change and to determine whether the global impact on the food processing sector is positive or negative. Research focused on the effects of climate change on the food industry and the value chain, from farm to fork, is an essential component to develop solutions capable to meet the current challenges in agriculture.

Addressing root causes would also provide us with a range of benefits for biodiversity, health, pollution and natural resource use. This would lead to a more sustainable, robust and dynamic food system where farmers benefit and receive a fair price for their produce; where local industries thrive all over the world and consumers are provided with healthier food and a better living environment. So what would contribute to this change? Friends of the Earth are researching some of the above areas in detail in order to provide possible long term solutions and interventions needed by Government and Industry (BFCC, 2007). But in the meantime there are some easy ways for everyone to do their bit as recommended below (Sustainability Ventures Group, 2014).

Recommended Actions for Consumer

- Buy better quality meat – organic, locally produced, free range and from your local butcher. They're much more likely to be able to tell you where the meat is sourced from than the supermarket.
- Buy organically produced dairy products.

- Check where your food has come from – the further away, the less likely you are to be able to verify how it is produced. And it will have clocked up more food miles.
- Shop local! Local shops tend to be within walking distance, reducing car travel and helps local food producers survive.
- Seek out seasonal produce.
- Watch out for processed food such as ready meals, sauces etc. Buy fresh produce as much as possible.
- Try and buy foods with less packaging.
- Try and reduce food wastage – if you're not going to be able to eat a whole bag of fruits, just buy a few.
- Change the balance of what you eat. Generally, eating lower down in the food chain has less climate impacts i.e. more cereals and vegetables and less meat and dairy.
- Accept more variability in the appearance of your food. Imperfectly round tomatoes can still taste perfect!
- Write to your supermarket as asking them to stop sourcing meat that uses GM and imported animal feed.

Recommended Actions for Government

- Identify new meat slaughter rules that foster local meat processing.
- Ensure local policies/laws support food processing activities.
- Research marketing quotas rules to support local processing.
- Encourage food buyers to buy from local processors.
- Give processors info on how to become wholesale ready.
- Provide business training/support for food entrepreneurs.
- Publish directory processors, buyers, and facilities.
- Develop agri-tourism plan that profiles processors.
- Establish shared use processing facility with test kitchen.
- Access existing shared processing equipment/storage.
- Create online food ordering website.
- Create food hub with processing, storage, distribution, and retail sales.
- Establish community agriculture training/incubator programme.
- Create a food production/processing consortium/peer group.
- Create a food innovation district.
- Enhance local financing for processors.

References

BFCC (Briefing Food and Climate Change). 2007. Friends of the earth. Underwood St., London. www.foe.co.uk

DEFRA (Department for Environment, Food and Rural Affairs). 2005. The Validity of Food Miles as an Indicator of Sustainable Development. Group on Energy and Climate Change.

DEFRA (Department for Environment, Food and Rural Affairs). 2006. Food Industry Sustainability Strategy. Group on Energy and Climate Change.

DEFRA (Department for Environment, Food and Rural Affairs). 2007. Final submission of the Food Industry Sustainability Strategy Champions' Group on Energy and Climate Change.

Delgado, C., Rosegrant, M., Steinfeld, H., Ehui, S., Courbois, C. 1999. Live stock to 2020. The Next Food Revolution. Food, Agriculture, and the Environment Discussion Paper.

FAO, 2006. Livestock's Long Shadow : Environmental Issues and Options. Food and Agriculture Organisation. pp. 4, 5, 43.

Francis, P. 2004. The Impact of UK Households on the Environment, Economic Trends. National Statistics. 611.

Garnett, T. 2000. Wise Moves: Exploring the relationship between food, transport and CO_2, Transport 2000.

Garnett, T. 2007, Food Refrigeration: What is the Contribution to Greenhouse Gas Emissions and How Might Emissions Be Reduced? Food Climate Research Network.

National Travel Survey (NTS). 2007. National Statistics. Department of Transport. Personal Travel Factsheet.

Nelson, G.C., M.W. Rosegrant, J. Koo, R. Robertson, T. Sulser, T. Zhu, C. Ringler, S. Msangi, A. Palazzo, M. Batka, M. Magalhaes, R. Valmonte-Santos, M. Ewing, and D. Lee. 2009. Climate Change: Impact on Agriculture and Costs of Adaptation. IFPRI Food Policy Report, Washington DC.

OECD (Organisation of Economic Co-operation and Development) 2011 Industrial Biotechnology and Climate Change: Challenges and Opportunities. Directorate for Science, Technology and Industry. pp. 8-12.

Ojo, O.O. 1998. Quantitative and Qualitative Analysis of Wastes from Nigeria's Brewery Industry, M. Eng. Project, Dept. of Agric. Engineering, FUT, Minna, Nigeria.

P. K. Shetty, S. Ayyappan and M. S. Swaminathan. 2013. *Climate Change and Sustainable Food Security*, ISBN: 978-81-87663-76-8, National Institute of Advanced Studies, Bangalore and Indian Council of Agricultural Research, New Delhi.

Pearson, S. F. 2003. How to improve energy efficiency in refrigerating equipment, International Institute of Refrigeration (IIR) Intergovernmental organisation for the development of refrigeration.

Perrings, Charles. 2010. Biodiversity, Ecosystem Services, and Climate Change: *The Economic Problem.* The World Bank Environment Department. Environmental Economics Series. pp. 7-8.

The Sustainability Ventures Group. 2014. Creating Climate Change Resilience: Enhancing Food Processing in the CVRD. Agriculture andAgri-Food Canada and BC Ministry of Agriculture. pp. iii-iv.

UNIDO (United Nations Industrial Development Organisation). 2009. Industrial Development Report. Vienna, Austria. United Nations Industrial Development Organisation. 143 p.

World Bank. 2008. World Development Report 2008: Agriculture for Development. Washington, D.C.: The World Bank.

World Bank. 2010. World Development Report 2010: Development and Climate Change. Washington, D.C.: The World Bank.

46

Plastic Packaging Waste Impact on Climate Change and its Mitigation

V. K. Chandegara, S. P. Cholera, J. N. Nandasana, M. T. Kumpavat and K. C. Patel

Abstract

Use of plastic in multiphase aspects of human life projects critical environmental challenges such as Greenhouse gas (GHG) emissions and depletion of fossil resources, mainly due to the consumption of fossil energy. This paper reviews the impact of plastic packaging waste on climate change due to impact on earth's environment and their emissions of Greenhouse Gases (GHG) into the atmosphere. In this paper assessment of the effects of packaging plastics recycling on GHG mitigation and resource saving was described. The strategy for reduction of plastic waste and its recycling was suggested to curb the problem of climate change. Recycling of plastic waste and innovative packaging is the alternative to drastically reduce the emission of such gases and vapours that would otherwise have had damaging effects on the environment.The paper implicates minimum plastic use strategies and management of plastic packaging wastes to mitigate impact of climate change at local, state and national level.

Keywords: *Packaging plastic, Change, climate, mitigating, recycling, waste.*

Introduction

The Earth has gone through many natural cycles of warming and cooling during droughts, flooding and extreme weather patterns. Scientists have confirmed that the Earth's atmosphere and oceans are warming gradually as a result of human activity (Intergovernmental Panel on Climate Change-IPCC, 2007). This warming will exacerbate climate variability and ultimately, adversely impact food and water security around the planet. Central to global warming and climate change is the "greenhouse effect". Carbon dioxide (CO_2), Nitrogen Oxides (NO_X), Sulphur dioxide (SO2), dioxins, fine particles and other greenhouse gases entering the Earth's atmosphere by activities of everyday energy use and the way of management of the environment still contribute to the build-up of Green House Gases (GHG), which are directly released into the atmosphere. Climate change impacts are only one of a number of environmenta limpacts that derive from solid waste management options.

Plastic is the major source of food packaging as its versatility in terms of cost, availability and comfort. We generate large quantities of waste in the form of plastics that remain in the environment for many years and cause damage. Packaging represents roughly one -third of municipal waste. Packaging used once and then promptly discarded, it seems like only an ephemeral presence in our lives as it rushes from factory to landfill. Consumers benefit from packages because they protect products as they travel, whether fast food or refrigerators, no matter how far they have to go. A well-designed package is attractive and appealing to consumers, and inspires confidence of product safety.

Potential sources of distribution of plastic chemicals to the environment include waste water, sludge from sewage treatment plants, leakage from landfills, incineration fumes, and local, regional and global transport of chemicals from plastic waste. Methane is also emitted from landfills and other waste dumps. If the waste is put into an incinerator or burnt in the open, carbon dioxide is emitted.

The potential material recovery from recycling amounted to 903 kg of plastic granules per tonne of plastic waste and thereby it is assumed that an equivalent amount of virgin plastic production could be eliminated. The estimated net GHG emissions and net fossil resource consumption amounted to 853 kg CO_2-eq and -1,374 kg crude oil-eq per tonne of recycled packaging plastic respectively. The resulting negative values indicate that packaging plastic recycling can make a significant contribution to GHG emissions mitigation and can significantly contribute to avoiding the depletion of fossil resources. (Menikpura *et.al.*, 2014)

It advocates that instead of production of materials from earth's virgin resources and disposal of wastes by burning or degradation, minimal production from virgin materials and total recycling of wastes drastically reduce the emission of such gases and vapours that would otherwise have had damaging effects on the environment. This was based on the European Commission Study on Environment's Report, (Smith *et al.*, 2001), which showed that reduced dependence on fresh production of goods and overall source segregation of Municipal Solid Waste (MSW), followed by recycling, gives the lowest net flux of greenhouse gases, compared with other options for industrial production processes and treatment of bulk MSW.

On an average, 134 tonnes/day of source separated packaging plastic waste is collected and transported by private companies using compactor trucks followed by baling. The baled plastic is then transported to several facilities, which are situated in different prefectures, for recycling. Recycling activities contribute to the global environmental challenges of GHG emissions and the depletion of fossil resources (Menikpura *et al.,* 2011). However, material recovery from plastic recycling can offset the GHG emissions and resource depletion that would otherwise occur through the production of virgin resin.

The objective of this paper is therefore to encourage and advocate minimum plastic use, zero-waste recycling, and innovative packaging in production of goods and services which leads less energy and less carbon emissions. Sustainable packaging and packaging waste reduction means to promote cleaner environment.

Environmental impacts of plastic packaging waste

Packaging is used in business to contain and promote finished products for consumer sales and also to transport them to the point of sale safely. The function of packaging includes protecting products in shipping, offering consumers information, providing a branding billboard and giving us a way to carry things home. The most commonly used packaging materials are paper, fiberboard, plastic, glass, steel and aluminum.

Pollution problems of plastics

Industrial practices in plastic manufacture can lead to polluting effluents and the use of toxic intermediates, the exposure to which can be hazardous. There is growing concern about the excess use of plastics, particularly in packaging. This has been done, in part, to avoid the theft of small objects. The use of plastics can be reduced through a better choice of container sizes and through the distribution of liquid products in more concentrated form. A concern is the proper disposal of waste plastics. Litter results from careless disposal, and decomposition rates in landfills can be extremely long. Marine pollution arising from disposal of plastics from ships or flow from storm sewers must be avoided.

Production and wastes contribute to climate change

Much of the plastic waste that is disposed of in landfills decompose, resulting in the release of both methane and carbon dioxide. In 2008, 20 million tonnes of CO_2 equivalent (eq.CO_2) were released from the disposal of solid waste on land (Gregory, 2010). Energy consumption contributes directly to climate change by adding carbon-based molecules to the atmosphere in excess of naturally occurring amounts. Carbon molecules, primarily carbon dioxide from burning petroleum products, trap radiant heat and keep it from escaping from the Earth's atmosphere. The resulting warming of the air has continued to change in global climate. Materials consumption contributes indirectly to climate change because it requires energy to mine, extract, harvest, process and transport raw materials; more energy to manufacture, transport and dispose of waste products.

A life cycle analysis study has indicated that the use of plastics leads to significantly less energy consumption and emissions of greenhouse gases than the use of alternative materials (Pilz *et al.*, 2010). In 2009, around 230 million tonnes of plastic were produced and around 25 per cent of these plastics were used in the EU (Mudgal *et al.*, 2011).

The disposal of solid waste produces GHG emissions in a number of ways.

- The anaerobic decomposition of waste in landfills produces methane (CH_4), a GHG *21 times more potent* than carbon dioxide. Landfills are the top human-caused source of methane.
- The incineration of waste that could be recycled produces carbon dioxide (CO_2) as a by-product.

Consumers are increasingly willing to buy concentrated products in light weight refill packs for dilution at home.

Plastic bags and their nuisances

Plastic bags are popular with consumers and retailers as they are a functional, light-weight, strong, cheap, and hygienic way to transport food and other products. Most of these go to landfill and garbage heaps after they are used, and some are recycled. Once littered plastic bags, it find their way on to our streets, parks and into our waterways.

Plastic bags create visual pollution problems and can have harmful effects on aquatic and terrestrial animals. Plastic bags are particularly noticeable components of the litter stream due to their size and can take a long time to fully break down. Many carrier bags end up as unsightly litter in trees, streets, parks and gardens which, besides being ugly, can kill birds, small mammals and other creatures. Bags that make it to the ocean may be eaten by sea turtles and marine mammals, who mistake them for jellyfish, with disastrous consequences. The biggest problem with plastic bags is that they do not readily break down in the environment. It has been found that, the average plastic carrier bag is used for five minutes, but takes 500 years to decompose.

Issues relating to plastics – There are about 50 different groups of plastics, with hundreds of different varieties. All types of plastic are recyclable. To make sorting and thus recycling easier, the American Society of Plastics

Industry developed a standard marking code to help consumers identify and sort the main types of plastic. Before recycling, plastics are sorted according to their resin identification code.

Plastic food containers recyclability

Plastic food containers cannot be recycled to make new food containers for sanitation reasons.

But plastics used in the food industry can be recycled for other uses if they can be separated easily. For example, recycled PET plastic is used for carpet backing, fiberfill for sleeping bags or ski jackets, fiberglass tubs and shower stalls, paint brush and appliance handles, floor tiles, and more.

Recycled HDPE plastic is used for such things as trashcans, flowerpots, traffic cones, and plastic "lumber" for park benches, railroad ties, boat docks, and fences. Polystyrene can be recycled, but systems for doing this are not well established.

Rigid plastic containers must be identified by code numbers to assist in sorting for recycling:

#1 = PET (polyethylene terephthalate),

#2 = HDPE (high density polyethylene),

#3 = PVC (polyvinyl chloride),

#4 = LDPE (low density polyethylene),

#5 = P/P (polypropylene),

#6 = P/S (polystyrene),

#7 = other, including multi-layer.

The type of plastics (as per the resin identification code) and their most common uses are given below:

(a) **Plastic process scrap recycling** - Currently most plastic recycling in of the developed countries are of 'process scrap' from industry, i.e. polymers left over from the production of plastics. This is relatively simple and economical to recycle, as there is a regular and reliable source and the material is relatively uncontaminated. This is usually described as reprocessing rather than recycling.

(b) **Post-use plastic recycling** – Post-use plastic can be described as plastic material arising from products that have undergone a first full service life prior to being recovered. Households are the biggest source of plastic waste, but recycling household plastics presents a number of challenges. One of these relates to collection.

(c) **Mechanical recycling** - Mechanical recycling of plastics refers to processes which involve the melting, shredding or granulation of waste plastics. Plastics must be sorted manually prior to mechanical recycling. Recently, technology is being introduced to sort plastics automatically, using various techniques such as X-ray fluorescence, infrared and near infrared spectroscopy, electrostatics and flotation. Following sorting, the plastic is either melted down directly and molded into a new shape, or melted down after being shredded into flakes and then processed into granules called re-granulate.

(d) **Chemical or feedstock recycling** – Feedstock recycling describes a range of plastic recovery techniques to make plastics, which break down polymers into their constituent monomers, which in turn can be used again in refineries, or petrochemical and chemical production. A range of feedstock recycling technologies is currently being explored. These include: (i) Pyrolysis, (ii) Hydrogenation, (iii) Gasification and (iv) Thermal cracking.

Feedstock recycling has a greater flexibility over composition and is more tolerant to impurities than mechanical recycling, although it is capital intensive and requires very large quantities of used plastic for reprocessing to be economically viable.

Indicators for the quantification of the impacts

Appropriate indicators were identified in order to assess the environmental implications of packaging plastic recycling activities. It was found that all kinds of recycling activities are associated with a significant amount of fossil energy consumption in the form of diesel fuel, grid electricity and thermal energy. This contributes to *Conference Proceeding, 3R International Scientific conference on Material Cycle and Waste Management, 10-12 March 2014, Kyoto, Japan* greenhouse gas (GHG) emissions as well as fossil resource depletion. In contrast, the materials recovered as a result of recycling enable to gain environmental benefits from the avoided virgin production of such materials and related GHG emissions. In order to quantify the overall environmental implications of recycling activities, net GHG emissions potential and net fossil fuel consumption potential were therefore identified as the most relevant environmental indicators. In order to estimate the climate impact, all the GHGs were considered, such as carbon dioxide (CO_2), methane (CH_4) and nitrous oxide (N_2O). The GHG mitigation potential is expressed in the unit of CO_2 equivalent. Fossil fuel consumption potential is expressed in crude oil equivalents (42 MJ/kg, in ground).

Strategies to mitigate climate change for plastic packaging waste

Although no packaging is the best choice of all, it is not always practical. The need for any packaging should be evaluated in the research, design and marketing stages of a product. The goal should always be to reduce unnecessary packaging. The bulk delivery of solids and liquids to food industries and bulk retail sales from bins (including hardware products, produce, housewares, toys and other items) eliminate unnecessary packaging. Where the need for packaging exists, packaging should follow the 3R's hierarchy. Waste minimisation, recycling and re-use represent an important and increasing potential for indirect reduction of GHG emissions through the conservation of raw materials, improved energy, resource efficiency and fossil fuel avoidance (Bogner *et al.*, 2007).

The concept of Reduce Reuse Recycle is applicable to any sphere of life – whether it is in home front or in industries such as energy and power.

1. Reduction of packaging waste

Packaging should be reduced prior to the manufacturing stage, by designing and marketing products for the first "R". This means reducing the number of layers, materials and toxins at source. If people refuse plastic as a packaging material, the industry will decrease production for that purpose, and the associated problems such as energy use, pollution, and adverse health effects will diminish.

In general order of hierarchy, reduction occurs by:

1. Using less packaging and by meeting all or most of the 3R's hierarchy, including reuse and recycle.
2. Minimising the number of materials used.
3. Minimising the weight and volume of materials used.
4. Employing bulk delivery systems.
5. Product concentration resulting in smaller packages.
6. Using fewer toxic chemicals in the product and its packaging.
7. Utilising modes of shipping requiring less packaging and use of repairable pallets by manufacturers.
8. Using multi-layered, multi-material packaging. However, this usually makes the product non-recyclable (i.e. composites, laminates).

Minimising Packaging and Disposables: Reasons for zero-waste recycling are saving in energy, preservation of the natural environment and its resources, reduction in air and water pollution, elimination of need for landfill sites and employment generation. TEPA′s approach to waste prevention put a strong emphasis on Extended Producer Responsibility(EPR)—making producers responsible for changes in design and production to reduce the waste generated by their products and packaging, by:

(a) Restricting the weight of boxes.

(b) Banning disposable tableware at schools and government agencies.

(c) Reducing plastic bags and plastic packaging.

(d) Encouraging a reduction in disposable chop-sticks.

(e) Reducing disposable cups.

2. Reuse of packaging waste

Packaging should be designed to be reusable, refillable, returnable and durable to the greatest extent possible. Source reduction, Retailers and consumers can select products that use little or no packaging. Select packaging materials that are recycled into new packaging – such as glass and paper. Since refillable plastic containers can be reused for many times, container reuse can lead to a substantial reduction in the demand for disposable plastic and reduced use of materials and energy, with the consequent reduced environmental impacts. Container designers will take into account the fate of the container beyond the point of sale and consider the service the container provides.

In general order of hierarchy, reuse is achieved by:

1. Reusing/refilling commercially and redistributing refilled products.
2. Refilling by the consumer through dispensing systems at retail outlets.
3. Reusing containers which have been standardised to assist in reuse applications.
4. Refilling via a second package (i.e. smaller, concentrated containers or larger family-size packages).
5. Reusing in the home - INFREQUENTLY purchased, durable and distinctive containers (i.e. teddy bear peanut butter jars that can later be used as cookie or candy jars).
6. Reusing in the home - FREQUENTLY purchased containers (i.e. margarine tubs).

3. Recycle of packaging waste

Packaging should be designed to be recyclable and/or made with recycled content.

A package or packaging material is considered to be "recyclable" if there is a widely available and economically viable collection, processing and marketing system for the product/material.

In general order of hierarchy, packaging may be recycled in the following ways:

1. Recycling over and over back into its original packaging type (also known as primary or "c losed loop" recycling.

2. Recycling back into another recyclable, useful package/marketable product.
3. Recycling back into another non-recyclable product (also known as "open loop" or tertiary recycling). Examples include:
4. Durable and marketable goods such as synthetic carpet.
5. "Cascaded" (delayed disposal), short-lived or single use marketable items such as seeding flowerpots
6. Recycling into "show piece" product that is not marketable in quantity such as park benches made from disposable diapers.

4. Extended producer responsibility

Extended Producer Responsibility (EPR) is a waste management framework that seeks to shift the responsibility for managing the end-of-life of a product from the government and taxpayer to those in charge of designing and producing the product. The theory is that if a producer is burdened with the cost of disposing a product at the end of its life, it has an incentive to design the product for recyclability or reusability as well as to reflect on the environmental cost of the disposal.

Get plastic manufacturers directly involved with plastic disposal and closing the material loop, which can stimulate them to consider the product's life cycle from cradle to grave. Make reprocessing easier by limiting the number of container types and shapes, using only one type of resin in each container, making collapsible containers, eliminating pigments, using water-dispersible adhesives for labels, and phasing out associated metals such as aluminum seals. Container and resin makers can help develop the reprocessing infrastructure by taking back plastic from consumers.

5. Legislatively require recycled content

Requiring that all containers be composed of a percentage of post-consumer material reduces the amount of virgin material consumed.

6. Standardise labeling and inform the public

Standardised labels for "recycled," "recyclable," and "made of plastic type X" must be developed for easy identification.

7. Innovative and sustainable packaging

Innovative packaging design can mean that the amount of materials used is reduced and that all packaging can be reused or at least recycled. In other words, it is nothing but creating more 'sustainable packaging'. Using the minimum and most efficient packaging will increase your competitiveness and save your money, as well as attracting environmentally-aware consumers.

Designers should expand and evolve creative processes to include strategies like

(a) optimising material and energy resources;

(b) sourcing materials that are produced responsibly with environmental best practices, fair labour and trade;

(c) choosing materials that are non-toxic, bio-based and made from renewable resources; and

(d) designing for resources recovery or reintegration back into nature after end use.

Sustainability criteria for packaging

1. Is beneficial, safe and healthy for individuals and communities throughout its lifecycle;
2. Meets market criteria for performance and cost;
3. Is sourced, manufactured, transported and recycled using renewable energy;
4. Maximises the use of renewable or recycled source materials;
5. Is manufactured using clean production technologies and best practices?
6. Is made from materials healthy in all probable end-of-life scenarios;
7. Is physically designed to optimise materials and energy;
8. Is effectively recovered and utilised in biological and/or industrial cradle-to-cradle cycles.

Degradable plastic packages

Degradable plastic packages decompose over time from exposure to light, hydrolysis, biological organisms such as fungi or bacteria, or some combination of environmental factors. Currently, degradable plastics are used in such non-food items as garbage bags and disposable diapers. A food package must be a sufficient barrier to prevent contamination from the surrounding environment during the intended shelf life of the product. Also, the degradation or decomposition process must not release toxic products that could migrate into the food, making it unsafe. Standards for measuring acceptability of degradable plastics for use in the food industry currently are in development, and must be approved by the Food and Drug Administration.

8. Online shopping

Online shopping has meant a rise in goods being delivered direct to customers' homes. This means that more packaging is needed to supply the goods. You can encourage customers to return your packaging to you for reuse by supplying them with a "freepost" return label.

Conclusion

Plastic packaging waste generated a great problem in environment lead climate change due to GHG production. Plastic packaging waste management should reduce emissions of carbon-dioxide (CO_2), nitrogen-oxides (NO_x) and sulfur-dioxide (SO_2). In view of the issues described in this paper, it is recommended that minimum use of plastic, waste byproduct utilisation, recycling of plastic waste and innovative plastic packaging to curb the problem of climate change. The governments should encourage the following practices at local, state and national levels to implement waste reduction practices like minimum use, reuse, recycle, use recycled content products and take back the waste of plastic packaging.

References

Anselm, E.O. E. and S. N. Oluigbo, (2012). Mitigating the Impact of Climate Change through Waste Recycling, Research Journal of Environmental and Earth Sciences 4(8): 776-781.

Bogner, J., A.M. Abdelrafie, C. Diaz, A. Faaij, Q. Gao, S. Hashimoto, K. Mareckova, R. Pipatti and T. Zhang, (2007). Waste Management. In: Metz, B., O.R. Davidson, P.R. Bosch, R. Dave and L.A. Meyer (Eds.), Climate Change: Mitigation. Contribution of Working Group III to the Fourth Assessment Report of the Intergovernmental Panel on Climate Change. Cambridge University Press, Cambridge and New York, Retrieved from: http://www. ipcc.ch/pdf/ assessment-report/ar-4/ wg3/ar4/-wg3-chapter10.pdf, (Accessed on: August 19, 2011).

Gregory, J., (2010). Climate Change and Waste-The Missing Link. Retrieved from: http://www.ceilap.org/pdf/ EPR. Climate change Brief. Pdf, (Accessed on: August 20, 2011).

Mudgal, S., Lyons, L., and J. Bain (2011) Plastic Waste in the Environment – Revised Final Report for European Commission DG Environment. Bio Intelligence Service. Downloadable from http://www.ec.europa.eu/environment/waste/studies/pdf/ plastics.pdf

Menikpura, S.N.M., Yoshiaki T. and S.A. Janya (2014). The Implications of Packaging Plastic Recycling on Climate Change Mitigation and Fossil Resource Savings – A Case Study in Japan Conference Proceeding, 3R International Scientific conference on Material Cycle and Waste Management, 10-12 March 2014, Kyoto, Japan.

Narayan, P. (2001). Analyzing Plastic Waste Management in India: Case study of Polybags and PET bottles published by IIIEE, Lund University, Sweden, pp. 24-25 accessed at http://www.iiiee.lu.se/information/library/publications/reports/2001/Priya-Narayan.pdf

Pilz, H., Brandt, B. and R. Fehringer (2010). The impact of plastics on life cycle energy consumption and greenhouse gas emissions in Europe. Summary report by Denkstatt for Plastics Europe.

47

Climate Change and Cement Manufacturing Industries

D. S. Thanki, V. K. Chandegara, G.S. Kharadi, G. D. Gohil and N.B. Parmar

Abstract

The production of one tonne of Portland cement requires 1.5 tonnes of raw material. The production of Portland cement is highly energy intensive, consuming four to seven MJ of fossil fuel energy per kg, and releases approximately one tonne of carbon dioxide for manufacture of each tonne of Portland cement. The production of cement contributes six per cent of the global greenhouse gas emissions. Utilisation of unused industrial by-products such as slag, which otherwise would contribute to land pollution, as Portland cement substitute in concrete, results in less energy for the manufacture of cement while also reducing the emissions arising from the burning of fossil fuels. The paper discusses the greenhouse gas emission reductions which can be achieved by the use of slag as a partial replacement of Portland cement (blended cements) and as a full replacement of Portland cement by the use of alkali activated slag concrete. The global cement industry contributes around six per cent of all man-made CO_2 emissions and is consequently responsible for around four per cent of man-made global warming. CO_2 emissions trading is likely to be of huge importance to the industry in the future.

Keywords: *greenhouse gas emissions; global warming; Portland cement; slag blended cement; alkali activated slag concrete.*

1. Introduction

Greenhouse gases are a natural part of the atmosphere. They trap the sun's warmth, and maintain the earth's surface temperature at a level necessary to support life. The problem we now face is that human actions are increasing the concentrations of these gases, creating the prospect of global climate change. This is the enhanced greenhouse effect. Water vapour is the most important greenhouse gas, but human activities make little direct impact on the amount in the atmosphere. Humans have most impact on carbon dioxide, methane and nitrous oxide. Also making a relatively small contribution to climate change are various artificial chemicals such as halocarbons. Carbon dioxide emissions resulting from direct human activities have perturbed the global carbon cycle. Thus the CO_2 concentration in the atmosphere is now about 30 per cent higher than 200 years ago. Burning of oil, coal and natural gas, and the clearing and burning of vegetation, are the main causes of the rise.This gas makes the biggest contribution to the enhanced greenhouse effect (about 70 per cent). The other gases, which contributes to 30 per cent of the enhanced greenhouse effect, include methane (CH_4), Nitrous oxide (N_2O), chlorofluorocarbons (CFCs) and perfluorocarbons (PFCs). The Intergovernmental Panel on Climate Change (IPCC) has found that global mean surface temperature has increased by 0.3 - 0.6°C since the late 19th century. The Panel concluded, "The balance of evidence suggests a discernible human influence on global climate" (The Australian Greenhouse Office).

If emissions continue to grow at current rates, it is almost certain that atmospheric levels of carbon dioxide will double from pre-industrial levels during the 21st century. If no steps are taken to slow greenhouse gas emissions, it is quite possible that levels will triple by the year 2100. The most direct result, says the scientific consensus, is likely to be a "global warming" of 1 to 3.5°C over the next 100 years. If the current predictions prove correct, the climatic changes over the coming century will be larger than any since the dawn of human civilisation. The effects of climate change will vary around the globe. Temperature rises are expected to be greater towards the Poles than near the Equator, and over land than at sea. While rainfall is expected to increase in some areas and decrease in others, likely changes in rainfall for particular regions are still highly uncertain. Given the wide range of climates experienced across Australia and the large natural variability of Australia and its climate, it is hard to predict what the impacts of climate change will be and where they will occur. It is possible that there will be positive and negative impacts. These impacts may affect the distribution of plants and animals; the frequency of storms and floods; and the spread of weeds, pests and diseases, which may influence agriculture and human health (United Nations Environment Programme).

2. Environmental Impact by Cement Industry

There are limited natural resources such as limestone, coal, and natural gas, for manufacture of ordinary Portland cement. The production of one tonne of Portland cement requires 1.5 tonnes of raw material (Swamy, R. N., 1998). Cement production is a highly energy intensive production process. The energy consumption by the cement industry is estimated at about two per cent of the global primary energy consumption, or almost five per cent of the total global industrial energy consumption (Davidovits, J. 1994). Due to the dominant use of carbon intensive fuels, e.g. coal, in clinker making, the cement industry is also a major emitter of CO_2 emissions. Besides energy consumption, the clinker making process also emits CO_2 due to the calcining process. Approximately one tonne of carbon dioxide is emitted for manufacture of each tonne of Portland cement (Mehta, P. K.1994). It was estimated that the one billion tonnes of Portland cement produced worldwide during 1998 accounted for one billion tonnes of carbon dioxide release, or fiver per cent of the world carbon dioxide emission (Davidovits, J., 1994).

Three production steps are distinguished in the description of the production of cement:

1. Preparing raw materials: Mixing/homogenising, grinding and preheating (drying) produces the raw meal.
2. Burning of raw meal to form cement clinker in the kiln: The components of the raw meal react at high temperatures (900-1500°C) in the precalciner and in the rotary kiln, to give cement clinker.
3. Finish grinding of clinker and mixing with additives: After cooling the clinker is ground together with additives.

2.1 Carbon Dioxide Emissions from Energy Use

The clinker making process (step 2), is the most highly energy consuming step in the manufacture of cement as it involves high temperature processing. This process removes water from the raw meal, calcines the limestone at temperatures between 900-1000°C and finally clinker the kiln material at about 1500 °C. The theoretical heat requirement for clinker making is calculated to be about 1.75 MJ/kg (Taylor, H.F.W. 1998). The actual heat requirement is higher, and depends on the type of process applied. Dry process is generally less energy consuming and typically in the range of 3.5 to 4.5 MJ/kg, whereas the wet process consumes about 6.2 MJ/kg (Hendriks, C A., *et al* 1998).The total carbon dioxide emission during the cement production process depends mainly on the type of production process (efficiency of the process and sub-processes) and the fuel used (coal, fuel oil, natural gas, petroleum coke, alternative fuels). The CO_2 emissions in kg per kg cement produced for dry and wet cement production process for various fuels are presented in Table 1.

Most of the cement manufactured is for concrete making. Typically cement forms 10 per cent to 20 per cent of the concrete (by weight). An analysis carried out to illustrate the energy consumption of an edge beam through all the life cycle phases is shown in Figure 1. The energy consumption of cement production make up more than 90 per cent of the total energy consumption of all constituent materials and approximately 1/3 of the total life cycle energy consumption.

Table 1: Fuel-Related Carbon Dioxide Emissions in kg per kg Ordinary Portland cement

Dry Process				Wet Process			
Coal	*Fuel Oil*	*Natural Gas*	*Waste*	*Coal*	*Fuel Oil*	*Natural Gas*	*Waste*
0.89	0.81	0.75	0.57	1.09	0.95	0.90	0.57

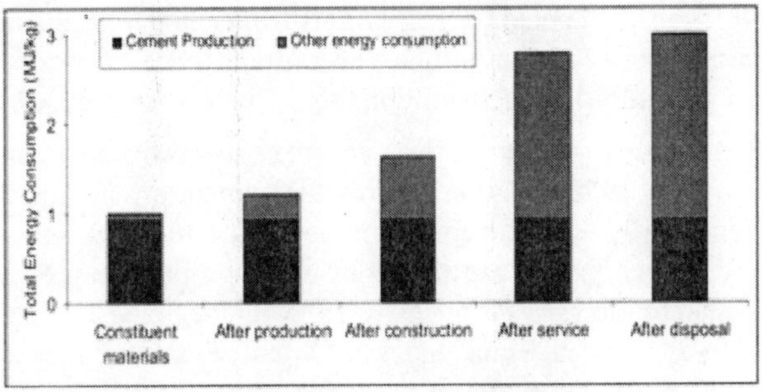

Fig. 1. Life Cycle Energy Consumption of a Concrete Edge Beam.

Carbon dioxide emissions in cement manufacturing also come directly from calcining the limestone in the raw mix. Only half of the emitted CO_2 originates from the fuel and the other half originates from the calcination of the raw material. Therefore, if the CO_2 emission were considered, nearly two-third of the total emission during the life cycle of the edge beam example would be attributed to cement production.

2.2 Carbon Dioxide Emissions from Calcination (Process Emissions)

Cement is produced from predominantly limestone, together with silica, iron oxide and alumina. The manufacture of cement involves pyro-processing the raw materials for long periods to produce clinker, which is then ground with gypsum to produce cement. During this process the calcium carbonate in the limestone is turned into calcium oxide, and this reaction is called calcinations. Process calcinations can be expressed by the following equation:

$$CaCO_3 = CaO + CO_2$$
$$(1 kg) = (0.56 kg) + (0.44 kg)$$

The share of CaO in clinker amounts to 64-67 per cent. The remaining part consists of iron oxides and aluminium oxides. CO_2 emissions from clinker production amounts therefore at about 0.5 kg/clinker. Ordinary Portland cement contains approximately 95 per cent clinker, contributing to approximately 0.48 kg emission for 1 kg manufacture of cement.

3. Global Trend in CO_2 Emissions from Cement Production

Global cement production grew from 594 million tonnes in 1970 to 1453 million tonnes in 1995 at an average annual rate of 3.6 per cent (Cembureau, 1995). The largest increase in cement production was recorded in China,where one-fifth of the world population contributes to one-third of the total cement world production. Estimated CO_2 emissions from cement production in 1994 were 1126 million tonnes, of which 587 million tonnes were from process emissions (calcination) and 539 million tonnes were from energy use.

These emissions account for five per cent of 1994 world carbon emissions (Marland, G *et al* 1998). Figure 2 shows the cement production from major world regions in 1994 and the corresponding CO_2 emissions from the cement manufacture. Depending on the efficiency of the production process, the CO_2 emission/cement ratio varies from 0.7 to 1.0, with the highest values estimated to be from North America, China, India and Middle East (Hendrics *et al.*, 1998).

For a long time, scientists have perceived the detrimental effects of increasing pollution on the atmosphere. The first international symposium began in Rio de Janeiro in 1992, followed by Kyoto (Japan) in 1997. At this conference many countries agreed to sign a protocol, from which they would commit to reduce their general greenhouse gas emission in such a way as to obtain a general five per cent minimum decrease to below the 1990 value during the commitment period, from 2008 to 2012.

Cement industries in many countries are already struggling to be competitive with huge new plants with minimised costs. The urbanisation also pushes old cement plants out of towns. The cement industries worldwide now have to face the concerns about emissions and new regulations. In many countries, the cement industries have made voluntary agreements to reduce the green house gas emissions. In Germany, the cement industry has signed an agreement to reduce the CO_2 emissions by 20 per cent from 1990 to 2008-2012. In France, an agreement has been signed by Ministry of Environmentand Cement industry to reduce 25 per cent CO_2 emission due to energy and fuels. In both countries, the industries has agreed yearly reporting of the progress, and consider the use of ground granulated blast furnace slag and fly ash as a way to achieve the goal (Marchal, G, 1998). In Denmark, the joint government and industry programme "Green Concrete in Denmark", is aimed to assist the industry in reducing the CO_2 emission by 25 per cent. Denmark produces 1.5 tonne of concrete per capita, which is about 50 per cent higher than the average per capita concrete consumption worldwide. However, only 2 per cent of Denmark's total CO_2 emission stems from cement and concrete production (Glavind, M, *et al* 1999).

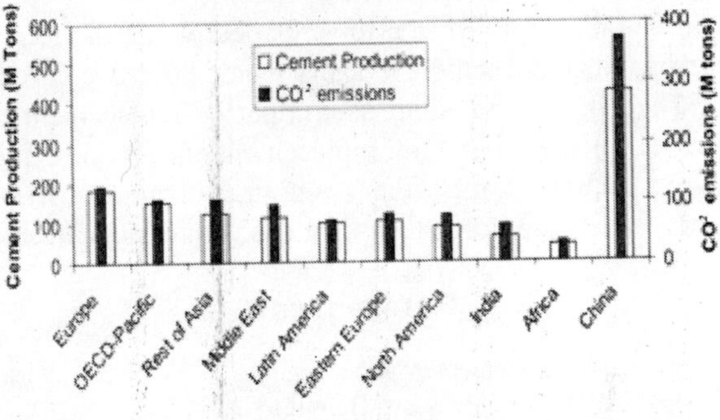

Fig. 2. Cement Production and CO² Emissions in Major World Regions.

4. Conclusions

The scientific studies show that if the current rate of greenhouse gas emissions continues then the resulting climatic changes over the coming century will be larger than any since the dawn of human civilisation. The manufacture of cement contributes about fiver per cent of the total greenhouse gas emissions. Significant reductions of these emissions can be achieved by replacing Portland cement with ground granulated blast furnace slag, which is an industrial by-product and would otherwise cause land pollution.

For a start, it is possible to decrease the proportion of calcined material in cement. This is an attractive option for cement manufacturers, since calcined material is expensive, and substitutes are very often cheaper. Such lower cost and low or zero-greenhouse-gas emission substitutes include slag, fly ash, limestone and the use of proprietary fillers such as Slagstar or Cemstar. All of these reduce the amount of clinker used to produce a tonne of cement, and so reduce the amount of CO_2 produced while producing a tonne of cement. For example, authors from Ecotrade have calculated that the total CO_2 emission per tonne of OPC produced is 970 kg, whereas for each tonne of ground granulated blast furnace slag produced, only 70 kg of CO_2 is produced. Substitution rates of over 33 per cent are achievable.

A second major step to decreasing the total amount of CO_2 produced by the global cement industry would be to decrease the proportion of cement in concrete. This may be possible through the use of higher-strength cement, such as Vicon (Sobolev and Podmore, 2001). This is a new type of cement produced with a siliceous admixture introduced during grinding, producing very high 1-day (44 MPa) and 90-day (96 MPa) strengths, allowing cement and concrete production with lower clinker content.

Although reducing the amount of cement in concrete might not be popular among cement producers, it is possible that they would be able to charge a premium for such 'supercement,' which would allow them to recoup their losses on volumes. A third step towards decreasing the total amount of CO_2 produced by the global cement industry would be to decrease the number of buildings. The slag can be either used a partial replacement of Portland cement (blended cements) or as a full replacement of Portland cements (alkali activated slag). In both forms, the resulting concretes offer sexcellent engineering properties, providing technical and economical incentives.

References

Cembureau (1995). "World Cement Market Figures 1913/1995", World Statistical Review No. 18, Cembureau, Brussels, Belgium.

Davidovits, J. (1994). "High-alkali cements for 21st century concretes", Proc. Mohan Malhotra Symposium on Concrete Technology: Past, Present, and Future, Mehta, P. K. (ed), ACI SP144, 383-398.

Glavind, M., Munch, C., Damtoft, J., and Berrig, A.,(1999). "Green Concrete in Denmark", Proceedings of the 19th Biennial Conference, Concrete 99 – Our Concrete Environment, Concrete Institute of Australia, May 1999, pp. 440 – 448.

Hendriks, C.A., Worrell, E., De J, D., Blok, K., and P. Riemer (1998). "Emission Reduction of Greenhouse Gases from the Cement Industry", International Energy Agency (IEA) Greenhouse Gas R&D Programme, Proceedings of the GHGT-4 in Interlaken, Switzerland, 30 Aug-2 Sep, 1998.

Marchal, G. "The eco-efficiency of blended cements", A study commissioned by Ecotrade, http://www.ecotrade.net/lavori/cement2_e.pdf

Marland, G., Boden, T. and Brenkert, A.(1998). "Revised Global CO_2 Emissions from Fossil Fuel Burning, Cement Manufacture, and Gas Flaring", pp. 1751-1995, Oak Ridge, TN:Carbon Dioxide Information and Analysis Center, Oak Ridge National Laboratory.

Mehta, P. K.(1994). "Mineral admixtures for concrete – an overview of recent developments", Proc. Sixth Engineering Foundation Conference Advances in Cement and Concrete, Durham, Grutzeck, M. W. and Sakar, S. L. (eds), American Society of Civil Engineers, 243-256.

Swamy, R. N.(1998). " Designing concrete and concrete structures for sustainable development", Proc. Sixth International Conference on Fly Ash, Slag, Silica Fume and Other Natural Pozzolans in Concrete", Bangkok, Thailand, Malhotra, V. M. (ed), ACI SP-178, 1, 245255.

Taylor, H.F.W. (1998), Cement Chemistry, Thomas Telford, 2nd Ed., p. 459.

The Australian Greenhouse Office, http://www.greenhouse.gov.au/

United Nations Environment Programme, Information Unit for Conventions (IUC), http://www.unep.ch/iuc/submenu/begin/beginner.htm

WEC (1995). "Efficient Use of Energy Utilizing High Technology: An Assessment of Energy Use in Industry and Buildings", World Energy Council, London, United Kingdom.

48

Effect of Climate Change on Import and Export of Food Products

V.A. Naliapara, M.H. Jethva and S. P. Cholera

Abstract

This paper discuss the effect of changing environmental condition on global trade of food products. Climate change will change recent cropping patterns and the market share of recent commodities related to countries. The severe effect will be first seen in equatorial countries like sub-Saharan countries and Latin America. The increased length of warmer region will show increase in cereal production but overall decline in crop yields. The world food system is linked system and one weak link will affect others. This may change food prices and GDP sharing and will affect food security in developing countries. Some recommendations are also reviewed at the end of paper.

Introduction

Climate change is a change in the statistical distribution of weather patterns when that change lasts for an extended period of time (i.e., decades to millions of years). Climate change may refer to a change in average weather conditions, or in the time variation of weather around longer-term average conditions (i.e., more or fewer extreme weather events). Climate change is caused by factors such as biotic processes, variations in solar radiation received by Earth, plate tectonics, and volcanic eruptions. Certain human activities have also been identified as significant causes of recent climate change, often referred to as "global warming" (ACC, 2010). This refers to the long-term effect on practices followed by human being such as agriculture, food habits, global trade etc.

Climate change is expected to influence crop and livestock production, hydrologic balances, input supplies and other components of agricultural systems. However, the nature of these biophysical effects and the human responses to them are complex and uncertain (Adams *et al.*, 1998).

Most studies and models on impacts of climate change on agricultural production indicate that there will be negative effects on crop yields over the next century. Some models predict that there will be an estimated 600 million additional people at risk of hunger if temperatures increase by 3° C (Warren *et al.*, 2006), particularly in developing countries where people are already at risk. On the other hand, models also predict that global cereal production will increase over the next century, even considering climate change, and appears to be sufficient to feed the projected world population of 9-15 billion in the year 2080 (Fischer *et al.*, 2005). Global cereal production under different climate change scenarios, however, is projected to be highly uneven between regions, with temperate regions on the northern hemisphere benefiting from increased temperatures and longer growing periods for moderate degrees of warming, while tropical regions will lose agricultural production potential.

This implies that fewer people will have the possibility to feed themselves (even if this is desirable) and, hence, that there will be greater reliance on markets and trade. Currently, some developing countries are particularly vulnerable to additional impacts of climate change on their ability to attain food security, due to their unfavourable positioning in international trade, compounded by poor development of domestic and regional markets. Being able to balance growing differences between food demand and production will mean paying greater attention over the next two decades to develop policies supporting trade, and putting in place the necessary infrastructure and institutions. The uncertainties with regard to climate change impacts on agriculture are considerable. Climate change impacts on international trade, markets and investments are even less clear. What can be anticipated is that climate change will affect the prices and volumes of goods traded between developed and developing countries, particularly agricultural raw materials and food, with wider macroeconomic consequences. Few studies so far have examined the effects of climate change on global trade patterns, but the consequences could be substantial, particularly for sea-borne trade and linked coastal manufacturing and refining activities (Stern, 2006).

Food import export and markets for growth and poverty reduction

Very few countries aim for, or achieve, food self-sufficiency; trade in food products is typical. Hence, the supply of food within a country is a function of: the volumes produced domestically, the price of imports (which in turn depends on global demand and supply), and the price of the exports used to generate foreign exchange. Climate change could affect all three of these variables.

The types of effect on import export

The price of imports can be affected both directly and indirectly by climate change. If increased production from the areas that will 'benefit' from climate change is smaller than the decline in those that will 'lose', global supply will fall and prices will rise. This direct impact would be additional to all of the others influences on global demand and supply (such as rising incomes and demand for meat in East Asia). It could also be affected indirectly by countries' responses to climate change. If, for example, concern over climate change leads to a large increase in the production of biofuels, this could, in turn, raise prices even further by diverting agricultural resources away from food production.

Changes to a country's export prices may reinforce or offset any change to prices of imports. There could be several – difficult to predict – indirect effects for non-agricultural exports. Will concern over climate change lead, for example, to a change in the pattern of manufacturing with increased use of some materials (with a lower carbon footprint) and declining use of others (with a higher one)? It would be easier to predict the potential changes affecting agricultural exports.

In cases where a country exports agricultural goods, the net impact of climate change will result from a combination of the impact on their production of exported crops and demand for these in the world. For example, if an exported crop/animal product is less susceptible to the adverse production effects of climate change in a particular country than those products destined for domestic consumption, the decline in domestically produced food supply may be offset by an increased ability to import (especially if world prices for the export crop also rise). By contrast, if the export crop is affected to the same (or a greater) extent than is domestic food production, the two adverse effects will reinforce each other.

Finally, as with imports, there is a range of potential indirect effects of climate change on exports that could result from the policy responses of different governments. Concern over 'food miles' could, for example, lead to a reduction in consumer demand (or the imposition of government controls) in export markets, thus, reducing the ability of developing countries to mitigate the adverse effects of climate change through trade.

These effects will bring the maximum population of world to poverty which is middle class and below middle class and biggest consumer of world commodities. The trade-induced poverty effects of climate change will thus be a combination of changes that:

- Alter the total supply of food within a country (from the combined effects of changes to domestic supply, to the volume of exports and to the relative price of imports and exports);

- Alter the balance of production between individuals who are currently net producers and those that are net consumers, between those that produce goods for the domestic market and for export, and between those who source their consumption domestically or from imports;
- Alter the structure of production between small and large farmers and between landed and landless labour.

Impacts of climate change on agricultural resources

Modelling of environmental constraints for agriculture shows that under the current climate about two-thirds of the global land surface – some 8.9 billion hectares – suffers severe constraints for crop cultivation: 13.2 per cent is too cold, 26.5 per cent is too dry, 4.6 per cent is too steep, 2.0 per cent is too wet, and 19.8 per cent has poor soils. Under climate change and by 2080, regional analyses predict an expansion of land with severe constraints. For southern Africa, these changes are particularly severe: an additional 11 per cent of the total land area of 265 million hectares will be at risk of being severely constrained for crop cultivation.

Especially in Africa, CO_2 fertilisation effects will be unable to counter the projected 20 per cent reduction in cereal yields (Parry *et al*, 2004). At country level, more mixed results emerge. A group of 40 countries worldwide will experience mean losses of 15 per cent of their cereal-productivity potential. By 2080, the total population of the currently over 80 food-insecure countries is projected to increase to about 6.8 billion (up from currently 4.2 billion). 20-40 poor and food-insecure countries may loose on average 10-20 per cent of their cereal-production potential under climate change. Simulations for sub-Saharan Africa predict that Sudan, Nigeria, Senegal, Mali, Burkina Faso, Somalia, Ethiopia, Zimbabwe, Chad, Sierra Leone, Angola, Mozambique and Niger would lose cereal-production potential by 2080 across all emission scenarios. The population of these countries currently is equivalent to 45 per cent of the total undernourished population in sub-Saharan Africa. Ten countries, on the other hand (Zaire, Tanzania, Kenya, Uganda, Madagascar, Cote d'Ivoire, Benin, Togo, Ghana and Guinea), with 38 per cent of the undernourished population of sub-Saharan Africa will gain cereal-production potential by 2080 (Fischer *et al*., 2005).

World food systems - linked food systems

While climate and farm management are key determinants of food production locally, agro economics and world trade combine to shape regional productivity significantly. The BLS (Basic Linked System) developed at IIASA (International Institute of Applied Systems Analysis) comprises of a series of national and regional agricultural models, providing for analysis of world food systems, where national agricultural economies interact with each other through international trade.

Global cereal production: Current world cereal production is modelled at 1.8 billion metric tonnes (G tonne). By 2080, BLS projects global cereal-production 4.8 G tonne. For developing countries, increases are projected to range from 2.3 - 3.2 G tonne – about a threefold increase in production from the 1990 baseline level (Fischer *et al.*, 2005). Africa is to gain even more as a consequence of the substantial economic development assumed in the scenarios. This global production forecast, excluding regional surpluses and deficits or necessary adjustments through trade or food aid, appears to be sufficient to feed the world population of about nine billion by 2080 (Fischer *et al.*, 2005).

Prices and agricultural GDP

Price changes resulting from climate change impacts are modest and range between two per cent and 20 per cent in the short to medium term. Higher output associated with a moderate increase in global mean temperatures will likely result in a small decline in real world food (cereals) prices, while global mean temperature changes in the range of 5.5°C or more could lead to a pronounced increase in food prices (Easterling*et al.*, 2007). Also, the impacts of climate change on agricultural GDP until 2080 are likely to be small at global level, and are estimated to range between -1.5 to +2.6 per cent, depending on the scenario. Agricultural GDP in mid to high latitude countries would benefit under climate change, whereas BLS results indicate decreases in agricultural GDP in low latitude regions, with the exception of Latin America (Fischer *et al.*, 2005).

Impacts on cereal trade

SRES (Special Report on Emissions Scenarios) simulations predict a growing dependence of developing countries on net cereal imports, totaling in 2080 between 170 million tonnes and 430 million tonnes, depending on scenario. The comparative advantage for producing cereals is predicted to shift towards developed countries, and net imports of developing countries increase by about 25 per cent, i.e. between 90 and 110 million tonnes of additional cereal imports. OECD countries, on the other hand, are projected to remain major cereal exporters, with net exports in 2080, 380 million ton (Fischer *et al.*, 2005).

Food security a major promise to developing countries

It is estimated that in 2080 about 768 million people (almost the same as in 1999, where the FAO estimated 776 million) will be undernourished (Fischer *et al.*, 2005). Regional differences between developed countries generally benefiting from climate change, and developing countries where crop production is projected to decline considerably are likely to grow stronger through time, leading to significant polarisation with a substantial increase in risk of hunger among the poorest countries (Parry *et al.*, 2004), especially in sub-Saharan Africa and South Asia, where still a large portion of the population will depend on agriculture, and where capacities (e.g. technologies, finances, investments, etc.), both at national and farm level to adapt to climate change, are lowest.

Climate change implications for food product import export and markets

Despite improvements in data availability and models, there are a number of uncertainties with regard to climate change impacts on agricultural productivity, especially in developing countries, where data is even scarcer. There are, however, a number of issues where general agreement exist (Peskett, 2007):

- The only certainties about the impact of climate change on agriculture are increasing uncertainty and variability and an increase in frequency and severity of extreme events (storms, hurricanes, droughts, etc.);
- There are opportunities for some countries under all but the most extreme scenarios (e.g. North America, Russia, China), leading to an expansion of potential agricultural crop land;
- These scenarios show relatively similar impacts on agriculture in the next one to two decades and predict impacts to be moderate during this period at global and regional scales. Impacts at smaller scales (intra-regional and within countries) may be much more severe but they are much harder to predict.

Recommendations

- Early and marked shift in public support to research and development, extension, market development, rural infrastructure, and services as to specifically, but not exclusively, benefit directly smaller farmers able to produce for export.
- Accompanying development of and support for safety net and social protection programmes for those smallholders unable to make necessary changes to their farm operations.
- Accompanying support to sustainable management of natural resources to make land resources more resilient against climate change impacts where feasible.

- Improve aid delivery mechanisms to increase the flexibility to incorporate climate change measures.
- Consideration of climate change in how such aid flows are used needs to become standard practice. Donors can support this process by developing and providing frameworks for the consideration of climate change and directing funds to support governments through this process.
- There is a need for more analysis of where investment related to climate change in agriculture should be targeted, the potential costs involved and the tradeoffs between mainstreamed versus non-mainstreamed approaches. Three areas are of particular importance in targeting: (1) supporting agriculture in areas where it is already promoting growth and where climate change is predicted to have potential positive effects; (2) supporting activities such as technology development that will help agriculture in all areas – both low and high potential; (3) supporting diversification out of agriculture and/or alternative forms of aid (such as insurance mechanisms and social protection) which will be especially important for economic growth and in situations where adaptation through agriculture is not an option.
- The level of funding available to agriculture in new funding mechanisms created to address climate change is unclear. Specific funding for agricultural adaptation needs to be identified and if necessary earmarked for access by developing countries.
- Agricultural markets and trade are likely to become more important under most climate change scenarios. Funds for adaptation should therefore be specifically targeted at areas that promote the development of a viable agricultural export sector to offset negative impacts of production for domestic markets, market development and improvement, and trade.
- The private sector will also play a crucial role in supporting agriculture under a changed climate and potentially offers new market opportunities, for example through carbon trading, the value of which could exceed ODA (Official development assistance). Donors can play a role in increasing opportunities for farmers to access this market through (1) engaging with international negotiations under the UNFCCC (The United Nations Framework Convention on Climate Change), and regional initiatives such as the EU (European Union) Emissions Trading System, to ensure that the interests of developing countries are taken into account in the design of these mechanisms and (2) help governments to support the development of the private sector to engage in regulated and voluntary carbon markets more effectively.

References

ACC. 2010. (America's Climate Choices: Panel on Advancing the Science of Climate Change; National Research Council). "Advancing the Science of Climate Change". Washington, D.C.: The National Academies Press. ISBN 0-309-14588-0, p. 1.

Adams, RM, McCarl, BA, Segerson, K, Rosenzweig, C, Bryant KJ, Dixon, BL, Conner, R, Evenson, RE, and Ojima, D. 1998. "The economic effects of climate change on U.S. agriculture, Chap 2." In: Mendelsohn R, Neumann J (eds) The economics of climate change. Cambridge University Press, Cambridge (in press).

Easterling, W.E., P.K. Aggarwal, P. Batima, K.M. Brander, L. Erda, S.M. Howden, A. Kirilenko, J. Morton, J.-F. Soussana, J. Schmidhuber and F.N. Tubiello, 2007. "Food, fibre and forest products.' Climate Change 2007: Impacts, Adaptation and Vulnerability. Contribution of Working Group II to the Fourth Assessment Report of the Intergovernmental Panel on Climate Change", M.L. Parry, O.F. Canziani, J.P. Palutikof, P.J. van der Linden and C.E. Hanson (Eds.). Cambridge University Press, Cambridge. pp. 273-313.

Fischer, G., M. Shah, F.N. Tubiello and H. van Velhuizen, 2005. "Socio-economic and climate change impacts on agriculture: an integrated assessment 1990 – 2080." Philosophical Transactions of the Royal Society, Vol. 360, pp. 2067-2083.

Parry, M., C. Rosenzweig, A. Iglesias, M. Livermore and G. Fischer 2004. "Effects of climate change on global food production under SRES emissions and socio-economic scenarios". Global Environmental Change14 (1): 53-67.

Peskett, L., 2007. "A rough guide to climate change and agriculture." Draft. ODI, London.

Richard, M. Adams, Brian H. Hurd, Stephanie Lenhart, Neil Leary.1998. "Effects of global climate change on agriculture: an interpretative review", Climate Research, Vol.11, pp. 19-30.

Stern, N. 2006. "Stern Review on the Economics of Climate Change". Her Majesty's Treasury and the Cabinet Office. London.

Warren, R., N. Arnell, R. Nicholls, P. Levy and J. Price, 2006. "Understanding the regional impacts of climate change." Report prepared for the Stern Review. Tyndall Centre Working Paper 90. Tyndall Centre, Norwich.

49

Pitcher Irrigation, the Cheapest, Easiest and Most Effective Indigenous Micro Irrigation Method for Orchard Crops and Summer Vegetables

Prafulla Kumar Mandal

Abstract

Western tract of the State of West Bengal is the extended part of the eastern Chhotanagpur plateau. In the up-lands of this vast area, fruit crops can be grown by proper technology with minimal irrigation. Pitcher Irrigation method was standardised through research for the fruit plants like Guava (Psidium guajava), Lemon (Citrus limonia). The main tool Pitcher was standardised. Paste made of clayey soil and sieved saw dust of 30—50 mesh at 60:40 ratio by volume mixed with which Pitcher of 10-12 lts water holding capacity was prepared and burnt hard. It was installed with lid cover 30 cm near the base of the pre-planted plant or at a time together and packed with the mixture of soil and decomposed Compost at 50:50 ratio. It was filled with un-turbid water covered with lid, when 60 per cent got exhausted, was refilled. It moistened the adjacent soil with capillary water. Active growing roots approached the Pitcher toward the moist soil for its geotropic and hydrotropic character and absorbed. Further, the method on the summer, Pumpkin and Tomato, was found equally effective. In summer Pumpkin Cucurbita moschata), seeds at four places around the Pitcher at 15 cm apart were sown /dibbled. In Tomato (Solanum lycopersicum) four seedlings were planted. It was also effective in coastal saline area of the State of West Bengal. The expenditure of installation per hectare for 205 no. of Guava (Var L-49) and 494 no. of Lemons (Var. Gandharaj) were Rs 8138 and Rs 19612 respectively while installed together pitcher and planting saplings. And while only pitcher was installed in the pre-planted saplings Rs 5379 and

Rs. 12962 respectively. (Estimates in Indian Rupee 2011). Method in comparison with (1) Flood, (2) Ring and (3) Basin in regards to water requirement and performance on the fruit plants of newly planted orchard to fourth year during February to May was for 205 Guava and 494 Lemon (1) Flood 7500000 and 7500000, (2) Basin 259490 and 628370, (3) Ring 577040 and 1397350, (4) Pitcher 48210 and 212480 liters respectively. In Pitcher it was least. Average no. of fruits per plant were in (1) Basin 220, drop 14.54 per cent, (2) Ring 239, drop 12.13 per cent, (3) Flood 160, drop 25.0 per cent and (4) Pitcher 342, drop 4.97 per cent; fruits drop were in (1) Basin 14.54 per cent, (2) Ring 12.13 per cent, (3) Flood 25.0 per cent and (4) Pitcher 4.97 per cent. Thus, fruit yield in Pitcher irrigation was highest and drop least. In adaptive trail, it was equally efficient in Pomegranate (Punicum granatum), Sweet Orange (Citrus sinensis), Lime (Citrus aurentifolia), Custard apple (Annona squamosa). Once Pitcher is installed it functions up to six years. Water was brought by Bullock cart placing two Barrels of 200 liters each and in each occasion and two unskilled labourers could do.

Introduction

In absence of rainfall, irrigation is required for growth, development and reproduction of fruit plants. Fruit plants are grown in uplands at wide gaps. The Western Tract of the State of West Bengal is the extended part of the eastern plateau of India, i.e. Chhotanagpur Plateau. It is called Red and Lateritic Tract/Western Tract of the State of West Bengal. The features are rainfall comparatively low than rest of the State, erratic, uncertainty of arrival–departure-intermittent cessation of monsoon, topography undulating, sloppy, rough, hillocks, gravelly coarse soil having low moisture holding capacity, high rate of run-off and high rate of soil erosion, extreme air temperature, high rate of Pan Evaporation and drought susceptible. About 60 per cent of the area comprises upland situation. Fruits and plantation crops can be grown in uplands by adoption of proper technology with irrigation in summer months.

Back ground:- Conventional methods of orchard irrigation- Mainly three conventional methods of surface irrigation, viz **(1) Flood, (2) Ring and (3) Basin** are adopted in the fruit orchard.

In the **Flood method,** entire field is irrigated with water of 5 – 7.5 cm depth in each irrigation. In **Ring method**, Rings around the base of the fruit plant of 1.0 – 1.2 meter internal dia. 20 cm width 25 cm depth is excavated which are connected by excavating shallow-depth and narrow-width surface channels for water transmission to the rings.

In **Basin method,** the Basin around the base of the fruit plant of 1.2- 1.5 meter dia and 7.5 -8 cm depth is constructed. Basins are connected similar to Rings. In these after each irrigation once enough water, follows to rapid depletion to least, prior to next irrigation in the field, plenty water is necessary, involves high cost but less irrigation efficiency, while water is precious. In uneven topography, uplands, sloppy lands, in hill flanks, these irrigation methods mostly do not fit, not possible, not feasible.

Other sophisticated methods are (1) Drip and (2) Sprinkler irrigation. In the Drip method, emitters drippers are fitted with laid pipes at the required interval/spacing so that, the water comes out, falls drop by drop at the base of the plant at uniform rate for desired period. In Sprinkler method, vertical standing pipes are fitted with the laid pipes and at the top of those, sprinklers are fitted. Water is sprayed out circulating around each sprinkler in the form of artificial rains for the desired quantity and period. These two methods Sprinkler and Drip are highly sophisticated, mechanised, and efficient; but involve high initial cost, need big sizes plot, continuous maintenance, pump set either electrically or fossil fuel operated in the water source site and need expert operator. In spite of all these, other constraints observed are that (1) difficult to lay pipes over uneven undulating sloppy topography, (2) damage of pipes by non-pest animals, (3) clotting of drippers/ emitters, (4) interrupted flow/exit of water, (5) out of total emitters/drippers while few dysfunctions, it is another difficult to apply water to those left outs, (6) the top soil is more moistened, roots grow superficial layers, (7) considerable evaporation of applied water takes place as the application is over the surface and in addition, in the vacant space is also irrigated, (2) length of spray of water is regulated by wind velocity and direction. Because of these practical constraints and limitations, till time these two sophisticated methods are yet to spread enough and adopt in large scale by the farmers.

The Research approach and effort for an alternative:- A suitable irrigation method that will be of low cost, easy, minimum water but maximum efficiency, least wastage, minimum device. In the fruit crops, the irrigation at the very first stage during summer months is very much important, crucial and essential. After three to four years. its root system grows deeper and spreads laterally when, it can collect soil moisture from greater sphere of the field that can withstand weather aberration to some extent. Then, it can meet additional water need by the life saving irrigation. A long-term research programme was launched by the author to fulfill all the criteria that were desired by the peasantry, mentioned in the introduction hereinbefore. It was **Pitcher Irrigation,** a Micro /In-situ irrigation for the fruit plants like Guava (Psidium guajava), Sweet Orange (Citrus sinensis). Generally the pitcher is prepared by the artisans i.e. Potters with the processed soil/mud on the wheel and burnt on the Potters furnace. But the essential **Pitcher** for the purpose should be somewhat different, that can fulfill the purpose.

Research programme, components and observation- The programme was long-term. Research location: was in between 22°20'- 22°30' North Latitude and 87° 20'- 87° 24' East Longitude, Soil Conservation Research Station, in the district Midnapore, State West Bengal. General weather-During at summer months hot wind blows with low Relative Humidity from the western parts of the plateau. High rate of evaporation. Water source-Open well. Representative area- Vast 2.2 million ha within the State of West Bengal in the Western Tract. Likewise further million hectares of area exists in other states also.

Programme consisted of (1) standardisation of Pitcher, (2) Guava and Lemon fruit crops, (3) comparison with other three conventional methods, (4) Growth and development of plants, (5) production in each methods. (6) its application in other fruit crops and summer Pumpkin and Tomato. Land Capability Classes- IV and VI. Soil - Red and Lateritic soil character and physical properties of the sites were determined by soil survey. Profile of the soil describing physico-chemical properties are cited below.

Standardisation of Pitcher. The Pitcher should be durable, hard, without leak, no crack, but should release water by auto control mode to moisten the adjacent soil by blotting action, so that the entire is Capillary water. After several searches, the charrable bio-material identified was the **Saw dust of wood**. Soil containing Clay 60 per cent + Silt 20 per cent + Sand 20 per cent (+ 5 per cent) i.e. the Clayey texture AND saw dust of 30-50 mesh by volume 60 per cent and 40 per cent respectively and some water at appropriate consistency are thoroughly mixed to prepare the paste (slury) to stand the figure of the Pitcher. The Pitcher of 10-12 ltrs water holding capacity 4-5 mm thick is prepared with the said mixture/paste on the Potter's wheel by the artisans Potter. The moist shape is air dried and burnt vary hard on the Potter's furnace. Leak-less, crack-less Pitchers are selected. For charring of the saw dust, its body contains numerous micro pores which permit out water from inside of the Pitcher when it is kept attached with the dry soil. The characteristics/ properties of the desired Pitcher, its mode of use and performance for the purpose observed are - (1) The raw Pitcher could get shape with the mud and saw dust mixture during making and before burning; (2) When the raw pitcher was burnt in to the furnace, the bio-material (vegetative) i.e. saw dust charred; (3) The strength of the burnt Pitcher was hard that gave metallic sound; (4) Due to charring of saw dust within the mud and saw dust mixture, the pitcher contained numerous micro pores, but no macro leak; (5) After filling the pitcher with water, no water came out in the form of drop or spring but got high rate of permeability; (6) Pit was excavated and the Pitcher was installed near the base of the pre-planted plant or both at a time together. After placement in to, was packed with the mixture of soil and Compost at 50:50 ratio; (7) Pitcher was filled with un-turbid water. When 60 per cent depth of water in the pitcher got exhausted, was refilled, which depended as per the dryness of the soil arisen out of air evaporation and absorption by the root system of the plant. The soil has the universal blotting action. It was observed that, the adjacent soil got moist and the depth of the water of the pitcher got decreased; (8) Leakless and uncracked hard Pitcher functioned very well for auto exit of the water by blotting action of the adjacent soil in the process of imbibitions. The rate of exit of water from the pitcher was according to the dryness of the soil; (9) The active growing roots approached onwards the Pitcher and also encircled it out side by advancing onward to the moist soil and absorbed the capillary water. It is mentionable that, about 20-25 per cent of the entire roots of the plant act for active absorption of soil moisture, whereas about 75-80 per cent acts for anchorage i.e. mechanical function; (10) Only one Pitcher within 30 cm of the plant can serve the purpose.; (11) Entire moisture status of the soil was always uniform. Active roots grew deep in to the soil as per its habit of geotropism and onwards to the source of moist soil as per its habit of hydrotropism and naturally encircled the Pitcher which moistened the adjacent soil. The soil held the capillary moisture 100 per cent of which is hold the capillary moisture 100 per cent of which is absorbable by the plant. As the fruit plant is non-aquatic, it always prefers **Capillary water** in the soil. (12) Growth and development of the plant influenced by the Pitcher irrigation were measured. Desiccation effect on leaves, flowering, flower drop, fertilisation of flowers, fruiting, fruit drop, fruit weight, total no. of fruits per plant, total fruit production per ha were studied. The response of irrigation on the plants were studied. Growth and development rate, leaf fall, flowering, flower fall, fruiting, fruit drop, weight of fruit, smoothness of the surface, total yield per plant etc. were studied to ascertain the efficiency of the irrigation method.

Comparison with -Conventional irrigation methods i.e. (1) Flood, (2) Ring and (3) Basin as to the economy, total requirement of water, utilisation of water and moisture fluctuation of the soil after irrigation at three depth classes (i) 0-15 cm, (ii) 15-30 cm , (iii) 30-45 cm were studied at depth.

Irrigation to the plant is given when moisture content of the soil depletes to such quantum that the root system becomes incapable to absorb. Other hand in presence of Gravitational water for consecutive days, the root system not only becomes defunct but ferments/rots and suffocates. Hence such quantity of water is useful which maintains Capillary water status of the soil. In the Pitcher irrigation, the soil always attains the Capillary water, not less not more.

Critical stage of plants when irrigation is essential. At the very beginning of planting of fruit plants, the moisture content of the soil is very crucial till the root system grows deep to deeper and spreads laterally. **Timings of initial installation of Pitcher**. Two moments are considered as appropriate (1) Together with planting of Sapling/Graft/ Gooti/ Cutting/Air laired of fruit plant i.e. during rainy season. This is shown in the diagram. (2) Planting of Sapling/Graft/Gooti /Cutting/Air laired of fruit plant alone during rainy season but installation of Pitcher subsequently before summer month.

The method of installation of pitcher- (1) Together with planting of Graft/ Air laired Gooti of fruit plant i.e. during rainy season. Vertical section if the view is shown in Fig 1. Pit of the dimension of top and corresponding bottom length 90 and 75 cm respectively; top and bottom width correspondingly 60 and 45 cm respectively, depth 60 cm is excavated. The excavated soil is sunned and grounded and mixed with decomposed organic manure/ Compost at 50:50 ratio and required fertilisers as well as termicide (insecticide against termites). The mixture should contain some moisture. Graft/ Air laired Gooti/Cutting of fruit plant is planted in one side of the pit along its length at required depth. The Pitcher is placed at 30 cm apart from the plant along the length of the pit and at such depth that its mouth level remains 7.5 cm above the ground level with lid. Pit is back filled with the mixed soil and both Pitcher and Graft/ Air laired Gooti /Cutting of fruit plant are packed firmly. The pitcher is filled with water. The one time expenditure involves as on 2011(up dated) for Guava plantation of 205 no. of Air laired Gooti per ha is Rs 8,138 and for Lemon plantation of 494 Saplings /Cuttings per ha is Rs 19612 inclusive of overhead Table 1(a).

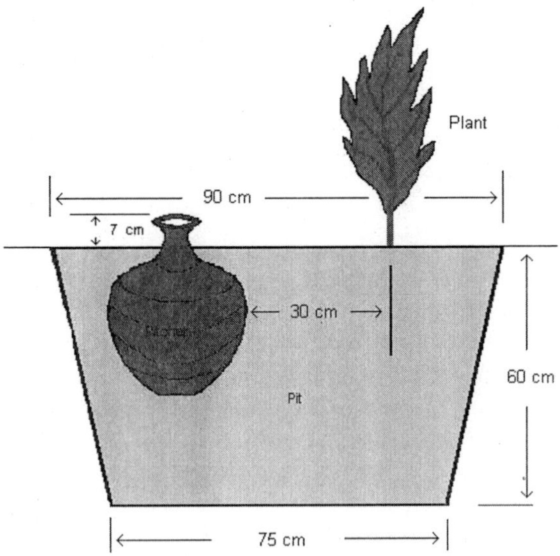

Fig. 1. Vertical Section of installation of special type of burnt Pitcher and sapling of fruit plants at a time [B.W.= Bottom width ; T.W.= Top width ; B.L = Bottom level; D=Depth].

In this expenditure, planting of sapling is added with that of installation of Pitcher as both are done together. (2) Planting of Graft/Air laired Gooti/ Cutting of fruit plant during rainy season and installation of Pitcher during subsequent summer month. Vegetative propagated (Graft/Air laired Gooti/Cutting) fruit plant is planted normally as usual. During the advance to subsequent summer season, the pit of the dimensions both top and corresponding bottom length and width cm and depth 50 cm is excavated. The Pitcher is placed at 30 cm apart from growing Plant within the pit and at such depth that its mouth level remains 7.5 cm above the ground level and covered with a lid. The pit is back filled as

Table 1(a): Expenditure per hectare for installation of Pitcher @ one per Sapling of fruit plant per pit together with planting at a time during rainy season. (Rate in Rs)

Sl.	Items	Rate	Guava	Lemon
1.	**Pitcher** of 10-12 ltr water holding capacity	12.00	2460	5928
2.	Excavation of pit for **Pitcher**.(Size –Top length 0.90 m X breadth X 0.60m ;Bottom length width 0.75 cm X 0.45 cm X depth 0.60 m = 0.26 cum@ Rs.51.66 per cum) (Rs.73.80 /cum - less 30 per cent due to no lead)	13.43	2753	6634
3.	Mixing of compost + in situ soil, placing of **Pitcher** in to the pit, backfilling with the mixture, compaction, initial filling with water etc. @Rs.47.60 per cum	12.38	2538	6116
	Sub Total(1+2+3)	-	7751	18678
4.	Add overhead 5 per cent (reserve)	-	387	934
5.	Grand Total Per hectare plantation	-	8138	19612

(Estimates in Indian Rupee. Standard Scheduled of Rates PWD, West Bengal- 2011)

Table 1(b): Expenditure per hectare for installation of pitcher @ one per sapling during summer after the plantation of sapling during preceding rainy season. (Rate in Rs.)

Sl.	Items	Unit	Rate	Guava	Lemon
1.	Special type of **Pitcher** of 10-12 ltr water holding capacity	Each	12.00	2460	5928
2.	Excavation of pit for **Pitcher**.(Size –Top length 0.50 m X breadth X 0.50m ; depth 0.50 m = 0.125 cum@Rs.51.66/ cum) (Rs.73.80 / cum - less 30% due to no lead and very little lift)	cum	6.46	1324	3191
3.	Mixing of compost + in situ soil, placing of **Pitcher** in to the pit, backfilling with the mixture, compaction, initial filling with water etc. all complete @Rs.47.60 per cum	Each	5.95	1220	2939
	Sub total(1+2+3)	-	-	5004	12058
4.	Add over head 7.5 per cent (reserve)	-	-	375	904
5.	Grand Total Per hectare plantation	-	-	5379	12962

(Estimates in Indian Rupee. Standard Scheduled of Rates PWD, West Bengal- 2011)

before. The Pitcher is filled with water. One time expenditure involves as per estimates of 2011 (up dated) for Guava plantation of 205 no. of Air laired Gooti/ per ha is Rs. 5379 and for Lemon plantation of 494 Saplings /Cuttings per ha is Rs 12962 inclusive of overhead. Table 1(b).

How many times and when to fill the Pitcher. It has been observed that exist of the water of the Pitcher is auto controlled. It depends on the dryness of the soil but the Pitcher has to be refilled when 75 per cent water got exhausted.

Comparison of water need of four irrigation methods. For appreciation, a comparison study of water need for irrigation during summer months *i.e.* February to May among the methods (1) Flood, (2) Basin, (3) Ring and (4) Pitcher was done. Irrigation was given to newly planted orchard and up to 3 years. The observation was taken in newly planted orchard and up to third year, presented in Table 2.

Table-2: Comparison of water need in four irrigation methods during the summer months.

Method	*Month*	*Nos. of occasion*	*Water needed (in liters) for the fruit plants*	
			Guava 204 nos.	*Lemon 494 nos.*
Flood	February	2	12,50,000	12,50,000
	March	3	18,75,000	18,75,000
	April	4	25,00,000	25,00,000
	May	3	18,75,000	18,75,000
	Total	**12**	**75,00,000**	**75,00,000**
Basin	February	2	43,250	1,04,370
	March	3	64,870	1,57,090
	April	4	86,500	2,09,460
	May	3	64,870	1,57,090
	Total	**12**	**2,59,490**	**6,28,370**
Ring	February	2	96,170	2,32,890
	March	3	1,44,260	3,49,340
	April	4	1,92,350	4,64,780
	May	3	1,44,260	3,49,340
	Total	**12**	**5,77,040**	**13,97,350**
Pitcher	February	4 - 5	9,940	24,850
	March	5 – 6	11,840	28,790
	April	6 – 7	13,650	35,370
	May	6 – 7	13,140	32,470
	Total	**21 - 25**	**48,210**	**1,21,480**

Thus, least quantity of water was needed in Pitcher Irrigation as against several more lakhs of liters of water in all other methods. Depletion of soil moisture after each open surface irrigation in the methods (1) Flood, (2) Basin, (3) Ring was measured to know moisture status at three depth (0-15 cm, >15-30 cm, >30-40 cm) i.e. root zone commencing two days after irrigation during summer month shown in Table 3.

It was observed that, after the open surface irrigation, moisture content of top sail was more that subsoil, whereas afterwards, the moisture content of

Table 3: The moisture content worked out by dry weight in per cent cited in the table below-

Method	*Depth (cm)*	*Moisture depletion per depth in cm below Ground level per cent by dry Weight days after irrigation*								*Irrigation per plant (In liters)*
		1	*2*	*3*	*4*	*6*	*8*	*11*	*13*	
Flood per 49 sqm	0-15	17.1	14.1	8.6	7.0	4.1	3.3	3.5	2.3	3064
	15-30	16.1	14.1	9.9	8.4	5.5	5.8	4.1	3.3	
	30-45	11.2	12.2	9.9	7.1	6.2	5.4	4.4	3.7	
Ring per 2.261 sqm	0-15	16.1	14.5	11.3	10.6	6.5	5.4	4.1	4.2	236
	15-30	16.3	15.3	12.5	11.7	8.4	6.3	5.0	4.2	
	30-45	14.3	13.7	12.0	11.7	8.6	6.8	5.6	4.9	
Basin per 3.5325 sqm	0-15	14.2	8.6	8.2	7.3	6.0	5.3	4.0	3.1	106
	15-30	15.3	12.1	10.1	8.4	6.6	5.8	4.9	3.3	
	30-45	12.8	11.1	10.6	10.0	6.8	6.5	5.8	3.7	

top soil was less than sub-soil. With the advancement of days from the date of open surface irrigation, the moisture content started reducing. Thus, no uniform moisture content existed up to the depth of 45 cm, once high, then low. Growth rate as the effect of Pitcher irrigation to ascertain the exit of the water from the pitcher, i.e. the water that was decreased, study was conducted for the summer months on the Guava and Lemon plants. Decrease of water from the Pitcher happened due dryness of the adjacent soil of the pitcher. The dryness was due to absorption of moisture from the soil. The average decrease of water from the Pitcher in Guava and Lemon given in the Table 4.

Table 4: Average decrease of water of the Pitcher

Month	*Guava 205 plants*			*Lemon 494 plants*		
	Per day per plant	*Per month per plant*	*Per ha.*	*Per day per plant*	*Per month per plant*	*Per ha.*
February	1.60	46.40	9,940	1,65	47.90	24,850
March	1.73	53.60	11,480	1.79	55.50	28,790
April	2.23	66.90	13,650	2.27	68.18	35,370
May	2.10	64.40	13,140	2.00	62.60	32,470
120 days	1.97	-	48,210	2.05	-	1,21,480

Maximum water decreases from the Pitcher per day was in Lemon 4.5 liters and in Guava 4.82 liters, during consecutive dry days coupled with high air temperature, low relative humidity and depletion of normal moisture content of the soil. Minimum water loss was 0.60 liter in Guava and 0.75 liter in Lemon per day in the rainy / moist day. The decrease of water denotes irrigation for absorption by plants. To ascertain the efficiency of the pitcher irrigation to moisten the adjacent soil of the Pitcher, study on moisture status of the Soil in control and influenced by Pitcher irrigation under the climatic situation during summer months was done. Average water application per plant per month for (1) Guava plant -61.5 ltrs, (2) Lemon plant- 59 ltrs. The influenced moisture content was measured by installing Gypsum Resistance Block in between the plant and the pitcher and in vacant opened place for comparison. The measurement by Resistance Block is done in terms of Atmospheric Moisture Tension Bar (AMTB).With the increasing of dryness of the soil, the reading scale in Electrical Moisture Meter also rises. The observations are mentioned in Table 5.

Table 5: Variation of moisture content of soil with and without Pitcher irrigation and control.

Day of reading	*Weather parameters*					*AMTB reading*			*Remarks*
	Temp. °C		*Rainfall (mm)*	*RH (per cent)*	*PE (mm)*	*Guava*	*Lemon*	*Con-trol*	
	Air	Soil							
1st	29.6	32.3	0	50.5	6.2	0.75	1.0	11	
2nd	30.5	32.7	0	46.5	8.0	0.60	0.75	13	
3rd	30.8	32.7	0	41.0	9.2	0.7	0.75	14	
6th	31.8	32.8	0	53.0	7.8	0.75	0.90	18	
14th	31.8	34.5	0	36.5	9.8	1.2	1.5	23	
15th	32.5	35.6	0	43.5	9.9	1.3	1.6	23	
16th	33.6	35.8	0	30.6	10.0	1.4	1.7	25	
33rd	26.6	33.6	7.2	56.0	8.8	0.80	0.90	13	
34th	28.4	30.8	13.6	57.0	8.2	0.70	0.80	10	Influence of rainfall
35th	28.9	31.1	0	69.5	5.2	0.70	0.80	8	
48th	-	-	42.6	-	-	-	-	-	
53rd	31.0	31.6	0	82.0	7.1	0.60	0.80	8	Influence of rainfall
55th	32.7	31.1	4.8	84.0	7.1	0.50	0.70	7	

Soil moisture were influences by the rainfall. It was observed that, the soil was moistened at uniform status with the Capillary water by pitcher irrigation all along the irrigation period. As to the efficiency of the Pitcher Irrigation, study was conducted to observe the growth and development of the plants after plantation and irrigation in the method. Among all the fruit plants, the observation cited are of the Lemon- Var. Gandharaj, broad leaved branchy type. Sapling prepared by Cutting of Stem (Vegetative propagation). And Guava- Var L-49, broad leaved branchy type. Sapling prepared by air lairing of the top of the branch. The observation presented up to two years from the date of plantation in the Table 6.

It was observed that the (1) Lemon grew 114 cm in height, 34 auxillary branches, base dia. 26 mm and (2) Guava grew 97 cm in heigh, 25 auxillary branches and base dia 20 mm respectively within 730 days. As the plants were vegetative propagated, there was flowering and fruiting within this period. But those were discarded till to full growth and development of the plants. This is appreciable and in response of the Pitcher irrigation. The ultimate aim is the production. Optimum growth and development, leaf hold, flowering and flower hold, fruit formation and fruit hold, maturity of fruits. Observation was taken on the efficiency of the four irrigation methods (1) Flood, (2) Basin, (3) Ring and (4) Pitcher for comparison. The observations are presented in Table 7.

Table-6: The growth and development of the plants by Pitcher irrigation

Fruit plant	Days After planting	Growth				Mode of planting of sapling (vegetative propagated) & installation of pitcher
		Height (cm)	Base Diam (mm)	Branch		
				no.	Length (cm)	
Lemon	0	11	9	0	0	Rainy season plantation and summer installation of Pitcher in to the pit. Pitcher irrigation during entire summer months in non-rainy days. Measurement of growth after passing two rainy seasons.
	108	50	+	8	15	
	365	100	20	20	25	
	730	125	35	34	45	
Guava	0	18	5	0	0	
	63	45	+	7	14	
	365	75	12	14	20	
	730	115	25	25	30	

Table: 7; Production of fruits from plants of 4 Irrigation methods in a hectare of Guava orchard (205 plants) by one time harvest as response of irrigation methods.

Method	*Plant size (m)*		*Average no. of fruits/plant*				*Yield of fruit per cent*
				Drop			
	Height	*Canopy*	*Total*	*No*	*Per cent*	*Net*	*Above(+), below(-)*
Basin	3.05	1.8	220	32	14.54	188	(-)50.00
Ring	2.15	2.0	239	29	12.13	208	(-)43.75
Flood	1.8	1.5	160	40	25.00	120	(-)68.75
Pitcher	3.65	3.1	342	17	4.97	325	100.00

From the result, it transpires that the fruit yield in **Pitcher** irrigation was significantly highest. In adaptive trials, this method was applied in Pomegranate, Lime, Sweet orange in which it was observed very much effective.

References

Mandal, P.K. Annual Reports of Soil and Water Conservation Research Stations, Midnapore, (1987-93), Directorate of Agriculture, Govt. of West Bengal.

Mandal, P.K. Pitcher Irrigation in orchard crops in red and laterite soils, National Conference on Challenge of our Watersheds,1993(SCSI).

Mandal, P.K. Pitcher Irrigation in fruit Plants and Summer Cucurbits Vegetables, 2000, SCR Stn.Midnapore, Directorate of Agriculture, G.O.W.B.

Mandal, P.K. Pitcher Irrigation Method for Fruit Plants and Vegetables, 2004, SCTC, Midnapore, Govt. of West Bengal.

50

Irrigation Management in Pointed Gourd Under South Gujarat Conditions

Savani, N. G., Solia, B. M., Patel R. B. and Patil, R. G.

Abstract

Pointed gourd is an important vegetable crop of this zone. This crop is very sensitive to excess water situation which causes wilt disease. The severity of wilt is accentuated in high clay containing soils and more so when conventional method of irrigation is followed. In order to generate information on water management aspects of pointed gourd, field experiments were conducted during 2007 to 2009 at Soil and Water Management Farm, NAU, Navsari (Gujarat). The experimental soil is clay in texture, low in available N, medium in available P and high in available K with high moisture retention capacity. The results of experiment revealed that the treatment with drip irrigation @ 0.6 PEF(I_2) recorded significantly higher length and average weight of fruit which were closely followed by surface irrigation at 1.0 IW/CPE ratio (I_3). Among the mulching treatments, no mulch (Mo) treatment recorded significantly higher volume of fruit (44.1 ml) as compared to trash mulch (M_T)(40.8 ml) and black plastic mulch(M_B)(38.8 ml). Here, the combination $I_2 M_T$ (5.1 per cent) and $I_3 M_B$ (5.1 per cent) recorded identical values of TSS and both these treatments were significantly higher than I_2M_0 (4.1 per cent), I_3M_T (4.4 per cent) and I_1M_T (4.3 per cent) treatments. Based on the yield results of the experiments indicated that irrigation through drip @ 0.6 PEF along with black plastic mulching in pointed gourd was found to be the best treatment for getting 47 per cent higher fruit yield and 42 per cent more net income than the surface control without mulch. This treatment could also lead to water saving to the extent of 37 per cent.

Key words: *Pointed gourd, Drip, Mulching.*

Introduction

South Gujarat is blessed with natural resources which are highly suitable for cultivation of agricultural as well as horticultural crops. In view of this,

Government of Gujarat has declared South Gujarat as special Agri-Export Zone for fruit and vegetable crops. As a result of this, farmers are going for cultivation of vegetable crops in a big way. There was demand from the farmers for appropriate method and schedule of irrigation particularly for gourd crops. Drip irrigation with its ability of small, frequent irrigation applications has created interest because of decreased water requirements and possible increased production (Singh and Singh, 1973). Further, mulching along with drip has been found to be more beneficial in increasing productivity (Parikh *et al.*,1996). There was a significant difference in length of vine of bottle gourd grown under micro and surface irrigation (Singh *et al.*, 2001). Pointed gourd is an important vegetable crop of this zone. This crop is very sensitive to excess water situation which causes wilt disease. The severity of wilt is accentuated in high clay containing soils and more so when conventional method of irrigation is followed. Farmers of this zone approaching us with wilt affected samples of pointed gourd. In view of paucity of information on water management aspects of this crop, it is not possible to give any advice to farmers coming with such problems. In order to generate information on water management aspects of pointed gourd, present study was conducted.

Material and Methods

This experiment was conducted at Soil and Water Management Farm, Navsari Agricultural University, Navsari (Gujarat). The location lies between 21° N latitude,72° E longitude and altitude of 10 m above mean sea level. The experimental soil belong to Vertic Ustochrepts with low in organic carbon, medium in available P and high in available K content. The soils are non-calcareous in nature with no problem of salinity and sodicity. In this experiment nine treatment combinations consisting of two levels of drip irrigation each of fraction two day, cumulative pan evaporation (I_1= 0.4 PEF and I_2= 0.6 PEF), one surface method control (I_3=IW/CPE 1.0, Depth: 60 mm) and three mulches (M_0: No mulch, M_T: sugarcane trash mulch @ 2.5 t/ha and M_B: black plastic mulch: 50 micron) were tried in RBD with four replications.The normal plant spacing of 100 x 200 cm was followed in this study. Recommended dose of fertiliser *i.e.* 90 : 60 : 40 NPK kg/ha were applied in both method. In case of surface method,100 per cent of PK and 25 per cent of N were applied as basal and remaining 75 per cent N was applied in two equal splits *i.e.*, before and after monsoon. But in case of drip method, 100 per cent of PK and 10 per cent of N were applied as basal and remaining 90 per cent N was applied in nine equal splits at an interval of 30 days (except monsoonic months). In drip irrigation system, lateral lines of 16 mm diameter with on line drippers of 4 Lph capacity were kept at 1 m spacing along with crop row. The system was operated at 1.2 kg/cm^2 pressure which gave uniformity co-efficient of about 89 to 91 per cent. Irrigation was applied every alternate day in drip irrigated crop based on two days cumulative pan evaporation. All the data recorded in these experiments were analyzed statistically and the results are presented here.

Results and Discussion

Yield attributes and quality parameter

The yield attributes *viz.,* length of fruit, girth of fruit, average weight of fruit and average volume of fruit along with quality parameter *viz.,* TSS content in fruit were determined during the year 2009-10. The length of fruit of pointed gourd was affected significantly only due to main effect of irrigation method (Table1). The treatment with drip irrigation @ 0.6 PEF recorded significantly higher fruit length which was closely followed by surface irrigation at 1.0 IW/CPE ratio. However, these treatments were significantly higher as compared to I_1, (drip irrigation @ 0.4 PEF). Similar pattern of treatment effect was also

Table 1: Yield attributes and quality parameters of pointed gourd under different treatments (2009-10)

Treatment	*Length of fruit (cm)*	*Girth of fruit (cm)*	*Average Weight of fruit (g)*	*Average volume of fruit (ml)*	*Average TSS (per cent)*
I_1 – 0.4 PEF	7.28	8.79	26.20	41.70	4.51
I_2 – 0.6 PEF	7.93	9.41	29.40	41.93	4.63
I_3 – 1.0 IW/CPE	7.53	9.31	27.69	40.04	4.84
CD @ 5 %	0.46	NS	2.3	NS	NS
Mo	7.52	9.21	27.82	44.13	4.53
M_T	7.72	9.20	28.21	40.78	4.58
M_B	7.48	9.11	27.30	38.75	4.87
SEm±	0.156	0.209	0.781	1.146	0.112
CD @ 5 %	NS	NS	NS	3.34	NS
I x M	NS	NS	NS	5.79	0.57
CV %	7	8	10	10	8

observed in the case of average weight of fruit as I_2 treatment recorded significantly heavier fruit (29.4 g) than I_1 (26.2 g) but was at par with I_3 (27.69 g). As far as girth of fruit is concerned, it was not affected significantly due to any of the factors or their interaction. The volume of fruit was influenced significantly due to main effect of mulching and its interaction with irrigation methods (Tables 1 and 2). Among the mulching treatments, no mulch (Mo) treatment recorded significantly higher volume of fruit (44.1 ml) as compared to M_T (40.8 ml) and M_B (38.8 ml). Among all the combinations, I_2Mo (46.5 ml), I_3Mo (46.0 ml) and I_1 M_T (46.0 ml) recorded significantly higher volume of fruits as compared to rest of the combinations, but these combinations were at par with each other. With respect to TSS content in fruit of pointed gourd, it was affected significantly only due to interaction effect of I x M (Table 3). The trend of treatment effect was on TSS content in fruit was similar as was observed in volume of fruit. Here, the combination $I_2 M_T$ (5.1 per cent) and $I_3 M_B$ (5.1 per cent) recorded identical values of TSS and both these treatments were significantly higher than I_2M_0 (4.1 %), I_3M_T (4.4%) and I_1M_T (4.3%) treatments.

Table 2: Effect of different treatments on volume of fruit (ml) (2009-10)

Treatments	*Mulching*			*Mean*
	Mo	*M_T(STM)*	*M_B(BPM)*	
I_1 – 0.4 PEF	39.9	46.0	39.3	41.7
I_2 – 0.6 PEF	46.5	38.7	40.6	41.9
I_3 – 1.0 IW/CPE	46.0	37.7	36.5	40.0
Mean	44.1	40.8	38.8	
	I	M	I x M	
SEm±	1.146	1.146	1.985	
C.D. at 5 %	NS	3.34	5.79	
CV %	10.0			

Table 3: Effect of different treatments on TSS content in fruit (%) (2009-10)

Treatments	Mulching			Mean
	Mo	M_T (STM)	M_B (BPM)	
I_1 – 0.4 PEF	4.6	4.3	4.7	4.5
I_2 – 0.6 PEF	4.1	5.1	4.8	4.6
I_3 – 1.0 IW/CPE	4.9	4.4	5.1	4.8
Mean	4.5	4.6	4.8	
	I	M	I x M	
SEm±	0.11	0.11	0.19	
C.D. at 5 %	NS	NS	0.57	
CV %	8.0			

Fruit yield

During the first year (2006-07), the fruit yield of pointed gourd was affected significantly due to I, M and their interaction (I x M) also (Table 4). Among the mulches, both *i.e.,* sugarcane trash (10.72 t/ha) and BPM (11.48 t/ha) were statistically at par with each other, but were significantly better than no mulch control (9.43 t/ha). Similarly, the treatment I_2 (11.12 t/ha) and I_3 (10.71 t/ha) were at par with each other, but were significantly higher in comparison to I_1 (9.86 t/ha). Among all the combinations, treatment I_2M_B recorded significantly higher fruit yield (11.95 t/ha) as compared to rest of the combinations. Of course, it was at par with some of the combinations, *viz.*, I_3M_B (11.82 t/ha), I_3M_T(11.72 t/ha) and I_2 Mo (11.01 t/ha). Almost similar pattern of treatment effect was observed on fruit yield recorded during 2008-09 and 2009-10. Only exception to this pattern was that during 2008-09, interaction I x M effect was not significant on fruit yield.

The trend of treatment effect of fruit yield observed during individual year was also reflected in pooled analysis (Table 4). Here also treatment I_2 (drip at 0.6 PEF) and M_B (black plastic mulch) recorded conspicuously higher fruit yield individually as well as jointly. The I_2M_B recorded significantly higher fruit yield of 12.13 t/ha as compared to rest of the combinations of I and M Water Use Efficiency (WUE).

Table 4: Effect of different treatments on fruit yield (t/ha) of pointed gourd

Treatments	*Mulching*			*Mean*
	Mo (No Mulch)	*M_T (Trash mulch)*	*M_B (Black plastic)*	
	2006-07			
I_1 – 0.4 PEF drip	8.86	10.05	10.68	9.86
I_2 – 0.6 PEF drip	11.01	10.40	11.95	11.12
I_3 – 1.0 IW/CPE surface	8.58	11.72	11.82	10.71
Mean	9.43	10.72	11.48	
	Irrigation (I)	**Mulch (M)**	**I x M**	
SEm±	0.305	0.305	0.528	
C.D. at 5 %	0.89	0.890	1.54	
CV %	10			
	2008-09			
I_1 – 0.4 PEF drip	9.03	8.82	8.65	8.83
I_2 – 0.6 PEF drip	9.15	11.10	12.98	11.08
I_3 – 1.0 IW/CPE surface	8.06	10.23	10.60	9.63
Mean	8.75	10.05	10.74	
	Irrigation (I)	**Mulch (M)**	**I x M**	
SEm±	0.407	0.407	0.704	
C.D. at 5 %	1.19	1.19	NS	
CV %	14			
	2009-10			
I_1 – 0.4 PEF drip	9.20	8.42	8.23	8.62
I_2 – 0.6 PEF drip	9.11	10.75	11.47	10.44
I_3 – 1.0 IW/CPE surface	8.13	9.89	10.36	9.46
Mean	8.81	9.69	10.02	
	Irrigation (I)	**Mulch (M)**	**I x M**	
SEm±	0.335	0.335	0.581	
C.D. at 5 %	0.98	0.98	1.69	
CV %	12			
	Pooled			
I_1 – 0.4 PEF drip	9.03	9.10	9.19	9.10
I_2 – 0.6 PEF drip	9.76	10.75	12.13	10.88
I_3 – 1.0 IW/CPE surface	8.26	10.61	10.93	9.93
Mean	9.01	10.15	10.75	
	Irrigation (I)	**Mulch (M)**	**I x M**	
SEm±	0.203	0.203	0.352	
C.D. at 5 %	0.57	0.57	0.99	
CV %	12			

Interactions: *Y × I: NS, Y × M : NS, Y×I×M : NS and Y : Sig.*

The WUE was calculated only for treatments involving irrigation levels/ method. Among the irrigation treatments, the WUE recorded with drip @ 0.4 PEF was maximum (15.9 kg/ha-mm) followed by drip @ 0.8 PEF (13.1 kg fruit / ha-mm) and least was with surface method of irrigation (7.52 kg / ha-mm). This implies that drip method of irrigation could almost double the WUE in comparison to surface method of irrigation (Table 5).

Table 5: WUE as influenced by different treatments of pointed gourd (Pooled)

Wreatments	*Mulching*			*Mean*
	Mo	*M_T (STM)*	*M_B (BPM)*	
I_1 – 0.4 PEF	15.76	15.88	16.04	**15.9**
I_2 – 0.6 PEF	11.77	12.97	14.63	**13.1**
I_3 – 1.0 IW/CPE	6.26	8.04	8.28	**7.52**
Mean	**11.26**	**12.30**	**12.98**	

Soil respiration (CO_2 content in soil air)

In order to understand the beneficial effects of mulching on crop yield, the CO_2 content in soil air was measured during March of 2009 and 2010 using LC pro-soil respiration chamber (ADC Bio scientific, Landon). The values of CO_2 content in soil air under different treatments are reported in Table 6. For clarity, the data were tabulated in two-way table. It is interesting to note that at I_1 and I_2, the net CO_2 content in soil air was almost same, but it declined at I_3 *i.e.,* 0.8 PEF. Similarly, among the mulching treatments, under black plastic mulch treatment, the CO_2 content in soil air was almost double as compared to no mulch and trash mulch treatments. This suggests that the soil microbial and root respiration activities were more profuse under black plastic mulching than no mulch and trash mulch treatments. The higher microbial and root respiration activities lead to higher nutrient availability in soil. This is the main reason for higher fruit yield of pointed gourd recorded with drip + mulch treatment.

Table 6 : The CO_2 content in soil air (difference between atmosphere and soil air) under different treatments (Pointed gourd)

Treatments	*Mulching*			*Mean*
	Mo (No Mulch)	*M_T (Trash mulch)*	*M_B (Black plastic)*	
I_1 – 0.4 PEF drip	107	142	234	161
I_2 – 0.6 PEF drip	123	138	245	169
I_3 – 1.0 IW/CPE surface	94	109	178	127
Mean	108	129	219	

Economics

As the treatment involving drip irrigation at 0.6 PEF with black plastic mulch out yielded rest of the treatments, economics was computed for this treatment along with surface control. The data presented in Table 7 revealed that treatment $I_2 M_B$ (Drip @ 0.6 PEF + BPM) recorded higher net income of Rs 1.5 lakh /ha along with water saving of 37 per cent over surface control. Even mulching in surface method also realised net income of Rs 1.38 lakh/ha which is more than unmulched surface irrigation control.

Table 7: Economics of different treatments

Sr. No.	*Particulars*	*Unit*	*Drip at 0.6 PEF with BPM*	*Surface irrigation with mulch*	*Surface irrigation without mulch*
1.	Fixed cost	'000 Rs/ ha	28.6	15.5	15.5
2.	Variable cost	'000 Rs/ ha	40.6	43.0	28.0
3.	Total cost	'000 Rs/ ha	69.2	58.5	43.5
4.	Yield	t/ha	12.13	10.93	8.26
5.	Income	'000 Rs/ ha	218.3	196.7	148.7
6.	Net income	'000 Rs/ ha	149.1	138.2	105.2
7.	Water applied	mm	829	1320	1320
8.	Water saving over surface	%	37		
9.	Net income over surface without mulch	%	42		

* Selling price of pointed gourd @ 18 Rs/kg

Conclusion

Based on the results of three years, it is concluded that irrigation through drip @ 0.6 PEF along with black plastic mulching in pointed gourd was found to be the best treatment for getting 47 per cent higher fruit yield and 42 per cent more net income than the surface control without mulch. This treatment could also lead to water saving to the extent of 37 per cent.

References

Singh, S. D. and Singh, Punjab (1973). Trickle irrigation ensure higher water use efficiency. *Indian Farming* 23 (7) : 14-17.

Parikh, M. M., Savani N. G., Shrivastava, P. K., Avadaria J. D., Thanki, J. D., Shah, G. B., Holer, G. H., Desai, S. G., Gohil, K. B., Patel, H. R. and Raman S. (1996). Response of various crops to micro-irrigation, mulching and fertigation. All India Seminar on Modern Irrigation Techniques organised by Institute of Engineers (Agric. Engg. Div.) held at Banglore during 26-27 June 1996 : 206-212.

Singh, K. L., Chauhan, H. S., Singh, P. K. and Shukla, K. N. (2001). Responses of biometric yield of bottle gourd under micro- irrigation, Micro irrigation, *Edition* : 387-392.

51

Integrated Water Management for Climate Smart Agriculture for Small Holders

Suhas P. Wani, Kesava Rao AVR and Pathak P.

Abstract

Sixty percent of agriculture (85.2 m ha) in India is rainfed supporting livelihoods of millions of small farmholder families for achieving food and nutritional security. An increasing trend in intense rainfall events over east-central India along with decreased number of rainy days and acidification of rainwater due to impacts of climate change further increases the vulnerability of rainfed agriculture.

During 1971-2004, increased semi-arid areasof 3.82 m ha in Madhya Pradesh and arid areas of 1.53 m ha in Rajasthan and 0.98 m ha in Gujarat are reported. More number of below average rainfall years affecting length of growing period and droughts are recorded in >90% of districts in Gujarat, making farmers vulnerable to impacts of climate change.

ICRISAT and partners have developed, validated and piloted integrated watershed management approach. The IWM brings together measures to increase efficiency of greenwater through landform management, augmenting blue water through rainwater harvesting for storage in tanks and increased groundwater recharge and conjunctive use through efficient irrigation methods and delivery systems for harnessing the potential of rainfed small farms. Integrated watershed management has adopted the approach of 4 ICEs i.e.

I	*C*	*Es*
Innovate	*Convergence*	*Efficiency*
Inclusive	*Collective*	*Equity*
Intensive	*Consortium*	*Environment protection*
Integrated	*Capacity building*	*Economic gain*

Using this approach, model watersheds in 13 states supported by MoA and MoRD showed that crop productivity increased by 43 to

67%, cropping intensity increased by 27 to 60%, reduced runoff by 38 to 59% and soil loss by 4.2 t ha^{-1} y^{-1} per watershed, and increased diversity of crops and livelihood systems.

Scaling out of sustainable intensification initiative viz., using soil health assessment as an entry point followed with improved cultivars, land, water, crop and pest management options, improved delivery systems, institutions, monitoring systems and policies increased crop yields by 20 to 66% over farmers' practice in Karnataka thru a mission project Bhoochetana. Area coverage in Bhoochetana increased from 0.2 m ha to 3.76 m ha in 2012 covering 4.3 million farmers. Accrued additional economic benefits during 2009 – 2012 were estimated as Rs.1267 crores (US$ 230 million).

Introduction

Global Warming is one of the most important issues prevailing in the world affecting food systems as well as livelihoods globally. Besides, the effects of Climate change can be seen in the unprecedented heatwaves, cyclones, floods, salinization of the coastline and impacts on agriculture, horticulture, livestock and fisheries. Sixty percent of agriculture in India is rainfed supporting millions of small farm holder families for achieving food and nutritional security.

The goal of Integrated Watershed Management is to improve livelihood security by minimising the adverse effects of climatic variability while protecting or enhancing the sustainability of the environment and the agricultural resource base through suitable adaptation strategies.ICRISAT in partnership with the NARSs has developed an innovative and up scalable consortium model for managing watersheds holistically for improving livelihoods through an integrated approach. This approach of integrated and participatory watershed development and management has emerged as the cornerstone of rural development in the SAT.

Climate change in India

India ranks first among the countries that practice rainfed agriculture in terms of both extent and value of production. The rainfed agro-ecologies cover about 60 per cent of the net sown area of 141 million ha and are widely distributed in the country (DOAC, 2011).

Alternating sequence of multi-decadal periods of thirty years having frequent droughts and flood years are observed in the all India monsoon rainfall data (Guhathakurta and Rajeevan, 2006). The decades 1961-70, 1971-80 and 1981-90 were dry periods. Though Indian monsoon rainfall as a whole does not show any significant trend, significant rainfall trends are observed over some specific areas. During the southwest monsoon season, three subdivisions viz. Jharkhand, Chhattisgarh, Kerala show significant decreasing trend and eight subdivisions **viz.** Gangetic WB, West UP, Jammu & Kashmir, Konkan & Goa, Madhya Maharashtra, Rayalaseema, Coastal AP and North Interior Karnataka show significant increasing trends. June rainfall is getting importance as its contribution to annual rainfall is increasing in almost 19 subdivisions while decreasing in the remaining 17 subdivisions. Contribution of July rainfall is decreasing in central and west peninsular India. But contribution of August rainfall is increasing in all these areas.

Goswami et al., (2006) analysed gridded rainfall data for the period 1951-2000 and found significant rising trends in the frequency and the magnitude of extreme rainfall events, and a significant decreasing trend in the frequency of moderate events over central India during the monsoon seasons. The seasonal mean rainfall does not show a significant trend, because the contribution from increasing heavy events is offset by decreasing moderate events. They concluded that a substantial increase in hazards related to heavy rainfall is expected over central India in the future. Increased frequency and intensity of extreme weather events in the past 15 years were also reported by Samra et al., (2003 and 2006).

Kumar et al (2012) have analyzed monsoon rainfall (June- September) in different parts of Gujarat in relation to ENSO events using 52 years data (1958-2009). The extent of impact has been found to vary from year to year as well as over regions to region. Data of four stations Anand (52 years) of middle Gujarat, Navsari (30 years) of South Gujarat, Junagadh (25 years) of Saurashtra and S K Nagar (25 years) of north Gujarat were used; out of 52 years of data 17 were El Nino or hot years, 11 were La Nina or cold years and rest 24 were neutral or normal years. Their analysis revealed that El Nino and La Nina events play significant role in seasonal rainfall variation in Gujarat.

Chinchorkar et al (2013) investigated monthly, seasonal and annual climatic variability in Anand City based on mean maximum, mean minimum and mean air temperatures. Their results confirmed a significant warming trend in average temperatures in Anand, of about 1.5°C in the past 30 years, concentrated in spring and summer months. Analysis of maximum and minimum temperatures revealed a warming trend for the annual and all seasonal series. Warming trend for the summer and winter seasons is statistically significant.

Changes in arid and semi-arid areas in India with associated shifts during 1971-2004 were earlier reported (Kesava Rao et al., 2013).Based on IMD's one degree gridded data of 34 years (1971-2004), climatic water balances were computed for 351 pixels in India and climates were classified in to six types following Thornthwaite's moisture regime classification; areas falling under different climatic zones delineated. Considerable change in the climates was observed between the two periods; 1971-90 and 1991-2004. Salient features are increase in the arid areas in Rajasthan (1.53 m ha) and Gujarat (0.98 m ha), and increase in semi-arid areas in Madhya Pradesh (3.82 m ha), Bihar (2.66 m ha) and Uttar Pradesh (1.57 m ha).Little increase in the areas under subhumid and humid climates is noticed in Himachal Pradesh, Uttarakhand and West Bengal.Results indicated that dryness and wetness are increasing in different parts of the country in the place of moderate climates existing earlier in these regions. They mentioned that ICRISAT's Hypothesis of Hope through future climate ready crops and Integrated Watershed Management could be a potential adaptation strategy to cope with the variability and changes anticipated.

Integrated Watershed Management

In rural areas, most of the poor make their livelihoods on the use of natural resources, which are degraded and inefficiently used. This is because of the inadequate traditional management practices of managing agriculture as well as the fact that resulting crop yields are much below the expected potential yields. IN addition per capita availability of water and land resources has declined sharply during last 50 years due to increased population. ICRISAT has adopted an integrated genetic and natural resource management (IGNRM) approach to enhance agricultural productivity in rainfed areas, which is a powerful integrative strategy of enhancing agricultural productivity. While addressing the core issue of rural development, this approach maximizes synergies among the disciplines of natural resource management, crop improvement, and social sciences, along with people's empowerment through capacity-building measures (Twomlow et al., 2008).Wani *et al.*, (2009) described the watershed scale as the "entry point" for effective management of smallholder agro-ecosystems for improving livelihoods.

ICRISAT in the early1970s initiated research on watersheds for integrated use of land, water, and crop management technologies for increasing crop production through efficient use of natural resources, especially rainfall that is highly variable in the SAT and is the main cause of year-to-year variation in crop production in India. Improved watershed management on Vertisols more than doubled crop productivity, and rainfall-use efficiency increased from 35% to 70% when compared with traditional technology (Singh et al., 2009). This watershed model is more holistic and puts rural communities and their collective actions at centre stage for implementing improved watershed technologies with technical backstopping and convergence by consortium partners.

Integrated watershed management comprises improvement of land and water management, integrated nutrient management including application of micronutrients, improved varieties and integrated pest and disease management is also recommended to bridge the yield gap (Wani *et al.*, 2012). Watershed management program is one of the most suitable options for increasing water use efficiency and also as an adaptive strategy to cope with climate change effect in rainfed areas (Wani *et al*, 2011).

ICRISAT's Integrated Watershed Management has adopted the approach of 4 ICEs i.e.

I	C	Es
Innovate	Convergence	Efficiency
Inclusive	Collective	Equity
Intensive	Consortium	Environment protection
Integrated	Capacity building	Economic gain

This approach has increased crop productivity and cropping intensity, reduced runoff and soil loss as well as increased diversity of crops and livelihood systems.

Surface and groundwater availability increased across the watersheds through in-situ water conservation and establishment of water harvesting and groundwater recharging structures. Results showed that open wells located near Water Harvesting Structures (WHS) have significantly higher water levels compared to those away from the WHS. Even after the rainy season, the water level in wells nearer to WHS sustained good groundwater yield. In case of Raisen watershed in Madhya Pradesh, treated area registered a groundwater level rise by 9.4 m. At Jamnagar watershed in Gujarat, the average rise in groundwater was at 6.1 m and the irrigated area increased from 171 ha to 390 ha. At Kothapally watershed in Telangana, the average groundwater level rise was at 4.2 m in open wells. With this improvement in groundwater availability, the supply of water for agriculture and clean drinking water was guaranteed.

In a model watershed project supported by Ministry of Agriculture and Department of Land Resources, Government of India, 13 model watersheds provisioning and regulatory ecosystem services were quantified. Additional surface water storage capacity of 88000 to 192000 m^3 was created in different watersheds. Watershed interventions reduced runoff by 38 to 59%, peak runoff rate by 42-63% and soil loss by 4.2 t ha^{-1}per year. Studied conducted by ICRISAT scientists have shown that the cost of water harvesting and groundwater recharging structures varies considerably with type of structures (Figure 1) and selection of appropriate location. Selection of appropriate location for structures also can play very important role in reducing the cost of structures (Figure 2).

In most situations, the small and medium water harvesting and groundwater recharging structures are highly cost effective as well as addresses the key issue of equity and long-term sustainability (Table 1). These small structures can be easily constructed and maintained by local community and provides good employment generation.

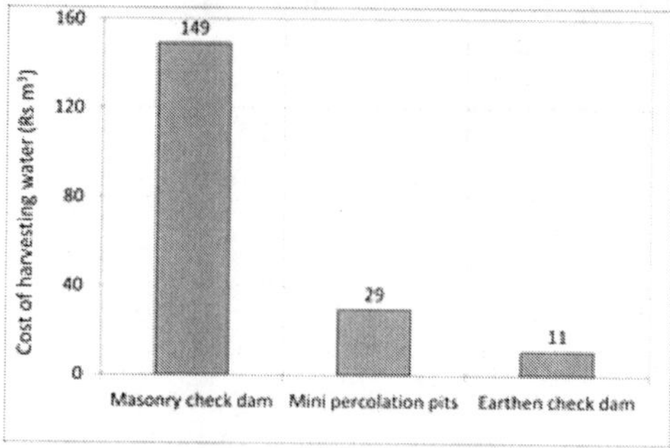

Figure 1. Cost of harvesting water in different structures at Kothapally watershed, AP, India (Source: Pathak et al., 2009)

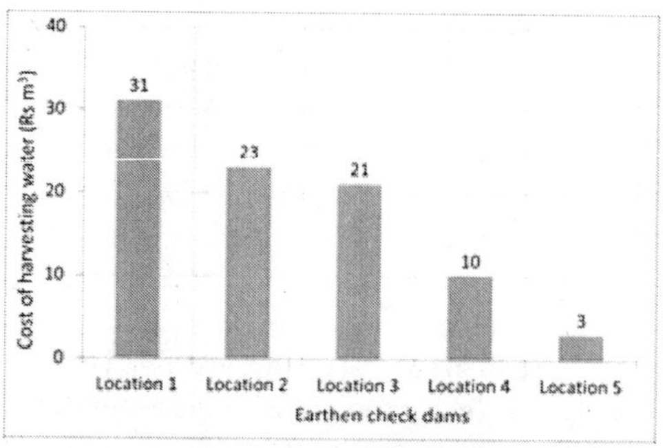

Figure 2. Cost of water harvesting at different locations in Lalatora watershed, MP, India (Source: Pathak et al., 2009)

The integrated watershed management along with improved Vertisol technology for managing rainy season fallow was found highly beneficial in increasing total annual crop yield compare to conventional farmer's practice (Table 2). In the soybean, chickpea sequential cropping system the improved technology gaveaverage total annual yield of 3.69 tha^{-1}compared to 1.77 tha^{-1} under the conventional farmers practice. The improved Vertisol technology was also found highly effective in increasing rainwater use efficiency from 35% under the rainy season fallow system to 79% under the double cropping.

Improved system of double cropping has been found highly effective in increasing the net annual benefit (Table 3). In Raisen district the improved system increased the net annual benefit from Rs.61575 tha^{-1}to Rs.121660 tha^{-1}(97.58% increase) in soybean, chickpea sequential cropping systems.

Increased and sustained crop productivity is found across the watersheds. Of the studied watersheds in India, implementing best-bet practices resulted in significant yield advantages in sorghum (35–270%), maize (30–174%), pearl

Table1. Comparison of different types of water harvesting structures in term of cost effectiveness, equity and employment generation, model watershed, Madhya Pradesh

Type of structure	*Total no. of structures / cost (Rs.)*	*Unit water storage cost (Rs m^{-3})*	*No. of farmers benefited*	*Total area benefited (ha)*	*Employment generation (man-days)*
Low cost small and medium structure	24 (19 lakhs)	7	139	204	8780
Large structure with RCC or cement stone	3 (13 lakhs)	94	30	61	1660

Table 2. Average crop yield and rainwater use efficiency in different management systems during 2010 -13.

District	*Crops/ Cropping systems*	*Mean Yield (t ha^{-1})*		*Increase in rainfall use efficiencies (%)**
		Double cropping with other improved practices	*Conventional farmers practice of rainy season fallow*	
Raisen	Soybean +	2.23	Nil	42
	Chickpea sequential system	1.67	1.73	
Sagar	Soybean +	1.86	Nil	46
	Chickpea sequential system	1.62	1.81	
Average annual crop yield		3.69	1.77	44

*Simulated Values

millet (72–242%), groundnut (28–179%), sole pigeonpea (97–204%) and as an intercrop (40–110%). In addition, cropping intensity increased by 32-57% along with higher crop diversity and livelihood systems.

Greater resilience of crop income in Adarsha Watershed, Kothapally (Rangareddy district, Andhra Pradesh) during the drought year 2002 was indeed due to watershed interventions. While the share of crops in household

Table 3. Mean economic return from the different management systems during 2010-13.

District	*Crops/ Cropping systems*	*Double cropping with other improved practices*		*Conventional farmers practice of rainy season fallow*		*Increase in Benefit (%)*
		Net benefit (Rs ha^{-1})	*B:C Ratio*	*Net benefit (Rs ha^{-1})*	*B:C Ratio*	
Raisen	Soybean +	62360	7.93	61575	8.08	
	Chickpea sequential system	59300	8.91			
	Total annual benefit	121660		61575		97.58
Sagar	Soybean +	50520	6.61	64775	8.50	
	Chickpea sequential system	57300	8.64			
	Total annual benefit	107820		64775		66.45

income declined from 44% to 12% in the non-watershed project villages, crop income remained largely unchanged from 36% to 37% in the watershed village (Wani et al., 2003).

Bundelkhand region of Central India is a hot spot of water scarcity, land degradation, poverty and poor socio-economic status. Impacts of integrated watershed development (IWD) interventions on water balanceand different ecosystem services are analyzed in one of the selected watershed of 850 ha in Bundelkhand region (Ramesh Singh et al., 2014) of India. Improved soil, water and crop management interventions in Garhkundar-Dabar (GKD) watershed enhanced ET to 64% as compared to 58% in untreated (control) watershed receiving 815 mm annual average rainfall. Reduced storm flow (21% vs. 34%) along with increased base flow (4.5% vs. 1.2%) and groundwater recharge (11% vs. 7%) of total rainfall received were recorded in treated watershed as compared to untreated control watershed. Economic Water productivity and total income increased from Rs.2.5 to 5.0 per m^3 and Rs.11,500 to 27,500 tha^{-1}y^{-1} after implementing integrated watershed development interventions in GKD watershed, respectively. Results demonstrated that integrated watershed management practices addressed issues of poverty in GKD watershed. Benefit to cost ratio of project interventions was found three and payback period within four years suggest economic feasibility to scale-up IWD interventions in Bundelkhand region.

Scaling up of sustainable intensification

Based on thelearning of integrated watershed management approach over the last four decades ICRISAT-led consortium along with Department of Agriculture, Government of Karnataka has undertaken a mission mode approach for increasing agricultural productivity of rainfed agriculture in 24 districts of Karnataka. In this approach soil health assessment was used as an entry point and based on the observed widespread deficiencies of multiple nutrients (macro as well as secondary and micro nutrients) taluk-wise fertilizer recommendations were developed, disseminated to the farmers ensuring availability of needed inputs thru incentives. Along with soil test based fertilisers improved cultivar seeds, seed treatment as well as use of soil and water management interventions increased crop yields by 20 to 66 per cent over the farmers' practice. During four years (2009 to 2012) the project covered 3.73 mha and 4.3 million farmers. During the four years individual farmers benefitted with a B:C ratio varying from 2.6 to 14:1 and total net value of increased production worked out to be US $ 240 million (Rs. 1267 crores) (Wani et al., 2014). The need to scale up the available technologies to enable the small farmers in developing countries to adopt climate smart agriculture to cope with the impacts of the climate change is immense and need to be taken up urgently. Most technologies are not reaching the small farm-holders and large yield gap exists between the current farmers' yields and the potential achievable yields. The science of delivery needs to be pursued vigorously to achieve the impacts through scaling up of technologies, products and knowledge to benefit billions of small farm holders across Asia and Africa.

Conclusions

ICRISAT and its consortium partners across years of experience have developed the integrated watershed management approach, which has demonstrated promising results in enhancing crop productivity and efficient use of natural resources under both on-station and on-farm watersheds with local community participation.

Integrated watershed management framework is needed to increase agricultural productivity and to cope up with climate change leading to resilient dryland agriculture; which can be operationalized through integrated genetic and natural resource management (IGNRM) approach. This approach includes implementation of both scientific and supporting solutions such as enabling policies, institutions, and socioeconomic aspects for enhancing impact and adoption of technologies and practices by farmers and the implementing agencies. ICRISAT in partnership with the National Agricultural

Research Systems has developed an innovative and up-scalable consortium model for managing watersheds holistically. This approach of integrated and participatory watershed development and management has emerged as the cornerstone of rural development in the SAT. It ties together the biophysical notion of a watershed as a hydrological unit with the social aspect of community and its institutions for sustainable management of land, water and other resources. During the last two decades this approach of watershed management had shown multi-faceted impacts on water availability, land degradation, agricultural production and productivity, environment, income and poverty.

References

Chinchorkar, S.S., Vaidya, V.B, Pandey, V. (2013) Monthly, Seasonal & Annual Air Temperature Variability and Trends – A Case Study to Assess Climate Change on Anand (Gujarat State). The International Daily Journal for Climate Change, Global Warming & Sustainability 1(1):20-25.

DOAC. (2011). Annual Report 2010-11. Department of Agriculture and Cooperation, Ministry of Agriculture, Government of India. http://agricoop.nic.in/Annual report2010-11/AR.pdf

Goswami, B.N., Venugopal, V., Sengupta, D., Madhusoodanan, M.S. and Xavier, P.K. (2006). Increasing trend of extreme rain events over India in a warming environment.*Science*, 2006, 314, 1442–1445.

Guhathakurta P and Rajeevan M. 2006.Trends in the rainfall pattern over India. National Climate Centre, India Meteorological Department, PUNE. INDIA. Research Report 2/2006.

Kesava Rao A.V.R., Suhas P Wani, K.K. Singh, M. Irshad Ahmed, K. Srinivas, Snehal D. Bairagi and O. Ramadevi. 2013. Increased arid and semi-arid areas in India with associated shifts during 1971-2004. Journal of Agrometeorology 15 (1): 11- 18 (June 2013).

Kumar, A., Pandey, V., Shekh, A.M., Lunagaria, M.M and Patel H.R. (2012).The impact of El Nino and La Nina (ENSO) on monsoon rainfall in Gujarat. Journal of Agrometeorology 14, Special Issue 151-156.

Pathak, P and Mishra, P.K and Rao, K.V and Wani, S.P and Sudi, R. (2009) *Best-bet Options on Soil and Water Conservation.* In: Best-bet Options for Integrated Watershed Management Proceedings of the Comprehensive Assessment of Watershed Programs in India, 25-27 July 2007, ICRISAT Patancheru, Andhra Pradesh, India.

Ramesh Singh, Kaushal K. Garg, Suhas P.Wani, R.K Tewari and S.K Dhyani. 2014. Impact of water management interventions on hydrology and ecosystem services in Garhkundar-Dabar watershed of Bundelkhandregion, Central India. Journal of Hydrology 509 (2014) 132–149.

Samra, J.S., Singh, G. and Ramakrishna, Y.S. (2003). Cold wave of 2002-03: Impact on agriculture. NRM Division, ICAR, KrishiAnusandhanBhawan-II, Pusa, New Delhi, India, 2003, pp. 1–49.

Samra, J.S., Ramakrishna, Y.S., Desai, S., SubbaRao, A.V.M., Rama Rao, C.A.Y.V.R., Rao G.G.S.N., Victor, U.S., Kumar, Vijaya, Lawande, K.E., Srivastava, K.L. and Krishna Prasad, V.S.R. (2006). Impact of excess rains on yield, market availability and prices of onion. Information bulletin Hyderabad. Central Research Institute for Dryland Agriculture (ICAR), 2006, p. 52.

Singh, Piara, Pathak, P., Wani, S.P and Sahrawat, K.L. (2009) 'Integrated Watershed Management for Increasing Productivity and Water-Use Efficiency in Semi-Arid Tropical India', Journal of Crop Improvement, 23:4, 402-429.

Twomlow, S., B. Shiferaw, P. Cooper, and J.D.H. Keatinge. 2008. Integrating genetics and natural resource management for technology targeting and greater impact of agricultural research in the semi-arid tropics. *Experimental Agric*. 44:1–22.

Wani S.P, Singh H.P, Sreedevi T.K, Pathak P, Rego T.J, Shiferaw B and IyerShailaja Rama. (2003). Farmer-participatory integrated watershed management: Adarsha watershed, Kothapally India, An innovative and upscalable approach. A Case Study. Pages 123-147*in* Research towards integrated natural resources management: Examples of research problems, approaches and partnerships in action in the CGIAR (Harwood R.R and Kassam A.H, eds.).

Interim Science Council, Consultative Group on International Agricultural Research. Washington, DC, USA.

Wani S.P, Sreedevi T.K, Rockström J and Ramakrishna Y.S.(2009). Rainfed agriculture - Past trend and future prospects. Pages 1-35 *in*Rainfed agriculture: Unlocking the Potential. Comprehensive Assessment of Water Management in Agriculture Series (Wani SP, Rockström J and Oweis T, eds.). CAB International, Wallingford, UK.

Wani, S.P, B Venkateswarlu., and V.N. Sharda. 2011. Watershed development for rainfed areas: Concept, principles, and approaches Chapter 3. In Integrated Watershed Management in Rainfed Agriculture (Suhas P Wani, J. Rockstrom and KL Sahrawated.). CRC Press, The Netherlands. pp 53-86.

Wani S.P, Garg K.K, Singh A.K and Rockstrom J. (2012). Sustainable Management of Scarce Water Resources in Tropical Rainfed Agriculture.In:Soil Water and Agronomic Productivity. Advances in Soil Science, CRC Press., pp. 347-408.

Wani S.P, William D. Dar, Dileep K Guntuku, Kaushal K. Garg and A.V.R. Kesava Rao. (2012). Improved livelihoods and building resilience in the semi-arid tropics: science-led, knowledge-based watershed management, Climate Exchange, The Global Framework for Climate Services, World Meteorological Organization, October 2012, Pages 69-71, ISBN 978-0-9568561-3-5. (http://www.wmo.int/pages/gfcs/tudor-rose/index.html).

Wani S.P, Sarvesh K.V, Sahrawat K.L, Krishnappa K, Dharmarajan B.K, Raju K.V, Kaushik Mukherjee and Dar W.D. 2014. Bhoochetana: Building Resilience and Livelihoods through Integrated Watershed Management. Resilient Dryland Systems Report No. 62. International Crops Research Institute for the Semi-Arid Tropics (ICRISAT) Patancheru 502 324, Andhra Pradesh,India. pp. 44.

52

Effect of Irrigation Scheduling on Crop Yield and Quality of Fenugreek (Trigonella *Foenum-Graecum* L.) under Changing Climate

P. K. Chovatia, B. S. Gohil, S. P. Kachhadiya and D. S. Hirpara

Abstract

An experiment was conducted to study the effect of irrigation schedules on the growth, yield attributes, yields and quality of fenugreek on medium black soil during the winterseason of the year 2006-07 and 2007-08 at Junagadh (Gujarat). The results revealed that significantly the highest seed yield, straw yield and test weight was observed when crop was irrigated five times, i.e. seedling, branching, flowering, pod formation and pod development stages. Based on Benefit Cost ratio, crop irrigated five times at different critical stages found to be better management practices in fenugreek. Treatment I_3 *(Irrigation at seedling, branching, flowering, pod formation and pod development stage), being statistically at par with* I_2 *(Irrigation at seedling, branching, flowering and pod formation stage) recorded significantly highest protein per cent, NPK uptake, content of N, P and K in seed as well as N and K content in straw along with appreciable saving of water under changing climate.*

Introduction

Fenugreek is one of the important minor spice crops. India is one of the major producer and exporter of fenugreek with an annual export of nearly 1500 tones. Fenugreek is raised in *rabi* season and fairly tolerant to frost and low temperature. It can be grown in all types of soil under irrigated conditions, but does best on loamy soils. Judicious use of water along with suitable agronomic techniques at appropriate crop growth stages would substantially increase both plant growth and yield. Increasing use of fertilizer and irrigation water would also increase manifolds weed problem. Therefore, application of irrigation water in proper amount and proper time will go a long way in arresting the problem created by weeds. Water is a scare commodity, which if used judiciously along with suitable agro-techniques at critical plant growth stages would substantially increase nutrient uptake by crop and ultimately yield. Keeping this in view, the present experiment was planned.

Materials and Methods

A field experiment was conducted during *rabi* seasons of 2006-07 and 2007-08 at the Instructional Farm, College of Agriculture, Junagadh Agricultural University, Junagadh. Soil of the experimental area was medium black in texture, low in available nitrogen and medium in available phosphorus and potassium with the pH of 8.05. The treatments consisted of three irrigation schedules. The experiment was laid out in split plot design with three replications. Irrigation schedules were allotted to main plot and weed management practices to sub-plots. Fenugreek variety GF-1 was sown by drilling on November 9, in 2006 and November 11, in 2007 at row spacing of 30 cm with seed rate of 25 kg ha^{-1}. Crop was fertilized with 20-40-0 NPK kg ha^{-1} as basal before sowing. The data were recorded for seed and straw yield on net plot basis and then converted on hectare basis. Chemical analysis of NPK content carried out by taking representative sample excluding root from each net plot at harvest of the crop. The samples were oven dried at 60 ^{0}c for 24 hours and then powered by using pestle and mortar. Finally, powdered sample were utilized for extraction of various elements. The protein (%) content of seeds worked out by multiplying nitrogen content with the factor of 6.25 (Angelo and Mann, 1973).

Results and Discussion

Effect on yield

Seed and straw yields of fenugreek were significantly affected by irrigation schedules (Table 1). In case of irrigation schedules, significantly higher seed and straw yields were recorded when the crop was irrigated five times, i.e. at seedling, branching, flowering, pod formation and pod development stage in both the years. This suggests that increasing irrigation numbers might have caused faster growth of fenugreek plants and inhibited weed seed germination and the growth of early emerged weeds, which in turn, reduced the weed-crop competition resulting into higher seed yield. Effect of irrigations and weed management practices was also observed on seed size (weight). Thousand seed weight was significantly higher when irrigation was applied at all five critical growth stages of crop during both the years. This might be due to better utilization of nutrients, moisture and sunlight by the fenugreek

Table 1. Effect of irrigations schedules on yield parameters, protein content and economics of fenugreek

Treatment	*Test weight (g)*		*Seed yield (kg ha^{-1})*			*Straw yield (kg ha^{-1})*		*Protein (%)*		*Net returns (Rs ha^{-1})*	*B:C ratio*
	2006-07	*2007-08*	*2006-07*	*2007-08*	*pooled*	*2006-07*	*2007-08*	*2006-07*	*2007-08*		
Irrigation schedules											
At seedling, branching and pod formation stages	12.74	13.07	1322	1406	1364	2003	2124	23.21	23.72	13282	2.65
At seedling, branching, flowering and pod formation stages	14.16	14.52	1512	1567	1539	2295	2418	24.38	25.60	15735	2.88
At seedling, branching, flowering, pod formation and pod development stages	16.17	16.58	1687	1720	1703	2603	2647	25.11	25.88	18009	3.07
C.D. (P=0.05)	1.55	1.61	171	152	94.80	234	217	1.27	1.43	1478	0.18

plants.These findings are in close vicinity of those of Kumar *et al.* (2000),Singh *et al.* (2001), Singh *et al.* (2002), Yadav and Dahama (2003), Patel *et al.*(2005), Bhunia *el al.* (2006), Datta and Chatterjee (2006) and Lakpale *et al.* (2007).

Effect on content and uptake of nutrients

N, P and K content and uptake of fenugreek were significantly influenced by irrigation treatments except P content in straw. Treatment I_3 (Irrigation at seedling, branching, flowering, pod formation and pod development stage), being statistically at par with I_2 (Irrigation at seedling, branching, flowering and pod formation stage) recorded significantly higher content of N, P and K in seed (Table 2) as well as N and K content in straw (Table 3). The treatment I_1 (Irrigation at seedling, branching and pod formation stage) registered significantly the lowest content of N, P and K in seed and N and K in straw. Phosphorus content in straw was not influenced significantly by various irrigation levels. Similarly, significantly the highest N, P and K uptake (Table 4) were observed under treatment I_3 (Irrigation at seedling, branching, flowering, pod formation and pod development stage), which remained at par with I_2 (Irrigation at seedling, branching, flowering and pod formation stage). Significantly the lowest uptake of N, P and K was observed under treatment I_1 (Irrigation at seedling, branching and pod formation stage). The increase in solubility of nutrients with increase in water content is also responsible for higher uptake of nutrients especially nitrogen. Secondly, the conductivity of the soil increases with the increase in moisture content of soil and vice-a-versa. Thus, nutrients which are moving along with the stream of moisture which is termed as mass flow transport of nutrient increases when conductivity of the soil is high. Therefore, it is obvious that when moisture content is more, the rate at which nutrients reach to root surface is high which in turn contributes to high N uptake. The higher uptake of nutrientswas due to high seed and straw production under different levels of irrigation. The findings are in accordance partly with those of Tanwar (1982) and Bhunia *el al*. (2006).

Effect on quality

As far as protein content of seed is concerned significantly highest protein content (25.50 %) was found under treatment I_3 (Irrigation at seedling, branching, flowering, pod formation and pod development stage), which remained statistically at par with treatments I_2 (Irrigation at seedling, branching, flowering and pod formation stage), Whereas, the lowest protein content (23.47%) was observed under treatment I_1 (Irrigation at seedling, branching and pod formation stage). This can be ascribed to increased uptake of nutrients and water, which in turn enhanced assimilation of amino acid leading to increased synthesis of protein (Table 1). Similar findings were observed by Tanwar (1982).

Table 2. Effect of irrigation schedules on N, P and K content in seed

Treatment	*N content in seed (%)*			*P content in seed (%)*			*K content in seed (%)*		
	2006-07	*2007-08*	*Pooled*	*2006-07*	*2007-08*	*Pooled*	*2006-07*	*2007-08*	*Pooled*
At seedling, branching and pod formation stages	3.714	3.795	3.754	0.381	0.377	0.379	0.597	0.602	0.599
At seedling, branching, flowering and pod formation stages	3.901	4.097	3.999	0.388	0.388	0.388	0.641	0.662	0.652
At seedling, branching, flowering pod formation and pod development stages	4.018	4.141	4.079	0.401	0.394	0.397	0.667	0.671	0.669
C.D. (P=0.05)	0.203	0.228	0.127	NS	NS	0.012	0.049	0.052	0.030

Table 3. Effect of irrigation schedules on N, P and K content in straw

Treatment	*N content in straw (%)*			*P content in straw (%)*			*K content in straw (%)*		
	2006-07	*2007-08*	*Pooled*	*2006-07*	*2007-08*	*Pooled*	*2006-07*	*2007-08*	*Pooled*
At seedling, branching and pod formation stages	2.034	1.986	2.010	0.211	0.210	0.211	0.515	0.506	0.510
At seedling, branching, flowering and pod formation stages	2.149	2.111	2.130	0.217	0.214	0.215	0.556	0.525	0.541
At seedling, branching, flowering pod formation and pod development stages	2.150	2.121	2.136	0.221	0.221	0.221	0.570	0.542	0.556
C.D. (P=0.05)	0.096	0.107	0.060	NS	NS	NS	0.037	0.025	0.018

Table 4. Effect of irrigation schedules on N, P and K uptake by fenugreek

Treatment	*N uptake (kg ha⁻¹)*			*P uptake (kg ha⁻¹)*			*K uptake (kg ha⁻¹)*		
	2006-07	*2007-08*	*Pooled*	*2006-07*	*2007-08*	*Pooled*	*2006-07*	*2007-08*	*Pooled*
At seedling, branching and pod formation stages	90.10	95.72	92.91	9.14	9.78	9.46	18.23	19.33	18.78
At seedling, branching, flowering and pod formation stages	108.33	115.34	111.83	10.82	11.28	11.05	22.45	23.26	22.86
At seedling, branching, flowering pod formation and pod development stages	124.02	127.66	125.84	12.50	12.67	12.59	26.22	25.93	26.07
C.D. (P=0.05)	7.36	8.18	4.48	0.82	1.23	0.55	2.04	2.86	1.42

Economics

In terms of monetary return, irrigation at five different stages: seedling, branching, flowering, pod formation and pod development stages recorded significantly the highest net return (Rs. 18009) and B:C ratio of Rs. 1:3.07(Table 1).These findings are in agreement with those of Patel et al (2005).Hence, the results of the present study revealed that irrigations at five critical stages, i.e. seedling, branching, flowering, pod formation and pod development, gave significantly higher seed yield. Thus, five irrigations found to be better for getting higher seed yield, nutrient uptakes and protein content in fenugreek under the changing climate.

References

Angelo, A.J. and Mann, G.E. (1973). Peanut culture and uses. Peanut Research Eduction Socoety, stone printing Corporation, Raonote USA, p. 561.

Bhunia, S.R., Chauhan, R.P.S., Yadav, B.S., and Bhati, A.S. (2006). Effect of phosphorus, irrigation and Rhizobium on productivity, water use and nutrient uptake in fenugreek (Trigonella foenum-graecum). *Indian. J. Agron.*51(3): 239-241.

Datta, S. and Chatterjee, R. (2006). Effect of irrigation regimes on moisture extraction pattern, evapotranspiration and yield of fenugreek (*Trigonella foenum-graecum* L.). *Journal of Spices and Aromatic Crops.*15(2): 125-129.

Kumar, V., Yadav, J.S., Singh, J. and Yadav, B.D. (2000). Irrigation and phosphorus requirement of fenugreek (*Trigonella foenum-graecum* L.) on light soil. *Indian J. Agic. Sci.* 70(8): 515-517.

Lakpale, R., Shrivastava, G.K., Tripathi, R.S. (2007). Effect of irrigation schedule on growth, yield and economics of spice crops. *Indian Journal of agric. Sci.*77(3): 170-173.

Patel, R. H., Oliver, J. and Usadadia, V. P. (2005). Influence of irrigation schedule and weed management practices on growth and yield of fenugreek (Trigonella foenum-graecum L.). *Indian J. weed Sci.*37(3 & 4): 237-239.

Singh, V.P., Yadav, A.C. and Thakral, K.K. (2001). Economics, consumptive use of water and water use efficiency of coriander seed crop as affected by various irrigation levels and weed control measures. *Haryana Agricultural University Journal of Research.*31(3/4): 105-110.

Singh, V.P., Yadav, A.C.and Thakral, K.K. (2002). Effect of irrigation levels and weed control treatments on yield and quality of coriander seed. *Seed Research.* 30(2): 328-330.

Tanwar, D.D. (1982). Effect of different levels of irrigation, nitrogen and method of weed control on growth, yield and quality of cumin. M.Sc.(Agri.) thesis (unpublished) submitted to University of Udaipur, Jobner.

Yadav, R.S. and Dahama, A.K. (2003).Effects of planting date, irrigation and weed control method on the yield and water use efficiency of cumin (*Cuminum cyminum*). *Indian Journal of agric. Sci.*73(9): 494-496.

List of Contributors

Authors	Page No.
A.J. Bhatt, College of Fisheries, Junagadh Agricultural University, Veraval. Gujarat	79, 220, 304
A.Y. Desai, College of Fisheries, Junagadh Agricultural University, Veraval. Gujarat	79, 304
Amit M. Pujar, Ph.D. Scholars, College of Agriculture, University of Agricultural Sciences, Dharwad-580005, Karnataka state, India	256, 266
Arti Joshi, College of Fisheries, Junagadh Agricultural University, Veraval. Gujarat	79
Arun Jhajharia, Department of Farm Engineering, Institute of Agricultural Sciences, Banaras Hindu University, Varanasi 221005 Uttar Pradesh	243
B. S. Gohil, Department of Agronomy, College of Agriculture, Junagadh Agricultural University, Junagadh-362001 (Gujarat)	461
B.B. Limbasiya, Assistant Professor, Polytechnic in Agril. Engg., Junagadh Agricultural University, Targhadia	35
B.B. Ramani, Office of Directorate of Research, Junagadh Agricultural University, Junagadh-362001, Gujarat	63
B.K. Patel, Soil and Water Management Research Unit, Navsari Agricultural University, Navsari- 396 450	183
B.K. Sagarka, Department of Agronomy, College of Agriculture, Junagadh Agricultural University, Junagadh-362 001 (Gujarat)	31, 47, 52, 128
B.M. Dabhi, Department of Agronomy, College of Agriculture, Junagadh Agricultural University, Junagadh-362 001 (Gujarat)	128
B.S. Gohil, Department of Agronomy, College of Agriculture, Junagadh Agricultural University, Junagadh-362001	56, 94, 148
Bhoomika B. Patel, Soil and Water Management Research Unit, Navsari Agricultural University, Navsari- 396 450	183
C.K. Patel, Main Cotton Research Station, Navsari Agricultural University, Athwa Farm, Surat-395007 (Gujarat) India.	237
C.M. Sutaria, Main Cotton Research Station, Navsari Agricultural University, Athwa Farm, Surat-395007 (Gujarat) India.	237
C.R. Bharodia, Post Graduate Institute of ABM,Junagadh agricultural University, Junagadh-362001	100
D S Thanki, Polytechnic in Agro-Processing, Junagadh Agricultural University, Junagadh	416
D. S. Hirpara, Department of Agronomy, College of Agriculture, Junagadh Agricultural University, Junagadh-362001 (Gujarat)	461

Authors	***Page No.***
Hanamant M. Halli, Ph.D. Scholar, University of Agricultural Sciences, Dharwad -580005, Karnataka state, India.	339
I.U.Dhruj, Directorate of Research office, Junagadh Agricultural University, Junagadh-362 001	68
J.B.Vasave, Ph. D. Scholar, Dept. of Agronomy, Navsari Agricultural University, Navsari	167
J.G. Patel, Main Cotton Research Station, Navsari Agricultural University, Athwa Farm, Surat-395007 (Gujarat) India.	237
J.H. Kelaiya, Senior Research Fellow,College of Agricultural Engineering and Technology, Junagadh Agricultural University, Junagadh-362 001	116, 227
J.M. Hemani, College of Agricultural Engineering and Technology, Junagadh Agricultural University, Junagadh-362 001	174
J.M. Modhawadia, Cotton Research Centre, Khapat Farm, Junagadh Agricultural University, Porbandar-360579 (Gujarat)	94, 148
J.M. Patel, Soil and Water Management Research Unit, Navsari Agricultural University, Navsari- 396 450	183
J.N. Nandasana, Assistant Professor, Polytechnic in Agro Processing, JAU, Junagadh,	392, 404
Joshi Viral P., Department of RE & RE, College of Agricultural Engineering and Technology, Junagadh Agricultural University, Junagadh, Gujarat, India.	359
K. Avil Kumar, Water Technology Centre, College of Agriculture Professor Jayashankar Telangana State Agricultural University Rajendranagar, Hyderabad -500 030.	24
K. Radhika, Water Technology Centre, College of Agriculture Professor Jayashankar Telangana State Agricultural University Rajendranagar, Hyderabad -500 030.	24
K. Ramaswamy, Agricultural Engineering College and Research Institute, Tamil Nadu Agricultural UniversityKumulur 621 712, Trichy District, Tamil Nadu, India	375
K.B. Sankat, Main Cotton Research Station, Navsari Agricultural University, Athwa Farm, Surat-395007 (Gujarat) India.	237
K.C. Patel, College of Agricultural Engineering and Technology, Junagadh Agricultural University, Junagadh-362 001	47, 174, 220, 350, 392, 404
K.N. Sondarva, Krushi Sala, Junagadh Agricultural University, Halvad	286
Kalpesh Kumar, Post Graduate Institute of ABM,Junagadh agricultural University, Junagadh-362001	100
Kesava Rao AVR, Scientist (Agroclimatology). ICRISAT Development Centre International Crops Research Institute for the Semi-Arid Tropics Patancheru- 502 324, Telangana, India	449
Keshavamurthy G. M., Ph.D Scholars,College of Agriculture, University of Agricultural Sciences, Dharwad-580 005, Karnataka, India	321, 331

Authors	*Page No.*
P.J. Prajapati, Agricultural Officer, Wheat Research Station, JAU, Junagadh	107, 167
P.M. Chauhan, Department of Renewable Energy and Rural Engineering, College of Agricultural Engineering and Technology, Junagadh Agricultural University, Junagadh, Gujarat, India.	359
P.M. Kyada, Subject Matter Specialist, Krishi Vigyan Kendra, Lokbharti Gramvidhyapith, Sanosara, Bhavnagar	63
P.M. Vaghasia, Main Oilseeds Research Station, Junagadh Agricultural University, Junagadh(Gujarat)	158
P.R. Mathukia, Department of Agronomy, College of Agriculture, Junagadh Agricultural University, Junagadh-362001	56
P.S. Jayswal, Assistant Professor, College of Agriculture, Junagadh Agricultural University, Amreli.	286
Parag. P, AICRP on Groundnut R.A.R.S. Polasa, Jagtial, Karim Nagar, Telangana	380
Patel R. B. Soil and Water Management Research Unit,Navsari Agricultural University, Navsari- 396450	442
Pathak P, ICRISAT Development Centre International Crops Research Institute for the Semi-Arid Tropics Patancheru- 502 324, Telangana, India	449
Patil, R. G. Soil and Water Management Research Unit,Navsari Agricultural University, Navsari- 396450	442
Prafulla Kumar Mandal, Rtd. Addl. Director of Agriculture, Govt. of West Bengal	431
Pramod Mohnot, Associate Director of Research, Junagadh Agricultural University, Junagadh -362001	311
R. Basavarajappa, Professor, College of Agriculture, University of Agricultural Sciences, Dharwad-580 005, Karnataka state, India	331
R. Raghavaiah, Water Technology Centre, College of Agriculture Professor Jayashankar Telangana State Agricultural University Rajendranagar, Hyderabad -500 030.	24
R. Subbaiah, Research Scientist (Agril. Engg), Centre of Excellence on Soil and Water Management, R.T.T.C, Junagadh Agricultural University, Junagadh-362 001 (Gujarat), India	201, 295
R.B. Thanki, Cotton Research Centre, Khapat Farm, Junagadh Agricultural University, Porbandar-360579 (Gujarat)	148
R.H. Patil, Agronomist, Department of Agronomy, College of Agriculture, University of Agricultural Sciences, Dharwad-580005, Karnataka state, India	256, 266, 321, 331, 339
R.J. Patel, Assistant Professor,College of Agricultural Engineering and Technology, Junagadh Agricultural University, Junagadh-362 001	116, 134, 227

Authors	***Page No.***
Suhas P Wani, Director,ICRISAT, Development Centre International Crops Research Institute for the Semi-Arid Tropics Patancheru - 502 324, Telangana, India	449
Sumeet Meshram, Department of Farm Engineering, Institute of Agricultural Sciences, Banaras Hindu University, Varanasi 221005 Uttar Pradesh	243
Tatte Sumathi, Department of Floriculture and Landscape Architecture, ACHF, NAU, Navsari	275, 280
V. Praveen Rao, Water Technology Centre, College of Agriculture Professor Jayashankar Telangana State Agricultural University Rajendranagar, Hyderabad -500 030.	24
V. Ramulu, Water Technology Centre, College of Agriculture Professor Jayashankar Telangana State Agricultural University Rajendranagar, Hyderabad -500 030.	24
V.A. Naliapara, M. Tech (Agril. Engg.) Student,Department of Processing and Food Engineering, College of Agricultural Engineering and Technology, Junagadh Agricultural University, Junagadh-362001, Gujarat	423
V.B. Bhalu, Main Oilseeds Research Station, Junagadh Agricultural University, Junagadh(Gujarat)	158
V.H. Kachhadiya, Main Oilseeds Research Station, Junagadh Agricultural University, Junagadh(Gujarat)	158
V.K. Chandegara, Polytechnic in Agro-Processing, Junagadh Agricultural University,Junagadh	392, 404, 416
V.K. Sardhara, Assistant Research Scientist, RTTC, Junagadh Agricultural University, Junagadh.	35, 210, 295
V.P. Usadadia, Main Cotton Research Station, Navsari Agricultural University, Athwa Farm, Surat-395007 (Gujarat) India.	237